Advances in Experimental Medicine and Biology

Volume 1494

Advances in Experimental Medicine and Biology provides a platform for scientific contributions in the main disciplines of the biomedicine and the life sciences. This series publishes thematic volumes on contemporary research in the areas of microbiology, immunology, neurosciences, biochemistry, biomedical engineering, genetics, physiology, and cancer research. Covering emerging topics and techniques in basic and clinical science, it brings together clinicians and researchers from various fields.

Advances in Experimental Medicine and Biology has been publishing exceptional works in the field for over 40 years, and is indexed in SCOPUS, Medline (PubMed), EMBASE, Reaxys, EMBiology, the Chemical Abstracts Service (CAS), and Pathway Studio.

2023 CiteScore: 5.9

Umut Toprak • Laura Musselman
Editors

Insect Lipid Metabolism

Springer

Editors
Umut Toprak
Department of Plant Protection
(Entomology), Faculty of Agriculture
Ankara University
Diskapi, Ankara, Türkiye

Laura Musselman
Department of Biological Sciences
Binghamton University
Binghamton, NY, USA

ISSN 0065-2598 ISSN 2214-8019 (electronic)
Advances in Experimental Medicine and Biology
ISBN 978-3-032-04841-7 ISBN 978-3-032-04842-4 (eBook)
https://doi.org/10.1007/978-3-032-04842-4

This Springer imprint is published by the registered company Springer Nature Switzerland AG
The registered company address is: Gewerbestrasse 11, 6330 Cham, Switzerland

Preface

The field of insect lipid metabolism—though foundational to our understanding of physiology, development, and ecological adaptation—has long remained scattered across studies, sub-disciplines, and fragmented reviews. This book was born out of a need for cohesion: to bring together the many voices of this intricate domain into a single, structured, and enduring scientific work.

This volume is the first comprehensive reference dedicated solely to insect lipid metabolism. Rather than a loosely compiled set of recent reviews, it offers a carefully curated, thematically unified, and academically structured collection of 21 chapters, authored by experts across multiple continents. Each chapter contributes to a coherent narrative arc—spanning basic lipid biochemistry to ecological function, from omics to endocrine control, from flight and diapause to vector biology, symbiosis, and even translational relevance to human health. The result is not just a resource—it is a reference: a book to return to, to teach from, and to build upon.

The idea first took shape when Lars Koerner of Springer reached out to us in November 2019, after noticing our symposium "From Fat to Fact: Recent Insights into Insect Lipid Metabolism at Entomology 2019 in St. Louis." His invitation was more than a publishing proposal—it was a recognition of a scientific void and an opportunity to fill it with depth and intention. We thank him sincerely for planting the seed.

Like any endeavor of worth, this project unfolded alongside life itself—with all its unpredictability, burdens, and shifting tides. The COVID-19 pandemic, personal health challenges, and global disruptions tested timelines, momentum, and morale. Yet as we often rediscover in science, adversity also refines. It demanded patience, compassion, and tenacity—not only from us but from every contributor who joined us in this journey. Bringing together such a diverse group of scholars, and ensuring that their chapters resonated not just individually but in harmony, was a task that required not only editorial rigor but also deep mutual respect.

This book is a reflection of what science truly is—and what it should remain. In an age of hyperproduction and superficiality, it reclaims the long path: knowledge acquired slowly, tested patiently, and presented with care. Scientific progress is not noise; it is an inheritance. Like precious metals in an alembic, real insight emerges not from haste but from distillation, synthesis, and intellectual lineage. This volume is part of that lineage. It stands as a humble continuation of the wisdom passed from mentors to students, from masters to apprentices, across generations and cultures. It is science with memory.

We dedicate this effort in spirit to all who have walked this path with integrity. To Rosalind Franklin, Marie Curie, and countless others whose names may never appear in headlines, yet whose devotion defines our disciplines. To the Isaac Asimovs who reminded us that "this world is big enough for all of us," and to the Craig Mellos whose humility dignifies discovery. Science, at its best, is built on values: patience, curiosity, honesty, and compassion. These values are etched into the pages of this book.

As we wrote and edited, we often returned to a simple but universal truth once spoken by Mustafa Kemal Atatürk—one of the twentieth century's most visionary statesmen and reformers—who stated, "In life, the truest guide is science." These words resonate beyond time or geography. They remind us that in moments of uncertainty, conflict, or confusion, the light we must seek is not found in slogans or pride, but in the quiet, persistent flame of genuine inquiry.

And perhaps no group of organisms better exemplifies the power of metabolic adaptation than insects. From polar extremes to desert dunes, from aquatic larvae to airborne foragers, insects navigate life with metabolic precision. They thrive on decaying wood, floral nectar, carrion, and blood—regulating lipids across all imaginable substrates. When we presented data on insect insulin signaling, many colleagues were stunned: "Insects have insulin?" Yes—and they deploy it elegantly.

This book is thus also a tribute to the insect world—the hidden masters of the biochemical economy. They succeed where we often struggle. And through their lipid machinery, we glimpse answers to broader biological puzzles: why tiny variations in regulation lead to immense phenotypic outcomes; how adaptation is etched in molecules; and what we, as humans, might learn from evolutionary design millions of years in the making.

This volume, we believe, offers more than knowledge. It offers perspective. It reminds us that science is not a commodity—it is a cultural heritage, passed down like an unwritten genetic code. With each chapter, we invite the reader not just to learn, but to listen—to the echoes of those who came before, and to the whispers of nature's most masterful survivors.

Diskapi, Ankara, Türkiye — Umut Toprak
Binghamton, NY USA — Laura Palanker Musselman

Contents

Editors and Contributors

About the Editors

Umut Toprak is Professor of Entomology at Ankara University, Faculty of Agriculture, Türkiye, where he also leads the Molecular Entomology (MOLEN) Laboratory (http://www.molenlab.com/). He completed his PhD at the University of Saskatchewan (Department of Biology, Canada), focusing on the molecular architecture and functions of the lepidopteran peritrophic matrix using genomics, proteomics, and RNA interference. His current research program integrates molecular insect physiology, biochemistry, and pathology with emphasis on lipid metabolism, diapause biology, host–pathogen interactions, and neuroendocrinology. He aims to develop next-generation pest control strategies targeting fundamental insect physiological systems using dsRNA-based platforms, venom-derived bioinsecticidal peptides, and microbial agents. In parallel, his laboratory uses insect models to decipher the mechanisms underlying human lipid metabolism disorders—including metabolic dysfunction-associated steatotic liver disease (MASLD) and obesity—toward the long-term goal of contributing to therapeutic development. He has secured more than $2.5 million in competitive research funding and authored over 70 peer-reviewed publications, which have been cited over 3000 times. He has delivered numerous keynote lectures and chaired major scientific events, including the International Molecular Plant Protection Congresses Series, the 5th International Entomopathogens and Microbial Control Congress, and the Physiology & Biochemistry sessions at the XII European Congress of Entomology. He is also an elected Presidium member of the European Congress of Entomology. In addition to organizing international symposia and scientific congresses, he actively serves on the editorial and executive boards of several leading journals in insect physiology, molecular entomology, and pest management, contributing as an associate editor, guest editor, and reviewer. He currently serves as the Physiology, Biochemistry, and Toxicology Section Representative to the Publication Council of the Entomological Society of America.

Laura Palanker Musselman is Associate Professor of Biological Sciences at Binghamton University, State University of New York. She earned her PhD under the mentorship of Dr. Carl Thummel at the University of Utah, where she used genomics and lipidomics to study the transcriptional regulation of lipid catabolism. As a postdoctoral fellow, she was mentored by Dr. Thomas Baranski at Washington University in St. Louis, where she studied dietary and genetic regulation of lipid biosynthesis. In her own lab, she continues to study the effect of high-calorie diets on physiology, metabolism, gene expression, and evolution.

About the Contributors

Perla Akiki is a researcher born and raised in Lebanon. She completed her bachelor's degree at the Lebanese University before moving to France to pursue graduate studies. She obtained her master's degree from the University of Rennes 1, where her interest in scientific research deepened and led her to complete a PhD at the Institut de Biologie Intégrative de la Cellule (I2BC, CNRS) in Jacques Montagne's laboratory. Her doctoral work investigated how social interactions influence tumor progression using *Drosophila melanogaster* as a model, leading to the publication of "Male manipulation impinges on social-dependent tumor suppression in *Drosophila melanogaster* females" in *Scientific Reports* (2024). Following her PhD, she pursued post-doctoral research in Tihana Jovanic's laboratory at NeuroPSI, focusing on neural circuits and behavior using electrophysiology, imaging, and behavioral approaches. She contributed to the publication "Feeding state-dependent neuropeptidergic modulation of reciprocally interconnected inhibitory neurons biases sensorimotor decisions in Drosophila" in *Nature Communications*. She is currently the Marketing and Supplier Relationship Manager at Cliniscience, where she bridges scientific innovation with industry needs. In this role, she oversees the scientific product portfolio and collaborates directly with international suppliers developing cutting-edge research tools.

Thomas Baranski is Professor of Medicine at Washington University School of Medicine. He received his MD/PhD from Washington University and completed residency and fellowship training at UC-San Francisco. He returned to join the faculty in the Division of Endocrinology, Metabolism, and Lipid Research. His research interests include mechanisms of insulin resistance and glucose toxicity. With his collaborator, Dr. Ross Cagan while at Washington University, he developed a model of high sugar diet-induced insulin resistance in fruit flies. Much of the metabolic characterization of the model system was pioneered with then postdoctoral fellow in his lab, Dr. Laura Musselman. He has also used *Drosophila* RNAi transgenic flies for functional genomic studies to evaluate candidates identified in genome-wide association studies of human obesity.

Gustavo M. Calderón-Fernández graduated in Biological Sciences from the Universidad Nacional de La Plata (UNLP) in 2003. He obtained a doctorate degree in Natural Sciences from the UNLP in 2009. He is Associate Professor of Animal Histology and Embryology at the Facultad de Ciencias

Naturales, UNLP, since 2011, and a member of the Consejo Nacional de Investigaciones Científicas y Técnicas (CONICET) research center since 2009 at the Instituto de Investigaciones Bioquímicas de La Plata (INIBIOLP). He has expertise in lipid analysis with research interests in biochemistry and molecular biology of cuticle lipids in insect vectors of Chagas disease.

Lilian Etelvina Canavoso is an Associate Professor in the Department of Clinical Biochemistry and the Center for Research in Clinical Biochemistry and Immunology (CIBICI), Faculty of Chemical Sciences, Universidad Nacional de Córdoba (Argentina). She graduated as a Biochemist and obtained her PhD degree in Chemical Sciences (discipline: Biochemistry) in 1995 at the Universidad Nacional de Córdoba. She performed her postdoctoral training at the Department of Biochemistry and the Center for Insect Science, University of Arizona, USA, from 1998 to 2000. Since 2004, she has been a member of the National Scientific and Technical Research Council staff in the Independent category. Current and previous fields of interest of Dr. Canavoso involve insect biochemistry and physiology, with a focus on lipid and protein metabolism, reproductive biology, and immunity.

Pierre Delamotte completed his scientific studies at Paris-Saclay University, where he also obtained a Magistère in Biology and Biotechnologies. He pursued his PhD at the Institute for Integrative Biology of the Cell (I2BC, Gif-sur-Yvette, France) under the supervision of Dr. Jacques Montagne. His research focused on deciphering the metabolic requirements in a *Drosophila* midgut tumor model, providing new insights into tumor formation dynamics and polyclonality. In the meantime, he investigated physiological responses to different social contexts, highlighting the involvement of the immune system and apoptosis in tumor progression. In parallel, he also contributed to the identification of a novel oenocyte subpopulation involved in egg activation in *D. melanogaster*. Having first discovered Heidelberg during his master's studies, where he was introduced to new experimental and collaborative approaches, Pierre Delamotte later returned there as a postdoctoral researcher. He currently works in Prof. Gilles Storelli's team at the Centre for Organismal Studies, focusing on *Drosophila* midgut biology, immunity, and microbiota. A long-standing fascination with insects—their astonishing diversity of forms, colors, anatomies, and adaptive strategies—continues to inspire his work. The opportunity to engage daily with *Drosophila*, creating new genetic combinations to address fundamental biological questions, remains the cornerstone of his scientific journey.

Justin DiAngelo is a Professor and Program Chair of Biochemistry and Molecular Biology at Penn State Berks, an undergraduate campus of Pennsylvania State University. He teaches undergraduate courses in Biology and Biochemistry and runs a small research lab, mentoring undergraduates on the regulation of metabolism using *Drosophila melanogaster* as the model system. He is also very active in the Genetics Society of America and has previously served as the Chair of the Education Committee and the Primarily

Undergraduate Institution Representative on the Drosophila Board of Directors.

Laurence Dinan Ecdysteroids have been a constant theme throughout Laurie DINAN's research career. His PhD research in Insect Biochemistry at the University of Liverpool (UK) concerned the chemistry, identification, and metabolism of ecdysteroids in insects. During a Royal Society Fellowship in Darmstadt (FRG) and a subsequent postdoctoral fellowship in Düsseldorf (FRG), he investigated the biochemical characterization of ecdysteroid receptors in *Drosophila melanogaster*. He then established an independent research group at the University of Exeter (UK), where the research concerned the mode of action of ecdysteroids in insects, the establishment of a cellular bioassay for the identification of ecdysteroid agonists and antagonists and its application to their QSAR and the distribution of phytoecdysteroids in the plant world. More recently, research (in Paris, in collaboration with Prof. René Lafont) has concerned the chemistry of ecdysteroids, aspects of phytoecdysteroid identification and distribution, and the study of the pharmacological effects of ecdysteroids in mammals.

Agnes Fekete has been the scientific manager of the metabolomics core unit at the University of Würzburg since 2011. She specializes in LC-MS-based analyses and is interested in performing metabolite fingerprinting and profiling to characterize changes in the lipidome.

Leonardo Luis Fruttero is an Adjunct Professor in the Department of Clinical Biochemistry and the Center for Research in Clinical Biochemistry and Immunology (CIBICI), Faculty of Chemical Sciences, Universidad Nacional de Córdoba (Argentina). He graduated as Biologist in 2005 and obtained his PhD degree in Chemical Sciences (discipline: Biochemistry) in 2011 at the Universidad Nacional de Córdoba. He performed his postdoctoral training at the Brain Institute, Pontifícia Universidade Católica do Rio Grande do Sul, Brazil, from 2014 to 2017. Since 2018, he has been a member of the National Scientific and Technical Research Council staff, currently in the Adjunct category. His current and previous fields of interest involve insect biochemistry and physiology, with a focus on lipid and protein metabolism, reproduction, and the effects of entomotoxic proteins on the immune system.

Juan R. Girotti graduated in Biochemistry from the Universidad Nacional de La Plata (UNLP) in 2005. He obtained a doctorate degree at the Facultad de Ciencias Exactas (UNLP) in 2010. He is Associate Professor of Organic Chemistry at the Facultad de Ciencias Naturales (UNLP) since 2012 and researcher at the Instituto de Investigaciones Bioquímicas de La Plata (INIBIOLP) as a member of the Consejo Nacional de Investigaciones Bioquímicas y Técnicas (CONICET) since 2013. He has expertise in lipid analysis with research interests in biochemistry of insects and pathogenic fungi.

Katia Calp Gondim is a Full Professor at the Instituto de Bioquímica Médica Leopoldo de Meis, Federal University of Rio de Janeiro (UFRJ), Brazil. She holds a PhD in Biological Sciences (Biophysics) from UFRJ and conducted postdoctoral research at the University of Arizona as a Pew Latin American Fellow. She leads a research project focused on the biochemistry and molecular biology of triatomines, and her work primarily investigates lipid transport and metabolism in the insect vector *Rhodnius prolixus*, a key model for understanding insect lipid physiology and a primary vector of Chagas disease. She was a member of the *R. prolixus* Genome Consortium and has contributed significantly to elucidating the roles of lipophorin, lipid metabolism pathways, and their regulation in insect vectors. With over 70 scientific publications and more than 2,000 citations, she serves as an editorial board member for *Insect Biochemistry and Molecular Biology* and as an Associate Editor for *Frontiers in Physiology (Invertebrate Physiology)*. She has supervised numerous master's and PhD students to completion and is an ad hoc consultant for national and international funding agencies.

Agata Kaczmarek is an Assistant Professor at the Museum and Institute of Zoology, Polish Academy of Sciences, Warsaw, Poland. Her research focuses on insect physiology, host–pathogen interactions, and the molecular mechanisms underlying innate immunity, particularly in the model organism *Galleria mellonella*. She employs biochemical, molecular, and microscopic techniques to investigate cuticular lipids, mitochondrial activity, and immune responses in insects exposed to microbial infections and environmental stressors. Her recent studies address the biological roles of free fatty acids (FFAs) and lipid metabolism in modulating immune function, energy balance, and stress adaptation. She also promotes the use of insects as alternative experimental models to mammals in biological and biomedical research, highlighting their ethical and methodological advantages. She obtained her PhD in Biology and previously worked at the Witold Stefański Institute of Parasitology, Polish Academy of Sciences. Her work integrates basic and applied approaches, contributing to sustainable biological control strategies and advancing understanding of insect–microbe interactions.

Natraj Krishnan is Professor of Insect Physiology and Graduate Coordinator at Mississippi State University, where he holds a dual appointment in the Department of Agricultural Sciences and Plant Protection and the Department of Biochemistry, Nutrition and Health Promotion. His interdisciplinary research bridges neuroendocrinology, physiology, biochemistry, and molecular biology, with a central focus on the endocrine regulation of oxidative stress responses, free radical biology, and antioxidant systems. He is particularly interested in how circadian rhythms modulate oxidative stress and contribute to aging-related physiological changes. His laboratory also explores the molecular basis of neurological disorders and disrupted sleep/wake cycles through the lens of tRNA modifications. Using the fruit fly *Drosophila melanogaster* as a model system, his team investigates conserved mechanisms that link cellular stress responses to neurodegeneration and circadian biology.

His work contributes to a deeper understanding of insect physiology while offering translational insights into human health and disease. As Graduate Coordinator, he is committed to mentoring emerging scientists and fostering collaborative, cross-disciplinary training environments that prepare students for leadership in agricultural and biomedical sciences.

René Lafont was born in 1946 in Aubervilliers (France). He was a student at the Ecole Normale Supérieure in Paris (1964–1968) where he graduated as a Master in Chemistry and Physiology. He received his PhD from the University Pierre and Marie Curie (Paris) in 1975 with a thesis entitled "Biochemical and endocrine aspects of the differentiation of wing imaginal discs in Lepidoptera." He was Assistant Professor at the ENS (1969–1985), then Professor at University P. & M. Curie (1985–2008). His teaching activities concerned mainly comparative animal physiology and biochemistry, analytical biochemistry, and insect–plant relationships. His research activities have been on ecdysteroids since 1972, first in insects and other arthropods (ticks, crustaceans), and then on non-arthropod invertebrates. His research later extended to plants (phytoecdysteroids) and, in recent years, focused on the pharmacological effects of ecdysteroids on mammals. This activity has resulted in more than 200 original articles and 60 reviews or book chapters.

Jimena Leyria is an Adjunct Professor at the Universidad Nacional de Córdoba (Argentina) and a member of the National Scientific and Technical Research Council (CONICET). She received her bachelor's degree in Biochemistry in 2008 and her PhD in Chemical Sciences in 2015, both at the Universidad Nacional de Córdoba. From 2018 to 2023, she performed her postdoctoral training as a fellow in the Department of Biology at the University of Toronto Mississauga (Canada). She was awarded two trainee research fellowships at the Universidade Federal do Rio Grande do Sul (Brazil) and another at the Institut de Biologia Evolutiva-UPF-CSIC (Spain). Her field of expertise is centered on insect reproductive physiology, mainly on how tissues communicate with one another to promote successful reproduction.

David Majerowicz is an Associate Professor at the Faculty of Pharmacy of the Federal University of Rio de Janeiro (UFRJ), Brazil. A biologist by training, he holds a master's and a PhD in Biological Chemistry, with his research focused on understanding lipid metabolism in insects. For over 15 years, he has extensively used models such as the kissing bug *Rhodnius prolixus* and the beetle *Tribolium castaneum* to investigate lipid digestion, transport, storage, and mobilization, as well as the role of nuclear receptors and transcription factors in metabolic regulation. His work aims to unravel the molecular basis of metabolic diseases by exploring the remarkable similarities between insect and mammalian metabolic pathways. A recipient of the prestigious "Jovem Cientista do Nosso Estado" fellowship from the Carlos Chagas Filho Foundation for Research Support in the State of Rio de Janeiro (FAPERJ), he leads a prolific research group and is a frequent reviewer for leading international journals in insect biochemistry, molecular biology, and physiology.

Jacques Montagne was a secondary school biology teacher in France when he decided to embark on research training in 1988. During his PhD at *École Normale Supérieure de Lyon* (France), he studied the transcriptional regulation of Human T-Cell Leukemia Virus Type I (HTLV-I) under the supervision of Dr. Pierre Jalinot. In 1992, he joined the team of Pr Walter Gehring at the *Biozentrum* in Basel (Switzerland) to study pattern formation using the powerful genetics of the *Drosophila melanogaster* model. In 1996, he moved to the Friedrich Miescher Institute (FMI) in Basel (Switzerland) to develop the *Drosophila melanogaster* system in the team of Dr. Georges Thomas that focused on the mTOR-dependent growth control. In 2004, he obtained a position at CNRS (Centre National de la Recherche Scientifique) to set his own research group at the *Centre de Génétique Moléculaire* (CGM) and the *Institut de Biologie intégrative de la cellule* (I2BC) in Gif-sur-Yvette (France). There, using the *Drosophila melanogaster* model, he focused on the integration of nutrition and metabolism at the physiological level of the entire organism. He published several fundamental studies investigating the metabolism of the oenocytes, the fat body, the digestive tract, and the prothoracic gland. Since 2023, he has been retired, but continues to work as an Emeritus Professor in the team of Dr. Francois Leulier at *École Normale Supérieure de Lyon* (France), where he investigates the inter-relation between microbiota and metabolism of *Drosophila melanogaster*.

Ralf Nauen is an insect toxicologist and biochemist working on functional (toxico)genomics, molecular entomology, fundamental and applied aspects of insecticide mode of action, selectivity, and detoxification, as well as biochemical and molecular mechanisms of insecticide resistance and its management. He received his PhD from the University of Portsmouth (UK), is a Bayer Distinguished Science Fellow, and is winner of the Otto Bayer Medal (2020) and Crop Science R&D Impact Award (2023). He is a Fellow of the Royal Entomological Society (London, UK), and in 2013, he was awarded Fellowship of the Entomological Society of America (ESA). In 2014, he received the prestigious American Chemical Society (ACS) International Award for Research in Agrochemicals, in recognition of his outstanding and influential research into insecticide and acaricide modes of action and resistance. In 2021, he received the ESA Physiology, Biochemistry and Toxicology (PBT) Section Award. He authored more than 300 scientific papers/book chapters with more than 37,000 citations (h-index 97). He received the Highly Cited Researcher award by the Web of Science (Clarivate) 2018–2025, awarded to scientists whose work is among the top 1% of the most cited papers in their field over a 10-year period. He is appointed Visiting Professor by the Chinese Academy of Agricultural Sciences (Beijing, China) and served as Lecturer/Adjunct Professor in agricultural entomology at the University of Bonn (Germany). Throughout his career he supervised more than 40 graduate students and is an active associate and executive Editorial Board member of several renowned journals in insect molecular sciences and pest management.

Mathilde Scheifler is an evolutionary ecologist and postdoctoral researcher at the University of Liège, Belgium. She obtained her master's degree in Ecology and Evolution at Sorbonne University (France), where she received the Jacques Lebbe Prize for the best Master 2 internship in theoretical or applied systematics in France in 2017. After completing her PhD at Sorbonne University (France) in 2021, she joined the University of Liège (Belgium) as a postdoctoral researcher of the National Fund for Scientific Research (FNRS). She then worked as a Temporary Teaching and Research Fellow from 2024 to 2025 at the École Normale Supérieure (PSL University, France) before returning to Liège to continue her career. Her main research line focuses on the study of microorganisms and their role in ecosystems. She is particularly interested in host–microbe and host–parasite interactions, and in the evolutionary relationships between symbiotic communities and host physiology, metabolism, and health in response to environmental factors.

Patrycja Staniszewska (PhD eng) is a bioengineering graduate of the University of Life Sciences in Lublin (Poland). Since the beginning of her academic career, she has been conducting research related to the biology and physiology of invertebrates, with particular emphasis on the immunology and ecophysiology of honeybees. She is the author or co-author of numerous publications in the field of: the effects of supplementations and environmental factors on immune processes occurring in key immunogenic tissues of insects, such as the hemolymph and fat body, along with an analysis of their physiology. She obtained her PhD in 2023 for her dissertation entitled "The Effect of Hemp Extract and CBD Oil on Physiological and Biochemical Parameters Related to Immunity in Honeybee Workers (*Apis mellifera*).

She is currently pursuing her scientific development at the University of Life Sciences in Lublin in the Department of Invertebrate Ecophysiology and Experimental Biology, under the supervision of Prof. Aneta Strachecka.

Nick Teets is an Associate Professor in the Department of Entomology at University of Kentucky. He completed his PhD in 2012 at Ohio State University of postdoctoral training at the University of Florida. His lab takes an integrative approach to identify mechanisms by which insects cope with temperature variation. Specific areas of interest include the genetic basic of thermal tolerance, seasonal biology and adaptations of invasive species, and mechanisms underlying the extreme stress tolerance of Antarctic insects. His lab also investigates the influence of environmental conditions on the efficacy of genetic technologies being developed for pest control. Outside of research, he teaches courses in general entomology and molecular genetics, and he serves as an Editor in Chief for *Physiological Entomology*.

Bertanne Visser is an evolutionary ecologist and Research Associate of the National Fund for Scientific Research at the University of Liège, Belgium. She did a master's in Ecology and Evolution at Leiden University (the Netherlands) and completed her PhD in 2012 at the VU University Amsterdam (the Netherlands). She then performed a short-term research project at the Netherlands Institute of Ecology (the Netherlands), after which she worked as a postdoctoral associate at the University of Florida (USA). She then obtained a competitive EU Marie Curie grant to work as a postdoctoral fellow at the University of Tours (France), followed by another postdoctoral fellowship of the National Fund for Scientific Research (FNRS) at UCLouvain (Belgium). In 2019, she became a Principal Investigator obtaining an FNRS Research Associate and Professorship position at UCLouvain. Her Evolution and Ecophysiology Group moved to the University of Liège (Belgium) in 2021. Her main research line focuses on studying plasticity of fat metabolism in parasitic insects, but she also studies behavioral plasticity, sexual selection, and life history evolution. She is further actively involved in fighting harassment, discrimination, and power abuse in the academic world through her international NGO.

Christian Wegener studied Biology at the Universities of Konstanz and Jena. In 2000, he received his PhD with honors, working on a new family of neuropeptides in cockroaches (periviscerokinins) at the Institute of General Zoology and Animal Physiology in Jena. After 3 years of postdoctoral training within a Human Frontiers Science project centered about insect clocks at the Zoological Department of Stockholm University, he started an independent Emmy-Noether-junior research group at the Department of Animal Physiology-Neurobiology, University of Marburg. He was also an EMBO short-term fellow at the Institute of Integrative Biology at Leeds University. Since 2011, he has been a Professor of Neurogenetics at the Biocenter, University of Würzburg. His research focuses on the biology of peptide hormones and the interplay between the neuroendocrine system and the circadian clock in insects, mainly the genetic model *Drosophila*. Using peptidomic techniques, his group has significantly contributed to chemically characterize insect neuropeptide hormones and brain-gut peptides and their processing. At the center of his current research is the role of neuroendocrine peptides in the circadian control of behavior, lipid metabolism and development, and its significance for the ability of insects to adapt to a daily changing environment.

Adv Exp Med Biol (2026) 1494: 1–45
https://doi.org/10.1007/5584_2024_805

Published online: 08 June 2024

Insect Lipids: Structure, Classification, and Function

Agata Kaczmarek, Anna Katarzyna Wrońska, Justyna Sobich, and Mieczysława Irena Boguś

Abstract

Lipids are a diverse group of compounds that play several important roles in insect physiology. Among biological lipids, the fundamental category comprises fatty acyl structures, with significant members being fatty acids (FAs). They play several crucial functions in insect physiology; they are used as the source of energy for flight and play key roles in the insect immune system. The FAs present in the insect cuticle are known to demonstrate antibacterial and antifungal activity and are considered as potential insecticides. The most abundant family of lipids are the glycerolipids, with numerous cellular functions including storage of energy, structural compartmentation of cells and organelles, and important signaling activities required for regulation of physiological processes (i.e., growth, development, reproduction, diapause, and overwintering). The phospholipids are also highly diversified key components of all cell membranes; they can modify cellular components in response to rapid cold-hardening (RCH), enhancing membrane fluidity and improving survival at low temperatures. The sphingolipids are important structural and signaling bioactive compounds, mostly detected in membranes.

Insects are sterol-auxotrophs: they do not have genes, which code enzymes converting farnesyl pyrophosphate to squalene. Similarly, to mammals, the production of steroids in insects is regulated by cytochrome P450 enzymes that convert sterols (mostly cholesterol) to hormonally active steroids. The major molting hormone in insects is 20-hydroxyecdysone, and cholesterol is the required precursor; however, several exemptions from this rule have been noted. This manuscript also reviews the roles of prenol lipids, isoprenoids, lipid vitamins, polyketides, and waxes in the vital processes of insects.

Keywords

Fatty acids · Glycerolipids · Isoprenoids · Phospholipids · Polyketides · Prenols · Sphingolipids · Steroids · Vitamins · Waxes

A. Kaczmarek (✉), A. K. Wrońska, and M. I. Boguś
Museum and Institute of Zoology, PAS, Warsaw, Poland

Witold Stefański Institute of Parasitology, PAS, Warsaw, Poland
e-mail: akaczmarek@miiz.waw.pl; awronska@miiz.waw.pl; mbogus@miiz.waw.pl

J. Sobich
Witold Stefański Institute of Parasitology, PAS, Warsaw, Poland

1 Introduction

The undeniable evolutionary success of insects allowing them to adapt to different climatic zones and ecological niches is their perfectly orchestrated and controlled lipid management.

Together with proteins, lipids form membranes compartmentalizing cells and subcellular elements, are a source of readily available energy used in life processes, and participate in cell signaling and defense mechanisms against pathogens; they hence serve as irreplaceable structural and functional components of the insect body and cells.

Lipids also constitute a source of easily available energy necessary for insect embryogenesis, growth, development, metamorphosis, diapause, reproduction, prolonged flight, and survival during winter and periods of food shortage. The center for lipid metabolism in insects is the fat body. Lipid metabolism, i.e., the breakdown and storage of fats and the synthesis of structural and functional lipids, takes place via the enzymatic digestion of the dietary lipids contained in the midgut, followed by transport of the hydrolyzed lipids into the fat body, muscles, and other target tissues by lipophorins, as well as cellular uptake and accumulation, all of these processes are controlled by the insect endocrine system (Wrońska et al. 2023).

A layer of lipids covers the cuticle; this helps maintain general homeostasis, and protects insects against wetting, water loss, predators, and some entomopathogens. Cuticular lipids also play important roles as sex pheromones, nestmate, and caste recognition cues in social insects.

This chapter focuses on the structure of major lipid classes and the roles they play in insect physiological processes essential for the preservation of the species and its successful functioning in the natural environment.

2 Fatty Acyls

The chemical structure of the fatty acyl serves as an essential element in the construction of complex lipids; it is therefore a fundamental category of biological lipids (Fahy et al. 2011). The fatty acyl group is created by a repeating series of methylene groups of varying length, degrees of unsaturation, branching, and various other features such as the presence of inter alia hydroxyl groups, cyclopentene, peroxy, and nitro-moieties. They are produced by chain extension of an acetyl-CoA primer by malonyl-CoA (or methylmalonyl-CoA) groups that can contain a cyclic functionality and/or are substituted with heteroatoms (Fahy et al. 2005).

The first subclass includes fatty acids (FA), which may be separated into few classes based on their structure, physiological role, and biological effects. FAs have a backbone with varying numbers of carbon atoms and double bonds. Short-chain FAs (SCFAs) are FAs with up to five carbon molecules, medium-chain FAs (MCFAs) have 6–12, long-chain FAs (LCFAs) 13 to 21, and very long-chain FAs (VLCFAs) have more than 22 carbon atoms. Other FAs classifications are based on the presence of double bonds in their carbon chain. FAs comprising only single bonds are called saturated FAs (SFAs), with one double bond—monounsaturated FAs (MUFAs) and with two or more double bonds—polyunsaturated FAs (PUFAs). Unsaturated FAs with one double bond can occur in two forms, depending on the position of hydrogen atoms in the molecule: cis (bent form), where the hydrogens are on the same side, or trans (straight form) with hydrogens on opposite sides. In nature, most unsaturated fatty acids occur in the cis form (CFAs), and natural trans acids (TFAs) occur only in ruminants (Kaczmarek and Boguś 2021a). Examples of fatty acids are presented in Table 1 and Fig. 1.

Lipids are an essential source of energy for flight and different insects use different types of fuel. *Drosophila melanogaster* firstly used the glycogen and then switched the fuel to FA to maintain long-term flight (Wigglesworth 1949), while aphids or the mountain pine beetle (*Dendroctonus ponderosae*) primarily use FAs (Wijerathna and Evenden 2019). In locusts, carbohydrate is the primary fuel source for the first 10–15 min of flight, during which they fly at high speed; however, after this period, locusts fly slowly and the flight muscles favorable the oxidizing the more energetic content found in lipids. In this cruising stage, 1 mg of lipid creates an equal amount of energy as 8 mg of glycogen (Minter et al. 2018).

On the first phase of flight the diacylglycerols (DAGs) are released from the fat body as the

Table 1 Classification of fatty acids with examples in insects

	Length	Function	Examples in insects	Literature data
Short-chain fatty acids (SCFAs)	Up to 5 atoms of carbon	Part of defense system	Stingless bees, Oxytrigona	Roubik et al. (1987)
			Carabid beetles	Rossini et al. (1997)
			Stingless formicine ants	Hefetz and Blum (1978)
			Moth larvae	Ricke et al. (2020)
		Aggregation factor	*Acheta domesticus*	McFarlane et al. (1983)
Medium-chain fatty acids (MCFAs)	6 to 12 atoms of carbon	Insecticidal activity	*Galleria mellonella*	Kaczmarek et al. (2022)
			Drosophila melanogaster	Legal et al. (1994)
		Sex pheromone component	*Vespa mandarinia*	Dong et al. (2022)
		Oviposition pheromone	*Lutzomyia longipalpis*	Dougherty and Hamilton (1997)
		Attractant for mosquito males	*Aedes albopictus*	Carraretto et al. (2022)
Long-chain fatty acids (LCFAs)	13 to 21atoms of carbon	Hydrocarbon synthesis	All insects	Blomquist and Bagnères (2010)
			Drosophila melanogaster	Wicker-Thomas et al. (2015) and Holze et al. (2021)
			Triatoma infestans	Juárez and Brenner (1989) and Juarez et al. (1996)
			Blattella germanica	Juárez et al. (1992)
		Inhibition of growth	*Gryllodes sigillatus*	McFarlane and Henneberry (1965)
		Sex pheromone biosynthesis	*Ctenopseustis herana*	Foster and Roelofs (1996)
			Chilecomadia valdiviana	Herrera et al. (2019)
		Storage of energy	All insects	Skowronek et al. (2021) and Arrese and Soulages (2010)
		Attractant	*Aulacophora foveicollis*	Mukherjee et al. (2014)
		Eicosanoid biosynthesis	All insects	Stanley and Kim (2014, 2019) and Stanley and Miller (2006)
		Cold tolerance (changes in membrane fluidity at low temperatures)	All insects	Clark and Worland (2008) and Sinclair and Marshall (2018)
		Fuel for flights	All insects	Haunerland (1997)
			Manduca sexta	Soulages and Wells (1994)
Very long-chain fatty acids (VLCFAs)	More than 22 atoms of carbon		*Sarcophaga argyrostoma*	Kaczmarek and Boguś (2021b)
		Part of cuticular barrier against fungal infection	All insects	Pedrini et al. (2007)
		Hydrocarbon synthesis	*Blattella germanica*	Pei et al. (2021)
		Compound of honeybee wax	*Apis mellifera*	Callow (1963 and Buchwald et al. (2009)

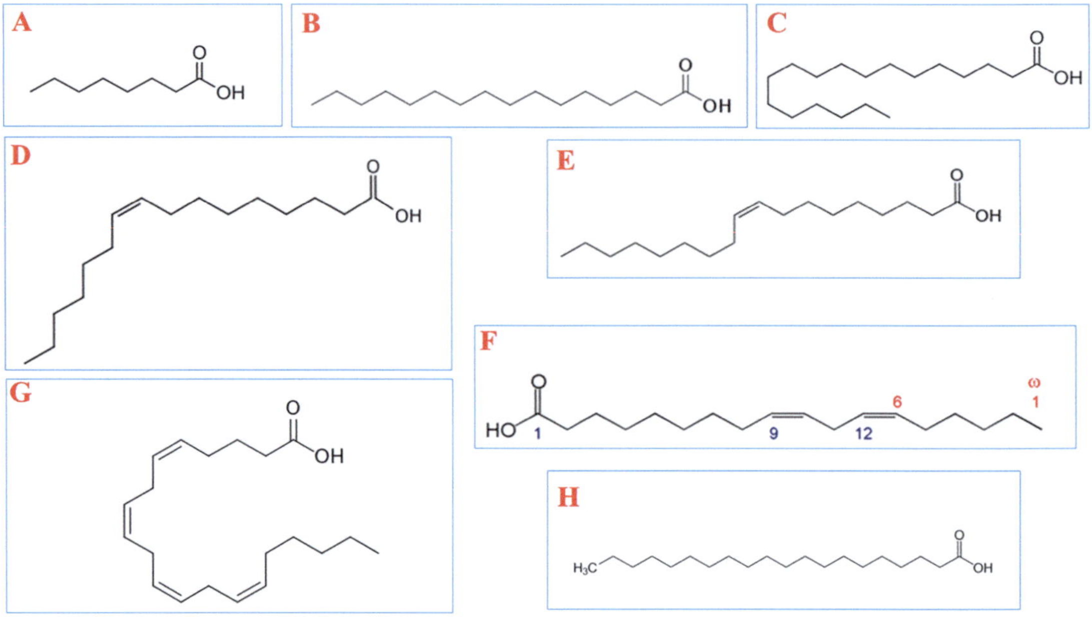

Fig. 1 The examples of important fatty acids (structure based on PubChem (National Center for Biotechnology Information 2023)). (**a**) Octanoic acid (caprylic acid; C8:0); (**b**) hexadecanoic acid (palmitic acid; C16:0); (**c**) octadecanoic acid (stearic acid; C18:0); (**d**) 9-cis-hexadecenoic acid (palmitoleic acid; C16:1 cis-9;); (**e**) cis-9-octadecenoic acid (oleic acid; 18:1 cis-9); (**e**) cis, cis-9,12-octadecadienoic acid (linoleic acid; 18: 2 cis-9,12); (**f**) cis,cis,cis-9,12,15-octadecatrienoic acid (α-linolenic acid, ALA; C18:3 cis-9,12,15); (**g**) cis-5,8,11,14-eicosatetraenoic acid (arachidonic acid, AA; 20:4 cis-5,8,11,14); (**h**) icosanoic acid (arachidic acid; C20:0)

result of octamine activity, next the adipokinetic hormone (AKH) activates the triacylglycerol (TAG) mobilization, that causes the increase of DAG concentration in the hemolymph, which is based for flighting activity (Toprak 2020). Shortly, during longer muscles activity the AKH is released from the corpus cardiacum and this hormone activates the mobilization of lipid content in adipose cells, which finally consequences in the release of DAG from TAG stores (Skowronek et al. 2021). In *Rhodnius prolixus*, turning off Adipokinetic Hormone Receptor (AKHR) results in TAG accumulation in the adipose tissue and flight muscles, and decreases the lipid content in hemolymph after starvation, thus highlighting the role played by AKHR in TAG mobilization (Zandawala et al. 2015). Lipophorin (Lp) transports the TAG to the muscles responsible for flight, where they are digested to FAs and oxidized for energy generation (Van der Horst et al. 2002; Jenni-Eiermann and Srygley 2018).

FAs are an important factor regulating the insect defense mechanism, both those that come from external sources (for example with food) as well as those synthesized inside the insect body via different pathways. The impact of FAs on different parts of the mammal's immunology is extensively described, there is only few literature data describing those processes in insects. Table 2 summarizes the impact of particular FAs on the elements of the defense system of various insect species.

Arachidonic acid ((5Z,8Z,11Z,14Z)-Icosa-5,8,11,14-tetraenoic acid) acts as a precursor to the synthesis of eicosanoids, the group of metabolites with immunomodulatory activity. The biosynthesis of eicosanoid from linoleic acid is well-described in insects, and is presented in Fig. 2 (Kim and Stanley 2021).

The effect of eicosanoids on the insect immune system is well described in the literature. PLA2B is a prostaglandin that is involved in important

Table 2 The influence of lipids on the mechanisms of the immune response of insects

Example of lipids (Systematic name; common name; chemical structure)	Insects	Effects on the immune system	References
All-cis-5,8,11,14-eicosatetraenoic acid; arachidonic acid; $C_{20}H_{32}O_2$	*Apis mellifera ligustica*	Reduced mortality after infection with *Escherichia coli;* Increased expression of immune genes (*defensin-2, toll, myd88*, and *dorsal*)	Yu et al. (2022)
	Spodoptera exigua	Induced immune response by exhibiting nodule formation (after injection to larvae)	Hasan et al. (2019)
		Reversing the inhibitory effects of phentolamine and ketanserin on the cellular immune response	Geun et al. (2009)
	Drosophila	Lack of IMD pathway activation after using as the sole agonist	Yajima et al. (2003)
	Galleria mellonella	Increased granulocytes adhesion, to the glass surface	Marin et al. (2005)
	Rhodnius prolixus	Reversing the inhibition of phagocytosis	Figueiredo et al. (2008)
	Bombyx mori	Induced expression of genes, which code the cecropin and lysozyme protein	Morishima et al. (1997)
(Z)-but-2-enedioic acid; maleic acid; $C_4H_4O_4$	*Manduca sexta*	No effect on the ability of insects to defeat *Serratia marcescens* infection	Stanley-Samuelson et al. (1991)
(2E)-10-Hydroxydec-2-enoic acid; the queen bee acid, 10-HAD; $C_{10}H_{18}O_3$	*Apis mellifera*	Inhibition of the extracellular polymeric substances' secretion, which caused decreased *Staphylococcus aureus* adhesion and aggregation, and disrupted biofilm architecture	Gao et al. (2022)
Octanoic acid; caprylic acid; C8:0	*Galleria mellonella*	Deformation of hemocytes, disordered networking, cell death (apoptosis, necrosis); the activation of apoptosis-related protein—caspases and increase of oxidative stress marker—8-hydroxy-2′-deoxyguanosine (8-OHdG) concentration in hemocytes	Kaczmarek et al. (2022)

physiological processes like development and immunity of *Spodoptera exigua.* The blockage of *PLA2* genes not only results in reduced PLA2 enzyme activity but also developmental retardation and immunosuppression. Previous studies have indicated a significant increase in lipid peroxidation despite no significant change being observed in the FA composition of fat body lipids (Wrońska et al. 2023; Sadekuzzaman et al. 2017). More detailed examples are summarized in Table 3.

Even though arachidonic acid plays an important role as a substrate to eicosanoids synthesis, its level in insects remains low. Stanley and Kim claim that terrestrial insects minimize the rate of this compound to minimize oxidative stress. Research has shown that insects have high metabolic levels in comparison to other animals, according to the effective tracheal system that efficiently transports oxygen to metabolically active tissues such as flight muscles. However, this may make insects susceptible to reactive oxygen species (ROS) produced by highly oxidative catabolism. Long-chain PUFAs, like arachidonic acid, mostly react with ROS and undergo peroxidation. Such peroxidized PUFAs are known to cause various cellular damage (Stanley and Kim 2020).

The antimicrobial activity of free FA (FFA) in insects has been described widely in the literature. As mentioned above, the biological action of each FFA is determined by its structure, in which the length of the carbon chain, the presence, number, orientation, and position of double bonds play a decisive role. The presence of a hydroxyl group in the FFA molecule predisposes to antibacterial

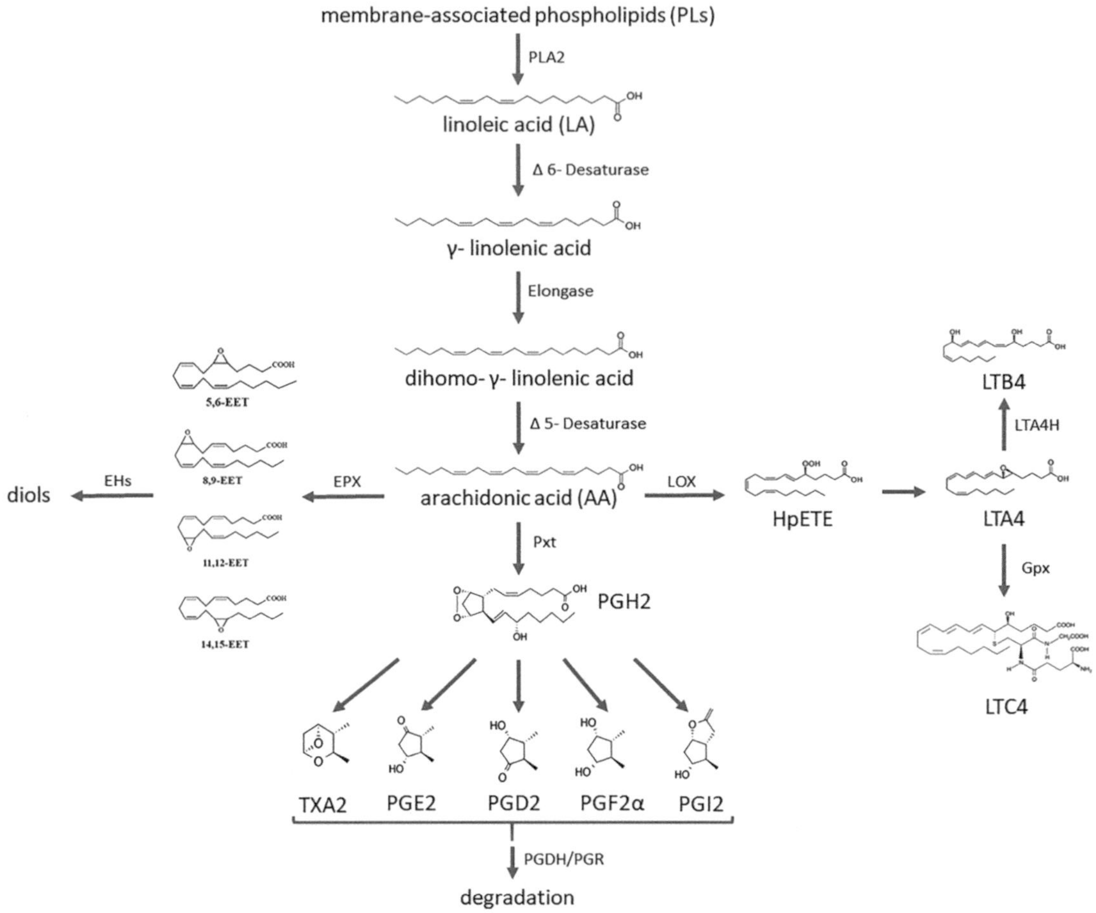

Fig. 2 Eicosanoid biosynthesis and degradation in insects. Phospholipase A2 (PLA2) is a catalysator for hydrolysis of linoleic acid (LA) from membrane-associated phospholipids (PLs). Δ 6-desaturase inserts a double bond in LA converting these fatty acids into γ-linolenic acid. After adding a two-carbon unit by elongase and an additional double bond by the action of Δ 5-desaturase, arachidonic acid (AA) is synthesized. Next, AA is oxygenated by epoxidase (EPX) into epoxyeicosatrienoic acid (EET), lipoxygenase (LOX) into leukotriene (LT), or cyclooxygenase-like peroxynectin (Pxt) to prostaglandin (PG). The EETs are reduced by soluble epoxide hydrolase (sEH). LTA4 is transformed from 5-hydroxyperoxide eicosatetraenoic acid (HpETE) and transformed into LTB4 by LTA4 hydrolase (LTA4H) or into LTC4 by glutathione peroxidase (Gpx). Lastly, several PGs are synthesized from PGH2 by cell-specific enzymes, thromboxane A2 (TXA2) synthase (TXAS), PGD2 synthase (PGDS), PGE2 synthase (PGES), and PGI2 synthase (PGIS); these PGs are reduced by PG dehydrogenase (PGDH) and PG reductase (based on (Kim and Stanley 2021); structural formulas of chemical compounds from PubChem (National Center for Biotechnology Information 2023))

activity, and unsaturated FFAs show higher antimicrobial activity than saturated ones with the same carbon chain length (Zheng et al. 2005). The interaction between FA and bacterial cell membrane seems to be the main mechanism of the antibacterial activities of FAs, and the bacterial cell membrane might be the primary target of lipid compounds (Batalha et al. 2020).

A mix of FFA extracted from biting midge *Forcipomyia nigra* (C14:0, C16:1, C16:0, C18:2, C18:1, and C18:0 FAs) have been shown to have strong antibacterial activity, especially toward *Bacillus cereus* and *Enterococcus faecalis*; however, the highest antibacterial activity was detected after application of C9:0, C10:0, and C16:1 (Urbanek et al. 2012). The FA extract

Table 3 Examples of the influence of selected eicosanoids on the mechanisms of the immune response of insects

Eicosanoids	Insects	Effect on immunity	References
Prostaglandins (PG) in general	*Drosophila*	Rearranging the cytoskeleton in hemocytes by activation of the fascin	Groen et al. (2012)
	Spodoptera exigua	The induction of antimicrobial ROS production by increasing Duox expression	Sajjadian and Kim (2020)
Prostaglandin E2 (PGE2)	*Spodoptera exigua*	Impairment cellular immune responses such as nodulation and phagocytosis	Ahmed and Kim (2019)
		PPO release from oenocytoids	Shrestha and Kim (2008) and Shrestha et al. (2015)
		Suppressed hemocyte-spreading behavior and nodule formation 65–66	Ahmed et al. (2019) and Park et al. (2014)
		Cytoskeletal rearrangement by activate aquaporin (mediated by hemocyte-spreading via cyclic adenosine monophosphate (cAMP)) signal and actin activation via Ca^{2+} signal	Ahmed and Kim (2021)
	Anopheles gambiae	Attraction hemocytes to the basal surfaces of midgut cells Activation of systemic cellular reactions to *Plasmodium berghei* infection	Barletta et al. (2019)
		Increased expression of two genes (*PPO1* and *PPO3*) crucial in immune response during Plasmodium infection Promotion of oenocytoid lysis	Kwon et al. (2021)
Prostaglandin F2 alpha (PGF2α)	*Pseudaletia unipuncta*	Mediator of insect cellular immunity during bacterial infection	Jurenka et al. (1999)
Prostacyklina (PGI2)	*Spodoptera exigua*	Suppression of nodule formation in response to bacterial challenge	Ahmed et al. (2021)
Thromboxane B2 (TXB2)	*Spodoptera exigua*	Increased hemocyte aggregation induced by the fungal infection via a Ca^{2+} signal through a PG receptor	Roy et al. (2021)
Epoxyoctadecamonoenoic acids (EpOMEs)	*Spodoptera exigua*	Suppression of the expression of antimicrobial peptide (AMP) genes Nodule formation interfered with hemocyte-spreading behavior Participant in facilitating anti-inflammatory responses	Vatanparast et al. (2020)
12-(3-adamantan-1-yl-ureido) dodecanoic acid (AUDA)	*Spodoptera exigua*	Suppressing immune responses	Vatanparast et al. (2020)
5(S)-hydroperoxyeicosa-6E,8Z,11Z,14Z-tetraenoic acid (5-HpETE)	*Manduca sexta*	The ability of reversing the nodulation inhibition caused by dexamethasone treatment	Lord et al. (2002)

of *Lucilla sericata* larvae has shown effective antibacterial activity against *Staphylococcus aureus* and *Streptococcus pneumoniae*. In addition, this FFA mixture has a remarked anti-biofilm action against these two bacteria (Liu et al. 2019). Cuticular FFA extract from the South American palm weevil *Rhynchophorus palmarum* demonstrated antimicrobial activity with minimum inhibitory concentrations (MIC) between 1.5 and 20 μg/ml against some species of Gram-positive and -negative bacteria and fungi (*Candida albicans, C. tropicalis, Enterococcus faecalis, E. coli, Klebsiella pneumonia, Pseudomonas aeruginosa, Staphylococcus epidermidis*) (Batalha et al. 2020).

The antifungal activity of cuticular FAs and FA methyl esters from *Calliphora vomitoria*, *C. vicina*, *Sarcophaga carnaria*, and *L. sericata* was detected against entomopathogenic fungi (Liu et al. 2019; Gołębiowski et al. 2012a, 2013a, 2013b, 2014a). The FFAs C14:0, 16:0, 16:1, 18:0, 18:1, 18:2, 18:3, 20:0, and 20:1 from *G. mellonella* show antifungal activity against *Conidiobolus coronatus* after in vitro treatment, causing the decrease of sporulation and hyphal biomass, reduction of the fungal infection potential against the wax moth larvae and decreased the toxicity of the metabolites released into the culture medium, without killing the fungus (Boguś et al. 2010).

FAs are also considered promising insecticides. The anti-insect activity of FAs is well-described in literature, with examples of action against aphids, mosquitoes, whitefly, caterpillars, and cockroaches (Sims et al. 2014; Mallavadhani et al. 2003; Javed 2002; Ribeiro-Neto et al. 2017). Octanoic acid (C8:0) efficiently and rapidly kills *G. mellonella* and *L. sericata* (Kaczmarek et al. 2022; Kaczmarek 2018). The application of C8:0 on the surface of exoskeleton of wax moth, in amounts corresponding to their LD50 and LD100 doses, had a significant effect on the two essential defense barriers involved in protecting the insect against pathogens, viz. the cuticular barrier and immune system; treatment caused important variations of concentration and occurrence of particular cuticular FFAs in both development stages and have destructive effect on insect immunocompetent cells. Cultures of *G. mellonella* immunocompetence cells after in vitro administration of C8:0 or collected from larvae treated in vivo with examined fatty acid have shown many changes in hemocyte morphology and physiology, like cells deformation, disordered cross-linking, hemocyte death (apoptosis and necrosis pathways), stimulation of caspases activity, and increase of the oxidative DNA damage marker (8-OHdG) concentration. Mentioned research also indicated that C8:0 has a cytotoxic effect on the *Spodoptera frugiperda* cell line (Sf9), confirmed by using WST-1 and LDH (lactate dehydrogenase) tests (Kaczmarek et al. 2022).

FA esters are another important lipid group in the fatty acyl category, including wax monoesters and diesters and the lactones (Fahy et al. 2005). The FA ester might be divided also into fatty acyl thioester-CoA derivatives, fatty acyl thioester-acyl carrier protein (ACP) derivatives, fatty acyl carnitines (esters of carnitine), and fatty adenylates, which are mixed anhydrides; there fatty compounds are significant biochemical intermediates (Donato et al. 2013). FA esters, such as methyl palmitate and methyl linolenate, play function of pheromones for the honeybee (Le Conte et al. 1990). Carnitine esters also play an important function in carbohydrate utilization in *Phormia regina* flight muscles while flying (Childress et al. 1967). Moreover, these compounds take part in FA transport through the outer and inner mitochondrial membranes in muscles during flight, e.g., in *Locusta migratoria* (Haunerland 1997). Adenylate compounds also form part of the luciferase enzyme, which regulates the oxidation of firefly luciferin with oxygen to release light in insects from the *Lampyridae* family (fireflies) (Inouye 2010). They also control mitochondrial glycerol 3-phosphate oxidation through allosteric controlling of cytochrome *c* oxidase (COX) activity in *Aedes aegypti* flight muscles (Gaviraghi et al. 2019).

The fatty alcohols and fatty aldehydes are recognized by terminal hydroxy and oxo groups, respectively (Snyder 1996). The fatty alcohols act as insect sex pheromones (Petkevicius et al. 2020; Hambalko et al. 2021). They form a significant part of the epicuticular lipids, and have been found in honeybees (Donzé et al. 1998), the kissing bug *Triatoma infestans* (Cocchiararo-Bastias et al. 2011), grasshoppers *Melanoplus sanguinipes* and *M. packardii* (Soliday et al. 1974), *Bemisia argentifolii* nymphs (Buckner et al. 1999), the tobacco budworm *Heliothis virescens* pupae and *Aleyrodes singularis* exuviae (Nelson et al. 1998; Buckner et al. 1996), and *C. vomitoria* (Gołębiowski et al. 2013b), among others. In addition, the alcohol mixtures present in the cuticular lipids of *Musca domestica* have demonstrated antibacterial and antifungal activities (against *Aspergillus niger, B. subtilis, C. albicans, C. lipolytica, C. tropicalis, E. coli, K. pneumoniae, Proteus vulgaris, Rhodococcus equi, S. aureus*) (Gołębiowski et al. 2012b).

The fatty alcohols detected in the cuticles of fly species resistant to *C. coronatus* (Gołębiowski et al. 2013b, 2012b, 2014b, 2015; Boguś et al. 2017) have been found to have diverse effects on *C. coronatus* growth, sporulation, and infection potential against the larvae of wax moth *G. mellonella* (Włóka et al. 2022). Also, non-FA cuticle lipids from four species of fly (*C. vicina, C. vomitoria, M. domestica*, and *S. carnaria*) have been found to demonstrate antifungal activity. Butyl stearate, 1-oleoyl glycerol, 9-tricosene, and dimethyl suberate demonstrated inhibitory effects against *Beauveria bassiana* (Dv-1/07), *B. bassiana* (Tve-N39), *Lecanicillium lecanii, Metarhizium anisopliae, Paecilomyces fumosoroseus*, and *P. lilacinus* (Gołębiowski et al. 2014b).

The fatty amides, including *N*-fatty acylated amines and unsubstituted amides, as well as various simple amides, also have fascinating biological activities in numerous animals (Farrell et al. 2012; Citoler et al. 2019). *N*-fatty acyl amides have been identified in *D. melanogaster, B. mori, A. mellifera*, and *Tribolium castaneum* (Tortoriello et al. 2013; Kamleh et al. 2009; Ni et al. 2021; Anderson et al. 2018; Jeffries et al. 2014; Dempsey et al. 2014). N-(17-hydroxylinolenoyl)-l-glutamine (volicitin) is an elicitor of plant volatiles that is also produced in insects (Alborn et al. 1997), and decylamine and dodecylamine, two fatty amines from *Monomorium minimum*, have demonstrated toxicity against *Solenopsis invicta* workers (Wang and Chen 2015).

Hydrocarbons are classed as FA derivatives as they correspond to six products formed by electron reduction of FAs; they are produced by the loss of the carboxylic acid from an FA or fatty acyl moiety. Long-chain hydrocarbons such as n-alkanes, n-alkenes, terminally branched monomethylalkanes, internally branched monomethylalkanes, dimethylalkanes, and trimethylalkanes are typical components of lipids covering the cuticle of many insects (Blomquist and Jackson 1979). The variety of hydrocarbons on insect cuticles might be responsible for chemical communication, where relatively non-volatile signals are obligatory (Howard and Blomquist 2005). This group of compounds plays also the function of insect pheromones to help in chemical composition and to facilitate reproduction, working as sex attractant, kairomones, species and sex recognition cues, nestmate recognition factors, dominance and fertility cues, chemical mimicry, primer pheromones, and even maternal cues for the offspring (Blomquist and Bagnères 2010; Fan et al. 2013; Van Oystaeyen et al. 2014; Le Conte and Hefetz 2008).

Insect cuticular lipids keep the aquatic stability and avoid drying, and play the function of signaling molecules in mate recognition and chemical communication as well; they also serve as a barrier against microbial or chemical penetration (Juárez 2006). In many insects, cuticular lipid components are hydrocarbons such as straight-chain, methyl-branched, and unsaturated components with 21–40 and more carbons (Blomquist and Jackson 1979). The surface of the lipid deposit has been known to show spatial heterogeneities such as tangled filaments, blooms, particles or tubular-shaped curls, islands or pillars (Nelson et al. 2000; Vrkoslav et al. 2010; Nguyen et al. 2013).

Several factors might be crucial in the spatial distribution of particular hydrocarbons in cuticle, like for example their physical activity (Howard and Blomquist 2005; Gibbs and Crowe 1991). For example, in the outer layer mostly the lipids with high value of melting point are detected, while in the lower layers the lipids with low melting point are predominant. The desert drosophilid, *D. mojavensis*, produces a higher than average proportion of longer-chain hydrocarbons and it is much more desiccation resistant than other fruit fly species (Matzkin et al. 2009; Etges and Jackson 2001). Moreover, in fruit flies, designated for desiccation resistance research, longer-chain hydrocarbons predominate in the contrary to control flies (Kwan and Rundle 2010). The level of cuticular might differ substantially. For instance, lipids of tenebrionid beetles, which live in deserts, are responsible for water loss reduction (Hadley 1994). Vast concentrations of lipids can also play a protective role against heat load by reflecting solar radiation (Gibbs and Rajpurohit 2010) or discouraging predators (Eigenbrode and Espelie 1995); hence, it might suggest that lipid blooms play key function in water maintenance. Cuticular lipids of grasshoppers (*Melanopus sanguinipes*) populations from lower latitudes have higher melting points resulting in slower water loss (Rourke 2000). Summer-collected tenebrionid beetles are characterized by minor water loss than insects collected in autumn and winter (Jackson et al. 1980). Analogues changes have been observed in the laboratory condition during insect acclimatization to high temperatures, which might confirm that seasonal temperature fluctuations in the environment may cause lipid variability and changes in water loss percentage as well (Gibbs and Rajpurohit 2010).

Lipids also influence the level of transpiration through the insect cuticle; when the temperature rises, the lipid layer indicates a rapid growth in permeability to water. This serious temperature differs between different insect species and even development stages as well; for example, while the impermeable film of lipid is wholly superficial in the newly formed puparium of *Calliphora*, like in other insects, the main waterproof layer is on the surface of the true pupa after pupation. The life-threatening temperature of the pupa is much higher than that of the puparium (Wigglesworth 1985, 1945).

The processes like diapause and aestivation require other water-proofing barriers (Benoit 2010). The two most common examples involve insects that go dormant during the egg or pupa stage. The better waterproof activity was observed in first-instar larvae within the chorion of the egg; the presence of an additional layer (the egg chorion and the larval integument) prevents the dehydration of insects. Such enormous resistance to loss of water has been shown in many insect eggs (Roberts 2004; Yoder et al. 2004) and is particularly prominent in diapausing mosquitoes (Sota and Mogi 1992). A similar situation has been detected in fly pupae, where the puparium plays a similar function as the egg chorion; it creates an extra layer that helps to suppress water loss. Additionally, the higher water resistance might be supported by the presence of additional lipids inside the puparium or chorion (Yoder and Denlinger 1992).

3 Glycerolipids

Glycerolipids make up the abundant family of lipids and have numerous cellular functions ranging from energy supply, to structural compartmentation of cells and organelles, to important signaling activities that are vital for regulating cell growth and development (Voelker 2013).

All eukaryotic glycerolipids are based on a glycerol backbone (Fig. 3a) (Cockcroft 2021). Glycerol is a symmetrical compound, and its two terminal $-CH_2OH$ groups (referred to as α and α') are stereochemically equivalent. The fundamental carbon C2, which is not asymmetric, is referred to as β. Each hydroxyl group of glycerol can react with the carboxylic group of an FA, leading to an ester derivative called glyceride, esters being formed by the condensation of a carboxylic acid and an alcohol (Fantini and Yahi 2015). Examples of glycerolipids are presented in Table 4 and Fig. 3.

In triacylglycerol (TAG), each of the hydroxyls of the glycerol backbone is exchanged by an FA esterified to the sn-3 location of glycerol. Similarly, structural differences take place when glycerol contains two FA (diacylglycerols; DAGs) or one FA (monoacylglycerol; MAG). The structures of glycerolipids confer significant physical properties upon the particles. The FA substituents comprise quite long hydrocarbon chains (typically 16–22 carbons, with 0–6 double bonds), which are non-polar and sequestered away from aqueous environments.

The primary lipid components of the diet are typically TAGs (Holtof et al. 2019). Lipids are usually digested and absorbed in the midgut, and digestion is believed to take place using numerous enzymes, such as TAG lipases, phospholipases, and cholesterol esterases, depending on the species (Abou El Asrar et al. 2020; Weidlich et al. 2015). The main products

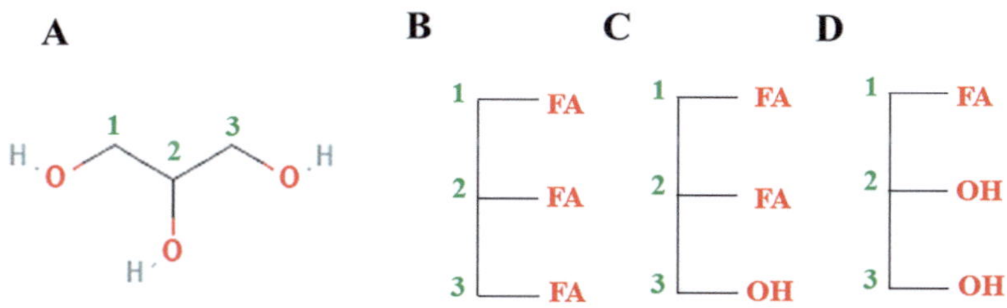

Fig. 3 The structures of glycerol (**a**), triacylglycerol (TAG; **b**), diacylglycerol (DAG, **c**), and monoacylglycerol (MAG; **d**) (structure based on PubChem (National Center for Biotechnology Information 2023)). *FA* fatty acid

Table 4 Classification of glycolipids with examples in insects

	Definition	Function	Examples in insects	Literature data
Triacylglycerols (TAGs)	Tri-esters of a glycerol skeleton bounding with three fatty acid molecules.	Energy supply for insect development, survival, and reproduction	*Rhodnius prolixus*	Pontes et al. (2008)
			Hermetia illucens	Matthäus et al. (2019)
			Panstrongylus megistus	Canavoso et al. (2004)
			Culex pipiens	Sawabe and Moribayashi (2000)
			Aedes aegypti	Silva et al. (2021)
			Ochlerotatus atropalpus	Telang and Wells (2004)
			Bombus lucorum	Kofroňová et al. (2009)
		The main lipid transported by lipophorin	*Aedes aegypti, Chaoborus maximus/minimus, Simulans vittatum*	Pennington and Wells (2002)
Diglycerides (DAGs)	Glyceride consisting of two fatty acid chains covalently bonded to a glycerol molecule through ester linkages	The main lipid transported by lipophorin	*Locusta migratoria*	Tiezt (1967)
			Rhodnius prolixus	Grillo et al. (2007)
			Hermetia illucens	Ushakova et al. (2016)
			Dysdercus koenigii	Mishra et al. (2004)
			Manduca sexta	Canavoso and Wells (2000)
Monoglyceride (MAGs)	Glycerol linked to a fatty acid via an ester bond.	Antimicrobial activity	*Hermetia illucens*	Borrelli et al. (2021)
		Synthesis of higher glycerides	*Galleria mellonella, Exeristes roborator*	Barlow and Jones (1981)
		Hydrolytic products of triglycerides	*Dysdercus koenigii*	Barlow and Jones (1981)

generated by this digestion, such as FFAs, non-esterified cholesterol and lysophospholipids, are next absorbed by the intestinal epithelial cells (Behmer and Vogel 2020; Huang et al. 2015), and these pre-digested lipid particles are used as substrates for the production of more complex structures that are transported to the other organs; FFAs are used for the synthesis of DAGs, phospholipids, and TAGs (Gondim et al. 2018). Lipid transport from enterocytes in insects varies from that observed in vertebrates in two significant ways: DAG is the key lipid transported and DAG release does not include the production of a lipoprotein element in the midgut, but rather the DAG is released straight to the existing lipoprotein (Lp) transporter in the hemolymph (Heier and Kühnlein 2018; Rodenburg and Van Der Horst 2005; Van Der Horst and Rodenburg

2010). Lp may carry lipids to the appropriate tissue and cells with higher energy demands, such as the brain, oocytes, oenocytes, and imaginal discs; but the majority of the transported DAG is directed to the primary energy store tissue, the fat body.

Lps are classified based on their density, viz. low-density Lp (LDLp), high-density Lp (HDLp), and very high-density Lp (vHDLp). New Lp is synthesized as an underloaded HDLp and is released from the fat body. When it reaches the midgut, it acquires fat from food. When DAG is loaded into the HDLps, it changes into an LDLp. Following this, the Lp moves back to the fat body, where it releases DAG, and then returns to the hemolymph as an HDLp (Wu et al. 2014; Pontes et al. 2002; Grillo et al. 2003).

Lp activity is controlled by the AKH (Van Der Horst 2003). They could transport particular lipids in two diverse manners dependent on the insect species. In one approach, it may be possible that the Lp is related to particular membrane receptors on the cells of adipocytes; the lipid transfer particle (LTP) may also take part in this transfer. Otherwise, Lp can be endocytosed by the adipose tissue (Rodríguez-Vázquez et al. 2015; Canavoso and Wells 2001; Matsuo et al. 2019).

The vast majority (over 90%) of the lipids accumulated in adipocyte cells are TAGs, which are encapsulated in cytoplasmic lipid droplets (LDs) surrounded by a hydrophilic phospholipid's monolayer and linked proteins (Wrońska et al. 2023). LDs are formed in the endoplasmic reticulum (ER) (Olarte et al. 2020). According to genome-wide RNAi screen in *Drosophila* S2 cells, approximately 1.5% of all genes were found to take part in the formation and regulation of LD processes (Guo et al. 2008). Fat body-specific knockdown of genes encoding adenosine diphosphate ribosylation factor/coat protein I (ARF/COPI) leads to large and diffuse LDs (Baumbach et al. 2014a). Correspondingly, the silencing of genes responsible for protein taking part in phospholipids production might be correlated with bigger LD creation (Guo et al. 2008). The transmembrane proteins Pex3, Pex13, and Pex14 (peroxisome complex, Peroxin, Pex) have been observed to surround newly formed LDs in *Drosophila*, and the lower level of these proteins in the fat body caused characteristic changes in larval fat body development and affected the level of surviving insect reared on lipid-rich or lipid-poor diet (Ueda et al. 2022). Reducing hnRNP-K (heterogeneous nuclear ribonucleoprotein) activity caused the lover TAG accumulation, while reducing Hrb27C and Hrb98DE led to improved triacylglycerol accumulation (Bhogal et al. 2020). However, other authors have pointed out the role of transportin-serine/arginine (Tnpo-SR) rich protein, suggesting its pivotal role in the appropriate accumulation of lipids compounds in the fat body of fruit flies (Nagle et al. 2022). The loss of tachykinin (Tk), neuropeptides that cause fast contraction of intestinal tissue and muscles, causes an increase in TAG concentration in the midgut and fat body as TKs act as negative regulators of lipid accumulation (Chowański et al. 2021).

Another important factor in insect lipogenesis is calcium signaling, as impaired SERCA (sarcoendoplasmic reticulum (SR) calcium transport ATPase) activity, which caused the reduced TAG accumulation in the fat body cells and LD size in fruit fly (Baumbach et al. 2014a, 2014b). Fat storage-inducing transmembrane proteins 1 and 2 (FIT1/FITM1 and FIT2/FITM2) are part of the unique family of evolutionarily conserved proteins localized to the ER that participate in the partitioning of cellular TAG into LDs, but not TAG biosynthesis (Gross et al. 2010). Thiel et al. have shown the presence of a new LD-associated protein family, represented by the fruit fly protein CG9186 which are lipid storage regulators (Thiel et al. 2013).

TAGs are collected to develop hunger resistance. Mild hunger may cause the storage of fat, depending on the insect species and its level of feeding (Lorenz 2001). As increased fat mass may enhance starvation resistance, it is possible that a correlation exists between lipid levels and resistance to food shortage (Parkash et al. 2014).

It is essential that insects are able to survive starvation, and the ratio of TAG to lipids in the body is crucial during this process (Renault et al. 2002; Ballard et al. 2008; Laparie et al. 2012). Different species of insects have been shown to express various lipid reactions to a lack of nutrition. For example, in *L. migratoria* the threefold increase in lipid concentration caused by the elevation of diglycerides was detected in locust hemolymph during starvation (Th et al. 1985), while in *M. sexta* only a twofold increase was observed (Ziegler 1991). The higher concentration of lipids has been shown also in in *B. mori* (Satake et al. 2000) under laboratory starvation conditions, while in *Pachnoda sinuata* the lipid concentration in the flight muscles and the adipocytes is meaningly reduced (Auerswald and Gäde 2000). In *Merizodus soledadinus*, the hydrolysis of TAGs significantly increases during food shortage and comes back to normal stages after the meal (Laparie et al. 2012). Lipolysis, the main biochemical process taking place during starvation, is regulated by perilipin, a lipid droplet-associated protein, which increases the lipid level during starvation and lowers it upon food intake. Perilipin is capable of both inhibiting and stimulating the hydrolysis and phosphorylation of TAGs; it can also stimulate phosphorylation and facilitate the translocation of TAGs from the cell cytosol to the surface of the fat droplets, where TAG are hydrolyzed (Skowronek et al. 2021).

The fat body plays an important storage role in holometabolous insects. Throughout the larval feeding stages, energy supplies are collected for use during metamorphosis and as storage used for the new adult, as insects need to supply at least a minimal level of nutrient accumulation to survive through metamorphosis (Arrese and Soulages 2010): in the southern house mosquito, *Culex quinquefasciatus*, lipids provide about 90% of the energy required for the development of the embryo (Wang et al. 2017). Research has shown that LDs extracted from whole *M. sexta* ovary contain great amounts of apoLp-I and apoLp-II. Since Lp transferences mostly DAG and because TAG is the main lipid accumulated in the LD in insect oocytes, the attachment of apolipophorins to the LD may facilitate the transition of DAG to TAG and/or the subsequent transfer of TAG to the LD base (Soulages et al. 2012).

TAG storage was also observed in the fat body of sugar-fed females from the genus *Aedes* and *Anopheles stephensis* (Sawabe and Moribayashi 2000; Ziegler and Ibrahim 2001). In blood-feeding the yellow fever mosquito, *Ae. aegypti*, the TAG storage in the adipocytes is limited, because of transport of lipid compounds to the growing oocytes (Ziegler and Ibrahim 2001). But, in *C. molestus*, TAG supplies are reduced even when females are rearing with sugar after emergence. This happens because the *Cx. molestus* is an example of autogenous, the oogenesis can start even when there is no blood meal (Sawabe and Moribayashi 2000). Hence, the larval diet is a key factor to ensuring the reproductive success of autogenous mosquitoes; in the mosquito *Ocheleratatus atropalpus*, newly emerged females fed a nutrient-rich diet during larval development have three times more TAGs accumulated in the fat body compared to mosquitoes reared on a poor diet (Telang and Wells 2004). In adults *An. gambiae* and *An. stephensi*, feeding a rich diet raises TAG amounts as well (Takken et al. 2013).

TAGs are also major storage lipids in insects preparing to enter to diapause process. During this process insects, change their lipid levels due to various processes, including the production of energy or cell membrane modifications. Insects collect lipids before diapause, and a failure to store suitable amounts of lipids causes incomplete diapause and probably death (Toprak et al. 2014). Lipids are commonly used for the storage of energy or nutrition during overwintering. The amount of lipids decreases during the diapause, indicating that it is the key source of nutrition (Hahn and Denlinger 2011). Stored lipids are converted to FFAs and used by mitochondria as an energy source (Zhang et al. 2019a).

After diapause, lipid storage is used for egg production (Zhou and Miesfeld 2009). Insulin/Foxo signaling is an important part of the pathway regulating primary diapause energetics in *C. pipiens*. The insulin signal is not triggered in the initial steps of diapause, thus blocking the suppression of the fork-head of the transcriptional factor (Foxo). The deactivation of Foxo causes the decrease of nutrient reserves, which proposes that the downstream genes targeted by Foxo are vital for glycogen and fat homeostasis in diapausing females.

Glycogen synthase is a key enzyme playing a crucial role in the regulation of glycogen and lipid metabolism during diapause in overwintering insects, responsible for overwintering surviving in *C. pipiens*. This enzyme changes glucose residues into a polymeric chain for packing as glycogen. Research has shown that *C. pipiens* females with glycogen synthase knockdown fed with 1% glucose indicate a 46% decrease in glycogen level and a 6% decrease in lipid accumulation. Moreover, suppression of glycogen synthase intensely decreases the viability of diapausing *C. pipiens* (by 88% vs control) (Olademehin et al. 2020; King et al. 2020).

Literature data has shown the overexpression of FA synthase during eggs production in diapause in adult females *A. albopictus* fed in blood compared to control one. Moreover, a comparison of the amount of lipids in diapausing and nondiapausing eggs showed 30% more lipids in diapausing larvae than in non-diapausing one (Reynolds et al. 2012). This is related to larger maternal lipid delivery to diapausing eggs (Huang et al. 2015). Furthermore, *A. albopictus* eggs demonstrated increased protein content in lipid storage droplets during diapause preparation and differential expression of twenty-three lipid metabolism-related genes during early diapause maintenance (Poelchau et al. 2013a, b). The increase in lipid abundance and changes in its saturation observed in *A. albopictus* during diapause has been attributed to the accumulation of DAGs and TAGs. An increased amount of unsaturated FA encourages lipid motion and membrane fluidity at cold, contributing to low-temperature tolerance (Clark and Worland 2008; Hazel 1995). In *A. albopictus*, increased expression of three genes encoding FA desaturases was demonstrated in adult females fed blood under diapause-inducing conditions, supporting increased maternal transfer of unsaturated FAs (Huang et al. 2015). Additionally, it is known that the gene encoding Δ(9)-desaturase, an enzyme that limits the rate of unsaturated FA synthesis, is significantly increased in expression during preparation for diapause (Reynolds et al. 2012). In this species, diapause moderately expands cold tolerance, but this effect is greatly enhanced by a period of prolonged exposure to sublethal low temperatures (Hanson and Craig Jnr 1994). Therefore, a more likely hypothesis is that in *A. albopictus*, fluctuations in lipid unsaturation may happen as a result of cold exposure rather than diapause itself (Batz and Armbruster 2018).

4 Glycerophospholipids

Phospholipids (PLs) are amphiphilic molecules built of a polar head group, two FAs, a phosphate moiety, and two non-polar acyl chains, and the PLs which contain an sn-glycerol-3-phosphate as a backbone are the glycerophospholipids (Ochiai and Komiya 2021). The sn-glycerol-3-phosphate molecule, ester of dibasic phosphoric acid (HOPO32-) and glycerol, is asymmetric and optically active as well thanks to the chemical structure where the OH group is presented on the left side of the molecule and absolute stereo configuration is 2-R-glycerol-3-phosphate (in Fischer projection (Voelker 2013). The structure of sn-glycerol-3-phosphate is presented in Fig. 4a.

The addition of different headgroups (choline, ethanolamine, serine, inositol, and glycerol) to the phosphate of PA can generate a diverse range of phospholipids. These can be sectioned into different classes based on the nature of the hydrophilic "head group" (Fahy et al. 2005). The whole phospholipids that are made from these compressions are phosphatidylcholine (PC), phosphatidylethanolamine (PE), phosphatidylserine (PS), phosphatidylinositol (PI), and phosphatidylglycerol (PGl). The last hydroxyl group of PG1 can equally form a phosphodiester with other phosphatidic acid molecules to generate diphosphatidylglycerol, a lipid known as *cardiolipin*. Additionally, these PLs can have different acyl chain compositions (Voelker 2013). Literature data described the occurrence of CDP-diacylglycerol, an additional glycerophospholipid linked glycan, found in many organisms as biosynthetic intermediate for phospholipid production, while phosphatidyl glucose, and the "glycosylphosphatidylinositol anchor" (GPI) is used to attach some proteins to membranes (Kenwood and Merrill 2016). Some examples of these compounds are presented in Table 5 and Fig. 4.

The inositol headgroup is built by three hydroxyls at the 3, 4, and 5 positions that can be phosphorylated independently and in combination,

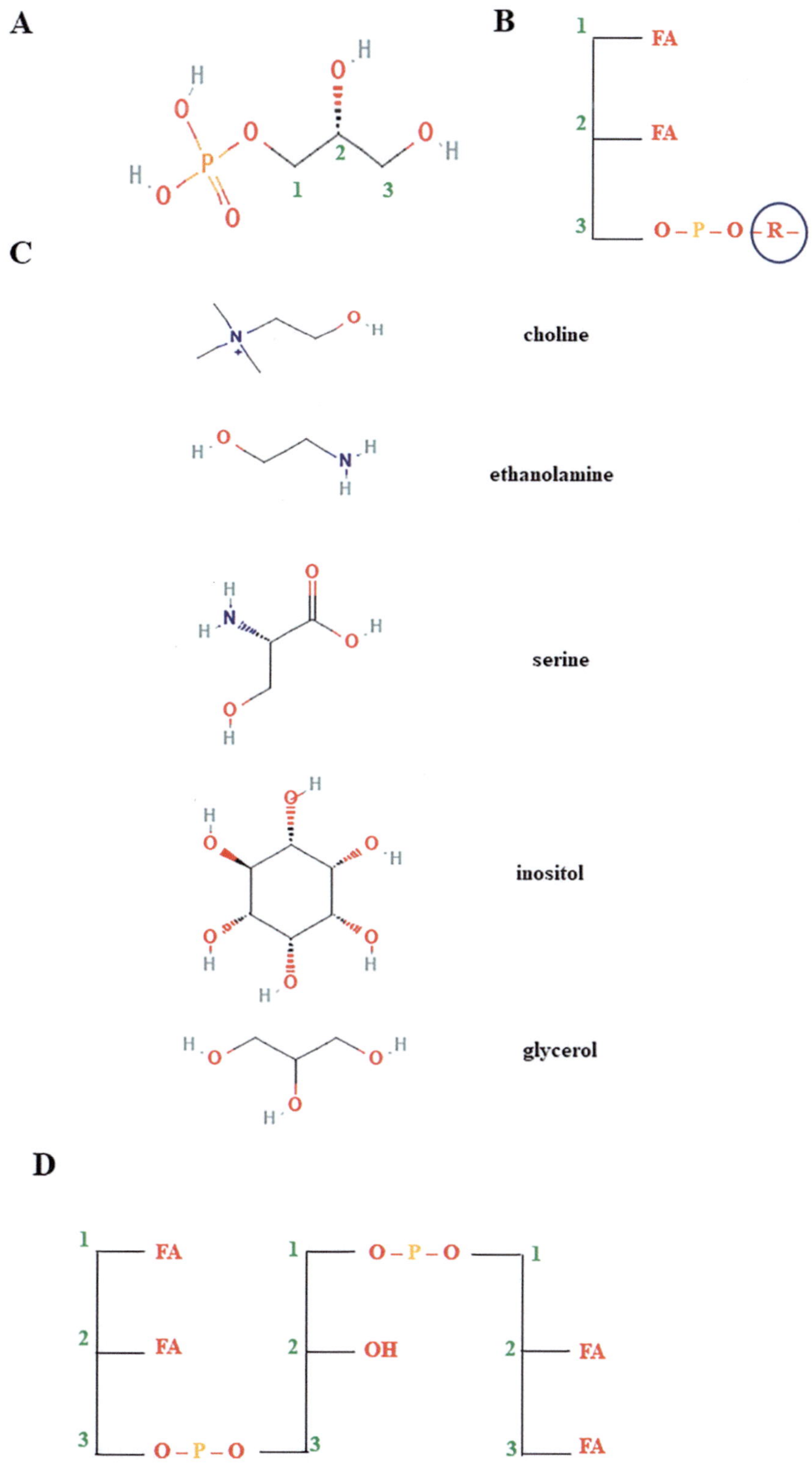

Fig. 4 The structure of sn-glycerol-3-phosphate (**a**), the general scheme of phospholipids (**b**), the structure of headgroups (**c**), and the general scheme of cardiolipin (**d**) (based on Pubchem (National Center for Biotechnology Information 2023)). *FA* fatty acid, *R* headgroup

Table 5 Classification of glycerophospholipids with examples in insects

	Definition	Function	Examples in insects	Literature data
Phosphatidylcholine (PE)	Phospholipids with choline as a headgroup	Major component of biological membranes	*Diamesa tonsa, Pseudodiamesa branickii*	Trenti et al. (2022)
			Sf9 cells line	Parathath et al. (2004)
			Bombyx mori	Nakahara et al. (2002)
			Phormia regina	Shelley and Hodgson (1970)
Phosphatidylserine (PS)	Glycerophospholipid—with serine attached through a phosphodiester linkage to the third carbon of the glycerol	Mediate phagocytosis	*Drosophila*	(Ji et al. 2022)
		Component of biological membranes	*Leptinotarsa decemlineata, Agelastica alni, Chrysolina polita*	Lehmann et al. (2020))
			Cydia pomonella	Forte et al. (2002)
Phosphatidylinositol (PI)	Glycerophospholipid—with inositol attached through a phosphodiester linkage to the third carbon of the glycerol	An ice nucleator site	*Tipula trivittata*	Neven et al. (1989)
		Component of biological membranes	*Pieris brassicae*	Turunen and Pärnänen (1987)
			Insects from the orders *Plecoptera, Epbemerella, Ecdyonurus v. and Corixidae*	Ghioni et al. (1996)
			Drosophila melanogaster	Hofbauer et al. (2021)
			Musca domestica	Hildenbrandt et al. (1971)
Phosphatidylglycerol (PGl)	Glycerophospholipid—with glycerol attached through a phosphodiester linkage to the third carbon of the glycerol	Component of biological membranes	*Drosophila melanogaster*	Colinet et al. (2016) and Ko et al. (2019)
			Pyrrhocoris apterus	Koštál et al. (2013)
			Diamesa tonsa, Pseudodiamesa branickii	Trenti et al. (2022)
			Aedes aegypti	Pinch et al. (2021)
			Hermetia illucens	Li et al. (2021)
Phosphatidylethanolamine (PE)	Glycerophospholipid—with ethylomanine attached through a phosphodiester linkage to the third carbon of the ethanolamine	Part of cells membrane	Sf9 cells line	Dawaliby et al. (2016)
			Drosophila melanogaster	Rawson (2003)
			Callosobruchus maculatus	Ximenes et al. (2008)
			Rhodnius prolixus	Majerowicz et al. (2013)
			Rhynchophorus phoenicis	Fogang Mba et al. (2017)
			Nilaparvata lugens	Wang et al. (2021b)

(continued)

Table 5 (continued)

	Definition	Function	Examples in insects	Literature data
Cardiolipin	Construct from two phosphatidic acid moieties linked with a glycerol backbone in the center to form a dimeric structure	Part of mitochondrial membrane	*Bombyx mori*	Ramos-Martín et al. (2022)
			Drosophila melanogaster	Martelli et al. (2022) and Malhotra et al. (2009)
			Sf9 cells line	Malhotra et al. (2009) and Schlame (2009) and Kumar et al. (2012)
			Rhodnius prolixus	Salerno et al. (2002)
			Musca domestica	Chan (1970)
			Glossina morsitans	D'Costa and Rutesasira (1973)

giving rise to seven different phosphorylated derivatives. Consequently, PI can be changed into PI4P, PI3P, PI5P, phosphatidylinositol 4,5 bisphosphate (PI(4,5)P2), PI(3,5)P2, PI(3,4)P2, and phosphatidylinositol 3,4,5-trisphosphate (PI (3,4,5)P3). Literature data confirmed the presence of various enzymes that regulate these conversions, which take place in diverse membrane sections. The localization of a particular phosphoinositide delivers identity to that compartment and acts as a lipid code (Cockcroft 2021).

Phospholipid diversity is not only caused by different headgroups but also by the nature of the acyl chains of the FAs; palmitic acid and stearic acid are the most common saturated FAs found in many phospholipids (Cockcroft 2021).

The glycerophospholipids represent the main lipid constituents of the membrane and are settled in a tail-to-tail conformation. The nonpolar base of the lipid bilayer of biological membranes is built by the FAs and provides the barrier function of biological membranes (Van Meer et al. 2008; Harayama and Riezman 2018).

In the plasma membranes of eukaryotes, PE and PS are detected in the inner leaflet of the bilayer membrane, while PC particles are predominant in its outer leaflet (Van Meer et al. 2008; Murate et al. 2015). On the contrary to the eucaryotic cells, the *Drosophila* phospholipids are equally spread between the bilayer leaflets of the insect plasma membrane, where phospholipid scramblase (XKR), which disrupts the lipid asymmetry, is constitutively active (Shiomi et al. 2021). In the Sf9 cell line, research indicates minimal lower sterol content and a four-times lower ratio of PE to PC compared to mammalian cells (Dawaliby et al. 2016). The research on two insect cell lines from *S. frugiperda* (Sf9) and *Trichoplusia ni* indicate that PE (diacyltype) and PE as key phospholipids, followed by PS and PI, and only little amount of ethanolamine plasmalogen. These lipid compounds are characterized by high amounts of monoenoic FAs, e.g., 16:1 and 18:1, which contain more than 70% of the total FAs; however, minor concentrations of PUFAs such as 18:2 and 20:4 are detected too. Research data indicated that the ratio cholesterol: phospholipid is lower in insect cells than in mammals (Marheineke et al. 1998; Yeh et al. 1997).

The chemical modifications of phospholipid head groups are a response to rapid cold-hardening (RCH) and might also enhance membrane fluidity (Michaud and Denlinger 2006), and a significant increase in the overall amount of unsaturated FAs might improve survival ability (Overgaard et al. 2005). Moreover, it has been proved that a rise of polyunsaturated linoleic acid and a decline of monounsaturated and saturated phospholipid FA is a common answer to RCH in insect membrane restructuring during RCH; indeed, studies on *D. melanogaster* found that linoleic, not oleic acid, is responsible for improved membrane fluidity (Kaczmarek and Boguś 2021a). Increased amounts of oleic acid have been detected in *Hepialus xiaojinensis* (Overgaard et al. 2005; Zhu et al. 2016). Membrane fluidity may also be influenced by switching the positions of the moieties within glycerophospholipids, and the polar head groups of phospholipids located in the membrane can also be restructured. In addition, the presence of glycerophosphoethanolamines may promote membrane disorders at low temperatures, by enhancing membrane fluidity and increasing the levels of unsaturated and short-chain FFAs (Teets and Denlinger 2013; Teets et al. 2020). It has also been demonstrated that the FA chains located in the membrane shorten during RCH, and this is accompanied by increased unsaturation of membrane FFAs (Overgaard et al. 2006; Colinet et al. 2016).

Similar to RCH, during diapause, the proportion of unsaturated FAs influence on cell membrane fluidity; in *Eurosta solidaginis* the amount of membrane FAs with double bonds rise by 50% during the fall (Bennett et al. 1997). As short-chain FAs have lower melting points, numerous insect species demonstrate a higher proportion of 16-carbon FAs relative to 18-carbon FA (Michaud and Denlinger 2006; Tomčala et al. 2006). Also, the place of the moieties within glycerophospholipids influences membrane fluidity: changing the *sn*-1 position to the *sn*-2 position allows them to penetrate deeper into the lipid bilayer, and this has a stronger influence on membrane fluidity (Overgaard et al. 2008). Moreover, restructuring the polar head groups of membrane phospholipids can also enhance membrane fluidity; glycerophosphoethanolamines, at the expense of glycerophosphocholines, promote membrane disorder during cold (Košťál et al. 2004). For example, in *S. crassipalpis* during diapause the proportion of glycerophosphoethanolamines rise by 16% (Michaud and Denlinger 2006).

5 Sphingolipids

The sphingolipids are a group of membrane lipids which are important structural and signaling biological active particles in organisms (Wang et al. 2021a). They are a complex group of lipid particles which have a similar structural base, a so-called sphingoid base backbone, which is composed of an alkyl chain with one to three hydroxyl groups and one amino group. The backbone is linked to an FA and a polar head group (Heaver et al. 2018). In mammals, C18 sphingoid bases serve as the major backbone, while shorter FA chains which are amide linked to the sphingoid bases predominate in insects (Rietveld et al. 1999); in *D. melanogaster* they are C14 and C16 (Jarne et al. 2018). The illustrations of sphingolipids diversity have been shown in Table 6 and Fig. 5.

The hydroxyl group on the first carbon can be substituted by various phosphosphingolipid headgroups, such as sphingomyelins, ceramide phosphoethanolamines, and ceramide-1-phosphates; they can also be fitted with neutral glycosphingolipids, in which the headgroup contains the uncharged sugars (glucose, galactose, glucosamine, galactosamine, and fructose, for mammals), as well as acidic glycosphingolipids, such as gangliosides (which contain the sialic acid) and sulfatides (sulfated glycans), among others (Kenwood and Merrill 2016). In their simplest forms, sphingosine, phytosphingosine, and dihydrosphingosine are the main core, from which more complexity is reached. For instance, phosphorylation of the C1 hydroxyl group degradation of end products and/or significant signaling molecules, for example: sphingosine-1-phosphate (S1P), phytosphingosine-1-phosphate, and dihydrosphingosine-1-phosphate (Quinville et al. 2021). In *Drosophila*, it has been shown that S1P mutants have indicated higher level of apoptosis, reduced muscle development and integrity, higher mortality, and reduced fecundity during embryogenesis (Herr et al. 2003). These reproductive system abnormalities have been shown to involve cell death, as indicated by increased levels of apoptosis in affected tissues, including the ovaries and testes

Table 6 Classification of sphingolipids with examples in insects

	Definition	Function	Examples in insects	Literature data
Sphingosine (2-amino-4-trans-octadecene-1,3-diol)	Amino alcohol (18-carbon) with an unsaturated hydrocarbon chain	Part of membrane	*Stenocoris furcifera, Nephotettix cincticeps, Laodelphax striatellus*	Zhang et al. (2021)
			Musca domestica	Robb et al. (1972)
			Sf9 cell line	Novak et al. (2020)
		Triggered the accumulation of hydrogen peroxide (H_2O_2)	*Drosophila*	Zhang et al. (2019b)
Sphingosine-1-phosphate	Sphingosine with a phospho group attached at position 1	Regulation of apoptosis during gametogenesis	*Drosophila*	Phan et al. (2007)
Ceramides	Sphingoid base derivatives with an amide-linked fatty acid	Reproduction	*Nilaparvata lugens*	Shi et al. (2018, 2021)
		Part of membrane	*Calliphora vicina*	Helling et al. (1991)
			Manduca sexta	Abeytunga et al. (2004)
			Tenebrio molitor	Tsochatzis et al. (2022)
			Bombyx mori	Itonori et al. (2018)
		Components of the defense gland	*Coptotermes formosanus*	Ohta et al. (2007)
Sphingomyelin	A sphingosine base with an 18-carbon chain and a double bond at position 4, attached to a phosphorylcholine fatty acid	Caste differentiation	*Cardiocondyla obscurior*	Schrader et al. (2015)
		Part of cell membrane compound	Sf21	Bartelsen et al. (1998)
			Stomoxys calcitran	Spates et al. (1990)
			Sf9	Ding et al. (2015)
Ceramide phosphoethanolamine	The amino group of a sphingoid base in amide linkage with one of several fatty acids, while the terminal hydroxy group of the sphingoid base is esterified to phosphorylethanolamine	Compound of membrane	*Drosophila*	Acharya and Acharya (2005) and Bhat et al. (2015)
			Leptinotarsa decemlineata	Panevska et al. (2019)
			Manduca sexta	Abeytunga et al. (2004)
Glycosphingolipids	Carbohydrate-containing derivative of a sphingoid or ceramide	Compound of membrane	*Drosophila*	Rietveld et al. (1999)
			Calliphora vicina	Dennis et al. (1985)
			Lucilia caesar	Itonori et al. (1991)
			Manduca sexta	Abeytunga et al. (2008)
			Bombyx mori	Itonori et al. (2018)

(Phan et al. 2007). Furthermore, loss of function of the sphingosine kinase 2 (Sk2) gene, which encodes one of two *Drosophila* sphingosine kinases whose main function is the conversion of the metabolite sphingosine to S1P, causes a 60% lower egg-laying capacity, characterized by an egg retention phenotype (Phan et al. 2007).

The acylation of sphingosine, phytosphingosine, or dihydrosphingosine with one of acyl CoA fragments through the action of distinct

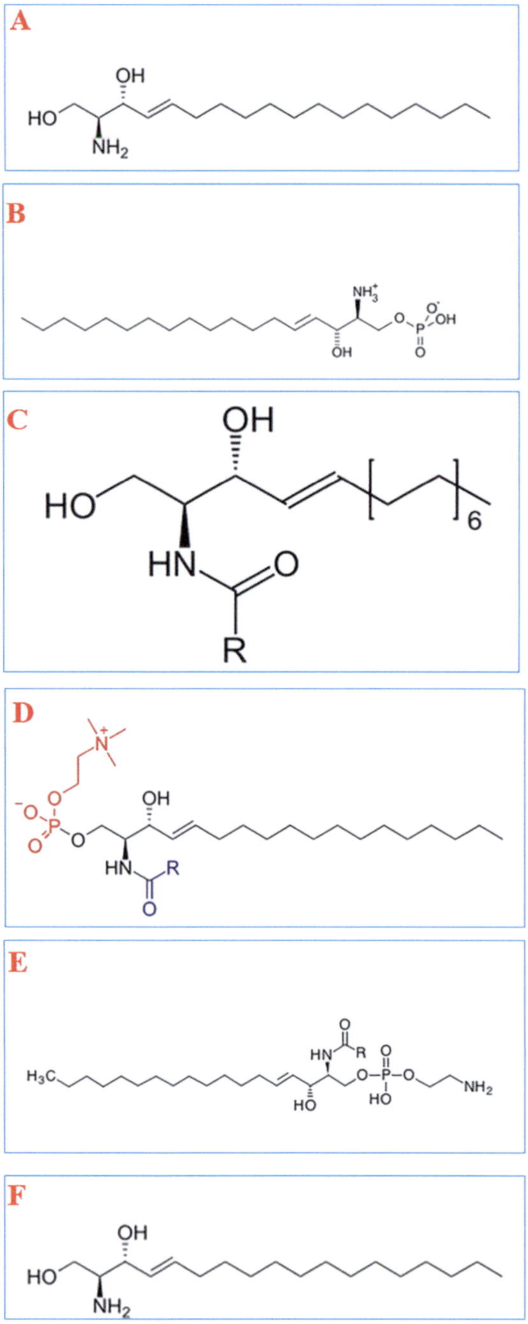

Fig. 5 The structure of main sphingolipids (based on PubChem (National Center for Biotechnology Information 2023)) sphingosine (**a**), sphingosine-1-phosphate (**b**), ceramide (**c**), sphingomyelin (**d**), glycosphingolipid (**e**)

ceramide synthases results in the production of ceramide, phytoceramide, or dihydroceramide. Ceramide is produced by the attachment of an FA through an amino group of sphingosine, and this reaction is catalyzed by ceramidases (CDases, EC 3.5.1.23) (Mao and Obeid 2008). The function of CDases in insects is not very known yet, and has been described in only a few insect species: *D. melanoganster* (Yoshimura et al. 2002), *T. castaneum* (Zhou et al. 2011), *Laodelphax striatellus* (Zhou et al. 2013), and *Nilaparvata lugens* (Shi et al. 2018). Inactivation of all CDases not only extended the development of fruit flies, but also enhanced its anti-oxidative stress activity (Yang et al. 2010). There are some literature data that has shown that CDase can be responsible for maintaining the photoreceptor homeostasis of fruit flies in a cell non-autonomous manner (Acharya et al. 2008). To sum up, these studies proposed the important roles of CDases in cellular signaling and biological processes in insects. Moreover, ceramides themselves play a key role in the development regulation of *N. lugens* reproductive system by controlling ceramide metabolism in the ovaries (Shi et al. 2021).

Ceramide is the lipid substrate used to synthesize sphingomyelin (SM) and glycosphingolipids (GSLs). Addition of the phosphocholine headgroup to the hydroxyl of ceramide makes the phospholipid, SM, the most popular sphingolipid in mammals (Cockcroft 2021). Insects (e.g., *Drosophila* and *Musca*) mostly synthesize ceramide phosphoethanolamine (CPE), not sphingomyelin, as the main sphingolipid in their cell membranes (Masood et al. 2010; Kraut 2011). While some insects (*Ae. aegypti*, *C. quinquefasciatus*, *M. sexta*, and *A. mellifera*) have both CPE and sphingomyelin as their bulk (Abeytunga et al. 2004), *N. lugens* (Shi et al. 2018), and three auchenorrhynchous insects, viz. *Sogatella furcifera*, *Nephotettix cincticeps*, and *L. striatellus*, were found to store sphingomyelin alone (Zhang et al. 2021). Sphingomyelin and ceramides seem to play a significant role in insect viral infections; the upregulation of sphingolipids and ceramides genes was observed in *A. cerana cerana* larvae during Chinese Sacbrood Virus (CSBV) infection (Dang et al. 2021), in *A. aegypti* mosquitoes during dengue virus infection (Chotiwan et al. 2018; Perera et al. 2012), and in most stages of *L. striatellus* during Rice Stripe Virus infection (Jiao et al. 2017). These compounds also play important roles in reproduction. The knocking-down of sphingomyelinases, i.e., enzymes that take part in the transformation of sphingomyelins into ceramides and phosphorylcholine, in *N. lugens* resulted in ovarian malformation (Shi et al. 2022). Literature data also suggest that sphingomyelin may play a role in caste differentiation in the ant (*Cardiocondyla obscurior*) (Schrader et al. 2015).

Glycosphingolipids (GSL) contain various sphingolipid classes with different orders and types of sugar residues attached to their main groups. Research has shown that the main glycosphingolipid base in *C. vicina* is tetradecasphing-4-enine (C14:1) constituting about 54%, and hexadecasphing-4-enine (C16:1) about 23%; and the remaining 23% contains C15 and C17 bases (Helling et al. 1991). Glycosphingolipids may show differences in the composition of their acyl chains, revealing an additional layer of differences because they are synthesized from ceramide precursors. The glycosphingolipids are generally classified into two classes: glycosphingolipids and galactosphingolipids. Glycosphingolipids initially depend on the enzyme glucosylceramide synthase (GCS), which functions to attach glucose as the first residue to the C1 hydroxyl position. On the other hand, galactosphingolipids are formed from galactosylceramide synthase (GalCerS), an enzyme evolutionarily distinct from GCS. Glycosphingolipids are next classified according to their change by several glycosyltransferases, which significantly raises the possible variance in lipid subtypes (Gault et al. 2010).

Insect glycosphingolipids have been known for many years (Wiegandt 1992). The glycosphingolipids in fruit flies are diverse groups that differ from occurring in mammals. The main core disaccharide in insects contains Manβ1–4Glcβ1-Cer, mactosyl-ceramide, but in mammals predominate the Galβ1–4Glcβ1-Cer (Seppo and Tiemeyer 2000). There are 12 main types of glycosphingolipid in insects, none of which are sialylated as in mammals, but many contain one or two PE-substituted-GlcNAc (Seppo et al. 2000). Glycosphingolipids can be extremely long in *Drosophila,* the longest species having 10 sugar residues (GlcNAc, GalNAc, galactose, or a terminal glucuronic acid) in a linear chain (Kraut 2011). In opposite to mammals, in insects sphingolipids also have shorter FA chains that are amide-linked to the sphingoid bases (Rietveld et al. 1999). Based on this, the main biological function of these structures can maintain the membrane fluidity at cold, which correlates with the optimal temperature for fruit fly survival (Rietveld et al. 1999).

6 Steroids (Sterols)

Steroids are multipart four-ringed organic compounds that play many functions in organisms (Cole et al. 2019). There are many literature data describing the role of cholesterol and its derivatives in mammalian systems, mostly as membrane lipids and important regulator factors (Kenwood and Merrill 2016; Bach and Wachtel 2003). Sterols are divided according to the number of carbons in the base skeleton; the steroids with 18 carbons in skeleton contain important estrogen family, while other hormones like androgens such as testosterone and androsterone are composed from skeleton with 19 carbons, or progestogens as well as the glucocorticoids and mineralocorticoids, which contain skeleton with 21 carbons and a two-carbon side chain at the C17 position. An important lipid vitamin, vitamin D, and its various forms included as secosteroids are categorized according to the cleavage of the B ring in the basic structure, later the Latin seco prefix (Jones et al. 1998). Another group of sterols are bile acids, which in mammals are formed mostly as derivatives of cholan-24 acid produced from cholesterol in the liver and their conjugates (including sulfuric acid, taurine, glycine, glucuronic acid). Further examples of sterols are presented in Table 7 and Fig. 6.

Unlike yeasts and vertebrates, insects are sterol-auxotrophs, as are all other arthropods (Behmer and David Nes 2003; Carvalho et al. 2010). This lack of ability of sterol synthesis is a result of deficiency of the genes coding the

Table 7 Classification of sterols with examples in insects

	Definition	Function	Examples in insects	Literature data
Cholesterol	A cholestanoid consisting of cholestane with a double bond at the 5,6-position and a 3β-hydroxy group	Part of cell membranes and factor regulating their fluidity	U4.4, C7–10, C6/36, Sf21cell lines	Hafer et al. (2009)
			Insects from the orders Coleoptera, Diptera and Lepidoptera	Terra et al. (2006)
			Spodoptera frugiperda	Capella et al. (1997)
		Production of hormones	All insects	Jing and Behmer (2020)and Svoboda and Feldlaufer (1991)
			Bombyx mori	Igarashi et al. (2018)
Campesterol	A steroid derivative, characterized by the hydroxyl group in position C-3 of the steroid skeleton, and saturated bonds throughout the sterol structure, with the exception of the 5–6 double bond in the B ring	Not known	*Dermestes ater, Dermestes maculatus*	Cerkowniak et al. (2020)
			Bombyx mori	Takeshima et al. (2019)
		Conversion into cholesterol	All insects	Svoboda and Feldlaufer (1991) and Maruyama et al. (1982)
			Drosophila	Feldlaufer et al. (1995)
β-Sitosterol	Member of the class of phytosterols that is stigmast-5-ene substituted by a β-hydroxy group at position 3.	Conversion into cholesterol	*Blattella germanica*	Robbins et al. (1962)
			Manduca sexta	Svoboda et al. (1971)
			Musca domestica	Kaplanis et al. (1965)
Stigmasterol	3β-sterol consisting of 3β-hydroxystigmastane having double bonds at the 5,6- and 22,23-positions	Conversion into cholesterol	*Bombyx mori*	Fujimoto et al. (1984)
		Conversion to cholestanol	*Epilachna varivestis*	Svoboda et al. (1986, 1975)

protein that regulates the reaction of changing the farnesyl pyrophosphate to squalene (Seegmiller et al. 2002; Vinci et al. 2008). A hypothesis has been developed that explains why the facility to synthesize sterols de novo has gone in insect; it is possible that sterol auxotrophy has an evolutionary advantage, given that insects have native restrictions in oxygen transport due to the structure of the respiratory system, which includes their blind-ended trachea (Harrison et al. 2010). The other hypothesis claims that the production of cholesterol is a process that requires large amounts of oxygen and is metabolically expensive. Insufficient oxygen amount can decrease sterol synthesis, which in turn can diminish delay insect development (Desmond and Gribaldo 2009; Dupont et al. 2012; Nguyen et al. 2007; Rosenfeld and Beauvoit 2003).

Steroids synthesis has a similar mechanism in different cells. Cholesterol or other sterol compounds are transformed into hormones released into the circulation and their role is to regulate numerous physiological and developmental processes. In both mammals and insects (like *D. melanogaster*) the synthesis of steroid lipids is regulated by a group of cytochrome

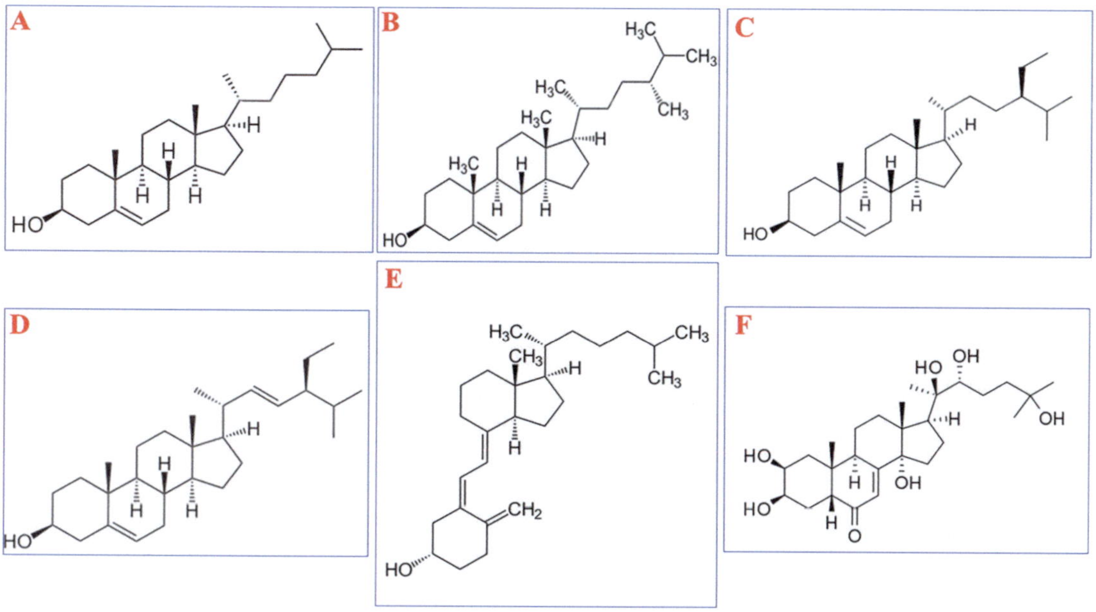

Fig. 6 The structure of cholesterol (**a**), campesterol (**b**), β-sitosterol (**c**), stigmasterol (**d**), vitamin D (**e**), 20-hydroxyecdysone (**f**) (based on PubChem (National Center for Biotechnology Information 2023))

P450 enzymes, which mediate the conversion of active compounds from cholesterol or other appropriate sterols (Carvalho et al. 2010; Lavrynenko et al. 2015; Rewitz et al. 2013).

Vitamin D is a lipid vitamin that is a precursor for the active signaling molecule—1-alpha, 25-dihydroxyvitamin D3 (1,25 VD3) and even though it is not a steroid, it is produced from a steroid compound (Wheeler and Nijhout 2003). Vitamin D3 concentration can be changed with UVB contact in some species of insect. The potential for de novo vitamin D3 synthesis was detected in *L. migratoria*, *Acheta domesticus*, and *Tenebrio molitor* after UVB exposure; in contrast, in the *Hermetia illucens* the concentration of vitamin D3 change after rearing, not after UVB application (Oonincx et al. 2018).

Cholesterol is used for the precursor of molting hormone (20-hydroxyecdysone; 20E) for the majority of insects. Ecdysteroids demonstrate great structural diversity in each insect species (Lavrynenko et al. 2015; Lafont and Koolman 2009). Arthropoda steroidogenesis has been described as cholesterol change into the ecdysone prohormone in the prothoracic glands (Huang et al. 2008). In the prothoracic glands, cholesterol is transformed into 7-dehydrocholesterol (7dC) and this reaction is controlled by Rieske-enzyme named Neverland, and lastly ecdysone is formed via several processes with ecdysone synthesizing proteins (Huang et al. 2008; Yoshiyama et al. 2006; Yoshiyama-Yanagawa et al. 2011). The secreted ecdysone is transformed into 20E, the bioactive form of ecdysteroid, in the peripheral tissues other than prothoracic glands by the P450 enzyme (CYP314A1) (Huang et al. 2008; Petryk et al. 2003).

Due to a lack of cellular sterol synthesis, cholesterol must be transported from hemolymph for the production of ecdysone. The lipid compounds are transported to each insect tissue via the Lp system in insect hemolymph (Kaczmarek and Boguś 2021a; Yun et al. 2002). When incubated in a medium containing Lp in vitro, the prothoracic glands of silkworm *B. mori* showed an increase in ecdysone production (Chino et al. 1974). Thus, cholesterol transported by Lp is probably a main component in ecdysone synthesis in the prothoracic glands (Igarashi et al. 2018). Some cholesterol transporters and enzymes are vital for its metabolism in the translation stage of the ecdysteroid production process also in *Drosophila* (Huang et al. 2008).

Even though the 20E is considered as a main molting hormone, some insects use other ecdysteroids. Examples of organisms that use makisterone as their primary molting hormone include plant-feeding heteropterans, leaf-cutter ants, and honey bees (Lafont et al. 2005). These organisms have the capability to dealkylate phytosterols and therefore they directly change campesterol into makisterone A (which has a methyl at the C24 position) or sitosterol into makisterone C (which has an ethyl at the C24 position). *Drosophila* uses cholesterol (*D. melanogaster*) or lathosterol (*Drosophila pachea*) and this diversity is dependent on different diets of the mentioned species. The limited dependence of *D. pachea* on using lathosterol as a precursor substrate is related to a change in four amino acids in the oxygenase-like protein encoded by the Neverland gene (Lang et al. 2012).

The most important sterol in honey bees is 24-methylenecholesterol (24MC). In honey bees, diet rich in 24MC caused the highest level of worker survival rate and the longest period of brood production among the tested substances (Chakrabarti et al. 2020). Further research has shown great concentrations of 24MC in honey bee pupae (Feldlaufer et al. 1986; Svoboda et al. 1980) and that higher dietary intake can raise the survival rate of bees and expand physiological indicators such as head protein and abdominal lipid concentration (Chakrabarti et al. 2021).

Sterols are also an important part of cell membranes. Being amphipathic compounds, they are built from two basic parts: polar (hydrophilic) and nonpolar (hydrophobic). The hydrophilic fragment, containing a C3 hydroxyl group, positions outward from the bilayer and could form hydrogen bonds with the hydrophilic phosphate head of a phospholipid particle. Contrary, the hydrophobic part is built from tetracyclic triterpenoid and an isooctyl hydrocarbon chain. This nonpolar fragment of a sterol covers inside and cooperates with phospholipid FA tails; precisely, the tetracyclic part and the side chain align adjacent to the FA chain (Hofsäß et al. 2003; Jing and Behmer 2020). In the membrane layers, cholesterol and phospholipids cooperate to improve the mechanical coherence of the membrane by lowering fluidity (cholesterol plays the function of stabilator when it is surrounded by phospholipid particles) and growing rigidity (the organization of the hydrocarbon chains). This suppresses passive permeability, which expands the cell's capacity to control the drive of several particles, across the membrane and into the cell, particularly polar ones (Mouritsen and Zuckermann 2004). The *lipid rafts*, cholesterol-rich area, are also present in the phospholipid bilayer and these regions with unique physical activity are different from those of the surrounding membrane area and are frequently connected with fundamental membrane proteins (like transmembrane proteins). Nevertheless, in these areas mostly predominant are sphingolipids rather than phospholipids (Guan et al. 2013; Zhuang et al. 2002). Currently, the knowledge of lipid raft dispersal, structure, and activity in cell membranes is elementary, particularly in insects.

The growth of *Drosophila* is dependent on the availability of dietary sterols. When fruit flies are rearing on diet poor in sterols, the larvae grow slowly and give rise to small adults. However, the increasing amount of sterol in the diet does not raise their concentration in the membrane, but lets the growth of bigger adults. However, other processes should be involved in the determination of the accumulation level of particular sterols in tissue and cell membrane. Remarkably, in the gut and central nervous system (CNS), the organs where the main accumulation of membrane sterols takes place, the predominant lipid fraction are sphingolipids (Carvalho et al. 2010). Due to the more favorable interaction of sterols with sphingolipids than with phospholipids (Sengupta et al. 2007), there is a chance that membrane containing higher level of sphingolipids has a bigger ability to retain sterols. Therefore, the higher sphingolipid production could collect more sterols in some tissues than others, even if cells do not have the ability of sterol production (Carvalho et al. 2010).

When sterols are removed from the food, fruit fly larvae stop growth and development and research data indicated the sixfold reduction of membrane sterol levels without changing the larval viability (Carvalho et al. 2010). To survive under these conditions, fruit flies increase sphingolipids synthesis, characterized by doubly desaturated long chain base and hydroxylated FA moieties. Sphingolipid hydroxylation indorses hydrogen bonding and tighter structure (Löfgren and Pascher 1977), emerging structures are predominant in membranes particularly focused on playing barrier roles, such as myelin and the gut apical membrane (Maldonado et al. 2008; Uchida et al. 2007). Therefore, it is possible that these lipids might be crucial to keep some biophysical activity of the membrane, such as impermeability. However, even if fluctuating of the sphingolipid arrangement can improve cell viability when membrane sterol concentrations are lower, it may not let finish the last development stage. Evidently, some roles of substance membrane sterol could not be substituted by sphingolipids (Carvalho et al. 2010).

There is still no answer to how fruit flies detect the concentration of membrane sterol and how they influence growth and sphingolipid biosynthesis. The sensing process in *Drosophila* is probably autonomous of the sterol regulatory element-binding proteins (SREBP) pathway, which prefers PE rather than sterols (Seegmiller et al. 2002; Dobrosotskaya et al. 2002; Kunte et al. 2006). The changes in membrane sterols may regulate the glycolysis and translation; the lower amount of sterol causes the transcriptional changes in genes. The insulin/Akt signaling control of the growth contains analogous transcriptional changes affecting glycolysis and translation, and happens via phosphorylation and nuclear translocation of the transcription factor Foxo (Jünger et al. 2003; Puig et al. 2003). Nevertheless, the lipid compound could not control the nuclear translocation of Foxo. In vertebrates, Foxo physically cooperates with nuclear hormone receptors, which have lipidic ligands (Dowell et al. 2003; Ganjam et al. 2009; Li et al. 2003; Qiuping et al. 2009). Additionally, sirtuins change the action and relations of nuclear Foxo via deacetylation (Frescas et al. 2005; Li et al. 2007; Wang and Tong 2009; Yang et al. 2005; Brunet et al. 2004). Unraveling such processes will offer new understandings of membrane homeostasis and growth control (Carvalho et al. 2010).

Literature data has shown that cholesterol taken from food controls intestinal epithelial cell differentiation from stem cells in the midgut of fruit fly adults by changing the level and persistence of Notch signaling, a crucial intercellular signaling pathway responsible for cell differentiation processes control during embryonic development and adult life (Obniski et al. 2018). Moreover, cholesterol participates in resistance to DCV (*Drosophila* C virus) infections in Wolbachia-infected flies (Caragata et al. 2013).

Cholesterol is also an important molecule that protects the insect from the negative impact of cold. Research has shown that after feeding with cholesterol-rich diet observed in insects not only increased tolerance for low temperatures but the ability of RCH to survive as well (Shreve et al. 2007). Also, cholesterol may play a role in diapause cold-hardening, as membrane-bound cholesterol may protect from coalescence of adjacent fatty acyl chains (Shreve et al. 2007).

7 Prenol Lipids

Prenols are produced from the five-carbon isopentenyl diphosphate precursors and dimethylallyl diphosphate, which are synthesized mostly via the mevalonic acid pathway (Fahy et al. 2005; Kuzuyama and Seto 2003). The simple isoprenoids, such as chain alcohols and diphosphates, are synthetized by the following addition of C5 terpene parts, and their classification is based on the number of pieces. The isoprenoid containing more than 40 carbons are described as polyterpenes. Carotenoids are examples of simple isoprenoids, with important biological activity, they work as antioxidants and are considered as vitamin A precursors.

The essential class of molecules, that are biologically active, are quinones and hydroquinones; they are built from isoprenoid tail attached to a quinonoid base of non-isoprenoid origin. Two good examples of this class are tocopherol (vitamin E), menadione (vitamin K), and ubiquinones (Fahy et al. 2005). Insect ubiquinones, which are part of the respiratory chain, are a mixture of three homologs with eight, nine, and ten isoprenoid units in the side chain (Van den Bergh 1967). Some examples of prenol lipids are presented in Table 8 and Fig. 7.

The juvenile hormone (JH) is isoprenoids that regulate a variety of insect processes during postembryonic development (larval development and metamorphosis) and adult reproduction; moreover, it takes part in the regulation of several physiological mechanisms, crucial for insect survival like diapause, migratory behavior, changes of wing length, develop of seasonal form in response to changes in the environment. Additionally in ants, bees, wasps, and termites mentioned hormone is responsible for caste determination, and in scarab beetles juvenile hormone regulates sexual dimorphism, by influencing on size and shape of female horns (Goodman and Cusson 2012). It is produced and secreted by the corpora allata (CA), a couple of endocrine glands with neural connections to the brain (Wyatt and

Table 8 Classification of prenols with examples in insects

	Definition	Function	Examples in insects	Literature data
Isoprenoids	Lipid formally derived from isoprene (2-methylbuta-1,3-diene), the skeleton of which can generally be discerned in repeated occurrence in the molecule	Not known	*Plodia interpunctella, Ephestia kuehniella, Cadra cautella, Sitophilus zeamais*	Tanaka et al. (2020)
			Hermetia illucens, Chilecomadia moorei, Blatta lateralis, Musca domestica	Finke (2013)
		Prohormones	*Locusta migratoria*	Darrouzet et al. (1998)
			Rhodnius prolixus	Nakamura et al. (2007)
		The compound of eyes	Insects from the orders *Lepidoptera, Diptera, Coleoptera, Neuroptera, Hemiptera and Odonata*	Seki et al. (1987)
Quinones and hydroquinones	A fully conjugated cyclic dione structure, such as that of benzoquinones, derived from aromatic compounds by conversion of an even number of –CH = groups into –C(=O)– groups with any necessary rearrangement of double bonds (polycyclic and heterocyclic analogs are included)	Not known	*Tribolium* spp.	St. Agatha Suter and Ladisch (1963) and Ladisch and McQue (1953)
			Conistra erythrocephala	Beedle et al. (1975)
			Bombyx mori	Sridhara and Bhat (1965)
		Immunity	All insects	Eleftherianos et al. (2021)
		Part of mitochondrial respiratory chain		Van den Bergh (1967)
		Growth and reproduction	*Aeshna affinis*	House (1965)
			Bactrocera tryoni	Moadeli et al. (2020)
Polyprenols	The general formula H-[CH_2C (Me) = $CHCH_2]_nOH$ in which the carbon skeleton is composed of more than one isoprene units	Not known	*Pieris rapae crucivora*	Suga and Shishibori (1979)
			Manduca sexta	Li et al. (1995)
Hopanoids	A triterpenoid based on a hopane skeleton	Not known	*Hydrotaea aenescens*	Frere et al. (2014)

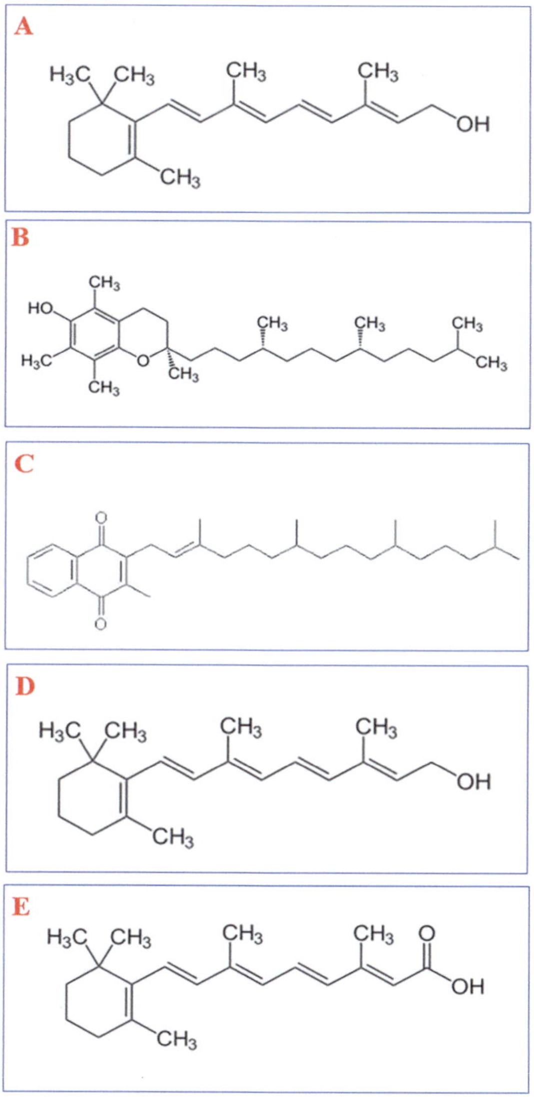

Fig. 7 The structure of the main prenol lipids (based on PubChem (National Center for Biotechnology Information 2023)) vitamin A (**a**), vitamin E (**b**), vitamin K (**c**), retinol (**d**) and retinoic acid (**e**)

Davey 1996). The biosynthesis of JHs is separated into two important steps (Bellés et al. 2005): (1) the synthesis of farnesyl pyrophosphate (FPP) from acetyl-CoA (mevalonate pathway) and (2) hydrolysis of FPP to farnesol (Cao et al. 2009), followed by oxidation to farnesal (Mayoral et al. 2009) and farnesoic acid (FrA) (Baker et al. 1983). In the end, FrA is converted to JH III by methyl transfer (Shinoda and Itoyama 2003) and epoxidation (Helvig et al. 2004).

Literature data has shown that the JH precursor released by the corpora allata during development might also play a function of prohormones in particular tissue, for example in the imaginal discs in tobacco hornworm (*M. sexta*) where the inactive precursor JH acid released by the corpora allata is changed to the active hormone (Sparagana and Bhaskaran 1985). If the isoprenoid released by the corpora allata did not play a role per se or only plays the function of prohormone, it could serve to non-activate hormones at the level of the producing gland, or it could act as a metabolic shunt that competes with JH for mevalonate requirement in the corpora allata cells (Darrouzet et al. 1998).

Lipid vitamins have huge biological activity in physiology, development, and gene expression without direct impact on the gene transcription process; moreover, some of them may act through nuclear receptors belonging to the steroid receptor superfamily. In insects vitamins A, D, and E play important functions. Vitamin A is a lipid compound that acts as a precursor for the active signaling molecule all-trans retinoic acid and the nuclear receptors of the mentioned vitamin are named the retinoic acid receptor. On the other hand, vitamin E, which is chemically similar to vitamins A and D, has the ability to work as an active signaling molecule without nuclear receptors or it is not known now (Carlberg 1999).

In general, vitamin A and its metabolites (retinoic acid for example) take part in neuronal differentiation, cell viability, and chemotaxis (Albalat 2009; Bui-Göbbels et al. 2015). Retinoids form the compound eyes of insects in the orders Lepidoptera, Diptera, Coleoptera, Neuroptera, Hemiptera, and Odonata (Seki et al. 1987). The application of retinoic acid to *Pyrrhocoris upterus, Dysdercus cingufutus,* and *T. mofitor* caused a juvenilizing effect in last-instar larvae or pupae, as well as embryonic malformations, and it was found to improve reproduction and vitellogenin synthesis, and increase the deposition of fertile eggs which provided normal progeny (after application to eggs) (Němec et al. 1993). Low concentrations of 9-cis retinoic acid demonstrated positive effects on cell viability and neurite cell development in cultures taken from the CNS of

L. migratoria embryos (Sukiban et al. 2014). In contrast, injection to the hemocoel resulted in abnormal larval development in the blood-sucking insect *R. prolixus* (Nakamura et al. 2007). The deficit of vitamin A in fruit flies has a negative impact on the visual system, by destroying the production of the retinal chromophore, disrupting the formation of the visual pigments (Rhodopsins), and destruction of the photoreceptors (Dewett et al. 2021a, 2021b). Moreover, retinol shows some antioxidant actions and extends the life span in SOD1-deficient *Drosophila*; however, research has shown that under hyperoxia, retinol exhibits a strong prooxidant activity and increases mortality of flies (Bahadorani et al. 2008).

Retinoic acid signal in vertebrates is transduced by two retinoid receptors: the retinoic acid receptor and the retinoid X receptor (RAR and RXR, correspondingly). In invertebrates, such as moths and flies, the role of RXR is played by its homolog ultraspiracle (USP) (Gutierrez-Mazariegos et al. 2014). In some insects, like *L. migratoria,* the RXR is more similar to human RXRs than USPs (Nowickyj et al. 2008). USP/RXR are members of the nuclear receptor superfamily (Sharma et al. 2022) and with the ecdysone receptor (EcR) create a crucial part of heterodimeric receptor complex. This complex plays a regulation role in the important physiological processes in insects, like growth, development, metamorphosis, and reproduction, according to the ability of ecdysteroids binding (Hult et al. 2011).

Research conducted by Mansfield and co-workers identified the *M. sexta* intracellular lipid-binding proteins (iLBPs), which are believed to serve as invertebrate cellular retinoic acid-binding proteins (CRABPs), suggesting that the vitamin A derivative retinoic acid, may play a possible role in invertebrates (Mansfield et al. 1998). The amino acid sequence of *M. sexta* CRABP demonstrates 71% homology overall with bovine/murine CRABP and 88% homology for the ligand binding pocket of the protein (Mansfield et al. 1998). However, this has not been confirmed in other studies, which found iLBPs to demonstrate high binding preference for FAs, particularly for unsaturated FAs (Folli et al. 2005).

Vitamin E is a generic term including α-, β-, γ-, and δ-tocopherols, and their equivalent four tocotrienols, together named tocochromanols (Niki and Traber 2012). Alpha-tocopherol has a very similar biological function to JH in the regulation of polyphenism and reproduction in insects (Wheeler and Nijhout 2003). Whereas the role and metabolism of those vitamins have been extensively studied in vertebrate organisms, relatively less information is available concerning invertebrates. Vitamin E was detected in *B. mori* pupae (Singh and Jayasomu 2002), *D. melanogaster* (Parker and McCormick 2005), and *S. cosmioides* cuticle (Fronza et al. 2013). The addition of vitamin E to the larval food of the parasitoid *Agria affinis* and *Pimpla turionellae* was found to be essential for the flies to produce viable offspring (Coşkun et al. 2005). Moreover, the dietary intake of this vitamin might protect *G. mellonella* larvae against selenium toxicity (Coskun et al. 2020).

The majority of pheromones are synthesized from the same pool of biochemical precursors including FAs and isoprenes in insects, which results in the range of chemical compounds being created by minor changes in a basic structure (Yew and Chung 2015). A number of coleopteran pheromones are isoprenoid in origin or derived from FAs and amino acids (Francke et al. 1995; Tillman et al. 1999). Cockroaches utilize both volatile and cuticular contact pheromones (Gemeno and Schal 2009). Beetle pheromone biosynthesis involves isoprenoid, FA, or amino acid (Jurenka 2004a). Sometimes compounds taken with food can be converted to pheromones; however, most of the beetle pheromones are synthesized de novo. The crucial enzymes in the primary steps of the isoprenoid pathway are HMG-CoA (3-hydroxy-3-methylglutaryl-coenzymeA) synthase and HMG-CoA reductase, which take part in the mevalonate formation pathway; however, in the final part of the process the key enzymes seem to be geranyl diphosphate synthase. Research data confirmed the presence of other pathways of isoprenoid compounds synthesis, which take part in the utilization of pyruvate and glyceraldehyde 3-phosphate to form 1-deoxy-xylulose,5-phosphate that is next transformed to isopentenyl diphosphate (Rohmer 1999). These primary non-mevalonate stages in the isoprenoids synthesis have been detected yet in bacteria and plants but not in insects (Jurenka 2004b).

8 Polyketides

Polyketides are a cluster of natural compounds, which are produced by derivation from acetyl or malonyl parts and their structural features (including polyphenols, macrolides, polyenes, enediynes, and polyethers); they are considered as lipids' according to their ability to dissolve in organic solvents (O'Hagan 1992). They are mostly divided based on the structure of their central functional group or kind of enzyme regulating the process of synthesis (polyketide synthase type). Generally, the biologically active polyketides are put into three structural classes. The first one includes polycyclic polyketides, which comprise the medically significant tetracyclines. The compounds with macrocyclic skeletons are considered as the next class, like the commercially available macrolide erythromycin. The polyketides that are not poly- or macrocyclic are categorized as the third group (Van Moll et al. 2021).

Carminic acid is an aromatic polyketide detected in *Dactylopius coccus* and other scale insects and is commonly used as a natural red dye (Flores-Alatorre et al. 2014). Literature data indicate the important function of polyketides as pheromone communication and as a defense against the enemy. Research data suggests that the polyketide chain is a precursor for the production of the (2E,4E,6E)-5-ethyl-3-methyl-2,4,6-nonatriene, a pheromone factor released by the *Carpophilus freeman* (Petroski et al. 1994; Bartelt and Weisleder 1996). Next potential polyketide substances are also detected in the sex pheromone blend of the female *Lasioderma serricorne* such as serricornin, anhydroserricornin, and 4,6-dimethylnonan-3,7-diol (Chuman et al. 1985, 1983).

Nevertheless, the biogenetic origin of polyketides in insects is frequently unidentified. Although polyketide biogenesis is regulated by polyketide synthases (PKS) in microorganisms (fungi, bacteria) and plants, there is a lack of research data confirming the presence of PKS, which take part in pheromones or defense compounds production. Polyketides detected in insects may also be biosynthesized by endosymbionts (Pankewitz and Hilker 2008) and literature data confirming the presence of at least 26 different active compounds from this group produced by symbionts (Van Moll et al. 2021). Research has indicated the production of the polyketide lagriamide by *Burkholderia gladioli*, i.e., symbionts of the invasive beetle *Lagria villosa*, and found it to demonstrate antifungal activity against the entomopathogen *Purpureocillium lilacinum* (Niehs et al. 2020). The diaphorin, a polyketide produced by *Profftella armatura*, microbial endosymbionts *of Diaphorina citri,* influences the insect protein expression during infection with *Liberibacter asiaticus*, the pathogen bacteria caused the citrus greening disease, transmitted by *D. citri*. They appear to influence the expression of protection and immunity compounds, particles responsible for energy storage and utilization, and proteins taking part in endocytosis and cellular adhesion, as well as those participated in cytoskeletal remodeling, which are related to microbial invasion of host cells (Ramsey et al. 2015).

9 Aliphatic Compounds: Waxes

Scale insects and honeybees generate higher mass of waxes ester for defense, brood raising, and honey storage. Insect waxes have been found to demonstrate good thermal stability; research found the melting point of insect waxes (shellac, coccerin, white, and beewaxes) to range from 70.3 to 90.4 °C whereas the melting point for paraffin wax (synthetic wax) was is lower and is 63.3 °C and the carnauba wax (vegetable wax) 87.9 °C (Zheng et al. 2011).

The *Ericerus pela* (Chavannes) male produces the wax, trees parasite (*Privet* or *Fraxinus*), which is as well used in traditional Chinese medicine as *white wax*. The white wax contains mostly

(93%–95% of the total amount) the ester mixtures of monoacids and monoalcohols, and the dominate the esters of hexacosanoic acid and octacosanoic acid. The insect waxes are also used for the preparation of policosanol *via* the process of reduction or saponification, and mostly contain hexacosanol and octacosanol (Hou et al. 2011; Ma et al. 2022).

Beeswax is a naturally occurring wax produced in hives by the honeybee *A. mellifera*. It is secreted by glands under the abdomen and *A. mellifera* is used to produce the honeycomb. Literature data confirmed the presence of eight glands in the bee abdominal segment of female worker bees that produce the wax. The beeswax is obtained as a by-product when honey is harvested and refined (Tinto et al. 2017). Wax glands on the underside of the abdomens of the young bees secrete small wax platelets, after rearing with royal jelly and are involved in the structure of the hive. These are scraped off by the bee, chewed, and masticated into pliable bits with the addition of saliva and a variety of enzymes. The composition of beeswax differs between geographic localization (Europe, Asia, or Africa), species of honeybee, and age of the wax. It has been shown that unhydrolyzed beeswax usually contain mostly esters (71%), other important compounds are hydrocarbons and free fatty acids (8%) (Tulloch 1970).

10 Conclusions and Future Directions

A wealth of lipid forms is known to be associated with insects, and a considerable body of research has detailed the wide-ranging functions they perform in their lives. However, many aspects of lipid metabolism in insects still remain unclear, and further in-depth research is needed to fully understand their influence on the mechanisms underlying vital processes.

Research on the composition of cuticular lipids present in scavenger insect species, which are able to remain resistant to infections despite being in close contact with a plethora of pathogenic microorganisms, appears very promising. The identification of lipids with antimicrobial properties followed by extensive screening tests might open the way to developing new targeted therapies and innovative methods of preventing bacterial and/or fungal infections in mammals. Indeed, lipidomic techniques have been extensively used to analyze lipid metabolism in mammals. Recently, insect models are become increasingly attractive in research; these might provide new insights into health and disease processes at significantly lower investigation costs while avoiding ethical problems.

References

Abeytunga DTU, Glick JJ, Gibson NJ, Oland LA, Somogyi A, Wysokid VH, Polt R (2004) Presence of unsaturated sphingomyelins and changes in their composition during the life cycle of the moth *Manduca sexta*. J Lipid Res 45:1221–1231

Abeytunga DTU, Oland L, Somogyi A, Polt R (2008) Structural studies on the neutral glycosphingolipids of *Manduca sexta*. Bioorg Chem 36:70–76

Abou El Asrar R, Cools D, Vanden Broeck J (2020) Role of peptide hormones in insect gut physiology. Curr Opin Insect Sci 41:71–78

Acharya U, Acharya JK (2005) Enzymes of sphingolipid metabolism in *Drosophila melanogaster*. Cell Mol Life Sci 62:128–142

Acharya JK, Dasgupta U, Rawat SS et al (2008) Cell-nonautonomous function of ceramidase in photoreceptor homeostasis. Neuron 57:69–79

Ahmed S, Kim Y (2019) PGE2 mediates cytoskeletal rearrangement of hemocytes *via* Cdc42, a small G protein, to activate actin-remodeling factors in *Spodoptera exigua* (Lepidoptera: Noctuidae). Arch Insect Biochem Physiol 102:e21607

Ahmed S, Kim Y (2021) PGE2 mediates hemocyte-spreading behavior by activating aquaporin *via* cAMP and rearranging actin cytoskeleton *via* Ca2+. Dev Comp Immunol 125:104230

Ahmed S, Hasan A, Kim Y (2019) Overexpression of PGE2 synthase by *in vivo* transient expression enhances immunocompetency along with fitness cost in a lepidopteran insect. J Exp Biol. https://doi.org/10.1242/jeb.207019

Ahmed S, Al Baki MA, Lee J, Seo DY, Lee D, Kim Y (2021) The first report of prostacyclin and its physiological roles in insects. Gen Comp Endocrinol 301:113659

Albalat R (2009) The retinoic acid machinery in invertebrates: ancestral elements and vertebrate innovations. Mol Cell Endocrinol 313:23–35

Alborn HT, Turlings TCJ, Jones TH, Stenhagen G, Loughrin JH, Tumlinson JH (1997) An elicitor of plant volatiles from beet armyworm oral secretion. Science (80–) 276:945–949

Anderson RL, Battistini MR, Wallis DJ, Shoji C, O'Flynn BG, Dillashaw JE, Merkler DJ (2018) Bm-iAANAT and its potential role in fatty acid amide biosynthesis in *Bombyx mori*. Prostaglandins Leukot Essent Fat Acids 135:10–17

Arrese EL, Soulages JL (2010) Insect fat body: energy, metabolism, and regulation. Annu Rev Entomol 55: 207–225

Auerswald L, Gäde G (2000) Metabolic changes in the African fruit beetle, *Pachnoda sinuata*, during starvation. J Insect Physiol 46:343–351

Bach D, Wachtel E (2003) Phospholipid/cholesterol model membranes: formation of cholesterol crystallites. Biochim Biophys Acta Biomembr 1610:187–197

Bahadorani S, Bahadorani P, Phillips JP, Hilliker AJ (2008) The effects of vitamin supplementation on *Drosophila* life span under normoxia and under oxidative stress. J Gerontol Ser A Biol Sci Med Sci 63:35–42

Baker FC, Mauchamp B, Tsai LW, Schooley DA (1983) Farnesol and farnesal dehydrogenase(s) in corpora allata of the tobacco hornworm moth, *Manduca sexta*. J Lipid Res 24:1586–1594

Ballard JWO, Melvin RG, Simpson SJ (2008) Starvation resistance is positively correlated with body lipid proportion in five wild caught *Drosophila simulans* populations. J Insect Physiol 54:1371–1376

Barletta ABF, Trisnadi N, Ramirez JL, Barillas-Mury C (2019) Mosquito midgut prostaglandin release establishes systemic immune priming. iScience 19: 54–62

Barlow JS, Jones D (1981) A comparative study of transacylation in three insect species. Can J Zool 59: 1141–1147

Bartelsen O, Lansmann S, Nettersheim M, Lemm T, Ferlinz K, Sandhoff K (1998) Expression of recombinant human acid sphingomyelinase in insect Sf21 cells: purification, processing and enzymatic characterization. J Biotechnol 63:29–40

Bartelt RJ, Weisleder D (1996) Polyketide origin of pheromones of *Carpophilus davidsoni* and *C. mutilatus* (Coleoptera: Nitidulidae). Bioorg Med Chem 4:429–438

Batalha M d MC, Goulart HF, Santana AEG, Barbosa LAO, Nascimento TG, da Silva MKH, Dornelas CB, Grillo LAM (2020) Chemical composition and antimicrobial activity of cuticular and internal lipids of the insect *Rhynchophorus palmarum*. Arch Insect Biochem Physiol. https://doi.org/10.1002/arch.21723

Batz ZA, Armbruster PA (2018) Diapause-associated changes in the lipid and metabolite profiles of the Asian tiger mosquito, *Aedes albopictus*. J Exp Biol. https://doi.org/10.1242/jeb.189480

Baumbach J, Hummel P, Bickmeyer I, Kowalczyk KM, Frank M, Knorr K, Hildebrandt A, Riedel D, Jäckle H, Kühnlein RP (2014a) A *Drosophila in vivo* screen identifies store-operated calcium entry as a key regulator of adiposity. Cell Metab 19:331–343

Baumbach J, Xu Y, Hehlert P, Kühnlein RP (2014b) Gαq, Gγ1 and Plc21C control *Drosophila* body fat storage. J Genet Genomics 41:283–292

Beedle AS, Walton MJ, Goodwin TW (1975) Isoprenoid biosynthesis in aseptic larvae of *Calliphora erythrocephala*. Insect Biochem 5:465–472

Behmer ST, David Nes W (2003) Insect sterol nutrition and nhysiology: a global overview. Adv Insect Phys 31:1–72

Behmer ST, Vogel H (2020) Editorial overview: molecular physiology of the multifunctional insect gut. Curr Opin Insect Sci 41:iv

Bellés X, Martín D, Piulachs MD (2005) The mevalonate pathway and the synthesis of juvenile hormone in insects. Annu Rev Entomol 50:181–199

Bennett VA, Pruitt NL, Lee RE (1997) Seasonal changes in fatty acid composition associated with cold-hardening in third instar larvae of *Eurosta solidaginis*. J Comp Physiol B Biochem Syst Environ Physiol 167: 249–255

Benoit JB (2010) Water management by dormant insects: comparisons between dehydration resistance during summer aestivation and winter diapause. Prog Mol Subcell Biol 49:209–229

Bhat HB, Ishitsuka R, Inaba T et al (2015) Evaluation of aegerolysins as novel tools to detect and visualize ceramide phosphoethanolamine, a major sphingolipid in invertebrates. FASEB J 29:3920–3934

Bhogal JK, Kanaskie JM, DiAngelo JR (2020) The role of the heterogeneous nuclear ribonucleoprotein (hnRNP) Hrb27C in regulating lipid storage in the *Drosophila* fat body. Biochem Biophys Res Commun 524:178–183

Blomquist GJ, Bagnères AG (2010) Insect hydrocarbons biology, biochemistry, and chemical ecology. Insect Hydrocarb Biol Biochem Chem Ecol. https://doi.org/10.1017/CBO9780511711909

Blomquist GJ, Jackson LL (1979) Chemistry and biochemistry of insect waxes. Prog Lipid Res 17:319–345

Boguś MI, Czygier M, Gołębiowski M, Kędra E, Kucińska J, Mazgajska J, Samborski J, Wieloch W, Włóka E (2010) Effects of insect cuticular fatty acids on *in vitro* growth and pathogenicity of the entomopathogenic fungus *Conidiobolus coronatus*. Exp Parasitol 125:400–408

Boguś MI, Włóka E, Wrońska A, Kaczmarek A, Kazek M, Zalewska K, Ligęza-Żuber M, Gołębiowski M (2017) Cuticle hydrolysis in four medically important fly species by enzymes of the entomopathogenic fungus *Conidiobolus coronatus*. Med Vet Entomol 31:23–35

Borrelli L, Varriale L, Dipineto L, Pace A, Menna LF, Fioretti A (2021) Insect derived lauric acid as promising alternative strategy to antibiotics in the antimicrobial resistance scenario. Front Microbiol. https://doi.org/10.3389/fmicb.2021.620798

Brunet A, Sweeney LB, Sturgill JF et al (2004) Stress-dependent regulation of FOXO transcription factors by the SIRT1 deacetylase. Science (80–) 303:2011–2015

Buchwald R, Breed MD, Bjostad L, Hibbard BE, Greenberg AR (2009) The role of fatty acids in the mechanical properties of beeswax. Apidologie 40: 585–594

Buckner JS, Mardaus MC, Nelson DR (1996) Curticular lipid composition of *Heliothis virescens* and *Helicoverpa zea* pupae. Comp Biochem Physiol B Biochem Mol Biol 114:207–216

Buckner JS, Hagen MM, Nelson DR (1999) The composition of the cuticular lipids from nymphs and exuviae of the silverleaf whitefly, *Bemisia argentifolii*. Comp Biochem Physiol B Biochem Mol Biol 124:201–207

Bui-Göbbels K, Quintela RM, Bräunig P, Mey J (2015) Is retinoic acid a signal for nerve regeneration in insects? Neural Regen Res 10:901–903

Callow RK (1963) Chemical and biochemical problems of beeswax. Bee World 44:95–101

Canavoso LE, Wells MA (2000) Metabolic pathways for diacylglycerol biosynthesis and release in the midgut of larval *Manduca sexta*. Insect Biochem Mol Biol 30: 1173–1180

Canavoso LE, Wells MA (2001) Role of lipid transfer particle in delivery of diacylglycerol from midgut to lipophorin in larval *Manduca sexta*. Insect Biochem Mol Biol 31:783–790

Canavoso LE, Frede S, Rubiolo ER (2004) Metabolic pathways for dietary lipids in the midgut of hematophagous *Panstrongylus megistus* (Hemiptera: Reduviidae). Insect Biochem Mol Biol 34:845–854

Cao L, Zhang P, Grant DF (2009) An insect farnesyl phosphatase homologous to the N-terminal domain of soluble epoxide hydrolase. Biochem Biophys Res Commun 380:188–192

Capella AN, Terra WR, Ribeiro AF, Ferreira C (1997) Cytoskeleton removal and characterization of the microvillar membranes isolated from two midgut regions of *Spodoptera frugiperda* (Lepidoptera). Insect Biochem Mol Biol 27:793–801

Caragata EP, Rancès E, Hedges LM, Gofton AW, Johnson KN, O'Neill SL, McGraw EA (2013) Dietary cholesterol modulates pathogen blocking by Wolbachia. PLoS Pathog. https://doi.org/10.1371/journal.ppat.1003459

Carlberg C (1999) Lipid soluble vitamins in gene regulation. Biofactors 10:91–97

Carraretto D, Soresinetti L, Rossi I, Malacrida AR, Gasperi G, Gomulski LM (2022) Behavioural responses of male *Aedes albopictus* to different volatile chemical compounds. Insects. https://doi.org/10.3390/insects13030290

Carvalho M, Schwudke D, Sampaio JL et al (2010) Survival strategies of a sterol auxotroph. Development 137:3675–3685

Cerkowniak M, Boguś MI, Włóka E, Stepnowski P, Gołębiowski M (2020) The composition of lipid profiles in different developmental stages of *Dermestes ater* and *Dermestes maculatus* and their susceptibility to the entomopathogenic fungus *Conidiobolus coronatus*. Phytoparasitica 48:247–260

Chakrabarti P, Lucas HM, Sagili RR (2020) Novel insights into dietary phytosterol utilization and its fate in honey bees (*Apis mellifera* L.). Molecules. https://doi.org/10.3390/molecules25030571

Chakrabarti P, Lucas HM, Sagili RR (2021) Evaluating effects of a critical micronutrient (24-methylenecholesterol) on honey bee physiology. Ann Entomol Soc Am 113:176–182

Chan SK (1970) Phospholipid composition in the mitochondria of the housefly, *Musca domestica*: a re-examination. J Insect Physiol 16:1575–1577

Childress CC, Sacktor B, Traynor DR (1967) Function of carnitine in the fatty acid oxidase-deficient insect flight muscle. J Biol Chem 242:754–760

Chino H, Sakurai S, Ohtaki T, Ikekawa N, Miyazaki H, Ishibashi M, Abuki H (1974) Biosynthesis of α-ecdysone by prothoracic glands *in vitro*. Science (80–) 183:529–530

Chotiwan N, Andre BG, Sanchez-Vargas I et al (2018) Dynamic remodeling of lipids coincides with dengue virus replication in the midgut of *Aedes aegypti* mosquitoes. PLoS Pathog 14:e1006853

Chowański S, Walkowiak-Nowicka K, Winkiel M, Marciniak P, Urbański A, Pacholska-Bogalska J (2021) Insulin-like peptides and cross-talk with other factors in the regulation of insect metabolism. Front Physiol 12:701203

Chuman T, Mochizuki K, Kato K, Ono M, Okubo A (1983) Serricorone and serricorole, new sex pheromone components of cigarette beetle. Agric Biol Chem 47:1413–1415

Chuman T, Mochizuki K, Mori M, Kohno M, Kato K, Noguchi M (1985) Lasioderma chemistry sex pheromone of cigarette beetle (*Lasioderma serricorne* F.). J Chem Ecol 11:417–434

Citoler J, Derrington SR, Galman JL, Bevinakatti H, Turner NJ (2019) A biocatalytic cascade for the conversion of fatty acids to fatty amines. Green Chem 21: 4932–4935

Clark MS, Worland MR (2008) How insects survive the cold: molecular mechanisms—a review. J Comp Physiol B Biochem Syst Environ Physiol 178:917–933

Cocchiararo-Bastias LM, Mijailovsky SJ, Calderon-Fernández GM, Figueiras ANL, Juárez MP (2011) Epicuticle lipids mediate mate recognition in *Triatoma infestans*. J Chem Ecol 37:246–252

Cockcroft S (2021) Mammalian lipids: structure, synthesis and function. Essays Biochem 65:813–845

Cole TJ, Short KL, Hooper SB (2019) The science of steroids. Semin Fetal Neonatal Med 24:170–175

Colinet H, Renault D, Javal M, Berková P, Šimek P, Koštál V (2016) Uncovering the benefits of fluctuating thermal regimes on cold tolerance of *Drosophila* flies by combined metabolomic and lipidomic approach. Biochim Biophys Acta Mol Cell Biol Lipids 1861: 1736–1745

Coşkun M, Özalp P, Emre I (2005) Effects of Vitamin E concentrations on sex ratio of *Pimpla turionellae*

(Hymenoptera: Ichneumonidae) adults. Ann Entomol Soc Am 98:336–339

Coskun M, Kayis T, Gulsu E, Alp E (2020) Effects of selenium and Vitamin E on enzymatic, biochemical, and immunological biomarkers in *Galleria mellonella* L. Sci Rep 10:9953

D'Costa MA, Rutesasira A (1973) Phospholipid composition of flight muscle sarcosomes from the tsetse fly, *Glossina morsitans*. Comp Biochem Physiol—Part B Biochem 45:491–498

Dang X, Li Y, Li X et al (2021) Lipidomic profiling reveals distinct differences in sphingolipids metabolic pathway between healthy apis cerana cerana larvae and Chinese sacbrood disease. Insects. https://doi.org/10.3390/insects12080703

Darrouzet E, Rossignol F, Couillaud F (1998) The release of isoprenoids by locust corpora allata *in vitro*. J Insect Physiol 44:103–111

Dawaliby R, Trubbia C, Delporte C, Noyon C, Ruysschaert JM, Van Antwerpen P, Govaerts C (2016) Phosphatidylethanolamine is a key regulator of membrane fluidity in eukaryotic cells. J Biol Chem 291:3658–3667

Dempsey DR, Jeffries KA, Anderson RL, Carpenter AM, Rodriquez Opsina S, Merkler DJ (2014) Identification of an arylalkylamine N-acyltransferase from *Drosophila melanogaster* that catalyzes the formation of long-chain N-acylserotonins. FEBS Lett 588:594–599

Dennis RD, Geyer R, Egge H, Menges H, Stirm S, Wiegandt H (1985) Glycosphingolipids in insects: chemical structures of ceramide monosaccharide, disaccharide, and trisaccharide from pupae of *Calliphora vicina* (Insecta: Diptera). Eur J Biochem 146:51–58

Desmond E, Gribaldo S (2009) Phylogenomics of sterol synthesis: insights into the origin, evolution, and diversity of a key eukaryotic feature. Genome Biol Evol 1: 364–381

Dewett D, Labaf M, Lam-Kamath K, Zarringhalam K, Rister J (2021a) Vitamin A deficiency affects gene expression in the *Drosophila melanogaster* head. G3 Genes Genomes Genet 11:jkab297

Dewett D, Lam-Kamath K, Poupault C, Khurana H, Rister J (2021b) Mechanisms of vitamin A metabolism and deficiency in the mammalian and fly visual system. Dev Biol 476:68–78

Ding T, Kabir I, Li Y et al (2015) All members in the sphingomyelin synthase gene family have ceramide phosphoethanolamine synthase activity. J Lipid Res 56:537–545

Dobrosotskaya IY, Seegmiller AC, Brown MS, Goldstein JL, Rawson RB (2002) Regulation of SREBP processing and membrane lipid production by phospholipids in *Drosophila*. Science (80–) 296: 879–883

Donato P, Dugo P, Mondello L (2013) Chapter 9—Separation of lipids. Elsevier Science

Dong S, Sun A, Tan K, Nieh JC (2022) Identification of giant hornet *Vespa mandarinia* queen sex pheromone components. Curr Biol 32:R211–R212

Donzé G, Schnyder-Candrian S, Bogdanov S, Diehl PA, Guerin PM, Kilchenman V, Monachon F (1998) Aliphatic alcohols and aldehydes of the honey bee cocoon induce arrestment behavior in *Varroa jacobsoni* (Acari: Mesostigmata), an ectoparasite of *Apis mellifera*. Arch Insect Biochem Physiol 37:129–145

Dougherty M, Hamilton G (1997) Dodecanoic acid is the oviposition pheromone of *Lutzomyia longipalpis*. J Chem Ecol 23:2657–2671

Dowell P, Otto TC, Adi S, Lane MD (2003) Convergence of peroxisome proliferator-activated receptor γ and Foxo1 Signaling pathways. J Biol Chem 278: 45485–45491

Dupont S, Lemetais G, Ferreira T, Cayot P, Gervais P, Beney L (2012) Ergosterol biosynthesis: a fungal pathway for life on land? Evolution (NY) 66:2961–2968

Eigenbrode SD, Espelie KE (1995) Effects of plant epicuticular lipids on insect herbivores. Annu Rev Entomol 40:171–194

Eleftherianos I, Heryanto C, Bassal T, Zhang W, Tettamanti G, Mohamed A (2021) Haemocyte-mediated immunity in insects: cells, processes and associated components in the fight against pathogens and parasites. Immunology. https://doi.org/10.1111/imm.13390

Etges WJ, Jackson LL (2001) Epicuticular hydrocarbon variation in *Drosophila mojavensis* cluster species. J Chem Ecol 27:2125–2149

Fahy E, Subramaniam S, Brown HA et al (2005) A comprehensive classification system for lipids. J Lipid Res 46:839–861

Fahy E, Cotter D, Sud M, Subramaniam S (2011) Lipid classification, structures and tools. Biochim Biophys Acta Mol Cell Biol Lipids 1811:637–647

Fan P, Manoli DS, Ahmed OM et al (2013) Genetic and neural mechanisms that inhibit *Drosophila* from mating with other species. Cell 154:89–102

Fantini J, Yahi N (2015) Chemical basis of lipid biochemistry. In: Fantini J, Yahi NBT-BL in SF and ND (eds) Brain lipids in synaptic function and neurological disease. Academic, San Diego, pp 1–28

Farrell EK, Chen Y, Barazanji M, Jeffries KA, Cameroamortegui F, Merkler DJ (2012) Primary fatty acid amide metabolism: conversion of fatty acids and an ethanolamine in N 18TG 2 and SCP cells. J Lipid Res 53:247–256

Feldlaufer MF, Herbert EW, Svoboda JA, Thompson MJ (1986) Biosynthesis of makisterone A and 20-hydroxyecdysone from labeled sterols by the honey bee, *Apis mellifera*. Arch Insect Biochem Physiol 3:415–421

Feldlaufer MF, Weirich GF, Imberski RB, Svoboda JA (1995) Ecdysteroid production in *Drosophila melanogaster* reared on defined diets. Insect Biochem Mol Biol 25:709–712

Figueiredo MB, Genta FA, Garcia ES, Azambuja P (2008) Lipid mediators and vector infection: *Trypanosoma rangeli* inhibits *Rhodnius prolixus* hemocyte phagocytosis by modulation of phospholipase A2 and PAF-acetylhydrolase activities. J Insect Physiol 54: 1528–1537

Finke MD (2013) Complete nutrient content of four species of feeder insects. Zoo Biol 32:27–36

Flores-Alatorre HL, Abrego-Reyes V, Reyes-Esparza JA, Angeles E, Alba-Hurtado F (2014) Variation in the concentration of carminic acid produced by *Dactylopius coccus* (Hemiptera: Dactylopidae) at various maturation stages. J Econ Entomol 107:1700–1705

Fogang Mba AR, Kansci G, Viau M, Hafnaoui N, Meynier A, Demmano G, Genot C (2017) Lipid and amino acid profiles support the potential of *Rhynchophorus phoenicis* larvae for human nutrition. J Food Compos Anal 60:64–73

Folli C, Ramazzina I, Percudani R, Berni R (2005) Ligand-binding specificity of an invertebrate (*Manduca sexta*) putative cellular retinoic acid binding protein. Biochim Biophys Acta Proteins Proteomics 1747:229–237

Forte SN, Ferrero AA, Alonso TS (2002) Content and composition of phosphoglycerols and neutral lipids at different developmental stages of the eggs of the codling moth, *Cydia pomonella* (Lepidoptera: Tortricidae). Arch Insect Biochem Physiol 50:121–130

Foster SP, Roelofs WL (1996) Sex pheromone biosynthesis in the tortricid moth, *Ctenopseustis herana* (Felder & Rogenhofer). Arch Insect Biochem Physiol 33:135–147

Francke W, Bartels J, Meyer H, Schröder F, Kohnle U, Baader E, Pierre Vité J (1995) Semiochemicals from bark beetles: new results, remarks, and reflections. J Chem Ecol 21:1043–1063

Frere B, Suchaud F, Bernier G, Cottin F, Vincent B, Dourel L, Lelong A, Arpino P (2014) GC-MS analysis of cuticular lipids in recent and older scavenger insect puparia. An approach to estimate the postmortem interval (PMI). Anal Bioanal Chem 406:1081–1088

Frescas D, Valenti L, Accili D (2005) Nuclear trapping of the forkhead transcription factor FoxO1 *via* sirt-dependent deacetylation promotes expression of glucogenetic genes. J Biol Chem 280:20589–20595

Fronza E, Migues I, Specht A, De Barros NM, Heinzen H (2013) Identification of -tocopherol and -tocopheryl acetate from the cuticle of soybean pods armyworm (*Spodoptera cosmioides*). Nat Prod Res 27:1808–1811

Fujimoto Y, Kimura M, Takasu A, Khalifa FAM, Morisaki M, Ikekawa N (1984) Mechanism of stigmasterol dealkylation in insect. Tetrahedron Lett 25:1501–1504

Ganjam GK, Dimova EY, Unterman TG, Kietzmann T (2009) FoxO1 and HNF-4 are involved in regulation of hepatic glucokinase gene expression by resveratrol. J Biol Chem 284:30783–30797

Gao K, Su B, Dai J, Li P, Wang R, Yang X (2022) Antibiofilm and anti-hemolysis activities of 10-hydroxy-2-decenoic acid against *Staphylococcus aureus*. Molecules. https://doi.org/10.3390/molecules27051485

Gault CR, Obeid LM, Hannun YA (2010) An overview of sphingolipid metabolism: from synthesis to breakdown. Adv Exp Med Biol 688:1–23

Gaviraghi A, Correa Soares JBR, Mignaco JA, Fontes CFL, Oliveira MF (2019) Mitochondrial glycerol phosphate oxidation is modulated by adenylates through allosteric regulation of cytochrome c oxidase activity in mosquito flight muscle. Insect Biochem Mol Biol 114:103226

Gemeno C, Schal C (2009) Sex pheromones of cockroaches. In: Millar JG, Cardé RT (eds) Advances insect chemical ecology. Cambridge University Press, Cambridge, pp 179–247

Geun SK, Nalini M, Kim Y, Lee DW (2009) Octopamine and 5-hydroxytryptamine mediate hemocytic phagocytosis and nodule formation *via* eicosanoids in the beet armyworm, *Spodoptera exigua*. Arch Insect Biochem Physiol 70:162–176

Ghioni C, Bell JG, Sargent JR (1996) Polyunsaturated fatty acids in neutral lipids and phospholipids of some freshwater insects. Comp Biochem Physiol B Biochem Mol Biol 114:161–170

Gibbs A, Crowe JH (1991) Intra-individual variation in cuticular lipids studied using Fourier transform infrared spectroscopy. J Insect Physiol 37:743–748

Gibbs AG, Rajpurohit S (2010) Cuticular lipids and water balance. In: Blomquist GJ, Bagnères A-G (eds) Insect Hydrocarbons: biochemistry and chemical ecology. Cambridge University Press, Cambridge, pp 100–120

Gołębiowski M, Boguś MI, Paszkiewicz M, Wieloch W, Włóka E, Stepnowski P (2012a) The composition of the cuticular and internal free fatty acids and alcohols from *Lucilia sericata* males and females. Lipids 47: 613–622

Gołębiowski M, Dawgul M, Kamysz W, Bogucœ MI, Wieloch W, Włóka E, Paszkiewicz M, Przybysz E, Stepnowski P (2012b) Antimicrobial activity of alcohols from *Musca domestica*. J Exp Biol. https://doi.org/10.1242/jeb.073155

Gołębiowski M, Cerkowniak M, Boguś MI, Włóka E, Dawgul M, Kamysz W, Stepnowski P (2013a) Free fatty acids in the cuticular and internal lipids of *Calliphora vomitoria* and their antimicrobial activity. J Insect Physiol 59:416–429

Gołębiowski M, Cerkowniak M, Dawgul M, Kamysz W, Boguś MI, Stepnowski P (2013b) The antifungal activity of the cuticular and internal fatty acid methyl esters and alcohols in *Calliphora vomitoria*. Parasitology 140:972–985

Gołębiowski M, Urbanek A, Oleszczak A, Dawgul M, Kamysz W, Boguś MI, Stepnowski P (2014a) The antifungal activity of fatty acids of all stages of *Sarcophaga carnaria* L. (Diptera: Sarcophagidae). Microbiol Res 169:279–286

Gołębiowski M, Cerkowniak M, Urbanek A, Dawgul M, Kamysz W, Boguś MI, Sosnowska D, Stepnowski P (2014b) Antimicrobial activity of untypical lipid compounds in the cuticular and internal lipids of four fly species. J Appl Microbiol 116:269–287

Gołębiowski M, Cerkowniak M, Urbanek A, Dawgul M, Kamysz W, Boguś MI, Stepnowski P (2015) Identification and antifungal activity of novel organic compounds found in cuticular and internal lipids of medically important flies. Microbiol Res 170:213–222
Gondim KC, Atella GC, Pontes EG, Majerowicz D (2018) Lipid metabolism in insect disease vectors. Insect Biochem Mol Biol 101:108–123
Goodman WG, Cusson M (2012) The juvenile hormones. In: Insect Endocrinology. Elsevier, pp 310–365
Grillo LAM, Pontes EG, Gondim KC (2003) Lipophorin interaction with the midgut of *Rhodnius prolixus*: characterization and changes in binding capacity. Insect Biochem Mol Biol 33:429–438
Grillo LAM, Majerowicz D, Gondim KC (2007) Lipid metabolism in *Rhodnius prolixus* (Hemiptera: Reduviidae): role of a midgut triacylglycerol-lipase. Insect Biochem Mol Biol 37:579–588
Groen CM, Spracklen AJ, Fagan TN, Tootle TL (2012) *Drosophila* Fascin is a novel downstream target of prostaglandin signaling during actin remodeling. Mol Biol Cell 23:4567–4578
Gross DA, Snapp EL, Silver DL (2010) Structural insights into triglyceride storage mediated by fat storage-inducing transmembrane (FIT) protein 2. PLoS One 5:e10796
Guan XL, Cestra G, Shui G et al (2013) Biochemical membrane Lipidomics during *Drosophila* development. Dev Cell 24:98–111
Guo Y, Walther TC, Rao M, Stuurman N, Goshima G, Terayama K, Wong JS, Vale RD, Walter P, Farese RV (2008) Functional genomic screen reveals genes involved in lipid-droplet formation and utilization. Nature 453:657–661
Gutierrez-Mazariegos J, Schubert M, Laudet V (2014) Evolution of retinoic acid receptors and retinoic acid signaling. In: Asson-Batres MA, Rochette-Egly C (eds) Subcellular biochemistry. Springer Netherlands, Dordrecht, pp 55–73
Hadley NF (1994) Water relations of terrestrial arthropods. CUP Archive
Hafer A, Whittlesey R, Brown DT, Hernandez R (2009) Differential incorporation of cholesterol by sindbis virus grown in mammalian or insect cells. J Virol 83: 9113–9121
Hahn DA, Denlinger DL (2011) Energetics of insect diapause. Annu Rev Entomol 56:103–121
Hambalko J, Gajdoš P, Nicaud JM, Ledesma-Amaro R, Tupec M, Pichová I, Čertík M (2021) Production of long chain fatty alcohols found in bumblebee pheromones by *Yarrowia lipolytica*. Front Bioeng Biotechnol. https://doi.org/10.3389/fbioe.2020.593419
Hanson SM, Craig Jnr GB (1994) Cold acclimation, diapause, and geographic origin affect cold hardiness in eggs of *Aedes albopictus* (Diptera: Culicidae). J Med Entomol 31:192–201
Harayama T, Riezman H (2018) Understanding the diversity of membrane lipid composition. Nat Rev Mol Cell Biol 19:281–296
Harrison JF, Kaiser A, Vanden Brooks JM (2010) Atmospheric oxygen level and the evolution of insect body size. Proc R Soc B Biol Sci 277:1937–1946
Hasan MA, Ahmed S, Kim Y (2019) Biosynthetic pathway of arachidonic acid in *Spodoptera exigua* in response to bacterial challenge. Insect Biochem Mol Biol. https://doi.org/10.1016/j.ibmb.2019.103179
Haunerland NH (1997) Transport and utilization of lipids in insect flight muscles. Comp Biochem Physiol B Biochem Mol Biol 117:475–482
Hazel JR (1995) Thermal adaptation in biological membranes: is homeoviscous adaptation the explanation? Annu Rev Physiol 57:19–42
Heaver SL, Johnson EL, Ley RE (2018) Sphingolipids in host–microbial interactions. Curr Opin Microbiol 43: 92–99
Hefetz A, Blum MS (1978) Biosynthesis of formic acid by the poison glands of formicine ants. BBA Gen Subj 543:484–496
Heier C, Kühnlein RP (2018) Triacylglycerol metabolism in *Drosophila melanogaster*. Genetics 210:1163–1184
Helling F, Dennis R, Weske B, Nores G, Peter-Katalinic J, Dabrowski U, Egge H, Wiegandt H (1991) Glycosphingolipids in insects: the amphoteric moiety, N-acetylglucosamine-linked phosphoethanolamine, distinguishes a group of ceramide oligosaccharides from the pupae of *Calliphora vicina* (Insecta: Diptera). Eur J Biochem 200:409–421
Helvig C, Koener JF, Unnithan GC, Feyereisen R (2004) CYP15A1, the cytochrome P450 that catalyzes epoxidation of methyl farnesoate to juvenile hormone III in cockroach corpora allata. Proc Natl Acad Sci USA 101:4024–4029
Herr DR, Fyrst H, Phan V, Heinecke K, Georges R, Harris GL, Saba JD (2003) Sply regulation of sphingolipid signaling molecules is essential for *Drosophila* development. Development 130:2443–2453
Herrera H, Barros-Parada W, Bergmann J (2019) Linoleic acid and stearic acid are biosynthetic precursors of (7Z,10Z)-7,10-hexadecadienal, the major component of the sex pheromone of *Chilecomadia valdiviana* (Lepidoptera: Cossidae). PLoS One. https://doi.org/10.1371/journal.pone.0215769
Hildenbrandt GR, Abraham T, Bieber LL (1971) Metabolism of ceramide phosphorylethanolamine, phosphatidylinositol, phosphatidylserine and phosphatidylglycerol by housefly larvae. Lipids 6:508–516
Hofbauer HF, Heier C, Sen Saji AK, Kühnlein RP (2021) Lipidome remodeling in aging normal and genetically obese *Drosophila* males. Insect Biochem Mol Biol 133:103498
Hofsäß C, Lindahl E, Edholm O (2003) Molecular dynamics simulations of phospholipid bilayers with cholesterol. Biophys J 84:2192–2206

Holtof M, Lenaerts C, Cullen D, Vanden Broeck J (2019) Extracellular nutrient digestion and absorption in the insect gut. Cell Tissue Res 377:397–414

Holze H, Schrader L, Buellesbach J (2021) Advances in deciphering the genetic basis of insect cuticular hydrocarbon biosynthesis and variation. Heredity (Edinb) 126:219–234

Hou XY, Cao MH, Gong J, Li N, Gao A, Jia X, Ni SF, Zhao T, Zheng H (2011) Overview of pharmacological research of insect wax. J Anhui Agric Sci 39:2817–2818

House HL (1965) Effects of vitamin A acetate and structurally related substances on growth and reproduction of *Agria affinis* (Fallén) (Diptera: Sarcophagidae). J Insect Physiol 11:1039–1045

Howard RW, Blomquist GJ (2005) Ecological, behavioral, and biochemical aspects of insect hydrocarbons. Annu Rev Entomol 50:371–393

Huang X, Warren JT, Gilbert LI (2008) New players in the regulation of ecdysone biosynthesis. J Genet Genomics 35:1–10

Huang JH, Jing X, Douglas AE (2015) The multi-tasking gut epithelium of insects. Insect Biochem Mol Biol 67: 15–20

Hult EF, Tobe SS, Chang BSW (2011) Molecular evolution of ultraspiracle protein (USP/RXR) in insects. PLoS One 6:e23416

Igarashi F, Ogihara MH, Iga M, Kataoka H (2018) Cholesterol internalization and metabolism in insect prothoracic gland, a steroidogenic organ, *via* lipoproteins. Steroids 134:110–116

Inouye S (2010) Firefly luciferase: an adenylate-forming enzyme for multicatalytic functions. Cell Mol Life Sci 67:387–404

Itonori S, Nishizawa M, Suzuki M, Inagaki F, Hori T, Sugita M (1991) Polar glycosphingolipids in insect: chemical structures of glycosphingolipid series containing 2′--aminoethylphosphoryl-(→6)-N-acetylglucosamine as a polar group from larvae of the green-bottle fly, Lucilia caesar. J Biochem 110:479–485

Itonori S, Hashimoto K, Nakagawa M, Harada M, Suzuki T, Kojima H, Ito M, Sugita M (2018) Structural analysis of neutral glycosphingolipids from the silkworm *Bombyx mori* and the difference in ceramide composition between larvae and pupae. J Biochem 163:201–214

Jackson LL, Hadley NF, Blomquist GJ (1980) Epicuticular lipids of the desert tenebrionid beetle, *Eleodes armata*: identification of the branched hydrocarbons. Insect Biochem 10:399–402

Jarne C, Savirón M, Pilar Lapieza M, Membrado L, Orduna J, Galbán J, Garriga R, Morlock GE, Cebolla VL (2018) High-performance thin-layer chromatography coupled with electrospray ionization tandem mass spectrometry for identifying neutral lipids and sphingolipids in complex samples. J AOAC Int 101:1993–2000

Javed MGA (2002) Bioresidual and integrated pest management status of a biorational agent and a novel insecticide against whitefly and its key parasitoids. Int J Pest Manag 48:13–17

Jeffries KA, Dempsey DR, Behari AL, Anderson RL, Merkler DJ (2014) Drosophila melanogaster as a model system to study long-chain fatty acid amide metabolism. FEBS Lett 588:1596–1602

Jenni-Eiermann S, Srygley RB (2018) Physiological aeroecology: anatomical and physiological adaptations for flight. In: Aeroecology. Springer, pp 87–118

Ji H, Wang B, Labib D, Lei J, Chen X, Sapar M, Boulanger A, Dura J-M, Han C (2022) The *Drosophila* chemokine-like Orion bridges phosphatidylserine and Draper in phagocytosis of neurons. bioRxiv 2022.02.11.480148

Jiao WJ, Li FQ, Bai YL, Shi XX, Zhu MF, Zhang MJ, Mao CG, Zhu ZR (2017) Rice stripe virus infection alters mRNA levels of sphingolipid-metabolizing enzymes and sphingolipids content in laodelphax striatellus. J Insect Sci 17:16

Jing X, Behmer ST (2020) Insect sterol nutrition: physiological mechanisms, ecology, and applications. Annu Rev Entomol 65:251–271

Jones G, Strugnell SA, DeLuca HF (1998) Current understanding of the molecular actions of vitamin D. Physiol Rev 78:1193–1231

Juárez MP (2006) Metabolism of insect cuticular lipids. In: Encyclopedia of entomology. Springer Netherlands, Dordrecht, pp 1381–1383

Juárez P, Brenner RR (1989) Fatty acid biosynthesis in the integument tissue of *Triatoma infestans*. Comp Biochem Physiol—Part B Biochem 93:763–772

Juárez P, Chase J, Blomquist GJ (1992) A microsomal fatty acid synthetase from the integument of *Blattella germanica* synthesizes methyl-branched fatty acids, precursors to hydrocarbon and contact sex pheromone. Arch Biochem Biophys 293:333–341

Juarez MP, Ayala S, Brenner RR (1996) Methyl-branched fatty acid biosynthesis in *Triatoma infestans*. Insect Biochem Mol Biol 26:599–605

Jünger MA, Rintelen F, Stocker H, Wasserman JD, Végh M, Radimerski T, Greenberg ME, Hafen E (2003) The *Drosophila* Forkhead transcription factor FOXO mediates the reduction in cell number associated with reduced insulin signaling. J Biol 2:20

Jurenka R (2004a) Insect pheromone biosynthesis. In: Schulz S (ed) The chemistry of pheromones and other semiochemicals I. Springer, Berlin/Heidelberg, pp 97–132

Jurenka R (2004b) Insect pheromone biosynthesis. In: Schulz S (ed) Insect pheromone biosynth, pp 97–132

Jurenka RA, Pedibhotla VK, Stanley DW (1999) Prostaglandin production in response to a bacterial infection in true armyworm larvae. Arch Insect Biochem Physiol 41:225–232

Kaczmarek A (2018) PhD thesis [Wpływ metabolitów entomopatogennego grzyba *Conidiobolus coronatus*: kwasu oktanowego, kwasu 2-oktenowego oraz 2,6-dimetylofenolu na wybrane procesy fizjologiczne

Galleria mellonella oraz *Lucilia sericata*]. Witold Stefanski Institute Of Parasitology

Kaczmarek A, Boguś M (2021a) The metabolism and role of free fatty acids in key physiological processes in insects of medical, veterinary and forensic importance. PeerJ 9:e12563

Kaczmarek A, Boguś MI (2021b) The impact of the entomopathogenic fungus *Conidiobolus coronatus* on the free fatty acid profile of the flesh fly *Sarcophaga argyrostoma*. Insects. https://doi.org/10.3390/insects12110970

Kaczmarek A, Wrońska AK, Kazek M, Boguś MI (2022) Octanoic acid—an insecticidal metabolite of *Conidiobolus coronatus* (Entomopthorales) that affects two majors antifungal protection systems in *Galleria mellonella* (Lepidoptera): cuticular lipids and hemocytes. Int J Mol Sci. https://doi.org/10.3390/ijms23095204

Kamleh MA, Hobani Y, Dow JAT, Zheng L, Watson DG (2009) Towards a platform for the metabonomic profiling of different strains of *Drosophila melanogaster* using liquid chromatography-Fourier transform mass spectrometry. FEBS J 276:6798–6809

Kaplanis JN, Robbins WE, Monroe RE, Shortino TJ, Thompson MJ (1965) The utilization and fate of β-sitosterol in the larva of the housefly, *Musca domestica* L. J Insect Physiol 11:251–258

Kenwood BM, Merrill AH (2016) Lipidomics. In: Bradshaw RA, Stahl PDBT-E of CB (eds) Encyclopedia of cell biology. Academic, Waltham, pp 147–159

Kim Y, Stanley D (2021) Eicosanoid signaling in insect immunology: new genes and unresolved issues. Genes (Basel) 12:1–15

King B, Li S, Liu C, Kim SJ, Sim C (2020) Suppression of glycogen synthase expression reduces glycogen and lipid storage during mosquito overwintering diapause. J Insect Physiol 120:103971

Ko L, Harshman L, Hangartner S, Hoffmann A, Kachman S, Black P (2019) Changes in lipid classes of *Drosophila melanogaster* in response to selection for three stress traits. J Insect Physiol 117:103890

Kofroňová E, Cvačka J, Jiroš P, Sýkora D, Valterová I (2009) Analysis of insect triacylglycerols using liquid chromatography-atmospheric pressure chemical ionization-mass spectrometry. Eur J Lipid Sci Technol 111:519–525

Koštál V, Vambera J, Bastl J (2004) On the nature of pre-freeze mortality in insects: water balance, ion homeostasis and energy charge in the adults of *Pyrrhocoris apterus*. J Exp Biol 207:1509–1521

Koštál V, Urban T, Řimnáčová L, Berková P, Šimek P (2013) Seasonal changes in minor membrane phospholipid classes, sterols and tocopherols in overwintering insect, Pyrrhocoris apterus. J Insect Physiol 59:934–941

Kraut R (2011) Roles of sphingolipids in *Drosophila* development and disease. J Neurochem 116:764–778

Kumar JS, Suman S, Singh V, Chandna S (2012) Radioresistant Sf9 insect cells display moderate resistance against cumene hydroperoxide. Mol Cell Biochem 367:141–151

Kunte AS, Matthews KA, Rawson RB (2006) Fatty acid auxotrophy in *Drosophila* larvae lacking SREBP. Cell Metab 3:439–448

Kuzuyama T, Seto H (2003) Diversity of the biosynthesis of the isoprene units. Nat Prod Rep 20:171–183

Kwan L, Rundle HD (2010) Adaptation to desiccation fails to generate pre- and postmating isolation in replicate *Drosophila melanogaster* laboratory populations. Evolution (NY) 64:710–723

Kwon H, Hall DR, Smith RC (2021) Prostaglandin E2 signaling mediates oenocytoid immune cell function and lysis, limiting bacteria and plasmodium oocyst survival in *Anopheles gambiae*. Front Immunol 12: 680020

Ladisch RK, McQue B (1953) Methods of obtaining quinones from flour beetles. Science (80–) 118:324–325

Lafont R, Koolman J (2009) Diversity of ecdysteroids in animal species. In: Ecdysone: structures and functions. Springer, pp 47–71

Lafont R, Dauphin-Villemant C, Warren JT, Rees H (2005) Ecdysteroid chemistry and biochemistry. In: Gilbert LIBT-CMIS (ed) Comprehensive molecular insect science. Elsevier, Amsterdam, pp 125–195

Lang M, Murat S, Clark AG et al (2012) Mutations in the neverland gene turned *Drosophila* pachea into an obligate specialist species. Science (80–) 337:1658–1661

Laparie M, Larvor V, Frenot Y, Renault D (2012) Starvation resistance and effects of diet on energy reserves in a predatory ground beetle (*Merizodus soledadinus*; Carabidae) invading the Kerguelen Islands. Comp Biochem Physiol A Mol Integr Physiol 161:122–129

Lavrynenko O, Rodenfels J, Carvalho M, Dye NA, Lafont R, Eaton S, Shevchenko A (2015) The ecdysteroidome of *Drosophila*: influence of diet and development. Dev 142:3758–3768

Le Conte Y, Hefetz A (2008) Primer pheromones in social hymenoptera. Annu Rev Entomol 53:523–542

Le Conte Y, Arnold G, Trouiller J, Masson C, Chappe B (1990) Identification of a brood pheromone in honeybees. Naturwissenschaften 77:334–336

Legal L, Chappe B, Jallon JM (1994) Molecular basis of *Morinda citrifolia* (L.): toxicity on *Drosophila*. J Chem Ecol. https://doi.org/10.1007/BF02066234

Lehmann P, Westberg M, Tang P, Lindström L, Käkelä R (2020) The diapause lipidomes of three closely related beetle species reveal mechanisms for tolerating energetic and cold stress in high-latitude seasonal environments. Front Physiol. https://doi.org/10.3389/fphys.2020.576617

Li H, Wang H, Hackett M, Schooley DA (1995) The structure of dolichols isolated from *Manduca sexta* larvae. Insect Biochem Mol Biol 25:1019–1026

Li P, Lee H, Guo S, Unterman TG, Jenster G, Bai W (2003) AKT-independent protection of prostate cancer cells from apoptosis mediated through complex formation between the androgen receptor and FKHR. Mol Cell Biol 23:104–118

Li X, Zhang S, Blander G, Tse JG, Krieger M, Guarente L (2007) SIRT1 deacetylates and positively regulates the nuclear receptor LXR. Mol Cell 28:91–106

Li M, Wang G, Shang R, Xu Q, Zhang J, Sun R, Li L (2021) Comparative lipid profile analysis of *Hermetia illucens* larvae fed food waste at different days of age using an LC-MS-based lipidomics approach. J Insect Sci 21:17

Liu J, Jiang J, Zong J, Li B, Pan T, Diao Y, Zhang Z, Zhang X, Lu M, Wang S (2019) Antibacterial and antibiofilm effects of fatty acids extract of dried *Lucilia sericata* larvae against *Staphylococcus aureus* and *Streptococcus pneumoniae in vitro*. Nat Prod Res. https://doi.org/10.1080/14786419.2019.1627353

Löfgren H, Pascher I (1977) Molecular arrangements of sphingolipids. The monolayer behaviour of ceramides. Chem Phys Lipids 20:273–284

Lord JC, Anderson S, Stanley DW (2002) Eicosanoids mediate *Manduca sexta* cellular response to the fungal pathogen *Beauveria bassiana*: a role for the lipoxygenase pathway. Arch Insect Biochem Physiol 51:46–54

Lorenz MW (2001) Synthesis of lipids in the fat body of *Gryllus bimaculatus*: age-dependency and regulation by adipokinetic hormone. Arch Insect Biochem Physiol 47:198–214

Ma J, Li K, Zhang W, Ma L, Xu J, Liu L, Chen X, Zhang H (2022) Acute toxicity and chromosomal aberration toxicity of insect wax and its policosanol. Food Sci Human Wellness 11:356–365

Majerowicz D, Cezimbra MP, Alves-Bezerra M, Entringer PF, Atella GC, Sola-Penna M, Meyer-Fernandes JR, Gondim KC (2013) *Rhodnius prolixus* lipophorin: lipid composition and effect of high temperature on physiological role. Arch Insect Biochem Physiol 82: 129–140

Maldonado EN, Alderson NL, Monje PV, Wood PM, Hama H (2008) FA2H is responsible for the formation of 2-hydroxy galactolipids in peripheral nervous system myelin. J Lipid Res 49:153–161

Malhotra A, Xu Y, Ren M, Schlame M (2009) Formation of molecular species of mitochondrial cardiolipin. 1. A novel transacylation mechanism to shuttle fatty acids between sn-1 and sn-2 positions of multiple phospholipid species. Biochim Biophys Acta Mol Cell Biol Lipids 1791:314–320

Mallavadhani UV, Mahapatra A, Raja SS, Manjula C (2003) Antifeedant activity of some pentacyclic triterpene acids and their fatty acid ester analogues. J Agric Food Chem 51:1952–1955

Mansfield SG, Cammer S, Alexander SC, Muehleisen DP, Gray RS, Tropsha A, Bollenbacher WE (1998) Molecular cloning and characterization of an invertebrate cellular retinoic acid binding protein. Proc Natl Acad Sci USA 95:6825–6830

Mao C, Obeid LM (2008) Ceramidases: regulators of cellular responses mediated by ceramide, sphingosine, and sphingosine-1-phosphate. Biochim Biophys Acta Mol Cell Biol Lipids 1781:424–434

Marheineke K, Grünewald S, Christie W, Reilander H (1998) Lipid composition of *Spodoptera frugiperda* (Sf9) and *Trichoplusia ni* (Tn) insect cells used for baculovirus infection. FEBS Lett 441:49–52

Marin D, Dunphy GB, Mandato CA (2005) Cyclic AMP affects the haemocyte responses of larval *Galleria mellonella* to selected antigens. J Insect Physiol 51: 575–586

Martelli F, Hernandes NH, Zuo Z et al (2022) Low doses of the organic insecticide spinosad trigger lysosomal defects, elevated ROS, lipid dysregulation, and neurodegeneration in flies. elife 11:e73812

Maruyama S, Fujimoto Y, Morisaki M, Ikekawa N (1982) Mechanism of campesterol demethylation in insect. Tetrahedron Lett 23:1701–1704

Masood MA, Yuan C, Acharya JK, Veenstra TD, Blonder J (2010) Quantitation of ceramide phosphorylethanolamines containing saturated and unsaturated sphingoid base cores. Anal Biochem 400: 259–269

Matsuo N, Nagao K, Suito T, Juni N, Kato U, Hara Y, Umeda M (2019) Different mechanisms for selective transport of fatty acids using a single class of lipoprotein in *Drosophila*. J Lipid Res 60:1199–1211

Matthäus B, Piofczyk T, Katz H, Pudel F (2019) Renewable resources from insects: exploitation, properties, and refining of fat obtained by cold-pressing from *Hermetia illucens* (Black soldier Fly) larvae. Eur J Lipid Sci Technol 121:1800376

Matzkin LM, Watts TD, Markow TA (2009) Evolution of stress resistance in *Drosophila*: interspecific variation in tolerance to desiccation and starvation. Funct Ecol 23:521–527

Mayoral JG, Nouzova M, Navare A, Noriega FG (2009) NADP+-dependent farnesol dehydrogenase, a corpora allata enzyme involved in juvenile hormone synthesis. Proc Natl Acad Sci USA 106:21091–21096

McFarlane JE, Henneberry GO (1965) Inhibition of the growth of an insect by fatty acids. J Insect Physiol 11: 1247–1252

McFarlane IE, Steeves E, Alli I (1983) Aggregation of larvae of the house cricket, acheta domesticus (L.), by propionic acid present in the excreta. J Chem Ecol 9: 1307–1315

Michaud MR, Denlinger DL (2006) Oleic acid is elevated in cell membranes during rapid cold-hardening and pupal diapause in the flesh fly, *Sarcophaga crassipalpis*. J Insect Physiol 52:1073–1082

Minter M, Pearson A, Lim KS, Wilson K, Chapman JW, Jones CM (2018) The tethered flight technique as a tool for studying life-history strategies associated with migration in insects. Ecol Entomol 43:397–411

Mishra A, Singh AK, Kumar D (2004) Stage- and sex-dependent changes in the lipid profile of *Dysdercus koenigii* (Heteroptera: Pyrrhocoridae) during development. Int J Trop Insect Sci 24:236–241

Moadeli T, Mainali B, Ponton F, Taylor PW (2020) Effects of fatty acids and vitamin E in larval diets on development and performance of Queensland fruit fly. J Insect Physiol 125:104058

Morishima I, Yamano Y, Inoue K, Matsuo N (1997) Eicosanoids mediate induction of immune genes in the fat body of the silkworm, *Bombyx mori*. FEBS Lett 419:83–86

Mouritsen OG, Zuckermann MJ (2004) What's so special about cholesterol? Lipids 39:1101–1113

Mukherjee A, Sarkar N, Barik A (2014) Long-chain free fatty acids from *Momordica cochinchinensis* leaves as attractants to its insect pest, *Aulacophora foveicollis* Lucas (Coleoptera: Chrysomelidae). J Asia Pac Entomol 17:229–234

Murate M, Abe M, Kasahara K, Iwabuchi K, Umeda M, Kobayashi T (2015) Transbilayer distribution of lipids at nano scale. J Cell Sci 128:1627–1638

Nagle C, Bhogal JK, Nagengast AA, DiAngelo JR (2022) Transportin-serine/arginine-rich (Tnpo-SR) proteins are necessary for proper lipid storage in the *Drosophila* fat body. Biochem Biophys Res Commun 596:1–5

Nakahara Y, Nakamura M, Hiraoka T, Iwabuchi K (2002) Insect lipophorin and vertebrate lipoproteins support larval development of the endoparasitoid venturia canescens (Hymenoptera: Ichneumonidae) as dietary lipid sources. Appl Entomol Zool 37:637–643

Nakamura A, Stiebler R, Fantappié MR, Fialho E, Masuda H, Oliveira MF (2007) Effects of retinoids and juvenoids on moult and on phenoloxidase activity in the blood-sucking insect *Rhodnius prolixus*. Acta Trop 103:222–230

National Center for Biotechnology Information (2023) National Center for Biotechnology Information; 2004–. PubChem compound summary. National Library of Medicine (US), Bethesda

Nelson DR, Guershon M, Gerling D (1998) The surface wax composition of the exuviae and adults of *Aleyrodes singularis*. Comp Biochem Physiol B Biochem Mol Biol 119:655–665

Nelson DR, Freeman TP, Buckner JS (2000) Waxes and lipids associated with the external waxy structures of nymphs and pupae of the giant whitefly, *Aleurodicus dugesii*. Comp Biochem Physiol B Biochem Mol Biol 125:265–278

Němec V, Kodrík D, Matolín S, Laufer H (1993) Juvenile hormone-like effects of retinoic acid in insect metamorphosis, embryogenesis and reproduction. J Insect Physiol 39:1083–1093

Neven LG, Duman JG, Low MG, Sehl LC, Castellino FJ (1989) Purification and characterization of an insect hemolymph lipoprotein ice nucleator: evidence for the importance of phosphatidylinositol and apolipoprotein in the ice nucleator activity. J Comp Physiol B 159:71–82

Nguyen AD, McDonald JG, Bruick RK, DeBose-Boyd RA (2007) *Hypoxia stimulates* degradation of 3-hydroxy-3-methylglutaryl-coenzyme A reductase through accumulation of lanosterol and hypoxia-inducible factor-mediated induction of insigs. J Biol Chem 282:27436–27446

Nguyen SHT, Webb HK, Hasan J, Tobin MJ, Crawford RJ, Ivanova EP (2013) Dual role of outer epicuticular lipids in determining the wettability of dragonfly wings. Colloids Surf B Biointerfaces 106:126–134

Ni R, Bhandari S, Mitchell PR, Suarez G, Patel NB, Lamb K, Bisht KS, Merkler DJ (2021) Synthesis, quantification, and characterization of fatty acid amides from *in vitro* and *in vivo* sources. Molecules. https://doi.org/10.3390/molecules26092543

Niehs SP, Kumpfmüller J, Dose B, Little RF, Ishida K, Flórez LV, Kaltenpoth M, Hertweck C (2020) Insect-associated bacteria assemble the antifungal butenolide gladiofungin by non-canonical polyketide chain termination. Angew Chemie Int Ed 59:23122–23126

Niki E, Traber MG (2012) A history of vitamin E. Ann Nutr Metab 61:207–212

Novak M, Krpan T, Panevska A, Shewell LK, Day CJ, Jennings MP, Guella G, Sepčić K (2020) Binding specificity of ostreolysin A6 towards Sf9 insect cell lipids. Biochim Biophys Acta Biomembr 1862: 183307

Nowickyj SM, Chithalen JV, Cameron D, Tyshenko MG, Petkovich M, Wyatt GR, Jones G, Walker VK (2008) Locust retinoid X receptors: 9-Cis-retinoic acid in embryos from a primitive insect. Proc Natl Acad Sci USA 105:9540–9545

O'Hagan D (1992) The polyketide metabolites. Gen Pharmacol Vasc Syst. https://doi.org/10.1016/0306-3623(92)90327-g

Obniski R, Sieber M, Spradling AC (2018) Dietary lipids modulate notch signaling and influence adult intestinal development and metabolism in *Drosophila*. Dev Cell 47:98–111.e5

Ochiai M, Komiya Y (2021) Detection of edible insect derived phospholipids with polyunsaturated fatty acids by thin-layer chromatography, gas chromatography, and enzymatic methods. J Food Compos Anal 99: 103869

Ohta M, Matsuura F, Henderson G, Laine RA (2007) Novel free ceramides as components of the soldier defense gland of the Formosan subterranean termite (Coptotermes formosanus). J Lipid Res 48:656–664

Olademehin OP, Liu C, Rimal B, Adegboyega NF, Chen F, Sim C, Kim SJ (2020) Dsi-RNA knockdown of genes regulated by Foxo reduces glycogen and lipid accumulations in diapausing *Culex pipiens*. Sci Rep 10:17201

Olarte MJ, Kim S, Sharp ME, Swanson JMJ, Farese RV, Walther TC (2020) Determinants of endoplasmic reticulum-to-lipid droplet protein targeting. Dev Cell 54:471–487.e7

Oonincx DGAB, Van Keulen P, Finke MD, Baines FM, Vermeulen M, Bosch G (2018) Evidence of vitamin D synthesis in insects exposed to UVb light. Sci Rep 8: 10807

Overgaard J, Sørensen JG, Petersen SO, Loeschcke V, Holmstrup M (2005) Changes in membrane lipid composition following rapid cold hardening in *Drosophila melanogaster*. J Insect Physiol 51:1173–1182

Overgaard J, Sørensen JG, Petersen SO, Loeschcke V, Holmstrup M (2006) Reorganization of membrane lipids during fast and slow cold hardening in *Drosophila melanogaster*. Physiol Entomol 31:328–335

Overgaard J, Tomčala A, Sørensen JG, Holmstrup M, Krogh PH, Šimek P, Koštál V (2008) Effects of acclimation temperature on thermal tolerance and membrane phospholipid composition in the fruit fly *Drosophila melanogaster*. J Insect Physiol 54: 619–629

Panevska A, Hodnik V, Skočaj M et al (2019) Pore-forming protein complexes from Pleurotus mushrooms kill western corn rootworm and Colorado potato beetle through targeting membrane ceramide phosphoethanolamine. Sci Rep 9:5073

Pankewitz F, Hilker M (2008) Polyketides in insects: ecological role of these widespread chemicals and evolutionary aspects of their biogenesis. Biol Rev 83:209–226

Parathath S, Connelly MA, Rieger RA, Klein SM, Abumrad NA, De La Llera-Moya M, Iden CR, Rothblat GH, Williams DL (2004) Changes in plasma membrane properties and phosphatidylcholine subspecies of insect Sf9 cells due to expression of scavenger receptor class B, type I, and CD36. J Biol Chem 279: 41310–41318

Park J, Stanley D, Kim Y (2014) Roles of peroxinectin in PGE2-mediated cellular immunity in *Spodoptera exigua*. PLoS One 9:e105717

Parkash R, Ranga P, Aggarwal DD (2014) Developmental acclimation to low or high humidity conditions affect starvation and heat resistance of *Drosophila melanogaster*. Comp Biochem Physiol Part A Mol Integr Physiol 175:46–56

Parker RS, McCormick CC (2005) Selective accumulation of α-tocopherol in *Drosophila* is associated with cytochrome P450 tocopherol-ω-hydroxylase activity but not α-tocopherol transfer protein. Biochem Biophys Res Commun 338:1537–1541

Pedrini N, Crespo R, Juárez MP (2007) Biochemistry of insect epicuticle degradation by entomopathogenic fungi. Comp Biochem Physiol C Toxicol Pharmacol 146:124–137

Pei X-J, Fan Y-L, Bai Y, Bai T-T, Schal C, Zhang Z-F, Chen N, Li S, Liu T-X (2021) Modulation of fatty acid elongation in cockroaches sustains sexually dimorphic hydrocarbons and female attractiveness. PLoS Biol 19: e3001330

Pennington JE, Wells MA (2002) Triacylglycerol-rich lipophorins are found in the dipteran infraorder Culicomorpha, not just in mosquitoes. J Insect Sci. https://doi.org/10.1093/jis/2.1.15

Perera R, Riley C, Isaac G, Hopf-Jannasch AS, Moore RJ, Weitz KW, Pasa-Tolic L, Metz TO, Adamec J, Kuhn RJ (2012) Dengue virus infection perturbs lipid homeostasis in infected mosquito cells. PLoS Pathog 8:e1002584

Petkevicius K, Löfstedt C, Borodina I (2020) Insect sex pheromone production in yeasts and plants. Curr Opin Biotechnol 65:259–267

Petroski RJ, Bartelt RJ, Weisleder D (1994) Biosynthesis of (2E,4E,6E)-5-ethyl-3-methyl-2,4,6-nonatriene: the aggregation pheromone of *Carpophilus freemani* (Coleoptera: Nitidulidae). Insect Biochem Mol Biol 24:69–78

Petryk A, Warren JT, Marqués G, Jarcho MP, Gilbert LI, Kahler J, Parvy JP, Li Y, Dauphin-Villemant C, O'Connor MB (2003) Shade is the *Drosophila* P450 enzyme that mediates the hydroxylation of ecdysone to the steroid insect molting hormone 20-hydroxyecdysone. Proc Natl Acad Sci USA 100: 13773–13778

Phan VH, Herr DR, Panton D, Fyrst H, Saba JD, Harris GL (2007) Disruption of sphingolipid metabolism elicits apoptosis-associated reproductive defects in *Drosophila*. Dev Biol 309:329–341

Pinch M, Mitra S, Rodriguez SD, Li Y, Kandel Y, Dungan B, Holguin FO, Attardo GM, Hansen IA (2021) Fat and happy: profiling mosquito fat body lipid storage and composition post-blood meal. Front Insect Sci. https://doi.org/10.3389/finsc.2021.693168

Poelchau MF, Reynolds JA, Elsik CG, Denlinger DL, Armbruster PA (2013a) Deep sequencing reveals complex mechanisms of diapause preparation in the invasive mosquito, *Aedes albopictus*. Proc R Soc B Biol Sci 280:20130143

Poelchau MF, Reynolds JA, Elsik CG, Denlinger DL, Armbruster PA (2013b) RNA-Seq reveals early distinctions and late convergence of gene expression between diapause and quiescence in the Asian tiger mosquito, *Aedes albopictus*. J Exp Biol 216:4082–4090

Pontes EG, Grillo LAM, Gondim KC (2002) Characterization of lipophorin binding to the fat body of *Rhodnius prolixus*. Insect Biochem Mol Biol 32: 1409–1417

Pontes EG, Leite P, Majerowicz D, Atella GC, Gondim KC (2008) Dynamics of lipid accumulation by the fat body of *Rhodnius prolixus*: the involvement of lipophorin binding sites. J Insect Physiol 54:790–797

Puig O, Marr MT, Ruhf ML, Tjian R (2003) Control of cell number by *Drosophila* FOXO: downstream and feedback regulation of the insulin receptor pathway. Genes Dev 17:2006–2020

Qiuping M, Fu W, Li P, Nicosia SV, Jenster G, Zhang X, Bai W (2009) FoxO1 mediates PTEN suppression of androgen receptor N- and C-terminal interactions and coactivator recruitment. Mol Endocrinol 23:213–225

Quinville BM, Deschenes NM, Ryckman AE, Walia JS (2021) A comprehensive review: sphingolipid metabolism and implications of disruption in sphingolipid homeostasis. Int J Mol Sci. https://doi.org/10.3390/ijms22115793

Ramos-Martín F, Herrera-León C, D'Amelio N (2022) *Bombyx mori* Cecropin D could trigger cancer cell apoptosis by interacting with mitochondrial cardiolipin. Biochim Biophys Acta Biomembr 1864: 184003

Ramsey JS, Johnson RS, Hoki JS et al (2015) Metabolic interplay between the asian citrus psyllid and its profftella symbiont: an achilles' heel of the citrus greening insect vector. PLoS One 10:e0140826

Rawson RB (2003) The SREBP pathway—insights from insigs and insects. Nat Rev Mol Cell Biol 4:631–640

Renault D, Hervant F, Vernon P (2002) Comparative study of the metabolic responses during food shortage and subsequent recovery at different temperatures in the adult lesser mealworm, *Alphitobius diaperinus* (Coleoptera: Tenebrionidae). Physiol Entomol 27:291–301

Rewitz KF, Yamanaka N, O'Connor MB (2013) Developmental checkpoints and feedback circuits time insect maturation. In: Shi Y-BBT-CT in DB (ed) Current topics in developmental biology. Academic, pp 1–33

Reynolds JA, Poelchau MF, Rahman Z, Armbruster PA, Denlinger DL (2012) Transcript profiling reveals mechanisms for lipid conservation during diapause in the mosquito, *Aedes albopictus*. J Insect Physiol 58: 966–973

Ribeiro-Neto JA, Pinto MEA, Ferreira VV, Tibúrcio JD, Varotti FP, Azevedo DO, Siqueira-Filho EP, Serrão JE, Lima LAS, Alves SN (2017) Larvicidal activity of vegetable oils and esterified compounds against *Culex quinquefasciatus* (Diptera: Culicidae). Ecotoxicol Environ Saf 143:57–61

Ricke SC, Dittoe DK, Richardson KE (2020) Formic acid as an antimicrobial for poultry production: a review. Front Vet Sci 7:563

Rietveld A, Neutz S, Simons K, Eaton S (1999) Association of sterol- and glycosylphosphatidylinositol-linked proteins with *Drosophila* raft lipid microdomains. J Biol Chem 274:12049–12054

Robb R, Hammond R, Bieber LL (1972) Temperature-dependent changes in sphingosine composition and composition of fatty acids of glycerophosphatides from *Musca domestica* larvae. Insect Biochem 2: 131–136

Robbins WE, Dutky RC, Monroe RE, Kaplanis JN (1962) The metabolism of H3-β-sitosterol by the german cockroach1. Ann Entomol Soc Am 55:102–104

Roberts D (2004) Prolonged survival of eggs of the rock-pool mosquito, *Aedes vittatus*, in the extreme heat of the Arabian peninsula. J Arid Environ 57:203–210

Rodenburg KW, Van Der Horst DJ (2005) Lipoprotein-mediated lipid transport in insects: analogy to the mammalian lipid carrier system and novel concepts for the functioning of LDL receptor family members. Biochim Biophys Acta Mol Cell Biol Lipids 1736:10–29

Rodríguez-Vázquez M, Vaquero D, Parra-Peralbo E, Mejía-Morales JE, Culi J (2015) *Drosophila* lipophorin receptors recruit the lipoprotein ltp to the plasma membrane to mediate lipid uptake. PLoS Genet 11:e1005356

Rohmer M (1999) The discovery of a mevalonate-independent pathway for isoprenoid biosynthesis in bacteria, algae and higher plants. Nat Prod Rep 16: 565–574

Rosenfeld E, Beauvoit B (2003) Role of the non-respiratory pathways in the utilization of molecular oxygen by *Saccharomyces cerevisiae*. Yeast 20:1115–1144

Rossini C, Attygalle AB, González A, Smedley SR, Eisner M, Meinwald J, Eisner T (1997) Defensive production of formic acid (80%) by a carabid beetle (*Galerita lecontei*). Proc Natl Acad Sci USA 94: 6792–6797

Roubik DW, Smith BH, Carlson RG (1987) Formic acid in caustic cephalic secretions of stingless bee, *Oxytrigona* (Hymenoptera: Apidae). J Chem Ecol 13:1079–1086

Rourke BC (2000) Geographic and altitudinal variation in water balance and metabolic rate in a California grasshopper, *Melanoplus sanguinipes*. J Exp Biol 203: 2699–2712

Roy MC, Nam K, Kim J, Stanley D, Kim Y (2021) Thromboxane mobilizes insect blood cells to infection foci. Front Immunol 12:791319

Sadekuzzaman M, Gautam N, Kim Y (2017) A novel calcium-independent phospholipase A2 and its physiological roles in development and immunity of a lepidopteran insect, *Spodoptera exigua*. Dev Comp Immunol 77:210–220

Sajjadian SM, Kim Y (2020) PGE 2 upregulates gene expression of dual oxidase in a lepidopteran insect midgut *via* cAMP signalling pathway. Open Biol 10: 200197

Salerno AP, Dansa-Petretski M, Silva-Neto MSL, Coelho HSL, Masuda H (2002) *Rhodnius prolixus* vitellin is composed of three different populations: comparison with vitellogenin. Insect Biochem Mol Biol 32:709–717

Satake S, Kawabe Y, Mizoguchi A (2000) Carbohydrate metabolism during starvation in the silkworm *Bombyx mori*. Arch Insect Biochem Physiol 44:90–98

Sawabe K, Moribayashi A (2000) Lipid utilization for ovarian development in an autogenous mosquito, *Culex pipiens* molestus (Diptera: Culicidae). J Med Entomol 37:726–731

Schlame M (2009) Formation of molecular species of mitochondrial cardiolipin. 2. A mathematical model of pattern formation by phospholipid transacylation. Biochim Biophys Acta Mol Cell Biol Lipids 1791: 321–325

Schrader L, Simola DF, Heinze J, Oettler J (2015) Sphingolipids, transcription factors, and conserved toolkit genes: developmental plasticity in the ant cardiocondyla obscurior. Mol Biol Evol 32:1474–1486

Seegmiller AC, Dobrosotskaya I, Goldstein JL, Ho YK, Brown MS, Rawson RB (2002) The SREBP pathway in *Drosophila*: regulation by palmitate, not sterols. Dev Cell 2:229–238

Seki T, Fujishita S, Ito M, Matsuoka N, Tsukida K (1987) Retinoid composition in the compound eyes of insects. Exp Biol 47:95–103

Sengupta P, Baird B, Holowka D (2007) Lipid rafts, fluid/fluid phase separation, and their relevance to plasma membrane structure and function. Semin Cell Dev Biol 18:583–590

Seppo A, Tiemeyer M (2000) Function and structure of *Drosophila* glycans. Glycobiology 10:751–760

Seppo A, Moreland M, Schweingruber H, Tiemeyer M (2000) Zwitterionic and acidic glycosphingolipids of the *Drosophila melanogaster* embryo. Eur J Biochem 267:3549–3558

Sharma S, Shen T, Chitranshi N et al (2022) Retinoid X receptor: cellular and biochemical roles of nuclear receptor with a focus on neuropathological involvement. Mol Neurobiol 59:2027–2050

Shelley RM, Hodgson E (1970) Biosynthesis of phosphatidylcholine in the fat body of Phormia regina larvae. J Insect Physiol 16:131–139

Shi XX, Huang YJ, Begum MA, Zhu MF, Li FQ, Zhang MJ, Zhou WW, Mao C, Zhu ZR (2018) A neutral ceramidase, NlnCDase, is involved in the stress responses of brown planthopper, *Nilaparvata lugens* (Stål). Sci Rep 8:1130

Shi XX, Zhu MF, Wang N et al (2021) Neutral ceramidase is required for the reproduction of brown planthopper, *Nilaparvata lugens* (Stål). Front Physiol. https://doi.org/10.3389/fphys.2021.629532

Shi XX, Zhang H, Quais MK, Chen M, Wang N, Zhang C, Mao C, Zhu ZR (2022) Knockdown of sphingomyelinase (NlSMase) causes ovarian malformation of brown planthopper, *Nilaparvata lugens* (Stål). Insect Mol Biol 31:391–402

Shinoda T, Itoyama K (2003) Juvenile hormone acid methyltransferase: a key regulatory enzyme for insect metamorphosis. Proc Natl Acad Sci USA 100:11986–11991

Shiomi A, Nagao K, Yokota N et al (2021) Extreme deformability of insect cell membranes is governed by phospholipid scrambling. Cell Rep 35:109219

Shrestha S, Kim Y (2008) Eicosanoids mediate prophenoloxidase release from oenocytoids in the beet armyworm *Spodoptera exigua*. Insect Biochem Mol Biol 38:99–112

Shrestha S, Park J, Ahn SJ, Kim Y (2015) Pge2 mediates oenocytoid cell lysis *via* a sodium-potassium-chloride cotransporter. Arch Insect Biochem Physiol 89:218–229

Shreve SM, Yi SX, Lee RE (2007) Increased dietary cholesterol enhances cold tolerance in *Drosophila melanogaster*. Cryo-Letters 28:33–37

Silva ERMN, Santos LV, Caiado MS, Hastenreiter LSN, Fonseca SRR, Carbajal-de-la-Fuente AL, Carvalho MG, Pontes EG (2021) The influence of larval density on triacylglycerol content in *Aedes aegypti* (Linnaeus) (Diptera: Culicidae). Arch Insect Biochem Physiol 106:e21757

Sims SR, Balusu RR, Ngumbi EN, Appel AG (2014) Topical and vapor toxicity of saturated fatty acids to the german cockroach (Dictyoptera: Blattellidae). J Econ Entomol 107:758–763

Sinclair BJ, Marshall KE (2018) The many roles of fats in overwintering insects. J Exp Biol. https://doi.org/10.1242/jeb.161836

Singh KP, Jayasomu RS (2002) *Bombyx mori*—a review of its potential as a medicinal insect. Pharm Biol 40:28–32

Skowronek P, Wójcik Ł, Strachecka A (2021) Fat body—multifunctional insect tissue. Insects. https://doi.org/10.3390/insects12060547

Snyder F (1996) Chapter 7 Ether-linked lipids and their bioactive species: occurrence, chemistry, metabolism, regulation, and function. In: Vance DE, Vance JEBT-NCB (eds) New comprehensive biochemistry. Elsevier, pp 183–210

Soliday CL, Blomquist GJ, Jackson LL (1974) Cuticular lipids of insects. VI. Cuticular lipids of the grasshoppers *Melanopulus sanguinipes* and *Melanoplus packardii*. J Lipid Res 15:399–405

Sota T, Mogi M (1992) Survival time and resistance to desiccation of diapause and non-diapause eggs of temperate *Aedes* (Stegomyia) mosquitoes. Entomol Exp Appl 63:155–161

Soulages JL, Wells MA (1994) Metabolic fate and turnover rate of hemolymph free fatty acids in adult *Manduca sexta*. Insect Biochem Mol Biol 24:79–86

Soulages JL, Firdaus SJ, Hartson S, Chen X, Howard AD, Arrese EL (2012) Developmental changes in the protein composition of *Manduca sexta* lipid droplets. Insect Biochem Mol Biol 42:305–320

Sparagana SP, Bhaskaran G, Barrera P (1985) Juvenile hormone acid methyltransferase activity in imaginal discs of *Manduca sexta* prepupae. Arch Insect Biochem Physiol 2:191–202

Spates GE, Bull DL, Chen AC (1990) Hydrolysis of sphingomyelin and phosphatidylcholine by midgut homogenates of the stable fly. Arch Insect Biochem Physiol 14:1–12

Sridhara S, Bhat JV (1965) Lipid composition of the silkworm *Bombyx mori* L. J Insect Physiol 11:449–462

St. Agatha Suter M, Ladisch RK (1963) Toxicity of insect quinones to mice. Proc Pennsylvania Acad Sci 37:137–141

Stanley D, Kim Y (2014) Eicosanoid signaling in insects: from discovery to plant protection. CRC Crit Rev Plant Sci 33:20–63

Stanley D, Kim Y (2019) Insect prostaglandins and other eicosanoids: from molecular to physiological actions. In: Jurenka RBT-A in IP (ed) Advances in insect physiology. Academic, pp 283–343

Stanley D, Kim Y (2020) Why most insects have very low proportions of C20 polyunsaturated fatty acids: the oxidative stress hypothesis. Arch Insect Biochem Physiol 103:e21622

Stanley DW, Miller JS (2006) Eicosanoid actions in insect cellular immune functions. Entomol Exp Appl 119:1–13

Stanley-Samuelson DW, Jensen E, Nickerson KW, Tiebel K, Ogg CL, Howard RW (1991) Insect immune response to bacterial infection is mediated by eicosanoids. Proc Natl Acad Sci USA 88:1064–1068

Suga T, Shishibori T (1979) Biotransformation of polyprenols by the larvae of *Pieris rapae crucivora*. Experientia 35:1423–1424

Sukiban J, Bräunig P, Mey J, Bui-Göbbels K (2014) Retinoic acid as a survival factor in neuronal development of the grasshopper, Locusta migratoria. Cell Tissue Res 358:303–312

Svoboda JA, Feldlaufer MF (1991) Neutral sterol metabolism in insects. Lipids 26:614–618

Svoboda JA, Thompson MJ, Robbins WE (1971) Identification of fucosterol as a metabolite and probable intermediate in conversion of ß-sitosterol to cholesterol in the tobacco hornworm. Nat New Biol 230: 57–58

Svoboda JA, Thompson MJ, Robbins WE, Elden TC (1975) Unique pathways of sterol metabolism in the Mexican bean beetle, a plant-feeding insect. Lipids 10: 524–527

Svoboda JA, Thompson MJ, Herbert EW, Shimanuki H (1980) Sterol utilization in honey bees fed a synthetic diet: analysis of prepupal sterols. J Insect Physiol 26:291–294

Svoboda JA, Cohen CF, Lusby WR, Thompson MJ (1986) New intermediates in the conversion of stigmasterol to cholestanol in the Mexican bean beetle. Lipids 21: 639–642

Takeshima M, Ogihara MH, Kataoka H (2019) Sterol characteristics in silkworm brain and various tissues characterized by precise sterol profiling using LC-MS/MS. Int J Mol Sci. https://doi.org/10.3390/ijms20194840

Takken W, Smallegange RC, Vigneau AJ, Johnston V, Brown M, Mordue-Luntz AJ, Billingsley PF (2013) Larval nutrition differentially affects adult fitness and Plasmodium development in the malaria vectors *Anopheles gambiae* and *Anopheles gambiae*. Parasites Vectors 6:1–10

Tanaka F, Magariyama Y, Miyanoshita A (2020) Volatile biomarkers for early-stage detection of insect-infested brown rice: Isopentenols and polysulfides. Food Chem 303:125381

Teets NM, Denlinger DL (2013) Physiological mechanisms of seasonal and rapid cold-hardening in insects. Physiol Entomol 38:105–116

Teets NM, Gantz JD, Kawarasaki Y (2020) Rapid cold hardening: ecological relevance, physiological mechanisms and new perspectives. J Exp Biol. https://doi.org/10.1242/jeb.203448

Telang A, Wells MA (2004) The effect of larval and adult nutrition on successful autogenous egg production by a mosquito. J Insect Physiol 50:677–685

Terra WR, Costa RH, Ferreira C (2006) Plasma membranes from insect midgut cells. An Acad Bras Cienc 78:255–269

Th M, Beenakkers A, Van der Horst DJ, Van Marrewijk WJA (1985) Insect lipids and lipoproteins, and their role in physiological processes. Prog Lipid Res 24:19–67

Thiel K, Heier C, Haberl V, Thul PJ, Oberer M, Lass A, Jäckle H, Beller M (2013) The evolutionarily conserved protein CG9186 is associated with lipid droplets, required for their positioning and for fat storage. J Cell Sci 126:2198–2212

Tiezt A (1967) Fat transport in the locust: the role of diglycerides. Eur J Biochem 2:236–242

Tillman JA, Seybold SJ, Jurenka RA, Blomquist GJ (1999) Insect pheromones—an overview of biosynthesis and endocrine regulation. Insect Biochem Mol Biol 29:481–514

Tinto WF, Elufioye TO, Roach J (2017) Waxes. In: Badal S, Delgoda RBT-P (eds) Pharmacognosy: fundamentals, applications and strategies. Academic, Boston, pp 443–455

Tomčala A, Tollarová M, Overgaard J, Šimek P, Koštál V (2006) Seasonal acquisition of chill tolerance and restructuring of membrane glycerophospholipids in an overwintering insect: triggering by low temperature, desiccation and diapause progression. J Exp Biol 209: 4102–4114

Toprak U (2020) The role of peptide hormones in insect lipid metabolism. Front Physiol. https://doi.org/10.3389/fphys.2020.00434

Toprak U, Guz N, Gurkan MO, Hegedus DD (2014) Identification and coordinated expression of perilipin genes in the biological cycle of sunn pest, Eurygaster maura (Hemiptera: Scutelleridae): implications for lipolysis and lipogenesis. Comp Biochem Physiol B Biochem Mol Biol 171:1–11

Tortoriello G, Rhodes BP, Takacs SM, Stuart JM, Basnet A, Raboune S, Widlanski TS, Doherty P, Harkany T, Bradshaw HB (2013) Targeted lipidomics in *Drosophila melanogaster* identifies novel 2-monoacylglycerols and N-acyl amides. PLoS One 8:e67865

Trenti F, Sandron T, Guella G, Lencioni V (2022) Insect cold-tolerance and lipidome: membrane lipid composition of two chironomid species differently adapted to cold. Cryobiology 106:84–90

Tsochatzis ED, Berggreen IE, Vidal NP, Roman L, Gika H, Corredig M (2022) Cellular lipids and protein alteration during biodegradation of expanded polystyrene by mealworm larvae under different feeding conditions. Chemosphere 300:134420

Tulloch AP (1970) The composition of beeswax and other waxes secreted by insects. Lipids 5:247–258

Turunen S, Pärnänen S (1987) Eicosapentaenoic acid: incorporation into phosphatidylinositol and other lipids in Pieris brassicae. Insect Biochem 17:891–895

Uchida Y, Hama H, Alderson NL, Douangpanya S, Wang Y, Crumrine DA, Elias PM, Holleran WM (2007) Fatty acid 2-hydroxylase, encoded by FA2H, accounts for differentiation-associated increase in 2-OH ceramides during keratinocyte differentiation. J Biol Chem 282:13211–13219

Ueda K, Anderson-Baron MN, Haskins J, Hughes SC, Simmonds AJ (2022) Recruitment of Peroxin 14 to lipid droplets affects lipid storage in *Drosophila*. J Cell Sci. https://doi.org/10.1242/jcs.259092

Urbanek A, Szadziewski R, Stepnowski P, Boros-Majewska J, Gabriel I, Dawgul M, Kamysz W, Sosnowska D, Gołębiowski M (2012) Composition and antimicrobial activity of fatty acids detected in the hygroscopic secretion collected from the secretory setae of larvae of the biting midge *Forcipomyia nigra* (Diptera: Ceratopogonidae). J Insect Physiol 58: 1265–1276

Ushakova NA, Brodskii ES, Kovalenko AA, Bastrakov AI, Kozlova AA, Pavlov DS (2016) Characteristics of lipid fractions of larvae of the black soldier fly *Hermetia illucens*. Dokl Biochem Biophys 468: 209–212

Van den Bergh SG (1967) [22] insect mitochondria. In: Methods in enzymology. Academic, pp 117–122

Van Der Horst DJ (2003) Insect adipokinetic hormones: release and integration of flight energy metabolism. Comp Biochem Physiol B Biochem Mol Biol 136: 217–226

Van Der Horst DJ, Rodenburg KW (2010) Lipoprotein assembly and function in an evolutionary perspective. Biomol Concepts 1:165–183

Van der Horst DJ, Van Hoof D, Van Marrewijk WJA, Rodenburg KW (2002) Alternative lipid mobilization: the insect shuttle system. Mol Cell Biochem 239: 113–119

Van Meer G, Voelker DR, Feigenson GW (2008) Membrane lipids: where they are and how they behave. Nat Rev Mol Cell Biol 9:112–124

Van Moll L, De Smet J, Cos P, Van Campenhout L (2021) Microbial symbionts of insects as a source of new antimicrobials: a review. Crit Rev Microbiol 47: 562–579

Van Oystaeyen A, Oliveira RC, Holman L et al (2014) Conserved class of queen pheromones stops social insect workers from reproducing. Science (80–) 343: 287–290

Vatanparast M, Ahmed S, Lee DH, Hwang SH, Hammock B, Kim Y (2020) EpOMEs act as immune suppressors in a lepidopteran insect, *Spodoptera exigua*. Sci Rep 10:20183

Vinci G, Xia X, Veitia RA (2008) Preservation of genes involved in sterol metabolism in cholesterol auxotrophs: facts and hypotheses. PLoS One 3:e2883

Voelker DR (2013) Glycerolipid structure, function, and synthesis in eukaryotes. In: Lennarz WJ, Lane MDBT-E of BC (eds) Encyclopedia of biological chemistry, 2nd edn. Academic, Waltham, pp 412–418

Vrkoslav V, Muck A, Cvačka J, Svatoš A (2010) MALDI imaging of neutral cuticular lipids in insects and plants. J Am Soc Mass Spectrom 21:220–231

Wang L, Chen J (2015) Fatty amines from little black ants, *Monomorium minimum*, and their biological activities against red imported fire ants, *Solenopsis invicta*. J Chem Ecol 41:708–715

Wang F, Tong Q (2009) SIRT2 suppresses adipocyte differentiation by deacetylating FOXO1 and enhancing FOXO1's repressive interaction with PPARγ. Mol Biol Cell 20:801–808

Wang X, Hou Y, Saha TT, Pei G, Raikhel AS, Zou Z (2017) Hormone and receptor interplay in the regulation of mosquito lipid metabolism. Proc Natl Acad Sci USA 114:E2709–E2718

Wang J, Chen YL, Li YK, Chen DK, He JF, Yao N (2021a) Functions of sphingolipids in pathogenesis during host–pathogen interactions. Front Microbiol. https://doi.org/10.3389/fmicb.2021.701041

Wang W, Ma Y, Yang RR, Cheng X, Huang HJ, Zhang CX, Bao YY (2021b) An MD-2-related lipid-recognition protein is required for insect reproduction and integument development. Open Biol 11:210170

Weidlich S, Hoffmann KH, Woodring J (2015) Secretion of lipases in the digestive tract of the cricket *Gryllus bimaculatus*. Arch Insect Biochem Physiol 90: 209–217

Wheeler DE, Nijhout HF (2003) A perspective for understanding the modes of juvenile hormone action as a lipid signaling system. BioEssays 25:994–1001

Wicker-Thomas C, Garrido D, Bontonou G, Napal L, Mazuras N, Denis B, Rubin T, Parvy JP, Montagne J (2015) Flexible origin of hydrocarbon/pheromone precursors in *Drosophila melanogaster*. J Lipid Res 56:2094–2101

Wiegandt H (1992) Insect glycolipids. Biochim Biophys Acta/Lipids Lipid Metab 1123:117–126

Wigglesworth VB (1945) Transpiration through the cuticle of insects. J Exp Biol 21:97–114

Wigglesworth VB (1949) The utilization of reserve substances in *Drosophila* during flight. J Exp Biol 26:150

Wigglesworth VB (1985) Sclerotin and lipid in the waterproofing of the insect cuticle. Tissue Cell 17:227–248

Wijerathna A, Evenden M (2019) Energy use by the mountain pine beetle (Coleoptera: Curculionidae: Scolytinae) for dispersal by flight. Physiol Entomol 44:200–208

Włóka E, Boguś MI, Wrońska AK, Drozdowski M, Kaczmarek A, Sobich J, Gołębiowski M (2022) Insect cuticular compounds affect *Conidiobolus coronatus* (Entomopthorales) sporulation and the activity of enzymes involved in fungal infection. Sci Rep 12: 13641

Wrońska AK, Kaczmarek A, Boguś MI, Kuna A (2023) Lipids as a key element of insect defense systems. Front Genet 14:1183659

Wu Z, Soulages JL, Joshi BD, Daniel SM, Hager ZJ, Arrese EL (2014) TGL-mediated lipolysis in *Manduca sexta* fat body: possible roles for lipoamide-dehydrogenase (LipDH) and high-density lipophorin (HDLp). Insect Biochem Mol Biol 45:58–68

Wyatt GR, Davey KG (1996) Cellular and molecular actions of juvenile hormone. ii. roles of juvenile hormone in adult insects. In: Evans PDBT-A in IP (ed) Advances in insect physiology. Academic, pp 1–155

Ximenes AA, Oliveira GA, Bittencourt-Cunha P, Tomokyo M, Leite DB, Folly E, Golodne DM, Atella GC (2008) Purification, partial characterization and role in lipid transport to developing oocytes of a novel lipophorin from the cowpea weevil, *Callosobruchus maculatus*. Braz J Med Biol Res 41: 18–25

Yajima M, Takada M, Takahashi N, Kikuchi H, Natori S, Oshima Y, Kurata S (2003) A newly established *in vitro* culture using transgenic *Drosophila* reveals functional coupling between the phospholipase A2-generated fatty acid cascade and lipopolysaccharide-dependent activation of the immune deficiency (IMD) pathway in insect immunity. Biochem J 371:205–210

Yang Y, Hou H, Haller EM, Nicosia SV, Bai W (2005) Suppression of FOXO1 activity by FHL2 through SIRT1-mediated deacetylation. EMBO J 24:1021–1032

Yang Q, Gong ZJ, Zhou Y et al (2010) Role of *Drosophila* alkaline ceramidase (Dacer) in *Drosophila* development and longevity. Cell Mol Life Sci 67:1477–1490

Yeh LHP, Bajpai RK, Sun GY (1997) Membrane lipid metabolism and phospholipase activity in insect *Spodoptera frugiperda* 9 ovarian cells. Lipids 32:481–487

Yew JY, Chung H (2015) Insect pheromones: an overview of function, form, and discovery. Prog Lipid Res 59: 88–105

Yoder JA, Denlinger DL (1992) Water vapour uptake by diapausing eggs of a tropical walking stick. Physiol Entomol 17:97–103

Yoder JA, Benoit JB, Opaluch AM (2004) Water relations in eggs of the lone star tick, *Amblyomma americanum*, with experimental work on the capacity for water vapor absorption. Exp Appl Acarol 33:235–242

Yoshimura Y, Okino N, Tani M, Ito M (2002) Molecular cloning and characterization of a secretory neutral ceramidase of *Drosophila melanogaster*. J Biochem 132:229–236

Yoshiyama T, Namiki T, Mita K, Kataoka H, Niwa R (2006) Neverland is an evolutionally conserved Rieske-domain protein that is essential for ecdysone synthesis and insect growth. Development 133:2565–2574

Yoshiyama-Yanagawa T, Enya S, Shimada-Niwa Y et al (2011) The conserved rieske oxygenase DAF-36/ Neverland is a novel cholesterol-metabolizing enzyme. J Biol Chem 286:25756–25762

Yu J, Zhang W, Chi X, Chen W, Li Z, Wang Y, Liu Z, Wang H, Xu B (2022) The dietary arachidonic acid improved growth and immunity of honey bee (*Apis mellifera* ligustica). Bull Entomol Res 112:261–270

Yun HK, Jouni ZE, Wells MA (2002) Characterization of cholesterol transport from midgut to fat body in *Manduca sexta* larvae. Insect Biochem Mol Biol 32: 1151–1158

Zandawala M, Hamoudi Z, Lange AB, Orchard I (2015) Adipokinetic hormone signalling system in the Chagas disease vector, *Rhodnius prolixus*. Insect Mol Biol 24: 264–276

Zhang C, Wei D, Shi G et al (2019a) Understanding the regulation of overwintering diapause molecular mechanisms in *Culex pipiens* pallens through comparative proteomics. Sci Rep. https://doi.org/10.1038/s41598-019-42961-w

Zhang CH, Zhang MJ, Shi XX, Mao C, Zhu ZR (2019b) Alkaline ceramidase mediates the oxidative stress response in *Drosophila melanogaster* through sphingosine. J Insect Sci 19:13

Zhang MJ, Shi XX, Bai YL, Zhou WW, Zhu ZR (2021) Sphingolipid composition and metabolism differ in three auchenorrhynchous pests of rice. J Asia Pac Entomol 24:772–779

Zheng CJ, Yoo JS, Lee TG, Cho HY, Kim YH, Kim WG (2005) Fatty acid synthesis is a target for antibacterial activity of unsaturated fatty acids. FEBS Lett 579: 5157–5162

Zheng H, Zhang R, Zhang H, Feng Y, Li K, Zhang W (2011) Thermal analysis of four insect waxes based on differential scanning calorimetry (DSC). Procedia Eng 18:101–106

Zhou G, Miesfeld RL (2009) Energy metabolism during diapause in *Culex pipiens* mosquitoes. J Insect Physiol 55:40–46

Zhou Y, Lin XW, Yang Q, Zhang YR, Yuan JQ, Da LX, Xu R, Cheng J, Mao C, Zhu ZR (2011) Molecular cloning and characterization of neutral ceramidase homologue from the red flour beetle, *Tribolium castaneum*. Biochimie 93:1124–1131

Zhou Y, Lin XW, Zhang YR et al (2013) Identification and biochemical characterization of Laodelphax striatellus neutral ceramidase. Insect Mol Biol 22:366–375

Zhu W, Zhang H, Li X et al (2016) Cold adaptation mechanisms in the ghost moth *Hepialus xiaojinensis*: metabolic regulation and thermal compensation. J Insect Physiol 85:76–85

Zhuang M, Oltean DI, Gómez I, Pullikuth AK, Soberón M, Bravo A, Gill SS (2002) *Heliothis virescens* and *Manduca sexta* lipid rafts are involved in Cry1A toxin binding to the midgut epithelium and subsequent pore formation. J Biol Chem 277:13863–13872

Ziegler R (1991) Changes in lipid and carbohydrate metabolism during starvation in adult *Manduca sexta*. J Comp Physiol B 161:125–131

Ziegler R, Ibrahim MM (2001) Formation of lipid reserves in fat body and eggs of the yellow fever mosquito, *Aedes aegypti*. J Insect Physiol 47:623–627

Adv Exp Med Biol (2026) 1494: 47–64
https://doi.org/10.1007/5584_2025_874

Published online: 25 July 2025

The Fat Body: Structure and Functions

Patrycja Staniszewska and Aneta Strachecka

Abstract

The fat body plays a vital role in the proper functioning of invertebrates, contributing to their metabolic processes and resilience. It serves as the central tissue for metabolism integrating signals, regulating molting and metamorphosis and producing hormones that govern the overall body function and immune system protein synthesis. Thanks to this multifunctionality, the fat body is responsible for the metabolism of basic compounds: lipids, carbohydrates and proteins, storing them in the form of reserves (protecting against weather conditions, starving, etc.) and "post-trials" metabolites. In this tissue the remodeling process takes place, which enables the metamorphosis of insects. Also, the fat body is the place of the synthesis of immune proteins, some hormones, pheromones and vitellogenin. Understanding its physiology has, therefore, become an important element of research on insects in the context of general health.

Keywords

Biochemical pathways · Hormones · Immunity · Lipids · Metabolism of insect · Oenocytes · Storage · Trophocytes · Urocytes

The importance of lipids in the life of insects is also confirmed by the presence of the key to their functioning, multifunctional tissue, i.e., the fat body. It is the center of the physiological and biochemical processes of insects, which, despite the common morphological and anatomical features, may differ depending on the type, species, caste and location in the body and may be dependent on the development phase or external factors modifying the structure. The fat body performs many roles: from the place of accumulation of substances, energy production, through the course of metabolic pathways related to the conversion of lipids, carbohydrates or proteins, to the synthesis of immune proteins, detoxification and regulation of hormonal balance and metamorphoses (Law and Wells 1989; Skowronek et al. 2021; Strachecka et al. 2022).

Knowledge about the processes taking place in the fat body is an important element of understanding the physiology of insects, especially in times of strongly departmental negative factors on their life and health. Information on the functioning of this tissue and its maintenance may turn out to be crucial for maintaining the

P. Staniszewska (✉) and A. Strachecka
Department of Invertebrate Ecophysiology and Experimental Biology, University of Life Sciences in Lublin, Lublin, Poland
e-mail: patrycja.staniszewska@up.lublin.pl; aneta.strachecka@up.lublin.pl

population of insects and their biodiversity (including utility ones such as pollinators) (Skowronek et al. 2021).

1 Structure

There are five main types of fat body cells that can occur in insects (the composition of the fat body varies between species in orders). These include (1) trophocytes—main fat body cells (also called adipocytes), (2) oenocytes, and in a smaller amount, (3) mycetocytes, (4) chromatocytes and 5) urocytes. The most abundant cells in most insect orders are trophocytes and right after them—oenocytes (Fig. 1). Size, arrangement, type and number of fat body cells depend on many factors, such as the type, species, caste, developmental stage of the insect and the location of the fat body. Each of these cells is characterized by a specific structure and functions in the fat body (Chobotow and Strachecka 2013; Cruz-Landim et al. 2006).

1.1 Cells

1.1.1 Trophocytes

Trophocytes are the most numerous and are a frequent object of scientific research. Their number/presence varies depending on the insect species and gender. Females have a higher trophocyte count than males. A higher number of these cells also occur during moulting (mesoderm differentiation). They are mainly responsible for the storage and secretion of substances and detoxification processes in the body of insects. In recent studies, it turns out that this is also a place where vitellogenin accumulates in female casts in honey bees (Strachecka et al. 2021). Due to the variety of consumed foods, physiological state and performed work requiring energy inputs,

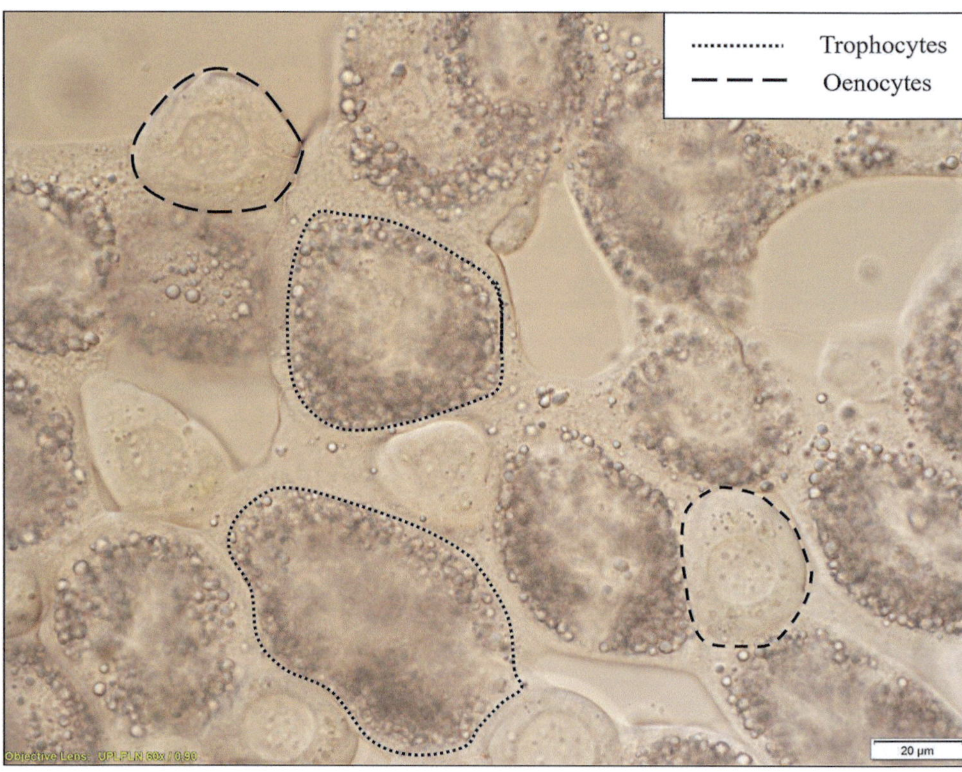

Fig. 1 *Apis mellifera* fat body's cells under a light microscope: trophocytes and oenocytes (copyright by the Department of Invertebrate Ecophysiology and Experimental Biology, University of Life Sciences in Lublin, Poland)

they may show a different physiological status (Cruz-Landim et al. 2006; Roma et al. 2010; Strachecka et al. 2021; Chapman 2022).

Trophocyte is surrounded by a thin membrane (basal lamina) and filled with cytoplasm. Inside, we can distinguish organelles such as nucleus, vacuoles, large vesicles, mitochondria, Golgi apparatus, endoplasmic reticulum (mainly rough), peroxisomes, lysosomes and storage structures like droplets/granules of protein, lipids and carbohydrates (Skowronek et al. 2021).

Cells also change their size depending on their life stage. This is due to changes in the size of the Golgi apparatus, the accumulation of spare substances (mainly the size of lipid droplets) and the swelling of the vacuole (or contraction during metamorphosis) (Chobotow and Strachecka 2013; Cruz-Landim et al. 2006; Roma et al. 2010).

There are several types of vacuoles in cells that accumulate appropriate substances. The main types include storage, digestive, condensation and superficial. Their size depends on the development stage (Locke 1984). During metamorphosis, vacuoles with the fat body are reduced, but between transitions to the next developmental stages, they usually increase in volume. Storage vacuoles accumulate spare substances in the form of amino acids (i.e., tyrosin), urates, proteins, lipid droplets and glycogen or trehaloze (along with symbiotes present in its structures) (Locke 1984).

Digestive vacuoles (autophagous, heterophagous) release spare substances, while the body needs large amounts of energy or during diapause. Inside, we can also find remains of digested organelles, which occur in cooperation with lysosomes. Condensation vacuoles are formed in the process of combining provacuoles. Their work is related to the operation of the Golgi apparatus and lysosomes (Skowronek et al. 2021; Cruz-Landim et al. 2006; Roma et al. 2010).

The form of organic substances accumulation depends on the species. In social insects, lipids occur in the form of large droplets. Protein can occur in the form of granules (Attini ants) or crystals (*Monomorium pharaonic* queens) of various shapes and sizes (Roma et al. 2008). Carbohydrates are usually found in free form as glycogen or in combination with other organic compounds (Roma et al. 2010).

Protein granules are also present in the trophocyte cytoplasm itself. They are usually derived from pinocytosis (absorbing protein from the hemolymph to the fat body cells). Proteins selectively absorbed are usually organism specific, and selectivity is based on the specificity of the connections between membrane receptors (defined during pinocytosis). In addition to the proteins absorbed and stored, the fat body is the main tissue in protein synthesis. Protein synthesis is dependent on the developmental stage. In *Heliothis zea*, protein synthesis is inhibited after the larval stage, while in honeybees, it can be continued in the prepupal stage, but the efficiency of synthesis is lower (Haunerland et al. 1990). The protein accumulated in this way is used as an energy source for the transformation of the pupa. Proteins are then transported and exchanged between the fat body and the hemolymph, which distributes them throughout the body (Cruz-Landim et al. 2006; Roma et al. 2010).

Lipids reserves in trophocytes consist mainly of triglycerides (diglycerides form mainly in hemolymph). Most lipids are used as an energy source (flight in Hymenoptera, Lepidoptera, Diptera) and additionally as structural material during oogenesis in females (Moribayashi et al. 1985).

Apart from changes in their number in the fat body, cells show internal changes in the morphology of the organelles. This is mainly due to the accumulation of nutrients (Kelley 1985). Depending on the substances consumed by insects and the time of their life in the cytoplasm, we can observe variability in the accumulation of protein (seasonal changes in *Bombus terrestris*), lipid droplets, the presence of a developed rough endoplasmic reticulum (with high storage of lipid droplets) and elongated mitochondria (Chapman 2022).

The size of the trophocytes also varies between social insect castes. For example honeybee queens and rebel workers had larger trophocytes in tergite no. III (vitellogenin store for sexually active / partially active animals) than in other tergital parts in the comparision to standard workers or drones (Strachecka et al. 2021).

Although trophocytes constitute the main and most numerous cells in most insects, some orders or species of insects may possess alternative or additional specialized cells (in Lepidoptera, Hymenoptera, Hemiptera). However, other cells can be observed in them that play a role similar to trophocytes. An alternative to trophocytes may be mycetocytes (cockroaches, Hemiptera—true bugs) and urocytes (cockroaches) (Cruz-Landim et al. 2006; Zülfikaroğlu et al. 2022).

1.1.2 Oenocytes

Usually, oenocytes are less common cells in the fat body than trophocytes. According to genetic studies on *Drosophila melanogaster* and *Tribolium castaneum,* origin of oenocytes is ectodermal, in comparision to trophocytes. These cells produce long-chain fatty acids responsible for sealing tissues (trachea, heart, spiracles). Oenocytes are important for metabolic pathways of lipids (Makki et al. 2014). They are responsible for maintaining the body's homeostasis, production of pheromones, protection against toxins, detoxification and the synthesis of carbohydrates that are transported between the fat body and the hemolymph (imago oenocytes). For sample, synthesize epithelial hydrocarbons responsible for resistance to drying and hardening of the epidermis. As in the case of trophocytes, they have basic organelles such as nucleus (centrally in cells), mitochondria, smooth endoplasmic reticulum, vacuoles with droplets/granules of lipids, proteins and glycogen. In a study of Wójcik et al. (2022) discover the presence of binuclear oenocytes in *Formica cinerea* ants. Usually, oenocytes can be distinguished due to their darker color compared to other structures from yellow to amber. In some cases, the dark color may be caused by the presence of hemoglobin observed in Diptera order. Hemoglobin in Chironomus (nonbiting midges) is used as an oxygen carrier, helping with respiration, particularly in aquatic habitats where oxygen availability can be limited (Bergtrom et al. 1976). The amount of fatty acids present in oenocytes depends on the neuropeptides active in the insect's organism, i.e., sulfacanines. The distribution of oenocytes can be random in the whole fat body, or we can observe them in clusters (the aforementioned surrounding tissues and organelles) (Chobotow and Strachecka 2013; Gutierrez et al. 2006; Vafopoulou-Mandalos and Laufer 1983; Wójcik et al. 2022; Szymczak-Cendlak et al. 2022; Des Marteaux et al. 2018).

In fruit flies (*Drosophila* spp.), the presence of oenocytes varies depending on developmental stages or specific environmental conditions (Makki et al. 2014).

1.1.3 Mycetocytes

Mycetocytes are a specialized cell type located in the fat body of certain insects, such as cockroaches and aphids. These are cells that mainly contain prokaryotic organisms that live in symbiosis with the host. The most research on mycetocytes is related to cockroaches and Hemiptera (true bugs). For Hemiptera, mycetocytes can may constitute the most numerous fraction of fat body cells. It has been shown that the number of these cells increases with the age of cockroaches (*Byrsotria fumigata, Blaberus craniifer, Blaberus discoidalis, Nauphoeta cinerea*) and that they contain bacterioids, which increase in size and lose their cell wall with changing physiological state. The size of mycetocytes may also be related to the temperature of the environment. Interestingly, at 35° Celsius, a decrease in the number of these cells was observed in relation to lower temperatures. The presence of yeast has also been observed in Hemiptera and Coleoptera insects. The presence of yeasts may be related not only to their symbiotic nature with the organism of insects but also to their symbiosis to some bacteria present in mycetocytes. Such a relationship exists between Sulcia and Nasuia bacteria, and *Mietchnikowia pimensis* in *Hishimonus phycitis*. In these cells, in addition to symbiotes, we also find lipid droplets and glycogen. Their presence may in some cases be associated with a poor diet of insects. In such cases, these organisms synthesize some amino acids (*Blattabacterium cuenoti*) and B vitamins. They cease to be active at the time of his death. In addition, it is believed that some of the symbionts present in these cells are involved in the recovery of nitrogen from uric acid, which makes it an important link in the nitrogen cycle. It also proves the relationship between mycetocyte and urocyte cells (Makki et al. 2014; Willott et al. 1988; Tunaz et al. 2003; Xianhui and Baoping 2006; Milburn 1966; Malassigné et al. 2021).

1.1.4 Chromatocytes

Chromatocytes are fat body cells characterized by specific pigmentation. One of these pigments is ommochromes, which play an important role in detoxification and protecting the insect organism from external factors. The presence of pigments is also related to another function of chromatocytes to accumulate the lipids needed for the insect metamorphosis (Martins et al. 2011). This function is crucial because it supports the energy requirements associated with the transformation from larval to imago stages. Their presence has been mainly noted in aquatic insects, indicating a potential adaptation to their environmental needs and life cycle requirements. Chromatocytes are flat and contain a centrally located nucleus and membrane. These cells are located in thin layers of the fat body. As with almost all cells of the fat body, they are of mesodermal origin (Chobotow and Strachecka 2013).

1.1.5 Urocytes

The role of urocytes is to store and mobilize reserves, derived from uric acid and polysaccharides, during starvation. Uric acid is stored as a result of the transformation of proteins and nucleic acids in the body of insects. This type of cell contains only some insects, i.e., cockroaches. In Lepidoptera order, these types of fat body cells are not observed. Uric acid is stored in the form of crystals in large vacuoles. In addition to the vacuole, cells have an endoplasmic reticulum, a small, spherical nucleus, a cytoplasm rich in sugars (polysaccharides), lipid droplets, protein and, in addition to urate granules—calcium salt granules. In honeybee, urocytes are present at immature stages of development and probably play the role of Malpighian tubules by accumulating metabolic compounds until the tubules are functionally active. Urocytes assist in controlling nitrogen levels in the insect's body, playing a role in maintaining metabolic balance. When the insect faces starvation or stress, the uric acid stored in urocytes can be used as a nitrogen source (Park et al. 2013).

They may or may not be present depending on the developmental stage of the insects. Therefore, they are observed in the vicinity of the trachea of insects (Chobotow and Strachecka 2013; Cruz-Landim et al. 2006; Strachecka et al. 2021; Furtado et al. 2013).

As with other cells, their occurrence depends on the species. Urocytes may be absent in Hemiptera or reduced as in Orthoptera. Many of the non-hematophagous Diptera insects do not have urocytes (Arrese and Soulages 2010).

1.2 Tissue Differentiation and Histotypes

1.2.1 Structural Diversity

Depending on the placement of the fat body, we distinguish two main types: 1) subcutaneous and 2) visceral/axial.

The subcutaneous fat body has a structure that can be formed into thin layers, tightly adjacent to each other—this kind of tissue is located just below the surface, attached to it. This fat body is made up of a layer(s) of lobes made up of rows of cells. In this tissue, the fat body layer is separated from the armor by a layer of muscles.

In addition to the subcutaneous fat body, we distinguish a deeper—visceral fat body with a lower density and regularity of cells that surround the organs and the intestine in the abdomen, thorax and head—perivascular (visceral) (Skowronek et al. 2021; Xianhui and Baoping 2006).

The arrangement of cells depends on the function performed. The fat body tissues, which are arranged closely, make contact very easily by signal transmission and transmit information about the physiological state between the segments. The proximity and continuous contact of the fat body layers with the hemolymph, gonads, nerve connections, trachea and Malpighian tubules are also important. In this way, hormonal/metabolic signals, metabolites, synthesized substances, etc. can be easily transported between the basic organs of insects inside the body. It would seem that visceral fat body is more metabolically active, but subcutaneous fat body is a storage of spare substances. However, according to Roma (2010), the metabolic activity is also noticed in the lobes under the body surface and is especially related with lipid metabolism (Roma et al. 2010; Zülfikaroğlu et al. 2022). It turns out that metabolic activity and storage functions in different types of fat body depend on the type of organic compound. The visceral part synthesizes and is a storage for protein, while the subcutaneous part has a greater role in lipid metabolism and lipid storage (Roma et al. 2010; Wójcik et al. 2022).

One of the new discoveries is the presence of the two-layer structure of the fat body and the presence of binuclear oenocytes, forming a layer separating the epidermis from the fat body composed of trophocytes and mononuclear oenocytes. The lobes may consist of a single cell type (trophocytes or oenocytes), e.g., in *Drosophila melanogaster* and *Formica cinerea*. The amount of lobes can be different and can be separated by free spaces between them. In other insects that are less primitive from the evolutionary point of view (honey bee), we observe the presence of mixed cells of trophocytes and oenocytes at one lobe/layer (Wójcik et al. 2022; Rizki and Rizki 1962; Krupp and Levine 2010).

As research shows, the fat body is distinguished by high adaptive plasticity and susceptibility to changes. Diversity may result from their evolution, nutritional strategies and other anatomical adaptations to environmental conditions and their functions. The research by Ł. Wójcik et al. (2022) showed that the size of fat body cells is species specific. In worker honeybees, trophocytes and oenocytes differed in size compared to the same cells in *Formica (Seviformica) cinerea* ants. In all cases, the ant trophocytes were larger and the oenocytes more numerous. In the case of trophocytes, they were more numerous in bees, and oenocytes had larger nuclei than in ants. Moreover, the number and size of cells may vary depending on the stage of life. It was noted in the studies that the largest cells are often found in the oldest stage larvae (Wójcik et al. 2022).

1.2.2 Regional Variation

In addition to the location-related differences in structure, the differentiation also occurs within a given type of fat body (visceral). A given type of fat body may have a different structure depending on its anatomical location. For example, a subtype fat body may differ in structure depending on whether it is located in tergite from no. I to no. III (upper side in the abdomen—the order is counted from the thorax or in the sternits of abdomen. The size of the cells present in the adjacent tergites and abdominal sternites differs from each other (Fig. 2). The largest cell diameters are assigned to the sternite part. The presence of cells with specific characteristics can also be noted in the tissue surrounding organs; for example, binuclear oenocytes often form clusters surrounding the trachea and heart. Probably a different structure is related to the performance of different functions depending on the place of occurrence. This is confirmed by the fact that the lobes/layers of the fat body depending on the location (a given tergite/sternite) are characterized by different biochemical activities, the content of specific compounds like pigments: xanthomatins or pteridines. Xanthomatins accumulate in the anterior three of the thorax, when pteridines accumulate in the last three abdominal segments. This may contribute to their multifunctionality and give them such great importance (Strachecka et al. 2021; Wójcik et al. 2022; Rizki and Rizki 1962; Haunerland and Shirk 2003).

1.2.3 Histotypes

In addition to the variability related to the location and functions of the fat body, it is possible to distinguish several main histotypes specific to given types and species of insects, as observed by the color of the tissues. The color depends on the structure, development stage, accumulation of

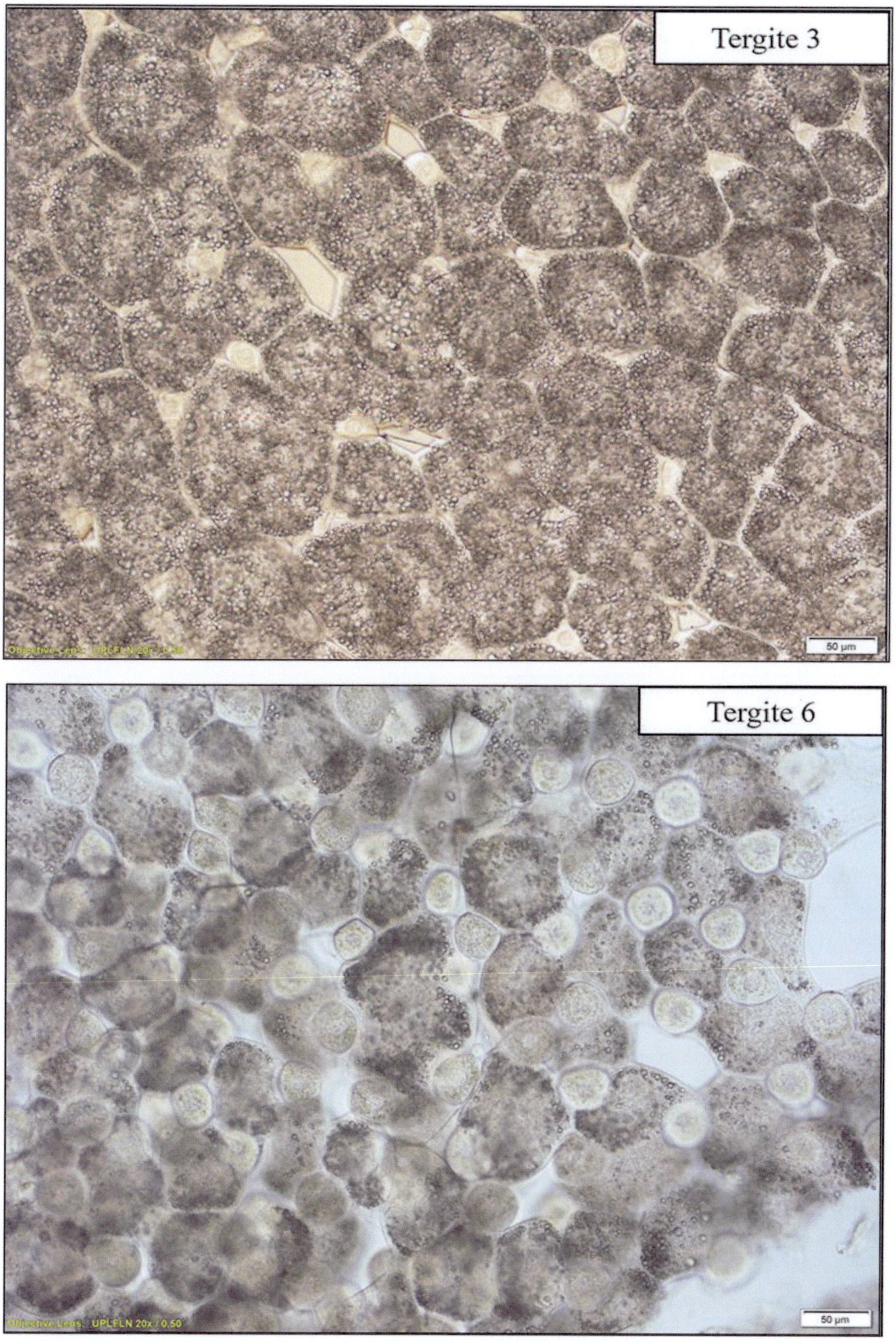

Fig. 2 Differentiation of *Apis mellifera* worker's fat body between tergite 3 and tergite 6 (copyright by the Department of Invertebrate Ecophysiology and Experimental Biology, University of Life Sciences in Lublin, Poland)

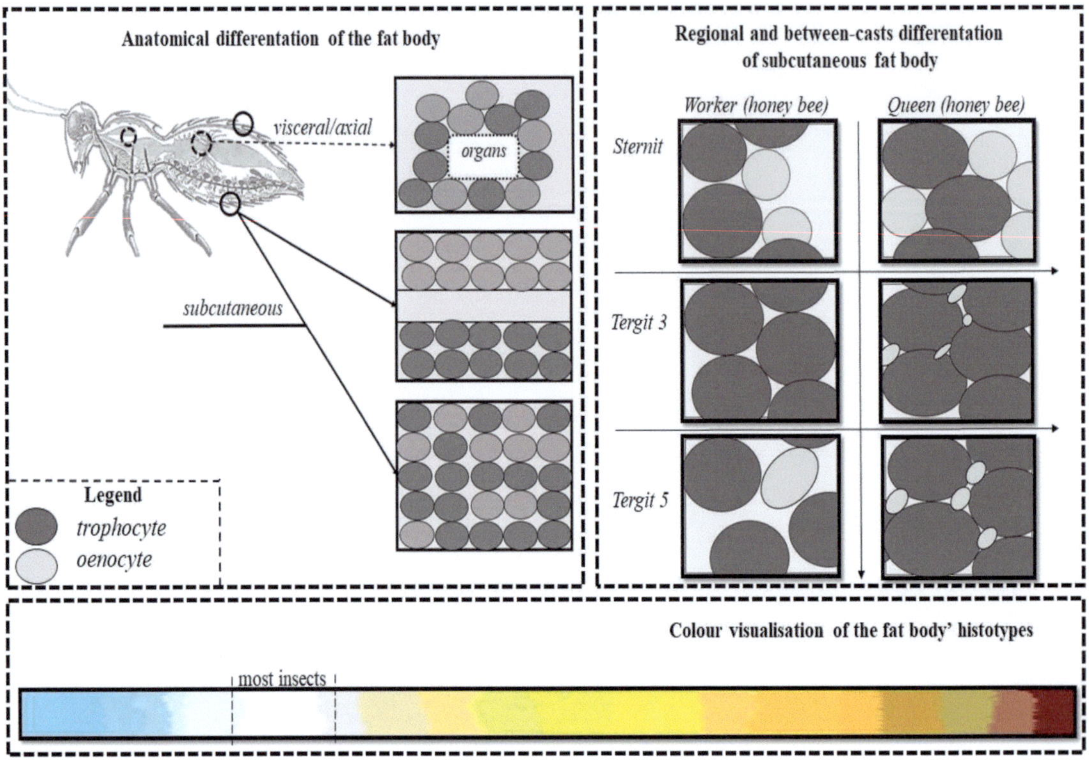

Fig. 3 Anatomical, regional, cast differentiation and histotype variation of the honey bee fat body. (Author: Patrycja Staniszewska)

nutritional resources such as protein, polysaccharides and the content of pigments (ammochrome, pteridines—eye pigment precursors in a fly). We distinguish between such histotypes as white in most insects, brown (Indian moth), colored/mixed (Lepidoptera), yellow and blue (Fig. 3) (Rizki and Rizki 1962; Haunerland and Shirk 2003).

The color differences in fat body cells of insects indicate their specific functions, such as energy storage or expenditure, much like the variations found in mammalian fat tissues. This flexibility enables insects to effectively regulate their energy resources based on environmental factors and physiological requirements.

1.2.4 Fat Body Microbiota

In addition to the mycetocyte cells themselves, microorganisms are present in the fat body. In the cockroach *Nilaparvata lugens* (brown planthopper), the studies showed the presence of Firmicutes, Proteobacteria, Actinobacteria, Bacteroidetes, Fusobacteria, Cyanobacteria, Chlamydiae, Deferribacteres and Saccharibacteria clusters. Firmicutes + Proteobacteria accounted for up to 99% of clusters. The types of microbes in the fat body belong to *Bacillus*, *Arsenophonus*, *Lactococcus*, *Asaia*, *Enterococcus*, *Streptococcus* and *Carnobacterium* (Zhang et al. 2019).

2 The Fat Body Functions

The insect fat body is a multifunctional tissue that plays a central role in physiological regulation and adaptation. Beyond its well-known role as a storage organ for lipids, carbohydrates and proteins, the fat body is actively involved in a wide range of vital processes, including metamorphosis, hormonal signaling, detoxification, reproduction and immune responses. These diverse functions are dynamically regulated across developmental stages and in response to internal and environmental cues. The following subsections provide an overview of the major metabolic and regulatory roles of the fat body in insect biology.

2.1 Metamorphosis and Fat Body Remodeling

The plasticity of the fat body and its extraordinary role are visible during metamorphosis in insects. During this process, the fatty body changes and enables insects to develop, grow and adapt to the insects' new physiology and environmental needs. The metamorphosis process takes place in the last stage of the larval stage and allows for the development of new functions (Zheng et al. 2016).

Each cell of the fat body has a common development point. Mature cells of the fat body reproduce by endoreplication. Initially, all types of fat body cells are derived from the same type of embryonic progenitor cells. Their formation takes place in the early stages of the insect's development. They differentiate in the later stages of growth after obtaining the appropriate number through mitotic processes. When the right moment comes in an insect's life, it undergoes metamorphosis. This process has a huge impact on the appearance of the body fat. Depending on the type of insect, metamorphosis can take place thanks to remodeling the cells of the fat body that is already present or due to complete histolysis and the production of a new fat body with its resources (holometabolous insects). Here are several signals to start/trigger metamorphosis. Adverse environmental conditions similar to hunger initiate autophagosome activation by activating pathways mediated by: the enzyme 5'adenosine monophosphate, activated protein kinase (AMPK) and a transducer of regulated activity of CREB 1 (TORC1). Biochemical stimulation of remodeling occurs by the stimulation of ecdysone/altraspiracle white receptor (nuclear receptor) of the transcriptional cascade by the effect of 20-hydroxyecdysone (20E). Then the transcription cascade is activated (Haunerland and Shirk 2003; Chang and Neufeld 2009; Delanoue et al. 2010; Lippai et al. 2008; Liu et al. 2013; Scott et al. 2004; Bond et al. 2011; Ryoo and Baehrecke 2010; Xie et al. 2016).

Histolysis begins with the process of cell dissociation, which results in cell splitting (20E and juvenile hormone affect the regulation of metalloproteinases: Mmps 1 is responsible for the splitting of closely located cells and Mmps 2 destroys the basal membrane covering the cells of the fat body). As a result of this stage, free cells, spherically floating in the hemocoel, are obtained. After further dissociation is inhibited (thanks to the action of hemocytes from the hemolymph), the main histolysis process begins by activating the programmed cell death. Histolysis induces hormones such as 20E, which affect the activation of the caspase cascade (CAC) that destroys cellular structures. Cysteine proteinase reaches the damaged cell and breaks down the membrane, causing the cells to break down. Ultimately, CAC and autophagy (ATG) 5–6 proteins cause cell death. Thousands of cells are completely histolyzed. After histolysis, a new fat body is reborn from the leftover cells. The body produces new cells by activating 10E, ecydone-induced protein 93F (E93) and the bric-a-brac-tramtrack-broad member Broad-Complex (*Br-C*) gene. As in the case of the formation of a fat body in young animals, also in this case, the cells undergo mitotic divisions until they reach the physiological specific number of cells for insects. After mitosis, cells undergo endoreplication (activation of Myc, IIS-TORC1 pathway and cyclin D/Cdk4 activating cyclin E/Ckd2 activating cyclone E). Within 3 days, the fat body matures up (even to about 18,000 cells), but the exact number of cells varies depending on species, developmental stage and environmental conditions. The new tissue is formed after the activation of the Kruppel homologue (Kr-h1) which, under the influence of juvenile hormone—Met receptor complex, inhibits E93 and Br-C expression (Bond et al. 2011; Ryoo and Baehrecke 2010; Xie et al. 2016; Li et al. 2019;

Jia et al. 2014; Agrawal et al. 2016; Yamanaka et al. 2013; Wang et al. 2021; Johnson and Butterworth 1985).

The second type of fat body cell replacement in insects begins as described above but does not involve apoptosis. There is also a stage of dissociation of cells into a loose structure (histolysis), but immediately after this stage, apoptosis does not occur, but cells aggregate and form theirs organelles. The final effect of an insect's metamorphosis (size, body location) is mainly determined by 20E and insulin/insulin-like growth factor signaling (IIS) (Zheng et al. 2016; Mirth and Riddiford 2007; Yuan et al. 2020).

2.2 Nutrient Storage and Metabolism

One of the roles of the fat body is to carry out the metabolic processes of basic organic compounds, i.e., lipids, carbohydrates and proteins in its tissues. In cells, the fat body accumulates the following substances in a percentage share: lipids over 50% (mainly triglycerides and diglycerides in droplets), carbohydrates below 50% (glucose in the form of glycogen) and proteins/amino acids below 50% (e.g., arylphorin, proline) are as reserves of substances with high energy potential (Arrese and Soulages 2010; Bi et al. 2012; Soulages and Wells 1994; Canavoso 2001; Patel et al. 2005; Brasaemle 2007).

- Lipids: More on lipid metabolism through lipogenesis and lipolysis in the following chapters.
- Carbohydrates: Glycogen as an easy-to-mobilize glucose reserve (built from glucose residues) is a good source of energy and a building material for the shell of insects (chitin), and its resources are usually found in muscles and eggs of insects. Sugar alcohols are substances that support the adaptation to harsh environmental conditions (temperature—osmolytes). Sugars are also used in lipid metabolism in as much as 50%. The pathways of sugar transformation include glycogen phosphorylation, glycogenesis, glycolysis and glucogenesis. Thanks to glycogenesis, we also obtain trehalose in the body of insects, which is used for high-energy tasks, such as flight (Cruz-Landim et al. 2006; Roma et al. 2010; Arrese and Soulages 2010; Siuda et al. 2014; Vardanis 1963; Steele 1982).
- Proteins and amino acids are the source of energy and are also involved in the transformation of sugars (proline is the main source of energy in the tse-tse flight). They are also a building element (when creating a new imago body), allow to survive the hunger period and its reserves are extremely important during the metamorphosis of insects. Therefore, its greatest synthesis occurs in the larval stages, in particular in the last stage through metamorphosis. Some species stop protein synthesis after the larval stage, but others (e.g., some Hymenoptera and Coleoptera) continue lower levels of synthesis during metamorphosis. Growth rate and time of reproduction depend on the amount of resources due to the presence of reserve sensors (expression of amino acid transporters) that determine the right moment of growth/metamorphosis/moulting over time and adequate accumulation of nutrients. Protein metabolism is controlled by 20E and juvenile hormone (JH), the same hormones as in metamorphosis (Cruz-Landim et al. 2006; Mirth and Riddiford 2007; Bursell 1981).

2.3 Metabolite Accumulation and Detoxification

The urea and fatty acids in vacuoles present in the cells of the fat body (content up to 75%) suggest that the fat body in insects can play a role of the equivalent of vertebrate liver. This usually plays a highly important role when Malpighian tubules are not active. During their life, insects accumulate metabolites in their vacuoles, and after tubular activation, they are discharged in the form of urea cassette ect. An example of this phenomenon is the occurrence of white frost on *Manduca sexta* at the time of tubular activation. The appearance of this activity results in a simultaneous reduction in urea concentration in the cell vacuoles (Haunerland and Shirk 2003; Nijhout and Williams 1974; Zhang et al. 2022).

2.4 Hormonal and Pheromonal Functions

Despite the fact that the fat body is influenced by many compounds, i.e., hormones, only a part is synthesized in its tissue. The hormones synthesized by the fat body include cytokines and the CCHa2 peptide (CCHamide-2). The cytokines synthesized in the fat body include mainly Unpaired 2 (Upd2), a leptin ortholog. Upd2 affects the activation of insulin pathways by affecting IPCs and ILPs (insulin-like peptides). Cytokines are similar to lectins. The hormone/peptide (CCHa2) also acts on insulin pathways as it does with the cytokine, affecting pupation time (lag) in insects with mutations of this hormone (Toprak 2020).

In addition to hormones, pheromones can be biosynthesized in the fat body, i.e., aggregation pheromone and sex pheromones. The aggregation pheromone is synthesized by de novo pathway, i.e., bark beetles. It is used by insects to stimulate attacks and aggregate colonies. Sex pheromones in *Blattella germanica*, composed mainly of ketones, are synthesized to a lesser extent. Its greater amounts are produced in the sternite part than in the tergite part (Keeling et al. 2021; Gu et al. 1995).

Only some species use the fat body for pheromone biosynthesis, while others rely on specialized exocrine glands (some of these are placed in fat body). These glands are responsible for the production of highly essential pheromones for the functioning of a honey bee colony. In queens (and rebels—cast of workers who have some features of queens), there are tergal Renner and Baumann glands, located on tergites no. II–IV, which are responsible for the production of pheromones (Strachecka et al. 2022; Strachecka et al. 2021).

2.5 Immune Function

The insect fat body plays a central role in innate immunity, functioning as the primary site for the synthesis of immune effectors and coordinating systemic responses to pathogenic threats. In close cooperation with hemocytes and the hemolymph, the fat body detects pathogen-associated signals and initiates the production of antimicrobial peptides, enzymes and signaling molecules. This section outlines the key immune functions of the fat body, including its role in producing antimicrobial peptides, lipid-mediated immune modulation, metalloprotease activity and interactions with fungal pathogens.

2.5.1 Antimicrobial Peptides and Their Regulation

The most significant antimicrobial proteins include lysozyme, lectins, attacins, defensin, cecropins, etc. Additionally, depending on the insect type, other compounds influencing the host's immunity may be present (Arrese and Soulages 2010; Eleftherianos et al. 2021; Zhang et al. 2021).

- *Lysozyme* catalyzes the hydrolysis of bonds present in the cell wall of gram-positive bacteria (presence of murein). It can function in various aerobic conditions and environments with different pH (acidic and alkaline). Insects have an inborn level of this protein, with the highest levels found in Diptera, Blattidea and Isoptera; intermediate levels in Lepidoptera; and the lowest levels in Hymenoptera (especially in Apidae). In healthy insects, the concentration of lysozyme varies from 5 to 1000 qg/ml of hemolymph (Gliński and Jarosz 2021; Strachecka et al. 2018).
- *Lectins* are usually present in the body of insects during the larval period. They recognize foreign substances in the body, cause aggregation of pathogenic cells (easier removal from the body), and support metamorphosis (facilitating phagocytosis of cell debris) (Gliński and Jarosz 2021).

- *Attacins* are found in Lepidoptera insects. We also distinguish other proteins with its activity in Diptera (diptericin) or Coleopter (coleoptericin) insects. They are mainly composed of aspartic acid, glycine, alanine, phenylalanine and threonine. It mainly affects the permeability of cell membranes in the division phase and shows activity against microorganisms, i.e., *Pseudomonas manthophilia*, *Escherichia coli* and *Acinetobacter colcoaceticus* (Eleftherianos et al. 2021; Gliński and Jarosz 2021; Liu et al. 2009).
- *Defensin* (in ants, e.g., formicin), may be active against gram-negative and gram-positive bacteria. However, its activity depends on the concentration of insects produced in the organism (Eleftherianos et al. 2021; Gliński and Jarosz 2021; Bulet et al. 1991).
- *Cecropins* are a multifunctional protein found mainly in holometabolic insects (Lepidoptera) with a broad antibacterial effect (works against gram-positive and gram-negative bacteria, combating, e.g., *Pseudomonas aeruginosa*, *Serratia marcescens* and *Xenorhabdus nematophilus*) (Eleftherianos et al. 2021; Gliński and Jarosz 2021).
- *Sarcotoxins II* (in *Sarcophag peregrina*) have a narrow range of action, mainly against gram-negative bacteria. They act mainly by inhibiting cell wall synthesis (Eleftherianos et al. 2021).
- *Gloverins* are so far known only in Lepidoptera and are biosynthesized as proboteins (Eleftherianos et al. 2021).
- *Hymenoptaecin* is found in the hemolymph of immature honey bees and has a wide range of antibacterial activity against gram-negative and gram-positive bacteria (Eleftherianos et al. 2021).
- *Coleoptericin* kills gram-negative bacteria rapidly within 12–24 h (Eleftherianos et al. 2021).
- *Tanatin* has a broad spectrum of activity: against gram-positive and gram-negative bacteria, filamentous fungi and yeasts (Eleftherianos et al. 2021).

The fat body also produces glycoproteins, i.e., *transferrin*. This glycoprotein affects increased survival after infections and reduces oxidative stress. Transferrin concentration increases after exposition on pathogens in the insect organism (Eleftherianos et al. 2021; Zhang et al. 2021).

The synthesis of these proteins in the case of an acquired response is often stimulated by hemokines, which carry information between the fat body and the hemolymph. When a pathogen enters the insect's organism, the pathways responsible for recognizing pathogenic organisms, i.e., bacteria, are activated. Gram-positive and gram-negative bacteria are recognized by the pattern recognition receptors (PRR) on the basis of the molecular patterns of the organism (PAMP). Diagnosis is based on the analysis of the components of the bacterial cell wall, i.e., peptidoglycan. Once the pathogens are recognized, the JAK/STAT (viruses), Toll (gram positive) or Imd (gram negative) pathway is activated. As a result of the activation of the indicated pathways, appropriate antimicrobial peptides are produced (Lindsay and Wasserman 2014; Manniello et al. 2021; Myllymäki et al. 2014; Nunes et al. 2020, 2021; Rao et al. 2018).

The cooperation of the fat body with the hemolymph is also evidenced by the combination of cellular and humoral immunity during the neutralization of pathogens. In the hemolymph, hemocytes are responsible for the processes of phagocytosis, nodulation or encapsulation, thanks to which the microorganism is closed in the follicle and rendered harmless by digestion. Then the rest substances from the neutralized organism go to the fat body (Gliński and Jarosz 2021; Stroschein-Stevenson et al. 2005).

2.5.2 Lipids in Immunity

In the case of an insect organism infection attempt by pathogenic fungi, the first significant immune response is the anatomical barriers of the body, i.e., the cuticle. However, as a result of evolution, some of these microorganisms have developed special mechanisms like the production of hydrolytic enzymes with chitin, lipo- and proteolytic activity, which can break through the insect cuticle and effectively penetrate its interior. In this case, lipids stored in the fat body not only provide energy for immune processes but also participate in the protection of anatomical barriers. The key factor in insect resistance to fungi turns out to be the presence of certain types of lipids, i.e., free fatty acids, waxes or hydrocarbons. These lipids can act favorably or inhibit the possibility of attachment of given fungal organisms to the cuticle (Wrońska et al. 2023).

2.5.3 Metalloproteases

Metalloproteases present in the insect fat body contribute to antifungal defense, although their role is generally supportive rather than central. These enzymes function in concert with core immune mechanisms such as hemocyte-mediated responses (e.g., phagocytosis, encapsulation) and the production of antimicrobial peptides. Experimental evidence from *Galleria mellonella* indicates that metalloproteases can influence susceptibility to fungal infection. In a study by Grizanova et al. (2021), RNAi-mediated silencing of metalloprotease genes in the fat body resulted in a $>10\%$ increase in mortality following fungal challenge, suggesting that these enzymes enhance overall immune competence (Grizanova et al. 2021).

Although not primary immune effectors, metalloproteases may facilitate tissue remodeling or modulate immune signaling cascades that support the antifungal response, making them a noteworthy component of the insect immune arsenal.

2.6 Vitellogenin Synthesis and Reproduction

Synthesis of vitellogenin takes place mainly in the fat body during a heterosynthetic process called vitellogenesis (orders: Lepidoptera and Hymenoptera). Diptera insects (fruit flies and houseflies) synthesized in follicular cells different substance with similar functions—yolk protein. The fat body produces a high molecular weight provitellogenin peptide (apo-proVg), which is then processed further. As a result of glycosylation, the provitellogenin peptide (proVg) is obtained, and then they are modified into the following peptides: vitellogenin 160 (Vg160) and vitellogenin 102 (Vg102), which are then transported to the hemolymph and in the final circulation in the oocytes (Wojchowski et al. 1986; Wu et al. 2021). Its large amount can be found in oocytes where it is deposited. Thanks to this precursor, in adult insects, it facilitates egg maturation, embryo growth (after eggs are laid) and oocyte nourishment. Additionally, thanks to the synthesis of vitellogenin, *Apis mellifera* queens are protected against oxidative stress, extend life and affect taste perception. It also influences the creation of intergenerational resistance in young bees as a receptor that recognizes the pathogen and presents its molecules to the progeny of *A. mellifera* (Valle 1993; Pan et al. 1969).

2.7 Aging and Functional Decline

As with other organisms, the production of substances and the rate of biochemical pathways change with age. This is due to the aging process, which is crucial for tissue function. In the case of the fat body, a phenomenon known as immunosenescence occurs, where tissue degradation leads to inflammation as a result of biochemical changes. Genetic factors drive these alterations in tissue function. In the fat body, the expression of *IIS-TORC1* genes, which are involved in metabolism, decreases over time. This decline can lead to changes in the lipid composition of the fat body, such as variations in the types and amounts of stored lipids, which can impact energy availability and metabolic functions. Lipid metabolism in the fat body is regulated by various hormones, including adipokinetic hormone and brain insulin-like peptides, which control responses to nutritional status and environmental factors. Hormonal changes during aging can influence lipid storage and mobilization, affecting overall energy balance. The efficiency of lipid metabolism in the fat body is vital for an insect's longevity and health, as it plays a crucial role in sustaining energy levels needed for survival, reproduction and other essential functions during senescence (Toprak 2020; Wrońska et al. 2023). Furthermore, the gradual decline of B-lamins triggers stem cell activation. Chronic inflammation results in the continuous, sterile activation of the Toll pathway. Immunosenescence often leads to tumor formation and shortened lifespan in insects. Signaling molecules like eicosanoids are commonly observed during aging, and eicosanoid pathways related to insect resistance are also linked to the prostaglandin receptor (PGE2). Prostaglandin E2 is an important element in modulating immunity in invertebrates via receptor-dependent signaling pathways (e.g., in the mosquito *Anopheles gambiae*), which influences the ability to regulate immune cell dynamics. PGE2 signaling with the receptor affects the increased expression of prophenooxidases, responsible for melanization, and pathogen encapsulation and increases the expression of AMPs (antimicrobial peptides like: cecropin, defensin 1), which translates directly into bactericidal processes (Kwon et al. 2021). While prostaglandins regulate numerous pro-immune processes, in excessive concentrations, they can also impair immune responses in invertebrates by inhibiting cellular activity during pathogenic infections. This occurs through the regulation of phospholipase A2, which facilitates the biosynthesis of prostaglandins from arachidonic acid in membrane phospholipids, thus maintaining homeostasis and lipid reserves. Studies on *Spodoptera exigua* have shown that prostaglandin can also negatively affect oocyte development and reduce insect fertility and at a concentration of 1–2 μM; it can also cause a decrease in the synthesis of AMPs, phenyloxidase and the lysis of oenocytoids (PGE-dependent central immune nodes). Oenocytoids link lipid signaling with humoral and cellular defense mechanisms (Kwon et al. 2021; Kim and Stanley 2021; Vatanparast et al. 2020; Ahmed et al. 2018, 2021; Kim et al. 2020; Aw et al. 2007; Barletta et al. 2020).

3 Conclusion

The insect fat body is a multifunctional tissue that plays a pivotal role in maintaining physiological homeostasis across developmental stages and environmental conditions. It serves not only as a central hub for nutrient storage and metabolism but also as an endocrine and immune-responsive organ. Its dynamic remodeling during metamorphosis, involvement in hormonal signaling, detoxification, reproductive processes and immune responses—including the synthesis of antimicrobial peptides and lipid-mediated pathogen resistance—highlight its integrative function in insect biology.

Advances in molecular and omics technologies continue to shed light on the complex regulatory networks within the fat body, revealing its essential contributions to insect survival, adaptation and aging. Understanding the diverse roles of this tissue offers valuable insights into insect physiology and holds potential applications in pest control, biotechnology and comparative physiology.

References

Agrawal N, Delanoue R, Mauri A, Basco D, Pasco M, Thorens B et al (2016) The Drosophila TNF Eiger is an Adipokine that acts on insulin-producing cells to mediate nutrient response. Cell Metab 23(4):675–684

Ahmed S, Stanley D, Kim Y (2018) An insect prostaglandin E2 synthase acts in immunity and reproduction. Front Physiol 9(Sep):1–13

Ahmed S, Al Baki MA, Lee J, Seo DY, Lee D, Kim Y (2021) The first report of prostacyclin and its physiological roles in insects. Gen Comp Endocrinol 301: 113659

Arrese EL, Soulages JL (2010) Insect fat body: energy, metabolism, and regulation. Annu Rev Entomol [Internet]. 55:207. [cited 2022 Oct 30]. Available from: https://pmc.ncbi.nlm.nih.gov/articles/PMC3075550/

Aw D, Silva AB, Palmer DB (2007) Immunosenescence: emerging challenges for an ageing population [internet]. Immunology 120:435–446. [cited 2021 Jul 1]. Available from: https://onlinelibrary.wiley.com/doi/full/10.1111/j.1365-2567.2007.02555.x

Barletta ABF, Alves E, Silva TL, Talyuli OAC, Luna-Gomes T, Sim S, Angleró-Rodríguez Y et al (2020)

Prostaglandins regulate humoral immune responses in *Aedes aegypti*. PLoS Negl Trop Dis [Internet] 14(10): e0008706. [cited 2025 Jan 31]. Available from: https://journals.plos.org/plosntds/article?id=10.1371/journal.pntd.0008706

Bergtrom G, Laufer H, Rogers R (1976) Fat body: a site of hemoglobin synthesis in *Chironomus thummi* (diptera). J Cell Biol [Internet] 69(2):264–274. [cited 2025 Jan 31]. Available from:. https://doi.org/10.1083/jcb.69.2.264

Bi J, Xiang Y, Chen H, Liu Z, Grönke S, Kühnlein RP et al (2012) Opposite and redundant roles of the two Drosophila: Perilipins in lipid mobilization. J Cell Sci [Internet] 125(15):3568–3577. [cited 2022 Oct 30]. Available from: https://journals.biologists.com/jcs/article/125/15/3568/32398/Opposite-and-redundant-roles-of-the-two-Drosophila

Bond ND, Nelliot A, Bernardo MK, Ayerh MA, Gorski KA, Hoshizaki DK et al (2011) ßFTZ-F1 and matrix metalloproteinase 2 are required for fat-body remodeling in Drosophila. Dev Biol 360(2):286–296

Brasaemle DL (2007) Thematic review series: adipocyte biology. The perilipin family of structural lipid droplet proteins: stabilization of lipid droplets and control of lipolysis. J Lipid Res [Internet] 48(12):2547–2559. [cited 2022 Oct 30]. Available from: http://www.jlr.org/article/S0022227520429050/fulltext

Bulet P, Cociancich S, Dimarcq JL, Lambert J, Reichhart JM, Hoffmann D et al (1991) Insect immunity. Isolation from a coleopteran insect of a novel inducible antibacterial peptide and of new members of the insect defensin family. J Biol Chem 266(36):24520–24525

Bursell E. The role of Proline in energy metabolism. Energy Metab Insects [Internet] 1981 135–154. [cited 2022 Oct 30]. Available from: https://link.springer.com/chapter/10.1007/978-1-4615-9221-1_5

Canavoso LE, Jouni ZE, Karnas KJ, Pennington JE, Wells MA. Fat metabolism in insects. Annu Rev Nutr. 2001;21:23-46. Available from: https://pubmed.ncbi.nlm.nih.gov/11375428/

Chang YY, Neufeld TP (2009) An Atg1/Atg13 complex with multiple roles in TOR-mediated autophagy regulation. Mol Biol Cell [Internet] 20(7):2004–2014. [cited 2022 Oct 30]. Available from: https://www.molbiolcell.org/doi/10.1091/mbc.e08-12-1250

Chapman RF. (2022) The insects: structure and function. Available from: https://books.google.pl/books?hl=pl&lr=&id=jHUCdbgW4MAC&oi=fnd&pg=PR11&dq=The+Insects+Structure+and+Function+chapman&ots=LQB9PZnZgk&sig=F8W5J6cq_Mvug5kapdQu9mEJBc&redir_esc=y#v=onepage&q=The%20Insects%20Structure%20and%20Function%20chapman&f=false

Chobotow J, Strachecka A (2013) Morfologia i funkcje ciała tłuszczowego owadów z uwzględnieniem pszczoły miodnej *Apis mellifera* L. Med Weter 69(12):712–715

Cruz-Landim C, Paes De Oliveira VT, Da Cruz-Landim C (2006) Morphology and function of insect fat body cells: a review. [cited 2022 Oct 30]. Available from: https://www.researchgate.net/publication/277226002

Delanoue R, Slaidina M, Léopold P (2010) The steroid hormone ecdysone controls systemic growth by repressing dMyc function in Drosophila fat cells. Dev Cell 18(6):1012–1021

Des Marteaux LE, Štětina T, Koštál V (2018) Insect fat body cell morphology and response to cold stress is modulated by acclimation. J Exp Biol [Internet]. 221(21) [cited 2022 Oct 30]. Available from: https://journals.biologists.com/jeb/article/221/21/jeb189647/33899/Insect-fat-body-cell-morphology-and-response-to

Eleftherianos I, Zhang W, Heryanto C, Mohamed A, Contreras G, Tettamanti G et al (2021) Diversity of insect antimicrobial peptides and proteins – a functional perspective: a review. Int J Biol Macromol 191:277–287

Furtado WCA, Azevedo DO, Martins GF, Zanuncio JC, Serrão JE (2013) Histochemistry and ultrastructure of Urocytes in the pupae of the stingless bee *Melipona quadrifasciata* (Hymenoptera: Meliponini). Microsc Microanal [Internet] 19(6):1502–1510. [cited 2022 Oct 30]. Available from: https://www.cambridge.org/core/journals/microscopy-and-microanalysis/article/abs/histochemistry-and-ultrastructure-of-urocytes-in-the-pupae-of-the-stingless-bee-melipona-quadrifasciata-hymenoptera-meliponini/A79DA76FBB1B997C7251F7DAE1963A9F

Gliński Z, Jarosz J (2021) Immunobiologia pszczoły miodnej [Internet]. [cited 2021 Jul 2]. Available from: https://vetbooks.pl/p/7/421/immunobiologia-pszczoly-miodnej-pszczelarstwo-hodowla-zwierzat.html

Grizanova EV, Coates CJ, Butt TM, Dubovskiy IM (2021) RNAi-mediated suppression of insect metalloprotease inhibitor (IMPI) enhances *Galleria mellonella* susceptibility to fungal infection. Dev Comp Immunol 122: 104126

Gu X, Quilici D, Juarez P, Blomquist GJ, Schal C (1995) Biosynthesis of hydrocarbons and contact sex pheromone and their transport by lipophorin in females of the German cockroach (*Blattella germanica*). J Insect Physiol 41(3):257–267

Gutierrez E, Wiggins D, Fielding B, Gould AP (2006) Specialized hepatocyte-like cells regulate Drosophila lipid metabolism. Nat 2006 [Internet]. 445(7125): 275–280. [cited 2022 Oct 30]. Available from: https://www.nature.com/articles/nature05382

Haunerland NH, Shirk PD (2003) Regional and functional differentiation in the insect fact body [Internet]. 40(1): 121–145. [cited 2022 Oct 30]. Available from: https://www.annualreviews.org/doi/abs/10.1146/annurev.en.40.010195.001005

Haunerland N, Nair K, Biochemistry WBI (1990) Fat body heterogeneity during development of *Heliothis zea*. Elsevier [Internet]. [cited 2025 Jan 31]. Available from: https://www.sciencedirect.com/science/article/pii/002017909090101Y

Jia Q, Liu Y, Liu H, Li S (2014) Mmp1 and Mmp2 cooperatively induce Drosophila fat body cell dissociation with distinct roles. Sci Reports 4(1):1–12. [cited 2022 Oct 30]. Available from: https://www.nature.com/articles/srep07535

Johnson MB, Butterworth FM (1985) Maturation and aging of adult fat body and oenocytes in Drosophila as revealed by light microscopic morphometry. J Morphol [Internet] 184(1):51–59. [cited 2022 Oct 30]. Available from: https://onlinelibrary.wiley.com/doi/full/10.1002/jmor.1051840106

Keeling C, Tittiger C, MacLean M, Blomquist GJ (2021 Jan) Pheromone production in bark beetles. Insect Pheromone Biochem Mol Biol 1:123–162

Kim Y, Stanley D (2021) Eicosanoid signaling in insect immunology: new genes and unresolved issues [Internet]. Genes 12(2):211. [cited 2022 Oct 30]. Available from: https://www.mdpi.com/2073-4425/12/2/211/htm

Kim Y, Ahmed S, Al Baki MA, Kumar S, Kim K, Park Y et al (2020) Deletion mutant of PGE2 receptor using CRISPR-Cas9 exhibits larval immunosuppression and adult infertility in a lepidopteran insect, *Spodoptera exigua*. Dev Comp Immunol [Internet] 111 (May):103743. Available from:. https://doi.org/10.1016/j.dci.2020.103743

Krupp JJ, Levine JD (2010) Dissection of oenocytes from adult *Drosophila melanogaster*. J Vis Exp [Internet] (41) [cited 2022 Oct 30]. Available from: https://www.jove.com/t/2242/dissection-of-oenocytes-from-adult-drosophila-melanogaster

Kwon H, Hall DR, Smith RC (2021) Prostaglandin E2 signaling mediates oenocytoid immune cell function and lysis, limiting bacteria and plasmodium oocyst survival in *Anopheles gambiae*. Front Immunol [Internet] 12:680020. [cited 2025 Apr 7]. Available from: www.frontiersin.org

Law JH, Wells MA (1989) Insects as biochemical models. J Biol Chem 264(28):16335–16338

Li S, Yu X, Feng Q (2019) Fat body biology in the last decade. Annu Rev Entomol 64:315–333

Lindsay SA, Wasserman SA (2014) Conventional and non-conventional Drosophila toll signaling. Dev Comp Immunol 42(1):16–24

Lippai M, Csikós G, Maróy P, Lukácsovich T, Juhász G, Sass M (2008) SNF4Aγ, the Drosophila AMPK γ subunit is required for regulation of developmental and stress-induced autophagy [Internet]. 4(4): 476–486. [cited 2022 Oct 30]. Available from: https://www.tandfonline.com/doi/abs/10.4161/auto.5719

Liu Y, Liu H, Liu S, Wang S, Jiang RJ, Li S (2009) Hormonal and nutritional regulation of insect fat body development and function. Arch Insect Biochem Physiol [Internet] 71(1):16–30. [cited 2022 Oct 30]. Available from: https://onlinelibrary.wiley.com/doi/full/10.1002/arch.20290

Liu H, Jia Q, Tettamanti G, Li S (2013) Balancing crosstalk between 20-hydroxyecdysone-induced autophagy and caspase activity in the fat body during Drosophila larval-prepupal transition. Insect Biochem Mol Biol 43(11):1068–1078

Locke M (1984) The structure and development of the vacuolar system in the fat body of insects. Insect Ultrastruct [Internet]. [cited 2025 Jan 31] 151–197. Available from: https://link.springer.com/chapter/10.1007/978-1-4613-2715-8_5

Makki R, Cinnamon E, Gould AP (2014) The development and functions of oenocytes. 101146/annurev-ento-011613-162056 [Internet] 59:405–425. [cited 2022 Oct 30]. Available from:. https://www.annualreviews.org/doi/abs/10.1146/annurev-ento-011613-162056

Malassigné S, Minard G, Vallon L, Martin E, Moro CV, Luis P (2021) Diversity and functions of yeast communities associated with insects. Microorganism [Internet] 9(8):1552. [cited 2022 Oct 30]. Available from: https://www.mdpi.com/2076-2607/9/8/1552/htm

Manniello MD, Moretta A, Salvia R, Scieuzo C, Lucchetti D, Vogel H et al (2021) Insect antimicrobial peptides: potential weapons to counteract the antibiotic resistance. Cell Mol Life Sci [Internet] 78(9): 4259–4282. [cited 2022 Oct 30]. Available from: https://link.springer.com/article/10.1007/s00018-021-03784-z

Martins GF, Serrão JES, Ramalho-Ortigão JM, Pimenta PFP (2011) A comparative study of fat body morphology in five mosquito species. Mem Inst Oswaldo Cruz [Internet] 106(6):742–747. [cited 2025 Jan 31]. Available from: https://www.scielo.br/j/mioc/a/JbvpJy7tJQbwRSdvVMt57RR

Milburn NS (1966) Fine structure of the pleomorphic bacteroids in the mycetocytes and ovaries of several genera of cockroaches. J Insect Physiol 12(10): 1245–1254

Mirth CK, Riddiford LM (2007) Size assessment and growth control: how adult size is determined in insects. BioEssays [Internet] 29(4):344–355. [cited 2022 Oct 30]. Available from: https://onlinelibrary.wiley.com/doi/full/10.1002/bies.20552

Moribayashi A, Kurahashi H, Agui N. (1985) Lipid metabolism. Compr Insect Physiol Biochem Pharmacol [Internet] 47(3):255–261. [cited 2025 Jan 31]. Available from: https://cir.nii.ac.jp/crid/1572543024782832256

Myllymäki H, Valanne S, Rämet M (2014) The Drosophila IMD signaling pathway. J Immunol [Internet] 192(8):3455–3462. [cited 2022 Oct 30]. Available from: https://www.jimmunol.org/content/192/8/3455

Nijhout HF, Williams CM (1974) Control of Moulting and metamorphosis in the tobacco hornworm, *Manduca sexta* (L.): cessation of juvenile hormone secretion as a trigger for pupation. J Exp Biol [Internet] 61(2): 493–501. [cited 2022 Oct 30]. Available from: https://journals.biologists.com/jeb/article/61/2/493/

22257/Control-of-Moulting-and-Metamorphosis-in-the

Nunes B, Simões MI, Navarro JC, Castro BB (2020) First ecotoxicological characterization of paraffin microparticles: a biomarker approach in a marine suspension-feeder, Mytilus sp. Environ Sci Pollut Res [Internet] 27(33):41946–41960. [cited 2021 Oct 7]. Available from: https://link.springer.com/article/10.1007/s11356-020-10055-0

Nunes C, Sucena É, Koyama T (2021) Endocrine regulation of immunity in insects. FEBS J [Internet] 288(13): 3928–3947. [cited 2022 Oct 30]. Available from: https://onlinelibrary.wiley.com/doi/full/10.1111/febs.15581

Pan ML, Bell WJ, Telfer WH. Vitellogenic blood protein synthesis by insect fat body Science (80–) [Internet]. 1969 165(3891):393–394. [cited 2022 Oct 30]. Available from: https://www.science.org/doi/10.1126/science.165.3891.393

Park MS, Park P, Takeda M (2013) Roles of fat body trophocytes, mycetocytes and urocytes in the American cockroach, *Periplaneta americana* under starvation conditions: an ultrastructural study. Arthropod Struct Dev [Internet] 42(4):287–295. [cited 2025 Jan 31]. Available from: https://pubmed.ncbi.nlm.nih.gov/23567491/

Patel RT, Soulages JL, Hariharasundaram B, Arrese EL (2005) Activation of the lipid droplet controls the rate of lipolysis of triglycerides in the insect fat body. J Biol Chem [Internet] 280(24):22624–22631. [cited 2022 Oct 30]. Available from: http://www.jbc.org/article/S0021925820614284/fulltext

Rao XJ, Zhan MY, Pan YM, Liu S, Yang PJ, Yang LL et al (2018) Immune functions of insect βGRPs and their potential application. Dev Comp Immunol 83:80–88

Rizki TM, Rizki RM (1962) Cytodifferentiation in the rosy mutant of *Drosophila melanogaster*. J Cell Biol [Internet] 12(1):149–157. [cited 2022 Oct 30]. Available from: http://rupress.org/jcb/article-pdf/12/1/149/1063601/149.pdf

Roma GC, Bueno OC, Camargo-Mathias MI (2010) Morpho-physiological analysis of the insect fat body: a review. Micron 41(5):395–401

Ryoo HD, Baehrecke EH (2010) Distinct death mechanisms in Drosophila development. Curr Opin Cell Biol 22(6):889–895

Scott RC, Schuldiner O, Neufeld TP (2004) Role and regulation of starvation-induced autophagy in the Drosophila fat body. Dev Cell 7(2):167–178

Siuda M, Wilde J, Woyke J, Jasiński Z, Madras-Majewska B (2014) Wintering reserve queens in mini-plus and 3-comb nuclei. J Apic Sci 58(1):61–68

Skowronek P, Wójcik Ł, Strachecka A, Brivio F (2021) Fat body-multifunctional insect tissue. Available from: https://doi.org/10.3390/insects12060547

Soulages JL, Wells MA (1994) Lipophorin: the structure of an insect lipoprotein and its role in lipid transport in insects. Adv Protein Chem 45:371–415

Steele JE (1982) Glycogen phosphorylase in insects. Insect Biochem 12(2):131–147

Strachecka A, Łoś A, Filipczuk J, Schulz M (2018) Individual and social immune mechanisms of the honey bee. Medycyna Weterynaryjna Polskie Towarzystwo Nauk Weterynaryjnych 74:426–433

Strachecka A, Olszewski K, Kuszewska K, Chobotow J, Wójcik Ł, Paleolog J et al (2021) Segmentation of the subcuticular fat body in *Apis mellifera* females with different reproductive potentials. Sci Reports [Internet] 11(1):1–11. [cited 2022 Jul 13]. Available from: https://www.nature.com/articles/s41598-021-93357-8

Strachecka A, Chobotow J, Kuszewska K, Olszewski K, Skowronek P, Bryś M et al (2022) Morphology of Nasonov and Tergal Glands in Apis mellifera rebels. Insects [Internet] 13:401. [cited 2022 Oct 30]. Available from: https://www.mdpi.com/2075-4450/13/5/401/htm

Stroschein-Stevenson SL, Foley E, O'Farrell PH, Johnson AD (2005) Identification of Drosophila gene products required for phagocytosis of *Candida albicans*. PLOS Biol [Internet] 4(1):e4. [cited 2022 Oct 30]. Available from: https://journals.plos.org/plosbiology/article?id=10.1371/journal.pbio.0040004

Szymczak-Cendlak M, Gołębiowski M, Chowański S, Pacholska-Bogalska J, Marciniak P, Rosiński G et al (2022) Sulfakinins influence lipid composition and insulin-like peptides level in oenocytes of *Zophobas atratus* beetles. J Comp Physiol B Biochem Syst Environ Physiol [Internet] 192(1):15–25. [cited 2022 Oct 30]. Available from: https://link.springer.com/article/10.1007/s00360-021-01398-2

Toprak U (2020) The role of peptide hormones in insect lipid metabolism. Front Physiol [Internet] 11 (May):1–33. [cited 2022 Oct 30]. Available from: www.frontiersin.org

Tunaz H, Park Y, Büyükgüzel K, Bedick JG, Aliza ARN, Stanley DW (2003) Eicosanoids in insect immunity: bacterial infection stimulates hemocytic phospholipase A2 activity in tobacco hornworms. Arch Insect Biochem Physiol [Internet] 52(1, 1):–6. [cited 2022 Oct 30]. Available from: https://onlinelibrary.wiley.com/doi/full/10.1002/arch.10056

Vafopoulou-Mandalos X, Laufer H (1983) Tissue-specificity of hemoglobin synthesis: localization of heme synthesis in the subepidermal fat body of *Chironomus thummi* (Diptera). Arch Insect Biochem Physiol [Internet] 1(2):191–197. [cited 2022 Oct 30]. Available from: https://onlinelibrary.wiley.com/doi/full/10.1002/arch.940010209

Valle D (1993) Vitellogenesis in insects and other groups: a review. Mem Inst Oswaldo Cruz [Internet] 88(1): 1–26. [cited 2022 Oct 30]. Available from: http://www.scielo.br/j/mioc/a/zkQ8mCMLB444cm8G5bRDrFp/abstract/?lang=en

Vardanis A (1963) Glycogen synthesis in the insect fat body. Biochim Biophys Acta Spec Sect Enzymol Subj 73(4):565–573

Vatanparast M, Lee DH, Kim Y (2020) Biosynthesis and immunity of epoxyeicosatrienoic acids in a lepidopteran insect, *Spodoptera exigua*. Dev Comp Immunol 107:103643

Wang X, Ding Y, Lu X, Geng D, Li S, Raikhel AS et al (2021) The ecdysone-induced protein 93 is a key factor regulating gonadotrophic cycles in the adult female mosquito *Aedes aegypti*. Proc Natl Acad Sci U S A [Internet] 118(8):e2021910118. [cited 2022 Oct 30]. Available from: https://www.pnas.org/doi/abs/10.1073/pnas.2021910118

Willott E, Bew LK, Nagle RB, Wells MA (1988) Sequential structural changes in the fat body of the tobacco hornworm, *Manduca sexta*, during the fifth larval stadium. Tissue Cell 20(4):635–643

Wojchowski DM, Parsons P, Nordin JH, Kunkel JG (1986) Processing of pro-vitellogenin in insect fat body: a role for high-mannose oligosaccharide. Dev Biol 116(2):422–430

Wójcik Ł, Chęć M, Skowronek P, Grabowski M, Persona K, Strachecka A (2022 Jul) Do the different life history strategies of ants and honeybees determine fat body morphology? Arthropod Struct Dev 1(69): 101186

Wrońska AK, Kaczmarek A, Boguś MI, Kuna A (2023) Lipids as a key element of insect defense systems. 14: 1183659. [cited 2025 Jan 31]. Available from: https://pubmed.ncbi.nlm.nih.gov/37359377/

Wu Z, Yang L, He Q, Zhou S (2021) Regulatory mechanisms of vitellogenesis in insects. Front Cell Dev Biol 8:1863

Xianhui L, Baoping L (2006) Effect of temperature on the abundance of mycetocytes in *Aphis fabae* Scopoli (Aphididae, Homoptera) and body size of their hosts Kun Chong xue bao. Acta Entomol Sin [Internet] 49(3):428–432. [cited 2022 Oct 30]. Available from: https://europepmc.org/article/cba/619337

Xie K, Tian L, Guo X, Li K, Li J, Deng X et al (2016) BmATG5 and bmATG6 mediate apoptosis following autophagy induced by 20-hydroxyecdysone or starvation. Autophagy [Internet] 12(2):381–396. [cited 2022 Oct 30]. Available from: https://www.tandfonline.com/doi/abs/10.1080/15548627.2015.1134079

Yamanaka N, Rewitz KF, O'Connor MB (2013) Ecdysone control of developmental transitions: lessons from Drosophila research [Internet]. 58:497–516. [cited 2022 Oct 30]. Available from: https://www.annualreviews.org/doi/abs/10.1146/annurev-ento-120811-153608

Yuan D, Zhou S, Liu S, Li K, Zhao H, Long S et al (2020) The AMPK-PP2A axis in insect fat body is activated by 20-hydroxyecdysone to antagonize insulin/IGF signaling and restrict growth rate. Proc Natl Acad Sci U S A [Internet] 117(17):9292–9301. [cited 2022 Oct 30]. Available from: https://www.pnas.org/doi/abs/10.1073/pnas.2000963117

Zhang JH, Yu N, Xu XX, Liu ZW (2019) Community structure, dispersal ability and functional profiling of microbiome existing in fat body and ovary of the brown planthopper, Nilaparvata lugens. Insect Sci [Internet] 26(4):683–694. [cited 2022 Oct 30]. Available from: https://onlinelibrary.wiley.com/doi/full/10.1111/1744-7917.12575

Zhang W, Tettamanti G, Bassal T, Heryanto C, Eleftherianos I, Mohamed A (2021) Regulators and signalling in insect antimicrobial innate immunity: functional molecules and cellular pathways. Cell Signal 83:110003

Zhang S-Y et al. (2022) 20-hydroxyecdysone-induced storage of uric acid in *Manduca sexta* fat body. [Internet]. [cited 2022 Oct 30]. Available from: https://agris.fao.org/agris-search/search.do?recordID=US19820745972

Zheng H, Yang X, Xi Y (2016) Fat body remodeling and homeostasis control in Drosophila. Life Sci 167:22–31

Zülfikaroğlu T, Turgay-İzzetoğlu G, Yikilmaz MS, İzzetoğlu S (2022) Demonstrating the general structure and cell types of the fat body in *Blatta orientalis* (oriental cockroach). Anat Histol Embryol [Internet] 51(1):23–35. [cited 2025 Jan 31]. Available from: https://pubmed.ncbi.nlm.nih.gov/34668578/

Adv Exp Med Biol (2026) 1494: 65–96
https://doi.org/10.1007/5584_2024_835

Published online: 21 November 2024

Dietary Lipids and Their Metabolism in the Midgut

Pierre Delamotte and Jacques Montagne

Abstract

Animals use dietary lipids to sustain their growth and survival. Insects can synthesize fatty acids (FAs) and are autotroph for a number of lipids, but auxotroph for specific lipids classes (e.g. sterols, polyunsaturated FAs). Once ingested, lipids are hydrolysed in the intestinal lumen and taken up into intestinal cells within specific regions of the insect digestive tract. These lipids can be either stored in the intestinal cells or exported through the haemolymph circulation to specific organs. In this chapter, we describe the various lipids provided by insect diets, their extracellular hydrolysis in the gut lumen and their intake and metabolic fate in the intestinal cells. This chapter emphasizes the critical role of the digestive tract and its regionalization in processing dietary lipids prior to their transfer to the requiring tissues.

Keywords

Columnar cells · Digestion · Emulsifier · Hydrolysis · Lipase · Lipid transporter · Membrane · Microbiota

P. Delamotte and J. Montagne (✉)
Institute for Integrative Biology of the Cell (I2BC), UMR 9198, CNRS, Université Paris-Sud, CEA, Gif-sur-Yvette, France
e-mail: Pierre-Delamotte@i2bc.paris-saclay.fr; Jacques.Montagne@i2bc.paris-saclay.fr; jacques.montagne@ens-lyon.fr

Insects colonize a great variety of ecological niches on Earth and, thus, their survival relies on diverse feeding diets. Dietary macromolecules provided by the environment are used to produce energy and building blocks required to sustain developmental growth and reproduction, and to maintain homeostasis. The digestive tract is the first organ exposed to these external macromolecules. It must take in charge several essential functions, including protection against pathogens, neutralization of toxic compounds, extra-cellular digestion of macromolecules into assimilable nutrients, and absorption and efflux of these nutrients into the haemolymph (the insect open circulatory system). In addition, intestinal cells can fulfil storage and further enzymatic modifications prior to the distribution of nutrients throughout the overall body to the needing organs. Therefore, the digestive tract plays a major role in coordinating local and systemic homeostasis (Zhao and Karpac 2020). Mechanistically, the food bolus progresses through specialized compartments that process specific nutrients. The diversity of insect species and their broad diet spectrum is associated with a large variety of digestive tract anatomical structures favouring nutrient breakdown and assimilation (Terra and Ferreira 2020). Nonetheless, despite this huge diversity, the nutrient requirements and the digestive tracts of insects exhibit conserved functional and structural properties.

Enzymatic digestions of carbohydrates, proteins and lipids have been investigated in several insect species for more than a century (Plateau 1876) to understand how the basic nutrients are produced and assimilated. Food provides essential and dispensable lipids, which are enzymatically processed in the gut lumen and taken up by the intestinal cells. In this chapter, we will first provide an overview of the insect digestive tract and discuss essential dietary lipids. Next, we will describe the lipid digestive processes in the gut lumen and the uptake of the resulting products by intestinal cells. Finally, we will describe the lipid metabolic processes that take place in midgut cells. Digestive processes are tightly regulated by systemic controls, including endocrine signals from the brain and the fat body, which will not be addressed here. In this chapter, we will focus on lipid digestion, the dynamics of this process along the whole midgut and its essential role prior to the transfer of basic dietary lipids into the body internal milieu.

1 The Insect Digestive Tract

1.1 Structure of the Digestive Tract

In general, the diversity in the morphology of the alimentary canal strongly depends on the order rather than on the insect feeding regimen (Santos et al. 2017; Terra and Ferreira 2020). Nonetheless, irrespective of the phylogeny, the insect midgut always comprises three major regions: the foregut, the midgut and the hindgut (Chapman et al. 2013) (Fig. 1a). The foregut and the hindgut are of ectodermal origin and are covered by a cuticle facing the intestinal lumen. The midgut is of endodermal origin and is protected by a thin mucus layer and, for most species, by another structure called the peritrophic matrix (PM, also called peritrophic membrane) (Terra 2001). From the anterior to the posterior ends, the foregut comprises the mouth, the pharynx connected to the salivary glands, the oesophagus, the crop in which the food bolus is transiently stored and the proventriculus that controls the entry of food into the midgut. The midgut is subdivided into an anterior, a median and a posterior region. Genomic and morphological analyses revealed that the midgut of *Drosophila melanogaster* contains five regions, which can be further divided into up to 14 sub-regions (Buchon et al. 2013; Dutta et al. 2015; Marianes and Spradling 2013). In most species, the anterior midgut is connected to gastric caeca (Terra and Ferreira 2020). The hindgut contributes to water reabsorption and faeces formation. It comprises the ileum connected to the Malpighian tubules, which fulfil excretory functions, the colon, the rectum and the anus.

The midgut is the main organ of food digestion and nutrient absorption, which are facilitated by the presence of a PM and the absence of an external cuticle. The PM is a chitin-based lattice of proteoglycans and proteins. Depending on the insect species, the PM is produced either by the proventriculus or by delamination from the intestinal cells along the whole midgut (Hegedus et al. 2009; Lehane 1997). Hemipterans and thysanopterans do not have a PM but a lipoproteic perimicrovillar membrane that fulfils a similar function {Terra, 2006 #1400}. The PM compartmentalizes the midgut lumen into an endoperitrophic space, the PM itself and an ectoperitrophic space (Fig. 1a). Depending on their physicochemical properties, the digestive enzymes are either attached to the epithelium membrane, free in the ectoperitrophic or endoperitrophic spaces, or framed within the PM (Hegedus et al. 2009; Hu et al. 2012). Analysis of the digestive enzymes that cross the larval PM of *Spodoptera frugiperda* suggested that the PM pore diameter is about 7.5–8 nm (Ferreira et al. 1994). As a result, macromolecules and undigested materials remain in the endoperitrophic space, whereas small molecules and nutrients can pass through, for eventual digestion and absorption by the intestinal cells (Bolognesi et al. 2008). The PM would allow the establishment of bidirectional fluxes along the midgut {Berridge, 1970 #1937}: contractions of the proventriculus push the food bolus into the midgut generating an anterior/posterior flux inside the endoperitrophic space (Peller et al. 2009; Woodring and Lorenz 2007), while fluid secretions by the Malpighian tubules exert a

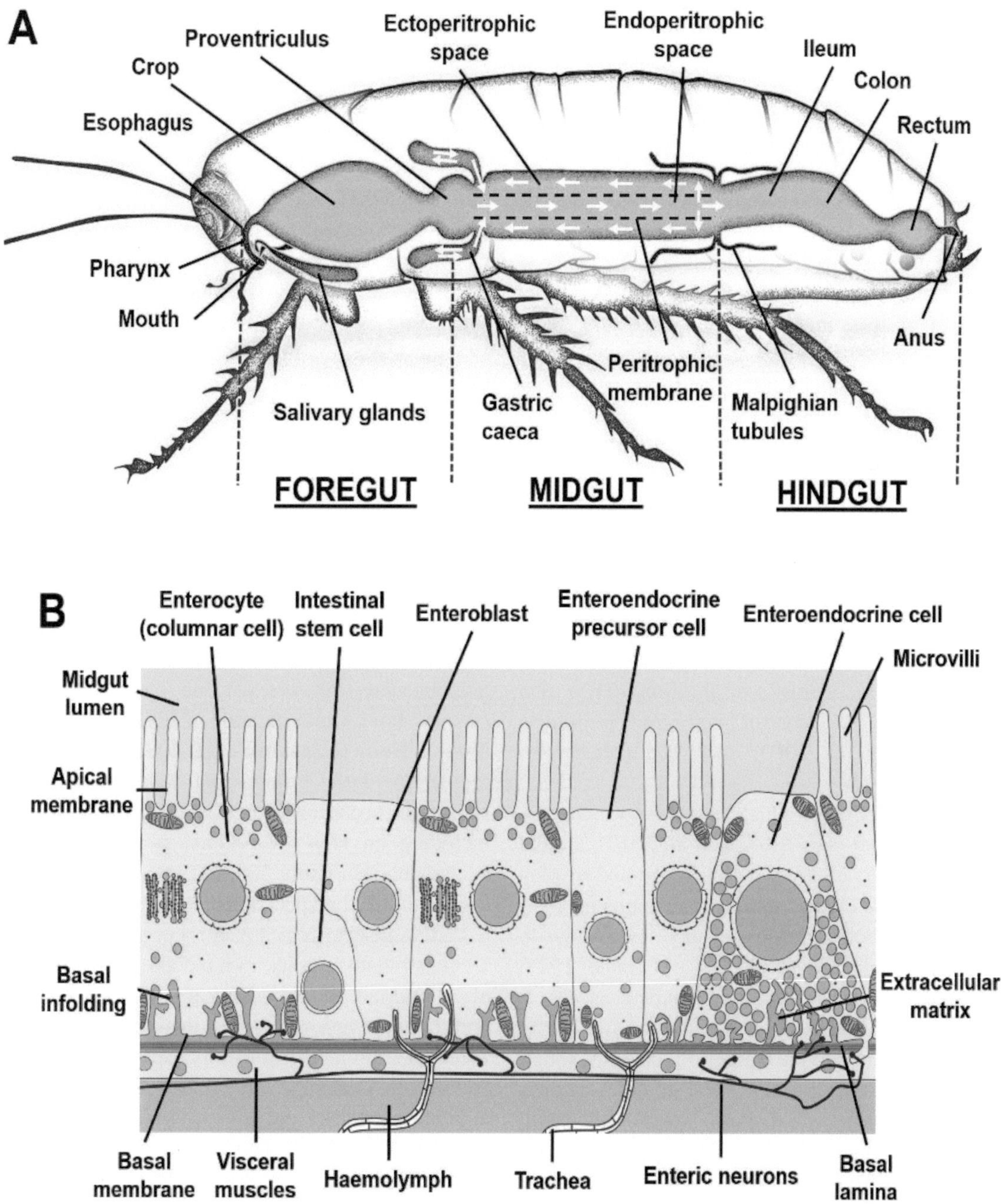

Fig. 1 The insect digestive tract. (**a**) Basic scheme of the digestive tract, showing the different gut sections and the bidirectional fluxes (white arrows) putatively generated by the PM (black dashed line). (**b**) The midgut epithelium showing three main cell types: intestinal stem cells, enteroendocrine cells and columnar cells. These schemes represent the typical structures and cells of a digestive tract but cannot be attributed to a given insect species

counter pressure within the ectoperitrophic space (Biagio et al. 2009) (Fig. 1a). Although not formally demonstrated, these opposing fluxes are expected to favour the mixing of foods to be digested and the recycling of digestive enzymes.

1.2 Cell Types and Their Functions

The midgut is a monolayer epithelium supported by a basal lamina with the apical cell side facing the intestinal lumen (Caccia et al. 2019). This epithelium is surrounded by striated muscles and connected to fine-branched tracheoles and nerves (Miguel-Aliaga et al. 2018) (Fig. 1b). Pluripotent intestinal stem cells (ISCs), located at the basis of the epithelium, ensure a constant homeostatic renewal of intestinal cells (Hakim et al. 2010; Zwick et al. 2019). ISCs give rise to either enteroblasts or enteroendocrine progenitors that differentiate into enterocytes or enteroendocrine cells, respectively (Zeng and Hou 2015).

The functions of enteroendocrine cells have been studied in several insect species (Reiher et al. 2011; Wegener and Veenstra 2015), but elucidation of their activity mostly relies on *D. melanogaster* studies (Issaq et al. 2009). These cells secrete various peptidic hormones that systemically control appetite and homeostasis (Veenstra 2009; Veenstra et al. 2008). They also act locally to control midgut peristalsis (LaJeunesse et al. 2010) and lipid metabolism (Song et al. 2014).

The enterocytes, which should formally be called columnar cells (Caccia et al. 2019), are the most abundant cell type in the midgut and display a characteristic organization that accounts for their secretive and/or absorptive functions (Azevedo et al. 2009; Fialho et al. 2013; Lehane 1977; Rost-Roszkowska et al. 2007; Santos et al. 2017; Teixeira et al. 2013). Columnar cells mainly specialized in enzyme secretion contain rough endoplasmic reticulum (ER) and granules in their apical cytosol and display apical microvilli (de Sousa et al. 2009). Release of vesicles containing digestive enzymes may occur through various types of secretion systems, namely holocrine, apocrine and merocrine (Fialho et al. 2013; Fuzita et al. 2019; Silva et al. 2013). During holocrine secretion, secretory vesicles and cytoplasm are released into the gut lumen by rupture of the plasma membrane, whereas in apocrine secretion, a variable part of the apical cytoplasm containing vesicles is released by membrane budding. During merocrine secretion, vesicles fuse with the apical membrane and release their content into the gut lumen without membrane and cytoplasm loss. Conversely, columnar cells mainly specialized in absorption exhibit a dense apical microvillar structure and basal membrane infolds and accumulate lipid droplets (LDs) depending on diet, gut region and insect species. These subcellular structures as well as the membrane composition greatly vary along the anterior-posterior axis of the midgut, suggesting different nutrient absorption and enzyme secretion abilities, although the existence of populations of columnar cells strictly specialized in either function has not been formally demonstrated (Caccia et al. 2019; Terra et al. 2006).

The pH greatly varies along the anteroposterior axis of the intestinal tract depending on gut regions and insect species (Harrison 2001; Terra and Ferreira 2020). In various species, restricted regions display extreme pH values; for instance, in the median region of the midgut of some Scarabaeidae, alkaline pH can reach 12, which favours lipid saponification (Lehane 1977) and lipase activity (Canavoso et al. 2001). The precise regulation of the luminal pH depends on active acid-base secretion processes across epithelia, including an H + -V-ATPase potassium/proton exchanger (Harrison 2001). Extreme pH values appear to rely on specialized cells that are specific to certain insect species (Caccia et al. 2019). Goblet cells mostly studied in lepidopteran larvae, but also reported in ephemeropterans, plecopterans and trichopterans, display an invagination of their apical membrane that forms a cavity (Cioffi 1984). Active ion exchanges in this cavity generate an alkaline pH in the gut lumen, which favours the neutralization of toxic compounds present in the plant material eaten by phytophagous lepidopteran larvae. Copper cells found in brachycerous dipterans generate an acidic pH in a restricted region of the median midgut, thereby favouring the killing of luminal bacteria (Dubreuil 2004; Storelli et al. 2018). In sum, the whole lumen environment depends on exogenous and endogenous parameters, which determine the final digestive abilities of each insect species.

2 Essential and Non-essential Dietary Lipids

2.1 Dietary Lipids

Insects find in their food a great variety of lipids, including those essential for survival and development that they are unable to synthesize. The main dietary lipids are triacylglycerols (TAGs), phospholipids, galactolipids, sulpholipids and sterols. TAGs and phospholipids are present in several types of foods. Galactolipids (mono- and di-galactosyl diacylglycerols, MGDG and DGDG respectively) (Fig. 2a) and sulpholipids (sulphated glycolipids), the most abundant membrane lipids of chloroplasts, represent up to 80% of total lipids of leaf cells consumed by folivorous insects (Sahaka et al. 2020). Depending on their diet, insects ingest sterols as phytosterol from plants, cholesterol from animals or ergosterol from fungi (Jing and Behmer 2020).

Only a few studies investigated the need for fat-soluble vitamins; they are not metabolized in the midgut but directly transferred to the requiring organs. Experimental strategies to determine the essential need of a given lipid are based on either controlling the food composition or feeding radiolabelled precursors to monitor their incorporation in final products (Canavoso et al. 2001). In this way, it has been shown that insects are unable to synthesize sterols, most of them being also disabled for polyunsaturated fatty acid (PUFAs) synthesis.

2.2 Essential Fatty Acids

Although most insect diets contain TAGs, insects are able to synthesize long-chain fatty acids (LCFAs), the monounsaturated palmitoleic (C16:1) and oleic (C18:1) acids, and TAGs (Garrido et al. 2015). In contrast, they are classically considered to be auxotrophs for the polyunsaturated linoleic (C18:2) and α-linolenic (C18:3) acids (Fig. 2b). Nevertheless, Blomquist and colleagues have shown that a few insect species, including the cockroach *Periplaneta americana* and the house cricket *Acheta domesticus*, are autotrophs for linoleic but not α-linolenic acid (Blomquist et al. 1991). Linoleic acid is the precursor of arachidonic acid (AA, C20:4), whereas α-linolenic acid is the precursor of docosahexanoic (DHA, C22:6) and eicosapentanoic (EPA, C20:5) acids (Fig. 2b), both pathways requiring the same metabolic enzymes (Stark et al. 2008). Surprisingly, *D. melanogaster* is disabled for AA, EPA and DHA synthesis (Shen et al. 2010), a default likely shared by mosquitoes that require dietary AA or EPA to sustain their larval development (Dadd and Kleinjan 1979; Merritt et al. 1992). Oxidation of AA directs the synthesis of eicosanoids (e.g. prostaglandins), appearing to be important for juvenile development, immune response and reproductive functions in insects (Kim and Stanley 2021). The specific requirements of EPA and DHA have been investigated in *D. melanogaster* and revealed that feeding an EPA/DHA mixture is neuroprotective against intoxication with the Paraquat insecticide (de Oliveira Souza et al. 2019).

The phenotypes induced by PUFA deficiency differ between insect species. In the locusts *Schistocerca gregaria* and *Locusta migratoria*, this deficiency results in impaired adult emergence (Dadd 1961), whereas in other species, it may result in milder phenotypes, including wing expansion defects and reduced reproductive capacity (Beenakkers et al. 1985). In some coleopteran, dipteran and lepidopteran species, phenotypes may appear only after two generations of deprivation, suggesting that the very low amounts provided by the oocyte stores are sufficient to sustain their critical functions (Canavoso et al. 2001; Stanley-Samuelson 1994).

2.3 Sterol Requirement

Insects are unable to synthesize sterols (Clark and Block 1959; Svoboda 1999), although for a few species, sterols are provided by intestinal symbionts (Clayton 1964; Jurzitza 1974). Cholesterol is essential i) as a membrane component present at high levels in lipid rafts (Carvalho et al. 2010; Rietveld et al. 1999), ii) as a precursor

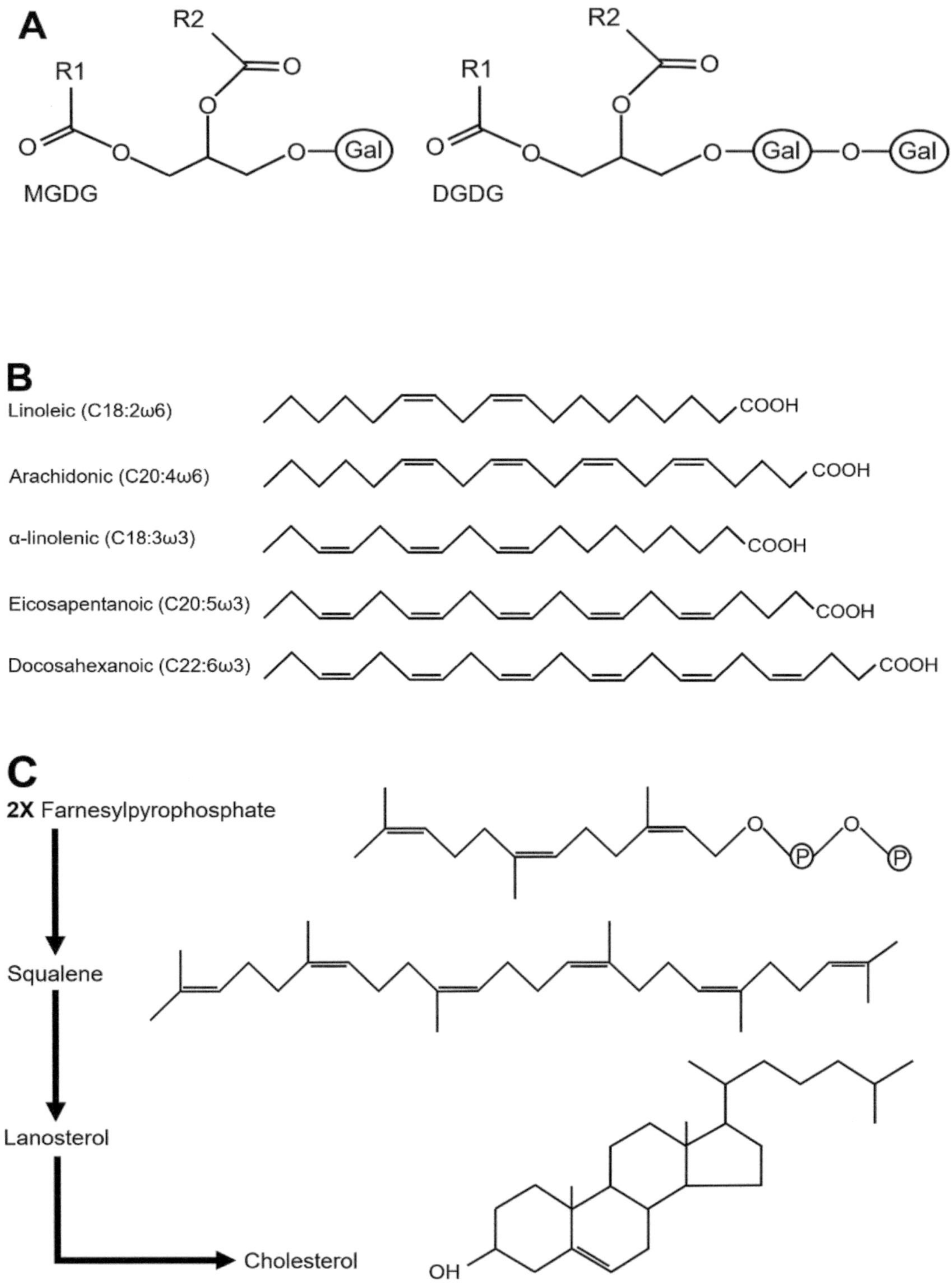

Fig. 2 Essential and non-essential lipids. (**a**) Galactolipids are the main components of membrane thylakoids; R1 and R2 represent the esterified FA moieties; Gal the galactose residues. (**b**) Essential and non-essential PUFAs. (**c**) The synthesis of cholesterol from farnesylpyrophsophate does not operate in insects

of the molting ecdysteroid hormones (Rewitz et al. 2006; Yamanaka et al. 2013) and iii) as a conjugate of the Hedgehog morphogen required for its signalling efficiency (Cooper et al. 2003). The need for dietary sterols for insect development was first discovered in the blow-fly *Lucilia sericata* (Hobson 1935); this requirement was next extended to all investigated insect species (Canavoso et al. 2001; Jing and Behmer 2020). Deficiency in dietary sterols results in defects in developmental growth, molting transitions and reproduction, although maternally provided sterol stores may be sufficient to support early developmental processes, as reported for the first and second larval stages of *D. melanogaster* (Fluegel et al. 2006).

Sterol synthesis requires the condensation of two farnesylpyrophsophate units and its reduction to squalene (Fig. 2c); next, cyclization and oxidation of squalene lead to the formation of lanosterol, which after further oxidation steps result in cholesterol. Insects can synthesize the sesquiterpene intermediate farnesylpyrophsophate in the biosynthetic pathway of the juvenile hormone (Noriega 2014). In contrast, insects lack the enzymes required for squalene, lanosterol and cholesterol synthesis (Zhang et al. 2019). These final metabolic steps demand a high number of oxygen molecules. Given their blind tracheal system limiting oxygen delivery to the needing tissues, it has been proposed that the loss of such a high oxygen-consuming pathway may provide an evolutionary advantage to insects (Harrison et al. 2010; Zhang et al. 2019).

2.4 Carotenoids and Fat-Soluble Vitamins

Carotenoids are lipophilic isoprenoids divided into carotenes and xanthophylls, the latter result from the oxidation of the former. They constitute pigments that colour the tegument and the eye photoreceptors of several insect species (Canavoso et al. 2001). Carotenoids were thought to be synthesized only by plants, fungi and microorganisms, but recent studies revealed that aphid and cecidomyiid genomes encode the enzymes needed for their synthesis (Cobbs et al. 2013; Moran and Jarvik 2010).

A few studies investigated the role of the fat-soluble vitamins A, D, E and K in insects. Vitamin A derives from β-carotene; its deprivation in *D. melanogaster* does not affect essential functions but results in blindness (Dewett et al. 2021). Three forms of vitamin K exist VK1, VK2 and VK3. It has been shown that a VK1 diet supplementation improves the developmental growth of the house cricket *A. domesticus* (McFarlane 1976). Studies in *D. melanogaster* revealed that VK2 acts as a mitochondrial electron carrier (Vos et al. 2012) and that feeding VK2 protects against mitochondrial dysfunctions in lymphoma and Alzheimer models (Dragh et al. 2017; Lin et al. 2021). The need for vitamin E (α-tocopherol) has been studied in a few insect species and shown to be essential for spermatogenesis in the house cricket *A. domesticus* and for female reproduction in *Agria affinis* (Fraenkel and Blewett 1946; House 1966; Meikle and McFarlane 1965). Vitamin D3 is a steroid whose deficiency increases DNA damage. In *D. melanogaster*, vitamin D3 supplementation reduces the formation of experimentally induced tumours (Vasconcelos et al. 2020). Characterization of the metabolic pathways in which these vitamins are involved may identify specific targets to develop novel strategies aimed at controlling insect pest propagation (Moadeli et al. 2020).

2.5 Microbiota and Dietary Lipids

The gut microbiota comprises all the microorganisms that colonize transiently or permanently the digestive tract. The gut microbiota has been shown to contribute to insect digestion, homeostasis, growth, detoxification and resistance to pathogens (Engel and Moran 2013; Grenier and Leulier 2020; Lee and Hase 2014; Schmidt and Engel 2021). The microbiota is essential for xylophagous insects, as described in termites, which—when deprived of their anaerobic protists—continue to feed on wood but die of starvation (Cleveland 1923). Lower termites are exclusive wood feeders, and their hindgut paunch hosts flagellates, bacteria and archaea, whereas higher termites have a more diverse diet but are devoid of flagellates (Brune and Dietrich 2015). Digestion of wood polysaccharides by these symbionts yields short-chain FAs (SCFAs), namely acetate, propionate and butyrate, which are taken up by the hindgut epithelium. Therefore, the hindgut paunch is referred to as a micro-bioreactor mostly providing acetate, which constitutes the major source of energy and anabolic metabolism of xylophagous termites (Brune 2014).

The digestive physiology of xylophagous but also of polyphagous cockroaches shares characteristic features with that of termites: i) their hindgut contains the highest bacterial density and is critical for metabolizing indigestible dietary compounds and ii) these biochemical activities result in the production of SCFAs, which are taken up by intestinal epithelial cells (Cruden and Markovetz 1987). Studies investigating the effect of diet in shaping the microbiota composition found variable results. One of those showed no significant change in the core microbial populations after shifting *P. americana* on a butter-based diet, suggesting that a selection pressure acts to maintain the specificity of the symbiosis between the insect and its microbiota (Tinker and Ottesen 2016). Volatile acids, namely acetate, propionate, butyrate, isobutyrate, valerate and isovalerate, are produced by microbiota-mediated digestion of cellulose; these compounds have been proposed to mediate social aggregation in several insect species (McFarlane and Alli 1985). In *Blattella germanica*, volatile FAs produced by the hindgut microbiota are excreted in the frass and are the main olfactory molecules mediating social aggregation (Wada-Katsumata et al. 2015).

In the honeybee *Apis mellifera*, social interactions elicit the colonization of the digestive tract by a conserved gut microbial community that comprises five core bacterial strains (Kwong et al. 2017). Pollen wall components resistant to midgut enzymes are metabolized to organic acids in the hindgut, where the bacterial density is the highest (Kesnerova et al. 2017). Honeybees devoid of microbiota exhibit reduced body and midgut weight through an insulin-dependent process (Zheng et al. 2017). These defects result from the lack of SCFAs produced from sugar fermentation by specialized bacteria in the hindgut.

The burying beetle *Nicrophorus vespilloides* (Coleoptera), which feeds its offspring on vertebrate carrions, excretes with their frass bacteria and yeasts that preserve carcasses from putrefaction, but also secrete lipases to operate extracellular lipid digestion (Shukla et al. 2018; Vogel et al. 2017a). Lipase activities have also been reported in the bacteria of the anal droplets from another coleopteran, the weevil *Cryptorhynchus lapathi* (Jing et al. 2020).

In the velvetbean caterpillar (Lepidoptera) *Anticarsia gemmatalis*, antibiotic (tetracycline) treatment to dampen the gut microbiota slightly affected survival and development, indicating that gut bacteria are not essential for this species survival in laboratory setting. On the other hand, this antibiotic treatment strongly suppressed intestinal lipase activity, suggesting that the microbiota contributes to lipid digestion in velvetbean caterpillars (Visotto et al. 2009). Analysis of the gut bacteria of three other lepidopteran species, *Antheraea assamensis*, *Helicoverpa armigera* and *Plutella xylostella*, revealed that among 71 isolates, 46 exhibited lipolytic activity, suggesting that their microbiota likely plays a role in lipid digestion (Gandotra et al. 2016).

The microbiota has been reported to modulate midgut lipid metabolism in several dipteran species. The Tsetse fly (*Glossina pallidipes*) and the sand fly (*Lutzomyia longipalpis*) are the vectors of *Trypanosoma brucei* and *Leishmania infantum*, respectively. In the Tsetse fly, midgut infection provokes a severe immune response to eliminate *Trypanosoma brucei*, while *Leishmania infantum* establishes a commensal association in the midgut sand fly. However, in both species, midgut infection induces lipid remodelling (Bateta et al. 2017; Coutinho-Abreu et al. 2020). Axenic mosquito *Aedes aegypti* larvae, lacking bacteria, cannot reach the critical size required for molting; this phenotype is associated with several metabolic perturbations, including changes in lipid metabolism (Valzania et al. 2018; Vogel et al. 2017b).

During the past two decades, the *D. melanogaster* model has been extensively used to investigate the physiological impact of gut microbiota, which, in the fruitfly, comprises only a few bacterial species. The two major strains *Acetobacter pomorum* and *Lactobacillus plantarum* cooperate in a metabolic dialogue that sustains larval growth under nutritional stress (Consuegra et al. 2020a; Consuegra et al. 2020b). Axenic *D. melanogaster* larvae are viable and develop to adulthood, but are sensitive to food scarcity (Shin et al. 2011; Storelli et al. 2011). This phenotype may in part rely on digestion efficiency, as the colonization of the midgut by commensal bacteria enhances, through the IMD/Relish pathway, the expression of digestive enzymes, including lipases (Erkosar and Leulier 2014). The mechanism of this regulation has been in part decrypted and initially involves the production of acetate by the commensal bacteria in the anterior midgut, which is the intestinal segment exhibiting the highest bacterial density (Jugder et al. 2021; Shin et al. 2011; Storelli et al. 2018). Acetate activates the IMD/Relish pathway in enteroendocrine cells, which in turn enhances the expression of tachykinin (Jugder et al. 2021; Kamareddine et al. 2018). Tachykinin is a secreted peptide that binds its cognate receptor at the membrane of columnar cells to regulate the intracellular lipid storage/catabolism rate (Kamareddine et al. 2018; Song et al. 2014).

In summary, the insect microbiota strongly affects host homeostasis by providing additional nutrients and optimizing the digestion process. As to lipid metabolism, the microbiota stimulates the expression of host lipases and produces energy metabolites, in particular SCFAs. Acetate appears to play a central role, both as a carbon source for energy metabolism and as a signalling molecule that triggers a lipid homeostasis response. In human, acetate produced by the colon microbiota provides the most abundant energy source for enterocytes and protects against infections (Fukuda et al. 2011; Lee and Hase 2014), a topic that deserves further investigations in insects.

3 Lipid Hydrolysis in the Gut Lumen

3.1 Lipase Characterization

Early studies on lipid digestion were based on the usage of radiolabelled molecules and the analyses of the end products after feeding or incubation with midgut extracts. Further experimentations were based on the separation of midgut sections and/or lumen gut fractions, and on the purification of enzymes. More recently, genomic and transcriptomic strategies allowed the identification of all the potential digestive enzymes and the analysis of their tissue-specific expression. Digestive enzymes have to hydrolyse dietary lipids, i.e. TAGs, phospholipids, galactolipids and sterols, the latter being possibly ingested as sterol conjugates. TAG hydrolases/lipases, commonly called lipases, cleave carboxylester bonds in TAGs, diacylglycerols (DAGs) and monoacylglycerols (MAGs), but may also be active on steryl-esters and phospholipids (Horne et al. 2009).

Lipases typically contain a Gly-X-Ser-X-Gly consensus sequence, where the Ser together with a His and an Asp/Glu residue form the tridimensional catalytic site (Derewenda 1994). When covered by a peptidic domain—called the lid—the catalytic site is not accessible to the substrate. Lipases are water soluble but poorly active in aqueous solution, whereas at the interface of lipid micelles, a conformational rearrangement opens the lid so that the catalytic site becomes accessible to the substrate (Holmquist 2000). Depending on their primary structures, lipases are classified as neutral, acid, lipase2 (not reported in insects), lipase3, GDSL, hormone sensitive lipase (HSL) and adipocyte triglyceride lipase (ATGL) (Horne et al. 2009). In mammals, digestive lipases correspond to the families of gastric/acid lipases active at low pH, and of neutral/pancreatic lipases active at neutral pH, the activity of the latter depending on colipase-specific binding at the C-terminal domain of the lipase (Miled et al. 2000). Congruent with the

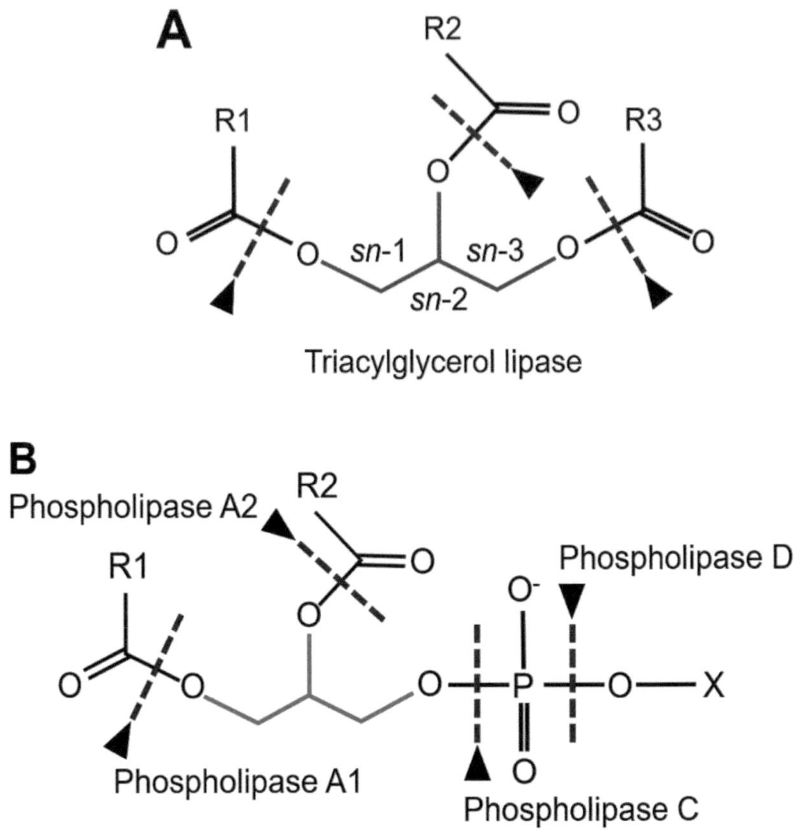

Fig. 3 Lipase activities. (**a**) TAG lipases hydrolyse the ester bounds (arrowheads) between glycerol and FA moieties (R1, R2, R3). (**b**) PLA1, PLA2, PLC and PLD cleave phospholipids at specific sites (arrowheads)

absence of colipase in insects, their neutral lipases lack such a C-terminal domain (Sahaka et al. 2020).

TAG lipases (EC 3.1.1.3) cleave ester bounds between FA carboxylic and glycerol-alcohol residues, although several of these lipases preferentially remove FA in position sn-1 and sn-3 of the glycerol, leading to sn-2 MAGs (Fig. 3a). Phospholipases are defined in respect of their specific cleavage site. Phospholipases A1 (PLA1, EC 3.1.1.32) and A2 (PLA2, EC 3.1.1.4) catalyse the formation of lysophospholipids by removing FA at the sn-1 and sn-2 positions of glycerol, respectively (Fig. 3b). Phospholipases B (PLB) cleave both sn-1 and sn-2 sites. Finally, phospholipases that hydrolyse the phosphate bonds located proximally and distally relative to the glycerol backbone are referred to as PLC (EC 3.1.4.3) and PLD (EC 3.1.4.4), respectively (Fig. 3b). The lid domain and a β-loop located near the catalytic site are involved in substrate recognition; both long domains are characteristic of TAG lipases, whereas shorter domains are characteristic of phospholipases (Horne et al. 2009).

3.2 Insect Digestive Lipases

Insect genomes encode a high number of lipases, most of them belonging to the acid and neutral lipase families (Table 1). Interestingly, these lipases are more abundant in insects than in mammals (Horne et al. 2009), which may be the consequence of a high level of redundancy and/or of a broader diversity of dietary lipids. The high number of lipases together with their substrate flexibility renders difficult the precise physiological assignment of a specific substrate to a given lipase. Consequently, investigations aimed at deciphering the lipid digestion processes hardly rely on genetic approaches but rather on monitoring enzymatic activities of midgut extracts or of

Table 1 Number of lipases per class in various insect species

	Neutral	Acid	Lipase 3	GDSL	HSL	ATGL	Total
Aedes aegypti	42	23	2	4	1	0	72
Anopheles gambiae	23	13	2	5	1	1	45
Apis mellifera	14	5	1	5	1	1	27
Bombyx mori	25	17	1	1	1	1	46
Cimex lectularius	14	12	2	1	1	2	32
Drosophila melanogaster	37	25	1	2	1	2	68
Locusta migratoria	22	14	2	3	1	1	43
Manduca sexta	45	28	1	1	1	1	77
Musca domestica	54	39	1	6	1	1	102
Rhodnius prolixus	18	6	3	2	1	0	30
Spodoptera frugiperda	57	22	1	2	1	0	83
Tenebrio molitor	28	25	2	1	1	1	58

These numbers have been established using Flybase (Flybase 2003) and insectBase (Mei et al. 2022), providing an estimate of the respective frequencies of each lipase class; they may be refined based on studies specific to each insect species. For instance, *R. prolixus* studies identified 21 neutral and 5 acidic lipases {Majerowicz, 2017 #1936} and one ATGL {Aredes, 2024 #1935}. Lipases 2 that are absent in insects and yolk proteins that contain inactive catalytic lipase sites and most phospholipases are not represented

purified enzymes, and on analysing gene expression in intestinal cells.

In the insect midgut, a few studies suggest that TAG hydrolysis preferentially removes FA from the sn-1 and sn-3 positions of the glycerol, leading to sn-2 MAGs (Bollade et al. 1970; Hoffman and Downer 1979; Treherne 1958), while others rather suggest complete digestion to glycerol and free FAs (FFAs) (Table 2) (Canavoso et al. 2004; Tsuchida and Wells 1988; Weintraub and Tietz 1973). Given that alkaline pH favours the complete hydrolysis to glycerol and FFAs (Canavoso et al. 2001), these discrepancies may be the consequence of species variability in intestinal luminal pH (Terra and Ferreira 2020). It may also depend on experimental conditions and/or lipase affinity for specific lipid classes. For instance, midgut lipases of *L. migratoria* hydrolyse triolein more efficiently than tripalmitin (Weintraub and Tietz 1973), suggesting a higher affinity for unsaturated FAs, as reported for mammalian pancreatic lipases.

Phospholipase activity was first described in the midgut of several species of coleopterans, lepidopterans and dipterans (Somerville and Pockett 1976; Turunen 1988a; Turunen and Chippendale 1989; Turunen and Kastari 1979). To date, PLA2 are the most studied phospholipases (Nor Aliza et al. 2018; Nor Aliza and Stanley 1998; Rana et al. 1997; Sajjadian et al. 2019; Stanley et al. 1998; Uscian et al. 1995; Weiher and Komnick 1997), while only a few studies focused on digestive phospholipases A1, C and D (references in Table 2). Further, some of the enzymes referred to as PLA1 are actually galactolipases that hydrolyse preferentially the acyl residue at the sn-1 position of the glycerol backbone of galactolipids (Christeller et al. 2011; Sahaka et al. 2020).

The removal of phosphate groups is catalysed by non-specific alkaline (EC 3.1.3.1) or acid (EC 3.1.3.2) phosphatases, defined by the optimal pH range of their activity. These phosphatases are active on a great variety of phosphomonoester compounds. Both acid and alkaline phosphatases have been described as soluble or membrane-bound forms in the midgut of several insect species (da Silva et al. 2019; Eguchi 1995; Matic et al. 2021; Terra et al. 1988).

Waxes are LCFA/long-chain alcohol esters, which are structural components of most insect cuticles and, thus, are present in the insect-based diet of carnivorous insects (Table 2). It has been suggested that wax hydrolysis might be accomplished by intestinal symbionts, but a recent study revealed that the honeycomb moth *Galleria mellonella* exhibits a microbiota-independent

Table 2 Lipase activities in the midgut lumen. A number of studies (column reference) have investigated enzymatic activities (column activity) in the midgut of various insect species (column species)

Activity	Species	Reference
TAG to sn-2-MAG	*Periplaneta americana; Blabera craniifer* *Pieris brassicae* *Periplaneta americana*	Bollade et al. (1970) Turunen (1975) Hoffman and Downer (1979)
TAG to FFAs+glycerol	*Locusta migratoria* *Periplaneta americana; Locusta migratoria* *Manduca sexta* *Panstrongylus megistus*	Weintraub and Tietz (1973) Male and Storey (1981) Tsuchida and Wells (1988) Canavoso et al. (2004)
Phospholipase A1	*Drosophila melanogaster* Lepidoptera species *Musca domestica*	Table 3 Canavoso et al. (2001) Barroso et al. (2021)
Phospholipase A2	*Cicindella circumpicta* *Manduca sexta* *Aedes aegypti* *Spodoptera exigua* *Musca domestica*	Uscian et al. (1995) Stanley et al. (1998) Nor Aliza and Stanley (1998) Sajjadian et al. (2019) Barroso et al. (2021)
Phospholipase B	*Drosophila melanogaster*	Table 3
Phospholipase C	*Stomoxys calcitrans* *Aeshna cyanea* *Musca domestica*	Spates et al. (1990) Weiher and Komnick (1997) Barroso et al. (2021)
Phospholipase D	*Pieris brassicae*	Turunen (1988a)
Galactolipase	*Epiphyas postvittana; Helicoverpa armigera*	Christeller et al. (2011)
Sterol dealkilation	*Bombyx mori*	Ciufo et al. (2011)
Sterol esterase	*Drosophila melanogaster*	Sieber and Thummel (2012)
Wax hydrolysis	*Aeshna cyanea* *Galleria mellonella*	Komnick and Bauerfeind (1991) Kong et al. (2019)
Lipid synthesis	*Locusta migratoria* *Pieris brassicae*	Weintraub and Tietz (1978) Turunen (1988b)

digestive enzyme activity that hydrolyses the wax ester bonds (Kong et al. 2019).

Dietary sterols are present in a free form or as steryl-ester or steryl-glycoside (Li and Jing 2020). Cholesterol ester hydrolase (EC 3.1.1.13) was first characterized in the beetle *Trogoderma granarium* (Agarwal and Nair 1976). In *D. melanogaster*, the lipase Magro acts as a sterol esterase in the midgut (Sieber and Thummel 2012). Phytosterols that contain an alkyl branching are not readily usable by phytophagous insects (Svoboda 1999). The efficiency of dealkylation varies, depending on the species (Jing and Behmer 2020). A few enzymes potentially required for phytosterol dealkylation have been identified in *Helicoverpa zea* (Jing et al. 2012), while a desmosterol reductase that catalyses the final step of this process has been characterized in *Bombyx mori* (Ciufo et al. 2011).

Database mining revealed that the genome of *D. melanogaster* encodes at least 90 putative lipases (Flybase 2003), 44 of which are significantly expressed in the midgut (Chintapalli et al. 2007). Among those 90 putative lipases, a few are inactive, such as the yolk proteins of higher dipterans that fall in a family of proteins containing sequences similarities to mammalian lipases, but fail to have lipase activity (Bownes 1992). Furthermore, given that lipid digestion mostly takes place in the gut lumen, the presence of a signal peptide as a prerequisite for protein secretion identified 29 of them as putative functional lipases. These enzymes are predicted (Flybase 2003) to fulfil most lipid-related enzymatic functions (Table 3), although specific activity has been formally demonstrated for only a few. Due to a high level of redundancy, the critical requirement of a given enzyme for a specific

Table 3 Enzymatic activities of *Drosophila melanogaster* lipases potentially involved in lipid digestion in the midgut lumen

	Genes														
Activity	CG1583	CG2772	CG3488	CG4267	CG5112	CG5162	CG5295	CG5932	CG5966	CG6113	CG6271	CG6277	CG6283	CG6295	CG6296
TAG Lipase 3.1.1.3	.	X	.	X	.	X	X	X	X	X	X	X	X	X	X
MAG Lipase 3.1.1.23	.	.	X	.	X	.	.	.	.	.	.	.	.	.	.
Sterol Esterase 3.1.1.13	.	.	.	.	.	.	.	X	.	.	.	.	.	.	.
PLA1 3.1.1.32	.	.	.	.	.	.	.	.	.	.	X	X	X	X	X
PLA2 3.1.1.4	X	.	.	.	.	.	.	.	.	.	.	.	.	.	.
PLB 3.1.1.5	.	.	.	.	.	.	.	.	.	.	.	.	.	.	.

	Genes													
Activity	CG6718	CG6847	CG7279	CG7900	CG8093	CG8823	CG10116	CG11029	CG15111	CG17035	CG17192	CG30503	CG31089	CG31091
TAG Lipase 3.1.1.3	.	X	X	.	X	X	X	.	.	.	X	.	X	X
MAG Lipase 3.1.1.23	.	.	.	X	.	.	.	.	.	.	.	.	.	.
Sterol Esterase 3.1.1.13	.	.	.	.	.	.	.	.	.	.	.	.	.	.
PLA1 3.1.1.32	.	.	.	.	.	.	.	.	.	.	X	.	.	.
PLA2 3.1.1.4	X	.	.	.	.	.	.	X	.	X	.	X	.	.
PLB 3.1.1.5	.	.	.	.	.	.	.	X	X	.	.	.	.	.

The list of genes and their activities (CG number) have been established mining Flybase (Flybase 2003) for lipases and phospholipases. The final selection retained the genes that were significantly expressed in the midgut based on FlyAtlas (Chintapalli et al. 2007) and contain a signal peptide based on signalP-4.1 (Petersen et al. 2011) dietary lipid substrate cannot be easily addressed by a systematic gene mutant approach.

3.3 Lipase Activity and Expression

In the mammalian pancreas, signal peptide cleavage converts several proenzymes into active secreted enzymes. This maturation process might be conserved among insect digestive lipases, as reported for a secreted PLA2 of *Spodoptera exigua* (Sajjadian et al. 2019; Vatanparast et al. 2018). Lipase activity mainly relies on the penetration of the enzyme at the interface of lipid micelles. In contrast to mammals, insects do not produce bile acids to emulsify dietary lipids. This process likely involves lysophospholipids, which derive from phospholipid cleavage by either PLA1 or PLA2 (Fig. 3b) (Ian De Veau and Schultz 1992; Sajjadian et al. 2019; Stanley 2006). In addition, the penetration of lipases into micelles is influenced by the acyl chain length, the pH and the concentration of several ions, in particular Ca^{2+} (Verger et al. 1973). For instance, the midgut lipase activity of the bloodsucking bug, *Rhodnius prolixus*, is optimal at neutral pH (Grillo et al. 2007), whereas that of PLA2 in *Cicindella circumpicta* (Uscian et al. 1995) and in *Nicrophorus marginatus* (Rana et al. 1997) is optimal at alkaline pH, all of them depending on Ca^{2+} concentration. However, a digestive PLA2 of *A. aegypti* has been shown to be Ca^{2+}-independent (Nor Aliza and Stanley 1998). In most insect species, the midgut lumen exhibits an alkaline pH (Harrison 2001), which favours the activity of neutral lipases. Conversely, a few species exhibit an acidic midgut region (Barroso et al. 2021; Bonelli et al. 2019; Dubreuil 2004), which is optimal to sustain the activity of acid lipases. Nonetheless, genes encoding acid lipases can be found in most insect species (Table 1). In *Manduca sexta*, acid lipases are expressed in the midgut, although devoid of acidic region. Given that three of these acid lipases are not detected in the lumen juice, it has been proposed that they act in acidic lysosomes to hydrolyse dietary lipids taken up by the columnar cells (Miao et al. 2020).

Early studies focusing on monitoring lipase activity of midgut juice revealed an effect of the nutritional status. In the hematophagous species *A. aegypti*, *Panstrongylus megistus* and *R. prolixus*, lipase activity is induced by the blood meal and reaches a maximum a few days later (Canavoso et al. 2004; Grillo et al. 2007; Rudin and Hecker 1979). In *M. sexta* (Stanley et al. 1998) and *Gryllus bimaculatus* (Weidlich et al. 2015), the activity of midgut lipases is higher in fed than in starved animals. Furthermore, a positive correlation between midgut lipase activity and dietary lipid content has been observed in *Oedaleus asiaticus* (Huang et al. 2017).

Conversely, FA supplementation in the diet of *Epiphyas pastvittana* represses lipase gene expression (Christeller et al. 2010), suggesting the existence of an FA-sensing mechanism that adjusts lipase levels to FFA amounts. These studies revealed that the expression and/or the secretion of lipases is controlled by the diet, a vision generally recognized for discontinuous insect feeders but not for continuous ones. However, transcriptomic analyses in *D. melanogaster* larvae—a continuous feeder—counteract this vision, in that starvation but also high dietary glucose levels repress the expression of digestive lipases (Chng et al. 2014; Mattila et al. 2015; Zinke et al. 2002). At the molecular scale, the nuclear receptor DHR96 controls midgut TAG and cholesterol metabolism through the expression of lipases (Sieber and Thummel 2009, 2012). Finally, the insect gut microbiota, which contributes to homeostasis by providing essential nutrients (Grenier and Leulier 2020), has also been shown to modulate enzyme expression. In *D. melanogaster*, midgut colonization by *Lactobacillus plantarum* enhances the expression of digestive lipases (Erkosar and Leulier 2014). In summary, these studies support the notion that gene expression and activity of digestive lipases depend on the nutrient and bacterial load in the midgut lumen to optimize the digestion process.

4 Lipid Uptake into Intestinal Cells

4.1 Lipid Emulsifiers

The molecular mechanism of lipid crossing cell membranes remains puzzling, albeit extensively studied. The idea that lipids—due to their biochemical nature—could easily cross cell membranes suggests that the lipid bilayer would fail to strictly maintain compartment integrity. Single lipid molecules are poorly soluble, if not insoluble, in aqueous solution. Lipids integrating a membrane bilayer tend to remain within the lipid leaflets, rather than to exit at the other membrane side in the cytosolic aqueous phase. Within a membrane leaflet, lipids are free in their lateral movements, while, unless catalysed by flippases (Montigny et al. 2016), spontaneous lipid flip-flops are rather rare events (Perez-Salas et al. 2021). Nonetheless, spontaneous flip-flops are favoured by an extreme membrane curvature, which may happen in the microvilli of columnar cells (Niot et al. 2009). Thus, a critical issue is lipid desorption, which likely requires specific cytosolic proteins. These proteins may also act as lipid sequesters, thereby preventing lipid transfer back into the gut lumen upon fasting.

Emulsification of lipids facilitates the access to water-soluble lipases (Sect. 3.1) and the tight interaction of lipid molecules with the hydrophilic side of the microvillar membrane prior to their absorption. Insects do not produce bile acid and the intestinal absorption of lipids is proposed to proceed through the formation of glycolipids and amino-lipid complexes (Chapman et al. 2013; Miguel-Aliaga et al. 2018). In lepidopterans, the alkaline midgut favours the formation of FA micelles (Canavoso et al. 2001), questioning whether the molecular nature of the lipids emulsifiers varies among insect species. Analysis of surface-active substances in *G. bimaculatus* revealed that FA-amino acid complexes (FACs) were potential emulsifiers (Collatz and Mommsen 1974). In phytophagous lepidopterans, FAC synthesis depends on the gut bacteria (Spiteller et al. 2000). FACs are regurgitated in oral secretions and act as potential emulsifiers to favour external lipolysis, yielding volatile compounds (Alborn et al. 1997; Halitschke et al. 2001; Spiteller and Boland 2003). These volatiles act at long distance to elicit a protective plant response (Dicke et al. 1990). Despite this adversative effect, the conservation of FAC synthesis and secretion suggests a benefit for the insect. Indeed, in *Spodoptera litura* and *Heliothis virescens*, these compounds are proposed to constitute nitrogen stores in the gut lumen, in the form of FA-glutamine complexes (Kuhns et al. 2012; Yoshinaga et al. 2008). As mentioned above (Sect. 3.3), lysophospholipids resulting from phospholipid hydrolysis are likely playing a critical digestive role as lipid emulsifiers (Ian De Veau and Schultz 1992; Sajjadian et al. 2019; Turunen and Kastari 1979). Finally, the MGDG galactolipid, highly abundant in chloroplast thylakoids (Sect. 2.1), has been shown to possess emulsifier properties (Chu et al. 2009). As for lysophospholipids, lysogalactolipid resulting from partial hydrolysis might act as intestinal emulsifiers in the midgut of phytophagous species. However, in spite of the characterization of all these putative lipid emulsifiers, their effective contribution to lipid uptake is not formally demonstrated and remains to be experimentally addressed in various insect species.

4.2 Lipid Transporters

The molecular mechanism of lipid uptake into the intestinal cells has been poorly studied in insects, with most data relying on mammalian investigations. In mammals, dietary lipids are emulsified by bile acids, which facilitate lipolysis and lipid absorption, mostly MAGs, FFAs, lysophospholipids and cholesterol (Ko et al. 2020). An early study in the rat intestine proposed that linoleic acid uptake proceeds through two distinct mechanisms. At low concentrations, the uptake is saturable, thus, potentially energy- and/or transporter dependent. In contrast, at higher concentrations, linoleic acid intake follows a linear plot, suggesting a concentration-dependent passive diffusion (Chow and Hollander 1979). This passive diffusion mechanism may result from a vectorial fusion/fission between micelles and membranes from the gut lumen towards the cytoplasm of intestinal cells, although thermodynamic considerations postulate that micelle-membrane fusion proceeds through unstable states of high-energy intermediates (Markvoort and Marrink 2011). Alternatively, it is tempting to speculate that high amounts of FAs might react with monovalent cations to form soaps, which could act as detergents, destabilizing cell membranes. Congruently, it has been shown that perfusion of the rat jejunum with emulsified lipids provokes membrane injuries in the brush border of the enterocytes, allowing lipid micelles intake (Kvietys et al. 1991). In sum, the molecular mechanism of this so-called passive diffusion remains largely unclear. The uptake of oleic acid has been shown to be more active than the one of palmitic or stearic acid in the midgut of the dragonfly *Aeshna cyanea* (Kirfel and Komnick 1999), suggesting either the existence of selective transporters or that passive diffusion is favoured for unsaturated FAs. In addition, the observation that the uptake of lipids is not uniform along the entire midgut but takes place within specific sub-regions (see below § 6.6.) further supports the notion that the principle of passive lipid diffusion must be considered with caution. Of note, while intestinal absorption of cholesterol and FAs was initially believed to proceed merely in this way, transporter-dependent intake is now characterized for either.

In mammals, several proteins have been proposed to contribute to FA absorption into the enterocytes, including fatty acid transporter protein 4 (FATP4), fatty acid–binding proteins (FABPs), FAT/CD36 and caveolin. FATP4 knockdown by antisense oligonucleotides reduces FA uptake by 50% in murine primary enterocytes (Stahl et al. 1999). However, a recent study indicated that FATP4 localizes at the ER to contribute to FA esterification, subsequently driving FA intake (Milger et al. 2006). Two other FA binding proteins, FABP1 and FABP2, despite their high intestinal expression, are also improbable FA transporters, rather acting to sequester FAs, thereby preventing lipotoxicity (Cifarelli and Abumrad 2018). The scavenger receptor CD36 is another candidate for mediating FA uptake, although such involvement in mammalian enterocytes remains controversial. In the mouse, one study reports the role of CD36 in FA absorption in the proximal, but not in the distal region of the small intestine (Nassir et al. 2007). However, two other studies rather suggest that CD36 plays a role in intracellular lipid trafficking and chylomicron formation (Goudriaan et al. 2002; Nauli et al. 2006). SR-B1, another scavenger receptor of the CD36 family highly expressed in the mouse intestine, is required for intestinal absorption of vitamin E (Reboul et al. 2006), but neither for FA nor for cholesterol intake (Cifarelli and Abumrad 2018). Caveolin1–3 are proteins associated with lipid rafts, which cover 50% of the membrane microvilli of the enterocytes (Kunding et al. 2010). Caveolins are proteins that channel lipid raft–derived endocytic vesicles to the ER. Caveolin-1 has been identified as an FA-binding protein (Trigatti et al. 1999), and its knockout in mice virtually abolishes the uptake of albumin-bound oleic acid (Siddiqi et al. 2013).

Most of these potential lipid transporters are conserved in insects (Table 4), although their roles in midgut lipid metabolism remain elusive. Two FABP homologues, MFB1 and MFB2 have been isolated from the midgut cytosol of *M. sexta* larvae (Smith et al. 1992). The *D. melanogaster* FATP4 homologue, Fatp1, has been the focus of interest in several tissues, although not in the midgut (Parvy et al. 2012). As reported for its mammalian counterpart, a study on *D. melanogaster* oenocytes suggests that it controls FA-CoA synthesis rather than FA uptake (Wicker-Thomas et al. 2015). A great number of scavenger receptors are encoded by insect genomes (Table 4); one of them, NinaD, has been shown to mediate carotenoid uptake in the midgut of *D. melanogaster* (Kiefer et al. 2002). To date, no caveolin genes have been found in insects. Nevertheless, two flotillin genes, encoding caveolae-associated proteins, have been described in *D. melanogaster* (Galbiati et al. 1998). Interestingly, ectopic expression of the murine flotillin-1 in Sf21 insect cells drives the formation of caveolae-like vesicles (Volonte et al. 1999).

Table 4 Number of receptors or transporters involved in lipid uptake or transfer in various insect species

	Scavenger receptor	FABP	FATP	Npc1a	Npc1b	Npc2	Flo
Aedes aegypti	3	1	2	0	1	4	2
Anopheles gambiae	4	2	2	1	1	4	3
Apis mellifera	1	2	3	1	1	2	2
Bombyx mori	1	5	5	1	1	1	3
Cimex lectularius	3	2	4	1	1	3	2
Drosophila melanogaster	3	1	3	1	1	8	2
Locusta migratoria	3	7	4	1	1	1	2
Manduca sexta	1	18	5	1	1	2	2
Musca domestica	4	2	2	1	1	9	3
Rhodnius prolixus	2	1	7	1	1	1	2
Spodoptera frugiperda	2	6	4	2	1	1	2
Tenebrio molitor	3	3	2	2	1	4	3

These numbers were established using Flybase (Flybase 2003) and InsectBase (Mei et al. 2022) and could potentially increase based on future studies

The mechanism of cholesterol uptake into intestinal cells has been elucidated thanks to mammalian and insect studies. The Niemann-Pick type C genes (*NPC1* or *NPC2*), which cause abnormal cholesterol accumulation, are conserved in insects (Table 4) (Zheng et al. 2018). The *D. melanogaster* Npc1a and Npc1b genes are the homologues of the mammalian *NPC1* and *NPC1*-like 1 (*NPC1L1*) genes, whose products fulfil intracellular sterol trafficking and intestinal sterol absorption, respectively (Huang et al. 2005; Voght et al. 2007). The expression of either NPC1L1 in mammals or Npc1b in *D. melanogaster* is restricted to the apical membrane of columnar cells (Jing and Behmer 2020). In a rat hepatic cell line, flotillins have been shown to associate with NPC1L1, thereby mediating micelle sterol uptake (Ge et al. 2011). Further, in *D. melanogaster*, Npc1b mutation results in a defect of intestinal sterol absorption (Voght et al. 2007). In vertebrates, the sterol carrier protein-2 (SCP-2) is a cytoplasmic protein involved in the ER to plasma transfer of cholesterol (Jing and Behmer 2020). SCP-2 homologues have been identified and shown to be highly expressed in the midgut of lepidopteran and dipteran species (Gong et al. 2006; Kim and Lan 2010; Lan and Wessely 2004). SCP-2 knockdown in *A. aegypti* and in *S. litura* decreases overall cholesterol levels, whereas its overexpression in cell lines improves cholesterol uptake (Blitzer et al. 2005; Guo et al. 2009; Lan and Massey 2004). These studies identified bona fide cholesterol transporters, acting on cholesterol uptake and transfer between subcellular compartments.

In summary, the mechanism of intestinal sterol uptake has been elucidated and depends on evolutionary conserved transporters. In contrast, the intestinal uptake of FA and phospholipids is lesser elucidated but likely involves specific transporters, which remain to be properly identified in insects.

5 Lipid Metabolism in Intestinal Cells

5.1 Lipid Usage to Sustain Intestinal Cell Activity

Once taken up, digested lipid products can be directly exported through the haemolymph, stored within specialized organelles (LDs, lysosomes) or used to sustain the energy and structural needs of the intestinal cells. The diet likely influences lipid metabolism in the midgut epithelium. For instance, in *D. melanogaster,* the FA composition of the diet significantly modifies the phospholipid composition of the different tissues, including the midgut (Carvalho et al. 2012). In addition, this lipidome profiling revealed that, compared to other tissues, the midgut contains a higher fraction of sphingolipids with two double bonds in the long-chain base and in the esterified fatty acyl chain.

Intestinal cells are highly active, requiring ATP as an energy source. The major ATP pool is produced via the respiratory chains coupled to the tricarboxylic acid cycle that relies on glycolysis, FA β-oxidation or amino acid catabolism. However, it has not been formally demonstrated which of these metabolic pathways is preferential in the intestinal cells; most likely it depends on the diet composition and thus on the insect species. For instance, in the stablefly *Stomoxys calcitrans*, a midgut posterior region (called lipoid zone) incorporates TAGs and phospholipids in the form of LDs often associated with mitochondria, suggesting that these lipid stores could be used for β-oxidation (Lehane 1977). In *A. cyanea*, peroxisome content increases in the apical region of the columnar cells of larvae fed on either natural lipid-rich food or high amounts of monounsaturated LCFA (Faber and Komnick 1989). Given that peroxisomes act to shorten and process FA not readily usable by mitochondria (Renne and Hariri 2021), these observations suggest that excess dietary LCFAs are first metabolized in peroxisomes to potentially sustain mitochondrial ATP production. In the midgut *R. prolixus*, a study reported high expression levels of β-oxidation enzymes (Ribeiro et al. 2014), whereas other studies reported moderate expression levels of the mitochondrial trifunctional protein A subunit HADHA {Aredes, 2022 #1939} and of CPT1, the gatekeeper of FA-CoA into the mitochondria {de Bruin, 2014 #871}. However, none of these studies investigated FA consumption levels through mitochondrial β-oxidation in the midgut of this species. Further, the microbiota may affect the midgut energy metabolism, through the production of short-chain FAs that may be consumable by the intestinal cells, as described for the digestion of wood by termites and of plant polymers from pollen by honeybees (Brune 2014; Zheng et al. 2017).

The secretion of enzymes and the absorption of nutrients by columnar cells suggest a high rate of microvillar membrane replacement. In *L. migratoria*, it has been shown that glycerol and FFAs hydrolysed from dietary triolein are found in the phospholipids of the intestinal wall (Weintraub and Tietz 1973). In *Pieris brassicae* larvae, the kinetics of phosphatidylcholine (PC) and phosphatidylethanolamine (PE) biosynthesis in the columnar cells, compared to that of DAG, suggests that these phospholipids are used for columnar cell membrane regeneration (Turunen 1993). In *R. prolixus*, the microvillar membrane contains high cholesterol levels (Albuquerque-Cunha et al. 2009), and the dynamics of cholesterol accumulation in the columnar cells further suggests the utilization for membrane regeneration (Entringer et al. 2021).

5.2 Lipid Metabolic Pathways in Intestinal Cells

In intestinal cells, dietary lipids can be converted to DAGs, TAGs or phospholipids. In *S. calcitrans*, it has been estimated that 60% of dietary FAs are converted to TAGs and 20% to phospholipids (Lehane 1977). In *M. sexta* fed radioactive triolein, labelling is eventually found in phospholipids and TAGs of the intestinal epithelium and in the circulating DAGs (Tsuchida and Wells 1988). DAGs are the prevalent form of lipid transport through the haemolymph and are also the precursors for phospholipid biogenesis (Vance 2015).

FFAs, in particular the saturated ones, are highly toxic (Listenberger et al. 2003). To prevent this deleterious effect, FAs can be either degraded through mitochondrial β-oxidation or esterified to TAGs and stored in LDs. The biosynthesis of TAGs may proceed through three distinct pathways, which begin by either acylation of glycerol-phosphate in the mitochondria or the ER, acylation of dihydroxyacetone in the peroxisomes, or acylation of MAGs in the ER; the subsequent steps taking place in the ER (Coleman and Mashek 2011; Ko et al. 2020). MAG acylation gives rise to DAG, whereas the glycerol-phosphate and dihydroxyacetone pathways result in phosphatidic acid (PA), which after dephosphorylation by lipin leads to

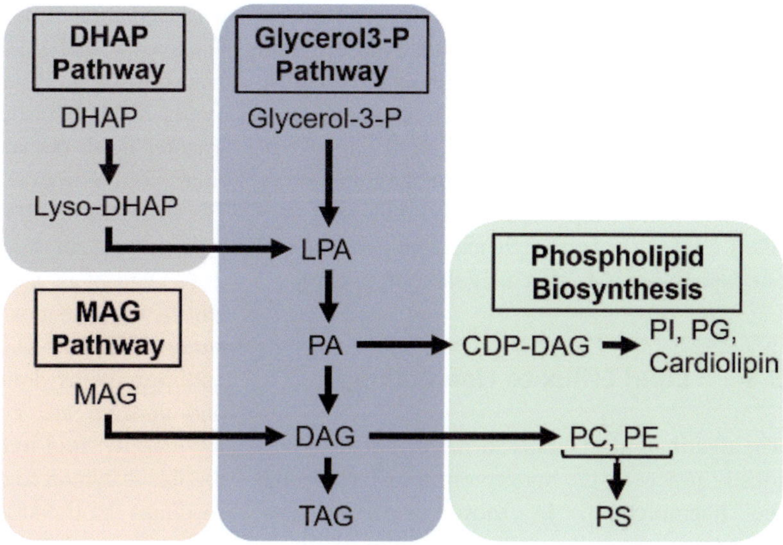

Fig. 4 Phospholipid and TAG biosynthesis. The DHAP, glycerol-3-phosphate and MAG pathways to TAG biosynthesis operate in an organ-specific manner. DAG is the precursor of the phospholipids PC, PE and PS, whereas PA is the precursor of PI, PG and cardiolipin.

DAG (Fig. 4). The glycerol-phosphate pathway also called the Kennedy pathway is the most universal. In mammals, it takes place in the liver and adipose tissue, while the conversion of MAGs—resulting from luminal TAG hydrolysis—into DAGs is prevalent in the small intestine. To date, besides the characterization of a putative MAG acyltransferase activity in the midgut of *P. americana* (Hoffman and Downer 1979), no evidence suggests that the MAG pathway operates in the insect midgut. In contrast, the PA pathway has been reported in the midgut of several insect species, including *P. brassicae* (Turunen 1993), *M. sexta* (Canavoso and Wells 2000), *P. megistus* (Canavoso et al. 2004) and *D. melanogaster* (Kuhnlein 2012).

DAGs are the precursors for the biogenesis of the most abundant phospholipids, phosphocholine (PC) and phosphoethanolamine (PE) (Fig. 4) (Vance 2015). First, phosphocholine and phosphoethanolamine are converted to CDP-choline and CDP-ethanolamine at the nuclear and ER membranes, respectively. The final step to PC and PE synthesis takes place at the ER membrane. Synthesis of phosphatidylserine may proceed by the exchange of the head group of PC or PE to a serine residue at the ER membrane. Synthesis of other phospholipids utilizes CTP and PA to form CDP-DAG; replacement of the CDP group by other functional groups results in the formation of phosphatidylinositol (PI), phosphatidylglycerol (PG) or cardiolipin (Fig. 4). The synthesis of sphingomyelin, the most abundant sphingolipid, occurs at the Golgi apparatus from ceramide and PC. Phospholipid biosynthesis and expression of the potentially required enzymes have been reported in the midgut of several insect species, including *S. calcitrans*, *M. sexta*, *R. prolixus* and *D. melanogaster* (Buchon et al. 2013; Grillo et al. 2007; Lehane 1977; Ribeiro et al. 2014; Tsuchida and Wells 1988). However, the functional characterization of the underlying enzymes in intestinal cells remains poorly investigated.

Excesses of dietary sterols can be temporarily stored as free or steryl-ester forms in lysosomes or LDs, or efflux back to the gut lumen likely via ABC transporters, as reported in mammal studies (Entringer et al. 2021; Jing et al. 2012; Sieber and Thummel 2012). In insects, Npc2 are soluble proteins encoded by multiple genes (Table 4), which are not equally expressed across different organs (Huang et al. 2007; Jing and Behmer 2020; Jupatanakul et al. 2014). Two of the *D. melanogaster* homologues, Npc2a and Npc2b, have a redundant function in controlling intracellular sterol distribution (Huang et al. 2007). Based on mammalian and insect studies, Npc2 homologues are required for intralysosomal sterol transfer to Npc1a (Huang et al. 2007; Jing and Behmer 2020). Loss of Npc1a in

D. melanogaster results in abnormal cholesterol retention in lysosome-like organelles in different tissues, including the midgut (Huang et al. 2005). Importantly, Npc1a and Npc1b are two large transmembrane proteins, which fulfil cholesterol intracellular trafficking and its uptake into columnar cells (Sect. 4.2), respectively. Although they likely evolved from a single ancestor gene, Npc1a and Npc1b are not interchangeable transports.

5.3 Lipid Efflux to Haemolymph

The distribution of lipids to the different tissues mostly proceeds via lipoprotein transfer through the haemolymph. In most insect species, lipoproteins typically carry sterols, phospholipids and DAGs (Canavoso et al. 2001), although DAGs are replaced by TAGs in some dipteran species (Ford and Van Heusden 1994; Pennington and Wells 2002). The transfer of lipids from the columnar cells to lipoproteins requires first the remobilization from the storage organelles. Given that intracellular LDs store TAGs in higher proportions than DAGs, the former must be hydrolysed to DAGs prior to their haemolymph transfer. In *D. melanogaster*, the adipocyte triglyceride lipase Brummer, which converts TAG to DAG, is expressed in the midgut and, thus, potentially works in the columnar cells (Gronke et al. 2005). Free sterols rather than steryl-esters are loaded on lipoproteins (Carvalho et al. 2012; Entringer et al. 2021; Jouni et al. 2002; Yun et al. 2002), suggesting that a steryl-esterase must be active in the columnar cells prior to haemolymph efflux (Sieber and Thummel 2012).

The lipid transfer from the columnar cells to the haemolymph requires a lipophorin receptor, a lipid transfer particle (LTP) and a circulating lipophorin (Lpp). The lipophorin receptor (s) acting to recruit both lipophorin and LTP at the basal membrane of the columnar cells have not been formally identified (Arrese et al. 2001; Rodriguez-Vazquez et al. 2015). A number of studies in several insect species have reported the purification of lipophorins, characterized their binding to the intestinal cell membrane and analysed their lipid content (Atella et al. 1995; Canavoso and Wells 2001; Ford and Van Heusden 1994; Fruttero et al. 2009; Gondim and Wells 2000; Grillo et al. 2003). LTP that acts to transfer lipids between intracellular and extracellular compartments in a bidirectional manner (Arrese et al. 2001) has been shown to interact with the midgut of *M. sexta* (Canavoso and Wells 2001) and of *D. melanogaster* (Palm et al. 2012; Rodriguez-Vazquez et al. 2015). An in-depth study of the *D. melanogaster* scaffolding apolipoproteins demonstrated that both apoLTP and apoLpp are synthesized in the fat body, matured by lipid loading and transferred through the haemolymph to reach the midgut, where LTP mediates the transfer of lipids from the columnar cells to Lpp (Palm et al. 2012).

6 The Dynamics of Lipid Digestion in Insects

6.1 An Attempt to Schematize the Lipid Digestive Process

Lipid digestion mostly proceeds in the midgut lumen (Chapman et al. 2013; Gazara et al. 2017; Toprak et al. 2020), although it can occur externally in a few species that regurgitate salivary gland enzymes to liquefy their food (Canton and Bonning 2020; Holtof et al. 2019; Rana et al. 1997). Lipid digestion may also start within the crop in species for which the countercurrent flux of the ectoperitrophic space dumps enzymes secreted from the anterior midgut into the foregut lumen (Holtof et al. 2019; Weidlich et al. 2015). For instance, in *P. americana*, the crop is the largest digestive organ, which contains high amounts of lipases (Tamaki et al. 2014) and can uptake FFAs (Hoffman and Downer 1976). Conversely, in xylophagous and phytophagous insects, the digestion of plant-derived macromolecules by symbionts and the uptake of the resulting SCFAs take place in the hindgut (Breznak 1982; Brune and Dietrich 2015).

Figure 5 summarizes the various processes of lipid digestion that occur in insect midguts. Most studies suggest that TAGs are completely

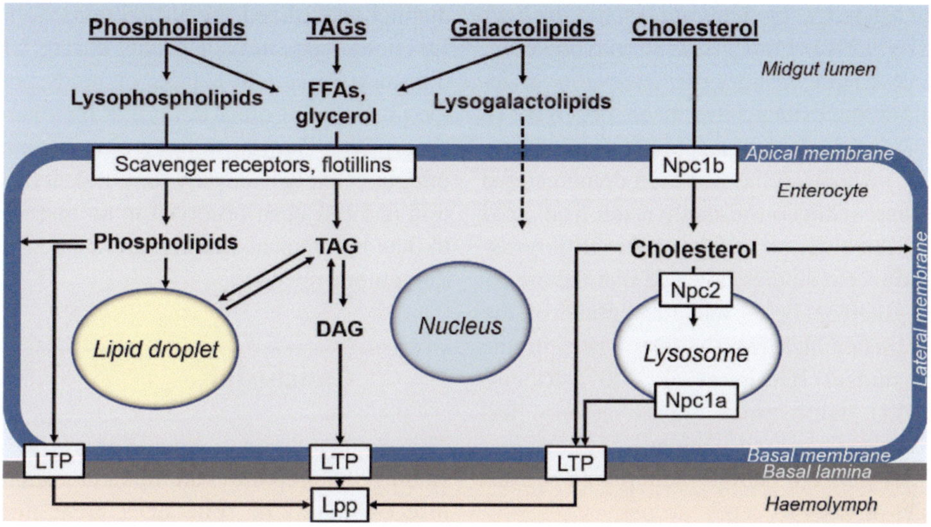

Fig. 5 Lipid digestive processes of insects. Enzymatic digestion of dietary lipids mostly takes place in the lumen. Although lipid uptake has been proposed to proceed passively, specific transporters have been reported. In the columnar cells, lipids may be used to sustain homeostasis, stored in specialized organelles or exported via Lpp via the haemolymph

digested to glycerol and FFAs, although MAGs may also be produced in some species (Hoffman and Downer 1979; Turunen 1975), but likely excreted within the frass. Phospholipids may be partially or completely hydrolysed; the resulting FAs, glycerol and choline, and the hydrolate intermediates, such as lysophosphatidylcholine, are absorbed by columnar cells (Turunen and Kastari 1979; Weiher and Komnick 1997). Galactolipids are hydrolysed to lysogalactolipids (Christeller et al. 2011), and likely through complete hydrolysis to galactose, glycerol and FAs, as described in mammals (Sahaka et al. 2020). Lipid uptake might proceed through passive diffusion, although not in all midgut sub-regions, suggesting that it depends on the molecular composition of the plasma membrane or on putative transporters (Sect. 4.2). Indeed, transporters for cholesterol have been identified in the midgut, whereas those for FAs and lysophospholipids have not been formally identified to date. To renew their plasma membrane, intestinal cells can incorporate dietary sterol and phospholipids derived from lipid digestive products (Fig. 5).

TAGs, steryl-esters and phospholipids can be stored in LDs, whereas lysosomes store free sterols. In most species, lipids are exported to the organs via the haemolymph in the form of DAGs, sterols and phospholipids loaded on lipophorins. This synthetic picture partly reflects the process of lipid digestion, since enzymatic activity, lipid emulsion and their absorption vary along specialized midgut sub-regions depending on the diet type of each insect species.

6.2 Midgut Regionalization of Lipid Digestion

The anterior midgut and the associated caeca play a critical role in lipid digestion (Toprak et al. 2020). For instance, the caeca of *G. bimaculatus* (Orthoptera) fulfil most of the total lipase activity (Weidlich et al. 2015). Studies in three lepidopteran species, *Diatraea grandiosella* (Turunen and Chippendale 1977), *P. brassicae* (Turunen 1990) and *M. sexta* (Smith et al. 1992), suggest that lipid absorption takes place in the anterior midgut, although secretion of PLA2 in the midgut of *M. sexta* is low in the anterior region and highest in the middle region (Rana and Stanley 1999).

In three species of hemipterans, *R. prolixus* (Billingsley 1990), *P. megistus* (Canavoso et al. 2004) and *Cimex hemipterus* (Azevedo et al. 2009), lipid inclusions have been observed in the anterior midgut after the blood meal. However, in *R. prolixus*, it has not been demonstrated whether these inclusions directly result from lipid uptake in the anterior midgut cells (Billingsley 1990), and recent studies revealed that the breakdown of complex lipids and the uptake of the resulting metabolites mostly take place in the posterior midgut (Grillo et al. 2007; Ribeiro et al. 2014). Furthermore, lipase activity has been observed in all midgut regions of *Brontocoris tabidus*, another hemipteran species (Guedes et al. 2007).

Studies on dipterans suggest that the posterior midgut is a critical region for lipid digestion. In *S. calcitrans*, enzymatic digestion and lipid absorption have been reported in the posterior midgut (Lehane 1977). In the midgut juice of *Hermatia illucens*, lipase activity is high in the anterior and posterior regions, but not in the middle region (Bonelli et al. 2019). Gene expression analyses from midgut sections provide considerable information about functional midgut regionalization. In *D. melanogaster*, lipid-related digestive enzymes are expressed in specific midgut regions, suggesting that each region fulfils distinct digestive functions (Buchon et al. 2013; Zinke et al. 2002). In *Anopheles gambiae*, several genes encoding lipid transporters are enriched in the gastric caeca and the anterior midgut, although a few appear to be also enriched in the posterior midgut, the Malpighian tubules and the hindgut (Neira Oviedo et al. 2008). In *Musca domestica*, the anterior midgut expresses PLA1 at high levels, whereas the middle midgut expresses high levels of acid lipases and the posterior midgut acid lipase as well as potential lipid transporters (Barroso et al. 2021). Based on these results, the authors propose that lysophospholipids resulting from PLA1 activity in the anterior midgut favour micelle formation in the middle midgut and the digestion of TAGs to FFAs, which are eventually absorbed in the posterior midgut (Barroso et al. 2021).

In summary, lipid digestion is a multiple-step process recruiting specific enzymes active in distinct specialized midgut sections, following a functional sequence of events during which the digestion of certain classes of lipids potentiates the digestion of other lipids and their subsequent absorption. The functional regionalization of the midgut is likely necessary for efficient lipid digestion and has been observed in all insects studied to date but does not exhibit an identical architecture among species.

7 Conclusion

This chapter aimed to provide an overview of the studies that reported on lipid digestion in the insect midgut. In spite of a great number of publications, data related to lipid digestion processes remain sparse and many questions are still opened. Not surprisingly, it is not possible to draw a uniform picture of insect digestion, because of the tremendous number of insect species and of the large variety of diet types. Nonetheless, the comprehension of the insect digestive processes is a crucial challenge for the future: i) to develop appropriate strategies against agriculture pests (e.g. the cotton bollworm *H. armigera*) and vectors of pathogens (e.g. the malaria vector *A. gambiae*), while maintaining biodiversity; ii) to improve the efficacy of insect-dependent industrial production (e.g. silk biogenesis by *B. mori*); iii) to set alternative models aimed at investigating human health and diseases (e.g. *D. melanogaster*). To this end, future investigations should evaluate the relative contribution of passive- versus transporter-dependent lipid uptake and precisely identify the lipases that are active in the midgut as well as their substrate specificity. More importantly, to provide an integrated view of lipid digestion for a given species, it will be essential to clarify the contribution of each midgut sub-region in the different digestive processes and to further determine to which extent the microbiota influences these processes. While setting a uniform picture of insect lipid digestion is a futile challenge, these studies will certainly improve our understanding of the digestive physiology of key insect species.

Acknowledgements We wish to thank the reviewers for their very constructive suggestions. We also apologize for several studies that were not cited here.

References

Agarwal HC, Nair AM (1976) Studies on the cholesterol ester hydrolase of trogoderma (coleoptera). Biochem J 157:111–116. https://doi.org/10.1042/bj1570111

Alborn HT, Turlings TCJ, Jones TH, Stenhagen G, Loughrin JH, Tumlison JH (1997) An elicitor of plant volatiles from beet armyworm oral secretion. Science 276:945–949

Albuquerque-Cunha JM, Gonzalez MS, Garcia ES, Mello CB, Azambuja P, Almeida JC, de Souza W, Nogueira NF (2009) Cytochemical characterization of microvillar and perimicrovillar membranes in the posterior midgut epithelium of *Rhodnius prolixus*. Arthropod Struct Dev 38:31–44. https://doi.org/10.1016/j.asd.2008.06.001

Arrese EL, Canavoso LE, Jouni ZE, Pennington JE, Tsuchida K, Wells MA (2001) Lipid storage and mobilization in insects: current status and future directions. Insect Biochem Mol Biol 31:7–17

Atella GC, Gondim C, Masuda H (1995) Loading of lipophorin particles with phospholipids at the midgut of *Rhodnius prolixus*. Arch Insect Biochem Physiol 30: 337–350. https://doi.org/10.1002/arch.940300404

Azevedo DO, Neves CA, Mallet JR, Goncalves TC, Zanuncio JC, Serrao JE (2009) Notes on midgut ultrastructure of *Cimex hemipterus* (hemiptera: Cimicidae). J Med Entomol 46:435–441. https://doi.org/10.1603/033.046.0304

Barroso IG, Cardoso C, Ferreira C, Terra WR (2021) Transcriptomic and proteomic analysis of the underlying mechanisms of digestion of triacylglycerols and phosphatides and absorption and fate of fatty acids along the midgut of *Musca domestica*. Comp Biochem Physiol Part D Genomics Proteomics 39:100826. https://doi.org/10.1016/j.cbd.2021.100826

Bateta R, Wang J, Wu Y, Weiss BL, Warren WC, Murilla GA, Aksoy S, Mireji PO (2017) Tsetse fly (*Glossina pallidipes*) midgut responses to *Trypanosoma brucei* challenge. Parasit Vectors 10:614. https://doi.org/10.1186/s13071-017-2569-7

Beenakkers AM, Van der Horst DJ, Van Marrewijk WJ (1985) Insect lipids and lipoproteins, and their role in physiological processes. Prog Lipid Res 24:19–67. https://doi.org/10.1016/0163-7827(85)90007-4

Biagio FP, Tamaki FK, Terra WR, Ribeiro AF (2009) Digestive morphophysiology of *Gryllodes sigillatus* (orthoptera: Gryllidae). J Insect Physiol 55:1125–1133. https://doi.org/10.1016/j.jinsphys.2009.08.015

Billingsley PF (1990) Blood digestion in the mosquito, *Anopheles stephensi* liston (diptera: Culicidae): partial characterization and post-feeding activity of midgut aminopeptidases. Arch Insect Biochem Physiol 15: 149–163. https://doi.org/10.1002/arch.940150304

Blitzer EJ, Vyazunova I, Lan Q (2005) Functional analysis of aescp-2 using gene expression knockdown in the yellow fever mosquito, *Aedes aegypti*. Insect Mol Biol 14:301–307. https://doi.org/10.1111/j.1365-2583.2005.00560.x

Blomquist G, Borgeson CE, Vundla M (1991) Polyunsaturated fatty acids and ecosanoids in insects. Insect Biochem. 21:99–106

Bollade D, Paris R, Moulins M (1970) Origine et mode d'action de la lipase intestinale chez les blattes. J Insect Physiol 16:45–53

Bolognesi R, Terra WR, Ferreira C (2008) Peritrophic membrane role in enhancing digestive efficiency. Theoretical and experimental models. J Insect Physiol 54:1413–1422. https://doi.org/10.1016/j.jinsphys.2008.08.002

Bonelli M, Bruno D, Caccia S, Sgambetterra G, Cappellozza S, Jucker C, Tettamanti G, Casartelli M (2019) Structural and functional characterization of *Hermetia illucens* larval midgut. Front Physiol 10: 204. https://doi.org/10.3389/fphys.2019.00204

Bownes M (1992) Why is there sequence similarity between insect yolk proteins and vertebrate lipases? J Lipid Res 33:777–790

Breznak JA (1982) Intestinal microbiota of termites and other xylophagous insects. Ann Rev Microbiol 36: 323–343. https://doi.org/10.1146/annurev.mi.36.100182.001543

Brune A (2014) Symbiotic digestion of lignocellulose in termite guts. Nat Rev Microbiol 12:168–180. https://doi.org/10.1038/nrmicro3182

Brune A, Dietrich C (2015) The gut microbiota of termites: digesting the diversity in the light of ecology and evolution. Ann Rev Microbiol 69:145–166. https://doi.org/10.1146/annurev-micro-092412-155715

Buchon N, Osman D, David FP, Fang HY, Boquete JP, Deplancke B, Lemaitre B (2013) Morphological and molecular characterization of adult midgut compartmentalization in drosophila. Cell Rep 3:1725–1738. https://doi.org/10.1016/j.celrep.2013.04.001

Caccia S, Casartelli M, Tettamanti G (2019) The amazing complexity of insect midgut cells: types, peculiarities, and functions. Cell Tissue Res 377:505–525. https://doi.org/10.1007/s00441-019-03076-w

Canavoso LE, Frede S, Rubiolo ER (2004) Metabolic pathways for dietary lipids in the midgut of hematophagous *Panstrongylus megistus* (hemiptera: Reduviidae). Insect Biochem Mol Biol 34:845–854. https://doi.org/10.1016/j.ibmb.2004.05.008

Canavoso LE, Jouni ZE, Karnas KJ, Pennington JE, Wells MA (2001) Fat metabolism in insects. Annu Rev Nutr 21:23–46

Canavoso LE, Wells MA (2000) Metabolic pathways for diacylglycerol biosynthesis and release in the midgut of larval *Manduca sexta*. Insect Biochem Mol Biol 30: 1173–1180. https://doi.org/10.1016/s0965-1748(00)00094-1

Canavoso LE, Wells MA (2001) Role of lipid transfer particle in delivery of diacylglycerol from midgut to lipophorin in larval *Manduca sexta*. Insect Biochem Mol Biol 31:783–790. https://doi.org/10.1016/s0965-1748(00)00183-1

Canton PE, Bonning BC (2020) Extraoral digestion: outsourcing the role of the hemipteran midgut. Curr Opin Insect Sci. 41:86–91. https://doi.org/10.1016/j.cois.2020.07.006

Carvalho M, Sampaio JL, Palm W, Brankatschk M, Eaton S, Shevchenko A (2012) Effects of diet and development on the drosophila lipidome. Mol Syst Biol 8:600. https://doi.org/10.1038/msb.2012.29

Carvalho M, Schwudke D, Sampaio JL, Palm W, Riezman I, Dey G, Gupta GD, Mayor S, Riezman H, Shevchenko A et al (2010) Survival strategies of a sterol auxotroph. Development 137:3675–3685

Chapman RF, Simpson SJ, Douglas AE. (2013). The insects. Structure and function

Chintapalli VR, Wang J, Dow JA (2007) Using flyatlas to identify better *Drosophila melanogaster* models of human disease. Nat Genet 39:715–720

Chng WA, Sleiman MSB, Schupfer F, Lemaitre B (2014) Transforming growth factor beta/activin signaling functions as a sugar-sensing feedback loop to regulate digestive enzyme expression. Cell Rep 9:336–348. https://doi.org/10.1016/j.celrep.2014.08.064

Chow SL, Hollander D (1979) A dual, concentration-dependent absorption mechanism of linoleic acid by rat jejunum in vitro. J Lipid Res 20:349–356

Christeller JT, Amara S, Carriere F (2011) Galactolipase, phospholipase and triacylglycerol lipase activities in the midgut of six species of lepidopteran larvae feeding on different lipid diets. J Insect Physiol 57:1232–1239. https://doi.org/10.1016/j.jinsphys.2011.05.012

Christeller JT, Poulton J, Markwick NM, Simpson RM (2010) The effect of diet on the expression of lipase genes in the midgut of the lightbrown apple moth (epiphyas postvittana walker; tortricidae). Insect Mol Biol 19:9–25. https://doi.org/10.1111/j.1365-2583.2009.00924.x

Chu BS, Rich GT, Ridout MJ, Faulks RM, Wickham MS, Wilde PJ (2009) Modulating pancreatic lipase activity with galactolipids: effects of emulsion interfacial composition. Langmuir 25:9352–9360. https://doi.org/10.1021/la9008174

Cifarelli V, Abumrad NA (2018) Intestinal cd36 and other key proteins of lipid utilization: role in absorption and gut homeostasis. Compr Physiol 8:493–507. https://doi.org/10.1002/cphy.c170026

Cioffi M (1984) Ultrastructure of arthropod transporting epithelia. Amer Zool 24:139–156

Ciufo LF, Murray PA, Thompson A, Rigden DJ, Rees HH (2011) Characterisation of a desmosterol reductase involved in phytosterol dealkylation in the silkworm, *Bombyx mori*. PLoS One 6:e21316. https://doi.org/10.1371/journal.pone.0021316

Clark AJ, Block K (1959) The absence of sterol synthesis in insects. J Biol Chem 234:2578–2582

Clayton RB (1964) The utilization of sterols by insects. J Lipid Res 5:3–19

Cleveland LR (1923) Symbiosis between termites and their intestinal protozoa. Proc Natl Acad Sci USA 9: 424–428. https://doi.org/10.1073/pnas.9.12.424

Cobbs C, Heath J, Stireman JO 3rd, Abbot P (2013) Carotenoids in unexpected places: gall midges, lateral gene transfer, and carotenoid biosynthesis in animals. Mol Phylogenet Evol 68:221–228. https://doi.org/10.1016/j.ympev.2013.03.012

Coleman RA, Mashek DG (2011) Mammalian triacylglycerol metabolism: synthesis, lipolysis, and signaling. Chem Rev 111:6359–6386. https://doi.org/10.1021/cr100404w

Collatz K-G, Mommsen T (1974) Die struktur der emulgierenden substanzen verschicdener invertebraten. J Comp Physiol 94:339–352

Consuegra J, Grenier T, Akherraz H, Rahioui I, Gervais H, da Silva P, Leulier F (2020a) Metabolic cooperation among commensal bacteria supports drosophila juvenile growth under nutritional stress. iScience 23: 101232. https://doi.org/10.1016/j.isci.2020.101232

Consuegra J, Grenier T, Baa-Puyoulet P, Rahioui I, Akherraz H, Gervais H, Parisot N, da Silva P, Charles H, Calevro F et al (2020b) Drosophila-associated bacteria differentially shape the nutritional requirements of their host during juvenile growth. PLoS Biol 18:e3000681. https://doi.org/10.1371/journal.pbio.3000681

Cooper MK, Wassif CA, Krakowiak PA, Taipale J, Gong R, Kelley RI, Porter FD, Beachy PA (2003) A defective response to hedgehog signaling in disorders of cholesterol biosynthesis. Nat Genet 33:508–513. https://doi.org/10.1038/ng1134

Coutinho-Abreu IV, Serafim TD, Meneses C, Kamhawi S, Oliveira F, Valenzuela JG (2020) Leishmania infection induces a limited differential gene expression in the sand fly midgut. BMC Genomics 21:608. https://doi.org/10.1186/s12864-020-07025-8

Cruden DL, Markovetz AJ (1987) Microbial ecology of the cockroach gut. Ann Rev Microbiol 41:617–643. https://doi.org/10.1146/annurev.mi.41.100187.003153

da Silva G, Costa Ramos LF, Dos Santos SH, Mendonca Gomes F, Reis Cortines J, Ramos I, Dinis Anobom C, de Alcantara ME, Perpetua de Oliveira DM (2019) Biochemical characterization of digestive membrane-associated alkaline phosphatase from the velvet bean caterpillar *Anticarsia gemmatalis*. Arch Insect Biochem Physiol 102:e21591. https://doi.org/10.1002/arch.21591

Dadd RH (1961) The nutritional requirements of locusts. J Ins Physiol 6:126–145

Dadd RH, Kleinjan JE (1979) Essential fatty acid for the mosquito *Culex pipiens*: arachidonic acid. J Insect Physiol 25:495–502. https://doi.org/10.1016/s0022-1910(79)80008-6

de Oliveira Souza A, Couto-Lima CA, Catalao CHR, Santos-Junior NN, Dos Santos JF, da Rocha MJA, Alberici LC (2019) Neuroprotective action of eicosapentaenoic (epa) and docosahexaenoic (dha) acids on paraquat intoxication in *Drosophila melanogaster*. Neurotoxicology 70:154–160. https://doi.org/10.1016/j.neuro.2018.11.013

de Sousa ME, Wanderley-Teixeira V, Teixeira AA, de Siqueira HA, Santos FA, Alves LC (2009) Ultrastructure of the *Alabama argillacea* (hubner) (lepidoptera: Noctuidae) midgut. Micron 40:743–749. https://doi.org/10.1016/j.micron.2009.04.008

Derewenda ZS (1994) Structure and function of lipases. Adv Protein Chem 45:1–52. https://doi.org/10.1016/s0065-3233(08)60637-3

Dewett D, Lam-Kamath K, Poupault C, Khurana H, Rister J (2021) Mechanisms of vitamin a metabolism and deficiency in the mammalian and fly visual system. Dev Biol 476:68–78. https://doi.org/10.1016/j.ydbio.2021.03.013

Dicke M, Sabelis MW, Takabayashi J, Bruin J, Posthumus MA (1990) Plant strategies of manipulating predatorprey interactions through allelochemicals: prospects for application in pest control. J Chem Ecol 16:3091–3118. https://doi.org/10.1007/BF00979614

Dragh MA, Xu Z, Al-Allak ZS, Hong L (2017) Vitamin k2 prevents lymphoma in drosophila. Sci Rep 7:17047. https://doi.org/10.1038/s41598-017-17270-9

Dubreuil RR (2004) Copper cells and stomach acid secretion in the drosophila midgut. Int J Biochem Cell Biol 36:745–752. https://doi.org/10.1016/j.biocel.2003.07.004

Dutta D, Dobson AJ, Houtz PL, Glasser C, Revah J, Korzelius J, Patel PH, Edgar BA, Buchon N (2015) Regional cell-specific transcriptome mapping reveals regulatory complexity in the adult drosophila midgut. Cell Rep 12:346–358. https://doi.org/10.1016/j.celrep.2015.06.009

Eguchi M (1995) Alkaline phosphatase isozymes in insects and comparison with mammalian enzyme. Comp Biochem Physiol B Biochem Mol Biol 111:151–162. https://doi.org/10.1016/0305-0491(94)00248-s

Engel P, Moran NA (2013) The gut microbiota of insects - diversity in structure and function. FEMS Microbiol Rev 37:699–735. https://doi.org/10.1111/1574-6976.12025

Entringer PF, Majerowicz D, Gondim KC (2021) The fate of dietary cholesterol in the kissing bug *Rhodnius prolixus*. Front Physiol 12:654565. https://doi.org/10.3389/fphys.2021.654565

Erkosar B, Leulier F (2014) Transient adult microbiota, gut homeostasis and longevity: novel insights from the drosophila model. FEBS Lett 588:4250–4257. https://doi.org/10.1016/j.febslet.2014.06.041

Faber V, Komnick H (1989) Peroxisomes of the midgut epithelium, Malpighian tubules and fat body of larvae of the dragonfly, *Aeshna cyanea*. Tissue Cell 21:917–924. https://doi.org/10.1016/0040-8166(89)90042-6

Ferreira C, Capella AN, Sitnik R, Terra WR (1994) Properties of the digestive enzymes and the permeability of the peritrophic membrane of *Spodoptera frugiperda* (lepidoptera) larvae. Comp Biochem Physiol 107A:631–640

Fialho MCQ, Terra WR, Moreira NR, Zanuncio JC, Serrao JE (2013) Ultrastructure and immunolocalization of digestive enzymes in the midgut of *Podisus nigrispinus* (heteroptera: Pentatomidae). Arthropod Struct Dev 42:277–285. https://doi.org/10.1016/j.asd.2013.03.002

Fluegel ML, Parker TJ, Pallanck LJ (2006) Mutations of a drosophila npc1 gene confer sterol and ecdysone metabolic defects. Genetics 172:185–196. https://doi.org/10.1534/genetics.105.046565

Flybase (2003) The flybase database of the drosophila genome projects and community literature. Nucleic Acids Res 31:172–175. https://doi.org/10.1093/nar/gkg094

Ford PS, Van Heusden MC (1994) Triglyceride-rich lipophorin in *Aedes aegypti* (diptera: Culicidae). J Med Entomol 31:435–441. https://doi.org/10.1093/jmedent/31.3.435

Fraenkel G, Blewett M (1946) Linoleic acid, vitamin e and other fat-soluble substances in the nutrition of certain insects, *Ephestia kuehniella*, *E. elutella*, *E. cautella* and *Plodia interpunctella* (lep.). J Exp Biol 22:172–190. https://doi.org/10.1242/jeb.22.3-4.172

Fruttero LL, Rubiolo ER, Canavoso LE (2009) Biochemical and cellular characterization of lipophorin-midgut interaction in the hematophagous *Panstrongylus megistus* (hemiptera: Reduviidae). Insect Biochem Mol Biol 39:322–331. https://doi.org/10.1016/j.ibmb.2009.01.009

Fukuda S, Toh H, Hase K, Oshima K, Nakanishi Y, Yoshimura K, Tobe T, Clarke JM, Topping DL, Suzuki T et al (2011) Bifidobacteria can protect from enteropathogenic infection through production of acetate. Nature 469:543–547. https://doi.org/10.1038/nature09646

Fuzita FJ, Pimenta DC, Palmisano G, Terra WR, Ferreira C (2019) Detergent-resistant domains in *Spodoptera frugiperda* midgut microvillar membranes and their relation to microapocrine secretion. Comp Biochem Physiol B Biochem Mol Biol 235:8–18. https://doi.org/10.1016/j.cbpb.2019.05.008

Galbiati F, Volonte D, Goltz JS, Steele Z, Sen J, Jurcsak J, Stein D, Stevens L, Lisanti MP (1998) Identification, sequence and developmental expression of invertebrate flotillins from *Drosophila melanogaster*. Gene 210:229–237. https://doi.org/10.1016/s0378-1119(98)00064-x

Gandotra S, Bhuyan PM, Gogoi DK, Kumar A, Subramanian S (2016) Screening of nutritionally important gut bacteria from the lepidopteran insects through qualitative enzyme assays. Proc Natl Acad Sci, India 88:329–337. https://doi.org/10.1007/s40011-016-0762-7

Garrido D, Rubin T, Poidevin M, Maroni B, Le Rouzic A, Parvy JP, Montagne J (2015) Fatty acid synthase cooperates with glyoxalase 1 to protect against sugar toxicity. PLoS Genet 11:e1004995. https://doi.org/10.1371/journal.pgen.1004995

Gazara RK, Cardoso C, Bellieny-Rabelo D, Ferreira C, Terra WR, Venancio TM (2017) De novo transcriptome sequencing and comparative analysis of midgut tissues of four non-model insects pertaining to hemiptera, coleoptera, diptera and lepidoptera. Gene 627:85–93. https://doi.org/10.1016/j.gene.2017.06.008

Ge L, Qi W, Wang LJ, Miao HH, Qu YX, Li BL, Song BL (2011) Flotillins play an essential role in niemann-pick c1-like 1-mediated cholesterol uptake. Proc Natl Acad Sci USA 108:551–556. https://doi.org/10.1073/pnas.1014434108

Gondim KC, Wells MA (2000) Characterization of lipophorin binding to the midgut of larval *Manduca sexta*. Insect Biochem Mol Biol 30:405–413. https://doi.org/10.1016/s0965-1748(00)00014-x

Gong J, Hou Y, Zha XF, Lu C, Zhu Y, Xia QY (2006) Molecular cloning and characterization of *Bombyx mori* sterol carrier protein x/sterol carrier protein 2 (scpx/scp2) gene. DNA Seq 17:326–333. https://doi.org/10.1080/10425170600886706

Goudriaan JR, Dahlmans VE, Febbraio M, Teusink B, Romijn JA, Havekes LM, Voshol PJ (2002) Intestinal lipid absorption is not affected in cd36 deficient mice. Mol Cell Biochem 239:199–202

Grenier T, Leulier F (2020) How commensal microbes shape the physiology of *Drosophila melanogaster*. Curr Opin Insect Sci. 41:92–99. https://doi.org/10.1016/j.cois.2020.08.002

Grillo LA, Majerowicz D, Gondim KC (2007) Lipid metabolism in *Rhodnius prolixus* (hemiptera: Reduviidae): role of a midgut triacylglycerol-lipase. Insect Biochem Mol Biol 37:579–588. https://doi.org/10.1016/j.ibmb.2007.03.002

Grillo LA, Pontes EG, Gondim KC (2003) Lipophorin interaction with the midgut of *Rhodnius prolixus*: characterization and changes in binding capacity. Insect Biochem Mol Biol 33:429–438. https://doi.org/10.1016/s0965-1748(03)00007-9

Gronke S, Mildner A, Fellert S, Tennagels N, Petry S, Muller G, Jackle H, Kuhnlein RP (2005) Brummer lipase is an evolutionary conserved fat storage regulator in drosophila. Cell Metab 1:323–330. https://doi.org/10.1016/j.cmet.2005.04.003

Guedes BAM, Zanuncio JC, Ramalho FS, Serrao JE (2007) Midgut morphology and enzymes of the obligate zoophytophagous stinkbug *Brontocoris tabidus* (signoret, 1963) (heteroptera: Pentatomidae). Pan Pac Entomol 83:66–74

Guo XR, Zheng SC, Liu L, Feng QL (2009) The sterol carrier protein 2/3-oxoacyl-coa thiolase (scpx) is involved in cholesterol uptake in the midgut of *Spodoptera litura*: gene cloning, expression, localization and functional analyses. BMC Mol Biol 10:102. https://doi.org/10.1186/1471-2199-10-102

Hakim RS, Baldwin K, Smagghe G (2010) Regulation of midgut growth, development, and metamorphosis. Annu Rev Entomol 55:593–608. https://doi.org/10.1146/annurev-ento-112408-085450

Halitschke R, Schittko U, Pohnert G, Boland W, Baldwin IT (2001) Molecular interactions between the specialist herbivore *Manduca sexta* (lepidoptera, sphingidae) and its natural host nicotiana attenuata. Iii. Fatty acid-amino acid conjugates in herbivore oral secretions are necessary and sufficient for herbivore-specific plant responses. Plant Physiol 125:711–717. https://doi.org/10.1104/pp.125.2.711

Harrison JF (2001) Insect acid-base physiology. Annu Rev Entomol 46:221–250. https://doi.org/10.1146/annurev.ento.46.1.221

Harrison JF, Kaiser A, VandenBrooks JM (2010) Atmospheric oxygen level and the evolution of insect body size. Proc Biol Sci 277:1937–1946. https://doi.org/10.1098/rspb.2010.0001

Hegedus D, Erlandson M, Gillott C, Toprak U (2009) New insights into peritrophic matrix synthesis, architecture, and function. Annu Rev Entomol 54:285–302. https://doi.org/10.1146/annurev.ento.54.110807.090559

Hobson RP (1935) On a fat-soluble growth factor required by blow-fly larvae: identity of the growth factor with cholesterol. Biochem J 29:2023–2026. https://doi.org/10.1042/bj0292023

Hoffman AGD, Downer RGH (1976) The crop as an organ of glyceride absorption in the American cockroach, *Periplaneta americana* l. Can J Zool 54:1165–1171

Hoffman AGD, Downer RGH (1979) End product specificity of triacylglycerol lipases from intestine, fat body, muscle and haemolymph of the American cockroach, *Periplaneta americana* l. Lipids 14:893–899

Holmquist M (2000) Alpha/beta-hydrolase fold enzymes: structures, functions and mechanisms. Curr Protein Pept Sci 1:209–235. https://doi.org/10.2174/1389203003381405

Holtof M, Lenaerts C, Cullen D, Vanden Broeck J (2019) Extracellular nutrient digestion and absorption in the insect gut. Cell Tissue Res 377:397–414. https://doi.org/10.1007/s00441-019-03031-9

Horne I, Haritos VS, Oakeshott JG (2009) Comparative and functional genomics of lipases in holometabolous insects. Insect Biochem Mol Biol 39:547–567. https://doi.org/10.1016/j.ibmb.2009.06.002

House HL (1966) Effects of vitamins e and a on growth and development, and the necessity of vitamin e for reproduction in the parasitoid *Agria affinis* (fallen) (diptera, sarcophagidae). J Insect Physiol 12:409–417. https://doi.org/10.1016/0022-1910(66)90004-7

Hu X, Chen L, Xiang X, Yang R, Yu S, Wu X (2012) Proteomic analysis of peritrophic membrane (pm) from the midgut of fifth-instar larvae, *Bombyx mori*. Mol Biol Rep 39:3427–3434. https://doi.org/10.1007/s11033-011-1114-6

Huang X, Ma J, Qin X, Tu X, Cao G, Wang G, Nong X, Zhang Z (2017) Biology, physiology and gene expression of grasshopper *Oedaleus asiaticus* exposed to diet stress from plant secondary compounds. Sci Rep 7:8655. https://doi.org/10.1038/s41598-017-09277-z

Huang X, Suyama K, Buchanan J, Zhu AJ, Scott MP (2005) A drosophila model of the niemann-pick type c lysosome storage disease: Dnpc1a is required for molting and sterol homeostasis. Development 132: 5115–5124. https://doi.org/10.1242/dev.02079

Huang X, Warren JT, Buchanan J, Gilbert LI, Scott MP (2007) Drosophila niemann-pick type c-2 genes control sterol homeostasis and steroid biosynthesis: a model of human neurodegenerative disease. Development 134:3733–3742

Ian De Veau EJ, Schultz JC (1992) Reassessment of interaction between gut detergents and tannins in lepidoptera and significance for gypsy moth larvae. J Chem Ecol 18:1437–1453. https://doi.org/10.1007/BF00994367

Issaq HJ, Chan KC, Blonder J, Ye X, Veenstra TD (2009) Separation, detection and quantitation of peptides by liquid chromatography and capillary electrochromatography. J Chromatogr A 1216:1825–1837. https://doi.org/10.1016/j.chroma.2008.12.052

Jing TZ, Qi FH, Wang ZY (2020) Most dominant roles of insect gut bacteria: digestion, detoxification, or essential nutrient provision? Microbiome 8:38. https://doi.org/10.1186/s40168-020-00823-y

Jing X, Behmer ST (2020) Insect sterol nutrition: physiological mechanisms, ecology, and applications. Annu Rev Entomol 65:251–271. https://doi.org/10.1146/annurev-ento-011019-025017

Jing X, Vogel H, Grebenok RJ, Zhu-Salzman K, Behmer ST (2012) Dietary sterols/steroids and the generalist caterpillar *Helicoverpa zea*: physiology, biochemistry and midgut gene expression. Insect Biochem Mol Biol 42:835–845. https://doi.org/10.1016/j.ibmb.2012.07.009

Jouni ZE, Zamora J, Wells MA (2002) Absorption and tissue distribution of cholesterol in *Manduca sexta*. Arch Insect Biochem Physiol 49:167–175. https://doi.org/10.1002/arch.10017

Jugder BE, Kamareddine L, Watnick PI (2021) Microbiota-derived acetate activates intestinal innate immunity via the tip60 histone acetyltransferase complex. Immunity 54:1683–1697. https://doi.org/10.1016/j.immuni.2021.05.017

Jupatanakul N, Sim S, Dimopoulos G (2014) *Aedes aegypti* ml and niemann-pick type c family members are agonists of dengue virus infection. Dev Comp Immunol 43:1–9. https://doi.org/10.1016/j.dci.2013.10.002

Jurzitza G (1974) Furnishing of sterol by the yeast-like endosymbionts of *Lasioderma serricorne* f. (coleoptera, anobiidae) and its ecological importance for the host. Oecologia 16:163–172. https://doi.org/10.1007/BF00345580

Kamareddine L, Robins WP, Berkey CD, Mekalanos JJ, Watnick PI (2018) The drosophila immune deficiency pathway modulates enteroendocrine function and host metabolism. Cell Metab 28:449–462. https://doi.org/10.1016/j.cmet.2018.05.026

Kesnerova L, Mars RAT, Ellegaard KM, Troilo M, Sauer U, Engel P (2017) Disentangling metabolic functions of bacteria in the honey bee gut. PLoS Biol 15:e2003467. https://doi.org/10.1371/journal.pbio.2003467

Kiefer C, Sumser E, Wernet MF, Von Lintig J (2002) A class b scavenger receptor mediates the cellular uptake of carotenoids in drosophila. Proc Natl Acad Sci USA 99:10581–10586. https://doi.org/10.1073/pnas.162182899

Kim MS, Lan Q (2010) Sterol carrier protein-x gene and effects of sterol carrier protein-2 inhibitors on lipid uptake in *Manduca sexta*. BMC Physiol 10:9. https://doi.org/10.1186/1472-6793-10-9

Kim Y, Stanley D (2021) Eicosanoid signaling in insect immunology: new genes and unresolved issues. Genes (Basel) 12. https://doi.org/10.3390/genes12020211

Kirfel G, Komnick H (1999) Differential absorption and esterification of dietary long-chain fatty acids by larvae of the dragonfly, *Aeshna cyanea*. Arch Insect Biochem Physiol 40:183–193. https://doi.org/10.1002/(SICI)1520-6327(1999)40:4<183::AID-ARCH3>3.0.CO;2-W

Ko CW, Qu J, Black DD, Tso P (2020) Regulation of intestinal lipid metabolism: current concepts and relevance to disease. Nat Rev Gastroenterol Hepatol 17: 169–183. https://doi.org/10.1038/s41575-019-0250-7

Komnick H, Bauerfeind R (1991) Intestinal absorption of defined lipids by the larval dragonfly *Aeshna cyanea* (insecta, odonata): wax ester and fatty alcohols. J Insect Physiol 37:179–191

Kong HG, Kim HH, Chung JH, Jun J, Lee S, Kim HM, Jeon S, Park SG, Bhak J, Ryu CM (2019) The galleria mellonella hologenome supports microbiota-independent metabolism of long-chain hydrocarbon beeswax. Cell Rep 26:2451–2464. https://doi.org/10.1016/j.celrep.2019.02.018

Kuhnlein RP (2012) Thematic review series: lipid droplet synthesis and metabolism: from yeast to man. Lipid droplet-based storage fat metabolism in drosophila. J Lipid Res 53:1430–1436

Kuhns EH, Seidl-Adams I, Tumlinson JH (2012) A lepidopteran aminoacylase (l-acy-1) in *Heliothis virescens* (lepidoptera: Noctuidae) gut lumen hydrolyzes fatty acid-amino acid conjugates, elicitors of plant defense. Insect Biochem Mol Biol 42:32–40. https://doi.org/10.1016/j.ibmb.2011.10.004

Kunding AH, Christensen SM, Danielsen EM, Hansen GH (2010) Domains of increased thickness in microvillar membranes of the small intestinal enterocyte. Mol Membr Biol 27:170–177. https://doi.org/10.3109/09687688.2010.494625

Kvietys PR, Specian RD, Grisham MB, Tso P (1991) Jejunal mucosal injury and restitution: role of hydrolytic products of food digestion. Am J Phys 261:G384–G391. https://doi.org/10.1152/ajpgi.1991.261.3.G384

Kwong WK, Medina LA, Koch H, Sing KW, Soh EJY, Ascher JS, Jaffe R, Moran NA (2017) Dynamic microbiome evolution in social bees. Sci Adv 3: e1600513. https://doi.org/10.1126/sciadv.1600513

LaJeunesse DR, Johnson B, Presnell JS, Catignas KK, Zapotoczny G (2010) Peristalsis in the junction region of the drosophila larval midgut is modulated by dh31 expressing enteroendocrine cells. BMC Physiol 10:14. https://doi.org/10.1186/1472-6793-10-14

Lan Q, Massey RJ (2004) Subcellular localization of the mosquito sterol carrier protein-2 and sterol carrier protein-x. J Lipid Res 45:1468–1474. https://doi.org/10.1194/jlr.M400003-JLR200

Lan Q, Wessely V (2004) Expression of a sterol carrier protein-x gene in the yellow fever mosquito, *Aedes aegypti*. Insect Mol Biol 13:519–529. https://doi.org/10.1111/j.0962-1075.2004.00510.x

Lee WJ, Hase K (2014) Gut microbiota-generated metabolites in animal health and disease. Nat Chem Biol 10:416–424. https://doi.org/10.1038/nchembio.1535

Lehane MJ (1977) Transcellular absorption of lipids in the midgut of the stablefly, Stomoxys calcitrans. J Insect Physiol 23:945–954. https://doi.org/10.1016/0022-1910(77)90121-4

Lehane MJ (1997) Peritrophic matrix structure and function. Annu Rev Entomol 42:525–550. https://doi.org/10.1146/annurev.ento.42.1.525

Li S, Jing X (2020) Fates of dietary sterols in the insect alimentary canal. Curr Opin Insect Sci. 41:106–111. https://doi.org/10.1016/j.cois.2020.08.001

Lin X, Wen X, Wei Z, Guo K, Shi F, Huang T, Wang W, Zheng J (2021) Vitamin k2 protects against abeta42-induced neurotoxicity by activating autophagy and improving mitochondrial function in drosophila. Neuroreport 32:431–437. https://doi.org/10.1097/WNR.0000000000001599

Listenberger LL, Han X, Lewis SE, Cases S, Farese RV Jr, Ory DS, Schaffer JE (2003) Triglyceride accumulation protects against fatty acid-induced lipotoxicity. Proc Natl Acad Sci USA 100:3077–3082. https://doi.org/10.1073/pnas.0630588100

Male KB, Storey KB (1981) Enzyme activities and isozyme composition of triglyceride, diglyceride and monoglyceride lipases in *Periplaneta americana*, *Locusta migratoria* and *Polia adjuncta*. Insect Biochem. 11:423–427

Marianes A, Spradling AC (2013) Physiological and stem cell compartmentalization within the drosophila midgut. elife 2:e00886. https://doi.org/10.7554/eLife.00886

Markvoort AJ, Marrink SJ (2011) Lipid acrobatics in the membrane fusion arena. Curr Top Membr 68:259–294. https://doi.org/10.1016/B978-0-12-385891-7.00011-8

Matic D, Vlahovic M, Ilijin L, Grcic A, Filipovic A, Todorovic D, Peric-Mataruga V (2021) Implications of long-term exposure of a *Lymantria dispar* l. population to pollution for the response of larval midgut proteases and acid phosphatases to chronic cadmium treatment. Comp Biochem Physiol C Toxicol Pharmacol 250:109172. https://doi.org/10.1016/j.cbpc.2021.109172

Mattila J, Havula E, Suominen E, Teesalu M, Surakka I, Hynynen R, Kilpinen H, Vaananen J, Hovatta I, Kakela R et al (2015) Mondo-mlx mediates organismal sugar sensing through the gli-similar transcription factor sugarbabe. Cell Rep 13:350–364

McFarlane JE (1976) Vitamin k: growth factor for the house cricket (orthoptera: Gryllidae). Can Entomol 108:391–394

McFarlane JE, Alli I (1985) Volatile fatty acids of frass of certain omnivorous insects. J Chem Ecol 11:59–63. https://doi.org/10.1007/BF00987605

Mei Y, Jing D, Tang S, Chen X, Chen H, Duanmu H, Cong Y, Chen M, Ye X, Zhou H et al (2022) Insectbase 2.0: a comprehensive gene resource for insects. Nucleic Acids Res 50:D1040–D1045. https://doi.org/10.1093/nar/gkab1090

Meikle JES, McFarlane JE (1965) The role of lipid in the nutrition of the house cricket, Acheta domesticus l. Can J Zool 43:87–98

Merritt RW, Dadd RH, Walker ED (1992) Feeding behavior, natural food, and nutritional relationships of larval mosquitoes. Annu Rev Entomol 37:349–376. https://doi.org/10.1146/annurev.en.37.010192.002025

Miao Z, Cao X, Jiang H (2020) Digestion-related proteins in the tobacco hornworm, *Manduca sexta*. Insect Biochem Mol Biol 126:103457. https://doi.org/10.1016/j.ibmb.2020.103457

Miguel-Aliaga I, Jasper H, Lemaitre B (2018) Anatomy and physiology of the digestive tract of *Drosophila melanogaster*. Genetics 210:357–396. https://doi.org/10.1534/genetics.118.300224

Miled N, Canaan S, Dupuis L, Roussel A, Riviere M, Carriere F, de Caro A, Cambillau C, Verger R (2000) Digestive lipases: from three-dimensional structure to physiology. Biochimie 82:973–986. https://doi.org/10.1016/s0300-9084(00)01179-2

Milger K, Herrmann T, Becker C, Gotthardt D, Zickwolf J, Ehehalt R, Watkins PA, Stremmel W, Fullekrug J (2006) Cellular uptake of fatty acids driven by the er-localized acyl-coa synthetase fatp4. J Cell Sci 119:4678–4688. https://doi.org/10.1242/jcs.03280

Moadeli T, Mainali B, Ponton F, Taylor PW (2020) Effects of fatty acids and vitamin e in larval diets on development and performance of Queensland fruit fly. J Insect Physiol 125:104058. https://doi.org/10.1016/j.jinsphys.2020.104058

Montigny C, Lyons J, Champeil P, Nissen P, Lenoir G (2016) On the molecular mechanism of flippase- and scramblase-mediated phospholipid transport. Biochim Biophys Acta 1861:767–783. https://doi.org/10.1016/j.bbalip.2015.12.020

Moran NA, Jarvik T (2010) Lateral transfer of genes from fungi underlies carotenoid production in aphids. Science 328:624–627. https://doi.org/10.1126/science.1187113

Nassir F, Wilson B, Han X, Gross RW, Abumrad NA (2007) Cd36 is important for fatty acid and cholesterol uptake by the proximal but not distal intestine. J Biol Chem 282:19493–19501. https://doi.org/10.1074/jbc.M703330200

Nauli AM, Nassir F, Zheng S, Yang Q, Lo CM, Vonlehmden SB, Lee D, Jandacek RJ, Abumrad NA,

Tso P (2006) Cd36 is important for chylomicron formation and secretion and may mediate cholesterol uptake in the proximal intestine. Gastroenterology 131:1197–1207. https://doi.org/10.1053/j.gastro.2006.08.012

Neira Oviedo M, Vanekeris L, Corena-McLeod MD, Linser PJ (2008) A microarray-based analysis of transcriptional compartmentalization in the alimentary canal of Anopheles gambiae (diptera: Culicidae) larvae. Insect Mol Biol 17:61–72. https://doi.org/10.1111/j.1365-2583.2008.00779.x

Niot I, Poirier H, Tran TT, Besnard P (2009) Intestinal absorption of long-chain fatty acids: evidence and uncertainties. Prog Lipid Res 48:101–115. https://doi.org/10.1016/j.plipres.2009.01.001

Nor Aliza AR, Othman M, Sabri M, Stanley DW (2018) A midgut digestive phospholipase a2 in larval mosquitoes, *Aedes albopictus* and *Culex quinquefasciatus*. Enzyme Res 2018:9703413. https://doi.org/10.1155/2018/9703413

Nor Aliza AR, Stanley D (1998) A digestive phospholipase a2 in larval mosquitoes, *Aedes aegypti*. Insect Biochem Mol Biol 28:561–569

Noriega FG (2014) Juvenile hormone biosynthesis in insects: what is new, what do we know, and what questions remain? Int Sch Res Notices 2014:967361. https://doi.org/10.1155/2014/967361

Palm W, Sampaio JL, Brankatschk M, Carvalho M, Mahmoud A, Shevchenko A, Eaton S (2012) Lipoproteins in *Drosophila melanogaster*–assembly, function, and influence on tissue lipid composition. PLoS Genet 8:e1002828. doi: https://doi.org/10.1371/journal.pgen.1002828

Parvy JP, Napal L, Rubin T, Poidevin M, Perrin L, Wicker-Thomas C, Montagne J (2012) *Drosophila melanogaster* acetyl-coa-carboxylase sustains a fatty acid-dependent remote signal to waterproof the respiratory system. PLoS Genet 8:e1002925. https://doi.org/10.1371/journal.pgen.1002925

Peller CR, Bacon EM, Bucheger JA, Blumenthal EM (2009) Defective gut function in drop-dead mutant drosophila. J Insect Physiol 55:834–839. https://doi.org/10.1016/j.jinsphys.2009.05.011

Pennington JE, Wells MA (2002) Triacylglycerol-rich lipophorins are found in the dipteran infraorder culicomorpha, not just in mosquitoes. J Insect Sci 2: 15. https://doi.org/10.1093/jis/2.1.15

Perez-Salas U, Garg S, Gerelli Y, Porcar L (2021) Deciphering lipid transfer between and within membranes with time-resolved small-angle neutron scattering. Curr Top Membr 88:359–412. https://doi.org/10.1016/bs.ctm.2021.10.004

Petersen TN, Brunak S, von Heijne G, Nielsen H (2011) Signalp 4.0: discriminating signal peptides from transmembrane regions. Nat Methods 8:785–786. https://doi.org/10.1038/nmeth.1701

Plateau F (1876). Les phenomenes de la digestion chez la blatte americaine (*Periplaneta americana*). In: Hayez F, editor.: Imprimeur de l'Académie Royale de Belgique

Rana RL, Hoback WW, Rahim NAR, Bedick J, Stanley DW (1997) Pre-oral digestion: a phospholipase a2 associated with oral secretions in adult burying beetles, *Nicrophorus marginatus*. Comp Biochem Physiol 118: 375–380

Rana RL, Stanley DW (1999) In vitro secretion of digestive phospholipase a(2) by midguts isolated from tobacco hornworms, *Manduca sexta*. Arch Insect Biochem Physiol 42:179–187. https://doi.org/10.1002/(SICI)1520-6327(199911)42:3<179::AID-ARCH2>3.0.CO;2-R

Reboul E, Klein A, Bietrix F, Gleize B, Malezet-Desmoulins C, Schneider M, Margotat A, Lagrost L, Collet X, Borel P (2006) Scavenger receptor class b type i (sr-bi) is involved in vitamin e transport across the enterocyte. J Biol Chem 281:4739–4745. https://doi.org/10.1074/jbc.M509042200

Reiher W, Shirras C, Kahnt J, Baumeister S, Isaac RE, Wegener C (2011) Peptidomics and peptide hormone processing in the drosophila midgut. J Proteome Res 10:1881–1892. https://doi.org/10.1021/pr101116g

Renne MF, Hariri H (2021) Lipid droplet-organelle contact sites as hubs for fatty acid metabolism, trafficking, and metabolic channeling. Front Cell Dev Biol 9: 726261. https://doi.org/10.3389/fcell.2021.726261

Rewitz KF, Rybczynski R, Warren JT, Gilbert LI (2006) The halloween genes code for cytochrome p450 enzymes mediating synthesis of the insect moulting hormone. Biochem Soc Trans 34:1256–1260

Ribeiro JM, Genta FA, Sorgine MH, Logullo R, Mesquita RD, Paiva-Silva GO, Majerowicz D, Medeiros M, Koerich L, Terra WR et al (2014) An insight into the transcriptome of the digestive tract of the bloodsucking bug, *Rhodnius prolixus*. PLoS Negl Trop Dis 8:e2594. https://doi.org/10.1371/journal.pntd.0002594

Rietveld A, Neutz S, Simons K, Eaton S (1999) Association of sterol- and glycosylphosphatidylinositol-linked proteins with drosophila raft lipid microdomains. J Biol Chem 274:12049–12054. https://doi.org/10.1074/jbc.274.17.12049

Rodriguez-Vazquez M, Vaquero D, Parra-Peralbo E, Mejia-Morales JE, Culi J (2015) Drosophila lipophorin receptors recruit the lipoprotein ltp to the plasma membrane to mediate lipid uptake. PLoS Genet 11:e1005356. https://doi.org/10.1371/journal.pgen.1005356

Rost-Roszkowska MM, Pilka M, Szymska R, Klag J (2007) Ultrastructural studies of midgut epithelium formation in *Lepisma saccharina* l. (insecta, zygentoma). J Morphol 268:224–231. https://doi.org/10.1002/jmor.10513

Rudin W, Hecker H (1979) Functional morphology of the midgut of *Aedes aegypti* l. (insecta, diptera) during blood digestion. Cell Tissue Res 200:193–203. https://doi.org/10.1007/BF00236412

Sahaka M, Amara S, Wattanakul J, Gedi MA, Aldai N, Parsiegla G, Lecomte J, Christeller JT, Gray D,

Gontero B et al (2020) The digestion of galactolipids and its ubiquitous function in nature for the uptake of the essential alpha-linolenic acid. Food Funct 11: 6710–6744. https://doi.org/10.1039/d0fo01040e

Sajjadian SM, Vatanparast M, Stanley D, Kim Y (2019) Secretion of secretory phospholipase a2 into *Spodoptera exigua* larval midgut lumen and its role in lipid digestion. Insect Mol Biol 28:773–784. https://doi.org/10.1111/imb.12588

Santos HP, Rost-Roszkowska M, Vilimova J, Serrao JE (2017) Ultrastructure of the midgut in heteroptera (hemiptera) with different feeding habits. Protoplasma 254:1743–1753. https://doi.org/10.1007/s00709-016-1051-2

Schmidt K, Engel P (2021) Mechanisms underlying gut microbiota-host interactions in insects. J Exp Biol 224. https://doi.org/10.1242/jeb.207696

Shen LR, Lai CQ, Feng X, Parnell LD, Wan JB, Wang JD, Li D, Ordovas JM, Kang JX (2010) Drosophila lacks c20 and c22 pufas. J Lipid Res 51:2985–2992. https://doi.org/10.1194/jlr.M008524

Shin SC, Kim SH, You H, Kim B, Kim AC, Lee KA, Yoon JH, Ryu JH, Lee WJ (2011) Drosophila microbiome modulates host developmental and metabolic homeostasis via insulin signaling. Science 334: 670–674. https://doi.org/10.1126/science.1212782

Shukla SP, Plata C, Reichelt M, Steiger S, Heckel DG, Kaltenpoth M, Vilcinskas A, Vogel H (2018) Microbiome-assisted carrion preservation aids larval development in a burying beetle. Proc Natl Acad Sci USA 115:11274–11279. https://doi.org/10.1073/pnas.1812808115

Siddiqi S, Sheth A, Patel F, Barnes M, Mansbach CM 2nd (2013) Intestinal caveolin-1 is important for dietary fatty acid absorption. Biochim Biophys Acta 1831: 1311–1321. https://doi.org/10.1016/j.bbalip.2013.05.001

Sieber MH, Thummel CS (2009) The dhr96 nuclear receptor controls triacylglycerol homeostasis in drosophila. Cell Metab 10:481–490

Sieber MH, Thummel CS (2012) Coordination of triacylglycerol and cholesterol homeostasis by dhr96 and the drosophila Lipa homolog magro. Cell Metab 15:122–127. https://doi.org/10.1016/j.cmet.2011.11.011

Silva W, Cardoso C, Ribeiro AF, Terra WR, Ferreira C (2013) Midgut proteins released by microapocrine secretion in *Spodoptera frugiperda*. J Insect Physiol 59:70–80. https://doi.org/10.1016/j.jinsphys.2012.10.015

Smith AF, Tsuchida K, Hanneman E, Suzuki TC, Wells MA (1992) Isolation, characterization, and cdna sequence of two fatty acid-binding proteins from the midgut of *Manduca sexta* larvae. J Biol Chem 267: 380–384

Somerville HJ, Pockett HV (1976) Phospholipase activity in gut juice of lepidopterous larvae. Insect Biochem. 6: 351–353

Song W, Veenstra JA, Perrimon N (2014) Control of lipid metabolism by tachykinin in drosophila. Cell Rep 9: 40–47. https://doi.org/10.1016/j.celrep.2014.08.060

Spates GE, Bull DL, Chen AC (1990) Hydrolysis of sphingomyelin and phosphatidylcholine by midgut homogenates of the stable fly. Arch Insect Biochem Physiol 14:1–12. https://doi.org/10.1002/arch.940140102

Spiteller D, Boland W (2003) N-(17-acyloxy-acyl)-glutamines: novel surfactants from oral secretions of lepidopteran larvae. J Org Chem 68:8743–8749. https://doi.org/10.1021/jo0342525

Spiteller D, Dettner K, Boland W (2000) Gut bacteria may be involved in interactions between plants, herbivores and their predators: microbial biosynthesis of n-acylglutamine surfactants as elicitors of plant volatiles. Biol Chem 381:755–762

Stahl A, Hirsch DJ, Gimeno RE, Punreddy S, Ge P, Watson N, Patel S, Kotler M, Raimondi A, Tartaglia LA et al (1999) Identification of the major intestinal fatty acid transport protein. Mol Cell 4:299–308. https://doi.org/10.1016/s1097-2765(00)80332-9

Stanley-Samuelson DW (1994) Prostaglandins and related eicosanoids in insects. Adv Insect Physiol 24:115–212

Stanley D (2006) The non-venom insect phospholipases a2. Biochim Biophys Acta 1761:1383–1390. https://doi.org/10.1016/j.bbalip.2006.05.011

Stanley DW, Sarath G, Rana RL (1998) A digestive phospholipase a(2) in midguts of tobacco hornworms, *Manduca sexta* l. J Insect Physiol 44:297–303. https://doi.org/10.1016/s0022-1910(97)00118-2

Stark AH, Crawford MA, Reifen R (2008) Update on alpha-linolenic acid. Nutr Rev 66:326–332. https://doi.org/10.1111/j.1753-4887.2008.00040.x

Storelli G, Defaye A, Erkosar B, Hols P, Royet J, Leulier F (2011) *Lactobacillus plantarum* promotes drosophila systemic growth by modulating hormonal signals through tor-dependent nutrient sensing. Cell Metab 14:403–414. https://doi.org/10.1016/j.cmet.2011.07.012

Storelli G, Strigini M, Grenier T, Bozonnet L, Schwarzer M, Daniel C, Matos R, Leulier F (2018) Drosophila perpetuates nutritional mutualism by promoting the fitness of its intestinal symbiont *Lactobacillus plantarum*. Cell Metab 27:362–377. https://doi.org/10.1016/j.cmet.2017.11.011

Svoboda JA (1999) Variability of metabolism and function of sterols in insects. Crit Rev Biochem Mol Biol 34: 49–57. https://doi.org/10.1080/10409239991209183

Tamaki FK, Pimentel AC, Dias AB, Cardoso C, Ribeiro AF, Ferreira C, Terra WR (2014) Physiology of digestion and the molecular characterization of the major digestive enzymes from *Periplaneta americana*. J Insect Physiol 70:22–35. https://doi.org/10.1016/j.jinsphys.2014.08.007

Teixeira A, Fialho Mdo C, Zanuncio JC, Ramalho Fde S, Serrao JE (2013) Degeneration and cell regeneration in the midgut of *Podisus nigrispinus* (heteroptera:

Pentatomidae) during post-embryonic development. Arthropod Struct Dev 42:237–246. https://doi.org/10.1016/j.asd.2013.02.004
Terra WR (2001) The origin and functions of the insect peritrophic membrane and peritrophic gel. Arch Insect Biochem Physiol 47:47–61. https://doi.org/10.1002/arch.1036
Terra WR, Costa RH, Ferreira C (2006) Plasma membranes from insect midgut cells. An Acad Bras Cienc 78:255–269. https://doi.org/10.1590/s0001-37652006000200007
Terra WR, Ferreira C (2020) Evolutionary trends of digestion and absorption in the major insect orders. Arthropod Struct Dev 56:100931. https://doi.org/10.1016/j.asd.2020.100931
Terra WR, Ferreira C, Garcia ES (1988) Origin, distribution, properties and functiosn of the major *Rhodnius prolixus* midgut hydrolases. Insect Biochem. 18:423–434
Tinker KA, Ottesen EA (2016) The core gut microbiome of the American cockroach, *Periplaneta americana*, is stable and resilient to dietary shifts. Appl Environ Microbiol 82:6603–6610. https://doi.org/10.1128/AEM.01837-16
Toprak U, Hegedus D, Dogan C, Guney G (2020) A journey into the world of insect lipid metabolism. Arch Insect Biochem Physiol 104:e21682. https://doi.org/10.1002/arch.21682
Treherne JE (1958) The digestion and absorption of tripalmitin in the cockroach, *periplanata americana* l. J Exp Biol 35:862–870
Trigatti BL, Anderson RG, Gerber GE (1999) Identification of caveolin-1 as a fatty acid binding protein. Biochem Biophys Res Commun 255:34–39. https://doi.org/10.1006/bbrc.1998.0123
Tsuchida K, Wells MA (1988) Digestion, absoprtion, transport and storage of fat during the last larval stadium of *Manduca sexta*. Changes in the role of lipophorin in the delivery of dietary lipid to the fat body. Insect Biochem. 18:263–268
Turunen S (1975) Absorption and transport of dietary lipid in *Pieris brassicae*. J Insect Physiol 21:1521–1529
Turunen S (1988a) Digestion and absorption of glycerophospholipid in *Pieris brassicae*. Comp Biochem Physiol 89A:19–24
Turunen S (1988b) Uptake of dietary lipids: a novel pathway in *Pieris brassicae*. Insect Biochem. 18:499–505
Turunen S (1990) Absorption of choline, myo-inositol, and oleic acid in th emidgut of *Pieris brassicae*: sectional differentiation and uptake into the haemolymph. J Insect Physiol 36:737–741
Turunen S (1993) Metabolic pathways in the midgut epithelium of *Pieris brassicae* during carbohydrate and lipid assimilation. Insect Biochem Molec Biol 23:681–689
Turunen S, Chippendale GM (1977) Lipid absorption and transport: sectional analysis of the larval midgut of the corn borer, *Diatraea grandiosella*. Insect Biochem. 7: 203–208
Turunen S, Chippendale GM (1989) Relationship between dietary lipids, midgut lipids, and lipid absorption in eight species of lepidoptera reared on artificial and natural diets. J Insect Physiol 35:627–633
Turunen S, Kastari T (1979) Digestion and absorption of lecithin in larvae of the cabbage butterfly, *Pieris brassicae*. Comp Biochem Physiol 62A:933–937
Uscian JM, Miller JS, Sarah G, Stanley-Samuelson DW (1995) A digestive phospholipase a2 in the tiger beetle *Cicindella circumpicta*. J Insect Physiol 41:135–141
Valzania L, Coon KL, Vogel KJ, Brown MR, Strand MR (2018) Hypoxia-induced transcription factor signaling is essential for larval growth of the mosquito *Aedes aegypti*. Proc Natl Acad Sci USA 115:457–465. https://doi.org/10.1073/pnas.1719063115
Vance JE (2015) Phospholipid synthesis and transport in mammalian cells. Traffic 16:1–18. https://doi.org/10.1111/tra.12230
Vasconcelos MA, Orsolin PC, Oliveira VC, Lima P, Naves MPC, de Morais CR, Nicolau-Junior N, Bonetti AM, Spano MA (2020) Modulating effect of vitamin d3 on the mutagenicity and carcinogenicity of doxorubicin in *Drosophila melanogaster* and in silico studies. Food Chem Toxicol 143:111549. https://doi.org/10.1016/j.fct.2020.111549
Vatanparast M, Ahmed S, Herrero S, Kim Y (2018) A non-venomous spla2 of a lepidopteran insect: its physiological functions in development and immunity. Dev Comp Immunol 89:83–92. https://doi.org/10.1016/j.dci.2018.08.008
Veenstra JA (2009) Peptidergic paracrine and endocrine cells in the midgut of the fruit fly maggot. Cell Tissue Res 336:309–323. https://doi.org/10.1007/s00441-009-0769-y
Veenstra JA, Agricola HJ, Sellami A (2008) Regulatory peptides in fruit fly midgut. Cell Tissue Res 334:499–516. https://doi.org/10.1007/s00441-008-0708-3
Verger R, Mieras MC, de Haas GH (1973) Action of phospholipase a at interfaces. J Biol Chem 248:4023–4034
Visotto LE, Oliveira MG, Guedes RN, Ribon AO, Good-God PI (2009) Contribution of gut bacteria to digestion and development of the velvetbean caterpillar, *Anticarsia gemmatalis*. J Insect Physiol 55:185–191. https://doi.org/10.1016/j.jinsphys.2008.10.017
Vogel H, Shukla SP, Engl T, Weiss B, Fischer R, Steiger S, Heckel DG, Kaltenpoth M, Vilcinskas A (2017a) The digestive and defensive basis of carcass utilization by the burying beetle and its microbiota. Nat Commun 8:15186. https://doi.org/10.1038/ncomms15186
Vogel KJ, Valzania L, Coon KL, Brown MR, Strand MR (2017b) Transcriptome sequencing reveals large-scale changes in axenic *Aedes aegypti* larvae. PLoS Negl Trop Dis 11:e0005273. https://doi.org/10.1371/journal.pntd.0005273
Voght SP, Fluegel ML, Andrews LA, Pallanck LJ (2007) Drosophila npc1b promotes an early step in sterol

absorption from the midgut epithelium. Cell Metab 5: 195–205. https://doi.org/10.1016/j.cmet.2007.01.011

Volonte D, Galbiati F, Li S, Nishiyama K, Okamoto T, Lisanti MP (1999) Flotillins/cavatellins are differentially expressed in cells and tissues and form a hetero-oligomeric complex with caveolins in vivo. Characterization and epitope-mapping of a novel flotillin-1 monoclonal antibody probe. J Biol Chem 274:12702–12709. https://doi.org/10.1074/jbc.274.18.12702

Vos M, Esposito G, Edirisinghe JN, Vilain S, Haddad DM, Slabbaert JR, Van Meensel S, Schaap O, De Strooper B, Meganathan R et al (2012) Vitamin k2 is a mitochondrial electron carrier that rescues pink1 deficiency. Science 336:1306–1310. https://doi.org/10.1126/science.1218632

Wada-Katsumata A, Zurek L, Nalyanya G, Roelofs WL, Zhang A, Schal C (2015) Gut bacteria mediate aggregation in the German cockroach. Proc Natl Acad Sci USA 112:15678–15683. https://doi.org/10.1073/pnas.1504031112

Wegener C, Veenstra JA (2015) Chemical identity, function and regulation of enteroendocrine peptides in insects. Curr Opin Insect Sci. 11:8–13. https://doi.org/10.1016/j.cois.2015.07.003

Weidlich S, Hoffmann KH, Woodring J (2015) Secretion of lipases in the digestive tract of the cricket *Gryllus bimaculatus*. Arch Insect Biochem Physiol 90:209–217. https://doi.org/10.1002/arch.21303

Weiher B, Komnick H (1997) Digestion of phosphatidylcholines, absorption, and esterification of lipolytic products by *Aeshna cyanea* larvae as studied in vivo and in vitro. Arch Insect Biochem Physiol 36: 273–293. https://doi.org/10.1002/(SICI)1520-6327(1997)36:4<273::AID-ARCH3>3.0.CO;2-T

Weintraub H, Tietz A (1973) Triglyceride digestion and absorption in the locust, *Locusta migratoria*. Biochim Biophys Acta 306:31–41. https://doi.org/10.1016/0005-2760(73)90205-1

Weintraub H, Tietz A (1978) Lipid absorption by isolated intestinal preparations. Insect Biochem 8:267–274

Wicker-Thomas C, Garrido D, Bontonou G, Napal L, Mazuras N, Denis B, Rubin T, Parvy JP, Montagne J (2015) Flexible origin of hydrocarbon/pheromone precursors in *Drosophila melanogaster*. J Lipid Res 56:2094–2101. https://doi.org/10.1194/jlr.M060368

Woodring J, Lorenz MW (2007) Feeding, nutrient flow, and functional gut morphology in the cricket *Gryllus bimaculatus*. J Morphol 268:815–825. https://doi.org/10.1002/jmor.10554

Yamanaka N, Rewitz KF, O'Connor MB (2013) Ecdysone control of developmental transitions: lessons from drosophila research. Annu Rev Entomol 58:497–516

Yoshinaga N, Aboshi T, Abe H, Nishida R, Alborn HT, Tumlinson JH, Mori N (2008) Active role of fatty acid amino acid conjugates in nitrogen metabolism in *Spodoptera litura* larvae. Proc Natl Acad Sci USA 105:18058–18063. https://doi.org/10.1073/pnas.0809623105

Yun HK, Jouni ZE, Wells MA (2002) Characterization of cholesterol transport from midgut to fat body in *Manduca sexta* larvae. Insect Biochem Mol Biol 32: 1151–1158. https://doi.org/10.1016/s0965-1748(02)00051-6

Zeng X, Hou SX (2015) Enteroendocrine cells are generated from stem cells through a distinct progenitor in the adult drosophila posterior midgut. Development 142:644–653. https://doi.org/10.1242/dev.113357

Zhang T, Yuan D, Xie J, Lei Y, Li J, Fang G, Tian L, Liu J, Cui Y, Zhang M et al (2019) Evolution of the cholesterol biosynthesis pathway in animals. Mol Biol Evol 36:2548. https://doi.org/10.1093/molbev/msz167

Zhao X, Karpac J (2020) The drosophila midgut and the systemic coordination of lipid-dependent energy homeostasis. Curr Opin Insect Sci 41:100–105. https://doi.org/10.1016/j.cois.2020.07.003

Zheng H, Powell JE, Steele MI, Dietrich C, Moran NA (2017) Honeybee gut microbiota promotes host weight gain via bacterial metabolism and hormonal signaling. Proc Natl Acad Sci USA 114:4775–4780. https://doi.org/10.1073/pnas.1701819114

Zheng JC, Sun SL, Yue XR, Liu TX, Jing X (2018) Phylogeny and evolution of the cholesterol transporter npc1 in insects. J Insect Physiol 107:157–166. https://doi.org/10.1016/j.jinsphys.2018.04.007

Zinke I, Schutz CS, Katzenberger JD, Bauer M, Pankratz MJ (2002) Nutrient control of gene expression in drosophila: microarray analysis of starvation and sugar-dependent response. EMBO J 21:6162–6173. https://doi.org/10.1093/emboj/cdf600

Zwick RK, Ohlstein B, Klein OD (2019) Intestinal renewal across the animal kingdom: comparing stem cell activity in mouse and drosophila. Am J Physiol Gastrointest Liver Physiol 316:G313–G322. https://doi.org/10.1152/ajpgi.00353.2018

Adv Exp Med Biol (2026) 1494: 97–114
https://doi.org/10.1007/5584_2024_806

Published online: 15 June 2024

Lipophorin: The Lipid Shuttle

Katia C. Gondim and David Majerowicz

Abstract

Insects need to transport lipids through the aqueous medium of the hemolymph to the organs in demand, after they are absorbed by the intestine or mobilized from the lipid-producing organs. Lipophorin is a lipoprotein present in insect hemolymph, and is responsible for this function. A single gene encodes an apolipoprotein that is cleaved to generate apolipophorin I and II. These are the essential protein constituents of lipophorin. In some physiological conditions, a third apolipoprotein of different origin may be present. In most insects, lipophorin transports mainly diacylglycerol and hydrocarbons, in addition to phospholipids. The fat body synthesizes and secretes lipophorin into the hemolymph, and several signals, such as nutritional, endocrine, or external agents, can regulate this process. However, the main characteristic of lipophorin is the fact that it acts as a reusable shuttle, distributing lipids between organs without being endocytosed or degraded in this process. Lipophorin interacts with tissues through specific receptors of the LDL receptor superfamily, although more recent results have shown that other proteins may also be involved. In this chapter, we describe the lipophorin structure in terms of proteins and lipids, in addition to reviewing what is known about lipoprotein synthesis and regulation. In addition, we reviewed the results investigating lipophorin's function in the movement of lipids between organs and the function of lipophorin receptors in this process.

Keywords

Apolipophorins · Lipid transport · Lipids · Lipophorin · Lipophorin receptor · Lipoprotein composition

1 Introduction

Lipids are essential as membrane structural molecules and as energy sources. Insects can obtain them from the diet or by de novo synthesis, although some are essential, such as sterols (Clark and Block 1959). The midgut absorbs lipids from the diet and must secrete them to be delivered to the fat body, where lipids are stored, and other sites (Zhao and Karpac 2020). De novo lipid synthesis occurs mainly in the fat body itself but

The Authors Katia C. Gondim and David Majerowicz have equally contributed to this chapter

K. C. Gondim (✉)
Instituto de Bioquímica Médica Leopoldo de Meis, Universidade Federal do Rio de Janeiro, Rio de Janeiro, Brazil
e-mail: katia@bioqmed.ufrj.br

D. Majerowicz (✉)
Departamento de Biotecnologia Farmacêutica, Faculdade de Farmácia, Universidade Federal do Rio de Janeiro, Rio de Janeiro, Brazil
e-mail: majerowicz@pharma.ufrj.br

also in the integument oenocytes and in the midgut cells. Furthermore, in energy-demanding actions such as flight or oogenesis, the fat body degrades its triacylglycerol (TAG) stores, and releases lipids which also need to be delivered to the muscles and ovaries (Gondim et al. 2018; Toprak et al. 2020).

However, a problem appeared with the coelom and the circulatory system in the evolutionary history of metazoans: how to transport lipids, hydrophobic molecules, between increasingly specialized tissues and organs through an aqueous medium? Lipoproteins emerged as a solution to this challenge. In mammals, at least three different lipoproteins, very low-density lipoprotein (VLDL), high-density lipoprotein (HDL), and chylomicrons are essential for the transport of lipids through the blood to different organs (Ying et al. 2021). First, enterocytes produce chylomicrons, which transport lipids absorbed from the intestine to peripheral tissues, and their remnants end up being endocytosed by the liver (Xiao et al. 2019). Next, the liver synthesizes VLDL, which maintains the supply of lipids to peripheral organs between meals. Then, the liver itself endocytoses and recycles the remnants of VLDL, low-density lipoprotein (LDL) (Tiwari and Siddiqi 2012). Finally, the liver and intestine produce apolipoprotein A-1, which receives cholesterol from peripheral tissues and macrophages, forming HDL. This lipoprotein makes the reverse cholesterol transport, carrying this lipid back to the liver (Ben-Aicha et al. 2020).

Insects developed a more straightforward and elegant solution: a single lipoprotein, called lipophorin (Lp), transports all the lipids between organs through the hemolymph, functioning as a reusable lipoprotein (Chino et al. 1981a; Van der Horst et al. 2002). Similarities with mammalian VLDL metabolism indicate a common evolutionary origin. However, there are notable differences as well.

This chapter describes Lp protein structure and lipid composition, and the mechanisms and regulations of its synthesis by the fat body. We also discuss the functioning of lipid transport through the hemolymph and Lp receptors and point out the gaps in our understanding and the field's future directions.

2 Protein Structure and Lipid Content

In all insects studied so far, Lp consists of two apolipoproteins, ApoLp-I and ApoLp-II (Robbs et al. 1985; Ochanda et al. 1991, 1992; Capurro et al. 1994; Trowell et al. 1994; Ford and Van Heusden 1994; Pho et al. 1996; Sevala et al. 1999; Schal et al. 2001; Pennington and Wells 2002; Lucas et al. 2004; Atella et al. 2006; Matsuoka et al. 2006). The apparent molecular weight of Lp ranges from 640 to 530 kDa (Robbs et al. 1985; Ochanda et al. 1991, 1992; Pho et al. 1996; Atella et al. 2006). However, when separated by denaturing electrophoresis, ApoLp-I has a molecular weight between 250 and 212 kDa, and ApoLp-II between 86 and 73 kDa (Robbs et al. 1985; Gondim et al. 1989a; Ochanda et al. 1991, 1992; Capurro et al. 1994; Ford and Van Heusden 1994; Trowell et al. 1994; Pho et al. 1996; Sevala et al. 1999; Schal et al. 2001; Pennington and Wells 2002; Lucas et al. 2004).

A third apolipoprotein with molecular weight around 20 kDa (Kawooya et al. 1984; Gondim et al. 1989a; Kanost et al. 1995), called apolipophorin-III (ApoLp-III), may be present in Lp, depending on the insect species studied and the physiological situation. The association of Lp and ApoLp-III is correlated with the quantity of diacylglycerol (DAG) in the particle (Nagao and Chino 1991; Soulages and Wells 1994a). ApoLp-III has a greater affinity for DAG than phospholipids (PL) (Demel et al. 1992). In the cricket *Gryllus bimaculatus*, ApoLp-III knockdown by RNA interference did not change the hemolymphatic lipid levels in animals with free access to food, but it prevented the increase in lipids the injection of adipokinetic hormone (AKH) would cause (Konuma et al. 2016). Thus, ApoLp-III seems essential in times of high lipid mobilization rates. The infection of the moth *Galleria mellonella* larvae by different pathogens altered the levels of ApoLp-III in the hemolymph, both in the free and Lp-associated forms (Stączek et al. 2018), and this result may indicate that the infection stimulates the mobilization of lipid reserves from the fat body, leading to a reduction in Lp density.

Other lipoproteins appear to be unique to some of the studied insects. For example, the mealybug *Dactylopius confusus* has two lipoproteins with 25 and 22 kDa molecular weight. They are structurally similar to ApoLp-III and carry mostly DAG (Ziegler et al. 1999). The fly *Musca domestica*, on the other hand, has a larval-specific lipoprotein that contains four apoproteins with molecular weights between 26 and 20 kDa, density of 1.134 g/mL, and carrying mainly hydrocarbons (HC) and PL (de Bianchi and Capurro 1991).

The primary structure of Lp indicates the existence of a lipid-binding domain, as shown for the silkworm *Bombyx mori*. Interestingly, azadirachtin, an insecticidal compound extracted from the *Azadirachta indica* tree, competes with PLs for binding to Lp (Thangaraj et al. 2018). This phenomenon may help explain the effects of azadirachtin on insects, as it inhibits PL transfer from Lp to the ovaries, affecting reproduction (Moreira et al. 1994).

Lp lipid composition shows some variation among insects. DAG and HC are commonly the main neutral lipids; PLs are also present in considerable amounts (Chino and Kitazawa 1981; Robbs et al. 1985; Prasad et al. 1986; Sevala et al. 1999; Young et al. 1999; Schal et al. 2001; Pennington and Wells 2002; Fan et al. 2004; Palm et al. 2012; Majerowicz et al. 2013). The exception is in the infraorder Culicomorpha, which includes mosquitoes and blackflies, in which Lp mainly transports TAG instead of DAG (Ford and Van Heusden 1994; Pennington and Wells 2002). Lp from the mosquito *Nephrotoma abbreviata* of the infraorder Tipulomorpha carries equal amounts of TAG and DAG (Pennington and Wells 2002). Lp also transports other lipids in smaller amounts, essential for biological processes, as cholesterol and free fatty acids (Chino and Kitazawa 1981; de Bianchi and Capurro 1991; Sevala et al. 1999; Majerowicz et al. 2013). In the moths *B. mori*, *Ascotis selenaria*, and *Lymantria dispar*, Lp transports trienes that are precursors of the pheromones of these species (Jurenka et al. 2003; Matsuoka et al. 2006). Lp binds to juvenile hormone (JH), an acyclic sesquiterpenoid with hydrophobic characteristics, in several insects (Trowell et al. 1994; Campbell et al. 1998; Engelmann and Mala 2000, 2005). Interestingly, this association between Lp and JH protects the hormone from degradation by esterases (Campbell et al. 1998). In this way, Lp can regulate the signaling levels of the JH pathway. *B. mori* Lp also plays a role in carotenoid transport (Tsuchida et al. 1998).

Lp density is a direct consequence of the amount of lipids present in the particle, and it can vary between insects and physiological conditions. In general terms, Lp density varies between 1.135 and 1.103 g/mL (Shapiro and Law 1983; Robbs et al. 1985; Prasad et al. 1986; Chino et al. 1986; Kawooya et al. 1988; Gondim et al. 1989a; Ochanda et al. 1991, 1992; Capurro et al. 1994; Trowell et al. 1994; Ford and Van Heusden 1994; Pho et al. 1996; Van Heusden et al. 1997; Sevala et al. 1999; Pennington and Wells 2002; Lucas et al. 2004; Atella et al. 2006) and can be classified as very high-density Lp (VHDLp), high-density Lp (HDLp), or low-density Lp (LDLp) (Beenakkers et al. 1988). The conversion from HDLp to LDLp doubles the particle size, increasing the amount of DAG (Chino et al. 1986; Nagao and Chino 1991; Soulages et al. 1996); in the moth *Acherontia atropos*, LDLp also carries more HC (Surholt et al. 1992). In the beetle *Callosobruchus maculatus*, the larval Lp is less dense, larger, and more lipid-laden compared to the Lp of adult insects (Ximenes et al. 2008). In *Anopheles gambiae*, the larval Lp is denser than in sugar-fed adult mosquitoes, but feeding females on blood also increases Lp density (Marinotti et al. 2006). The same pattern occurs in another mosquito, *Aedes aegypti* (Van Heusden et al. 1997). In the kissing-bug *Rhodnius prolixus*, Lp density also varies during the feeding cycle; it decreases after blood feeding and increases during ovogenesis, but always in the HDLp range (Coelho et al. 1997). Interestingly, the blood meal has opposite effects on Lp density in kissing-bugs and mosquitoes, although the reasons are unclear. In the moth *Manduca sexta*, Lp density is lower in molting larvae, and ApoLp-III associates with it (Ziegler et al. 1995). In wandering larvae and pupae, the density of Lp is increased, with a reduction in the amount of HC and an increase in carotenes (Prasad et al. 1986; Engler et al. 1996). Furthermore, larvae reduce Lp density when fasting, changing the distribution of Lp populations, and hemolymph lipids increase in fasted animals when compared to fed controls. These results indicate a mobilization of lipid reserves. Interestingly, these effects are independent of brain/head factors, as abdominally ligated fasting larvae also showed a decrease in Lp density (Gondim et al. 2013).

Flight has a high energy demand and, in *R. prolixus*, this activity increased the density of Lp and reduced its molecular weight, but without changes in the interaction with ApoLp-III (Oliveira et al. 2006). On the other hand, in the kissing-bug *Panstrongylus megistus*, flight induced the formation of a larger LDLp (Canavoso et al. 2003). The differences observed in these two evolutionarily close models are unclear, but they could stand in the used protocols for flight induction. In the locust *Locusta migratoria*, flight increases the amount of ApoLp-III in Lp (Surholt et al. 1992). The *corpora cardiaca* releases AKH during flight activity, and this hormone induces lipolysis of TAG reserves in the fat body, increasing the amount of lipids available for transport by Lp (Van der Horst and Rodenburg 2010). AKH injection reduces Lp density in several models (Shapiro and Law 1983; Chino et al. 1986; Canavoso and Rubiolo 1994; Kanost et al. 1995; Engler et al. 1996), which is accompanied by the association of ApoLp-III with the Lp particle (Chino et al. 1989; Surholt et al. 1992; Ziegler et al. 1995).

In *G. mellonella* larvae, the injection of dead bacteria induced the formation of LDLp (Dettloff et al. 2001), indicating that stimulation of the immune system generates an energy demand. Furthermore, some studies indicate that Lp is directly involved with immune processes, as described later in this chapter.

The structure of Lp varies little among the studied insects. In general, Lp is spherical and has on its surface PL, cholesterol, part of the DAG, and ApoLp-I and ApoLp-III, when present (Katagiri et al. 1987, 1991; Kawooya et al. 1991; Soulages and Brenner 1991; Rimoldi et al. 1996). ApoLp-II dips into the particle and is present on the surface and in the middle layer, composed mainly by DAG and TAG (Katagiri et al. 1987, 1991; Kawooya et al. 1991; Soulages and Brenner 1991; Soulages et al. 1994; Wang et al. 1995; Rimoldi et al. 1996). HC makes up the interior core of the particle (Katagiri et al. 1987, 1991; Soulages and Brenner 1991) (Fig. 1). LDLp is heterogeneous in size, whereas HDLp is more homogeneous (Chino et al. 1986; Nagao and Chino 1991). Structural studies with Lp showed that this lipoprotein is very resistant to changes in the physical environment. Lp from *R. prolixus* is remarkably resistant to high temperatures, and to hydrostatic pressure as well (Maya-Monteiro et al. 1997; Majerowicz et al. 2013).

Although they are present on the surface of the particles and are considered crucial to the interaction with the aqueous medium, PLs are not essential for the structural stability of Lp (Kawooya et al. 1991; Rimoldi et al. 1996; Maya-Monteiro et al. 1997), or for its solubility or association with ApoLp-III (Gondim et al. 1992). It is important to note that in the case of *R. prolixus*, Lp carries a significant amount of PL (Gondim et al. 1989a; Coelho et al. 1997; Majerowicz et al. 2013), and PL depletion increases Lp density (Gondim et al. 1992). The removal of DAG from Lp using lipases did not cause ApoLp-III dissociation (Hiraoka and Katagiri 1992). However, the lipids present in Lp particle protect the apolipoproteins from proteolysis (Ryan et al. 1992) and, when both PL and DAG were removed, Lp precipitated in an aqueous medium (Kawooya et al. 1991). Even when depleted of PL or DAG, Lp maintained its capacity to receive lipids from the fat body; this maintenance of Lp structure is essential for its functioning as a reusable lipoprotein (Kawooya et al. 1991; van Heusden et al. 1991; Gondim et al. 1992).

In addition to its action on lipid transport, several research groups have shown that Lp interacts with other hemolymphatic proteins and participates in other physiological roles. In the

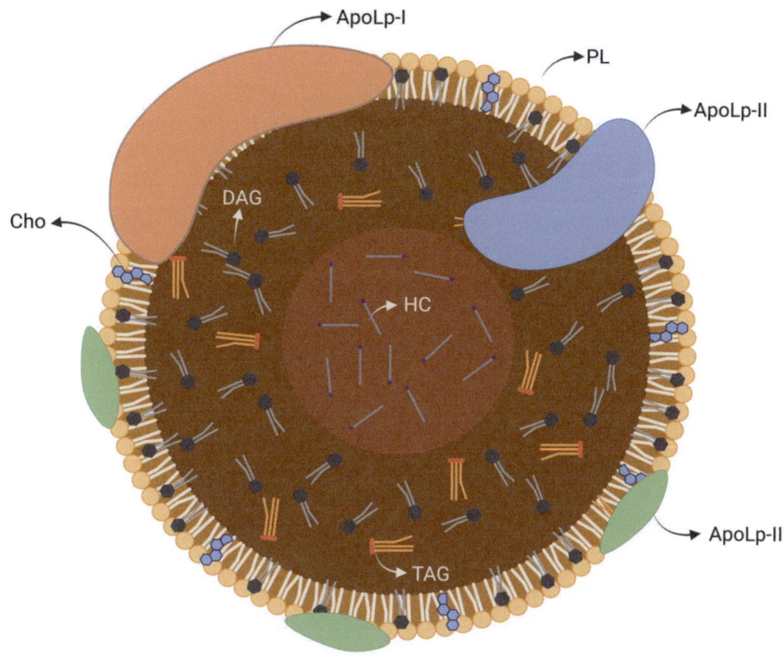

Fig. 1 Schematic representation of lipophorin structure. One molecule each of ApoLp-I and ApoLp-II is always present, and ApoLp-III may be present or not, and in variable number, according to the insect species and physiological condition. Lipids found: *Cho* cholesterol, *DAG* diacylglycerol, *HC* hydrocarbon, *PL* phospholipid, *TAG* triacylglycerol

kissing bug *Dipetalogaster maxima*, Lp acts as a cathepsin D carrier for developing oocytes, where it functions as a yolk protein (Leyria et al. 2018). In the fruit fly *Drosophila melanogaster*, Lp acts as a carrier of signaling proteins from the Hedgehog family, responsible for controlling development and homeostasis (Palm et al. 2013). Lp even crosses the blood-brain barrier and can deliver these signaling proteins to the brain (Brankatschk and Eaton 2010). Lp also binds to hexamerin (Magee et al. 1994) and ApoLp-III to arylphorin and to another hexamerin (Stączek et al. 2018), although the effects of these interactions are unknown. Hexamerins are a class of hexameric storage proteins, present in the hemolymph, which are stored in the fat body before pupation to be used during metamorphosis in holometabolous insects. Some hexamerins contain a high content of aromatic amino acids, and are called arylphorins (Haunerland 1996; Burmester 2002). Lp is also involved in insect resistance to external aggression. For example, Lp from the moths *Helicoverpa armigera* and *Ephestia kuehniella* binds to the Cry1Ac protein toxin from the bacterium *Bacillus thuringeiensis* in in vitro (Ma et al. 2012) and in vivo (Rahman et al. 2007) assays, respectively. The Cry1Ac toxin is one of the modifications introduced in transgenic plants of agricultural interest to prevent them from being attacked by pest insects and, therefore, Lp can be an essential factor for the emergence of resistant insect populations. In *A. gambiae*, ApoLp-I is present in clot formation (Agianian et al. 2007), and Lp of wasps from the genus *Venturia* and *D. melanogaster* binds to hemomucin, which is vital for coagulation (Schmidt and Theopold 1997; Kinuthia et al. 1999). In *E. kuehniella*, Lp associates with pattern-recognition protein, prophenoloxidase-regulatory protein, and lectins (Rahman et al. 2006). Lp also binds and forms aggregates with lipopolysaccharides (LPS) of gram-negative bacteria, reducing the toxicity of these compounds (Kato et al. 1994; Ma et al. 2006). Lp can also form dimers with the CrV1 protein of polydnaviruses present in the parasitoid wasp *Cotesia rubecula*, facilitating the uptake of this protein by the host hemocytes (Schmidt et al. 2005). These results indicate that Lp can participate in wound healing processes, encapsulation of invading foreign bodies in hemolymph, and pathogen recognition and combat.

3 Lipophorin Synthesis

The fat body is the organ responsible for the synthesis of Lp (Gellissen and Wyatt 1981; Glinka and Triseleva 1989; Weers et al. 1992; Trowell et al. 1994; Van Heusden et al. 1998; Sun et al. 2000; Marinotti et al. 2006; Benoit et al. 2011) and ApoLp-III (Halwani et al. 2001). Although two apoproteins are present in the Lp particle, a single gene encodes them (Weers et al. 1993; Van Heusden et al. 1998; Marinotti et al. 2006). A furin-like protease cleaves the propeptide, thus forming ApoLp-I and ApoLp-II. However, apolipophorin II/I cleavage seems to be not essential for particle assembly as inhibition of furin activity did not prevent Lp secretion as an HDLp (Smolenaars et al. 2005). The secondary structure of Lp contains alpha-helix and amphipathic beta-sheets, and the formation of HDLp depends on amphipathic beta-sheets (Smolenaars et al. 2007). The Lp assembly and secretion processes depend on microsomal triglyceride transfer protein (MTP), a protein also from the lipoprotein family (Palm et al. 2012). In fact, in in vitro experiments, MTP expression increased Lp production by the fat body and reduced lipoprotein density (Smolenaars et al. 2007). The fat body secretes a PL-rich HDLp (Weers et al. 1992, 1993; Palm et al. 2012).

The regulation of Lp secretion is still far from being fully understood, and we only know a few pieces of the puzzle. For example, in *A. aegypti*, microRNA-8 is necessary for Lp secretion by the fat body (Lucas et al. 2015), and in *D. melanogaster*, the overexpression of beta-spectrin, a cell cytoskeleton protein, prevents secretion of Lp (Diaconeasa et al. 2013).

Interestingly, Lp is present in the saliva of several insects of the Order Hemiptera (Nicholson et al. 2012; Nicholson and Puterka 2014; Liu et al. 2016; Huang et al. 2018). It is unclear whether salivary gland cells produce Lp or whether it is secreted by the fat body to the hemolymph and somehow reaches the salivary gland lumen. Which functions Lp would have in the saliva of these phytophagous insects is also a mystery. In the bee *Apis mellifera*, the reproductive system of drones expresses Lp, and this expression decreases as the insect ages (Colonello-Frattini et al. 2010). However, we do not know whether this Lp expression has any role in the reproduction of bees. In *L. migratoria*, glial cells in the brain also express Lp (Bogerd et al. 2000), whereas in *D. melanogaster*, embryos express Lp during their development (Kutty et al. 1996). Finally, the gut lumen of *E. kuehniella* contains Lp (Rahman et al. 2007), although the origin of the lipoprotein is not known.

Different physiological signals regulate the transcription of the *Lp* gene in the fat body, including nutritional ones. In several hematophagous insects, blood-feeding induces Lp expression at both the mRNA and protein level (Capurro et al. 1994; Van Heusden et al. 1997, 1998; Sun et al. 2000; Marinotti et al. 2006; Ramos et al. 2021). Interestingly, in *A. aegypti*, this regulation is independent of the amount of lipids present in the blood (Van Heusden et al. 1997), indicating that other factors mediate the regulation of Lp expression. In *A. mellifera*, the diet does not affect the expression of ApoLp-I or ApoLp-III (Lourenço et al. 2012) but, in drones, Lp expression decreases with age (Colonello-Frattini et al. 2010). In *M. sexta*, the amount of Lp in the hemolymph increases during larval development (Beckage and Kanost 1993; Gondim et al. 2013). Interestingly, the amount of Lp in the hemolymph correlates with lipase activity in the fat body (Arrese et al. 1996). In the fly *Glossina morsitans*, Lp expression increases with larval development, reaching a peak just before parturition. In adult flies, Lp expression is constitutive (Benoit et al. 2011). In the cockroach *Leucophaea maderae*, Lp in the hemolymph increases in the middle of the vitellogenic period (Engelmann and Mala 2005). In another cockroach (*Blattella germanica*), the Lp titer in the hemolymph increases during the development of last-stage nymphs (Young et al. 1999). In the leafhopper *Nilaparvata lugens*, fatty acid desaturase knockdown by RNA interference reduced Lp expression (Ye et al. 2020). However, it is unclear whether this effect was direct on the Lp gene or indirect, caused by a change in the fatty acid saturation profile or the amount of fatty acids or TAG.

Studies with different models showed that endocrine signals also regulate Lp expression. In *D. maxima*, JH signaling mediates at least part of the effect of feeding on Lp expression (Ramos et al. 2021). Similarly, in *L. maderae* and *L. migratoria*, treatment with JH increased Lp concentration in the hemolymph (Gellissen and Wyatt 1981; Engelmann and Mala 2000). On the other hand, in the stink bug *Dysdercus cingulatus*, treatment with methoprene, a JH analog, reduced Lp expression (Mohan and Muraleedharan 2005). The promoter of the Lp gene from *A. gambiae* has response elements to ecdysone (Marinotti et al. 2006), and treatment of *D. cingulatus* and *A. aegypti* with ecdysone induced the expression of this gene (Sun et al. 2000; Mohan and Muraleedharan 2005). Other but unknown endocrine signals regulate Lp synthesis. In *M. sexta*, the injection of trehalose, the major circulating sugar in the hemolymph, increased the amount of Lp in the hemolymph. On the other hand, decapitation of larvae reduced the amount of Lp. Nevertheless, the injection of trehalose into decapitated larvae did not change the amount of Lp. These results show that carbohydrate levels in the hemolymph regulate Lp expression indirectly, through a compound produced by the head but still unknown (Arrese et al. 1996).

External agents can also alter Lp expression in some insects. In the leafhopper *Laodelphax striatellus*, treatment with the insecticide buprofezin, which acts by inhibiting chitin synthesis, reduced Lp expression (Fang et al. 2020), although the mechanisms involved are not precise. Infection with pathogens or parasitism also regulates Lp expression. *Pieris rapae* larvae, when parasitized by *C. rubecula*, showed a reduction in Lp expression (Asgari and Schmidt 2004). In contrast, parasitization by the wasp *Cotesia congregata* did not change the amount of Lp in *M. sexta* larvae (Beckage and Kanost 1993). In *A. mellifera*, infection with the bacteria *Serratia marcescens* slightly reduced the amount of ApoLp-I in the hemolymph (Lourenço et al. 2012). In mosquitoes, infection with several types of pathogens caused an increase in Lp expression (Cheon et al. 2006; Araujo et al. 2011; Kumar and Paily 2011). Specifically in *A. aegypti*, the Toll pathway was responsible for this induction (Cheon et al. 2006).

The apoLp-III expression can also vary. In *G. mellonella*, ApoLp-III increases during the larval feeding stage (Halwani et al. 2001). However, injection of LPS did not induce ApoLp-III expression (Niere et al. 1999), although ApoLp-III appears to play a role in the immune response. In *B. mori*, the embryo expresses ApoLp-III, which associates with LDLp (Tsuchida et al. 2010).

4 Lipid Transport

The insect lipid transport system has the particular feature of having a single lipoprotein, Lp, that transports the various lipid classes in the hemolymph, between all the insect organs, under the most diverse physiological conditions. Although it is a very intriguing system, the details about how lipids are incorporated by Lp and delivered to the organs, and how these events are regulated, remain mostly unknown.

Lp was initially identified as a DAG-carrying lipoprotein in insect hemolymph, loaded at the fat body, and only later it was shown to also receive lipids from the midgut (Chino et al. 1969, 1981b). Because it was soon recognized that Lp had a central role in the supply of DAG to the flight muscles, this process was studied in detail in insects that use lipids to fuel flight under the control of the AKH, as *L. migratoria* and *M. sexta*, and which have DAG as a main component of Lp lipids (Mwangi and Goldsworthy 1981; Wells et al. 1987; Van der Horst and Rodenburg 2010). However, it was clear soon that other lipids that were present in Lp particle were also transported and delivered at diverse sites by the lipoprotein (Fig. 2).

As insects cannot synthesize cholesterol de novo from acetate (Li and Jing 2020), sterol acquisition from the diet is critical and, in fact, it was first shown in *Philosamia cynthia* larvae that the midgut absorbed and released cholesterol to

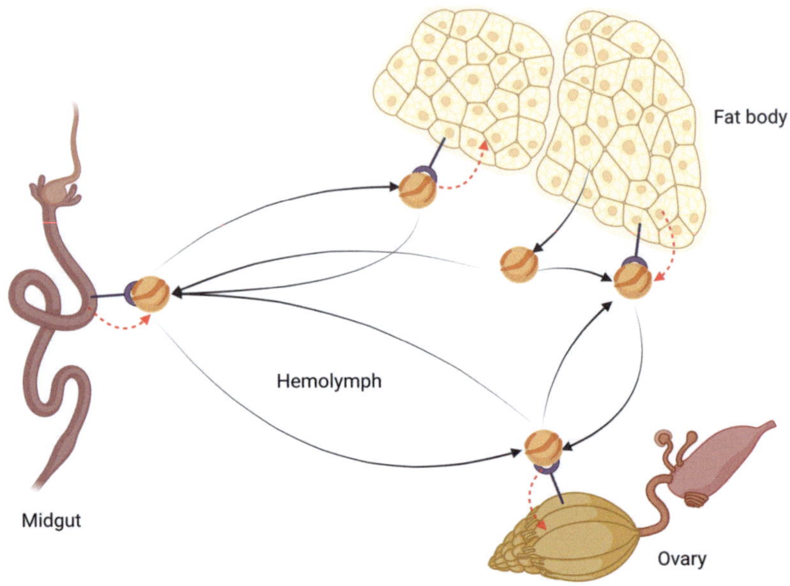

Fig. 2 Lipid transport in insects. Lipophorin is synthesized by the fat body, and secreted to the hemolymph. It acquires lipids at the midgut and the fat body, and specifically unloads its cargo at multiple sites, as the ovaries and the fat body, where they are stored, and others

circulating Lp (Chino and Gilbert 1971). It was later confirmed in other insect species that Lp takes up cholesterol from the midgut, mostly in the free form, and transports it to the other organs, as the fat body, where it seems to be transiently stored, and the ovaries (Jouni et al. 2002a, b, 2003; Entringer et al. 2021).

Hydrocarbons are very abundant in Lp from many species (Chino and Kitazawa 1981; Sevala et al. 1999; Fan et al. 2004; Majerowicz et al. 2013). These lipids are de novo synthesized by the oenocytes, transferred to Lp present in the hemolymph and unloaded at the cuticle for deposition, where they are important for water proofing, and at internal organs, as well (Katase and Chino 1982; Juárez and Brenner 1989; Gu et al. 1995). In *B. germanica*, Lp transfers hydrocarbons to the ovaries during vitellogenesis, which will be deposited into an egg case during oviposition (Fan et al. 2002).

Although the presence of phospholipids in Lp particle has been early acknowledged, their role in the lipid transport system was not so clear. It was shown that phospholipids were released from the fat body of *Hyalophora cecropia* to lipoproteins in the hemolymph (Thomas and Gilbert 1967). They are located at the lipoprotein surface, and it was initially proposed that they were not transferred between Lp and the organs and were only a component of Lp particle (Katagiri 1985). It was then demonstrated that in *R. prolixus* phospholipids are transferred to the vitellogenic ovaries by Lp (Gondim et al. 1989b), and that following the blood meal, it is loaded at the midgut and the fat body (Atella et al. 1992, 1995; Coelho et al. 1997). Moreover, after enzymatic removal of phospholipids from *R. prolixus* Lp, it remained structurally stable and retained the ability to interact with and receive lipids from the fat body, reinforcing phospholipids as a Lp cargo, similar to other transported lipids (Gondim et al. 1992; Maya-Monteiro et al. 1997).

Lps from many insects have a rich yellow color because of the abundance of carotenoids, which are acquired from the diet and absorbed by the midgut (Chino et al. 1969). In *B. mori* larvae, attention has been given to the transport of these lipids, and it was shown that the carotenoids are taken up from the midgut by Lp, that delivers them at the silk gland, where they are accumulated into the cocoon (Jouni and Wells 1996; Tsuchida et al. 2004).

Insects do not have a specific albumin-like protein to transport free fatty acids in the hemolymph, and they are also carried by Lp. The role of Lp in the transport of fatty acids was demonstrated in a few insects, and in all of them the determined circulating fatty acid half-life was very short, around 2–5 min (Soulages et al. 1988; Soulages and Wells 1994b; Atella et al. 2000).

Soon after the identification of Lp as the insect lipid carrier in the hemolymph, it was shown in *L. migratoria* that the DAG content of Lp had a much higher turnover rate than its protein moiety, indicating that this lipoprotein could transfer lipids between the insect organs without degradation of its apoproteins, acting as a lipid shuttle (Downer and Chino 1985). This was confirmed by in vitro assays that demonstrated that, upon stimulation, Lp received DAG from the fat body, ApoLp-III associated with the particle which acquired a lower density, and then these lipids were delivered to the flight muscles, ApoLp-III was released in the medium, and then the Lp particle could be loaded again at the fat body (Van Heusden et al. 1987). In the same way, in *R. prolixus* lipids are transported by Lp to the vitellogenic ovaries, but apoproteins are not incorporated (Gondim et al. 1989b; Machado et al. 1996; Santos et al. 2011; Entringer et al. 2013). In accordance with the shuttle hypothesis, in this insect, circulating Lp can acquire its lipid cargo at the fat body, the integument, and the midgut (Atella et al. 1992, 1995; Coelho et al. 1997). Similarly, in *D. melanogaster*, neutral lipid uptake by the oocytes does not require Lp internalization (Parra-Peralbo and Culi 2011).

In *M. sexta*, LDLp present in the hemolymph transfers DAG to the ovaries during oogenesis, and its protein component is not accumulated (Kawooya and Law 1988). However, HDLp which is also present in the hemolymph, is endocytosed by the oocytes, stripped of its lipids being converted to VHDLp, also contributing to the egg lipid content (Kawooya et al. 1988). Although LDLp is the major lipid supplier to the ovary, HDLp uptake is clearly relevant as well, and these results showed that both mechanisms could simultaneously exist. In line with these observations, in the triatomines *P. megistus* and *D. maxima*, Lp supplies lipids for the ovaries during vitellogenesis by both mechanisms. Lp endocytosis by the oocytes occurs, but lipids are mainly transferred by Lp by a non-endocytic pathway (Fruttero et al. 2011; Leyria et al. 2014; Ramos et al. 2021). Lp was also found in the yolk granules of ovarian follicles of the mosquitoes *A. aegypti* and *A. gambiae* (Sun et al. 2000; Atella and Shahabuddin 2002), but it was not investigated whether a non-endocytic lipid delivery to the ovaries also occurs in these insects. Likewise, in *L. migratoria,* the fat bodies of young adults have the capacity to endocytose Lp (Dantuma et al. 1997). Thus, although there is only one lipoprotein in charge of lipid transport in insects, the mechanisms responsible for lipid transfer between Lp and organs may present considerable variation, and are still poorly understood.

5 Lipophorin Receptor

In order to receive or deliver lipids, Lp needs to properly interact with tissues, under diverse conditions. Biochemical properties of Lp interaction with binding sites at cell membrane were studied in membrane preparations obtained from organs of some insect species, as the fat body, midgut and ovary of *R. prolixus*, the midgut of *P. megistus*, the fat body of *L. migratoria*, the fat body and midgut of *M. sexta* (Tsuchida and Wells 1990; Dantuma et al. 1996; Gondim and Wells 2000; Pontes et al. 2002; Grillo et al. 2003; Fruttero et al. 2009; Entringer et al. 2013). Lp was shown to bind to cell membranes specifically and with high affinity, as expected for specific receptors. This interaction was affected by pH and it was maximal around pH 6.0–7.0, although some variation occurred depending on the insect tissue or species, and it did not depend on Ca^{2+}, with the exception of the *M. sexta* fat body. Lp binding to cell surface is modulated by the insect's physiological condition, and this can be clearly observed in hematophagous insects, which have reproduction and development triggered and controlled by the blood meal. The Lp binding capacity in the midgut of *R. prolixus* peaked 2 days following a blood meal, period of the most efficient transfer of lipids by the organ to Lp (Coelho et al. 1997; Grillo et al. 2003), indicating that the abundance of receptors at the cell surface may control lipid flux. Similarly, in *P. megistus*, the midgut of unfed insects showed lower Lp binding levels than that of fed insects (Fruttero et al. 2009). The diet quality may also interfere with Lp binding to the organs. In *R. prolixus*, when compared to the insects fed whole blood, the binding of fluorescently labeled Lp to vitellogenic oocytes was decreased in females fed blood plasma (Entringer et al. 2013).

Lp receptors (LpRs), present at the plasma membrane of various organs, were identified in multiple species, and shown to be members of the superfamily of low-density lipoprotein receptors (LDLR) (Dantuma et al. 1999). As a general characteristic, the members of this receptor family contain five domains: an optional O-linked glycosylation domain, an N-terminal ligand-binding domain with cysteine-rich repeats, an EGF precursor homology domain, a single transmembrane domain, and a C-terminal cytoplasmic domain with at least one copy of the sequence motif NPXY (Rodenburg et al. 2006; Tufail and Takeda 2009). Interestingly, phylogenetical analyses placed LpRs in a unique group, that seems to have evolved independently from the other family groups and that shows a variability degree that may help to explain the different routes of lipid uptake and delivery that occur (Rodenburg et al. 2006; Tufail and Takeda 2009).

In order to study Lp intracellular fate, as it was shown to be endocytosed by the *L. migratoria* fat body under some conditions, chinese hamster ovary (CHO) cells were transfected with LpR and, after incubation with Lp, the lipoprotein seemed to be recycled, and re-secreted to the medium (Van Hoof et al. 2002). However, these results were not confirmed when insect cells, either *Drosophila* Schneider line 2 (S2) or *Spodoptera frugiperda* Sf9 cells were transfected with LpR, and Lp was retained inside these cells upon its uptake (Van Hoof et al. 2005). In this way, although it is clear that in some cases Lp is endocytosed, its route inside cells is still unclear.

Multiple factors probably take part in the adjustment of the insect lipid transport system to specific conditions, and hormones are certainly involved. In *A. aegypti*, LpR expression is affected by JH titers, possibly related to the previtellogenic nutritional condition (Clifton and Noriega 2012). In line with these observations, in the kissing bug *D. maxima*, JH titer increases after a blood meal, and stimulates LpR transcription in the ovaries, as well as Lp synthesis in the fat body (Ramos et al. 2021).

Ecdysone, which has long been known to take part in the control of insect development and reproduction, seems to act in the regulation of LpR expression. In *G. mellonella* larvae, LpR abundance in the fat body was higher in the prepupal stage, indicating a possible relationship with ecdysone signal variation during development. In fact, when larvae were treated with 20E (20-hydroxyecdysone), there was a huge increase in the amount of LpR in the insect's whole body (Lee et al. 2003). In the brown planthopper, *N. lugens*, LpR is important to guarantee lipid accumulation and successful ovarian development. In this insect, ecdysone was shown to act in the control of LpR transcription, since knockdown of either ecdysone receptor or USP (ultraspiracle) decreased LpR expression, and topic application of 20E increased it (Lu et al. 2018). In *D. melanogaster*, it was demonstrated that ecdysone stimulates LpR expression in the ovary through the activation of the sterol regulatory element-binding protein, or SREBP, a lipogenic transcription factor, as part of a general strategy for lipid accumulation by the follicles during oogenesis (Sieber and Spradling 2015).

Appropriate nutritional condition is critical for energy storage and reproductive success, and the insulin-like peptides (ILPs), as nutritional sensors, have central roles in these processes (Smykal and Raikhel 2015). Females of *R. prolixus* that were knockdown for the insulin receptor indicated a reduction in the fat body's TAG buildup after a blood meal due, at least in part, to the inhibition of LpR gene expression and consequent impairment of lipid incorporation from the hemolymph (Silva-Oliveira et al. 2021). Overall, these results clearly reveal that LpR expression is under the control of diverse agents, and is finely tuned to modulate lipid transport in insects.

In some insects, alternative splicing of the *LpR* was detected. In *A. aepypti* and *G. mellonella*, two transcripts originating from a single *LpR* gene were identified (Lee et al. 2003; Seo et al. 2003). In the mosquito, one of the variants was expressed exclusively in the ovaries, during the pre-vitellogenic and vitellogenic stages, while the other one was detected in the fat body, but its expression was restricted to the postvitellogenic period, when the fat body stores lipids and other molecules (Seo et al. 2003). Four LpR isoforms exist in *B. mori*, originated from a single gene, which were characterized and shown to be expressed according to the tissue and developmental stage (Gopalapillai et al. 2006). A more complex system was described in *D. melanogaster*, which has two *LpR* genes that are transcribed into multiple isoforms (Parra-Peralbo and Culi 2011). In this way, *LpR* splice variants seem to have their expression controlled by diverse conditions according to specific roles.

6 Other Proteins Involved in Lipid Transport

Although it is well accepted that LpR has a central role in lipid transfer between Lp and organs, the insect lipid transport system is complex, variable, and not fully understood. In fact, other proteins were identified that also take part in lipid delivery and/or interaction of Lp with tissues. In *B. mori*, Cameo 2, a protein that is related to CD-36, plays a role in transporting carotenoids from the midgut to the silk gland (Sakudoh et al. 2010). Interestingly, an ectopic β-chain of ATP synthase interacts with Lp at the midgut, ovary, and fat body in the triatomine *P. megistus* and, when it was blocked by an antibody anti-β-ATPase, the release of DAG from the midgut to Lp was impaired. At the ovary and fat body, β-ATPase blocking inhibited fatty acids incorporation from Lp but not its endocytosis by the oocytes (Fruttero et al. 2014, 2017, 2019). These results reinforced the occurrence of diverse mechanisms involved in lipid transfer in insects, and the participation of other components in Lp-tissue interaction.

The lipid transfer particle (LTP) is another hemolymphatic lipoprotein that is also implicated in the insect lipid transport system. It is a lipoprotein with a very high density and molecular mass, first described in the hemolymph of *M. sexta*, and which mediates the exchange of lipids between Lp populations (Ryan et al. 1986; Takeuchi and Chino 1993; Golodne et al. 2001). The participation of LTP in lipid transfer between Lp and tissues has been reported, although the mechanisms involved in this interaction are still unclear (Van Heusden and Law 1989; Rodríguez-Vázquez et al. 2015).

7 Conclusions and Future Directions

Insects have a very peculiar system for lipid transport in the hemolymph, between the various organs. A single lipoprotein, Lp, selectively transports the specific lipid classes to and from the diverse cell types according to each physiological condition. Lp lipid composition varies among insect species, as would be expected from the multiple feeding habits and developmental features of these organisms. Moreover, there are different mechanisms that respond for lipid incorporation by organs, sometimes simultaneously, that somehow evolved to fulfill specific insect needs, and the molecular components and control that take part of lipid delivery in each case are still largely unknown. Lp lipid transport system needs to interact with the intracellular metabolic pathways and organelles, as lipid droplets, so that lipid homeostasis is properly achieved, and basically nothing is understood about this putative crosstalk. In this way, although knowledge about insect lipid metabolism advanced a lot since Lp identification, many points still deserve to be investigated and many questions are still unanswered.

References

Agianian B, Lesch C, Loseva O, Dushay MS (2007) Preliminary characterization of hemolymph coagulation in *Anopheles gambiae* larvae. Dev Comp Immunol 31:879–888. https://doi.org/10.1016/j.dci.2006.12.006

Araujo RV, Maciel C, Hartfelder K, Capurro ML (2011) Effects of *plasmodium gallinaceum* on hemolymph physiology of *Aedes aegypti* during parasite development. J Insect Physiol 77:265–273

Arrese EL, Rojas-Rivas BI, Wells MA (1996) The use of decapitated insects to study lipid mobilization in adult *Manduca sexta*: effects of adipokinetic hormone and trehalose on fat body lipase activity. Insect Biochem Mol Biol 26:775–782

Asgari S, Schmidt O (2004) Isolation of an imaginal disc growth factor homologue from *Pieris rapae* and its expression following parasitization by *Cotesia rubecula*. J Insect Physiol 50:687–694. https://doi.org/10.1016/j.jinsphys.2004.05.003

Atella GC, Shahabuddin M (2002) Differential partitioning of maternal fatty acid and phospholipid in neonate mosquito larvae. J Exp Biol 205:3623–3630

Atella GC, Gondim KC, Masuda H (1992) Transfer of phospholipids from fat body to lipophorin in *Rhodnius prolixus*. Arch Insect Biochem Physiol 19:133–144. https://doi.org/10.1002/arch.940190206

Atella GC, Gondim KC, Masuda H (1995) Loading of lipophorin particles with phospholipids at the midgut of *Rhodnius prolixus*. Arch Insect Biochem Physiol 30: 337–350. https://doi.org/10.1002/arch.940300404

Atella GC, Arruda MA, Masuda H, Gondim KC (2000) Fatty acid incorporation by *Rhodnius prolixus* midgut. Arch Insect Biochem Physiol 43:99–107

Atella GC, Silva-Neto MAC, Golodne DM et al (2006) *Anopheles gambiae* lipophorin: characterization and role in lipid transport to developing oocyte. Insect Biochem Mol Biol 36:375–386. https://doi.org/10.1016/j.ibmb.2006.01.019

Beckage NE, Kanost MR (1993) Effects of parasitism by the braconid wasp *Cotesia congregata* on host hemolymph proteins of the tobacco hornworm, *Manduca sexta*. Insect Biochem Mol Biol 23:643–653. https://doi.org/10.1016/0965-1748(93)90038-t

Beenakkers AMT, Chino H, Law JH (1988) Lipophorin nomenclature. Insect Biochem 18:1–2

Ben-Aicha S, Badimon L, Vilahur G (2020) Advances in HDL: much more than lipid transporters. Int J Mol Sci 21:732. https://doi.org/10.3390/ijms21030732

Benoit JB, Yang G, Krause TB et al (2011) Lipophorin acts as a shuttle of lipids to the milk gland during tsetse fly pregnancy. J Insect Physiol 57:1553–1561. https://doi.org/10.1016/j.jinsphys.2011.08.009

Bogerd J, Babin PJ, Kooiman FP et al (2000) Molecular characterization and gene expression in the eye of the apolipophorin II/I precursor from *Locusta migratoria*. J Comp Neurol 427:546–558

Brankatschk M, Eaton S (2010) Lipoprotein particles cross the blood-brain barrier in *Drosophila*. J Neurosci 30: 10441–10447. https://doi.org/10.1523/JNEUROSCI.5943-09.2010

Burmester T (2002) Origin and evolution of arthropod hemocyanins and related proteins. J Comp Physiol B 172:95–107. https://doi.org/10.1007/s00360-001-0247-7

Campbell PM, Oakeshott JG, Healy MJ (1998) Purification and kinetic characterisation of juvenile hormone esterase from *Drosophila melanogaster*. Insect Biochem Mol Biol 28:501–515. https://doi.org/10.1016/s0965-1748(98)00037-x

Canavoso LE, Rubiolo ER (1994) Adipokinetic hormone causes formation of low-density lipophorin in the hemolymph of *Triatoma infestans*. Mem Inst Oswaldo Cruz 89:233–234

Canavoso LE, Stariolo R, Rubiolo ER (2003) Flight metabolism in *Panstrongylus megistus* (Hemiptera: Reduviidae): the role of carbohydrates and lipids. Mem Inst Oswaldo Cruz 98:909–914. https://doi.org/10.1590/S0074-02762003000700009

Capurro ML, de Bianchi AG, Marinotti O (1994) *Aedes aegypti* lipophorin. Comp Biochem Physiol Biochem Mol Biol 108:35–39

Cheon HM, Shin SW, Bian G et al (2006) Regulation of lipid metabolism genes, lipid carrier protein lipophorin, and its receptor during immune challenge in the mosquito *Aedes aegypti*. J Biol Chem 281:8426–8435. https://doi.org/10.1074/jbc.M510957200

Chino H, Gilbert LI (1971) Uptake and transport of cholesterol by haemolymph lipoproteins. Insect Biochem 1:337–347

Chino H, Kitazawa K (1981) Diacylglycerol-carrying lipoprotein of hemolymph of the locust and some insects. J Lipid Res 22:1042–1052

Chino H, Murakami S, Harashima K (1969) Diglyceride-carrying lipoproteins in insect hemolymph isolation, purification and properties. Biochim Biophys Acta 176:1–26

Chino H, Downer RGH, Wyatt GR, Gilbert LI (1981a) Lipophorins, a major class of lipoproteins of insect hemolymph. Insect Biochem 11:491

Chino H, Katase H, Downer RGH, Takahashi K (1981b) Diacylglycerol-carrying lipoprotein of hemolymph of the American cockroach—purification, characterization, and function. J Lipid Res 22:7–15

Chino H, Downer RGH, Takahashi K (1986) Effect of adipokinetic hormone on the structure and properties of lipophorin in locusts. J Lipid Res 27:21–29

Chino H, Kiyomoto Y, Takahashi K (1989) *In vitro* study of the action of adipokinetic hormone in locusts. J Lipid Res 30:571–578

Clark AJ, Block K (1959) The absence of sterol synthesis in insects. J Biol Chem 234:2578–2582

Clifton ME, Noriega FG (2012) The fate of follicles after a blood meal is dependent on previtellogenic nutrition and juvenile hormone in *Aedes aegypti*. J Insect Physiol 58:1007–1019. https://doi.org/10.1016/j.jinsphys.2012.05.005

Coelho HSL, Atella GC, Moreira MF et al (1997) Lipophorin density variation during oogenesis in *Rhodnius prolixus*. Arch Insect Biochem Physiol 35: 301–313

Colonello-Frattini NA, Guidugli-Lazzarini KR, Simões ZL, Hartfelder K (2010) Mars is close to venus—female reproductive proteins are expressed in the fat body and reproductive tract of honey bee (*Apis mellifera* L.) drones. J Insect Physiol 56:1638–1644. https://doi.org/10.1016/j.jinsphys.2010.06.008

Dantuma NP, Van Marrewijk WJ, Wynne HJ, Van Der Horst DJ (1996) Interaction of an insect lipoprotein with its binding site at the fat body. J Lipid Res 37: 1345–1355

Dantuma NP, Pijnenburg MA, Diederen JH, Van der Horst DJ (1997) Developmental down-regulation of receptor-mediated endocytosis of an insect lipoprotein. J Lipid Res 38:254–265

Dantuma NP, Potters M, De Winther MP et al (1999) An insect homolog of the vertebrate very low density lipoprotein receptor mediates endocytosis of lipophorins. J Lipid Res 40:973–978

de Bianchi AG, Capurro ML (1991) *Musca domestica* larval lipoprotein. Arch Insect Biochem Physiol 17: 15–27

Demel RA, Van Doorn JM, Van der Horst DJ (1992) Insect apolipophorin III: interaction of locust apolipophorin III with diacylglycerol. Biochim Biophys Acta 1124:151–158

Dettloff M, Wittwer D, Weise C, Wiesner A (2001) Lipophorin of lower density is formed during immune responses in the lepidopteran insect *Galleria mellonella*. Cell Tissue Res 306:449–458. https://doi.org/10.1007/s00441-001-0468-9

Diaconeasa B, Mazock GH, Mahowald AP, Dubreuil RR (2013) Genetic studies of spectrin in the larval fat body of *Drosophila melanogaster*: evidence for a novel lipid uptake apparatus. Genetics 195:871–881. https://doi.org/10.1534/genetics.113.155192

Downer RGH, Chino H (1985) Turnover of protein and diacylglycerol components of lipophorin in insect haemolymph. Insect Biochem 15:627–630. https://doi.org/10.1016/0020-1790(85)90124-6

Engelmann F, Mala J (2000) The interactions between juvenile hormone (JH), lipophorin, vitellogenin, and JH esterases in two cockroach species. Insect Biochem Mol Biol 30:793–803

Engelmann F, Mala J (2005) The cockroach *Leucophaea maderae* needs more than juvenile hormone, vitellogenin and reserves to make a yolky egg. J Insect Physiol 51:465–472. https://doi.org/10.1016/j.jinsphys.2005.01.003

Engler DL, Willingham LA, Ziegler R (1996) Variations in the density of lipophorins during late larval and early pupal development of *Manduca sexta*. Arch Insect Biochem Physiol 33:53–61

Entringer PF, Grillo LAM, Pontes EG et al (2013) Interaction of lipophorin with *Rhodnius prolixus* oocytes: biochemical properties and the importance of blood feeding. Mem Inst Oswaldo Cruz 108:836–844. https://doi.org/10.1590/0074-0276130129

Entringer PF, Majerowicz D, Gondim KC (2021) The fate of dietary cholesterol in the kissing bug *Rhodnius prolixus*. Front Physiol 12:654565

Fan Y, Chase J, Sevala VL, Schal C (2002) Lipophorin-facilitated hydrocarbon uptake by oocytes in the German cockroach *Blattella germanica* (L.). J Exp Biol 205:781–790

Fan Y, Schal C, Vargo EL, Bagnères AG (2004) Characterization of termite lipophorin and its involvement in hydrocarbon transport. J Insect Physiol 50:609–620

Fang Y, Park MG, Choi JY et al (2020) Insecticidal and synergistic activity of dsRNAs targeting buprofezin-specific genes against the small brown planthopper, *Laodelphax striatellus*. Arch Insect Biochem Physiol 105:e21739. https://doi.org/10.1002/arch.21739

Ford PS, Van Heusden MC (1994) Triglyceride-rich lipophorin in *Aedes aegypti* (Diptera: Culicidae). J Med Entomol 31:435–441

Fruttero LL, Rubiolo ER, Canavoso LE (2009) Biochemical and cellular characterization of lipophorin-midgut interaction in the hematophagous *Panstrongylus megistus* (Hemiptera: Reduviidae). Insect Biochem Mol Biol 39:322–331. https://doi.org/10.1016/j.ibmb.2009.01.009

Fruttero LL, Frede S, Rubiolo ER, Canavoso LE (2011) The storage of nutritional resources during vitellogenesis of *Panstrongylus megistus* (Hemiptera: Reduviidae): the pathways of lipophorin in lipid delivery to developing oocytes. J Insect Physiol 57:475–486. https://doi.org/10.1016/j.jinsphys.2011.01.009

Fruttero LL, Demartini DR, Rubiolo ER et al (2014) β-Chain of ATP synthase as a lipophorin binding protein and its role in lipid transfer in the midgut of *Panstrongylus megistus* (Hemiptera: Reduviidae). Insect Biochem Mol Biol 52:1–12. https://doi.org/10.1016/j.ibmb.2014.06.002

Fruttero LL, Leyria J, Ramos FO et al (2017) The process of lipid storage in insect oocytes: the involvement of β-chain of ATP synthase in lipophorin-mediated lipid transfer in the chagas' disease vector *Panstrongylus megistus* (Hemiptera: Reduviidae). J Insect Physiol 96:82–92. https://doi.org/10.1016/j.jinsphys.2016.10.014

Fruttero LL, Leyria J, Moyetta NR et al (2019) The fat body of the hematophagous insect, *Panstrongylus megistus* (Hemiptera: Reduviidae): histological features and participation of the β-chain of ATP synthase in the Lipophorin-mediated lipid transfer. J Insect Sci 19:16

Gellissen G, Wyatt GR (1981) Production of lipophorin in the fat body of adult *Locusta migratoria*: comparison with vitellogenin. Can J Biochem 59:648–654

Glinka AV, Triseleva TA (1989) Differential regulation of protein biosynthesis at translational level in the fat body cells of adult *Locusta migratoria*. Mol Reprod Dev 1:278–282

Golodne DM, Van Heusden MC, Gondim KC et al (2001) Purification and characterization of a lipid transfer particle in *Rhodnius prolixus*: phospholipid transfer. Insect Biochem Molec Biol 31:563–571

Gondim KC, Wells MA (2000) Characterization of lipophorin binding to the midgut of larval *Manduca sexta*. Insect Biochem Mol Biol 30:405–413. https://doi.org/10.1016/S0965-1748(00)00014-X

Gondim KC, Oliveira PL, Coelho HSL, Masuda H (1989a) Lipophorin from *Rhodnius prolixus*: purification and partial characterization. Insect Biochem 19:153–161

Gondim KC, Oliveira PL, Masuda H (1989b) Lipophorin and oogenesis in *Rhodnius prolixus*: transfer of phospholipids. J Insect Physiol 35:19–27

Gondim KC, Atella GC, Kawooya JK, Masuda H (1992) Role of phospholipids in the Lipophorin particles of Rhodnius prolixus. Arch Insect Biochem Physiol 20: 303–314

Gondim KC, Pennington JE, Meyer-Fernandes JR et al (2013) Effect of starvation on lipophorin density in fifth instar larval *Manduca sexta*. Arch Insect Biochem Physiol 84:145–156. https://doi.org/10.1002/arch.21133

Gondim KC, Atella GC, Pontes EG, Majerowicz D (2018) Lipid metabolism in insect disease vectors. Insect Biochem Mol Biol 101:108–123. https://doi.org/10.1016/j.ibmb.2018.08.005

Gopalapillai R, Kadono-Okuda K, Tsuchida K et al (2006) Lipophorin receptor of *Bombyx mori*: cDNA cloning, genomic structure, alternative splicing, and isolation of a new isoform. J Lipid Res 47:1005–1013. https://doi.org/10.1194/jlr.M500462-JLR200

Grillo LAM, Pontes EG, Gondim KC (2003) Lipophorin interaction with the midgut of *Rhodnius prolixus*: characterization and changes in binding capacity. Insect Biochem Mol Biol 33:429–438. https://doi.org/10.1016/S0965-1748(03)00007-9

Gu X, Quilici D, Juarez P et al (1995) Biosynthesis of hydrocarbons and contact sex pheromone and their transport by lipophorin in females of the German cockroach (*Blattella germanica*). J Insect Physiol 41: 257–267

Halwani AE, Niven DF, Dunphy GB (2001) Apolipophorin-III in the greater wax moth, *Galleria mellonella* (Lepidoptera: Pyralidae). Arch Insect Biochem Physiol 48:135–143. https://doi.org/10.1002/arch.1066

Haunerland NH (1996) Insect storage proteins: gene families and receptors. Insect Biochem Molec Biol 26:755 765. https://doi.org/10.1016/s0965-1748(96)00035-5

Hiraoka T, Katagiri C (1992) Treatment of low density lipophorin with lipoprotein lipase: diacylglycerol content has no effect on dissociation of apolipophorin III from low-density lipophorin. J Biochem 112:689–693

Huang HJ, Lu JB, Li Q et al (2018) Combined transcriptomic/proteomic analysis of salivary gland and secreted saliva in three planthopper species. J Proteome 172:25–35. https://doi.org/10.1016/j.jprot.2017.11.003

Jouni ZE, Wells MA (1996) Purification and partial characterization of a lutein-binding protein from the midgut of the silkworm Bombyx mori. J Biol Chem 271: 14722–14726

Jouni ZE, Yun HK, Wells MA (2002a) Cholesterol efflux from larval *Manduca sexta* fat body in vitro: high-density lipophorin as the acceptor. J Insect Physiol 48:609–618

Jouni ZE, Zamora J, Wells MA (2002b) Absorption and tissue distribution of cholesterol in Manduca sexta. Arch Insect Biochem Physiol 49:167–175

Jouni ZE, Takada N, Gazard J et al (2003) Transfer of cholesterol and diacylglycerol from lipophorin to *Bombyx mori* ovarioles in vitro: role of the lipid transfer particle. Insect Biochem Mol Biol 33:145–153

Juárez P, Brenner RR (1989) Fatty acid biosynthesis in the integument tissue of *Triatoma infestans*. Comp Biochem Physiol: Part B Biochem 93:763–772. https://doi.org/10.1016/0305-0491(89)90043-6

Jurenka RA, Subchev M, Abad JL et al (2003) Sex pheromone biosynthetic pathway for disparlure in the gypsy moth, *Lymantria dispar*. Proc Natl Acad Sci USA 100: 809–814. https://doi.org/10.1073/pnas.0236060100

Kanost MR, Sparks KA, Wells MA (1995) Isolation and characterization of Apolipophorin-III from the Giant water bug (Lethocerus-Medius). Insect Biochem Mol Biol 25:759–764

Katagiri C (1985) Structure of lipophorin in insect blood: location of phospholipid. Biochim Biophys Acta 834: 139–143

Katagiri C, Sato M, Tanaka N (1987) Small-angle x-ray scattering study of insect lipophorin. J Biol Chem 262: 15857–15861

Katagiri C, Sato M, de Kort S, Katsube Y (1991) Structure of Colorado potato beetle lipophorin: differential scanning calorimetric and small-angle X-ray scattering studies. Biochemistry 30:9675–9681

Katase H, Chino H (1982) Transport of hydrocarbons by the lipophorin of insect hemolymph. Biochim Biophys Acta 710:341–348

Kato Y, Motoi Y, Taniai K et al (1994) Lipopolysaccharide-lipophorin complex formation in insect hemolymph: a common pathway of lipopolysaccharide detoxification both in insects and in mammals. Insect Biochem Mol Biol 24:547–555

Kawooya JK, Law JH (1988) Role of lipophorin in lipid transport to the insect egg. J Biol Chem 263:8748–8753

Kawooya JK, Keim PS, Ryan RO et al (1984) Insect apolipophorin III: purification and properties. J Biol Chem 259:733–737

Kawooya JK, Osir EO, Law JH (1988) Uptake of the major hemolymph liporprotein and its transformation in the insect egg. J Biol Chem 263:8740–8747

Kawooya JK, van der Horst DJ, van Heusden MC et al (1991) Lipophorin structure analyzed by *in vitro* treatment with lipases. J Lipid Res 32:1781–1788

Kinuthia W, Li D, Schmidt O, Theopold U (1999) Is the surface of endoparasitic wasp eggs and larvae covered by a limited coagulation reaction? J Insect Physiol 45: 501–506. https://doi.org/10.1016/s0022-1910(98)00164-4

Konuma T, Tsukamoto Y, Nagasawa H, Nagata S (2016) Imbalanced Hemolymph lipid levels affect feeding motivation in the two-spotted cricket, *Gryllus bimaculatus*. PLoS One 11:e0154841. https://doi.org/10.1371/journal.pone.0154841

Kumar BA, Paily KP (2011) Up-regulation of lipophorin (Lp) and lipophorin receptor (LpR) gene in the mosquito, Culex quinquefasciatus (Diptera: Culicidae), infected with the filarial parasite, Wuchereria bancrofti (Spirurida: Onchocercidae). Parasitol Res 108:377–381. https://doi.org/10.1007/s00436-010-2075-8

Kutty RK, Kutty G, Kambadur R et al (1996) Molecular characterization and developmental expression of a retinoid- and fatty acid-binding glycoprotein from *Drosophila* A putative lipophorin. J Biol Chem 271: 20641–20649. https://doi.org/10.1074/jbc.271.34.20641

Lee CS, Han JH, Kim BS et al (2003) Wax moth, *Galleria mellonella*, high density lipophorin receptor: alternative splicing, tissue-specific expression, and developmental regulation. Insect Biochem Mol Biol 33:761–771. https://doi.org/10.1016/S0965-1748(03)00066-3

Leyria J, Fruttero LL, Aguirre SA, Canavoso LE (2014) Ovarian nutritional resources during the reproductive cycle of the hematophagous *Dipetalogaster maxima* (Hemiptera: Reduviidae): focus on lipid metabolism. Arch Insect Biochem Physiol 87:178–163. https://doi.org/10.1002/arch.21186

Leyria J, Fruttero LL, Ligabue-Braun R et al (2018) DmCatD, a cathepsin D-like peptidase of the hematophagous insect *Dipetalogaster maxima* (Hemiptera: Reduviidae): purification, bioinformatic analyses and the significance of its interaction with lipophorin in the internalization by developing oocyte. J Insect Physiol 105:28–39. https://doi.org/10.1016/j.jinsphys.2018.01.002

Li S, Jing X (2020) Fates of dietary sterols in the insect alimentary canal. Curr Opin Insect Sci 41:106–111. https://doi.org/10.1016/j.cois.2020.08.001

Liu X, Zhou H, Zhao J et al (2016) Identification of the secreted watery saliva proteins of the rice brown planthopper, *Nilaparvata lugens* (Stål) by transcriptome and shotgun LC-MS/MS approach. J Insect Physiol 89:60–69. https://doi.org/10.1016/j.jinsphys.2016.04.002

Lourenço AP, Martins JR, Guidugli-Lazzarini KR et al (2012) Potential costs of bacterial infection on storage protein gene expression and reproduction in queenless *Apis mellifera* worker bees on distinct dietary regimes. J Insect Physiol 58:1217–1225. https://doi.org/10.1016/j.jinsphys.2012.06.006

Lu K, Chen X, Li Y et al (2018) Lipophorin receptor regulates *Nilaparvata lugens* fecundity by promoting lipid accumulation and vitellogenin biosynthesis. Comp Biochem Physiol A Mol Integr Physiol 219–220:28–37. https://doi.org/10.1016/j.cbpa.2018.02.008

Lucas C, Pho DB, Fresneau D, Jallon JM (2004) Hydrocarbon circulation and colonial signature in *Pachycondyla villosa*. J Insect Physiol 50:595–607

Lucas KJ, Roy S, Ha J et al (2015) MicroRNA-8 targets the wingless signaling pathway in the female mosquito fat body to regulate reproductive processes. Proc Natl Acad Sci USA 112:1440–1445. https://doi.org/10.1073/pnas.1424408112

Ma G, Hay D, Li D et al (2006) Recognition and inactivation of LPS by lipophorin particles. Dev Comp Immunol 30:619–626

Ma G, Rahman MM, Grant W et al (2012) Insect tolerance to the crystal toxins Cry1Ac and Cry2Ab is mediated by the binding of monomeric toxin to lipophorin glycolipids causing oligomerization and sequestration reactions. Dev Comp Immunol 37:184–192. https://doi.org/10.1016/j.dci.2011.08.017

Machado EA, Atella GC, Gondim KC et al (1996) Characterization and immunocytochemical localization of lipophorin binding sites in the oocytes of *Rhodnius prolixus*. Arch Insect Biochem Physiol 31:185–196

Magee J, Kraynack N, Massey HC, Telfer WH (1994) Properties and significance of a riboflavin-binding hexamerin in the hemolymph of *Hyalophora cecropia*. Arch Insect Biochem Physiol 25:137–157

Majerowicz D, Cezimbra MP, Alves-Bezerra M et al (2013) *Rhodnius prolixus* lipophorin: lipid composition and effect of high temperature on physiological role. Arch Insect Biochem Physiol 82:129–140. https://doi.org/10.1002/arch.21080

Marinotti O, de Capurro ML, Nirmala X et al (2006) Structure and expression of the lipophorin-encoding gene of the malaria vector, *Anopheles gambiae*. Comp Biochem Physiol B Biochem Mol Biol 144: 101–109. https://doi.org/10.1016/j.cbpb.2006.01.012

Matsuoka K, Tabunoki H, Kawai T et al (2006) Transport of a hydrophobic biosynthetic precursor by lipophorin in the hemolymph of a geometrid female moth which secretes an epoxyalkenyl sex pheromone. Insect Biochem Mol Biol 36:576–583

Maya-Monteiro C, Oliveira PL, Gondim KC et al (1997) Role of phospholipids in the protein stability of an insect lipoprotein, lipophorin from *Rhodnius prolixus*. Biochemistry 36:11216–11222

Mohan KG, Muraleedharan D (2005) Modulatory influence of juvenile hormone analogue (JHa) and 20-hydroxyecdysone on lipophorin synthesis in red cotton bug, *Dysdercus cingulatus* Fabr. Indian J Exp Biol 43:1176–1181

Moreira MF, Garcia ES, Masuda H (1994) Inhibition by azadirachtin of phospholipid transfer from lipophorin to the oocytes in *Rhodnius prolixus*. Arch Insect Biochem Physiol 27:287–299

Mwangi RW, Goldsworthy GJ (1981) Diacylglycerol-transporting lipoproteins and flight in *Locusta*. J Insect Physiol 27:47–50

Nagao E, Chino H (1991) Further characterization of locust low density lipophorin induced by adipokinetic hormone. J Lipid Res 32:417–422

Nicholson SJ, Puterka GJ (2014) Variation in the salivary proteomes of differentially virulent greenbug (*Schizaphis graminum* Rondani) biotypes. J Proteome 105:186–203. https://doi.org/10.1016/j.jprot.2013.12.005

Nicholson SJ, Hartson SD, Puterka GJ (2012) Proteomic analysis of secreted saliva from Russian wheat aphid (*Diuraphis noxia* Kurd.) biotypes that differ in virulence to wheat. J Proteome 75:2252–2268. https://doi.org/10.1016/j.jprot.2012.01.031

Niere M, Meisslitzer C, Dettloff M et al (1999) Insect immune activation by recombinant *Galleria mellonella* apolipophorin III. Biochim Biophys Acta 1433:12–26. https://doi.org/10.1016/s0167-4838(99)00148-x

Ochanda JO, Osir EO, Nguu EK, Olembo NK (1991) Lipophorin from the tsetse fly, *Glossina morsitans morsitans*. Comp Biochem Physiol B 99:811–814

Ochanda JO, Osir EO, Nguu EK, Olembo NK (1992) Isolation and properties of 600-kDa and 23-kDa haemolymph proteins from the tsetse fly, *Glossina morsitans*: their possible role as biological insecticides. Scand J Immunol Suppl 11:41–47

Oliveira GA, Baptista DL, Guimaraes-Motta H et al (2006) Flight-oogenesis syndrome in a blood-sucking bug: biochemical aspects of lipid metabolism. Arch Insect Biochem Physiol 62:164–175. https://doi.org/10.1002/arch.20132

Palm W, Sampaio JL, Brankatschk M et al (2012) Lipoproteins in *Drosophila melanogaster*—assembly, function, and influence on tissue lipid composition. PLoS Genet 8:e1002828. https://doi.org/10.1371/journal.pgen.1002828

Palm W, Swierczynska MM, Kumari V et al (2013) Secretion and signaling activities of lipoprotein-associated hedgehog and non-sterol-modified hedgehog in flies and mammals. PLoS Biol 11:e1001505. https://doi.org/10.1371/journal.pbio.1001505

Parra-Peralbo E, Culi J (2011) *Drosophila* lipophorin receptors mediate the uptake of neutral lipids in oocytes and imaginal disc cells by an endocytosis-independent mechanism. PLoS Genet 7:e1001297. https://doi.org/10.1371/journal.pgen.1001297

Pennington JE, Wells MA (2002) Triacylglycerol-rich lipophorins are found in the dipteran infraorder Culicomorpha, not just in mosquitoes. J Insect Sci 2:15

Pho DB, Pennanech M, Jallon JM (1996) Purification of adult *Drosophila melanogaster* lipophorin and its role in hydrocarbon transport. Arch Insect Biochem Physiol 31:289–303

Pontes EG, Grillo LAM, Gondim KC (2002) Characterization of lipophorin binding to the fat body of *Rhodnius prolixus*. Insect Biochem Mol Biol 32: 1409–1417. https://doi.org/10.1016/S0965-1748(02)00061-9

Prasad SV, Ryan RO, Law JH, Wells MA (1986) Changes in lipoprotein composition during larval-pupal metamorphosis of an insect, *Manduca sexta*. J Biol Chem 261:558–562

Rahman MM, Ma G, Roberts HL, Schmidt O (2006) Cell-free immune reactions in insects. J Insect Physiol 52: 754–762. https://doi.org/10.1016/j.jinsphys.2006.04.003

Rahman MM, Roberts HL, Schmidt O (2007) Tolerance to bacillus thuringiensis endotoxin in immune-suppressed larvae of the flour moth Ephestia kuehniella. J Invertebr Pathol 96:125–132

Ramos FO, Leyria J, Nouzova M et al (2021) Juvenile hormone mediates lipid storage in the oocytes of *Dipetalogaster maxima*. Insect Biochem Mol Biol 133:103499. https://doi.org/10.1016/j.ibmb.2020.103499

Rimoldi OJ, Garda HA, Brenner RR (1996) Effect of phospholipids on the structure of *Triatoma infestans* lipophorin studied by fluorescence methods. J Lipid Res 37:2125–2135

Robbs SL, Ryan RO, Schmidt JO et al (1985) Lipophorin of the larval honeybee, *Apis mellifera* L. J Lipid Res 26:241–247

Rodenburg KW, Smolenaars MM, Van Hoof D, van der Horst DJ (2006) Sequence analysis of the non-recurring C-terminal domains shows that insect lipoprotein receptors constitute a distinct group of LDL receptor family members. Insect Biochem Mol Biol 36:250–263

Rodríguez-Vázquez M, Vaquero D, Parra-Peralbo E et al (2015) *Drosophila* Lipophorin receptors recruit the lipoprotein LTP to the plasma membrane to mediate lipid uptake. PLoS Genet 11:e1005356. https://doi.org/10.1371/journal.pgen.1005356

Ryan RO, Wells MA, Law JH (1986) Lipid transfer protein from *Manduca sexta* hemolymph. Biochem Biophys Res Commun 136:260–265

Ryan RO, Kay CM, Oikawa K et al (1992) Effect of particle lipid content on the structure of insect lipophorins. J Lipid Res 33:55–63

Sakudoh T, Iizuka T, Narukawa J et al (2010) A CD36-related transmembrane protein is coordinated with an intracellular lipid-binding protein in selective carotenoid transport for cocoon coloration. J Biol Chem 285: 7739–7751. https://doi.org/10.1074/jbc.M109.074435

Santos R, Rosas-Oliveira R, Saraiva FB et al (2011) Lipid accumulation and utilization by oocytes and eggs of *Rhodnius prolixus*. Arch Insect Biochem Physiol 77:1–16. https://doi.org/10.1002/arch.20414

Schal C, Sevala V, Capurro ML et al (2001) Tissue distribution and lipophorin transport of hydrocarbons and sex pheromones in the house fly, *Musca domestica*. J Insect Sci 1:12

Schmidt O, Theopold U (1997) *Helix pomatia* lectin and annexin V, two molecular probes for insect microparticles: possible involvement in hemolymph coagulation. J Insect Physiol 43:667–674. https://doi.org/10.1016/s0022-1910(97)00013-9

Schmidt O, Glatz RV, Asgari S, Roberts HL (2005) Are insect immune suppressors driving cellular uptake reactions? Arch Insect Biochem Physiol 60:153–158. https://doi.org/10.1002/arch.20111

Seo SJ, Cheon HM, Sun J et al (2003) Tissue- and stage-specific expression of two Lipophorin receptor variants with seven and eight ligand-binding repeats in the adult mosquito. J Biol Chem 278:41954–41962. https://doi.org/10.1074/jbc.M308200200

Sevala V, Shu S, Ramaswamy SB, Schal C (1999) Lipophorin of female *Blattella germanica* (L.): characterization and relation to hemolymph titers of juvenile hormone and hydrocarbons. J Insect Physiol 45: 431–441

Shapiro JP, Law JH (1983) Locust adipokinetic hormone stimulates lipid mobilization in *Manduca sexta*. Biochem Biophys Res Commun 115:924–931

Sieber MHMH, Spradling ACAC (2015) Steroid signaling establishes a female metabolic state and regulates

SREBP to control oocyte lipid accumulation. Curr Biol 25:993–1004. https://doi.org/10.1016/j.cub.2015.02.019

Silva-Oliveira G, De Paula IF, Medina JM et al (2021) Insulin receptor deficiency reduces lipid synthesis and reproductive function in the insect *Rhodnius prolixus*. Biochim Biophys Acta Mol Cell Biol Lipids 1866: 158851

Smolenaars MM, Kasperaitis MA, Richardson PE et al (2005) Biosynthesis and secretion of insect lipoprotein: involvement of furin in cleavage of the apoB homolog, apolipophorin-II/I. J Lipid Res 46:412–421. https://doi.org/10.1194/jlr.M400374-JLR200

Smolenaars MM, de Morrée A, Kerver J et al (2007) Insect lipoprotein biogenesis depends on an amphipathic beta cluster in apolipophorin II/I and is stimulated by microsomal triglyceride transfer protein. J Lipid Res 48:1955–1965. https://doi.org/10.1194/jlr.M600434-JLR200

Smykal V, Raikhel AS (2015) Nutritional control of insect reproduction. Curr Opin Insect Sci 11:31–38

Soulages JL, Brenner RR (1991) Study on the composition-structure relationship of lipophorins. J Lipid Res 32: 407–415

Soulages JL, Wells MA (1994a) Effect of diacylglycerol content on some physicochemical properties of the insect lipoprotein, lipophorin: correlation with the binding of apolipophorin-III. Biochemistry 33:2356–2362

Soulages JL, Wells MA (1994b) Metabolic fate and turnover rate of hemolymph free fatty acids in adult Manduca sexta. Insect Biochem Mol Biol 24:79–86

Soulages JL, Rimoldi OJ, Peluffo OR, Brenner RR (1988) Transport and utilization of free fatty acids in Triatoma infestans. Biochem Biophys Res Commun 157:465–471

Soulages JL, Rivera M, Walker FA, Wells MA (1994) Hydration and localization of diacylglycerol in the insect lipoprotein lipophorin. A ^{13}C-NMR study. Biochemistry 33:3245–3251

Soulages JL, Van Antwerpen R, Wells MA (1996) Role of diacylglycerol and Apolipophorin-III in regulation of the physicochemical properties of the lipophorin surface: metabolic implications. Biochemistry 35: 5191–5198

Stączek S, Zdybicka-Barabas A, Mak P et al (2018) Studies on localization and protein ligands of *Galleria mellonella* apolipophorin III during immune response against different pathogens. J Insect Physiol 105:18–27. https://doi.org/10.1016/j.jinsphys.2017.12.009

Sun J, Hiraoka T, Dittmer NT et al (2000) Lipophorin as a yolk protein precursor in the mosquito, *Aedes aegypti*. Insect Biochem Molec Biol 30:1161–1171

Surholt B, Van Doorn JM, Goldberg J, Van der Horst DJ (1992) Insect lipophorin conversions. Compositional analysis of high- and low-density lipophorin of *Acherontia atropos* and *Locusta migratoria*. Biol Chem Hoppe Seyler 373:10–13

Takeuchi N, Chino H (1993) Lipid transfer particle in the hemolymph of the american cockroach: evidence for its capacity to transfer hydrocarbons between lipophorin particles. J Lipid Res 34:543–551

Thangaraj P, Neelamegam RK, Nagarajan K, Muthukalingan K (2018) Interaction of azadirachtin with the lipid-binding domain: suppression of lipid transportation in the silkworm, *Bombyx mori*. Pestic Biochem Physiol 152:62–68. https://doi.org/10.1016/j.pestbp.2018.09.001

Thomas KK, Gilbert LI (1967) In vitro studies on the release and transport of phospholipids. J Insect Physiol 13:963–980

Tiwari S, Siddiqi SA (2012) Intracellular trafficking and secretion of VLDL. Arterioscler Thromb Vasc Biol 32: 1079–1086. https://doi.org/10.1161/ATVBAHA.111.241471

Toprak U, Hegedus D, Doğan C, Güney G (2020) A journey into the world of insect lipid metabolism. Arch Insect Biochem Physiol 104:e21682. https://doi.org/10.1002/arch.21682

Trowell SC, Hines ER, Herlt AJ, Rickards RW (1994) Characterization of a juvenile hormone binding lipophorin from the blowfly *Lucilia cuprina*. Comp Biochem Physiol Part B, Biochem Mol Biol 109: 339–357

Tsuchida K, Wells MA (1990) Isolation and characterization of a lipoprotein receptor from the fat body of an insect, *Manduca sexta*. J Biol Chem 265:5761–5767

Tsuchida K, Arai M, Tanaka Y et al (1998) Lipid transfer particle catalyzes transfer of carotenoids between lipophorins of *Bombyx mori*. Insect Biochem Mol Biol 28:927–934

Tsuchida K, Katagiri C, Tanaka Y et al (2004) The basis for colorless hemolymph and cocoons in the Y-gene recessive Bombyx mori mutants: a defect in the cellular uptake of carotenoids. J Insect Physiol 50:975–983

Tsuchida K, Yokoyama T, Sakudoh T et al (2010) Apolipophorin-III expression and low density lipophorin formation during embryonic development of the silkworm, *Bombyx mori*. Comp Biochem Physiol B Biochem Mol Biol 155:363–370. https://doi.org/10.1016/j.cbpb.2009.12.006

Tufail M, Takeda M (2009) Insect vitellogenin/lipophorin receptors: molecular structures, role in oogenesis, and regulatory mechanisms. J Insect Physiol 55:87–103. https://doi.org/10.1016/j.jinsphys.2008.11.007

Van der Horst DJ, Rodenburg KW (2010) Locust flight activity as a model for hormonal regulation of lipid mobilization and transport. J Insect Physiol 56:844–853. https://doi.org/10.1016/j.jinsphys.2010.02.015

Van der Horst DJ, Van Hoof D, Van Marrewijk WJ, Rodenburg KW (2002) Alternative lipid mobilization: the insect shuttle system. Mol Cell Biochem 239: 113–119

Van Heusden MC, Law JH (1989) An insect lipid transfer particle promotes lipid loading from fat body to lipoprotein. J Biol Chem 264:17287–17292

Van Heusden MC, Van der Horst DJ, Voshol J, Beenakkers AMT (1987) The recycling of protein components of the flight-specific lipophorin in Locusta migratoria. Insect Biochem 17:771–776

van Heusden MC, van der Horst DJ, Kawooya JK, Law JH (1991) *In vivo* and *in vitro* loading of lipid by artificially lipid-depleted lipophorins: evidence for the role of lipophorin as a reusable lipid shuttle. J Lipid Res 32:1789–1794

Van Heusden MC, Erickson BA, Pennington JE (1997) Lipophorin levels in the yellow fever mosquito, *Aedes aegypti*, and the effect of feeding. Arch Insect Biochem Physiol 34:301–312

Van Heusden MC, Thompson F, Dennis J (1998) Biosynthesis of Aedes aegypti lipophorin and gene expression of its apolipoproteins. Insect Biochem Mol Biol 28: 733–738

Van Hoof D, Rodenburg KW, Van der Horst DJ (2002) Insect lipoprotein follows a transferrin-like recycling pathway that is mediated by the insect LDL receptor homologue. J Cell Sci 115:4001–4012

Van Hoof D, Rodenburg KW, Van der Horst DJ (2005) Receptor-mediated endocytosis and intracellular trafficking of lipoproteins and transferrin in insect cells. Insect Biochem Mol Biol 35:117–128. https://doi.org/10.1016/j.ibmb.2004.09.009

Wang J, Liu H, Sykes BD, Ryan RO (1995) Identification and localization of two distinct microenvironments for the diacylglycerol component of lipophorin particles by 13C NMR. Biochemistry 34:6755–6761

Weers PMM, Vanderhorst DJ, Vanmarrewijk WJA et al (1992) Biosynthesis and secretion of insect lipoprotein. J Lipid Res 33:485–491

Weers PMM, Vanmarrewijk WJA, Beenakkers AMT, Vanderhorst DJ (1993) Biosynthesis of locust lipophorin. Apolipophorin I and II originate from a common precursor. J Biol Chem 268:4300–4303

Wells MA, Ryan RO, Kawooya JK, Law JH (1987) The role of Apolipophorin-III in *in vivo* lipoprotein interconversions in adult *Manduca sexta*. J Biol Chem 262:4172–4176

Xiao C, Stahel P, Lewis GF (2019) Regulation of chylomicron secretion: focus on post-assembly mechanisms. Cell Mol Gastroenterol Hepatol 7:487–501. https://doi.org/10.1016/j.jcmgh.2018.10.015

Ximenes AA, Oliveira GA, Bittencourt-Cunha P et al (2008) Purification, partial characterization and role in lipid transport to developing oocytes of a novel lipophorin from the cowpea weevil, Callosobruchus maculatus. Braz J Med Biol Res 41:18–25

Ye W, Zeng J, Hu W et al (2020) The desaturase gene *Nlug-desatA2* regulates the performance of the Brown Planthopper *Nilaparvata lugens* and its relationship with Rice. Int J Mol Sci 21:4143. https://doi.org/10.3390/ijms21114143

Ying Q, Chan DC, Barrett PHR, Watts GF (2021) Unravelling lipoprotein metabolism with stable isotopes: tracing the flow. Metabolism 124:154887. https://doi.org/10.1016/j.metabol.2021.154887

Young HP, Bachmann JA, Sevala V, Schal C (1999) Site of synthesis, tissue distribution, and lipophorin transport of hydrocarbons in *Blattella germanica* (L.) nymphs. J Insect Physiol 45:305–315

Zhao X, Karpac J (2020) The *Drosophila* midgut and the systemic coordination of lipid-dependent energy homeostasis. Curr Opin Insect Sci 41:100–105. https://doi.org/10.1016/j.cois.2020.07.003

Ziegler R, Willingham LA, Sanders SJ et al (1995) Apolipophorin-III and adipokinetic hormone in lipid metabolism of larval *Manduca sexta*. Insect Biochem Mol Biol 25:101–108

Ziegler R, Willingham LA, Engler DL et al (1999) A novel lipoprotein from the hemolymph of the cochineal insect, *Dactylopius confusus*. Eur J Biochem 261: 285–290. https://doi.org/10.1046/j.1432-1327.1999.00276.x

Adv Exp Med Biol (2026) 1494: 115–162
https://doi.org/10.1007/5584_2024_823

Published online: 10 October 2024

Insect Sterols and Steroids

René Lafont and Laurence Dinan

Abstract

Insects are incapable of biosynthesising sterols de novo so they need to obtain them from their diets or, in certain cases, from symbiotic microorganisms. Sterols serve a structural role in cellular membranes and act as precursors for signalling molecules and defence compounds. Many phytophagous insects dealkylate phytosterols to yield primarily cholesterol, which is also the main sterol that carnivorous and omnivorous insects obtain in their diets. Some phytophagous species have secondarily lost the capacity to dealkylate and consequently use phytosterols for structural and functional roles. The polyhydroxylated steroid hormones of insects, the ecdysteroids, are derived from cholesterol (or phytosterols in non-dealkylating phytophagous species) and regulate many crucial aspects of insect development and reproduction by means of precisely regulated titres resulting from controlled synthesis, storage and further metabolism/excretion. Ecdysteroids differ significantly from vertebrate steroid hormones in their chemical, biochemical and biological properties. Defensive steroids (cardenolides, bufadienolides, cucurbitacins and ecdysteroids) can be accumulated from host plants or biosynthesised within the insect, depending on species, stored in significant amounts in the insect and released when it is attacked. Other allelochemical steroids serve as pheromones. Vertebrate-type steroids have also been conclusively identified from insect sources, but debate continues about their significance. Side chain dealkylation of phytosterols, ecdysteroid metabolism and ecdysteroid mode of action are targets of potential insect control strategies.

Keywords

Allelochemical · Bufadienolide · Cardenolide · Cucurbitacin · Defensive steroids · Dietary choice · Ecdysteroid · Feeding deterrent · Hormonal steroid · Insect pest control · Membrane sterol · Pheromonal steroid · Phytophagy · Steroid metabolism · Sterol dealkylation · Vertebrate steroid

R. Lafont (✉) and L. Dinan
BIOSIPE, Sorbonne Université, Paris, France
e-mail: rene.lafont@sorbonne-universite.fr

Definitions

Steroids: *A class of natural or synthetic organic lipidic compounds characterised by a structure of 17 carbon atoms arranged in four fused rings (a cyclopentanoperhydrophenanthrene ring system; rings A-D). Steroids can possess extra C-atoms in the form of methyl groups at C-10 and C-13 and a side chain at C-17 ($C_{19} \rightarrow C_{29}$) and various oxygenated substituents located around the molecule. Some steroids have fewer than 19 C-atoms through later removal of specific C-atoms during* *biosynthesis* (e.g. *oestradiol, which lacks the methyl at C-10 and is a* C_{18}*-steroid). Steroids include sterols, steroid hormones, bile acids, steroidal alkaloids and certain medicinal steroids.*

Sterols*: Sterols are a subgroup of steroids with a β-hydroxyl group at C-3 of the A-ring and a side chain of 8–10 C-atoms located at the 17β-position. Consequently, sterols typically possess 27–29 carbons and are amphipathic alcohols. Cholesterol is a representative sterol.*

1 Purpose

In this chapter, we will first consider insect sterols (origin, diversity and roles), then hormonal steroids (biosynthesis, roles, mode of action and further metabolism) and finally those diverse steroids produced internally or derived from the food, which are involved in insect communication with other species as repellents, toxins or pheromones (Fig. 1).

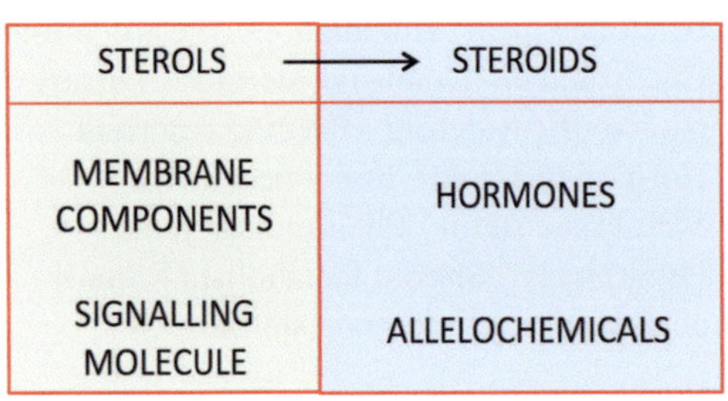

Fig. 1 Sterols and steroids and their biological roles

2 Insect Sterols

2.1 Introduction

Sterols represent a class of non-hydrolysable lipids. The classical animal representative is cholesterol (Fig. 2), but there are numerous analogues in animals, plants and fungi. The major sterol present in most, but not all, insects is cholesterol.

The cholesterol molecule as such is a structural component of cell membranes, where it modifies the fluidity/rigidity of the phospholipid bilayer. It is also a component of a signalling molecule when covalently bound to the protein hedgehog (in invertebrates, or its vertebrate counterparts), which controls embryonic patterning (Ingham 2022). Cholesterol has also been implicated in pathogen and parasitoid blocking in *Drosophila melanogaster* (Caragata et al. 2013; Parades et al. 2016).

Sterol structures differ between animals, plants and fungi, thus insects are faced with very different situations according to the nature of their diet, as cholesterol is mainly present in animals, whereas plants contain phytosterols (e.g. campesterol and sitosterol), while fungi possess mycosterols (e.g. ergosterol), which differ primarily in the number of carbon atoms (Fig. 3). The additional (one or two) carbons are attached at position C-24 on the side chain. This introduces a chiral centre in this position, the stereochemistry of which differs between plants and fungi (24*S* in plants and 24*R* in fungi). Sterols also differ in the position of double bonds, e.g. $\Delta^{5,6}$ for cholesterol and $\Delta^{7,8}$ for lathosterol; additional

Fig. 2 Structure of cholesterol showing the numbering of the carbon atoms, the designation of the rings (A-D) in the sterol and the stereochemistry

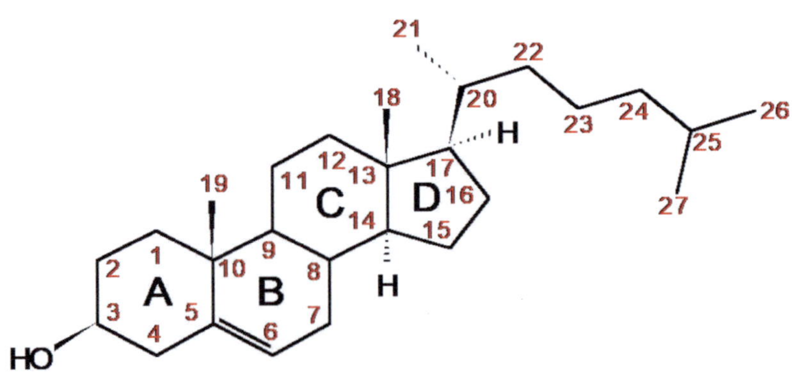

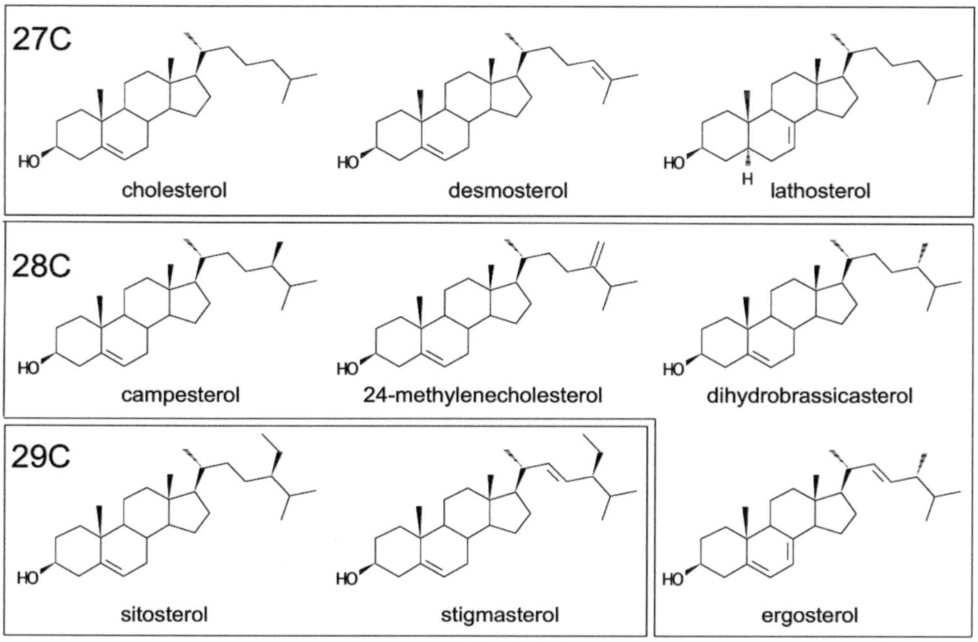

Fig. 3 Structure of representative sterols. (Redrawn from Lafont et al. 2017) arranged in relation to the total number of C-atoms

unsaturations may also exist, e.g. $\Delta^{22,23}$ (stigmasterol, ergosterol), $\Delta^{24,25}$ (desmosterol) or $\Delta^{24,28}$ (24-methylenecholesterol) (Fig. 3; Lafont et al. 2017).

Sterols are converted by animals into a wide array of metabolites, the steroids, which are either key hormones controlling primarily development and reproduction (e.g. ecdysteroids in insects) or which may also be used as allelochemicals involved in interspecific relationships (e.g. as toxicants or repellents). Sterols are not catabolised to generate energy in insects (Lafont et al. 2017).

The ecdysteroids biosynthesised and used hormonally by insects are related to the insect's dietary sterols and whether the insect is able to dealkylate phytosterols. As explained in Sect. 3, some plant-feeding heteropterans and hymenopterans (e.g. milkweed bugs, leaf-cutting ants and honey bees) have secondarily lost the capability to dealkylate, and they convert campesterol to makisterone A (24-methyl-20-hydroxyecdysone) or sitosterol to makisterone C (24-ethyl-20-hydroxyecdysone) and are able to use these in place of 20-hydroxyecdysone (Kaplanis et al. 1975; Aldrich et al. 1982; Feldlaufer et al. 1986, 1991; Maurer et al. 1993).

Drosophila pachea is specialised on the senita cactus (*Lophocerus schottii*) growing in the Sonoran Desert, which contains exceedingly low levels of cholesterol, but which is rich in lathosterol (Fig. 3). An early study (Heed and Kircher 1965) showed that lathosterol supported development and reproduction in *D. pachae*, but cholesterol did not, even though cholesterol is acceptable to other drosophilids. A more recent study (Lang et al. 2012) has clarified the basis for this since the *neverland* gene product, which converts cholesterol to 7-dehydrocholesterol at the start of ecdysteroid biosynthesis, was found to be modified by four amino acid changes such that it will no longer accept cholesterol as a substrate, but does accept lathosterol; the neverland enzyme of *D. melanogaster* accepts both cholesterol and lathosterol as substrates (Kopp 2012).

Half of all insects (especially Orthoptera, Hemiptera, Coleoptera, Diptera, Lepidoptera and Hymenoptera) consume terrestrial plants, while the other half are omnivorous/carnivorous or, to a much lesser extent, eat fungi, lichens and algae.

The most common phytosterols are sitosterol, campesterol and stigmasterol, whereas cholesterol is a minor component in most plants. Consequently, phytophagous insects, in contrast to omnivorous/carnivorous insects, rarely encounter enough cholesterol in their diets. Lepidopteran insects dealkylate phytosterols very efficiently to cholesterol. Orthopterans are able to convert campesterol and sitosterol, but not stigmasterol, to cholesterol. Aphids can dealkylate, but not very efficiently, with only 40% of the consumed sitosterol being converted to cholesterol. Dipterans have a limited capacity to metabolise phytosterols, but are able to use non-cholesterol sterols effectively as membrane components (Jing and Behmer 2020).

2.2 Origin and Uptake of Sterols in Insects

Unlike vertebrates, insects are unable to produce sterols de novo because they lack the enzymes for the conversion of farnesyl pyrophosphate to squalene and subsequent steps in cholesterol biosynthesis. Consequently, insects must obtain sterols from their diet (Behmer and Nes 2003; Jing and Behmer 2020). This auxotrophy for sterols is not restricted to insects but is shared with many other invertebrates (e.g. crustaceans (Clayton 1964), molluscs (Giner et al. 2016) or nematodes (Shamsuzzama et al. 2020).

In insects, sterols taken up from food can be either free or esterified to fatty acids or, less commonly, conjugated to sugars. Sterol esters are hydrolysed by the enzyme magro (homologue of mammalian lipase A), a specific lipase secreted by enterocytes and adsorbed on the chitinous peritrophic membrane (matrix), then free sterols enter midgut enterocytes through a membrane transporter (*NPC1b*) (Li and Jing 2020). Phytosterols reach the endoplasmic reticulum where they usually undergo a well-described multi-step dealkylation to desmosterol and finally to cholesterol in phytophagous insects (Svoboda et al. 1982; Svoboda and Weirich 1995; Fujimoto et al. 1985; Ciufo et al. 2011). Cholesterol (and remaining phytosterols) is then transferred to a haemolymph carrier protein (lipophorin: Lp) at the basal face of enterocytes and distributed through the haemolymph to the different tissues. Studies on *Bombyx mori* have shown that the brain and prothoracic glands are relatively cholesterol-rich while the midgut and fat body are phytosterol-rich (Takeshima et al. 2019, 2022) Cholesterol is transferred into insect tissues by lipophorin receptor (LpR)-mediated endocytosis; LpR is ubiquitously expressed in insect tissues. Phytosterol uptake involves variants of the scavenger receptor SRB1, which are expressed in the midgut and fat body and appear to be more or less specific for particular phytosterols (Takeshima et al. 2022). All these mechanisms are summarised in Fig. 4.

2.3 Dealkylation of C-24 Alkyl Sterols

The dealkylation process requires 4–5 steps to produce cholesterol, and the sequence is similar whatever the starting phytosterol (Fig. 5; Ikekawa et al. 1993; Svoboda and Weirich 1995). The pathways were elucidated using a combination of radiolabelled phytosterols and inhibitors of desmosterol reductase (the last, and common, step in the branched pathway) which resulted in the build-up of intermediates (e.g. Svoboda et al. 1989). The only enzyme in the pathway which has so far been extensively characterised is desmosterol reductase (from *Bombyx mori*; Ciufo et al. 2011). It is not currently clear if each branch of the pathway uses the same enzymes or whether phytosterol-specific enzymes exist; the finding that certain insects can dealkylate some phytosterols, but not others (e.g. grasshoppers (Behmer and Elias 2000), which are able to dealkylate sitosterol and campesterol, but not stigmasterol), may indicate that the latter is the case.

Dealkylation is usually not complete, with the result that tissues of phytophagous insects contain a mixture of sterols, where cholesterol is usually the major component (Takeshima et al. 2019). The sterols are not equally distributed between tissues, with the highest percentages of

Fig. 4 Absorption and transformation of dietary sterols. (Composite scheme after Jing and Behmer 2020; Li and Jing 2020; Takeshima et al. 2019, 2022. *Lp* lipophorin, *Lp-R* Lipophorin receptor, *SRB1* scavenger receptor. The black dot represents a sterol molecule)

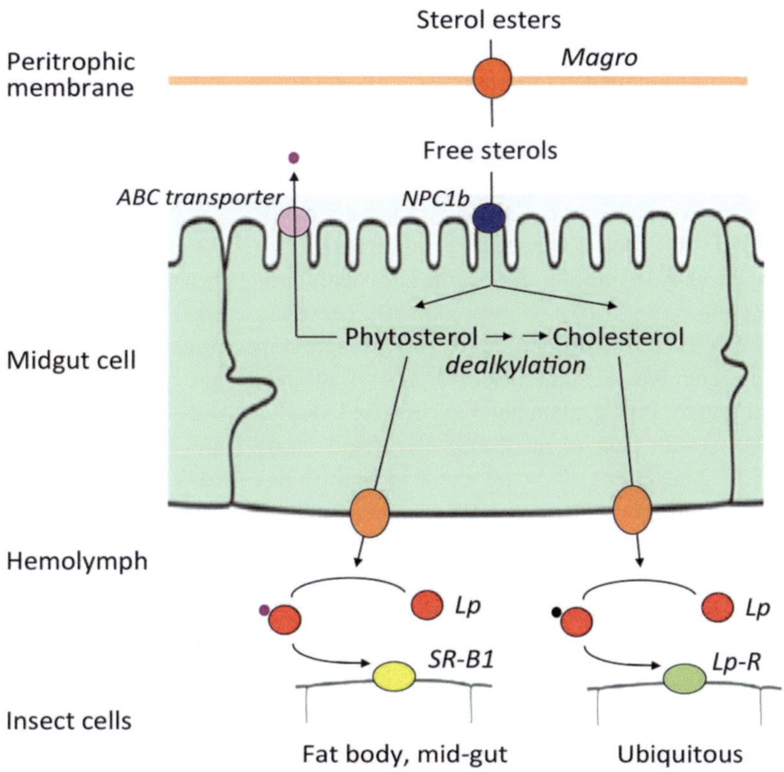

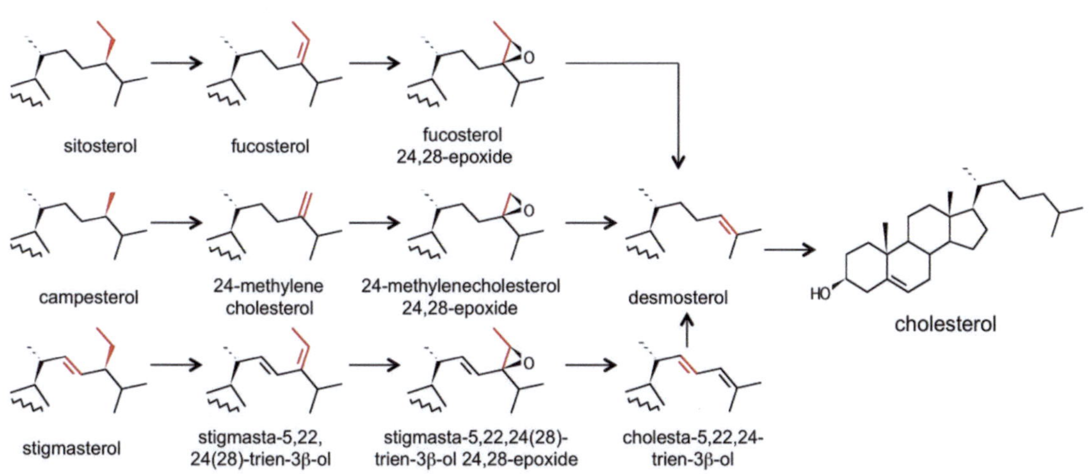

Fig. 5 The dealkylation process of the phytosterols sitosterol (C_{29}), campesterol (C_{28}) and stigmasterol (C_{29}, Δ^{22}). (Redrawn from Ikekawa et al. 2013)

cholesterol being found in the brain and the prothoracic glands, which suggests the presence of specific cholesterol uptake mechanisms (Takeshima et al. 2019). Indeed, it has been shown that prothoracic glands have a specific process for cholesterol internalisation (Blaszek and Mala 1978; Igarashi et al. 2018).

Some (phytophagous) insects (e.g. cockroaches and bees) lack the dealkylation enzymes, and they have to cope with the sterols they find in their diet, e.g. 24-methylenecholesterol (Chakrabarti et al. 2020; Tsuruda et al. 2021). This has limited consequences, at least as concerns the structural role of sterols in cell membranes (see below), but it may impact ecdysteroids (see Sect. 3).

Finally, it appears that not all insects are auxotrophic for sterols. Indeed, some insects can obtain sterols through endosymbiosis with prokaryotic or yeast-like microorganisms. These are present in specialised structures (mycetocytes) in representatives of some insect Orders (Blattodea, Hemiptera, Phthiraptera and Coleoptera: Douglas 1988, 1989; Suh et al. 2001; Nasir and Noda 2003). These species are all characterised by living on nutritionally poor diets (e.g. sap-sucking insects such as aphids and planthoppers) which provide low levels of sterols. The symbionts produce mainly ergosterol, which is converted by insects to 7-dehydrocholesterol or cholesterol (Nasir and Noda 2003). Sterols could be produced by the bacterial symbionts as well as the yeast-like symbionts (Wei et al. 2016).

2.4 Dietary Choice of Sterols

As explained above, dietary sterols fulfil structural and metabolic roles in insects. Initial studies (Clayton et al. 1964; Lasser et al. 1966; Dutky et al. 1967) led to the hypothesis that most insects have the capacity to spare certain sterols, whereby the first call on the available cholesterol would be for metabolic purposes (e.g. conversion to ecdysteroids), whereas phytosterols and remaining cholesterol would fulfil the structural functions. More recent studies (summarised in Behmer 2017) have revealed that the situation is more complex, since feeding insects on defined mixed sterol diets has shown that how well the insects thrive depends not only on the identities and absolute amounts of sterols but also on the proportions of the sterols in the diet. For example (Behmer et al. 1999; Behmer and Elias 1999, 2000), nine generalist grasshopper species were reared on diets containing sitosterol (able to be dealkylated) and stigmasterol (unable to be dealkylated) in defined proportions and amounts. On the sitosterol-only diet, the insects completed development, while on the stigmasterol-only diet, they died in the third or fourth instar. When fed sitosterol: stigmasterol at 3:1, the insects developed into adults as well as they did on the sitosterol-only diet. At 2:1, the survival rate was ca. 50%, and at 1:1, the mortality was >90%. As generalist feeders, the grasshoppers can be expected to encounter both sitosterol and stigmasterol in significant amounts, and the above findings help to explain why one of the grasshopper species, *Schistocerca americana*, is unable to develop on spinach where the dominant sterols are spinasterol ($\Delta^{7,22}$) and dihydrospinasterol (Δ^{7}) (Behmer and Elias 2000). Thus, the omnivory of the grasshoppers can be limited by the sterol composition of the plants they encounter and not just by the ability to derive adequate cholesterol and enough total sterol.

Insects seem to have the ability to detect sterols (Ugine et al. 2021). Moreover, the nature of dietary sterols may influence their fitness and even modify their feeding behaviour (Ugine et al. 2019), as indicated by experiments with ladybird beetles (*Coccinella septempunctata*). The authors observed that insects consuming an all-prey diet (i.e. aphids, which are sterol-poor) had normal growth and development, but did not reproduce (owing to the lack of spermatogenesis). However, reproduction was restored by the administration of phytosterols or cholesterol or herbivory, in order to compensate for the sterol-poor diet. Very interestingly, experiments show that a sterol appetite is able to turn a predator into omnivory by extending their diet to include plant material (Ugine et al. 2019). Similarly, a search for suitable sterol sources operates in bees by selecting an adequate range of plant pollens in an environment which changes with seasons (Vanderplanck et al. 2019).

2.5 Cholesterol as a Component of a Signalling Molecule (Hedgehog)

Sterols may display unexpected functions even in the absence of any conversion to steroids. The most investigated one is for cholesterol when bound to hedgehog (Hh) protein (reviewed in Ingham 2022; Sigafoos et al. 2021; Kandel and Wang 2022), which has a role in the development of most metazoans, including insects. In fact, the importance of Hh was first identified from a mutation in *Drosophila melanogaster* affecting embryological development. The hedgehog precursor protein comprises three regions, an N-terminal signalling sequence (SS) which is responsible for targeting the protein to the endoplasmic reticulum, the main Hh signalling domain (HhN) and a C-terminal auto-processing domain (HhC), which also incorporates a sterol recognition region (SRR). The precursor undergoes cleavage of the SS in the ER and then autoprocessing and maturation to produce the active signalling molecule, which involves internal splicing of the peptide chain, covalent attachment of a cholesterol molecule to the new C-terminal amino acid (Gly) and palmitoylation of the N-terminal Cys. This double lipidation is essential for both the correct transport and signalling potency of HhN. The precursor molecule interacts with the lipid membrane to abstract a molecule of cholesterol from it (by a currently unknown mechanism) and binds it non-covalently to the SRR. The splicing activity inherent in the HhC domain creates an internal thioester to what will become the terminal Gly of HhN and then cholesterolysis of this thioester bond with the cholesterol molecule bound to SRR. The sterol specificity is not absolute for cholesterol, as various native and non-natural sterols and steroids can act as substrates, but there is a clear requirement for a 3β-OH. The auto-processing of Hh is inhibited by Zn^{2+}-ions. The covalently bound cholesterol serves as a lipidic anchor to bind mature HhN to cell membranes (Kandel and Wang 2022). The membrane anchoring is reinforced by the palmitoylation of the NH_2-terminus (Ingham 2022). The hedgehog signalling pathway is very complex and involves a large set of factors, the details of which are outside the scope of this chapter. Defects in hedgehog signalling have also been implicated in the development of various types of cancer in humans (Sigafoos et al. 2021).

2.6 Ergosterol in Antibacterial Protection

Another unexpected function was recently discovered for ergosterol: this fungal sterol was shown able to protect *Drosophila* against bacterial infection (Deng et al. 2022). These authors showed that ergosterol was required for host defence; haemolymph levels of ergosterol increased after bacterial infection, and ergosterol supplementation enhanced host survival. Understanding the detailed mechanism still requires additional experiments.

2.7 Insect Pest Control Strategies Involving Modified Sterol Profiles

The nutritional difference between insects and mammals in how sterols are obtained and metabolised is biochemically and physiologically very significant. The absence of de novo biosynthesis of cholesterol in insects, which means that most phytophagous insects are dependent on the dealkylation of phytosterols, and their further metabolism to the ecdysteroids, a class of steroid hormone which is chemically distinct from vertebrate steroid hormones, provide three targets, i.e. phytosterol dealkylation, ecdysteroid biosynthesis/metabolism and ecdysteroid mode of action which are a priori interesting for the development of more specific pest control agents. The first of these will be summarised below, while the other two will be briefly covered in Sect. 3.

The first development in this area concerned the finding that certain agents (triparanol and 22,25-diazacholesterol), which had a hypocholesterolaemic effect in mammals by inhibiting the Δ^{24}-reductase that converts desmosterol to cholesterol, also inhibited dealkylation in *Manduca sexta* and other phytophagous insects by inhibiting the same enzyme activity. SAR studies identified two further azasteroids (25-azacholestane and 25-azacoprostane), which were 200-fold more active than 22,25-diazacholesterol and were inhibitory at dietary concentrations of 1 ppm against

M. sexta. Unfortunately, these could not be developed further because the target enzyme is also present in vertebrates and because of the cost of chemical synthesis of agents containing a steroid nucleus. Inexpensive non-steroidal analogues were identified which had low vertebrate toxicity, but these were less potent against dealkylating insects and were also phytotoxic (reviewed in Svoboda and Weirich 1995).

29-Fluorophytosterols have been proposed as potential pro-insecticides to control insect pest species which dealkylate phytosterols (Prestwich et al. 1983, 1984). C-24-dealkylation of such phytosterol derivatives results in the release of fluoroacetate, which is converted to fluorocitrate by the TCA cycle and is toxic to the insect, causing reductions in growth, weight and survival. 29-Fluorostigmasterol was found to be 100-fold more active than 29-fluorositosterol in *Manduca sexta*, owing to the presence of a 22(23)-double bond.

Also, we may wonder whether a modification of phytosterol profiles of cultivated plants could be able to make them unsuitable for supporting phytophagous insect development. Costet et al. (1987) reared *Locusta migratoria* adults on wheat seedlings treated with the fungicide fenpropimorph. The seedlings accumulated 9β,19-cyclopropylsterols (95% of total sterols) in the place of the usual phytosterols. These modified sterols are unsuitable as precursors for insect ecdysteroid production, and the eggs laid by *Locusta* females contained lower ecdysteroid levels and failed to develop normally. Control experiments showed that fenpropimorph was not directly toxic to the locusts and that the effect could be reversed by the application of exogenous cholesterol to fenpropimorph-treated wheat seedlings. Genetic modification of tobacco plants such that they accumulate the ketosteroids cholestan-3-one and cholest-4-en-3-one in the phloem sap makes them resistant to the aphid *Myzus persicae* (Behmer et al. 2011; Bouvaine et al. 2014). Whether such strategies could be developed practically remains to be demonstrated.

3 Ecdysteroids

3.1 Introduction

Ecdysteroids are the steroid hormones of insects and, more generally, of all arthropods (Lafont et al. 2017). They have multiple functions, the primary ones being the control of development (Kozlova and Thummel 2000), including embryonic development (Sonobe and Yamada 2004), metamorphosis (Truman and Riddiford 2002; Belles 2020), developmental arrests (diapauses; Du et al. 2022; Hutfilz 2022) and reproduction (Roy et al. 2018). Their structures differ significantly from those of vertebrate steroid hormones, as they have conserved the full carbon skeleton of their sterol precursors and differ in their greater polarity and the 3D shape of the steroid ring structure. They represent a unique system as (i) the same molecules are produced by different organs in larvae (moulting glands) and adults (gonads and accessory glands) and (ii) they have pleiotropic functions. The major ecdysteroid hormone is 20-hydroxyecdysone (20E), but some variants do exist in relation to the major sterol precursors present in insects from which they derive (Fig. 6).

Many natural ecdysteroidal compounds have been isolated and identified, comprising major and minor compounds from invertebrates, plants and fungi; biosynthetic intermediates; and further metabolites and deterrent compounds (see the compilation in Ecdybase; Lafont et al. 2002) Owing to the wide structural diversity, ecdysteroids have been classified into true ecdysteroids and ecdysteroid-like compounds (Lafont and Horn 1989). The true ecdysteroids are compounds where the steroid nucleus possesses a *cis*-fused A/B-ring junction, a 7-en-6-one chromophore and a 14α-hydroxyl group. The

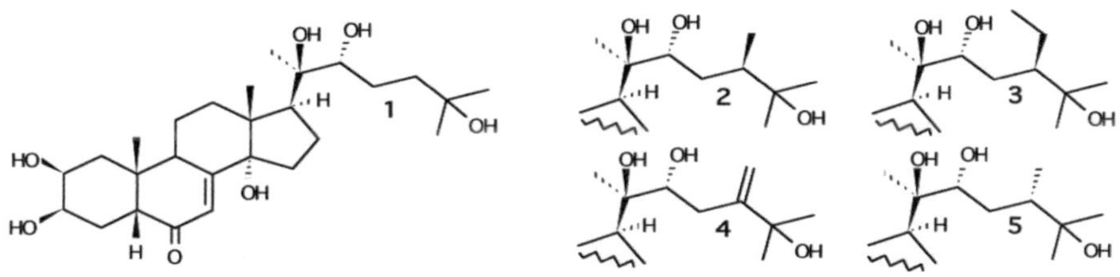

Fig. 6 Structure of 20-hydroxyecdysone (1) and related moulting hormones: makisterone A (2), makisterone C (3), 24,28-dehydromakisterone A (4) and 24-*epi*-makisterone A (5)

ecdysteroid-related compounds are analogues which do not fulfil all of these structural criteria.

3.2 Zooecdysteroids

Ecdysteroids alkylated at C-24 are found in phytophagous or mycophagous insects unable to dealkylate sterols (Fig. 6). Thus, makisterone A was first identified in milkweed bug embryos (Kaplanis et al. 1975), then in other bugs (Aldrich et al. 1982), honey bees (Feldlaufer et al. 1986) and *Drosophila* (Redfern 1984; Feldlaufer et al. 1995). Makisterone C was identified in embryos of the cotton-stainer bug (Feldlaufer et al. 1991). 24-*Epi*-makisterone A was identified in a leaf-cutting ant (Maurer et al. 1993), and 24,28-dihydro-makisterone A was identified in *Drosophila* (Lavrynenko et al. 2015). Interestingly, it was shown that *Drosophila* is able to adapt steroidogenesis to the sterols found in its diet and accordingly produce 20E and/or the three C-28 moulting hormones as shown in Fig. 6 (Redfern 1986; Feldlaufer et al. 1995; Lavrynenko et al. 2015).

3.3 Ecdysteroid Biosynthesis in Insects

Steroidogenesis takes place in moulting glands (and possibly in the epidermis in some cases) of insect larvae and in gonads of adults (follicle cells of ovaries or interstitial tissue of testes) (Lafont et al. 2017). In male adults of the mosquito *Anopheles gambiae*, however, ecdysteroid production takes place in accessory glands (Pondeville et al. 2008). In all cases, steroidogenic cells produce either ecdysone (or a C24-alkylated analogue) or 3-dehydro-ecdysone, i.e. precursors of the active hormones, that will require one or two further steps to be converted to the active 20-hydroxyecdysone. Those reactions take place in haemolymph or peripheral tissues, e.g. fat body, epidermis or Malpighian tubules.

Ecdysteroids are polyhydroxylated molecules and are rather water soluble (up to 8 mg/mL for 20E) and consequently do not readily cross cell membranes. Thus, ecdysone secretion by moulting gland cells takes place by exocytosis (Birkenbeil 1983; Yamanaka et al. 2015; Pan et al. 2021) and, conversely, 20E enters into target cells by means of the membrane ecdysteroid importer EcI, which belongs to the organic anion transporter (OATP) family (Okamoto et al. 2018; Okamoto and Yamanaka 2020; Masterson et al. 2022; Samantsidis et al. 2022). The situation is in fact still more complex, since three additional *EcI* genes (EcI 2–4) have been described, which may have different species-specific roles to control development and reproduction (Hun et al. 2022).

Ecdysteroid biosynthesis is a multi-step process not yet fully elucidated (Fig. 7); the early steps still resist elucidation and constitute the so-called black box (Warren et al. 2009; Niwa and Niwa 2014; Pan et al. 2021). It is assumed that a similar pathway operates for 24-alkylated ecdysteroids. The elucidation of the black box is under active investigation. It is a difficult topic as early intermediates do not accumulate and are possibly unstable and short-lived; they are still a matter of controversy (Shimmel and O'Connor 2022).

The early experiments aimed at elucidating the biosynthetic pathway were mainly based on

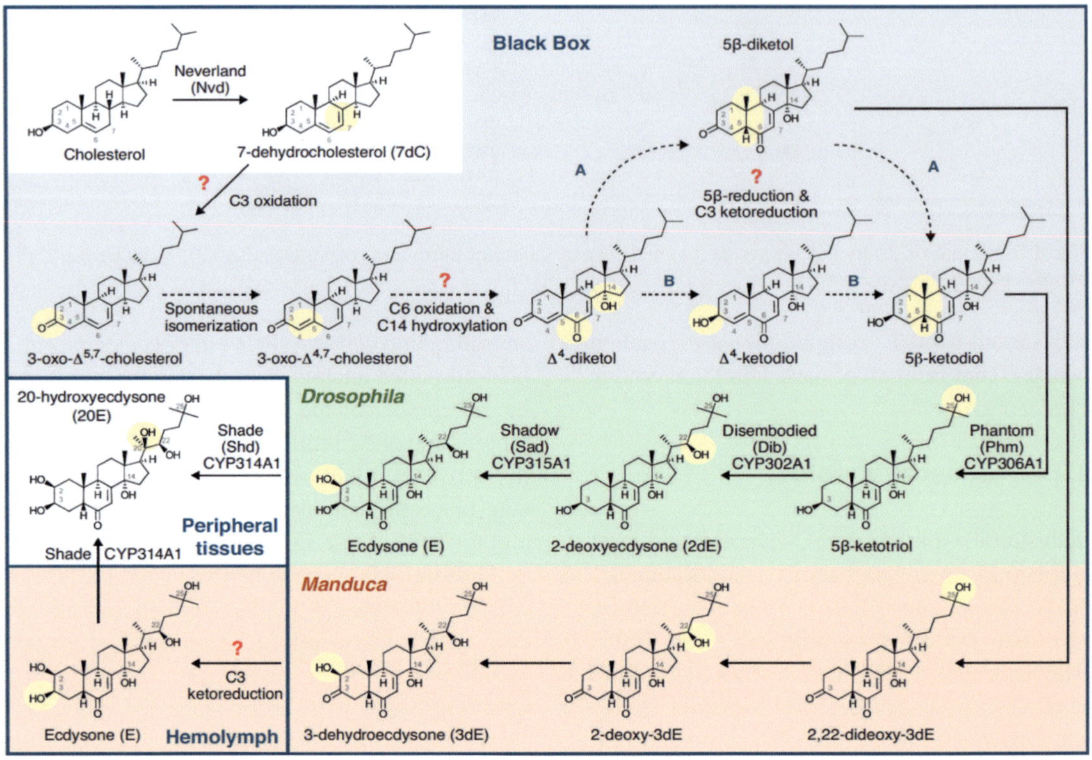

Fig. 7 Summary of the biosynthetic pathway for 20-hydroxyecdysone in *Drosophila melanogaster* (Diptera) and *Manduca sexta* (Lepidoptera). The early part of the pathway takes place in the ring gland (*Drosophila*) or prothoracic gland (*Manduca*), and the prohormones (3dE and/or E) are released into the haemolymph and converted to 20E (the major, hormonally active ecdysteroid) in peripheral tissues. The C6-oxidation and 14-α-hydroxylation reactions are as yet uncharacterised (the "Black Box"). (Figure reproduced from Pan et al. 2021 with permission)

chemistry, i.e. the synthesis of (radioactive) putative intermediates and the determination of their metabolic fate in vitro or in vivo. This approach allowed the identification of the sequence from ketodiol (2,22,25-trideoxyecdysone) to ecdysone (e.g. Meister et al. 1985) and to investigate potential early intermediates. Then the enzymes involved in these reactions were biochemically characterised and their subcellular locations were determined (Warren and Hétru 1990). A key advance then came from the study of Halloween developmental mutants of *Drosophila melanogaster*, which allowed the identification of the genes encoding most of the enzymes involved in 20-hydroxyecdysone biosynthesis (Gilbert 2004; Rewitz et al. 2006). This knowledge is of particular importance for the study of the regulation of ecdysteroid production and to identify the different factors involved in the control of ecdysteroid titres.

3.4 Further Metabolism of Ecdysteroids in Insects

The correct modulation of hormonal titres depends not only on the rates of synthesis and release of the hormone but also on the appropriate further metabolism to less active compounds and excretion. Thus, for example, during larval development in insects, the rising peak of 20E at the end of the interphase will induce apolysis (separation of the old cuticle from the new one), but ecdysis itself (shedding of the old cuticle) takes place effectively only during a falling 20E titre.

It is convenient to divide the further metabolism of ecdysteroids into three areas depending on the purpose of the metabolism: (i) deactivation of hormonal activity, (ii) creation of storage forms of ecdysteroids and (iii) detoxification of ingested phytoecdysteroids.

3.4.1 Deactivation of Hormonal Activity

The pathways and enzymes of the further metabolism of ecdysone have been reviewed previously (Lafont and Connat 1989; Weirich 1989; Rees 1995; Lafont et al. 2017). The types of reactions involved in the further metabolism of ecdysone and 20-hydroxyecdysone are summarised in Fig. 8.

Some of these reactions (20-hydroxylation, 26-hydroxylation, formation of 3-dehydroecdysteroids and ecdysonoic acids) appear to be widespread in insects occurring in many organs, while others are restricted to specific tissues (e.g. 3-*epi*-ecdysteroid formation in the midgut and conjugation in the gut, Malphigian tubules and fat body). Over 50 ecdysteroid metabolites have been identified to date (see Lafont et al. 2002 for structures, physico-chemical data and relevant references). From an analytical and chemical point of view, the metabolites can be divided into three classes: (i) ionic and non-ionic polar (ecdysonoic acids, phosphate conjugates and glucosides), (ii) medium polarity (3-dehydroecdysteroids, 3-*epi*-ecdysteroids, 26-hydroxyecdysteroids, 14-deoxyecdysteroids and side chain cleavage products) and (iii) non-polar (acetyl- and acyl [fatty-acid]-derivatives). There is variation in the metabolic routes depending on species, developmental stage and the tissues involved. The 3-dehydroecdysteroids may be activation or deactivation (in their own right or as intermediates in the formation of 3-*epi*-ecdysteroids) products, since, depending on the assay system they may be more (Sommé-Martin et al. 1988, 1990) or less (Dinan 2003) active than the parent 3β-hydroxyl ecdysteroids. 14-Deoxyecdysteroids, which have been identified as metabolites after injection or ingestion of ecdysone and 20-hyroxyecdysone in the cricket *Gryllus bimaculatus* (Hoffmann et al. 1990), are believed to be products of the gut

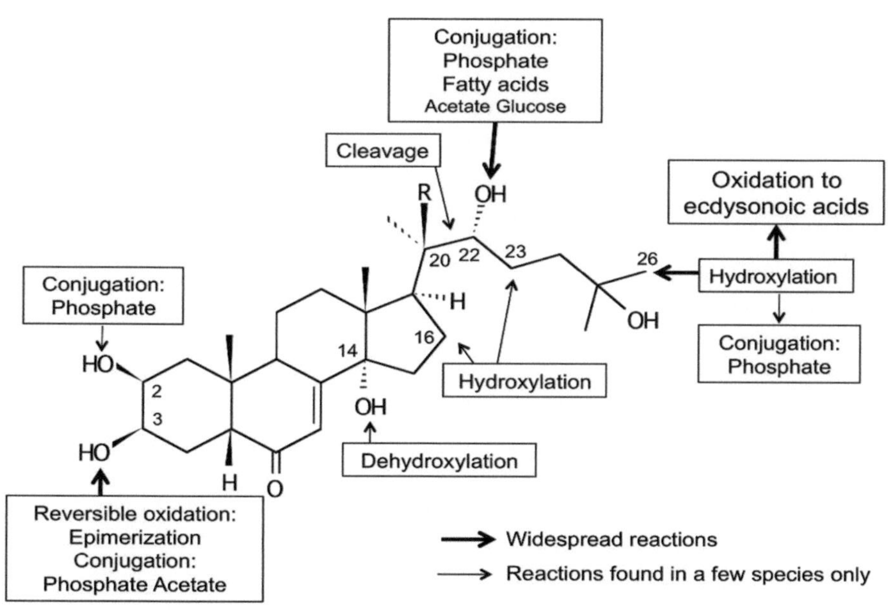

Fig. 8 Summary of the routes for further metabolism of ecdysone (R = –H) and 20-hydroxyecdysone (R = –OH) in insects. (Reproduced and modified from Lafont et al. 2017)

flora, rather than being produced by insect tissues. The true significance of side chain cleavage metabolites (e.g. poststerone and rubrosterone) in insects is currently difficult to assess properly, because most metabolic studies have been performed with side chain radiolabelled E or 20E, with the result that the radioactive labelling would be lost during the formation of any side chain cleavage products. However, poststerone has been identified as a metabolite of injected 20E in the blowfly *Calliphora stygia* (Galbraith et al. 1969) and the silkworm *Bombyx mori* (Hikino et al. 1975).

3.4.2 Creation of Storage Forms of Ecdysteroids

Relevant previous reviews can be found (Hoffmann and Lagueux 1985; Isaac and Slinger 1989; Lafont et al. 2017). The highest concentrations of ecdysteroids in insects are often associated with maturing ovaries and the eggs derived from them. This was first investigated extensively for *Bombyx mori* (Mizuno and Ohnishi 1975) and then in the locusts *Locusta migratoria* (Lagueux et al. 1984) and *Schistocerca gregaria* (Dinan and Rees 1981). The majority of the ecdysteroids are present in the form of polar 22-phospho-conjugates. The ecdysteroids are synthesised by the follicle cells surrounding the oöcytes and passed, along with a small amount of free hormone (mainly E and 2dE), into the developing egg to be associated with the yolk. It appears that the conjugates serve as a hormonal storage form which can be hydrolysed by the developing embryo to release free active ecdysteroids until such time as the embryo is itself capable of biosynthesising ecdysteroids. As the egg is a closed system, all the incorporated maternal ecdysteroids remain in the system, and its metabolism can be followed until hatching. The 22-phosphotransferase from *S. gregaria* follicle cells has been characterised (Kabbouh and Rees 1991). There are peaks of free ecdysteroid which coincide with the formation of several membranes and cuticles during embryogenesis (Sall et al. 1983) and an ecdysteroid phosphate phosphatase is present, the activity of which increases at the time of greater utilisation of the ecdysteroid 22-phosphate store (Isaac et al. 1983). A similar enzyme has been purified and cloned from *B. mori* (Yamada and Sonobe 2003). As embryogenesis in locusts proceeds, other (presumably inactivation) metabolites accumulate; ecdysteroid 2-phosphate 3 acetate derivatives, 3-*epi*-2-deoxyecdysone 3-phosphate and ecdysteroid 26-oic acids (Rees and Isaac 1984). At blastokinesis, the ecdysteroids become associated with the embryo (Scalia et al. 1987). In certain insects, embryonic synthesis of ecdysteroids occurs in the latter stages of embryogenesis once the embryonic prothoracic glands have developed. Similarly, an extensive range of ecdysteroid metabolites has been identified during egg development in *B. mori* (Sonobe and Yamada 2004). In *Manduca sexta*, the major conjugate in newly laid eggs is 26-hydroxyecdysone 26-phosphate, which is hydrolysed and 20-hydroxylated to give 20,26-didhydroxyecdysone as the free (active) hormone, which in turn is then inactivated by 22-glucosylation, 3-epimer formation and oxidation to 26-oic acids (Warren et al. 1986; Thompson et al. 1988).

3.4.3 Detoxification of Ingested Phytoecdysteroids

This involves many of the reactions mentioned above, but will not be covered here as it is the subject of a recent review (Dinan and Lafont 2022).

3.5 Physiological Effects and Functions of Ecdysteroids in Insects

Ecdysteroids undergo large variations in their haemolymph titres and show periodic peaks in larvae, which precede/induce each larval moult, and usually a large peak during metamorphosis. Ecdysteroids produced in adults play diverse roles relative to reproduction. Ovarian ecdysteroids fulfil different functions in relation to the duration of insect adult life and either single or multiple oögenesis cycles (Swevers 2019). Ovarian ecdysteroids may either circulate and induce vitellogenin synthesis in fat body, as in mosquitoes (Hagedorn et al. 1975), induce meiotic reinitiation (Lanot et al. 1988), or be stored as inactive conjugates in maturing oöcytes (Tsoupras et al. 1982) to be later hydrolysed and used during early embryonic development, until endogenous synthesis becomes active (Yamada and Sonobe 2003; Fujinaga et al. 2020). The roles of ecdysteroids in vitellogenesis and meiosis control are reminiscent of those of ovarian steroids in vertebrates (e.g. amphibians).

Ecdysteroids may have additional roles in insect reproduction: in some insects, they control the production of sex pheromone (Blomquist et al. 1995; Chiang et al. 2016), or the behavioural response to sex pheromone (Abrieux et al. 2013). *Anopheles gambiae* males transfer ecdysteroids to females during mating, and this triggers vitellogenesis (Pondeville et al. 2008); moreover, this transfer turns off female receptivity after mating (Gabrieli et al. 2014; Peng et al. 2022).

Ecdysteroids are also involved in the regulation of insect immune defence. They activate the production of antimicrobial peptides in *Helicoverpa armigera* fat body (Wang et al. 2014), of immulectin-1 (a defence against entomogenous fungi) in *Locusta migratoria* (Han et al. 2020) or of lysozyme in *Bombyx mori* (Ma et al. 2019).

3.6 Modes of Action of Ecdysteroids in Insects

The mechanism of action of ecdysteroids illustrates the key role they have in the development of our general understanding of how steroid hormones control gene activity. At first, the elucidation took advantage of the giant polytenic chromosomes present in the salivary glands of chironomid larvae: these chromosomes present a banding which allows their mapping, appearing like a barcode, and local expansions (puffs) where chromatin is decondensed and which represent sites of active transcription. The in vitro induction of "puffing" by ecdysteroids in salivary glands allowed the visualisation of a direct relationship between hormones and gene activity (Clever and Karlson 1960) and the conceptualisation of a mechanism similar to the regulation of the lactose operon in *Escherichia coli* according to Jacob and Monod (1961). It was observed that ecdysteroids induced a characteristic sequence of puffs, which allowed the distinction of early and late effects, which were interpreted originally by Ashburner et al. (Ashburner et al. 1974) (Fig. 9).

Ecdysteroid receptors (EcR) belong to the family of nuclear receptors, and upon ligand binding,

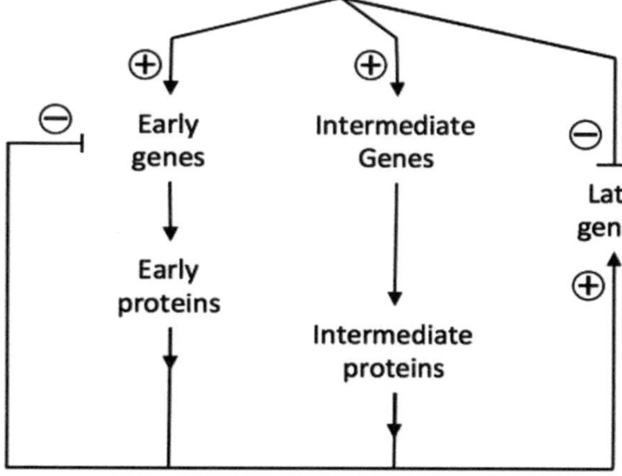

Fig. 9 20E evokes sequential gene activation (early genes, intermediate genes, then late genes); the early gene products are transcription factors necessary to induce the late effects. Note that 20E removal is also required for late gene activation

they heterodimerise with the ultraspiracle (USP/RXR) protein (Hill et al. 2013). The active complex will bind to specific sequences of DNA (ecdysteroid-response elements or ECREs), which control the transcription of specific genes. Interestingly, both EcR and USP exist as multiple isoforms, which allow several active complexes to be formed. Different isoforms are expressed in different tissues, and this explains why the prepupal hormonal peak induces apoptosis of larval tissues and proliferation of imaginal cells (Talbot et al. 1993; Morrow and Mirth 2024).

We may wonder whether in a given insect species the ligand-specificity of EcR is consistent with the nature of ecdysteroids produced. Indeed, in female milkweed bugs, makisterone A, the endogenous hormone, is tenfold more active than 20E in inducing cuticulogenesis and inhibiting vitellogenesis (Aldrich et al. 1981), which may mean that some adaptation in the EcR protein ligand-binding site has taken place. The situation in *Drosophila melanogaster* is of special interest, as this insect produces four different ecdysteroids according to sterol supply. *Drosophila* receptor affinity for different moulting hormones produced is 24,28-dehydromakA (K_D 4.0 x 10^{-9} M) > 20E (7.6 x 10^{-9} M) > MakA (1.3 × 10^{-8} M) > > 24-epimakA (2.2 × 10^{-7} M) (Dinan and Hormann 2005). Interestingly, it has been shown that *Drosophila* can selectively use some sterols, even if very minor, to produce ecdysteroids (i.e. traces of ergosterol precursors to make 24,28-dehydromakisterone A and traces of cholesterol to synthesise 20-hydroxyecdysone, i.e. the two best ligands of EcR). It is notable that 24-*epi*-makisterone A, which has the lowest affinity for EcR, supports only larval development, whereas the other ecdysteroids allow full adult development (Lavrynenko et al. 2015).

According to Taylor and Heyland (2022), there are three major classes of steroid signalling: (1) genomic signalling via the nuclear receptor, (2) non-genomic signalling via other receptors, and (3) non-genomic signalling via the nuclear receptor. Indeed, rapid non-genomic effects have been described for ecdysteroids, and even the puffing activity itself received an alternative explanation as being a consequence of changing ionic environment ([Na^+]/[K^+]) on chromatin (Kroeger and Lezzi 1966). Applebaum and Gilbert (1972) described the stimulation of adenyl cyclase in pupal wing epidermis by 20E, and a lot of other rapid effects have been reported (Schlattner et al. 2006). At that time, it was difficult to decide whether these effects involved EcR itself or another receptor, but later on a true membrane receptor for 20E was identified, in *Drosophila*, belonging to the GPCR family of receptors (Srivastava et al. 2005). This receptor was also able to bind dopamine and was named DopEcR. Interestingly, its ecdysteroid binding activity is very different from that of EcR and it does not bind makisterone A (Srivastava et al. 2005). Further experiments on this receptor explored its distribution (mainly in the nervous system) and its roles in sex pheromone perception in *Agrotis ipsilon* (Abrieux et al. 2013, 2014) and feeding cessation before metamorphosis in *Helicoverpa armigera* (Kang et al. 2019).

Other scientists working on *Helicoverpa armigera* isolated two other 20E-responsive GPCRs designated as ErGPCR-1 and ErGPCR-2 (Cai et al. 2014; Wang et al. 2015). These are different from DopEcR as they belong to another subfamily of GPCRs, and their activation rapidly increases intracellular Ca^{2+} concentrations. More recently, a third receptor, ErGPCR-3, was identified in the same species, the activation of which evokes rapid increases in calcium, cAMP and protein phosphorylation (Kang et al. 2021). Interestingly, upon 20E binding, this receptor undergoes homotetramerisation and promotes 20E entry into the cells.

Thus, membrane actions of 20E appear very complex, with several receptors activating different signalling cascades and being required for normal development and controlling behaviour (reviewed by Zhao 2020). Moreover, this list is perhaps not complete, if we consider 20E detection by gustatory receptors (GRs), since GRs do not seem to correspond to canonical GPCRs, but to transmembrane ligand-gated channels which are inverted with regard to GPCRs (i.e. the N-terminus in intracellular and the C-terminus is extracellular, or both termini are either intracellular or extracellular (Zhang et al. 2011; Xu 2020; Morinaga et al. 2022).

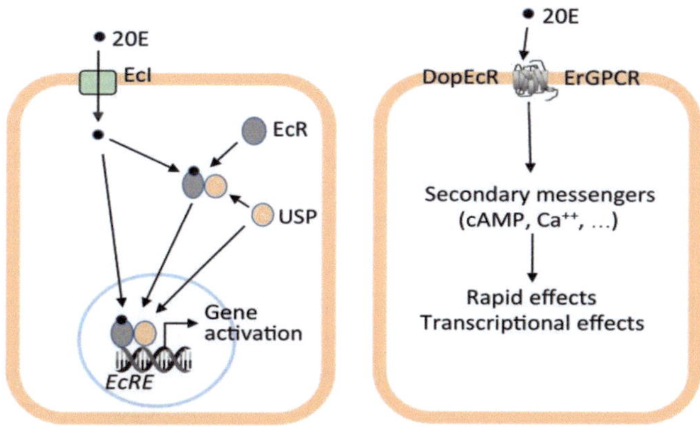

Fig. 10 Genomic and non-genomic effects of ecdysteroids (inspired by Taylor and Heyland 2022); *20E* 20-hydroxyecdysone, *EcI* ecdysteroid importer, *EcR* ecdysteroid receptor, *USP* ultraspiracle/retinoid X receptor, *EcRE* ecdysteroid response element, *DopEcR* dopamine/ecdysteroid receptor, *ErGPCR* 20E-activated G-protein-coupled receptor

Taylor and Heyland (2022) describe a third type of mechanism of action, non-genomic nuclear receptor signalling. It proceeds via multiple mechanisms that often involve phosphorylation cascades via MAPK or PI3K. It often involves membrane-associated forms of nuclear receptors formed by variant splicing (truncated forms) anchored to the membrane by scaffolding with GPCRs or anchored post-translational fatty acid coupling. Such systems are well known for vertebrate steroid hormones (Pedram et al. 2007). Whether it exists also in insects remains to be investigated. Our present knowledge of 20E actions is summarised in Fig. 10.

3.7 Non-hormonal Ecdysteroids

We also wish to mention briefly the role of exogenous ecdysteroids. These molecules are also produced by many plants (phytoecdysteroids) and thus represent potential endocrine disrupters for phytophagous insects (Dinan 2009). Some insects have developed efficient detoxification mechanisms in order to protect themselves against ingested ecdysteroids but, more interestingly, many are able to detect these molecules thanks to bitter taste receptors on their mouth parts, which will evoke anti-appetite effects and deter them from eating ecdysteroid-containing foods (Marion-Poll and Descoins 2002). Similarly, taste receptors present on legs may prevent egg-laying on ecdysteroid-containing plants (Calas et al. 2006).

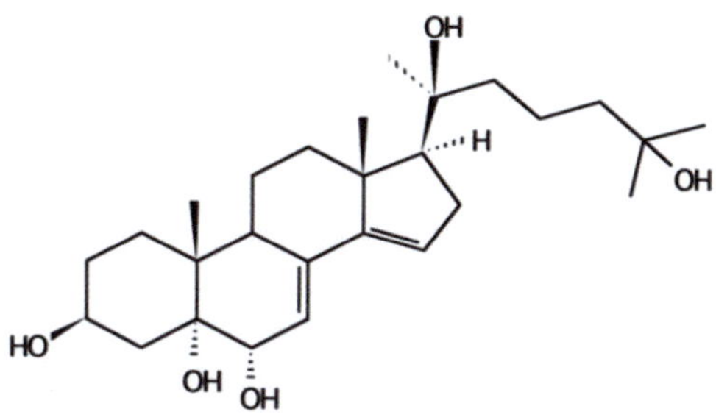

Fig. 11 The structure of bombycosterol

Finally, it is worth mentioning that the leaf beetle *Chrysolina carnifex* (Coleoptera: Chrysomelidae) possesses a defence gland secreting ecdysteroids, which are released upon attack and would protect it against spiders (Laurent et al. 2003). This situation is reminiscent of the pycnogonid (sea spider) *Pycnogonum litorale*, whose epidermis accumulates and releases ecdysteroids to deter crabs from eating them (Tomaschko 1994).

3.8 Bombycosterol

Bombycosterol (Fig. 11) is an unusual ecdysteroid-like compound which has been isolated so far only from mature ovaries of *Bombyx mori* (Fujimoto et al. 1985; Hiramoto et al. 1988; Ohnishi et al. 1989). It is also present as the 3-phosphate conjugate. Bombycosterol has low biological activity in the *D. melanogaster* BII cell bioassay (reflecting its affinity for DmEcR/DmUSP receptor complex), but its function (if any) in *B. mori* is currently unknown.

3.9 Insect Pest Control through Disruption of Ecdysteroid Metabolism and Mode of Action

The structural differences between ecdysteroids and vertebrate steroid hormones and the reliance of insects on a sole class of steroid hormones make the metabolism and mode of action of ecdysteroids interesting targets for the control of insect pest species as one might expect such agents to be more specific than conventional insecticides.

3.9.1 Enzyme Inhibitors

A series of acetylenic and allenic cholesteryl derivatives were prepared (Burger et al. 1988; 1989; Mauvais et al. 1993, 1994), which were shown to specifically inhibit the C-22- and C-25-hydroxylases of ecdysteroid biosynthesis in ovaries and prothoracic glands of *Locusta migratoria* by acting as suicide substrates for the $CytP_{450}$ hydroxylases. The acetylenic or allenic functional groups of the active compounds are located at C-20 or C-22 for the 22-hydroxylase (Disembodied: Dib) inhibitors and at C-24 or C-25 for the 25-hydroxylase (Phantom: Phm) inhibitors, but the nature of the steroid nucleus did not appear to be critical for inhibitory activity (Roussel 1994). These compounds also inhibited ecdysteroid synthesis in the abdominal integument of adult female *Gryllus bimaculatus* in vitro (Lorenz et al. 1995).

3.9.2 Ecdysteroid Receptor Agonists

Almost all natural phytoecdysteroids act as ecdysteroid receptor agonists, but they differ significantly in their potencies with EC_{50} values in the *D. melanogaster* BII cell bioassay varying from 10^{-10} M to $>10^{-4}$ M (Dinan 2003). This structurally diverse range of analogues has permitted an extensive SAR and molecular modelling of ecdysteroid binding (Dinan et al. 1999; Hormann et al. 2003; Dinan and Hormann 2005). However, steroidal ecdysteroid analogues per se are not suitable as insect pest control agents as they are chemically too complex to synthesise efficiently and economically, and the compounds are too unstable under field conditions, not to mention problems of penetration and ready metabolism in the pest insects themselves. Fortuitously, a class of non-steroidal compounds, the bis(di)acylhydrazines, were discovered (Wing 1988; Wing et al. 1988) which are chemically much simpler, can be readily synthesised and are effective against a range of lepidopteran and coleopteran pest species. It has been unambiguously demonstrated that the active compounds interact with the ligand-binding site of the EcR/RXR(USP) complexes and induce developmentally inappropriate responses, resulting in malformations and death (Dhadialla et al. 1998; Smagghe et al. 2012). Several members of this class have been commercialised as insecticides.

3.9.3 Ecdysteroid Receptor Antagonists

One phytoecdysteroid (24ξ-hydroxycathamasterone) has been identified which acts as an ecdysteroid receptor antagonist (Dinan 2003). A screen of plants and purified natural products identified several cucurbitacins and withanolides which act as receptor antagonists in the μM range (Dinan 2003). Molecular modelling showed structural similarity between the cucurbitacins and 20-hydroxyecdysone, and it is believed that the cucurbitacins act as Michael acceptors to covalently interact with the ligand-binding domain (LBD), thereby blocking access to 20-hydroxyecdysone, which interacts non-covalently, but with higher affinity, with the LBD (Dinan et al. 1997).

3.9.4 Other Compounds

Azadirachtin is a tetranorterpenoid natural product isolated from the neem tree which has powerful moult inhibitory activity (amongst many other effects on insect development; reviewed in Mordue and Blackwell 1993). Evidence (Lorenz et al. 1995) indicates that the compound inhibits ecdysteroid synthesis in *Gryllus bimaculatus*. Its biochemical mode of action, however, remains obscure.

The synthetic compound 1-benzyl-5-[(*E*)-2,6-dimethyl-1,5-heptadienyl]imidazole (KK-42) inhibits ecdysone synthesis in the prothoracic glands of *Bombyx mori* larvae, bringing about premature pupation (Kadono-Okuda et al. 1987). Secretion of ecdysone by prothoracic glands in vitro was inhibited by KK-42 in a dose-dependent manner, but the mode of action of the compound has not been fully elucidated although evidence has been presented that KK-42 preferentially inhibits the 22-hydroxylase of ecdysteroid biosynthesis in follicle cells of *Locusta migratoria* (Jarvis et al. 1994).

4 Vertebrate Steroids in Insects

4.1 Introduction

Debate concerning the significance of the presence of vertebrate-type steroids in insects goes back a long way, as the first reports for vertebrate steroids in insects were made about 90 years ago using heterologous assays of insect extracts on castrated mammals (Steidle 1930; Loewe 1931; Loewe et al. 1932). Since that time, many reports have concerned the identification and isolation of vertebrate-type steroids in insects and other invertebrates. There is convincing evidence that such compounds are present (Table 1), as their identification relies not only on immunoassays but also on GC-MS data or even isolation and

Table 1 Vertebrate-type steroids identified in insects (recreated and updated from Lafont 1991)

Species	Biological material	Method	Steroids identified[a]	References
Abedus herberti (Hemiptera)	Cephalic glands	Isolation	DOC, Δ5P, P, 3α-hydroxypregn-5-en-20-one	Lokensgard et al. (1993)
Acheta domesticus (Orthoptera)	Eggs	TLC-RIA, RPHPLC-RIA	T	Bradbrook et al. (1990)
Agabus seriatus (Coleoptera)	Defensive glands	GC, MS	DOC	Miller and Mumma (1973)
Apis mellifera (hymenoptera)	Royal jelly	GC RIA	T	Vittek and Slomyani (1984)
Attacus atlas (Lepidoptera)	Ovaries	Bioassay	Oestrogen-like	Loewe (1931)
Bombyx mori (Lepidoptera)	Haemolymph	RIA	T-like	Nagashima et al. (1983)
	Ovaries	RIA, GC-MS	E2	Ohnishi et al. (1985)
Cybister spp. (Coleoptera)	Prothoracic defensive secretion	Isolation	12OHΔ4P3,20dione + 8 other C21 steroids	Chadha et al. (1970), Sipahimalani et al. (1970), Baumgarten (1995)
Various dytiscid beetles (Coleoptera)	Prothoracic defensive glands	Isolation	61 × C21, C19 & C18 steroids, including E1, E2, T, P, D5P, DOC & C	Reviewed in Schildknecht (1970), Dettner (2014)
Graphoderus cinereus (Coleoptera)	Prothoracic defensive glands	GC-MS	3α,11α-Dihydroxy-5β-pregnan-20-one	Schaaf et al. (2000)
Graphoderus liberus (Coleoptera)	Defensive glands	GC, MS	DOC	Miller and Mumma (1973)
Ilybius fenestratus (Coleoptera)	Defensive glands	Isolation	T	Schildknecht et al. (1967a, b)

(continued)

Table 1 (continued)

Species	Biological material	Method	Steroids identified[a]	References
Ips typographus (Coleoptera)	Whole body	GC-MS	17β-Hydroxyandrosta-1,4–3-one benzoate, cholecalciferol (vit D3)	Kacprzyk et al. (2021)
Laccophilus minutus (Coleoptera)	Prothoracic defensive glands	GC-MS	3α,12α-Dihydroxy-5β-pregnan-20-one	Schaaf et al. (2000)
Leptinotarsa decemlineata (Coleoptera)	Haemolymph	GC-MS	T, DHEA, 5αDHT, 11KT, 11βOHT, P, Δ4, 17αOHP, Δ5P, 17,20βP	De Clerck et al. (1988)
Locusta migratoria (Orthoptera)	Haemolymph	RIA	Δ5P, P, T, 5αDHT, E1, E2	Novak et al. (1987)
	Gonads	RIA	P, Δ5P, T, 5αDHT, E1, E2	Novak et al. (1990)
	Ovaries	GC-MS	Δ5P, T, E2	Novak and Lambert (1989)
	Testes	GC-MS	Δ5P, T, E2	Novak and Lambert (1989)
Manduca sexta (Lepidoptera)	Ovaries, eggs	Isolation	Pregnen-2β,20β-diol glucoside	Thompson et al. (1985)
Necrophila (Silpha) Americana (Coleoptera)	Rectal glands	Isolation	15βOHP + 6 minor related pregnanes	Meinwald et al. (1985)
Necrophila (Silpha) novaboracensis (Coleoptera)	Rectal glands	Isolation	3α,7β-Dihydroxy-14β-pregn-4-en-15,20-dione, 3α,7β,20ξ-trihydroxy-14β-pregn-4-en-15-one	Meinwald et al. (1987)
Nevrorthus fallax (Neuroptera)	Adult defensive glands	Isolation	DOC	Dettner (2019)
Periplaneta americana (Blattoidea)	Internal organs	EIA	Progesterone-like	Darvas et al. (1997)
Pityogenes chalcographus (Coleoptera)	Whole body	GC-MS	17β-Hydroxyandrosta-1,4–3-one benzoate, cholecalciferol (vit D3)	Kacprzyk et al. (2021)
Sarcophaga bullata (Diptera)	Whole body	GC-MS	E2, E3	Mechoulam et al. (1984)
	Haemolymph	GC-MS	P, T, Δ4, 5αDHT, 5α-androstane-3β,17β-diol, 5β-androstane-3β,17β-diol, androst-5-ene-3β,17β-diol androstenedione,, 11KT, 11βOHT, 17αOHP, 17α,20βP	De Clerck et al. (1983, 1984)
Sarcophaga crassipalpis (Diptera)	Whole body	RIA	Oestrogen-like, testosterone-like	Yocum et al. (1987)
Thermonectus marmoratus (Coleoptera)	Prothoracic defensive glands	Isolation	Mirasorvone, cybisterone	Meinwald et al. (1998)
Tribolium confusum (Coleoptera)	Whole body	^{14}C-chol label	Δ5P, P, DHEA	Smissmann et al. (1964)

[a]**Steroids**: *Δ4* androstenedione, *DHEA* dehydroepiandrosterone, *5αDHT* 5α-dihydrotestostereone, *DOC* deoxycorticosterone, *C* corticosterone, *E1* oestrone, *E2* oestradiol, *E3* oestriol, *12OHΔ4P3,20dione* 12β-hydroxypregn-Δ4-en-3,20-dione, *11KT* 11-ketotestosterone, 11β-hydroxytestosterone, *Δ5P* pregnenolone, *P progesterone*, *17αOHP* 17α-hydroxyprogesterone, *15βOHP* 15β-hydroxyprogesterone, *17α,20βP* 17α,20β-dihydroprogesterone, *T* testosterone

full chemical identification. The function of the steroids is clear when they serve a defensive role in predaceous diving (Dytiscidae) and carrion (Silphidae) beetles, but in most other cases, the functions (if any; Scott 2018) remain enigmatic. The literature on this topic has been reviewed previously (Schildknecht 1970; Lehoux and Sandor 1970; Lafont 1991; Swevers et al. 1991a; b; Lafont and Mathieu 2007).

Several approaches have been applied in this research area: (1) isolation and spectroscopic identification when large amounts of the steroids are present, as in beetle defensive secretions, (2) when levels are lower, detecting and quantifying vertebrate-type steroids in whole insect or isolated organs by means of immunoassays, or more specifically by GC-MS (a typical "reverse endocrinology" approach (Lafont 1991) which assumes the steroids play similar roles in vertebrates and insects), (3) studies with radiolabelled sterols to demonstrate biosynthesis and radiolabelled steroids to identify metabolites and which enzyme activities are present and (4) attempts to correlate steroid levels with stages of development or reproduction in order to identify potential functions.

4.2 Presence and Identity of Molecules

Table 1 provides a summary of the vertebrate-type steroids which have been identified from insect sources. The reports can be divided into two categories. The first is where the steroids are components of the defensive (allomonal) secretions of water or carrion beetles (Fig. 12). These steroids are present in large amounts and could be readily quantified and unambiguously identified. A compilation of 61 steroids is identified, and their occurrence in the defensive secretions of 37 species of dytiscid predaceous diving beetles is available in Dettner (2014). C21, C19 and C18 steroids have been identified, including compounds which are identical to the classical mammalian steroid hormones (progestagens, corticosteroids, androgens and oestrogens; Fig. 12), but also a diverse array of analogues. The second category of reports concerns other types of insects, where the levels are much lower, and their functions are uncertain. Here the range of structures identified is much more limited, but this is in large part certainly

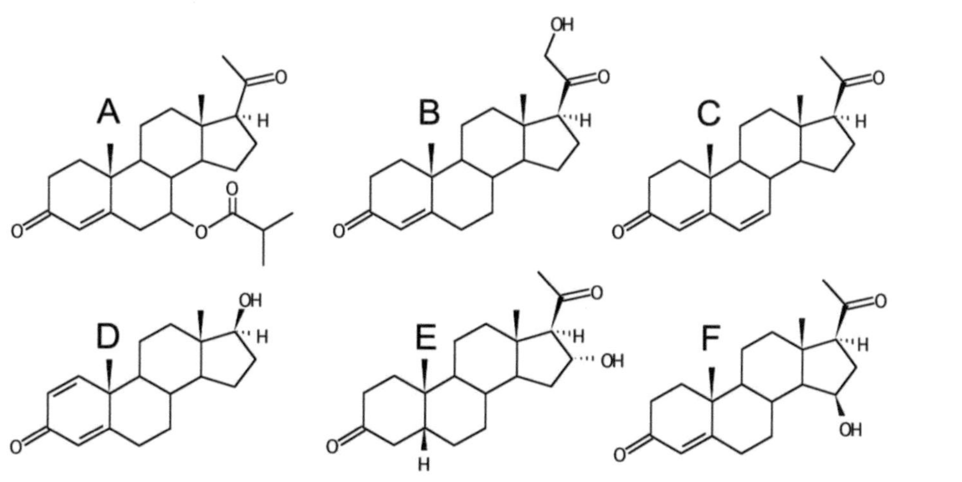

Fig. 12 Representative examples of defensive steroids from dytiscid water beetles. (**a**) 4-pregnen-7-ol-3,20-dione isobutyrate; (**b**) 11-deoxycorticosterone; (**c**): cybisterone; (**d**): $\Delta^{1,2}$-testosterone; (**e**): 5β-pregnan-16α-ol-3,20-dione; (**f**): 15β-hydroxyprogesterone

a consequence of methodological difficulties. Vitamin D3 (a seco-steroid hormone) has recently been detected in bark beetles (Kacprzyk et al. 2021). This molecule is formed from 7-dehydrocholesterol under UV radiation (thus a non-enzymatic reaction) and has no identified function in insects. A UV treatment has been proposed to enhance vitamin D3 levels in insects raised to use as animal (or human) food (Oonincx et al. 2018).

4.3 Sources

The presence of vertebrate-type steroids in insects does not necessarily imply that they are biosynthesised within insect tissues, nor that they have functions (hormonal or otherwise) in the insects within which they occur. The vertebrate steroids could be taken in with the food or produced/modified by microbes associated with the insects, since steroids or, in a wider sense, polyoxygenated sterols are found in organisms from bacteria to plants and mammals, where they have a variety of hormonal and non-hormonal functions (Ji et al. 1990). Vertebrate-type steroids are also present in plants (Heftmann 1975; Sláma 1979; Kovganko and Achrem 1990; Janeczko and Skoczowski 2005). Hypothetically, even if vertebrate steroids are not endogenous to insects, they might have a significant (and hence vitamin-like) role in insects. Thus, it is necessary to pose three questions: where do the vertebrate-type steroids originate? What evidence is there that insects biosynthesise/modify vertebrate-type steroids? Is there convincing evidence that vertebrate-type steroids have identifiable physiological functions in insects? Ultimately, it is necessary to demonstrate that the effects are specific, induced by circulating concentrations, and to clarify the mode of action.

Are Vertebrate-Type Steroids Biosynthesised/ Metabolised in Insects?

In specific cases, it has been possible to demonstrate that vertebrate-type steroids are endogenously biosynthesised in insects by the demonstration of the conversion of injected radioactive cholesterol. This has been possible only in the special cases where large amounts of the steroids are produced, as in the defensive secretions of diving beetles (Coleoptera: Dytiscidae), carrion beetles (Coleoptera: Silphidae), giant water bugs (Hemiptera: Belastomatidae) and ancient lacewings (Neuroptera: Nevrorthidae), which produce 50–1,000 μg defensive steroids/ insect (Schildknecht 1970; Dettner 2014; Dettner 2019). For example, in the water beetle *Acilius sulcatus* (Chapman et al. 1977), it was possible not only to demonstrate the conversion of labelled cholesterol into the defensive steroids (cortexone, 20α-hydroxypregn-4-en-3-one, 6,7-dehydrocortexone, 20α-hydroxypregna-4,6-dien-3-one and pregna-4,6-dien-3,20-dione) but also, by using cholesterol tritiated at specific positions, to reach conclusions about the mechanisms involved in biosynthesis.

The levels of vertebrate-type steroids found in other investigated species are far lower (50–250 ng/g). In certain cases, e.g. progesterone in the confused flour beetle (*Tribolium confusum*; Dubé and Lemonde 1970a, b) it has been concluded that the levels found in the insect can be accounted for by the levels found in the food. It is also possible that a steroidal precursor could be present in the food and converted in the insect (or by the gut microbes) to the detected steroid. A number of studies have concerned the metabolism of injected radiolabelled steroids in insects (Table 2).

4.4 Presence of Specific Enzyme Activities?

The presence of vertebrate-type enzyme activities in insects can be demonstrated by injection into intact animals in vivo or incubation of tissues (intact/ minced/homogenised) in vitro with various radiolabelled known precursors (e.g. cholesterol, pregnenolone and testosterone). This makes the assumption that the same pathways would operate as in vertebrates (Fig. 13), and the case is stronger if whole pathways, or at least large sections of them, can be shown to be operational, rather than isolated reactions which might be a result of non-specific conversions. This is a

Table 2 Products of metabolism of various steroids in insects after application of the steroid in vivo or in vitro

Species	Biological material	Conditions	Substrate applied	Products identified	References
Acheta (Gryllus) domesticus (Orthoptera)	Testes	Homogenised in vitro	Testosterone	Androstenedione	Lehoux and Sandor (1969)
Acilius sulcatus (Coleoptera)	Adult insects	Injected in vivo	Cholesterol	Cortexone, 6,7-dehydrocortexone, cybisterone, 6,7-dehydrocybisterone, 6.7-dehydroprogesterone	Schildknecht (1970)
			Progesterone	Cortexone, 6,7-dehydrocybisterone	
Acilius sulcatus (Coleoptera)	Adult insects	Injected in vivo	Cholesterol	Cortexone, 20α-hydroxypregn-4-en-3-one, 6,7-dehydrocortexone, 20α-hydroxypregna-4,6-dien-3-one, pregna-4,6-dien-3,20-dione	Chapman et al. (1977)
Byrsotria fumigata (Blattoidea)	Testes	Sliced in vitro	Pregnenolone	Progesterone	Lehoux et al. (1970)
Gromphadorina portentosa (Blattoidea)	Testes	Minced in vitro	Progesterone	20β-Hydroxypregn-4-en-3-one	Lehoux et al. (1968)
		Homogenised in vitro		20α- Hydroxypregn-4-en-3-one, 20β-hydroxypregn-4-en-3-one	
	Leg muscles	Homogenised in vitro		20α- Hydroxypregn-4-en-3-one, 20β-hydroxypregn-4-en-3-one	
Gromphadorina portentosa (Blattoidea)	Testes	Sliced in vitro	Pregnenolone	Progesterone	Lehoux et al. (1970)
Gryllus assimilis (Orthoptera)	Testes	Minced in vitro	Progesterone	20α- Hydroxypregn-4-en-3-one, 20β-hydroxypregn-4-en-3-one	Lehoux et al. (1968)
		Homogenised in vitro		20α- Hydroxypregn-4-en-3-one	
	Leg muscles	Homogenised in vitro		20α- Hydroxypregn-4-en-3-one	
Gryllus assimilis (Orthoptera)	Testes	Sliced in vitro	Pregnenolone	Progesterone	Lehoux et al. (1970)
Locusta migratoria (Orthoptera)	Faeces	After feeding	Pregnenolone	20ξ-Dihydropregnenolone, pregnenolone apolar ester, 20ξ-dihydropregnenolone polar conjugate, pregnenolone polar conjugate	Sweevers et al. (1992)
		After feeding	Testosterone	Androstenedione, polar and apolar conjugates of testosterone	
		After injection	Testosterone	Androstenedione, polar and apolar conjugates of testosterone	
		After feeding	Oestradiol	Oestrone, polar conjugates of oestradiol and oestrone	
Schistocerca gregaria (Orthoptera)	Ovaries	Incubation in vitro	Progesterone	20α-Dihydroprogesterone, 20β-dihydroprogesterone	Dubé et al. (1968)

(continued)

Table 2 (continued)

Species	Biological material	Conditions	Substrate applied	Products identified	References
Schistocerca gregaria (Orthoptera)	Intact adult females	Injected in vivo	Progesterone	20α-Hydroxypregn-4-en-3-one	Dubé and Lemonde (1970a, b)
			17α-progesterone	17α,20ξ-Dihydroxypregn-4-en-3-one	
			Androstenedione	Testosterone	
			Oestrone	Oestradiol	
			Dehydroepiandrosterone	3β,17-dihydroxyandrost-5-ene	
			Oestradiol	Oestrone	
			Testosterone	Androst-4-ene-3,17-dione	
Manduca sexta (Lepidoptera)	Female pupae	Injected in vivo	Cholesterol	Pregn-5-en-3β,20β-diol triglucoside	Thompson et al. (1985)
Sarcophaga bullata (Diptera)	Intact larvae	Injected in vivo	Progesterone	17a-Hydroxyprogesterone	De Clerck et al. (1987)
			17α-Hydroxypregnenolone	17α-Hydroxyprogesterone	
			Androstenedione	Testosterone	
			17α-Hydroxyprogesterone	Testosterone	
			Dehydroepiandrosterone	Testosterone	

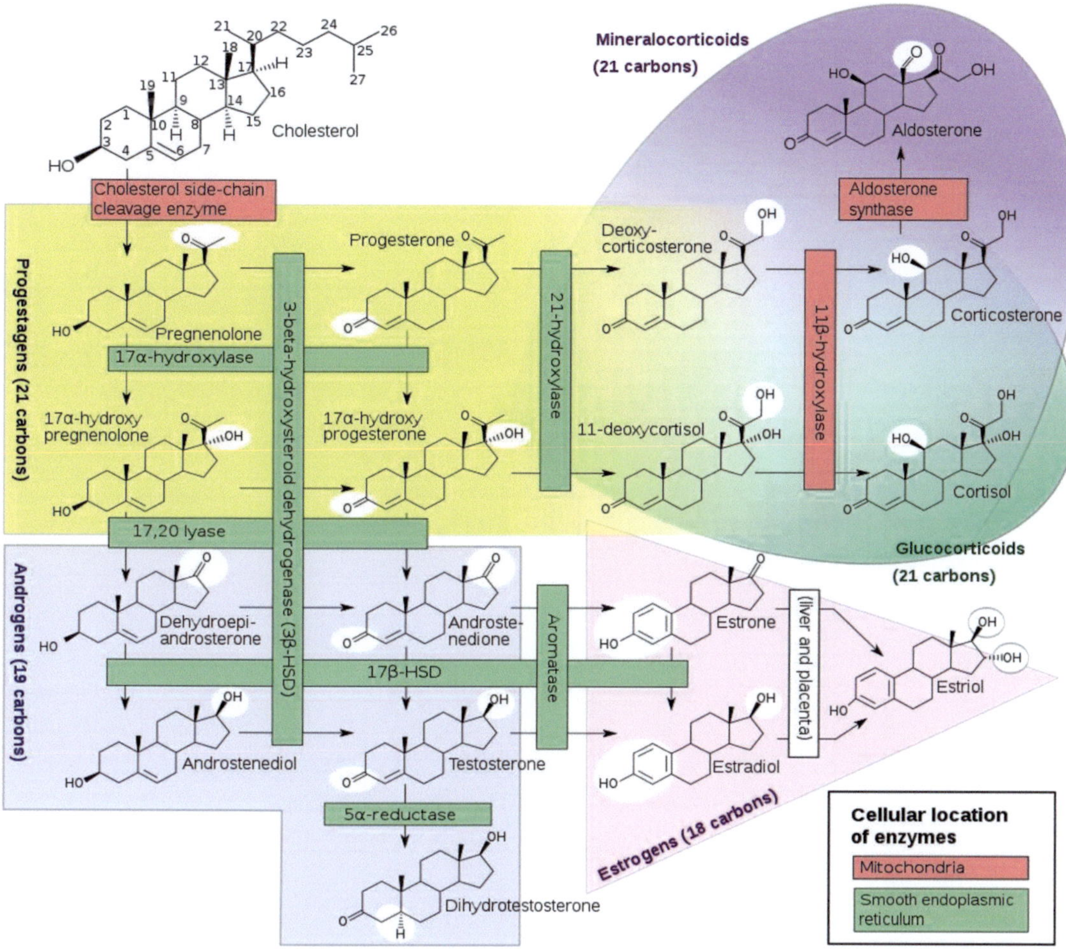

Fig. 13 Summary of the biosynthesis of vertebrate steroids from cholesterol (Häggström and Richfield 2014, https://commons.wikimedia.org/w/index.php?curid=4494399)

complementary approach to the isolation and identification of endogenous steroidal end products and intermediates, as it is highly desirable that in vitro findings should be compatible with the in vivo data. There are a number of other caveats to this approach as precursors may be delivered incorrectly and the enzyme activities might require different reaction conditions in insect systems than in vertebrates. Also, when using a general precursor such as cholesterol, the radiolabelled molecules may be so diluted by endogenous cholesterol that it is difficult to detect the sought-for radiolabelled metabolites, or the products of more major metabolic pathways are preferentially detected. However, a number of studies have been performed using variants of this approach (summarised in Table 2).

A summary of which enzyme activities associated with vertebrate steroid synthesis have been detected in insects is presented in Table 3. This provides a fragmentary picture because of the many caveats associated with this type of approach; so it cannot be expected to provide definite information on which enzyme activities are present and which are not, and virtually nothing is known about the substrate specificities or kinetic parameters of the enzymes responsible for these activities. It can, however, provide consensus information which can help to indicate which sections of vertebrate steroid biosynthesis might

Table 3 Summary of the steroidogenic enzyme activities detected in insects (recreated and updated from Lafont 1991)

		Enzyme activity[a]																								
Species	Method/organ	1	2	3	4	5	6	7	8	9	10	11	12	13	14	15	16	17	18	19	20	21	22	23	24	25	References
Acilius sulcatus (Coleoptera)	In vivo labelling	✓	✓	✓				✓				✓										✓		✓			Schildknecht (1970)
Acheta (Gryllus) domesticus (Orthoptera)	Testes									✓																	Lehoux and Sandor (1969)
Apis mellifera (Hymenoptera)	In vivo labelling		✓	✓	✓	✓							✓	✓													Veith et al. (1974)
Bombyx mori (Lepidoptera)	Ovary																					✓ (3-&17-glc)					Ogiso and Ohnishi (1986)
Byrsotria fumigata (Blattoidea)	Testes		✓	✓																							Lehoux et al. (1970)
Gromphadarina portentosa (Blattoidea)	Testes		✓	✓								✓	✓														Lehoux et al. (1968, 1970)
	Muscles											✓															Lehoux et al. (1968)
Gryllus assimilis (Orthoptera)	Testes		✓	✓						✓		✓	✓														Lehoux et al. (1968, 1970)
	Muscles											✓															Lehoux et al. (1968, 1970)
Leptinotarsa decemlineata (Coleoptera)	In vivo labelling											✓	✓									✓					Swevers et al. (1991b)
Manduca sexta (Lepidoptera)	In vivo labelling	✓											✓									✓ (20-glc)					Thompson et al. (1985)
Necrophila (Silpha) Americana (Coleoptera)	Deduced from isolated compounds	✓	✓	✓											✓	✓			✓				✓		✓		Meinwald et al. (1985)

Samia (Philosamia) ricini (Lepidoptera)	Larval testes/ histochemistry		✓							✓	✓																Hurkadli et al. (1980)
Sarcophaga bullata (Diptera)	In vivo labelling		(✓)	(✓)	✓	✓				✓		✓	✓														De Clerck et al. (1987)
	In vivo labelling									✓												✓					Swevers et al. (1991b)
	Ovary, fat body, CNS								✓																		Denlinger et al. (1987)
Schistocerca gregaria (Orthoptera)	Ovary in vitro											✓	✓														Dube et al. (1968)
	In vivo injected females		✓	✓						✓		✓	✓														Dube and Lemonde (Dubé and Lemonde 1970a)
Tribolium confusum (Coleoptera)	In vivo labelling	✓	✓	✓	✓	✓																					Smissmann et al. (1964)

[a]**Enzyme activities**: (1) 20,22-Desmolase (= 20- + 22-hydroxylases +20,22-lyase); (2) 3β-hydroxysteroid oxidoreductase; (3) Δ^{5-4}-3-oxosteroid isomerase; (4) 17-hydroxylase; (5) 17,20-desmolase; (6) 11β-hydroxylase; (7) 21-hydroxylase; (8) aromatase (=19-hydroxylase +19-hydroxysteroid oxidoreductase+10,19-desmolase); (9) 17β-hydroxysteroid oxidoreductase; (10) 11β-hydroxysteroid oxidoreductase; (11) 20α-hydroxysteroid oxidoreductase; (12) 20β-hydroxysteroid oxidoreductase; (13) Δ^4–3-oxosteroid 5α-reductase; (14) Δ^4–3-oxosteroid 5β-reductase; (15) 3α-hydroxysteroid oxidoreductase; (16) 6α-hydroxylase; (17) 6β-hydroxylase; (18) 16α-hydroxylase; (19) 18-hydroxylase; (20) 18-hydroxysteroid oxidoreductase; (21) conjugating enzyme(s) (sulfotransferase, glucosyl-transferase, palmitoyl-transferase etc.); (22) 7β-hydroxylase; (23) 15α-hydroxylase; (24) 15β-hydroxylase; (25) 17α-hydroxysteroid oxidoreductase

be operational in insects. 20–22-Desmolase (20- and 22-hydroxylases plus 20,22-lyase activities, converting cholesterol to pregnenolone) activity has been detected in several insect systems and might be expected as a similar reaction is found in the conversion of 20-hydroxyecdysone to its metabolite poststerone, although the steroid nucleus part of the molecule is rather different. 3β-Hydroxysteroid dehydrogenase/Δ^{5-4} isomerase activities (converting pregnenolone to progesterone) appear to occur widely in insects. The enzyme activities converting pregnenolone to testosterone (17α-hydroxylase, 17,20-desmolase and 17β-hydroxysteroid dehydrogenase) have been detected. Aromatase activity (converting testosterone to oestradiol) appears to be rare (only detected in *Sarcophaga bullata*; Denlinger et al. 1987) and the significance and reliability of this solitary finding have been questioned (Swevers et al. 1991b). The 17- and 21-hydroxylase activities which are involved in the conversion of progesterone to corticosteroids have been detected in insects, if somewhat rarely, but 11β-hydroxylase activity which contributes to more potent glucocorticoid activity in mammals has not. The mammalian mineralocorticoid aldosterone has apparently never been detected in an insect source, although the C18-oxygenated C21-steroid, mirasorvone ([1*R*,2*S*,5*S*,6*S*,9*R*,12*S*,13*R*]-6-hydroxy-6,13-dimethyl-7-oxapentacyclo[10.8.0.$0^{2,9}$.$0^{5,9}$.$0^{13,18}$]icosa-17,19-dien-16-one), has been found in the defensive secretion of the dytiscid beetle *Thermonectes marmoratus* (Meinwald et al. 1998).

4.5 Roles of Vertebrate-Type Steroids in Insects

Although the presence of vertebrate-type steroids seems to be widespread in insects, substantial evidence for functions for them in insects is only available for a few of them, and these roles seem to be specific to the species or family of insects and cannot be more generalised. In other cases, attempts have been made to correlate fluctuations in the levels of the vertebrate-type steroids with physiological events occurring during the development of the insect, but without demonstrating cause and effect. In the majority of cases, however, it is still true to say that the presence of the vertebrate-type steroids has been reported without evidence for endogenous biosynthesis or physiological, or even pharmacological effects. The evidence for the most substantiated functions are summarised below.

4.5.1 Defensive Secretions in Beetles Released from Prothoracic or Rectal Glands

Water beetles (Coleoptera: Dystiscidae), owing to being exposed on the water surface of ponds or lakes, are susceptible to attack by fish, amphibians and birds. They have evolved to release secretions from prothoracic defensive glands, the contents of which are repulsive/toxic to predators. The defensive chemicals differ widely in their chemistry, but many species produce and release significant (mg) amounts of defensive steroids. Most of the steroids identified to date are a wide diversity of C21 pregane-type steroids (Fig. 12; reviewed in Schildknecht 1970; Dettner 2014). It has been demonstrated that these molecules are biosynthesised from cholesterol (e.g. Chapman et al. 1977), although it is possible that bacteria in the foregut of the insect may contribute to the process (Schaaf and Dettner 2000a; b). The main site of cholesterol uptake has been shown to be the foregut in omnivorous and carnivorous insects, including *Dysticus marginalis*, whereas the midgut is the predominant site of (chole)sterol uptake in phytophagous insects, because of the need of phytophagous insects to dealkylate phytosterols (Joshi and Agarwal 1977). The steroids have been shown to possess anaesthetic activity and be toxic towards minnows (*Pimephales promelas*), with pregn-4-en-3-one being the most active with concentrations of 10^{-6}–10^{-5} M causing anaesthesia and 10^{-4}–10^{-3} M causing death (Miller and Mumma 1973, 1976a, b).

Carrion beetles (Coleoptera; Silphidae) also incorporate vertebrate-type steroids into their defensive secretions. Although these are stored and released from the rectal glands, they serve similar allomonal functions. Leaf beetles of the family Chrysomelidae also secrete a wide range of defensive substances and one section of the family (Chrysolinina) is specialised to release cardenolides or polyoxygenated steroids (reviewed in Pasteels et al. 1994). In one case, *Chrysolina carniflex* (Laurent et al. 2003), the defensive steroid, has been identified as the ecdysteroid 22-acetyl-20-hydroxyecdysone.

4.5.2 Oestradiol in Bombyx mori

Pupae and the fifth instar larvae of *Bombyx mori* possess oestrogenic activity (Yang et al. 2010; Lee et al. 2022) and adult ovaries of *B. mori* were found to contain oestradiol (E2; Table 1; Ohnishi et al. 1985), which was detected by RIA (176 pg/g) and GC-MS (63 pg/g). Subsequently (Ogiso and Ohnishi 1986), the effects of exogenous oestradiol, injected into pupae on ovarian development, oviposition rate, embryological development and determination of diapause, were examined. Only the rate of oviposition was affected by the injection of high doses of E2; none of the other parameters were significantly affected. Topical application of E2 at the start of the fifth instar was, however, found to have a dose-dependent effect on the size of the silk gland and its fibroin content, which could be counteracted by ICI-182780 (fulvestrant, an oestrogen-receptor antagonist in mammals) (Keshan and Ray 1998). Application of E2 at 1 μg/g improves cocoon weight and silk quality without affecting fecundity or hatchability (Keshan and Ray 2000), which is reminiscent of the effects of exogenous 20-hydroxyecdysone on silk production in *B. mori* (Ninagi and Maruyama 1996), and may have a sericultural application. Specific binding sites for E2 have been identified in the posterior silk gland with a K_d of 9 nM and B_{max} of 56 fmol/mg (Keshan and Ray 2001) and female fat body of the fifth instar larvae with a K_d of 13 nM (Roy et al. 2007). No homologues of oestrogen receptors (ERα and ERβ) are found in insects, but an orthologue of oestrogen-related receptor (ERR: a further nuclear receptor, but not particularly closely evolutionarily related to ER) has been identified in *Drosophila melanogster* (Maglich et al. 2001; Thornton et al. 2003; Tennessen et al. 2011) and, more recently, in *B. mori* (Wu et al. 2022). Interestingly, expression of BmERR is upregulated by 20-hydroxyecdysone, owing to the presence of several putative ecdysteroid response elements (binding sites for 20E/EcR/USP complexes) in the promoter region of BmERR (Wu et al. 2022). Shen et al. (2015) presented evidence that E2 promotes the release of vitellogenin from the fat body of *B. mori* and its accumulation in the haemolymph, ovaries and eggs and that titres of E2 during the fifth instar larvae and pupae fluctuate, implying that *B. mori* can modulate E2 levels. Although these findings do not yet form a fully coherent story, it looks as though E2 may fulfil a physiological (and perhaps even hormonal) function in *B. mori*.

The origin of E2 in *B. mori* is a puzzle, because, as explained above, it appears that insects do not possess the aromatase activity necessary to to form oestrogens. *B. mori* is monophagous, feeding only on leaves of mulberry (*Morus* spp., with *M. alba* being the preferred food plant). Mulberry is known to possess oestrogenic activity (Cohen 2005; Guzmán 2005), but this is normally attributed to the presence of phyto-oestrogenic flavonoids (Wei et al. 2016). E2, or other steroidal oestrogens, has not as yet been isolated from the plant (Rohela et al. 2020).

4.5.3 Dihydroxypregn-5-ene Triglucoside in Manduca Sexta Pupae and Eggs

In an experiment on radiolabel ecdysteroids from ^{14}C-cholesterol injected into female pupae of *M. sexta* (tobacco hornworm) and extraction of either ovaries or eggs deriving from the emerged adults, it was found that the major labelled meta bolite was pregn-5-en-3β-20β-diol triglucoside (Thompson et al. 1985). The yield was ca. 0.02% of the fresh weight, which was threefold higher than the amount of ecdysteroids present. Thus, the endogenous synthesis of a conjugated vertebrate-type steroid present in significant amounts has been clearly demonstrated, but, unfortunately, the role (if any) of this steroid conjugate remains unknown. The compound has not, as yet, been detected in any other insect species.

4.5.4 Maturation of Reproductive Capacity in Rabbit Fleas

The rabbit flea, *Spilopsyllus cuniculi*, is specific for its host, and reproduction of the flea takes place in synchrony with that of the pregnant doe rabbit (Mead-Briggs and Rudge 1960). The maturation of the oöcytes in adult female fleas is induced by oestrogens and corticosteroids present in the host blood prior to parturition in the rabbit. The fleas move to the new-born kits and feed on them for several days, mate with male adult fleas and deposit eggs on the kits and in the nest, before moving back to the lactating doe, where the progesterone-rich postpartum blood induces ovarian regression (oöcyte resorption) and cessation of acceptance of copulation (Rothschild and Ford 1964a, b, 1966). The involvement of the steroids was elucidated by the application of exogenous steroids to the rabbits or fleas, but the identities and levels of the active steroids in the rat fleas have not been determined. The mode of action of the steroids in the rabbit fleas is also unknown. This developmental coordination seems to be specific to the rabbit fleas and its host, as a similar hormonal dependence is not observed between rat fleas (*Xenopsylla cheopis* and *X. astia*) and their host (Prasad 1976).

4.5.5 Correlation of Vertebrate-Type Steroid Titres with Physiological Events

Levels of vertebrate-type steroids have been determined, usually by immunoassay (RIA or EIA), during phases of development in a few insects. In *Sarcophaga crassipalpis* (flesh fly; Yocum et al. 1987), oestrogen- and androgen-like immunoreactivity was detected throughout development with male and female insects containing almost equal levels and with a sharp increase in both corresponding to mating. High levels were also found in ovoviparous females during the second half of the reproductive cycle, which the authors attributed to the developing embryos because high levels were also found in first instar larvae. In *Leptinotarsa decemlineata* (Colorado potato beetle), levels of testosterone and 5α-dihydrotestosterone were shown to fluctuate during the larval and pupal stages. Titres of pregnenolone, progesterone, testosterone, 5-α-dihydrotestosterone, oestrone and oestradiol were measured by RIAs in haemolymph (Novak et al. 1987) and gonads (Novak et al. 1990) of larval and adult *Locusta migratoria* (migratory locust). Measured levels of immunoreactive materials were significantly lower (five- to ten-fold) in the haemolymph than in tissues of the insect. Haemolymph titres of most of the steroids did not fluctuate during development, with only oestrone showing peaks in the middle of the fifth instar (males and females) and 3-week-old adult males. In the gonad study, the higher levels of immunoreactive material allowed levels to be determined in samples from individual insects, which revealed considerable variation between samples of the same developmental stage for progesterone-like and pregnenolone-like quantifications. Levels of oestrone-like and oestradiol-like materials were low throughout development, while testosterone-like material peaked in gonads of aged (60 days after the adult moult) male and female locusts. In the adult cockroach *Nauphoeta cinerea* (speckled cockroach; Takáč et al. 1993), progesterone-like immunoreactive material was detected in the intestine, gonads and, to a much lesser extent, haemolymph. When the insects were raised at 29 °C, there was a significant peak in progesterone-like material (rising from 50 to 150–250 pmol progesterone equivalents/g) in the intestines and gonads of both males and females, but the authors were not able to correlate this to a physiological event in the gonotrophic cycle. In male and female adults of *Periplaneta americana* (American cockroach; Darvas et al. 1997), there was found to be a peak (150 ng progesterone-equivalents/g) of progesterone-like material (determined by an EIA which cross-reacts with pregnenolone) coinciding with sclerotisation of the cuticle after the adult moult and most of this (>95%) were associated with the digestive system. The intake of progesterone-like material in the food was 120 pg progesterone equivalents/g adult/day (>1,000-fold lower than the peak value in the insect). Progesterone-like immunoreactivity was only present during the first 2 days after the adult moult and remained low (1–10 ng progesterone equivalents/g) throughout the rest of the adult stage; no peaks were seen for oestradiol-like or testosterone-like materials which were assessed in separate EIAs. When the progesterone-like peak material was reassessed by the more specific method RP-HPLC/RIA, the level of RIA-positive material dropped to 16 ng progesterone equivalents/g, which the authors interpreted as the separation off of progesterone-like molecules through the chromatographic separation, but could also be a result of separation of non-specific interference or greater specificity of the RIA over the EIA. A peak of 3β-hydroxy steroid dehydrogenase Δ^5/Δ^4-isomerase (converting pregnenolone to progesterone) activity in homogenates of whole insects was found to coincide with the peak of progesterone-like material and had dropped significantly by 48 h after the adult moult, but the localisation of this enzyme activity in the insect was not determined. Unfortunately, the level of pregnenolone in the food was not determined, so what may have been observed could correspond to the conversion of dietary pregnenolone to progesterone by gut microbes during peri-ecdysis when food does not pass through the gut so rapidly, or increased activity of the enzyme in the temporarily more aerobic environment of the gut when air is taken in to expand the new cuticle after the moult, which allows the reduced cofactor product (NADH) of the enzyme to be reoxidised, and hence recycled, more readily.

The above studies show the difficulties of trying to correlate titres with physiological events since interpretation can be confused by methodological difficulties, lack of specificity of the detection method, variability and lack of synchrony in insects and correlation not being the same as cause and effect.

5 Cardenolides, Bufadienolides and Other Steroids

5.1 Introduction

The cardenolides and bufadienolides (Fig. 14) are sterol derivatives which are collectively referred to as cardiac glycosides because they increase the strength of heart contractions in mammals by inhibiting Na^+/K^+-ATPase (Botelho et al. 2019). The multiplicity of substituents on the steroid nucleus and the diversity of glycosyl moieties contribute to a very large number of natural cardiac glycosides (Steyn and van Heerden 1998; Bejček et al. 2021).

These molecules are generally toxic at high doses, but some of them (e.g. the cardenolides digoxin and digitoxin) are widely used at low doses for treating heart failure and irregular heartbeat, and the plants containing them are used as traditional medicines.

Remarkably, they are found not only in several plant families but also in some animals (Table 4;

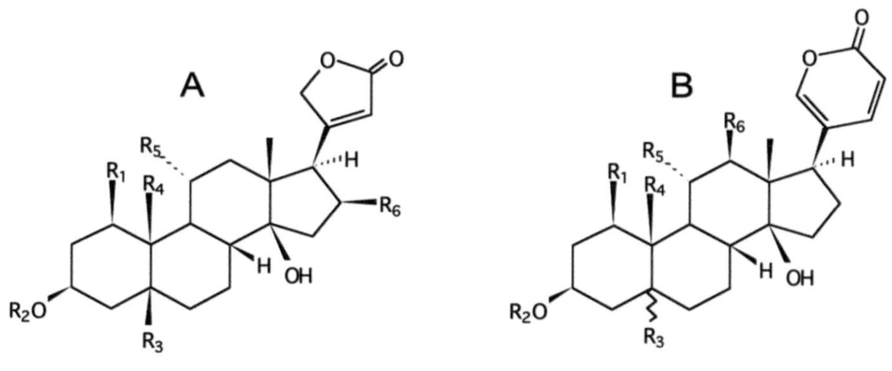

Fig. 14 Generalised structures of cardenolides (**a**) and bufadienolides (**b**). (**a**) R_1 = H/OH; R_2 = H/glycosyl; R_3 = H/OH; R_4 = –CH_3/CH_2OH/CHO; R_5 = H/OH; R_6 = H/OH. (**b**) R_1 = H/OH; R_2 = H/glycosyl; R_3 = α/β H/OH, R_4 = –CH_3/CHO; R_5 = H/OH; R_6 = H/OH/=O

Table 4 Sources of cardiac glycosides

Compounds	Occurrence		Examples of genera
Bufadienolides	Plants	Crassulaceae	*Kalanchoe*
		Hyacinthaceae	*Urginea, Bowiea*
		Iridaceae	*Homeria, Moraea*
		Melianthaceae	*Bersama, Mellanthus*
		Ranunculaceae	*Helleborus*
		Santalaceae	*Thesium*
	Animals	Toads	*Bufo*
		Snakes	*Rhabdophis*
		Insects (fireflies)	*Photinus*
Cardenolides	Plants	Apocynaceae	*Strophantus, Nerium*
		Fabaceae	*Coronilla*
		Liliaceae	*Convallaria*
		Scrophulariaceae	*Digitalis*
	Animals	Mammals	*Rattus, homo, Canis*
		Insects	*Danaus, Oncopeltus*

Compiled from Steyn and van Heerden (1998), Bejček et al. (2021)

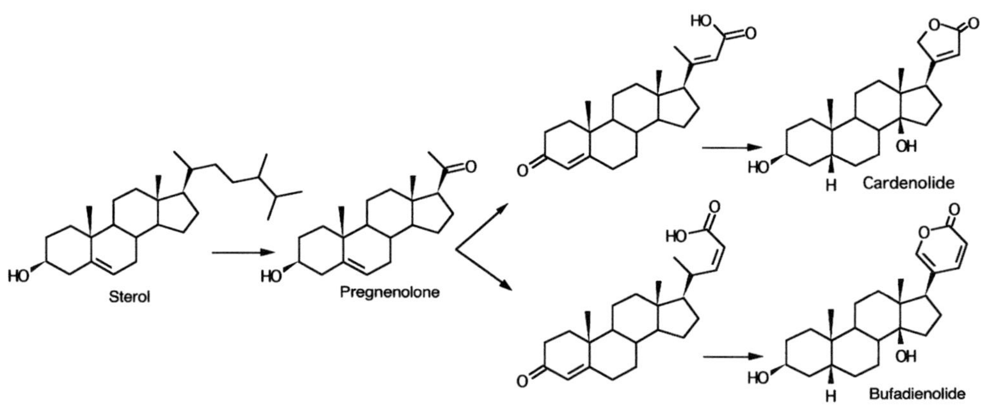

Fig. 15 Simplified biosynthetic pathway of cardiac glycosides in plants

Steyn and van Heerden 1998; Bejček et al. 2021). Their presence results from the endogenous synthesis in plants, but it appears that some animals are also able to synthesise them, whereas others sequester them from their diet.

Their multi-step biosynthesis in plants (Steyn and van Heerden 1998; Kreis et al. 1998; Bejček et al. 2021) uses pregnenolone as an intermediate (Fig. 15), and a short side chain is rebuilt by coupling with either acetic acid (cardenolides) or oxaloacetic acid (bufadienolides). An alternative pathway using a limited side chain cleavage of cholesterol, similar to that used to generate bile acids, seems also possible (Agrawal et al. 2012). Diversification then takes place mainly through various hydroxylations and conjugation at C-3 with more or less complex glycosyl moieties (Steyn and van Heerden 1998).

Experiments with insects have provided evidence that they generally do not synthesise cardenolides, which originate and are sequestered from their food. However, certain chrysomelid beetles produce defensive secretions which contain cardenolides, and there is good evidence that the insects synthesise the compounds themselves and by a pathway from cholesterol which is similar to that occurring in plants (Pasteels and Daloze 1977; Daloze and Pasteels 1979; van Oycke et al.

1987; Pasteels et al. 1989). On the other hand, some insect species (fireflies) appear to produce their own bufadienolides, termed lucibufagins (Eisner et al. 1978), although it is not known if the pathway is the same as those found in plants or toads (Steyn and van Heerden 1998; Kreis et al. 1998).

5.2 Cardenolides

Phytophagous insects of various Orders (e.g. *Aphis nerii* [Hemiptera; Sternorrhyncha], *Oncopeltus fasciatus* [Hemiptera: Heteroptera] and *Poikilocerus bufonius* (Orthoptera); van Oycke et al. 1987) may sequester dietary cardenolides. The most famous examples are the milkweed butterflies (Lepidoptera: Danainae, e.g. monarch butterfly *Danaus plexippus* (Parsons 1965)), queen butterfly *Danaus gilippus* and the tropical milkweed butterfly *Lycorea cleobaea)* which feed on milkweeds (*Asclepias* spp.) and accumulate cardenolides from them. Sequestration of this class of compounds has also been observed in other insect species (bugs, beetles, aphids; see Dobler et al. 2011).

Lack of toxicity in accumulating species results from modifications of Na^+/K^+-ATPase by a single amino acid substitution (N122H) in the α-subunit, which reduces its sensitivity to toxins by two orders of magnitude (Holzinger et al. 1992; Aardema et al. 2012; Agrawal et al. 2012; El-Mallakh et al. 2019), and from a concentration of the toxins in remote compartments of the insect, i.e. storage in the midgut in *D. plexippus* (Petschenka and Agrawal 2015) or in a dorsolateral space delimited by a double epidermis in the milkweed bug *Oncopeltus fasciatus* (Duffey et al. 1978).

The presence of these toxins in insects protects them significantly from predation, by being toxic, distasteful or emetic depending on their potential aggressors (e.g. spiders and birds), and, in the case of *D. plexippus*, the toxin provides resistance against the protozoan parasite *Ophryocystis elektroscirrha* (Tan et al. 2018). The toxic insects often display aposematic colours, which facilitate predators learning to avoid them. In the case of *Oncopeltus*, the cuticle presents some weak areas that are easily broken when the insect is squeezed, thus allowing the stored cardenolide-rich solution to be released onto the insect body surface and made detectable to predator taste receptors (Scudder et al. 1986). Thus, cardenolides protect insects which store them in the same way as they may protect plants from being eaten by non-adapted phytophagous insects (e.g. Dimock et al. 1991). The complexity of the interactions between cardenolide-accumulating prey species and their potential predators and the diversity of cardenolide structures present is shown by a recent study (Pokharel et al., 2020) which compared the resistance of the palearctic milkweed bug (*Lygaeus equestris*, Heteroptera), which sequesters cardenolides from pheasant's eye *Adonis vernalis* (Ranunculaceae) and its close relative Horvath's knight bug *Horvathiolus superbus*, which obtains its cardenolides from the common foxglove *Digitalis purpurae* (Plantaginaceae), so the two insect species have different profiles of protective cardenolides. Both insect species were protected against predation by birds, but only *H. superbus* was protected against predation by larvae of the green lacewing *Chrysoperla carnea*. Protection against lacewings could be conferred on *L. equestris* by rearing it on *D. purpurea*.

This protection is, however, not absolute: in the case of *Danaus plexippus*, some rodents are not repelled and eat the insects (Weinstein and Dearing 2022), while Chinese mantids (*Tenodera sinensis*) carefully remove the insect midgut before eating the rest of the caterpillar's body, thereby reducing the toxin intake by one half (Mebs et al. 2017), but the gutting behaviour is innate and not dependent on the presence of cardenolides in the prey (Rafter et al. 2017).

Leaf beetles (Coleoptera: Chrysomelidae) are mainly phytophagous insects, usually monophagous or oligophagous, which form large clusters on their host plants. They possess an impressive array of defence strategies against predators, and this has contributed to several species becoming serious agricultural pests (Pasteels et al. 1982). Certain chrysomelid beetles have the capacity to biosynthesise cardenolides de novo from sterol (van Oycke et al. 1987). Examination of a significant number of chrysomelid species (Daloze and Pasteels 1979; van Oycke et al. 1988; Pasteels et al. 1989; Eggenberger et al. 1992; Daloze et al. 1995) indicated that cardenolides were only found in the genera *Chrysolina*, *Chrysochloa*, *Dlochrysa* and *Oreina*, but not in all species of these genera. The cardenolide profiles, which vary from species to species, were independent of the food plant and consistent over generations implying that synthesis occurs within the insect and is genetically determined, especially as the food plants of these species do not themselves accumulate cardenolides. Further, cardenolides are present throughout the life cycle and increase in amount as the insect ages (Daloze and Pasteels 1979; Hilker et al. 1992).

Polyoxygenated Steroids in the Defensive Secretions of Certain Chrysomelid Beetles

Although the steroidal components of the defensive secretions of chrysomelid leaf beetles are characteristically cardenolides, certain species are devoid of them. Amongst these are five species (*Chrysolina hyperici*, *C. varians*, *C. brunsvicensis*, *C. germinata* and *C. quadrigemina*), which feed on *Hypericum perforatum* (St. John's wort) and accumulate instead polyoxygenated steroid glycosides which are biosynthesised in the insects de novo from plant-derived sterols (Daloze et al. 1985, 1991; Randoux et al. 1990). The structures all possess a full sterol side chain, an oxo-group at C-6 and a C-20 hydroxyl group with further oxygenation at other positions around the molecule (including C-25). Intriguingly, the genins of these compounds show structural similarities to ecdysteroids, and one of them is even 2,14,22-trideoxy-20-hydroxyecdysone (Fig. 16 and see Sect. 3). Thus, chrysomelid beetles are able to generate a range of steroidal defensive compounds from cardenolides through polyoxygenated steroids to ecdysteroids (Pasteels et al. 1989; Randoux et al. 1990; Laurent et al. 2003).

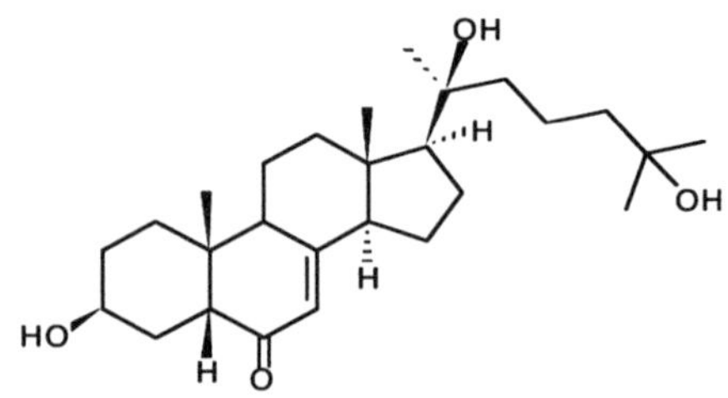

Fig. 16 The structure of 2,14,22-trideoxy-20-hydroxyecdysone. This compound isolated from *Chrysolina varians* was the 3-*O*-β-sophoroside derivative. (Randoux et al. 1990)

5.3 Bufadienolides

Compounds in a sub-category of bufadienolides (lucibufagins) were originally found in fireflies of the genus *Photinus* (*P. pyralis*, *P. ignitus* and *P. marginellus*) and the species *Lampyris noctiluca* and *Ellychnia corrusca* (Coleoptera: Lampyridae; Eisner et al. 1978; Meinwald et al. 1979; Goetz et al. 1979, 1981; Tyler et al. 2008; Smedley et al. 2017), where they are believed to be biosynthesised within the insect, although this does not appear to have been directly demonstrated. A differing mixture of lucibufagins is present in each species; for example, *P. pyralis* produces the parent compound 2β,5β,11α-trihydroxy-12-oxobufalin and its 5 acetyl, ethyl and isopropyl ester derivatives, of which the 2β-acetoxy-3-*O*-acetyl derivative makes up 60% of the mixture, but the smaller amount of the parent compound accounts for all the deterrent activities (Fig. 17; Meinwald et al. 1979). Lucibufagins are present at all stages of the life cycle (Smedley et al. 2017) in those species which are believed to biosynthesise them for themselves, and this makes the insects strongly repellent to most potential predators including ants, spiders, lizards, birds and bats (Leavell et al. 2018). The lucibufagins are stored in the haemolymph and, when an adult insect is attacked, it undergoes reflex bleeding along the pronotal and elytral margins and at the antennal sockets to produce a noxious and unpalatable exudate on the surface of the insect (Blum and Sannasi 1974).

Over 2000 species of Lampyridae (fireflies) are known. A recent survey of samples of 59 previously uninvestigated species found lucibufagins to be present in 21 of them. The authors extrapolated their findings to suggest that all species in the genera *Lampyris* and *Nyctophila* will possess

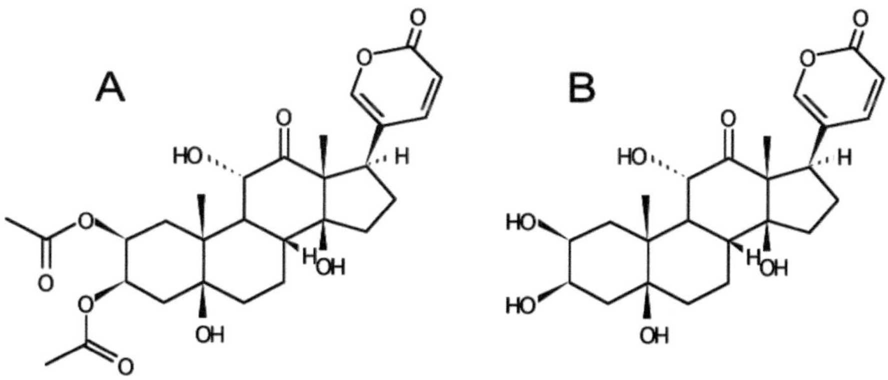

Fig. 17 Structures of the main (**a**) and most active (**b**) lucibufagins from *Photinus pyralis*. (**a**): 2β-acetoxy-3-*O*-acetyl-5β,11α-dihydroxy-12-oxo-bufalin; (**b**): 2β,5β,11α-trihydroxy-12-oxo-bufalin

the compounds, while the genera *Lamprorhiza*, *Lampyroidea* and *Luciola* do not (Berger et al. 2021). Combining their findings with those of others allowed the authors to propose that fireflies within the sub-family Lampyrinae (the genera *Nyctophila*, *Pelania*, *Phosphaenus, Photinus* and *Lampyris*) contain, and probably biosynthesise, lucibufagins, while those within the Photurinae (genus *Photuris*) can sequester the compounds from their diet, and the other sub-families do not accumulate this class of compound (Berger et al. 2021).

The occurrence of lucibufagins in *Photuris* (Photurinae) is an interesting case as it does not produce them, but can nevertheless store them if present in the diet (Eisner et al. 1997). Male and female fireflies communicate through specific light signals for mating. The females of *Photuris* are able to lure *Photinus* (Lampyrinae) males by emitting light signals similar to those that *Photinus* females use to attract *Photinus* males. In this way, *Photuris* "femmes fatales" can catch *Photinus* males, eat them and accumulate their toxins (Eisner et al. 1997). *Photuris* females partially modify the sequestered lucibfagins, as the profile of compounds is different from that of adult *Photinus* males (González et al. 1999). Interestingly, *Photuris* males also contain lucibufagins, albeit at lower levels than the females, and it is currently not clear how these are obtained (Eisner et al. 1997).

The occurrence of lucibufagins seems limited to some species of fireflies (Berger et al. 2021). Their presence in other insect groups has not been reported. Such a finding is consistent with experiments of Jones et al. (1983) on *Oncopeltus fasciatus* showing no sequestration of bufadienolides, whereas cardenolides were significantly accumulated. Given the chemical similarity of both categories of toxins, this deserves further investigation to find a proper explanation.

5.4 Other Steroids

5.4.1 Cockroach Aggregation Pheromone

The arrestant component of the aggregation pheromone present in the faeces of the German cockroach, *Blattella germanica*, has been found to contain two unusual (non-volatile) chlorinated sterol glucosides, blattellastanosides A and B (Fig. 18; Sakuma and Fukamai 1993a, b; Mori 2005), of which blattellastanoside A is the significantly more (70-fold) active compound. The attractant components are volatile and non-steroidal, attracting conspecifics, while the arrestant components are contact chemicals providing a durable signal at the nest site of conspecifics. The effectiveness of blatellostanoside A has been questioned by other authors (Scherkenbeck et al. 1999).

5.4.2 Caterpillar Trail Pheromone

5β-Cholestane-3,24-dione (Fig. 19) has been identified as an active component of the trail

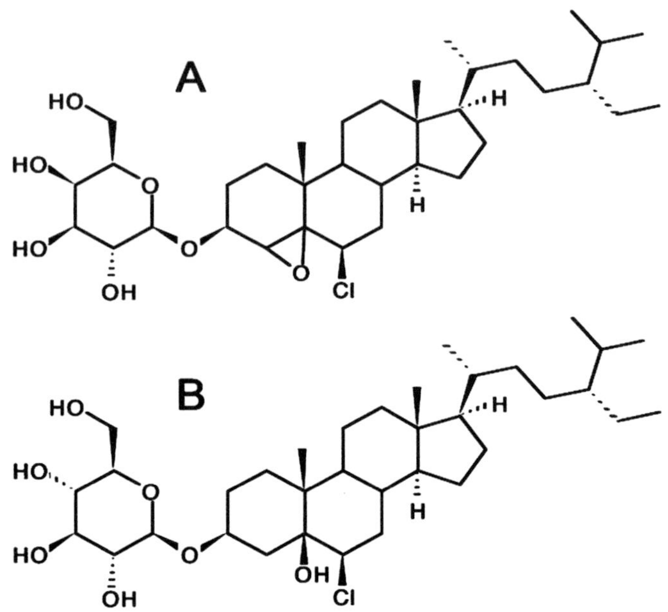

Fig. 18 Structures of the steroids of the arrestant component of the aggregation pheromone of *Blattella germanica*. (**a**): Blattellastanoside A; (**b**): Blattellastanoside B

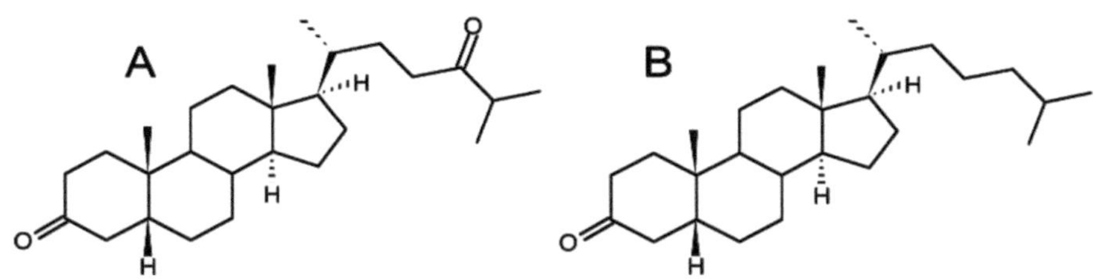

Fig. 19 Steroid components of the larval trail pheromone of *Malacosoma americanum*. (**a**): 5β-cholestane-3,24-dione; (**b**): 5β-cholestan-3-one

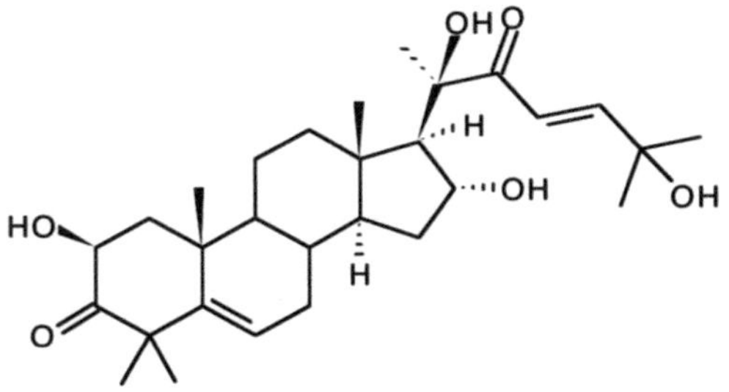

Fig. 20 The structure of cucurbitacin D

pheromone of larvae of the eastern tent caterpillar, *Malacosoma americanum* (Crump et al. 1987; Peterson and Fitzgerald 1991). It is secreted from the abdomen and used to mark trails leading from the communal nest tent to the feeding sites on plum and apple trees. 5β-Cholestan-3-one elicits the trail-following response even more strongly (Fitzgerald 1993).

5.5 Cucurbitacins

Although not strictly steroids, it is perhaps worth mentioning that the triterpenoid cucurbitacins can be taken up from their host plants in the Cucurbitaceae and accumulated in the haemolymph and tissues of leaf beetles of the Galerucinae subfamily within the Chrysomelidae. Studies on *Diabrotica undecimpunctata* (spotted cucumber beetle) and *Diabrotica virgifera* (western corn rootworm) have shown that they are able to sequester plant cucurbitacins (e.g. cucurbitacin D; Fig. 20) and partially modify them in vivo by glucosylation at C-3, hydrogenation at the Δ^{22}-double bond, desaturation at C-1-C-2 and acetylation at 26-OH (Andersen et al. 1988). Cucurbitacins are not only exceedingly bitter-tasting and deterrent to most predators, but they also act as ecdysteroid receptor antagonists (Dinan et al. 1997).

6 Summary and Conclusions

6.1 Summary

6.1.1 Sterols

Insects are cannot biosynthesise sterols de novo, which means they have to obtain them in their diets to provide adequate amounts to fulfil the structural function of modulating membrane fluidity and permeability, as precursors for ecdysteroids and other steroids and as signalling molecules when bound to the protein hedgehog. Carnivorous and omnivorous insects obtain enough cholesterol in their food to fulfil all these roles. Phytophagous insects are faced with the problem that plants generally contain very low levels of cholesterol (ca. 1% of the phytosterols), the majority of the sterols being C24-alkylated sterols (e.g. campesterol and sitosterol) and perhaps also esterified to fatty acids at C-3. Sterol esters are hydrolysed in the gut by a specific lipase to release free sterols which are taken up by intestinal enterocytes. The majority of phytophagous species dealkylate the phytosterols in the midgut to yield cholesterol. However, some species use the C24-alkylated sterols for both the plasma membrane structural function and, more surprisingly, as precursors for ecdysteroids, since this necessitates modification of ecdysteroid receptor specificity. Other non-dealkylating insects appear to be flexible as to whether they use C24-alkylated sterols or cholesterol, depending on what their dietary intake provides. Yet other insects obtain sterols through endosymbiosis with yeast-like microorganisms present in specialised mycetocytes located in some fat body cells or aggregated as a specialised organ.

6.1.2 Ecdysteroids

Zooecdysteroids are a class of steroid hormones which differ significantly in their structure from the classical vertebrate steroid hormones, in that they are polyhydroxylated and typically retain the full sterol side chain. In insects, they are biosynthesised from cholesterol (or C24-alkylated sterols) in a pathway which has not yet been completely elucidated. Synthesis occurs in larval prothoracic glands, adult female ovaries and adult male testes, and it may occur in other tissues and organs as well. Ecdysteroids regulate development, moulting, reproduction and diapause through fluctuations in haemolymph ecdysteroid titres which arise from changes in the rates of biosynthesis and metabolism/excretion. In insects, 20-hydroxyecdysone is regarded as the major biologically active ecdysteroid. The immediate precursors are ecdysone or 3-dehydroecdysone which are released from the PGs and converted to 20-hydroxyecdysone in the haemolymph and peripheral target tissues in larval insects. 20E brings about developmental and physiological changes either by interacting with intracellular receptor complexes (EcR/USP[RXR]), which are members of the canonical nuclear receptor superfamily, and result in long-term changes in gene expression, or by interacting with membrane-bound receptors (e.g. DopEcR), which bring about rapid changes in the physiological state of the target cell. Proper development progression in insects depends on the correct spatiotemporal sequence of rises and declines in ecdysteroid levels which results from a complex interaction of external signals, peptidic factors and juvenile hormones. Ecdysteroids may also play a defensive role in insects, since at least one species of leaf beetle incorporates large amounts of 20-hydroxyecdysone 22-acetate into its defensive secretion, which is believed to protect against predatory spiders.

6.1.3 Vertebrate-Type Steroids

Vertebrate-type steroids are detectable and widespread in insects. Their presence does not necessarily imply endogenous synthesis or a physiological function. In the cases of predatory diving beetles and carrion beetles, the defensive secretions of which contain high levels of a wide range of C21, C19 and C18 steroids, biosynthesis from cholesterol in the insect has been demonstrated and the deterrent function is clear and effective. A C21 steroid glucoside is synthesised in pupae of *Manduca sexta*, but the function of this compound is unknown. It remains unclear whether endogenous synthesis of vertebrate-type steroids in the majority of other insects is possible. Enzyme activities for various steroid interconversions appear to be present in insect tissues. Only in very specific cases (e.g. silkworm and rabbit fleas) has it been possible to make some progress in clarifying the physiological functions of vertebrate-type steroids in the control of development.

6.1.4 Cardenolides and Bufadienolides

The classes of cardiac glycosides are accumulated as protective agents in certain insects: fireflies of the genus *Photinus* (bufadienolides), *Danaus plexippus* and *Oncopeltus fasciatus* (cardenolides). Both classes are synthesised as defensive secondary products in certain plant species. It appears that insects sequester cardenolides from their plant food sources, while *Photinus* spp. have to capacity to synthesise bufadienolide lucibufagins themselves. These compounds are well known as powerful inhibitors of Na^+/K^+-ATPase. *D. plexippus* protects itself from the sequestered cardenolides by storing them in the midgut and by possessing an Na^+/K^+-ATPase which is significantly less sensitive to cardenolides. In *O. fasciatus*, the cardenolides are stored in a dorso-lateral space and released to the outer body surface when the insect is attacked. The stored bufadienolides in *Photinus* spp. are also released when the insects are attacked.

6.2 Conclusions

Although insects are incapable of de novo biosynthesis of cholesterol and other sterols, there is a huge diversity of sterol and steroid molecules present in insects where they fulfil a wide array of different functions, ranging from a structural role in membranes to precursors for ecdysteroids and various defensive steroids and signalling molecules. Sterols are taken up from the diet by all insects and may also be generated by symbiotic organisms. Other steroids present in insects are sequestered and perhaps modified from the food. Given the enormous diversity within the Class Insecta, their habitats, their food sources and the predators and parasites which need to be resisted, the biochemical and metabolic diversity of insects should perhaps not be surprising, but it is nevertheless still amazing. It almost seems that any category of secondary compound or steroid hormone found in the natural world, can, if one looks hard enough, also be found in one group of insects or another, not only by sequestration but also by endogenous biosynthesis. Current exceptions to this are the mineralocorticoid aldosterone and the plant brassinosteroid hormones, but the proportion of all insects which has been examined to date is tiny, so who knows what the future will reveal?

References

Aardema ML, Zhen Y, Andolfatto P (2012) The evolution of cardenolide-resistant forms of Na^+,K^+-ATPase in Danainae butterflies. Mol Ecol 21:340–349. https://doi.org/10.1111/j.1365-294X.2011.05379.x

Abrieux A, Debernard S, Maria A, Gaertner C, Anton S, Gadenne C et al (2013) Involvement of the G-protein-coupled dopamine/ecdysteroid receptor DopEcR in the behavioral response in sex pheromone in an insect. PLoS One 8(9):e72785. https://doi.org/10.1371/journal.pone.0072785

Abrieux A, Duportets L, Debernard S, Gadenne C, Anton S (2014) The GPCR membrane receptor, DopEcR, mediates the actions of both dopamine and ecdysone to control sex pheromone perception in an insect. Front Behavl Neurosci 8:312. https://doi.org/10.3389/fnbeh.2014.00312

Agrawal AA, Petschenka G, Bingham RA, Weber MG, Rasmann S (2012) Toxic cardenolides: chemical ecology and coevolution of specialized plant-herbivore interactions. New Phytol 194(1):28–45. https://doi.org/10.1111/j.1469-8137.2011.04049.x

Aldrich JR, Svoboda JA, Thompson MJ (1981) Cuticle synthesis and inhibition of vitellogenesis: makisterone A is more active than 20-hydroxyecdysone in female milkweed bugs. J Exp Zool 218:133–137. https://doi.org/10.1002/jez.1402180205

Aldrich JR, Kelly TJ, Woods CW (1982) Larval moulting hormone of Trichophoran Hemiptera-Heteroptera: makisterone A, not 20-hydroxyecdysone. J Insect Physiol 28(10):857–861

Andersen JF, Plattner RD, Weisleder D (1988) Metabolic transformations of cucurbitacins by *Diabrotica virgifera virgifera* Leconte and *D. undecimpunctata howardi* Barber. Insect Biochem 18:71–77. https://doi.org/10.1016/0020-1790(88)90038-8

Applebaum SW, Gilbert LI (1972) Stimulation of adenyl cyclase in pupal wing epidermis by α-ecdysone. Dev Biol 27(2):165–175. https://doi.org/10.1016/0012-1606(72)90095-4

Ashburner M, Chihara C, Meltzer P, Richards G (1974) Temporal control of puffing activity in polytene chromosomes. Cold Spring Habor Symp Quant Biol 38:655–662. https://doi.org/10.1101/sqb.1974.038.01.070

Baumgarten J (1995) Vergleichende chemische Untersuchungen der Naturstoffe aus den Prothorakalwehrdrüsen der Laccophilinae und Hydroporinae (Dytiscidae, Coleoptera). Diploma thesis, University of Bayreuth, p 484

Behmer ST (2017) Overturning dogma: tolerance of insects to mixed-sterol diets is not universal. Curr Opin Insect Sci 23:89–95. https://doi.org/10.1016/j.cois.2017.08.001

Behmer ST, Elias DO (1999) The nutritional significance of of sterol metabolic constraints in the generalist grasshopper *Schistocerca americana*. J Insect Physiol 45:339–348. https://doi.org/10.1016/s0022-1910(98)00131-0

Behmer ST, Elias DO (2000) Sterol metabolic constraints as a factor contributing to the maintenance of diet mixing in grasshoppers (Orthoptera: Acrididae). Physiol Biochem Zool 73:219–230. https://doi.org/10.1086/316728

Behmer ST, Nes WD (2003) Insect sterol nutrition and physiology: a global overview. Adv Insect Physiol 31:1–72. https://doi.org/10.1016/S0065-2806(03)31001-X

Behmer ST, Elias DO, Grebenok HJ (1999) Phytosterol metabolism and absorption in the generalist grasshopper, *Schistocerca americana* (Orthoptera: Acrididae). Arch Insect Biochem Physiol 42:13–25

Behmer ST, Grebenok RJ, Douglas AE (2011) Plant sterols and host plant suitability for a phloem-feeding insect. Funct Ecol 25:484–491. https://doi.org/10.3389/fpls.2013.00370

Bejček J, Jurášek M, Spiwok V, Rimpelová S (2021) *Quo vadis* cardiac glycoside research? Toxins 13(344). https://doi.org/10.3390/toxins13050344

Belles X (2020) Insect metamorphosis. From natural history to regulation of development and evolution. Academic, p 290

Berger A, Petschenka G, Degenkolb T, Geisthardt M, Vilcinskas A (2021) Insect collections as an untapped source of bioactive compounds – fireflies (Coleoptera: Lampyridae) and cardiotonic steroids as a proof of concept. Insects 12:689. https://doi.org/10.3390/insects12080689

Birkenbeil H (1983) Ultrastructural and immunochemical investigation of ecdysteroid secretion by the prothoracic gland of the waxmoth *Galleria mellonella*. Cell Tissue Res 229:433–441. https://doi.org/10.1007/BF00214984

Blaszek I, Mala J (1978) Steroid transport through the surface of the prothoracic gland cells in *Galleria mellonella*. Cell Tissue Res 187(3):507–513. https://doi.org/10.1007/BF00229615

Blomquist GJ, Tillman JA, Reed JR, Gu P, Vanderwel D, Choi S et al (1995) Regulation of enzymatic activity involved in sex pheromone production in the housefly, *Musca domestica*. Insect Biochem Mol Biol 25(6):751–757. https://doi.org/10.1016/0965-1748(95)00015-n

Blum MS, Sannasi A (1974) Reflex bleeding in the lampyrid *Photinus pyralis*: defensive function. J Insect Physiol 20:451–460. https://doi.org/10.1016/0022-1910(74)90153-X

Botelho AFM, Pierezan F, Soto-Blanco B, Melo MM (2019) A review of cardiac glycosides: structure, toxicokinetics, clinical signs, diagnosis and antineoplastic potential. Toxicon 158:63–68. https://doi.org/10.1016/j.toxicon.2018.11.429

Bouvaine S, Faure M-L, Grebenok RJ, Behmer ST, Douglas AE (2014) A dietary test of putative deleterious sterols for the aphid *Myzus persicae*. PLoS One 9:e86256. https://doi.org/10.1371/journal.pone.0086256

Bradbrook DA, Clément CY, Cook B, Dinan L (1990) The occurrence of vertebrate-type steroids in insects and a comparison with ecdysteroid levels. Comp Biochem Physiol 95B:365–374. https://doi.org/10.1016/0305-0491(90)90089-C

Burger A, Colobert F, Hétru C, Luu B (1988) Acetylenic cholesterol derivatives as irreversible inhibitors of ecdysone biosynthesis. Tetrahedron 44:1141–1152. https://doi.org/10.1016/S0040-4020(01)85894-4

Burger A, Hétru C, Luu B (1989) Allenic cholesteryl derivatives as inhibitors of ecdysone biosynthesis. Tetrahedron 45:155–164. https://doi.org/10.1016/0040-4020(89)80042-0

Cai MJ, Dong DJ, Wang Y, Liu PC, Liu W, Wang JX et al (2014) G-protein-coupled receptor participates in 20-hydroxyecdysone signaling on the plasma membrane. Cell Commun Signal 12:9. https://doi.org/10.1186/1478-811X-12-9

Calas D, Thiéry D, Marion-Poll F (2006) 20-Hydroxyecdysone deters oviposition and larval feeding in the European grapevine moth, *Lobesia botrana*. J Chem Ecol 32(11):2443–2454. https://doi.org/10.1007/s10886-006-9155-3

Caragata EP, Rancès E, Hedges LM, Gofton AW, Johnson KN, O'Neill SL et al (2013) Dietary cholesterol modulates pathogen blocking by *Wolbachia*. PLoS Pathog 9(6):e1003459. https://doi.org/10.1371/journal.ppat.1003459

Chadha MS, Joshi NK, Mamdapur VR, Sipahimalani AT (1970) C-21 steroids in the defensive secretions of some Indian water beetles–II. Tetrahedron 26:2061–2064. https://doi.org/10.1016/S0040-4020(01)92782-6

Chakrabarti P, Lucas HM, Sagili RR (2020) Novel insights into dietary phytosterol utilization and its fate in honey bees (*Apis mellifera* L.). Molecules 25:571. https://doi.org/10.3390/molecules25030571

Chapman JC, Lockley WJS, Rees HH, Goodwin TW (1977) Stereochemistry of olefinic bond formation in defensive steroids of *Acilius sulcatus* (Dytiscidae). Eur J Biochem 81:293–298. https://doi.org/10.1111/j.1432-1033.1977.tb11951.x

Chiang YN, Tan KJ, Chung H, Lavrynenko O, Shevchenko A, Yew JY (2016) Steroid hormone signaling is essential for pheromone production and oenocyte survival. PLoS Genet 12(6):e1006126. https://doi.org/10.1371/journal.pgen.1006126

Ciufo LF, Murray PA, Thompson A, Rigden DJ, Rees HH (2011) Characterization of a desmosterol reductase involved is phytosterol dealkylation in the silkworm, *Bombyx mori*. PLoS One 6(6):e21316. https://doi.org/10.1371/journal.pone.0021316

Clayton RB (1964) The utilization of sterols by insects. J Lipid Res 5:3–19. https://doi.org/10.1016/S0022-2275(20)40254-8

Clayton RB, Edwards AM, Hinkle PC, Smith DA (1964) The intestinal absorption of cholesterol, its esters and some related sterols and analogues in the roach, *Eurycotis floridana*. Comp Biochem Physiol 11:333–350. https://doi.org/10.1016/0010-406x(64)90001-5

Clever U, Karlson P (1960) Induktion von Puff-Veränderungen in den Speicheldrüsenchromosomen von *Chironomus tentans* durch Ecdyson. Exp Cell Res 20:623–626. https://doi.org/10.1016/0014-4827(60)90141-5

Cohen I (2005) Estrogenic extracts of *Morus alba* and uses thereof. Australian patent; AU 2005316833 B2

Costet MF, El Achouri M, Charlet M, Lanot R, Benveniste P, Hoffmann JA (1987) Ecdysteroid biosynthesis and embryonic development are disturbed in insects (*Locusta migratoria*) reared on plant diet (*Triticum sativum*) with a selectively modified sterol profile. Proc Natl Acad Sci USA 84:643–647. https://doi.org/10.1073/pnas.84.3.643

Crump D, Silverstein RM, Williams HJ, Fitzgerald TD (1987) Identification of trail pheromone of larva of eastern tent caterpillar *Malacosoma americanum* (Lepidoptera: Lasiocampidae). J Chem Ecol 13(3):397–402. https://doi.org/10.1007/BF01880088

Daloze D, Pasteels JM (1979) Production of cardiac glycosides by chrysomelid beetles and larvae. J Chem Ecol 5:63–77. https://doi.org/10.1007/BF00987688

Daloze D, Braekman JC, Pasteels JM (1985) New polyoxygenated steroidal glycosides from *Chrysolina hyperici*. Tetrahedron Letts 26:2311–2314

Daloze D, Braekman JC, Delbrassine A, Pasteels JM (1991) Polyoxygenated steroid sophorosides from the defense glands of *Chrysolina quadrigemina*. J Nat Prod 54:1553–1557. https://doi.org/10.1021/np50078a009

Daloze D, Broeders F, Braekman J-C, Araujot J, Pasteels JM (1995) New cardiac glycosides containing 2-deoxyhexoses from the defensive secretion of adult *Chrysolina banski* (Coleoptera: Chrysomelidae). Biochem System Ecol 23:113–119. https://doi.org/10.1016/0305-1978(94)00085-U

Darvas B, Székács A, Fónagy A, Szécsi M, Tóth I (1997) Progesterone in *Periplaneta americana* and *Neobellieria bullata* adults from the procuticle phase until first progeny production. Gen Comp Endocr 107:450–460. https://doi.org/10.1006/gcen.1997.6952

De Clerck D, Eechaute W, Leusen I, Diederik H, De Loof A (1983) Identification of testosterone and progesterone in hemolymph of larvae of the fleshfly *Sarcophaga bullata*. Gen Comp Endocrinol 52:368–378. https://doi.org/10.1016/0016-6480(83)90175-2

De Clerck D, Diederik H, De Loof A (1984) Identification by capillary gas chromatography-mass spectrometry of eleven non-ecdysteroid steroids in the haemolymph of larvae of *Sarcophaga bullata*. Insect Biochem 14:199–208. https://doi.org/10.1016/0020-1790(84)90030-1

De Clerck D, Eechaute W, Leusen I, De Loof A (1987) Study of the metabolism of steroids in larvae of the fleshfly *Sarcophaga bullata*. Comp Biochem Physiol B 97:821–826. https://doi.org/10.1016/0305-0491(87)90395-6

De Clerck D, Diederik H, Paesen G, De Loof A (1988) Identification and quantification of C21 and C19 steroids in the haemolymph of *Leptinotarsa decemlineata*, a phytophagous insect. Insect Biochem 18:93–99. https://doi.org/10.1016/0020-1790(88)90041-8

Deng Z, Yang Y, Luo J, Zhang B, Liu J, Shui G et al (2022) An integrated transcriptomics and lipidomics analysis reveals that ergosterol is required for host defense against bacterial infection in *Drosophila*. Front Immunol 13:933137. https://doi.org/10.3389/fimmu.2022.933137

Denlinger DL, Brueggemeier RW, Mechoulam R, Katlic N, Yocum LB, Yocum GD (1987) Estrogens and androgens in insects. In: Law JH (ed) Molecular entomology. Liss, New York, pp 189–199

Dettner K (2014) Chemical ecology and biochemistry of Dysticidae. In: Yee DA (ed) Ecology, systematics, and the natural history of predaceous diving beetles (Coleoptera: Dytiscidae). Springer Science+Business Media BV, pp 235–323. https://doi.org/10.1007/978-94-017-9109-0_6

Dettner K (2019) Defenses of water insects. In: Del-Klaro K, Guillermo R (eds) Aquatic insects. Springer Nature, Switzerland AG, pp 191–262. https://doi.org/10.1007/978-3-030-16327-3_9

Dhadialla TS, Carlson GR, Le DP (1998) New insecticides with ecdysteroidal and juvenile hormone activity. Annu Rev Entomol 43:545–569. https://doi.org/10.1146/annurev.ento.43.1.545

Dimock MB, Renwick JA, Radke CD, Sachdev-Gupta K (1991) Chemical constituents of an unacceptable crucifer, *Erysimum cheiranthoides*, deter feeding by *Pieris rapae*. J Chem Ecol 17(3):525–533. https://doi.org/10.1007/BF00982123

Dinan L (2003) Ecdysteroid structure-activity relationships. In: Atta-ur-Rahman (ed) Studies in natural products research, vol 29. Elsevier, Amsterdam, pp 3–71. https://doi.org/10.1016/S1572-5995(03)80004-2

Dinan L (2009) The Karlson lecture. Phytoecdysteroids: what use are they? Arch Insect Biochem Physiol 72:126–141. https://doi.org/10.1002/arch.20334

Dinan L, Hormann RE (2005) Ecdysteroid agonists and antagonists. In: Gilbert LI, Iatrou K, Gill S (eds) Comprehensive molecular insect science, vol 3. Elsevier, pp 197–242

Dinan L, Lafont R (2022) Ecdysteroids as defensive chemicals. Adv Insect Physiol 63:107–174. https://doi.org/10.1016/bs.aiip.2022.09.001

Dinan L, Rees HH (1981) The identification and titres of conjugated and free ecdysteroids in developing ovaries and newly-laid eggs of *Schistocerca gregaria*. J Insect Physiol 27:51–58. https://doi.org/10.1016/0022-1910(81)90032-9

Dinan L, Whiting P, Girault JP, Lafont R, Dhadialla TS, Cress DE et al (1997) Cucurbitacins are insect steroid hormone antagonists acting at the ecdysteroid receptor. Biochem J 327:643–650. https://doi.org/10.1042/bj3270643

Dinan L, Hormann RE, Fujimoto T (1999) An extensive ecdysteroid COMFA. J Comp Aid Mol Des 13:185–207. https://doi.org/10.1023/a:1008052320014

Dobler S, Petschenka G, Pankoke H (2011) Coping with toxic plant compounds – the insect's perspective on iridoid glycosides and cardenolides. Phytochemistry 72(13):1593–1604. https://doi.org/10.1016/j.phytochem.2011.04.015

Douglas AE (1988) On the source of sterols in the green peach aphid, *Myzus persicae*, reared on holidic diets. J Insect Physiol 34(5):403–408. https://doi.org/10.1016/0022-1910(88)90110-2

Douglas AE (1989) Mycetocyte symbiosis in insects. Biol Rev 64:409–434. https://doi.org/10.1111/j.1469-185X.1989.tb00682.x

Du J, Zhao P, Wang J, Ma S, Yao L, Zhu X et al (2022) Pupal diapause termination and transcriptional response of *Antheraea pernyi* (Lepidoptera: Saturniidae) triggered by 20-hydroxyecdysone. Front Physiol 13:888643. https://doi.org/10.3389/fphys.2022.888643

Dubé J, Lemonde A (1970a) Transformations des stéroïdes par la femelle adulte d'un insecte orthoptère, *Schistocerca gregaria* Forskal. Gen Comp Endocrinol 15:158–164. https://doi.org/10.1016/0016-6480(70)90106-1

Dubé J, Lemonde A (1970b) The origin of progesterone in the confused flour beetle (*Tribolium confusum*). Experientia 26:543–544. https://doi.org/10.1007/BF01898500

Dubé J, Villeneuve J-L, Lemonde A (1968) Métabolisme *in vitro* de la progestérone 7α-^{3}H dans l'ovaire de *Schistocerca gregaria* (Orthoptère). Arch Int Physiol Biochim 76:64–70

Duffey SS, Blum MS, Isman MB, Scudder GGE (1978) Cardiac glycosides: a physical system for their sequestration by the milkweed bug. J Insect Physiol 24:639–645. https://doi.org/10.1016/0022-1910(78)90127-0

Dutky RC, Robbins WE, Shortino TJ, Kaplanis JN, Vroman HE (1967) Conversion of cholestanone to cholestanol by housefly *Musca domestica* L. J Insect Physiol 13:1501–1510. https://doi.org/10.1016/0022-1910(67)90175-8

Eggenberger F, Daloze D, Pasteels JM, Rowell-Rahier M (1992) Identification and seasonal quantification of defensive secretion components of *Oreina gloriosa* (Coleoptera: Chrysomelidae). Experientia 48:1173–1179. https://doi.org/10.1007/BF01948018

Eisner T, Wiemer DF, Haynes LW, Meinwald J (1978) Lucibufagins: defensive steroids from the fireflies *Photinus ignitus* and *P. marginellus* (Coleoptera: Lampyridae). Proc Natl Acad Sci USA 75(2):905–908. https://doi.org/10.1073/pnas.75.2.905

Eisner T, Goetz MA, Hill DE, Smedley SR, Meinwald J (1997) Firefly "femmes fatales" acquire defensive steroids (lucibufagins) from their firefly prey. Proc Natl Acad Sci USA 94:9723–9728. https://doi.org/10.1073/pnas.94.18.9723

El-Mallakh RS, Brar KS, Yeruva RR (2019) Cardiac glycosides in human physiology and disease: update for entomologists. Insects 10:102. https://doi.org/10.3390/insects10040102

Feldlaufer MF, Herbert EW, Svoboda JA, Thompson MJ (1986) Biosynthesis of makisterone A and 20-hydroxyecdysone from labeled sterols by the honey bee, *Apis mellifera*. Arch Insect Biochem Physiol 3:415–421. https://doi.org/10.1002/arch.940030502

Feldlaufer MF, Weirich GF, Lusby WR, Svoboda JA (1991) Makisterone C: a 29-carbon ecdysteroid from developing embryos of the cotton stainer bug, *Dysdercus fasciatus*. Arch Insect Biochem Physiol 18:71–79. https://doi.org/10.1002/arch.940180202

Feldlaufer MF, Weirich GF, Imberski RB, Svoboda JA (1995) Ecdysteroid production in *Drosophila melanogaster* reared on defined diets. Insect Biochem Molec Biol 25(6):709–712. https://doi.org/10.1016/0965-1748(95)00009-k

Fitzgerald TD (1993) Trail following and recruitment: response of eastern tent caterpillar *Malacosoma americanum* to 5β-cholestane-3,24-dione and 5-β-cholestan-3-one. J Chem Ecol 19(3):449–457. https://doi.org/10.1007/BF00994318

Fujimoto Y, Morisaki M, Ikekawa N (1985) Enzymatic dealkylation of phytosterols in insects. Methods Enzymol 111:346–352. https://doi.org/10.1016/s0076-6879(85)11021-9

Fujimoto Y, Miyasaka S, Ikeda T, Ikekawa N, Ohnishi E, Mizuno T, Watanabe K (1985) An unusual ecdysteroid, (20*S*)-cholesta-7,14-diene-3β,5α,6α,20,25-pentaol (bombycosterol) from the ovaries of the silkworm, *Bombyx mori*. J Chem Soc Chem Comm:10–12. https://doi.org/10.1039/C39850000010

Fujinaga D, Gu J, Kawahara H, Ogihara MH, Kojima I, Takeshima M et al (2020) Twenty-hydroxyecdysone produced by dephosphorylation and ecdysteroidogenesis regulates early embryonic development in the silkmoth, *Bombyx mori*. Insect Biochem Mol Biol 127:103491. https://doi.org/10.3389/fphys.2022.888643

Gabrieli P, Kakani EG, Mitchell SN, Mameli E, Want EJ, Mariezcurrena Anton A et al (2014) Sexual transfer of the steroid hormone 20E induces the postmating switch in *Anopheles gambiae*. Proc Natl Acad Sci USA 111(46):16353–16358. https://doi.org/10.1073/pnas.1410488111

Galbraith MN, Horn DHS, Middleton EJ, Thompson JA, Siddall JB, Hafferl W (1969) The catabolism of crustecdysone in the blowfly *Calliphora stygia*. J Chem Soc D: Chem Comm:1134–1135. https://doi.org/10.1039/C29690001134

Gilbert LI (2004) Halloween genes encode P450 enzymes that mediate steroid hormone biosynthesis in *Drosophila melanogaster*. Mol Cell Endocrinol 215(1–2):1–10. https://doi.org/10.1016/j.mce.2003.11.003

Giner J-L, Zhao H, Dixon MS, Wikfors GH (2016) Bioconversion of ^{13}C-labeled microalgal phytosterols to cholesterol by the Northern Bay scallop, *Argopecten irradians irradians*. Comp Biochem Physiol B 192:1–8. https://doi.org/10.1016/j.cbpb.2015.11.003

Goetz M, Wiemer DF, Haynes LW, Meinwald J (1979) Lucibufagines. Partie III. Oxo-11- et oxo-12-bufalines, stéroïdes défensifs des lampyres *Photinus ignitus* et *P. Marginellus* (Coleoptera: Lampyridae). Helv Chim Acta 62:1396–1400. https://doi.org/10.1002/hlca.19790620503

Goetz MA, Meinwald J, Eisner T (1981) Lucibufagins, IV. New defensive steroids and a pterin from the firefly, *Photinus pyralis* (Coleoptera: Lampyridae). Experientia 37:679–680. https://doi.org/10.1007/BF01967916

González A, Schroeder FC, Attygalle AB, Svatos A, Meinwald J, EisnerT. (1999) Metabolic transformations of acquired lucibufagins by firefly 'femmes fatales'. Chemoecol 9:105–112. https://doi.org/10.1007/s000490050040

Guzmán I (2005) Isolation and characterization of mulberry phytoestrogens. MSc Thesis, Texas Woman's University, Denton, Texas, USA

Hagedorn HH, O'Connor JD, Fuchs MS, Sage B, Schlaeger DA, Bohm MK (1975) The ovary as a source of alpha-ecdysone in an adult mosquito. Proc Natl Acad Sci USA 72(8):3255–3259. https://doi.org/10.1073/pnas.72.8.3255

Häggström M, Richfield D (2014) Diagram of the pathways of human steroidogenesis. Wiki J Med 1(1). https://doi.org/10.15347/wjm/2014.005

Han P, Han J, Zhang M, Fan J, Gong Q, Ma E et al (2020) 20-Hydroxyecdysone enhances immulectin-1 mediated immune response against entomogenous fungus in *Locusta migratoria*. Pest Manag Sci 76(1): 304–313. https://doi.org/10.1002/ps.5515

Heed WB, Kircher HW (1965) Unique sterol in the ecology and nutrition of *Drosophila pachea*. Science 149:758–761. https://doi.org/10.1126/science.149.3685.758

Heftmann E (1975) Steroid hormones in plants. J Nat Products: Lloydia 38:195–209

Hikino H, Ohizumi Y, Takemoto T (1975) Steroid metabolism in *Bombyx mori*. I. Catabolism of ponasterone A and ecdysterone in *Bombyx mori*. Hoppe Seylers Z Physiol Chem 356:309–314. https://doi.org/10.1515/bchm2.1975.356.1.309

Hilker M, Daloze D, Pasteels JM (1992) Cardenolide glycosides from the adults and eggs of *Chrysolina fuliginosa* (Coleoptera: Chrysomelidae). Experientia 48:1023–1027. https://doi.org/10.1007/BF01919158

Hill RJ, Billas IML, Bonneton F, Graham LD, Lawrence MC (2013) Ecdysone receptors: from the Ashburner model to structural biology. Annu Rev Entomol 58: 251–271

Hiramoto M, Fujimoto Y, Kakinuma K, Ikekawa N (1988) Ecdysteroid conjugates in the ovaries of the silkworm, *Bombyx mori*: 3-phosphates of 2,22-dideoxy-20-hydroxyecdysone and of bombycosterol. Experientia 44:623–624. https://doi.org/10.1007/BF01953319

Hoffmann JA, Lagueux M (1985) Endocrine aspects of embryonic development. In: Kerkut GA, Gilbert LI (eds) Comprehensive insect physiology, biochemistry and pharmacology, vol 1. Pergamon Press, Oxford, pp 435–460

Hoffmann K, Thiry E, Lafont R (1990) 14-Deoxyecdysteroids in an insect (*Gryllus bimaculatus*). Z Naturforsch C 45(6):703–708. https://doi.org/10.1515/znc-1990-0622

Holzinger F, Frick C, Wink M (1992) Molecular basis for the insensitivity of the monarch (*Danaus plexippus*) to cardiac glycosides. FEBS Lett 314(3):477–480. https://doi.org/10.1016/0014-5793(92)81530-y

Hormann RE, Dinan L, Whiting P (2003) Superimposition evaluation of ecdysteroid agonist chemotypes through multidimensional QSAR. J Comp Aid Mol Des 17: 135–153. https://doi.org/10.1023/a:1025365722403

Hun LV, Okamoto N, Imura E, Maxson R, Bittar R, Yamanaka N (2022) Essential functions of mosquito ecdysone importers in development and reproduction. Proc Natl Acad Sci USA 119(25):e2202932119. https://doi.org/10.1073/pnas.2202932119

Hurkadli HK, Hooli MA, Nadkarni VB (1980) Steroidogenic cells in the testis of eri silkmoth, *Philosamia ricini:* a biochemical study. In: Clark WH, Adams TS (eds) Advances in invertebrate reproduction. Elsevier, Amsterdam, p 378

Hutfilz C (2022) Endocrine regulation of lifespan in insect diapause. Front Physiol 13:825057. https://doi.org/10.3389/fphys.2022.825057

Igarashi F, Ogihara MH, Iga M, Kataoka H (2018) Cholesterol internalisation and metabolism in insect prothoracic gland, a steroidogenic organ, via lipoproteins. Steroids 134:110–116. https://doi.org/10.1016/j.steroids.2018.01.012

Ikekawa N, Morisaki M, Fujimoto Y (1993) Sterol metabolism in insects: dealkylation of phytosterol to cholesterol. Acc Chem Res 26:139–146. https://doi.org/10.1021/ar00028a002

Ikekawa N, Fujimoto Y, Ishiguro M (2013) Reminiscences of research on the chemistry and biology of natural sterols in insects, plants and humans. Proc Jpn Acad Ser B 89(8):349–369. https://doi.org/10.2183/pjab.89.349

Ingham PW (2022) Chapter one – hedgehog signaling. Curr topics. Dev Biol 149:1–58. https://doi.org/10.1016/bs.ctdb.2022.04.003

Isaac RE, Slinger AJ (1989) Storage and excretion of ecdysteroids. In: Koolman J (ed) Ecdysone: from chemistry to mode of action. Thieme Verlag, Stuttgart, pp 250–253

Isaac RE, Sweeney FP, Rees HH (1983) Enzymic hydrolysis of ecdysteroid phosphate during embryogenesis in the desert locust (*Schistocerca gregaria*). Biochem Soc Trans 11:379–380. https://doi.org/10.1042/bst0110379

Jacob F, Monod J (1961) Genetic regulatory mechanisms in the synthesis of proteins. J Mol Biol 3:318–356. https://doi.org/10.1016/s0022-2836(61)80072-7

Janeczko A, Skoczowski A (2005) Mammalian sex hormones in plants. Folia Histochem Cytobiol 43: 71–79

Jarvis TD, Earley FGP, Rees HH (1994) Inhibition of the ecdysteoid biosynthetic pathway in ovarian follicle cells of *Locusta migratoria*. Pestic Biochem Physiol 48:153–162. https://doi.org/10.1006/pest.1994.1016

Ji HY, Moog C, Schmitt G, Luu B (1990) Polyoxygenated sterols and triterpenes: chemical structures and biological activities. J Steroid Biochem 6:741–744. https://doi.org/10.1016/0022-4731(90)90317-l

Jing X, Behmer ST (2020) Insect sterol nutrition: physiological mechanisms, ecology, and applications. Annu Rev Entomol 65:251–271. https://doi.org/10.1146/annurev-ento-011019-025017

Jones CG, Hoggard M, Blum MS (1983) Is sequestration structure-specific in the milkweed bug, *Oncopeltus fasciatus*. Comp Biochem Physiol 76C(2):283–284. https://doi.org/10.1016/0742-8413(83)90079-8

Joshi M, Agarwal HC (1977) Site of cholesterol absorption in some insects. J Insect Physiol 23:403–404. https://doi.org/10.1016/0022-1910(77)90280-3

Kabbouh M, Rees HH (1991) Characterization of the ATP:2-deoxyecdysone 22-phosphotransferase (2-deoxyecdysone 22-kinase) in the follicle cells of *Schistocerca gregaria*. Insect Biochem 21:607–613. https://doi.org/10.1016/0020-1790(91)90064-L

Kacprzyk M, Bednarz B, Choczyński M (2021) Attempt to identify sex hormones in the bodies of selected Norway spruce bark beetles. Forests 12:536. https://doi.org/10.3390/f12050536

Kadono-Okuda K, Kuwanto E, Eto M, Yamashita O (1987) Inhibitory action of an imidazole compound on ecdysone synthesis in prothoracic glands of the silkworm, *Bombyx mori*. Develop Growth Differ 29:527–533. https://doi.org/10.1111/j.1440-169X.1987.00527.x

Kandel N, Wang C (2022) Hedgehog autoprocessing: from structural mechanisms to drug discovery. Front Mol Biosci 9:900560. https://doi.org/10.3389/fmolb.2022.900560

Kang XL, Zhang JY, Wang D, Zhao YM, Han XL, Wang JX et al (2019) The steroid hormone 20-hydroyecdysone binds to dopamine receptor to repress lepidopteran insect feeding and promote pupation. PLoS Genet 15(8):e1008331. https://doi.org/10.1371/journal.pgen.1008331

Kang XL, Li YX, Li YL, Wang JX, Zhao XF (2021) The homotetramerization of a GPCR transmits the 20-hydroxyecdysone signal and increases its entry into cells for insect metamorphosis. Development 148:dev196667. https://doi.org/10.1242/dev.196667

Kaplanis JN, Dutky SR, Robbins WE, Thompson MJ, Lindquist EL, Horn DH et al (1975) Makisterone A: a 28-carbon hexahydroxy molting hormone from the embryo of the milkweed bug. Science 190(4215): 681–682. https://doi.org/10.1126/science.1237931

Keshan B, Ray AK (1998) Action of estradiol-17β on the synthetic activity of the silk gland in *Bombyx mori* L. J Insect Physiol 44:491–498. https://doi.org/10.1016/s0022-1910(97)00141-8

Keshan B, Ray AK (2000) Estradiol-17β in *Bombyx mori*: possible significance and it effect on silk production. J Insect Physiol 46:1061–1068. https://doi.org/10.1016/s0022-1910(99)00218-8

Keshan B, Ray AK (2001) The presence of estradiol-17β and its specific binding sites in posterior silk gland of *Bombyx mori*. Gen Comp Endocrinol 123:23–30. https://doi.org/10.1006/gcen.2001.7642

Kopp A (2012) Evolutionary genetics: no coming back from Neverland. Curr Biol 22:R1004–R1006. https://doi.org/10.1016/j.cub.2012.10.040

Kovganko NV, Achrem AA (1990) Steroids: ecological functions. Nauka i Technika, Minsk, p 223

Kozlova T, Thummel CS (2000) Steroid regulation of postembryonic development and reproduction in *Drosophila*. Trends Endocrinol Metab 11(7):276–280. https://doi.org/10.1016/s1043-2760(00)00282-4

Kreis W, Hensel A, Stuhlemmer U (1998) Cardenolide biosynthesis in foxglove. Planta Med 64:491–499. https://doi.org/10.1055/s-2006-957500

Kroeger H, Lezzi M (1966) Regulation of gene action in insect development. Annu Rev Entomol 11(1–22) https://doi.org/10.1146/annurev.en.11.010166.000245

Lafont R (1991) Reverse endocrinology, or “hormones” seeking functions. Insect Biochem 21(7):697–721. https://doi.org/10.1016/0020-1790(91)90112-R

Lafont R, Connat J-L (1989) Pathways of ecdysone metabolism. In: Koolman J (ed) Ecdysone: from chemistry to mode of action. Thieme Verlag, Stuttgart, pp 167–173

Lafont R, Horn DHS (1989) Phytoecdysteroids: structures and occurrence. In: Koolman J (ed) Ecdysone: from chemistry to mode of action. Thieme Verlag, Stuttgart, pp 39–64

Lafont R, Mathieu M (2007) Steroids in aquatic invertebrates. Ecotoxicology 16:109–130. https://doi.org/10.1007/s10646-006-0113-1

Lafont R, Harmatha J, Marion-Poll F, Dinan L, Wilson ID (2002) Ecdybase: the Ecdysone Handbook, 3rd edn. https://ecdybase.org

Lafont R, Dauphin-Villemant C, Warren JT, Rees HH (2017) Ecdysteroid chemistry and biochemistry. reference module in life sciences. Elsevier. https://doi.org/10.1016/B978-0-12-809633-8.04026-7

Lagueux M, Hoffmann JA, Goltzené F, Kappler C, Tsoupras G, Hetru C et al (1984) Ecdysteroids in ovaries and embryos of *Locusta migratoria*. In: Hoffmann JA, Porchet M (eds) Biosynthesis, metabolism and mode of action of invertebrate hormones. Springer Verlag, Berlin, pp 168–189. https://doi.org/10.1007/978-3-642-69922-1_18

Lang M, Murat S, Clark AG, Gouppil G, Blais C, Matzkin LM et al (2012) Mutations in the neverland gene turned *Drosophila pachea* into an obligate specialist species. Science 337(6102):1658–1661. https://doi.org/10.1126/science.1224829

Lanot R, Thiebold J, Costet-Coriot MF, Benveniste P, Hoffmann JA (1988) Further experimental evidence for the involvement of ecdysone in the control of meiotic reinitiation in oocytes of *Locusta migratoria* (Insecta, Orthoptera). Dev Biol 126(1):212–214. https://doi.org/10.1016/0012-1606(88)90255-2

Lasser NL, Edwards AM, Clayton RB (1966) Distribution and dynamic state of sterols and steroids in tissues of an insect, the roach *Eurycotis floridana*. J Lipid Res 7: 403–412

Laurent P, Braekman JC, Daloze D, Pasteels JM (2003) An ecdysteroid (22-acetyl-20-hydroxyecdysone) from the defense gland secretion of an insect: *Chrysolina carnifex* (Coleoptera; Chrysomelidae). Chemoecology 13:109–111. https://doi.org/10.1007/s00049-003-0235-3

Lavrynenko O, Rodenfels J, Carvalho M, Dye NA, Lafont R, Eaton S et al (2015) The ecdysteroidome of *Drosophila*: influence of diet and development. Development 142:3758–3768. https://doi.org/10.1242/dev.124982

Leavell BC, Rubin JJ, McClure CJW, Miner KA, Branham MA, Barber JR (2018) Fireflies thwart bat attack with multisensory warnings. Sci Adv 4: eaat6601. https://doi.org/10.1126/sciadv.aat6601

Lee S, Nguyen QN, Kim SJ, Lee J, Shin MS (2022) Estrogenic activity of freeze-dried silkworm extracts through the activation of estrogen receptors in MCF-7 cells. App Biol Chem 65:43. https://doi.org/10.1186/s13765-022-00714w

Lehoux J-G, Sandor T (1969) Conversion of testosterone to Δ^4-androstene-3,17-dione by house cricket (*Gryllus domesticus*) male gonad preparations *in vitro*. Endocrinology 84:652–657. https://doi.org/10.1210/endo-84-3-652

Lehoux J-G, Sandor T (1970) The occurrence of steroids and steroid metabolizing enzyme systems in invertebrates. A review. Steroids 16:141–171. https://doi.org/10.1016/S0039-128X(70)80102-7

Lehoux J-G, Sandor T, Lanthier A, Lusis O (1968) Metabolism of exogenous progesterone by insect tissue preparations *in vitro*. Gen Comp Endocrinol 11:481–488. https://doi.org/10.1016/0016-6480(68)90063-4

Lehoux J-C, Chapdelaine A, Sandor T (1970) L'Oxydation de la prégnènolone en progésterone par des préparations de tissus des Orthoptères *in vitro*. Can J Biochem 48:407–411. https://doi.org/10.1139/o70-066

Li S, Jing X (2020) Fates of dietary sterols in the insect alimentary canal. Curr Opin Insect Sci 41:106–111. https://doi.org/10.1016/j.cois.2020.08.001

Loewe S (1931) Hormone und Sexualität bei den Schmetterlingen. Naturwissenschaften 19:775. https://doi.org/10.1007/BF01520524

Loewe S, Raudenbusch W, Loss HE, Van Heurn JWC (1932) Nachweis des Sexualhormon-Vorkommens bei Schmetterlingen. Biochem Z 224:347–456

Lokensgard J, Smith RL, Eisner T, Meinwald J (1993) Pregnanes from defensive glands of a belastomid bug. Experientia 49(2):175–176. https://doi.org/10.1007/BF01989425

Lorenz J, Lenz M, Hoffmann KH (1995) Effects of pharmacological agents on ecdysteroid synthesis *in vitro* in ovaries and abdominal integument from female adult crickets, *Gryllus bimaculatus* de geer (Ensifera, Gryllidae). Z Naturforsch 50c:286–293. https://doi.org/10.1515/znc-1995-3-419

Ma H, Abbas MN, Zhang K, Hu X, Xu M, Liang H et al (2019) 20-Hydroxyecdysone regulates the transcription of the lysozyme via broad-complex Z2 gene in silkworm *Bombyx mori*. Dev Comp Immunol 94:66–72. https://doi.org/10.1016/j.dci.2019.01.014

Maglich JM, Sluder A, Guan X, Shi Y, McKee DD, Carrick K et al (2001) Comparison of complete nuclear receptor sets from the human, *Caenorhabditis elegans* and *Drosophila* genomes. Genome Biol 2(8): RESEARCH0029. https://doi.org/10.1186/gb-2001-2-8-research0029

Marion-Poll F, Descoins C (2002) Taste detection of phytoecdysteroids in larvae of *Bombyx mori*, *Spodoptera littoralis* and *Ostrinia nubilalis*. J Insect Physiol 48:467–476. https://doi.org/10.1016/s0022-1910(02)00068-9

Masterson M, Bittar R, Chu H, Yamanaka N, Haga-Yamanaka S (2022) Rapid assessment of insect steroid hormone entry into cultured cells. Front Physiol 12: 816058. https://doi.org/10.3389/fphys.2021.816058

Maurer P, Girault JP, Larchevêque M, Lafont R (1993) 24-Epi-makisterone A (not makisterone A) is the major ecdysteroid in the leaf-cutting ant *Acromyrmex octospinosus* (Reich). Arch Insect Biochem Physiol 23:29–35. https://doi.org/10.1002/arch.940230105

Mauvais A, Hétru C, Luu B (1993) Synthesis of ecdysteroid inhibitors of ecdysone biosynthesis. Inhibition of the 25-hydroxylation. Bioorg Med Chem Lett 3(10):1951–1952. https://doi.org/10.1016/S0960-894X(01)80993-8

Mauvais A, Burger A, Roussel JP, Hétru C, Luu B (1994) Acetylenic inhibitors of C-22-hydroxylase of ecdysone biosynthesis. Bioorg Chem 22:36–50. https://doi.org/10.1006/bioo.1994.1003

Mead-Briggs AR, Rudge AJB (1960) Breeding of the rabbit flea, *Spilopsyllus cuniculi* (Dale): requirement of a 'factor' from a pregnant rabbit for ovarian maturation. Nature 187:1136–1137. https://doi.org/10.1038/1871136a0

Mebs D, Wunder C, Pogoda W, Toennes SW (2017) Feeding on toxic prey. The praying mantis (Mantodea) as predator of poisonous butterfly and moth (Lepidoptera) caterpillars. Toxicon 131:16–19. https://doi.org/10.1016/j.toxicon.2017.03.010

Mechoulam R, Brueggemeier RW, Denlinger DL (1984) Estrogens in insects. Experientia 40:942–944. https://doi.org/10.1007/BF01946450

Meinwald J, Wiemer DF, Eisner T (1979) Lucibufagins 2. Esters of 12-oxo-2β,5β,11α-tridydroxybufalin, the major defensive steroids of the firefly *Photinus pyralis* (Coleoptera: Lampyridae). J Am Chem Soc 101:3055–3060. https://doi.org/10.1021/JA00505A037

Meinwald J, Roach B, Hicks K, Alsop D, Eisner T (1985) Defensive steroids from a carrion beetle (*Silpha americana*). Experientia 41(4):516–519. https://doi.org/10.1007/BF01966178

Meinwald J, Roach B, Hicks K, Alsopp D, Eisner T (1987) Defensive steroids from a carrion beetle (*Silpha novaboracensis*). J Chem Ecol 13:35–38. https://doi.org/10.1007/BF01020349

Meinwald J, Huang Q, Vrkoč J, Herath KB, Yang Z-C, Schröder F et al (1998) Mirasorvone: a masked 20-ketopregnane from the defensive secretion of a diving beetle (*Thermonectus marmoratus*). Proc Natl Acad Sci USA 95:2733–2737. https://doi.org/10.1073/pnas.95.6.2733

Meister MF, Dimarcq JL, Kappler C, Hetru C, Lagueux M, Lanot R et al (1985) Conversion of a radiolabelled ecdysone precursor, 2,22,25-trideoxyecdysone, by embryonic and larval tissues of *Locusta migratoria*. Mol Cell Endocrinol 41(1):27–44. https://doi.org/10.1016/0303-7207(85)90140-6

Miller JR, Mumma RO (1973) Defensive agents of the American water beetles *Agabus seriatus* and *Graphoderus liberus*. J Insect Physiol 19:917–925. https://doi.org/10.1016/0022-1910(73)90161-3

Miller JR, Mumma RO (1976a) Physiological activity of water beetle defensive agents. I. Toxicity and anaesthetic activity of steroids and norsesquiterpenes administered in solution to the minnow *Pimephales promelas*. J Chem Ecol 2:115–130. https://doi.org/10.1007/BF00987737

Miller JR, Mumma RO (1976b) Physiological activity of water beetle defensive agents. II. Absorption of selected anesthetic steroids and norsesquiterpenes across gill membranes of the minnow *Pimephales promelas* Raf. J Chem Ecol 2:131–146. https://doi.org/10.1007/BF00987737

Mizuno T, Ohnishi E (1975) Conjugated ecdysone in the eggs of the silkworm, *Bombyx mori*. Devel Growth Differ 17:219–225. https://doi.org/10.1111/j.1440-169X.1975.00219.x

Mordue AJ, Blackwell A (1993) Azadirachtin: an update. J Insect Physiol 39:903–924. https://doi.org/10.1016/0022-1910(93)90001-8

Mori K (2005) Synthetic examination of incorrectly proposed structures of biomolecules. Chem Rec 5:1–16

Morinaga S, Nagata K, Ihara S, Yumita T, Nimura Y, Sato K et al (2022) Structural model for ligand binding and channel opening of an insect gustatory receptor. J Biol Chem 298(11):102573. https://doi.org/10.1016/j.jbc.2022.102573

Morrow H, Mirth CK (2024) Timing *Drosophila* development through steroid hormone action. Curr Op Genetics Devel 84:102148. https://doi.org/10.1016/j.gde.2023.102148

Nagashima E, Nakagaki M, Sato M, Wakabayashi H, Banno Y, Xiang Z (1983) Testosterone-like immunoreactivities in the haemolymph of the silkworm, *Bombyx mori*. J Ser Sci Japan 52:243–244

Nasir H, Noda H (2003) Yeast-like symbiotes as a sterol source in Anobiid beetles (Coleoptera, Anobiidae): possible metabolic pathways from fungal sterols to 7-dehydrocholesterol. Arch Insect Biochem Physiol 52:175–182. https://doi.org/10.1002/arch.10079

Ninagi O, Maruyama M (1996) Utilization of 20-hydroxyecdysone extracted from a plant in sericulture. JARQ 30:123–128

Niwa R, Niwa YS (2014) Enzymes for ecdysteroid biosynthesis: their biological functions in insects and beyond. Biosci biotech. Biochemist 78(8):1283–1292. https://doi.org/10.1080/09168451.2014.942250

Novak FJS, Lambert JGD (1989) Pregnenolone, testosterone and estradiol in the migratory locust, *Locusta migratoria*: a gas chromatographical-mass spectrometrical study. Endocr 76:73–82. https://doi.org/10.1016/0016-6480(89)90034-8

Novak FJS, Paesen G, De Loof A (1987) C18, C19 and C21 steroids in hemolymph of *Locusta migratoria migratorioides* R & F. Gen Comp Endocrinol 66:143

Novak F, De Clerck D, Paesen G et al (1990) Radioimmunological quantification of C_{21}, C_{19} and C_{18} steroids in the gonads of *Locusta migratoria migratorioides* R and F. Comp Biochem Physiol 97:727–733. https://doi.org/10.1016/0305-0491(90)90114-9

Ogiso M, Ohnishi E (1986) Does estradiol play a role in ovarian maturation or embryonic development of the silkworm? Gen Comp Endocrinol 61:82–86. https://doi.org/10.1016/0016-6480(86)90252-2

Ohnishi E, Ogiso M, Wakabayashi K, Fujimoto Y, Ikekawa N (1985) Identification of estradiol in the ovaries of the silkworm, *Bombyx mor*. Gen Comp Endocrinol 60:35–38. https://doi.org/10.1016/0016-6480(85)90289-8

Ohnishi E, Hiramoto M, Fujimoto Y, Kakinuma K, Ikekawa N (1989) Isolation and identification of major ecdysteroid conjugates from the ovaries of *Bombyx mori*. Insect Biochem 19:95–101. https://doi.org/10.1016/0020-1790(89)90014-0

Okamoto N, Yamanaka N (2020) Steroid hormone entry into the brain requires a membrane transporter in *Drosophila*. Curr Biol 30:359–366. https://doi.org/10.1016/j.cub.2019.11.085

Okamoto N, Viswathana R, Bittar R, Li Z, Haga-Yamanaka S, Perrimon N et al (2018) A membrane transporter is required for steroid hormone uptake in *Drosophila*. Develop Cell 47:294–305. https://doi.org/10.1016/j.devcel.2018.09.012

Oonincx DGAB, van Keuken P, Finke MD, Baines FM, Vermeulen M, Bosch G (2018) Evidence of vitamin D synthesis in insects exposed to UVb light. Sci Rep 8:10807. https://doi.org/10.1038/s41598-018-29232-w

Pan X, Connacher RP, O'Connor MB (2021) Control of the insect metamorphic transition by ecdysteroid production and secretion. Curr Opin Insect Sci 43:11–20. https://doi.org/10.1016/j.cois.2020.09.004

Parades JC, Herren JK, Schupfer F, Lemaitre B (2016) The role of lipid composition for endosymbiont-mediated protection against parasitoid wasps in *Drosophila*. MBio 7(4):e01006-16. https://doi.org/10.1128/mbio.01006-16

Parsons JA (1965) A digitalis-like toxin in the monarch butterfly, *Danaus plexippus*. J Physiol 178(2):290–304. https://doi.org/10.1113/jphysiol.1965.sp007628

Pasteels JM, Daloze D (1977) Cardiac glycosides in the defensive secretion of chrysomelid beetles: evidence for their production by the insects. Science 197:70–72. https://doi.org/10.1126/science.867051

Pasteels JM, Braekman JC, Daloze D (1982) Chemical defence in chrysomelid larvae and adults. Tetrahedron 38:1891–1897. https://doi.org/10.1016/0040-4020(82)80038-0

Pasteels JM, Rowell-Rahier M, Braekman JC, Daloze D, Duffey S (1989) Evolution of exocrine chemical defense in leaf beetles (Coleoptera: Chrysomelidae). Cell Mol Life Sci 45:295–300. https://doi.org/10.1007/BF01951815

Pasteels JM, Rowell-Rahier M, Braekman JC, Daloze D (1994) Chemical defense of adult leaf beetles updated. In: Jolivet PH, Cox ML, Petitpierre E (eds) Novel aspects of the biology of Chrysomelidae. Kluwer, Dordrecht, pp 289–301. https://doi.org/10.1007/978-94-011-1781-4_22

Pedram A, Razandi M, Sainson RCA, Kim JK, Hughes CC, Levin ER (2007) A conserved mechanism for steroid receptor translocation to the plasma membrane. J Biol Chem 282(31):22278–22288. https://doi.org/10.1074/jbc.M611877200

Peng D, Kakani EG, Mameli E, Vidoudez C, Mitchell SN, Merrihew GE et al (2022) A male steroid controls female sexual behaviour in the malaria mosquito. Nature 608(7921):93–97. https://doi.org/10.1038/s41586-022-04908-6

Peterson SC, Fitzgerald TD (1991) Chemoorientation of eastern tent caterpillars to trail pheromone, 5-β-cholestane-3,24-dione. J Chem Ecol 17(10):1963–1972. https://doi.org/10.1007/BF00992581

Petschenka G, Agrawal AA (2015) Milkweed butterfly resistance to plant toxins is linked to sequestration, not coping with a toxic diet. Proc R Soc B 282:20151865. https://doi.org/10.1098/rspb.2015.1865

Pokharel P, Sippel M, Vilcinskas A, Petschenka G (2020) Defense of milkweed bugs (Heteroptera: Lygaeinae) against predatory lacewing larvae depends on structural differences of sequestered cardenolides. Insects 11:485. https://doi.org/10.3390/insects11080485

Pondeville E, Maria A, Jacques C, Bourgouin C, Dauphin-Villemant C (2008) *Anopheles gambiae* males produce and transfer the vitellogenic steroid hormone 20-hydroxyecdysone to females during mating. Proc Natl Acad Sci USA 105(50):19631–19636. https://doi.org/10.1073/pnas.0809264105

Prasad RS (1976) Studies on host-flea relationship IV. Progesterone and cortisone do not influence the reproductive potential of rat fleas *Xenopsylla cheopsis* (Rothschild) and *X. astia* (Rothschild). Z Parasitenk 50:81–86. https://doi.org/10.1007/BF00389935

Prestwich GD, Gayen AK, Phirwa S, Kline TB (1983) 29-Fluorophytosterols: novel pro-insecticides which cause death by dealkylation. Nature. Bio/Technology 1:62–65. https://doi.org/10.1038/nbt0383-62

Prestwich GD, Yamaoka R, Phirwa S, DePalma A (1984) Isolation of 2-fluoro citrate produced by *in vivo* dealkylation of 29-fluorostigmasterol in an insect. J Biol Chem 259:11022–11026. https://doi.org/10.1016/S0021-9258(18)90616-2

Rafter JL, Vendettuoli JF, Gonda-King L, Niesen D, Seeram NP, Rigsby CM et al (2017) Pretty picky of a generalist: impacts of toxicity and nutritional quality on mantid prey processing. Environ Entomol 48:626–632. https://doi.org/10.1093/ee/nvx038

Randoux T, Braekman JC, Daloze D, Pasteels JM, Riccio R (1990) New polyoxygenated steroid glycosides from the defence glands of several species of *Chrysolinina* beetles (Coleoptera: Chrysomelidae). Tetrahedron 46:3879–3888. https://doi.org/10.1016/S0040-4020(01)90523-X

Redfern CPF (1984) Evidence for the presence of makisterone A in *Drosophila* larvae and the secretion of 20-deoxymakisterone A by the ring gland. Proc Natl Acad Sci USA 82:5643–5647. https://doi.org/10.1073/pnas.81.18.5643

Redfern CPF (1986) Changes in patterns of ecdysteroid secretion by the ring gland of *Drosophila* in relation to the sterol composition of the diet. Experientia 42:307–309. https://doi.org/10.1007/BF01942516

Rees HH (1995) Ecdysteroid biosynthesis and inactivation in relation to function. Eur J Entomol 92:9–39

Rees HH, Isaac RE (1984) Biosynthesis of ovarian ecdysteroid phosphates and their metabolic fate during embryogenesis in *Schistocerca gregaria*. In: Hoffmann JA, Porchet M (eds) Biosynthesis, metabolism and mode of action of invertebrate hormones. Springer Verlag, Berlin, pp 181–195. https://doi.org/10.1007/978-3-642-69922-1_19

Rewitz KT, Rybczynski H, Warren JT, Gilbert LI (2006) The Halloween genes code for cytochrome P450 enzymes mediating synthesis of the insect moulting hormone. Biochem Soc Trans 34(6):1256–1260. https://doi.org/10.1042/BST0341256

Rohela GK, Shukla P, Muttanna KR, Chowdhury SR (2020) Mulberry (*Morus* spp.): an ideal plant for sustainable development. Trees, Forests and People 2: 100011. https://doi.org/10.1016/j.tfp.2020.100011

Rothschild M, Ford B (1964a) Breeding of the rabbit flea [*Spilopsyllus cuniculi* (Dale)] controlled by the reproductive hormones of the host. Nature 201:103–104

Rothschild M, Ford B (1964b) Maturation and egg-laying of the rabbit flea (*Spilopsyllus cuniculi* Dale) induced by the external application of hydrocortisone. Nature 203:210–211. https://doi.org/10.1038/203210a0

Rothschild M, Ford B (1966) Hormones of the vertebrate host controlling ovarian regression and copulation of the rabbit flea. Nature 211:261–266. https://doi.org/10.1038/211261a0

Roussel JP (1994) Inhibition de la biosynthèse de l'ecdysone par des molécules synthétiques. Arch Int Physiol Biochim Biophys 102:297–310. https://doi.org/10.3109/13813459409007550

Roy S, De J, Kundu S, Biswas A, Pramanik M, Ray AK (2007) Estradiol-17β: tracing its metabolic significance in female fatbody of fifth instar larvae of silkworm, *Bombyx mori* L (race: Nistari). Life Sci 80:446–453. https://doi.org/10.1016/j.lfs.2006.09.032

Roy S, Saha TT, Zou Z, Raikhel AS (2018) Regulatory pathways controlling female insect reproduction. Annu Rev Entomol 63:489–511. https://doi.org/10.1146/annurev-ento-020117-043258

Sakuma M, Fukamai H (1993a) Aggregation arrestant pheromone of the German cockroach, *Blattella germanica* (L.) (Dictyoptera: Blattellidae): isolation and structure elucidation of blattellastanoside-A and -B. J Chem Ecol 19:2521–2541

Sakuma M, Fukamai H (1993b) Novel steroid glycosides as aggregation pheromones of the German cockroach. Tetrahedron Letts 34:6059–6062. https://doi.org/10.1016/S0040-4039(00)61727-6

Sall C, Tsoupras G, Kappler C, Lagueux M, Zachary D, Luu B et al (1983) Fate of maternal conjugated ecdysteroids during embryonic development in *Locusta migratoria*. J Insect Physiol 29:491–507. https://doi.org/10.1016/0022-1910(83)90080-X

Samantsidis GR, Fotiadou M, Tzavellas S, Geibel S, Nauen R, Swevers L et al (2022) Functional characterization of putative ecdysone transporters in lepidopteran pests. Insect Biochem Mol Biol 151:103830. https://doi.org/10.1016/j.ibmb.2022.103830

Scalia S, Sbrenna-Miacciarelli A, Sbrena G, Morgan ED (1987) Ecdysteroid titres and location in developing eggs of *Schistocerca gregaria*. Insect Biochem 17: 227–236. https://doi.org/10.1016/0020-1790(87)90164-8

Schaaf O, Dettner K (2000a) Polyunsaturated monoglycerides and a pregnadiene in defensive glands of the water beetle *Agabus affinis*. Lipids 35(5): 543–550. https://doi.org/10.1007/s11745-000-554-6

Schaaf O, Dettner K (2000b) Transformation of steroids by *Bacillus* strains isolated from the foregut of water beetles (Coleoptera: Dytiscidae): II. Metabolism of 3-β-hydroxypregn-5-en-20-one (pregnenolone). J Steroid Biochem Mol Biol 75:187–199. https://doi.org/10.1016/S0960-0760(00)00166-7

Schaaf O, Baumgarten J, Dettner K (2000) Identification and function of prothoracic exocrine gland steroids of the Dytiscid beetles *Graphoderus cinereus* and *Laccophilus minutus*. J Chem Ecol 26(10):2291–2305. https://doi.org/10.1023/A:1005570609835

Scherkenbeck J, Nentwig G, Justus K, Lenz J, Gondol D, Wendler G et al (1999) Aggregation agents in German cockroach *Blattella germanica:* examination of efficacy. J Chem Ecol 25:1105–1119. https://doi.org/10.1023/A:1020885926578

Schildknecht H (1970) The defensive chemistry of land and water beetles. Angew Chem Int Edit 9(1):1–9. https://doi.org/10.1002/anie.197000011

Schildknecht H, Birringer H, Maschwitz U (1967a) Testosterone as protective agent of the water beetle *Ilybius*. Angew Chen Int Ed 6:558–559

Schildknecht H, Hotz D, Maschwitz U (1967b) On arthropod defense substances. XXVII. The C21-steroids from the prothoracic defense glands of *Acilius sulcatus*. Z Naturforsch B 22(9):938–944

Schlattner U, Vafopoulou X, Steel CG, Hormann RE, Lezzi M (2006) Non-genomic ecdysone effects and the invertebrate nuclear steroid hormone receptor EcR – new role for an "old" receptor? Mol Cell Endocrinol 247(1–2):64–72. https://doi.org/10.1016/j.mce.2005.12.051

Scott AP (2018) Is there any value in measuring vertebrate steroids in invertebrates? Gen Comp Endocr 265:77–82. https://doi.org/10.1016/j.ygcen.2018.04.005

Scudder GGE, Moore LV, Isman MB (1986) Sequestration of cardenolides in *Oncopeltus fasciatus*: morphological and physiological adaptations. J Chem Ecol 12(5):1171–1187. https://doi.org/10.1007/BF016390003

Shamsuzzama LR, Trabelcy B, Goncalves IL, Gerchman Y, Sapir A (2020) Metabolic reconfiguration in *C. elegans* suggests a pathway for widespread sterol auxotrophy in the animal kingdom. Curr Biol 30: 3031–3038. https://doi.org/10.1016/j.cub.2020.05.070

Shen G, Lin Y, Yang C, Xing R, Zhang H, Chen E, Han C et al (2015) Vertebrate estrogen regulates the development of female characteristics in silkworm, *Bombyx mori*. Gen Comp Endocrinol 210:30–37. https://doi.org/10.1016/j.ygcen.2014.09.016

Shimmel M, O'Connor MB (2022) The cytochrome P450 Cyp6t3 is not required for ecdysone biosynthesis in *Drosophila melanogaster*. microPubl Biol. https://doi.org/10.17912/micropub.biology.000611

Sigafoos AN, Paradise BD, Fernandez-Zapico ME (2021) Hedgehog/GLI signaling pathway: transduction, regulation, and implications for disease. Cancers 13:3410. https://doi.org/10.3390/cancers13143410

Sipahimalani AT, Mamdapur VR, Joshi NK, Chadha MS (1970) Steroids in the defensive secretion of the water beetle, *Cybister limbatus*. I Naturwiss 57(1):40. https://doi.org/10.1007/BF00593559

Sláma K (1979) Insect hormones and antihormones in plants. In: Rosenthal GA, Janzen DH (eds) Herbivores:

their interaction with secondary plant metabolism. Academic Press, New York, pp 683–700

Smagghe G, Gomez LE, Dhadialla TS (2012) Bisacylhydrazine insecticides for selective pest control. Adv Insect Physiol 43:163–249. https://doi.org/10.1016/B978-0-12-391500-9.00002-4

Smedley SR, Risteen RG, Tonyai KK, Pitino JC, Hu Y, Ahmed ZB et al (2017) Bufadienolides (lucibufagins) from an ecologically aberrant firefly (*Ellychnia corrusca*). Chemoecol 27:141–153. https://doi.org/10.1007/s00049-017-0240-6

Smissmann EE, Jenny NA, Beck SD (1964) Sterol metabolism in larvae of the confused flour beetle, *Tribolium confusum*. J Pharm Sci 53:1515–1519. https://doi.org/10.1002/jps.2600531218

Sommé-Martin G, Colardeau J, Lafont R (1988) Conversion of ecdysone and 20-hydroxyecdysone into 3-dehydroecdysteroids is a major pathway in third instar *Drosophila melanogaster* larvae. Insect Biochem 18:729–734. https://doi.org/10.1016/0020-1790(88)90082-0

Sommé-Martin G, Colardeau J, Beydon P, Blais C, Lepesant JA, Lafont R (1990) P1 gene expression in Drosophila larval fat body: regulation by various ecdysteroids. Arch Insect Biochem Physiol 15:43–56. https://doi.org/10.1002/arch.940150105

Sonobe H, Yamada R (2004) Ecdysteroids during early embryonic development in silkworm *Bombyx mori*. Zool Sci 21(5):503–516. https://doi.org/10.2108/zsj.21.503

Srivastava DP, Yu EJ, Kennedy K, Chatwin H, Reale V, Hamon M et al (2005) Rapid, nongenomic responses to ecdysteroids and catecholamines mediated by a novel *Drosophila* G-protein-coupled receptor. J Neurosci 25(26):6145–6155. https://doi.org/10.1523/JNEUROSCI.1005-05.2005

Steidle H (1930) Über die Verbreitung des weiblichen Sexualhormons. Arch Exp Pathol Pharmakol 157:89. https://doi.org/10.1007/BF01972095

Steyn PS, van Heerden FR (1998) Bufadienolides of plant and animal origin. Nat Prod Rep 15(4):397–413. https://doi.org/10.1039/a815397y

Suh SO, Noda H, Blackwell M (2001) Insect symbiosis: derivation of yeast-like endosymbionts within an entomopathogenic filamentous lineage. Mol Biol Evol 18(3):995–1000. https://doi.org/10.1093/oxfordjournals.molbev.a003901

Svoboda JA, Weirich GF (1995) Sterol metabolism in the tobacco hornworm, *Manduca sexta* – a review. Lipids 30:263–267. https://doi.org/10.1007/BF02537831

Svoboda JA, Thompson MA, Herbert EW, Shortino TJ, Szczepanik-Vanleeuwen PA (1982) Utilization and metabolism of dietary sterols in the honey bee and the yellow fever mosquito. Lipids 17:220–225. https://doi.org/10.1007/BF02535107

Svoboda JA, Rees HH, Thompson MJ, Hoggard N (1989) Intermediates of stigmasterol metabolism in *Spodoptera littoralis*. Steroids 53:329–343. https://doi.org/10.1016/0039-128x(89)90018-4

Swevers L (2019) An update on ecdysone signaling during insect oogenesis. Curr Opin Insect Sci 31:8–13. https://doi.org/10.1016/j.cois.2018.07.003

Swevers L, Lambert JG, De Loof A (1991a) Synthesis and metabolism of vertebrate-type steroids by tissues of insects: a critical evaluation. Experientia 47(7):687–698. https://doi.org/10.1007/BF01958817

Swevers L, Lambert JG, Novak F, Paesen G, De Loof A (1991b) Lack of essential enzymes for the biosynthesis of C19 and C18 steroids in gonads of the migratory locust, *Locusta migratoria*. Gen Comp Endocr 84(2):237–248. https://doi.org/10.1016/0016-6480(91)90047-a

Swevers L, Lambert JG, De Loof A (1992) On the origin of vertebrate-type steroids present in *Locusta migratoria*: do they originate from the food? Comp Biochem Physiol B 102(3):591–600. https://doi.org/10.1016/0305-0491(92)90052-s

Takáč P, Kozánek M, Somogyiová E, V□boh P (1993) Immunoreactive progesterone concentration in some tissues of the cockroach *Nauphoeta cinerea* (Blattoidea: Panchloridae) during development. Eur J Entomol 90:37–44

Takeshima M, Ogihara MH, Kataoka H (2019) Sterol characteristics in silkworm brain and various tissues characterized by precise sterol profiling using LC-MS/MS. Int J Mol Sci 20:4840. https://doi.org/10.3390/ijms20194840

Takeshima M, Ogihara MH, Kataoka H (2022) Characterization and functional analysis of *Bm*SR-B1 for phytosterol uptake. Steroids 184:109039. https://doi.org/10.1016/j.steroids.2022.109039

Talbot WS, Swyryd EA, Hogness DS (1993) Drosophila tissues with different metamorphic responses to ecdysone express different ecdysone receptor isoforms. Cell 73:1323–1337. https://doi.org/10.1016/0092-8674(93)90359-x

Tan W-H, Tan L, Hoang KM, Hunter MD, de Roode JC (2018) The effects of milkweed induced defense on parasite resistance in monarch butterflies, *Danaus plexippus*. J Chem Ecol 44:1040–1044. https://doi.org/10.1007/s10886-018-1007-4

Taylor E, Heyland A (2022) Evolution of non-genomic receptor function. Mol Cell Endocrinol 539:111468. https://doi.org/10.1016/j.mce.2021.111468

Tennessen JM, Baker KD, Lam G, Evans J, Thummel CS (2011) The *Drosophila* estrogen-related receptor directs a metabolic switch that supports developmental growth. Cell Metab 13(2):139–148. https://doi.org/10.1016/j.cmet.2011.01.005

Thompson MJ, Svoboda JA, Lusby WR, Rees HH, Oliver JE, Weirich GF et al (1985) Biosynthesis of a C21 steroid conjugate in an insect. The conversion of [^{14}C]cholesterol to [^{14}C]pregnen-3β,20β-diol glucoside in the tobacco hornworm, *Manduca sexta*. J Biol Chem 260(29):15410–15412

Thompson MJ, Svoboda JA, Lozano R, Wilzer KR (1988) Profile of free and conjugated ecdysteroids and ecdysteroid acids during embryonic development of *Manduca sexta* (L.) following maternal incorporation of [^{14}C]cholesterol. Arch Insect Biochem Physiol 7:157–172. https://doi.org/10.1002/arch.940070303

Thornton JW, Need E, Crews D (2003) Resurrecting the ancestral steroid receptor: ancient origin of estrogen signaling. Science 301:1714–1717. https://doi.org/10.1126/science.1086185

Tomaschko KH (1994) Ecdysteroids from *Pycnogonum litorale* (Arthropoda, Pantopoda) act as chemical defense against *Carcinus maenas* (Crustacea, Decapoda). J Chem Ecol 20:1445–1455. https://doi.org/10.1007/BF02059872

Truman JW, Riddiford LM (2002) Endocrine insights into the evolution of metamorphosis in insects. Annu Rev Entomol 47:467–500. https://doi.org/10.1146/annurev.ento.47.091201.145230

Tsoupras G, Luu B, Hoffmann JA (1982) Isolation and identification of three ecdysteroid conjugates with a C-20 hydroxy group in eggs of *Locusta migratoria*. Steroids 40(5):551–560. https://doi.org/10.1016/0039-128x(82)90075-7

Tsuruda JM, Chakrabarti P, Sagili RR (2021) Honey bee nutrition. Vet Clin Food Anim 37:505–519. https://doi.org/10.1016/j.cvfa.2021.06.006

Tyler J, McKinnon W, Lord GA, Hilton PJ (2008) A defensive steroidal pyrone in the glow-worm *Lampyris noctiluca* L. (Coleoptera: Lampyridae). Physiol Entomol 33:167–170. https://doi.org/10.1111/j.1365-3032.2007.00610.x

Ugine TA, Krasnoff SB, Grebenok RJ, Behmer ST, Losey JE (2019) Prey nutrient content creates omnivores out of predators. Ecol Lett 22:275–283. https://doi.org/10.1111/ele.13186

Ugine TA, Gill HK, Hernandez N, Grebenok RJ, Behmer ST, Losey JE (2021) Predator performance and fitness is dictated by herbivore prey type plus indirect effects of their host plant. J Chem Ecol 47(10–11):877–888. https://doi.org/10.1007/s10886-021-01251-4

van Oycke S, Braekman JC, Daloze D, Pasteels JM (1987) Cardenolide biosynthesis in chrysomelid beetles. Experientia 43:460–462. https://doi.org/10.1007/BF01940455

van Oycke S, Randoux T, Braekman JC, Pasteels JM (1988) New cardenolide glycosides from the defence glands of *Chrysolinina* beetles (Coleoptera: Chrysomelidae). Bull Soc Chim Belg 97:297–311. https://doi.org/10.1002/bscb.19880970409

Vanderplanck M, Zerck PL, Lognay G, Michez D (2019) Generalized host-plant feeding can hide sterol-specialized foraging behaviors in bee-plant interactions. Ecol Evol 10:150–162. https://doi.org/10.1002/ece3.5868

Veith HJ, Barbier M, Pain J, Roger B (1974) Transformation des stéroïdes par l'abeille, *Apis mellifera* L. Comp Biochem Physiol 47:459–472. https://doi.org/10.1016/0305-0491(74)90076-5

Vittek J, Slomyani BL (1984) Testosterone in royal jelly. Experientia 40:104–106

Wang JL, Chen L, Tang L, Zhao HB, Liu XS, Wang YF (2014) 20-Hydroxyecdysone transcriptionally regulates humoral immunity in the fat body of *Helicoverpa armigera*. Insect Mol Biol 23(6): 842–856. https://doi.org/10.1111/imb.12131

Wang D, Zhao WL, Cai MJ, Wang JX, Zhao XF (2015) G-protein-coupled receptor controls steroid hormone signaling in cell membrane. Sci Rep 5:8675. https://doi.org/10.1038/srep08675

Warren JT, Hétru C (1990) Ecdysone biosynthesis: pathways, enzymes and the early step problem. Invert Reprod Dev 18(1–2):91–99. https://doi.org/10.1080/07924259.1990.9672131

Warren JT, Steiner B, Dorn A, Pak M, Gilbert LI (1986) Metabolism of ecdysteroids during the embryogenesis of *Manduca sexta*. J Liquid Chromatogr 9(8):1759–1782. https://doi.org/10.1080/01483918608076716

Warren JT, O'Connor MB, Gilbert LI (2009) Studies on the black box: incorporation of 3-oxo-7-dehydrocholesterol into ecdysteroids by *Drosophila melanogaster* and *Manduca sexta*. Insect Biochem Mol Biol 39:677–687. https://doi.org/10.1016/j.ibmb.2009.08.004

Wei JH, Yin X, Welander PV (2016) Sterol synthesis in diverse bacteria. Front Microbiol 7:990. https://doi.org/10.3389/fmicb.2016.00990

Weinstein SB, Dearing MD (2022) Harvest mice (*Reithropontomys megalotis*) consume monarch butterflies (*Danaus plexippus*). Ecology 103:e3607. https://doi.org/10.1002/ecy.3607

Weirich GF (1989) Enzymes involved in ecdysone metabolism. In: Koolman J (ed) Ecdysone: from chemistry to mode of action. Thieme Verlag, Stuttgart, pp 174–180

Wing KD (1988) RH5849, a nonsteroidal ecdysone agonist: effects on a *Drosophila* cell line. Science 241: 467–469. https://doi.org/10.1126/science.3393913

Wing KD, Slawecki RA, Carlson G (1988) RH 5849, a nonsteroidal ecdysone agonist: effects on larval Lepidoptera. Science 241:470–472. https://doi.org/10.1126/science.241.4864.470

Wu J, Shen G, Liu D, Xu H, Jiao M, Zhang Y et al (2022) The response of the estrogen-related receptor to 20-hydroxyecdysone in *Bombyx mori*: insight into the function of estrogen-related receptor in insect 20-hydroxyecdysone signaling pathway. Front Physiol 12:785637. https://doi.org/10.3389/fphys.2021.785637

Xu W (2020) How do moth and butterfly taste? Molecular basis of gustatory receptors in Lepidoptera. Insect Sci 27:1148–1157. https://doi.org/10.1111/1744-7917.12718

Yamada R, Sonobe H (2003) Purification, kinetic characterization, and molecular cloning of a novel enzyme ecdysteroid-phosphate phosphatase. J Biol Chem 278(29):26365–26373. https://doi.org/10.1074/jbc.M304158200

Yamanaka N, Marqués G, O'Connor MB (2015) Vesicle-mediated steroid hormone secretion in *Drosophila melanogaster*. Cell 163(4):907–919. https://doi.org/10.1016/j.cell.2015.10.022

Yang H-J, Lee J-W, Lee S-H, Ryu JS, Kwak DH, Nam KS et al (2010) Estrogenic activity produced by aqueous extracts of silkworm (*Bombyx mori*) pupae in ovariectomized rats. Am J Clin Med 38:89–97. https://doi.org/10.1142/S0192415X10007683

Yocum LB, Denlinger DL, Katlic NE, Brueggemeier RW, Mechoulam R (1987) A developmental profile of estrogen and androgen radioimmunoactive substances in the flesh fly, *Sarcophaga crassipalpis*. Insect Biochem 17:1149–1153. https://doi.org/10.1016/0020-1790(87)90087-4

Zhang H-J, Anderson AR, Trowell SC, Luo A-R, Xiang Z-H, Xia Q-Y (2011) Topological and functional characterization of an insect gustatory receptor. PLoS One 6(8):e24111. https://doi.org/10.1371/journal.pone.0024111

Zhao XF (2020) G protein-coupled receptors function as cell membrane receptors for the steroid hormone 20-hydroxyecdysone. Cell Commun Signal 18(1):146. https://doi.org/10.1186/s12964-020-00620-y

Adv Exp Med Biol (2026) 1494: 163–200
https://doi.org/10.1007/5584_2024_808

Published online: 24 May 2024

Transcriptional Control of Lipid Metabolism

Laura Palanker Musselman, Huy G. Truong, and Justin R. DiAngelo

Abstract

Transcriptional control of lipid metabolism uses a framework that parallels the control of lipid metabolism at the protein or enzyme level, via feedback and feed-forward mechanisms. Increasing the substrates for an enzyme often increases enzyme gene expression, for example. A paucity of product can likewise potentiate transcription or stability of the mRNA encoding the enzyme or enzymes needed to produce it. In addition, changes in second messengers or cellular energy charge can act as on/off switches for transcriptional regulators to control transcript (and protein) abundance. Insects use a wide range of DNA-binding transcription factors (TFs) that sense changes in the cell and its environment to produce the appropriate change in transcription at gene promoters. These TFs work together with histones, spliceosomes, and additional RNA processing factors to ultimately regulate lipid metabolism. In this chapter, we will first focus on the important TFs that control lipid metabolism in insects. Next, we will describe non-TF regulators of insect lipid metabolism such as enzymes that modify acetylation and methylation status, transcriptional coactivators, splicing factors, and microRNAs. To conclude, we consider future goals for studying the mechanisms underlying the control of lipid metabolism in insects.

Keywords

Insect · Lipid · Lipogenesis · Lipolysis · Metabolism · microRNA · Splicing · Transcription factor · Transcriptional coactivator · Triglyceride

The Authors Laura Palanker Musselman and Justin R. DiAngelo have equally contributed to this chapter

L. P. Musselman
Department of Biological Sciences, Binghamton University, State University of New York, Binghamton, NY, USA

H. G. Truong and J. R. DiAngelo (✉)
Division of Science, Pennsylvania State University, Berks Campus, Reading, PA, USA
e-mail: Jrd5671@psu.edu

1 Transcription Factors that Control Lipid Metabolism

1.1 Transcription Factors that Respond to High Sugar

In the early 2000s, Michael Pankratz's group showed that altering the nutrient environment of developing *Drosophila melanogaster* (called *Drosophila* in this chapter) larvae (either removing food or feeding them a high-sugar diet) had profound changes in the expression of a number of lipid metabolic genes (Zinke et al. 2002). However, the full complement of transcription factors

(TFs) responsible for these changes in gene expression were not completely understood. Over the next two decades, much research has led to the identification and characterization of TFs that respond to changing nutrient conditions (especially high sugar) and we will describe these factors in this section. These TFs serve as central regulators of these global changes in gene expression in response to nutrient availability as described by the Pankratz group (Zinke et al. 2002).

1.1.1 Mondo (dChREBP/Mio) and Bigmax (dMlx)

A very well-characterized nutrient-responsive transcription factor is Mondo, the *Drosophila* homolog of the mammalian carbohydrate response element binding protein (ChREBP). A high-sugar diet promotes lipogenesis by increasing the expression of a number of lipid metabolic enzyme genes important for the synthesis of fatty acids and triglycerides (TAGs) such as fatty acid synthase (known as FAS, FASN, or FASN1 in different insects; we will refer to this enzyme as FAS throughout this chapter), acetyl-CoA carboxylase (ACC) and ATP citrate lyase (ATPCL) (Zinke et al. 2002; Musselman et al. 2011). *Mondo* levels are also increased in larvae fed a high-sugar diet (Musselman et al. 2013) and the high-sugar induction of expression of *FAS*, *ACC*, and *ATPCL* genes is blunted in *Mondo* mutants (Sassu et al. 2012), indicating a central role of Mondo in regulating high-sugar-induced expression of lipid metabolic genes. Mondo is also important in *Drosophila* development as mutants in *Mondo* arrest in pupal development (Fig. 1; (Billin and Ayer 2006)). However, whether the developmental defects observed in *Mondo* mutants are due to a decrease in lipid stores throughout the larval stage leading to not having enough energy to support metamorphosis or whether there are lipid-independent functions of *Mondo* important for insect development is unknown.

Mondo is required to survive high sugar feeding and functions in the *Drosophila* fat body to regulate lipid metabolism as fat body-specific knockdown of *Mondo* using RNA interference (RNAi) results in less TAGs in both the third instar larval and adult stages (Musselman et al. 2013; Sassu et al. 2012). Whereas feeding a diet of 1 M sucrose is lethal to *Mondo*-deficient larvae, feeding a high-sugar diet of 0.7 M or 20% sucrose results in decreased expression of *FAS* as well as *ATPCL* (Musselman et al. 2013; Sassu et al. 2012). In addition, fat bodies from fat body-specific *Mondo-RNAi* larvae fed 0.7 M sucrose have longer and more saturated lipids as well as decreased ketones; however, lipolysis was unchanged, despite increased expression of a number of lipases (Musselman et al. 2013). Interestingly, adult female *Drosophila* with fat body-specific decreases in *Mondo* eat less, consistent with the decreased lipid levels in these flies (Sassu et al. 2012) and suggesting a nonautonomous role of *Mondo* in the fat body to regulate food consumption. Together, these data indicate that *Mondo* plays an important role in the insect fat body to regulate lipid storage and metabolism under normal and nutrient-rich environments.

An ortholog of mammalian *ChREBP* has also been identified in termites. Termite *ChREBP* is expressed in the fat body and ovaries of reproductive females from eight different termite species and its expression is induced under high carbohydrate conditions (Sillam-Dusses et al. 2016). The targets of ChREBP, *FAS* and *ACC*, are both expressed in a similar pattern as *ChREBP* in termites (Sillam-Dusses et al. 2016), suggesting that ChREBP may regulate similar transcriptional targets in termites as in fruit flies. In termites, fatty acids are used by the queen to produce pheromones that signal fertility and her presence. Without this pheromone being produced, prospective queens will change their behavior increasing head-butting (Penick et al. 2013). Interestingly, inhibiting fatty acid synthesis by pharmacologically targeting ACC increased head-butting behavior among worker termites suggesting that ChREBP and its targets may be involved in regulating queen fertility, but additional experimentation is needed to support this claim (Sillam-Dusses et al. 2016).

Mondo acts as a heterodimer with bigmax, the *Drosophila* homolog of the mammalian Max-like protein X (Mlx) (Havula et al. 2013).

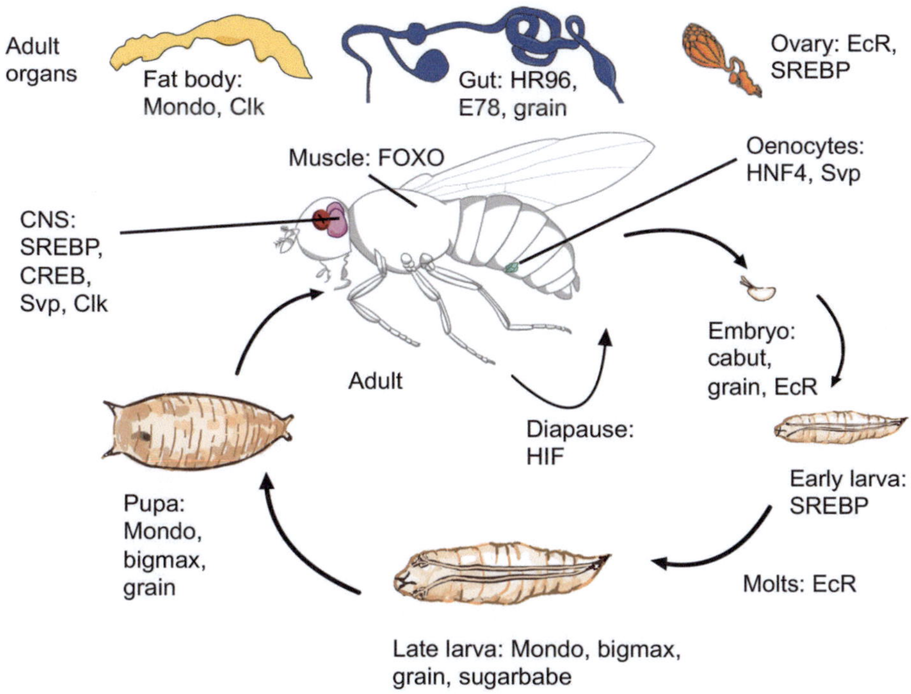

Fig. 1 Transcriptional regulation of lipid metabolism during insect development
Lipids are used to build membranes and as hormone signals during development. Loss-of-function phenotypes in TFs that regulate lipid metabolism often include lethality or reduced fitness in specific organs or developmental stages. The sugar-responsive TFs **cabut** and **grain** and the ecdysone receptor **EcR** are required for embryogenesis. **EcR** is also required to respond to pulses of its lipid ligand to promote the expression of genes that advance the insect through the transitions to its next stage. Throughout the larval stages, **SREBP** is required to build the fats needed for membrane biosynthesis and lipid stores. In addition, the sugar-responsive TFs **Mondo**, **bigmax**, **sugarbabe**, and **grain** are all necessary for larval TAG storage. The sugar-responsive TFs **Mondo**, **bigmax** and **grain** are all necessary for normal pupal development. In adulthood, **HR96** and **E78** promote gut-to-fat body lipid flux while grain regulates the expression of lipogenic enzymes in the gut. **EcR** and **SREBP** also function in the adult ovary to promote lipid homeostasis. The TFs **Mondo** and **Clk** function in the adult fat body to regulate TAG storage, while **SREBP**, **CREB**, **Clk**, and **Svp** function in the adult brain and/or eye. Also shown: other adult tissues where transcriptional regulation of lipid homeostasis has been shown include the adult oenocytes, where **HNF4** waterproofs the cuticle and makes lipid pheromones used in mating. **Svp** in oenocytes helps promote ovarian function and offspring production. **FOXO** functions in the adult flight muscle to remotely control corpora cardiaca secretory activity and ultimately to mobilize stored fat. The oxygen sensor TF **HIF** functions in lipid oxidation during adult diapause. In addition, two large RNAi screens for regulators of TAG content in whole *Drosophila* identified a number of TFs that are yet to be validated (Baumbach et al. 2014a; Pospisilik et al. 2010). See Sects. 1.1–1.5 for more details and citations

Similar to Mondo, bigmax plays an important role in the transcriptional response to high sugar as 325 genes that were upregulated in sugar-fed controls were decreased in *bigmax* mutants (Mattila et al. 2015). Mutants in *bigmax* are also developmentally delayed and arrest in pupal development and this lethality is exacerbated by adding excess sugar to the diet (Havula et al. 2013). The developmental phenotypes can be rescued by adding bigmax back specifically to the fat body but not muscle (Havula et al. 2013), suggesting a functional role of bigmax in fat body tissue. *Bigmax* mutants have altered lipid levels; phosphatidylethanolamine and lysophosphatidylcholine levels are decreased as well as medium and long fatty acids such as myristoleic acid, lauric acid and ceramides are increased (Havula et al. 2013). Consistent with these results, fat body-specific knockdown of *bigmax* in the larval fat body results in decreased TAG storage (Sassu et al. 2012). *Bigmax* mutants have decreased expression of lipogenic enzyme genes such as *FAS* and *ACC*, similar to *Mondo* mutants (Sassu et al. 2012; Havula et al. 2013). When fed a high-sugar diet, the expression of a similar set of lipogenic enzyme genes was blunted (i.e., *FAS*, *ACC*, *ATPCL*); however, additional genes like the perilipin-like genes *Lsd-1* and *Lsd-2* were also identified to be *bigmax* dependent (Mattila et al. 2015). The similarity of the *Mondo* and *bigmax* mutant phenotypes suggests that these two proteins work together in the fat body and throughout the fly to regulate lipid storage and metabolism in response to nutrient excess. The Mondo/bigmax heterodimer acts to regulate the expression of three additional TFs sugarbabe, cabut, and grain to control lipid metabolism under high sugar conditions and the functions of these three target genes will be discussed next.

1.1.2 Sugarbabe

A major target of the Mondo-bigmax heterodimer is the gene encoding the TF sugarbabe (Mattila et al. 2015). Sugarbabe was originally identified as a gene highly induced under high sugar conditions (Zinke et al. 2002) and this induction is dependent on *bigmax* (Mattila et al. 2015). While a direct transcriptional target of *bigmax*, *sugarbabe* expression is also dependent on the activin/bone morphogenic protein (BMP) ligand dawdle and the Smad TF, Smox, suggesting a synergy between Mondo/bigmax and activin signaling (Mattila et al. 2015). Sugarbabe is also important for the expression of lipogenic enzyme genes as *sugarbabe* mutants have decreased levels of *FAS*, *ACC*, and *ATPCL* as well as a fatty acid desaturase gene when fed high sugar. Consistent with regulating the production of lipogenic enzymes, *sugarbabe* mutant larvae have decreased TAG storage under normal food conditions as well as blunted palmitate desaturation under high-sugar conditions (Mattila et al. 2015). These data position sugarbabe alongside Mondo and bigmax to promote lipid storage during high-sugar feeding.

1.1.3 Cabut

Another target of the Mondo-bigmax heterodimer is the gene encoding the TF cabut. *Cabut* is a direct bigmax target and decreasing *cabut* levels in larvae results in developmental delay and impairment of survival which is enhanced by feeding the flies a high-sugar diet, a similar phenotype to *bigmax* mutants (Havula et al. 2013; Bartok et al. 2015). While *cabut* mutants are embryonic lethal (Munoz-Descalzo et al. 2005), adults were generated by using a *cabut-RNAi* transgene using *timeless*-gal4, whose expression is restricted to many of the cells in the fly head (Bartok et al. 2015). Cabut serves as a transcriptional repressor with its targets involved in lipolysis (i.e., the intracellular TAG lipase encoded by *brummer* (*bmm*)) and glucose and glycerol metabolism (i.e., glycerol kinase) when fed a high-sugar diet (Bartok et al. 2015). The gene encoding the gluconeogenic enzyme phosphoenolpyruvate carboxykinase (Pepck) is also highly repressed by cabut and interestingly, *Pepck* mutants have decreased free glycerol and TAGs (Bartok et al. 2015). These results indicate that cabut is a member of the Mondo-bigmax network, serving as the repressive arm of the Mondo-Mlx sugar sensing pathway. This positions cabut to play an important role in coordinating the expression of target genes involved in glucose metabolism, gluconeogenesis, and lipid metabolism under high-sugar conditions.

1.1.4 Grain

A third target of the Mondo-bigmax heterodimer is the gene encoding the TF grain, a member of the GATA TF family known for binding to the nucleotide motif "GATA" in target gene promoters. GATA TF binding motifs, specifically the motif to which grain binds, were identified as being highly enriched in sugar-regulated genes (Kokki et al. 2021). Like *cabut* and *sugarbabe*, the expression of *grain* is induced under high-sugar conditions and this induction is dependent on bigmax (Kokki et al. 2021). While *grain* mutants are embryonic lethal (Brown and Castelli-Gair Hombria 2000), *grain* is essential for normal development on a high-sugar diet as *grain* knockdown animals fed high sugar have decreased pupal volume and increased pupal lethality, similar to bigmax and cabut (Kokki et al. 2021). *Grain* mutants fed a low-sugar diet also had decreased pupal volume, suggesting a more general developmental role for *grain* (Kokki et al. 2021). The expression of *cabut* and *Smox* were both decreased in *grain-RNAi* larvae fed normal sugar levels and sugarbabe target genes like *FAS*, *ACC*, and *ATPCL* were also decreased in *grain* knockdown animals (Kokki et al. 2021) suggesting that grain, cabut, sugarbabe, and Mondo-bigmax regulate overlapping transcriptional targets to control lipid metabolism. Consistent with this decreased expression of lipogenic enzyme genes, *grain* knockdown animals also had less TAGs and smaller lipid droplets (Kokki et al. 2021). Interestingly, *grain* expression is induced in the fat body but not the midgut when animals are fed high-sugar diets. In the fat body, knockdown of *grain* blunted high-sugar-induced expression of lipogenic enzyme genes like *FAS*, *ACC*, and *ATPCL*, but this was not the case in the gut. However, the expression of these lipogenic enzyme genes was blocked in the midgut under low-sugar conditions (Kokki et al. 2021). Together, these data indicate the presence of a complex network of TFs with overlapping targets that function in a tissue-specific manner to regulate lipid metabolism under normal and high-sugar conditions.

1.2 SREBP

Whereas Mondo/ChREBP regulation is similar between insects and mammals, the insect ortholog of the cholesterol-dependent mammalian sterol response element binding protein (SREBP) TF appears to be cholesterol-independent (Dobrosotskaya et al. 2003; Rawson 2003). Rather, insect SREBP is a polar lipid sensor and regulator of fatty acid, phospholipid, and glycerolipid metabolism (Dobrosotskaya et al. 2002; Kunte et al. 2006; Seegmiller et al. 2002). This TF's main role is to promote lipogenesis when lipid levels are low, supplying fatty acid substrates for a variety of cellular processes, including cell growth and stem cell maintenance (Li et al. 2022; Porstmann et al. 2008; Senos Demarco et al. 2019). Accordingly, insect SREBP plays critical roles not only in so-called "metabolic" organs like the gut, ovary, and fat body, but also in neurons, where it acts as a transcriptional regulator of lipids that maintain vesicle pools and synapse strength.

First cloned and named for its helix-loop-helix structure, HLH106 (called SREBP from now on) was identified as a membrane-resident protein with two isoforms that had high sequence similarity to mammalian SREBP (Theopold et al. 1996). As in mammals, the insect TF undergoes a lipid-dependent cleavage event that allows it to translocate from the endoplasmic reticulum (ER) membrane to the nucleus, activating gene expression. Studies in cultured cells found that SREBP was able to bind human SREBP promoters and drive expression of these promoters in *Drosophila* S2 cells (Rosenfeld and Osborne 1998). In *Drosophila* cells, cleavage of the inactive precursor to form the active nuclear form of SREBP was decreased upon addition of fatty acids (Seegmiller et al. 2002), phosphatidylethanolamines, or ceramides (Dobrosotskaya et al. 2002), but not cholesterol, a sterol that suppresses SREBP's cleavage in mammals. Accordingly, *Drosophila* SREBP positively regulates the expression of enzymes in fatty acid synthesis, including the Mondo targets *FAS* and *ACC*, but not those in sterol metabolism (Dobrosotskaya et al. 2002). Knockdown of *SREBP* reduced the concentration of phosphatidylethanolamines in an S2-derived cell line and also affected ER structure and function during stress (Sanchez-Alvarez et al. 2014). This is of particular interest, as the ER is not only the location of inactive SREBP but is also the organelle from which new lipid droplets arise. Taken together, these findings are consistent with a feedback mechanism where insect SREBP increases the expression of lipogenesis genes to support cellular demands and is inactive until another bolus of acyl chains is needed.

In accordance with the biochemical data from cell culture, *SREBP* null mutants have reduced expression of lipogenesis enzymes and reduced fatty acid content, failing to grow, and dying in the late second instar (Kunte et al. 2006). Lethality of *SREBP* mutants could be rescued by dietary supplementation with a mixture of soy lipids or with individual 14-, 16-, or 18-carbon long-chain free fatty acids during larval development (Fig. 1; Kunte et al. 2006). Once *SREBP* mutants reach adulthood, this TF is dispensable, at least under normal husbandry conditions (Kunte et al. 2006). Efficient processing of SREBP from its inactive, ER-associated form to the transcriptionally active nuclear form occurs with the help of the SREBP-binding protein Scap (Seegmiller et al. 2002) which translocates SREBP to the Golgi when phosphatidylethanolamine concentrations are low (Dobrosotskaya et al. 2003). In the Golgi, processing into the mature, active form requires two proteases, the site one and site two proteases (S1P and S2P) (Matthews et al. 2009) although the caspase Drice can compensate for a lack of S2P and cleave SREBP in its absence (Amarneh et al. 2009).

SREBP functions in a number of different organs in insects. A reporter of SREBP activity that expresses GFP upon proteolytic cleavage was observed in the canonical "metabolic" tissues, including the gut, fat body, and lipid-secreting cells called oenocytes (Kunte et al. 2006). Fat body expression alone only very weakly rescued the larval lethality of *SREBP* mutants (8% of controls), but expression in the gut and fat body was able to rescue most (91%) of these animals to adulthood (Kunte et al. 2006). Studies using tissue-specific approaches have found evidence for transcriptional regulation of lipid metabolism by SREBP in a wide range of insect cell types. The fat body, the major lipid metabolic organ and storage depot, is a key site of SREBP activity, where it promotes lipid storage (Zhao et al. 2012). The ceramide synthase Schlank is essential in the fat body and promotes lipogenesis as well as expression of *SREBP* and its target genes (Bauer et al. 2009). Like *SREBP* mutants, larval lethality in *schlank* mutants could be rescued to adulthood by supplementing the diet with soy lipids (Bauer et al. 2009). Other pathways that regulate lipid metabolism via SREBP in the fat body include cyclin-dependent kinase 8, which phosphorylates and inactivates SREBP; *CDK8* mutants have increased expression of SREBP target genes and increased TAG content (Zhao et al. 2012). The actomyosin contractile effector GTPase RhoA also inactivates SREBP, as reducing this protein in the fat body promoted SREBP translocation to the nucleus and increased lipid storage in larval fat body (Bertolio et al. 2019).

Neurons also require SREBP for synapse maintenance and refinement. As a major regulator of phospholipid concentrations, SREBP helps supply the lipid bilayers needed for vesicle biogenesis and plasma membrane expansion, two active processes in neuron growth and plasticity. In *Drosophila* larvae, knockdown of SREBP in peripheral sensory neurons reduces dendritic growth and branching while, surprisingly, rendering the neuron field hypersensitive (Ziegler et al. 2017). Loss of *SREBP* reduces brain expression of lipogenic genes and phospholipid content (Ziegler et al. 2017). Reducing SREBP targets *FAS* or *ACC* and reduces dendrite arborization in this model (Ziegler et al. 2017). Interestingly, reducing the phosphatidylethanolamine biosynthetic enzyme easily shocked (*eas*) reduces phosphatidylethanolamine (PE) and increases SREBP activity, but also reduces arborization of these neurons (Meltzer et al. 2017). Similar results were observed when SREBP was reduced or constitutively activated in amacrine neurons of the *Drosophila* optic neuropil, both of which produced reductions in dendrite field size (Luo et al. 2020). A genetic screen for visual defects identified a mutant with reduced phosphatidylethanolamine concentrations that diminished vesicle pools and increased photoreceptor neurodegeneration by increasing the nuclear translocation and transcriptional activity of SREBP (Tsai et al. 2019). Neuron-supporting glia in the adult *Drosophila* ommatidium also activates SREBP and lipid storage in the process of neurodegeneration, which can be rescued when one copy of the *SREBP* gene is removed (Liu et al. 2015). Therefore, SREBP seems to help optimize lipid homeostasis to enable synaptogenesis and synapse maintenance in *Drosophila*.

Due to the metabolic demands for reproduction, it is not surprising that mating increases SREBP proteolytic cleavage (Landis et al. 2021) and target gene expression (Reiff et al. 2015) in *Drosophila*. In the oocyte, SREBP supports healthy lipogenesis via transcriptional upregulation of the *LpR2* gene, which encodes a lipoprotein receptor that enables lipid uptake by the germline (Sieber and Spradling 2015). In the germline of the testis, however, increasing SREBP activation and target gene expression produces germline stem cell loss (Senos Demarco et al. 2019). These studies both discovered new regulators of SREBP in addition to the downstream effects on gonads, namely the nuclear receptor EcR (Sieber and Spradling 2015) and mitofusin, a GTPase that regulates the fusion of the mitochondrial outer membrane (Senos Demarco et al. 2019). In addition to direct effects in reproductive organs, SREBP also supports cardiac, gut, and immune function. Cardiac function was impaired by steatosis when SREBP activity increased or phosphatidylethanolamine metabolism was targeted in *eas* mutants (Lim et al. 2011). Gut SREBP activity in absorptive enterocytes promotes an expansion of the posterior midgut after mating that fuels lipogenesis and reproduction by increasing the expression of fatty acid metabolic genes including those encoding long-chain fatty acid CoA ligases and SREBP itself (Reiff et al. 2015). Oral introduction of some, but not all, bacteria to *Drosophila* females can also promote lipogenesis by activating fat body SREBP in an insulin-dependent manner, which is antagonized in the presence of the TF NFκB. This is consistent with a potential role for SREBP in the allocation of resources needed to modulate infectious and/or commensal bacterial loads (Charroux and Royet 2022). In *Aedes aegypti*, gut SREBP promotes Dengue virus infection and increases the viral load (Raquin et al. 2017). Together, these studies support a central role for SREBP in the transcriptional control of lipid metabolism in insects during development, reproduction, and infection and position it as a master regulator of membrane structure and function.

1.3 Nuclear Receptors

Nuclear receptors (or NRs, formerly called nuclear hormone receptors) are a highly conserved family of ligand- and cofactor-regulated TFs that can bind, and regulate the metabolism of, fatty acids, phospholipids, and steroid hormones (King-Jones and Thummel 2005). Ligand binding to these TFs can lead to activation or repression via control of cofactor recruitment, dimerization, and receptor localization and, in turn, DNA binding and transcription of target genes by the NR (Fahrbach et al. 2012). As a result, NR agonists have been useful as insecticides and in inducible expression systems. The most well-known of the insect NRs is the ecdysone receptor (EcR) which enables transcriptional regulation by the classical steroid insect molting hormones, ecdysteroids (Lafont et al. 2021), and is therefore required for insect development (Bender et al. 1997; Kozlova and Thummel 2003; Schubiger et al. 1998; Wang et al. 2018).

1.3.1 HNF4

The insect NR that seems to have the biggest role in the transcriptional control of lipid homeostasis is HNF4. Activity of this NR increases in the larval fat body upon starvation and promotes the expression of genes encoding enzymes that promote fatty acid catabolism (lipases, carnitine synthesis and transacylase enzymes, enzymes of the mitochondrial beta-oxidation pathway) and oxidative phosphorylation (complexes 1–4 and mitochondrial ATPases) (Barry and Thummel 2016; Palanker et al. 2009). A GAL4-HNF4 reporter can be activated by culturing *Drosophila* tissues with free fatty acids, suggesting a feed-forward mechanism where HNF4 helps to both liberate stored fatty acids and promote their use as an energy source. Transcriptional changes in *HNF4* mutant *Drosophila* larvae correlated with evidence for reduced fatty acid catabolism, as removing *HNF4* increased fatty acid content and prevented mobilization of stored lipids during starvation (Palanker et al. 2009). Accordingly, an overexpression study in adult *Drosophila* found that increasing HNF4 in the fat body was able to reduce lipid storage and oogenesis (Almeida-Oliveira et al. 2021). In the same study, *HNF4* overexpression in the ovary increased oogenesis and egg-laying, yet reduced the number of offspring produced per female, suggesting a role for HNF4-dependent fatty acids or energy in the control of reproduction (Almeida-Oliveira et al. 2021).

An *HNF4-RNAi* study in the brown planthopper *Nilaparvata lugens* also found a role for HNF4 in embryogenesis along with decreased expression of fatty acid catabolism and oxidative phosphorylation genes in the ovaries of RNAi-treated females (Cheng et al. 2020). Similar target genes were also identified in female *Aedes aegypti* mosquitoes where *HNF4* RNAi increased stored TAG and reduced the expression of lipid catabolism genes (Wang et al. 2017). The Chagas disease vector, *Rhodnius prolixus*, expresses two *HNF4* isoforms, *HNF4-1* and *HNF4-2*, both of which are expressed in the adult gut, fat body, muscle, ovary, and testis. Isoform 2, which is more highly expressed, is most similar to other insect HNF4 orthologs and is more abundant than isoform 1 in most tissues, whereas isoform 1 predominates in the ovary (Nascimento et al. 2021). It remains to be seen what the roles of these isoforms may be in this species.

HNF4 is downstream of the deacetylase Sir2 (Palu and Thummel 2016) and the peroxisomal import and biogenesis protein Pex19 (Bulow et al. 2018). Loss of Pex19 reduces peroxisomes, presumably increasing the load on mitochondrial fatty acid oxidation. Increased HNF4 target gene expression in *Pex19* mutants produces lipotoxicity, whereas removing one copy of *HNF4* can rescue *Pex19* mutant phenotypes, including lethality (Bulow et al. 2018). Sir2, on the other hand, acts to deacetylate and stabilize HNF4 and therefore Sir2 loss-of-function, like HNF4 loss-of-function, is characterized by increased TAG content (Palu and Thummel 2016; Reis et al. 2010).

HNF4 not only promotes the catabolism of long-chain fatty acids, but it also promotes the expression of fatty acid elongases and associated enzymes that help to produce very long-chain fatty acids and other hydrocarbons for waterproofing the adult cuticle (Storelli et al. 2019). Mutants lacking this receptor survive adulthood only under humidified conditions, a phenotype which can be rescued by expressing HNF4 in oenocytes, the specialized population of cells that secretes lipids to coat the insect cuticle (Storelli et al. 2019). Therefore, this receptor seems to play a critical role in fatty acid homeostasis across a range of different lipids in a variety of tissues during both insect development and adult physiology.

1.3.2 HR96

Another NR, HR96, can bind a subset of EcR response elements in the promoters of target genes (Fisk and Thummel 1995) and is expressed in insect organs active in lipid metabolism, including the gut, fat body, and epidermis (Giraudo et al. 2013; Liu et al. 2022). Like *Drosophila HNF4* mutants, *Drosophila HR96* mutants are sensitive to starvation and exhibit impaired lipolysis. Furthermore, *HR96* mutant animals are lean and resistant to diet-induced obesity due to impaired acyl chain homeostasis in the midgut (Sieber and Thummel 2009). Loss of *HR96* also impairs intestinal cholesterol homeostasis and is required for survival on a low-cholesterol diet (Bujold et al. 2010; Horner et al. 2009). This NR has a key role at the interface between cholesterol and triglyceride metabolism via transcriptional regulation of the intestinal lumen resident TAG lipase and cholesterol esterase magro (also called LipA) (Sieber and Thummel 2012). This conserved enzyme promotes TAG turnover in the intestinal lumen via expression in the lipid-rich anterior gut region known as the proventriculus, whereas it promotes cholesterol turnover in midgut cells themselves, rather than the lumen (Sieber and Thummel 2012). Cholesterol binds HR96 and its targets include the *Niemann-Pick type C-1b* gene, which is required for the uptake of dietary cholesterol (Horner et al. 2009).

Interestingly, HR96 not only regulates intestinal lipid homeostasis but also promotes the differentiation of enteroendocrine cells via inhibiting Notch signaling, although the direct transcriptional targets that underlie this function of HR96 have not yet been described (Obniski et al. 2018). In addition, HR96 is also important for the detoxification of xenobiotics in *Drosophila* (King-Jones et al. 2006) and other insect species (Gao et al. 2022) via transcriptional regulation of cytochrome p450 enzymes, although the links between xenobiotics and lipid metabolism are also not yet well characterized. This NR therefore serves as an intriguing basis for studies in which we might better understand the interactions between differentiation, detoxification, and lipid metabolism.

1.3.3 EcR

The EcR is both a NR and a master regulator of NR expression. With its heterodimeric partner NR, ultraspiracle (Usp), EcR also promotes the expression of genes that regulate the major developmental transitions between insect life stages (Hill et al. 2013). Preceding each molt, there is a brief increase in the circulating EcR ligand, after which the animal stops feeding and becomes dormant for a short time, then undergoes morphological changes to enter the next stage and continue growing (Nijhout et al. 2014). The most active form of the ecdysteroid family, the steroid hormone 20-hydroxyecdysone (20E), binds and activates EcR (Kamiyama and Niwa 2022). This triggers a cascade of transcriptional and other changes that include many downstream targets, including those involved in lipid metabolism (explained in more detail in Chap. 15).

As might be suggested by its activation when feeding is low, activating EcR typically reduces lipogenesis and supports lipolysis. In *Bombyx mori* larvae, a 20E injection reduced feeding and upregulated the expression of TAG lipase (Wang et al. 2010) although the effect seemed to be indirect, via the "starvation" TF FOXO, which is described in Sect. 1.4 (Hossain et al. 2013). Another study found that 20E injection increased the expression of fatty acid beta-oxidation genes and activated the lipolytic NR HNF4 in *Aedes aegypti* females (Wang et al. 2017). RNAi targeting *EcR* in these mosquitos produced an increase in TAG as well as stored and circulating sugars, consistent with a role for this NR in metabolism (Hou et al. 2015).

In *Drosophila*, a more complex picture of EcR's effect on lipid metabolism has emerged. As in the aforementioned studies, some *Drosophila* studies show a negative relationship between EcR and lipogenesis. Reducing *EcR* in the larval fat body leads to increased TAG storage, whereas culturing larval guts or fat bodies with the synthetic EcR ligand muristerone-A increases expression of the adipose gene, a negative regulator of fat accumulation (Kamoshida et al. 2012; Suh et al. 2007). By contrast, another larval study using whole-animal expression of a dominant negative EcR found reduced TAG storage (Francis et al. 2010). A third developmental study found an inverse correlation between EcR and SREBP transactivation during development and starvation; lipogenesis mRNAs were reduced when EcR targets were most strongly expressed (Xie et al. 2015). Neither of the last two studies identified direct targets of EcR that function in lipogenesis; rather, the latter study suggests that common upstream regulators in the Mediator complex coactivate EcR and inhibit SREBP simultaneously and in parallel. One potential mechanism by which EcR may regulate lipogenesis at the transcriptional level is via the TF FOXO, as 20E antagonizes insulin signaling via promoting the nuclear translocation and activation of FOXO in the larval fat body (Colombani et al. 2005).

While EcR is able to negatively regulate lipogenesis during *Drosophila* development, the opposite is true in female reproduction, where EcR *promotes* lipid deposition during oogenesis via SREBP (Sieber and Spradling 2015). *EcR* mutants expressed lower levels of lipogenesis genes in the developing somatic egg chamber and had reduced lipid storage there. To support energy demand, EcR also functions in the CNS to increase feeding, which in turn increases whole-body TAG in adult female *Drosophila* (Sieber and Spradling 2015). Similarly, 20E injection into *Rhodnius prolixus* females increased lipogenesis (Nascimento et al. 2021) as did topical application in *Nilaparvata lugens* females (Lu et al. 2018). Genes that encode the lipophorin receptor LpR are transcriptional targets of EcR that mediate lipid uptake into the germline in both model systems. Taken together, these and other studies suggest that there are important links between EcR and lipid metabolism that await discovery in specific developmental or environmental contexts that have not yet been tested.

In addition to the large pools of acyl chains used in energy metabolism, EcR exerts transcriptional control of ecdysteroid metabolism in a well-characterized pathway via the so-called "halloween" genes, which encode ecdysteroid biosynthetic enzymes under the regulation of numerous TFs (reviewed in (Kamiyama and Niwa 2022)). The ligand 20E binds the EcR half of the EcR-Usp heterodimer to promote the expression of so-called "early" and "late" genes that enable synthesis and secretion of the hormone, then turn off signaling after a short time several times during development (Truman 2019). Ecdysteroid metabolism itself will be the focus of Chap. 15 and is not covered in further detail here.

1.3.4 Other NRs: Seven-Up, E78, HR38, and ERR

The NR Seven-up (Svp) was named for its mutant phenotype in the developing *Drosophila* eye, where photoreceptors are transformed to adopt a cell 7 fate (Mlodzik et al. 1990). Subsequent studies have identified a role for this NR in a number of developmental processes, including neuroblast (Kanai et al. 2005; Ren et al. 2017), cardioblast (Lo and Frasch 2001), and Malpighian tubule tip cell (Kerber et al. 1998) differentiation. Because Svp is expressed in most organs (Leader et al. 2018) and interacts with growth-promoting pathways such as Ras signaling (Begemann et al. 1995), it is not surprising that it plays a role in lipid homeostasis. *Svp* is highly expressed in the lipid processing cells known as oenocytes, which play critical roles in insect lipid metabolism (Gutierrez et al. 2007). An RNAi screen in oenocytes identified *svp* as a regulator of the cuticular hydrocarbon profile in adult *Drosophila* females (Chiang et al. 2016). Another RNAi screen in the larval fat body found that fat body *svp-RNAi* led to increased hemolymph glucose and acyl-carnitine concentrations and increased overall TAG content and fat body lipid droplet size (Musselman et al. 2018). Expression profiling of late larval fat bodies identified differentially expressed lipid metabolism and immunity genes, as well as a large number of EcR targets (Musselman et al. 2018). Another adult-specific RNAi screen targeting NRs identified *svp* as a positive regulator of egg production (Weaver and Drummond-Barbosa 2019). When targeted in somatic cells, *svp-RNAi* led to a smaller germarium and a reduced number of both germline cells and eggs laid. Tissue-specific *svp* knockdown supports a role for this NR in both the fat body and oenocyte in reproductive output (Weaver and Drummond-Barbosa 2019). Interestingly, there did not appear to be changes in lipid storage when *svp* was targeted in these adult females (Weaver and Drummond-Barbosa 2019), and there were no effects on lipid metabolism gene expression when *svp* was decreased in adult fat body, although there were changes in immunity and detoxification genes (Weaver and Drummond-Barbosa 2020). Therefore, more work is needed to understand if Svp exerts its non-autonomous regulatory effects on reproduction via transcriptional control of lipid metabolism in adults.

The NR E78 is encoded by an ecdysone-inducible gene that encodes two isoforms, one of which (E78B) lacks a DNA-binding domain (Stone and Thummel 1993). Like Svp, E78 is required for ovarian follicle development and fertility in *Drosophila* (Ables et al. 2015), although, to our knowledge, no role for E78 in ovarian lipid metabolism has yet been reported. An RNAi study in the red flour beetle found that *E78* was also required for sperm production (Xu et al. 2012a). Consistent with these findings, recent studies of *Drosophila E78A* mutants found no phenotypes during development, but mutants had reduced TAG in both young adult males and females (Praggastis et al. 2021). Differentially expressed genes in *E78A* mutants included those expressed highly in the midgut and testis, leading to the identification of the intestinal lipase encoded by *CG17192* as an E78A target gene that maintains whole-animal TAG levels during early adulthood (Praggastis et al. 2021). Thus, like EcR and Svp, E78 may also regulate lipid homeostasis during reproduction.

Another ecdysteroid-responsive NR, HR38, regulates transcription of genes encoding lipid metabolic enzymes such as lipases, acyl transferases, and fatty acid translocases (Ruaud et al. 2011). This NR also regulates EcR target genes (Baker et al. 2003; Kozlova et al. 2009) but has not yet been shown to affect fatty acid or ecdysteroid metabolism itself. In addition, HR38 heterodimerizes with EcR's binding partner, ultraspiracle (Sutherland et al. 1995), and also controls carbohydrate metabolism genes and glycogen storage in *Drosophila* larvae (Ruaud et al. 2011). Interestingly, the closest mammalian HR38 ortholog, Nurr1, binds fatty acids in its LBD and is associated with changes in lipid metabolism (de Vera et al. 2016; Pearen and Muscat 2010). Therefore, HR38 may regulate lipid metabolism in a context that has not yet been explored or reported.

Like HR38, ERR exerts transcriptional control of both carbohydrate metabolism and lipid metabolism (Beebe et al. 2020; Li et al. 2013; Tennessen et al. 2011). This NR acts at the end of embryogenesis to reprogram metabolic gene expression and is required for the expression of the glycolytic and citric acid cycle genes needed to process carbohydrates. These *ERR* mutants accumulate trehalose and alter the metabolism of amino acids and Krebs cycle intermediates along with a decreased ability to grow and store energy as TAG, leading to a model where this NR enables the provision of dietary carbons for a broad range of metabolic processes (Tennessen et al. 2011). During larval development, *Drosophila* ERR drives the expression of genes with roles in aerobic glycolysis and the pentose phosphate pathway (Tennessen et al. 2011). In adulthood, the conditional loss of *ERR* results in reduced fatty acids and TAG, likely due to a reduction in the expression of glycolytic, NADPH-producing, and lipogenic genes (Beebe et al. 2020). As with HR38, insect ERR studies have focused largely on glucose metabolism (Long et al. 2020; Park et al. 2021) and clear roles for ERR in lipid metabolism may become apparent in future work.

Considering the extensive mammalian literature on the roles of NRs in lipid metabolism with the highly conserved nature of this family of TFs, it seems that researchers are likely to uncover additional roles of NRs in insect lipid metabolism over the coming years.

1.4 FOXO and Other AMP Kinase-Regulated TFs

Insect insulin-like peptides and numerous other endocrine factors control the insect's response to poor (or rich) nutrition. Insulin signaling is a positive regulator of lipogenesis that can act through a number of TFs during development, adult maturation, and the subsequent maintenance of metabolic homeostasis. Insulin-like peptides are negative regulators of the AMP-regulated kinase AMPK, which phosphorylates and activates several TFs to drive the expression of target genes under low nutrient conditions. Next, we focus on the highly conserved forkhead box O (FOXO) and other TFs that are considered "classic" AMPK-dependent downstream regulators of energy balance and insulin signaling. These TFs are highly conserved across taxa and function in the production, metabolism, and distribution of lipids.

1.4.1 FOXO

The TF FOXO is typically activated by starvation, when it is phosphorylated and enters the nucleus to drive the expression of genes that help animals survive starvation. Although FOXO null mutants are viable and normal in size (Junger et al. 2003), a number of essential pathways can act through FOXO, including the insulin signaling and TOR pathways. Briefly, insulin signaling inhibits FOXO and is active during feeding. When nutrients are low, AMP levels increase and activate AMPK, which in turn phosphorylates FOXO, increasing its transcriptional activity to drive the expression of genes that encode proteins with beneficial roles during starvation, including those in lipid metabolism. When nutrition is good, or InR is constitutively activated, FOXO is phosphorylated by the insulin-dependent kinase Akt and migrates out of the nucleus (DiAngelo et al. 2009; Gershman et al. 2007; Luong et al. 2006; Nassel and Vanden Broeck 2016; Puig and Mattila 2011).

Many studies have shown a role for FOXO in the maintenance of lipid homeostasis across insect species, particularly during starvation. As a master regulator of the starvation response, FOXO typically helps the organism mobilize stored lipids from the fat body and can also modulate the expression of the gastric lipases that recover lipids from the intestinal lumen. One direct transcriptional target of *Drosophila* FOXO is the gene encoding the mitochondrial acyl coA synthase Pudgy, which is required for lipolysis and fatty acid beta-oxidation upon starvation (Xu et al. 2012b). The gene encoding the *Drosophila* lipid droplet-associated lipase dLip4 is also a direct FOXO target (Vihervaara and Puig 2008). FOXO directly activates the expression of two fat body TAG lipases in the silkworm *Bombyx mori* in response to 20E (Hossain et al. 2013). Expression of the gene encoding the TAG lipase bmm is also FOXO-dependent upon starvation in the cockroach *Blattella germanica* (Suren-Castillo et al. 2014) and in reproductively active tsetse fly females (Baumann et al. 2013). *Drosophila* FOXO promotes *bmm* expression and reduces TAG during starvation or when the TF *Kruppel-h1* is mutated, which increases FOXO's transactivation ability (Kang et al. 2017). These two TFs have a direct physical interaction at the *bmm* promoter (Kang et al. 2017). The TF Atf3 is upstream of FOXO in the *Drosophila* larval gut, where it promotes expression of the gastric lipase Lip3 and turnover of glycerolipids, which may represent an attempt to absorb fatty acids from the gut when energy is low (Rynes et al. 2012). Sirtuins and HDAC4 deacetylate and activate *Drosophila* FOXO to promote lipolysis, which can be a disadvantage and produces starvation sensitivity, due to inadequate lipid stores (Wang et al. 2011). Another role for FOXO is in *Drosophila* adult flight muscle, where it inhibits expression of the leptin-like cytokine Unpaired2 and, in turn, inhibits adipokinetic hormone secretion and lipid mobilization during the day (Zhao and Karpac 2017).

Like EcR, FOXO's effects on lipid homeostasis appear to be context-dependent and can vary based on the tissue, stage, genotype, and environmental conditions. Inducing FOXO in adult female *Drosophila* using an RU486-inducible promoter produced only a few overlapping differentially expressed genes between the gut and fat body (Alic et al. 2014). Gastric lipases, which are extracellular and act on dietary glycerolipids, typically promote lipid storage, whereas intracellular lipases in tissues like the fat body typically reduce lipid storage. Accordingly, expression of two gastric lipase-encoding genes, *magro* and *CG6295*, was decreased by overexpressing FOXO via the transcriptional repressor Aop in the gut, which corresponded to an overall decrease in whole-animal stored TAG in these flies (Alic et al. 2014). Fat body lipolysis activated by FOXO in starved *Drosophila* adults can be antagonized when the immunity-promoting NF-kB family TF Relish is present, giving the animal an additional mechanism to carefully control lipid metabolism, one that could be very useful to conserve energy during infection (Molaei et al. 2019). Consistent with a role for FOXO in the regulation of energy balance and immune function, fat body-specific knockdown of the transcriptional co-activator *fs(1)h* led to increased expression of antimicrobial peptides and insulin signaling accompanied by an inability of flies to mobilize stored TAG during starvation. Both the gene expression and metabolic changes depended on FOXO, suggesting that fs(1)h knockdown phenotypes result from FOXO hyperactivation (Sharrock et al. 2019). Overexpressing a constitutively active FOXO (CA-FOXO) in the *Drosophila* adult fat body did not affect overall TAG content (DiAngelo and Birnbaum 2009) although another study found an increase in lipid droplet size when the same transgene was expressed in the fat body that surrounds the brain in the adult head (Hwangbo et al. 2004). Such studies are presumably complicated by simultaneous presence of active insulin signaling, which would normally suppress FOXO activity. Ubiquitous overexpression of CA-FOXO in feeding *Drosophila* larvae, when FOXO is not typically active, also increased TAG (Luong et al. 2006). After oxidative stress, activated FOXO in *Drosophila* imaginal discs and other tissues drives expression of the gene encoding Sestrin, a conserved protein that promotes lipolysis over lipogenesis (Lee et al. 2010). Finally, in the *Drosophila* heart, FOXO also promotes lipolysis, which improves cardiac function during overnutrition, as might be expected (Birse et al. 2010).

In the *Drosophila* adult midgut, FOXO inhibits lipase gene expression and lipolysis after high-sugar feeding and during aging, including expression of the extracellular gastric lipase *magro* (Karpac et al. 2013). In this context, animals can use sugar to synthesize fat and energy, making the absorption of dietary lipids less important. By contrast, another study found that FOXO promoted expression of the gastric lipases encoded by *CG6277* and *CG8093* in lean *ceramide kinase* mutant guts, a context where energy recovery is needed (Nirala et al. 2013). An RNAi injection approach to decrease *FOXO* in the mosquito *Culex pipiens* reduced lipid storage in the fat bodies of diapausing females, although no transcriptional targets were determined in the study (Sim and Denlinger 2008). A ChIP-seq study comparing wing hinges in the dimorphic-winged pest *Sogatella furcifera* found a number of lipid metabolism genes associated with FOXO (Gao et al. 2019). Another study in the Asian agricultural pest *Galeruca daurica* found that knockdown of *FOXO* led to an overall decrease in TAG with an apparent decrease in fat body size and an increase in ovary size (Duan et al. 2022). This is an interesting finding in light of this and other genotypes where changes in lipid metabolism in one organ are able to produce endocrine effects on other organs via inter-organ communication (Agrawal et al. 2016; Chatterjee et al. 2014; Rajan and Perrimon 2012).

FOXO seems to play key roles in the crosstalk between insect organs during the regulation of lipid metabolism, especially that between the muscle and fat body. Targeting FOXO in the muscle led to increased muscle lipogenesis, but decreased fat body lipogenesis and lipid content, consistent with a model where FOXO in the muscle is able to control fatty acid flux between tissues and therefore systemic lipid homeostasis (Zhao and Karpac 2017). In a complementary fat body-to-muscle paradigm, the deacetylase Sir2 works cooperatively with FOXO in the *Drosophila* fat body to promote lipolysis and hyperlipidemia, leading to insulin resistance in the muscle (Banerjee et al. 2013). Several other tissues require FOXO in the transcriptional control of cell-autonomous and non-autonomous lipid metabolism. In the developing *Drosophila* wing disc, the gene encoding the lipid storage droplet protein Lsd2 genetically interacts with *FOXO*, and dsRNA targeting *Lsd2* led to increased *FOXO* expression in third instar wing imaginal discs (Binh et al. 2019). In *Drosophila* perineural glia, FOXO controls sphingolipid metabolism and growth of adjacent peripheral neurons (Dahlgaard et al. 2012). *FOXO* interacts genetically with *EcR* in the *Drosophila* prothoracic gland to control ecdysteroid metabolism and organismal growth (Koyama et al. 2014). Injecting dsRNA targeting *FOXO* into *Tribolium castaneum* larvae impaired ecdysone release and metamorphosis, consistent with what has been observed in *Drosophila* (Lin et al. 2018). The EcR coactivator DOR, described in more detail below, is a direct target of FOXO that transcriptionally regulates lipid metabolism (Francis et al. 2010). In summary, therefore, FOXO plays a range of distinct roles in the transcriptional regulation of both tissue-autonomous and systemic lipid homeostasis.

1.4.2 The CREB Family and CRTC

The cAMP response element-binding proteins, or Crebs, are members of the conserved family of CREB basic leucine zipper TFs that typically act downstream of AMPK to promote lipid catabolism during fasting in humans and insects via regulation of fatty acid oxidation and oxidative stress genes (Wang et al. 2008). Two *Drosophila* orthologs, CrebA and CrebB, have been functionally characterized and play a variety of roles. Loss-of-function in *CrebA* produces dorsal-ventral patterning (Andrew et al. 1997), immune (Troha et al. 2018), and secretory (Abrams and Andrew 2005; Fox et al. 2010) defects that may reflect a role for this TF in lipid metabolism, and the CrebA human ortholog Creb-H promotes expression of lipid metabolic genes and apolipoprotein secretion (Barbosa et al. 2013), but there is currently no evidence to support a role for this isoform in insect lipid metabolism and therefore we shall focus on the role of CrebB.

Activation of adenylate cyclase by G-protein coupled receptors such as the adipokinetic hormone receptor increases intracellular cAMP concentrations and CrebB activity when insects need to mobilize stored energy (Baumbach et al. 2014b). Reducing CrebB activity in the *Drosophila* fat body using a dominant negative approach led to an obese phenotype and was associated with increased food intake (Iijima et al. 2009). Similarly, RNAi targeting *CREB* increased fat body lipid droplet size and TAG content in the insect *Bactrocera dorsalis* and led to the identification of the microRNA *miR-2b-2-5p* as a negative regulator of CREB expression (Zhang et al. 2023). Surprisingly, loss of *CrebB* or the gene encoding the CREB-regulated transcriptional coactivator (CRTC) produced lean flies with gut permeability defects, reduced starvation survival, and increased infection susceptibility despite increased antimicrobial peptide expression (Shen et al. 2016).

Flies required neuronal CrebB activity to promote short neuropeptide F (sNPF) secretion, which in turn acted on gut sNPF receptors to enhance epithelial integrity (Shen et al. 2016). Expressing a dominant-negative CrebB in the *Drosophila* brain reduced lipid storage (Iijima et al. 2009) and neuronal RNAi targeting *CrebB* reduced the survival of starved flies (Choi et al. 2011). The latter study placed CRTC and CrebB downstream of the AMPK-related salt-inducible kinase 2 (Sik2), which is also known to function in metabolic homeostasis during fasting (Choi et al. 2011). Multiple isoforms of CrebB with different functions are encoded by the *Drosophila* genome (Yin et al. 1995). An in vitro study in *Drosophila* S2 cells found that human CREB could co-immunoprecipitate with and increase promoter binding of SREBP to drive expression of a downstream reporter construct, suggesting another potential mechanism by which Crebs regulate lipid metabolism (Ngo et al. 2002). Therefore, these TFs have a complex role in the regulation of lipid homeostasis that is likely to depend on the cell type or context.

A third, more distantly related Creb TF, *Drosophila* Atf-2, promotes glycerogenesis, and *atf2* RNAi in the fat body reduced TAG storage and starvation survival, possibly via reduction of *acyl coA synthetase* and/or *Pepck* expression (Okamura et al. 2007). Overexpression of Atf-2 also seems to increase fat body lipogenesis, at least in part, via the sphingolipid pathway enzyme glucosylceramide synthase (Kohyama-Koganeya et al. 2011). In the *Drosophila* intestine, Atf-2 is phosphorylated by the kinase p38c during an enteric infection and, in turn, regulates the immune and oxidative stress responses (Chakrabarti et al. 2014). In this paradigm, p38c interacts with a related TF, Atf3, to control intestinal lipid metabolism. In contrast with Atf-2, *Drosophila Atf3* mutants display a larval obesity phenotype, with lipid accumulation in both the fat body and intestine that is accompanied by changes in the expression of a number of gastric lipases (Rynes et al. 2012). Taken together, the CREB family of TFs plays numerous roles in the control of insect lipid metabolism and may serve to link lipid metabolism to infection susceptibility and redox status.

1.4.3 MEF2

The myocyte enhancer factor 2 MEF2 is another AMPK-dependent TF that was identified in a *Drosophila* RNAi screen to identify genotypes exhibiting differences in lipid droplet morphology in the indirect flight muscle (Zhao et al. 2020). Reduction of *MEF2* in this metabolically active tissue, found in the insect thorax, increased lipid droplet size as well as the ratio of cholesterol esters to TAG in *Drosophila* fed a high-calorie diet (Zhao et al. 2020). Note that in the indirect flight muscles from this RNAi genotype, increased lipid droplet size corresponded with *reduced* TAG on a high-calorie diet. Interestingly, the differentially expressed genes had roles throughout metabolism and *Mef2* knockdown also produced reductions in flight muscle glucose and lactate levels. There was no effect of

Mef2-RNAi on lipogenesis from radiolabeled glucose, so this TF may regulate the flux of carbons from intermediary metabolites into droplets, or perhaps lipid turnover, rather than the de novo synthesis of fatty acids destined for storage (Zhao et al. 2020). Overexpression of *Mef2* did increase both the expression of genes encoding lipid droplet proteins Lsd-2 and DGAT/Midway and lipid droplet size in *Drosophila* imaginal epithelial tumors, consistent with a role for MEF2 in controlling the balance between cholesterol and glycerol esterification (Harsh and Eleftherianos 2020).

Interestingly, MEF2 also plays a role in lipid metabolism in the nervous system. A forward genetic screen identified an allele of *frataxin* (*fh^{1}*) that led to neurodegeneration in *Drosophila* (Yamamoto et al. 2014). These flies exhibit a sphingolipid-dependent loss of photoreceptors that occurs via the activation of MEF2 and its target genes in the CNS. Dihydrosphingosine accumulates and activates MEF2 transcriptional activity in a phosphoinositide-dependent kinase-dependent manner, placing this TF in the AMPK and lipid signaling pathways in the nervous system (Chen et al. 2016). How sphingolipids activate MEF2 remains to be determined but is likely to help researchers better understand the molecular links between lipids and transcriptional regulation of lipid metabolism.

1.4.4 HIF

Another AMPK-dependent TF, hypoxia-inducible factor (HIF), works in parallel with FOXO and is repressed by AMPK and stimulated by low oxygen and insulin (Barretto et al. 2020; Dekanty et al. 2005). The oxygen-regulated HIF1α subunit controls the expression of a number of lipid metabolic genes in *Drosophila*, including those that encode the lipases Lip1 and Lip2, the serine palmitoyltransferases lace and Spt1, and a number of other glycerolipid metabolic enzymes including glycerol kinase, the acyl coA ligase bubblegum, the acetyl coA synthase AcCoAS, and the lipid phosphate phosphatase wunen (Li et al. 2013). Mitochondrial antioxidant feeding improves gut integrity and survival on a high-fat diet in a HIF-dependent manner, consistent with a model where HIF controls flux of fatty acids through the electron transport chain (Watson et al. 2021). Hypoxia is associated with increased fat storage in *Drosophila* larvae (Lee et al. 2019), which might be expected, considering the requirement for oxygen as electron acceptor during the mitochondrial respiratory chain. By contrast, HIF is required for glycogen catabolism during hypoxia, a process augmented by ERR-dependent glycolytic gene expression, which may act to compensate for the reduced ability to use lipids as an energy source under low oxygen conditions (Li et al. 2013). These two TFs physically interact, providing a potential mechanistic link between lipid and carbohydrate metabolism at the transcriptional level (Li et al. 2013). Larval hypoxia reduces adult size and starvation survival (Polan et al. 2020), supporting a model where hypoxia and its downstream effectors, such as HIF, control the balance between lipid and carbohydrate storage and catabolism in *Drosophila*. This TF does regulate lipid metabolic gene expression in *C. elegans* (Xie and Roy 2012) and in several insects (below) and may, like CREB, integrate redox status with lipid metabolism.

One remarkable study done in the mosquito *Aedes aegypti* found that pharmacological stabilization of HIF reduced neutral lipids in the larval gut and increased lipids in the fat body, while destabilizing HIF reduced fat body lipid accumulation and increased lipids in the gut. The expression of the lipophorin Lpp, which carries acyl chains from the gut to the fat body, was decreased by HIF destabilization, suggesting a model where a deficit in lipid transport underlies the reduced fat body neutral lipid storage seen when HIF is inactive. Axenic culture in these larvae reduces growth (Valzania et al. 2018a) because bacteria-induced hypoxia in the developing gut serves as an upstream growth-promoting signal that increases lipid transport to the fat body via transcriptional regulation of lipid metabolism by HIF (Valzania et al. 2018b).

A second study explored the role of HIF during arousal from diapause in the flesh fly *Sarcophaga crassipalpis*, which entails an increase in metabolic activity via the TCA cycle along with a decrease in hypoxia signaling (Chen et al. 2021). Paradoxically, HIF protein increased at the onset of arousal, yet injecting a HIF inhibitor into diapausing flesh flies led to the acceleration of arousal and increased CO_2 production. This may result from the inhibition of the HIF transcriptional target *pyruvate dehydrogenase kinase*, which encodes the PDK enzyme that inactivates pyruvate dehydrogenase to suppress aerobic respiration; a PDK inhibitor produced a similar acceleration in arousal and respiratory activity. One by-product of fatty acid oxidation is oxidative stress and accordingly, reactive oxygen species were increased during arousal along with fatty acid oxidation, acylcarnitines, and *carnitine palmitoyltransferase* (*CPT*) expression. Isotope tracer experiments support a model where reducing HIF may act to promote arousal by controlling the flux of carbons between glycolysis, the TCA cycle, and fatty acid metabolism via pyruvate (Chen et al. 2021). In accordance with a role for HIF in controlling lipid metabolism during diapause, a study in the cotton bollworm *Helicoverpa armigera* found increased HIF and reduced mitochondrial oxidative activity during diapause (Lin et al. 2016). Therefore, HIF may serve to help balance oxidative stress with lipid catabolism during development, diapause, and in response to changes in environmental conditions.

1.5 Circadian Transcription Factors: Clock and Cycle

Many behaviors and physiological processes are regulated in a circadian fashion, meaning that their activities cycle throughout a 24-h day. Metabolism is well known to be controlled by the circadian clock in both mammals and insects, with many important metabolites and metabolic hormones cycling throughout the day (Bass and Takahashi 2010; Sehgal 2016). The circadian clock mechanism has been well characterized in *Drosophila* and other insects and is controlled by four major proteins: Clock (Clk), cycle (cyc), period (per), and timeless (tim), with Clk and cyc being TFs (Patke et al. 2020). While TAG levels don't cycle throughout the day in *Drosophila* fat tissue (DiAngelo et al. 2011; Seay and Thummel 2011), acylcarnitines do cycle throughout the day under normal feeding conditions (Schabler et al. 2020) and TAGs with medium chain fatty acids do cycle under conditions of dietary restriction (Katewa et al. 2016) suggesting an integration of diet and the circadian clock. Cycling of acylcarnitines and medium-chain fatty acids during dietary restriction are abrogated in *per* and *tim* mutants, respectively, providing further evidence that the metabolism of acyl chains is under circadian control. In addition, TAG, diglyceride (DAG) and acylcarnitine levels are blunted in the heads of *per* mutants while only DAGs and acylcarnitines are decreased in *per* mutant bodies (Schabler et al. 2020). Moreover, *tim* mutants have decreased TAG breakdown and synthesis during dietary restriction, suggesting a role for the circadian clock in regulating glycerolipid homeostasis (Katewa et al. 2016). Consistent with work in *Drosophila*, cabbage beetles (*Colaphellus bowringi*) with the levels of *tim* or *per* decreased via RNAi have less TAG storage during the diapause preparation period (Zhu et al. 2019). This change in TAGs may be due to less expression of the fatty acid binding protein, *FABP*, and increased expression of the TAG lipase, *TGL1* (Zhu et al. 2019). In addition, *Drosophila* lipid storage is controlled by both peripheral and central clocks as expressing a dominant-negative form of *Clk* in the fat body (a peripheral clock) results in decreased TAGs (Xu et al. 2008) while expressing the same dominant-negative construct in pigment-dispersing factor (PDF) neurons in the fly brain (a central clock) results in increased TAG levels (DiAngelo et al. 2011). These data highlight the complexity of how lipid homeostasis is maintained, and that this regulation is integrated with the time of day to make lipid metabolism most efficient. Taken together, the functions of all of the TFs described in this section give us a model of the transcriptional regulation of lipid synthesis and breakdown in insects (Figs. 2 and 3).

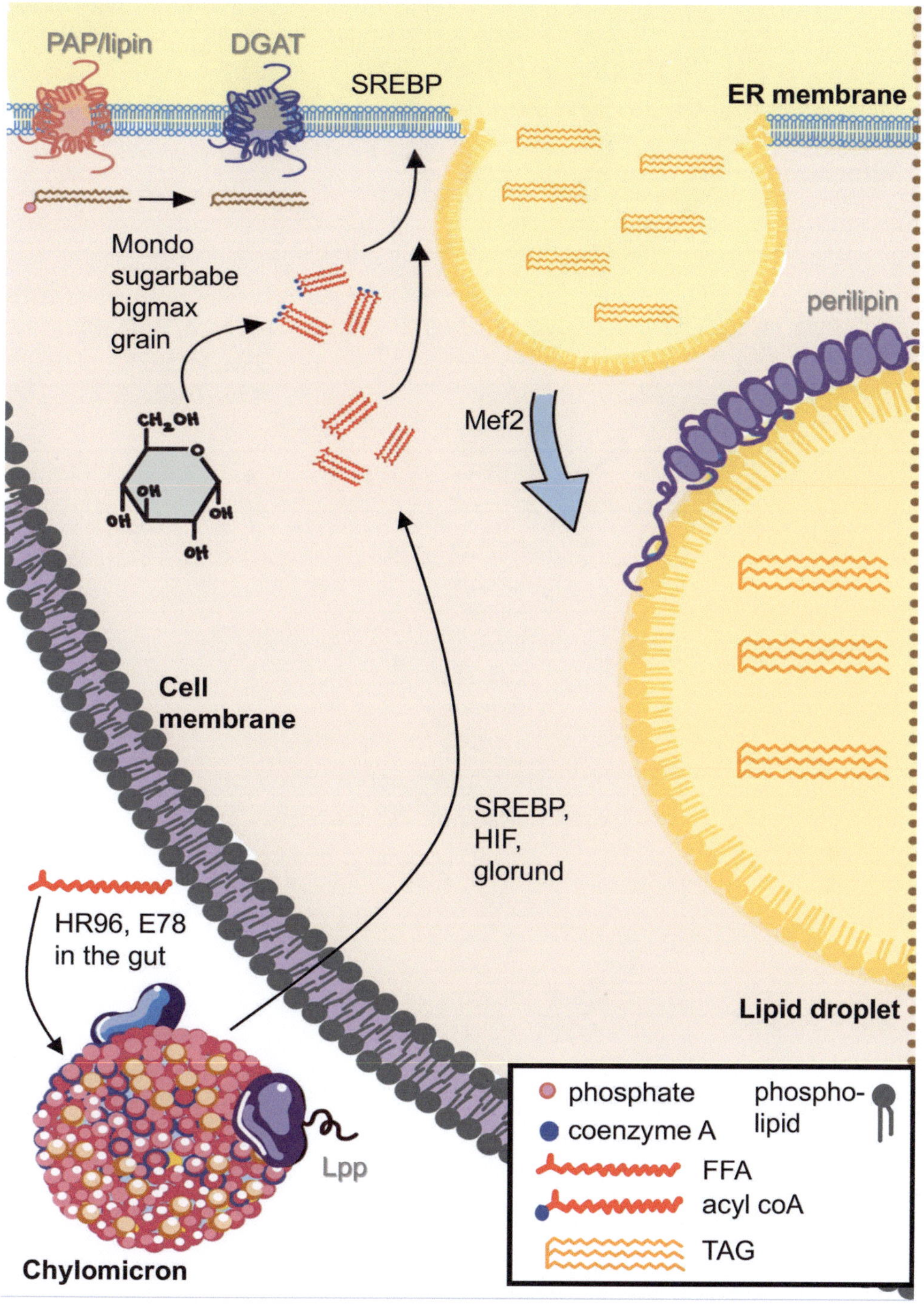

Fig. 2 Transcriptional regulation of lipid biosynthesis in insects

Lipid biosynthesis is favored when the cellular energy charge is high and the substrates needed to store lipids are abundant, either from synthetic pathways in the cell or via transport through the hemolymph. High-sugar feeding activates a group of TFs that promote both glycolysis and fatty acid synthesis and esterification as TAG, including **Mondo**, **bigmax**, **grain**, and **sugarbabe**. Changes in the ER membrane's phospholipid composition can increase transcriptional activation by **SREBP**, which also increases fatty acid synthesis and esterification as TAG to promote lipid droplet expansion. Genes encoding lipoproteins (Lpps) and their receptors are activated by **SREBP, HIF,** and the ribonucleoprotein **glorund**. The loading of lipoproteins from the gut, which are required to fuel lipid storage in the fat body, depends on gut-encoded lipases that are transcriptionally regulated by **HR96** and **E78**. **Mef2** functions in adult flight muscle to control lipid droplet size and composition. See Sects. 1.1–1.5 and 2.3 for more details and citations

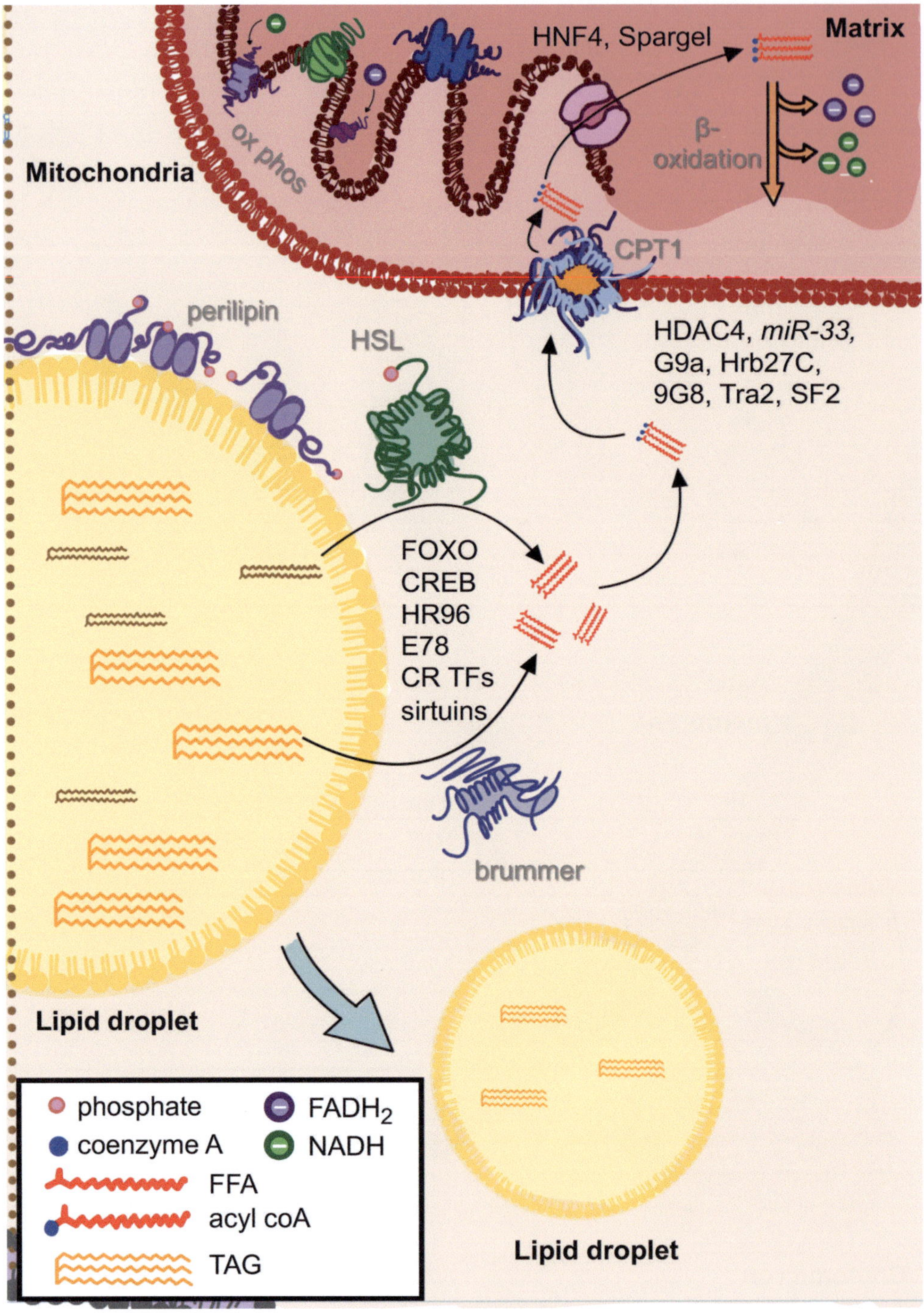

Fig. 3 Transcriptional regulation of lipid catabolism in insects

Lipid catabolism helps the cell mobilize energy stored in lipid droplets for use by the insect to fuel flight, reproduction, and other energetically demanding processes. The TF **FOXO** typically activates expression of lipases to release fatty acids from glycerolipids in the lipid droplet when nutrients are low. In the fat body, **CREB** also promotes lipid catabolism, although its roles in other organs, such as the brain, are more complex. The circadian rhythm (**CR**) **TFs** tim and per also regulate lipase expression and therefore TAG in a periodic manner. The nuclear receptors **HR96** and **E78** promote lipase expression in the gut to produce an overall increase in stored TAG. The **sirtuins** group of deacetylases also regulate lipase expression. During starvation, **HNF4** drives the expression of fatty acid beta-oxidation and oxidative phosphorylation genes, which are also targets of the transcriptional coactivator **Spargel**. A major target gene for non-TF regulatory mechanisms encodes the carnitine palmitoyl transferase CPT1, which imports fatty acids into the mitochondrion for catabolism. See Sects. 1.1–1.5 and 2.1–2.4 for more details and citations

2 Additional Transcriptional and Post-transcriptional Mechanisms that Control Lipid Metabolism

While so far in this chapter we have focused on the TFs that regulate lipid metabolism in insects, we haven't discussed the mechanisms that allow for these TFs to become activated and access promoters or other regulatory mechanisms that alter mRNAs that encode lipid metabolic enzymes. In this section, we will explore the transcriptional control mechanisms used by insects to regulate the production and processing of mRNAs and proteins important for controlling lipid homeostasis (Fig. 4).

2.1 Acetylation and Methylation

2.1.1 Sirtuins

The sirtuins are a class of NAD^+-dependent deacetylases that regulate aging and metabolism in a number of organisms (Chang and Guarente 2014). The *Drosophila* genome contains five sirtuin genes (*Sirt1*, *Sirt2*, *Sirt4*, *Sirt6*, and *Sirt7*) but only *Sirt1*, *Sirt4*, and *Sirt6* regulate lipid metabolism. *Sirt1* (also known as *dSir2*) was identified in a larval buoyancy screen as being a potential obesity-causing gene as mutants had increased buoyancy and body fat (Reis et al. 2010). Sirt1 acts in the fat body to regulate lipid metabolism as *Sirt1-RNAi* flies have increased TAG storage (Reis et al. 2010; Banerjee et al.

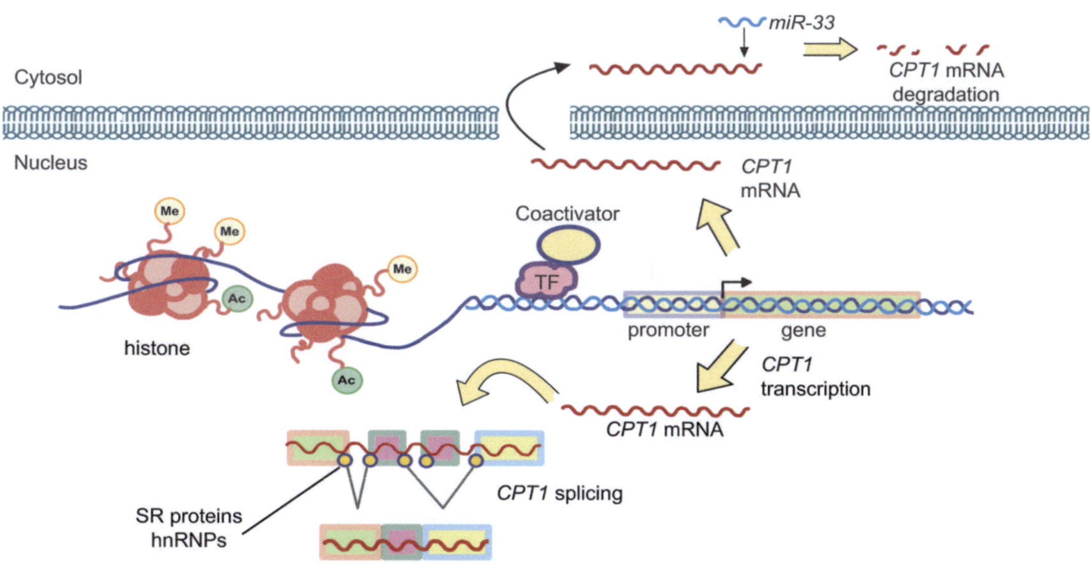

Fig. 4 Transcriptional and post-transcriptional mechanisms that control lipid metabolism in insects Transcription of genes important for lipid metabolism such as the lipase *brummer*, the phosphatidic phosphatase *Lipin*, and the beta-oxidation gene *CPT1* can be regulated by adding methyl groups (**Me**) or acetyl groups (**Ac**) to histone tails, ultimately altering chromatin structure and gene expression. Once the upstream region of a gene is accessible, **TFs** can bind to the promoter and recruit **coactivators** such as spargel and Lipin to help increase transcription. As transcription is occurring, pre-mRNAs are being processed through the process of alternative splicing where serine/arginine-rich domain-containing proteins (SR proteins) and heterogeneous nuclear ribonucleoproteins (hnRNPs) control the usage of splice sites to allow different transcripts to be generated from a single gene. The *CPT1* transcript can be alternatively spliced by **SR proteins** like 9G8, Tra2, and SF2 or **hnRNPs** like Hrb27C to form two different isoforms that differ by a single exon and produce proteins that have different enzymatic activity. In addition, levels of mRNAs encoding important lipid metabolic enzymes such as CPT1, the DAG acyltransferase midway, and acetyl coenzyme A synthase (AcCoAS) can also be decreased by the presence of microRNAs (i.e., ***miR-33*** targets the *CPT1* mRNA) that will bind to target mRNAs and cause their destruction. See Sects. 2.1–2.4 for more details and citations

2012). Fat body-specific *Sirt1-RNAi* flies also have decreased expression of the lipases *Lip3* and *bmm* and increased expression of *FAS* (Banerjee et al. 2012), suggesting that Sirt1 regulates both lipid breakdown and synthesis to control TAG storage. Decreasing *Sirt1* in the fat body also increases the expression of the insulin-like peptide, *ilp5*, and this increase in systemic insulin signaling may account for the increased TAGs in these flies (Banerjee et al. 2012). This increase in *ilp5* levels may also be compensatory as insulin-stimulated Akt phosphorylation is decreased in *Sirt1* mutants and fat body-specific *Sirt1-RNAi* flies (Palu and Thummel 2016). In addition, Sirt1 may be acting through HNF4 to regulate lipid storage as *HNF4* overexpression in *Sirt1* mutants can rescue the Akt phosphorylation defects found in these flies (Palu and Thummel 2016).

Sirt1 also acts in the heart to regulate lipid storage as well as heart function. Decreasing *Sirt1* levels specifically in the heart using RNAi results in increased cardiac TAG levels as well as decreased heart function (Wen et al. 2019). Conversely, overexpression of *Sirt1* specifically in the fly heart decreases cardiac TAGs and increases heart function (Wen et al. 2019). These changes in cardiac TAG may be due to *Sirt1* regulating lipid breakdown as cardiac-specific *Sirt1-RNAi* flies have decreased expression of the lipase *bmm* while cardiac *Sirt1* overexpressing flies have increased *bmm* levels (Wen et al. 2019). In addition, *Sirt1* overexpression specifically in the heart can blunt the increased cardiac TAGs and lipotoxic cardiomyopathy caused by feeding a high-fat diet (Wen et al. 2019). Overall, these data indicate that *Sirt1* functions in multiple insect tissues to control lipid metabolism.

Sirt4 and *Sirt6* also regulate insect lipid homeostasis. Sirt4 is a mitochondrially localized sirtuin that promotes lifespan (Wood et al. 2018). *Sirt4* mutants are sensitive to starvation while ubiquitous overexpression of *Sirt4* results in starvation resistance (Wood et al. 2018). Interestingly, *Sirt4* mutants have increased TAGs under fed conditions as well as during starvation, suggesting that the starvation sensitivity in these flies is due to an inability to break down lipids (Wood et al. 2018). Similar to *Sirt4* mutants, *Sirt6* mutants also have decreased lifespan and increased TAGs under fed conditions with TAGs remaining high throughout a 48-h starvation period (Shukla and Kolthur-Seetharam 2022). *Sirt6* mutants also have decreased expression of two lipases, *Lip3* and *bmm*, during starvation which may account for the increased TAGs observed under these conditions (Shukla and Kolthur-Seetharam 2022). These phenotypes suggest that *Sirt4* and *Sirt6* may regulate the expression of important lipid catabolism genes to regulate lipid storage and lifespan.

2.1.2 Histone Acetyltransferases and Deacetylases

Adding acetyl groups to histones can alter chromatin structure and gene expression generally in cells. If the chromatin structure of critical lipid metabolic enzyme genes was specifically altered, organismal lipid storage could also be impacted. N(alpha)-acetyltransferase 40 (Naa40) is a histone acetyltransferase that adds acetyl groups on histones H2A and H4 (Hole et al. 2011). When *Naa40* levels are decreased specifically in the larval *Drosophila* fat body using RNAi, overall organismal triglyceride accumulates and the number of lipid droplets found in each fat cell increases (Charidemou et al. 2022). In addition, *Naa40-RNAi* larvae have increased expression of the phosphatidic acid phosphatase, *Lipin*, suggesting that increased TAG esterification may account for the augmented TAG storage in these animals.

The fruit fly genome contains four histone deacetylase (HDAC) genes: *HDAC 1, 3, 4*, and *6* (Cho et al. 2005). While HDAC3 does not seem to play a role in regulating lipid storage, HDAC1, HDAC4, and HDAC6 have lipid metabolic functions. HDAC1 (also known as Rpd3) was identified in an adult *Drosophila* screen as an anti-obesity gene as flies with *HDAC1* decreased in fat body and midgut conditionally in adulthood had increased TAGs (Baumbach et al. 2014a). In addition, when *HDAC1* levels were decreased specifically in the fat body throughout development, increased lipid breakdown during a 20-h fast was observed (Molaei et al. 2019). The expression of the lipase *bmm* was also increased during a 20-h fast, which may be responsible for the increase in lipid breakdown observed (Molaei et al. 2019). Together, these data suggest that HDAC1 regulates lipid metabolism differently depending on the stage of development or nutritional status of the animal.

HDAC4 was identified as functioning downstream of salt-inducible kinase 3 (Sik3), a kinase in the AMP protein kinase family, to promote lipid mobilization. *Sik3* mutant animals have a decreased TAG phenotype that can be suppressed by a *HDAC4* mutant allele (Wang et al. 2011). While fat body-specific depletion of *HDAC4* using RNAi has no effect on lipid levels, overexpressing *HDAC4* in the fat body increases expression of the lipase encoded by *bmm* and the rate-limiting enzyme in mitochondrial fatty acid import, encoded by *CptI* (Wang et al. 2011; Choi et al. 2015). *HDAC4* acts downstream of the kinases LKB1 and Sik3 to regulate lipid levels as expression of a nonphosphorylatable mutant of *HDAC4* specifically in the fat body results in less TAG storage (similar to *Sik3* mutants) and increased *bmm* levels may explain this phenotype (Choi et al. 2015). During feeding, Sik3 phosphorylates HDAC4, leading to its export from the nucleus to the cytoplasm, favoring lipid storage. During fasting, the glucagon-like molecule adipokinetic hormone (AKH) inhibits LKB1 and Sik3, causing HDAC4 to remain in the nucleus and activate transcription of *bmm*, favoring lipid breakdown. In this way, HDAC4 acts in maintaining overall lipid homeostasis (Choi et al. 2015).

Unlike the other HDACs, HDAC6 functions in the cytoplasm to regulate lipid metabolism. HDAC6 functions in the *Drosophila* oenocyte to regulate lipid levels during starvation. During starvation, lipids are broken down from the fat body and transported to the oenocyte for further processing (Gutierrez et al. 2007). In the oenocyte, HDAC6 promotes the transfer of lipid droplets to autophagic vesicles for degradation as lipid droplets accumulate in the oenocytes of *HDAC6* mutants (Yan et al. 2019). HDAC6 also regulates ectopic fat accumulation in non-adipose tissue in aging flies. Lipid droplets accumulate in tubular jump muscles of aging flies, and this is augmented in *HDAC6* mutants (Yan et al. 2017). This lipid droplet accumulation phenotype in *HDAC6* mutants can be suppressed by mutating perilipin-2 (*Lsd-2*) (Yan et al. 2017). The ubiquitination of Lsd-2 is increased in old flies and augmented in *HDAC6* mutants (Yan et al. 2017) suggesting that HDAC6 controls ectopic lipid levels in tubular jump muscle by regulating the levels of the lipid droplet-associated protein, Lsd-2.

2.1.3 Nucleic Acid and Histone Methyltransferases

Similar to adding acetyl groups to histones, adding methyl groups to either DNA/RNA or histones can also regulate gene expression. Mutants in *Mt2*, a nucleic acid methyltransferase in *Drosophila*, have decreased TAGs, but increased sphingosine 1-phosphate in 15-day-old flies as shown by thin-layer chromatography (Abhyankar et al. 2018). *Mt2* mutants also have decreased survival after bacterial infection and decreased expression of antimicrobial peptides, suggesting a potential link between the lipid storage defects and the immune response in these flies (Abhyankar et al. 2018). In addition, the histone methyltransferase G9a regulates lipid levels under baseline and oxidative stress conditions. Heads from *G9a* mutants have increased TAGs under basal conditions that continue to stay high during treatment with the oxidative stress inducer paraquat (Riahi et al. 2019). Consistent with the augmented TAG levels, lipid droplet diameter was also increased in *G9a* mutant head fat body (Riahi et al. 2019). This suggests that G9a is necessary for lipid catabolism and perhaps regulates the expression of important lipid breakdown genes like the lipase *bmm* or the beta-oxidation gene *CptI* as we described above for HDAC4.

2.2 Transcriptional Coactivators

2.2.1 Spargel (PGC-1)

TFs many times will act with additional proteins called coactivators that recruit the transcriptional machinery and help to activate transcription (Spiegelman and Heinrich 2004). One very well-characterized transcriptional coactivator with metabolic functions is PPAR-gamma coactivator-1 (PGC-1 in mammals and spargel in *Drosophila*). Mutants in the *spargel* gene weigh less and have decreased TAG stores compared to controls and also have impeded growth (Tiefenbock et al. 2010). Consistent with this lean phenotype, *spargel* mutants have increased expression of genes involved in beta-oxidation and lipid catabolism and decreased expression of major TCA cycle genes (Tiefenbock et al. 2010). In addition, *spargel* mutants have decreased oxygen consumption when electrons enter the electron transport chain through Complex II, but not Complex I, indicating mitochondrial functions of this gene (Tiefenbock et al. 2010). *Spargel* expression is increased by activating insulin signaling and is necessary for insulin-mediated growth effects in developing flies (Tiefenbock et al. 2010). However, whether spargel and FOXO act together or in parallel to regulate lipid metabolism is not yet understood.

In addition to regulating lipid metabolism throughout development, spargel also regulates lipid storage in *Drosophila* adults fed a high-fat diet. Flies fed a high-fat diet have decreased *spargel* transcript levels and have increased lipid stores; this excess storage of TAGs is blunted when *spargel* is overexpressed (Diop et al. 2015). Consistent with this finding, ubiquitous *spargel* RNAi flies store more TAGs when fed a high-fat diet than wildtype flies fed the same diet (Diop et al. 2015). This result is in contrast to the lean phenotype observed in *spargel* mutants fed a normal diet (Tiefenbock et al. 2010), suggesting that *spargel* may function differently depending on nutrient availability. Moreover, high-fat diet feeding or ubiquitously decreasing *spargel* levels causes lipids to accumulate in the heart, leading to cardiac lipotoxicity (Diop et al. 2015). In addition, spargel protein is ubiquitinated and degraded by the E3 ubiquitin ligase, dRNF34, and decreasing *dRNF34* specifically in all muscles using RNAi decreases high-fat-induced TAG accumulation in a spargel-dependent manner (Wei et al. 2018). Together, these data support a role of spargel in regulating lipid metabolism throughout the fly life cycle and in a tissue-specific manner.

2.2.2 Lipin

Lipin is a highly conserved protein that serves as both a phosphatidic acid phosphatase and a transcriptional coactivator in insects (Lehmann 2021). In *Drosophila*, the *Lipin* gene generates three major protein-coding isoforms by alternative splicing that are expressed differentially throughout development (Valente et al. 2010). Mutants in the *Lipin* gene in *Drosophila* fail to produce viable adults, but mutant larvae have less stored TAGs and smaller lipid droplets in their fat bodies (Ugrankar et al. 2011). Consistent with the decreased TAG levels, *Lipin* mutants also have increased levels of ketone bodies, suggesting excess fatty acid oxidation, and fat body-specific *Lipin-RNAi* causes starvation sensitivity (Ugrankar et al. 2011). Deleting the nuclear localization sequence (NLS) in the Lipin protein, thus blocking the transcription functions of Lipin, has no effect on viability in fed *Lipin* mutants, suggesting that the nuclear functions of Lipin are dispensable for normal fly development (Hood et al. 2020). Interestingly, flies with the NLS deleted from Lipin have no defects in TAG storage but are still sensitive to starvation and have decreased metabolic rate (Hood et al. 2020). Consistent with the increased sensitivity to starvation, the expression of a number of lipases was increased while fatty acid synthesis enzyme genes were decreased in flies with the NLS deleted from Lipin suggesting that perhaps these flies use up their TAG stores more quickly during starvation causing them to die sooner (Hood et al. 2020).

Lipin has also been characterized in *Rhodnius prolixus*, the insect vector of Chagas disease, and the cabbage beetle, *Colaphellus bowringi* (Alves-Bezerra and Gondim 2012; Guo et al. 2021). In *Rhodnius prolixus*, *Lipin* is expressed highly in the anterior midgut and at lower levels in the fat body and ovary (Alves-Bezerra and Gondim 2012). *Lipin* is also expressed in the midgut, fat body, and ovary of the cabbage beetle and decreasing *Lipin* levels in females using RNAi results in defective ovaries suggesting a role for Lipin in regulating insect reproduction (Guo et al. 2021). Interestingly, *Lipin-RNAi* cabbage beetles have increased TAG levels and larger lipid droplets compared to controls (Guo et al. 2021), which is different from the phenotypes observed in *Drosophila Lipin* mutants as described above. Consistent with the increased TAGs, *Lipin-RNAi* cabbage beetles also have increased expression of *FAS* and *fatty acid transporter* genes and decreased expression of two lipases (Guo et al. 2021). In contrast to the *Drosophila Lipin* mutants, *Lipin* levels were only decreased in cabbage beetle adults, suggesting that Lipin has pleiotropic effects throughout insect development. Together, these results suggest that Lipin regulates lipid storage and metabolism in complex ways throughout the entire lifecycle of the insect and additional experiments are needed to increase our understanding of Lipin function.

2.2.3 Diabetes and Obesity Regulated (DOR)

Diabetes and obesity regulated (DOR) was originally identified as being downregulated in diabetic rats and subsequently shown to serve as a coactivator for the thyroid hormone receptor in mammals (Baumgartner et al. 2007). In *Drosophila*, *DOR* mutants have reduced viability and decreased TAG storage (Francis et al. 2010). Lowering *DOR* levels ubiquitously using RNAi also decreased autophagy in the fat body (Mauvezin et al. 2010). Interestingly, DOR can bind directly to EcR and inhibiting EcR results in lean flies, indicating that DOR acts with EcR to regulate lipid metabolism (Francis et al. 2010). In addition, *DOR* transcript levels are decreased in the presence of insulin as well as in *FOXO* mutants (Francis et al. 2010), contributing to the mounting evidence that insulin and ecdysone signaling interact to regulate lipid metabolism.

2.2.4 CREB-Regulated Transcription Coactivator (CRTC or TORC)

During fasting, production of cyclic AMP causes protein kinase A to phosphorylate cyclic AMP response element binding protein (CREB) to activate the transcription of a number of metabolic genes (Altarejos and Montminy 2011). A family of coactivators (CRTCs also known as TORCs) shuttle into the nucleus and bind to CREB to help stimulate its activity and increase transcription (Altarejos and Montminy 2011). CRTC nuclear localization is inhibited by phosphorylation by Sik2 (Altarejos and Montminy 2011). In *Drosophila*, CRTC levels are increased in response to oxidative stress and starvation (Wang et al. 2008). *CTRC* mutants are sensitive to starvation and consistent with this phenotype have decreased lipid stores (Wang et al. 2008). Interestingly, ectopic *CRTC* expression in all neurons can rescue the starvation sensitivity and lean phenotypes in *CRTC* mutants (Wang et al. 2008). Activating the insulin signaling pathway increases CRTC phosphorylation resulting in decreased CRTC activity (Wang et al. 2008). Consistent with an inhibitory role of Sik2 on CRTC in mammals (Altarejos and Montminy 2011), decreasing *Sik2* levels in flies results in increased fat stores and resistance to starvation, phenotypes opposite of *CRTC* mutants (Wang et al. 2008; Choi et al. 2011). Consistent with the function of CRTC in neurons to regulate lipid storage and starvation resistance, ectopic expression of *Sik2* in neurons can rescue the lipid accumulation and starvation resistance phenotypes observed in *Sik2* mutants (Wang et al. 2008; Choi et al. 2011). Together, these data indicate that Sik2 regulates CRTC during fasting to control overall lipid homeostasis.

2.3 Splicing

As transcription occurs, the growing mRNA must be processed by adding a guanosine nucleotide cap to the 5′ end of the transcript, adding a poly (adenosine) tail to the 3′ end of the transcript, and removing non-coding intron sequences in the process called splicing. This removal of intron sequences is dependent on small nuclear ribonucleoproteins (U1, U2, U4, U5, and U6) that form the spliceosome and work together to recognize intronic regions of mRNAs and catalyze intron removal (Kornblihtt et al. 2013). Different combinations of exons can also be included in growing mRNAs to generate multiple proteins from a single gene in a process known as alternative splicing that is dependent on the function of serine-arginine (SR) domain-containing proteins and heterogeneous nuclear ribonucleoproteins (hnRNPs) that help identify intron-exon borders (Kornblihtt et al. 2013). The role of alternative splicing in regulating lipid metabolism in insects was first appreciated in the late 2000s when two independent genome-wide RNAi screens performed in *Drosophila* cells uncovered splicing factors as being important for the formation of lipid droplets as knocking down members of the spliceosome resulted in the production of fewer lipid droplets (Beller et al. 2008; Guo et al. 2008). Consistent with this result, decreasing the expression of the constitutive splicing factors *snRNP-U1-70K* and *U2AF38*, members of the U1 and U2 snRNPs, respectively, in the *Drosophila* fat body results in blunted TAG storage (Gingras et al. 2014; Shabar and DiAngelo 2022). Interestingly, decreasing the levels of the SR proteins *9G8*, *Tra2, RBP1* and *SF2* in the *Drosophila* fat body using RNAi results in an obese phenotype indicating different roles of proteins involved in constitutive and alternative splicing in regulating lipid storage in flies (Gingras et al. 2014; Bennick et al. 2019; Mikoluk et al. 2018). Consistent with a role of SR proteins in regulating lipid storage, blocking the transport of SR proteins in and out of the nucleus also results in excess TAG storage (Nagle et al. 2022).

A potential alternatively spliced gene in insects involved in regulating lipid metabolism is the gene encoding the enzyme *Cpt1* (encoded by *whd* in *Drosophila*). CPT1 adds a carnitine group to a fatty acyl chain controlling the rate of fatty acid entry into the mitochondrion for beta-oxidation. In *Drosophila*, transcripts of the *Cpt1* gene are spliced to contain one of two alternative exons (exon 6A or 6B) encoding two protein isoforms, and this arrangement is also observed in the genomes of the mosquitos *A. gambiae* and *A. aegypti*, the wasp *N. vitripennis*, the beetle *T. castaneum*, and the honey bee *A. mellifera* (Price et al. 2010). CPT1 enzymes produced from transcripts including exon 6A are more enzymatically active than enzymes produced from transcripts including exon 6B (Price et al. 2010). Fat body-specific *9G8-RNAi*, *Tra2-RNAi*, and *SF2-RNAi* flies have altered *Cpt1* splicing such that they produce more transcripts containing exon 6B suggesting that less beta-oxidation may account for the TAG accumulation observed in these flies (Gingras et al. 2014; Bennick et al. 2019; Mikoluk et al. 2018). Blocking SR protein transport into the nucleus also results in the production of more *Cpt1* transcripts that contain exon 6B providing additional evidence for SR proteins regulating *Cpt1* splicing (Nagle et al. 2022). In addition, the splicing of the NADPH-producing enzyme genes glucose 6-phosphate dehydrogenase (known as *Zw* in *Drosophila*) and malic enzyme (known as *Men* in *Drosophila*) is also altered in fat body-specific *9G8-RNAi* flies (Weidman et al. 2022). Since NADPH is a necessary coenzyme for the synthesis of fatty acids, it is possible that alternative splicing of transcripts of these genes is also important for the regulation of lipid metabolism in insects, but additional experimentation is necessary to test this hypothesis.

In addition to SR proteins, a number of hnRNPs have been implicated in the regulation of fat storage in *Drosophila* (Bhogal et al. 2020; Kolasa et al. 2021). Decreasing the expression of *Hrb27C* specifically in the adult *Drosophila* fat body results in the accumulation of TAGs and this may be due to decreased expression of the *bmm* lipase (Bhogal et al. 2020). Conversely, decreasing the expression of the hnRNP *glorund* (*glo*) results in a lean phenotype (Kolasa et al. 2021). While whole *glo-RNAi* animals have decreased TAGs, fat bodies dissected from *glo-RNAi* animals have excess TAG storage, suggesting a defect in transporting lipids from the fat body to peripheral tissues. Consistent with this hypothesis, *glo-RNAi* fat bodies have decreased expression of the apoB-containing lipoprotein gene *apolpp*, the apolipoprotein lipid transfer particle gene *apoltp* and the microsomal triacylglycerol transfer protein gene *mtp*, indicating that fewer lipids are being packaged into lipoproteins in the fat body and transported through the hemolymph to other tissues throughout the fly (Kolasa et al. 2021). Together, these studies have identified a number of genes whose expression and splicing are regulated to control insect lipid metabolism and provide support for continued efforts toward identifying additional genes that may be regulated in a similar fashion to control overall lipid homeostasis.

Another pathway that relies on mRNA splicing to regulate development and physiology is the sex determination pathway. Sex-lethal (Sxl) is an RNA-binding protein that when present promotes female development and the generation of female-specific traits in flies (Salz and Erickson 2010). Sxl acts on the gene *transformer* (*tra*) to regulate its splicing to form a functional protein that promotes female fly development (Salz and Erickson 2010). In addition, tra functions with another RNA-binding protein, transformer 2 (tra2), to regulate the splicing of the *doublesex* gene which results in the sexual dimorphic phenotypes observed in adult flies (Sciabica and Hertel 2006). The storage of triglycerides is a sexually dimorphic trait with adult female flies storing more triglycerides than adult males, while male larvae store more triglycerides than female larvae (Diaz et al. 2023; Wat et al. 2020). The RNA-binding protein and *Sxl* regulator spenito acts in the larval fat body to regulate the sexual dimorphism of triglyceride storage in *Drosophila* larvae as decreasing *spenito* in larval fat tissue results in similar fat storage between male and female larvae (Diaz et al. 2023). In addition, ectopic expression of *tra* in larval fat tissue also results in similar fat levels in male and female larvae suggesting a role for tra in regulating sexually dimorphic larval lipid storage (Diaz et al. 2023). In adults, deleting *tra* from female flies decreases the percent body fat in those animals and conversely, ectopically expressing *tra* in male flies increases their percent body fat (Wat et al. 2021). Moreover, *tra* regulation of triglyceride storage in adults can be localized to the fly nervous system as ectopic expression of *tra* localized specifically to the brain increases the percent body fat in males (Wat et al. 2021). Interestingly, decreasing *tra2* expression in the adult female fat body results in the accumulation of triglycerides (Mikoluk et al. 2018), suggesting that tra and tra2 may act differently or in different cell types to regulate lipid storage in female flies. Additional experiments designed to tease out the developmental- and tissue-specific roles of sex determination pathway genes are essential to better understand the regulation of the sexually dimorphic storage of lipids in flies throughout development.

2.4 microRNAs (miRNAs)

microRNAs (miRNAs) are small interfering RNAs common in insects that regulate their growth and development, behavior, and immunity (Lucas et al. 2015). miRNAs are generated first in a longer primary form in the nucleus and these primary miRNAs are exported to the cytoplasm where they are cleaved into small approximately 22 bp mature miRNAs that bind to target mRNAs inhibiting their translation, promoting their degradation, or altering their stability (Lucas et al. 2015). A number of miRNAs have been discovered that control lipid metabolism in insects (Redmond et al. 2019) and we will describe some of the major ones here.

2.4.1 miRNAs That Promote Lipid Storage

The miRNA *miR-33* promotes lipid storage in both mammals and insects (Gerin et al. 2010). Ubiquitous and fat body-specific overexpression of *miR-33* results in TAG accumulation (Redmond et al. 2019; Gerin et al. 2010; Davalos et al. 2011). *miR-33* can bind to and decrease the expression of the enzyme CPT1, which may account for the obese phenotype observed in *miR-33* overexpressing flies (Redmond et al. 2019; Gerin et al. 2010; Davalos et al. 2011). Interestingly, *miR-33* mutants also have an obese phenotype that is diet-dependent and are starvation resistant (Clerbaux et al. 2021). *miR-33* also targets the insulin signaling kinase gene *Akt1* as well as the DAG acyltransferase gene *midway* and lipogenic enzyme *ATPCL* (Clerbaux et al. 2021). Since genes involved in lipid synthesis (i.e., *midway* and *ATPCL*) and lipid breakdown (*whd*) are all targeted by *miR-33*, this might explain the inconsistencies in the overexpression and mutant phenotypes. Regardless, these data suggest that the expression of *miR-33* must be finely controlled in order to achieve lipid homeostasis.

A second miRNA that promotes lipid storage is *miR-278*. *miR-278* was initially identified in a *Drosophila* overexpression screen for growth modulators (Teleman et al. 2006). *miR-278* is highly expressed in the fly fat body and *miR-278* mutant adults are lean, suggesting that *miR-278* plays a role in promoting lipid storage (Redmond et al. 2019; Teleman et al. 2006). Interestingly, the expression of insulin-like peptides 2, 3, and 5 (*ilp2, ilp3, ilp5)* is increased in *miR-278* mutants (Teleman et al. 2006) raising the possibility that this microRNA may target the insulin signaling pathway to regulate lipid metabolism. Consistent with this hypothesis, the translational inhibitor, *4EBP*, and the insulin receptor, *dInR*, both insulin signaling-FOXO targets, were upregulated in *miR-278* mutants, indicating increased insulin signaling activity (Teleman et al. 2006). In addition, inhibiting insulin signaling by mutating the catalytic subunit of phosphatidylinositol 3-kinase, *dp110*, can partially rescue the lean phenotype of *miR-278* mutants (Teleman et al. 2006). Interestingly, *miR-278* was shown to target the gene *expanded*, a growth regulator of the hippo pathway (Boedigheimer et al. 1997). *expanded* mRNA levels were increased in *miR-278* mutants and overexpressing *expanded* using *miR-278-Gal4* phenocopies the lean phenotype observed in *miR-278* mutants (Teleman et al. 2006). Together these data suggest that *miR-278* promotes lipid storage by targeting the gene *expanded* and modulating insulin signaling activity.

Another miRNA that targets insulin pathway members to control lipid metabolism is *miR-277* in the mosquito *Aedes aegypti*. *miR-277* was deleted from the mosquito using CRISPR-Cas9 technology and this resulted in smaller lipid droplets and an overall decrease in TAG levels 24 h after a blood meal, indicating that *miR-277* promotes lipid storage (Ling et al. 2017). *miR-277* also targets insulin signaling to regulate lipid metabolism as FOXO accumulates in the cytoplasm in *miR-277*-depleted animals. Consistent with FOXO's cytoplasmic localization, two insulin-like peptides, *ilp7* and *ilp8*, have increased expression in *miR-277*-depleted animals (Ling et al. 2017). *miR-277*-depleted mosquitos also have ovarian defects and small primary follicles (Ling et al. 2017) indicating coordinated control of lipid metabolism and reproduction by *miR-277* in an insulin-signaling dependent manner.

2.4.2 miRNAs That Inhibit Lipid Storage

To identify miRNAs that control adiposity in *Drosophila*, an overexpression and loss-of-function screen was performed (Redmond et al. 2019). The microRNA *miR-969* was identified from this microRNA screen. Ubiquitous expression of *miR-969* resulted in blunted TAG storage while *miR-969* mutants have increased adiposity (Redmond et al. 2019). Gustatory receptor 47b (*Gr47b*) was identified as a target gene of *miR-969* as *Gr47b* levels were reduced when *miR-969* was overexpressed while *Gr47b* levels increased in *miR-969* mutants (Redmond et al. 2019). Interestingly, inducing RNAi to reduce *Gr47b* in neurons had no effect on TAG storage, but fat body-specific induction of *Gr47b-RNAi* resulted in a lean phenotype (Redmond et al. 2019). This suggests that *miR-969* targets *Gr47b* specifically in the fat body to regulate organismal adiposity.

A second miRNA that inhibits insect lipid storage is *miR-14*. *miR-14* was originally identified in a *Drosophila* screen for apoptosis modulators and while the investigators were analyzing *miR-14* mutants for cell death phenotypes, they noticed that these mutants had large lipid droplets and accumulate TAGs and DAGs while increasing *miR-14* copy number decreases TAGs and DAGs (Xu et al. 2003). However, the genes that *miR-14* targets to regulate lipid storage were not identified. Years later, another group continued studying *miR-14* and its effects on lipid metabolism. Interestingly, expressing *miR-14* in the fly fat body cannot rescue the obesity phenotype in *miR-14* mutants, but pan-neuronal expression of *miR-14* or even more selective expression in the insulin-producing cells (IPCs) can rescue the obesity phenotype (Varghese et al. 2010). The expression of the three insulin-like peptides that are expressed in the IPCs (*ilp2, ilp3,* and *ilp5*) is decreased in *miR-14* mutants and overexpression of *ilp3* in the IPCs can rescue the TAG accumulation phenotype observed in *miR-14* mutants (Varghese et al. 2010) indicating that *miR-14* regulates insulin levels to control lipid storage. In addition, the nutrient-responsive TF sugarbabe was identified as a *miR-14* target in the fly IPCs. *sugarbabe* mRNA levels were increased in *miR-14* mutants and decreasing *sugarbabe* specifically in the IPCs using RNAi partially rescues the obesity phenotype of the *miR-14* mutants (Varghese et al. 2010). In addition, overexpressing *sugarbabe* in the IPCs results in TAG accumulation and blunting of *ilp3* and *ilp5* expression (Varghese et al. 2010). Together, these data suggest that *miR-14* regulates lipid storage by modulating *sugarbabe* expression specifically in the IPCs, which then affects insulin signaling activity in the fly.

Another miRNA that acts in the insect brain to control lipid metabolism is *miR-210*. *miR-210* is expressed in the *Drosophila* eye and other sensory organs and *miR-210* mutants display retinal degeneration (Lyu et al. 2021). Interestingly, loss of *miR-210* results in lipid droplet accumulation in the retina and accumulation of TAGs and phospholipids in the head while DAGs decreased (Lyu et al. 2021). The retinal lipid droplet accumulation was dependent on SREBP as SREBP protein is cleaved and active in *miR-210* mutants and an *SREBP* mutant allele can partially suppress the lipid droplet and retinal degeneration phenotypes (Lyu et al. 2021). While a number of lipid metabolic enzyme genes (i.e., *FAS* and *ACC*) had increased expression in *miR-210* mutant retinas, *Acetyl Coenzyme A synthase* (*AcCoAS*) was shown to be a direct target of *miR-210*. *AcCoAS* levels increased in *miR-210* mutants and an *AcCoAS* mutant allele can partially suppress the retinal degeneration phenotype (but not the lipid droplet accumulation phenotype) (Lyu et al. 2021). These results highlight the role of *miR-210* in regulating lipid storage in the insect retina and ultimately retinal function.

The miRNA *mir-2765-3p* inhibits lipid storage in the Asian pest *Galeruca daurica*. Overexpression of *mir-2765-3p* decreases lipid storage and this may be due to decreased expression of the fatty acid synthase, *FAS*, and increased expression of the TAG lipase, *TGL1* (Duan et al. 2022). Conversely, depleting *mir-2765-3p* results in lipid accumulation that may be due to increased *FAS* and decreased *TGL1* expression (Duan et al. 2022). These changes in TAG storage and expression of lipid metabolic enzyme genes are due to *mir-2765-3p* directly targeting the insulin-responsive transcription factor, FOXO (Duan et al. 2022). In addition, increasing *mir-2765-3p* activity increased ovary size and delayed entry into diapause indicating coordinated regulation of lipid metabolism and reproduction by this microRNA (Duan et al. 2022).

3 Future Goals and Considerations

While much research has been performed to study the transcriptional control of lipid metabolism in insects, there is still much to be discovered. For example, in young *Drosophila* adults, many genes that encode for enzymes that act in the Kreb's cycle and the electron transport chain are upregulated (Beebe et al. 2020), but how this is regulated at the transcriptional level is not well understood. In addition, it is well-documented that the gut microbiome plays a role in regulating lipid storage in *Drosophila* and other insects (Didion et al. 2021; Kang and Douglas 2020); however, the role of transcriptional control in this microbiome-lipid storage interaction is poorly understood. Another field where links to lipid metabolism are being uncovered is immunity. In *Drosophila*, innate immune signaling has been shown to regulate lipid metabolism and altering phospholipid metabolism can affect an animal's ability to mount an immune response (Molaei et al. 2019; Martinez et al. 2020). The transcriptional control of these processes is only beginning to be elucidated. In addition, the expression of genes with roles in lipid metabolism can be altered in insects by changes in temperature and geographic location (Brown et al. 2018; Chang et al. 2020; Zhu et al. 2016), but how this is coupled to physiological changes is not yet well-established. Finally, there is great interest in understanding how individual organs such as the brain, fat body, muscle, and gut interact and communicate to regulate lipid metabolism in insects (Zhao and Karpac 2017; Al-Anzi et al. 2009; Palermo et al. 2022; Scopelliti et al. 2019; Zhao and Karpac 2020). However, to what extent this inter-organ communication is mediated by transcriptional mechanisms is not fully understood. These fields and others provide ripe areas for further inquiry into the transcriptional control of lipid metabolism in insects.

References

Abhyankar V, Kaduskar B, Kamat SS, Deobagkar D, Ratnaparkhi GS (2018) Drosophila DNA/RNA methyltransferase contributes to robust host defense in aging animals by regulating sphingolipid metabolism. J Exp Biol 221. https://doi.org/10.1242/jeb.187989

Ables ET, Bois KE, Garcia CA, Drummond-Barbosa D (2015) Ecdysone response gene E78 controls ovarian germline stem cell niche formation and follicle survival in Drosophila. Dev Biol 400:33–42. https://doi.org/10.1016/j.ydbio.2015.01.013

Abrams EW, Andrew DJ (2005) CrebA regulates secretory activity in the Drosophila salivary gland and epidermis. Development 132:2743–2758. https://doi.org/10.1242/dev.01863

Agrawal N, Delanoue R, Mauri A, Basco D, Pasco M, Thorens B, Leopold P (2016) The drosophila TNF Eiger is an Adipokine that acts on insulin-producing cells to mediate nutrient response. Cell Metab 23:675–684. https://doi.org/10.1016/j.cmet.2016.03.003

Al-Anzi B, Sapin V, Waters C, Zinn K, Wyman RJ, Benzer S (2009) Obesity-blocking neurons in Drosophila. Neuron 63:329–341. https://doi.org/10.1016/j.neuron.2009.07.021

Alic N, Giannakou ME, Papatheodorou I, Hoddinott MP, Andrews TD, Bolukbasi E, Partridge L (2014) Interplay of dFOXO and two ETS-family transcription factors determines lifespan in Drosophila melanogaster. PLoS Genet 10:e1004619. https://doi.org/10.1371/journal.pgen.1004619

Almeida-Oliveira F, Tuthill BF 2nd, Gondim KC, Majerowicz D, Musselman LP (2021) dHNF4 regulates lipid homeostasis and oogenesis in Drosophila melanogaster. Insect Biochem Mol Biol 133:103569. https://doi.org/10.1016/j.ibmb.2021.103569

Altarejos JY, Montminy M (2011) CREB and the CRTC co-activators: sensors for hormonal and metabolic signals. Nat Rev Mol Cell Biol 12:141–151. https://doi.org/10.1038/nrm3072

Alves-Bezerra M, Gondim KC (2012) Triacylglycerol biosynthesis occurs via the glycerol-3-phosphate pathway in the insect Rhodnius prolixus. Biochim Biophys Acta 1821:1462–1471. https://doi.org/10.1016/j.bbalip.2012.08.002

Amarneh B, Matthews KA, Rawson RB (2009) Activation of sterol regulatory element-binding protein by the caspase Drice in Drosophila larvae. J Biol Chem 284:9674–9682. https://doi.org/10.1074/jbc.M900346200

Andrew DJ, Baig A, Bhanot P, Smolik SM, Henderson KD (1997) The Drosophila dCREB-A gene is required for dorsal/ventral patterning of the larval cuticle. Development 124:181–193. https://doi.org/10.1242/dev.124.1.181

Baker KD, Shewchuk LM, Kozlova T, Makishima M, Hassell A, Wisely B, Caravella JA, Lambert MH, Reinking JL, Krause H, Thummel CS, Willson TM, Mangelsdorf DJ (2003) The Drosophila orphan nuclear receptor DHR38 mediates an atypical ecdysteroid signaling pathway. Cell 113:731–742. https://doi.org/10.1016/s0092-8674(03)00420-3

Banerjee KK, Ayyub C, Sengupta S, Kolthur-Seetharam U (2012) dSir2 deficiency in the fatbody, but not muscles, affects systemic insulin signaling, fat mobilization and starvation survival in flies. Aging (Albany NY) 4:206–223. https://doi.org/10.18632/aging.100435

Banerjee KK, Ayyub C, Sengupta S, Kolthur-Seetharam U (2013) Fat body dSir2 regulates muscle mitochondrial physiology and energy homeostasis nonautonomously and mimics the autonomous functions of dSir2 in muscles. Mol Cell Biol 33:252–264. https://doi.org/10.1128/MCB.00976-12

Barbosa S, Fasanella G, Carreira S, Llarena M, Fox R, Barreca C, Andrew D, O'Hare P (2013) An orchestrated program regulating secretory pathway genes and cargos by the transmembrane transcription factor CREB-H. Traffic 14:382–398. https://doi.org/10.1111/tra.12038

Barretto EC, Polan DM, Beevor-Potts AN, Lee B, Grewal SS (2020) Tolerance to hypoxia is promoted by FOXO regulation of the innate immunity transcription factor NF-kappaB/relish in Drosophila. Genetics 215:1013–1025. https://doi.org/10.1534/genetics.120.303219

Barry WE, Thummel CS (2016) The Drosophila HNF4 nuclear receptor promotes glucose-stimulated insulin secretion and mitochondrial function in adults. elife 5. https://doi.org/10.7554/eLife.11183

Bartok O, Teesalu M, Ashwall-Fluss R, Pandey V, Hanan M, Rovenko BM, Poukkula M, Havula E, Moussaieff A, Vodala S, Nahmias Y, Kadener S, Hietakangas V (2015) The transcription factor Cabut coordinates energy metabolism and the circadian clock in response to sugar sensing. EMBO J 34:1538–1553. https://doi.org/10.15252/embj.201591385

Bass J, Takahashi JS (2010) Circadian integration of metabolism and energetics. Science 330:1349–1354. https://doi.org/10.1126/science.1195027

Bauer R, Voelzmann A, Breiden B, Schepers U, Farwanah H, Hahn I, Eckardt F, Sandhoff K, Hoch M (2009) Schlank, a member of the ceramide synthase family controls growth and body fat in Drosophila. EMBO J 28:3706–3716. https://doi.org/10.1038/emboj.2009.305

Baumann AA, Benoit JB, Michalkova V, Mireji P, Attardo GM, Moulton JK, Wilson TG, Aksoy S (2013) Juvenile hormone and insulin suppress lipolysis between periods of lactation during tsetse fly pregnancy. Mol Cell Endocrinol 372:30–41. https://doi.org/10.1016/j.mce.2013.02.019

Baumbach J, Hummel P, Bickmeyer I, Kowalczyk KM, Frank M, Knorr K, Hildebrandt A, Riedel D, Jackle H, Kuhnlein RP (2014a) A Drosophila in vivo screen identifies store-operated calcium entry as a key regulator of adiposity. Cell Metab 19:331–343. https://doi.org/10.1016/j.cmet.2013.12.004

Baumbach J, Xu Y, Hehlert P, Kuhnlein RP (2014b) Galphaq, Ggamma1 and Plc21C control Drosophila body fat storage. J Genet Genomics 41:283–292. https://doi.org/10.1016/j.jgg.2014.03.005

Baumgartner BG, Orpinell M, Duran J, Ribas V, Burghardt HE, Bach D, Villar AV, Paz JC, Gonzalez M, Camps M, Oriola J, Rivera F, Palacin M, Zorzano A (2007) Identification of a novel modulator of thyroid hormone receptor-mediated action. PLoS One 2:e1183. https://doi.org/10.1371/journal.pone.0001183

Beebe K, Robins MM, Hernandez EJ, Lam G, Horner MA, Thummel CS (2020) Drosophila estrogen-related receptor directs a transcriptional switch that supports adult glycolysis and lipogenesis. Genes Dev 34:701–714. https://doi.org/10.1101/gad.335281.119

Begemann G, Michon AM, vd Voorn L, Wepf R, Mlodzik M (1995) The Drosophila orphan nuclear receptor seven-up requires the Ras pathway for its function in photoreceptor determination. Development 121:225–235. https://doi.org/10.1242/dev.121.1.225

Beller M, Sztalryd C, Southall N, Bell M, Jackle H, Auld DS, Oliver B (2008) COPI complex is a regulator of lipid homeostasis. PLoS Biol 6:e292. https://doi.org/10.1371/journal.pbio.0060292

Bender M, Imam FB, Talbot WS, Ganetzky B, Hogness DS (1997) Drosophila ecdysone receptor mutations reveal functional differences among receptor isoforms. Cell 91:777–788. https://doi.org/10.1016/s0092-8674(00)80466-3

Bennick RA, Nagengast AA, DiAngelo JR (2019) The SR proteins SF2 and RBP1 regulate triglyceride storage in the fat body of Drosophila. Biochem Biophys Res Commun 516:928–933. https://doi.org/10.1016/j.bbrc.2019.06.151

Bertolio R, Napoletano F, Mano M, Maurer-Stroh S, Fantuz M, Zannini A, Bicciato S, Sorrentino G, Del Sal G (2019) Sterol regulatory element binding protein 1 couples mechanical cues and lipid metabolism. Nat Commun 10:1326. https://doi.org/10.1038/s41467-019-09152-7

Bhogal JK, Kanaskie JM, DiAngelo JR (2020) The role of the heterogeneous nuclear ribonucleoprotein (hnRNP) Hrb27C in regulating lipid storage in the Drosophila fat body. Biochem Biophys Res Commun 524:178–183. https://doi.org/10.1016/j.bbrc.2020.01.064

Billin AN, Ayer DE (2006) The Mlx network: evidence for a parallel max-like transcriptional network that regulates energy metabolism. Curr Top Microbiol Immunol 302:255–278. https://doi.org/10.1007/3-540-32952-8_10

Binh TD, Pham TLA, Men TT, Dang TTP, Kamei K (2019) LSD-2 dysfunction induces dFoxO-dependent cell death in the wing of Drosophila melanogaster. Biochem Biophys Res Commun 509:491–497. https://doi.org/10.1016/j.bbrc.2018.12.132

Birse RT, Choi J, Reardon K, Rodriguez J, Graham S, Diop S, Ocorr K, Bodmer R, Oldham S (2010) High-fat-diet-induced obesity and heart dysfunction are regulated by the TOR pathway in Drosophila. Cell Metab 12:533–544. https://doi.org/10.1016/j.cmet.2010.09.014

Boedigheimer MJ, Nguyen KP, Bryant PJ (1997) Expanded functions in the apical cell domain to regulate the growth rate of imaginal discs. Dev Genet 20:103–110

Brown S, Castelli-Gair Hombria J (2000) Drosophila grain encodes a GATA transcription factor required for cell rearrangement during morphogenesis. Development 127:4867–4876. https://doi.org/10.1242/dev.127.22.4867

Brown EB, Torres J, Bennick RA, Rozzo V, Kerbs A, DiAngelo JR, Keene AC (2018) Variation in sleep and metabolic function is associated with latitude and average temperature in Drosophila melanogaster. Ecol Evol 8:4084–4097. https://doi.org/10.1002/ece3.3963

Bujold M, Gopalakrishnan A, Nally E, King-Jones K (2010) Nuclear receptor DHR96 acts as a sentinel for low cholesterol concentrations in Drosophila melanogaster. Mol Cell Biol 30:793–805. https://doi.org/10.1128/MCB.01327-09

Bulow MH, Wingen C, Senyilmaz D, Gosejacob D, Sociale M, Bauer R, Schulze H, Sandhoff K, Teleman AA, Hoch M, Sellin J (2018) Unbalanced lipolysis results in lipotoxicity and mitochondrial damage in peroxisome-deficient Pex19 mutants. Mol Biol Cell 29:396–407. https://doi.org/10.1091/mbc.E17-08-0535

Chakrabarti S, Poidevin M, Lemaitre B (2014) The Drosophila MAPK p38c regulates oxidative stress and lipid homeostasis in the intestine. PLoS Genet 10:e1004659. https://doi.org/10.1371/journal.pgen.1004659

Chang HC, Guarente L (2014) SIRT1 and other sirtuins in metabolism. Trends Endocrinol Metab 25:138–145. https://doi.org/10.1016/j.tem.2013.12.001

Chang YW, Zhang XX, Lu MX, Gong WR, Du YZ (2020) Transcriptome analysis of Liriomyza trifolii (Diptera: Agromyzidae) in response to temperature stress. Comp Biochem Physiol Part D Genomics Proteomics 34:100677. https://doi.org/10.1016/j.cbd.2020.100677

Charidemou E, Tsiarli MA, Theophanous A, Yilmaz V, Pitsouli C, Strati K, Griffin JL, Kirmizis A (2022)

Histone acetyltransferase NAA40 modulates acetyl-CoA levels and lipid synthesis. BMC Biol 20:22. https://doi.org/10.1186/s12915-021-01225-8

Charroux B, Royet J (2022) Gut-derived peptidoglycan remotely inhibits bacteria dependent activation of SREBP by Drosophila adipocytes. PLoS Genet 18: e1010098. https://doi.org/10.1371/journal.pgen.1010098

Chatterjee D, Katewa SD, Qi Y, Jackson SA, Kapahi P, Jasper H (2014) Control of metabolic adaptation to fasting by dILP6-induced insulin signaling in Drosophila oenocytes. Proc Natl Acad Sci USA 111: 17959–17964. https://doi.org/10.1073/pnas.1409241111

Chen K, Lin G, Haelterman NA, Ho TS, Li T, Li Z, Duraine L, Graham BH, Jaiswal M, Yamamoto S, Rasband MN, Bellen HJ (2016) Loss of Frataxin induces iron toxicity, sphingolipid synthesis, and Pdk1/Mef2 activation, leading to neurodegeneration. elife 5. https://doi.org/10.7554/eLife.16043

Chen C, Mahar R, Merritt ME, Denlinger DL, Hahn DA (2021) ROS and hypoxia signaling regulate periodic metabolic arousal during insect dormancy to coordinate glucose, amino acid, and lipid metabolism. Proc Natl Acad Sci USA 118. https://doi.org/10.1073/pnas.2017603118

Cheng Y, Li Y, Li W, Song Y, Zeng R, Lu K (2020) Effect of hepatocyte nuclear factor 4 on the fecundity of Nilaparvata lugens: insights from RNA interference combined with transcriptomic analysis. Genomics 112:4585–4594. https://doi.org/10.1016/j.ygeno.2020.08.002

Chiang YN, Tan KJ, Chung H, Lavrynenko O, Shevchenko A, Yew JY (2016) Steroid hormone signaling is essential for pheromone production and Oenocyte survival. PLoS Genet 12:e1006126. https://doi.org/10.1371/journal.pgen.1006126

Cho Y, Griswold A, Campbell C, Min KT (2005) Individual histone deacetylases in Drosophila modulate transcription of distinct genes. Genomics 86:606–617. https://doi.org/10.1016/j.ygeno.2005.07.007

Choi S, Kim W, Chung J (2011) Drosophila salt-inducible kinase (SIK) regulates starvation resistance through cAMP-response element-binding protein (CREB)-regulated transcription coactivator (CRTC). J Biol Chem 286:2658–2664. https://doi.org/10.1074/jbc.C110.119222

Choi S, Lim DS, Chung J (2015) Feeding and fasting signals converge on the LKB1-SIK3 pathway to regulate lipid metabolism in Drosophila. PLoS Genet 11: e1005263. https://doi.org/10.1371/journal.pgen.1005263

Clerbaux LA, Schultz H, Roman-Holba S, Ruan DF, Yu R, Lamb AM, Bommer GT, Kennell JA (2021) The microRNA miR-33 is a pleiotropic regulator of metabolic and developmental processes in Drosophila melanogaster. Dev Dyn 250:1634–1650. https://doi.org/10.1002/dvdy.344

Colombani J, Bianchini L, Layalle S, Pondeville E, Dauphin-Villemant C, Antoniewski C, Carre C, Noselli S, Leopold P (2005) Antagonistic actions of ecdysone and insulins determine final size in Drosophila. Science 310:667–670. https://doi.org/10.1126/science.1119432

Dahlgaard K, Jung A, Qvortrup K, Clausen H, Kjaerulff O, Wandall HH (2012) Neurofibromatosis-like phenotype in Drosophila caused by lack of glucosylceramide extension. Proc Natl Acad Sci USA 109:6987–6992. https://doi.org/10.1073/pnas.1115453109

Davalos A, Goedeke L, Smibert P, Ramirez CM, Warrier NP, Andreo U, Cirera-Salinas D, Rayner K, Suresh U, Pastor-Pareja JC, Esplugues E, Fisher EA, Penalva LO, Moore KJ, Suarez Y, Lai EC, Fernandez-Hernando C (2011) miR-33a/b contribute to the regulation of fatty acid metabolism and insulin signaling. Proc Natl Acad Sci USA 108:9232–9237. https://doi.org/10.1073/pnas.1102281108

de Vera IM, Giri PK, Munoz-Tello P, Brust R, Fuhrmann J, Matta-Camacho E, Shang J, Campbell S, Wilson HD, Granados J, Gardner WJ Jr, Creamer TP, Solt LA, Kojetin DJ (2016) Identification of a binding site for unsaturated fatty acids in the orphan nuclear receptor Nurr1. ACS Chem Biol 11:1795–1799. https://doi.org/10.1021/acschembio.6b00037

Dekanty A, Lavista-Llanos S, Irisarri M, Oldham S, Wappner P (2005) The insulin-PI3K/TOR pathway induces a HIF-dependent transcriptional response in Drosophila by promoting nuclear localization of HIF-alpha/Sima. J Cell Sci 118:5431–5441. https://doi.org/10.1242/jcs.02648

DiAngelo JR, Birnbaum MJ (2009) Regulation of fat cell mass by insulin in Drosophila melanogaster. Mol Cell Biol 29:6341–6352. https://doi.org/10.1128/MCB.00675-09

DiAngelo JR, Bland ML, Bambina S, Cherry S, Birnbaum MJ (2009) The immune response attenuates growth and nutrient storage in Drosophila by reducing insulin signaling. Proc Natl Acad Sci USA 106:20853–20858. https://doi.org/10.1073/pnas.0906749106

DiAngelo JR, Erion R, Crocker A, Sehgal A (2011) The central clock neurons regulate lipid storage in Drosophila. PLoS One 6:e19921. https://doi.org/10.1371/journal.pone.0019921

Diaz AV, Stephenson D, Nemkov T, D'Alessandro A, Reis T (2023) Spenito-dependent metabolic sexual dimorphism intrinsic to fat storage cells. Genetics 225. https://doi.org/10.1093/genetics/iyad164

Didion EM, Sabree ZL, Kenyon L, Nine G, Hagan RW, Osman S, Benoit JB (2021) Microbiome reduction prevents lipid accumulation during early diapause in the northern house mosquito, Culex pipiens pipiens. J Insect Physiol 134:104295. https://doi.org/10.1016/j.jinsphys.2021.104295

Diop SB, Bisharat-Kernizan J, Birse RT, Oldham S, Ocorr K, Bodmer R (2015) PGC-1/Spargel counteracts high-fat-diet-induced obesity and cardiac lipotoxicity downstream of TOR and Brummer ATGL lipase. Cell Rep 10:1572–1584. https://doi.org/10.1016/j.celrep.2015.02.022

Dobrosotskaya IY, Seegmiller AC, Brown MS, Goldstein JL, Rawson RB (2002) Regulation of SREBP processing and membrane lipid production by phospholipids in Drosophila. Science 296:879–883. https://doi.org/10.1126/science.1071124

Dobrosotskaya IY, Goldstein JL, Brown MS, Rawson RB (2003) Reconstitution of sterol-regulated endoplasmic reticulum-to-Golgi transport of SREBP-2 in insect cells by co-expression of mammalian SCAP and Insigs. J Biol Chem 278:35837–35843. https://doi.org/10.1074/jbc.M306476200

Duan TF, Li L, Wang HC, Pang BP (2022) MicroRNA miR-2765-3p regulates reproductive diapause by targeting FoxO in Galeruca daurica. Insect Sci 30: 279. https://doi.org/10.1111/1744-7917.13089

Fahrbach SE, Smagghe G, Velarde RA (2012) Insect nuclear receptors. Annu Rev Entomol 57:83–106. https://doi.org/10.1146/annurev-ento-120710-100607

Fisk GJ, Thummel CS (1995) Isolation, regulation, and DNA-binding properties of three Drosophila nuclear hormone receptor superfamily members. Proc Natl Acad Sci USA 92:10604–10608. https://doi.org/10.1073/pnas.92.23.10604

Fox RM, Hanlon CD, Andrew DJ (2010) The CrebA/Creb3-like transcription factors are major and direct regulators of secretory capacity. J Cell Biol 191:479–492. https://doi.org/10.1083/jcb.201004062

Francis VA, Zorzano A, Teleman AA (2010) dDOR is an EcR coactivator that forms a feed-forward loop connecting insulin and ecdysone signaling. Curr Biol 20:1799–1808. https://doi.org/10.1016/j.cub.2010.08.055

Gao X, Fu Y, Ajayi OE, Guo D, Zhang L, Wu Q (2019) Identification of genes underlying phenotypic plasticity of wing size via insulin signaling pathway by network-based analysis in Sogatella furcifera. BMC Genomics 20:396. https://doi.org/10.1186/s12864-019-5793-z

Gao L, Qiao H, Wei P, Moussian B, Wang Y (2022) Xenobiotic responses in insects. Arch Insect Biochem Physiol 109:e21869. https://doi.org/10.1002/arch.21869

Gerin I, Clerbaux LA, Haumont O, Lanthier N, Das AK, Burant CF, Leclercq IA, MacDougald OA, Bommer GT (2010) Expression of miR-33 from an SREBP2 intron inhibits cholesterol export and fatty acid oxidation. J Biol Chem 285:33652–33661. https://doi.org/10.1074/jbc.M110.152090

Gershman B, Puig O, Hang L, Peitzsch RM, Tatar M, Garofalo RS (2007) High-resolution dynamics of the transcriptional response to nutrition in Drosophila: a key role for dFOXO. Physiol Genomics 29:24–34. https://doi.org/10.1152/physiolgenomics.00061.2006

Gingras RM, Warren ME, Nagengast AA, Diangelo JR (2014) The control of lipid metabolism by mRNA splicing in Drosophila. Biochem Biophys Res Commun 443:672–676. https://doi.org/10.1016/j.bbrc.2013.12.027

Giraudo M, Audant P, Feyereisen R, Le Goff G (2013) Nuclear receptors HR96 and ultraspiracle from the fall armyworm (Spodoptera frugiperda), developmental expression and induction by xenobiotics. J Insect Physiol 59:560–568. https://doi.org/10.1016/j.jinsphys.2013.03.003

Guo Y, Walther TC, Rao M, Stuurman N, Goshima G, Terayama K, Wong JS, Vale RD, Walter P, Farese RV (2008) Functional genomic screen reveals genes involved in lipid-droplet formation and utilization. Nature 453:657–661. https://doi.org/10.1038/nature06928

Guo S, Tian Z, Zhu F, Liu W, Wang XP (2021) Lipin modulates lipid metabolism during reproduction in the cabbage beetle. Insect Biochem Mol Biol 139:103668. https://doi.org/10.1016/j.ibmb.2021.103668

Gutierrez E, Wiggins D, Fielding B, Gould AP (2007) Specialized hepatocyte-like cells regulate Drosophila lipid metabolism. Nature 445:275–280. https://doi.org/10.1038/nature05382

Harsh S, Eleftherianos I (2020) Tumor induction in Drosophila imaginal epithelia triggers modulation of fat body lipid droplets. Biochimie 179:65–68. https://doi.org/10.1016/j.biochi.2020.09.011

Havula E, Teesalu M, Hyotylainen T, Seppala H, Hasygar K, Auvinen P, Oresic M, Sandmann T, Hietakangas V (2013) Mondo/ChREBP-Mlx-regulated transcriptional network is essential for dietary sugar tolerance in Drosophila. PLoS Genet 9:e1003438. https://doi.org/10.1371/journal.pgen.1003438

Hill RJ, Billas IM, Bonneton F, Graham LD, Lawrence MC (2013) Ecdysone receptors: from the Ashburner model to structural biology. Annu Rev Entomol 58: 251–271. https://doi.org/10.1146/annurev-ento-120811-153610

Hole K, Van Damme P, Dalva M, Aksnes H, Glomnes N, Varhaug JE, Lillehaug JR, Gevaert K, Arnesen T (2011) The human N-alpha-acetyltransferase 40 (hNaa40p/hNatD) is conserved from yeast and N-terminally acetylates histones H2A and H4. PLoS One 6:e24713. https://doi.org/10.1371/journal.pone.0024713

Hood SE, Kofler XV, Chen Q, Scott J, Ortega J, Lehmann M (2020) Nuclear translocation ability of Lipin differentially affects gene expression and survival in fed and fasting Drosophila. J Lipid Res 61:1720–1732. https://doi.org/10.1194/jlr.RA120001051

Horner MA, Pardee K, Liu S, King-Jones K, Lajoie G, Edwards A, Krause HM, Thummel CS (2009) The Drosophila DHR96 nuclear receptor binds cholesterol and regulates cholesterol homeostasis. Genes Dev 23: 2711–2716. https://doi.org/10.1101/gad.1833609

Hossain MS, Liu Y, Zhou S, Li K, Tian L, Li S (2013) 20-Hydroxyecdysone-induced transcriptional activity of FoxO upregulates brummer and acid lipase-1 and promotes lipolysis in Bombyx fat body. Insect Biochem Mol Biol 43:829–838. https://doi.org/10.1016/j.ibmb.2013.06.007

Hou Y, Wang XL, Saha TT, Roy S, Zhao B, Raikhel AS, Zou Z (2015) Temporal coordination of carbohydrate metabolism during mosquito reproduction. PLoS Genet 11:e1005309. https://doi.org/10.1371/journal.pgen.1005309

Hwangbo DS, Gershman B, Tu MP, Palmer M, Tatar M (2004) Drosophila dFOXO controls lifespan and regulates insulin signalling in brain and fat body. Nature 429:562–566. https://doi.org/10.1038/nature02549

Iijima K, Zhao L, Shenton C, Iijima-Ando K (2009) Regulation of energy stores and feeding by neuronal and peripheral CREB activity in Drosophila. PLoS One 4:e8498. https://doi.org/10.1371/journal.pone.0008498

Junger MA, Rintelen F, Stocker H, Wasserman JD, Vegh M, Radimerski T, Greenberg ME, Hafen E (2003) The Drosophila forkhead transcription factor FOXO mediates the reduction in cell number associated with reduced insulin signaling. J Biol 2:20. https://doi.org/10.1186/1475-4924-2-20

Kamiyama T, Niwa R (2022) Transcriptional regulators of Ecdysteroid biosynthetic enzymes and their roles in insect development. Front Physiol 13:823418. https://doi.org/10.3389/fphys.2022.823418

Kamoshida Y, Fujiyama-Nakamura S, Kimura S, Suzuki E, Lim J, Shiozaki-Sato Y, Kato S, Takeyama K (2012) Ecdysone receptor (EcR) suppresses lipid accumulation in the Drosophila fat body via transcription control. Biochem Biophys Res Commun 421:203–207. https://doi.org/10.1016/j.bbrc.2012.03.135

Kanai MI, Okabe M, Hiromi Y (2005) Seven-up controls switching of transcription factors that specify temporal identities of Drosophila neuroblasts. Dev Cell 8:203–213. https://doi.org/10.1016/j.devcel.2004.12.014

Kang D, Douglas AE (2020) Functional traits of the gut microbiome correlated with host lipid content in a natural population of Drosophila melanogaster. Biol Lett 16:20190803. https://doi.org/10.1098/rsbl.2019.0803

Kang P, Chang K, Liu Y, Bouska M, Birnbaum A, Karashchuk G, Thakore R, Zheng W, Post S, Brent CS, Li S, Tatar M, Bai H (2017) Drosophila Kruppel homolog 1 represses lipolysis through interaction with dFOXO. Sci Rep 7:16369. https://doi.org/10.1038/s41598-017-16638-1

Karpac J, Biteau B, Jasper H (2013) Misregulation of an adaptive metabolic response contributes to the age-related disruption of lipid homeostasis in Drosophila. Cell Rep 4:1250–1261. https://doi.org/10.1016/j.celrep.2013.08.004

Katewa SD, Akagi K, Bose N, Rakshit K, Camarella T, Zheng X, Hall D, Davis S, Nelson CS, Brem RB, Ramanathan A, Sehgal A, Giebultowicz JM, Kapahi P (2016) Peripheral circadian clocks mediate dietary restriction-dependent changes in lifespan and fat metabolism in Drosophila. Cell Metab 23:143–154. https://doi.org/10.1016/j.cmet.2015.10.014

Kerber B, Fellert S, Hoch M (1998) Seven-up, the Drosophila homolog of the COUP-TF orphan receptors, controls cell proliferation in the insect kidney. Genes Dev 12:1781–1786. https://doi.org/10.1101/gad.12.12.1781

King-Jones K, Thummel CS (2005) Nuclear receptors—a perspective from Drosophila. Nat Rev Genet 6:311–323. https://doi.org/10.1038/nrg1581

King-Jones K, Horner MA, Lam G, Thummel CS (2006) The DHR96 nuclear receptor regulates xenobiotic responses in Drosophila. Cell Metab 4:37–48. https://doi.org/10.1016/j.cmet.2006.06.006

Kohyama-Koganeya A, Nabetani T, Miura M, Hirabayashi Y (2011) Glucosylceramide synthase in the fat body controls energy metabolism in Drosophila. J Lipid Res 52:1392–1399. https://doi.org/10.1194/jlr.M014266

Kokki K, Lamichane N, Nieminen AI, Ruhanen H, Morikka J, Robciuc M, Rovenko BM, Havula E, Kakela R, Hietakangas V (2021) Metabolic gene regulation by Drosophila GATA transcription factor grain. PLoS Genet 17:e1009855. https://doi.org/10.1371/journal.pgen.1009855

Kolasa AM, Bhogal JK, DiAngelo JR (2021) The heterogeneous nuclear ribonucleoprotein (hnRNP) glorund functions in the Drosophila fat body to regulate lipid storage and transport. Biochem Biophys Rep 25:100919. https://doi.org/10.1016/j.bbrep.2021.100919

Kornblihtt AR, Schor IE, Allo M, Dujardin G, Petrillo E, Munoz MJ (2013) Alternative splicing: a pivotal step between eukaryotic transcription and translation. Nat Rev Mol Cell Biol 14:153–165. https://doi.org/10.1038/nrm3525

Koyama T, Rodrigues MA, Athanasiadis A, Shingleton AW, Mirth CK (2014) Nutritional control of body size through FoxO-Ultraspiracle mediated ecdysone biosynthesis. elife 3. https://doi.org/10.7554/eLife.03091

Kozlova T, Thummel CS (2003) Essential roles for ecdysone signaling during Drosophila mid-embryonic development. Science 301:1911–1914. https://doi.org/10.1126/science.1087419

Kozlova T, Lam G, Thummel CS (2009) Drosophila DHR38 nuclear receptor is required for adult cuticle integrity at eclosion. Dev Dyn 238:701–707. https://doi.org/10.1002/dvdy.21860

Kunte AS, Matthews KA, Rawson RB (2006) Fatty acid auxotrophy in Drosophila larvae lacking SREBP. Cell Metab 3:439–448. https://doi.org/10.1016/j.cmet.2006.04.011

Lafont R, Balducci C, Dinan L (2021) Ecdysteroids. Encyclopedia 1:1267–1302

Landis GN, Hilsabeck TAU, Bell HS, Ronnen-Oron T, Wang L, Doherty DV, Tejawinata FI, Erickson K, Vu W, Promislow DEL, Kapahi P, Tower J (2021) Mifepristone increases life span of virgin female Drosophila on regular and high-fat diet without reducing food intake. Front Genet 12:751647. https://doi.org/10.3389/fgene.2021.751647

Leader DP, Krause SA, Pandit A, Davies SA, Dow JAT (2018) FlyAtlas 2: a new version of the Drosophila melanogaster expression atlas with RNA-Seq, miRNA-Seq and sex-specific data. Nucleic Acids Res 46:D809–D815. https://doi.org/10.1093/nar/gkx976

Lee JH, Budanov AV, Park EJ, Birse R, Kim TE, Perkins GA, Ocorr K, Ellisman MH, Bodmer R, Bier E, Karin M (2010) Sestrin as a feedback inhibitor of TOR that prevents age-related pathologies. Science 327:1223–1228. https://doi.org/10.1126/science.1182228

Lee B, Barretto EC, Grewal SS (2019) TORC1 modulation in adipose tissue is required for organismal adaptation to hypoxia in Drosophila. Nat Commun 10:1878. https://doi.org/10.1038/s41467-019-09643-7

Lehmann M (2021) Diverse roles of phosphatidate phosphatases in insect development and metabolism. Insect Biochem Mol Biol 133:103469. https://doi.org/10.1016/j.ibmb.2020.103469

Li Y, Padmanabha D, Gentile LB, Dumur CI, Beckstead RB, Baker KD (2013) HIF- and non-HIF-regulated hypoxic responses require the estrogen-related receptor in Drosophila melanogaster. PLoS Genet 9:e1003230. https://doi.org/10.1371/journal.pgen.1003230

Li X, Zhang M, Liu M, Liu TH, Hemba-Waduge RU, Ji JY (2022) Cdk8 attenuates lipogenesis by inhibiting SREBP-dependent transcription in Drosophila. Dis Model Mech 15. https://doi.org/10.1242/dmm.049650

Lim HY, Wang W, Wessells RJ, Ocorr K, Bodmer R (2011) Phospholipid homeostasis regulates lipid metabolism and cardiac function through SREBP signaling in Drosophila. Genes Dev 25:189–200. https://doi.org/10.1101/gad.1992411

Lin XW, Tang L, Yang J, Xu WH (2016) HIF-1 regulates insect lifespan extension by inhibiting c-Myc-TFAM signaling and mitochondrial biogenesis. Biochim Biophys Acta 1863:2594–2603. https://doi.org/10.1016/j.bbamcr.2016.07.007

Lin X, Yu N, Smagghe G (2018) FoxO mediates the timing of pupation through regulating ecdysteroid biosynthesis in the red flour beetle, Tribolium castaneum. Gen Comp Endocrinol 258:149–156. https://doi.org/10.1016/j.ygcen.2017.05.012

Ling L, Kokoza VA, Zhang C, Aksoy E, Raikhel AS (2017) MicroRNA-277 targets insulin-like peptides 7 and 8 to control lipid metabolism and reproduction in Aedes aegypti mosquitoes. Proc Natl Acad Sci USA 114:E8017–E8024. https://doi.org/10.1073/pnas.1710970114

Liu L, Zhang K, Sandoval H, Yamamoto S, Jaiswal M, Sanz E, Li Z, Hui J, Graham BH, Quintana A, Bellen HJ (2015) Glial lipid droplets and ROS induced by mitochondrial defects promote neurodegeneration. Cell 160:177–190. https://doi.org/10.1016/j.cell.2014.12.019

Liu T, Xu X, An F, Zhu W, Luo D, Liu S, Wei G, Wang L (2022) Functional analysis of nuclear receptor HR96 gene in Bombyx mori exposed to phoxim. Arch Insect Biochem Physiol 111:e21910. https://doi.org/10.1002/arch.21910

Lo PC, Frasch M (2001) A role for the COUP-TF-related gene seven-up in the diversification of cardioblast identities in the dorsal vessel of Drosophila. Mech Dev 104:49–60. https://doi.org/10.1016/s0925-4773(01)00361-6

Long W, Wu J, Shen G, Zhang H, Liu H, Xu Y, Gu J, Jia L, Lin Y, Xia Q (2020) Estrogen-related receptor participates in regulating glycolysis and influences embryonic development in silkworm Bombyx mori. Insect Mol Biol 29:160–169. https://doi.org/10.1111/imb.12619

Lu K, Chen X, Li Y, Li W, Zhou Q (2018) Lipophorin receptor regulates Nilaparvata lugens fecundity by promoting lipid accumulation and vitellogenin biosynthesis. Comp Biochem Physiol A Mol Integr Physiol 219-220:28–37. https://doi.org/10.1016/j.cbpa.2018.02.008

Lucas KJ, Zhao B, Liu S, Raikhel AS (2015) Regulation of physiological processes by microRNAs in insects. Curr Opin Insect Sci 11:1–7. https://doi.org/10.1016/j.cois.2015.06.004

Luo J, Ting CY, Li Y, McQueen P, Lin TY, Hsu CP, Lee CH (2020) Antagonistic regulation by insulin-like peptide and activin ensures the elaboration of appropriate dendritic field sizes of amacrine neurons. elife 9. https://doi.org/10.7554/eLife.50568

Luong N, Davies CR, Wessells RJ, Graham SM, King MT, Veech R, Bodmer R, Oldham SM (2006) Activated FOXO-mediated insulin resistance is blocked by reduction of TOR activity. Cell Metab 4:133–142. https://doi.org/10.1016/j.cmet.2006.05.013

Lyu J, Chen Y, Yang W, Guo T, Xu X, Xi Y, Yang X, Ge W (2021) The conserved microRNA miR-210 regulates lipid metabolism and photoreceptor maintenance in the Drosophila retina. Cell Death Differ 28:764–779. https://doi.org/10.1038/s41418-020-00622-w

Martinez BA, Hoyle RG, Yeudall S, Granade ME, Harris TE, Castle JD, Leitinger N, Bland ML (2020) Innate immune signaling in Drosophila shifts anabolic lipid metabolism from triglyceride storage to phospholipid synthesis to support immune function. PLoS Genet 16:e1009192. https://doi.org/10.1371/journal.pgen.1009192

Matthews KA, Kunte AS, Tambe-Ebot E, Rawson RB (2009) Alternative processing of sterol regulatory element binding protein during larval development in Drosophila melanogaster. Genetics 181:119–128. https://doi.org/10.1534/genetics.108.093450

Mattila J, Havula E, Suominen E, Teesalu M, Surakka I, Hynynen R, Kilpinen H, Vaananen J, Hovatta I, Kakela R, Ripatti S, Sandmann T, Hietakangas V (2015) Mondo-Mlx mediates organismal sugar sensing through the Gli-similar transcription factor Sugarbabe. Cell Rep 13:350–364. https://doi.org/10.1016/j.celrep.2015.08.081

Mauvezin C, Orpinell M, Francis VA, Mansilla F, Duran J, Ribas V, Palacin M, Boya P, Teleman AA, Zorzano A (2010) The nuclear cofactor DOR regulates autophagy in mammalian and Drosophila cells. EMBO Rep 11: 37–44. https://doi.org/10.1038/embor.2009.242

Meltzer S, Bagley JA, Perez GL, O'Brien CE, DeVault L, Guo Y, Jan LY, Jan YN (2017) Phospholipid homeostasis regulates dendrite morphogenesis in Drosophila sensory neurons. Cell Rep 21:859–866. https://doi.org/10.1016/j.celrep.2017.09.089

Mikoluk C, Nagengast AA, DiAngelo JR (2018) The splicing factor transformer2 (tra2) functions in the Drosophila fat body to regulate lipid storage. Biochem Biophys Res Commun 495:1528–1533. https://doi.org/10.1016/j.bbrc.2017.12.002

Mlodzik M, Hiromi Y, Weber U, Goodman CS, Rubin GM (1990) The Drosophila seven-up gene, a member of the steroid receptor gene superfamily, controls photoreceptor cell fates. Cell 60:211–224. https://doi.org/10.1016/0092-8674(90)90737-y

Molaei M, Vandehoef C, Karpac J (2019) NF-kappaB shapes metabolic adaptation by attenuating Foxo-mediated lipolysis in Drosophila. Dev Cell 49:802–810. e806. https://doi.org/10.1016/j.devcel.2019.04.009

Munoz-Descalzo S, Terol J, Paricio N (2005) Cabut, a C2H2 zinc finger transcription factor, is required during Drosophila dorsal closure downstream of JNK signaling. Dev Biol 287:168–179. https://doi.org/10.1016/j.ydbio.2005.08.048

Musselman LP, Fink JL, Narzinski K, Ramachandran PV, Hathiramani SS, Cagan RL, Baranski TJ (2011) A high-sugar diet produces obesity and insulin resistance in wild-type Drosophila. Dis Model Mech 4:842–849. https://doi.org/10.1242/dmm.007948

Musselman LP, Fink JL, Ramachandran PV, Patterson BW, Okunade AL, Maier E, Brent MR, Turk J, Baranski TJ (2013) Role of fat body lipogenesis in protection against the effects of caloric overload in Drosophila. J Biol Chem 288:8028–8042. https://doi.org/10.1074/jbc.M112.371047

Musselman LP, Fink JL, Maier EJ, Gatto JA, Brent MR, Baranski TJ (2018) Seven-up is a novel regulator of insulin signaling. Genetics 208:1643–1656. https://doi.org/10.1534/genetics.118.300770

Nagle C, Bhogal JK, Nagengast AA, DiAngelo JR (2022) Transportin-serine/arginine-rich (Tnpo-SR) proteins are necessary for proper lipid storage in the Drosophila fat body. Biochem Biophys Res Commun 596:1–5. https://doi.org/10.1016/j.bbrc.2022.01.087

Nascimento PVP, Almeida-Oliveira F, Macedo-Silva A, Ausina P, Motinha C, Sola-Penna M, Majerowicz D (2021) Gene annotation of nuclear receptor superfamily genes in the kissing bug Rhodnius prolixus and the effects of 20-hydroxyecdysone on lipid metabolism. Insect Mol Biol 30:297–314. https://doi.org/10.1111/imb.12696

Nassel DR, Vanden Broeck J (2016) Insulin/IGF signaling in Drosophila and other insects: factors that regulate production, release and post-release action of the insulin-like peptides. Cell Mol Life Sci 73:271–290. https://doi.org/10.1007/s00018-015-2063-3

Ngo TT, Bennett MK, Bourgeois AL, Toth JI, Osborne TF (2002) A role for cyclic AMP response element-binding protein (CREB) but not the highly similar ATF-2 protein in sterol regulation of the promoter for 3-hydroxy-3-methylglutaryl coenzyme a reductase. J Biol Chem 277:33901–33905. https://doi.org/10.1074/jbc.M202135200

Nijhout HF, Riddiford LM, Mirth C, Shingleton AW, Suzuki Y, Callier V (2014) The developmental control of size in insects. Wiley Interdiscip Rev Dev Biol 3: 113–134. https://doi.org/10.1002/wdev.124

Nirala NK, Rahman M, Walls SM, Singh A, Zhu LJ, Bamba T, Fukusaki E, Srideshikan SM, Harris GL, Ip YT, Bodmer R, Acharya UR (2013) Survival response to increased ceramide involves metabolic adaptation through novel regulators of glycolysis and lipolysis. PLoS Genet 9:e1003556. https://doi.org/10.1371/journal.pgen.1003556

Obniski R, Sieber M, Spradling AC (2018) Dietary lipids modulate notch signaling and influence adult intestinal development and metabolism in Drosophila. Dev Cell 47:98–111. e115. https://doi.org/10.1016/j.devcel.2018.08.013

Okamura T, Shimizu H, Nagao T, Ueda R, Ishii S (2007) ATF-2 regulates fat metabolism in Drosophila. Mol Biol Cell 18:1519–1529. https://doi.org/10.1091/mbc.e06-10-0909

Palanker L, Tennessen JM, Lam G, Thummel CS (2009) Drosophila HNF4 regulates lipid mobilization and beta-oxidation. Cell Metab 9:228–239. https://doi.org/10.1016/j.cmet.2009.01.009

Palermo J, Keene AC, DiAngelo JR (2022) Expression of a constitutively active insulin receptor in drosulfakinin (Dsk) neurons regulates metabolism and sleep in Drosophila. Biochem Biophys Rep 30:101280. https://doi.org/10.1016/j.bbrep.2022.101280

Palu RA, Thummel CS (2016) Sir2 acts through hepatocyte nuclear factor 4 to maintain insulin signaling and metabolic homeostasis in Drosophila. PLoS Genet 12: e1005978. https://doi.org/10.1371/journal.pgen.1005978

Park WR, Lim DJ, Sang H, Kim E, Moon JH, Choi HS, Kim IS, Kim DK (2021) Aphid estrogen-related receptor controls glycolytic gene expression and fecundity. Insect Biochem Mol Biol 130:103529. https://doi.org/10.1016/j.ibmb.2021.103529

Patke A, Young MW, Axelrod S (2020) Molecular mechanisms and physiological importance of circadian rhythms. Nat Rev Mol Cell Biol 21:67–84. https://doi.org/10.1038/s41580-019-0179-2

Pearen MA, Muscat GE (2010) Minireview: nuclear hormone receptor 4A signaling: implications for metabolic disease. Mol Endocrinol 24:1891–1903. https://doi.org/10.1210/me.2010-0015

Penick CA, Trobaugh B, Brent CS, Liebig J (2013) Head-butting as an early indicator of reproductive disinhibition in the termite Zootermopsis nevadensis. J Insect

Behav 26:23–34. https://doi.org/10.1007/s10905-012-9332-x

Polan DM, Alansari M, Lee B, Grewal SS (2020) Early-life hypoxia alters adult physiology and reduces stress resistance and lifespan in Drosophila. J Exp Biol 223. https://doi.org/10.1242/jeb.226027

Porstmann T, Santos CR, Griffiths B, Cully M, Wu M, Leevers S, Griffiths JR, Chung YL, Schulze A (2008) SREBP activity is regulated by mTORC1 and contributes to Akt-dependent cell growth. Cell Metab 8:224–236. https://doi.org/10.1016/j.cmet.2008.07.007

Pospisilik JA, Schramek D, Schnidar H, Cronin SJ, Nehme NT, Zhang X, Knauf C, Cani PD, Aumayr K, Todoric J, Bayer M, Haschemi A, Puviindran V, Tar K, Orthofer M, Neely GG, Dietzl G, Manoukian A, Funovics M, Prager G, Wagner O, Ferrandon D, Aberger F, Hui CC, Esterbauer H, Penninger JM (2010) Drosophila genome-wide obesity screen reveals hedgehog as a determinant of brown versus white adipose cell fate. Cell 140:148–160. https://doi.org/10.1016/j.cell.2009.12.027

Praggastis SA, Lam G, Horner MA, Nam HJ, Thummel CS (2021) The Drosophila E78 nuclear receptor regulates dietary triglyceride uptake and systemic lipid levels. Dev Dyn 250:640–651. https://doi.org/10.1002/dvdy.287

Price NT, Jackson VN, Muller J, Moffat K, Matthews KL, Orton T, Zammit VA (2010) Alternative exon usage in the single CPT1 gene of Drosophila generates functional diversity in the kinetic properties of the enzyme: differential expression of alternatively spliced variants in Drosophila tissues. J Biol Chem 285:7857–7865. https://doi.org/10.1074/jbc.M109.072892

Puig O, Mattila J (2011) Understanding Forkhead box class O function: lessons from Drosophila melanogaster. Antioxid Redox Signal 14:635–647. https://doi.org/10.1089/ars.2010.3407

Rajan A, Perrimon N (2012) Drosophila cytokine unpaired 2 regulates physiological homeostasis by remotely controlling insulin secretion. Cell 151:123–137. https://doi.org/10.1016/j.cell.2012.08.019

Raquin V, Merkling SH, Gausson V, Moltini-Conclois I, Frangeul L, Varet H, Dillies MA, Saleh MC, Lambrechts L (2017) Individual co-variation between viral RNA load and gene expression reveals novel host factors during early dengue virus infection of the Aedes aegypti midgut. PLoS Negl Trop Dis 11: e0006152. https://doi.org/10.1371/journal.pntd.0006152

Rawson RB (2003) The SREBP pathway—insights from Insigs and insects. Nat Rev Mol Cell Biol 4:631–640. https://doi.org/10.1038/nrm1174

Redmond W, Allen D, Elledge MC, Arellanes R, Redmond L, Yeahquo J, Zhang S, Youngblood M, Reiner A, Seo J (2019) Screening of microRNAs controlling body fat in Drosophila melanogaster and identification of miR-969 and its target, Gr47b. PLoS One 14:e0219707. https://doi.org/10.1371/journal.pone.0219707

Reiff T, Jacobson J, Cognigni P, Antonello Z, Ballesta E, Tan KJ, Yew JY, Dominguez M, Miguel-Aliaga I (2015) Endocrine remodelling of the adult intestine sustains reproduction in Drosophila. elife 4:e06930. https://doi.org/10.7554/eLife.06930

Reis T, Van Gilst MR, Hariharan IK (2010) A buoyancy-based screen of Drosophila larvae for fat-storage mutants reveals a role for Sir2 in coupling fat storage to nutrient availability. PLoS Genet 6:e1001206. https://doi.org/10.1371/journal.pgen.1001206

Ren Q, Yang CP, Liu Z, Sugino K, Mok K, He Y, Ito M, Nern A, Otsuna H, Lee T (2017) Stem cell-intrinsic, seven-up-triggered temporal factor gradients diversify intermediate neural progenitors. Curr Biol 27:1303–1313. https://doi.org/10.1016/j.cub.2017.03.047

Riahi H, Brekelmans C, Foriel S, Merkling SH, Lyons TA, Itskov PM, Kleefstra T, Ribeiro C, van Rij RP, Kramer JM, Schenck A (2019) The histone methyltransferase G9a regulates tolerance to oxidative stress-induced energy consumption. PLoS Biol 17:e2006146. https://doi.org/10.1371/journal.pbio.2006146

Rosenfeld JM, Osborne TF (1998) HLH106, a Drosophila sterol regulatory element-binding protein in a natural cholesterol auxotroph. J Biol Chem 273:16112–16121. https://doi.org/10.1074/jbc.273.26.16112

Ruaud AF, Lam G, Thummel CS (2011) The Drosophila NR4A nuclear receptor DHR38 regulates carbohydrate metabolism and glycogen storage. Mol Endocrinol 25: 83–91. https://doi.org/10.1210/me.2010-0337

Rynes J, Donohoe CD, Frommolt P, Brodesser S, Jindra M, Uhlirova M (2012) Activating transcription factor 3 regulates immune and metabolic homeostasis. Mol Cell Biol 32:3949–3962. https://doi.org/10.1128/MCB.00429-12

Salz HK, Erickson JW (2010) Sex determination in Drosophila: the view from the top. Fly (Austin) 4:60–70. https://doi.org/10.4161/fly.4.1.11277

Sanchez-Alvarez M, Finger F, Arias-Garcia Mdel M, Bousgouni V, Pascual-Vargas P, Bakal C (2014) Signaling networks converge on TORC1-SREBP activity to promote endoplasmic reticulum homeostasis. PLoS One 9:e101164. https://doi.org/10.1371/journal.pone.0101164

Sassu ED, McDermott JE, Keys BJ, Esmaeili M, Keene AC, Birnbaum MJ, DiAngelo JR (2012) Mio/dChREBP coordinately increases fat mass by regulating lipid synthesis and feeding behavior in Drosophila. Biochem Biophys Res Commun 426:43–48. https://doi.org/10.1016/j.bbrc.2012.08.028

Schabler S, Amatobi KM, Horn M, Rieger D, Helfrich-Forster C, Mueller MJ, Wegener C, Fekete A (2020) Loss of function in the Drosophila clock gene period results in altered intermediary lipid metabolism and increased susceptibility to starvation. Cell Mol Life Sci 77:4939–4956. https://doi.org/10.1007/s00018-019-03441-6

Schubiger M, Wade AA, Carney GE, Truman JW, Bender M (1998) Drosophila EcR-B ecdysone receptor isoforms are required for larval molting and for neuron remodeling during metamorphosis. Development 125: 2053–2062. https://doi.org/10.1242/dev.125.11.2053

Sciabica KS, Hertel KJ (2006) The splicing regulators Tra and Tra2 are unusually potent activators of pre-mRNA splicing. Nucleic Acids Res 34:6612–6620. https://doi.org/10.1093/nar/gkl984

Scopelliti A, Bauer C, Yu Y, Zhang T, Kruspig B, Murphy DJ, Vidal M, Maddocks ODK, Cordero JB (2019) A neuronal relay mediates a nutrient responsive gut/fat body Axis regulating energy homeostasis in adult Drosophila. Cell Metab 29:269–284. e210. https://doi.org/10.1016/j.cmet.2018.09.021

Seay DJ, Thummel CS (2011) The circadian clock, light, and cryptochrome regulate feeding and metabolism in Drosophila. J Biol Rhythm 26:497–506. https://doi.org/10.1177/0748730411420080

Seegmiller AC, Dobrosotskaya I, Goldstein JL, Ho YK, Brown MS, Rawson RB (2002) The SREBP pathway in Drosophila: regulation by palmitate, not sterols. Dev Cell 2:229–238. https://doi.org/10.1016/s1534-5807(01)00119-8

Sehgal A (2016) Control of metabolism by central and peripheral clocks in Drosophila. In: Sassone-Corsi P, Christen Y (eds) A time for metabolism and hormones. Cham, pp 33–40

Senos Demarco R, Uyemura BS, D'Alterio C, Jones DL (2019) Mitochondrial fusion regulates lipid homeostasis and stem cell maintenance in the Drosophila testis. Nat Cell Biol 21:710–720. https://doi.org/10.1038/s41556-019-0332-3

Shabar H, DiAngelo JR (2022) The regulation of lipid and carbohydrate storage by the splicing factor gene snRNP-U1-70K in the Drosophila fat body. MicroPubl Biol 2022. https://doi.org/10.17912/micropub.biology.000580

Sharrock J, Estacio-Gomez A, Jacobson J, Kierdorf K, Southall TD, Dionne MS (2019) fs(1)h controls metabolic and immune function and enhances survival via AKT and FOXO in Drosophila. Dis Model Mech 12. https://doi.org/10.1242/dmm.037259

Shen R, Wang B, Giribaldi MG, Ayres J, Thomas JB, Montminy M (2016) Neuronal energy-sensing pathway promotes energy balance by modulating disease tolerance. Proc Natl Acad Sci USA 113:E3307–E3314. https://doi.org/10.1073/pnas.1606106113

Shukla N, Kolthur-Seetharam U (2022) Drosophila Sirtuin 6 mediates developmental diet-dependent programming of adult physiology and survival. Aging Cell 21:e13576. https://doi.org/10.1111/acel.13576

Sieber MH, Spradling AC (2015) Steroid signaling establishes a female metabolic state and regulates SREBP to control oocyte lipid accumulation. Curr Biol 25:993–1004. https://doi.org/10.1016/j.cub.2015.02.019

Sieber MH, Thummel CS (2009) The DHR96 nuclear receptor controls triacylglycerol homeostasis in Drosophila. Cell Metab 10:481–490. https://doi.org/10.1016/j.cmet.2009.10.010

Sieber MH, Thummel CS (2012) Coordination of triacylglycerol and cholesterol homeostasis by DHR96 and the Drosophila LipA homolog magro. Cell Metab 15:122–127. https://doi.org/10.1016/j.cmet.2011.11.011

Sillam-Dusses D, Hanus R, Poulsen M, Roy V, Favier M, Vasseur-Cognet M (2016) The role of the glucose-sensing transcription factor carbohydrate-responsive element-binding protein pathway in termite queen fertility. Open Biol 6. https://doi.org/10.1098/rsob.160080

Sim C, Denlinger DL (2008) Insulin signaling and FOXO regulate the overwintering diapause of the mosquito Culex pipiens. Proc Natl Acad Sci USA 105:6777–6781. https://doi.org/10.1073/pnas.0802067105

Spiegelman BM, Heinrich R (2004) Biological control through regulated transcriptional coactivators. Cell 119:157–167. https://doi.org/10.1016/j.cell.2004.09.037

Stone BL, Thummel CS (1993) The Drosophila 78C early late puff contains E78, an ecdysone-inducible gene that encodes a novel member of the nuclear hormone receptor superfamily. Cell 75:307–320. https://doi.org/10.1016/0092-8674(93)80072-m

Storelli G, Nam HJ, Simcox J, Villanueva CJ, Thummel CS (2019) Drosophila HNF4 directs a switch in lipid metabolism that supports the transition to adulthood. Dev Cell 48:200–214. e206. https://doi.org/10.1016/j.devcel.2018.11.030

Suh JM, Zeve D, McKay R, Seo J, Salo Z, Li R, Wang M, Graff JM (2007) Adipose is a conserved dosage-sensitive antiobesity gene. Cell Metab 6:195–207. https://doi.org/10.1016/j.cmet.2007.08.001

Suren-Castillo S, Abrisqueta M, Maestro JL (2014) FoxO is required for the activation of hypertrehalosemic hormone expression in cockroaches. Biochim Biophys Acta 1840:86–94. https://doi.org/10.1016/j.bbagen.2013.08.015

Sutherland JD, Kozlova T, Tzertzinis G, Kafatos FC (1995) Drosophila hormone receptor 38: a second partner for Drosophila USP suggests an unexpected role for nuclear receptors of the nerve growth factor-induced protein B type. Proc Natl Acad Sci USA 92: 7966–7970. https://doi.org/10.1073/pnas.92.17.7966

Teleman AA, Maitra S, Cohen SM (2006) Drosophila lacking microRNA miR-278 are defective in energy homeostasis. Genes Dev 20:417–422. https://doi.org/10.1101/gad.374406

Tennessen JM, Baker KD, Lam G, Evans J, Thummel CS (2011) The Drosophila estrogen-related receptor directs a metabolic switch that supports developmental growth. Cell Metab 13:139–148. https://doi.org/10.1016/j.cmet.2011.01.005

Theopold U, Ekengren S, Hultmark D (1996) HLH106, a Drosophila transcription factor with similarity to the vertebrate sterol responsive element binding protein.

Proc Natl Acad Sci USA 93:1195–1199. https://doi.org/10.1073/pnas.93.3.1195

Tiefenbock SK, Baltzer C, Egli NA, Frei C (2010) The Drosophila PGC-1 homologue Spargel coordinates mitochondrial activity to insulin signalling. EMBO J 29:171–183. https://doi.org/10.1038/emboj.2009.330

Troha K, Im JH, Revah J, Lazzaro BP, Buchon N (2018) Comparative transcriptomics reveals CrebA as a novel regulator of infection tolerance in D. melanogaster. PLoS Pathog 14:e1006847. https://doi.org/10.1371/journal.ppat.1006847

Truman JW (2019) The evolution of insect metamorphosis. Curr Biol 29:R1252–R1268. https://doi.org/10.1016/j.cub.2019.10.009

Tsai JW, Kostyleva R, Chen PL, Rivas-Serna IM, Clandinin MT, Meinertzhagen IA, Clandinin TR (2019) Transcriptional feedback links lipid synthesis to synaptic vesicle pools in drosophila photoreceptors. Neuron 101:721–737. e724. https://doi.org/10.1016/j.neuron.2019.01.015

Ugrankar R, Liu Y, Provaznik J, Schmitt S, Lehmann M (2011) Lipin is a central regulator of adipose tissue development and function in Drosophila melanogaster. Mol Cell Biol 31:1646–1656. https://doi.org/10.1128/MCB.01335-10

Valente V, Maia RM, Vianna MC, Paco-Larson ML (2010) Drosophila melanogaster lipins are tissue-regulated and developmentally regulated and present specific subcellular distributions. FEBS J 277:4775–4788. https://doi.org/10.1111/j.1742-4658.2010.07883.x

Valzania L, Martinson VG, Harrison RE, Boyd BM, Coon KL, Brown MR, Strand MR (2018a) Both living bacteria and eukaryotes in the mosquito gut promote growth of larvae. PLoS Negl Trop Dis 12:e0006638. https://doi.org/10.1371/journal.pntd.0006638

Valzania L, Coon KL, Vogel KJ, Brown MR, Strand MR (2018b) Hypoxia-induced transcription factor signaling is essential for larval growth of the mosquito Aedes aegypti. Proc Natl Acad Sci USA 115:457–465. https://doi.org/10.1073/pnas.1719063115

Varghese J, Lim SF, Cohen SM (2010) Drosophila miR-14 regulates insulin production and metabolism through its target, sugarbabe. Genes Dev 24:2748–2753. https://doi.org/10.1101/gad.1995910

Vihervaara T, Puig O (2008) dFOXO regulates transcription of a Drosophila acid lipase. J Mol Biol 376:1215–1223. https://doi.org/10.1016/j.jmb.2007.12.042

Wang B, Goode J, Best J, Meltzer J, Schilman PE, Chen J, Garza D, Thomas JB, Montminy M (2008) The insulin-regulated CREB coactivator TORC promotes stress resistance in Drosophila. Cell Metab 7:434–444. https://doi.org/10.1016/j.cmet.2008.02.010

Wang S, Liu S, Liu H, Wang J, Zhou S, Jiang RJ, Bendena WG, Li S (2010) 20-hydroxyecdysone reduces insect food consumption resulting in fat body lipolysis during molting and pupation. J Mol Cell Biol 2:128–138. https://doi.org/10.1093/jmcb/mjq006

Wang B, Moya N, Niessen S, Hoover H, Mihaylova MM, Shaw RJ, Yates JR 3rd, Fischer WH, Thomas JB, Montminy M (2011) A hormone-dependent module regulating energy balance. Cell 145:596–606. https://doi.org/10.1016/j.cell.2011.04.013

Wang X, Hou Y, Saha TT, Pei G, Raikhel AS, Zou Z (2017) Hormone and receptor interplay in the regulation of mosquito lipid metabolism. Proc Natl Acad Sci USA 114:E2709–E2718. https://doi.org/10.1073/pnas.1619326114

Wang ZZ, Ye XQ, Shi M, Li F, Wang ZH, Zhou YN, Gu QJ, Wu XT, Yin CL, Guo DH, Hu RM, Hu NN, Chen T, Zheng BY, Zou JN, Zhan LQ, Wei SJ, Wang YP, Huang JH, Fang XD, Strand MR, Chen XX (2018) Parasitic insect-derived miRNAs modulate host development. Nat Commun 9:2205. https://doi.org/10.1038/s41467-018-04504-1

Wat LW, Chao C, Bartlett R, Buchanan JL, Millington JW, Chih HJ, Chowdhury ZS, Biswas P, Huang V, Shin LJ, Wang LC, Gauthier ML, Barone MC, Montooth KL, Welte MA, Rideout EJ (2020) A role for triglyceride lipase brummer in the regulation of sex differences in Drosophila fat storage and breakdown. PLoS Biol 18:e3000595. https://doi.org/10.1371/journal.pbio.3000595

Wat LW, Chowdhury ZS, Millington JW, Biswas P, Rideout EJ (2021) Sex determination gene transformer regulates the male-female difference in Drosophila fat storage via the adipokinetic hormone pathway. elife 10. https://doi.org/10.7554/eLife.72350

Watson MA, Pattavina B, Hilsabeck TAU, Lopez-Dominguez J, Kapahi P, Brand MD (2021) S3QELs protect against diet-induced intestinal barrier dysfunction. Aging Cell 20:e13476. https://doi.org/10.1111/acel.13476

Weaver LN, Drummond-Barbosa D (2019) The nuclear receptor seven up functions in adipocytes and oenocytes to control distinct steps of Drosophila oogenesis. Dev Biol 456:179–189. https://doi.org/10.1016/j.ydbio.2019.08.015

Weaver LN, Drummond-Barbosa D (2020) The nuclear receptor seven up regulates genes involved in immunity and xenobiotic response in the adult drosophila female fat body. G3 (Bethesda) 10:4625–4635. https://doi.org/10.1534/g3.120.401745

Wei P, Guo J, Xue W, Zhao Y, Yang J, Wang J (2018) RNF34 modulates the mitochondrial biogenesis and exercise capacity in muscle and lipid metabolism through ubiquitination of PGC-1 in Drosophila. Acta Biochim Biophys Sin Shanghai 50:1038–1046. https://doi.org/10.1093/abbs/gmy106

Weidman T, Nagengast AA, DiAngelo JR (2022) The splicing factor 9G8 regulates the expression of NADPH-producing enzyme genes in Drosophila. Biochem Biophys Res Commun 620:92–97. https://doi.org/10.1016/j.bbrc.2022.06.073

Wen DT, Zheng L, Li JX, Lu K, Hou WQ (2019) The activation of cardiac dSir2-related pathways mediates physical exercise resistance to heart aging in old Drosophila. Aging (Albany NY) 11:7274–7293. https://doi.org/10.18632/aging.102261

Wood JG, Schwer B, Wickremesinghe PC, Hartnett DA, Burhenn L, Garcia M, Li M, Verdin E, Helfand SL (2018) Sirt4 is a mitochondrial regulator of metabolism and lifespan in Drosophila melanogaster. Proc Natl Acad Sci USA 115:1564–1569. https://doi.org/10.1073/pnas.1720673115

Xie M, Roy R (2012) Increased levels of hydrogen peroxide induce a HIF-1-dependent modification of lipid metabolism in AMPK compromised C. elegans dauer larvae. Cell Metab 16:322–335. https://doi.org/10.1016/j.cmet.2012.07.016

Xie XJ, Hsu FN, Gao X, Xu W, Ni JQ, Xing Y, Huang L, Hsiao HC, Zheng H, Wang C, Zheng Y, Xiaoli AM, Yang F, Bondos SE, Ji JY (2015) CDK8-cyclin C mediates nutritional regulation of developmental transitions through the ecdysone receptor in Drosophila. PLoS Biol 13:e1002207. https://doi.org/10.1371/journal.pbio.1002207

Xu P, Vernooy SY, Guo M, Hay BA (2003) The Drosophila microRNA Mir-14 suppresses cell death and is required for normal fat metabolism. Curr Biol 13:790–795. https://doi.org/10.1016/s0960-9822(03)00250-1

Xu K, Zheng X, Sehgal A (2008) Regulation of feeding and metabolism by neuronal and peripheral clocks in Drosophila. Cell Metab 8:289–300. https://doi.org/10.1016/j.cmet.2008.09.006

Xu J, Raman C, Zhu F, Tan A, Palli SR (2012a) Identification of nuclear receptors involved in regulation of male reproduction in the red flour beetle, Tribolium castaneum. J Insect Physiol 58:710–717. https://doi.org/10.1016/j.jinsphys.2012.02.006

Xu X, Gopalacharyulu P, Seppanen-Laakso T, Ruskeepaa AL, Aye CC, Carson BP, Mora S, Oresic M, Teleman AA (2012b) Insulin signaling regulates fatty acid catabolism at the level of CoA activation. PLoS Genet 8:e1002478. https://doi.org/10.1371/journal.pgen.1002478

Yamamoto S, Jaiswal M, Charng WL, Gambin T, Karaca E, Mirzaa G, Wiszniewski W, Sandoval H, Haelterman NA, Xiong B, Zhang K, Bayat V, David G, Li T, Chen K, Gala U, Harel T, Pehlivan D, Penney S, Vissers L, de Ligt J, Jhangiani SN, Xie Y, Tsang SH, Parman Y, Sivaci M, Battaloglu E, Muzny D, Wan YW, Liu Z, Lin-Moore AT, Clark RD, Curry CJ, Link N, Schulze KL, Boerwinkle E, Dobyns WB, Allikmets R, Gibbs RA, Chen R, Lupski JR, Wangler MF, Bellen HJ (2014) A drosophila genetic resource of mutants to study mechanisms underlying human genetic diseases. Cell 159:200–214. https://doi.org/10.1016/j.cell.2014.09.002

Yan Y, Wang H, Hu M, Jiang L, Wang Y, Liu P, Liang X, Liu J, Li C, Lindstrom-Battle A, Lam SM, Shui G, Deng WM, Jiao R (2017) HDAC6 suppresses age-dependent ectopic fat accumulation by maintaining the Proteostasis of PLIN2 in Drosophila. Dev Cell 43:99–111. e115. https://doi.org/10.1016/j.devcel.2017.09.001

Yan Y, Wang H, Wei C, Xiang Y, Liang X, Phang CW, Jiao R (2019) HDAC6 regulates lipid droplet turnover in response to nutrient deprivation via p62-mediated selective autophagy. J Genet Genomics 46:221–229. https://doi.org/10.1016/j.jgg.2019.03.008

Yin JC, Wallach JS, Wilder EL, Klingensmith J, Dang D, Perrimon N, Zhou H, Tully T, Quinn WG (1995) A Drosophila CREB/CREM homolog encodes multiple isoforms, including a cyclic AMP-dependent protein kinase-responsive transcriptional activator and antagonist. Mol Cell Biol 15:5123–5130. https://doi.org/10.1128/MCB.15.9.5123

Zhang S, Xie J, Luo R, Zhang H, Zheng W (2023) MiR-2b-2-5p regulates lipid metabolism and reproduction by targeting CREB in Bactrocera dorsalis. RNA Biol 20:164–176. https://doi.org/10.1080/15476286.2023.2204579

Zhao X, Karpac J (2017) Muscle directs diurnal energy homeostasis through a Myokine-dependent hormone module in Drosophila. Curr Biol 27:1941–1955. e1946. https://doi.org/10.1016/j.cub.2017.06.004

Zhao X, Karpac J (2020) The drosophila midgut and the systemic coordination of lipid-dependent energy homeostasis. Curr Opin Insect Sci 41:100–105. https://doi.org/10.1016/j.cois.2020.07.003

Zhao X, Feng D, Wang Q, Abdulla A, Xie XJ, Zhou J, Sun Y, Yang ES, Liu LP, Vaitheesvaran B, Bridges L, Kurland IJ, Strich R, Ni JQ, Wang C, Ericsson J, Pessin JE, Ji JY, Yang F (2012) Regulation of lipogenesis by cyclin-dependent kinase 8-mediated control of SREBP-1. J Clin Invest 122:2417–2427. https://doi.org/10.1172/JCI61462

Zhao X, Li X, Shi X, Karpac J (2020) Diet-MEF2 interactions shape lipid droplet diversification in muscle to influence Drosophila lifespan. Aging Cell 19:e13172. https://doi.org/10.1111/acel.13172

Zhu W, Zhang H, Li X, Meng Q, Shu R, Wang M, Zhou G, Wang H, Miao L, Zhang J, Qin Q (2016) Cold adaptation mechanisms in the ghost moth Hepialus xiaojinensis: metabolic regulation and thermal compensation. J Insect Physiol 85:76–85. https://doi.org/10.1016/j.jinsphys.2015.11.008

Zhu L, Tian Z, Guo S, Liu W, Zhu F, Wang XP (2019) Circadian clock genes link photoperiodic signals to lipid accumulation during diapause preparation in the diapause-destined female cabbage beetles Colaphellus bowringi. Insect Biochem Mol Biol 104:1–10. https://doi.org/10.1016/j.ibmb.2018.11.001

Ziegler AB, Thiele C, Tenedini F, Richard M, Leyendecker P, Hoermann A, Soba P, Tavosanis G (2017) Cell-autonomous control of neuronal dendrite expansion via the fatty acid synthesis regulator SREBP. Cell Rep 21:3346–3353. https://doi.org/10.1016/j.celrep.2017.11.069

Zinke I, Schutz CS, Katzenberger JD, Bauer M, Pankratz MJ (2002) Nutrient control of gene expression in Drosophila: microarray analysis of starvation and sugar-dependent response. EMBO J 21:6162–6173. https://doi.org/10.1093/emboj/cdf600

Adv Exp Med Biol (2026) 1494: 201–216
https://doi.org/10.1007/5584_2024_807

Published online: 24 May 2024

Endocrine Control of Lipid Metabolism

Natraj Krishnan

Abstract

Lipids are essential in insects and play pleiotropic roles in energy storage, serving as a fuel for energy-driven processes such as reproduction, growth, development, locomotion, flight, starvation response, and diapause induction, maintenance, and termination. Lipids also play fundamental roles in signal transduction, hormone synthesis, forming components of the cell membrane, and thus are essential for maintenance of normal life functions. In insects, the neuroendocrine system serves as a master regulator of most life activities, including growth and development. It is thus important to pay particular attention to the regulation of lipid metabolism through the endocrine system, especially when considering the involvement of peptide hormones in the processes of lipogenesis and lipolysis. In insects, there are several lipogenic and lipolytic hormones that are involved in lipid metabolism such as insulin-like peptides (ILPs), adipokinetic hormone (AKH), 20-hydroxyecdysone (20-HE), juvenile hormone (JH), and serotonin. Other neuropeptides such as diapause hormone-pheromone biosynthesis activating neuropeptide (DH-PBAN), CCHamide-2, short neuropeptide F, and the cytokines Unpaired 1 and 2 may play a role in inducing lipogenesis. On the other hand, neuropeptides such as neuropeptide F, allatostatin-A, corazonin, leukokinin, tachykinins, limostatins, and insulin-like growth factor (ILP6) stimulate lipolysis. This chapter briefly discusses the current knowledge of the endocrine regulation of lipid metabolism in insects that could be utilized to reveal differences between insects and mammalian lipid metabolism which may help understand human diseases associated with dysregulation of lipid metabolism. Physiological similarities of insects to mammals make them valuable model systems for studying human diseases characterized by disrupted lipid metabolism, including conditions like diabetes, obesity, arteriosclerosis, and various metabolic syndromes.

Keywords

Insulin-like peptides · Lipid metabolism · Lipogenesis · Lipolysis · Neuroendocrine system · Peptidergic hormones

N. Krishnan (✉)
Department of Biochemistry, Molecular Biology, Entomology and Plant Pathology, Mississippi State University, Mississippi State, MS, USA
e-mail: nkrishnan@entomology.msstate.edu; nk260@msstate.edu

1 Introduction

In insects, the main energy sources essential to perform all life activities are carbohydrates, lipids, and proteins similar to other animals. In particular, lipids serve as the primary storage molecule for fueling energetic processes such as reproduction, growth, development, locomotion, flight, response to starvation, and also in preparing for diapause. In insects, storage of lipids occurs in the fat body which is analogous to the vertebrate liver and adipose tissue (Law and Wells 1989). However, unlike the solid structure of the liver, the insect fat body has a loose tissue structure. This arrangement of the fat body in insects facilitates its close contact with the hemolymph and forms a system that is uniquely adapted to the physiology of insects (Keeley 1985). The insect fat body cells play multiple roles, not only managing the creation and utilization of energy stores like fat and glycogen but also participating in the production of a majority of hemolymph proteins and circulating metabolites. In general, the fat body comprises five primary cell types: trophocytes (formerly referred to as adipocytes), oenocytes, mycetocytes, chromatocytes, and urocytes. However, not all insects have all five cell types and some cell types occur only in specific insect groups (Roma et al. 2010; Skowronek et al. 2021). These cell types can exhibit differences in size, composition, and function based on the insect's growth stage (Roma et al. 2010; Skowronek et al. 2021).

Lipid compounds are not only stored but also released and processed in the cells of the fat body in the form of droplets and various granules. In the hemolymph of insects, lipids are found as diglycerides (DGs) whereas in cells, fats make up $>50\%$ of all components and mainly as triglycerides (TGs, anhydrous $>90\%$ of all fats), along with hydrated glycogen (Gilby 1965; Canavoso et al. 2001). The accumulated TGs usually are seen to exist as fat droplets (Patel et al. 2005; Arrese and Soulages 2010). In insects, the midgut serves as the primary location for lipid digestion and absorption. Enzymes like TG lipases, phospholipases, and cholesterol esterases are responsible for breaking down lipids (Palm et al. 2012; Kühnlein 2012; Grillo et al. 2007; Heier and Kühnlein 2018; Christeller et al. 2011; Toprak et al. 2020). Intestinal epithelial cells take up free fatty acids (FFAs), non-esterified cholesterol, and lysophospholipids, although the exact process of how each of these compounds is assimilated into the midgut remains largely unclear (Grillo et al. 2007; Turunen 1993; Juoni et al. 2002; Canavaso et al. 2004; Guo et al. 2009). Following absorption, these previously digested lipid molecules act as building blocks for the creation of more complex structures, which are intended for transportation to various organ systems within the insect. FFAs serve as precursors for synthesis of DGs, phospholipids (PLs), and TGs. Glycerolipids in insects appear to be synthesized via the glycerol-3-phosphate (G3P) pathway (Grillo et al. 2007; Canavaso et al. 2004). In this process, the main steps involve successive acylations of G3P until the formation of TG takes place, including the generation of phosphatidic acid (PA), DG, and other intermediary molecules, similar to what occurs in mammalian systems (Coleman and Mashek 2011). A clue to this process in insects comes from studies on phosphatidic acid phosphatase (PAP) activity in the midgut of the kissing bug (*Panstrongylus megistus*) (Canavaso et al. 2004) and PAP expression described in *Drosophila melanogaster* midgut (Valente et al. 2010). The lipids which are produced by this process are then loaded onto lipophorins (Lp), the principal lipid-transporting protein in insect hemolymph. The Lp then directs the lipids to storage organs such as the fat body, ovaries, and muscles. Lps can transport lipid cargo to specific tissues through two mechanisms. Firstly, they can employ specific membrane receptors (e.g., low-density lipoprotein receptor (LDLR) family LpR1 and LpR2 encoded by *lpr1* and *lpr2* in *Drosophila*) on fat body cells, a process akin to what happens in the midgut (Palm et al. 2012; Van der Horst and Ryan 2012). Secondly, Lps can undergo endocytosis by fat body cells (Van Hoof et al. 2003). Subsequently, the Lp, along with its receptor, is recycled and released back into the hemolymph. This reusable shuttle makes the Lp capable of unloading its cargo in one target organ and returning to another to reload and supply another target organ (Downer and Chino 1985; Van Hoof et al. 2005). Along with the lipids delivered by Lp, the fat body can also create fatty acids using acetate and amino acids as sources (Lorenz and Anand 2004; Anand and Lorenz 2008; Alabaster et al. 2011). This process is a highly intricate one, driven by several enzymes, with fatty acid synthases (FASs) standing out as pivotal and multifunctional players in both fatty acid synthesis and lipid metabolism. Triglycerides (TG) constitute the primary component of lipids, and FAS plays a crucial role in synthesizing new lipids by transforming acetyl-CoA into palmitate, ultimately resulting in the generation and retention of TGs (Menendez and Lupu 2007). Moreover, the pathway of fatty acid synthesis includes other enzymes like acetyl-CoA carboxylase (ACC), desaturases (Desat), and lipases. Nevertheless, it's worth noting that research on FAS genes and their functions has been limited to a very select number of insect species (Alabaster et al. 2011; Parvy et al. 2012; Wicker-Thomas et al. 2015; Song et al. 2022). Further exploration in this area is required for a more comprehensive knowledge of the role of FAS and related enzymes in insect lipid metabolism.

To meet the diverse metabolic requirements that arise as insects undergo growth and development, the fat body must possess the ability to harmonize signals received from various organs. These functions are primarily under hormonal control, making the fat body a key target for a range of hormones (Arrese and Soulages 2010; Gade 2004; Schooley et al. 2005). Many of these hormones, which influence lipid metabolism, are primarily synthesized by neurosecretory cells and endocrine glands like the *corpora cardiaca*, *corpora allata*, and prothoracic glands. The fat body's role in integrating these hormonal signals is pivotal in orchestrating the insect's metabolic responses to changing physiological conditions. These hormones act as messengers, transmitting vital information to the fat body, which then adjusts its metabolic activities accordingly. This intricate interplay between hormonal regulation and the fat body's functions is essential for the insect's growth, development, and overall metabolic homeostasis. Furthermore, organs such as fat body, ovaries, and midgut are also considered as endocrine glands because of their ability to synthesize hormones. The ensuing sections will be devoted to discussing the overall mechanism of endocrine regulation of lipid metabolism in insects.

2 Regulation of Insect Lipid Metabolism by Insect Hormones

2.1 Lipogenic Pathway

Lipogenesis is the process leading to creation of TGs. Hence, TGs are the most important substrate as well as product of the lipogenic pathway. TGs are generally found in the circulating hemolymph, where very low-density lipoprotein (VLDL) are hydrolyzed by lipoprotein lipase to fatty acids and glycerol. Insulin along with other VLDL compounds stimulate the process of lipogenesis. Glycerol is subsequently converted into glucose. Acetyl-CoA synthetase is synthesized at the cell surface through a complex process that involves the utilization of fatty acids and G3P, derived from dihydroxyacetone phosphate. This compound is a product of the conversion of glycerol into glucose, facilitated by the action of acetyl-CoA synthetase. The G3P, crucial in this context, can be obtained through two distinct pathways. Firstly, it can arise from the glycolytic pathway, where glucose serves as the primary substrate. Alternatively, G3P can be generated from glycogenic substrates, such as pyruvate or lactate, through a process known as glycerogenesis. This dual origin of G3P highlights its significance as a versatile molecule, essential for various cellular processes, including lipid metabolism and energy production. The intricate interplay of these pathways underscores the adaptability of cellular metabolism to different substrates and metabolic demands. About 50% of the obtained glucose is utilized for the synthesis of fats and stimulation of lipogenic processes (Arrese and Soulages 2010; Mead 1986). The G3P obtained from the TGs undergoes alcohol esterification by the action of enzyme acetyltransferase G3P, which results in lysophosphatidic acid. This is further esterified by acetyltransferase 1-acyl G3P resulting in diacylglycerol which is then acted upon by diacylglycerol acetyltransferase-acetyl-CoA to form TG, the end product which then becomes the starting substrate for the lipogenic pathway. A brief account of hormones involved in the lipogenic pathway is provided below.

2.1.1 Insulin-Like Peptides (ILPs)

Insulin-like peptides (ILPs) are synthesized within the insulin-producing cells (IPCs), situated in both the medial and lateral regions of the brain (Ikeya et al. 2002). These cells produce ILP 2, ILP3, and ILP5 all of which are involved in lipogenesis. ILP2 and ILP5 can also be produced from salivary glands, ovaries, malpighian tubules, larval midgut, and mesoderm (Toprak 2020). In *Drosophila melanogaster* eight ILPs have been described (DILP1-8) (Kannan and Fridell 2013). These hormones mainly exert their regulatory influence on carbohydrate levels (Kannan and Fridell 2013; Brown et al. 2008). Given the high degree of conservation in the insulin signaling pathway, it is anticipated that ILPs in insects function in a comparable manner (Teleman 2009).

Like mammalian insulin, insect ILPs are hormones involved in controlling carbohydrate and lipid metabolism, tissue growth, and the duration of lifespan (Kannan and Fridell 2013). Some ILPs in insects act mainly on lipid transformations in lipogenic pathways (LIP2, 3, and 5) and in lipolytic pathways (such as the insulin-like growth factor, ILP6). The ILPs play a multifaceted role in regulating various physiological processes in insects. Notably, they have the capacity to inhibit the activity of the forkhead box class O (FoxO) transcription factor, while concurrently activating sterol regulation and governing lipid accumulation, particularly during diapause—a state of suspended development and metabolic dormancy. ILP6, on the other hand, plays a distinct role by suppressing the activity of ILPs and enhancing resistance to hunger (Toprak 2020). The FoxO protein family of transcription factors plays a pivotal role in orchestrating a wide array of metabolic processes, as well as being involved in embryonic development. This multifunctionality stems from the fact that FoxO acts as the ultimate target for several critical signaling pathways, including the insulin/insulin growth factor (IGF) pathway, the mitogen-activated protein kinase (MAPK) pathway, and the Wnt/β-catenin pathway (Benayoun et al. 2011). Of particular significance is the close relationship between members of the FoxO class O family and insulin signaling, which has direct implications for lipid metabolism. While mammalian systems possess multiple genes encoding FoxO proteins, *D. melanogaster*, for example, has only one ortholog, known as *dfoxo* (Teleman 2009; Kramer et al. 2003; Puig et al. 2003). The interplay between FoxO and the Sterol Regulatory Element Binding Protein (SREBP) within the insulin signaling pathway is noteworthy. FoxO activates genes responsible for β-oxidation, a process involved in breaking down fatty acids for energy production, while SREBP induces the expression of genes related to lipogenesis, the synthesis of lipids (Seegmiller et al. 2002; Gershman et al. 2007). Consequently, ILPs inhibit FoxO activity while activating SREBP, ultimately leading to the accumulation of lipid reserves, i.e., lipogenesis. This intricate regulation of lipid metabolism showcases the intricate role of ILPs in maintaining metabolic homeostasis in insects.

In *D. melanogaster*, SREBP is not regulated by cholesterol levels unlike what occurs in mammals (Yang et al. 2002) but by the levels of phosphatidylethanolamine (PE) (Dobrosotskaya et al. 2002). The role of SREBP has only been investigated in *Drosophila*. Flies which are mutant for the gene do not survive beyond the second instar (Kunte et al. 2006). Activation of the target of rapamycin (TOR) pathway has a notable impact on the function of SREBP (Portsmann et al. 2008). When the insulin signaling pathway is silenced using RNA interference (RNAi), it results in a decrease in the expression of genes targeted by SREBP. Conversely, when the pathway is activated by inhibiting *dp110*, a gene involved in suppressing the insulin pathway, it leads to an increase in the expression of SREBP target genes (Portsmann et al. 2008). This observation highlights the interaction between insulin signaling and the regulation of SREBP activity, ultimately resulting in the expression of genes associated with lipogenesis, mirroring findings in mammals (Jeon and Osborne 2012). Moreover, ILPs are known to modulate the expression of genes through other transcription factors, including the cAMP response element-binding protein (CREB), by affecting the activity of its coactivator known as TORC (TOR complex). In the context of *D. melanogaster*, ILPs signaling exerts an inhibitory effect on TORC through phosphorylation, a process mediated by salt-inducible kinase 2 (SIK2) (Wang et al. 2008). Consequently, this results in a decrease in the activation of specific genes, including the acetyl-CoA binding protein (ACBP) (Wang et al. 2008).This functions in several aspects of lipid metabolism (Knudsen et al. 2000). These findings underscore the intricate regulatory network through which ILPs govern various aspects of gene expression, contributing to the modulation of lipid metabolism and other critical cellular processes.

To summarize, when nutrient availability is plentiful, ILPs are secreted from the brain and interact with insulin receptors (InR) located on the cell membrane (Fernandez et al. 1995; Chen et al. 1996). This interaction triggers the recruitment of the InR substrate, Chico (Böhni et al. 1999), which, in turn, activates phosphoinositide-3-kinase (PI3K) to phosphorylate phosphatidylinositol 4,5-bisphosphate (PIP2) into phosphatidylinositol 3,4,5-trisphosphate (PIP3) (Brown and Auger 2011). PIP3 recruits protein kinase B (AKT) to membrane which inhibits FoxO activity by phosphorylation (Puig et al. 2003; Wang et al. 2008). In contrast when insulin signaling is reduced, as in situations of starvation, dephosphorylation of FoxO promotes its nuclear translocation and upregulation of genes encoding lipases in TAG hydrolysis (Wang et al. 2011), supplying energy to the insect.

2.1.2 Ecdysteroids

Ecdysteroids, which are lipophilic steroid hormones, play a vital role in the molting process during insect growth and metamorphosis. The primary ecdysteroid, ecdysone, is produced from cholesterol in the prothoracic gland and transformed into its active form, 20-hydroxyecdysone (20-HE), which serves as the molting hormone. However, 20-HE has also been associated with other regulatory functions, including its involvement in lipid metabolism, particularly lipogenesis. 20-HE binds to a nuclear receptor, the ecdysone receptor (EcR). Subsequently, this EcR forms a heterodimer with ultraspiracle (USP), another nuclear receptor (Yao et al. 1992, 1993; Thomas et al. 1993). Together, this heterodimeric transcription factor influences the expression of numerous genes, most of which are categorized as "early genes."

Among these initial genes, E75 stands out as noteworthy because its transcriptional activity is regulated by the interaction of gases like carbon monoxide or nitric oxide with its heme prosthetic group (Reinking et al. 2005). Notably, when researchers employed RNA interference (RNAi) to inhibit the expression of nitric oxide synthase (NOS) in the prothoracic gland of *D. melanogaster* larvae, they observed a significant accumulation of lipids in fat body cells (Cáceres et al. 2011). Furthermore, the activity of EcR has been linked to the accumulation of lipids in fat body cells, which is attributed to the inhibition of NOS associated with E75 activity (Colombani et al. 2005). This relationship was further substantiated when the inhibition of EcR expression via RNAi led to the accumulation of TGs in the fat body (Kamoshida et al. 2012). Additionally, fly larvae with a null mutation in the e75 gene exhibit a phenotype characterized by an accumulation of fat (Reis et al. 2010). These findings suggest an opposing association between the lipogenic and lipolytic functions of EcR, suggesting that its roles in lipid metabolism may be stage-specific to the insect.

Within the ecdysone cascade, several other early genes are known to regulate genes associated with lipid metabolism. For example, the broad (*br*) gene is responsible for regulating the expression of *Niemann-Pick type C1* (*ncp1*), a gene involved in cholesterol metabolism and ecdysone biosynthesis in *D. melanogaster* (Xiang et al. 2010). In *Aedes aegypti*, an increase in Lp expression was observed in the fat body during vitellogenesis, a period when circulating 20-HE levels in the hemolymph are also elevated (Sun et al. 2000; Margam et al. 2006). There is also a link between ecdysteroid signaling via EcR and insulin signaling pathway—the action of this receptor is mediated through several co-activators/repressors. Certain genes within the ecdysone cascade, such as *Broad*, *HR3*, and *βFTZ-F1*, have the potential to influence insulin signaling and consequently impact lipid metabolism by acting as transcription factors (Toprak et al. 2020). Ecdysteroids exhibit the capability to impede TOR signaling within the fat body of feeding *Drosophila* larvae. This inhibition results in the generation of a signal, the nature of which remains unknown, that intricately governs the production of insulin-like peptides (ILPs) in the insulin-producing cells (IPCs), thereby exerting control over systemic growth (Colombani et al. 2005). The intricate interplay between the insulin/TOR and ecdysteroid signaling pathways provides a mechanism for precise regulation of body and organ growth, intricately intertwined with nutrition-dependent responses and lipid metabolism. These collective findings strongly imply that 20-hydroxyecdysone (20-HE) may assume a pivotal role in finely regulating lipid metabolism and its transport within the hemolymph.

2.1.3 Diapause Hormone-Pheromone Biosynthesis Activating Neuropeptide (DH-PBAN)

To overcome periods of seasonal challenges including adverse weather conditions and food shortages, insects resort to a developmental arrest that can be either termed facultative or obligatory diapause. This is essentially a period of metabolic depression. Although hormones such as ILPs and juvenile hormone (JH) are recognized for their involvement in coordinating diapause, certain insects like the silkworm (*Bombyx mori*) and the cotton bollworm (*Helicoverpa armigera*) possess a distinct peptide hormone known as the diapause hormone (DH) to govern the diapause process (Xu et al. 1995; Zhang et al. 2004a). Uniquely, while DH induces embryonic diapause in *B. mori* (Xu et al. 1995), it has the opposite effect on *H. armigera*, where it rekindles the metabolic activity of pupae, effectively resulting in the termination of diapause (Zhang et al. 2004a, b). DH-PBAn has also been shown to affect lipid metabolism, e.g., elevation of lipid content of eggs by DH-PBAN (Hasegawa and Yamashita 1965) and regulation of genes involved in lipid metabolism in *H. armigera* (Liu et al. 2005).

2.1.4 Serotonin

Serotonin, which acts as a neurohormone and plays a role in various physiological processes, including blood feeding, diuresis, saliva secretion in several insects, and particularly in *Rhodnius prolixus*, has also been shown to modulate gene expression (Orchard 2006). An increase in the expression of *RpCBP-1* in the posterior midgut of *R. prolixus* is observed following blood feeding, or after the injection of serotonin. Whereas, a reduction in *RpCBP-1* expression occurs following the injection of spiperone, a serotonin antagonist. These findings strongly indicate the involvement of serotonin in regulating lipid uptake in the midgut of hematophagous insects (Alves-Bezerra et al. 2010). Likewise, in *Aedes aegypti*, the upregulation of various genes associated with lipid metabolism and absorption in the intestine occurs following a blood meal. These genes encompass the FATP gene, responsible for fatty acid transport, as well as those involved in fatty acid synthesis (such as FAS and FAD) and the synthesis of phosphatidic acid (including G3P acyltransferase and AGPAT) (Sanders et al. 2003). Taken together, these findings strongly indicate that serotonin might assume a substantial role in lipid metabolism, particularly in the intricate process of lipogenesis.

2.1.5 Short Neuropeptide F and CCHamide-2

Short neuropeptide F (sNPF) is normally expressed by lateral neurosecretory cells in the brain, as well as by endocrine cells in the midgut (Reiher et al. 2011), the hindgut (Caers et al. 2016), the antennae, Malpighian tubules, and ovaries (Jiang et al. 2017). sNPFs bind to receptors (the sNPFR) which belong to the GPCR superfamily as other neuropeptide receptors. One important function of sNPFs is to regulate feeding behavior (Lee et al. 2004; Dillen et al. 2013) in addition to other physiological roles such as learning and memory, behavioral timing, sensitization to odors, etc. Overexpression of *sNPF* produces larger flies

(Lee et al. 2004) and also induces feeding in *B. mori* (Nagata et al. 2011). However, there are other reports that suggest an inhibitory effect of sNPFs on feeding (Dillen et al. 2013; Nässel et al. 2008; Christ et al. 2018). sNPFs have also been shown to influence lipid metabolism, e.g., downregulation of sNPF reduces food intake (Lee et al. 2004), increases starvation sensitivity (Kahsai et al. 2010), and results in lean fly phenotype in *Drosophila* (Baumbach et al. 2014a). A direct involvement of sNPF is evident from experiments where the silencing of sNPF in circadian clock neurons resulted in a decrease in TG levels. This observation further indicates a role for sNPF in promoting lipogenesis (Toprak et al. 2020).

The peptide hormone CCHamide2 is synthesized in the brain, fat body, and intestine in *Drosophila* (Ren et al. 2015; Sano et al. 2015). When flies lack CCHamide2, they exhibit diminished feeding activity, indicating that CCHamide2 likely has a stimulating influence on feeding behavior (Ren et al. 2015). It binds to the CCHamide-2 receptor (CCHamide2R) in the IPCs to stimulate insulin signaling (Ren et al. 2015; Sano et al. 2015). Disruption of CCHamide2R has been associated with reduced ILP concentrations and larval growth (Sano et al. 2015). The effects of CCHamide2 on lipid metabolism might thus be lipogenic and it exerts its effects indirectly through its interaction with the insulin signaling pathway.

2.2 Lipolytic Pathway

This is the pathway that participates in the breakdown of lipids. In this process, the TGs are hydrolyzed to fatty acids. The initial step in the lipolytic process involves the hydrolysis of TGs to produce DGs and monoglycerides (MGs). Subsequent transformations result in the production of three fatty acid molecules along with one glycerol molecule. The regulation of the lipolytic pathway is governed by lipid storage droplet-1 protein (Lsd1, perilipin), which can either stimulate or inhibit the hydrolysis or phosphorylation of TGs.

In insects, perilipin's activity has been shown to increase during periods of starvation and due to the action of glucocorticoids, but it decreases after food consumption (Arrese and Soulages 2010). Importantly, perilipin plays a pivotal role in promoting the phosphorylation process and facilitating the translocation of lipase enzymes from the cytosol to the surface of fat droplets. This translocation is essential for initiating TG hydrolysis at the fat droplet surface (Arrese and Soulages 2010; Sztalryd et al. 2013). The primary catalysts in driving the lipolytic process include Lsd1 and Brummer lipase, along with acid lipase and hormone-sensitive lipase (HSL). The activation of Brummer lipase is further influenced by Lsd2, which undergoes phosphorylation mediated by protein kinase A (PKA). Notably, PKA activation is triggered by cyclic adenosine monophosphate (cAMP), a molecule produced in response to the binding of adipokinetic hormone (AKH) to its receptor (AKHR) (Arrese and Soulages 2010). Thus, lipolysis can be promoted in response to nutritional cues, such as hunger.

An additional mechanism triggering lipolysis involves hedgehog signaling (Hh), which diminishes TG levels and hampers the formation of lipids (Jimenez-Sanchez et al. 2012). These intricate regulatory processes highlight the multifaceted nature of lipid metabolism and how it can be influenced by various signals and pathways, ultimately serving the metabolic needs of the organism. The principal hormones regulating the lipolytic pathway are discussed below.

2.2.1 Adipokinetic Hormone (AKH)

The adipokinetic hormone (AKH) is a neuropeptide produced by the neurosecretory cells located in the *corpora cardiaca* (Beenakkers et al. 1985; Goldsworthy and Mordue 1989; Goldsworthy et al. 1997; Van der Horst 2003). It is a relatively short peptide consisting of 8–10 amino acids and shares similarities in its action with glucagon (Beenakkers et al. 1984; Kim and Rulifson 2004). Initially, the hormone is synthesized as a prohormone, and its active form is released upon cleavage of AKH from the adipokinetic hormone precursor-related peptide (APRP). Besides its production in neurosecretory cells, AKH is also expressed in various tissues, including the ovaries, midgut, fat body, and muscle tissues (Abdel-latief and Hoffmann 2007; Kaufmann et al. 2009). It must be mentioned here that this evidence is primarily based on mRNA expression and not on the presence of the actual peptide. While there are pleotropic roles attributed to AKH, its best characterized role is primarily to initiate the mobilization of lipids and carbohydrates from the fat body (Van der Horst et al. 2001). Lipid mobilization occurs via the action of AKH on triglyceride-lipase (TGL) (Arrese and Soulages 2010; Arrese et al. 2006). The AKH system functions in response to rapid changes in lipid demands/energetic demands.

AKH exerts its effects by binding to membrane receptors known as adipokinetic hormone receptors (AKHR), which are G-protein coupled receptors (GPCRs) (Canavoso and Wells 2001; Park et al. 2002; Staubli et al. 2002; Lindemans et al. 2009). AKHR is found in various tissues, including the fat body, midgut, muscles, brain, and reproductive organs (Kaufmann and Brown 2006; Ziegler et al. 2011; Zandawala et al. 2015; Hou et al. 2017). Upon binding to AKHR, AKH can activate two distinct signaling pathways:

(a) In one pathway, AKH binding to AKHR triggers the activation of the G_q protein, which subsequently stimulates phospholipase C (PLC). PLC cleaves the membrane lipid phosphatidylinositol 4,5-diphosphate (PIP2) into inositol-1,4,5 triphosphate (IP3) and diacylglycerol (DAG). The release of IP3 leads to the release of intracellular calcium ions (Ca^{2+}) from the endoplasmic reticulum, serving as a signal for lipid mobilization.
(b) In the other pathway, AKH increases adenylate cyclase activity, leading to elevated intracellular levels of cyclic adenosine monophosphate (cAMP). This, in turn, activates protein kinase A (PKA), which promotes the phosphorylation of downstream targets such as lipid droplets (LDs), TGL, and Perilipin 1/Lipid Storage Droplet-1 protein (PLIN1/LSD1) (Arrese and Wells 1994; Arrese et al. 1999, 2008).

The activation of both pathways by AKH has also been reported in response to stress situations (Bednářová et al. 2013). Additionally, AKH may also have a direct effect on gene expression, e.g., increased cAMP and PKA activity could trigger CREB (Altarejos and Montminy 2011). There is a direct association between AKH signaling and the activation of CREB, as demonstrated in *Drosophila melanogaster* (Iijima et al. 2009). Furthermore, the activation of protein kinase A (PKA) can induce the phosphorylation and subsequent inhibition of proteins belonging to the salt-inducible kinase 3 (SIK3) family (Toprak 2020). This subsequently triggers the translocation of histone deacetylase 4 (HDAC4) to the nucleus where it activates the FoxO transcription factor (Toprak 2020; Choi et al. 2015). This leads to activation of *Brummer* leading to lipolysis and fatty acid oxidation (Choi et al. 2015). However, the mechanisms behind *Brummer* lipase regulation are still unclear. *AKHR* mutants or larvae that are mutant in AKH in *Drosophila* have greater activation and expression of Brummer (Grönke et al. 2007), while overexpression of AKH leads to inhibition of lipase expression (Grönke et al. 2007). On the other hand, null mutations in *AKH* or *AKHR* result in an obese phenotype in *Drosophila* adults, while their overexpression has been shown to drastically reduce lipid stores (Grönke et al. 2007; Bharucha et al. 2008; Baumbach et al. 2014b; Gáliková et al. 2015). Recent findings have intriguingly unveiled a crucial role for the sex determination gene transformer (*tra*) in orchestrating the male-female disparity in fat storage (Wat et al. 2021). The involvement of Tra in promoting fat storage has been traced to its specific function in neurons, particularly in the adipokinetic hormone (Akh)-producing cells (APCs). The study brought to light a male bias in APC activity and Akh pathway function, where this sex-dependent regulation significantly influenced the difference in fat storage by curbing TG accumulation in males. The loss of Tra in females led to an elevation in Akh pathway activity, and the manipulation of the Akh pathway at a genetic level successfully mitigated the Tra-dependent effects on fat storage (Wat et al. 2021). These findings not only underscore the importance of Tra in shaping sex differences in fat storage but also highlight the existence of sex-specific nuances in lipid metabolism.

2.2.2 Juvenile Hormone (JH)

Juvenile hormone (JH) is a sesquiterpenoid that is produced by the cells of the *corpora allata* and serves a role in many processes related to insect development from larvae to imago and also in the process of vitellogenesis and maturation of eggs (Jindra et al. 2013). JH prompts the activation of a lipase in the cotton bollworm *H. armigera*, whereas 20-HE leads to its suppression. This suggests that JH plays an antagonistic role in relation to ecdysone concerning the regulation of the lipolytic pathway (Sui et al. 2008). There are other results that seem to indicate a role for JH in lipid metabolism. For example, JH has been shown to mediate lipid storage in the oocytes of the hematophagous triatomine *Dipetalogaster maxima* (Ramos et al. 2021). In mosquito *A. aegypti AeSCP-2 L1* (a sterol carrier protein encoding gene) is enhanced by JH, whereas *AeSCP-2 L2* does not respond to JH, but the latter is induced by 20-HE (Vyazunova et al. 2007). JH has been demonstrated to regulate the presence of protein granules, and the absence of JH serves as a signal for metamorphosis, triggering the breakdown of the larval fat body through cytolysis. Additionally, JH-1 analog causes the vacuolization of old trophocytes (De Oliveira and Cruz-Landim 2003; Liu et al. 2009).

2.3 Other Players in Lipid Metabolism

In addition to SNPF, there are several other peptides which play a role in lipid metabolism such as neuropeptide F (NPF), allatostatin-A, corazonin, leukokinin, tachykinins, limostatins, and cytokines.

There is limited evidence for involvement of NPF—activation of *NPF*-expressing neurons in adult *Drosophila* has been linked to a reduction in TG levels (Chung et al. 2017). Conversely, inhibiting the activity of *NPF*-expressing neurons and downregulating *NPF* has been shown to reverse this effect (Gendron et al. 2014). Allatostatin-A (AstA) has primarily been shown to be expressed in the brain and gut of insects (Veenstra et al. 2008; Hentze et al. 2015). Its activity has been shown to regulate other hormones such as AKH and ILPs (Veenstra et al. 2008) through its receptor *Drosophila* Allatostatin Receptor 2 (DAR2). AstA stimulates both insulin and AKH signaling pathways by binding to the DAR2 receptor. Silencing of DAR2 leads to a decrease in both ILPs and AKH, resulting in increased resistance to starvation (Hentze et al. 2015; Buch et al. 2008; Nässel and Vanden 2015). In flies devoid of AstA, a significant buildup of lipids in the fat body is evident, suggesting that diminished AKH and ILP signaling also contributes to lipid accumulation in this tissue (Hentze et al. 2015). Conversely, AstA plays a role in the response to starvation by inhibiting starvation-induced feeding behavior, ultimately promoting lipolysis (Hergarden et al. 2012).

Corazonin (Crz) is a neuropeptide synthesized by the lateral neurosecretory cells in the brain (Duan-Sahbaz and Ýyison 2018). It is evolutionarily related to AKH (Veenstra 1994), and its expression is reduced by approximately 50% in AKH-mutants (Veenstra 1994). Crz has been shown to have an impact on both lipid and carbohydrate metabolism (Kapan et al. 2012; Kubrak et al. 2016; Gáliková et al. 2017). Changes in *Crz* expression have been demonstrated to affect trehalose levels and alter the concentrations of lipids and carbohydrates (Kapan et al. 2012; Kubrak et al. 2016; Zhao et al. 2010). It is likely that Crz's effect on lipid metabolism is indirect, possibly mediated through insulin signaling or other pathways, though further research is needed to clarify this relationship (Toprak 2020).

Leukokinin (Lk) is a myotropic neuropeptide produced in various tissues, including the brain, ventral ganglia, and ILP-producing cells. It primarily plays a crucial role in regulating ion and water balance in the Malpighian tubules and hindgut (Holman and Cook 1983; Schoofs et al. 1992). Lk's influence on lipid metabolism is likely indirect and occurs through its interaction with the insulin signaling pathway (Zandawala et al. 2018). Lk is likely involved in the lipid metabolism-related starvation response. The impact of leukokinin (Lk) on the starvation response appears to be mediated through its interaction with insulin signaling. Flies subjected to targeted knockdown of Lk receptor (LkR) in IPCs, as well as those with mutations in both *Lk* and *LkR*, exhibited elevated expression of *DILP2* and *DILP3* in IPCs, leading to increased resistance to starvation (Toprak 2020).

Tachykinins (Tk) is a neuropeptide produced by both the central nervous system and the intestine. When gut Tk is absent, TG levels have been found to significantly increase in the midgut, fat body, and hemolymph. Silencing the receptor for tachykinins (TkR) in *Drosophila* results in the increased expression of *ILP2* and *ILP3* in well-fed flies. Intriguingly, while *ILP2* levels were elevated in starved flies, the expression of *ILP3* decreased under starvation conditions. This implies that Tachykinins likely exert an influence on insulin signaling, suggesting a potential role in modulating lipid metabolism in the midgut (Song et al. 2014; Birse et al. 2011).

Limostatins (Lst) are expressed in the AKH-producing cells of the *corpora cardiaca* and fat body in situations of starvation (Birse et al. 2011). Increased levels of *DILP2*, *DILP3*, and *DILP5* mRNA were found in *Lst*-mutant flies which were hypoglycemic. In an AKH-deficient background, the Lst levels were decreased, whereas overexpression of AKH did not affect *Lst* expression levels (Alfa et al. 2015). Lst action is lipolytic and this occurs through suppression of insulin signaling.

Cytokines or adipokines are secreted by the fat body trophocytes (adipocytes) (Unpaired 2) and by the brain (Unpaired 1). Flies in which *Upd2* is knocked down in the fat body show elevated accumulation of ILPs in the brain when they are in a fed state (Rajan and Perrimon 2012). This implies that Upd2 acts on IPCs, affecting the release of ILPs and the activation of insulin signals. This connects the fat body to the brain in regulating lipid metabolism. Notably, it has been demonstrated that the adipokine leptin/Upd2 plays a role in reducing inhibitory signals to IPCs (Brent and Rajan 2020). In situations of nutrient abundance, Upd2 regulates the reorganization of actin-based synapses, leading to a decrease in bouton number within an inhibitory circuit. This establishes a neural tone conducive to insulin release (Brent and Rajan 2020). This underscores the growing understanding that inter-tissue communication is crucial for regulating organismal energy homeostasis. Muscle tissue is emerging as a key player in this homeostatic control by consuming lipids and governing systemic signaling networks. In *Drosophila*, a study has uncovered that the FOXO transcription factor regulates the expression of *Upd2* in skeletal muscle, acting as a myokine to control AKH secretion. Consequently, this regulation can influence lipid homeostasis in the fat body/adipose tissue and the intestine (Zhao and Karpac 2017).

3 Conclusions and Future Perspectives

The study of lipid metabolism in insects offers a window of opportunity to study dysregulation of lipid metabolism related to human disorders. While much of the information on lipid metabolism in insects stems from work on the fruit fly *D. melanogaster*, it is important to extend such studies on other insects along with exploring specific signaling mechanisms and pathways involved. The fat body is the target of most hormones in the metabolism of lipids but other organs such as midgut, Malpighian tubules, and ovaries are also impacted. There are several outstanding questions with regard to endocrine control of lipid metabolism such as what pathways are the direct targets of AKH and conversely what triggers AKH release? While SREBP is involved in the regulation of lipid metabolism, does it also function in lipid digestion? Are there other players in the regulation of lipid metabolism? What are the molecular underpinnings of sex differences in lipid metabolism in insects? Finally, it can be expected that none of the pathways discussed above work in isolation, what are the discrete pathways whereby these signaling systems influence each other, i.e., crosstalk? Future work in these and other areas of lipid metabolism and its regulation by the endocrine systems in insects may offer clues on analogous processes that could be operative in humans.

References

Abdel-latief M, Hoffmann KH (2007) The adipokinetic hormones in the fall armyworm, *Spodoptera frugiperda*: cDNA cloning, quantitative real time RT-PCR analysis, and gene specific localization. Insect Biochem Mol Biol 37:999–1014

Alabaster A, Isoe J, Zhou G, Lee A, Murhy A, Day WA, Miesfeld RL (2011) Deficiencies in acetyl-CoA carboxylase and fatty acid synthase 1 differentially affect eggshell formation and blood meal digestion in *Aedes aegypti*. Insect Biochem Mol Biol 41(12):946–955

Alfa RW, Park S, Skelly KR, Poffenberger G, Jain N, Gu X, Kockel L, Wang J, Liu Y, Powers AC, Kim SK (2015) Suppression of insulin production and secretion by a decretin hormone. Cell Metab 21:323–333

Altarejos JY, Montminy M (2011) CREB and CRTC co-activators: sensors for hormonal and metabolic signals. Nat Rev Mol Cell Biol 12(3):141–151

Alves-Bezerra M, Majerowicz D, Grillo LA, Tremonte H, Almeida CB, Braz GR, Sola-Penna M, Paiva-Silva GO, Gondim KC (2010) Serotonin regulates an acyl-CoA-binding protein (ACBP) gene expression in the midgut of *Rhodnius prolixus*. Insect Biochem Mol Biol 40:119–125

Anand AN, Lorenz MW (2008) Age-dependent changes of fat body stores and the regulation of fat body lipid synthesis and mobilization by adipokinetic hormone in the last larval instar of the cricket, *Gryllus bimaculatus*. J Insect Physiol 54(10–11):1404–1412

Arrese EL, Soulages JL (2010) Insect fat body: energy, metabolism, and regulation. Annu Rev Entomol 55: 207–225

Arrese EL, Wells MA (1994) Purification and properties of a phosphorylatable triacylglycerol lipase from the fat body of an insect, *Manduca sexta*. J Lipid Res 35: 1652–1660

Arrese EL, Flowers MT, Gazard JL, Wells MA (1999) Calcium and cAMP are second messengers in the adipokinetic hormone-induced lipolysis of triacylglycerols in *Manduca sexta* fat body. J Lipid Res 40:556–564

Arrese EL, Patel RT, Soulages JL (2006) The main triglyceride-lipase from the insect fat body is an active phospholipase A(1): identification and characterization. J Lipid Res 47:2657–2667

Arrese EL, Mirza S, Rivera L, Howard AD, Chetty PS, Soulages JL (2008) Expression of lipid storage droplet protein-1 may define the role of AKH as a lipid mobilizing hormone in *Manduca sexta*. Insect Biochem Mol Biol 38:993–1000

Baumbach J, Hummel P, Bickmeyer I, Kowalczyk KM, Frank M, Knorr K, Hildebrandt A, Riedel D, Jäckle H, Kühnlein RP (2014a) A *Drosophila* in vivo screen identifies store-operated calcium entry as a key regulator of adiposity. Cell Metabol 19:331–343

Baumbach J, Xu Y, Hehlert P, Kühnlein RP (2014b) Gaq, Gg1 and Plc21C control *Drosophila* body fat storage. J Genet Genomics 41:283–292

Bednářová A, Kodrík D, Krishnan N (2013) Adipokinetic hormone exerts its anti-oxidative effects using a conserved signal-transduction mechanism involving both PKC and cAMP by mobilizing extra- and intracellular Ca^{2+} stores. Comp Biochem Physiol C Toxicol Pharmacol 158:142–149

Beenakkers A, Vanderhorst D, Vanmarrewijk W (1984) Insect flight metabolism. Insect Biochem 14:243–260

Beenakkers AMT, Van der Horst DJ, Van Marrewijk WJA (1985) Biochemical processes directed to flight muscle metabolism. In: Kerkut GA, Gilbert LL (eds) Comprehensive insect physiology, biochemistry and pharmacology. Pergamon Press, Oxford, pp 451–486

Benayoun BA, Caburet S, Veitia RA (2011) Forkhead transcription factors: key players in health and disease. Trends Genet 27(6):233–240

Bharucha KN, Tarr P, Zipursky SL (2008) A glucagon-like endocrine pathway in *Drosophila* modulates both lipid and carbohydrate homeostasis. J Exp Biol 211: 3103–3110

Birse RT, Soderberg JA, Luo J, Winther AM, Nässel DR (2011) Regulation of insulin-producing cells in the adult *Drosophila* brain via the tachykinin peptide receptor DTKR. J Exp Biol 214:4201–4208

Böhni R, Riesgo-Escovar J, Oldham S, Brogiolo W, Stocker H, Andruss BF, Beckingham K, Hafen E (1999) Autonomous control of cell and organ size by CHICO, a *Drosophila* homolog of vertebrate IRS1-4. Cell 97:865–875

Brent AE, Rajan A (2020) Insulin and leptin/Upd2 exert opposing influences on synapse number in fat-sensing neurons. Cell Metab 32(5):786–800.e7

Brown JR, Auger KR (2011) Phylogenomics of phosphoinositide lipid kinases: perspectives on the evolution of second messenger signaling and drug discovery. BMC Evol Biol 11:4

Brown MR, Clark KD, Gulia M, Zhao Z, Garczynski SF, Crim JW, Suderman RJ, Strand MR (2008) An insulin-like peptide regulates egg maturation and metabolism in the mosquito *Aedes aegypti*. Proc Natl Acad Sci USA 105:5716–5721

Buch S, Melcher C, Bauer M, Katzenberger J, Pankratz MJ (2008) Opposing effects of dietary protein and sugar regulate a transcriptional target of *Drosophila* insulin-like peptide signaling. Cell Metab 7:321–332

Cáceres L, Necakov AS, Schwartz C, Kimber A, Roberts IJ, Krause HM (2011) Nitric oxide coordinates metabolism, growth and development via the nuclear receptor E75. Genes Dev 25(14):1476–1485

Caers J, Peymen K, Van Hiel MB, Van Rompay L, Van Den Abbeele J, Schoofs L, Beets I (2016) Molecular characterization of a short neuropeptide F signaling system in the tsetse fly, *Glossina morsitans morsitans*. Gen Comp Endocrinol 235:142–149

Canavaso LE, Frede S, Rubiolo ER (2004) Metabolic pathways for dietary lipids in the midgut of hematophagous *Panstrongylus megistus* (Heliptera: Reduviidae). Insect Biochem Mol Biol 34(8):845–854

Canavoso LE, Wells MA (2001) Role of lipid transfer particle in delivery of diacylglycerol from midgut to lipophorin in larval *Manduca sexta*. Insect Biochem Mol Biol 31:783–790

Canavoso L, Jouni Z, Karnas K, Pennington J, Wells M (2001) Fat metabolism in insects. Annu Rev Nutr 21: 23–46

Chen C, Jack J, Garofalo RS (1996) The *Drosophila* insulin receptor is required for normal growth. Endocrinology 137:846–856

Choi S, Lim DS, Chung J (2015) Feeding and fasting signals converge on the LKB1-SIK3 pathway to regulate lipid metabolism in *Drosophila*. PLoS Genet 11: e1005263

Christ P, Hill SR, Schachtner J, Hauser F, Ignell R (2018) Functional characterization of mosquito short neuropeptide F receptors. Peptides 103:31–39

Christeller JT, Amara S, Carrière F (2011) Galactolipase, phospholipase and triacylglycerol lipase activities in the midgut of six species of lepidopteran larvae feeding on different lipid diets. J Insect Physiol 57(9): 1232–1239

Chung BY, Ro J, Hutter SA, Miller KM, Guduguntla LS, Kondo S, Pletcher SD (2017) *Drosophila* neuropeptide F signaling independently regulates feeding and sleep-wake behavior. Cell Rep 19:2441–2450

Coleman RA, Mashek DG (2011) Mammalian triacylglycerol metabolism: synthesis, lipolysis, and signaling. Chem Rev 111(10):6359–6386

Colombani J, Bianchini L, Layalle S, Pondeville E, Dauphin-Villemant C, Antoniewski C, Carré C, Noselli S, Léopold P (2005) Antagonistic actions of ecdysone and insulins determine final size in *Drosophila*. Science 310:667–670

De Oliveira VP, Cruz-Landim C (2003) Morphology and function of insect fat body cells: a review. Biociencias 11:195–205

Dillen S, Zels S, Verlinden H, Spit J, van Wielendaele P, Vanden BJ (2013) Functional characterization of the short neuropeptide F receptor in the desert locust, *Schistocerca gregaria*. PLoS One 8:e53604

Dobrosotskaya IY, Seegmiller AC, Brown MS, Goldstein JL, Rawson RB (2002) Regulation of SREBP processing and membrane lipid production by phospholipids in *Drosophila*. Science 296(5569):879–883

Downer RGH, Chino H (1985) Turnover of protein and diacylglycerol components of lipophorin in insect hemolymph. Insect Biochem 15(5):627–630

Duan-Sahbaz B, Ýyison NB (2018) Neuropeptides as ligands of GPCRs. In: Saravanan C, Biswas CB (eds) Ligand. InTech, London, pp 77–101

Fernandez R, Tabarini D, Azpiazu N, Frasch M, Schlessinger J (1995) The *Drosophila* insulin receptor homolog: a gene essential for embryonic development encodes two receptor isoforms with different signaling potential. EMBO J 14:3373–3384

Gade G (2004) Regulation of intermediary metabolism and water balance of insects by neuropeptides. Annu Rev Entomol 49:93–113

Gáliková M, Diesner M, Klepsatel P, Hehlert P, Xu Y, Bickmeyer I, Predel R, Kühnlein RP (2015) Energy homeostasis control in *Drosophila* adipokinetic hormone mutants. Genetics 201:665–683

Gáliková M, Klepsatel P, Xu Y, Kühnlein RP (2017) The obesity related adipokinetic hormone controls feeding and expression of neuropeptide regulators of *Drosophila* metabolism. Eur J Lipid Sci Technol 119:1600138

Gendron CM, Kuo TH, Harvanek ZM, Chung BY, Yew JY, Dierick HA, Pletcher SD (2014) *Drosophila* life span and physiology are modulated by sexual perception and reward. Science 343:544–548

Gershman B, Puig O, Hang L, Peitzsch RM, Tatar M, Garofalo RS (2007) High-resolution dynamics of the transcriptional response to nutrition in *Drosophila*: a key role for dFOXO. Physiol Genomics 29(1):24–34

Gilby A (1965) Lipids and their metabolism in insects. Annu Rev Entomol 10:141–160

Goldsworthy G, Mordue W (1989) Adipokinetic hormones: functions and structures. Biol Bull 177: 218–224

Goldsworthy GJ, Lee MJ, Luswata R, Drake AF, Hyde D (1997) Structures assays and receptors for adipokinetic hormones. Comp Biochem Physiol B 117:483–496

Grillo LAM, Majerowicz D, Gondim KC (2007) Lipid metabolism in *Rhodnius prolixus* (Hemiptera: Rediviidae): role of a midgut triacylglycerol-lipase. Insect Biochem Mol Biol 37(6):579–588

Grönke S, Müller G, Hirsch J, Fellert S, Andreou A, Haase T, Jäckle H, Kühnlein RP (2007) Dual lipolytic control of body fat storage and mobilization in *Drosophila*. PLoS Biol 5:e137

Guo XR, Zheng SC, Liu L, Feng QL (2009) The sterol carrier protein 2/3-oxoacyl-CoA thiolase (SCPx) is involved in cholesterol uptake in the midgut of *Spodoptera litura*: gene cloning, expression, localization and functional analysis. BMC Mol Biol 10:102

Hasegawa K, Yamashita O (1965) Studies on the mode of action of the diapause hormone in the silkworm, *Bombyx mori* L.-VI. The target organ of the diapause hormone. J Exp Biol 43:271–277

Heier C, Kühnlein RP (2018) Triacylglycerol metabolism in *Drosophila melanogaster*. Genetics 210:1163–1184

Hentze JL, Carlsson MA, Kondo S, Nässel DR, Rewitz KF (2015) The neuropeptide Allatostatin a regulates metabolism and feeding decisions in *Drosophila*. Sci Rep 5:11680

Hergarden AC, Tayler TD, Anderson DJ (2012) Allatostatin-a neurons inhibit feeding behavior in adult *Drosophila*. Proc Natl Acad Sci USA 109:3967–3972

Holman GM, Cook BJ (1983) Isolation and partial characterization of a second myotropic peptide from the hindgut of the cockroach, *Leucophaea maderae*. Comp Biochem Physiol C 76:39–43

Hou QL, Chen EH, Jiang HB, Wei DD, Gui SH, Wang JJ, Smagghe G (2017) Adipokinetic hormone receptor gene identification and its role in triacylglycerol mobilization and sexual behavior in the oriental fruit fly (*Bactrocera dorsalis*). Insect Biochem Mol Biol 90: 1–13

Iijima K, Zhao L, Shenton C, Iijima-Ando K (2009) Regulation of energy stores and feeding by neuronal and peripheral CREB activity in *Drosophila*. PLoS One 4: e8498

Ikeya T, Galic M, Belawat P, Nairz K, Hafen E (2002) Nutrient dependent expression of insulin-like peptides from neuroendocrine cells in the CNS contributes to growth regulation in *Drosophila*. Curr Biol 12:1293–1300

Jeon TI, Osborne TF (2012) SREBPs: metabolic integrators in physiology and metabolism. Trends Endocrinol Metab 23(2):65–72

Jiang HB, Gui SH, Xu L, Pei YX, Smagghe G, Wang JJ (2017) The short neuropeptide F modulates olfactory sensitivity of *Bactrocera dorsalis* upon starvation. J Insect Physiol 99:78–85

Jimenez-Sanchez M, Menzies FM, Chang YY, Simecek N, Neufeld TP, Rubinsztein DC (2012) The hedgehog signalling pathway regulates autophagy. Nat Commun 3:1200

Jindra M, Palli SR, Riddiford LM (2013) The juvenile hormone signaling pathway in insect development. Annu Rev Entomol 58:181–204

Juoni ZE, Zamora J, Wells MA (2002) Absorption and tissue distribution of cholesterol in *Manduca sexta*. Arch Insect Biochem Physiol 49(3):167–175

Kahsai L, Kapan N, Dircksen H, Winther AM, Nässel DR (2010) Metabolic stress responses in *Drosophila* are modulated by brain neurosecretory cells that produce multiple neuropeptides. PLoS One 5:e11480

Kamoshida Y, Fujiyama-Nakamura S, Kimura S, Suzuki E, Lim J, Shiozaki-Sato Y, Kato S, Takeyama K (2012) Ecdysone receptor (EcR) suppresses lipid accumulation in the *Drosophila* fat body via transcriptional control. Biochem Biophys Res Commun 421(2):203–207

Kannan K, Fridell YWC (2013) Functional implications of Drosophila insulin-like peptides in metabolism, aging, and dietary restriction. Front Physiol 4:288

Kapan N, Lushchak OV, Luo J, Nässel DR (2012) Identified peptidergic neurons in the *Drosophila* brain regulate insulin-producing cells, stress responses and metabolism by coexpressed short neuropeptide F and corazonin. Cell Mol Life Sci 69:4051–4066

Kaufmann C, Brown MR (2006) Adipokinetic hormones in the African malaria mosquito, Anopheles gambiae: identification and expression of genes for two peptides and a putative receptor. Insect Biochem Mol Biol 36: 466–481

Kaufmann C, Merzendorfer H, Gäde G (2009) The adipokinetic hormone system in Culicinae (Diptera: Culicidae): molecular identification and characterization of two adipokinetic hormone (AKH) precursors from *Aedes aegypti* and *Culex pipiens* and two putative AKH receptor variants from *A. aegypti*. Insect Biochem Mol Biol 39:770–781

Keeley LL (1985) Biochemistry and physiology of the insect fat body. In: Kerkut GA, Gilbert LI (eds) Comprehensive insect physiology, biochemistry and pharmacology, vol 3. Pergamon, New York, pp 211–228

Kim SK, Rulifson EJ (2004) Conserved mechanisms of glucose sensing and regulation by Drosophila *corpora cardiaca* cells. Nature 431:316–320

Knudsen J, Neergaard TBF, Gaigg B, Jensen MV, Hansen JK (2000) Role of acyl-CoA binding protein in acyl-CoA metabolism and acyl-CoA-mediated cell signaling. J Nutr 130(2):294S–298S

Kramer JM, Davidge JT, Lockyer JM, Staveley BE (2003) Expression of Drosophila FOXO regulaes growth and can phenocopy starvation. BMC Dev Biol 3:5

Kubrak OI, Lushchak OV, Zandawala M, Nässel DR (2016) Systemic corazonin signalling modulates stress responses and metabolism in *Drosophila*. Open Biol 6: 160152

Kühnlein RP (2012) Lipid droplet-based storage fat metabolism in Drosophila. J Lipid Res 53:1430–1436

Kunte AS, Matthews KA, Rawson RB (2006) Fatty acid auxotrophy in *Drosophila* larvae lacking SREBP. Cell Metab 3(6):439–448

Law JH, Wells MA (1989) Insects as biochemical models. J Biol Chem 264:16335–16338

Lee KS, You KH, Choo JK, Han YM, Yu K (2004) *Drosophila* short neuropeptide F regulates food intake and body size. J Biol Chem 279:50781–50789

Lindemans M, Liu F, Janssen T, Husson SJ, Mertens I, Gaede G, Schoofs L (2009) Adipokinetic hormone signaling through the gonadotropin releasing hormone receptor modulates egg-laying in *Caenorhabditis elegans*. Proc Natl Acad Sci USA 106:1642–1647

Liu M, Zhang TY, Xu WH (2005) A cDNA encoding diazepam-binding inhibitor/acyl-CoA-binding protein in *Helicoverpa armigera*: molecular characterization and expression analysis associated with pupal diapause. Comp Biochem Physiol C 141(2):168–176

Liu Y, Liu H, Liu S, Wang S, Jiang RJ, Li S (2009) Hormonal and nutritional regulation of insect fat body development and function. Arch Insect Biochem Physiol 71:16–30

Lorenz MW, Anand AN (2004) Changes in the biochemical composition of fat body stores during adult development of female crickets, *Gryllus bimaculatus*. Arch Insect Biochem Physiol 56(3):110–119

Margam VM, Gelman DB, Palli SR (2006) Ecdysteroid titers and developmental expression of ecdysteroid-regulated genes during metamorphosis of the yellow fever mosquito, *Aedes aegypti* (Diptera:Culicidae). J Insect Physiol 52(6):558–568

Mead JF (1986) Lipids: chemistry, biochemistry, and nutrition. Plenum Press, New York

Menendez JA, Lupu R (2007) Fatty acid synthase and the lipogenic phenotype in cancer pathogenesis. Nat Rev Cancer 7:763–777

Nagata S, Morooka N, Matsumoto S, Kawai T, Nagasawa H (2011) Effects of neuropeptides on feeding initiation in larvae of the silkworm, *Bombyx mori*. Gen Comp Endocrinol 172:90–99

Nässel DR, Vanden BJ (2015) Insulin/IGF signaling in *Drosophila* and other insects: factors that regulate production, release and post-release action of the insulin-like peptides. Cell Mol Life Sci 73:271–290

Nässel DR, Enell LE, Santos JG, Wegener C, Johard HA (2008) A large population of diverse neurons in the *Drosophila* central nervous system expresses short neuropeptide F, suggesting multiple distributed peptide functions. BMC Neurosci 9:90

Orchard I (2006) Serotonin: a coordinator of feeding-related physiological events in the blood-gorging bug, *Rhodnius prolixus*. Comp Biochem Physiol A 144(3): 316–324

Palm W, Sampaio JL, Brankatschk M, Carvalho M, Mahmoud A, Shevchenko A, Eaton S (2012) Lipoproteins in *Drosophila melanogaster*—assembly, function, and influence on tissue lipid composition. PLoS Genet 8(7):e1002828

Park Y, Kim YJ, Adams ME (2002) Identification of G protein-coupled receptors for *Drosophila* PRXamide peptides, CCAP, corazonin, and AKH supports a theory of ligand-receptor coevolution. Proc Natl Acad Sci USA 99:11423–11428

Parvy J-P, Napal L, Rubin T, Poidevin M, Perrin L, Wicker-Thomas C, Montagne J (2012) *Drosophila melanogaster* acetyl-CoA-carboxylase sustains a fatty acid-dependent remote signal to waterproof the respiratory system. PLoS Genet 8:e1002925

Patel RT, Soulages JL, Hariharasundaram B, Arrese EL (2005) Activation of the lipid droplet controls the rate of lipolysis of triglycerides in the insect fat body. J Biol Chem 280:22624–22631

Portsmann T, Santos CR, Griffiths B, Cully M, Wu M, Leevers S, Griffiths JR, Cung Y-L, Schulze A (2008) SREBP activity is regulated by mTORC1 and contributes to Akt-dependent cell growth. Cell Metab 8(3):224–236

Puig O, Marr MT, Ruhf ML, Tjian R (2003) Control of cell number by Drosophila FOXO: downstream and feedback regulation of insulin receptor pathway. Genes Dev 17(16):2006–2020

Rajan A, Perrimon N (2012) *Drosophila* cytokine unpaired 2 regulates physiological homeostasis by remotely controlling insulin secretion. Cell 151:123–137

Ramos FO, Leyria J, Nouzova M, Fruttero LL, Noriega FG, Canavoso LE (2021) Juvenile hormone mediates lipid storage in the oocytes of *Dipetalogaster maxima*. Insect Biochem Mol Biol 133:103499

Reiher W, Shirras C, Kahnt J, Baumeister S, Isaac RE, Wegener C (2011) Peptidomics and peptide hormone processing in the *Drosophila midgut*. J Proteome Res 10:1881–1892

Reinking J, Lam MM, Pardee K, Sampson HM, Liu S, Yang P, Williams S, White W, Lajoie G, Edwards A, Krause HM (2005) The Drosophila nuclear receptor e75 contains heme and is gas responsive. Cell 122(2): 195–207

Reis T, Van Gilst MR, Hariharan IK (2010) A buoyancy-based screen of *Drosophila* larvae for fat-storage mutants reveals a role for Sir2 in coupling fat storage to nutrient availability. PLoS Genet 6(11):e1001206

Ren GR, Hauser F, Rewitz KF, Kondo S, Engelbrecht AF, Didriksen AK, Schjøtt SR, Sembach FE, Li S, Søgaard KC, Søndergaard L, Grimmelikhuijzen CJP (2015) CCHamide-2 is an orexigenic brain-gut peptide in *Drosophila*. PLoS One 10:e0133017

Roma GC, Bueno OC, Camargo-Mathias MI (2010) Morpho-physiological analysis of the insect fat body: a review. Micron 41:395–401

Sanders HR, Evans AM, Ross LS, Gill SS (2003) Blood meal induces global changes in midgut gene expression in the disease vector, *Aedes aegypti*. Insect Biochem Mol Biol 33(11):1105–1122

Sano H, Nakamura A, Texada MJ, Truman JW, Ishimoto H, Kamikouchi A, Nibu Y, Kume K, Ida T, Kojima M (2015) The nutrient-responsive hormone CCHamide-2 controls growth by regulating insulin-like peptides in the brain of *Drosophila melanogaster*. PLoS Genet 11:e1005209

Schoofs L, Holman GM, Proost P, Van Damme J, Hayes TK, De Loof A (1992) Locustakinin, a novel myotropic peptide from *Locusta migratoria*, isolation, primary structure and synthesis. Regul Pept 37:49–57

Schooley DA, Horodyski FM, Coast GM (2005) Hormones controlling homeostasis in insects: endocrinology. In: Gilbert LI, Iatrou K, Gill S (eds) Comprehensive molecular insect science, vol 3. Elsevier, Amsterdam, pp 493–550

Seegmiller AC, Dobrosotskaya I, Goldstein JL, Ho YK, Brown MS, Rawson RB (2002) The SREBP pathway in *Drosophila*: regulation by palmitate, not sterols. Dev Cell 2(2):229–238

Skowronek P, Wójcik L, Strachecka A (2021) Fat body—multifunctional tissue insects. Insects 12:547. https://doi.org/10.3390/insects12060547

Song W, Veenstra JA, Perrimon N (2014) Control of lipid metabolism by tachykinin in *Drosophila*. Cell Rep 9: 40–47

Song Y, Gu F, Liu Z, Li Z, Wu F, Sheng S (2022) The key role of fatty acid synthase in lipid metabolism and metamorphic development in a destructive insect pest *Spodoptera litura* (Lepidoptera: Noctuidae). Int J Mol Sci 23:9064

Staubli F, Jørgensen TJD, Cazzamali G, Williamson M, Lenz C, Søndergaard L, Roepstorf P, Grimmelikhuijzen CJP (2002) Molecular identification

of the insect adipokinetic hormone receptors. Proc Natl Acad Sci USA 99:3446–3451
Sui YP, Wang JX, Zhao XF (2008) Effects of classical insect hormones on the expression profiles of a lipase gene from the cotton bollworm (*Helicoverpa armigera*). Insect Mol Biol 17(5):523–529
Sun JX, Hiraoka T, Dittmer NT, Cho KH, Raikhel AS (2000) Lipophorin as a yolk protein precursor in the mosquito, *Aedes aegypti*. Insect Biochem Mol Biol 30(12):1161–1171
Sztalryd C, Xu G, Dorward H, Tansey JT, Contreras JA (2013) Perilipin A is essential for the translocation of hormone-sensitive lipase during lipolytic activation. J Cell Biol 161:1093–1103
Teleman AA (2009) Molecular mechanisms of metabolic regulation by insulin in Drosophila. Biochem J 425(1): 13–26
Thomas HE, Stunnenberg HG, Stewart AF (1993) Heterodimerization of the *Drosophila* ecdysone receptor with retinoid X receptor and ultraspiracle. Nature 362(6419):471–475
Toprak U (2020) The role of peptide hormones in insect lipid metabolism. Front Physiol 11:434
Toprak U, Hegedus D, Dogan C, Guney G (2020) A journey into the world of insect lipid metabolism. Arch Insect Biochem Physiol 104:e21682
Turunen S (1993) Metabolic pathways in the midgut epithelium of *Pieris brassicae* during carbohydrate and lipid assimilation. Insect Biochem Mol Biol 23(6):681–689
Valente V, Maia RM, Vianna MC, Paçó-Larson ML (2010) *Drosophila melanogaster* lipins are tissue-regulated and developmentally regulated and present in specific subcellular distributions. FEBS J 277(22): 4775–4788
Van der Horst DJ (2003) Insect adipokinetic hormones: release and integration of flight energy metabolism. Comp Biochem Physiol B 136:217–226
Van der Horst DJ, Ryan R (2012) Lipid transport. In: Insect molecular biology and biochemistry. Academic Press Ltd-Elsevier Science Ltd, pp 317–345
Van der Horst DJ, van Marrewijk WJ, Diederen JH (2001) Adipokinetic hormones of insect: release, signal transduction, and responses. Int Rev Cytol 211:179–240
Van Hoof D, Rodenburg KW, Van der Horst DJ (2003) Lipophorin receptor-mediated lipoprotein endocytosis in insect fat body cells. J Lipid Res 44(8):1431–1440
Van Hoof D, Rodenburg KW, Van der Horst DJ (2005) Receptor-mediated endocytosis and intracellular trafficking of lipoproteins and transferrin in insect cells. Insect Biochem Mol Biol 35(2):117–128
Veenstra JA (1994) Isolation and structure of the *Drosophila* corazonin gene. Biochem Biophys Res Commun 204:292–296
Veenstra JA, Agricola HJ, Sellami A (2008) Regulatory peptides in fruit fly midgut. Cell Tissue Res 334:499–516
Vyazunova I, Wessley V, Kim M, Lan Q (2007) Identification of two sterol carrier protein-2 like genes in the yellow fever mosquito, *Aedes aegypti*. Insect Mol Biol 16(3):305–314
Wang B, Goode J, Best J, Meltzer J, Schilman PE, Chen J, Garza D, Thomas JB, Montminy M (2008) The insulin-regulated CREB coactivator TORC promotes stress resistance in *Drosophila*. Cell Metab 7(5): 434–444
Wang B, Moya N, Niessen S, Hoover H, Mihaylova MM, Shaw RJ, Yates JR 3rd, Fischer WH, Thomas JB, Montminy M (2011) A hormone-dependent module regulating energy balance. Cell 145:596–606
Wat LW, Chowdhury ZS, Millington JW, Biswas P, Rideout EJ (2021) Sex determination gene transformer regulates the male-female difference in Drosophila fat storage via the adipokinetic hormone pathway. elife 10:e72350
Wicker-Thomas C, Garrido D, Bontonou G, Napal L, Mazuras N, Denis B, Rubin T, Parvy J-P, Montagne J (2015) Flexible origin of hydrocarbon/pheromone precursors in *Drosophila melanogaster*. J Lipid Res 56:2094–2101
Xiang Y, Liu Z, Huang X (2010) *br* regulates the expression of the ecdysone biosynthesis gene *npc1*. Dev Biol 344(2):800–808
Xu WH, Sato Y, Ikeda M, Yamashita O (1995) Molecular characterization of the gene encoding the precursor protein of diapause hormone and pheromone biosynthesis activating neuropeptide (DH-PBAN) of the silkworm *Bombyx mori* and its distribution in some insects. Biochim Biophys Acta 1261(1):83–89
Yang T, Espenshade PJ, Wright ME, Yabe D, Gong Y, Aebersold R, Goldstein JL, Brown MS (2002) Crucial step in cholesterol homeostasis: sterols promote binding of SCAP to INSIG-1, a membrane protein that facilitates retention of SREBPs in ER. Cell 110(4): 489–500
Yao TP, Seagraves WA, Oro AE, McKeown M, Evans RM (1992) *Drosophila* ultraspiracle modulates ecdysone receptor function via heterodimer formation. Cell 71(2):63–72
Yao TP, Forman BM, Jiang Z, Cherbas L, Chen JD, McKeown M, Cherbas P, Evans RM (1993) Functional ecdysone receptor is the product of *EcR* and *Ultraspiracle* genes. Nature 366(6454):476–479
Zandawala M, Hamoudi Z, Lange AB, Orchard I (2015) Adipokinetic hormone signalling system in the Chagas disease vector, *Rhodnius prolixus*. Insect Mol Biol 24: 264–276
Zandawala M, Yurgel ME, Texada MJ, Liao S, Rewitz KF, Keene AC, Nässel DR (2018) Modulation of *Drosophila* post-feeding physiology and behavior by the neuropeptide leucokinin. PLoS Genet 14: e1007767
Zhang TY, Sun JS, Zhang LB, Shen JL, Xu WH (2004a) Cloning and expression of the cDNA encoding the FXPRL family of peptides and a functional analysis of their effect on breaking pupal diapause in *Helicoverpa armigera*. J Insect Physiol 50(1):25–33

Zhang TY, Sun JS, Zhang QR, Xu J, Jiang RJ, Xu WH (2004b) The diapause hormone-pheromone biosynthesis activating neuropeptide gene of *Helicoverpa armigera* encodes multiple peptides that break rather than induce diapause. J Insect Physiol 50(6):547–554

Zhao X, Karpac J (2017) Muscle directs diurnal energy homeostasis through a myokine-dependent hormone module in Drosophila. Curr Biol 27(13):1941–1955.e6

Zhao Y, Bretz CA, Hawksworth SA, Hirsh J, Johnson EC (2010) Corazonin neurons function in sexually dimorphic circuitry that shape behavioral responses to stress in *Drosophila*. PLoS One 5:e9141

Ziegler R, Isoe J, Moore W, Riehle MA, Wells MA (2011) The putative AKH receptor of the tobacco hornworm, *Manduca sexta*, and its expression. J Insect Sci 11:40

Adv Exp Med Biol (2026) 1494: 217–237
https://doi.org/10.1007/5584_2025_875

Published online: 05 July 2025

Lipid Metabolism in Relation to Calcium Homeostasis

Umut Toprak

Abstract

Calcium (Ca^{2+}) homeostasis is a critical regulator of insect cellular functions, influencing neurotransmission, muscle contraction, hormone signaling, and lipid metabolism. This chapter explores the intricate relationship between Ca^{2+} signaling and lipid metabolism, emphasizing key molecular components that mediate this interaction. Store-operated calcium entry (SOCE) mechanisms, involving sarco/endoplasmic reticulum Ca^{2+}-ATPase (SERCA), inositol 1,4,5-trisphosphate receptor (IP_3R), ryanodine receptor (RyR), stromal interaction molecule (STIM), and Orai1, coordinate intracellular Ca^{2+} fluxes that regulate lipid storage, mobilization, and utilization. Other Ca^{2+}-binding proteins, such as calmodulin (CaM), calcineurin (CaN), regucalcin (RgN), calreticulin (CrT), and calnexin (CnX), further modulate Ca^{2+} homeostasis and impact lipid metabolism by influencing lipolysis, lipogenesis, and lipid droplet dynamics. This chapter also highlights the role of hepatocyte-like oenocytes in lipid metabolism. These cells, analogous to mammalian hepatocytes, regulate lipid processing and mobilization during fasting, forming a metabolic axis with fat body adipocytes. While Ca^{2+} signaling is well characterized in adipocytes, its role in oenocyte lipid metabolism remains largely unexplored. However, Ca^{2+}-dependent regulation of lipid metabolism in mammalian hepatocytes suggests a similar involvement in insect oenocytes. A central theme is the bidirectional relationship between Ca^{2+} homeostasis and lipid metabolism. While Ca^{2+} signaling regulates lipid accumulation and hydrolysis, impaired lipid metabolism can disrupt Ca^{2+} homeostasis. For instance, *Drosophila melanogaster seipin* mutants with defective lipid storage exhibit reduced SERCA activity, leading to lower ER and mitochondrial Ca^{2+} levels, which impair lipogenesis. Additionally, CaN promotes lipogenesis, whereas STIM and IP_3R serve as lipolytic regulators. This metabolic feedback loop is essential for maintaining energy balance. Understanding the Ca^{2+}-lipid interplay in insects provides insights into metabolic regulation, with implications for pest management and metabolic disease research. Future studies should further investigate Ca^{2+}-dependent mechanisms governing oenocyte function and systemic lipid homeostasis.

Keywords

Lipid · Calcium · SOCE · SERCA · Ip_3r · STIM · ORAI

U. Toprak (✉)
Molecular Entomology Laboratory, Faculty of Agriculture, Department of Plant Protection Ankara, Ankara University, Ankara, Türkiye
e-mail: utoprak@agri.ankara.edu.tr

1 Introduction

Calcium (Ca^{2+}) is the fifth most abundant element in the Earth's crust and derives its name from the Latin term *calx*, meaning "lime." Since limestone primarily comprises calcium carbonate ($CaCO_3$), the term eventually became synonymous with Ca^{2+}. However, the element was first isolated in 1808 by Sir Humphry Davy in England through the electrolysis of a lime and mercuric oxide mixture.

Calcium is essential for life and is critical in nearly all cell types across diverse organisms. It is best known for its importance in bone and dental health in humans, but its biological functions extend far beyond structural support. Calcium regulates numerous cellular and physiological processes, including cell motility, gene expression, vesicle fusion, nuclear pore function, apoptosis, muscle contraction, hormone biosynthesis, and posttranslational modifications (Berridge et al. 2000). Additionally, Ca^{2+} is a key signaling molecule in the nervous system, regulating neuronal excitability and neurotransmission (Ghosh and Greenberg 1995). Like its role in vertebrates, Ca^{2+} is crucial for insect physiology and development, regulating muscle function, neuronal activity, and metabolic processes.

Despite the central importance of Ca^{2+}, research on its roles in insects remains disproportionately limited compared to vertebrates. However, existing studies have demonstrated the involvement of Ca^{2+} in various physiological and metabolic processes, including development (Kaneuchi et al. 2015), molting (Gu et al. 1998; De Loof and Schoofs 2019), muscle activity and flight (Domingo et al. 1998; Gordon and Dickinson 2006), neurotransmitter release (Bronk et al. 2018), nephrocyte function (Sivakumar et al. 2022), cold sensing (Teets et al. 2013), reproduction (Takeo et al. 2006), olfactory responses (Bahk and Jones 2016), sex pheromone synthesis (Yoshiga et al. 2002), diapause maintenance (Doğan et al. 2021b), and carbohydrate metabolism (Pallen and Steele 1988). Notably, Ca^{2+} also plays a critical role in lipid metabolism (Doğan et al. 2021c).

Research on the interaction between lipid metabolism and Ca^{2+} in mammals has laid the groundwork for similar studies in insects. Initial investigations focused on the effects of dietary Ca^{2+} on lipid metabolism, demonstrating its influence on adipocyte function and fat storage across various models, including mammals and insects (Shi et al. 2001; Zemel 2002; Jacqmain et al. 2003; Alomaim et al. 2019; Doğan et al. 2021a). For instance, dietary Ca^{2+} inhibits adipocyte lipid accumulation and body fat gain in mouse models and human beings (Zemel et al. 2000; Shi et al. 2001; Sun and Zemel 2004; Alomaim et al. 2019). Similarly, *Spodoptera littoralis* larvae fed a Ca^{2+}-rich diet exhibited significantly lower triacylglycerol (TAG) levels and a reduced number and size of lipid droplets compared to those on control, high-fat, and high-sugar diets (Doğan et al. 2021a). Various other studies explored the role of Ca^{2+} in lipolytic and lipogenic pathways, revealing that Ca^{2+} regulates key metabolic processes governing lipid breakdown and synthesis (Allen and Beck 1986; Arruda and Hotamisligil 2015; Arruda et al. 2017; Maus et al. 2017; Doğan et al. 2021b, c). Before exploring the specifics of Ca^{2+} homeostasis and related pathways in insect lipid metabolism, the following section will provide an overview of Ca^{2+} signaling and its core genetic components.

2 Calcium Signaling in Insects

Calcium homeostasis is maintained through the coordinated action of Ca^{2+} channels, transporters, pumps, sensors, and binding proteins in the plasma membrane, endoplasmic reticulum (ER), mitochondria, Golgi apparatus, peroxisomes, and lysosomes (Chorna and Hasan 2012). The ER and mitochondria are the primary intracellular Ca^{2+} reservoirs among these organelles. Calcium flux occurs from the hemolymph to the cytosol and between intracellular organelles, including specialized contact sites between the ER and mitochondria. Under resting conditions, cytosolic free Ca^{2+} concentrations are typically low (~50–100 nM), while extracellular Ca^{2+} concentrations are significantly higher

(1–2 mM) (Fill and Copello 2002; Panda et al. 2021). The ER maintains a Ca^{2+} concentration of 50–500 μM, whereas mitochondrial Ca^{2+} levels are usually ~ 100 nM at rest but can rise to 10–500 μM in response to cellular signaling (Fill and Copello 2002; Megha and Hasan 2017a; Fernandez-Sanz and Sheu 2019). This steep Ca^{2+} gradient between extracellular and intracellular compartments is essential for Ca^{2+}-dependent signaling. When cytosolic Ca^{2+} levels drop, Ca^{2+} influx occurs primarily through plasma membrane channels, while additional Ca^{2+} release from intracellular stores (ER and mitochondria) contributes to cytosolic Ca^{2+} levels via organelle-associated channel and sensor proteins. Conversely, upon Ca^{2+} stimulation, excess cytosolic Ca^{2+} is rapidly sequestered into the ER, mitochondria, and other organelles to prevent cytotoxicity. Calcium binds to various effectors as a secondary messenger, activating a wide range of cellular pathways (Costas-Ferreira and Faro 2021). These pathways regulate critical physiological processes, including neurotransmitter release, muscle contraction, gene expression, and lipid metabolism. Additionally, Ca^{2+} serves as a crucial cofactor for ER chaperones, facilitating the proper folding of newly synthesized proteins (Corbett and Michalak 2000). A fundamental understanding of Ca^{2+} homeostasis and signaling is essential for exploring its role in lipid metabolism, which will be discussed in the subsequent sections.

Calcium homeostasis in the cell is regulated through a process known as store-operated calcium entry (SOCE), first described in rat exocrine gland cells by Putney (1986) and later refined in 1990 (Putney 1990). Insects, particularly *Drosophila melanogaster*, have been a primary model for studying SOCE, as they share the same core mammalian orthologs but with fewer genes (Baumbach et al. 2014a). The key components of SOCE include i) sarco/endoplasmic reticulum Ca^{2+}-ATPase (SERCA)—responsible for pumping Ca^{2+} back into the ER, ii) inositol 1,4,5-trisphosphate receptor (IP_3R) and ryanodine receptor (RyR) - mediate Ca^{2+} release from the ER, iii) stromal interaction molecule (STIM)-Orai1 complex—the core machinery of SOCE, where STIM senses ER Ca^{2+} depletion and activates Orai1 channels for Ca^{2+} influx, and iv) other Ca^{2+} signaling and binding molecules, including calmodulin (CaM), calcineurin (CaN), and regucalcin (RgN), which modulate various Ca^{2+}-dependent processes (Table 1).

2.1 SERCA and IP_3R/RyR

The two major components of SOCE are SERCA and the IP_3R or RyR. SERCA actively pumps Ca^{2+} from the cytosol into the ER lumen, maintaining intracellular Ca^{2+} homeostasis. In contrast, IP_3R and RyR mediate Ca^{2+} release from the ER into the cytosol or mitochondria, facilitating Ca^{2+}-dependent signaling. Notably, these receptors are also present in mitochondria and at ER-mitochondrial contact sites, where they regulate Ca^{2+} transfer between organelles (Ryu et al. 2010; Jakob et al. 2014; Bartok et al. 2019). While IP_3R and RyR share structural and functional similarities, their regulatory mechanisms and expression patterns differ. Thus, IP_3R is activated by the second messenger inositol 1,4,5-trisphosphate (IP_3) and Ca^{2+} (Paknejad and Hite 2018), while RyR is primarily activated by Ca^{2+} itself, amplifying Ca^{2+} signaling through "calcium-induced calcium release (CICR)" (Hamilton and Serysheva 2009; Lanner et al. 2010).

Insect genomes contain a single gene encoding SERCA, IP_3R, and RyR, whereas mammalian genomes typically encode three genes for SERCA, IP_3R, and RyR (Toprak et al. 2020, 2021) (Table 1).

2.2 STIM-Orai1 Complex

Another component of the SOCE is the STIM-Orai1 complex. STIM is predominantly located in the ER membrane and contains an N-terminal Ca^{2+}-binding EF-hand domain, which acts as a sensor for luminal Ca^{2+} levels (Roos et al. 2005). When ER Ca^{2+} levels drop, such as during the release of Ca^{2+} from the ER, STIM becomes activated and migrates to ER-plasma membrane (ER-PM) contact sites, where it rapidly

Table 1 Primary components of calcium signaling and their roles in lipid metabolism in adipocytes

Calcium signaling component	Function in calcium homeostasis	Role in lipid metabolism	Presence in insects vs. mammals	Key references
Sarco/endoplasmic reticulum Ca^{2+}-ATPase (SERCA)	Pumps Ca^{2+} into ER, maintaining intracellular Ca^{2+} balance	Promotes lipogenesis; dysfunction leads to lipid storage defects	Single gene in insects, three genes in mammals	Baumbach et al. (2014a) and Bi et al. (2014)
Inositol 1,4,5-trisphosphate receptor (IP_3R)	Releases Ca^{2+} from ER into cytosol, activated by IP_3	Regulates lipolysis and lipid mobilization	Single gene in insects, three genes in mammals	Subramanian et al. (2013b) and Megha and Hasan (2017a)
Ryanodine receptor (RyR)	Releases Ca^{2+} from ER into cytosol, activated by Ca^{2+}	Regulates lipolysis; knockdown leads to lipid accumulation	Single gene in insects, three genes in mammals	Bi et al. (2014)
Stromal interaction molecule (STIM)	Senses ER Ca^{2+} depletion and activates Orai1 for Ca^{2+} influx	Enhances lipolysis, reduces adiposity	Single gene in insects, two genes in mammals	Baumbach et al. (2014a) and Xu et al. (2019)
Orai1	Forms Ca^{2+} channels, mediating store-operated calcium entry	Facilitates SOCE, indirectly regulating lipid metabolism	Single gene in insects, three genes in mammals	Liou et al. (2005) and Parekh (2017)
Calmodulin (CaM)	Binds Ca^{2+} to regulate various signaling pathways	Modulates lipolytic enzyme activation and lipid mobilization	A highly conserved, single gene both in insects and mammals	Doyle et al. (1990) Wang et al. (2003) and Baumbach et al. (2014b)
Calcineurin (CaN)	Ca^{2+}/CaM-dependent phosphatase, involved in multiple pathways	Regulates lipid synthesis, influences fat accumulation	Single gene in insects and two calcineurin catalytic subunit genes and one regulatory subunit gene in mammals	Pfluger et al. (2015) and Doğan et al. (2021c)
Regucalcin (RgN)	Regulates intracellular Ca^{2+} levels, suppressing excess Ca^{2+} signaling	Regulates fatty acid synthesis and lipid storage	A highly conserved, single gene both in insects and mammals	Vesala et al. (2012) and Salminen et al. (2015)
Calreticulin (CrT)	ER chaperone involved in Ca^{2+} storage and protein folding	Involved in lipid droplet formation and stress response	A highly conserved, single gene both in insects and mammals	Somogyi et al. (2003) and Gold et al. (2010)
Calnexin (CnX)	Membrane-bound ER chaperone, regulating Ca^{2+} homeostasis and protein folding	Regulates glycoprotein quality control, possibly linked to lipid metabolism	A highly conserved, single gene both in insects and mammals	Xiao et al. (2017) and Doğan et al. (2021a)

oligomerizes and directly interacts with the Orai1 Ca^{2+} channel in the plasma membrane (Liou et al. 2005). Notably, "Orai" is named not based on an acronym, but rather as a reference to its role as a calcium gatekeeper, inspired by Greek mythology. Thus, "Orai" is the Greek goddess of fate and guardian of the gates of heaven. The STIM-Orai coupling activates the Orai1 channel, a key component of Ca^{2+} release-activated Ca^{2+} (CRAC) channels, mediating SOCE-dependent calcium influx. Once activated, Orai1 allows extracellular Ca^{2+} to enter the cytosol, and SERCA subsequently transports Ca^{2+} back into the ER, restoring ER Ca^{2+} stores (Parekh and Putney 2005; Prakriya and Lewis 2015). Once ER Ca^{2+} levels are replenished, CRAC channels become inactivated, leading to STIM binding to the Ca^{2+}-binding protein CaM, and dissociation of STIM from Orai1, returning both proteins to their resting states (Parekh 2017; Li et al. 2017).

Insect genomes contain a single STIM and Orai gene, whereas mammalian genomes encode two and three paralogs of STIM and Orai, respectively (Table 1).

2.3 Other Calcium-Binding Proteins

Several Ca^{2+}-binding proteins, including CaM, CaN, RgN, calreticulin (CrT), and calnexin (CnX), play crucial roles in regulating Ca^{2+} homeostasis, signaling, and storage within cells. These proteins also act as key intermediaries in the SOCE process, modulating Ca^{2+}-dependent cellular functions.

2.3.1 Calmodulin (CaM)

Calmodulin is a highly conserved Ca^{2+}-binding protein that functions as a universal Ca^{2+} sensor in eukaryotic cells. The name "calmodulin" comes from a combination of "calcium" and "modulating," reflecting its function as a calcium-binding regulatory protein that modulates various cellular processes by interacting with target proteins. Structurally, CaM contains four EF-hand motifs, each capable of binding one Ca^{2+} ion.

In low Ca^{2+} levels, CaM remains inactive, while in high Ca^{2+} levels, Ca^{2+} binds to CaM, inducing a structural change that allows it to bind and regulate downstream proteins. CaM is directly involved in SOCE regulation by interacting with STIM1 and Orai1, as well as modulating Ca^{2+}-dependent molecules such as CaN and SERCA. Beyond SOCE, CaM contributes to various physiological functions, including development, immune responses, muscle contraction, apoptosis, gene transcription, ion channel regulation, neuronal growth, and metabolism (Haeseleer et al. 2002).

Calmodulin is indispensable for various physiological processes in insects. Mutations in *CaM* result in severe developmental defects such as pupal lethality, head eversion failure, ectopic wing veins, and chromosomal anomalies during embryogenesis (Nelson et al. 1997). Calmodulin also functions as a Ca^{2+} sensor in gustatory receptors, modulating taste responses to sugars, salts, and bitter compounds (Seno et al. 2005), and regulates Ca^{2+}-activated ion channels in sensory neurons. It influences vitellogenin synthesis, transport, and uptake by modulating hormonal signaling and oocyte maturation (Brown et al. 2010). In olfaction, CaM interacts with the olfactory co-receptor Orco to adjust odorant responses (Mukunda et al. 2014; Bahk and Jones 2016; Jain et al. 2021). It is also critical for presynaptic excitability, neurotransmitter release, and synaptic plasticity (Bronk et al. 2018). In muscle physiology, CaM regulates contraction dynamics in flight, leg, and visceral muscles. Silencing CaM is lethal in *Leptinotarsa decemlineata* larvae (Doğan et al. 2021c) and causes movement defects in *Drosophila melanogaster* larvae, which can be partially rescued by neuronal expression, indicating both neural and non-neural roles for CaM (Heiman et al. 1996; Wang et al. 2003). In *Drosophila melanogaster*, a single *CaM* gene is present in the genome, like in mammals (Doyle et al. 1990) (Table 1).

2.3.2 Calcineurin (CaN)

Calcineurin is a Ca^{2+}/CaM-dependent serine/threonine phosphatase that also plays a crucial role in Ca^{2+} signaling (Rusnak and Mertz 2000). The name "calcineurin" is derived from a combination of "calcium" and "neurin,"which refers to its role in the nervous system as it was initially discovered in neural tissues, leading to the inclusion of "-neurin" in its name (Klee et al. 1979). However, its functions extend beyond the nervous system, influencing immune responses, metabolism, and development. Calcineurin is activated by Ca^{2+} and CaM, allowing it to regulate multiple cellular processes. Calcineurin functions as a heterodimer, consisting of calcineurin A (CNA), the catalytic subunit, which contains the active site responsible for dephosphorylation, and calcineurin B (CNB), the regulatory subunit, which binds Ca^{2+} and is essential for enzyme activation (Yoshiga et al. 2002). Upon Ca^{2+} influx, CaM binds to CaN, leading to its activation. Once activated, CaN dephosphorylates key transcription factors and proteins, modulating cell growth, differentiation, immune responses, and neuronal signaling.

Calcineurin acts as an inhibitor of the epidermal growth factor receptor (EGFR) pathway, a key regulator of cell growth, survival, and differentiation (O'Keefe et al. 1992). By downregulating EGFR signaling, CaN may prevent excessive cell proliferation, influencing development, immunity, and potentially tumor progression. Calcineurin plays a role in ion balance and is essential for myogenesis, regulating muscle fiber formation and differentiation (Rusnak and Mertz 2000; Sullivan and Rubin 2002). Calcineurin is critical for immune responses, including T cell activation and antimicrobial defense mechanisms (Dijkers and O'Farrell 2007; Yu et al. 2020; Wei et al. 2021). Additionally, CaN influences cellular stress responses and contributes to the fine-tuning of intracellular signaling pathways critical for growth and adaptation.

In insects, CaN is implicated in sleep homeostasis (Tomita et al. 2011; Yu et al. 2025), muscle function (Gajewski et al. 2003), learning and memory (Tomita et al. 2011), immunity (Dijkers and O'Farrell 2007; Yu et al. 2020; Wei et al. 2021), development and molting (Wei et al. 2019), reproduction (Wang et al. 2022), and sex pheromone biosynthesis (Du et al. 2017). It also contributes to ion balance and stress responses. In *Bombyx mori*, CaN activity is low in diapausing eggs but increases during development; CNA remains high, while CNB rises in developing eggs, suggesting a role in diapause regulation (Hsieh and Gu 2019). Similarly, in *Leptinotarsa decemlineata*, CNB is higher in non-diapause males but downregulated during and after diapause (Doğan et al. 2021c). Insects have a single CaN gene, unlike mammals, which have two catalytic and one regulatory subunit gene (Table 1).

2.3.3 Regucalcin (RgN)

Regucalcin is a highly conserved Ca^{2+}-binding protein that plays a critical role in Ca^{2+} homeostasis, intracellular signaling, and metabolic regulation across various organisms, including insects. The name "regucalcin" is derived from "regulating calcium," reflecting its role in calcium homeostasis and regulation. It was first identified in the liver (Yamaguchi and Yamamoto 1978) and later termed senescence marker protein-30 (SMP-30) due to its decline in expression with aging in rat liver (Fujita et al. 1992). Regucalcin is a multifunctional regulatory protein involved in intracellular Ca^{2+} balance, gene expression regulation, apoptosis modulation, energy homeostasis, and oxidative stress protection (Marques et al. 2014). Unlike many other Ca^{2+}-binding proteins, such as CaM, RgN lacks the characteristic EF-hand Ca^{2+}-binding motif (Shimokawa and Yamaguchi 1993). Instead, it interacts with Ca^{2+} transporters such as SERCA to maintain Ca^{2+} balance.

Studies on RgN in insects are limited, but existing research suggests its role in physiological adaptations. In *Drosophila montana*, and *Drosophila americana*, RgN has been implicated in cold adaptation, where its expression may contribute to cellular protection against low-temperature stress (Reis et al. 2011; Vesala et al. 2012). Additionally, RgN is thought to play a role in diapause regulation by suppressing metabolic and physiological processes, thereby conserving energy and maintaining dormancy (Salminen et al. 2015). These findings suggest that RgN contributes to insect stress resilience by modulating Ca^{2+}-dependent cellular functions. Both mammals and insects possess a conserved, single RgN gene (Table 1).

2.3.4 Calreticulin (CrT)

Calreticulin (CrT) is a multifunctional Ca^{2+}-binding protein primarily located in the ER. Therefore, its name "calreticulin" comes from a combination of "calcium" and "reticulum," reflecting its primary location, the ER (Ostwald and MacLennan 1974). Calreticulin is also found in the cytoplasm and on the cell surface. It plays a crucial role in Ca^{2+} homeostasis, protein folding, and quality control by ensuring the proper folding of newly synthesized glycoproteins (Somogyi et al. 2003; Gold et al. 2010). Beyond its role in intracellular Ca^{2+} regulation, CrT is involved in cell adhesion, immune responses, and stress signaling.

In insects, CrT has been linked to various physiological processes in addition to its primary roles above. These include odor-guided behavior (Stoltzfus et al. 2003), anesthetic sensitivity (Gamo et al. 2003), and the phagocytosis of apoptotic cells (Kuraishi et al. 2007). It also plays a role in development and immune function (Choi et al. 2002), diapause regulation (Doğan et al. 2021a), and interactions with pathogens, highlighting its importance as a regulatory protein in both cellular processes and organismal physiology. Both mammals and insects possess a conserved, single CrT gene (Table 1).

2.3.5 Calnexin (CnX)

Calnexin (CnX) is a Ca^{2+}-binding molecular chaperone located in the ER, where it works alongside CrT to facilitate protein folding and glycoprotein quality control.

The name "calnexin" comes from a combination of "calcium" and "nexus," meaning "connection" or "link," referring to its role in connecting newly synthesized glycoproteins with the quality control system of the ER (Wada et al. 1991). It is a key component of the CrT/CnX cycle, ensuring newly synthesized proteins achieve their proper conformation before being transported to their final cellular destinations. Unlike CrT, which is soluble, CnX is a membrane-bound protein, making it particularly important for membrane protein maturation. Despite this structural difference, both proteins serve as key regulators of protein quality control and Ca^{2+}-dependent signaling in eukaryotic cells, including those in insects.

Although still under investigation, CnX appears to contribute to immune signaling by aiding the maturation of glycoproteins like pattern recognition receptors and antimicrobial peptides. It also regulates the voltage-gated sodium channel Para in *Drosophila melanogaster*, crucial for neuronal function and locomotion (Xiao et al. 2017). Calnexin may support development by ensuring proper folding of hormone receptors, a function especially important under stress conditions like diapause and starvation. In *L. decemlineata*, both CnX and CrT have been linked to diapause regulation (Doğan et al. 2021b). Insects and mammals possess a single, conserved *CnX* gene (Table 1).

3 Lipid Metabolism in Relation to Calcium

Studies in mammalian models have demonstrated that Ca^{2+} influences lipid metabolism (Arruda et al. 2017; Maus et al. 2017). Disruptions in Ca^{2+} homeostasis can lead to lipid droplet accumulation in the liver, heart, and skeletal muscle of mice, highlighting its regulatory role in lipid storage and utilization (Maus et al. 2017). Similarly, in insects, Ca^{2+} plays a crucial role in lipid metabolism, affecting lipolysis, lipid storage, and energy mobilization.

Most research on Ca^{2+}-lipid interactions in insects has been conducted in *Drosophila melanogaster*, providing insights into key genetic regulators. However, the molecular mechanisms and primary genetic components may not be conserved across all insect species. In addition to *Drosophila melanogaster*, several studies in non-drosophilid insects have also revealed the importance of Ca^{2+} in lipid metabolism. These include investigations in the tobacco hornworm *Manduca sexta* (Lepidoptera: Sphingidae) (Arrese et al. 1999), the migratory locust *Locusta migratoria* (Orthoptera: Acrididae) (Auerswald and Gäde 2006), the desert locust *Schistocerca gregaria* (Orthoptera: Acrididae) (Ogoyi et al. 1998), and *L. decemlineata* (Doğan et al. 2021c). These studies collectively emphasize the conserved yet species-specific roles of Ca^{2+} in lipid metabolism across different insect taxa.

3.1 Primary Actors of SOCE in Relation to Lipid Metabolism

SERCA plays a crucial role in lipid metabolism by maintaining intracellular calcium homeostasis (Fig. 1, Table 1). Knocking down *SERCA* in *Drosophila melanogaster* adipose tissue increases cytosolic Ca^{2+} levels, leading to reduced body fat and smaller lipid droplets, indicating its role as a lipogenic factor (Baumbach et al. 2014a; Bi et al. 2014). This is supported by the downregulation of key lipogenic genes, including fatty acid synthase (*FAS*), acetyl-CoA carboxylase (*ACC*), and acyl-CoA synthetase (*ACS*), following *SERCA* silencing, which suppresses lipogenesis while enhancing fatty acid β-oxidation in adipocytes (Bi et al. 2014). Additionally, SERCA influences the expression of midway (*mdy*), encoding diacylglycerol acyltransferase 1 (DGAT1), a crucial enzyme in TAG synthesis. Elevated cytosolic Ca^{2+} levels suppress *mdy* expression, whereas *CaM* silencing upregulates it (Baumbach et al. 2014a; Bi et al. 2014). This aligns with the role of adipokinetic hormone (AKH) (the functional equivalent of mammalian glucagon), which also suppresses *mdy* expression to promote lipid mobilization. Conversely, reducing cytosolic Ca^{2+} leads to increased *mdy* expression, further linking Ca^{2+} homeostasis to lipid storage (Baumbach et al. 2014a; Xu et al. 2019). On the other hand, since SERCA controls ER Ca^{2+} levels, its activity also influences seipin—an ER-resident protein essential for the proper formation and maturation of lipid droplets in eukaryotic cells (Bi et al. 2014; Ding et al. 2018) (for further discussion, see section "A chicken-egg question: Does calcium homeostasis affect lipid metabolism or does lipid metabolism affect calcium homeostasis?"

IP_3R and RyR, the two intracellular Ca^{2+} release channels primarily located on the ER membrane, also influence lipid metabolism (Fig. 1, Table 1). Hypomorphic *Drosophila melanogaster* *IP_3R* mutants exhibit increased TAG storage, larger lipid droplets, and obesity on a normal diet (Subramanian et al. 2013b; Megha and Hasan 2017a). Consistently, *IP_3R* knockdown in the fat body leads to enlarged lipid droplets and increased TAG accumulation (Bi et al. 2014). Similarly, fat body–specific *RyR* knockdown results in increased lipid droplet size and TAG levels, suggesting that RyR also has a lipolytic role (Bi et al. 2014). Additionally, IP_3R mutants display hyperphagia, consuming three times more food than wild-type flies, with upregulated midgut lipase activity, enhancing dietary lipid digestion and absorption and contributing to obesity (Subramanian et al. 2013b). These mutants also exhibit starvation resistance, retaining stored lipids longer than wild-type flies, and survive significantly longer under starvation conditions (Subramanian et al. 2013b). On the other hand, IP_3R also influences insulin signaling, as mutants show significantly reduced insulin-like peptide 2 (DILP2) levels, suggesting impaired insulin signaling. However, increasing insulin signaling only partially rescues the obesity phenotype, reducing TAG storage but failing to fully restore normal metabolism (Subramanian et al. 2013b; Megha and Hasan 2017b). This suggests a functional interplay between Ca^{2+} and insulin signaling, with defective Ca^{2+} homeostasis playing a direct role in obesity rather than insulin dysfunction alone (Subramanian et al. 2013b). Mass spectrometric analysis revealed altered lipid composition in *IP_3R* mutants, characterized by a deficiency in long-chain fatty acid mobilization, resulting in persistent lipid storage (Subramanian et al. 2013b). Further research into peptidergic neurons has provided additional insights into IP_3R's role in lipid metabolism. Knocking down *IP_3R* specifically in peptidergic neurons resulted in obesity, increased TAG levels, hyperphagia, starvation resistance, and upregulated midgut lipase (*Magro*) expression (Subramanian et al. 2013a; Jayakumar and Hasan, 2018). Additionally, *IP_3R* loss suppresses target of rapamycin (TOR) signaling, a key regulator of lipid storage and protein synthesis, linking Ca^{2+}-dependent IP_3R activity to lipid metabolism via AMPK-TOR cross-talk (Megha and Hasan 2017a). As a result, *IP_3R* loss increases AMPK activity, promoting lipolysis and reducing lipid biosynthesis (Megha and

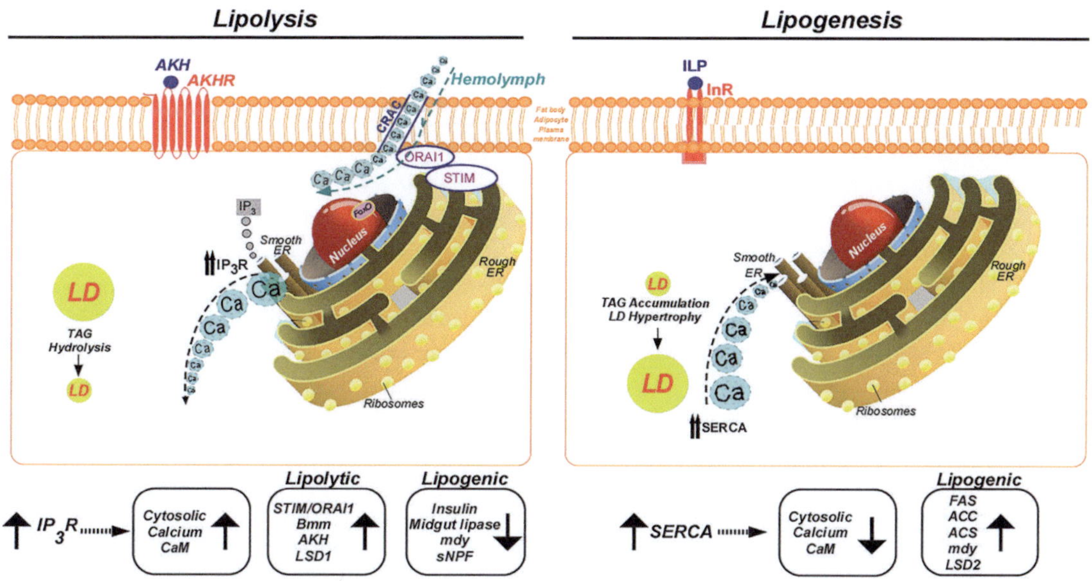

Fig. 1 Calcium-dependent regulation of lipolysis and lipogenesis in insect adipocytes. Left panel: Lipolysis, adipokinetic hormone (AKH) binds to its receptor (AKHR) on the adipocyte plasma membrane, triggering intracellular signaling cascades that increase cytosolic calcium (Ca^{2+}). The inositol 1,4,5-trisphosphate receptor (IP_3R) on the smooth endoplasmic reticulum (ER) facilitates Ca^{2+} release into the cytosol, where it activates calmodulin (CaM) and other lipolytic regulators, including stromal interaction molecule (STIM), Orai1, Brummer lipase (Bmm), adipokinetic hormone (AKH), and lipid storage droplet-1 (LSD1). These factors promote triacylglycerol (TAG) hydrolysis and mobilization of stored lipids. Right panel: Lipogenesis, insulin-like peptides (ILP) activate the insulin receptor (InR), leading to the stimulation of sarco/endoplasmic reticulum Ca^{2+}-ATPase (SERCA), which pumps Ca^{2+} into the ER, thereby lowering cytosolic Ca^{2+}. Reduced cytosolic Ca^{2+} promotes lipogenesis through key metabolic regulators, including fatty acid synthase (FAS), acetyl-CoA carboxylase (ACC), acyl-CoA synthetase (ACS), midway (mdy) (encoding diacylglycerol acyltransferase 1, DGAT1), and lipid storage droplet-2 (LSD2). This results in TAG accumulation and lipid droplet (LD) hypertrophy

Hasan 2017a). These findings indicate that IP_3R functions as a leanness gene by controlling AMPK activation and TOR signaling, confirming that low cytosolic Ca^{2+} levels in adipocytes promote lipid accumulation.

STIM, another key component of SOCE, is also crucial for adiposity regulation (Fig. 1, Table 1). Overexpressing *STIM* in adipose tissue elevates cytosolic Ca^{2+} and reduces body fat, whereas *STIM* knockdown decreases cytosolic Ca^{2+} and results in massive fat accumulation (Baumbach et al. 2014a; Xu et al. 2019). This phenotype is associated with the upregulation of the lipid synthesis gene *mdy* and downregulation of the major lipase *brummer* (*bmm*)/adipose triacylglycerol lipase (ATGL) (Baumbach et al. 2014a; Xu et al. 2019). Additionally, *STIM* knockdown increases the expression of short neuropeptide F (*sNPF*), a neuropeptide homologous to mammalian NPY, leading to hyperphagia and further fat accumulation (Baumbach et al. 2014a; Jayakumar and Hasan 2018; Xu et al. 2019). Supporting the systemic impact of neuronal calcium signaling, pan-neuronal expression of mutant Huntingtin protein in *Drosophila* causes calcium dyshomeostasis in the fat body, alters the expression of lipolytic (*bmm*) and lipogenic (*lipin*) genes, impairs mitochondrial function, and induces apoptosis, ultimately resulting in lipid depletion despite adequate caloric intake (Singh and Agrawal 2021). A similar neuroendocrine regulatory axis was revealed by Kumar et al. (2025), who showed that the TRPγ Ca^{2+} channel functions in Dh44-expressing neuroendocrine cells to regulate systemic lipid homeostasis. Loss of *TRPγ* resulted in impaired lipolysis, enlarged lipid droplets, and reduced *bmm* expression—phenocopying adiposity and starvation sensitivity. Restoration of TRPγ in Dh44 neurons rescued these defects, confirming a calcium-dependent neuroendocrine control of lipid metabolism. Further supporting this notion, mushroom body (MB) neurons in *Drosophila*, especially the α'β' lobes, have been shown to regulate fat storage via distinct output pathways. Silencing of these neurons reduces de novo fatty acid synthesis and leads to leanness, while their hyperactivation induces obesity by increasing food intake and fat accumulation (Al-Anzi and Zinn 2018). These MB neurons modulate fat metabolism and feeding behavior through separate cholinergic and glutamatergic circuits, underscoring the neural compartmentalization of systemic energy regulation. In parallel, acute *STIM* knockdown impairs lipid mobilization in response to AKH, while chronic *STIM* knockdown induces full AKH resistance, evidenced by downregulation of lipolytic genes and failure of lipid mobilization under starvation (Xu et al. 2019). Chronic STIM dysfunction also disrupts insulin signaling, leading to hyperglycemia (~1.9× increase in circulating sugar), further exacerbating fat accumulation (Xu et al. 2019). Together, these findings highlight SOCE as a critical regulator of adiposity, with high cytosolic Ca^{2+} promoting leanness and low cytosolic Ca^{2+} promoting fat accumulation (Baumbach et al. 2014a; Xu et al. 2019). Thus, SERCA acts as a lipogenic gene that promotes obesity, whereas STIM/Orai functions as lipolytic regulators that induce leanness.

Lipid mobilization in insects is regulated by AKH, which utilizes the cAMP-PKA and phospholipase C (PLC)-IP_3 pathways (Fig. 2). When AKH is secreted by the corpora cardiaca, it binds to AKH receptors (AKHR) on adipocyte plasma membranes, activating a G-protein-coupled receptor complex consisting of Gαq and Gγ1 (Toprak 2020). This interaction stimulates adenylyl cyclase, which converts ATP into cAMP, activating PKA. PKA, in turn, phosphorylates lipases and the lipolytic transcription factor FoxO, promoting TAG breakdown in fat body cells (Baumbach et al. 2014a; Gáliková et al. 2015; Toprak et al. 2020). In parallel, AKH binding to AKHR activates PLC, which hydrolyzes phosphatidylinositol 4,5-bisphosphate (PIP_2) into IP_3 and diacylglycerol (DAG) (Toprak et al. 2021). IP_3 binds to IP_3R, triggering Ca^{2+} release from the ER into the cytoplasm, activating Ca^{2+}-dependent enzymes and lipolytic pathways, while DAG activates PKC, which phosphorylates target proteins involved in lipid metabolism (Bednářová et al. 2013). PKC inhibition blocks Ca^{2+} influx, preventing lipid breakdown in *Periplaneta americana* (Sun and Steele 2001). Finally, *AKH*-deficient *Drosophila melanogaster* develop adult-onset obesity, accumulating excess TAGs even under prolonged fasting (Gáliková et al. 2015). Interestingly, AKH-induced lipid mobilization occurs only in adults, as larval fat reserves are unaffected by AKH signaling (Baumbach et al. 2014a; Gáliková et al. 2015). Similarly, involvement of Ca^{2+} in AKH-induced lipid hydrolysis has also been reported in *Manduca sexta* adult fat body (Arrese et al. 1999). On the other hand, knockdown of the genes encoding Gαq, Gγ1, and PLC resulted in lower cytosolic Ca^{2+} levels in fat body cells, obesity phenotype with a significant increase in body fat, increased lipid droplet size, and lipid accumulation in *Drosophila melanogaster* (Baumbach et al. 2014b). Conversely, overexpression of *Gαq* and *PLC* elevated cytosolic Ca^{2+} levels and reduced body fat, leading to a lean phenotype. It is also noteworthy that dietary amino acids such as threonine and methionine trigger AKH release from neurosecretory cells, promoting fat loss by activating Ca^{2+} waves in adipose tissue (Ahmad et al. 2025). By contrast, branched-chain amino acids (e.g., leucine) fail to activate AKH (Ahmad et al. 2025). Consequently, altering cytosolic Ca^{2+} levels in the larval fat body does not significantly impact whole-body TAG levels, aligning with findings that AKH does not influence larval fat reserves (Baumbach et al. 2014a; Gáliková et al. 2015). In summary, AKH regulates lipid mobilization via cAMP-PKA and PLC-IP_3 pathways, with PKA activating lipases and IP_3-induced Ca^{2+} release driving lipolysis.

The involvement of lipid hormones such as juvenile hormone (JH) and ecdysteroids in

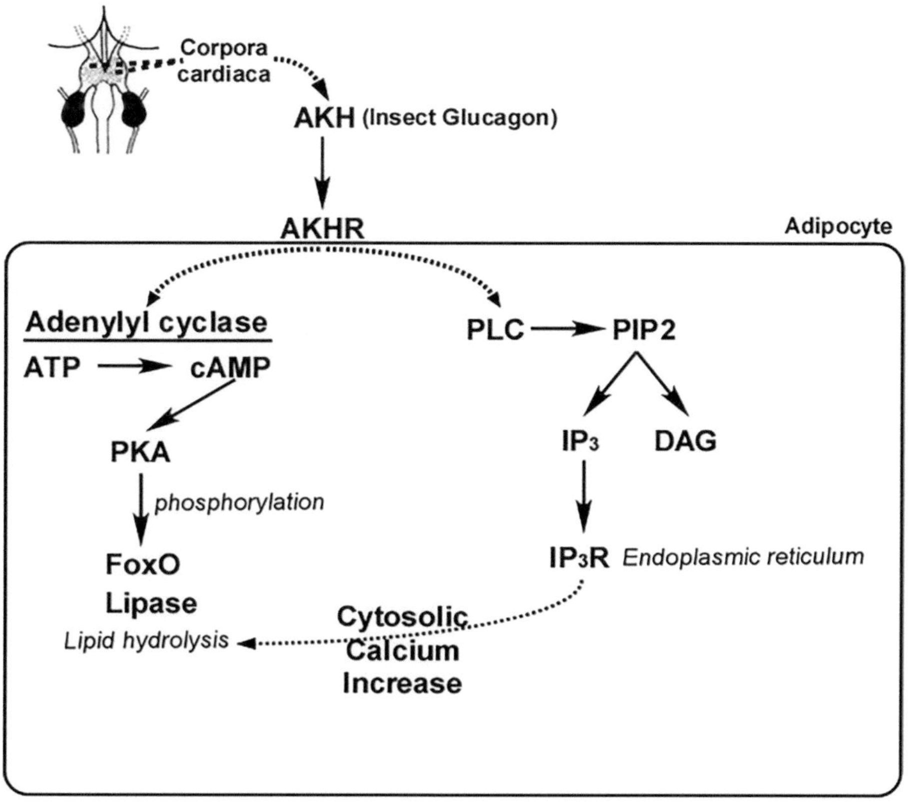

Fig. 2 Regulation of lipid mobilization in insect adipocytes by adipokinetic hormone via cAMP-PKA and PLC-IP_3 pathways. *AKH* adipokinetic hormone, *AKHR* adipokinetic hormone receptor, *ATP* adenosine triphosphate, *cAMP* cyclic adenosine monophosphate, *DAG* diacylglycerol, *FoxO* forkhead box protein O, *IP_3* inositol 1,4,5-trisphosphate, *IP_3R* inositol 1,4,5-trisphosphate receptor, *PIP_2* phosphatidylinositol 4,5-bisphosphate, *PKA* protein kinase A, *PLC* phospholipase C

lipid homeostasis is a critical but often underrepresented facet of insect physiology. Emerging evidence highlights the tight interplay between Ca^{2+} signaling and the biosynthesis, release, and function of these hormones. In cockroaches, JH biosynthesis by the corpora allata (CA) is strictly Ca^{2+} dependent, with JH production nearly abolished in Ca^{2+}-free media and significantly modulated by Ca^{2+} channel blockers or ionophores (Kikukawa et al., 1987). Similarly, Zhu et al. (2025) demonstrated that Ca^{2+} influx via voltage-gated Ca^{2+} channels is essential for JH-mediated signaling cascades in *Aedes aegypti*, further emphasizing a conserved regulatory mechanism across species. Beyond biosynthesis, Ca^{2+} also influences JH function. For instance, in *Choristoneura fumiferana*, treatment with the JH analog fenoxycarb impaired lipid biosynthesis and β-oxidation in the fat body, demonstrating the hormone's role in lipid metabolism (Mulye and Gordon, 1993).

Ecdysteroids are likewise regulated by Ca^{2+} dynamics. Classic work by Kikukawa et al. (1987) and more recent studies in *Drosophila* have shown that prothoracicotropic hormone (PTTH)-stimulated ecdysteroidogenesis in prothoracic glands is mediated through intracellular Ca^{2+} release and Ca^{2+}-dependent kinases (Yamanaka et al., 2015; Ramos et al., 2021). Moreover, Zang et al. (2024) revealed that STIM- and ORAI-mediated SOCE is required for proper ecdysone biosynthesis during molting and metamorphosis. Calcium also plays a role downstream of these hormones, modulating their metabolic outputs. For example, ecdysteroid signaling influences TAG hydrolysis and lipid mobilization, partly through Ca^{2+}-dependent transcriptional regulation (Wang et al. 2010). Taken together, these findings suggest that calcium is not only a cofactor in hormonal biosynthesis but also a key player in hormone-mediated lipid metabolic processes. Integrating this understanding with AKH and ILP signaling, which similarly relies on Ca^{2+} as a second messenger, highlights the central role of Ca^{2+} in orchestrating endocrine regulation of lipid metabolism in insects.

3.2 Other Calcium Signaling Molecules in Lipid Metabolism

As previously discussed, CaM is the primary Ca^{2+}-binding protein in the cytosol and is essential for Ca^{2+}-mediated activation of lipid metabolism. Inhibiting CaM disrupts Ca^{2+} homeostasis and prevents lipid breakdown, confirming its lipolytic role (Sun and Steele 2001). Additionally, *CaM* silencing downregulates *bmm* lipase gene, leading to more than a 50% increase in whole-body TAG levels in *Drosophila melanogaster* (Baumbach et al. 2014b). These findings reinforce the role of CaM as a regulator of lipolysis (Table 1).

Calcineurin also plays a crucial role in lipid metabolism by modulating key signaling pathways. Loss of CaN function in *Drosophila melanogaster* significantly reduces whole-body TAG levels, indicating its lipogenic role (Pfluger et al. 2015) (Table 1). Similarly, *CaN*-deficient mice exhibit a marked reduction in fat mass, suggesting that its role in lipid accumulation is conserved across species (Pfluger et al. 2015). Calcineurin exerts its effects primarily through the cAMP response element-binding protein (CREB)/CREB-regulated transcription coactivator (CRTC) pathway, essential for lipid homeostasis. Calcineurin dephosphorylates CRTC, leading to its activation and subsequent enhancement of CREB-dependent gene expression, which promotes lipid storage (Screaton et al. 2004; Koo et al. 2005). Conversely, insulin signaling counteracts this pathway by activating AMPK, which phosphorylates CRTC, leading to its degradation during feeding (Horiuchi et al. 2004; Wang et al. 2008). Inhibition of *CaN* disrupts this signaling axis, reducing lipid accumulation. Supporting this, RNAi-mediated silencing of *CaNB*, a key CaN subunit, significantly decreases total fat content in *L. decemlineata* larvae, further confirming its role in lipid metabolism (Doğan et al. 2021c). Beyond its direct role in lipid storage, CaN also influences lipid biosynthesis and oxidation through its interaction with AMPK. Under normal conditions, AMPK suppresses lipogenesis and promotes fatty acid oxidation. However, CaN negatively regulates AMPK, shifting the balance toward lipid synthesis and accumulation. Additionally, CaN activity correlates with increased SERCA function, reducing cytosolic Ca^{2+} levels, influencing lipid metabolic pathways (Toprak et al. 2020, 2021). Together, these findings suggest that CaN serves as a central regulator of lipid metabolism, promoting lipogenesis by activating CREB-CRTC signaling, inhibiting AMPK, and modulating Ca^{2+} homeostasis. Its role is closely tied to metabolic state and nutrient availability, highlighting its importance in energy storage and utilization.

Regucalcin is also linked to lipid metabolism, influencing fatty acid synthesis and lipid storage. It plays a role in energy regulation, particularly in cells with high metabolic demands, such as flight muscles and reproductive tissues in insects (Doğan et al. 2021c; Toprak et al. 2021). RNAi-mediated silencing of *RgN* in *L. decemlineata* larvae resulted in a significant reduction in total fat content, highlighting its role in lipid accumulation (Doğan et al. 2021c) (Table 1).

These findings collectively suggest that Ca^{2+} signaling regulates lipid metabolism not only through cell-autonomous mechanisms within the fat body but also via top-down neuroendocrine pathways involving peptidergic neurons and hormone-secreting neurosecretory cells. These findings are further supported by circuit-level evidence that neuronal IP_3R-mediated calcium signaling acts as an integrator of nutrient status and systemic energy homeostasis (Jayakumar and Hasan 2018). In parallel, a daily neuron-glia lipid metabolic cycle supports brain energy homeostasis, whereby neuronal fatty acid transporters such as NLaz mediate lipid transfer to glial cells for oxidation and clearance during sleep (Haynes et al. 2024). Glial mitochondrial β-oxidation, rather than neuronal metabolism, appears essential for sleep-dependent lipid clearance, reinforcing the functional division between lipid production in neurons and lipid catabolism in glia. Notably, Drp1, a mitochondrial fission protein activated by calcineurin-mediated dephosphorylation, is required in both neurons and glia to maintain proper lipid dynamics and sleep architecture (Haynes et al. 2024).

4 Lipid Metabolism in Hepatocyte-Like Cells "Oenocytes" in Relation to Calcium

In mammals, adipocytes primarily serve as lipid storage cells, while hepatocytes regulate lipid metabolism by synthesizing, processing, and mobilizing lipids to maintain systemic energy balance (Azzu et al. 2020). Although both adipocytes and hepatocytes store and hydrolyze lipids, their metabolic roles differ (Frayn et al. 2006). Adipocytes function as energy reservoirs, storing excess lipids as TAGs and releasing them when needed. In contrast, hepatocytes are metabolically dynamic, synthesizing lipids from excess carbohydrates, regulating lipoprotein metabolism, and playing a key role in fatty acid oxidation and ketogenesis.

Given these distinct roles, insect oenocytes, the hepatocyte-like cells in the fat body, should be considered separately from lipid-storing adipocytes. Research in *Drosophila* provides strong evidence that oenocytes function similarly to mammalian hepatocytes, particularly during fasting (Gutierrez et al. 2007). While insect adipocytes store lipids, oenocytes actively regulate lipid mobilization, fatty acid modification, and detoxification (Huang et al. 2022). A hallmark observation in *Drosophila melanogaster* larvae is that lipid droplets accumulate in oenocytes during fasting, a phenomenon also observed in mammalian hepatocytes undergoing hepatic steatosis (Gutierrez et al. 2007). During starvation, fat body–derived insulin-like peptide 6 (dILP6) activates insulin receptor signaling in oenocytes, triggering lipid accumulation and mobilization (Chatterjee et al. 2014). Additionally, ablation of oenocytes prevents normal lipid depletion from the fat body during fasting, impairing lipid mobilization. Oenocyte-deficient mutant larvae retain significantly more TAGs than controls under starvation, suggesting that oenocytes function as lipid-processing centers, capturing and modifying fatty acids released from the fat body (Gutierrez et al. 2007; Martins and Ramalho-Ortigão 2012; Huang et al. 2022). Rather than acting independently, oenocytes form a bidirectional lipid-metabolic axis with the fat body, mirroring the adipose-liver axis in mammals.

The fat body communicates lipid status to oenocytes via nutrient-sensing pathways, such as Slimfast (Slif), an amino acid transporter that modulates lipid release. Disrupting *Slif* expression in the fat body results in lipid droplet accumulation in oenocytes, even under fed conditions, suggesting that amino acid and lipid signaling pathways regulate lipid exchange between the fat body and oenocytes (Gutierrez et al. 2007). Additionally, bmm and lipid storage droplet-2 (LSD2), a perilipin homolog, regulate lipid mobilization in oenocytes. Overexpression of *bmm* in the fat body increases lipid droplet accumulation in oenocytes, whereas overexpression of *LSD2* reduces oenocyte lipid storage, indicating that fat body lipolysis directly influences oenocyte lipid accumulation (Gutierrez et al. 2007). Thus, insects exhibit a functional partitioning of lipid metabolism, where adipocytes primarily store lipids, while oenocytes process and modify them for systemic energy homeostasis, analogous to mammalian adipocytes and hepatocytes.

Cytosolic Ca^{2+} may regulate lipid metabolism in oenocytes, but this remains largely unexplored in insects. In mammals, SERCA dysfunction in hepatocytes is associated with hepatic steatosis, insulin resistance, and glucose intolerance, whereas increasing SERCA activity enhances lipid metabolism, reduces TAG accumulation, and improves systemic glucose homeostasis (Fu et al. 2011). SERCA reduces lipid accumulation in hepatocytes when functional, whereas its inhibition promotes lipid accumulation and contributes to hepatic steatosis. Interestingly, this contrasts with findings in adipocytes, where SERCA activation promotes lipid accumulation. Despite advances in understanding oenocyte function in lipid metabolism, the role of Ca^{2+} signaling in regulating lipid dynamics remains unclear. Given the strong links between Ca^{2+} homeostasis and lipid metabolism in adipocytes and hepatocytes, it is likely that Ca^{2+} also plays a role in oenocyte function. Additionally, the well-developed smooth ER in oenocytes suggests involvement in Ca^{2+}-dependent lipid processing (Martins and Ramalho-Ortigão 2012). Further research is needed to determine whether Ca^{2+} signaling directly regulates lipid accumulation or hydrolysis in oenocytes, which could provide new insights into their metabolic function in insects.

5 A Chicken-and-Egg Question: Does Calcium Homeostasis Affect Lipid Metabolism, or Does Lipid Metabolism Affect Calcium Homeostasis?

The interaction between Ca^{2+} homeostasis and lipid metabolism discussed in this chapter has primarily focused on how impaired Ca^{2+} signaling affects lipid metabolism. Given the scope of this book, this approach was necessary to highlight the importance of Ca^{2+} regulation in maintaining lipid balance in insects. However, from a broader biological perspective, it is equally important to consider the reverse relationship—how disruptions in lipid metabolism impact Ca^{2+} homeostasis. Addressing this question is crucial for understanding the reciprocal nature of metabolic regulation, as lipid metabolism is deeply intertwined with Ca^{2+} signaling. Although studies on this bidirectional relationship remain limited, analyzing the functional roles of key lipid metabolism genes provides insight into how lipid processing can directly influence Ca^{2+} dynamics.

One of the most essential lipid metabolism genes is seipin, which encodes an ER-localized protein required for proper lipid droplet formation and maturation in eukaryotic cells, including insects. In *Drosophila melanogaster*, *seipin* mutants exhibit defective lipid storage and lipodystrophy, emphasizing its critical role in lipid metabolism (Bi et al. 2014; Ding et al. 2018). Interestingly, impaired lipid metabolism in these mutants also disrupts Ca^{2+} homeostasis, demonstrating a direct link between these processes. Seipin physically interacts with SERCA, and *seipin* mutant *Drosophila melanogaster* show reduced SERCA activity, leading to decreased ER Ca^{2+} levels and subsequent reductions in mitochondrial Ca^{2+} (Bi et al. 2014; Ding et al. 2018). This reduction in mitochondrial Ca^{2+} has profound metabolic consequences. Mitochondrial Ca^{2+} plays a crucial role in regulating the tricarboxylic acid cycle, and its depletion impairs citrate production, a key precursor for lipogenesis. Consequently, lower citrate levels suppress lipid biosynthesis, reducing TAG accumulation and lipid droplet formation in adipose tissue (Ding et al. 2018). Supporting this, *seipin* mutants exhibit downregulation of key lipogenic genes, including ATP citrate lyase (*ATPCL*) and *ACC*, further reinforcing the link between Ca^{2+} homeostasis and lipid metabolism (Bi et al. 2014; Ding et al. 2018). Notably, the defects in lipid storage and lipid droplet size reduction in *seipin* mutants are directly linked to impaired SERCA activity and mitochondrial dysfunction (Tian et al. 2011; Bi et al. 2014; Ding et al. 2018). Overexpression of *SERCA* in the fat body of *Drosophila melanogaster seipin* mutants successfully rescues the lipodystrophy phenotype, restoring lipid storage (Bi et al. 2014). However, overexpression of *seipin* does not compensate for the lipid droplet defects caused by *SERCA* silencing, suggesting that SERCA functions downstream of seipin in promoting lipid storage (Bi et al. 2014). These findings indicate that seipin is crucial for maintaining ER Ca^{2+} levels by regulating SERCA activity and that defective lipid metabolism—via seipin dysfunction—directly impairs Ca^{2+} homeostasis. Further supporting this idea, decreased SERCA activity in *seipin* mutants disrupts ER Ca^{2+} transport, leading to reduced ER Ca^{2+} levels and increased cytosolic Ca^{2+} concentrations (Bi et al. 2014). In line with this, SOCE pathways are activated, as indicated by the upregulation of *STIM* in *seipin* mutant fat cells, further suggesting that cytosolic Ca^{2+} levels are elevated. Interestingly, adipose tissue–specific knockdown of *RyR* partially rescues fat storage defects in *seipin* mutants (Bi et al. 2014). However, silencing *IP_3R* does not restore lipid accumulation, highlighting a specific role for RyR-mediated Ca^{2+} release in lipid storage regulation (Bi et al. 2014). These findings suggest that a complex interplay between Ca^{2+} transporters and lipid metabolism regulators orchestrates adipocyte fat storage.

In summary, seipin and SERCA function together to maintain ER Ca^{2+} homeostasis, and disruptions in lipid metabolism can have cascading effects on Ca^{2+} signaling. When ER Ca^{2+} levels drop due to impaired SERCA activity, lipogenesis is downregulated, leading to reduced lipid storage in fat body cells. Conversely, restoring ER Ca^{2+} homeostasis by modulating Ca^{2+} transport pathways can partially rescue lipid metabolism defects. Therefore, rather than a one-directional relationship, lipid metabolism and Ca^{2+} homeostasis are reciprocally linked, forming a tightly regulated feedback loop that ensures metabolic balance in adipose tissue.

6 Conclusion

The intricate relationship between Ca^{2+} homeostasis and lipid metabolism in insects is an emerging area of research with profound implications for understanding metabolic regulation. The findings discussed in this chapter highlight the crucial roles of SOCE, SERCA, IP_3R, RyR, STIM, and Orai1 in modulating lipid storage, mobilization, and utilization. Additionally, Ca^{2+}-binding proteins such as CaM, CaN, RgN, CrT, and CnX contribute significantly to the fine-tuned regulation of lipid metabolic pathways.

The bidirectional nature of Ca^{2+}-lipid interactions is of particular interest. While Ca^{2+} signaling influences lipid metabolism through mechanisms such as lipogenesis and lipolysis, disruptions in lipid metabolism also reciprocally impact Ca^{2+} homeostasis. The case of *seipin* mutants in *Drosophila melanogaster*, where impaired lipid storage leads to decreased SERCA activity and disrupted ER and mitochondrial Ca^{2+} levels, serves as a compelling example of this feedback loop. These findings underscore the necessity of a holistic approach in metabolic studies, where lipid and Ca^{2+} homeostasis must be examined as interdependent regulatory systems.

A particularly underexplored avenue in insect physiology is the role of Ca^{2+} in hepatocyte-like oenocytes. These cells, which mirror mammalian hepatocytes in function, play a pivotal role in lipid mobilization and metabolic homeostasis. The evidence suggests that Ca^{2+} signaling may be as crucial in oenocytes as it is in adipocytes; yet little is known about how Ca^{2+} channels and sensors operate within these specialized cells. Future studies should aim to elucidate the molecular pathways governing Ca^{2+}-dependent lipid metabolism in oenocytes, as this could bridge a significant knowledge gap in insect physiology.

Moreover, increasing evidence suggests that calcium signaling contributes to lipid homeostasis not only through cell-autonomous mechanisms in adipose and hepatocyte-like tissues but also via systemic neuroendocrine and intercellular pathways. Calcium-dependent dynamics in specific neuronal and glial circuits appear to coordinate hormonal output, energy balance, and lipid flux across tissues. These circuits may involve calcium-modulated mitochondrial activity, lipid transfer processes, and sleep-linked metabolic cycles, reinforcing the integrative role of calcium in orchestrating both systemic and brain-specific aspects of lipid metabolism. Future research should aim to disentangle the interplay between these neural, hormonal, and metabolic pathways under varying physiological and environmental conditions.

From an applied perspective, elucidating the interplay between Ca^{2+} homeostasis and lipid metabolism presents promising opportunities for both pest management and biomedical research. Disrupting Ca^{2+}-dependent metabolic pathways in pest insects could serve as an effective strategy to impair energy balance, ultimately reducing their survival and reproductive capacity. Beyond pest control, insect models offer valuable insights into human metabolic disorders, including obesity, metabolic dysfunction-associated steatotic liver disease (MASLD) (formerly nonalcoholic fatty liver disease, NAFLD), and diabetes. In particular, hibernating insect species such as *L. decemlineata* provide unique models for studying lipid-calcium interactions and their implications for these conditions. Comparative studies on insects that enter diapause or undergo prolonged metabolic suppression could shed light on how Ca^{2+} homeostasis regulates lipid storage and mobilization under extreme physiological states. These models may help bridge critical gaps in our understanding of lipid metabolism disorders in humans and inspire novel therapeutic strategies.

In conclusion, studying Ca^{2+} homeostasis in relation to lipid metabolism in insects represents a frontier in entomological and metabolic research. A deeper understanding of these processes will not only advance fundamental knowledge in insect physiology but also provide translational opportunities in biotechnology, pest control, and human health. Future studies should focus on the precise mechanisms by which Ca^{2+} signaling influences lipid metabolism in oenocytes, hormonal regulation of Ca^{2+}-dependent metabolic pathways, and potential for exploiting these interactions in applied sciences.

References

Ahmad M, Wu S, Luo S, Shi W, Guo X, Cao Y, Perrimon N, He L (2025) Dietary amino acids promote glucagon-like hormone release to generate global calcium waves in adipose tissues in *Drosophila*. Nat Commun 16(1):247. https://doi.org/10.1038/s41467-024-55371-y

Al-Anzi B, Zinn K (2018) Identification and characterization of mushroom body neurons that regulate fat storage in *Drosophila*. Neural Dev 13:18. https://doi.org/10.1186/s13064-018-0116-7

Allen DO, Beck RR (1986) Role of calcium ion in hormone-stimulated lipolysis. Biochem Pharmacol 35:767–772. https://doi.org/10.1016/0006-2952(86)90244-3

Alomaim H, Griffin P, Swist E, Plouffe LJ, Vandeloo M, Demonty I, Kumar A, Bertinato J (2019) Dietary calcium affects body composition and lipid metabolism in rats. PLoS One 14:e0210760. https://doi.org/10.1371/journal.pone.0210760

Arrese EL, Flowers MT, Gazard JL, Wells MA (1999) Calcium and cAMP are second messengers in the adipokinetic hormone-induced lipolysis of triacylglycerols in *Manduca sexta* fat body. *J Lipid Res* 40(4):556–564. https://doi.org/10.1194/jlr.M900437-JLR200

Arruda AP, Hotamisligil GS (2015) Calcium homeostasis and organelle function in the pathogenesis of obesity and diabetes. Cell Metab 22:381–397. https://doi.org/10.1016/j.cmet.2015.06.010

Arruda AP, Pers BM, Parlakgul G, Guney E, Goh T, Cagampan E, Lee GY, Goncalves RL, Hotamisligil GS (2017) Defective STIM-mediated store operated Ca^{2+} entry in hepatocytes leads to metabolic dysfunction in obesity. eLife 6:e29968. https://doi.org/10.7554/eLife.29968

Auerswald L, Gäde G (2006) Endocrine control of TAG lipase in the fat body of the migratory locust, *Locusta migratoria*. Insect Biochem Mol Biol 36(10):759–768. https://doi.org/10.1016/j.ibmb.2006.07.004

Azzu V, Vacca M, Virtue S, Allison M, Vidal-Puig A (2020) Adipose tissue-liver cross talk in the control of whole-body metabolism: implications in nonalcoholic fatty liver disease. Gastroenterology 158(7):1899–1912. https://doi.org/10.1053/j.gastro.2019.12.054

Bahk S, Jones WD (2016) Insect odorant receptor trafficking requires calmodulin. BMC Biol 14:83. https://doi.org/10.1186/s12915-016-0306-x

Bartok A, Weaver D, Golenár T, Nichtová Z, Katona M, Bánsághi S, Alzayady KJ, Thomas VK, Ando H, Mikoshiba K, Joseph SK, Yule DI, Hajnóczky G (2019) IP_3 receptor isoforms differently regulate ER-mitochondrial contacts and local calcium transfer. Nat Commun 10:3726. https://doi.org/10.1038/s41467-019-11646-3

Baumbach J, Hummel P, Bickmeyer I, Kowalczyk KM, Frank M, Knorr K, Hildebrandt A, Riedel D, Jäckle H, Kühnlein RP (2014a) A Drosophila in vivo screen identifies store-operated calcium entry as a key regulator of adiposity. Cell Metab 19(3):331–343. https://doi.org/10.1016/j.cmet.2013.12.004

Baumbach J, Xu Y, Hehlert P, Kühnlein RP (2014b) Gaq, Gg1 and Plc21C control Drosophila body fat storage. J Genet Genomics 41(5):283–292. https://doi.org/10.1016/j.jgg.2014.03.005

Bednářová A, Kodrík D, Krishnan N (2013) Adipokinetic hormone exerts its anti-oxidative effects using a conserved signal-transduction mechanism involving both PKC and cAMP by mobilizing extra- and intracellular Ca^{2+} stores. Comp Biochem Physiol C 158: 142–149. https://doi.org/10.1016/j.cbpc.2013.07.002

Berridge MJ, Lipp P, Bootman MD (2000) The versatility and universality of calcium signalling. Nat Rev Mol Cell Biol 1:11–21. https://doi.org/10.1038/35036035

Bi J, Wang W, Liu Z, Huang X, Jiang Q, Liu G, Wang Y, Huang X (2014) Seipin promotes adipose tissue fat storage through the ER Ca^{2+}-ATPase SERCA. Cell Metab 19(5):861–871. https://doi.org/10.1016/j.cmet.2014.03.028

Bronk P, Kuklin EA, Gorur-Shandilya S, Liu C, Wiggin TD, Reed ML, Marder E, Griffith LC (2018) Regulation of Eag by Ca^{2+}/calmodulin controls presynaptic excitability in *Drosophila*. J Neurophysiol 119(5): 1665–1680. https://doi.org/10.1152/jn.00820.2017

Brown PT, Herbert P, Woodruff RI (2010) Vitellogenesis in *Oncopeltus fasciatus*: PLC/IP_3, DAG/PK-C pathway triggered by CaM. J Insect Physiol 56(9): 1300–1305. https://doi.org/10.1016/j.jinsphys.2010.04.006

Chatterjee D, Katewa SD, Qi Y, Jackson SA, Kapahi P, Jasper H (2014) Control of metabolic adaptation to fasting by dILP6-induced insulin signaling in *Drosophila* oenocytes. Proc Natl Acad Sci USA 111: 17959–17964. https://doi.org/10.1073/pnas.1409241111

Choi JY, Jang HH, Kim I, Je YH, Roh JY (2002) Calreticulin enriched as an early-stage encapsulation protein in *galleria mellonella*. Insect Biochem Mol Biol 32(12):1717–1723. https://doi.org/10.1016/S0965-1748(02)00153-1

Chorna T, Hasan G (2012) The genetics of calcium signaling in *Drosophila melanogaster*. Biochim Biophys Acta 1820(8):1269–1282. https://doi.org/10.1016/j.bbagen.2011.11.002

Corbett EF, Michalak M (2000) Calcium, a signaling molecule in the endoplasmic reticulum? Trends Biochem Sci 25(7):307–311. https://doi.org/10.1016/s0968-0004(00)01588-7

Costas-Ferreira C, Faro LRF (2021) Systematic review of calcium channels and intracellular calcium signaling: relevance to pesticide neurotoxicity. Int J Mol Sci 22(24):13376. https://doi.org/10.3390/ijms222413376

De Loof A, Schoofs L (2019) Intraluminal farnesol and farnesal in the mealworm's alimentary canal: an unusual storage site uncovering hidden eukaryote Ca^{2+}-homeostasis-dependent "Golgicrine" activities. Front Endocrinol 10:885. https://doi.org/10.3389/fendo.2019.00885

Dijkers PF, O'Farrell PH (2007) *Drosophila* calcineurin promotes induction of innate immune responses. Curr Biol 17(23):2087–2093. https://doi.org/10.1016/j.cub.2007.11.001

Ding L, Yang X, Tian H, Liang J, Zhang F, Wang G, Wang Y, Ding M, Shui G, Huang X (2018) Seipin regulates lipid homeostasis by ensuring calcium-dependent mitochondrial metabolism. EMBO J 37: e97572. https://doi.org/10.15252/embj.201797572

Doğan C, Güney G, Güzel KK, Can A, Hegedus DD, Toprak U (2021a) What you eat matters: nutrient inputs alter the metabolism and neuropeptide expression in Egyptian cotton leaf worm, *Spodoptera littoralis* (Lepidoptera: Noctuidae). Front Physiol 12: 773688. https://doi.org/10.3389/fphys.2021.773688

Doğan C, Hänniger S, Heckel DG, Coutu C, Hegedus DD, Crubaugh L, Groves RL, Bayram Ş, Toprak U (2021b) Two calcium-binding chaperones from the fat body of the Colorado potato beetle, *Leptinotarsa decemlineata* (Coleoptera: Chrysomelidae) involved in diapause. Arch Insect Biochem Physiol 106(1):e21755. https://doi.org/10.1002/arch.21755

Doğan C, Hänniger S, Heckel DG, Coutu C, Hegedus DD, Crubaugh L, Groves RL, Mutlu DA, Suludere Z, Bayram Ş, Toprak U (2021c) Characterization of calcium signaling proteins from the fat body of the Colorado potato beetle, *Leptinotarsa decemlineata* (Coleoptera: Chrysomelidae): implications for diapause and lipid metabolism. Insect Biochem Mol Biol 133:103549. https://doi.org/10.1016/j.ibmb.2021.103549

Domingo A, Gonzalez-Jurado J, Maroto M, Diaz C, Vinos J, Carrasco C, Cervera M, Marco R (1998) Troponin-T is a calcium-binding protein in insect muscle: in vivo phosphorylation, muscle-specific isoforms, and developmental profile in *Drosophila melanogaster*. J Muscle Res Cell Motil 19:393–403. https://doi.org/10.1023/a:1005349704790

Doyle KE, Kovalick GE, Lee E, Beckingham K (1990) *Drosophila melanogaster* contains a single calmodulin gene. Further structure and expression studies. J Mol Biol 213(4):599–605. https://doi.org/10.1016/S0022-2836(05)80245-1

Du M, Liu X, Ma N, Liu X, Wei J, Yin X, Zhou S, Rafaeli A, Song Q, An S (2017) Calcineurin-mediated dephosphorylation of acetyl-CoA carboxylase is required for pheromone biosynthesis activating neuropeptide (PBAN)-induced sex pheromone biosynthesis in *Helicoverpa armigera*. Mol Cell Proteomics 16(12): 2138–2152. https://doi.org/10.1074/mcp.RA117.000065

Fernandez-Sanz C, Sheu S (2019) Mitochondrial Ca^{2+} concentrations in live cells: quantification methods and discrepancies. FEBS Lett 593(13):1528–1541. https://doi.org/10.1002/1873-3468.13427

Fill M, Copello JA (2002) Ryanodine receptor calcium release channels. Physiol Rev 82:893–922. https://doi.org/10.1152/physrev.00013.2002

Frayn KN, Arner P, Yki-Järvinen H (2006) Fatty acid metabolism in adipose tissue, muscle and liver in health and disease. Essays Biochem 42:89–103. https://doi.org/10.1042/bse0420089

Fu S, Yang L, Li P, Hofmann O, Dicker LH, Hide WA, Hotamisligil GS (2011) Aberrant lipid metabolism disrupts calcium homeostasis causing liver endoplasmic reticulum stress in obesity. Nature 473:528–531. https://doi.org/10.1038/nature09968

Fujita T, Shirasawa T, Uchida K, Maruyama N (1992) Isolation of cDNA clone encoding rat senescence marker protein-30 (SMP30) and its tissue distribution. Biochim Biophys Acta 1132(3):297–305. https://doi.org/10.1016/0167-4781(92)90164-u

Gajewski K, Wang J, Molkentin JD, Chen EH, Olson EN, Schulz RA (2003) Requirement of the calcineurin subunit gene *canB2* for indirect flight muscle formation in *Drosophila*. Proc Natl Acad Sci USA 100:1040–1045. https://doi.org/10.1073/pnas.0337662100

Gáliková M, Diesner M, Klepsatel P, Hehlert P, Xu Y, Bickmeyer I, Predel R, Kühnlein RP (2015) Energy homeostasis control in *Drosophila* adipokinetic hormone mutants. Genetics 201(2):665–683. https://doi.org/10.1534/genetics.115.178897

Gamo S, Tomida J, Dodo K, Keyakidani D, Matakatsu H, Yamamoto D, Tanaka Y (2003) Calreticulin mediates anesthetic sensitivity in *Drosophila melanogaster*. Anesthesiology 99(4):867–875. https://doi.org/10.1097/00000542-200310000-00022

Ghosh A, Greenberg ME (1995) Calcium signaling in neurons: molecular mechanisms and cellular consequences. Science 268(5208):239–247. https://doi.org/10.1126/science.7716515

Gold LI, Eggleton P, Sweetwyne MT, Van Duyn LB, Greives MR, Naylor SM, Michalak M, Murphy-Ullrich JE (2010) Calreticulin: non-endoplasmic reticulum functions in physiology and disease. FASEB J 24(3):665–683. https://doi.org/10.1096/fj.09-145482

Gordon S, Dickinson MH (2006) Role of calcium in the regulation of mechanical power in insect flight. Proc Natl Acad Sci USA 103(11):4311–4315. https://doi.org/10.1073/pnas.0510109103

Gu SH, Chow YS, O'Reilly DR (1998) Role of calcium in the stimulation of ecdysteroidogenesis by recombinant prothoracicotropic hormone in the prothoracic glands of the silkworm, *Bombyx mori*. Insect Biochem Mol Biol 28:861–867. https://doi.org/10.1016/S0965-1748(98)00074-5

Gutierrez E, Wiggins D, Fielding B, Gould AP (2007) Specialized hepatocyte-like cells regulate Drosophila lipid metabolism. Nature 445(7125):275–280. https://doi.org/10.1038/nature05382

Haeseleer F, Imanishi Y, Sokal I, Filipek S, Palczewski K (2002) Calcium-binding proteins: intracellular sensors from the calmodulin superfamily. Biochem Biophys Res Commun 290(2):615–623. https://doi.org/10.1006/bbrc.2001.6228

Hamilton SL, Serysheva II (2009) Ryanodine receptor structure: progress and challenges. J Biol Chem 284(7):4047–4051. https://doi.org/10.1074/jbc.R800054200

Haynes PR, Pyfrom ES, Li Y, Stein C, Cuddapah VA, Jacobs JA, Yue Z, Sehgal A (2024) A neuron–glia lipid metabolic cycle couples daily sleep to mitochondrial homeostasis. Nat Neurosci 27(4):666–678. https://doi.org/10.1038/s41593-023-01568-1

Heiman RG, Atkinson RC, Andruss BF, Bolduc C, Kovalick GE, Beckingham K (1996) Spontaneous avoidance behavior in *Drosophila* null for calmodulin expression. Proc Natl Acad Sci USA 93(6):2420–2425. https://doi.org/10.1073/pnas.93.6.2420

Horiuchi J, Jiang W, Zhou H, Wu P, Yin JC (2004) Phosphorylation of conserved casein kinase sites regulates cAMP-response element-binding protein DNA binding in *Drosophila*. J Biol Chem 279(13):12117–12125. https://doi.org/10.1074/jbc.M212839200

Hsieh HY, Gu SH (2019) Expression of calcineurin in relation to the embryonic diapause process in the silkworm, *Bombyx mori*. Comp Biochem Physiol A 228:35–42. https://doi.org/10.1016/j.cbpa.2018.10.013

Huang K, Liu Y, Perrimon N (2022) Roles of insect oenocytes in physiology and their relevance to human metabolic diseases. Front Insect Sci 2:859847. https://doi.org/10.3389/finsc.2022.859847

Jacqmain M, Doucet E, Després JP, Bouchard C, Tremblay A (2003) Calcium intake, body composition, and lipoprotein-lipid concentrations in adults. Am J Clin Nutr 77:1448–1452. https://doi.org/10.1093/ajcn/77.6.1448

Jain K, Lavista-Llanos S, Grabe V, Hansson BS, Wicher D (2021) Calmodulin regulates the olfactory performance in *Drosophila melanogaster*. Sci Rep 11(1):3747. https://doi.org/10.1038/s41598-021-83296-9

Jakob R, Beutner G, Sharma VK, Duan Y, Gross RA, Hurst S, Jhun BS, Uchi J, Sheu SS (2014) Molecular and functional identification of a mitochondrial ryanodine receptor in neurons. Neurosci Lett 575:7–12. https://doi.org/10.1016/j.neulet.2014.05.026

Jayakumar S, Hasan G (2018) Neuronal calcium signaling in metabolic regulation and adaptation to nutrient stress. Front Neural Circuits 12:25. https://doi.org/10.3389/fncir.2018.00025

Kaneuchi T, Sartain CV, Takeo S, Horner VL, Buehner NA, Aigaki T, Wolfner MF (2015) Calcium waves occur as *Drosophila* oocytes activate. Proc Natl Acad Sci USA 112(3):791–796. https://doi.org/10.1073/pnas.1420589112

Kikukawa T, Tobe I, Solowiej S, Rankin SM, Stay B (1987) Calcium as a regulator of juvenile hormone biosynthesis and release in the cockroach *Diploptera punctata*. Insect Biochem 17:179–187. https://doi.org/10.1016/0020-1790(87)90158-2

Klee CB, Crouch TH, Krinks MH (1979) Calcineurin: a calcium- and calmodulin-binding protein of the nervous system. Proc Natl Acad Sci USA 76(12):6270–6273. https://doi.org/10.1073/pnas.76.12.6270

Koo SH, Flechner L, Qi L, Zhang X, Screaton RA, Jeffries S, Hedrick S, Xu W, Boussouar F, Brindle P, Takemori H, Montminy M (2005) The CREB coactivator TORC2 is a key regulator of fasting glucose metabolism. Nature 437(7062):1109–1111. https://doi.org/10.1038/nature03967

Kumar DKN, Dhakal S, Lee Y (2025) TRPγ regulates lipid metabolism through Dh44 neuroendocrine cells. eLife 14:e99258. https://doi.org/10.7554/eLife.99258.2

Kuraishi T, Manaka J, Kono M, Ishii H, Yamamoto N, Koizumi K, Shiratsuchi A, Lee BL, Higashida H, Nakanishi Y (2007) Identification of calreticulin as a marker for phagocytosis of apoptotic cells in *Drosophila*. Exp Cell Res 313(3):500–510. https://doi.org/10.1016/j.yexcr.2006.10.027

Lanner JT, Georgiou DK, Joshi AD, Hamilton SL (2010) Ryanodine receptors: structure, expression, molecular details, and function in calcium release. Cold Spring Harb Perspect Biol 2(11):a003996. https://doi.org/10.1101/cshperspect.a003996

Li X, Wu G, Yang Y, Fu S, Liu X, Kang H, Yang X, Su XC, Shen Y (2017) Calmodulin dissociates the STIM1-Orai1 complex and STIM1 oligomers. Nat Commun 8:1042. https://doi.org/10.1038/s41467-017-01135-w

Liou J, Kim ML, Heo WD, Jones JT, Myers JW, Ferrell JE Jr, Meyer T (2005) STIM is a Ca^{2+} sensor essential for Ca^{2+}-store-depletion-triggered Ca^{2+} influx. Curr Biol 15(13):1235–1241. https://doi.org/10.1016/j.cub.2005.05.055

Marques R, Maia CJ, Vaz C, Correia S, Socorro S (2014) The diverse roles of calcium-binding protein regucalcin in cell biology: from tissue expression and signalling to disease. Cell Mol Life Sci 71(1):93–111. https://doi.org/10.1007/s00018-013-1323-3

Martins GF, Ramalho-Ortigão JM (2012) Oenocytes in insects. Invertebr Surviv J 9(1):139–152

Maus M, Cuk M, Patel B, Lian J, Ouimet M, Kaufmann U, Yang J, Horvath R, Hornig-Do HT, Chrzanowska-Lightowlers ZM, Moore KJ, Cuervo AM, Feske S (2017) Store-operated Ca^{2+} entry controls induction of lipolysis and the transcriptional reprogramming to lipid metabolism. Cell Metab 25(3):698–712. https://doi.org/10.1016/j.cmet.2016.12.021

Megha, Hasan G (2017a) Control of protein translation by IP_3R-mediated Ca^{2+} release in *Drosophila* neuroendocrine cells. Fly 11(4):290–296. https://doi.org/10.1080/19336934.2017.1384103

Megha, Hasan G (2017b) IP_3R-mediated Ca^{2+} release regulates protein metabolism in *Drosophila* neuroendocrine cells: implications for development under nutrient stress. Development 144(8):1484–1489. https://doi.org/10.1242/dev.145235

Mukunda L, Miazzi F, Kaltofen S, Hansson BS, Wicher D (2014) Calmodulin modulates insect odorant receptor function. Cell Calcium 55(4):191–199. https://doi.org/10.1016/j.ceca.2014.02.013

Mulye H, Gordon R (1993) Effects of fenoxycarb, a juvenile hormone analogue, on lipid metabolism of the eastern spruce budworm, *Choristoneura fumiferana*. J Insect Physiol 39:721–727. https://doi.org/10.1016/0022-1910(93)90046-T

Nelson HB, Heiman RG, Bolduc C, Kovalick GE, Whitley P, Stern M, Beckingham K (1997) Calmodulin point mutations affect *Drosophila* development and behavior. Genetics 147(4):1783–1798. https://doi.org/10.1093/genetics/147.4.1783

Ogoyi DO, Osir EO, Olembo NK (1998) Fat body triacylglycerol lipase in solitary and gregarious phases of *Schistocerca gregaria* (Forskal) (Orthoptera: Acrididae). Comp Biochem Physiol B Biochem Mol Biol 119(1):163–167. https://doi.org/10.1016/S0305-0491(97)00300-3

O'Keefe SJ, Tamura J, Kincaid RL, Tocci MJ, O'Neill EA (1992) FK-506- and CsA-sensitive activation of the interleukin-2 promoter by calcineurin. Nature 357(6380):692–694. https://doi.org/10.1038/357692a0

Ostwald TJ, MacLennan DH (1974) Isolation of a high affinity calcium-binding protein from sarcoplasmic reticulum. J Biol Chem 249(3):974–979

Paknejad N, Hite RK (2018) Structural basis for the regulation of inositol trisphosphate receptors by Ca^{2+} and IP_3. Nat Struct Mol Biol 25(8):660–668. https://doi.org/10.1038/s41594-018-0089-6

Pallen CJ, Steele JE (1988) A putative role for calmodulin in corpus cardiacum-stimulated trehalose synthesis in fat body of the American cockroach (*Periplaneta americana*). Insect Biochem 18:577–584. https://doi.org/10.1016/0020-1790(88)90009-1

Panda S, Behera S, Alam MF, Syed GH (2021) Endoplasmic reticulum & mitochondrial calcium homeostasis: the interplay with viruses. Mitochondrion 58:227–242. https://doi.org/10.1016/j.mito.2021.03.008

Parekh AB (2017) Regulation of CRAC channels by Ca^{2+}-dependent inactivation. Cell Calcium 63:20–23. https://doi.org/10.1016/j.ceca.2016.12.003

Parekh AB, Putney JW Jr (2005) Store-operated calcium channels. Physiol Rev 85:757–810. https://doi.org/10.1152/physrev.00057.2003

Pfluger PT, Kabra DG, Aichler M, Schriever SC, Pfuhlmann K, Casquero García V, Lehti M, Weber J, Kutschke M, Rozman J, Elrod JW, Hevener AL, Feuchtinger A, Hrabě de Angelis M, Walch A, Rollmann SM, Aronow BJ, Müller TD, Perez-Tilve D, Jastroch M, De Luca M, Molkentin JD, Tschöp MH (2015) Calcineurin links mitochondrial elongation with energy metabolism. Cell Metab 22(5):838–850. https://doi.org/10.1016/j.cmet.2015.08.022

Prakriya M, Lewis RS (2015) Store-operated calcium channels. Physiol Rev 95:1383–1436. https://doi.org/10.1152/physrev.00020.2014

Putney JW Jr (1986) A model for receptor-regulated calcium entry. Cell Calcium 7(1):1–12. https://doi.org/10.1016/0143-4160(86)90026-6

Putney JW Jr (1990) Capacitative calcium entry revisited. Cell Calcium 11(10):611–624. https://doi.org/10.1016/0143-4160(90)90016-n

Ramos FO, Leyria J, Nouzova M, Fruttero LL, Noriega FG, Canavoso LE (2021) Juvenile hormone mediates lipid storage in the oocytes of *Dipetalogaster maxima*. Insect Biochem Mol Biol 133:103499. https://doi.org/10.1016/j.ibmb.2020.103499

Reis M, Vieira CP, Morales-Hojas R, Aguiar B, Rocha H et al (2011) A comparative study of the short term cold resistance response in distantly related Drosophila species: the role of regucalcin and frost. PLoS One 6(10): e25520. https://doi.org/10.1371/journal.pone.0025520

Roos J, DiGregorio PJ, Yeromin AV, Ohlsen K, Lioudyno M, Zhang S, Safrina O, Kozak JA, Wagner SL, Cahalan MD, Veliçelebi G, Stauderman KA (2005) STIM1, an essential and conserved component of store-operated Ca^{2+} channel function. J Cell Biol 169(3):435–445. https://doi.org/10.1083/jcb.200502019

Rusnak F, Mertz P (2000) Calcineurin: form and function. Physiol Rev 80(4):1483–1521. https://doi.org/10.1152/physrev.2000.80.4.1483

Ryu SY, Beutner G, Dirksen RT, Kinnally KW, Sheu SS (2010) Mitochondrial ryanodine receptors and other mitochondrial Ca^{2+} permeable channels. FEBS Lett 584(10):1948–1955. https://doi.org/10.1016/j.febslet.2010.01.032

Salminen TS, Vesala L, Laiho A, Merisalo M, Hoikkala A, Kankare M (2015) Seasonal gene expression kinetics between diapause phases in *Drosophila virilis* group species and overwintering differences between diapausing and non-diapausing females. *Sci Rep* 5: 11197. https://doi.org/10.1038/srep11197

Screaton RA, Conkright MD, Katoh Y, Best JL, Canettieri G, Jeffries S, Guzman E, Niessen S, Yates JR 3rd, Takemori H, Okamoto M, Montminy M (2004) The CREB coactivator TORC2 functions as a calcium- and cAMP-sensitive coincidence detector. Cell 119(1): 61–74. https://doi.org/10.1016/j.cell.2004.09.015

Seno K, Nakamura T, Ozaki M (2005) Biochemical and physiological evidence that calmodulin is involved in the taste response of the sugar receptor cells of the blowfly, *Phormia regina*. Chem Senses 30(6): 497–504. https://doi.org/10.1093/chemse/bji042

Shi H, Dirienzo D, Zemel MB (2001) Effects of dietary calcium on adipocyte lipid metabolism and body weight regulation in energy-restricted aP2-agouti transgenic mice. FASEB J 15:291–293. https://doi.org/10.1096/fj.00-0584fje

Shimokawa N, Yamaguchi M (1993) Molecular cloning and sequencing of the cDNA coding for a calcium-binding protein regucalcin from rat liver. FEBS Lett 327(3):251–255. https://doi.org/10.1016/0014-5793(93)80998-a

Singh A, Agrawal N (2021) Deciphering the key mechanisms leading to alteration of lipid metabolism in *Drosophila* model of Huntington's disease. Biochim Biophys Acta (BBA) Mol Basis Dis 1867(7):166127. https://doi.org/10.1016/j.bbadis.2021.166127

Sivakumar S, Miellet S, Clarke C, Hartley PS (2022) Insect nephrocyte function is regulated by a store-operated calcium entry mechanism controlling endocytosis and Amnionless turnover. J Insect Physiol 143: 104453. https://doi.org/10.1016/j.jinsphys.2022.104453

Somogyi E, Petersson U, Hultenby K, Wendel M (2003) Calreticulin—an endoplasmic reticulum protein with calcium-binding activity is also found in the extracellular matrix. Matrix Biol 22(2):179–191. https://doi.org/10.1016/s0945-053x(02)00117-8

Stoltzfus JR, Horton WJ, Grotewiel MS (2003) Odor-guided behavior in *Drosophila* requires calreticulin. J Comp Physiol A Neuroethol Sens Neural Behav Physiol 189(6):471–483. https://doi.org/10.1007/s00359-003-0425-z

Subramanian M, Jayakumar S, Richhariya S, Hasan G (2013a) Loss of IP3 receptor function in neuropeptide-secreting neurons leads to obesity in adult *Drosophila*. BMC Neurosci 14:157. https://doi.org/10.1186/1471-2202-14-157

Subramanian M, Metya SK, Sadaf S, Kumar S, Schwudke D, Hasan G (2013b) Altered lipid homeostasis in *Drosophila* InsP3 receptor mutants leads to obesity and hyperphagia. Dis Model Mech 6(3): 734–744. https://doi.org/10.1242/dmm.010017

Sullivan KM, Rubin GM (2002) The Ca^{2+}-calmodulin-activated protein phosphatase calcineurin negatively regulates EGF receptor signaling in *Drosophila* development. Genetics 161(1):183–193. https://doi.org/10.1093/genetics/161.1.183

Sun D, Steele JE (2001) Regulation of intracellular calcium in dispersed fat body trophocytes of the cockroach, Periplaneta americana, by hypertrehalosemic hormone. J Insect Physiol 47(12):1399–1408. https://doi.org/10.1016/S0022-1910(01)00130-5

Sun X, Zemel MB (2004) Calcium and dairy products inhibit weight and fat regain during ad libitum consumption following energy restriction in Ap2-agouti transgenic mice. J Nutr 134:3054–3060. https://doi.org/10.1093/jn/134.11.3054

Takeo S, Tsuda M, Akahori S, Matsuo T, Aigaki T (2006) The calcineurin regulator *sra* plays an essential role in female meiosis in *Drosophila*. Curr Biol 16:1435–1440. https://doi.org/10.1016/j.cub.2006.05.058

Teets NM, Yi SX, Lee RE Jr, Denlinger DL (2013) Calcium signaling mediates cold sensing in insect tissues. Proc Natl Acad Sci USA 110(22):9154–9159. https://doi.org/10.1073/pnas.1306705110

Tian Y, Bi J, Shui G, Liu Z, Xiang Y, Liu Y, Wenk MR, Yang H, Huang X (2011) Tissue-autonomous function of *Drosophila* Seipin in preventing ectopic lipid droplet formation. PLoS Genet 7(4):e1001364. https://doi.org/10.1371/journal.pgen.1001364

Tomita J, Mitsuyoshi M, Ueno T, Aso Y, Tanimoto H, Nakai Y, Aigaki T, Kume S, Kume K (2011) Pan-neuronal knockdown of calcineurin reduces sleep in the fruit fly, *Drosophila melanogaster*. J Neurosci 31(37):13137–13146. https://doi.org/10.1523/JNEUROSCI.5860-10.2011

Toprak U (2020) The role of peptide hormones in insect lipid metabolism. Front Physiol 11:434. https://doi.org/10.3389/fphys.2020.00434

Toprak U, Hegedus D, Doğan C, Güney G (2020) A journey into the world of insect lipid metabolism. Arch Insect Biochem Physiol 104:e21682. https://doi.org/10.1002/arch.21682

Toprak U, Doğan C, Hegedus D (2021) A comparative perspective on functionally-related, intracellular calcium channels: the insect ryanodine and inositol 1,4,5-trisphosphate receptors. Biomol Ther 11:1031. https://doi.org/10.3390/biom11071031

Vesala L, Salminen TS, Kankare M, Hoikkala A (2012) Photoperiodic regulation of cold tolerance and expression levels of regucalcin gene in *Drosophila Montana*. J Insect Physiol 58(5):704–709. https://doi.org/10.1016/j.jinsphys.2012.02.004

Wada I, Rindress D, Cameron PH, Ou WJ, Doherty JJ 2nd, Louvard D, Bell AW, Dignard D, Thomas DY, Bergeron JJ (1991) SSR alpha and associated calnexin are major calcium binding proteins of the endoplasmic reticulum membrane. J Biol Chem 266(29): 19599–19610

Wang B, Sullivan KM, Beckingham K (2003) *Drosophila* calmodulin mutants with specific defects in the musculature or in the nervous system. Genetics 165(3): 1255–1268. https://doi.org/10.1093/genetics/165.3.1255

Wang B, Goode J, Best J, Meltzer J, Schilman P, Chen J, Montminy M (2008) The insulin-regulated CREB coactivator TORC promotes stress resistance in *Drosophila*. Cell Metab 7(5):434–444. https://doi.org/10.1016/j.cmet.2008.02.010

Wang S, Liu S, Liu H, Wang J, Zhou S, Jiang R-J, Bendena WG, Li S (2010) 20-hydroxyecdysone reduces insect food consumption resulting in fat body lipolysis during molting and pupation. J Mol Cell Biol 2:128–138. https://doi.org/10.1093/jmcb/mjq006

Wang W, Zhu T, Wan P, Wei Q, He J, Lai F, Fu Q (2022) Cloning and functional analysis of calcineurin subunits a and B in development and fecundity of *Nilaparvata lugens* (Stål). Rice Sci 29(2):143–154. https://doi.org/10.1016/j.rsci.2022.01.003

Wei J, Li L, Yao S, Yang S, Zhou S, Liu X, Du M, An S (2019) Calcineurin-modulated antimicrobial peptide expression is required for the development of *Helicoverpa armigera*. Front Physiol 10:1312. https://doi.org/10.3389/fphys.2019.01312

Wei J, Yao X, Yang S, Liu S, Zhou S, Cen J, Liu X, Du M, Tang Q, An S (2021) Suppression of calcineurin enhances the toxicity of Cry1Ac to *Helicoverpa armigera*. Front Microbiol 12:634619. https://doi.org/10.3389/fmicb.2021.634619

Xiao X, Chen C, Yu TM, Ou J, Rui M, Zhai Y, He Y, Xue L, Ho MS (2017) Molecular chaperone calnexin regulates the function of Drosophila sodium channel paralytic. Front Mol Neurosci 10:57. https://doi.org/10.3389/fnmol.2017.00057

Xu Y, Borcherding AF, Heier C, Tian G, Roeder T, Kühnlein RP (2019) Chronic dysfunction of stromal interaction molecule by pulsed RNAi induction in fat tissue impairs organismal energy homeostasis in Drosophila. Sci Rep 9:6989. https://doi.org/10.1038/s41598-019-43327-y

Yamaguchi M, Yamamoto T (1978) Purification of calcium binding substance from soluble fraction of normal rat liver. Chem Pharm Bull (Tokyo) 26(6): 1915–1918. https://doi.org/10.1248/cpb.26.1915

Yamanaka N, Marqués G, O'Connor MB (2015) Vesicle-mediated steroid hormone secretion in *Drosophila melanogaster*. Cell 163:907–919. https://doi.org/10.1016/j.cell.2015.10.022

Yoshiga T, Yokoyama N, Imai N, Ohnishi A, Moto K, Matsumoto S (2002) cDNA cloning of calcineurin heterosubunits from the pheromone gland of the silkmoth, *Bombyx mori*. Insect Biochem Mol Biol 32: 477–486. https://doi.org/10.1016/S0965-1748(01)00125-4

Yu H, He L, Li ZQ, Li N, Ou-Yang YY, Huang GH (2020) Altering of host larval *Spodoptera exigua* calcineurin activity in response to ascovirus infection. *Pest Manag Sci* 76(3):1048–1059. https://doi.org/10.1002/ps.5615

Yu J, Liu H, Gao R, Wang TV, Li C, Liu Y, Yang L, Xu Y, Cui Y, Jia C, Huang J, Chen PR, Rao Y (2025) Calcineurin: an essential regulator of sleep revealed by biochemical, chemical biological, and genetic approaches. Cell Chem Biol 32(1):157–173.e7. https://doi.org/10.1016/j.chembiol.2024.12.003

Zang S, Wang R, Liu Y, Zhao S, Su L, Dai X, Chen H, Yin Z, Zheng L, Liu Q et al (2024) (2024) insulin signaling pathway mediates foxo–pepck axis regulation of glucose homeostasis in *Drosophila suzukii*. Int J Mol Sci 25:10441. https://doi.org/10.3390/ijms251910441

Zemel MB (2002) Regulation of adiposity and obesity risk by dietary calcium: mechanisms and implications. J Am Coll Nutr 21:146S–151S. https://doi.org/10.1080/07315724.2002.10719212

Zemel MB, Shi H, Greer B, Dirienzo D, Zemel PC (2000) Regulation of adiposity by dietary calcium. FASEB J 14:1132–1138. https://doi.org/10.1096/fasebj.14.9.1132

Zhu S, Liu F, Chen X, Xia S, Wu Y, Tang W, Ren C, Wang J, Li S (2025) Inter-organelle communication dynamically orchestrates juvenile hormone biosynthesis and female reproduction. Natl Sci Rev 12(3): nwaf022. https://doi.org/10.1093/nsr/nwaf022

Adv Exp Med Biol (2026) 1494: 239–259
https://doi.org/10.1007/5584_2024_821

Published online: 28 August 2024

Lipid Metabolism in Relation to Carbohydrate Metabolism

Perla Akiki, Pierre Delamotte, and Jacques Montagne

Abstract

Carbohydrates and lipids integrate into a complex metabolic network that is essential to maintain homeostasis. In insects, as in most metazoans, dietary carbohydrates are taken up as monosaccharides whose excess is toxic, even at relatively low concentrations. To cope with this toxicity, monosaccharides are stored either as glycogen or neutral lipids, the latter constituting a quasi-unlimited energy store. Breakdown of these stores in response to energy demand depends on insect species and on several physiological parameters. In this chapter, we review the multiple metabolic pathways and strategies linking carbohydrates and lipids that insects utilize to respond to nutrient availability, food scarcity or physiological activities.

Keywords

Fat body · Glycolysis · Haemolymph · Homeostasis · Lipogenesis · Midgut · Monosaccharides · Muscles · Oenocytes · Triacylglycerol

P. Akiki, P. Delamotte, and J. Montagne (✉)
Institute for Integrative Biology of the Cell (I2BC), Gif-sur-Yvette, France
e-mail: Perla.akiki@cnrs.fr; Pierre.DELAMOTTE@i2bc.paris-saclay.fr; Jacques.Montagne@i2bc.paris-saclay.fr

1 Introduction

Insects colonize a huge variety of ecological niches on Earth. Despite a great diversity of diets, their food always comprises carbohydrate biomolecules. Carbohydrates are organic compounds containing carbon, oxygen and hydrogen atoms, most frequently with a 2:1 hydrogen-oxygen ratio. They are divided into three main classes ranging from monosaccharide and disaccharide simple sugars (e.g. glucose, maltose) to complex polymers (e.g. glycogen, starch) and glycoconjugates (e.g. glycolipids) (Chandel 2022). The proportion of each class of dietary carbohydrates depends on the insect feeding source. Since, in most animal species, carbohydrates are commonly taken up by enterocytes as monosaccharides, the dietary oligosaccharides, polysaccharides and glycoconjugates require enzymatic hydrolysis in the digestive tract lumen (Terra and Ferreira 2012; Chapman et al. 2013; Sahaka et al. 2020). In insects, the major monosaccharides resulting from enzymatic digestion in the gut lumen are glucose, fructose and galactose, whereas trehalose is the most abundant circulating sugar in their haemolymph (Wyatt and Kale 1957). Trehalose is a non-reductive disaccharide linking two glucose units in a 1,1-glycosidic bound, which impedes the reductive toxicity of glucose (Mattila and Hietakangas 2017).

For most animal cells, monosaccharides and their metabolic by-products are toxic compounds, even at low concentrations (Rabbani and Thornalley 2013). To prevent this toxicity once taken up into animal cells, glucose is catabolized through glycolysis to produce energy or stored as glycogen granules. However, the capacity for glycogen storage is limited, so that additional storage strategies have been selected throughout evolution to provide energy resources in case of food scarcity (Steele 1982). Lipogenesis leading to the production of triacylglycerols (TAGs) is directly connected to carbohydrate catabolism through a metabolic axis comprising successively glycolysis, tricarboxylic acid (TCA) cycle and fatty acid (FA) synthesis. TAG storage in lipid droplets is, to a certain extent, unlimited, making it the essential strategy for energy storage (Brookheart et al. 2009; Patel and Abate 2013). In contrast, glycolipids do not constitute an energy storage form; they are mainly found at the extracellular membrane layer, playing critical roles in numerous cellular functions (Nishihara 2020). Cholesterol is critical for membrane fluidity and steroid hormone synthesis. Cholesterol biogenesis requires the condensation of two farnesylpyrophosphate units. Insects lack the enzymes for farnesylpyrophosphate condensation to sterol synthesis (Zhang et al. 2019). Therefore, cholesterol is not connected to carbohydrate metabolism and must be dietary provided (Clark and Block 1959).

Lipid catabolism in living organisms is not expected to reverse to carbohydrate biogenesis; thus, the carbohydrate to lipid metabolic axis is considered unidirectional. The fate of glucose to energy production or to glycogen/TAG storage is tightly controlled. In numerous animal species, this metabolic pathway involves the Mondo/Mlx transcription factor that responds to glucose metabolites to regulate several metabolic routes, including FA synthesis (Mattila and Hietakangas 2017; Richards et al. 2017; Havula and Hietakangas 2018). In this chapter, focusing on insect metabolism, we review the physiological links that connect the carbohydrate to lipid metabolism at the whole body level. To this end, we first summarize the metabolic routes operating in animal cells to direct dietary carbohydrates to lipid synthesis and storage. Next, we describe the roles of the various insect organs in these processes. Finally, we discuss the relevance of this vectorial metabolic axis in maintaining energy homeostasis.

2 The Metabolic Network Linking Carbohydrates to Lipids

Unless when referring to particular species, the metabolic pathways described in this section are expected to work in most metazoan organisms and often refer to publications based on mammalian metabolism.

2.1 Monosaccharide Fates

Monosaccharides do not accumulate in the intracellular compartment. They are either stored as glycogen or metabolized through catabolic pathways, including glycolysis or the pentose-phosphate pathway (Fig. 1).

Glycogen is a polymer of glucose units linked by $\alpha(1 \rightarrow 4)$ bonds, forming chains branched by $\alpha(1 \rightarrow 6)$ bonds (Adeva-Andany et al. 2016a; Chandel 2022). Glycogen synthesis comprises the intermediates glucose-1-phosphate (glucose-1P) and UDP-glucose and sequentially requires phosphoglucomutase, UDP-glucose phosphorylase, glycogen synthase and a branching enzyme (Fig. 1). Breakdown of glycogen is catalysed in the cytoplasm by glycogen phosphorylase and a debranching enzyme, and in lysosome by an acid α-glucosidase (Roach et al. 2012; Adeva-Andany et al. 2016a; Chandel 2022).

Glycolysis first proceeds in a preparatory phase that consumes two ATP units, leading to the formation of the trioses glyceraldehyde-3phosphate (GA3P) and dihydroxyacetone-phosphate (DHAP) (Fig. 1). In the glycolysis pay-off phase, each GA3P unit is oxidized to pyruvate, yielding one NADH and two ATP molecules.

The pentose-phosphate pathway (PPP) (Fig. 1) comprises a first oxidative phase, where glucose-6P is oxidized to produce NADPH and ribose-5P and a second non-oxidative phase that regenerates a pool of glucose-6P (Horecker 2002).

Fructose can be converted to glucose through the intermediate sorbitol (Fig. 1) (Krause and

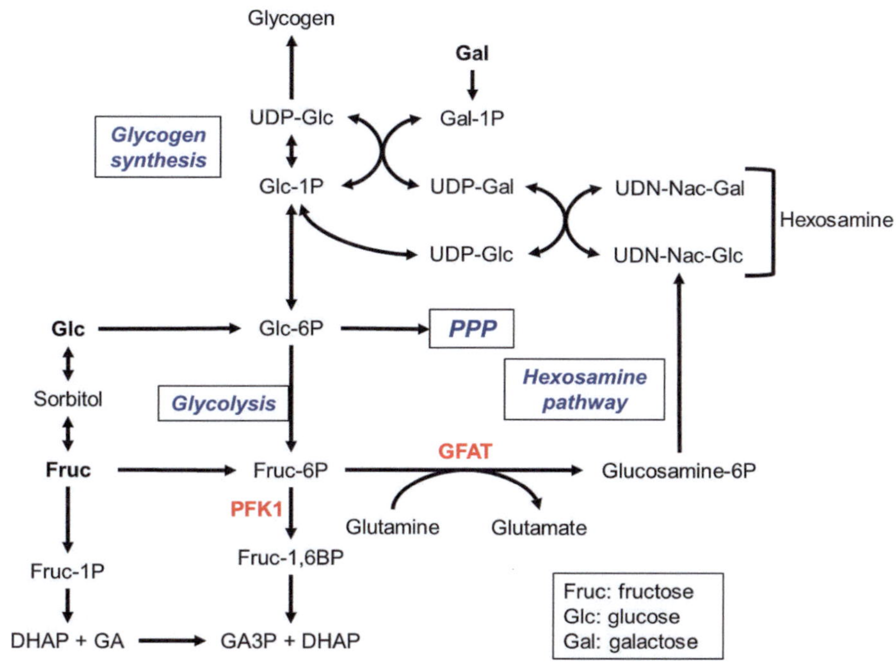

Fig. 1 Monosaccharide fates. Glucose, fructose and galactose are the main monosaccharides taken by enterocytes after hydrolysis of dietary carbohydrates. Glucose is metabolized to glycogen or through glycolysis. Fructose can enter glycolysis at various steps. Galactose enters this metabolic network as galactose-1P. Glycolytic intermediates may be used by the pentose phosphate pathway (PPP) or the hexosamine pathway

Wegner 2020). A few studies reveal that sorbitol plays a role in immune response, in learning and in the seasonal physiological turnover of overwintering insects (Yang et al. 2019; Sano et al. 2022; Weiglein et al. 2019; Mustard et al. 2018; Drahun et al. 2023). Nonetheless, the relevance of sorbitol in the fructose to glucose conversion to feed glycolysis has been sparsely evaluated in insects. In metazoan, to enter glycolysis (Fig. 1), fructose is phosphorylated either by a hexokinase to fructose-6P or by a ketohexokinase to fructose-1P (Johnson et al. 2020). The former enters glycolysis, whereas the latter is cleaved in DAHP and glyceraldehyde (GA), which is phosphorylated to GA3P by a triose kinase (Fig. 1). Phosphorylation of fructose-6P to fructose-1,6-bisphosphate by PFK1 is the rate-limiting step of glucose consumption through glycolysis (Mor et al. 2011). The fructose-1P pathway bypasses this key regulatory step and thus enhances lipogenesis, thereby, favouring obesity and metabolic syndrome in human patients (Krause and Wegner 2020).

The Leloir pathway is the main galactose metabolic pathway leading to the formation of UDP-galactose and glucose-1P, thereby (i) contributing to glycogen synthesis, (ii) entering glycolysis or (iii) initiating the hexosamine pathway required for glycoproteins and glycolipids biosynthesis (Adeva-Andany et al. 2016a) (Fig. 1). Nonetheless, the glycolytic metabolism of galactose is slower compared to that of glucose and fructose (Chandel 2022). Galactose-linked metabolic pathways are conserved in insects, such as *Drosophila melanogaster*, which can be used as a model system of galactosemia (Daenzer and Fridovich-Keil 2017).

The hexosamine pathway (Fig. 1) starts with the rate-limiting step catalysed by glutamine fructose 6-P amidotransferase (GFAT) to finally result in the production of UDP-N-acetylglucosamine (Chandel 2022). The UDP-galactose epimerase (GALE) catalyses the reversible interconversion of UDP-galactose to UDP-glucose, but also of UDP-N-acetylglucosamine to N-acetylgalactosamine, all these sugar derivatives being potential substrates for lipid glycosylation. Hexosamine-linked metabolic pathways are conserved in *Drosophila melanogaster* and likely in most insect species (Daenzer and Fridovich-Keil 2017; Nishihara 2020).

Trehalose is the main circulating sugar in insects. Its synthesis results from the condensation of glucose-6-phosphate and UDP-glucose to trehalose-6-phosphate catalysed by trehalose phosphate synthase and its dephosphorylation by trehalose phosphatase (Thompson 2003).

2.2 Glycolytic Products and By-products

The pay-off phase of glycolysis results in the production of one NADH and two ATP molecules through the conversion of each GA3P unit into pyruvate (Fig. 2). In the cytoplasm, the reversible conversion of one pyruvate to one lactate molecule is catalysed by lactate dehydrogenase (LDH), which concurrently oxidizes one NADH coenzyme to NAD+. In the mitochondria, the oxidation of pyruvate to acetyl-CoA by the pyruvate dehydrogenase enzymatic complex is coupled to the reduction of NAD+ to NADH. Acetyl-CoA reacts with oxaloacetate to form citrate, thereby initiating the TCA cycle. Citrate oxidation to oxaloacetate proceeds through seven enzymatic reactions coupled to the formation of one FAD and three NADH units. Regeneration of these coenzymes by the respiratory chains at the mitochondrial inner membrane yields high amounts of ATP (Towarnicki and Ballard 2020).

The lactate resulting from anaerobic fermentation in muscles is secreted and can be taken up by other tissues. For instance, in the mammalian liver, lactate from muscles is converted by LDH to pyruvate that is carboxylated to oxaloacetate by pyruvate carboxylase to fuel gluconeogenesis (Adeva-Andany et al. 2016b). To initiate gluconeogenesis, the conversion of oxaloacetate to phosphoenolpyruvate by the phosphoenolpyruvate carboxykinase (PEPCK) bypasses the irreversible phosphoenolpyruvate to pyruvate final reaction of glycolysis. When initiated from TCA cycle intermediates, gluconeogenesis depletes oxaloacetate, which must be compensated by anaplerotic reactions. Anaplerosis is commonly considered to depend on amino acid metabolism, in particular via transaminase-catalysed reactions that convert

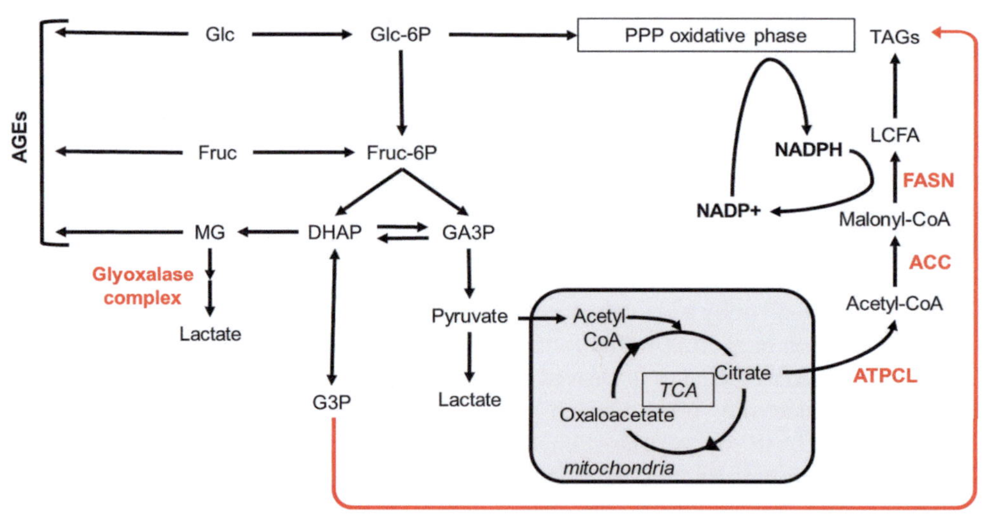

Fig. 2 Connecting glycolysis to lipogenesis. The final glycolytic product pyruvate is either fermented to lactate or oxidized to acetyl-CoA in the mitochondria. Excess of citrate is transferred back to the cytoplasm and used for FA synthesis. The NADPH required for this FA synthesis may be produced by the PPP oxidative phase. Glucose, fructose and trioses phosphate may lead to α-oxoaldehydes that react with amine groups to form advanced-glycation-end-products (AGEs). The toxicity of methylglyoxal (MG) can be neutralized by the glyoxalase complex. G3P is required for TAG synthesis through the Kennedy pathway

aspartate to oxaloacetate and glutamate to α-ketoglutarate (Brooks 1987; Inigo et al. 2021).

The trioses DHAP and GA3P can be interconverted by the triose phosphate isomerase (Fig. 2). GA3P fuels glycolysis, whereas DHAP can be reduced to glycerol 3-phosphate (G3P) that is used for TAG biogenesis through the Kennedy pathway (Coleman and Mashek 2011; Ko et al. 2020). In addition, DHAP and GA3P may spontaneously be oxidized to methylglyoxal (MG), which is an α-oxoaldehyde that strongly reacts with the amine groups of proteins, DNA or nucleotides to form advanced-glycation-end-products (AGEs). Several studies support the notion that AGEs—in particular, glycated proteins that become resistant to proteasome degradation—are responsible for the toxic effects of sugars (Annandale et al. 2021). Nonetheless, to prevent this toxicity, MG can be metabolized to lactate by the glyoxalase system composed of glyoxalase 1 (Glo1) and glyoxalase 2 enzymes (Thornalley 1993) (Fig. 2). This pathway is conserved in *Drosophila*, where it has been shown to play a critical role in preventing dietary sugar toxicity (Garrido et al. 2015).

2.3 Lipid Metabolism

Excess of citrate from the TCA cycle is transferred back to the cytoplasm, where it is converted to oxaloacetate and acetyl-CoA; the latter is used for FA synthesis that proceeds in two main enzymatic steps (Fig. 2). First, acetyl-CoA-carboxylase (ACC) catalyses the carboxylation of acetyl-CoA into malonyl-CoA that is the rate-limiting step for FA synthesis (Barber et al. 2005). Next, fatty acid synthase (FASN) sequentially incorporates malonyl-CoA units to one acetyl-CoA primer to produce long chain fatty acids (LCFAs) (Smith et al. 2003; Maier et al. 2010). The biochemical reactions catalysed by FASN require two NADPH coenzymes for the incorporation of each malonyl-CoA unit (Fig. 2). The NADPH may be synthesized through (i) the pentose phosphate pathway, (ii) the reduction of malate or (iii) the one-carbon metabolism (Fan et al. 2014).

Newly synthesized LCFA are used for the synthesis of phospholipids or stored as TAGs in lipid droplets (Coleman and Mashek 2011; Ko et al. 2020) (Fig. 3). Synthesis of TAGs that may proceed through three distinct pathways, as well as phospholipid biogenesis, are described in chapter "Dietary Lipids and Their Metabolism in the Midgut". In the context of energy demand, the breakdown of TAGs releases LCFAs used for mitochondrial β-oxidation. The resulting acetyl-CoA either feeds the TCA cycle or is converted to ketone bodies that efflux into the circulatory system to sustain the energy demand of other organs. In insects, ketone bodies transit through the haemolymph (Bailey and Horne 1972; Huang et al. 2019). Nonetheless, this catabolic route does not happen in the context of FA synthesis; in mammals, it has been shown that malonyl-CoA produced by ACC inhibits the CPT-1 mediated transfer of LCFAs into the mitochondria (Saggerson 2008). Notably, ketogenic—high fat, low carbohydrate—diet may have anti-epileptic effect in human; in *Drosophila melanogaster*, ketogenic diet has been shown to ameliorate seizure activity in a mitochondrial encephalomyopathy model (Fogle et al. 2019). Importantly, β-oxidation of even-chain LCFA cannot fuel gluconeogenesis but rather fosters anaplerotic

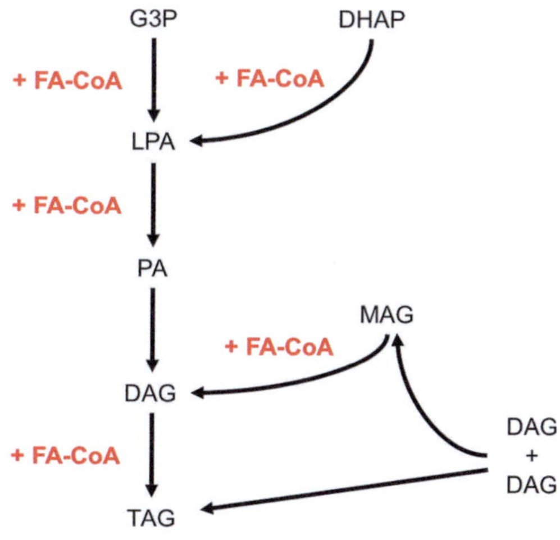

Fig. 3 TAG synthesis pathways. This synthesis may originate from G3P (Kennedy pathway), DHAP, MAG or transacylation of DAGs. Note that newly synthesized FAs are always required

reactions to increase the oxaloacetate pool required to metabolize the resulting high amounts of acetyl-CoA. Even in the case of β-oxidation of odd-chain LCFA, whose final product propionyl-CoA can be converted into succinyl-CoA (Inigo et al. 2021), the need in oxaloacetate substrates to metabolize acetyl-CoA impedes gluconeogenesis. Therefore, the metabolic vectorial axis from carbohydrate catabolism to lipid synthesis via the propionyl-CoA resulting from odd-chain LCFA oxidation is unlikely to reverse to carbohydrate biosynthesis.

Glycolipids are cell membrane components acting in signalling. Their synthesis is catalysed by glycosyltransferases that covalently attach sugar moieties to the hydroxyl group of either a glycerolipid or a sphingolipid backbone. Glycolipid synthesis takes place at the Golgi apparatus, so that glycolipids are presented at the cell surface exposing their sugar moiety in the extracellular milieu. The *Drosophila* genetics has proven a convenient model system to investigate glycosylation-related physiopathology (Nishihara 2020).

3 Major Organs Implicated

3.1 Digestive Tract

The type of dietary carbohydrates directly depends on the insect feeding source (Table 1) (Patrick et al. 2013). Given that monosaccharides are the main form of sugar absorbed by the enterocytes, dietary oligosaccharides, polysaccharides and glycoconjugates require enzymatic hydrolysis of glycosylic bonds in the midgut lumen prior to intestinal uptake. A number of carbohydrase activities have been identified in the salivary glands and in the midgut of several phytophagous bugs, suggesting that carbohydrate hydrolysis starts upon food ingestion (Hori 1975) or prior to it, as reported for external digestion of starch in *Drosophila melanogaster* (Haj-Ahmad and Hickey 1982; Boer and Hickey 1986; Szyszka and Galizia 2018). Insect enterocytes synthesize amylases that hydrolyse starch and glycogen, and maltases that hydrolyse maltose (Benkel and Hickey 1986; Terra and Ferreira 1994; Zinke et al. 2002; Mattila et al. 2015; Pimentel et al. 2018). Conversely, cellulose and pectin digestion relies in most species on enzymes synthesized by enteric microorganisms (Watanabe and Tokuda 2010; Ransom-Jones et al. 2012; Calderon-Cortes et al. 2012). Release of sugar moieties from glycolipids involves specific hydrolases secreted by the enterocytes, as reported for the folivorous lepidopteran *Epiphyas postvittana* and *Helicoverpa armigera* (Christeller et al. 2011; Sahaka et al. 2020). The origin of a given carbohydrate hydrolase cannot, however, be formally assigned to the genome of either the enterocyte

Table 1 Dietary carbohydrates of insects

Carbohydrate	Type	Main dietary source
Glucose	Monosaccharide $C_6H_{12}O_6$	Universal
Fructose	Monosaccharide $C_6H_{12}O_6$	Fruits
Galactose	Monosaccharide $C_6H_{12}O_6$	Milk, honey, plant galactolipids
Sucrose	1 glucose +1 fructose	Plant roots, fruits, sap
Maltose	2 glucose	Germinating seeds
Trehalose	2 glucose	Insect haemolymph
Cellobiose	2 glucose	Cellulose digestion
Raffinose	1 galactose +1 glucose +1 fructose	Plant seeds
Melezitose	2 glucose +1 fructose	Plant sap, honey
Glycogen	Branched glucose polymer	Animal stores
Starch	Branched glucose polymer	Plant stores
Inulin	Branched fructose polymer	Plant stores
Cellulose	Linear glucose polymer	Plant cell walls
Hemicellulose	Branched monosaccharide polymer	Plant cell walls
Pectin	Galacturonic acid polymer	Plant cell walls
Glycolipid	Monosaccharide + diacylglycerol	Cell membranes
Galactolipid	Galactose + diacylglycerol	Thylakoid membranes

or the microbiota, but varies depending on the insect species. For instance, cellulases and/or pectinases have been reported as encoded by the genome of the termite *Reticulitermes speratus*, the hemipteran *Lygus lineolaris* and the coleopteran *Phaedon cochleariae*, *Tribolium castaneum* and *Sphenophorus levis* (Watanabe et al. 1998; Girard and Jouanin 1999; Allen and Mertens 2008; Willis et al. 2011; Evangelista et al. 2015).

The physiology of aphids soaking up an unbalanced diet from plant phloem, containing high amounts of carbohydrates—mostly sucrose—with low amino acid and lipid levels, illustrates the tight metabolic link between carbohydrates and lipids through a unidirectional axis. It has been shown in the sap exclusive feeder *Acyrthosiphon pisum* that most of the dietary sucrose is used for lipid synthesis, mostly neutral lipids (Febvay et al. 1992). Consequently, TAGs are in high levels in the fat body but also in cornicle secretion that also contains high amounts of sugars (Rahbe et al. 1994). Their midgut symbionts are essential for their growth and their reproductive functions. However, antibiotic-induced depletion of their intestinal symbionts affects amino acid and glycogen synthesis but conversely increases lipid levels (Febvay et al. 1999; Lv et al. 2018), suggesting that these symbionts are required for amino acid but not lipid metabolism. The respective contributions of midgut and fat body cells in this lipogenic process in aphids have not been addressed to date.

Sugar transporters are members of the solute carrier proteins, including the major facilitators GLUT/SLC2, the sodium-driven glucose symporters SGLTs/SLC5 and the plant-specific SWEET/SLC50 (Deng and Yan 2016). In mammalian enterocytes, glucose uptake proceeds through SGLT1 located at the brush border, whereas passive reversible export at the basal membrane mainly depends on GLUT2 for glucose and galactose and on GLUT5 for fructose (Wright et al. 2007; Gonzalez and Betts 2019; Taskinen et al. 2019). In insects, the midgut is the main organ where monosaccharide absorption takes place, although this may also occur at the epidermis of endophagous parasitoids as reported for the larvae of the wasp *Aphidius ervi* (Caccia et al. 2005). Genomic analyses led to the identification of sugar transporters in several insect species (Kikuta et al. 2015; Moreira et al. 2017; Govindaraj et al. 2016; Pimentel et al. 2018; Wang and Wang 2020). Utilization of chemical inhibitors revealed a powerful approach to confirm sugar transporter specificity. In this way, expression of a potential hexose transporter (NlHT1) has been reported in the midgut of the hemipteran *Nilaparvata lugens* that feeds on sucrose-enriched phloem sap (Price et al. 2007). NlHT1 expression in vesicles mediates glucose but not fructose transport. This transport is inhibited by cytochalasin B, suggesting that NlHT1 is a functional GLUT2 homologue. Homologues of GLUTs and SGLTs have been found in the anterior midgut of *Dysdercus peruvianus* (Bifano et al. 2010). In this hemipteran, glucose transport is activated by K2SO4 and inhibited by the GLUT inhibitor phloretin and by the SGLT inhibitor phlorizin, supporting that GLUT- and SGLT-like functions are conserved in insects. In *D. melanogaster*, the soluble carrier dSLC5A5 has been shown to be the bona fide SGLT1 homologue required for apical endocytosis into the enterocytes (Li et al. 2021). Expression of a trehalose transporter (TreT) has been found in the midgut of *Anopheles gambiae* and *Musca domestica* (Liu et al. 2013; Pimentel et al. 2018). However, given the high luminal trehalase levels (Becker et al. 1996), TreT is unlikely fulfiling the uptake of dietary trehalose, but rather buffering haemolymph trehalose levels to maintain homeostasis.

In mammals, the classical view is that nutrients taken up from the intestinal lumen are directly routed to the liver, which is the major site of intermediary metabolism. In insect midgut cells, monosaccharides resulting from carbohydrate hydrolysis can either efflux into the haemolymph or be converted to FAs via neolipogenesis as reported in *Pieris brassicae* and *D. melanogaster* (Turunen 1993; Kokki et al. 2021). In the enterocytes, FAs are transiently esterified to TAGs and stored as lipid droplets, from where they are progressively exported in the form of diacylglycerol (DAG) loaded lipoproteins through the haemolymph to the requiring organs (Ford and

Van Heusden 1994; Pennington and Wells 2002; Grillo et al. 2003; Palm et al. 2012; Heier and Kuhnlein 2018).

3.2 Fat Body and Oenocytes

Fat body cells are in tight contact with oenocytes in a number of insect species. The fat body is an insect organ that combines hepatic and adipose functions related to energy homeostasis, nutrient storage and detoxification (Arrese and Soulages 2010; Li et al. 2019). In contrast, the oenocytes fulfil catabolic hepatic-like functions related to starvation (Gutierrez et al. 2007; Moraes and Montagne 2021; Huang et al. 2022). So far, investigations of the lipid anabolic pathways in the oenocytes of *D. melanogaster* did not reveal hepatic-like functions; instead, oenocytes produce FAs required to sustain the production of cuticular hydrocarbons (Wicker-Thomas et al. 2015; Storelli et al. 2019; Montagne and Wicker-Thomas 2020; Blomquist and Ginzel 2021; Huang et al. 2022) and to maintain tracheal water-tightness (Parvy et al. 2012).

The fat body accumulates high amounts of TAGs in the form of lipid droplets, which may involve import of dietary lipids or neolipogenesis. Whether the fat body is solely a storage organ or also contributes to intermediary metabolism has been addressed in a number of studies. Incorporation of dietary radiolabelled lipids in the fat fraction of the fat body of the lepidopteran *P. brassicae* and of the cockroach *Periplaneta americana* has been reported (Turunen 1975; Turunen and Kastari 1979; Chino and Downer 1979). Conversely, the fat body of the locusts *Schistocerca gregaria* and *Locusta migratoria* has been shown to perform intermediary metabolisms, including TCA cycle, respiration and fatty acid synthesis and to incorporate radiolabelled elements from glucose and amino acids into TAG-containing lipid droplets (Clements 1959; Tietz 1961; Hines and Smith 1963). These studies indicate that acyl groups of the TAG stores may have different origins (Coleman 2019; Olzmann and Carvalho 2019; Heier et al. 2021). TAG synthesis through the Kennedy pathway begins with the acylation of a glycerol-3-phosphate backbone in mitochondria or endoplasmic reticulum (ER), whereas the dihydroxyacetone-phosphate (DHAP) pathway begins with DHAP acylation in the peroxisome. Both pathways result in the production of lysophosphatidic acid, which after two acylation reactions in the ER leads successively to phosphatidic acid, DAG and TAG. Acylation of monoacylglycerol (MAG) in the ER is a third pathway for TAG biogenesis. A fourth pathway, which also takes place in the ER, is the transacylation of two DAG units that results in the formation of one TAG and one MAG molecules; the latter may follow the MAG pathway to TAG. In sum, any of these pathways requires additional FA units (Fig. 3). The TAG biogenesis pathway that takes place in the fat body has been sparsely addressed and likely depends on the insect species and/or its specific diet type (Kastari and Turunen 1977). For instance, in *Rhodnius prolixus*, the Kennedy pathway has been shown to be the only pathway for TAG synthesis (Alves-Bezerra and Gondim 2012). Nonetheless, given that in most species DAGs but not TAGs are transferred from the midgut to the fat body through the haemolymph, FA synthesis must operate in the fat body to sustain TAG biogenesis. Consistently, in *D. melanogaster*, fat body knockdown of ACC or FASN results in an almost complete depletion of overall TAGs and of fat body lipid droplets (Parvy et al. 2012; Garrido et al. 2015) and incorporation of isotope-labelled dietary glucose has been reported in total lipids (Musselman et al. 2013).

Glycogen has been reported to accumulate in the fat body of several insect species, including *S. gregaria*, *Bombyx mori*, *P. americana*, *Culex pipiens* and *D. melanogaster* (Clements 1959; Inagaki and Yamashita 1986; Veenstra 1989; King et al. 2020; Garrido et al. 2015; Yamada et al. 2018; Yamada et al. 2019). The fat body also synthesizes and accumulates trehalose and is considered to be the only organ where this synthesis takes place (Thompson 2003; Mattila and Hietakangas 2017). High levels of trehalose synthesis and the expression of trehalose-6-phosphate synthase have been reported in the fat body of *B. mori*, *L. migratoria* and *Mythimna separata* (Murphy and Wyatt 1965; Agbanyo and Taylor 1986; Yang et al. 2023). In *D. melanogaster*, starvation-induced glycogen breakdown of fat body stores is crucial to maintain homeostasis of circulating trehalose (Yamada et al. 2018).

In the context of starvation, TAGs are hydrolysed in glycerol and FAs to feed β-oxidation, which ends up with the formation of acetyl-CoA units. Acetyl-CoA may feed the TCA cycle or generate ketone bodies, which are transferred to the requiring organs (Bailey and Horne 1972). Studies in *D. melanogaster* suggest that the oenocytes perform β-oxidation upon fasting, as do mammalian hepatocytes that accumulate lipid droplets after the remobilization of adipocyte lipid stores. Likewise, fasting or genetically provoked breakdown of fat body TAGs induces lipid droplet accumulation in the oenocytes (Gutierrez et al. 2007). Congruently, enzymes responsible for ketone body formation are highly expressed in the oenocytes of *D. melanogaster* (Huang et al. 2019). Oenocyte-ablated *D. melanogaster* larvae are almost unable to burn their TAG stores in the context of fasting, which might suggest that the oenocyte are the only organs to perform FA β-oxidation (Gutierrez et al. 2007). Nonetheless, since the oenocytes produce an FA required for the watertightness of the tracheal system, oenocyte-ablated *D. melanogaster* larvae are in anoxic condition and thereby cannot perform FA β-oxidation (Parvy et al. 2012). Of note, FA consumption through β-oxidation has been reported in the flight muscles of a number of lepidopteran and orthopteran species (see below). Therefore, it is unlikely that the oenocytes are the only organs able to perform FA β-oxidation in the context of starvation and the respective contributions of the fat body and of the muscles remain to be precisely established.

3.3 Muscles

Muscles are necessary for locomotion dispersion in the environment (skeletal muscles) or for internal physiological functions (peristalsis of intestinal visceral muscles; heart activity). The metabolic pathway to provide energy depends on the species and on the muscle type. Metabolism of visceral muscles has been sparsely studied in insects. In *D. melanogaster*, a study focusing on the energy sensor AMP-activated protein kinase (AMPK) revealed that it is essential for intestinal peristalsis and that its mutation impaired nutrient absorption and glucose incorporation in glycogen and neutral lipids (Bland et al. 2010). Conversely, muscle metabolism has been extensively studied in insect skeletal flight muscles, which are the most energy demanding per tissue weight (Wegener 1996). The immediate on/off state transition at flight onset requires a rapid and efficient production of energy, which is considered to rely on carbohydrate consumption through glycolysis (Auerswald and Gade 1995; Canavoso et al. 2003; Suarez et al. 2005; Zhang et al. 2011; Gaviraghi et al. 2019). However, proline oxidation by the proline dehydrogenase at the mitochondrial respiratory chains is proposed to be a sparker metabolite at flight onset in several insect species (Bursell 1981; Auerswald and Gade 1995; Gade and Auerswald 2002; Soares et al. 2015; Gaviraghi et al. 2019; Stec et al. 2021).

Glucose may originate from muscle glycogen stores, whose breakdown is catalysed by the debranching enzyme AGL and by the glycogen phosphorylase that releases glucose-1P units (Childress and Sacktor 1970; Wegener 1996; Yamada et al. 2018). Glycogen storage and glycogen phosphorylase have been described in the muscles of numerous insect species (Childress et al. 1970; Downer and Matthews 1976; Turturro and Shafiq 1979; Steele 1982; Lorenz 2007; Zhang et al. 2011; Garrido et al. 2015; Yamada et al. 2019). Remobilization of glycogen from the fat body resulting in the synthesis of trehalose, which is transported through the haemolymph, is the other major source of glycolytic substrate in muscles after trehalose uptake and Trehalase hydrolysis to glucose (Kanamori et al. 2010; Thompson 2003; Murphy and Wyatt 1965; Agbanyo and Taylor 1986; Shukla et al. 2015; Mattila and Hietakangas 2017; Yang et al. 2023).

Consistently, glycogen phosphorylase has been reported in the fat body of several insect species including *D. melanogaster* and the cockroaches *Nauphoeta cinerea* and *P .americana* (Steele 1982; Gade 1991; Yamada et al. 2018; Yamada et al. 2019). Glucose consumption through glycolysis produces at the pay-off phase a pool of reduced NADH that can be regenerated to NAD+ in the anaerobic reduction of pyruvate to lactate (Fig. 4). Slower working skeletal muscles (e.g. leg muscles) may oxidize NADH to NAD+ by pyruvate to lactate, as described in the jumping muscles of *Schistocerca americana* (Kirkton and Tyler 2021). Conversely, the high glycolytic rate of flight muscles would produce dramatic amounts of lactate that might be toxic. To cope with this deleterious effect, flight muscles consume pyruvate through the TCA cycle, resulting in the lack of NAD+ to sustain the pay-off phase of glycolysis (Fig. 4). The fastest metabolic pathway to regenerate the pool of NAD+ proceeds through the G3P shuttle (Mracek et al. 2013; Gaviraghi et al. 2019). G3P is produced from DHAP reduction oxidizing NADH to NAD+. While G3P is a precursor for TAG synthesis through the Kennedy pathway in fat body cells, in muscles, it is regenerated at the mitochondrial inner membrane to DHAP by reducing the FAD coenzyme, which subsequently results in ATP production by the respiratory chains. Therefore, the G3P shuttle in flight muscle cells reduces the cytosolic pool of G3P, which would have been used for TAG or phospholipid synthesis in other tissues. Interestingly, a study on the bumblebee *Bombus terrestris*, which can fly at low ambient temperatures, suggests that the G3P shuttle plays a thermogenic role in pre-warming flight muscles (Masson et al. 2017).

The production of energy for flight muscles relies on carbohydrates for short-term and medium-term flights in several insect species (Beenakkers 1969), but may also depend on lipid consumption (Li et al. 2023), as reported for *Manduca sexta*, *L. migratoria* and *Gryllus*

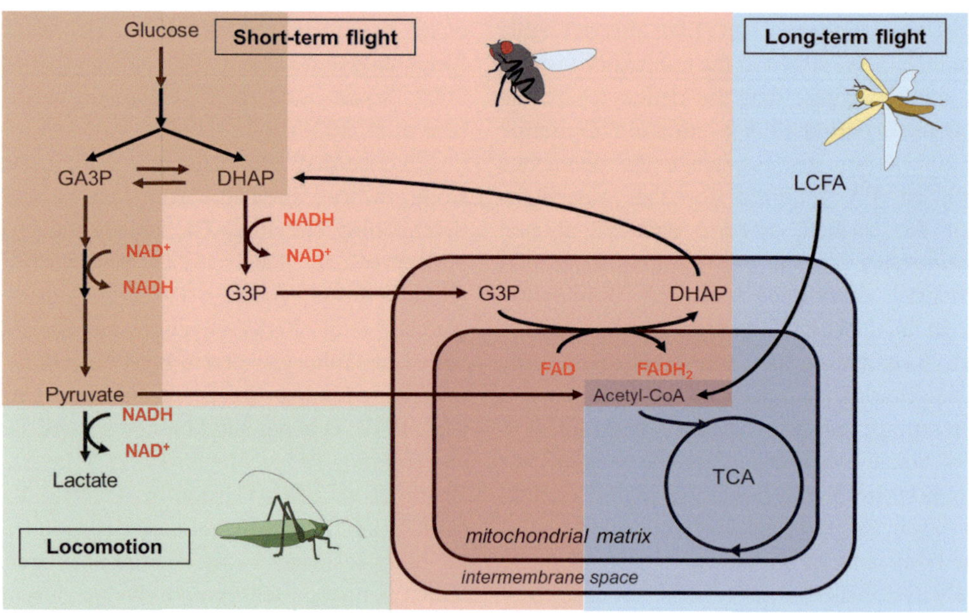

Fig. 4 Metabolism of insect flight muscles. Glucose consumption in flight muscles follows the aerobic glycolytic pathway through mitochondrial respiration; regeneration of the cytoplasmic NAD+ cofactor proceeds through the G3P shuttle. Energy production through anaerobic glycolysis yields high amounts of lactate that is toxic in flight muscles, but is used in locomotion muscles. Long-term flight usually consumes TAG stores through FA β-oxidation

bimaculatus (Ziegler and Schulz 1986; van der Horst et al. 1993; Lorenz 2007) (Table 2). Given that neutral lipids represent an enormous energy store, consumption of remobilized lipids represents an advantage for the long-term flight of non-feeding adult insects (Anand and Lorenz 2008). In *L. migratoria*, carbohydrate is the predominant energy source during the first 30 min of flight, while lipid remobilization from fat body stores in the form of circulating DAGs progressively increases to become the major energy source for the following 30 min of flight (Jutsum and Goldsworthy 1976). In *Panstrongylus megistus*, haemolymph carbohydrates and muscle glycogen progressively drop during 45 min flight, while fat body glycogen slightly decreases, suggesting that the latter do not sustain energy during flight. In contrast, between 15 and 45 min of flight, lipid stores from muscles and fat body strongly decrease, while those in the haemolymph increase, indicating that lipids become the predominant energy source during extended flight duration (Canavoso et al. 2003).

Studies to identify the mechanism of store remobilization to provide energy to flight muscles revealed the key role of neuropeptides (Li et al. 2023), including the functional glucagon homologue adipokinetic hormone (AKH). AKHs are small peptides synthesized in the corpora cardiaca and transferred through the haemolymph to the fat body, where they bind their cognate G-protein-coupled receptors (Gade and Auerswald 2003; Gade 2009). AKH-dependent remobilization is not store specific. In *L. migratoria*, AKH controls the remobilization of lipid and carbohydrate stores from the fat body to feed flight muscle activity (Robinson and Goldsworthy 1977). In *M. sexta*, AKH controls lipid mobilization to sustain adult flight, but glycogen phosphorylase in starving larvae (Ziegler et al. 1990). In *D. melanogaster*, AKH deficiency results in lipid consumption defects and obese flies, which surprisingly are not flight deficient, although this study assayed only the immediate flying performance, but not long-term flying ability (Galikova et al. 2015). In *A. gambia*, injection of a synthetic AKH increases flight performances and haemolymph carbohydrate but not lipids levels (Kaufmann and Brown 2008). In several beetle species that use proline to pre-warm flight

Table 2 Metabolites for flight muscles in various insect species

Insect species	Flight muscle metabolism requirements	Reference
A. aegypti	Proline, pyruvate and G3P	Gaviraghi et al. (2019)
A. gambia	AKH-I-dependent haemolymph trehalose	Kaufmann and Brown (2008)
Bombus impatiens	Proline at flight onset shift to carbohydrate	Stec et al. (2021)
B. terrestris	G3P shuttle to pre-warm flight muscle	Masson et al. (2017)
D. melanogaster	Glycolytic metabolites AKH-dependent lipid metabolism Peroxisome-dependent lipid metabolism Carbohydrate	Eanes et al. (2006) Katewa et al. (2012) Faust et al. (2014) Zhao et al. (2020)
Decapotoma lunata	Haemolymph carbohydrate	Auerswald and Gade (1995)
G. molesta	Carbohydrate	Su et al. (2022)
G. firmus	Fat body glycogen, haemolymph trehalose	Zhang et al. (2011)
L. migratoria	Trehalose and lipids Lipids, high FABP levels in flight muscles	Robinson and Goldsworthy (1977) van der Horst et al. (1993)
M. sexta	Haemolymph lipids FA oxidation shifts to glycolysis with ageing	Ziegler and Schulz (1986) Wone et al. (2018)
Pachnoda sinuata	Proline and haemolymph carbohydrates	Auerswald et al. (1998)
P. megistus	Carbohydrate first, shift to lipids	Canavoso et al. (2003)
P. americana	Glycogen High expression of glycogen phosphorylase	Downer and Matthews (1976) Steele (1982)
R. prolixus	Lipids	Braz et al. (2023)
S. gregaria	Carbohydrate first, shift to lipids	Rajapakse et al. (2019)
Orchid bees	Carbohydrate	Suarez et al. (2005)

muscles but also for flight, AKH controls haemolymph proline levels (Gade and Auerswald 2002). Likely because of its role in remobilizing fat body stores, a number of studies focused on the fat body AKH receptor, although it has been detected in a number of distinct cell types (Galikova and Klepsatel 2023). In *R. prolixus*, the AKH receptor is highly expressed in fat body and flight muscles, and its knockdown results in lipid accumulation in both organs (Alves-Bezerra et al. 2016). In sum, the insulin/glucagon balance to control homeostatic storage and remobilization of carbohydrates and lipids is tightly conserved in insects (Chowanski et al. 2021) (see also chapter "Dietary Lipids and Their Metabolism in the Midgut").

4 Integration of Carbohydrate/Lipid Metabolism at the Body Level

4.1 Carbohydrate and Lipid Homeostasis

Systemic regulators of metabolic homeostasis conserved throughout evolution have been identified in insects. In feeding state, dietary sugar and amino acids act through the fat body to trigger insulin-like peptide (ILP) secretion from neurosecretory cells, which in turn mediate several aspects of body growth and metabolic homeostatic response, in particular sugar storage in the form of glycogen and TAGs (Wu and Brown 2006; Toprak 2020; Chowanski et al. 2021). Conversely, in the case of high energy demand, neuropeptides such as AKH, AKH/Corazonin-related peptide and octopamine stimulate store breakdown to maintain homeostasis (Oguri and Steele 2003; Roeder 2005; Michitsch and Steele 2008; Zhou et al. 2018; Toprak 2020; Galikova and Klepsatel 2023). Neutral lipids are always the major components of fat body stores. However, the TAG/glycogen storage ratio as well as their consumption depends on the insect species and on several parameters, including behaviour, physiological status, ageing and energy level requirement (Lorenz and Anand 2004; Anand and Lorenz 2008; Wang et al. 2016). This section focuses on the relationships between carbohydrate and lipid metabolism per se, whereas the mechanisms of endocrine regulation of lipid metabolism are extensively documented in other chapters.

A number of insect-based studies emphasize the metabolic links between carbohydrates and lipids. As mentioned above, this metabolic link is nicely illustrated by the phloem exclusive feeder *A. pisum*, producing lipids in excess that accumulate in fat body but also in cornicle secretions (Febvay et al. 1992; Rahbe et al. 1994). In *R. prolixus*, it has been shown that deficiency in insulin receptor signalling strongly drops TAG synthesis in the fat body after blood meal (Silva-Oliveira et al. 2021). In *Blattella germanica*, high sucrose diet increases the expression of ACC—the rate-limiting enzyme of fatty acid synthesis—while ACC-knockdown decreases TAG and cuticular hydrocarbons levels and increases circulating sugar and glycogen storage (Pei et al. 2023). It has been shown in *P. americana* that FAs stimulate trehalose synthesis in isolated FB cells and that trehalose efflux into the haemolymph depends on arachidonic acid-derived metabolites (Ali et al. 1998; Ali and Steele 1997). Furthermore, a co-dependence in lipid and glycogen synthesis has been reported in *C. pipiens* where inhibition of either results in decreased biogenesis of both glycogen and TAGs (Olademehin et al. 2020). In *Aedes aegypti* females, where lipid stores are mobilized for egg production, a constant carbohydrate dietary source is required to sustain the accomplishment of the gonotrophic cycles (Briegel et al. 2002; Zhou et al. 2004). In *Grapholita molesta*, while flight muscle energy relies on carbohydrate consumption, lipid stores constitute the main energy source in starvation condition(s) (Su et al. 2022). The flying morph of *Gryllus firmus* female consumes carbohydrates for flight, while TAG consumption is likely related to reproductive behaviour in the non-flying reproductive morph (Zhang et al. 2011). A metabolic-based conflict between flight ability and egg production has been reported in several insect species. For instance, in *G. bimaculatus* adult females, carbohydrate and lipid stores are directed to egg production, while flight muscles are concurrently hydrolysed (Lorenz 2007). A screening-based study in *D. melanogaster* identified sirtuin 1 as a gene whose mutation results in fat accumulation but improved survival on a feeding media depleted in amino acid but supplemented with sugar (Reis et al. 2010). Another study investigating the relationship between dietary sugar and fat storage revealed that fructose was a more potent obesity-inducer than glucose (Rovenko et al. 2015). This effect likely results from the fructose-1P glycolytic route (§ 2.1), which bypasses the PFK1 rate-limiting reaction of glycolysis (Mor et al. 2011; Krause and Wegner 2020). Dietary sugar excesses are harmful in humans but also in insects. In *D. melanogaster*, high sugar diet (HSD) has been shown to provoke fat accumulation in the cardiac tube and heart dysfunction, which in part depends on hexosamine metabolism (Na et al. 2013). These studies strengthen the notion that carbohydrate and lipid metabolisms are inseparable processes to maintain homeostasis and might suggest that high amounts of dietary glucose are directly routed to TAG storage through the glycolytic/TCA/Lipogenic axis. However, a *D. melanogaster* study indicates that under chronic HSD, mitochondrial acetyl-coA could be recycled through an anaplerotic/gluconeogenic pathway before being re-metabolized and incorporated into TAGs (Musselman et al. 2013). Further, a study focusing on muscle ACC in *D. melanogaster* supports that lifespan extension upon diet restriction depends on a recycling loop of fat synthesis/oxidation (Katewa et al. 2012). In sum, the carbohydrate/lipid connection does not necessarily follow a unique direct axis but likely integrates within a complex metabolic network.

4.2 Sugar Diet and Type 2 Diabetic-like Syndrome in *D. Melanogaster*

The *D. melanogaster* model has been of powerful help in investigating the harmful effect of HSD. The target of rapamycin (TOR) signalling network plays a critical role in metabolic-related diseases (Saxton and Sabatini 2017; Moraes and Montagne 2021). TOR is present in two distinct complexes, TORC1 and TORC2, which operate via two parallel branches in *D. melanogaster* (Radimerski et al. 2002). TORC1 integrates nutrient availability through the tumour suppressor TSC and the small GTPase Rheb, while TORC2 resides in the insulin signalling branch that includes the tumour suppressor PTEN and the kinase AKT. The metabolism of *D. melanogaster* larvae over-activating either the TORC1 or the insulin response branch exhibits a significant decrease in overall glycogen and trehalose levels (Devilliers et al. 2021). Nonetheless, insulin—but not TORC1-induced cell-autonomous overgrowth—is significantly decreased by glycolysis or FA synthesis defect, suggesting that alternative metabolic strategies operate for TORC1, but not insulin-dependent growth (Devilliers et al. 2021). In mammals, mTOR-dependent metabolic regulations are in part mediated by the NAD^+-dependent deacetylases Sirtuins, as shown for SIRT4 (Csibi et al. 2013). In *D. melanogaster*, mutation of *Sirt4* results in shorter lifespan and increased TAG stores; these mutants are sensitive to starvation and weakly consume their carbohydrate and lipid stores upon fasting (Wood et al. 2018). In *D. melanogaster*, while amino acid controls the production of ilp2 (Colombani et al. 2003), circulating sugar promotes selective release of Dilp3 from insulin-producing cells (Kim and Neufeld 2015).

To characterize the genomic response to dietary sugar, a microarray analysis identified the sucrose-induced transcription factor encoded by *sugarbabe* and showed that it can repress genes required for dietary lipid absorption and fat breakdown (Zinke et al. 2002). Further studies revealed that HSD fed larvae decrease food intake, were hyperglycemic and developmentally delayed compared to larvae fed control diet. HSD-fed larvae accumulate more TAGs, express higher ILPs both at mRNA and protein levels and exhibit decreased insulin response (Musselman et al. 2011; Pasco and Leopold 2012; Musselman et al. 2013). In addition, larvae fed chronic HSD up-regulate genes of glycolysis, lipogenesis and trehalose synthesis, but also of gluconeogenesis and FA β-oxidation. Interestingly, obesity, ketogenesis, insulin resistance and increased gluconeogenesis are typical hallmarks of type 2 diabetes (Hatting et al. 2018). Congruent with a type 2 diabetic-like syndrome (Back and Kaufman 2012), genes involved in endoplasmic reticulum and oxidative stresses are also misregulated in larvae fed chronic HSD (Musselman et al. 2013).

The E-box transcription factor Mondo—homologue to the mammalian carbohydrate responsive element binding protein (ChREBP)—is up-regulated in HSD-fed larvae (Musselman et al. 2011). In mammals, ChREBP activity is stimulated by glycolytic metabolites to direct glucose usage (Richards et al. 2017; Havula and Hietakangas 2018). In *D. melanogaster, Mondo* knockdown and mutation of its partner encoded by *Mlx* result in high sugar sensibility and in the down-regulation of glycolytic and lipogenic genes in the fat body (Havula et al. 2013; Musselman et al. 2013). Consequently, Mondo and Mlx deficient larvae exhibited an increase in haemolymph sugar and a decrease in TAG and phospholipid levels. Finally, it has been shown that in response to dietary sugar, the Mondo/Mlx complex stimulates glycolysis and lipogenesis but also the pentose phosphate pathway and amino acid biosynthesis while repressing glucose intake and gluconeogenesis (Bartok et al. 2015; Mattila et al. 2015). These regulatory processes in part depend on Sugarbabe and on the TGF-β/Activin ligand Dawdle, either of them being metabolic regulators. On one hand, Dawdle has been shown to control the expression of carbohydrases and lipases in the enterocytes (Chng et al. 2014) and *Dawdle* mutant larvae exhibit elevated haemolymph sugars, TAG and glycogen stores (Ghosh and O'Connor 2014). On the other hand, over-expressing Sugarbabe in insulin-producing cells represses ilp3 expression, which in turn results in overall elevated TAG levels (Varghese et al. 2010). Finally, Sugarbabe has been shown to reside within an ERK7/TOR/PWP1 signalling axis to control fat body adiposity, where ERK7 is an anti-anabolic atypical MAP kinase and PWP1 a chromatin binding protein (Hasygar et al. 2021). Taken together, these studies demonstrate that Mondo and TOR signalling integrate in a complex network to regulate lipid metabolism in response to carbohydrate load.

4.3 Sugar Toxicity and Lipogenesis

The metabolic links between obesity and carbohydrate support the notion that lipogenesis protects against excess dietary sugar. This lipogenic-dependent protection is strongly supported by the phenotype induced by the mutation in the *FASN1* gene that encodes the ubiquitous FASN enzyme in *D. melanogaster* (Garrido et al. 2015). These mutants are early larval lethal but can be rescued by an appropriate mixture of dietary lipids. These rescued mutants decrease their food intake and TAG stores and increase their glycogen stores and haemolymph trehalose, but fail to survive on sucrose-supplemented media. Surprisingly, this sensitivity to dietary sugar also operates at the cell-autonomous level, since *FASN1* mutant clones in the fat body are strongly reduced in size in HSD-, but not standard, fed larvae (Garrido et al. 2015; Devilliers et al. 2021). An increase in AGEs (§ 2.2) is directly associated with high levels of circulating sugar in type 2 diabetic patients and is responsible for the harmful effects (Liu et al. 2023). Methylglyoxal that formed spontaneously from the glycolytic trioses phosphate GA3P and DHAP (Fig. 2) is a very active glycation reagent, which is detoxified by the glyoxalase complex (Rabbani and Thornalley 2012). Genetic interaction between *FASN1* mutation and *glyoxalase1* knockdown in *D. melanogaster* revealed that the toxic effect of sugar also depends on α-oxoaldehyde reagents, both at the organismal and cell-autonomous levels (Garrido et al. 2015). In addition, mutants for *glyoxalase 1*—encoding the first enzyme of the glyoxalase complex—are hyperglycemic, insulin resistant and exhibit elevated methylglyoxal levels and FASN1 activity (Moraru et al. 2018).

In conclusion, a critical function of lipogenesis is to protect against excesses of dietary carbohydrates. At first glance, it may be surprising that the basic enzymatic pathway of FA synthesis is conserved throughout evolution, suggesting that TAG synthesis is more important than the essential-FA metabolites for animal life.

Therefore, it is tempting to surmise that the protection against dietary carbohydrates is more critical than the essential FA requirement. Evolution is driven by adaptation to environmental constraints. To survive variations in food availability and scarcity episodes, metabolic pathways have been selected in a way that carbohydrates, which are abundant but potentially toxic, can be stored in high amounts in the form of neutral lipids. Excess of food may induce metabolic syndrome and insulin resistance, in part because of lipotoxicity. However, the strong gene conservation of the glycolytic/TCA/lipogenic axis suggests that lipotoxicity is not as harmful as glucotoxicity.

References

Adeva-Andany MM, Gonzalez-Lucan M, Donapetry-Garcia C, Fernandez-Fernandez C, Ameneiros-Rodriguez E (2016a) Glycogen metabolism in humans. BBA Clin 5:85–100

Adeva-Andany MM, Perez-Felpete N, Fernandez-Fernandez C, Donapetry-Garcia C, Pazos-Garcia C (2016b) Liver glucose metabolism in humans. Biosci Rep 36(6):e00416

Agbanyo M, Taylor NF (1986) Incorporation of 3-deoxy-3-fluoro-D-glucose into glycogen and trehalose in fat body and flight muscle in *Locusta migratoria*. Biosci Rep 6(3):309–316

Ali I, Steele JE (1997) Fatty acids stimulate trehalose synthesis in trophocytes of the cockroach (*Periplaneta americana*) fat body. Gen Comp Endocrinol 108(2): 290–297

Ali I, Finley C, Steele JE (1998) Evidence for the participation of arachidonic acid metabolites in trehalose efflux from the hormone activated fat body of the cockroach (*Periplaneta americana*). J Insect Physiol 44(11):1119–1126

Allen ML, Mertens JA (2008) Molecular cloning and expression of three polygalacturonase cDNAs from the tarnished plant bug, *Lygus lineolaris*. J Insect Sci 8:1–14

Alves-Bezerra M, Gondim KC (2012) Triacylglycerol biosynthesis occurs via the glycerol-3-phosphate pathway in the insect *Rhodnius prolixus*. Biochim Biophys Acta 1821(12):1462–1471

Alves-Bezerra M, De Paula IF, Medina JM, Silva-Oliveira G, Medeiros JS, Gade G, Gondim KC (2016) Adipokinetic hormone receptor gene identification and its role in triacylglycerol metabolism in the blood-sucking insect *Rhodnius prolixus*. Insect Biochem Mol Biol 69:51–60

Anand AN, Lorenz MW (2008) Age-dependent changes of fat body stores and the regulation of fat body lipid synthesis and mobilisation by adipokinetic hormone in the last larval instar of the cricket, Gryllus bimaculatus. J Insect Physiol 54(10–11):1404–1412

Annandale M, Daniels LJ, Li X, Neale JPH, Chau AHL, Ambalawanar HA, James SL, Koutsifeli P, Delbridge LMD, Mellor KM (2021) Fructose metabolism and cardiac metabolic stress. Front Pharmacol 12:695486

Arrese EL, Soulages JL (2010) Insect fat body: energy, metabolism, and regulation. Annu Rev Entomol 55: 207–225

Auerswald L, Gade G (1995) Energy substrates for flight in the blister beetle *Decapotoma lunata* (Meloidae). J Exp Biol 198(Pt 6):1423–1431

Auerswald L, Schneider P, Gade G (1998) Utilisation of substrates during tethered flight with and without lift generation in the African fruit beetle *Pachnoda sinuata* (Cetoniinae). J Exp Biol 201(Pt 15):2333–2342

Back SH, Kaufman RJ (2012) Endoplasmic reticulum stress and type 2 diabetes. Annu Rev Biochem 81: 767–793

Bailey E, Horne JA (1972) The role of ketone bodies in the metabolism of the adult desert locust. Biochem J 128(3):79P

Barber MC, Price NT, Travers MT (2005) Structure and regulation of acetyl-CoA carboxylase genes of metazoa. Biochim Biophys Acta 1733(1):1–28

Bartok O, Teesalu M, Ashwall-Fluss R, Pandey V, Hanan M, Rovenko BM, Poukkula M, Havula E, Moussaieff A, Vodala S, Nahmias Y, Kadener S, Hietakangas V (2015) The transcription factor Cabut coordinates energy metabolism and the circadian clock in response to sugar sensing. EMBO J 34(11): 1538–1553

Becker A, Schloder P, Steele JE, Wegener G (1996) The regulation of trehalose metabolism in insects. Experientia 52(5):433–439

Beenakkers AM (1969) Carbohydrate and fat as a fuel for insect flight. A comparative study. J Insect Physiol 15(3):353–361

Benkel BF, Hickey DA (1986) Glucose repression of amylase gene expression in*Drosophila melanogaster*. Genetics 114(1):137–144

Bifano TD, Alegria TG, Terra WR (2010) Transporters involved in glucose and water absorption in the *Dysdercus peruvianus* (Hemiptera: Pyrrhocoridae) anterior midgut. Comp Biochem Physiol B Biochem Mol Biol 157(1):1–9

Bland ML, Lee RJ, Magallanes JM, Foskett JK, Birnbaum MJ (2010) AMPK supports growth in Drosophila by regulating muscle activity and nutrient uptake in the gut. Dev Biol 344(1):293–303

Blomquist GJ, Ginzel MD (2021) Chemical ecology, biochemistry, and molecular biology of insect hydrocarbons. Annu Rev Entomol 66:45–60

Boer PH, Hickey DA (1986) The alpha-amylase gene in *Drosophila melanogaster*: nucleotide sequence, gene structure and expression motifs. Nucleic Acids Res 14(21):8399–8411

Braz V, Selim L, Gomes G, Costa ML, Mermelstein C, Gondim KC (2023) Blood meal digestion and changes in lipid reserves are associated with the post-ecdysis

development of the flight muscle and ovary in young adults of *Rhodnius prolixus*. J Insect Physiol 146:104492
Briegel H, Hefti M, DiMarco E (2002) Lipid metabolism during sequential gonotrophic cycles in large and small female *Aedes aegypti*. J Insect Physiol 48(5):547–554
Brookheart RT, Michel CI, Schaffer JE (2009) As a matter of fat. Cell Metab 10(1):9–12
Brooks GA (1987) Amino acid and protein metabolism during exercise and recovery. Med Sci Sports Exerc 19(5 Suppl):S150–S156
Bursell E (1981) The role of proline in energy metabolism. In: Downer RGH (ed) Energy metabolism in insects. Plenum Press, New York
Caccia S, Leonardi MG, Casartelli M, Grimaldi A, de Eguileor M, Pennacchio F, Giordana B (2005) Nutrient absorption by *Aphidius ervi* larvae. J Insect Physiol 51(11):1183–1192
Calderon-Cortes N, Quesada M, Watanabe H, Cano-Camacho H, Oyama K (2012) Endogenous plant cell wall digestion: a key mechanism in insect evolution. Annu Rev Ecol Evol Syst 43:45–71
Canavoso LE, Stariolo R, Rubiolo ER (2003) Flight metabolism in *Panstrongylus megistus* (Hemiptera: Reduviidae): the role of carbohydrates and lipids. Mem Inst Oswaldo Cruz 98(7):909–914
Chandel NS (2022) Carbohydrate metabolism. Cold Spring Harb Perspect Biol 13:a040568
Chapman RF, Simpson SJ, Douglas AE (2013) The insects. In: Structure and Function. Cambridge University Press, New York
Childress CC, Sacktor B (1970) Regulation of glycogen metabolism in insect flight muscle. Purification and properties of phosphorylases in vitro and in vivo. J Biol Chem 245(11):2927–2936
Childress CC, Sacktor B, Grossman IW, Bueding E (1970) Isolation, ultrastructure, and biochemal characterization of glycogen in insect flight muscle. J Cell Biol 45(1):83–90
Chino H, Downer RGH (1979) The role of diacylglycerol in absorption of dietary glyceride in the American cockroach *Periplanata americana* L. Insect Biochem 9:379–382
Chng WA, Sleiman MSB, Schupfer F, Lemaitre B (2014) Transforming growth factor beta/activin signaling functions as a sugar-sensing feedback loop to regulate digestive enzyme expression. Cell Rep 9(1):336–348
Chowanski S, Walkowiak-Nowicka K, Winkiel M, Marciniak P, Urbanski A, Pacholska-Bogalska J (2021) Insulin-like peptides and cross-talk with other factors in the regulation of insect metabolism. Front Physiol 12:701203
Christeller JT, Amara S, Carriere F (2011) Galactolipase, phospholipase and triacylglycerol lipase activities in the midgut of six species of lepidopteran larvae feeding on different lipid diets. J Insect Physiol 57(9): 1232–1239
Clark AJ, Block K (1959) The absence of sterol synthesis in insects. J Biol Chem 234:2578–2582
Clements AN (1959) Studies on the metabolism of locust fat body. J Exptl Biol 36:665–675
Coleman RA (2019) It takes a village: channeling fatty acid metabolism and triacylglycerol formation via protein interactomes. J Lipid Res 60(3):490–497
Coleman RA, Mashek DG (2011) Mammalian triacylglycerol metabolism: synthesis, lipolysis, and signaling. Chem Rev 111(10):6359–6386
Colombani J, Raisin S, Pantalacci S, Radimerski T, Montagne J, Leopold P (2003) A nutrient sensor mechanism controls Drosophila growth. Cell 114(6): 739–749
Csibi A, Fendt SM, Li C, Poulogiannis G, Choo AY, Chapski DJ, Jeong SM, Dempsey JM, Parkhitko A, Morrison T, Henske EP, Haigis MC, Cantley LC, Stephanopoulos G, Yu J, Blenis J (2013) The mTORC1 pathway stimulates glutamine metabolism and cell proliferation by repressing SIRT4. Cell 153(4):840–854
Daenzer JM, Fridovich-Keil JL (2017) *Drosophila melanogaster* models of Galactosemia. Curr Top Dev Biol 121:377–395
Deng D, Yan N (2016) GLUT, SGLT, and SWEET: structural and mechanistic investigations of the glucose transporters. Protein Sci 25(3):546–558
Devilliers M, Garrido D, Poidevin M, Rubin T, Le Rouzic A, Montagne J (2021) Differential metabolic sensitivity of insulin-like-response- and mTORC1-dependent overgrowth in Drosophila fat cells. Genetics 217(1):1–12
Downer RG, Matthews JR (1976) Glycogen depletion of thoracic musculature during flight in the American cockroach, *Periplaneta americana* L. Comp Biochem Physiol B 55(4):501–502
Drahun I, Poole EA, Hunt KA, van Herk WG, LeMoine CM, Cassone BJ (2023) Seasonal turnover and insights into the overwintering biology of wireworms (Coleoptera: Elateridae) in the Canadian prairies. Pest Manag Sci 79(2):526–536
Eanes WF, Merritt TJ, Flowers JM, Kumagai S, Sezgin E, Zhu CT (2006) Flux control and excess capacity in the enzymes of glycolysis and their relationship to flight metabolism in *Drosophila melanogaster*. Proc Natl Acad Sci USA 103(51):19413–19418
Evangelista DE, de Paula FF, Rodrigues A, Henrique-Silva F (2015) Pectinases from *Sphenophorus levis* Vaurie, 1978 (Coleoptera: Curculionidae): putative accessory digestive enzymes. J Insect Sci 15:168
Fan J, Ye J, Kamphorst JJ, Shlomi T, Thompson CB, Rabinowitz JD (2014) Quantitative flux analysis reveals folate-dependent NADPH production. Nature 510(7504):298–302
Faust JE, Manisundaram A, Ivanova PT, Milne SB, Summerville JB, Brown HA, Wangler M, Stern M, McNew JA (2014) Peroxisomes are required for lipid metabolism and muscle function in *Drosophila melanogaster*. PLoS One 9(6):e100213
Febvay G, Pageaux J, Bonnot G (1992) Lipid composition of the pea aphid, *Acyrthosiphon pisum* (Harris)

(hornoptera: Aphididae), reared on host plant and on artificial media. Arch Insect Biochem Physiol 21: 103–118

Febvay G, Rahbe Y, Rynkiewicz M, Guillaud J, Bonnot G (1999) Fate of dietary sucrose and neosynthesis of amino acids in the pea aphid, acyrthosiphon pisum, reared on different diets. J Exp Biol 202(Pt 19): 2639–2652

Fogle KJ, Smith AR, Satterfield SL, Gutierrez AC, Hertzler JI, McCardell CS, Shon JH, Barile ZJ, Novak MO, Palladino MJ (2019) Ketogenic and anaplerotic dietary modifications ameliorate seizure activity in Drosophila models of mitochondrial encephalomyopathy and glycolytic enzymopathy. Mol Genet Metab 126(4):439–447

Ford PS, Van Heusden MC (1994) Triglyceride-rich lipophorin in *Aedes aegypti* (Diptera: Culicidae). J Med Entomol 31(3):435–441

Gade G (1991) Glycogen phosphorylase in the fat body of two cockroach species, *Periplaneta americana* and *Nauphoeta cinerea*: isolation, partial characterization of three forms and activation by hypertrehalosaemic hormones. Z Naturforsch C J Biosci 46(1–2):149–162

Gade G (2009) Peptides of the adipokinetic hormone/red pigment-concentrating hormone family: a new take on biodiversity. Ann N Y Acad Sci 1163:125–136

Gade G, Auerswald L (2002) Beetles' choice–proline for energy output: control by AKHs. Comp Biochem Physiol B Biochem Mol Biol 132(1):117–129

Gade G, Auerswald L (2003) Mode of action of neuropeptides from the adipokinetic hormone family. Gen Comp Endocrinol 132(1):10–20

Galikova M, Klepsatel P (2023) Endocrine control of glycogen and triacylglycerol breakdown in the fly model. Semin Cell Dev Biol 138:104–116

Galikova M, Diesner M, Klepsatel P, Hehlert P, Xu Y, Bickmeyer I, Predel R, Kuhnlein RP (2015) Energy homeostasis control in Drosophila Adipokinetic hormone mutants. Genetics 201(2):665–683

Garrido D, Rubin T, Poidevin M, Maroni B, Le Rouzic A, Parvy JP, Montagne J (2015) Fatty acid synthase cooperates with glyoxalase 1 to protect against sugar toxicity. PLoS Genet 11(2):e1004995

Gaviraghi A, Correa Soares JBR, Mignaco JA, Fontes CFL, Oliveira MF (2019) Mitochondrial glycerol phosphate oxidation is modulated by adenylates through allosteric regulation of cytochrome c oxidase activity in mosquito flight muscle. Insect Biochem Mol Biol 114:103226

Ghosh AC, O'Connor MB (2014) Systemic Activin signaling independently regulates sugar homeostasis, cellular metabolism, and pH balance in *Drosophila melanogaster*. Proc Natl Acad Sci USA 111(15): 5729–5734

Girard C, Jouanin L (1999) Molecular cloning of cDNAs encoding a range of digestive enzymes from a phytophagous beetle, *Phaedon cochleariae*. Insect Biochem Mol Biol 29(12):1129–1142

Gonzalez JT, Betts JA (2019) Dietary sugars, exercise and hepatic carbohydrate metabolism. Proc Nutr Soc 78(2): 246–256

Govindaraj L, Gupta T, Esvaran VG, Awasthi AK, Ponnuvel KM (2016) Genome-wide identification, characterization of sugar transporter genes in the silkworm *Bombyx mori* and role in *Bombyx mori* nucleopolyhedrovirus (BmNPV) infection. Gene 579(2):162–171

Grillo LA, Pontes EG, Gondim KC (2003) Lipophorin interaction with the midgut of *Rhodnius prolixus*: characterization and changes in binding capacity. Insect Biochem Mol Biol 33(4):429–438

Gutierrez E, Wiggins D, Fielding B, Gould AP (2007) Specialized hepatocyte-like cells regulate Drosophila lipid metabolism. Nature 445(7125):275–280

Haj-Ahmad Y, Hickey DA (1982) A molecular explanation of frequency-dependent selection in Drosophila. Nature 299(5881):350–352

Hasygar K, Deniz O, Liu Y, Gullmets J, Hynynen R, Ruhanen H, Kokki K, Kakela R, Hietakangas V (2021) Coordinated control of adiposity and growth by anti-anabolic kinase ERK7. EMBO Rep 22(2): e49602

Hatting M, Tavares CDJ, Sharabi K, Rines AK, Puigserver P (2018) Insulin regulation of gluconeogenesis. Ann N Y Acad Sci 1411(1):21–35

Havula E, Hietakangas V (2018) Sugar sensing by ChREBP/Mondo-mlx-new insight into downstream regulatory networks and integration of nutrient-derived signals. Curr Opin Cell Biol 51:89–96

Havula E, Teesalu M, Hyotylainen T, Seppala H, Hasygar K, Auvinen P, Oresic M, Sandmann T, Hietakangas V (2013) Mondo/ChREBP-mlx-regulated transcriptional network is essential for dietary sugar tolerance in Drosophila. PLoS Genet 9(4):e1003438

Heier C, Kuhnlein RP (2018) Triacylglycerol metabolism in *Drosophila melanogaster*. Genetics 210(4):1163–1184

Heier C, Klishch S, Stilbytska O, Semaniuk U, Lushchak O (2021) The Drosophila model to interrogate triacylglycerol biology. Biochim Biophys Acta Mol Cell Biol Lipids 1866(6):158924

Hines WJW, Smith MJH (1963) Some aspects of intermediary metabolism in the desert locust (*Schistocerca gregaria* Forskal). J Ins Physiol 9:463–468

Horecker BL (2002) The pentose phosphate pathway. J Biol Chem 277(50):47965–47971

Hori K (1975) Digestive carbohydrases in the salivary gland and midgut of several phytophagous bugs. Comp Biochem Physiol B 50(1):145–151

Huang K, Chen W, Zhu F, Li PW, Kapahi P, Bai H (2019) RiboTag translatomic profiling of Drosophila oenocytes under aging and induced oxidative stress. BMC Genomics 20(1):50

Huang K, Liu Y, Perrimon N (2022) Roles of insect oenocytes in physiology and their relevance to human metabolic diseases. Front Insect Sci 2:859847

Inagaki S, Yamashita O (1986) Metabolic shift from lipogenesis to glycogenesis in the last instar larval fat body of the silkworm, Bombyx mori. Insect Biochem 16(2): 327–331

Inigo M, Deja S, Burgess SC (2021) Ins and outs of the TCA cycle: the central role of Anaplerosis. Annu Rev Nutr 41:19–47

Johnson RJ, Stenvinkel P, Andrews P, Sanchez-Lozada LG, Nakagawa T, Gaucher E, Andres-Hernando A, Rodriguez-Iturbe B, Jimenez CR, Garcia G, Kang DH, Tolan DR, Lanaspa MA (2020) Fructose metabolism as a common evolutionary pathway of survival associated with climate change, food shortage and droughts. J Intern Med 287(3):252–262

Jutsum AR, Goldsworthy GJ (1976) Fuels for flight in Locusta. J Insect Physiol 22:243–249

Kanamori Y, Saito A, Hagiwara-Komoda Y, Tanaka D, Mitsumasu K, Kikuta S, Watanabe M, Cornette R, Kikawada T, Okuda T (2010) The trehalose transporter 1 gene sequence is conserved in insects and encodes proteins with different kinetic properties involved in trehalose import into peripheral tissues. Insect Biochem Mol Biol 40(1):30–37

Kastari T, Turunen S (1977) Lipid utilization in *Pieris brassicae* reared on meridic and natural diest: implication for dietary improvement. Ent Exp Appl 22:71–80

Katewa SD, Demontis F, Kolipinski M, Hubbard A, Gill MS, Perrimon N, Melov S, Kapahi P (2012) Intramyocellular fatty-acid metabolism plays a critical role in mediating responses to dietary restriction in *Drosophila melanogaster*. Cell Metab 16(1):97–103

Kaufmann C, Brown MR (2008) Regulation of carbohydrate metabolism and flight performance by a hypertrehalosaemic hormone in the mosquito *Anopheles gambiae*. J Insect Physiol 54(2):367–377

Kikuta S, Nakamura Y, Hattori M, Sato R, Kikawada T, Noda H (2015) Herbivory-induced glucose transporter gene expression in the brown planthopper, *Nilaparvata lugens*. Insect Biochem Mol Biol 64:60–67

Kim J, Neufeld TP (2015) Dietary sugar promotes systemic TOR activation in Drosophila through AKH-dependent selective secretion of Dilp3. Nat Commun 6:6846

King B, Li S, Liu C, Kim SJ, Sim C (2020) Suppression of glycogen synthase expression reduces glycogen and lipid storage during mosquito overwintering diapause. J Insect Physiol 120:103971

Kirkton SD, Tyler SK (2021) American locust (*Schistocerca americana*) post-exercise lactate fate dataset. Data Brief 37:107263

Ko CW, Qu J, Black DD, Tso P (2020) Regulation of intestinal lipid metabolism: current concepts and relevance to disease. Nat Rev Gastroenterol Hepatol 17(3): 169–183

Kokki K, Lamichane N, Nieminen AI, Ruhanen H, Morikka J, Robciuc M, Rovenko BM, Havula E, Kakela R, Hietakangas V (2021) Metabolic gene regulation by Drosophila GATA transcription factor grain. PLoS Genet 17(10):e1009855

Krause N, Wegner A (2020) Fructose metabolism in cancer. Cells 9(12):2635

Li S, Yu X, Feng Q (2019) Fat body biology in the last decade. Annu Rev Entomol 64:315–333

Li Y, Wang W, Lim HY (2021) Drosophila solute carrier 5A5 regulates systemic glucose homeostasis by mediating glucose absorption in the midgut. Int J Mol Sci 22(22):12424

Li X, Zhou Y, Wu K (2023) Biological characteristics and energy metabolism of migrating insects. Meta 13(3): 439

Liu K, Dong Y, Huang Y, Rasgon JL, Agre P (2013) Impact of trehalose transporter knockdown on *Anopheles gambiae* stress adaptation and susceptibility to *Plasmodium falciparum* infection. Proc Natl Acad Sci USA 110(43):17504–17509

Liu J, Pan S, Wang X, Liu Z, Zhang Y (2023) Role of advanced glycation end products in diabetic vascular injury: molecular mechanisms and therapeutic perspectives. Eur J Med Res 28(1):553

Lorenz MW (2007) Oogenesis-flight syndrome in crickets: age-dependent egg production, flight performance, and biochemical composition of the flight muscles in adult female *Gryllus bimaculatus*. J Insect Physiol 53(8): 819–832

Lorenz MW, Anand AN (2004) Changes in the biochemical composition of fat body stores during adult development of female crickets, *Gryllus bimaculatus*. Arch Insect Biochem Physiol 56(3):110–119

Lv N, Wang L, Sang W, Liu CZ, Qiu BL (2018) Effects of endosymbiont disruption on the nutritional dynamics of the pea aphid *Acyrthosiphon pisum*. Insects 9(4):161

Maier T, Leibundgut M, Boehringer D, Ban N (2010) Structure and function of eukaryotic fatty acid synthases. Q Rev Biophys 43(3):373–422

Masson SWC, Hedges CP, Devaux JBL, James CS, Hickey AJR (2017) Mitochondrial glycerol 3-phosphate facilitates bumblebee pre-flight thermogenesis. Sci Rep 7(1):13107

Mattila J, Hietakangas V (2017) Regulation of carbohydrate energy metabolism in *Drosophila melanogaster*. Genetics 207(4):1231–1253

Mattila J, Havula E, Suominen E, Teesalu M, Surakka I, Hynynen R, Kilpinen H, Vaananen J, Hovatta I, Kakela R, Ripatti S, Sandmann T, Hietakangas V (2015) Mondo-Mlx mediates organismal sugar sensing through the Gli-similar transcription factor Sugarbabe. Cell Rep 13(2):350–364

Michitsch J, Steele JE (2008) Carbohydrate and lipid metabolism in cockroach (*Periplaneta americana*) fat body are both activated by low and similar concentrations of Peram-AKH II. Peptides 29(2): 226–234

Montagne J, Wicker-Thomas C (2020) Drosophila pheromone production. In: Blomquist G, Vogt RG (eds) Insect pheromone biochemistry and molecular biology. Elsevier Academic Press, London

Mor I, Cheung EC, Vousden KH (2011) Control of glycolysis through regulation of PFK1: old friends and recent additions, vol 76. Cold Spring Harb Symp Quant Biol, pp 211–216

Moraes KCM, Montagne J (2021) *Drosophila melanogaster*: a powerful tiny animal model for the study of metabolic hepatic diseases. Front Physiol 12: 728407

Moraru A, Wiederstein J, Pfaff D, Fleming T, Miller AK, Nawroth P, Teleman AA (2018) Elevated levels of the reactive metabolite methylglyoxal recapitulate progression of type 2 diabetes. Cell Metab 27(4): 926–934 e8

Moreira NR, Cardoso C, Dias RO, Ferreira C, Terra WR (2017) A physiologically-oriented transcriptomic analysis of the midgut of *Tenebrio molitor*. J Insect Physiol 99:58–66

Mracek T, Drahota Z, Houstek J (2013) The function and the role of the mitochondrial glycerol-3-phosphate dehydrogenase in mammalian tissues. Biochim Biophys Acta 1827(3):401–410

Murphy TA, Wyatt GR (1965) The enzymes of glycogen and Trehalose synthesis in silk moth fat body. J Biol Chem 240:1500–1508

Musselman LP, Fink JL, Narzinski K, Ramachandran PV, Hathiramani SS, Cagan RL, Baranski TJ (2011) A high-sugar diet produces obesity and insulin resistance in wild-type Drosophila. Dis Model Mech 4(6): 842–849

Musselman LP, Fink JL, Ramachandran PV, Patterson BW, Okunade AL, Maier E, Brent MR, Turk J, Baranski TJ (2013) Role of fat body lipogenesis in protection against the effects of caloric overload in Drosophila. J Biol Chem 288(12):8028–8042

Mustard JA, Alvarez V, Barocio S, Mathews J, Stoker A, Malik K (2018) Nutritional value and taste play different roles in learning and memory in the honey bee (*Apis mellifera*). J Insect Physiol 107:250–256

Na J, Musselman LP, Pendse J, Baranski TJ, Bodmer R, Ocorr K, Cagan R (2013) A Drosophila model of high sugar diet-induced cardiomyopathy. PLoS Genet 9(1): e1003175

Nishihara S (2020) Functional analysis of glycosylation using *Drosophila melanogaster*. Glycoconj J 37(1): 1–14

Oguri E, Steele JE (2003) A novel function of cockroach (*Periplaneta americana*) hypertrehalosemic hormone: translocation of lipid from hemolymph to fat body. Gen Comp Endocrinol 132(1):46–54

Olademehin OP, Liu C, Rimal B, Adegboyega NF, Chen F, Sim C, Kim SJ (2020) Dsi-RNA knockdown of genes regulated by Foxo reduces glycogen and lipid accumulations in diapausing *Culex pipiens*. Sci Rep 10(1):17201

Olzmann JA, Carvalho P (2019) Dynamics and functions of lipid droplets. Nat Rev Mol Cell Biol 20(3):137–155

Palm W, Sampaio JL, Brankatschk M, Carvalho M, Mahmoud A, Shevchenko A, Eaton S (2012) Lipoproteins in *Drosophila melanogaster*--assembly, function, and influence on tissue lipid composition. PLoS Genet 8(7):e1002828

Parvy JP, Napal L, Rubin T, Poidevin M, Perrin L, Wicker-Thomas C, Montagne J (2012) *Drosophila melanogaster* acetyl-CoA-carboxylase sustains a fatty acid-dependent remote signal to waterproof the respiratory system. PLoS Genet 8(8):e1002925

Pasco MY, Leopold P (2012) High sugar-induced insulin resistance in Drosophila relies on the lipocalin neural Lazarillo. PLoS One 7(5):e36583

Patel P, Abate N (2013) Role of subcutaneous adipose tissue in the pathogenesis of insulin resistance. J Obes 2013:489187

Patrick JW, Botha FC, Birch RG (2013) Metabolic engineering of sugars and simple sugar derivatives in plants. Plant Biotechnol J 11(2):142–156

Pei X, Bai T, Luo Y, Zhang Z, Li S, Fan Y, Liu TX (2023) Acetyl coenzyme a carboxylase modulates lipogenesis and sugar homeostasis in *Blattella germanica*. Insect Sci 31(2):387–404

Pennington JE, Wells MA (2002) Triacylglycerol-rich lipophorins are found in the dipteran infraorder Culicomorpha, not just in mosquitoes. J Insect Sci 2:15

Pimentel AC, Barroso IG, Ferreira JMJ, Dias RO, Ferreira C, Terra WR (2018) Molecular machinery of starch digestion and glucose absorption along the midgut of *Musca domestica*. J Insect Physiol 109: 11–20

Price DR, Wilkinson HS, Gatehouse JA (2007) Functional expression and characterisation of a gut facilitative glucose transporter, NlHT1, from the phloem-feeding insect *Nilaparvata lugens* (rice brown planthopper). Insect Biochem Mol Biol 37(11):1138–1148

Rabbani N, Thornalley PJ (2012) Methylglyoxal, glyoxalase 1 and the dicarbonyl proteome. Amino Acids 42(4):1133–1142

Rabbani N, Thornalley PJ (2013) Glyoxalase in diabetes, obesity and related disorders. Semin Cell Dev Biol 22(3):309–317

Radimerski T, Montagne J, Hemmings-Mieszczak M, Thomas G (2002) Lethality of Drosophila lacking TSC tumor suppressor function rescued by reducing dS6K signaling. Genes Dev 16(20):2627–2632

Rahbe Y, Delobel B, Febvay G, Chantegrel B (1994) Aphid-specific triglycerides in symbiotic and aposymbiotic *Acyrthosiphon pisum*. Insect Biochem Molec Biol 24(1):95–101

Rajapakse S, Qu D, Sayed Ahmed A, Rickers-Haunerland-J, Haunerland NH (2019) Effects of FABP knockdown on flight performance of the desert locust, *Schistocerca gregaria*. J Exp Biol 222(Pt 21):jeb203455

Ransom-Jones E, Jones DL, McCarthy AJ, McDonald JE (2012) The Fibrobacteres: an important phylum of cellulose-degrading bacteria. Microb Ecol 63(2): 267–281

Reis T, Van Gilst MR, Hariharan IK (2010) A buoyancy-based screen of Drosophila larvae for fat-storage mutants reveals a role for Sir2 in coupling fat storage to nutrient availability. PLoS Genet 6(11):e1001206

Richards P, Ourabah S, Montagne J, Burnol AF, Postic C, Guilmeau S (2017) MondoA/ChREBP: the usual suspects of transcriptional glucose sensing; implication in pathophysiology. Metabolism 70:133–151

Roach PJ, Depaoli-Roach AA, Hurley TD, Tagliabracci VS (2012) Glycogen and its metabolism: some new developments and old themes. Biochem J 441(3): 763–787

Robinson NL, Goldsworthy GJ (1977) Adipokinetic hormone and the regulation of carbohydrate and lipid metabolism in a working flight muscle preparation. J Insect Physiol 23(1):9–16

Roeder T (2005) Tyramine and octopamine: ruling behavior and metabolism. Annu Rev Entomol 50:447–477

Rovenko BM, Perkhulyn NV, Gospodaryov DV, Sanz A, Lushchak OV, Lushchak VI (2015) High consumption of fructose rather than glucose promotes a diet-induced obese phenotype in *Drosophila melanogaster*. Comp Biochem Physiol A Mol Integr Physiol 180:75–85

Saggerson D (2008) Malonyl-CoA, a key signaling molecule in mammalian cells. Annu Rev Nutr 28:253–272

Sahaka M, Amara S, Wattanakul J, Gedi MA, Aldai N, Parsiegla G, Lecomte J, Christeller JT, Gray D, Gontero B, Villeneuve P, Carriere F (2020) The digestion of galactolipids and its ubiquitous function in nature for the uptake of the essential alpha-linolenic acid. Food Funct 11(8):6710–6744

Sano H, Nakamura A, Yamane M, Niwa H, Nishimura T, Araki K, Takemoto K, Ishiguro KI, Aoki H, Kato Y, Kojima M (2022) The polyol pathway is an evolutionarily conserved system for sensing glucose uptake. PLoS Biol 20(6):e3001678

Saxton RA, Sabatini DM (2017) mTOR signaling in growth, metabolism, and disease. Cell 168(6):960–976

Shukla E, Thorat LJ, Nath BB, Gaikwad SM (2015) Insect trehalase: physiological significance and potential applications. Glycobiology 25(4):357–367

Silva-Oliveira G, De Paula IF, Medina JM, Alves-Bezerra-M, Gondim KC (2021) Insulin receptor deficiency reduces lipid synthesis and reproductive function in the insect *Rhodnius prolixus*. Biochim Biophys Acta Mol Cell Biol Lipids 1866(2):158851

Smith S, Witkowski A, Joshi AK (2003) Structural and functional organization of the animal fatty acid synthase. Prog Lipid Res 42(4):289–317

Soares JB, Gaviraghi A, Oliveira MF (2015) Mitochondrial physiology in the major arbovirus vector *Aedes aegypti*: substrate preferences and sexual differences define respiratory capacity and superoxide production. PLoS One 10(3):e0120600

Stec N, Saleem A, Darveau CA (2021) Proline as a sparker metabolite of oxidative metabolism during the flight of the bumblebee, *Bombus impatiens*. Meta 11(8):511

Steele JE (1982) Glycogen phosphorylase in insects. Insect Biochem 12(2):131–147

Storelli G, Nam HJ, Simcox J, Villanueva CJ, Thummel CS (2019) Drosophila HNF4 directs a switch in lipid metabolism that supports the transition to adulthood. Dev Cell 48(2):200–214 e6

Su S, Zhang X, Jian C, Huang B, Peng X, Vreysen MJB, Chen M (2022) Effects of adult feeding treatments on longevity, fecundity, flight ability, and energy metabolism enzymes of Grapholita molesta moths. Insects 13(8):725

Suarez RK, Darveau CA, Welch KC Jr, O'Brien DM, Roubik DW, Hochachka PW (2005) Energy metabolism in orchid bee flight muscles: carbohydrate fuels all. J Exp Biol 208(Pt 18):3573–3579

Szyszka P, Galizia CG (2018) The role of the sucrose-responsive IR60b neuron for *Drosophila melanogaster*: a hypothesis. Chem Senses 43(5):311–312

Taskinen MR, Packard CJ, Boren J (2019) Dietary fructose and the metabolic syndrome. Nutrients 11(9):1987

Terra WR, Ferreira C (1994) Insect digestive enzymes: properties, compartmentalization and function. Comp Biochem Physiol 109B(1):1–62

Terra WR, Ferreira C (2012) Biochemistry and molecular biology of digestion. In: Gilbert LI (ed) Insect molecular biology and biochemistry. Academic Press/ Elsevier, London

Thompson SN (2003) Trehalose – the insect 'blood' sugar. Adv Insect Physiol 31:205–285

Thornalley PJ (1993) The glyoxalase system in health and disease. Mol Asp Med 14(4):287–371

Tietz A (1961) Fat synthesis in cell-free preparations of the locust fat-body. J Lipid Res 2:182–187

Toprak U (2020) The role of peptide hormones in insect lipid metabolism. Front Physiol 11:434

Towarnicki SG, Ballard JWO (2020) Towards understanding the evolutionary dynamics of mtDNA. Mitochondrial DNA A DNA Mapp Seq Anal 31(8):355–364

Turturro A, Shafiq SA (1979) Quantitative morphological analysis of age-related changes in flight muscle of *Musca domestica* L. J Gerontol 34(6):823–833

Turunen S (1975) Absorption and transport of dietary lipid in *Pieris brassicae*. J Insect Physiol 21:1521–1529

Turunen S (1993) Metabolic pathways in the midgut epithelium of *Pieris brassicae* during carbohydrate and lipid assimilation. Insect Biochem Molec Biol 23(6): 681–689

Turunen S, Kastari T (1979) Digestion and absorption of lecithin in larvae of the cabbage butterfly, *Pieris brassicae*. Comp Biochem Physiol 62A:933–937

van der Horst DJ, van Doorn JM, Passier PC, Vork MM, Glatz JF (1993) Role of fatty acid-binding protein in lipid metabolism of insect flight muscle. Mol Cell Biochem 123(1–2):145–152

Varghese J, Lim SF, Cohen SM (2010) Drosophila miR-14 regulates insulin production and metabolism through its target, sugarbabe. Genes Dev 24(24):2748–2753

Veenstra JA (1989) Simulation of the activation of fat body glycogen phosphorylase and trehalose synthesis by peptide hormones in the American cockroach. Biosystems 23(1):31–40

Wang M, Wang J (2020) Glucose transporter GLUT1 influences *Plasmodium berghei* infection in *Anopheles stephensi*. Parasit Vectors 13(1):285

Wang Y, Campbell JB, Kaftanoglu O, Page RE Jr, Amdam GV, Harrison JF (2016) Larval starvation improves metabolic response to adult starvation in honey bees (*Apis mellifera* L.). J Exp Biol 219(Pt 7):960–968

Watanabe H, Tokuda G (2010) Cellulolytic systems in insects. Annu Rev Entomol 55:609–632

Watanabe H, Noda H, Tokuda G, Lo N (1998) A cellulase gene of termite origin. Nature 394(6691):330–331

Wegener G (1996) Flying insects: model systems in exercise physiology. Experientia 52(5):404–412

Weiglein A, Gerstner F, Mancini N, Schleyer M, Gerber B (2019) One-trial learning in larval Drosophila. Learn Mem 26(4):109–120

Wicker-Thomas C, Garrido D, Bontonou G, Napal L, Mazuras N, Denis B, Rubin T, Parvy JP, Montagne J (2015) Flexible origin of hydrocarbon/pheromone precursors in *Drosophila melanogaster*. J Lipid Res 56(11):2094–2101

Willis JD, Oppert B, Oppert C, Klingeman WE, Jurat-Fuentes JL (2011) Identification, cloning, and expression of a GHF9 cellulase from *Tribolium castaneum* (Coleoptera: Tenebrionidae). J Insect Physiol 57(2): 300–306

Wone BWM, Kinchen JM, Kaup ER, Wone B (2018) A procession of metabolic alterations accompanying muscle senescence in *Manduca sexta*. Sci Rep 8(1): 1006

Wood JG, Schwer B, Wickremesinghe PC, Hartnett DA, Burhenn L, Garcia M, Li M, Verdin E, Helfand SL (2018) Sirt4 is a mitochondrial regulator of metabolism and lifespan in *Drosophila melanogaster*. Proc Natl Acad Sci USA 115(7):1564–1569

Wright EM, Hirayama BA, Loo DF (2007) Active sugar transport in health and disease. J Intern Med 261(1): 32–43

Wu Q, Brown MR (2006) Signaling and function of insulin-like peptides in insects. Annu Rev Entomol 51:1–24

Wyatt GR, Kale GF (1957) The chemistry of insect hemolymph. II. Trehalose and other carbohydrates. J Gen Physiol 40(6):833–847

Yamada T, Habara O, Kubo H, Nishimura T (2018) Fat body glycogen serves as a metabolic safeguard for the maintenance of sugar levels in Drosophila. Development 145(6):dev158865

Yamada T, Habara O, Yoshii Y, Matsushita R, Kubo H, Nojima Y, Nishimura T (2019) The role of glycogen in development and adult fitness in Drosophila. Development 146(8):dev176149

Yang S, Zhao Y, Yu J, Fan Z, Gong ST, Tang H, Pan L (2019) Sugar alcohols of polyol pathway serve as Alarmins to mediate local-systemic innate immune communication in Drosophila. Cell Host Microbe 26(2):240–251 e8

Yang HJ, Cui MY, Zhao XH, Zhang CY, Hu YS, Fan D (2023) Trehalose-6-phosphate synthase regulates chitin synthesis in *Mythimna separata*. Front Physiol 14: 1109661

Zhang BC, Jiang CJ, An CJ, Zhang QW, Zhao ZW (2011) Variations in fuel use in the flight muscles of wing-dimorphic *Gryllus firmus* and implications for morph-specific dispersal. Environ Entomol 40(6):1566–1571

Zhang T, Yuan D, Xie J, Lei Y, Li J, Fang G, Tian L, Liu J, Cui Y, Zhang M, Xiao Y, Xu Y, Zhang J, Zhu M, Zhan S, Li S (2019) Evolution of the cholesterol biosynthesis pathway in animals. Mol Biol Evol 36(11): 2548–2556

Zhao X, Li X, Shi X, Karpac J (2020) Diet-MEF2 interactions shape lipid droplet diversification in muscle to influence Drosophila lifespan. Aging Cell 19(7):e13172

Zhou G, Pennington JE, Wells MA (2004) Utilization of pre-existing energy stores of female *Aedes aegypti* mosquitoes during the first gonotrophic cycle. Insect Biochem Mol Biol 34(9):919–925

Zhou YJ, Fukumura K, Nagata S (2018) Effects of adipokinetic hormone and its related peptide on maintaining hemolymph carbohydrate and lipid levels in the two-spotted cricket, Gryllus bimaculatus. Biosci Biotechnol Biochem 82(2):274–284

Ziegler R, Schulz M (1986) Regulation of lipid metabolism during flight in *Manduca sexta*. J Insect Physiol 32(10):903–908

Ziegler R, Eckart K, Law JH (1990) Adipokinetic hormone controls lipid metabolism in adults and carbohydrate metabolism in larvae of *Manduca sexta*. Peptides 11(5):1037–1040

Zinke I, Schutz CS, Katzenberger JD, Bauer M, Pankratz MJ (2002) Nutrient control of gene expression in Drosophila: microarray analysis of starvation and sugar-dependent response. EMBO J 21(22):6162–6173

Adv Exp Med Biol (2026) 1494: 261–291
https://doi.org/10.1007/5584_2025_878

Published online: 09 August 2025

Insect Lipidomics: Advances, Applications, and Physiological Insights

Laura Palanker Musselman, Doga Cedden, Gözde Güney, and Umut Toprak

Abstract

Lipidomics, a specialized branch of metabolomics, investigates the diversity and functionality of lipids in biological systems. Lipids serve crucial roles in energy storage, membrane composition, and environmental acclimation in insects, underpinning processes such as development and stress responses. Advances in analytical technologies, such as liquid chromatography-mass spectrometry (LC-MS), have enabled precise identification and quantification of lipid species, providing unprecedented insights into lipid metabolism and dynamics. Key lipid classes, including triacylglycerols and phospholipids, exhibit structural and functional versatility, adapting to environmental pressures through mechanisms like homeoviscous adaptation. These dynamic lipid responses are essential for maintaining cellular and cuticular integrity and functionality under stress. By exploring lipid diversity and adaptations, lipidomics offers valuable perspectives on insect physiology, survival strategies, and evolutionary ecology. This chapter summarizes methods used to study insect lipidomes and highlights comparative lipidomic studies that have advanced our understanding of insect biology.

Keywords

Cuticular hydrocarbons · Diapause · Ecdysteroids · Insect lipidomics · Mass spectrometry

L. P. Musselman (✉)
Department of Biological Sciences, Binghamton University, Binghamton, NY, USA
e-mail: lmusselm@binghamton.edu

D. Cedden
Department of Evolutionary Developmental Genetics, Johann-Friedrich-Blumenbach Institute, GZMB, University of Göttingen, Göttingen, Germany
e-mail: doga.cedden@biologie.uni-goettingen.de

G. Güney
Agricultural Entomology, Department of Crop Sciences, University of Göttingen, Göttingen, Germany
e-mail: ggueney@uni-goettingen.de

U. Toprak (✉)
Faculty of Agriculture, Molecular Entomology Lab, Department of Plant Protection, Ankara University, Ankara, Turkey
e-mail: utoprak@agri.ankara.edu.tr

1 Introduction

Lipidomics, a subfield of metabolomics, investigates the total lipid profiles and their roles in cellular and physiological processes, including structural, storage, and signaling functions (Han 2016). Lipids are essential cellular components, serving as energy reserves, structural barriers, and dynamic signaling molecules (Han 2016; Harayama and Riezman 2018). In insects, these versatile molecules play important roles in energy storage, membrane composition, and acclimation to environmental and physiological stresses. Processes like diapause and migration are particularly reliant on lipid-based energy reserves and membrane stability, highlighting the evolutionary significance of lipids in insect survival (Lehmann et al. 2020; Cedden et al. 2024).

The field of lipidomics has been revolutionized by advances in analytical technologies, such as liquid chromatography coupled with mass spectrometry (LC-MS) and high-performance liquid chromatography (HPLC). These tools enable precise identification and quantification of complex lipid species with high sensitivity and specificity. Such advancements allow researchers to dissect lipid composition and metabolism at an unprecedented level of detail. This has provided insights into lipid dynamics, membrane integrity, energy metabolism, and intracellular communication, especially under conditions of environmental or physiological stress. Insect lipidomics reveals the remarkable functional and structural diversity of lipids, ranging from energy-rich triacylglycerols (TAGs), also called triglycerides (TGs), to membrane-stabilizing phospholipids, such as phosphatidylcholine (PC) and phosphatidylethanolamine (PE) (Guan et al. 2013). Structural lipids, particularly glycerophospholipids and sphingolipids, are key components of insect membranes, influencing their fluidity, phase behavior, and functionality. These lipids dynamically adapt to environmental conditions, such as temperature fluctuations, through mechanisms like homeoviscous adaptation (Kayukawa et al. 2007). Specific physiological processes, such as waterproofing and diapause, further illustrate the critical role of lipids in insect survival (Fig. 1). These processes maintain membrane integrity and cellular functionality, enabling insects to thrive or survive under unfavorable environmental conditions. Using cutting-edge lipidomic methods, researchers can now analyze lipid metabolism at a molecular level. This has identified important physiological and behavioral strategies in insects, offering broader insights into lipid function across insect species. Understanding lipid dynamics not only advances our knowledge of metabolic regulation but may also provide practical applications, such as improving pest control strategies and enhancing the nutritional quality of edible insects (Toprak et al. 2020). This chapter begins by providing a historical overview of lipidomics and its methodological advancements. It then explores how the study of lipidomes has advanced our understanding of insect physiology, metabolism, and behavior. By examining the diversity and distribution of lipids, this chapter highlights the critical role of lipidomics in advancing our understanding of insect biology and its broader implications in ecology, evolution, and agriculture.

2 Insect Lipidomics in the Twentieth Century

Early lipidomic studies in insects discovered interesting cuticular hydrocarbons (CHCs) that acted as mating pheromones as well as defensive lipids and steroid hormones found across this animal class (Howard and Blomquist 1982). In addition to these specialized lipids, insects also contain the phospholipids and glycerolipids common to all kingdoms. Insect lipids comprise a large group of hydrophobic and amphipathic compounds and have been studied in earnest since the 1970s. Mass spectrometry approaches developed starting in the 1970s have enabled investigators to identify and quantify insect hormones and pheromones, among other lipids. These early studies gave rise to whole fields, including those of courtship behavior and developmental biology, which will not be covered here (reviewed in Jezovit et al. 2017). Insect lipidomics lends itself to a wide range of applications that have been used to identify unique features of insect biochemistry. For instance, early isotope tracer and gas chromatography-mass spectrometry (GC-MS) experiments in the beetle *Epilachna varivestis* found that serine and the long-chain fatty acid oleic acid were substrates for the synthesis of a defensive alkaloid found on pupae (Attygalle et al. 1994). Moreover, the introduction of bacteria to the armyworm *Pseudaletia unipuncta* led to prostaglandin biosynthesis that could be

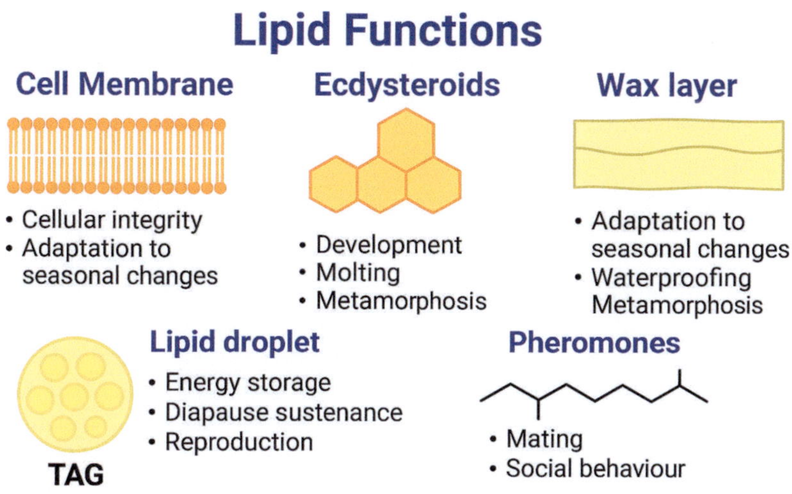

Fig. 1 Functional roles of key lipid classes and derivatives in insects. This infographic illustrates the major classes of lipids and their biological roles in insect physiology. Phospholipids, sterols, and sphingolipids contribute to membrane integrity, development, and environmental acclimation. Ecdysteroids act as steroid hormones that regulate molting, metamorphosis, and reproduction. Triacylglycerols (TAGs) serve as the primary energy reserve within lipid droplets, supporting diapause and reproduction. Cuticular hydrocarbons form a wax layer that enables waterproofing and protection against desiccation. Pheromones, composed of lipid-derived molecules, mediate mating and social behaviors. Together, these lipid classes orchestrate key developmental, metabolic, and behavioral processes across insect taxa

enhanced via supplementation with the fatty acid arachidonic acid (Jurenka et al. 1999). Another example of how GC-MS has contributed to insect lipid biology comes from studies on immune responses, where the eicosanoid pathway was found to be required for the response to infection in a range of insect species (Miller et al. 1999; Stanley et al. 1999; Bedick et al. 2000; Miller and Stanley 2001). Another membrane-specific approach used both thin layer chromatography (TLC) and liquid chromatography with tandem mass spectrometry (LC-MS-MS) to characterize the lipidome and was among the first to characterize lipid rafts and their resident glycosylphosphatidylinositol (GPI)-linked proteins in insects (Rietveld et al. 1999). This study identified the cholesterol-linked protein Hedgehog as a raft-associated protein and provided key insights into how this important insect morphogen is spatially regulated during development (Rietveld et al. 1999). As lipidomic methods were just developing in the twentieth century, most research in this field is relatively recent. However, it is important to acknowledge that foundational advances were already made in the twentieth century. Early studies using gas chromatography, thin-layer chromatography, and radiolabel tracing techniques identified key lipid species such as CHCs, ecdysteroids, and fatty acids in insect species from diverse insect taxa, including the tobacco hornworm *Manduca sexta*, the migratory locust *Locusta migratoria*, the American cockroach *Periplaneta americana*, and the black fly *Simulium damnosum* (Thompson et al. 1971; McDonough et al. 1972; Jacobson et al. 1973; Hetru et al. 1982). These efforts not only enabled biochemical characterization of lipid species but also laid the groundwork for understanding insect behavior, development, and environmental acclimation. These historical contributions provide essential context for the more comprehensive, high-resolution studies that define modern insect lipidomics. Many of these early approaches were limited in scope, especially considering the broad range of lipids that would ultimately be discovered and those yet to be discovered. Hence, the rest of this chapter focuses on twenty-first-century developments in insect lipidomics after summarizing methods used in this field.

3 Major Lipids and Lipid Derivatives in Insects

Insect lipids serve essential roles in energy storage, membrane structure, signaling, and environmental interaction. The most abundant subclasses include phospholipids, glycerolipids, and sterols, accompanied by minor but functionally significant classes such as sphingolipids, CHCs, and lipid-derived signaling molecules. Although many methodologies in lipidomics are adapted from mammalian systems, insects display unique lipid compositions reflective of their physiology and life history traits.

The following subsections covering fatty acids, storage lipids (TAGs and DAGs), membrane lipids (phospholipids and sphingolipids), sterols and ecdysteroids, and cuticular hydrocarbons are organized based on molecular structure and metabolic context, progressing from basic building blocks to complex derivatives. In contrast, Table 1 presents lipid classes grouped by their biological functions. Additional lipid categories, such as acyl-carnitines, cardiolipins (CLs), and lipid-derived signaling molecules (e.g., eicosanoids) are also discussed throughout this chapter and summarized in Table 1.

3.1 Fatty Acids

Fatty acids form the basic structural units of most complex lipids. In insects, medium and long-chain fatty acids dominate phospholipid and glycerolipid profiles (Tose et al. 2021). While very long-chain polyunsaturated fatty acids (≥C20) are rare (Shen et al. 2010), short-chain fatty acids have been observed in certain contexts, such as in *Drosophila* TAGs (Marshall et al. 2016). These fatty acids are either synthesized de novo or derived from the diet, and their composition influences membrane fluidity, energy storage, and signaling (Cedden et al. 2024). Notably, insects possess conserved orthologs of the core genes involved in fatty acid biosynthesis and metabolism, indicating that the enzymatic toolkit for fatty acid processing is broadly conserved across animal taxa.

3.2 Storage Lipids: Triacylglycerides and Diacylglycerols

Insects store neutral lipids predominantly as TAGs, which are sequestered in lipid droplets, primarily in the fat body. Triacylglycerides are mobilized during energy-demanding activities such as flight or starvation. Diacylglycerols, by contrast, serve as both structural intermediates and principal circulating lipids in the hemolymph, transported via lipophorin. TAGs in insects contain saturated and unsaturated fatty acyl chains ranging from C12 to C24 (Soliday et al. 1974; Sawabe and Moribayashi 2000).

3.3 Membrane Lipids: Phospholipids and Sphingolipids

Phospholipids constitute the primary structural components of insect membranes (Wegener et al. 2022). The most prevalent classes include phosphatidylcholine (PC), phosphatidylethanolamine (PE), phosphatidylglycerol (PG), phosphatidylserine (PS), and phosphatidylinositol (PI) (Brenner and Bernasconi 1987; Jurenka et al. 1987; Shipley et al. 1993; Gamo et al. 1999). Insect phospholipids often display shorter acyl chains (≤C18) than vertebrate equivalents. Sphingolipids are also present in various insect tissues and contribute to membrane stability and signal transduction (Rietveld et al. 1999).

3.4 Sterols and Ecdysteroids

Sterols are vital for insect physiology but cannot be synthesized de novo and must be acquired through the diet. Insects convert dietary sterols, such as cholesterol, into ecdysteroids—steroid hormones that regulate molting and development (Lafont et al. 2012). Techniques such as GC-MS and LC-MS have enabled the detection of ecdysteroids like 20-hydroxyecdysone in diverse insects including *Manduca sexta* (Thompson et al. 1971; Kaplanis et al. 1972; Kiyota et al. 2011), locusts (Hetru et al. 1982; Isaac et al. 1983; Tsoupras et al. 1983), and other species (Beydon

Table 1 Key lipids and lipid derivatives in insects (grouped by function)

Lipid class	Function	Methods	References
Membrane lipids			
Phospholipids [e.g., phosphatidylcholines (PC), phosphatidylethanolamines (PE)]	Major components of cell membranes; dynamically remodeled in response to diapause, development, thermal stress, and infection. Acyl chain diversity contributes to membrane fluidity and signaling capacity	LC-MS/MS, IMS, TLC	Batz and Armbruster (2018), Carvalho et al. (2012), Guan et al. (2013), and Matyash et al. (2008)
Sphingolipids (e.g., sphingomyelin)	Structural membrane components; involved in cell signaling and apoptosis	LC-MS/MS, shotgun lipidomics	Kraut (2011) and Merrill (2002)
Sterols (e.g., cholesterol, ecdysteroids)	Essential for molting and hormone signaling; cholesterol serves as a precursor for ecdysteroid biosynthesis	GC-MS, HPLC, RIA	Gilbert et al. (2002) and Lafont et al. (2012)
Cardiolipins (CL)	Mitochondrial-specific phospholipids crucial for respiratory chain function and mitochondrial integrity	LC-MS/MS, shotgun lipidomics	Schlame and Greenberg (2017) and Paradies et al. (2014)
Energy storage lipids			
Triacylglycerols (TAGs)	Main energy storage lipids in insects; stored in fat body and mobilized during stress, flight, reproduction, and metamorphosis	GC-MS, LC-MS/MS, TLC	Arrese and Soulages (2010) and Canavoso et al. (2001)
Diacylglycerols (DAGs)	Circulating intermediates in lipid metabolism that can function as minor storage lipids. Although not abundant in lipid droplets, they contribute to energy mobilization alongside cholesterol esters	LC-MS/MS, GC-MS	Arrese and Soulages (2010) and Canavoso et al. (2001)
Signaling lipids			
Ecdysteroids	Steroid hormones essential for molting and metamorphosis. Synthesized from dietary cholesterol and act via nuclear receptors to regulate gene expression	GC-MS, HPLC, RIA, LC-MS/MS	Gilbert et al. (2002), Lafont et al. (2012), and Rees and Isaac (1985)
Signaling lipids (e.g., eicosanoids, diacylglycerols)	Mediate immune responses, reproduction, and neuronal signaling in insects	LC-MS/MS, enzyme immunoassay	Stanley (2006) and Jurenka (2015)
Acyl-carnitines	Intermediates in fatty acid transport into mitochondria for β-oxidation; markers of lipid metabolic flux	LC-MS/MS, FIA-MS	Carvalho et al. (2012), and Chintapalli et al. (2013)
Structural/barrier lipids			
Cuticular hydrocarbons (CHCs)	Long-chain saturated and unsaturated hydrocarbons forming a waxy layer on the cuticle. They prevent desiccation and serve as pheromones for mating, recognition, and territorial behavior Composition varies with species, sex, life stage, and environmental cues	GC-MS, SPME-GC, MALDI-MSI	Blomquist and Bagnères (2010), Howard and Blomquist (2005), and Yew et al. (2008)
Wax esters	Long-chain esters found in the cuticle; contribute to water retention and surface hydrophobicity	GC-MS, TLC	Blomquist and Bagnères (2010) and Nelson et al. (1972)
Behavioral/pheromonal lipids			
Pheromones	Volatile or contact-active lipids involved in intraspecific communication, mate attraction, and aggregation behavior	GC-MS, SPME, EAG	Howard and Blomquist (2005) and Jurenka (2004)

GC gas chromatography, *MS* mass spectrometry, *LC* liquid chromatography, *TLC* thin-layer chromatography, *HPLC* high-performance liquid chromatography, *SPE* solid phase extraction, *DART* direct analysis in real time, *MALDI* matrix-assisted laser desorption/ionization, *IMS* imaging mass spectrometry, *SIMS* secondary ion mass spectrometry, *UHPLC* ultrahigh-performance liquid chromatography, *RIA* radioimmunoassay, *TAG* triacylglycerol, *DAG* diacylglycerol, *PC* phosphatidylcholine, *PE* phosphatidylethanolamine, *TCA* tricarboxylic acid et al. 1981; Grau and Lafont 1994; Blais et al. 2010; Miyashita et al. 2011; Lavrynenko et al. 2013, 2015).

Beyond development, ecdysteroids also play crucial roles in adult reproduction and mating behavior. In *D. melanogaster*, they modulate female postmating responses, including increased egg production and altered sexual receptivity, by acting on both the ovary and the central nervous system (Ameku and Niwa 2016).

3.5 Cuticular Hydrocarbons

Cuticular hydrocarbons are nonpolar lipids secreted by oenocytes and deposited on the insect cuticle. These molecules function in waterproofing, chemical signaling, and sexual communication (Wicker-Thomas 2007; Khallaf et al. 2021). Cuticular hydrocarbon biosynthesis originates from fatty acids and involves elongation and decarboxylation steps and can be hormonally regulated by peptides like PBAN (Pheromone Biosynthesis Activating Neuropeptide) (Martinez et al. 1990). Analytical approaches typically involve hexane extraction followed by gas chromatography-mass spectrometry (GC-MS), which has enabled detailed CHC profiling across species (Blomquist et al. 1976; Juárez et al. 1992).

Cuticular hydrocarbon profiles vary with sex, age, species, and environmental conditions and have been used to distinguish species (Carlson et al. 1997; Soliday et al. 1974), populations (Kruger and Pappas 1993; Byrne et al. 1995), infection status (Bernier et al. 1998; Buckner et al. 2000), and developmental stages (Liebig et al. 2000; Genin et al. 1986). Some CHCs are also found in the hemolymph and exhibit circadian variation (Turner and Acree 1967; Pho et al. 1996), highlighting their systemic distribution. This structural and temporal plasticity underscores the broad ecological and physiological relevance of CHCs across insect taxa (Blomquist and Ginzel 2021).

4 Methodological Approaches in Insect Lipidomics

4.1 Sampling and Extraction

Mass spectrometric approaches benefit from fractionation, where lipids are separated into simpler groups before analysis, improving signal-to-noise ratios and spectral resolution. Therefore, organic extraction of samples is the norm in insect lipidomics. It is standard practice to spike the starting material with a known quantity of one or more lipids to control for extraction efficiency. This can be an odd chain or other lipid that is easy to identify and quantify and is not typically observed in the samples (Han 2016). As different lipid classes vary in their susceptibility to oxidation or degradation, there is a need to work quickly and uniformly during sample preparation. As there is no signal amplification in lipidomics as in genomics, the quantity of material that can be isolated is, therefore, a key determinant of success. Balancing the need for substantial quantities of material (mg if possible) with the need to work quickly is a primary challenge of insect lipidomics.

Heat treatment has been reported to stabilize phospholipids, as their catabolism was higher in flash-frozen house crickets (*Acheta domesticus*) compared with those heated for just 25 s at 95 °C, suggesting that heating reduced phospholipase activity (Koelmel et al. 2018). Insects, of course, are segmented, and approaches to separately analyze the head, thorax, and/or abdomen are amenable for species of a sufficient size, such as *M. sexta* (Nelson et al. 1972). Quantifying cuticular pheromone concentrations is essential for understanding insect communication and behavior. These concentrations can be measured in washed abdomens or in situ on the cuticle's surface using direct real-time analysis, or direct analysis in real-time mass spectrometry (DART-MS) (Jennings et al. 2014). Similarly, hemolymph, excreta, and individual organs can be isolated, and the lipids can be extracted and quantified.

Hemolymph can be isolated by centrifugation (Lung and Wolfner 1999) or capillary action. Excreta can be collected from surfaces, including filter paper (Mendki et al. 2014) and plastic (Bush et al. 2012; Keesey et al. 2016; Fink et al. 2017; Wu et al. 2020), or obtained by placing larvae in solution (Hoi et al. 2014). Tissues and organs are typically isolated by fine dissection under the stereomicroscope.

The lipids of interest determine the chemistry of the extraction procedure. The Folch et al. (1957) or Bligh and Dyer (1959) extraction approaches, which use chloroform, have been the method of choice for most lipids. Few lipids partition into the aqueous (methanol-water) phase of these preparations, whereas the organic (chloroform) phase contains glycerolipids, sphingolipids, phospholipids, and steroids. However, caution is warranted when using the Folch method, as glycerol can co-extract into the organic phase and potentially confound gravimetric lipid quantification. This limitation is particularly relevant in studies relying solely on mass-based estimates of total lipids. Lipidomics-based workflows, by contrast, circumvent this issue through molecular-level identification and quantification (Williams et al. 2011). In addition to chloroform, one can also extract insect lipids with tert-butyl methyl ether (TBME), which also uses methanol and produces similar yields to chloroform-methanol extraction for major lipid classes (Matyash et al. 2008; Reis et al. 2013; Delgado et al. 2014; Salem et al. 2017).

4.2 Separation

A typical approach for analyzing insect lipids is to collect material and extract the organic phase that contains the lipids of interest. In this case, the material could be whole insects or spatially or temporally distinct lipid pools. Although spatial investigations are challenged by the small size of insects, advances such as fine dissection, cell sorting, laser capture microdissection, and imaging mass spectrometry have improved spatial resolution in lipidomic studies.

Separation of lipid classes is a crucial step in lipidomics. In addition to Folch or Bligh and Dyer extraction, techniques like solid-phase extraction (SPE), thin-layer chromatography (TLC), and gas chromatography (GC) have been used. However, high-performance liquid chromatography (HPLC) coupled with mass spectrometry (MS) is now favored for its ability to handle complex lipid profiles (Gao et al. 2018). Depending on the lipid type, normal-phase or reversed-phase HPLC is utilized, often employing C18 columns for efficient separation. For specific applications, such as analyzing phospholipids or neutral lipids, tailored HPLC methodologies are applied, enabling precise detection and quantification (Gao et al. 2018).

TLC is a classical technique that allows the separation and identification of lipids. Despite its early establishment, it is still routinely used for the analysis of lipids. TLC is a simple and effective method for analyzing lipids based on their polarity and mobility on a silica gel plate. Lipid extracts are applied as small spots on the TLC plate, which is then developed in a solvent system that separates lipid classes. By comparing the migration of unknown lipids to known standards, different lipid types such as TAGs, phospholipids, and fatty acids can be identified. TLC is widely used for qualitative analysis, lipid profile comparisons, and monitoring lipid purity or changes in lipid composition. Due to its simplicity and effectiveness, it has also been widely used in the study of insect lipidomes. For instance, TLC can be used to purify polar and neutral lipids from *Drosophila melanogaster* for further analysis (Jones et al. 1992).

4.3 Mass Spectrometry in Lipidomics

Lipidomic workflows can be broadly categorized into targeted and untargeted approaches. Targeted lipidomics involves the quantification of predefined lipid species using internal standards, enabling precise and sensitive measurements suitable for hypothesis-driven studies. In contrast, untargeted lipidomics aims to detect and annotate as many lipid molecules as possible without prior selection, providing a global view of lipidomic variation but requiring more advanced data processing and annotation. Both approaches have been used in insect studies depending on the research objective and sample complexity.

Mass spectrometry (MS) forms the cornerstone of modern lipidomics, offering high sensitivity, resolution, and selectivity for the detection and characterization of lipids. The fundamental principle involves ionizing lipid molecules and measuring their mass-to-charge (m/z) ratios. Early lipidomic studies relied on single-stage MS (non-tandem MS), which enabled identification of major lipid classes and quantification of abundant species (Frere et al. 2014; Gołębiowski et al. 2016). These foundational approaches paved the way for more advanced techniques by establishing basic lipid fingerprints in various biological matrices, including insect tissues. However, single-stage MS has limitations in resolving isobaric species and lacks detailed structural information. To overcome this, tandem mass spectrometry (MS/MS) was developed. This technique involves two stages of mass spectrometry separated by a fragmentation step, enabling the dissection of lipid ions into their structural components. MS/MS has proven essential for accurately identifying fatty acid chain length, degree of unsaturation, headgroup identity, and other fine structural details (Köfeler et al. 2021). It significantly enhances sensitivity and specificity, allowing the detection of low-abundance lipids in complex mixtures. MS/MS-based techniques have become standard in insect lipidomics, including for model organisms like *Drosophila melanogaster* (Gamo et al. 1999). Coupled with various ionization methods—such as electrospray ionization (ESI) or matrix-assisted laser desorption/ionization (MALDI)—and separation systems (e.g., HPLC or UHPLC), tandem MS enables both targeted and untargeted lipidomic profiling. These advances have expanded the capacity to study lipid remodeling under various biological and environmental conditions.

4.4 Specialized and Innovative Techniques

The unique developmental stages, morphological features, and life history adaptations of insects serve as platforms in which to characterize how the lipidome correlates with organismal biology. A growing number of insect lipidomic studies have employed innovative approaches—such as imaging mass spectrometry, laser capture microdissection, and in situ derivatization—to resolve temporally or spatially restricted features, offering insight into potential mechanisms by which lipids regulate fitness.

One study used high-resolution shotgun lipidomics—a direct-infusion mass spectrometry technique that analyzes complex lipid extracts without prior chromatographic separation—to quantify 250 lipid species in six lipid pools (gut, fat body, brain, salivary gland, wing disc, and lipoproteins from hemolymph) isolated by fine dissection from feeding third-instar *Drosophila* larvae (Carvalho et al. 2012). Phosphatidylethanolamine was the most abundant phospholipid across all sample types, whereas diacylglycerols (DAGs) were the most abundant glycerolipid in circulating lipoproteins, and TAGs predominated in tissues such as the fat body, wing disc, gut, and salivary gland (Carvalho et al. 2012). This study also found that the influence of dietary composition was dependent on the tissue analyzed. The nature of sterols and acyl chains differed among lipid depots, consistent with a model where both systemic and local lipid transport and metabolism control the lipidome in each tissue. Tissue lipids typically reflected the dietary lipid composition, although clear differences could be observed across tissues, especially for sterols. Low-abundance lipids such as sterol esters, CLs, and lysophospholipids were not observed due to the small size of the samples inherent in such a study (Carvalho et al. 2012).

A similar study in adult *Drosophila* quantified over 300 lipid species in 10 tissues isolated by fine dissection (Chintapalli et al. 2013). These authors focused on the gut (crop, midgut, and hindgut); the ovary, testis, and male accessory gland; the Malpighian tubules (anterior and posterior); the cuticle; and the head, all of which were homogenized. This LC-MS-based analysis revealed striking tissue-specific lipid signatures, including the enrichment of acyl-carnitines in the gut—reflecting high β-oxidation and metabolic activity—and the accumulation of ether lipids in the head, which are typically associated with neuronal membranes and oxidative protection. Additionally, several lipids were specifically enriched in gonads, consistent with their roles in reproduction. These findings emphasize the degree of lipid specialization across insect tissues and provide molecular insight into functional compartmentalization (Chintapalli et al. 2013). Another study compared the heads and genital tracts of lab-reared and wild-caught adult males and females over different diets and temperatures. Two different sterol sources, plant and yeast, produced the respective sterol profiles in fly heads and organs, and these profiles shifted within 7 days after the diets were swapped, an effect that was more pronounced in the female genital tract than in the head or male genital tract (Knittelfelder et al. 2020). Another lipid pool that can be accessed easily is the cuticular lipids, especially CHCs, which can be extracted directly from the surface by washing the cuticle in solvents and separating the extract (Jennings et al. 2014). Several interesting discoveries about the biochemistry and function of pheromones have been made using cuticular washing approaches.

A clever approach was developed using ambient mass spectrometry (Cooks et al. 2006) to analyze cuticular lipids in situ using DART-MS (Yew et al. 2008). The method can be used on appendages, dissected organs, across species, and even in living specimens. A gentle touch by stainless steel fine forceps or a metal insect pin can be used to sample quite effectively and to transfer the lipidome via helium gas to an ionization source. Lipid profiles from DART in *Drosophila* are consistent with previous findings using hexane-extracted cuticular lipids (Yew et al. 2008). Pioneering studies using this approach allowed observations such as the real-time transfer of male pheromones, including cis-vaccenic acid, to females after courtship (Yew et al. 2008). More recently, DART has been used in a genetic screen to identify Spidey as a novel regulator of CHC and steroid homeostasis as well as oenocyte survival (Chiang et al. 2016). This approach is also suitable for single-organ analyses, as shown recently in the male *Drosophila* pheromone-producing ejaculatory bulb, and may serve as a way to differentiate closely related species by their lipidome (Cetraro and Yew 2022).

A related procedure, ultraviolet laser desorption/ionization orthogonal time-of-flight (UV-LDI-o-TOF) MS, has higher resolution and the ability to detect more polar CHCs than DART (Yew et al. 2009, 2011b). This technique has allowed researchers to identify a new sex-dependent lipid and to characterize novel spatially restricted patterns of lipid abundance. In adult *Drosophila*, the male leg lipidome differed from the female leg, and cis-vaccenic acid was most concentrated in the male anogenital region and could be observed on the female cuticle only after mating. This study also identified a novel, larger male-specific pheromone with the molecular formula $C_{30}H_{56}O3$, which was also transferred to females during mating and persisted for 10 days postmating (Yew et al. 2009). Similar male-female CHC transfer phenomena were observed in the arid-adapted species *Drosophila arizonae* and *Drosophila mojavensis*, where a role for diet on CHC composition was also noted (Yew et al. 2011a). This study found that the leg and proboscis lipidomes were most affected in response to changes in the diet, consistent with the known roles of these regions in taste and feeding. Other insects have also been characterized using UV-LDI-o-TOF, where a broad range of spectra was acquired from the bee, wasp, and mosquito, in addition to spatially restricted spectra from the mandible and wing (Yew et al. 2011b). Transfer of osmeterium hydrocarbons from male leg segments to female wings was observed in the fly species *Themira superba* using UV-LDI-o-TOF (Araujo et al. 2014). With the high resolution and small size of insects, these in situ approaches are an exciting addition to the toolkit used for insect lipidomics that enable the assessment of live flies.

4.5 Imaging MS and MALDI-IMS

A technique with even higher spatial resolution is known as imaging mass spectrometry (IMS) or, alternately, mass spectrometric imaging (MSI). Typically, IMS requires substrates like body parts or organs to be immobilized on a slide and sprayed with a matrix that acts as a substrate from which lipids can be ionized into a matrix-assisted laser desorption/ionization (MALDI) detector mounted on a microscope. Initial studies in wings from adult *Drosophila* and the blow fly *Neobellieria bullata* were able to produce spectra containing a range of hydrocarbons and wax esters using a lithium salt matrix (Vrkoslav et al. 2010). Follow-up studies used additional matrices such as 2,5-dihydroxybenzoic acid (DHB), innovative scaffolds, thin sectioning, and fine dissection to explore additional anatomical regions and identify lipid species with higher sensitivity. For whole animal IMS in *Drosophila*, bespoke aluminum MALDI plates were designed to hold males or females so that the abdominal cuticle could be accessed by the laser (Kaftan et al. 2014). This approach detected cuticular TAG and localized cis-VA to the male genitalia (Kaftan et al. 2014). Another MALDI-IMS study benchmarked a number of approaches to profile the lipidome in cryosections from adult *Drosophila* (Niehoff et al. 2014). Using a sucrose infiltration medium followed by gelatin embedding and sectioning, the authors used positive and negative ion modes to detect phospholipids, TAG, and CHCs, comprising 92 lipid species in six classes (Niehoff et al. 2014). Although spatial differences could be observed across sections, assignment of lipids to specific organs was not possible at the resolution used (Niehoff et al. 2014).

Higher spatial resolution has provided additional insights. One example was achieved using MALDI-IMS with a neodymium-doped yttrium aluminum garnet laser in adult *Drosophila* Malpighian tubules (Yang et al. 2019). Extended washing, embedding, and matrix protocols were tested to optimize lipidomic analysis of dissected tubules and sectioned flies. Tubule-specific and organism-specific distribution patterns were observed, primarily for phospholipids. This group obtained accurate results using optimal cutting temperature (OCT) to embed organs for cryosectioning and found that 1,5-diaminonapthalene (DAN) was the best matrix for resolution in the negative ion mode.

A DAN matrix was used in a MALDI-IMS exploration of a *Drosophila* genetic model for amyotrophic lateral sclerosis (Jang et al. 2021). Thin sections of brains were analyzed after embedding in gelatin with a clever collar that fixes the head orientation in a reproducible manner (Jang et al. 2021). Brain-specific overexpression of a dipeptide-repeat region associated with the disease in humans produces dramatic changes in phospholipid content, compared to control brains, and regional patterns of lipid deposition changed over time, similar to findings in humans affected with this disease (Jang et al. 2021). A *Drosophila* disease model using high-sugar feeding to generate obesity and cardiac dysfunction was used to characterize the lipidome in whole-mounted adult fat bodies and hearts via MALDI-IMS (Tuthill et al. 2020). Organs were isolated by fine dissection and laid on glass slides before applying a DHB matrix, followed by acquisition of position-specific mass spectra (Tuthill et al. 2020). Using this approach to compare flies aged on control or high-sugar diets, 147 lipid features were putatively identified in one or both organs, including glycerolipids, phospholipids, and ether lipids, some of which differed significantly between diets (Tuthill et al. 2020). Heads from wild-type or lipoprotein receptor LpR1 mutants embedded in OCT were sectioned for MALDI-IMS and revealed reduced lipid content in mutant brains, consistent with reduced ability to traffic lipids in these animals (Chen et al. 2022).

Spatial resolution can be increased using atmospheric pressure high-resolution scanning microprobe MALDI, or atmospheric pressure (AP)-SMALDI-IMS, which was used to identify 67 lipids in sections from *Anopheles stephensi* mosquitoes embedded in carboxymethylcellulose

(Khalil et al. 2015). Although legs and wings were removed for this work, there were spatially restricted lipidomic signatures for phospholipids, glycerolipids, and sphingolipids in the head, thorax, and gut/abdomen (Khalil et al. 2015). In more detail, organ-specific lipid features were characterized using AP-SMALDI-IMS in tragacanth gum-embedded specimens of the rove beetle *Paederus riparius* (Bhandari et al. 2015). Exact mass was used to putatively identify 86 metabolites, most of which were lipids, including glycerolipids, sphingolipids, phospholipids, lysophospholipids, and acyl-carnitines. Many of these lipids exhibited spatially restricted pools in the brain, gut, fat body, ovary, and/or cuticle (Bhandari et al. 2015).

Secondary ion mass spectrometry (SIMS) is yet another in situ lipidomic method that was used successfully in frozen brain sections from *D. melanogaster* adults (Phan et al. 2017). This technique separates secondary ions ejected from a sample after treatment with an argon gas cluster ion beam to produce mass spectra. Using SIMS-IMS, the authors could putatively identify specific phospholipids and glycerolipids distributed throughout brain sections (Phan et al. 2017).

Going one step further by combining IMS and isotope tracer approaches (described in more detail later in the chapter), scientists have been able to produce data that are both spatially and temporally informative using multi-isotope IMS (MIMS). Cellular and subcellular features can be studied using MIMS following the carbons from dietary ^{13}C-palmitate, which could be observed in lipid droplets within the anterior midgut and fat body of *D. melanogaster* larvae (Steinhauser et al. 2012). These studies used a pulse-chase approach to detect ^{13}C enrichment in enterocyte lipid droplets within 1 h of feeding, with an estimated half-life of 9 h (Steinhauser et al. 2012). A subsequent study used isotopically labeled dietary glucose, acetate, and linoleate to follow ^{13}C incorporation into glial lipid droplets in the larval CNS of *D. melanogaster* during hypoxia (Bailey et al. 2015). This work revealed a requirement for glial lipogenesis in the maintenance of neuroblast proliferation during hypoxia, as glial lipid droplet TAG helps buffer the oxidative stress that occurs in the CNS during hypoxia (Bailey et al. 2015).

The "best" methods for using IMS in insects remain unclear, and the advantages of matrices, dissection, or live animal imaging, etc., vary depending on the desired application. The use of this technique is in its infancy, and few systematic approaches have been taken. One comprehensive approach in four insects [honeybees (*Apis cerana*), oriental fruit flies (*Bactrocera dorsalis*), fall armyworms (*Spodoptera frugiperda*), and diamondback moths (*Plutella xylostella*)] used thin sectioning to compare MALDI-IMS methods (Ma et al. 2022). This study found that there was more background with OCT as the embedding agent, and excellent results using gelatin, regardless of the matrix applied (DHB, DAN, and others) (Ma et al. 2022). IMS approaches are promising for the spatial and temporal characterization of lipids and other metabolites across species, genotypes, and environmental conditions.

Although it would be impossible to cover the entire range of innovative lipidomic analyses across insect genera, many intriguing examples from the literature exemplify the utility of these approaches. One study of *Drosophila* harvested from different geographic locations and reared in the laboratory for many generations identified distinct lipid profiles that were associated with the original habitat (Scheitz et al. 2013). This work also described lipid profiles arising from the endosymbiotic bacterium *Wolbachia*, which helps provision odd-chain lipids to the fly (Scheitz et al. 2013). Another study in larval *Drosophila* tissues took a computational approach to predict and confirm a role for dietary yeast in lipid metabolism by unbiased clustering lipids into biochemically relevant functional categories (Marella et al. 2015).

Additional lipidomic studies have revealed interesting spatial and temporal features at the tissue, cellular, and subcellular levels. In the mandibular gland of *Apis mellifera* worker bees, for example, a developmental time course was used to characterize shifts in the TAG composition over time (Hu et al. 2021). Exercise's effects on the flight muscle lipidome in butterflies (*Melitaea cinxia*) and moths (*Helicoverpa armigera*) also revealed potential roles for lipid species in flight capacity and the resulting stress (Pekny et al. 2018; Wang et al. 2020). Some researchers have gone beyond dissecting individual organs, including laser-capture microdissection of individual neuronal cell bodies or synaptic densities, followed by LC-MS in the *Drosophila* brain (Hebbar et al. 2014). A biochemical approach even enabled the purification of autophagosomes from within insect cells and found a role for the *Drosophila* protein Atg2 in the maintenance of lipid homeostasis in this organelle (Laczkó-Dobos et al. 2021). These and other studies set the stage for a wide range of future work exploiting lipidomics to discover new biology within insect cells and organs over time, diets, and ecological niches.

4.6 Isotopic Tracers

Isotopic tracer approaches use isotopic isomer feeding, injection, topical application, or in vitro cultures to follow the fate of isotopically labeled atoms into lipids using lipidomics. Depending on the tracer used and the outcomes to be measured, experimental design can be quite complex. Both the insect cuticle and gut are permeable to many metabolites, and most stages and species can be injected or dissected for organ-specific approaches where desired. Stable isotopes (^{13}C, ^{2}H, and ^{15}N) are those that do not decay rapidly and are, therefore, less dangerous than unstable (radioactive) isotopes, so these should be considered where possible. Stable isotopes contain an extra neutron and are somewhat abundant, making up a small fraction of the total atoms present in nature. Radioactive (^{14}C, ^{3}H, and ^{32}P) atoms are very rare in nature and easy to measure, so they provide greater sensitivity, but pose hazards in the laboratory.

Labeled acetate is a carbon source that is easily incorporated into lipids. Early studies used acetate feeding and injection to follow the fate of stable and radioactive carbon into lipids via GC/MS (Barkawi et al. 2003; Buckner and Hagen 2003). Labeled hydrogen atoms, either deuterium (2H) or tritium (3H), have also been useful. An ex vivo incubation of the Mediterranean fruit fly *Ceratitis capitata* corpus allatum with tritiated 3H-methyl-methionine results in transfer of the labeled methyl group to 3H-methyl-palmitate (Moshitzky et al. 2003). In *Helicoverpa* species, ^{2}H-labeled long chain fatty acids were incubated with pheromone glands, where they were incorporated into pheromones that could be detected by GC/MS (Choi et al. 2005). Topical application of ^{13}C fatty alcohols to the pheromone gland of the fall webworm, *Hyphantria cunea* Drury, also produced labeled pheromones observed by GC/MS (Kiyota et al. 2011). Isotope tracer studies using ^{13}C ethanolamine and choline supplementation have also been used to characterize phospholipid biosynthesis in *Aedes aegypti* mosquito cells after Dengue virus infection, which led to increased phospholipid remodeling and reduced de novo phospholipid biosynthesis (Vial et al. 2020). Thus, a range of substrates and delivery methods can be used to follow atoms when studying insects.

Feeding is a common way to study the metabolism of dietary sources via isotope tracers. Although most isotope feeding approaches in insects have used synthetic diets with specific added isotopically labeled components, others have exploited the natural differences in isotopomer abundance (Wessels et al. 2010). A 2010 study in the flesh fly *Sarcophaga crassipalpis* exploited the differing natural abundance of ^{13}C in sugar beets and cane as well as cattle that had grazed at different altitudes, both of which affect ^{13}C content of the fly's sugar- and liver-based diet (Wessels et al. 2010). These authors described the deposition of ^{13}C from the parental diet into neutral and polar lipids in both the maternal soma and in her eggs over time (Wessels et al. 2010). Another study in *Anopheles arabiensis* mosquitoes compared dietary and tissue trace element composition and used GC/MS to determine that the natural isotopomer distribution in tissues was highly correlated with the diet, and that tissues could synthesize a number of fatty acids from dietary substrates (Hood-Nowotny et al. 2012).

Table 2 Summary of key methodological approaches in insect lipidomics

Technique	Description	Advantages	Disadvantages	Typical applications
Folch/Bligh and Dyer extraction	Organic solvent-based total lipid extraction methods	High recovery of lipids, widely used, compatible with downstream MS	Uses toxic solvents (chloroform); poor recovery of polar metabolites	Bulk lipid extraction from whole insects/ tissues
TBME extraction	Methyl-tert-butyl ether-based lipid extraction	Less toxic than chloroform; similar efficiency	Phase separation requires careful handling	Whole-body lipidomics, compatible with MS
Solid phase extraction (SPE)	Fractionation of lipid classes on silica columns	Enables targeted fractionation (e.g., polar vs neutral lipids)	Limited loading capacity; possible loss of minor lipids	Class-specific lipid profiling
Thin-layer chromatography (TLC)	Separation based on lipid polarity on silica plates	Inexpensive, simple, visual, qualitative	Low resolution, semi-quantitative	Initial screening; lipid class separation
Gas chromatography (GC, GC-MS)	Separation of volatile derivatized lipids	High resolution; excellent for fatty acid profiling and CHCs	Requires derivatization; not suitable for large polar lipids	Cuticular hydrocarbons, fatty acids, sterols
High-performance liquid chromatography (HPLC-MS)	Separation of lipid classes by polarity or chain length	High resolution; suitable for complex mixtures	Requires method optimization per lipid class	Global and targeted lipidomics
Shotgun lipidomics	Direct infusion of total lipid extracts into MS	Fast, minimal sample prep; covers many lipid classes	Ion suppression; lower quantification accuracy	High-throughput lipid screening
Tandem MS (LC-MS/MS, MS/MS)	Two-stage mass spec with fragmentation for lipid ID	High specificity and sensitivity; detects low-abundance lipids	Requires expensive instrumentation; complex data analysis	Structural identification and quantification
Ambient MS (e.g., DART-MS)	Real-time, direct ionization from insect surface	Minimal sample prep, suitable for in situ cuticle/pheromone profiling	Limited depth; matrix effects; lower resolution than MALDI	Cuticular hydrocarbons, pheromonal lipids
Imaging MS (MALDI-IMS, SIMS)	Spatially resolved detection of lipids in tissues or whole insects	Enables tissue/organ-specific lipid mapping, subcellular localization	Complex setup, matrix interference, limited standardization	Brain, fat body, Malpighian tubules, cuticle
Isotopic tracer studies	Use of stable or radioactive labeled substrates (e.g., 13C-acetate)	Tracks lipid metabolism, synthesis, and flux over time	Requires careful design and isotopic purity; safety concerns for radioisotopes	Lipid synthesis, turnover, dietary incorporation studies
Laser capture microdissection + LC-MS	Collection of specific cell types or regions for MS analysis	Cell-specific lipidomics; reduces background	Technically challenging, sample limitation	Neuronal or organ-specific lipid studies

To facilitate comparison and provide a practical resource for researchers entering the field, Table 2 summarizes the primary methodological approaches used in insect lipidomics. Each technique is briefly described and evaluated in terms of its advantages, disadvantages, and common applications. This overview complements the detailed narrative in Sect. 3 and may serve as a quick reference for selecting appropriate lipidomic strategies based on experimental needs.

5 Comparative and Functional Lipidomics in Insects

5.1 Environmental Effects and Diapause

Lipidomic approaches using diapausing insects have revealed novel insights into lipid biochemistry during development. Most diapausing insects do not eat and must also survive changes in temperature, often overwintering until they exit diapause the following spring. Experimental temperature manipulation can modify diapause duration to allow innovative and precise studies in many diapausing insects, including the butterfly *Pieris napi*. The "intensity" of diapause in this system had the most significant effects on lipid composition in adults (Lehmann et al. 2016). As expected, prolonged cold temperatures that induced diapause also increased unsaturated fatty acids in membrane phospholipids, conferring fluidity (Lehmann et al. 2016). There were also surprising increases in acyl chain length in adult sphingolipids after diapause in these butterflies.

Overwintering larvae of the Asiatic rice borer *Chilo suppressalis* undergo significant changes in fatty acid composition to enhance cold tolerance (Atapour et al. 2007). From August to March, there is a marked increase in unsaturated fatty acids, particularly palmitoleic and oleic acids, and a concurrent decrease in saturated fatty acids like palmitic and stearic acids. This shift raises the unsaturation-to-saturation ratio, improving membrane fluidity and cold hardiness. These changes align with diapause initiation in October and deep diapause in January–February, enabling larvae to survive freezing temperatures as low as −25 °C. These adaptations exemplify homeoviscous adaptation, a strategy critical for maintaining membrane functionality under extreme conditions (Atapour et al. 2007).

Such changes are not limited to Lepidoptera; other insect orders exhibit similar but mechanistically diverse lipidomic responses. A particularly well-characterized example from Diptera is the mosquito *Aedes albopictus*, where embryonic diapause involves coordinated lipid and metabolite reprogramming. During embryonic diapause, there is marked accumulation of TAGs and DAGs, which support energy storage. Concurrently, alterations in PC and carnitine levels indicate a reduction in lipid catabolism and lipolysis. Additional shifts include decreased levels of developmentally active metabolites such as dopamine and octopamine, which may modulate hormonal pathways like insulin signaling. Changes in lipid unsaturation levels suggest membrane remodeling for enhanced stability at low temperatures. Furthermore, the reduced abundance of energy-intensive metabolites reflects a broader metabolic downregulation aimed at conserving energy during developmental arrest. Intriguingly, diapause-specific steroid-like metabolites hint at possible involvement of ecdysteroids. Together, these findings reveal a coordinated biochemical strategy to optimize survival under unfavorable conditions and exemplify shared physiological principles of diapause across insect taxa while retaining species-specific nuances (Batz and Armbruster 2018).

Three beetle species showed metabolic suppression and lipid store utilization during diapause, with lipid profiles adapting to ensure membrane fluidity at low temperatures (Lehmann et al. 2020). Native beetles exhibited more stable lipidomic responses than the invasive *Leptinotarsa decemlineata*, which may limit its adaptation to colder climates. Lipidomic adaptations included shifts in storage lipids and phospholipid species, highlighting mechanisms for energy efficiency and structural integrity during overwintering (Lehmann et al. 2020).

The metabolic flexibility seen in beetle diapause is highlighted by similar observations in other insects. For instance, research on the stink bug *Eysarcoris inconspicuous* demonstrated seasonal changes in the composition of phospholipids and TAGs to maintain cellular membrane fluidity and ensure survival under low-temperature conditions (Çakmak 2010). Key adaptations included a significant increase in the unsaturated fatty acid-to-saturated fatty acid ratio during colder months, particularly within phospholipid fractions. This shift, driven by elevated levels of oleic and linolenic acids, highlights the critical role of lipid remodeling in coping with thermal stress. These findings further reinforce the notion that lipidomic adjustments are not merely passive responses but are actively regulated to enhance overwintering success.

Early lipidomic studies in *Drosophila* found the effects of the rearing temperature on fatty acid composition (Geer et al. 1979; Ohtsu et al. 1993; Overgaard et al. 2005). In addition, chill coma recovery time (Goto et al. 2010), overwintering (Sinclair and Marshall 2018), and cold acclimation (Overgaard et al. 2008; Košťál et al. 2011) are all associated with lipidome remodeling. Researchers have observed the effects of temperature on the lipidome in a range of species, including *Drosophila suzukii* (Enriquez and Colinet 2019), the soil dweller *Orchesella cincta* (van Dooremalen and Ellers 2010), and the moth *Phthorimaea operculella* (Egbelakin et al. 2011), primarily in phospholipids, which determine membrane fluidity. Colder temperatures typically increase double bond content in these insects, a feature that counterbalances the decrease in membrane fluidity that occurs at lower temperatures by providing more flexible unsaturated acyl chains.

Overall, these studies support the idea that environmental conditions and diapause may have major impacts on the insect lipidome, as extensively reviewed in *Low Temperature Biology of Insects* (Denlinger and Lee 2010). Cold stress adaptations are not limited to changes in fatty acid unsaturation within membrane phospholipids: insects also remodel lipid stores, alter molecular species composition (e.g., shifts in PE/PC species), change lipid saturation ratios, and adjust storage lipid accessibility to maintain energetic reserves during winter diapause (Kostál et al. 2003). Such structural and compositional lipid remodeling ensures both membrane functionality and metabolic readiness under prolonged cold exposure.

5.2 Dietary Influences

The diet also has dramatic effects on the lipidome. Triacylglycerides are the most dramatically affected class, especially when the variable is caloric intake. As in other animals, starvation reduces fat storage, whereas overnutrition increases lipid storage across insect genera, including *A. mellifera* (Wang et al. 2021b), the diamondback moth *Plutella xylostella* (Warbrick-Smith et al. 2006), the black field cricket *Teleogryllus commodus* (Rapkin et al. 2018), the brown planthopper *Nilaparvata lugens* (Zhou et al. 2018), and *Drosophila* (Grönke et al. 2005; Wat et al. 2020). Varying the dietary content has effects not just on glycerolipids like TAG but also on phospholipids, sphingolipids, and sterols (Carvalho et al. 2012; Knittelfelder et al. 2020). The effects of diet on the insect lipidome are not restricted to the fat body, the site where the bulk of lipid metabolism occurs, but can be observed in the heart (Tuthill et al. 2020), hemolymph (Tuthill et al. 2020), salivary glands (Carvalho et al. 2012), wing discs (Carvalho et al. 2012), and head, where dietary lipids regulate photoreceptor function via polyunsaturated fatty acids (PUFAs) (Stark et al. 1993; Randall et al. 2015).

The lipidome of the house cricket (*Acheta domesticus*) responds distinctly to diet composition and sex, but not to the social environment (Gutiérrez et al. 2022). Female crickets fed a balanced diet displayed higher lipid storage, linked to longevity, while protein-rich diets correlated with higher fecundity despite reduced lipid reserves. Male lipid profiles, however, remained unaffected by dietary variations. The study highlights sex-specific lipid allocation strategies, with females prioritizing lipid storage for reproduction and males investing more in signaling-related lipids. These findings underscore the metabolic plasticity of crickets and provide insights into the interplay between nutrition, lipidomics, and life-history strategies in insects (Gutiérrez et al. 2022).

Diet composition directly influences the *Drosophila* lipidome, with specific dietary fatty acids significantly altering the fatty acid composition of membrane lipids in all tissues (Carvalho et al. 2012). A plant-based diet rich in long unsaturated fatty acids increases the length and unsaturation of tissue phospholipid fatty acids compared to a yeast-based diet. Notably, these changes occur in all phospholipid classes across tissues, demonstrating a striking flexibility in lipid incorporation. These shifts suggest that dietary fatty acids contribute directly to cellular lipid composition, offering insight into the adaptability of membrane properties in response to nutritional inputs (Carvalho et al. 2012).

5.3 Genetic and Hormonal Regulation of the Insect Lipidome

The effects of genotype on glycerolipid and phospholipid lipidomics have been most thoroughly studied in *Drosophila*, owing to its array of genetic tools. Reducing lipid flux through conserved pathways has both expected and unexpected effects on lipid profiles. For example, the deletion of two fatty acid synthases reduced TAG content as expected but also affected the acyl composition of sterol esters and TAGs, modifying chain length and saturation, without affecting phospholipid acyl chains (Garrido et al. 2015). Lipidomics in flies with knockdown of the stearoyl-CoA desaturase Desat1 increased fatty acid saturation in glycerolipids with negligible effects on phospholipids (Tuthill et al. 2021). Likewise, impairment of the Mondo-Mlx transcriptional complex, functionally analogous to mammalian ChREBP, led to reductions in TAG levels and altered fatty acid chain length and saturation (Musselman et al. 2013; Havula et al. 2013).

Building on these findings, more recent studies have used lipidomic profiling to uncover the roles of specific transcription factors. For example, the transcription factor Grain, a sugar-responsive regulator like ChREBP, promotes lipogenesis. RNAi knockdown of Grain reduced TAG storage and increased acyl chain length (Musselman et al. 2013; Havula et al. 2013). Lipidomics combined with loss- and gain-of-function genetics showed that another transcription factor, sugarbabe, promotes TAG synthesis and acyl chain desaturation (Mattila et al. 2015; Tuthill et al. 2021). These studies highlight the potential of combining lipidomic and genomic approaches to reveal interesting molecular mechanisms; however, such studies remain scarce in non-model insects.

Hormonal regulation is equally critical. The adipokinetic hormone (AKH) functions analogously to mammalian glucagon, triggering lipid mobilization during fasting or activity. In *Zophobas atratus*, Tenmo-AKH regulates not only major fatty acids but also sterols and aromatic lipid species in a developmental stage-dependent manner (Gołębiowski et al. 2014). Other comparative lipidomic studies indicate that peptide diversity in AKH family members may underlie species-specific lipidomic responses (Marco et al. 2022).

5.4 Developmental and Aging-Related Lipid Dynamics

As the need for lipids in growth and energy varies across insect developmental stages, so does the lipidome. The bulk of fatty acids can be considered destined to become membrane lipids (largely phospholipids) or storage lipids (mostly TAGs). Due to its extensive genetic toolkit and short generation time, *D. melanogaster* has served as a leading model for developmental lipidomics. Both membrane and storage lipids increase during *Drosophila* development, although the ratio favors TAG in the third instar, when larvae prepare for metamorphosis (Carvalho et al. 2012). In third instar larvae, levels of phosphatidic acids declined, followed by a reduction in DAGs during metamorphosis, while TAGs remained relatively stable. This pattern indicates that TAGs are not significantly mobilized as an energy source during metamorphosis, contrary to what might be expected in such a high-energy-demand phase (Carvalho et al. 2012). This was accompanied by a shift in acyl chain length from medium to long chains in DAG at the onset of metamorphosis. Numerous other dynamic patterns were observed for individual lipid classes in this study. The authors also found several sex-specific effects in adults, including fewer sterols and ceramides in males and an increase in ether lipids (Carvalho et al. 2012).

The aging lipidome may also provide information on the lipids that contribute to the lifespan and/or healthspan. Numerous lipid-associated diseases can be modeled in insects, making them an innovative and affordable model for studying the links between lipidome and fitness (Han 2016). A comparison of two different *Drosophila* genotypes, the short-lived Dahomey and the long-lived Oregon R, found increased acyl chain double bond content in total lipids from females from the short-lived population, consistent with a role for oxidative stress in lifespan (Gubina et al. 2019). Another study in long-lived female flies that had undergone laboratory selection found only minor differences in phospholipid content, compared with nonselected controls (Moghadam et al. 2013). Likewise, there were only small changes in acyl chain length and saturation from three to 19 days of adulthood in this study (Moghadam et al. 2013). Another lipidomic study in aging adult *Drosophila* males found a number of glycerolipid classes that changed over time (Hofbauer et al. 2021). In particular, the acyl chain length and double bond content both decreased over time in TAGs, whereas changes in phospholipid substituents were less pronounced between early and middle age. Interestingly, most lipid classes were stable across time points in old age, a feature that was observed across a number of genotypes, suggesting that there is a complex relationship between lipid content and aging (Hofbauer et al. 2021).

Several longitudinal and comparative studies in model insects have deepened our understanding of age-related lipid dynamics. A 2013 LC-MS/MS timecourse study characterized the lipidome from embryogenesis to adulthood and revealed dynamic variation across multiple lipid classes, including phospholipids, sphingolipids, and sterols (Guan et al. 2013). Phosphatidylethanolamine (PE) was the most abundant phospholipid, followed by phosphatidylcholine (PC), and triacylglycerols (TAGs) were more abundant than sterol esters throughout development (Goh and Guan 2021). TAG levels peaked during metamorphosis, while adulthood brought sex-specific differences—adult females exhibited higher TAG and sterol abundance than males, along with increased polyunsaturated acyl chains. Notably, phosphoceramides showed greater hydroxylation in females. These sex-specific findings led to the study of the *Ghiberti* mutant, which encodes a subunit of serine palmitoyl transferase, the first enzyme in sphingolipid biosynthesis from palmitate (Giansanti et al. 2004). These data illustrate that even low-abundance lipid species can have biologically significant and dynamic roles. Developmental lipidomics has also illuminated the biosynthesis and metabolism of ecdysteroids (Lavrynenko et al. 2015) and revealed spatially and temporally restricted lipid patterns in *Drosophila* tissues (Carvalho et al. 2012). Beyond flies, stage-specific changes in total lipid content and acyl chain properties have been documented in *Bemisia argentifolii* nymphs and adults using GC-MS, showing that lipid saturation and esterification are tightly coupled to development in other insect taxa as well (Buckner and Hagen 2003).

Beyond ontogenetic changes, lipidomics has provided insights into social differentiation within insect colonies. In *A. mellifera*, adult worker bees have high polyunsaturated fatty acid content compared to queens, which may be a result of the worker's pollen diet. Higher PUFA content has been proposed to reduce the oxidative stress buffering capacity of workers compared to queens, who can live many times longer than worker bees of the same genotype (Haddad et al. 2007). Interestingly, lipid content was similar across workers, drones, and queens during larval and pupal stages but changed after emergence, once the diets of pollen (for workers) or royal jelly (for queens) had been consumed (Martin et al. 2019). A recent lipidomic study defined a "lipotype" that characterizes honeybee caste by acyl chain composition and CL content (Wegener et al. 2022). An LC-MS/MS study in the social termite *Macrotermes natalensis* found decreases in many TAGs and lysophosphatidylcholines, accompanied by increases in DAGs, PC, and PE in hemolymph from mature queens, compared with field worker termites (Séité et al. 2022).

The development of mandibular glands in *A. mellifera ligustica* is marked by significant changes in lipid composition and gene expression

(Hu et al. 2021). Newly emerged bees show initial reservoirs of lipids, predominantly PC and TAG, which undergo dynamic remodeling as the bees mature into nurses and foragers. TAG unsaturation increases during development, enhancing energy availability and membrane fluidity. These lipid changes correlate with shifts in secretory activity, highlighted by increased production of 10-hydroxydecanoic acid (10-HDA), a key lipid involved in larval nutrition and colony communication. This lipidomic and transcriptomic study provides insights into the molecular basis of mandibular gland maturation and its functional implications for social behavior and survival in honeybee colonies (Hu et al. 2021).

The lipidome of *A. mellifera* workers reveals distinct "lipotypes" associated with behavioral roles such as cell heating, foraging, and overwintering. Cardiolipins, key components of the mitochondrial inner membrane, with short acyl chains dominate in cell heater bees, supporting their high energy turnover for thermoregulation. Conversely, foragers and drones exhibit more unsaturated PC, enhancing membrane fluidity needed for flight at low temperatures. Dietary fatty acids, such as linolenic acid (C18:3) from pollen, influence membrane lipid unsaturation, while lower rearing temperatures suppress heater-specific lipid patterns. These findings highlight the role of lipidomic plasticity in energy metabolism and environmental acclimation within honeybee colonies (Wegener et al. 2022).

These studies highlight that lipidomes can drastically differ between developmental stages and aging and can even be used to define lipotypes differentiating social insect castes.

5.5 Lipidomics for Waterproofing

The insect's cuticular wax layer acts as an osmotic barrier that waterproofs the animal, maintaining the internal milieu in the face of various environmental and chemical stressors. Accordingly, a number of CHCs, including branched and linear alkanes and alkenes, were associated with salinity tolerance across two genera of aquatic beetles, *Nebrioporus* and *Enochrus*, in a GC-MS/MS lipidomic study (Botella-Cruz et al. 2019). In a follow-up lipidomic study of the response to sublethal desiccation stress, both *Enochrus jesusarribasi* and *Nebrioporus baeticus*, modified their CHC profiles, albeit in different ways (Botella-Cruz et al. 2021). Upon a second desiccation challenge, *E. jesusarribasi* was protected against water loss, whereas *N. baeticus* was not. Furthermore, *E. jesusarribasi* increased chain length and branching, whereas *N. baeticus* reduced overall CHC abundance without changes in the nature of substituents after the latter bout of desiccation stress. Therefore, changes in CHC profile appear to be protective against desiccation in these saline-tolerant species (Botella-Cruz et al. 2021).

Genetic loss-of-function approaches have been combined with lipidomics to identify lipid metabolic genes that are essential for waterproofing in several insect species. Some studies have used RNAi to target the lipidome in other insects. In the kissing bug *Rhodnius prolixus*, RNAi targeting an integumental fatty acid synthase impaired cuticular hydrocarbon biosynthesis and waterproofing (Moriconi et al. 2019). Also, *N. lugens* waterproofing requires the fatty acyl-CoA reductase Far7 for waterproofing, which was also discovered through RNAi (Cui et al. 2024).

A GC-MS/MS study in *Drosophila* demonstrated that the nuclear receptor dHNF4 (hepatocyte nuclear factor 4) regulates the rapid conversion of lipid stores into CHCs at adult emergence, a key process for waterproofing and desiccation resistance (Storelli et al. 2019). dHNF4 expression in oenocytes is essential for the transcription of genes involved in very long chain fatty acid (VLCFA) and hydrocarbon biosynthesis, and loss of its function results in reduced hydrocarbon levels, impaired cuticular hydrophobicity, and increased water loss (Storelli et al. 2019). This pathway is evolutionarily conserved, as HNF4α similarly regulates epidermal waterproofing in mice.

Fruitless (*fru*), a transcription factor best known for its role in regulating male courtship behavior in *D. melanogaster*, has also been implicated in lipid metabolism. Recent studies revealed that the non-sex-specific isoform Fru^COM promotes HNF4 expression in both male and female oenocytes, thereby facilitating CHC biosynthesis and cuticular waterproofing (Sun et al. 2023). Oenocyte-specific knockdown of *fru* led to reduced long-chain alkane levels, impaired cuticular hydrophobicity in both sexes, and increased susceptibility to desiccation (Sun et al. 2023).

These findings underscore how reverse genetics and lipidomics-based approaches can uncover molecular mechanisms underlying insect waterproofing, with potential for cross-species insights.

5.6 Effects of Infection and Parasitism

Pathogens and parasites can also affect the insect lipidome. Lipidomics has provided insights into the biology of infection and stress responses in insects. Pathogens require lipid-derived energy and membrane lipids to reproduce, so it makes sense that they exert dramatic effects on the lipidome. For this reason, the effects of infection and parasitism on insect lipidomes have been explored via lipidomics.

The Chinese sacbrood virus (CSBV), a major pathogen in Asian apiculture, infects both larval and adult stages of the pollinator *Apis cerana*, often resulting in colony collapse. Lipidomic profiling using UHPLC-MS/MS during early larval infection (24–28 hpi) revealed significant remodeling of host lipid metabolism. Among 2164 detected lipid species, 266 were differentially regulated, including increases in glycerolipids, phospholipids, and sphingolipids—particularly TAG 39:6, which increased over tenfold and may serve as a potential early biomarker of infection. Elevated expression of acyl-transferases and sphingolipid-metabolizing enzymes suggests active viral manipulation of host lipid pathways to facilitate replication. These findings underscore the central role of lipid metabolism in host-pathogen dynamics and highlight lipid pathways as promising targets for disease monitoring and intervention strategies (Dang et al. 2021). Another UHPLC-MS/MS lipidomic approach was used to compare nuclear polyhedrovirus infection-susceptible and infection-resistant strains of the domestic silk moth *Bombyx mori* before and after infection, which is a concern for the silk industry. In this case, the authors observed that increased lipid content, especially ceramides and DAG, was associated with higher viral infection susceptibility (Zhu et al. 2021).

Drosophila is a common model to study the response of the lipidome to harmful or commensal infections. The common insect symbiont *Spiroplasma poulsonii* grows best and reduces adult lifespan in this species most dramatically when flies are fed a rich medium (Herren et al. 2014). This bacterium uses circulating hemolymph lipids as an energy source, particularly lipoprotein-bound DAG from the hemolymph (Herren et al. 2014). Interestingly, MALDI-Tof (time-of-flight)-MS lipidomics of fly hemolymph support a model where long acyl chains from these glycerolipids are used for cardiolipin production by Spiroplasma (Herren et al. 2014). Likewise, infecting *Drosophila* adults with *Mycobacterium marinarum* (Dionne et al. 2006) or *Listeria monocytogenes* (Chambers et al. 2012) led to metabolic “wasting” accompanied by reduced TAG content. Lipidomics in *L. monocytogenes*-infected whole flies using GC and LC-MS found decreases in a number of free and esterified fatty acids, consistent with the use of acyl chains for bacterial proliferation in vivo (Chambers et al. 2012).

A particularly intriguing *Drosophila*-based screening approach combined genetics and lipidomics using the oral diarrheal pathogen *Vibrio cholerae* to identify a role for lipogenesis in the fly gut (Vanhove et al. 2017). This pathogen channels acetate into long chain fatty acids stored in gut enterocyte lipid droplets. Untargeted LC-MS/MS lipidomics found that, in contrast to most studies, which focus on lipid droplet TAG catabolism, *V. cholerae* uses medium-chain PC and PE as an acyl chain source to support infectivity (Vanhove et al. 2017). Lipidomics using another *Drosophila* oral pathogen, *Erwinia carotovora* (*Ecc15*, also known as

Pectobacterium carotovorum), has been combined with the powerful genetic approaches available in *Drosophila* to discover the underlying mechanisms by which lipids support bacterial growth. Hemolymph lipidomics used UHPLC-MS to discover increases in both DAG and ergosterol after *Ecc15* infections in *Drosophila* (Deng et al. 2022). Ergosterol appeared to be protective, as supplementation of the fly media with ergosterol improved host survival after infection. It may be that cholesterol esters are critical for *Ecc15* biology, as adult-specific loss of the acyl CoA synthetase ACSL was also protective for infected flies (Deng et al. 2022). Another combined genomic and lipidomic study using cultured cells, flies, and mice discovered a key role in peroxisomal lipid metabolism in *Ecc15* pathogenesis (Nath et al. 2022).

Viral replication in insect vectors also has effects on the lipidome that regulates viral loads. This process has been studied in the most detail in mosquito models. A broad-spectrum LC-MS/MS lipidomic strategy in the *Ae. aegypti* midgut after Dengue infection identified changes in phospholipids, glycerolipids, and sphingolipids, among other species (Chotiwan et al. 2018). Biochemical or genetic manipulation of sphingolipid desaturation reduced viral titers in this model, consistent with a role for lipidome remodeling in viral fitness (Chotiwan et al. 2018). A more recent study used untargeted LC-MS/MS lipidomics to compare Dengue-resistant vs susceptible *Ae. aegypti* midguts and discovered increased midgut ceramides in the guts of the refractory strain, consistent with increased apoptosis in these guts, which could function as a mechanism to limit viral replication (Elliott et al. 2023). A third study in *Ae. aegypti* using LC-HRMS lipidomics after Dengue viral infection helped identify several phospholipids, which increased early (1–2 days postinfection) and decreased toward the end of the viral replication cycle (7–14 days postinfection) (Vial et al. 2019). This shift occurred at least in part via the phosphatidic acid biosynthetic enzyme acylglycerolphosphate acyltransferase 1 (AGPAT1), which was decreased at the mRNA level upon infection. Knockdown of AGPAT1 increased viral load and led to an apparent increase in the use of PC and PE for viral replication in this model (Vial et al. 2019).

Insect parasitoids also exert control over host lipid metabolism, an essential feature especially because many parasitoids lack the ability to effectively metabolize lipids (Visser et al. 2017). Parasitism of the cotton aphid *Aphis gossypii* by the parasitoid wasp *Lysiphlebia japonica* induces substantial remodeling of the host's lipid metabolism. Lipidomic analysis using UHPLC-MS/MS revealed increased levels of triacylglycerols (TAGs), particularly those with unsaturated long and very-long acyl chains, alongside reduced phospholipids such as phosphatidylcholine (PC), phosphatidylethanolamine (PE), and sphingomyelin. Conversely, ceramide levels were elevated in parasitized aphids. These lipidomic shifts were accompanied by transcriptional upregulation of key genes in TAG biosynthesis (e.g., *Gpat1*, *Ppap2a*) and activation of energy-generating pathways such as the TCA cycle and glycolysis. Fatty acid profiling indicated increased C18:1 and C18:2 in TAGs, suggesting preferential mobilization of energy-rich lipids. These findings support a model in which parasitoids exploit host lipid reserves to fuel their own development and reproduction (XueKe et al. 2017). A follow-up study used lipidomics to explore the larval and pupal lipidomes of *Lysiphlebia japonica* parasitoids and found similar changes to those in the host, consistent with a model where host lipids are provisioned to parasitoids during growth (Gao et al. 2018). A UHPLC-MS-based lipidomic approach in the cabbage pest *Pieris rapae* infected by the parasitoid *Pteromalus puparum* found that although this parasitoid wasp is able to synthesize some lipids, it, nonetheless, appears to mobilize stored acyl chains from the host fat body into the hemolymph for use during infection (Dang et al. 2021). Taken together, these studies reveal that pathogens and parasitoids profoundly reshape host lipidomes—not only to support their own replication and development but also with potential downstream effects on host physiology and behavior.

5.7 Lipidomics and Behavior

Lipidomics has significantly advanced our understanding of insect behavior, particularly by analyzing CHCs. These hydrophobic molecules, present on the insect exoskeleton, play crucial roles in communication, mating, and social organization.

In a study on *Drosophila montana*, Ala-Honkola et al. (2020) observed that diapausing individuals were less attractive to the opposite sex compared to nondiapausing individuals. This reduced attractiveness was attributed to an increase in long-chain hydrocarbons and a decrease in short-chain hydrocarbons on the cuticle of diapausing individuals.

In mosquitoes of the genus *Anopheles*, lipidomic analyses have revealed that the desaturation of CHCs is intricately linked to mating behaviors. This process is regulated by the circadian clock genes *period* (*per*) and *timeless* (*tim*), whose protein products influence the rhythmic expression of downstream genes, thereby coordinating pheromone synthesis with swarming and mating activities (Wang et al. 2021a). Such synchronization ensures that mating behaviors occur at optimal times, enhancing reproductive success.

In the confused flour beetle, *Tribolium confusum*, specific CHCs, particularly odd-chain n-alkanes, function as kairomones—chemical cues that attract parasitic wasps (Awater-Salendo et al. 2021). Interestingly, these hydrocarbons undergo a phase transition over time, solidifying into inert filaments. This change reduces the kairomonal activity of trails, potentially serving as a defense mechanism to decrease the likelihood of parasitism as the chemical cues become less detectable to the wasps (Awater-Salendo et al. 2021).

In eusocial insects like ants, queen-specific CHCs elicit distinct electrophysiological responses in worker ants' antennae. Antennal odorant receptors are finely tuned to detect and differentiate these hydrocarbons, enabling workers to recognize and respond to their queen appropriately (Sharma et al. 2015). This chemical communication is vital for maintaining colony hierarchy and social cohesion (Sharma et al. 2015). Studies on both eusocial and solitary bee species have demonstrated that variations in CHC profiles are associated with cuticle maturation and foraging behaviors (Falcon et al. 2019). Lipidomic analyses have shown that their CHC composition changes as bees mature, correlating with the onset of foraging activities. These findings suggest that CHC profiles may serve as indicators of age and task allocation within bee populations.

Collectively, these studies highlight the importance of lipidomics in understanding the biochemical underpinnings of insect behavior. By mapping the diversity and function of lipids like CHCs, one can gain deeper insights into the molecular mechanisms that drive communication, mating, and social organization in insects. However, more elaborate experimental approaches will be necessary to firmly link lipidomes changes to specific insect behaviors.

5.8 Lipidomics for Edible Insects

Edible insects have emerged as a promising dietary alternative that is rich in protein and lipids. Recently, researchers began to use lipidomics to better understand the lipidome of edible insect species to assess and find ways to improve their quality as food sources. For instance, the lipid composition of four edible cricket species—house cricket (*Acheta domesticus*), banded cricket (*Gryllodes sigillatus*), Jamaican field cricket (*Gryllus assimilis*), and two-spotted cricket (*Gryllus bimaculatus*)—varies significantly, despite identical rearing conditions (Tzompa-Sosa et al. 2021). Two-spotted and banded crickets had the highest fat content, with lipids rich in unsaturated FAs (>63%) and TAGs (>98%), making them liquid at room temperature. Cholesterol dominated the sterol profile, raising nutritional concerns, while phytosterols like campesterol and β-sitosterol added potential health benefits. These findings highlight cricket lipids as viable alternatives for food applications, with further research needed to optimize their nutritional and functional properties (Tzompa-Sosa et al. 2021).

The lipid content and fatty acid profiles of several orthopteran insects, namely, the house cricket *Acheta domesticus,* long-winged conehead *Conocephalus discolor,* the meadow grasshopper *Chorthippus parallelus*, and the coleopteran, yellow mealworm beetle *Tenebrio molitor* larvae, were analyzed to assess their potential as food lipid sources (Paul et al. 2017). *Tenebrio molitor* had the highest lipid content (32% dry weight), while Orthopterans ranged between 10% and 16%. Orthopterans exhibited higher levels of essential fatty acids like linoleic and α-linolenic acids, which are essential for human health. Differences in fatty acid profiles were influenced by diet, environment, and species-specific metabolic pathways. These findings suggest that insect lipids, especially from Orthopterans, could serve as sustainable dietary supplements due to their high nutritional value (Paul et al. 2017).

Lauric acid (C12:0) is the dominant fatty acid in the black soldier fly *Hermetia illucens* larvae, accounting for up to 41% of the total fatty acids after biological decomposition (Rabani et al. 2019). This fatty acid is recognized for its antimicrobial properties, particularly against Gram-positive bacteria, and holds potential applications in pharmaceuticals and cosmetics as a natural additive. Comparatively, the blowfly *Lucilia sericata* larvae exhibit a more diverse fatty acid profile, with oleic acid (C18:1 n-9) and palmitic acid (C16:0) as the primary components, although lauric acid levels also increase significantly post-decomposition (Rabani et al. 2019). These lipid profiles underline the potential of insect fats as sustainable alternatives in nonfood industries, such as cosmetics and animal feed additives, especially given their antimicrobial and functional properties.

6 Gaps and Future Directions

Despite significant advancements in lipidomics, several knowledge and technical gaps still limit our understanding of insect lipid biology. One major limitation is the current inability to resolve lipid dynamics at the single-cell level. While bulk lipidomic analyses are informative, they mask cellular heterogeneity within complex tissues such as the fat body, where lipid functions may vary dramatically across cell types. Emerging techniques in single-cell lipidomics, analogous to single-cell transcriptomics, have the potential to reveal such heterogeneity, offering insights into cell-specific lipid storage, mobilization, and signaling (Zhang et al. 2023). Further development of high-resolution mass spectrometry platforms capable of handling single-cell lipid measurements will be crucial for this goal.

Another critical challenge is the limited throughput and coverage of existing MS-based lipidomic workflows. The vast diversity of lipid species, spanning structural isomers and low-abundance signaling molecules, can lead to incomplete lipidome characterization. Improving mass accuracy, expanding ion libraries, and developing faster acquisition protocols are necessary to achieve more comprehensive and comparative insect lipidomics, including across species and environmental conditions.

A further conceptual challenge lies in our incomplete understanding of the upstream regulatory networks that control lipid metabolism and remodeling across developmental stages and environmental contexts. While lipidomic datasets offer rich snapshots of lipid abundance and composition, they often lack explanatory power about how and why such changes occur. Integrative multi-omics approaches combining lipidomics with transcriptomics and proteomics are essential to link observed lipid patterns to their regulatory drivers, such as transcription factors, enzymes, and hormonal signals. Such integrative approaches may be especially informative in non-model insects and under stress conditions like starvation, infection, or temperature extremes.

Together, these technological and conceptual advances will help bridge current gaps, accelerating discoveries in insect physiology and enabling practical applications in fields such as pest control, disease vector management, and the sustainable use of insects as nutritional resources.

References

Ala-Honkola O, Kauranen H, Tyukmaeva V, Boetzl FA, Hoikkala A, Schmitt T (2020) Diapause affects cuticular hydrocarbon composition and mating behavior of both sexes in *Drosophila montana*. Insect Sci 27:304–316. https://doi.org/10.1111/1744-7917.12639

Ameku T, Niwa R (2016) Mating-induced increase in germline stem cells via the neuroendocrine system in female Drosophila. PLoS Genet 12(6):e1006123. https://doi.org/10.1371/journal.pgen.1006123

Araujo DP, Tuan MJM, Yew JY, Meier R (2014) Analysing small insect glands with UV-LDI MS: high-resolution spatial analysis reveals the chemical composition and use of the osmeterium secretion in *Themira superba* (Sepsidae: Diptera). J Evol Biol 27:1744–1750. https://doi.org/10.1111/jeb.12420

Arrese EL, Soulages JL (2010) Insect fat body: energy, metabolism, and regulation. Annu Rev Entomol 55:207–225. https://doi.org/10.1146/annurev-ento-112408-085356. PMID: 19725772; PMCID: PMC3075550

Atapour M, Moharramipour S, Barzegar M (2007) Seasonal changes of fatty acid compositions in overwintering larvae of rice stem borer, *Chilo suppressalis* (Lepidoptera: Pyralidae). J Asia Pac Entomol 10:33–38. https://doi.org/10.1016/S1226-8615(08)60328-3

Attygalle AB, Blankespoor CL, Eisner T, Meinwald J (1994) Biosynthesis of a defensive insect alkaloid: epilachnene from oleic acid and serine. Proc Natl Acad Sci USA 91:12790–12793. https://doi.org/10.1073/pnas.91.26.12790

Awater-Salendo S, Voigt D, Hilker M, Fürstenau B (2021) Cuticular hydrocarbon trails released by host larvae lose their kairomonal activity for parasitoids by solidification. J Chem Ecol 47:998–1013. https://doi.org/10.1007/s10886-021-01310-w

Bailey AP, Koster G, Guillermier C, Hirst EMA, MacRae JI, Lechene CP, Postle AD, Gould AP (2015) Antioxidant role for lipid droplets in a stem cell niche of Drosophila. Cell 163:340–353. https://doi.org/10.1016/j.cell.2015.09.020

Barkawi LS, Francke W, Blomquist GJ, Seybold SJ (2003) Frontalin: de novo biosynthesis of an aggregation pheromone component by Dendroctonus spp. bark beetles (Coleoptera: Scolytidae). Insect Biochem Mol Biol 33:773–788. https://doi.org/10.1016/s0965-1748(03)00069-9

Batz ZA, Armbruster PA (2018) Diapause-associated changes in the lipid and metabolite profiles of the Asian tiger mosquito, *Aedes albopictus*. J Exp Biol 221:jeb189480. https://doi.org/10.1242/jeb.189480

Bedick JC, Pardy RL, Howard RW, Stanley DW (2000) Insect cellular reactions to the lipopolysaccharide component of the bacterium *Serratia marcescens* are mediated by eicosanoids. J Insect Physiol 46:1481–1487. https://doi.org/10.1016/s0022-1910(00)00073-1

Bernier UR, Carlson DA, Geden CJ (1998) Gas chromatography/mass spectrometry analysis of the cuticular hydrocarbons from parasitic wasps of the genus Muscidifurax. J Am Soc Mass Spectrom 9:320–332. https://doi.org/10.1016/S1044-0305(97)00288-2

Beydon P, Claret J, Porcheron P, Lafont R (1981) Biosynthesis and inactivation of ecdysone during the pupal-adult development of the cabbage butterfly, *Pieris brassicae* L. Steroids 38:633–650. https://doi.org/10.1016/0039-128x(81)90083-0

Bhandari DR, Schott M, Römpp A, Vilcinskas A, Spengler B (2015) Metabolite localization by atmospheric pressure high-resolution scanning microprobe matrix-assisted laser desorption/ionization mass spectrometry imaging in whole-body sections and individual organs of the rove beetle *Paederus riparius*. Anal Bioanal Chem 407:2189–2201. https://doi.org/10.1007/s00216-014-8327-1

Blais C, Blasco T, Maria A, Dauphin-Villemant C, Lafont R (2010) Characterization of ecdysteroids in *Drosophila melanogaster* by enzyme immunoassay and nano-liquid chromatography-tandem mass spectrometry. J Chromatogr B Analyt Technol Biomed Life Sci 878:925–932. https://doi.org/10.1016/j.jchromb.2010.02.018

Bligh EG, Dyer WJ (1959) A rapid method of total lipid extraction and purification. Can J Biochem Physiol 37:911–917. https://doi.org/10.1139/o59-099

Blomquist GJ, Bagneres AG (2010) Insect hydrocarbons: biology, biochemistry, and chemical ecology. Cambridge University Press, New York

Blomquist GJ, Ginzel MD (2021) Chemical ecology, biochemistry, and molecular biology of insect hydrocarbons. Annu Rev Entomol 66:45–60. https://doi.org/10.1146/annurev-ento-031620-071754

Blomquist GJ, Blailock TT, Scheetz RW, Jackson LL (1976) Cuticular lipids of insects-VII. Cuticular hydrocarbons of the crickets Acheta a domesticus, gryllus pennsylvanicus and Nemobius fasciatus. Comp Biochem Physiol B 54(3):381–386. https://doi.org/10.1016/0305-0491(76)90261-3

Botella-Cruz M, Pallarés S, Millán A, Velasco J (2019) Role of cuticle hydrocarbons composition in the salinity tolerance of aquatic beetles. J Insect Physiol 117:103899. https://doi.org/10.1016/j.jinsphys.2019.103899

Botella-Cruz M, Velasco J, Millán A, Hetz S, Pallarés S (2021) Cuticle hydrocarbons show plastic variation under desiccation in saline aquatic beetles. Insects 12:285. https://doi.org/10.3390/insects12040285

Brenner RR, Bernasconi AM (1987) The occurrence of arachidonic acid in *Triatoma infestans*. Comp Biochem Physiol B 87:815–819. https://doi.org/10.1016/0305-0491(87)90394-4

Buckner JS, Hagen MM (2003) Triacylglycerol and phospholipid fatty acids of the silverleaf whitefly: composition and biosynthesis. Arch Insect Biochem Physiol 53:66–79. https://doi.org/10.1002/arch.10086

Buckner JS, Poprawski TJ, Jones WA, Nelson DR (2000) Effect of whitefly parasitoids on the cuticular lipid composition of *Bemisia argentifolii* (Homoptera: aleyrodidae) nymphs. Arch Insect Biochem Physiol 44:82–89. https://doi.org/10.1002/1520-6327(200006)44:2<82::AID-ARCH3>3.0.CO;2-4

Bush DR, Wysocki VH, Scaraffia PY (2012) Study of the fragmentation of arginine isobutyl ester applied to arginine quantification in *Aedes aegypti* mosquito excreta. J Mass Spectrom 47:1364–1371. https://doi.org/10.1002/jms.3063

Byrne AL, Camann MA, Cyr TL, Catts EP, Espelie KE (1995) Forensic implications of biochemical differences among geographic populations of the black blow fly, *Phormia regina* (Meigen). J Forensic Sci 40:372–377

Çakmak O (2010) Seasonal changes in fatty acid composition of Eysarcoris inconspicuous (Herrich-Schaffer, 1844) (Heteroptera: Pentatomidae) adults. Turk J Entomol 34(1):15–27

Canavoso LE, Jouni ZE, Karnas KJ, Pennington JE, Wells MA (2001) Fat metabolism in insects. Annu Rev Nutr 21:23–46. https://doi.org/10.1146/annurev.nutr.21.1.23

Carlson DA, Reinert JF, Bernier UR, Sutton BD, Seawright JA (1997) Analysis of the cuticular hydrocarbons among species of the *Anopheles quadrimaculatus* complex (Diptera: Culicidae). J Am Mosq Control Assoc 13(Suppl):103–111

Carvalho M, Sampaio JL, Palm W, Brankatschk M, Eaton S, Shevchenko A (2012) Effects of diet and development on the Drosophila lipidome. Mol Syst Biol 8:600. https://doi.org/10.1038/msb.2012.29. PMID: 22864382; PMCID: PMC3421444

Cedden D, Güney G, Toprak U (2024) The integral role of de novo lipogenesis in the preparation for seasonal dormancy. Proc Natl Acad Sci 121:e2406194121. https://doi.org/10.1073/pnas.2406194121

Cetraro N, Yew JY (2022) In situ lipid profiling of insect pheromone glands by direct analysis in real time mass spectrometry. Analyst 147:3276–3284. https://doi.org/10.1039/d2an00840h

Chambers MC, Song KH, Schneider DS (2012) Listeria monocytogenes infection causes metabolic shifts in *Drosophila melanogaster*. PLoS One 7:e50679. https://doi.org/10.1371/journal.pone.0050679

Chen YX, Veerasammy K, Yin J, Choetso T, Zhong T, Choudhury MA, Weng C, Xu E, Hein MA, Abzalimov R, He Y (2022) Sample preparation for rapid lipid analysis in Drosophila brain using matrix-assisted laser desorption/ionization mass spectrometry imaging. J Vis Exp. https://doi.org/10.3791/63930

Chiang YN, Tan KJ, Chung H, Lavrynenko O, Shevchenko A, Yew JY (2016) Steroid hormone signaling is essential for pheromone production and Oenocyte survival. PLoS Genet 12:e1006126. https://doi.org/10.1371/journal.pgen.1006126

Chintapalli VR, Al Bratty M, Korzekwa D, Watson DG, Dow JAT (2013) Mapping an atlas of tissue-specific *Drosophila melanogaster* metabolomes by high resolution mass spectrometry. PLoS One 8:e78066. https://doi.org/10.1371/journal.pone.0078066. PMID: 24205093; PMCID: PMC3812166

Choi M-Y, Groot A, Jurenka RA (2005) Pheromone biosynthetic pathways in the moths *Heliothis subflexa* and *Heliothis virescens*. Arch Insect Biochem Physiol 59:53–58. https://doi.org/10.1002/arch.20051

Chotiwan N, Andre BG, Sanchez-Vargas I, Islam MN, Grabowski JM, Hopf-Jannasch A, Gough E, Nakayasu E, Blair CD, Belisle JT, Hill CA, Kuhn RJ, Perera R (2018) Dynamic remodeling of lipids coincides with dengue virus replication in the midgut of *Aedes aegypti* mosquitoes. PLoS Pathog 14:e1006853. https://doi.org/10.1371/journal.ppat.1006853

Cooks RG, Ouyang Z, Takats Z, Wiseman JM (2006) Detection technologies. Ambient mass spectrometry. Science 311:1566–1570. https://doi.org/10.1126/science.1119426

Cui Y-L, Guo J-S, Zhang C-X, Yu X-P, Li D-T (2024) Silencing NlFAR7 destroyed the pore canals and related structures of the brown planthopper. Insect Mol Biol 33:350–361. https://doi.org/10.1111/imb.12903

Dang X, Li Y, Li X, Wang C, Ma Z, Wang L, Fan X, Li Z, Huang D, Xu J, Zhou Z (2021) Lipidomic profiling reveals distinct differences in sphingolipids metabolic pathway between healthy *Apis cerana cerana* larvae and Chinese Sacbrood disease. Insects 12:703. https://doi.org/10.3390/insects12080703

Delgado R, Muñoz Y, Peña-Cortés H, Giavalisco P, Bacigalupo J (2014) Diacylglycerol activates the light-dependent channel TRP in the photosensitive microvilli of *Drosophila melanogaster* photoreceptors. J Neurosci 34:6679–6686. https://doi.org/10.1523/JNEUROSCI.0513-14.2014

Deng Z, Yang Y, Luo J, Zhang B, Liu J, Shui G, Jiao R, Wei C (2022) An integrated transcriptomics and lipidomics analysis reveals that ergosterol is required for host defense against bacterial infection in Drosophila. Front Immunol 13:933137. https://doi.org/10.3389/fimmu.2022.933137

Denlinger DL, Lee RE Jr (2010) Low temperature biology of insects. Cambridge University Press. https://doi.org/10.1017/CBO9780511675997

Dionne MS, Pham LN, Shirasu-Hiza M, Schneider DS (2006) Akt and FOXO dysregulation contribute to infection-induced wasting in Drosophila. Curr Biol 16:1977–1985. https://doi.org/10.1016/j.cub.2006.08.052

Egbelakin A, Ferguson MJ, MacGill EA, Lehmann AS, Topletz AR, Quinney SK, Li L, McCammack KC, Hall SD, Renbarger JL (2011) Increased risk of vincristine neurotoxicity associated with low CYP3A5 expression genotype in children with acute lymphoblastic leukemia. Pediatr Blood Cancer 56:361–367. https://doi.org/10.1002/pbc.22845

Elliott K, Caicedo PA, Haunerland NH, Lowenberger C (2023) Profiling lipidomic changes in dengue-resistant and dengue-susceptible strains of Colombian *Aedes aegypti* after dengue virus challenge. PLoS Negl Trop Dis 17:e0011676. https://doi.org/10.1371/journal.pntd.0011676

Enriquez T, Colinet H (2019) Cold acclimation triggers lipidomic and metabolic adjustments in the spotted wing drosophila *Drosophila suzukii* (Matsumara). Am J Physiol Regul Integr Comp Physiol 316:R751–R763. https://doi.org/10.1152/ajpregu.00370.2018

Falcon T, Pinheiro DG, Ferreira-Caliman MJ, Turatti ICC, de Abreu FCP, Galaschi-Teixeira JS, Martins JR, Elias-Neto M, Soares MPM, Laure MB, Figueiredo VLC, Lopes NP, Simões ZLP, Garófalo CA, Bitondi MMG (2019) Exploring integument transcriptomes, cuticle ultrastructure, and cuticular hydrocarbons profiles in eusocial and solitary bee species displaying heterochronic adult cuticle maturation. PLoS One 14: e0213796. https://doi.org/10.1371/journal.pone.0213796

Fink C, Von Frieling J, Knop M, Roeder T (2017) Drosophila fecal sampling. Bio Protoc 7:e2547. https://doi.org/10.21769/BioProtoc.2547

Folch J, Lees M, Sloane Stanley GH (1957) A simple method for the isolation and purification of total lipides from animal tissues. J Biol Chem 226:497–509

Frere B, Suchaud F, Bernier G, Cottin F, Vincent B, Dourel L, Lelong A, Arpino P (2014) GC-MS analysis of cuticular lipids in recent and older scavenger insect puparia. An approach to estimate the postmortem interval (PMI). Anal Bioanal Chem 406(4):1081–1088. https://doi.org/10.1007/s00216-013-7184-7

Gamo S, Kawabe A, Kohara H, Yamaguchi H, Tanaka Y, Yagi S (1999) Fast atom bombardment tandem mass spectrometric analysis of phospholipids in *Drosophila melanogaster*. J Mass Spectrom 34:590–600. https://doi.org/10.1002/(SICI)1096-9888(199906)34:6<590::AID-JMS810>3.0.CO;2-8

Gao X, Luo J, Lü L, Zhang L, Zhang S, Cui J (2018) RNA-Seq and UHPLC-Q-TOF/MS based lipidomics study in *Lysiphlebia japonica*. Sci Rep 8:7802. https://doi.org/10.1038/s41598-018-26139-4

Garrido D, Rubin T, Poidevin M, Maroni B, Le Rouzic A, Parvy J-P, Montagne J (2015) Fatty acid synthase cooperates with glyoxalase 1 to protect against sugar toxicity. PLoS Genet 11:e1004995. https://doi.org/10.1371/journal.pgen.1004995

Geer BW, Lindel DL, Lindel DM (1979) Relationship of the oxidative pentose shunt pathway to lipid synthesis in *Drosophila melanogaster*. Biochem Genet 17:881–895. https://doi.org/10.1007/BF00504310

Genin E, Jullien R, Perez F, Fuzeau-Braesch S (1986) Cuticular hydrocarbons of gregarious and solitary locusts *Locusta migratoria* cinerascens. J Chem Ecol 12:1213–1238. https://doi.org/10.1007/BF01012343

Giansanti MG, Farkas RM, Bonaccorsi S, Lindsley DL, Wakimoto BT, Fuller MT, Gatti M (2004) Genetic dissection of meiotic cytokinesis in Drosophila males. Mol Biol Cell 15:2509–2522. https://doi.org/10.1091/mbc.e03-08-0603

Gilbert LI, Rybczynski R, Warren JT (2002) Control and biochemical nature of the ecdysteroidogenic pathway. Annu Rev Entomol 47:883–916. https://doi.org/10.1146/annurev.ento.47.091201.145302

Goh EXY, Guan XL (2021) Targeted lipidomics of *Drosophila melanogaster* during development. In: Hsu F-F (ed) Mass spectrometry-based lipidomics: methods and protocols. Springer, New York, pp 187–213. https://doi.org/10.1007/978-1-0716-1410-5_13

Gołębiowski M, Cerkowniak M, Urbanek A, Słocińska M, Rosiński G, Stepnowski P (2014) Adipokinetic hormone induces changes in the fat body lipid composition of the beetle *Zophobas atratus*. Peptides 58:65–73. https://doi.org/10.1016/j.peptides.2014.05.013

Gołębiowski M, Cerkowniak M, Ostachowska A, Boguś MI, Stepnowski P (2016) Determination of cuticular and internal fatty acids of *Chorthippus brunneus* males and females using HPLC-LLSD and GC-MS. Biomed Chromatogr 30(8):1318–1323. https://doi.org/10.1002/bmc.3688

Goto SG, Udaka H, Ueda C, Katagiri C (2010) Fatty acids of membrane phospholipids in *Drosophila melanogaster* lines showing rapid and slow recovery from chill coma. Biochem Biophys Res Commun 391: 1251–1254. https://doi.org/10.1016/j.bbrc.2009.12.053

Grau V, Lafont R (1994) Metabolism of ecdysone and 20-hydroxyecdysone in adult *Drosophila melanogaster*. Insect Biochem Mol Biol 24:49–58. https://doi.org/10.1016/0965-1748(94)90122-8

Grönke S, Mildner A, Fellert S, Tennagels N, Petry S, Müller G, Jäckle H, Kühnlein RP (2005) Brummer lipase is an evolutionary conserved fat storage regulator in Drosophila. Cell Metab 1:323–330. https://doi.org/10.1016/j.cmet.2005.04.003

Guan XL, Cestra G, Shui G, Kuhrs A, Schittenhelm RB, Hafen E, van der Goot FG, Robinett CC, Gatti M, Gonzalez-Gaitan M, Wenk MR (2013) Biochemical membrane lipidomics during Drosophila development. Dev Cell 24:98–111. https://doi.org/10.1016/j.devcel.2012.11.012

Gubina N, Naudi A, Stefanatos R, Jove M, Scialo F, Fernandez-Ayala DJ, Rantapero T, Yurkevych I, Portero-Otin M, Nykter M, Lushchak O, Navas P, Pamplona R, Sanz A (2019) Essential physiological differences characterize short- and long-lived strains of *Drosophila melanogaster*. J Gerontol A Biol Sci Med Sci 74:1835–1843. https://doi.org/10.1093/gerona/gly143

Gutiérrez Y, Fresch M, Scherber C, Brockmeyer J (2022) The lipidome of an omnivorous insect responds to diet composition and social environment. Ecol Evol 12: e9497. https://doi.org/10.1002/ece3.9497

Haddad LS, Kelbert L, Hulbert AJ (2007) Extended longevity of queen honey bees compared to workers is associated with peroxidation-resistant membranes. Exp

Gerontol 42:601–609. https://doi.org/10.1016/j.exger.2007.02.008

Han X (2016) Lipidomics for studying metabolism. Nat Rev Endocrinol 12:668–679. https://doi.org/10.1038/nrendo.2016.98

Harayama T, Riezman H (2018) Understanding the diversity of membrane lipid composition. Nat Rev Mol Cell Biol 19:281–296. https://doi.org/10.1038/nrm.2017.138

Havula E, Teesalu M, Hyötyläinen T, Seppälä H, Hasygar K, Auvinen P, Orešič M, Sandmann T, Hietakangas V (2013) Mondo/ChREBP-Mlx-regulated transcriptional network is essential for dietary sugar tolerance in Drosophila. PLoS Genet 9(4):e1003438. https://doi.org/10.1371/journal.pgen.1003438

Hebbar S, Schulz WD, Sauer U, Schwudke D (2014) Laser capture microdissection coupled with on-column extraction LC-MS(n) enables lipidomics of fluorescently labeled Drosophila neurons. Anal Chem 86:5345–5352. https://doi.org/10.1021/ac500276r

Herren JK, Paredes JC, Schüpfer F, Arafah K, Bulet P, Lemaitre B (2014) Insect endosymbiont proliferation is limited by lipid availability. elife 3:e02964. https://doi.org/10.7554/eLife.02964

Hetru CC, Kappler C, Hoffmann JA, Nearn R, Bang L, Horn DH (1982) The biosynthetic pathway of ecdysone: studies with vitellogenic ovaries of *Locusta migratoria* (Orthoptera). Mol Cell Endocrinol 26:51–80. https://doi.org/10.1016/0303-7207(82)90006-5

Hofbauer HF, Heier C, Sen Saji AK, Kühnlein RP (2021) Lipidome remodeling in aging normal and genetically obese Drosophila males. Insect Biochem Mol Biol 133:103498. https://doi.org/10.1016/j.ibmb.2020.103498

Hoi KK, Daborn PJ, Battlay P, Robin C, Batterham P, O'Hair RAJ, Donald WA (2014) Dissecting the insect metabolic machinery using twin ion mass spectrometry: a single P450 enzyme metabolizing the insecticide imidacloprid in vivo. Anal Chem 86:3525–3532. https://doi.org/10.1021/ac404188g

Hood-Nowotny R, Schwarzinger B, Schwarzinger C, Soliban S, Madakacherry O, Aigner M, Watzka M, Gilles J (2012) An analysis of diet quality, how it controls fatty acid profiles, isotope signatures and stoichiometry in the malaria mosquito *Anopheles arabiensis*. PLoS One 7:e45222. https://doi.org/10.1371/journal.pone.0045222

Howard RW, Blomquist GJ (1982) Chemical ecology and biochemistry of insect hydrocarbons. https://doi.org/10.1146/annurev.en.27.010182.001053

Howard RW, Blomquist GJ (2005) Ecological, behavioral, and biochemical aspects of insect hydrocarbons. Annu Rev Entomol. 50:371–393. https://doi.org/10.1146/annurev.ento.50.071803.130359

Hu X, Zhang W, Chi X, Wang H, Liu Z, Wang Y, Ma L, Xu B (2021) Non-targeted lipidomics and transcriptomics analysis reveal the molecular underpinnings of mandibular gland development in *Apis mellifera ligustica*. Dev Biol 479:23–36. https://doi.org/10.1016/j.ydbio.2021.07.016

Isaac RE, Rose ME, Rees HH, Goodwin TW (1983) Identification of the 22-phosphate esters of ecdysone, 2-deoxyecdysone, 20-hydroxyecdysone and 2-deoxy-20-hydroxyecdysone from newly laid eggs of the desert locust, *Schistocerca gregaria*. Biochem J 213:533–541. https://doi.org/10.1042/bj2130533

Jacobson M, Ohinata K, Chambers DL, Jones WA, Fujimoto MS (1973) Insect sex attractants. 13. Isolation, identification, and synthesis of sex pheromones of the male Mediterranean fruit fly. J Med Chem 16:248–251. https://doi.org/10.1021/jm00261a018

Jang HJ, Le MUT, Park JH, Chung CG, Shon JG, Lee GS, Moon JH, Lee SB, Choi JS, Lee TG, Yoon S (2021) Matrix-assisted laser desorption/ionization mass spectrometry imaging of phospholipid changes in a Drosophila model of early amyotrophic lateral sclerosis. J Am Soc Mass Spectrom 32:2536–2545. https://doi.org/10.1021/jasms.1c00167

Jennings JH, Etges WJ, Schmitt T, Hoikkala A (2014) Cuticular hydrocarbons of *Drosophila montana*: geographic variation, sexual dimorphism and potential roles as pheromones. J Insect Physiol 61:16–24. https://doi.org/10.1016/j.jinsphys.2013.12.004

Jezovit JA, Levine JD, Schneider J (2017) Phylogeny, environment and sexual communication across the Drosophila genus. J Exp Biol 220:42–52. https://doi.org/10.1242/jeb.143008

Jones HE, Harwood JL, Bowen ID, Griffiths G (1992) Lipid composition of subcellular membranes from larvae and prepupae of *Drosophila melanogaster*. Lipids 27:984–987. https://doi.org/10.1007/BF02535576

Juárez P, Chase J, Blomquist GJ (1992) A microsomal fatty acid synthetase from the integument of *Blattella germanica* synthesizes methyl-branched fatty acids, precursors to hydrocarbon and contact sex pheromone. Arch Biochem Biophys 293:333–341. https://doi.org/10.1016/0003-9861(92)90403-j

Jurenka R (2004) Insect pheromone biosynthesis. In: Schulz S (ed) The chemistry of pheromones and other Semiochemicals I. Topics in current chemistry, vol 239. Springer, Berlin/Heidelberg. https://doi.org/10.1007/b95450

Jurenka R (2015) The PRXamide neuropeptide signalling system. Elsevier, pp 123–170. https://doi.org/10.1016/bs.aiip.2015.07.001

Jurenka RA, de Renobales M, Blomquist GJ (1987) De novo biosynthesis of polyunsaturated fatty acids in the cockroach *Periplaneta americana*. Arch Biochem Biophys 255:184–193. https://doi.org/10.1016/0003-9861(87)90309-2

Jurenka RA, Pedibhotla VK, Stanley DW (1999) Prostaglandin production in response to a bacterial infection in true armyworm larvae. Arch Insect Biochem Physiol 41:225–232. https://doi.org/10.1002/(SICI)1520-6327(1999)41:4<225::AID-ARCH6>3.0.CO;2-0

Kaftan F, Vrkoslav V, Kynast P, Kulkarni P, Böcker S, Cvačka J, Knaden M, Svatoš A (2014) Mass spectrometry imaging of surface lipids on intact *Drosophila melanogaster* flies. J Mass Spectrom 49:223–232. https://doi.org/10.1002/jms.3331

Kaplanis JN, Thompson MJ, Dutky SR, Robbins WE, Lindquist EL (1972) The metabolism of 4-14 C-22, 25-bisdeoxyecdysone during larval development in the tobacco hornworm, *Manduca sexta* (L.). Steroids 20: 105–120. https://doi.org/10.1016/0039-128x(72)90122-5

Kayukawa T, Chen B, Hoshizaki S, Ishikawa Y (2007) Upregulation of a desaturase is associated with the enhancement of cold hardiness in the onion maggot, *Delia antiqua*. Insect Biochem Mol Biol 37:1160–1167. https://doi.org/10.1016/j.ibmb.2007.07.007

Keesey IW, Koerte S, Retzke T, Haverkamp A, Hansson BS, Knaden M (2016) Adult frass provides a pheromone signature for Drosophila feeding and aggregation. J Chem Ecol 42:739–747. https://doi.org/10.1007/s10886-016-0737-4

Khalil SM, Römpp A, Pretzel J, Becker K, Spengler B (2015) Phospholipid topography of whole-body sections of the Anopheles stephensi mosquito, characterized by high-resolution atmospheric-pressure scanning microprobe matrix-assisted laser desorption/ionization mass spectrometry imaging. Anal Chem 87: 11309–11316. https://doi.org/10.1021/acs.analchem.5b02781

Khallaf MA, Cui R, Weißflog J, Erdogmus M, Svatoš A, Dweck HKM, Valenzano DR, Hansson BS, Knaden M (2021) Large-scale characterization of sex pheromone communication systems in Drosophila. Nat Commun 12:4165. https://doi.org/10.1038/s41467-021-24395-z

Kiyota R, Arakawa M, Yamakawa R, Yasmin A, Ando T (2011) Biosynthetic pathways of the sex pheromone components and substrate selectivity of the oxidation enzymes working in pheromone glands of the fall webworm, *Hyphantria cunea*. Insect Biochem Mol Biol 41:362–369. https://doi.org/10.1016/j.ibmb.2011.02.004

Knittelfelder O, Prince E, Sales S, Fritzsche E, Wöhner T, Brankatschk M, Shevchenko A (2020) Sterols as dietary markers for *Drosophila melanogaster*. Biochim Biophys Acta Mol Cell Biol Lipids 1865:158683. https://doi.org/10.1016/j.bbalip.2020.158683

Koelmel JP, Jones CM, Ulmer CZ, Garrett TJ, Yost RA, Schock TB, Bowden JA (2018) Examining heat treatment for stabilization of the lipidome. Bioanalysis 10: 291–305. https://doi.org/10.4155/bio-2017-0209

Köfeler HC, Ahrends R, Baker ES, Ekroos K, Han X, Hoffmann N, Holčapek M, Wenk MR, Liebisch G (2021) Recommendations for good practice in MS-based lipidomics. J Lipid Res 62:100138. https://doi.org/10.1016/j.jlr.2021.100138

Kostál V, Berková P, Simek P (2003) Remodelling of membrane phospholipids during transition to diapause and cold-acclimation in the larvae of *Chymomyza costata* (Drosophilidae). Comp Biochem Physiol B Biochem Mol Biol 135(3):407–419. https://doi.org/10.1016/s1096-4959(03)00117-9

Koštál V, Korbelová J, Rozsypal J, Zahradníčková H, Cimlová J, Tomčala A, Šimek P (2011) Long-term cold acclimation extends survival time at 0 °C and modifies the metabolomic profiles of the larvae of the fruit fly *Drosophila melanogaster*. PLoS One 6: e25025. https://doi.org/10.1371/journal.pone.0025025

Kraut R (2011) Roles of sphingolipids in Drosophila development and disease. J Neurochem 116(5):764–778. https://doi.org/10.1111/j.1471-4159.2010.07022.x. Epub 2011 Jan 7

Kruger EL, Pappas CD (1993) Geographic variation of cuticular hydrocarbons among fourteen populations of *Aedes albopictus* (Diptera: Culicidae). J Med Entomol 30:544–548. https://doi.org/10.1093/jmedent/30.3.544

Laczkó-Dobos H, Maddali AK, Jipa A, Bhattacharjee A, Végh AG, Juhász G (2021) Lipid profiles of autophagic structures isolated from wild type and Atg2 mutant Drosophila. Biochim Biophys Acta Mol Cell Biol Lipids 1866:158868. https://doi.org/10.1016/j.bbalip.2020.158868

Lafont R, Dauphin-Villemant C, Warren JT, Rees H (2012) 4 – Ecdysteroid chemistry and biochemistry. In: Gilbert LI (ed) Insect endocrinology. Academic Press, San Diego, pp 106–176. https://doi.org/10.1016/B978-0-12-384749-2.10004-4

Lavrynenko O, Nedielkov R, Möller HM, Shevchenko A (2013) Girard derivatization for LC-MS/MS profiling of endogenous ecdysteroids in Drosophila. J Lipid Res 54:2265–2272. https://doi.org/10.1194/jlr.D035949

Lavrynenko O, Rodenfels J, Carvalho M, Dye NA, Lafont R, Eaton S, Shevchenko A (2015) The ecdysteroidome of Drosophila: influence of diet and development. Development 142:3758–3768. https://doi.org/10.1242/dev.124982

Lehmann P, Pruisscher P, Posledovich D, Carlsson M, Käkelä R, Tang P, Nylin S, Wheat CW, Wiklund C, Gotthard K (2016) Energy and lipid metabolism during direct and diapause development in a pierid butterfly. J Exp Biol 219:3049–3060. https://doi.org/10.1242/jeb.142687

Lehmann P, Westberg M, Tang P, Lindström L, Käkelä R (2020) The diapause lipidomes of three closely related beetle species reveal mechanisms for tolerating energetic and cold stress in high-latitude seasonal environments. Front Physiol 11:576617. https://doi.org/10.3389/fphys.2020.576617

Liebig J, Peeters C, Oldham NJ, Markstädter C, Hölldobler B (2000) Are variations in cuticular hydrocarbons of queens and workers a reliable signal of fertility in the ant *Harpegnathos saltator*? Proc Natl Acad Sci USA 97:4124–4131. https://doi.org/10.1073/pnas.97.8.4124

Lung O, Wolfner MF (1999) *Drosophila* seminal fluid proteins enter the circulatory system of the mated female fly by crossing the posterior vaginal wall. Insect Biochem Mol Biol 29:1043–1052. https://doi.org/10.1016/S0965-1748(99)00078-8

Ma L, Xie Q, Du M, Huang Y, Chen Y, Chen D, Xu Y, Xu H, Wu X, Yin Z (2022) Sample preparation optimization of insects and zebrafish for whole-body mass spectrometry imaging. Anal Bioanal Chem 414:4777–4790. https://doi.org/10.1007/s00216-022-04102-7

Marco HG, König S, Gäde G (2022) Mass spectrometric proof of predicted peptides: novel adipokinetic hormones in insects. Molecules 27(19):6469. https://doi.org/10.3390/molecules27196469

Marella C, Torda AE, Schwudke D (2015) The LUX score: a metric for lipidome homology. PLoS Comput Biol 11:e1004511. https://doi.org/10.1371/journal.pcbi.1004511

Marshall DL, Pham HT, Bhujel M, Chin JSR, Yew JY, Mori K, Mitchell TW, Blanksby SJ (2016) Sequential collision- and ozone-induced dissociation enables assignment of relative acyl chain position in triacylglycerols. Anal Chem 88:2685–2692. https://doi.org/10.1021/acs.analchem.5b04001

Martin N, Hulbert AJ, Brenner GC, Brown SHJ, Mitchell TW, Else PL (2019) Honey bee caste lipidomics in relation to life-history stage and the long life of the queen. J Exp Biol 222:jeb207043. https://doi.org/10.1242/jeb.207043

Martinez T, Fabriás G, Camps F (1990) Sex pheromone biosynthetic pathway in *Spodoptera littoralis* and its activation by a neurohormone. J Biol Chem 265:1381–1387

Mattila J, Havula E, Suominen E, Teesalu M, Surakka I, Hynynen R, Kilpinen H, Väänänen J, Hovatta I, Käkelä R, Ripatti S, Sandmann T, Hietakangas V (2015) Mondo-Mlx mediates organismal sugar sensing through the gli-similar transcription factor sugarbabe. Cell Rep 13:350–364. https://doi.org/10.1016/j.celrep.2015.08.081

Matyash V, Liebisch G, Kurzchalia TV, Shevchenko A, Schwudke D (2008) Lipid extraction by methyl-tert-butyl ether for high-throughput lipidomics. J Lipid Res 49:1137–1146. https://doi.org/10.1194/jlr.D700041-JLR200

McDonough LM, George DA, Butt BA, Ruth JM, Hill KR (1972) Sex pheromone of the codling moth: structure and synthesis. Science 177:177–178. https://doi.org/10.1126/science.177.4044.177

Mendki MJ, Ganesan K, Parashar BD, Sukumaran D, Prakash S (2014) Aggregation responses of *Cimex hemipterus* F. to semiochemicals identified from their excreta. J Vector Borne Dis 51:224–229

Merrill AH Jr (2002) De novo sphingolipid biosynthesis: a necessary, but dangerous, pathway. J Biol Chem 277 (29):25843–25846. https://doi.org/10.1074/jbc.R200009200. Epub 2002 May 13

Miller JS, Stanley DW (2001) Eicosanoids mediate microaggregation reactions to bacterial challenge in isolated insect hemocyte preparations. J Insect Physiol 47:1409–1417. https://doi.org/10.1016/s0022-1910(01)00131-7

Miller JS, Howard RW, Rana RL, Tunaz H, Stanley DW (1999) Eicosanoids mediate nodulation reactions to bacterial infections in adults of the cricket, *Gryllus assimilis*. J Insect Physiol 45:75–83. https://doi.org/10.1016/s0022-1910(98)00089-4

Miyashita M, Matsushita K, Nakamura S, Akahane S, Nakagawa Y, Miyagawa H (2011) LC/MS/MS identification of 20-hydroxyecdysone in a scorpion (*Liocheles australasiae*) and its binding affinity to in vitro-translated molting hormone receptors. Insect Biochem Mol Biol 41:932–937. https://doi.org/10.1016/j.ibmb.2011.09.002

Moghadam NN, Holmstrup M, Pertoldi C, Loeschcke V (2013) Age-induced perturbation in cell membrane phospholipid fatty acid profile of longevity-selected *Drosophila melanogaster* and corresponding control lines. Exp Gerontol 48:1362–1368. https://doi.org/10.1016/j.exger.2013.08.018

Moriconi DE, Dulbecco AB, Juárez MP, Calderón-Fernández GM (2019) A fatty acid synthase gene (FASN3) from the integument tissue of *Rhodnius prolixus* contributes to cuticle water loss regulation. Insect Mol Biol 28:850–861. https://doi.org/10.1111/imb.12600

Moshitzky P, Miloslavski I, Aizenshtat Z, Applebaum SW (2003) Methyl palmitate: a novel product of the medfly (*Ceratitis capitata*) corpus allatum. Insect Biochem Mol Biol 33:1299–1306. https://doi.org/10.1016/j.ibmb.2003.06.008

Musselman LP, Fink JL, Ramachandran PV, Patterson BW, Okunade AL, Maier E, Brent MR, Turk J, Baranski TJ (2013) Role of fat body lipogenesis in protection against the effects of caloric overload in Drosophila. J Biol Chem 288(12):8028–8042. https://doi.org/10.1074/jbc.M112.371047. Epub 2013 Jan 25. PMID: 23355467; PMCID: PMC3605622

Nath AS, Parsons BD, Makdissi S, Chilvers RL, Mu Y, Weaver CM, Euodia I, Fitze KA, Long J, Scur M, Mackenzie DP, Makrigiannis AP, Pichaud N, Boudreau LH, Simmonds AJ, Webber CA, Derfalvi B, Hammon Y, Rachubinski RA, Di Cara F (2022) Modulation of the cell membrane lipid milieu by peroxisomal β-oxidation induces Rho1 signaling to trigger inflammatory responses. Cell Rep 38:110433. https://doi.org/10.1016/j.celrep.2022.110433

Nelson DR, Sukkestad DR, Zaylskie RG (1972) Mass spectra of methyl-branched hydrocarbons from eggs of the tobacco hornworm. J Lipid Res 13:413–421

Niehoff A-C, Kettling H, Pirkl A, Chiang YN, Dreisewerd K, Yew JY (2014) Analysis of Drosophila lipids by matrix-assisted laser desorption/ionization mass spectrometric imaging. Anal Chem 86:11086–11092. https://doi.org/10.1021/ac503171f

Ohtsu T, Katagiri C, Kimura MT, Hori SH (1993) Cold adaptations in Drosophila. Qualitative changes of triacylglycerols with relation to overwintering. J Biol Chem 268:1830–1834

Overgaard J, Sørensen JG, Petersen SO, Loeschcke V, Holmstrup M (2005) Changes in membrane lipid composition following rapid cold hardening in *Drosophila melanogaster*. J Insect Physiol 51:1173–1182. https://doi.org/10.1016/j.jinsphys.2005.06.007

Overgaard J, Tomcala A, Sørensen JG, Holmstrup M, Krogh PH, Simek P, Kostál V (2008) Effects of acclimation temperature on thermal tolerance and membrane phospholipid composition in the fruit fly *Drosophila melanogaster*. J Insect Physiol 54:619–629. https://doi.org/10.1016/j.jinsphys.2007.12.011

Paradies G, Paradies V, Ruggiero FM, Petrosillo G (2014) Oxidative stress, cardiolipin and mitochondrial dysfunction in nonalcoholic fatty liver disease. World J Gastroenterol 20(39):14205–14218. https://doi.org/10.3748/wjg.v20.i39.14205. PMID: 25339807; PMCID: PMC4202349

Paul A, Frederich M, Megido RC, Alabi T, Malik P, Uyttenbroeck R, Francis F, Blecker C, Haubruge E, Lognay G, Danthine S (2017) Insect fatty acids: a comparison of lipids from three orthopterans and *Tenebrio molitor* L. larvae. J Asia Pac Entomol 20: 337–340. https://doi.org/10.1016/j.aspen.2017.02.001

Pekny JE, Smith PB, Marden JH (2018) Enzyme polymorphism, oxygen and injury: a lipidomic analysis of flight-induced oxidative damage in a succinate dehydrogenase d (Sdhd)-polymorphic insect. J Exp Biol 221:jeb171009. https://doi.org/10.1242/jeb.171009

Phan NTN, Munem M, Ewing AG, Fletcher JS (2017) MS/MS analysis and imaging of lipids across Drosophila brain using secondary ion mass spectrometry. Anal Bioanal Chem 409:3923–3932. https://doi.org/10.1007/s00216-017-0336-4

Pho DB, Pennanec'h M, Jallon JM (1996) Purification of adult *Drosophila melanogaster* lipophorin and its role in hydrocarbon transport. Arch Insect Biochem Physiol 31:289–303. https://doi.org/10.1002/(SICI)1520-6327(1996)31:3<289::AID-ARCH4>3.0.CO;2-T

Rabani V, Cheatsazan H, Davani S (2019) Proteomics and lipidomics of black soldier fly (Diptera: Stratiomyidae) and blow fly (Diptera: Calliphoridae) larvae. J Insect Sci 19:29. https://doi.org/10.1093/jisesa/iez050

Randall AS, Liu C-H, Chu B, Zhang Q, Dongre SA, Juusola M, Franze K, Wakelam MJO, Hardie RC (2015) Speed and sensitivity of phototransduction in Drosophila depend on degree of saturation of membrane phospholipids. J Neurosci 35:2731–2746. https://doi.org/10.1523/JNEUROSCI.1150-14.2015

Rapkin J, Jensen K, House CM, Wilson AJ, Hunt J (2018) Genotype-by-sex-by-diet interactions for nutritional preference, dietary consumption, and lipid deposition in a field cricket. Heredity (Edinb) 121:361–373. https://doi.org/10.1038/s41437-018-0130-x

Rees HH, Isaac RE (1985) Biosynthesis and metabolism of ecdysteroids and methods of isolation and identification of the free and conjugated compounds. Methods Enzymol 111:377–410. https://doi.org/10.1016/s0076-6879(85)11024-4. PMID: 3897779

Reis A, Rudnitskaya A, Blackburn GJ, Mohd Fauzi N, Pitt AR, Spickett CM (2013) A comparison of five lipid extraction solvent systems for lipidomic studies of human LDL. J Lipid Res 54:1812–1824. https://doi.org/10.1194/jlr.M034330

Rietveld A, Neutz S, Simons K, Eaton S (1999) Association of sterol- and glycosylphosphatidylinositol-linked proteins with Drosophila raft lipid microdomains. J Biol Chem 274:12049–12054. https://doi.org/10.1074/jbc.274.17.12049

Salem M, Bernach M, Bajdzienko K, Giavalisco P (2017) A simple fractionated extraction method for the comprehensive analysis of metabolites, lipids, and proteins from a single sample. J Vis Exp:55802. https://doi.org/10.3791/55802

Sawabe K, Moribayashi A (2000) Lipid utilization for ovarian development in an autogenous mosquito, *Culex pipiens* molestus (Diptera: Culicidae). J Med Entomol 37:726–731. https://doi.org/10.1603/0022-2585-37.5.726

Schlame M, Greenberg ML (2017) Biosynthesis, remodeling and turnover of mitochondrial cardiolipin. Biochim Biophys Acta Mol Cell Biol Lipids 1862(1):3–7. https://doi.org/10.1016/j.bbalip.2016.08.010. Epub 2016 Aug 21. PMID: 27556952; PMCID: PMC5125896

Scheitz CJF, Guo Y, Early AM, Harshman LG, Clark AG (2013) Heritability and inter-population differences in lipid profiles of *Drosophila melanogaster*. PLoS One 8:e72726. https://doi.org/10.1371/journal.pone.0072726

Séité S, Harrison MC, Sillam-Dussès D, Lupoli R, Van Dooren TJM, Robert A, Poissonnier L-A, Lemainque A, Renault D, Acket S, Andrieu M, Viscarra J, Sul HS, de Beer ZW, Bornberg-Bauer E, Vasseur-Cognet M (2022) Lifespan prolonging mechanisms and insulin upregulation without fat accumulation in long-lived reproductives of a higher termite. Commun Biol 5:44. https://doi.org/10.1038/s42003-021-02974-6

Sharma KR, Enzmann BL, Schmidt Y, Moore D, Jones GR, Parker J, Berger SL, Reinberg D, Zwiebel LJ, Breit B, Liebig J, Ray A (2015) Cuticular hydrocarbon pheromones for social behavior and their coding in the ant antenna. Cell Rep 12:1261–1271. https://doi.org/10.1016/j.celrep.2015.07.031

Shen LR, Lai CQ, Feng X, Parnell LD, Wan JB, Wang JD, Li D, Ordovas JM, Kang JX (2010) Drosophila lacks C20 and C22 PUFAs. J Lipid Res 51:2985–2992. https://doi.org/10.1194/jlr.M008524

Shipley MM, Dillwith JW, Essenberg RC, Howard RW, Sauer JR (1993) Analysis of lipids in the salivary glands of *Amblyomma americanum* (L.): detection of a high level of arachidonic acid. Arch Insect Biochem Physiol 23:37–52. https://doi.org/10.1002/arch.940230106

Sinclair BJ, Marshall KE (2018) The many roles of fats in overwintering insects. J Exp Biol 221:jeb161836. https://doi.org/10.1242/jeb.161836

Soliday CL, Blomquist GJ, Jackson LL (1974) Cuticular lipids of insects. VI. Cuticular lipids of the grasshoppers *Melanoplus sanguinipes* and *Melanoplus packardii*. J Lipid Res 15:399–405

Stanley D (2006) Prostaglandins and other eicosanoids in insects: biological significance. Annu Rev Entomol

51:25–44. https://doi.org/10.1146/annurev.ento.51.110104.151021

Stanley DW, Hoback WW, Bedick JC, Tunaz H, Rana RL, Nor Aliza AR, Miller JS (1999) Eicosanoids mediate nodulation reactions to bacterial infections in larvae of the butterfly, *Colias eurytheme*. Comp Biochem Physiol C Pharmacol Toxicol Endocrinol 123:217–223. https://doi.org/10.1016/s0742-8413(99)00021-3

Stark WS, Lin TN, Brackhahn D, Christianson JS, Sun GY (1993) Fatty acids in the lipids of Drosophila heads: effects of visual mutants, carotenoid deprivation and dietary fatty acids. Lipids 28:345–350. https://doi.org/10.1007/BF02536321

Steinhauser ML, Bailey AP, Senyo SE, Guillermier C, Perlstein TS, Gould AP, Lee RT, Lechene CP (2012) Multi-isotope imaging mass spectrometry quantifies stem cell division and metabolism. Nature 481:516–519. https://doi.org/10.1038/nature10734

Storelli G, Nam HJ, Simcox J, Villanueva CJ, Thummel CS (2019) Drosophila HNF4 directs a switch in lipid metabolism that supports the transition to adulthood. Dev Cell 48(2):200–214.e6. https://doi.org/10.1016/j.devcel.2018.11.030

Sun J, Liu W-K, Ellsworth C, Sun Q, Pan Y, Huang Y-C, Deng W-M (2023) Integrating lipid metabolism, pheromone production and perception by fruitless and hepatocyte nuclear factor 4. Sci Adv 9:eadf6254. https://doi.org/10.1126/sciadv.adf6254

Thompson MJ, Robbins WE, Cohen CF, Kaplanis JN, Dutky SR, Hutchins FN (1971) Synthesis and biological activity of 5 beta-hydroxy analogs of alpha-ecdysone. Steroids 17:399–409. https://doi.org/10.1016/s0039-128x(71)80143-5

Toprak U, Hegedus D, Doğan C, Güney G (2020) A journey into the world of insect lipid metabolism. Arch Insect Biochem Physiol 104:e21682. https://doi.org/10.1002/arch.21682

Tose LV, Weisbrod CR, Michalkova V, Nouzova M, Noriega FG, Fernandez-Lima F (2021) Following de novo triglyceride dynamics in ovaries of *Aedes aegypti* during the previtellogenic stage. Sci Rep 11:9636. https://doi.org/10.1038/s41598-021-89025-6

Tsoupras G, Luu B, Hoffmann JA (1983) A cytokinin (isopentenyl-adenosyl-mononucleotide) linked to ecdysone in newly laid eggs of *Locusta migratoria*. Science 220:507–509. https://doi.org/10.1126/science.220.4596.507

Turner RB, Acree F (1967) The effect of photoperiod on the daily fluctuation of haemolymph hydrocarbons in the American cockroach. J Insect Physiol 13:519–522. https://doi.org/10.1016/0022-1910(67)90063-7

Tuthill BF 2nd, Searcy LA, Yost RA, Musselman LP (2020) Tissue-specific analysis of lipid species in Drosophila during overnutrition by UHPLC-MS/MS and MALDI-MSI. J Lipid Res 61(3):275–290. https://doi.org/10.1194/jlr.RA119000198

Tuthill BF 2nd, Quaglia CJ, O'Hara E, Musselman LP (2021) Loss of stearoyl-CoA desaturase 1 leads to cardiac dysfunction and lipotoxicity. J Exp Biol 224:jeb240432. https://doi.org/10.1242/jeb.240432

Tzompa-Sosa DA, Dewettinck K, Provijn P, Brouwers JF, de Meulenaer B, Oonincx DGAB (2021) Lipidome of cricket species used as food. Food Chem 349:129077. https://doi.org/10.1016/j.foodchem.2021.129077

van Dooremalen C, Ellers J (2010) A moderate change in temperature induces changes in fatty acid composition of storage and membrane lipids in a soil arthropod. J Insect Physiol 56:178–184. https://doi.org/10.1016/j.jinsphys.2009.10.002

Vanhove AS, Hang S, Vijayakumar V, Wong AC, Asara JM, Watnick PI (2017) Vibrio cholerae ensures function of host proteins required for virulence through consumption of luminal methionine sulfoxide. PLoS Pathog 13:e1006428. https://doi.org/10.1371/journal.ppat.1006428

Vial T, Tan W-L, Wong Wei Xiang B, Missé D, Deharo E, Marti G, Pompon J (2019) Dengue virus reduces AGPAT1 expression to alter phospholipids and enhance infection in *Aedes aegypti*. PLoS Pathog 15:e1008199. https://doi.org/10.1371/journal.ppat.1008199

Vial T, Tan W-L, Deharo E, Missé D, Marti G, Pompon J (2020) Mosquito metabolomics reveal that dengue virus replication requires phospholipid reconfiguration via the remodeling cycle. Proc Natl Acad Sci USA 117:27627–27636. https://doi.org/10.1073/pnas.2015095117

Visser B, Willett DS, Harvey JA, Alborn HT (2017) Concurrence in the ability for lipid synthesis between life stages in insects. R Soc Open Sci 4:160815. https://doi.org/10.1098/rsos.160815

Vrkoslav V, Muck A, Cvacka J, Svatos A (2010) MALDI imaging of neutral cuticular lipids in insects and plants. J Am Soc Mass Spectrom 21:220–231. https://doi.org/10.1016/j.jasms.2009.10.003

Wang S, Minter M, Homem RA, Michaelson LV, Venthur H, Lim KS, Withers A, Xi J, Jones CM, Zhou J-J (2020) Odorant binding proteins promote flight activity in the migratory insect, *Helicoverpa armigera*. Mol Ecol 29:3795–3808. https://doi.org/10.1111/mec.15556

Wang G, Vega-Rodríguez J, Diabate A, Liu J, Cui C, Nignan C, Dong L, Li F, Ouedrago CO, Bandaogo AM, Sawadogo PS, Maiga H, Alves e Silva TL, Pascini TV, Wang S, Jacobs-Lorena M (2021a) Clock genes and environmental cues coordinate anopheles pheromone synthesis, swarming, and mating. Science 371:411–415. https://doi.org/10.1126/science.abd4359

Wang X, Zhong Z, Chen X, Hong Z, Lin W, Mu X, Hu X, Zheng H (2021b) High-fat diets with differential fatty acids induce obesity and perturb gut microbiota in honey bee. Int J Mol Sci 22:834. https://doi.org/10.3390/ijms22020834

Warbrick-Smith J, Behmer ST, Lee KP, Raubenheimer D, Simpson SJ (2006) Evolving resistance to obesity in an insect. Proc Natl Acad Sci USA 103:14045–14049. https://doi.org/10.1073/pnas.0605225103

Wat LW, Chao C, Bartlett R, Buchanan JL, Millington JW, Chih HJ, Chowdhury ZS, Biswas P, Huang V, Shin LJ, Wang LC, Gauthier M-PL, Barone MC, Montooth KL, Welte MA, Rideout EJ (2020) A role for triglyceride lipase brummer in the regulation of sex differences in Drosophila fat storage and breakdown. PLoS Biol 18:e3000595. https://doi.org/10.1371/journal.pbio.3000595

Wegener J, Krause S, Parafianczuk V, Chaniotakis I, Schiller J, Dannenberger D, Engel KM (2022) Lipidomic specializations of honeybee (*Apis mellifera*) castes and ethotypes. J Insect Physiol 142:104439. https://doi.org/10.1016/j.jinsphys.2022.104439

Wessels FJ, Jordan DC, Hahn DA (2010) Allocation from capital and income sources to reproduction shift from first to second clutch in the flesh fly, *Sarcophaga crassipalpis*. J Insect Physiol 56:1269–1274. https://doi.org/10.1016/j.jinsphys.2010.03.033

Wicker-Thomas C (2007) Pheromonal communication involved in courtship behavior in Diptera. J Insect Physiol 53:1089–1100. https://doi.org/10.1016/j.jinsphys.2007.07.003

Williams CM, Thomas RH, MacMillan HA, Marshall KE, Sinclair BJ (2011) Triacylglyceride measurement in small quantities of homogenised insect tissue: comparisons and caveats. J Insect Physiol 57(12): 1602–1613. https://doi.org/10.1016/j.jinsphys.2011.08.008

Wu Q, Yu G, Park SJ, Gao Y, Ja WW, Yang M (2020) Excreta quantification (EX-Q) for longitudinal measurements of food intake in Drosophila. iScience 23:100776. https://doi.org/10.1016/j.isci.2019.100776

XueKe G, Shuai Z, JunYu L, LiMin L, LiJuan Z, JinJie C (2017) Lipidomics and RNA-Seq study of lipid regulation in *Aphis gossypii* parasitized by *Lysiphlebia japonica*. Sci Rep 7:1364. https://doi.org/10.1038/s41598-017-01546-1

Yang E, Gamberi C, Chaurand P (2019) Mapping the fly Malpighian tubule lipidome by imaging mass spectrometry. J Mass Spectrom 54:557–566. https://doi.org/10.1002/jms.4366

Yew JY, Cody RB, Kravitz EA (2008) Cuticular hydrocarbon analysis of an awake behaving fly using direct analysis in real-time time-of-flight mass spectrometry. Proc Natl Acad Sci USA 105:7135–7140. https://doi.org/10.1073/pnas.0802692105

Yew JY, Dreisewerd K, Luftmann H, Müthing J, Pohlentz G, Kravitz EA (2009) A new male sex pheromone and novel cuticular cues for chemical communication in Drosophila. Curr Biol 19:1245–1254. https://doi.org/10.1016/j.cub.2009.06.037

Yew JY, Dreisewerd K, de Oliveira CC, Etges WJ (2011a) Male-specific transfer and fine scale spatial differences of newly identified cuticular hydrocarbons and triacylglycerides in a Drosophila species pair. PLoS One 6:e16898. https://doi.org/10.1371/journal.pone.0016898

Yew JY, Soltwisch J, Pirkl A, Dreisewerd K (2011b) Direct laser desorption ionization of endogenous and exogenous compounds from insect cuticles: practical and methodologic aspects. J Am Soc Mass Spectrom 22:1273–1284. https://doi.org/10.1007/s13361-011-0110-3

Zhang H, Liu Y, Fields L, Shi X, Huang P, Lu H, Schneider AJ, Tang X, Puglielli L, Welham NV, Li L (2023) Single-cell lipidomics enabled by dual-polarity ionization and ion mobility-mass spectrometry imaging. Nat Commun 14:5185. https://doi.org/10.1038/s41467-023-40512-6

Zhou J, Chen X, Yan J, You K, Yuan Z, Zhou Q, Lu K (2018) Brummer-dependent lipid mobilization regulates starvation resistance in *Nilaparvata lugens*. Arch Insect Biochem Physiol 99:e21481. https://doi.org/10.1002/arch.21481

Zhu F, Song D, Chen H, Tang Q, Huo S, Liu X, Chen K (2021) A lipidome map of the silkworm *Bombyx mori*: influences of viral infection. J Proteome Res 20:695–703. https://doi.org/10.1021/acs.jproteome.0c00608

Adv Exp Med Biol (2026) 1494: 293–312
https://doi.org/10.1007/5584_2025_850

Published online: 19 March 2025

Lipid Metabolism in Diapause

Umut Toprak, Nicholas M. Teets, Doga Cedden, and Gözde Güney

Abstract

Organisms living in temperate and polar environments encounter seasonal fluctuations that entail changes in temperature, resource availability, and biotic interactions. Thus, adaptations for synchronizing the life cycle with essential resources and persisting through unfavorable conditions are critical. Diapause, a programmed period of developmental arrest and metabolic depression, is widely used by insects to survive winter and synchronize the life cycle. In some cases, insects spend over half the year (or in some cases, multiple years) in a nonfeeding diapause state. Thus, diapause is energetically challenging, and insects accumulate surplus energy stores and/or suppress metabolism to make it through the winter. As the most energy-dense, and often most abundant, energy reserve in insects, lipids play a central role in diapause energetics. In this chapter, we provide an overview of lipid metabolism in the context of diapause. First, as this is the only chapter in this book that covers diapause, we present some of the general features of diapause. We then discuss the role of lipids as an essential energy store during diapause, focusing on patterns of lipid accumulation before diapause and patterns of utilization during diapause. In the next section, we outline some other roles of lipids during diapause in addition to their role as an energy store. Finally, we end the chapter by discussing the molecular regulation of lipid metabolism in diapause, which has received increased attention in recent years.

U. Toprak (✉)
Molecular Entomology Laboratory, Department of Plant Protection, Faculty of Agriculture, Ankara University, Ankara, Türkiye
e-mail: utoprak@agri.ankara.edu.tr

N. M. Teets (✉)
Department of Entomology, University of Kentucky, Lexington, KY, USA
e-mail: n.teets@uky.edu

D. Cedden
Department of Evolutionary Developmental Genetics, Johann-Friedrich-Blumenbach Institute, GZMB, University of Göttingen, Göttingen, Germany
e-mail: doga.cedden@biologie.uni-goettingen.de

G. Güney
Agricultural Entomology, Department of Crop Sciences, University of Göttingen, Göttingen, Germany
e-mail: ggueney@uni-goettingen.de

Keywords

Diapause · Hibernation · Lipid metabolism · Lipogenesis · Lipolysis

1 General Characteristics of Diapause

Diapause is a programmed period of developmental arrest, metabolic suppression, and enhanced stress tolerance that begins in anticipation of adverse environmental conditions experienced during winter (Denlinger 2022). (Note: parallel adaptations for coping with seasonally dry conditions, i.e., aestivation, will not be covered in this chapter, but see Storey and Storey 2012 for a review). While diapause can occur at any life stage across insect species, each species can typically only diapause at a single life stage. Diapause appears to have multiple evolutionary origins. For example, among mosquitoes, there are species that undergo egg, larval, and adult diapauses, and even some mosquito genera have species that diapause at all three stages (Denlinger and Armbruster 2014). In addition to the life stage of diapause, another important feature is whether it is *facultative* or *obligate.* Multivoltine species undergo facultative diapause, triggered by token environmental stimuli (see below) before winter. In contrast, univoltine species typically undergo obligatory diapause, which is a genetically programmed diapause that occurs at a specific developmental stage regardless of developmental conditions. It is worth noting that not all insects undergo diapause, and some rely on quiescence, a period of dormancy that occurs in direct response to suboptimal environmental conditions, to survive the winter (Košťál 2006).

For facultative diapause, photoperiod is the most commonly reported token stimulus for triggering diapause development (Tauber and Tauber 1976). Insects with photoperiodic diapause have a photosensitive stage to cue diapause induction occurring before the diapausing stage that detects photoperiodic cues indicative of winter's approach. Most photoperiodic diapauses are "all or nothing" in that when a specific critical photoperiod is reached, all or nearly all of the population will enter diapause. This critical photoperiod typically varies with latitude, with species or populations at higher latitudes having a longer critical photoperiod (Bradshaw et al. 2003). In many cases, temperature can modulate the photoperiodic response, such that at a given photoperiod, insects are more likely to enter diapause at a lower temperature (Bradshaw et al. 2003). While photoperiod and temperature are the two most common cues that modulate diapause incidence, in some cases, other environmental factors, such as resource availability, can further influence diapause (Michaud and Qureshi 2005). Once the appropriate environmental signals to program diapause are obtained, dormancy is typically generated and maintained internally by endocrine signals. While the endocrine regulation of diapause is complex, in general, diapause in immature stages (embryo, larva, pupa) is maintained by low ecdysone titers, while a shutdown of juvenile hormone (JH) signaling maintains adult reproductive diapause. For a more in-depth review of the endocrine regulation of diapause, see Denlinger et al. (2005).

While diapause was originally considered a period of stasis, it is clear that diapause is a dynamic developmental program with distinct phases (Tauber and Tauber 1976; Košťál 2006). The distinct phases of diapause defined by Košťál (2006) are now widely used in the field. Prior to diapause, there is an induction phase in which the insect is sensitive to the token stimuli that triggers diapause, and this phase is followed by the preparation phase, in which development continues, but preparations for diapause may take place. Once diapause begins and development stops, there is an initiation phase in which energy storage can increase, metabolism is suppressed, and other physiological preparations occur. During the subsequent maintenance phase of diapause, developmental arrest continues, even if conditions are favorable, and the metabolic rate is typically low and constant. Finally, during the termination phase, changes in the external or internal environment cause diapause intensity to decrease, and by the end of this stage, the insect has the capacity to resume development. However, in many cases, after diapause termination, conditions remain unfavorable for development, and insects remain in post-diapause quiescence until conditions improve. While these phases were defined in eco-physiological terms, recent studies indicate that these ecophysiological transitions are also accompanied by distinct molecular shifts (Ragland et al. 2010; Košťál et al. 2017; Nadeau et al. 2022), and throughout the remainder of the chapter, we will discuss lipid metabolism in the context of these distinct phases of diapause.

2 Lipid as an Energy Source During Diapause

As briefly discussed above, maintaining energy balance is one of the most significant challenges during diapause (Hahn and Denlinger 2007, 2011). Many insects diapause in an immobile stage, and species that diapause as larvae or adults may not have access to food during the winter or be unable to move due to low temperatures. Further, while growth and development cease during diapause, the physiological and biochemical changes needed to cope with low temperature and other harsh abiotic conditions (e.g., production of cryoprotectants and cytoprotective proteins) can be energetically costly (Teets et al. 2023). Some species also rely on energy stored during diapause to fuel development and/or reproduction in the spring. Thus, efficiently storing and using energy throughout diapause is critical.

For most overwintering insects, lipids are the major source of energy. Lipids can generate more ATP energy per unit mass than other energy stores, and they do not require large amounts of water for storage, with water often being limited in winter (Hahn and Denlinger 2011). As discussed elsewhere in this book, triacylglycerides (or triglyceride, TAG) are the primary storage form for lipids, and most are stored in lipid droplets in the fat body. When lipids are transported to other tissues, a fatty acid is cleaved from triacylglycerol to form diacylglycerol that is then carried in the hemolymph by lipophorin proteins (Chino and Downer 1982). In some diapausing eggs, diacylglycerols are also abundant and accumulate during preparation for winter (Batz and Armbruster 2018). It is worth noting that some species primarily use carbohydrates as an overwintering energy source, although these cases appear to be the exception (Sinclair and Marshall 2018).

Many diapause-destined individuals are reported to have a higher lipid content than non-diapause-destined individuals (Enriquez and Visser 2023; see Fig. 1a, for an example). Further, lipid accumulation is prevalent across all diapausing life stages; for example, elevated lipid content is observed at the onset of diapause for eggs of the yellow fever mosquito *Aedes aegypti* (Batz and Armbruster 2018), larvae of the five-spot burnet *Zygaena trifolii* (Wipking et al. 1995), pupae of the flesh fly *Sarcophaga crassipalpis* (Adedokun and Denlinger 1985; Fig. 1a), and adults of the northern house mosquito, *Culex pipiens* (Sim and Denlinger 2008). It is worth noting that there are exceptions to increased lipid content in diapausing compared to non-diapausing insects (see Hahn and Denlinger (2007) and Short and Hahn (2023)). For instance, some insects primarily rely on energy conservation, rather than accumulation, to survive winter (see Fig. 1b in this chapter for an example of the stink bug *Halyomorpha halys* from Ciancio et al. (2021) and Table 1 in Enriquez and Visser (2023) for a summary). In many cases, these increases in fat stores can be quite dramatic, with diapausing individuals having more than twice the lipid content as non-diapause individuals. A notable example of this is adult Colorado potato beetles that accumulate approximately threefold higher lipid reserves when raised under diapause-inducing conditions (Güney et al. 2021). Lipid accumulation during diapause preparation is typically followed by significant metabolic depression during diapause (Hahn and Denlinger 2011; Fig. 1c), and the combination of these two increases the probability that insects can survive the winter without running out of energy.

With lipids serving as the primary energy source during diapause, lipids are expected to be depleted during diapause. Indeed, for many insects, lipid content decreases throughout diapause (Enriquez and Visser 2023). For example, in the pupal diapause of *S. crassipalpis,* lipid content decreases nearly fourfold across the ~90 day diapause period despite a marked reduction in metabolic rate (Adedokun and Denlinger 1985; Fig. 1a, c). However, it is worth noting that these experiments were conducted at a relatively high temperature (20 °C), so whether such dramatic reduction occurs at ecologically relevant winter temperatures is uncertain. While many insects deplete lipids in the winter, some

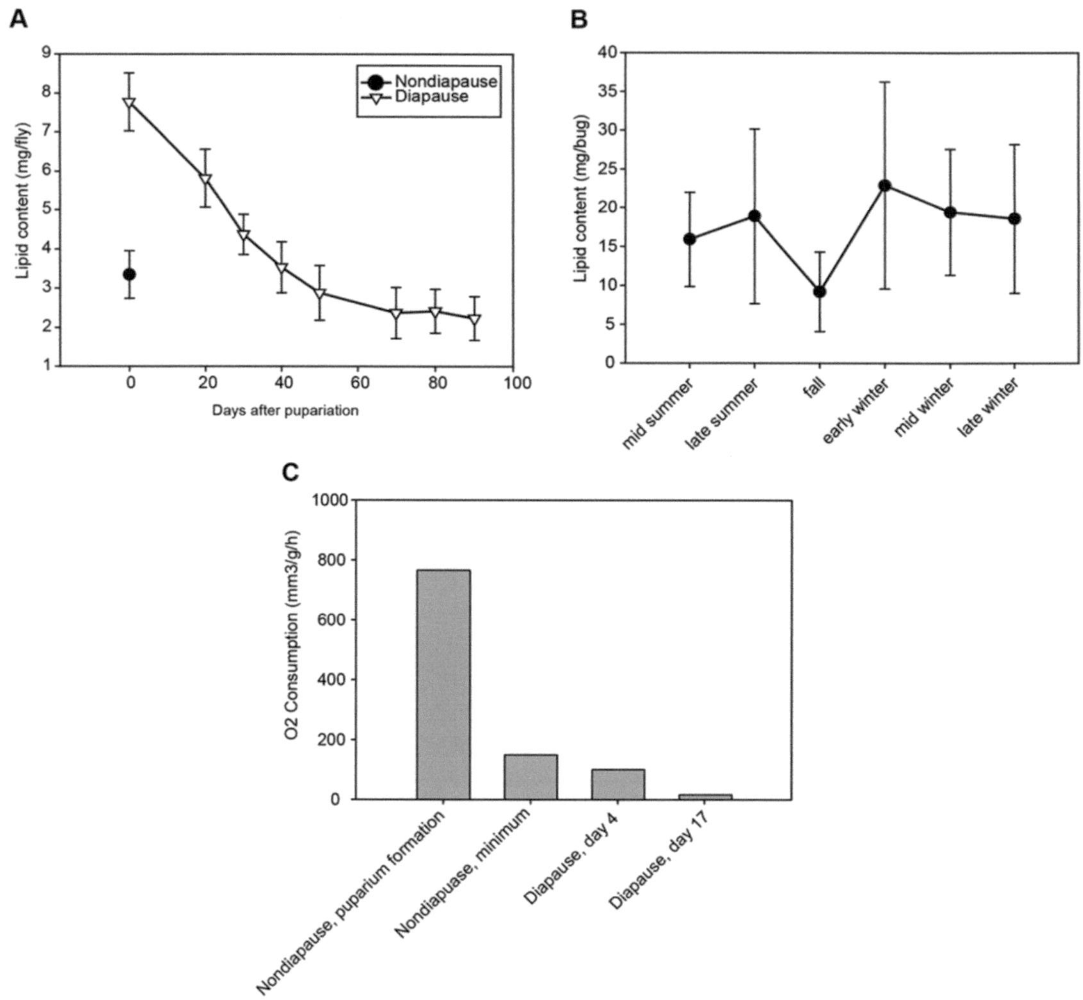

Fig. 1 Select examples of lipid stores and metabolic rate during diapause. (**a**) Lipid content between diapausing and non-diapausing flesh flies (*Sarcophaga crassipalpis*) from Adedokun and Denlinger (1985), showing an example of a species that has higher levels of lipid content at the onset of diapause but depletes these lipids throughout diapause. (**b**) Lipid content of brown marmorated stink bug (*Halyomorpha halys*) collected at different times of year from Ciancio et al. (2021), showing an example of a species that preserves lipid content during diapause. (**c**) Metabolic rate of specific stages of non-diapause and diapausing pupae from *S. crassipalpis* from Denlinger et al. (1972), showing a classic example of metabolic rate depression during diapause). In (**a**) and (**b**), error bars represent standard deviation

remarkably preserve lipid levels throughout the winter. For example, the stink bug *H. halys*, despite likely being unable to feed during its adult diapause, reduces its metabolic rate and experiences little to no reduction in lipid content through the winter (Ciancio et al. 2021; Fig. 1b). This ability to maintain lipid stores during winter may be critical for survival in some species because of energy requirement for post-diapause activities such as reproduction and development. In *C. pipiens*, cold tolerance is more than sufficient to survive temperatures present in underground hibernacula, but the duration of winter survival is limited by the amount of lipid stores (Rozsypal et al. 2021). Also, in some species, such as *Aedes aegypti*, local adaptation to temperate environments has favored increased lipid accumulation prior to diapause, suggesting that energy reserves in ancestral tropical populations are insufficient for longer temperate winters (Mensch et al. 2021). Similarly, high-latitude beetle species are adept at conserving lipid stores during winter (Lehmann et al. 2020), suggesting that the ability to conserve energy may be a critical adaptation for species expanding to colder habitats. Further, adaptations for conserving energy in winter may become even more critical as winters get warmer. Warmer winter environments can lead to higher rates of lipid depletion (Williams et al. 2012, 2015; Roberts et al. 2021), which may limit the survival of some species, especially species with prolonged diapause and limited opportunities to feed during winter.

3 Other Roles of Lipids in Diapause

3.1 Cuticular Waterproofing

The cuticle has a major role in ensuring water homeostasis in terrestrial arthropods. Cuticular permeability can be altered by adjusting the amount and type of cuticular hydrocarbons (CHCs) deposited on the epicuticle, mainly composed of n-alkanes, methyl-branched alkanes, and unsaturated hydrocarbons (Yang et al. 2020). The composition of CHCs varies between species, developmental stages, and in response to seasonal challenges. The latter suggests that changes in CHC composition might be a critical adaptive part of the diapause toolkit in insects (Yoder et al. 1992; Yoder et al. 1995; Gibbs et al. 1991; Gibbs 1998; Blomquist and Bagnères 2010; Holze et al. 2021). Indeed, studies on diverse insects indicate dynamic changes in CHC composition in diapause, especially in insects facing seasonal desiccation risk. In several *Drosophila* species, diapause induction leads to an increase in the proportion of long-chain hydrocarbons that are more effective in reducing water permeability compared to shorter-chain hydrocarbons due to their biophysical properties (Ala-Honkola et al. 2020; Kárpáti et al. 2023). These changes in CHC profiles are associated with modifications in gene expression related to hydrocarbon biosynthesis and metabolism, indicating a genetically regulated mechanism aimed at augmenting cuticular barrier function during diapause. For example, the upregulation of enzymes that modify hydrocarbons, including desaturase and elongase, suggested that the production of longer-chain hydrocarbons is regulated at the gene expression level to maintain water balance under diapause conditions (Li et al. 2019; Zhao et al. 2020; Holze et al. 2021). Similarly, the CHC profiles of both sexes shifted toward having increased long-chain and reduced short-chain CHCs in diapausing *Drosophila montana* adults (Ala-Honkola et al. 2020).

In mosquitoes, particularly *C. pipiens*, diapause results in significant differences in water balance properties (Benoit and Denlinger 2007). Diapausing females have a lower percentage of water content yet possess higher initial and dry masses compared to non-diapausing females. This physiological adjustment in diapausing females is associated with lower rate of water loss compared to non-diapausing females. Further experiments revealed that diapausing females do not utilize common polyols for water retention. Instead, they have twice the amount of cuticular hydrocarbons and bigger in size compared to non-diapausing females. These findings highlight the critical role of body size and cuticular hydrocarbons in enhancing water conservation during diapause, highlighting the complex interplay of physiological adaptations that enable *C. pipiens* to survive the winter months (Benoit and Denlinger 2007).

As in *C. pipens*, diapause programming affects CHC profiles in the spotted amber ladybird, *Hippodamia variegate*. This species enters diapause as an adult in winter and has distinct CHC profiles in diapausing versus non-diapausing adults (Khabir et al. 2024). Specifically, a gas chromatography-mass spectrometry (GC-MS) study in *H. variegate* identified 13 CHCs exclusive to diapausing adults, while 5 were exclusive to non-diapausing adults. Notably, the CHC profiles of diapausing adults had a significant bias in favor of longer chains, suggesting an adaptation to the cold winter conditions. Long and methyl-branched CHCs were shown to be primary molecules conferring resistance to desiccation in *Drosophila* species (Wang et al. 2022), further supporting the survival advantage of longer CHCs.

Like many other biochemical changes during diapause, the above changes in CHC composition are not universal. For example, in flesh flies (*Sarcophaga crassipalpis*), hydrocarbons in the puparia of diapausing and non-diapausing individuals have similar compositions of saturated hydrocarbons with lengths from 25 to 33 carbons. The analysis showed a mix of straight-chain, monomethyl-, and dimethyl-alkanes, with diapausing pupae comprising 39.4% n-alkanes, 46.5% monomethyl-alkanes, and 9.5% dimethyl-alkanes, and non-diapausing ones having slightly different percentages. However, diapausing individuals exhibited double the total hydrocarbon content compared to non-diapausing individuals, suggesting that the overall quantity of hydrocarbons, rather than their specific composition, is important for reducing water loss in diapausing pupae (Yoder et al. 1995). In a study examining *Musca autumnalis* adult flies caught in the wild, researchers analyzed and compared the hydrocarbon profiles of diapausing and reproductive adult males and females (Jurenka et al. 1998). Hydrocarbon profiles were largely similar between the sexes, with minimal qualitative differences. In contrast, diapause-related variations were observed in the percentage composition of various hydrocarbons: monounsaturated alkenes (4% in diapausing vs. 45% in reproductive), n-alkanes (24% vs. 37%), monomethyl-alkanes (57% vs. 15%), and dimethyl-alkanes (15% vs. 2%) in female flies. Moreover, a slight difference was detected in the total hydrocarbon content, with diapausing females having 7.3 ± 0.6 μg/fly and reproductive females having 9.7 ± 1.1 μg/fly. Male flies also showed a similar pattern of change in hydrocarbon profile and total amount between their diapausing and reproductive phases (Jurenka et al. 1998). These studies highlight that the changes in the hydrocarbon composition of diapausing individuals can have different quantities of CHCs of the same compositions, different compositions at the same total quantity of CHCs, or a combination of both depending on the species.

The literature indicates that diapause in various insects is associated with changes in CHC profiles, suggesting adaptation to seasonal conditions. Through the modulation of their cuticular hydrocarbon profiles, diapausing insects can enhance their desiccation resistance, thermal tolerance, and overall survivability. Moreover, CHCs have been shown to act as signaling molecules that facilitate intraspecific communication. It remains to be elucidated if specific CHC profiles convey information regarding the diapausing state to conspecific insects. In a study on *Drosophila montana*, the authors observed that diapausing individuals were less attractive to the opposite sex compared to the non-diapausing individuals. The authors linked this change in attractiveness to more long-chain and less short-chain hydrocarbons on the cuticle of diapausing individuals compared to non-diapausing individuals. Although this suggested a possible mechanism where the diapause state is conveyed through CHCs, direct experimental evidence is lacking (Ala-Honkola et al. 2020). Furthermore, variation in hydrocarbon profiles has been shown to influence thermal tolerance, predation risk, and interspecific interactions (Palazzo and Setzer 2009; Michelutti et al. 2018; Rajpurohit et al. 2020; Joel et al. 2022; Kárpáti et al. 2023; Li et al. 2024), albeit there are no comprehensive studies that implicate these functions of CHCs in diapause.

3.2 Homeoviscous Acclimation in Diapause

Insects, as ectothermic organisms, need to modulate their cellular membrane fluidity in response to temperature fluctuations (Sinensky 1974; Hazel 1995), which is covered in "Lipid Properties and Metabolism in Response to Cold" in this book. Homeoviscous acclimation or homeoviscous adaptation in insects refers to a physiological mechanism that allows insects to maintain the fluidity and functionality of their cell membranes across a range of environmental temperatures. This process ensures that the membrane's lipid composition is adjusted in response to temperature changes, such that the membrane remains in an optimal state for cellular processes. Homeoviscous acclimation is prevalent in insects, and recent studies are beginning to untangle the molecular pathways that govern it (also see Chapter 17 for a discussion of these molecular mechanisms) and its importance in insect life cycle.

Homeoviscous acclimation is critical to the success of winter diapause in many insect species (Michaud and Denlinger 2006; Vukašinović et al. 2015; Lehmann et al. 2020). A recent study systematically described the lipidomes of three beetle species with similar winter diapause responses, namely the Colorado potato beetle *Leptinotarsa decemlineata*, the alder leaf beetle *Agelastica alni,* and the knotgrass leaf beetle *Chrysolina polita* (Lehmann et al. 2020). Notably, the membrane lipid profiles of all three diapausing beetles exhibited significant changes compared to their non-diapausing counterparts. One of the common changes in all three beetles was the increase in the unsaturation levels of acidic phospholipid classes (PS + PI) during diapause (Lehmann et al. 2020). This is consistent with the notion that an increase in unsaturated fatty acids increases membrane fluidity under cold temperatures (Nozawa 2011; Wu et al. 2003). However, the study also revealed that the changes in other membrane lipid profiles were more complex. For instance, as beetles moved from the beginning stages of diapause to its maintenance phase across various species, a notable qualitative transformation occurred within the dynamics of phospholipids, specifically characterized by an increased heterogeneity in the acyl chain lengths of phosphatidylcholine (PC) species. During the transition into diapause, there was a preferential increase in the abundance of PC species possessing either short (32- and 34-carbon) or long (38-carbon) acyl chains, while those with intermediate lengths (36-carbon) acyl chains were comparatively diminished. Hence, it seems both canonical homeoviscous acclimation and other changes might be involved in ensuring a successful winter diapause in beetles (Lehmann et al. 2020).

The significance of homeoviscous acclimation in diapause is also highlighted through research on the European corn borer (*Ostrinia nubilalis*) (Vukašinović et al. 2015). In this species, diapause leads to an increase in the unsaturated to saturated fatty acids (UFAs/SFAs) ratio, driven by an increase in monounsaturated fatty acids (MUFAs) levels, alongside a reduction in both polyunsaturated fatty acids (PUFAs) and saturated fatty acids (SFAs) levels. The adjustments in phospholipid composition were relatively low, while changes in free FAs led to an overall increase in unsaturation FA levels. The study also showed that the melting transition temperatures of total lipids from whole larvae were notably reduced throughout diapause in contrast to non-diapause periods. This trend was linked to the FA makeup of triacylglycerols, presumably facilitating their operational integrity during winter low temperatures. Hence, the European corn borer *O. nubilalis* acclimates its FA profiles within both energy-storing triacylglycerols and structural phospholipids as part of its hibernation strategy, potentially enhancing its capacity for cold tolerance during diapause (Vukašinović et al. 2015).

An early study on the onion maggot, *Delia antiqua*, pupae that were either cold-acclimated during non-diapause conditions or in diapause during summer and winter, exhibited significantly greater cold tolerance compared to their non-acclimated, non-diapause counterparts. Further investigations highlighted the importance of homeoviscous acclimation, where cellular membranes adjust to maintain fluidity under cold conditions, with a focus on the presumed critical role of Δ9-acyl-CoA desaturase. An increase in Δ9-acyl-CoA desaturase enzyme mRNA abundance was observed in the brain, midgut, and Malpighian tubules of both cold-acclimated and diapause-programmed pupae, ranging from two to ten times higher than controls, which correlated strongly with their increased cold resilience. In pupae demonstrating enhanced cold hardiness, there was a notable rise in the levels of palmitoleic and oleic acids within the phospholipids, identified as probable products of desaturase activity. This evidence supports the notion that elevated Δ9-acyl-CoA desaturase enzyme expression is involved in the cold hardiness in *D. antiqua* by enhancing the production of these unsaturated fatty acids (Kayukawa et al. 2007).

The collective findings from diapause research in insects strongly suggest a role for homeoviscous acclimation in enabling insects to navigate survive winter cold. It is also apparent that noncanonical changes in the membrane compositions occur in diapausing insects, although it is not yet clear whether they are genuine examples of homeoviscous acclimation or other diapause-related changes. One of the gaps in the literature on homeoviscous acclimation in diapause is the lack of direct evidence implicating the suggested adaptations to a survival advantage during diapause. This gap may be addressed through techniques such as CRISPR-Cas9 and RNA interference targeting genes involved in membrane composition acclimation. For instance, studies on model organism *Drosophila melanogaster* have suggested a role for specific desaturase enzymes, including Δ12 fatty acid desaturase, in regulating fatty acid saturation levels, and hence membrane fluidity (Suito et al. 2020). Such enzymes could be good candidates for molecular studies in insects showing robust winter diapause phenotype.

4 Molecular Regulation of Lipid Metabolism in Diapause

The molecular regulation of lipid metabolism during insect diapause involves complex changes in gene expression, enzyme activity, and metabolic pathways that enable insects to conserve energy and survive adverse environmental conditions. During diapause, lipid stores, mainly triglycerides, are mobilized for energy homeostasis. Key enzymes such as lipases, fatty acid synthases, and desaturases are modulated to regulate lipid synthesis, breakdown, and remodeling. Additionally, hormonal control, especially through juvenile hormones and ecdysone, plays a significant role in coordinating lipid metabolism by influencing the expression of genes involved in fat storage and utilization. These molecular adaptations help insects efficiently manage their lipid reserves during diapause, ensuring survival until favorable conditions return. We summarize the key finding on the molecular regulation of lipid metabolism in diapause below.

4.1 Gene Regulation of Lipid Metabolism Throughout Phases of Diapause

Different phases of diapause are usually associated with distinct modes of lipid metabolism. The most basic and important distinction is whether lipids are consumed to produce energy (lipolysis) or are actively being accumulated (lipogenesis). These two opposite metabolic activities, in turn, are enabled by a specific set of genes whose expressions are dynamically regulated throughout the diapause (Hahn and Denlinger 2007, 2011; Arrese and Soulages 2010; Toprak et al. 2020). As a general rule, lipogenesis genes are upregulated during the pre-diapause, whereas the lipolysis genes are upregulated during the diapause and its termination (Sim and Denlinger 2009; Batz and Armbruster 2018; Hussein et al. 2020).

To support the synthesis of fatty acids, expression of acetyl-CoA carboxylase (ACC) and *FAS* genes peaks in the pre-diapause period in many insects, including the cabbage beetle *Colaphellus bowringi* (Tan et al. 2017b), *L. decemlineata* (Güney et al. 2021; Cedden et al. 2024), and green Lacewing, *Chrysoperla nipponensis* (Liu et al. 2024) (Fig. 2). The accumulation of adequate lipid reserves during pre-diapause appears crucial for diapause regulation, as knockdown of *ACC* or *FAS* prolonged the diapause preparatory phase in *L. decemlineata*, presumably allowing the insects more time to accumulate sufficient lipid intermediates for diapause entry through de novo lipogenesis (Cedden et al. 2024). Long-chain fatty acid-CoA ligase (ACSL), elongase of

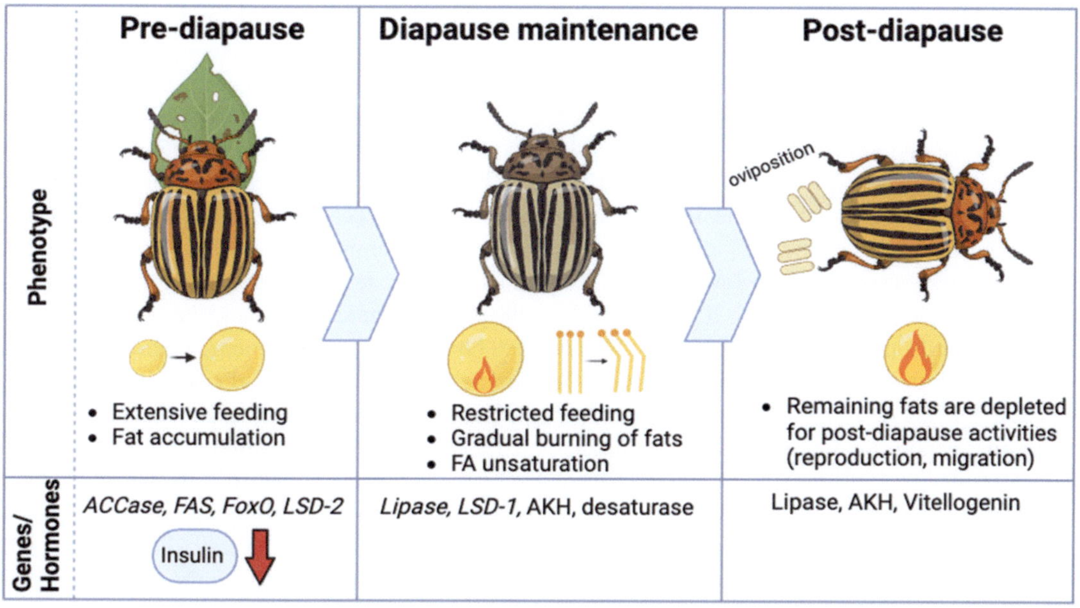

Fig. 2 Overview of lipid metabolism-related phenotypes and key associated genes/hormones in diapause-destined insects. A typical adult diapause is depicted. Pre-diapause (or diapause preparatory phase) is associated with extensive feeding and fat accumulation, which is enabled by the regulation of lipogenic genes such as *ACC* and *FOXO*, as well as hormonal regulation, including reduction in insulin signaling. In general, diapause maintenance is associated with limited or complete absence of feeding behavior, which requires the burning of accumulated fat stores to sustain the basal metabolism. The burning of fats is usually gradual and even not fully depleted at the termination of diapause. Lipases and adipokinetic hormone (AKH) may be involved in the burning of fat stores for sustenance. Moreover, non-metabolism-related changes in lipids, such as fatty acid unsaturation, may enable increased resistance to harsh environmental conditions, including low temperatures. The remaining fat stores are important to support post-diapause activities (e.g., reaching sexual maturity in adult diapause and continuation of development in larval or pupal diapause) in some insects

very-long-chain fatty acids (ELO), and very-long-chain 3-oxoacyl-CoA reductase (KAR) are also highly expressed during pre-diapause, but not in diapause, in the ladybird *Coccinella septempunctata* (Xiang et al. 2021). Functional studies using RNAi showed that these lipid genes were required for adequate lipid accumulation in the pre-diapause of *C. septempunctata*. Similarly, genes that are involved in the synthesis of TAGs (long-term lipid storage molecules), including monoacylglycerol acyltransferase (MGAT) and diacylglycerol acyltransferase (DGAT), are also upregulated during the pre-diapause in the mosquito *C. pipiens* (Sim and Denlinger 2009) and in the copepod (a group of small crustaceans) *Calanus finmarchicus*, like many other copepods (Tarrant et al. 2016).

The main gene family that ultimately initiates the catabolism of storage or dietary lipids for energy are lipases, especially the adipose triacylglycerol lipase homolog brummer lipase (Grönke et al. 2005, 2007; Kühnlein 2012; Wat et al. 2020). The catabolism of storage lipids is of importance in diapause maintenance because diapausing insects usually do not feed (Hahn and Denlinger 2011; Reynolds et al. 2012; Batz and Armbruster 2018) (Fig. 2). Lipases enable the provision of energy during diapause maintenance and termination, the latter being especially important for embryonic or pupal diapause due to the need for high amounts of energy for developmental purposes (Reynolds et al. 2012; Jia and Li 2023). Nevertheless, post-diapause development of insects undergoing reproductive (adult) diapause can also depend on the action of lipases.

Notably, lipases might be playing a prominent role, especially in the latter part of the diapause maintenance, as the primary energy source in animals is sugars. This is supported by the fact that lipase is upregulated in late rather than early diapause in *C. pipiens* (Sim and Denlinger 2009).

A primary characteristic of fat body cells in diapausing insects is enlarged lipid droplets that mainly contain accumulated TAGs (Sim and Denlinger 2013b; Zhang et al. 2019) (Fig. 2). Lipid storage droplet 1 (LSD1) 1 and 2 (LSD2) proteins, homologous to perilipins in mammalians, are the major proteins that ensure the formation and regulation of the lipid droplets (Miura et al. 2002; Grönke et al. 2003; Arrese et al. 2008a, b, 2014). LSD2 is upregulated in many pre-diapausing insects such as the Colorado potato beetle *L. decemlineata* (Güney et al. 2021) and the tortoise bug *Eurygaster maura* (Hemiptera: Scutelleridae) (Toprak et al. 2014) to allow the storage of the increasing TAG levels. The silencing of *LSD2* genes, in turn, leads to a decrease in TAG levels, suggesting that they are protecting the TAGs against lipolytic activity during diapause (Güney et al. 2021). An alternative explanation is that insufficient perilipins, such as LSD2, lead to the exclusion of TAGs from lipid droplets because droplet size is limited by perilipin availability, albeit there are no experimental studies for elucidating the exact mechanism. LSD1, in contrast, is known to induce lipolysis after its phosphorylation (Grönke et al. 2007; Arrese et al. 2008b; Beller et al. 2010; Bi et al. 2012). Consequently, *LSD1* genes are upregulated in diapause, presumably to allow the storage TAG to be activated for lipase activity (Toprak et al. 2014; Güney et al. 2021; Chen et al. 2023b). The mobilization and accumulation of lipid reserves also depend on fatty acid–binding proteins (FABP) (van der Horst et al. 1993; Febbraio et al. 1999; Sim and Denlinger 2009; Gondim et al. 2018). FABPs play a key role in the trafficking of fatty acids and their derivatives throughout cellular compartments, making them important for lipid metabolism (van der Horst et al. 1993). In the leaf beetle *C. bowringi*, the gene encoding the fatty acid–binding protein is highly expressed in the diapause-destined beetles, and its silencing inhibited lipid accumulation in the fat body, emphasizing its role in lipid accumulation via the transport of dietary or de novo–produced fatty acids to cellular locations of TAG synthesis and storage (Tan et al. 2017a).

Beyond the above canonical changes in lipid metabolism genes, recent work has demonstrated that more complex molecular interactions are at play to regulate lipid metabolism in diapause. In the pupae of lepidopteran and dipteran insects, the fat body undergoes cell dissociation, a process crucial for the activation of lipid metabolism and subsequent pupal development (Jia and Li 2023). This dissociation was driven by cathepsin L, which activates matrix metalloproteinases responsible for degrading extracellular matrix proteins, thus facilitating cell dissociation and, in turn, lipolysis required for the development of pupa. Contrastingly, in diapause pupae, the suppression of cathepsin L and Mmps activity was suggested to prevent dissociation, maintaining the insect in a state of developmental arrest. The study not only demonstrates the critical role of dynamic changes in lipid metabolism for the termination of pupal diapause but also suggests that the molecular mechanisms regulating the lipid metabolism of diapausing insects are not required to be canonical metabolism genes. However, the study is limited by the lack of proper developmental staging of the pupae, making it unclear whether the suppression of cathepsin L and Mmps activity is directly related to diapause termination or simply disrupts development. Therefore, further studies with precise developmental staging are necessary to firmly implicate fat body dissociation in diapause termination.

4.2 Hormonal Regulation of Lipid Metabolism

It has been long established that diapause is regulated at the hormonal level. As discussed above, lipid metabolism should be dynamically adjusted based on the stage in the diapause program, suggesting hormonal regulation is involved (Yamashita 1996; Lee et al. 1997; Sim and Denlinger 2008; Denlinger et al. 2012; Sinclair and Marshall 2018; Batz et al. 2019; Toprak 2020; Chen et al. 2021; Karp 2021; Ma et al. 2021; Hutfilz 2022). Indeed, accumulating evidence implicates a suite of hormones in the regulation of lipid metabolism throughout diapause.

4.2.1 Adipokinetic Hormone

One of the most well-characterized group of signals in the regulation of lipid metabolism is adipokinetic hormones (AKHs). AKHs are small peptides in insects and essential for mobilizing lipids, glycogen, and amino acids (Gäde and Auerswald 2003; Van der Horst 2003; Gäde and Marco 2022). Initially recognized for their role in fueling insect flight, AKHs also play a significant part in nutrient homeostasis in species such as the fruit fly, *Drosophila melanogaster*, and the domestic silk moth *Bombyx mori*, suggesting a potential role in regulating diapause lipid metabolism (Staubli et al. 2002; Sehadova et al. 2020; Hughson et al. 2021). As discussed previously, diapausing insects are usually in a state of lipolysis to maintain their basal metabolism due to restricted food intake. Hence, AKHs may be necessary to induce the mobilization of lipid stores in the fat body to other tissues during diapause.

Indeed, it appears that AKH may also play a role in diapause. For instance, in the firebug *Pyrrhocoris apterus* (Heteroptera), diapausing adults exhibit a heightened response to AKH, mobilizing significantly more lipid reserves than their non-diapausing counterparts (Socha and Kodrík 1999). This enhanced sensitivity to AKH during diapause points to a potential adaptive mechanism for energy conservation (Hahn and Denlinger 2011).

Investigations into the specific actions of AKH across life stages in non-diapause insects have revealed that the hormone's impact can vary, such as promoting glycogen mobilization in larvae while focusing on lipid mobilization in adults (Van der Horst 2003; Lorenz and Gäde 2009). This versatility is exemplified in the migratory locust, *Locusta migratoria*, which produces multiple AKH peptides, each with slightly different effects on nutrient mobilization (Van der Horst 2003). In *L. migratoria*, AKH-I is the most potent lipid mobilizer, whereas AKH-II and AKH-III exhibit comparatively lower effects on lipid mobilization. Moreover, in the hawk moth *Manduca sexta*, AKH stimulates the phosphorylation of key proteins involved in lipid mobilization, providing a mechanistic link between AKH and lipid mobilization (Lay and Dugail 2009; Arrese and Soulages 2010). These findings suggest that diverse AKH peptides might fine-tune the lipid metabolism to the requirements of different phases of the diapause program (e.g., initiation vs. maintenance phases). However, functional evidence in this regard is currently lacking.

On the other hand, low levels of AKH might be necessary during specific diapause stages to maintain energy balance throughout diapause by allowing gradual consumption of fat reserves. This argument is supported by a recent study where diapausing adult ladybeetle, *C. septempunctata*, injected with AKH experienced a reduction in fat stores, which may compromise the diapause energy balance in the long term (Shen et al. 2024). In contrast, the knockdown of AKH or its receptor mimicked the diapause induction phenotype by increasing fat stores under non-diapause-inducing conditions in the same beetle. Although these findings suggest AKHs could be involved in the regulation of lipid metabolism in diapause, we lack strong evidence for the role of AKH in diapause energy balance.

4.2.2 Insulin

Insulin signaling is considered a potential regulator of diapause due to its well-understood role in energy homeostasis (Sim and Denlinger 2013a). Research has demonstrated that the suppression of the insulin signaling pathway induces diapause in the mosquito *C. pipiens*, characterized by the activation of the transcription factor forkhead box (FOXO) and fat hypertrophy (Sim and Denlinger 2008). The role of insulin in diapause extends to reproductive functions, where the inhibition of the insulin receptor (InR) in female insects not undergoing diapause can induce conditions resembling diapause, such as halted ovary development (Sim and Denlinger 2008). Insulin's effects are predominantly observed during feeding periods, which precede diapause initiation, i.e., diapause preparatory phase. This is exemplified in *C. pipiens* females, which show a preference for sugar over blood meals in the pre-diapause phase, resulting in significantly higher lipid reserves than those seen in non-diapausing insects (Robich and Denlinger 2005). The involvement of insulin-like peptides (ILPs), particularly ILP1, has been identified as a key factor in regulating diapause-associated lipid accumulation in *C. pipiens*. Furthermore, as discussed previously, interactions between insulin signaling and hormonal pathways involving juvenile hormone (JH) and ecdysone have been noted to play a part in the termination of diapause, indicating a complex network of hormonal regulation governing this physiological state (Tatar and Yin 2001; Denlinger 2002; Robich and Denlinger 2005; Baker and Thummel 2007; Sim and Denlinger 2008; Denlinger et al. 2012).

4.2.3 Neuropeptide F and Short Neuropeptide F (sNPF)

Neuropeptide F (NPF) and short neuropeptide F (sNPF) are recognized for their roles in modulating feeding behavior, nutritional balance, and interactions with insulin signaling that are critical for lipid metabolism and potentially influence diapause regulation. *Drosophila* larvae that overexpress NPF exhibit increased feeding duration, which in turn activates insulin signaling pathways facilitating lipid accumulation in preparation for diapause, whereas larvae with mutations in the *NPF* gene demonstrate reduced feeding (Wu et al. 2003). This linkage between feeding behaviors modulated by NPF and the regulation of energy reserves underscores the potential importance of NPF in diapause preparation (Wu et al. 2003; Chung et al. 2017; Fadda et al. 2019). Similarly, sNPF has been associated with feeding patterns, evidenced in studies on the Colorado potato beetle, where sNPF presence correlates with active feeding in adults, contrasting with its absence in diapausing beetles. This absence suggests a potential involvement of sNPF in the transition to diapause, likely affecting the pre-diapause dietary shift toward increased lipid storage (Huybrechts et al. 2004). These findings highlight the complex interplay between neuropeptides, feeding behavior, and the physiological preparation for diapause through lipid management.

4.2.4 Case Study in Culex: Role of JH, Insulin, and FOXO in Regulating Fat Body Hypertrophy in Diapause

Culex pipens is a major vector of the West Nile virus and facultatively enters diapause as nonreproductive adults in winter (Sanburg and Larsen 1973; Sim and Denlinger 2008). This insect has served as a premier model for understanding the key mechanisms underlying the regulation of fat body hypertrophy in diapause-destined individuals (Zhou and Miesfeld 2009; Meuti et al. 2018). Earlier studies on energy metabolism during diapause preparation in *C. pipiens* mosquitoes demonstrated that diapause-destined mosquitoes significantly increased lipid synthesis from glucose during the diapause-preparation phase, compared to non-diapausing, reproductive counterparts (Mitchell and Briegel 1989; Robich and Denlinger 2005; Sim and Denlinger 2009; Sim et al. 2015; Zhang et al. 2019). This lipid accumulation serves as a crucial energy reserve throughout the diapause, supporting survival during this period of inactivity (Robich and Denlinger 2005; King et al. 2020). Interestingly, a portion of these lipids accumulated in the diapause-preparatory phase is later utilized for egg production once diapause concludes, highlighting the essential role of lipid reserves in both survival and reproductive success post-diapause (Hahn and Denlinger 2007; Dworschak et al. 2014; Margus and Lindström 2020). Given that *C. pipiens* exhibits a canonical lipid metabolism dynamic throughout its diapause program, it served as a great model for understanding the mechanism of managing energy reserves to facilitate overwintering and contribute to post-diapause reproductive success (Sim and Denlinger 2009; Zhou and Miesfeld 2009).

Subsequent studies in *C. pipiens* by Sim and Denlinger (2008) revealed that appropriate lipid metabolism, depending on the phase of the diapause program, is regulated via an intricate interplay between key hormones and transcription factors, including JH, Insulin, and forkhead box transcription factors (FOXO). Under non-diapause conditions, the insulin-producing cells in the mosquito secrete insulin, which binds to the insulin receptor (InR). This activates the PI3-kinase/Akt signaling pathway, leading to two main outcomes: the synthesis of juvenile hormone (JH) and the phosphorylation of the transcription factor FOXO. Meanwhile, phosphorylation of FOXO prevents its translocation into the nucleus, thereby inhibiting the expression of genes involved in lipid accumulation. Conversely, under diapause-inducing conditions, i.e., short day lengths, the production of insulin in the mosquito is suppressed. This leads to a lack of FOXO phosphorylation, allowing FOXO to translocate into the nucleus. Once inside the nucleus, FOXO can initiate the transcription of various genes, including those responsible for lipid accumulation, such as fatty acid synthase (FAS). This process results in the accumulation of lipid reserves, mainly in the form of TAG, and the suppression of ovarian development, preparing the mosquito for the diapause phase. Hence, the presence or absence of insulin leads to differential outcomes in lipid metabolism, primarily mediated through the actions of JH and FOXO. This switch is a critical adaptation for diapause, and this core mechanism seems to be largely maintained in many other insects from diverse orders (Sim and Denlinger 2008, 2013a, b; Sim et al. 2015). Interestingly, a more recent study in mice also showed that FOXO1 is involved in the regulation of lipid metabolism in embryos that become dormant after fertilization upon implantation (van der Weijden et al. 2024). This suggests the findings in *C. pipiens* may be widely conserved.

4.3 Emerging Research Directions

4.3.1 Symbiosis

Insect symbionts are involved in many metabolic processes, and recent evidence suggests that they might also be involved in the regulation of lipid metabolism in diapause. In a study exploring how the microbiome affects winter survival in *C. pipiens* mosquitoes, the composition of gut bacteria varies widely among individuals, regardless of whether they are preparing to enter diapause (a dormant state) or remain active (Didion et al. 2021). When mosquitoes entering diapause were raised in a germ-free environment, their bacterial levels dropped dramatically, nearly a thousand times lower than usual. This significant reduction in gut bacteria was followed by a notable decrease in survival rates two weeks into diapause, underscoring the critical role the microbiome plays in helping these mosquitoes survive the winter (Didion et al. 2021). These bacteria-depleted mosquitoes also showed an inability to convert sugars into fats—a crucial process for building the energy reserves needed for overwintering. However, the study found no significant differences in the activity of genes linked to fat metabolism between these mosquitoes and those with a normal microbiome. This suggests that the presence of a diverse set of gut bacteria might be important for processing nutrients and accumulating the fat reserves essential for the mosquitoes' diapause success, rather than any direct genetic influence on lipid storage.

Similarly, a study in cabbage beetle, *C. bowringi* found quantitative differences in the gut microbiota between diapausing and non-diapausing females (Liu et al. 2016). The phyla that strongly correlated with diapause included Firmicutes, Proteobacteria, and Bacteroidetes. Interestingly, most of these phyla were previously linked to human obesity (Shin et al. 2015; Liu et al. 2016), and the core diapause in insects resembles human obesity, due to high levels of fat stores with restricted catabolism (Güney et al. 2021). We speculate that similar mechanisms, where the microbiota regulates the host lipid metabolism, might be present across diverse animals and insects. These mechanisms, in turn, might be ensuring successful energetics during the different phases of diapause. One of the challenges in this line of research is that the differences seem to be largely quantitative rather than qualitative at the phyla or genus levels, requiring complex manipulative study designs to conclude causation.

4.3.2 Posttranscriptional Regulation of Lipid Metabolism

The regulation of lipid metabolism during diapause appears to be complex and is likely governed at multiple levels, including posttranscriptional regulation. The microRNA (miRNA) pathway is one of the key mechanisms involved in posttranscriptional regulation. miRNAs are crucial for the fine-tuning of gene expression and protein synthesis, acting by degrading or blocking the translation of target mRNAs that are partially complementary (Ying et al. 2008; Fernández-Hernando et al. 2011). There is growing evidence that insects exploit this mechanism to control lipid metabolism during diapause. In *C. pipiens,* several miRNAs, such as miR-8-3p, miR-13b-3p, miR-14-3p, miR-275-3p, and miR-305-5p, have reduced expression in diapause-destined females immediately after they emerge as adults and on day 5 (Meuti et al. 2018). This suggests these miRNAs are involved in triggering diapause and switching the developmental pathway from a non-diapause to a diapause state. Specifically, miR-14-3p, miR-277-3p, and miR-305-5p, which are implicated in fat metabolism and insulin signaling, are downregulated in mosquitoes entering diapause, coinciding with rapid fat accumulation. Particularly, miR-277-3p serves a dual function in controlling insulin signaling and the metabolism of branched-chain amino acids, which might influence the regulation of lipid metabolism. Likewise, miR-305-5p and miR-277-3p play significant roles in managing lipid storage by affecting genes involved in the pathways of fatty acid elongation and degradation. These findings underscore the pivotal function of these miRNAs in the enhanced lipid storage seen in diapausing mosquitoes (Meuti et al. 2018). While the research in insects has so far only implicated miRNAs in the regulation of lipid metabolism in diapause, the modulation of lipid metabolism in hibernating mammals by miRNAs and other noncoding RNAs (Lang-Ouellette et al. 2014) suggests further research in insects might be promising.

Another recently explored mechanism for potential posttranscriptional regulation of lipid metabolism in diapausing insects is N6-adenosine methylation (m6A). Methylation of adenosine, catalyzed by m6A methyltransferase complex, is a commonly observed mRNA modification in most eukaryotes, including insects. Depending on the site of the m6A modification on RNA, it can cause diverse alterations in RNA function, including differential splicing and nuclear transport (Liu et al. 2023). In the silkworm, *B. mori*, m6A levels varied in genes critical to lipid metabolism, specifically dihydroxyacetone phosphate acyltransferase (DHAPAT) and phosphatidate phosphatase (PAP) (Y.-H. Chen et al. 2023a). The research highlighted significant epigenetic differences between silkworm strains cultivated from eggs at a standard temperature, which led to diapause egg production (QFHT: diapause egg producer), and those incubated at lower temperatures, resulting in non-diapause eggs (QFLT: non-diapause egg producer). A key experiment within the study involved knocking down DHAPAT in the pupal stage of the QFLT group, which intriguingly resulted in the production of diapausing eggs from what were initially non-diapause destined eggs. Conversely, knocking down PAP led to a change in egg color from light yellow to pink three days post-oviposition, though these eggs eventually hatched normally. Further validation showed that YTHDF3 binds to m6A-modified DHAPAT and PAP mRNAs, enhancing their stability and translation. These findings underscore the role of RNA m6A methylation in the regulation of diapause in silkworms by altering the expression levels of critical genes involved in lipid metabolism and further suggest the presence of a complex interplay between epigenetic modifications and metabolic pathways in insect diapause.

5 Summary

Diapause is a dynamic, alternative developmental trajectory that helps diverse insects to survive winter. Diapause is often accompanied by fat hypertrophy and metabolic suppression to meet the energetic demands of overwintering. Triacylglycerides are the most widely used energy storage during diapause, and while diapause is often accompanied by significant lipid depletion, many insects have a remarkable ability to preserve lipid stores through the winter and use these banked lipids to contribute to spring reproduction. Beyond serving as an energy source, lipids also help conserve water during diapause by waterproofing the cuticle, while modifications to cell membrane lipids (i.e., homeoviscous acclimation) maintain cellular functions at low temperatures. While the molecular regulation of lipid metabolism during diapause has only been investigated in select species, it appears that insulin signaling plays an important role in many species. Specifically, shutdown of insulin signaling during diapause results in FOXO-dependent lipid accumulation. The patterns of lipid accumulation and depletion during diapause are in turn regulated by a cascade of gene expression changes involving lipid synthetic enzymes, lipases, and lipid transport proteins, and some of these changes are regulated by epigenetic modulators. The molecular regulation of lipid metabolism during diapause is an active, rapidly evolving field of research, but it holds promise for elucidating the evolutionary physiology of diapause and providing novel tools to manage beneficial and pest insects.

References

Adedokun TA, Denlinger DL (1985) Metabolic reserves associated with pupal diapause in the flesh fly, *Sarcophaga crassipalpis*. J Insect Physiol 31:229–233. https://doi.org/10.1016/0022-1910(85)90124-6

Ala-Honkola O, Kauranen H, Tyukmaeva V, Boetzl FA, Hoikkala A, Schmitt T (2020) Diapause affects cuticular hydrocarbon composition and mating behavior of both sexes in *Drosophila montana*. Insect Sci 27:304–316. https://doi.org/10.1111/1744-7917.12639

Arrese EL, Soulages JL (2010) Insect fat body: Energy, metabolism, and regulation. Annu Rev Entomol 55: 207–225. https://doi.org/10.1146/annurev-ento-112408-085356

Arrese EL, Mirza S, Rivera L, Howard AD, Chetty PS, Soulages JL (2008a) Expression of lipid storage droplet protein-1 may define the role of AKH as a lipid mobilizing hormone in *Manduca sexta*. Insect Biochem Mol Biol 38:993–1000. https://doi.org/10.1016/j.ibmb.2008.08.004

Arrese EL, Rivera L, Hamada M, Soulages JL (2008b) Purification and Characterization of recombinant lipid storage protein-2 from *Drosophila melanogaster*. Protein Pept Lett 15:1027–1032

Arrese EL, Saudale FZ, Soulages JL (2014) Lipid droplets as signaling platforms linking metabolic and cellular functions. Lipid Insights 7:7–16. https://doi.org/10.4137/LPI.S11128

Baker KD, Thummel CS (2007) Diabetic larvae and obese flies—emerging studies of metabolism in *Drosophila*. Cell Metab 6:257–266. https://doi.org/10.1016/j.cmet.2007.09.002

Batz ZA, Armbruster PA (2018) Diapause-associated changes in the lipid and metabolite profiles of the Asian tiger mosquito, *Aedes albopictus*. J Exp Biol 221:jeb189480. https://doi.org/10.1242/jeb.189480

Batz ZA, Brent CS, Armbruster PA (2019) Juvenile hormone III but not 20-hydroxyecdysone regulates the embryonic diapause of *Aedes albopictus*. Front Physiol 10. https://doi.org/10.3389/fphys.2019.01352

Beller M, Bulankina AV, Hsiao HH, Urlaub H, Jäckle H, Kühnlein RP (2010) Perilipin-dependent control oflipid droplet structure and fat storage in Drosophila. Cell Metabol 12(5), 521–532. https://doi.org/10.1016/j.cmet.2010.10.001

Benoit JB, Denlinger DL (2007) Suppression of water loss during adult diapause in the northern house mosquito, *Culex pipiens*. J Exp Biol 210:217–226. https://doi.org/10.1242/jeb.02630

Blomquist GJ, Bagnères A-G (2010) Insect hydrocarbons: biology, biochemistry, and chemical ecology. Cambridge University Press

Bi J, Xiang Y, Chen H, Liu Z, Grönke S, Kühnlein RP, Huang X (2012) Opposite and redundant roles of thetwo Drosophila perilipins in lipid mobilization. J Cell Sci 125(15), 3568–3577. https://doi.org/10.1242/jcs.101329

Bradshaw WE, Quebodeaux MC, Holzapfel CM (2003) The contribution of an hourglass timer to the evolution of photoperiodic response in the pitcher-plant mosquito, *Wyeomyia smithii*. Evolution 57:2342–2349

Cedden D, Güney G, Toprak U (2024) The integral role of de novo lipogenesis in the preparation for seasonal dormancy. Proc Natl Acad Sci 121:e2406194121. https://doi.org/10.1073/pnas.2406194121

Chen H, Wu G, Zhou H, Dai X, Steeghs NWF, Dong X, Zheng L, Zhai Y (2021) Hormonal regulation of reproductive diapause that occurs in the year-round mass rearing of *Bombus terrestris* Queens. J Proteome Res 20:2240–2250. https://doi.org/10.1021/acs.jproteome.0c00776

Chen Y-H, Jiang T, Yasen A, Fan B-Y, Zhu J, Wang M-X, Qian P, Shen X-J (2023a) m6A-dependent mevalonate kinase in juvenile hormone synthesis pathway regulates the diapause process of bivoltine silkworm

(*Bombyx mori*). Mol Biol Rep 50:5295–5306. https://doi.org/10.1007/s11033-023-08489-z

Chen Z-Z, Wang X, Kong X, Zhao Y-M, Xu M-H, Gao Y-Q, Huang H-Y, Liu F-H, Wang S, Xu Y-Y, Kang Z-W (2023b) Quantitative transcriptomic and proteomic analyses reveal the potential maintenance mechanism of female adult reproductive diapause in *Chrysoperla nipponensis*. Pest Manag Sci 79:1897–1911. https://doi.org/10.1002/ps.7375

Chino H, Downer RG (1982) Insect hemolymph lipophorin: a mechanism of lipid transport in insects. Adv Biophys 15:67–92. https://doi.org/10.1016/0065-227x(82)90005-3

Chung BY, Ro J, Hutter SA, Miller KM, Guduguntla LS, Kondo S, Pletcher SD (2017) *Drosophila* neuropeptide f signaling independently regulates feeding and sleep-wake behavior. Cell Rep 19:2441–2450. https://doi.org/10.1016/j.celrep.2017.05.085

Ciancio JJ, Turnbull KF, Gariepy TD, Sinclair BJ (2021) Cold tolerance, water balance, energetics, gas exchange, and diapause in overwintering brown marmorated stink bugs. J Insect Physiol 128:104171. https://doi.org/10.1016/j.jinsphys.2020.104171

Denlinger DL (2002) Regulation of diapause. Annu Rev Entomol 47:93–122. https://doi.org/10.1146/annurev.ento.47.091201.145137

Denlinger DL (2022) Insect diapause. Cambridge University Press. https://doi.org/10.1017/9781108609364

Denlinger DL, Armbruster PA (2014) Mosquito diapause. Annu Rev Entomol 59:73–93. https://doi.org/10.1146/annurev-ento-011613-162023

Denlinger DL, Willis JH, Fraenkel G (1972) Rates and cycles of oxygen consumption during pupal diapause in Sarcophaga flesh flies. J Insect Physiol 18:871–882. https://doi.org/10.1016/0022-1910(72)90026-1

Denlinger DL, Yocum GD, Rinehart JP (2005) 3.12 - Hormonal control of diapause. In: Gilbert LI (ed) Comprehensive molecular insect science. Elsevier, pp 615–650. https://doi.org/10.1016/B0-44-451924-6/00043-0

Denlinger DL, Yocum GD, Rinehart JP (2012) 10 - Hormonal control of diapause. In: Gilbert LI (ed) Insect endocrinology. Academic, San Diego, pp 430–463. https://doi.org/10.1016/B978-0-12-384749-2.10010-X

Didion EM, Sabree ZL, Kenyon L, Nine G, Hagan RW, Osman S, Benoit JB (2021) Microbiome reduction prevents lipid accumulation during early diapause in the northern house mosquito, *Culex pipiens pipiens*. J Insect Physiol 134:104295. https://doi.org/10.1016/j.jinsphys.2021.104295

Dworschak K, Gruppe A, Schopf R (2014) Survivability and post-diapause fitness in a scolytid beetle as a function of overwintering developmental stage and the implications for population dynamics. Ecol Entomol 39:519–526. https://doi.org/10.1111/een.12127

Enriquez T, Visser B (2023) The importance of fat accumulation and reserves for insect overwintering. Curr Opin Insect Sci 60:101118. https://doi.org/10.1016/j.cois.2023.101118

Fadda M, Hasakiogullari I, Temmerman L, Beets I, Zels S, Schoofs L (2019) Regulation of feeding and metabolism by neuropeptide f and short neuropeptide f in invertebrates. Front Endocrinol 10:64. https://doi.org/10.3389/fendo.2019.00064

Febbraio M, Abumrad NA, Hajjar DP, Sharma K, Cheng W, Pearce SFA, Silverstein RL (1999) A null mutation in murine cd36 reveals an important role in fatty acid and lipoprotein metabolism. J Biol Chem 274:19055–19062. https://doi.org/10.1074/jbc.274.27.19055

Fernández-Hernando C, Suárez Y, Rayner KJ, Moore KJ (2011) MicroRNAs in lipid metabolism. Curr Opin Lipidol 22:86–92. https://doi.org/10.1097/MOL.0b013e3283428d9d

Gäde G, Auerswald L (2003) Mode of action of neuropeptides from the adipokinetic hormone family. Gen Comp Endocrinol 132:10–20. https://doi.org/10.1016/S0016-6480(03)00159-X

Gäde G, Marco HG (2022) The adipokinetic peptides of hemiptera: structure, function, and evolutionary trends. Front Insect Sci 2:891615. https://doi.org/10.3389/finsc.2022.891615

Gibbs AG (1998) Water-proofing properties of cuticular lipids. Am Zool 38:471–482. https://doi.org/10.1093/icb/38.3.471

Gibbs A, Mousseau TA, Crowe JH (1991) Genetic and acclimatory variation in biophysical properties of insect cuticle lipids. Proc Natl Acad Sci 88:7257–7260. https://doi.org/10.1073/pnas.88.16.7257

Gondim KC, Atella GC, Pontes EG, Majerowicz D (2018) Lipid metabolism in insect disease vectors. Insect Biochem Mol Biol 101:108–123. https://doi.org/10.1016/j.ibmb.2018.08.005

Grönke S, Beller M, Fellert S, Ramakrishnan H, Jäckle H, Kühnlein RP (2003) Control of Fat Storage by a *Drosophila* PAT Domain Protein. Curr Biol 13:603–606. https://doi.org/10.1016/S0960-9822(03)00175-1

Grönke S, Mildner A, Fellert S, Tennagels N, Petry S, Müller G, Jäckle H, Kühnlein RP (2005) Brummer lipase is an evolutionary conserved fat storage regulator in *Drosophila*. Cell Metab 1:323–330. https://doi.org/10.1016/j.cmet.2005.04.003

Grönke S, Müller G, Hirsch J, Fellert S, Andreou A, Haase T, Jäckle H, Kühnlein RP (2007) Dual lipolytic control of body fat storage and mobilization in *Drosophila*. PLoS Biol 5:e137. https://doi.org/10.1371/journal.pbio.0050137

Güney G, Toprak U, Hegedus DD, Bayram Ş, Coutu C, Bekkaoui D, Baldwin D, Heckel DG, Hänniger S, Cedden D, Mutlu DA, Suludere Z (2021) A look into Colorado potato beetle lipid metabolism through the lens of lipid storage droplet proteins. Insect Biochem Mol Biol 133:103473. https://doi.org/10.1016/j.ibmb.2020.103473

Hahn DA, Denlinger DL (2007) Meeting the energetic demands of insect diapause: nutrient storage and utilization. J Insect Physiol 53:760–773. https://doi.org/10.1016/j.jinsphys.2007.03.018

Hahn DA, Denlinger DL (2011) Energetics of insect diapause. Annu Rev Entomol 56:103–121. https://doi.org/10.1146/annurev-ento-112408-085436

Hazel JR (1995) Thermal adaptation in biological membranes: is homeoviscous adaptation the explanation? Annu Rev Physiol 57:19–42. https://doi.org/10.1146/annurev.ph.57.030195.000315

Holze H, Schrader L, Buellesbach J (2021) Advances in deciphering the genetic basis of insect cuticular hydrocarbon biosynthesis and variation. Heredity 126:219–234. https://doi.org/10.1038/s41437-020-00380-y

Hughson BN, Shimell M, O'Connor MB (2021) AKH signaling in *D. melanogaster* alters larval development in a nutrient-dependent manner that influences adult metabolism. Front Physiol 12:619219. https://doi.org/10.3389/fphys.2021.619219

Hussein AM, Wang Y, Mathieu J, Margaretha L, Song C, Jones DC, Cavanaugh C, Miklas JW, Mahen E, Showalter MR, Ruzzo WL, Fiehn O, Ware CB, Blau CA, Ruohola-Baker H (2020) Metabolic control over mTOR-dependent diapause-like state. Dev Cell 52:236–250.e7. https://doi.org/10.1016/j.devcel.2019.12.018

Hutfilz C (2022) Endocrine regulation of lifespan in insect diapause. Front Physiol 13:825057. https://doi.org/10.3389/fphys.2022.825057

Huybrechts J, De Loof A, Schoofs L (2004) Diapausing Colorado potato beetles are devoid of short neuropeptide F I and II. Biochem Biophys Res Commun 317:909–916. https://doi.org/10.1016/j.bbrc.2004.03.136

Jia Q, Li S (2023) Mmp-induced fat body cell dissociation promotes pupal development and moderately averts pupal diapause by activating lipid metabolism. Proc Natl Acad Sci 120:e2215214120. https://doi.org/10.1073/pnas.2215214120

Joel A-C, Schmitt D, Baumgart L, Menzel F (2022) Insect cuticular hydrocarbon composition influences their interaction with spider capture threads. J Exp Biol 225:jeb242514. https://doi.org/10.1242/jeb.242514

Jurenka RA, Holland D, Krafsur ES (1998) Hydrocarbon profiles of diapausing and reproductive adult face flies (*Musca autumnalis*). Arch Insect Biochem Physiol 37:206–214. https://doi.org/10.1002/(SICI)1520-6327(1998)37:3<206::AID-ARCH3>3.0.CO;2-Q

Karp X (2021) Hormonal regulation of diapause and development in nematodes, insects, and fishes. Front Ecol Evol 9:735924. https://doi.org/10.3389/fevo.2021.735924

Kárpáti Z, Deutsch F, Kiss B, Schmitt T (2023) Seasonal changes in photoperiod and temperature lead to changes in cuticular hydrocarbon profiles and affect mating success in *Drosophila suzukii*. Sci Rep 13:5649. https://doi.org/10.1038/s41598-023-32652-y

Kayukawa T, Chen B, Hoshizaki S, Ishikawa Y (2007) Upregulation of a desaturase is associated with the enhancement of cold hardiness in the onion maggot, *Delia antiqua*. Insect Biochem Mol Biol 37:1160–1167. https://doi.org/10.1016/j.ibmb.2007.07.007

Khabir M, Izadi H, Mahdian K (2024) Impact of diapause on the cuticular hydrocarbons and physiological status of *Hippodamia variegata*. J Insect Sci 24:11. https://doi.org/10.1093/jisesa/ieae005

King B, Li S, Liu C, Kim SJ, Sim C (2020) Suppression of glycogen synthase expression reduces glycogen and lipid storage during mosquito overwintering diapause. J Insect Physiol 120:103971. https://doi.org/10.1016/j.jinsphys.2019.103971

Koštál V (2006) Eco-physiological phases of insect diapause. J Insect Physiol 52:113–127. https://doi.org/10.1016/j.jinsphys.2005.09.008

Koštál V, Štětina T, Poupardin R, Korbelová J, Bruce AW (2017) Conceptual framework of the eco-physiological phases of insect diapause development justified by transcriptomics profiling. Proc Natl Acad Sci USA 114:8532–8537. https://doi.org/10.1073/pnas.1707281114

Kühnlein RP (2012) Lipid droplet-based storage fat metabolism in *Drosophila*. J Lipid Res 53:1430–1436. https://doi.org/10.1194/jlr.R024299

Lang-Ouellette D, Richard TG, Morin P (2014) Mammalian hibernation and regulation of lipid metabolism: a focus on non-coding RNAs. Biochemistry (Moscow) 79:1161–1171. https://doi.org/10.1134/S0006297914110030

Lay SL, Dugail I (2009) Connecting lipid droplet biology and the metabolic syndrome. Prog Lipid Res 48:191–195. https://doi.org/10.1016/j.plipres.2009.03.001

Lee K-Y, Valaitis AP, Denlinger DL (1997) Further evidence that diapause in the gypsy moth, *Lymantria dispar*, is regulated by ecdysteroids: a comparison of diapause and non-diapause strains. J Insect Physiol 43:897–903. https://doi.org/10.1016/S0022-1910(97)00054-1

Lehmann P, Westberg M, Tang P, Lindström L, Käkelä R (2020) The Diapause lipidomes of three closely related beetle species reveal mechanisms for tolerating energetic and cold stress in high-latitude seasonal environments. Front Physiol 11:576617. https://doi.org/10.3389/fphys.2020.576617

Li D-T, Chen X, Wang X-Q, Moussian B, Zhang C-X (2019) The fatty acid elongase gene family in the brown planthopper, *Nilaparvata lugens*. Insect Biochem Mol Biol 108:32–43. https://doi.org/10.1016/j.ibmb.2019.03.005

Li Y, Parven N, Akimoto S (2024) Interspecific recognition based on cuticular hydrocarbons mediates reproduction control in aphids. Sci Rep 14:4079. https://doi.org/10.1038/s41598-024-54019-7

Liu W, Li Y, Guo S, Yin H, Lei C-L, Wang X-P (2016) Association between gut microbiota and diapause preparation in the cabbage beetle: a new perspective for studying insect diapause. Sci Rep 6:38900. https://doi.org/10.1038/srep38900

Liu Z, Gao L, Cheng L, Lv G, Sun B, Wang G, Tang Q (2023) The roles of N6-methyladenosine and its target regulatory noncoding RNAs in tumors: classification, mechanisms, and potential therapeutic implications. Exp Mol Med 55:487–501. https://doi.org/10.1038/s12276-023-00944-y

Liu S, Gao Y, Shi R, Huang H, Xu Y, Chen Z (2024) Transcriptomics provide insights into the photoperiodic regulation of reproductive diapause in the green lacewing, *Chrysoperla nipponensis* (Okamoto) (Neuroptera: Chrysopidae). Insects 15:136. https://doi.org/10.3390/insects15020136

Lorenz MW, Gäde G (2009) Hormonal regulation of energy metabolism in insects as a driving force for performance. Integr Comp Biol 49:380–392. https://doi.org/10.1093/icb/icp019

Ma H-Y, Li Y-Y, Li L, Tan Y, Pang B-P (2021) Regulation of juvenile hormone on summer diapause of *Geleruca daurica* and its pathway analysis. Insects 12:237. https://doi.org/10.3390/insects12030237

Margus A, Lindström L (2020) Prolonged diapause has sex-specific fertility and fitness costs. Evol Ecol 34:41–57. https://doi.org/10.1007/s10682-019-10024-1

Mensch J, Di Battista C, De Majo MS, Campos RE, Fischer S (2021) Increased size and energy reserves in diapausing eggs of temperate *Aedes aegypti* populations. J Insect Physiol 131:104232. https://doi.org/10.1016/j.jinsphys.2021.104232

Meuti ME, Bautista-Jimenez R, Reynolds JA (2018) Evidence that microRNAs are part of the molecular toolkit regulating adult reproductive diapause in the mosquito, *Culex pipiens*. PLoS One 13:e0203015. https://doi.org/10.1371/journal.pone.0203015

Michaud MR, Denlinger DL (2006) Oleic acid is elevated in cell membranes during rapid cold-hardening and pupal diapause in the flesh fly, *Sarcophaga crassipalpis*. J Insect Physiol 52:1073–1082. https://doi.org/10.1016/j.jinsphys.2006.07.005

Michaud JP, Qureshi JA (2005) Induction of reproductive diapause in *Hippodamia convergens* (Coleoptera: Coccinellidae) hinges on prey quality and availability. Eur J Entomol 102:483–487

Michelutti KB, Soares ERP, Sguarizi-Antonio D, Piva RC, Súarez YR, Cardoso CAL, Antonialli-Junior WF (2018) Influence of temperature on survival and cuticular chemical profile of social wasps. J Therm Biol 71:221–231. https://doi.org/10.1016/j.jtherbio.2017.11.019

Mitchell CJ, Briegel H (1989) Inability of diapausing *Culex pipiens* (Diptera: Culicidae) to use blood for producing lipid reserves for overwinter survival. J Med Entomol 26:318–326. https://doi.org/10.1093/jmedent/26.4.318

Miura S, Gan J-W, Brzostowski J, Parisi MJ, Schultz CJ, Londos C, Oliver B, Kimmel AR (2002) Functional Conservation for Lipid Storage Droplet Association among Perilipin, ADRP, and TIP47 (PAT)-related Proteins in Mammals, *Drosophila*, and Dictyostelium. J Biol Chem 277:32253–32257. https://doi.org/10.1074/jbc.M204410200

Nadeau EAW, Lecheta MC, Obrycki JJ, Teets NM (2022) Transcriptional regulation of reproductive diapause in the convergent lady beetle, *Hippodamia convergens*. Insects 13:343. https://doi.org/10.3390/insects13040343

Nozawa Y (2011) Adaptive regulation of membrane lipids and fluidity during thermal acclimation in Tetrahymena. Proc Jpn Acad Ser B Phys Biol Sci 87:450–462. https://doi.org/10.2183/pjab.87.450

Palazzo MC, Setzer WN (2009) Monoterpene hydrocarbons may serve as antipredation defensive compounds in *Boisea Trivittata*, the boxelder bug. Nat Prod Commun 4:1934578X0900400401. https://doi.org/10.1177/1934578X0900400401

Ragland GJ, Denlinger DL, Hahn DA (2010) Mechanisms of suspended animation are revealed by transcript profiling of diapausein the flesh fly. Proc Natl Acad Sci 107:14909–14914. https://doi.org/10.1073/pnas.1007075107

Rajpurohit S, Vrkoslav V, Hanus R, Gibbs AG, Cvačka J, Schmidt PS (2020) Post-eclosion temperature effects on insect cuticular hydrocarbon profiles. Ecol Evol 11:352–364. https://doi.org/10.1002/ece3.7050

Reynolds JA, Poelchau MF, Rahman Z, Armbruster PA, Denlinger DL (2012) Transcript profiling reveals mechanisms for lipid conservation during diapause in the mosquito, *Aedes albopictus*. J Insect Physiol 58:966–973. https://doi.org/10.1016/j.jinsphys.2012.04.013

Roberts KT, Rank NE, Dahlhoff EP, Stillman JH, Williams CM (2021) Snow modulates winter energy use and cold exposure across an elevation gradient in a montane ectotherm. Glob Change Biol 27:6103–6116. https://doi.org/10.1111/gcb.15912

Robich RM, Denlinger DL (2005) Diapause in the mosquito *Culex pipiens* evokes a metabolic switch from blood feeding to sugar gluttony. Proc Natl Acad Sci 102:15912–15917. https://doi.org/10.1073/pnas.0507958102

Rozsypal J, Moos M, Rudolf I, Košťál V (2021) Do energy reserves and cold hardiness limit winter survival of *Culex pipiens*? Comp Biochem Physiol A Mol Integr Physiol 255:110912. https://doi.org/10.1016/j.cbpa.2021.110912

Sanburg LL, Larsen JR (1973) Effect of photoperiod and temperature on ovarian development in *Culex pipiens pipiens*. J Insect Physiol 19:1173–1190. https://doi.org/10.1016/0022-1910(73)90202-3

Sehadova H, Takasu Y, Zaloudikova A, Lin Y-H, Sauman I, Sezutsu H, Rouhova L, Kodrik D, Zurovec M (2020) Functional analysis of adipokinetic hormone signaling in *Bombyx mori*. Cells 9:2667. https://doi.org/10.3390/cells9122667

Shen Z, Liu Z, Chen J, Li Y, Mao J, Wang M, Zhang L (2024) Functional analysis of adipokinetic hormone signaling in reproductive diapause *of Coccinella septempuctata*. Pest Manag Sci 80:3665–3674. https://doi.org/10.1002/ps.8070

Shin N-R, Whon TW, Bae J-W (2015) Proteobacteria: microbial signature of dysbiosis in gut microbiota. Trends Biotechnol 33:496–503. https://doi.org/10.1016/j.tibtech.2015.06.011

Short CA, Hahn DA (2023) Fat enough for the winter? Does nutritional status affect diapause? J Insect Physiol 145:104488. https://doi.org/10.1016/j.jinsphys.2023.104488

Sim C, Denlinger DL (2008) Insulin signaling and FOXO regulate the overwintering diapause of the mosquito *Culex pipiens*. Proc Natl Acad Sci 105:6777–6781. https://doi.org/10.1073/pnas.0802067105

Sim C, Denlinger DL (2009) Transcription profiling and regulation of fat metabolism genes in diapausing adults of the mosquito *Culex pipiens*. Physiol Genomics 39: 202–209. https://doi.org/10.1152/physiolgenomics.00095.2009

Sim C, Denlinger DL (2013a) Insulin signaling and the regulation of insect diapause. Front Physiol 4:54183. https://doi.org/10.3389/fphys.2013.00189

Sim C, Denlinger DL (2013b) Juvenile hormone III suppresses forkhead of transcription factor in the fat body and reduces fat accumulation in the diapausing mosquito, *Culex pipiens*. Insect Mol Biol 22:1–11. https://doi.org/10.1111/j.1365-2583.2012.01166.x

Sim C, Kang DS, Kim S, Bai X, Denlinger DL (2015) Identification of FOXO targets that generate diverse features of the diapause phenotype in the mosquito *Culex pipiens*. Proc Natl Acad Sci 112:3811–3816. https://doi.org/10.1073/pnas.1502751112

Sinclair BJ, Marshall KE (2018) The many roles of fats in overwintering insects. J Exp Biol 221:jeb161836. https://doi.org/10.1242/jeb.161836

Sinensky M (1974) Homeoviscous adaptation—a homeostatic process that regulates the viscosity of membrane lipids in *Escherichia coli*. Proc Natl Acad Sci 71:522–525. https://doi.org/10.1073/pnas.71.2.522

Socha R, Kodrík D (1999) Differences in adipokinetic response of *Pyrrhocoris apterus* (Heteroptera) in relation to wing dimorphism and diapause. Physiol Entomol 24:278–284. https://doi.org/10.1046/j.1365-3032.1999.00143.x

Staubli F, Jørgensen TJD, Cazzamali G, Williamson M, Lenz C, Søndergaard L, Roepstorff P, Grimmelikhuijzen CJP (2002) Molecular identification of the insect adipokinetic hormone receptors. Proc Natl Acad Sci 99:3446–3451. https://doi.org/10.1073/pnas.052556499

Storey KB, Storey JM (2012) Aestivation: signaling and hypometabolism. J Exp Biol 215:1425–1433. https://doi.org/10.1242/jeb.054403

Suito T, Nagao K, Takeuchi K, Juni N, Hara Y, Umeda M (2020) Functional expression of Δ12 fatty acid desaturase modulates thermoregulatory behavior in *Drosophila*. Sci Rep 10:11798. https://doi.org/10.1038/s41598-020-68601-2

Tan Q-Q, Liu W, Zhu F, Lei C-L, Hahn DA, Wang X-P (2017a) Describing the diapause-preparatory proteome of the beetle *Colaphellus bowringi* and identifying candidates affecting lipid accumulation using isobaric tags for mass spectrometry-based proteome quantification (iTRAQ). Front Physiol 8:251. https://doi.org/10.3389/fphys.2017.00251

Tan Q-Q, Liu W, Zhu F, Lei C-L, Wang X-P (2017b) Fatty acid synthase 2 contributes to diapause preparation in a beetle by regulating lipid accumulation and stress tolerance genes expression. Sci Rep 7:40509. https://doi.org/10.1038/srep40509

Tarrant AM, Baumgartner MF, Lysiak NSJ, Altin D, Størseth TR, Hansen BH (2016) Transcriptional profiling of metabolic transitions during development and diapause preparation in the copepod *Calanus finmarchicus*. Integr Comp Biol 56:1157–1169. https://doi.org/10.1093/icb/icw060

Tatar M, Yin C-M (2001) Slow aging during insect reproductive diapause: why butterflies, grasshoppers and flies are like worms. Exp Gerontol 36:723–738. https://doi.org/10.1016/S0531-5565(00)00238-2

Tauber MJ, Tauber CA (1976) Insect seasonality: diapause maintenance, termination, and postdiapause development. Annu Rev Entomol 21:81–107

Teets NM, Marshall KE, Reynolds JA (2023) Molecular mechanisms of winter survival. Annu Rev Entomol 68: 319–339. https://doi.org/10.1146/annurev-ento-120120-095233

Toprak U (2020) The role of peptide hormones in insect lipid metabolism. Front Physiol 11:434. https://doi.org/10.3389/fphys.2020.00434

Toprak U, Guz N, Gurkan MO, Hegedus DD (2014) Identification and coordinated expression of perilipin genes in the biological cycle of sunn pest, *Eurygaster maura* (Hemiptera: Scutelleridae): implications for lipolysis and lipogenesis. Comp Biochem Physiol B Biochem Mol Biol 171:1–11. https://doi.org/10.1016/j.cbpb.2014.02.001

Toprak U, Hegedus D, Doğan C, Güney G (2020) A journey into the world of insect lipid metabolism. Arch Insect Biochem Physiol 104:e21682. https://doi.org/10.1002/arch.21682

Van der Horst DJ (2003) Insect adipokinetic hormones: release and integration of flight energy metabolism. Comp Biochem Physiol B Biochem Mol Biol 136: 217–226. https://doi.org/10.1016/S1096-4959(03)00151-9

van der Horst DJ, van Doorn JM, Passier PC, Vork MM, Glatz JF (1993) Role of fatty acid-binding protein in lipid metabolism of insect flight muscle. Mol Cell Biochem 123:145–152. https://doi.org/10.1007/BF01076486

van der Weijden VA, Stötzel M, Iyer DP, Fauler B, Gralinska E, Shahraz M, Meierhofer D, Vingron M, Rulands S, Alexandrov T, Mielke T, Bulut-Karslioglu A (2024) FOXO1-mediated lipid metabolism maintains mammalian embryos in dormancy. Nat Cell Biol 26:181–193. https://doi.org/10.1038/s41556-023-01325-3

Vukašinović EL, Pond DW, Worland MR, Kojić D, Purać J, Popović ŽD, Grubor-Lajšić G (2015)

Diapause induces remodeling of the fatty acid composition of membrane and storage lipids in overwintering larvae of *Ostrinia nubilalis*, Hubn. (Lepidoptera: Crambidae). Comp Biochem Physiol B Biochem Mol Biol 184:36–43. https://doi.org/10.1016/j.cbpb.2015.02.003

Wang Z, Receveur JP, Pu J, Cong H, Richards C, Liang M, Chung H (2022) Desiccation resistance differences in *Drosophila* species can be largely explained by variations in cuticular hydrocarbons. eLife 11: e80859. https://doi.org/10.7554/eLife.80859

Wat LW, Chao C, Bartlett R, Buchanan JL, Millington JW, Chih HJ, Chowdhury ZS, Biswas P, Huang V, Shin LJ, Wang LC, Gauthier M-PL, Barone MC, Montooth KL, Welte MA, Rideout EJ (2020) A role for triglyceride lipase brummer in the regulation of sex differences in *Drosophila* fat storage and breakdown. PLoS Biol 18:e3000595. https://doi.org/10.1371/journal.pbio.3000595

Williams CM, Marshall KE, MacMillan HA, Dzurisin JDK, Hellmann JJ, Sinclair BJ (2012) Thermal variability increases the impact of autumnal warming and drives metabolic depression in an overwintering butterfly. PLoS One 7:e34470. https://doi.org/10.1371/journal.pone.0034470

Williams CM, Chick WD, Sinclair BJ (2015) A cross–seasonal perspective on local adaptation: metabolic plasticity mediates responses to winter in a thermal–generalist moth. Funct Ecol 29:549–561. https://doi.org/10.1111/1365-2435.12360

Wipking W, Viebahn M, Neumann D (1995) Oxygen consumption, water, lipid and glycogen content of early and late diapause and non-diapause larvae of the burnet moth *Zygaena trifolii*. J Insect Physiol 41: 47–56. https://doi.org/10.1016/0022-1910(94)00079-V

Wu Q, Wen T, Lee G, Park JH, Cai HN, Shen P (2003) Developmental control of foraging and social behavior by the *Drosophila* neuropeptide y-like system. Neuron 39:147–161. https://doi.org/10.1016/S0896-6273(03)00396-9

Xiang M, Zhang H-Z, Jing X-Y, Wang M-Q, Mao J-J, Li Y-Y, Zang L-S, Zhang L-S (2021) Sequencing, expression, and functional analyses of four genes related to fatty acid biosynthesis during the diapause process in the female ladybird, *Coccinella septempunctata* L. Front Physiol 12:706032. https://doi.org/10.3389/fphys.2021.706032

Yamashita O (1996) Diapause hormone of the silkworm, *Bombyx mori*: structure, gene expression and function. J Insect Physiol 42:669–679. https://doi.org/10.1016/0022-1910(96)00003-0

Yang Y, Zhao X, Niu N, Zhao Y, Liu W, Moussian B, Zhang J (2020) Two fatty acid synthase genes from the integument contribute to cuticular hydrocarbon biosynthesis and cuticle permeability in *Locusta migratoria*. Insect Mol Biol 29:555–568. https://doi.org/10.1111/imb.12665

Ying S-Y, Chang DC, Lin S-L (2008) The MicroRNA (miRNA): overview of the RNA genes that modulate gene function. Mol Biotechnol 38:257. https://doi.org/10.1007/s12033-007-9013-8

Yoder JA, Denlinger DL, Dennis MW, Kolattukudy PE (1992) Enhancement of diapausing flesh fly puparia with additional hydrocarbons and evidence for alkane biosynthesis by a decarbonylation mechanism. Insect Biochem Mol Biol 22:237–243. https://doi.org/10.1016/0965-1748(92)90060-R

Yoder JA, Blomquist GJ, Denlinger DL (1995) Hydrocarbon profiles from puparia of diapausing and nondiapausing flesh flies (*Sarcophaga crassipalpis*) reflect quantitative rather than qualitative differences. Arch Insect Biochem Physiol 28:377–385. https://doi.org/10.1002/arch.940280407

Zhang C, Wei D, Shi G, Huang X, Cheng P, Liu G, Guo X, Liu L, Wang H, Miao F, Gong M (2019) Understanding the regulation of overwintering diapause molecular mechanisms in *Culex pipiens* pallens through comparative proteomics. Sci Rep 9:6485. https://doi.org/10.1038/s41598-019-42961-w

Zhao X, Yang Y, Niu N, Zhao Y, Liu W, Ma E, Moussian B, Zhang J (2020) The fatty acid elongase gene *LmELO7* is required for hydrocarbon biosynthesis and cuticle permeability in the migratory locust, *Locusta migratoria*. J Insect Physiol 123:104052. https://doi.org/10.1016/j.jinsphys.2020.104052

Zhou G, Miesfeld RL (2009) Energy metabolism during diapause in *Culex pipiens* mosquitoes. J Insect Physiol 55:40–46. https://doi.org/10.1016/j.jinsphys.2008.10.002

Adv Exp Med Biol (2026) 1494: 313–344
https://doi.org/10.1007/5584_2024_809

Published online: 15 June 2024

Lipids in Insect Reproduction: Where, How, and Why

Jimena Leyria, Leonardo L. Fruttero, and Lilián E. Canavoso

Abstract

Modern insects have inhabited the earth for hundreds of millions of years, and part of their successful adaptation lies in their many reproductive strategies. Insect reproduction is linked to a high metabolic rate that provides viable eggs in a relatively short time. In this context, an accurate interplay between the endocrine system and the nutrients synthetized and metabolized is essential to produce healthy offspring. Lipids guarantee the metabolic energy needed for egg formation and represent the main energy source consumed during embryogenesis. Lipids availability is tightly regulated by a complex network of endocrine signals primarily controlled by the central nervous system (CNS) and associated endocrine glands, the *corpora allata* (CA) and *corpora cardiaca* (CC). This endocrine axis provides hormones and neuropeptides that significatively affect tissues closely involved in successful reproduction: the fat body, which is the metabolic center supplying the lipid resources and energy demanded in egg formation, and the ovaries, where the developing oocytes recruit lipids that will be used for optimal embryogenesis. The post-genomic era and the availability of modern experimental approaches have advanced our understanding of many processes involved in lipid homeostasis; therefore, it is crucial to integrate the findings of recent years into the knowledge already acquired in the last decades. The present chapter is devoted to reviewing major recent contributions made in elucidating the impact of the CNS/CA/CC-fat body-ovary axis on lipid metabolism in the context of insect reproduction, highlighting areas of fruitful research.

Keywords

Corpora allata · *Corpora cardiaca* · Ecdysteroids · Embryogenesis · Fat body · Insect reproduction · Juvenile hormones · Lipid metabolism · Peptide hormones · Vitellogenesis

1 Introduction

A successful reproductive event, including the optimal development of embryogenesis, requires high energy expenditure. The energy is obtained via the transformation of lipids into other energy forms, which can generate up to 9 kcal/g (Arrese and Soulages 2010). Thus, lipid metabolism

J. Leyria, L. L. Fruttero, and L. E. Canavoso (✉)
Departamento de Bioquímica Clínica, Facultad de Ciencias Químicas, Universidad Nacional de Córdoba, Córdoba, CP 5000, Argentina

Centro de Investigaciones en Bioquímica Clínica e Inmunología (CIBICI), Consejo Nacional de Investigaciones Científicas y Técnicas (CONICET), Córdoba, Argentina
e-mail: jimena.leyria@unc.edu.ar; lfruttero@unc.edu.ar; lilian.canavoso@unc.edu.ar

provides the most potent source of energy. Some insect species that feed on diets containing high levels of carbohydrates and low levels of lipid can convert dietary sugars into lipid reserves using them to support egg production, e.g., mosquitoes (Ziegler and Ibrahim 2001). During embryogenesis, lipids are used mainly as a source of energy but also as precursors for the synthesis of new cellular membranes and in signaling pathways (Downer and Matthews 1976). Therefore, the importance of lipid metabolism during reproduction has been increasingly recognized (Arrese and Soulages 2010; Fruttero et al. 2017a; Gondim et al. 2018; Hansen et al. 2013; Leyria et al. 2020a).

During vitellogenesis, high quantities of lipids, along with other nutrients, are taken up by developing oocytes and stored in specialized structures. This process is referred to as heterosynthetic since the majority of these nutrients are synthesized externally to the ovarian tissue. The primary tissue providing such a lipid source is the fat body via the hemolymph. The fat body works as a metabolic center to fulfill the energy demands of reproduction, metamorphosis, and flight, among other processes (Arrese and Soulages 2010).

Lipid mobilization is tightly regulated by a complex endocrine system, which promotes lipid synthesis and hydrolysis by the fat body, as well as lipid uptake and storage by oocytes. Insects precisely integrate the metabolism and the endocrine system to respond efficiently to different physiological requirements. The central nervous system (CNS) and associated endocrine glands, *corpora allata* (CA) and *corpora cardiaca* (CC), are the main sources providing hormones and neuropeptides. One well-known action of endocrine factors on lipid metabolism during reproduction is promoting the synthesis, release, and uptake of lipid carrier proteins, such as lipophorin (Lp), the main hemolymphatic lipoprotein; vitellogenin (Vg), a female-specific protein found in most insect species; and yolk proteins (YPs), found in dipterans such as the fruit fly, housefly, and fleshfly (Toprak 2020; Canavoso et al. 2001; Ziegler and Van Antwerpen 2006).

Lipid uptake is performed mainly through non-endocytic pathways, in which the lipids are mostly supplied by Lp via its classic docking mechanism, and by endocytosis of lipoproteins. The latter mechanism is mediated by receptors of the gene superfamily of low-density lipoprotein receptor (LDLR), particularly the Vg receptor, VgR, and Lp receptor, LpR (Fruttero et al. 2017a; Van der Horst et al. 2009), which will be discussed in detail later.

In this context, the coordination of the CNS/CA/CC-fat body-ovary axis seems to be essential for insects in order to produce viable eggs, explaining why it has been the topic of numerous reports. The post-genomic era has greatly advanced our comprehension of numerous processes involved in lipid homeostasis; therefore, is crucial to integrate the new findings of recent years into the knowledge already acquired in the last decades. In the current chapter, we focused on recent reports to approach lipid metabolism in the context of insect reproduction and its impact on the CNS/CA/CC-fat body-ovary axis interaction.

2 The Midgut as a Key Site for Dietary Lipid Dynamics in Successful Insect Reproduction

The production of viable eggs requires a substantial intake of nutritionally and energy-rich resources from food. In the context of increased egg production, such as after mating, the midgut plays a crucial role in processing available food to provide the female with the necessary nutrients to sustain oogenesis (White et al. 2021; Leyria et al. 2023). The midgut is the region of the alimentary canal where takes place most of the digestion and absorption of the lipids and other nutrients contained in the food. Interestingly, the spatial arrangement of the midgut in proximity to other metabolic organs like the fat body forms the basis for its involvement in inter-organ communication (Zhao and Karpac 2020). Morphologically, the midgut is a tube not covered by cuticle, from which blind sacs can branch. In the majority of insect species, the midgut surface is covered by a film-like acellular structure known as peritrophic membrane, that separates the luminal content, aids in the digestion process, and serves as protection from chemical and mechanical damage (Terra et al. 2019). At cellular level, the midgut consists of a single layer of epithelial cells, comprising both columnar cells or enterocytes, and endocrine cells. Enterocytes provided with microvilli on the apical membrane play a crucial role in the production of digestive enzymes and the absorption of nutrients, while endocrine cells are responsible for producing and releasing bioactive peptides. The epithelial layer is supported by a basal lamina and is surrounded by longitudinal and circular muscles. These muscles, upon contraction, generate peristaltic waves that propel the luminal contents (Terra et al. 2019).

Lipids from luminal hydrolysis are taken up by the midgut cells in a poorly characterized process in which membrane transfer proteins are possibly involved. In the enterocytes, the absorbed lipid precursors are involved in the synthesis of more complex lipids following different metabolic pathways. The transient lipid pools conformed in the enterocytes are mobilized to the hemolymph upon demand via Lp, and, once in circulation, these lipids are transported and transferred mainly to the fat body for storage. During oogenesis, fat body lipids are used as essential source of energy and mobilized to the ovarian tissue, as discussed in the next sections. For more in-depth information about lipid metabolism in the insect midgut, we refer the reader to a specific chapter of this book.

3 The Fat Body: A Multifunctional Organ Sensing Hormonal and Nutritional Signals for Egg Production

The fat body is analog to the liver and adipose tissue of the vertebrates (Skowronek et al. 2021). This organ is located underneath the tegument and surrounding the internal organs, arranged as loosely bound cell strings suspended in the hemocoel. This organization allows the tissue to be completely bathed by the hemolymph and in close contact with the organs, thus facilitating its main function of integrating signals from the whole body (Arrese and Soulages 2010). Its primary cell type, the trophocyte or adipocyte, is characterized by abundant lipid droplets, also termed adiposomes (Roma et al. 2010), which can undergo dynamic changes in terms of volume and lipid composition during vitellogenesis (Pinch et al. 2021). The fat body also participates in immunity, recognizing pathogens and synthesizing antimicrobial proteins and polypeptides in response to foreign insults (Wojda et al. 2020). The fat body structure, function, and main features are thoroughly described in a specific chapter of this book.

The fat body senses hormonal and nutritional stimuli initiating major biochemical and ultrastructural tissue changes according to insect demands during development. In the context of lipid metabolism and during vitellogenesis, this tissue synthesizes and secretes lipids and yolk protein precursors, such as Vg and Lp, which are then taken up by the oocyte and stored in yolk granules to be used by the developing embryo (Ziegler and Van Antwerpen 2006).

The lipid stored in the fat body as triacylglycerol (TAG) arises from circulation and *de novo* synthesis from acetate and amino acids, which has been well documented (Majerowicz and Gondim 2013). Signaling pathways involved in lipid homeostasis and successful reproduction have also been reported. The lipin family of proteins works as phosphatidate phosphatases (PAP), yielding diacylglycerol (DAG), a substrate for the *de novo* synthesis of TAG, and inorganic phosphate. In addition, lipin proteins can function as transcriptional regulators that promote fatty acid oxidation (Chen et al. 2015). In *Drosophila melanogaster*, lipin is shown to have cytoplasmic and nuclear localization, not only in the fat body but also in the ovaries. The presence of lipin in the nuclei of ovarian nurse and follicle cells suggests that this protein has gene regulatory functions that could be important for oogenesis. Meanwhile, the oocyte cytoplasm localization indicates that lipin is provided maternally to support embryogenesis

(Ugrankar et al. 2011). In the cabbage beetle *Colaphellus bowringi*, the lipin protein possesses a putative nuclear localization signal, likely participating in gene regulation, as well as a plasma membrane localization motif (Guo et al. 2021a). Lipin is highly expressed in the fat body and ovaries, two main tissues involved in insect reproduction. Low *lipin* gene expression not only leads to a decrease in *Vg* transcript expression in the fat body and *VgR* in the ovaries but also results in a rapid expansion of lipid stores in the fat body and promotes the expression of genes related to lipid synthesis, thereby blocking normal oogenesis (Guo et al. 2021a). Thus, the findings indicate that a lipolytic system mediated by lipin is necessary for preserving lipid homeostasis during reproduction.

4 Lipid Transport from the Fat Body to The Ovaries: Role of Circulating Lipoproteins

Lipoproteins are macromolecular complexes composed of a hydrophobic core containing neutral lipids and an outer amphipathic shell of phospholipids, free cholesterol, and apolipoproteins. This outer shell allows the lipoproteins to transport their hydrophobic cargo in the aqueous environment of the circulatory system of vertebrates and invertebrates (Soulages and Wells 1994). The structural and non-exchangeable apolipoproteins of Lp and Vg belong to the large lipid transfer protein (LLTP) superfamily, which also contains the mammalian apolipoprotein B (apoB) and microsomal triglyceride transfer proteins (MTP) (Wu et al. 2013). The apolipoprotein fraction aids solubility and interacts with receptors and other molecules. The principal lipoprotein in the circulation of insects is Lp, whose most thoroughly studied function is the transport and transfer of lipids and other lipophilic compounds among target tissues (Van der Horst et al. 2009). In resting insects, Lp has a density of ~1.15 g/ml, thus being in the range of high-density lipoproteins, a molecular weight of ~600–800 kDa, and a lipid content of 30–50%. Lipophorin is synthesized in the fat body and contains the apolipoproteins apolipophorin I (apoLp-I, Mr ~230–250 kDa) and apolipophorin II (apoLp-II, Mr ~70–85 kDa), which arise through furin protease-mediated cleavage of their common precursor apolipophorin II/I. This multifunctional lipoprotein carries different lipid classes, mainly DAG, although, in mosquitoes and insects belonging to the infraorder Culicomorpha, TAG is the predominant neutral lipid carried by Lp (Pennington and Wells 2002).

In blood-feeding mosquito females, studies have shown that the expression of *Lp* transcripts and protein synthesis is upregulated during vitellogenesis (Van Heusden et al. 1997; Sun et al. 2000; Stryapunina et al. 2024). Additionally, in another blood-feeding insect, *Rhodnius prolixus*, although *Lp* transcript level increases in early vitellogenesis, the Lp titer in the hemolymph remains quite stable (Leyria et al. 2022; Gondim et al. 1989). However, Lp density varies after feeding. In the initial three days following the blood meal, Lp density decreases as it receives lipids at the highest rates from the midgut. Subsequently, its density increases, likely due to the significant lipid unloading at the vitellogenic ovaries (Coelho et al. 1997).

Although Lp occurs in all insect stages and in both sexes (Canavoso et al. 2001), currently, Lp-mediated lipid transfer to specific tissues is not completely understood. The majority of lipid transfer occurs upon interaction with binding sites without lipoprotein internalization, but endocytosis of Lp has also been documented in the fat body and the ovarian tissue (Van Der Horst and Rodenburg 2010). In addition, it has been hypothesized that the anatomical proximity between the posterior region of the midgut and the ovary could enhance lipid transfer locally, thus maximizing its loading into the developing egg (Reiff et al. 2015). For more information refer the chapter on lipoproteins found in this book.

Vitellogenin (a lipoprotein of ~200–600 kDa) and YPs (three yolk polypeptides of ~44 to 49.5 kDa) are primarily synthesized in the fat body of females in response to endocrine and nutritional stimuli. These main yolk protein precursors provide lipids, protein, carbohydrates, and other nutrients for the development of the embryo (Raikhel et al. 2005). The Vg family of proteins resembles the mammalian ApoB, displaying similarity of regions involved in lipid binding. In contrast, YPs have conserved sequences similar to those surrounding the catalytic site of the triacylglycerol lipase and lipoprotein lipase from vertebrates, where the enzyme-lipid interaction occurs. However, YPs have no lipase activity and only seem to bind steroid hormones in a storage form conjugated to lipids, which then will be used during embryogenesis (Bownes 1992). Vitellogenin possesses a lower lipid content than Lp that ranges from 8 to 15% and presents a density of ~1.28 g/ml, thus being in the range of very high-density lipoproteins. In the fat body, the Vg particles are cleaved by proteases and co- and post-translationally modified in a well-characterized process. Moreover, in some insect species the ovary can also serve as a source of Vg (Melo et al. 2000). The consensus cleavage site for the *Vg* gene product is conserved near the N-terminus. Vitellogenin possesses polyserine residues, serving as phosphorylation sites; however, the function of Vg phosphorylation remains unclear. Glycosylation of Vg plays an essential role in folding and subunit assembly (Tufail and Takeda 2008). The structural conservation observed in Vg molecules across different taxa is likely due to the common nutritional demands of developing embryos in all oviparous species (Tufail et al. 2014). The number of Vgs varies significantly among different species, for instance, *D. melanogaster*, *Aedes aegypti*, and *Plautia stali* each exhibit three Vgs, while *R. prolixus* and *Periplaneta americana* have two Vgs (Melo et al. 2000; Tarone et al. 2012; Isoe and Hagedorn 2007; Lee et al. 2000; Tufail et al. 2001). It has been proposed that *Vg* genes may display different functions (Tufail and Takeda 2008, 2012).

Patency is a process regulated by juvenile hormones (JHs) in which the cells of the follicular epithelium shrink and form intercellular channels that facilitate the passage of molecules from the hemolymph to the surface of the oocyte (Raikhel and Dhadialla 1992; Wu et al. 2021). Once secreted to the hemolymph, Vg reaches the oocyte where it is taken up by receptor-mediated endocytosis and stored in yolk granules as vitellin (Tufail and Takeda 2008).

5 The Central Nervous System/Corpora Allata/Corpora Cardiaca-Fat Body-Ovary Axis and Its Implication in the Endocrine Regulation of Lipid Metabolism in Reproduction

Most of the processes involved in lipid metabolism are controlled by hormones and neuropeptides. The endocrine control of insect lipid metabolism during several physiological events has been recently reviewed by Toprak (2020). The main hormones and neuropeptides involved in lipogenesis and lipolysis that highly affect the hemolymph lipid levels and the lipids synthesized or stored in tissues are insulin-like peptides (ILPs), adipokinetic hormones (AKHs), JHs, and ecdysteroids (Table 1). In addition, there are some amine-type hormones with less influence on lipid metabolism, such as the biogenic monoamine serotonin (5-hydroxytryptamine, 5-HT) and octopamine, among others. The common characteristic of all these endocrine factors is that their titers in hemolymph strongly depend on the nutritional state. Indeed, most adult insects use dietary supplies to improve their energetic status, particularly to regulate the onset of reproductive maturity and egg production. Thus, a single physiological process as reproduction is controlled by multiple hormones and neuropeptides that play distinct roles on lipid metabolism in a spatial and temporal dynamic way, regulating events by acting on different targets through binding with their specific receptors. The majority of hormones and neuropeptides and numerous biological neuromodulators of insect metabolism are synthesized in the CNS, CA, and CC. It is important to point out that the fat body and the ovaries are also endocrine tissues as they can synthesize several hormones and neuropeptides (Toprak 2020).

The CNS in insects is relatively simple from a macroscopic morphological point of view. It is composed of a dorsal brain lying above the anterior end of the stomodeum and connected to a ventral nerve cord consisting of paired segmental ganglia extended along the ventral midline of the

Table 1 Endocrine control of lipid metabolism to successful reproduction. Main findings

Hormone/ Neuropeptide	Insect	Effect(s) confirmed or suggested	Reference(s)
AKH signaling			
	Gryllus bimaculatus	Stimulates catabolic reactions to promote lipid mobilization in fat body via activation of triacylglycerol lipase Inhibits lipid synthesis by the fat body	Lorenz and Anand (2004)
	Glossina morsitans	Regulates lipolysis activity through Brummer lipase to provide lipids for milk production	Attardo et al. (2012)
	Aedes aegypti	Potential involvement in metabolic control and liberation of energy stores in the form of lipids	Oryan et al. (2018)
	Bactrocera dorsalis	Promotes lipid mobilization in fat body to fuel for oocyte maturation	Hou et al. (2017)
Insulin signaling			
	Rhodnius prolixus	Signaling activation in fat body and ovaries after feeding Promotes expression of the lipid carrier protein Vg Involvement in *de novo* lipogenesis activity and capacity to utilize hemolymph-derived fatty acids Potential involvement in SREBP expression for lipid accumulation by oocytes	Silva-Oliveira et al. (2021), Leyria et al. (2020a, 2021)
	A. aegypti	Promotes metabolic activity after feeding by elevating lipid storage for egg maturation Modulation of ilp7 and ilp8 by the microRNA miR-277 is required for lipid metabolism and ovarian development	Brown et al. (2008), Ling et al. (2017)
	G. morsitans	Suppresses lipolysis between periods of lactation during tsetse fly pregnancy	Baumann et al. (2013)
	Drosophila melanogaster	Stimulates expression of Lsd2, in turn promoting the enlargement of the lipid droplets	Vereshchagina and Wilson (2006)
Juvenile hormones			
	Mostly in Orthoptera, Blattodea, and Hemiptera insect orders	Facilitate the accumulation of lipids into the oocyte by stimulating the synthesis and release of lipid carrier proteins from the fat body, mainly Vg and Lp Promote patency in the follicular epithelium	Reviewed by Raikhel and Dhadialla (1992), Wu et al. (2021), Khalid et al. (2021)
	Nilaparvata lugens	Increase both VgR mRNA and protein levels Lead to an increase in LpR transcript expression Bmm-mediated lipolysis interacts with JH signaling pathway to activate vitellogenesis and oocyte maturation (Lu, Chen, et al., 2018)	Lu et al. (2015), Lu et al. (2018a, b)
	Dipetalogaster maxima	Induce the expression of Lp/LpR genes in fat body and ovary, respectively, and prompt the uptake of Lp and Vg by oocytes	Ramos et al. (2021)

(continued)

Table 1 (continued)

Hormone/ Neuropeptide	Insect	Effect(s) confirmed or suggested	Reference(s)
	A. aegypti	Stimulate the expression of LpR to promote lipid accumulation in previtellogenic ovaries	Clifton and Noriega (2012)
	Blattella germanica	Regulate the expression of LpR post-transcriptionally	Ciudad et al. (2007)
	Colaphellus bowringi	Facilitate development of the reproductive system and inhibit diapause by suppressing lipid synthesis Kr-h1 acts as key mediator triggering the expression of triacylglycerol lipases to suppress lipid storage and block diapause	Liu et al. (2016, 2019), Tian et al. (2021), Guo et al. (2021)
	Locusta migratoria	Kr-h1 stimulates lipid accumulation in the primary oocytes as well as oocyte maturation and ovarian growth	Song et al. (2014)
	G. morsitans	Induce lipid synthesis and reduce Bmm expression through suppression of FOXO via insulin signaling activation	Baumann et al. (2013)
Ecdysteroids			
	A variety of insects, including dipterans, hymenopterans, and lepidopterans	Mediate the accumulation of lipids into the oocyte by stimulating the synthesis and release of lipid carrier proteins from the fat body, mainly Vg or YPs	Reviewed by Wu et al. (2021), Khalid et al. (2021)
	A. aegypti	Activate HNF4 expression, which in turn regulates of gene encoding for TAG catabolism and β-oxidation enzymes	Wang et al. (2017)
	D. melanogaster	Ec locally produced by ovaries promotes SREBP-mediated lipid uptake by regulating the expression of LpR In the CNS, stimulate female feeding promoting whole-body TAG storage at high levels	Sieber and Spradling (2015)
	Anopheles gambiae	When transferred by males during copulation along with MISO protein ecdysteroids increase the expression of Lp, facilitating oocyte development via the accumulation of lipids	Baldini et al. (2013)

Legend: *SREBP* Sterol regulatory-element binding proteins, *HNF4* nuclear receptor hepatocyte nuclear factor 4, *Vg* vitellogenin, *Lp* lipophorin, *YPs* yolk proteins, *LpR* lipophorin receptor, *Ec* ecdysone, *Kr-h1* Krupper homolog 1, *FOXO* Forkhead Box Sub Group O, *MISO* female mating-induced stimulator of oogenesis

thorax and abdomen that supply nerves to different body segments. In contrast to this apparent simplicity, the CNS works complexly and sophisticatedly. Within the CNS, diverse neurons can synthesize peptide hormones, which will become active neuropeptides after processing. Processing starts at the rough endoplasmic reticulum and Golgi apparatus, where preprohormones are converted to prohormones by the proteolytical removal of the N-terminal signal sequence and other post-translational modifications, thus resulting in mature peptide hormones. Finally, they are released within the CNS, working as neuromodulators or neurotransmitters, or into the circulation, acting as neurohormones (Orchard and Lange 2024). Other endocrine factors are synthesized after sequential enzymatic reactions, e.g., JHs by the CA and Ecdysone (Ec) by the larval prothoracic gland or the ovaries in adult insects. In CC and CA, the critical factors synthesized and involved in lipid mobilization are AKH and JH, respectively. Although hormonal control of insect reproduction has been widely studied and reported (Roy et al. 2018; Khalid et al. 2021), the knowledge on the endocrine control of lipid metabolism during reproduction or oogenesis still remains incomplete. The endocrine control of lipid metabolism is reviewed in a specific chapter of this book. Here, we discuss the primary endocrine factors involved in maintaining lipid homeostasis for successful reproduction.

5.1 Endocrine Control of Lipid Metabolism by Peptide Hormones

The members of the AKHs/red pigment concentrating hormones (RPCH) family are the main peptide hormones involved in both carbohydrate and lipid mobilization from the fat body, acting as functional homologs of glucagon from vertebrates (Van der Horst et al. 2001). Adipokinetic hormones represent a family of nearly 60 isoforms spanning across almost all major insect orders, all of them with a primary sequence ranging from 8 to 11 amino acids with highly conserved residues, which are essential for their biological activity (Gäde et al. 2013; Gäde and Marco 2013); the functional differences among isoforms are currently being investigated. The AKH is primarily synthesized and stored in endogenous neurosecretory cells of the CC. After release, AKH signaling begins by binding of AKH to G protein-coupled receptors (GPCRs), leading to the expression of second messengers in the cytoplasm that amplify the signal (Gäde and Auerswald 2003). The corazonin (CZN) receptor and the AKH/corazonin-related peptide (ACP) receptor are two insect molecules structurally related to the insect AKH receptor (AKHR), whose ligands are CZN and ACP, respectively. The CZN and ACP peptides share a number of conserved amino acid residues with AKH (Cazzamali et al. 2002; Hansen et al. 2010); however, they might not share the physiological effects of their signaling pathways. For example, in the blood-sucking bug, *R. prolixus*, AKH can elevate hemolymphatic lipid levels in adult male insects, whereas CZN increases the heartbeat frequency and seems not to affect lipid metabolism. The ACP does not affect either hemolymph lipid levels or heartbeat frequency (Patel et al. 2014).

As stated above, AKH represents one of the cases in which the ovaries may be acting as an endocrine tissue; the presence of the AKH precursor and its receptor in the ovaries of *R. prolixus* suggests a role for this neuropeptide in egg production, egg-laying behavior, or both, via an autocrine signaling pathway (Leyria et al. 2020a). The research on AKH has been focused on its role as a lipolytic hormone during flight (Van Der Horst and Rodenburg 2010) and on starvation survival (Arrese and Soulages 2010; Toprak 2020; Gäde 2004; Lorenz and Gäde 2009). However, AKH not only stimulates catabolic reactions but also prevents anabolic ones, including the synthesis of proteins, lipids, and glycogen (Lorenz and Gäde 2009). Interestingly, in *Gryllus bimaculatus* female crickets, AKH activates catabolic reactions that promote the mobilization of lipid stores in the fat body during vitellogenesis via triacylglycerol lipase activation. Moreover, elevated titers of AKH in the hemolymph would, at the same time, inhibit lipid synthesis in the fat body (Lorenz and Anand 2004). Consequently, it was presumed that the endocrine regulation of anabolic and catabolic reactions during insect reproduction has to be precisely coordinated by AKH. In the tsetse fly *Glossina morsitans*, adult females carry their progeny in their utero during embryonic and larval development. The mother provides nourishment in the form of milk secretions from a modified accessory gland. The milk is not synthesized during oogenesis and embryogenesis; thus, the lipolysis is decreased, and dietary nutrients are shifted to the fat body for storage of lipids. However, Attardo et al. (Attardo et al. 2012) have reported that, during lactation, the lipolytic activity of Brummer lipase (Bmm), an insect homolog of mammalian lipases, is regulated by AKH/AKHR signaling to provide lipids to the accessory gland for milk production. Remarkably, in the mosquito *A. aegypti*, the *AKHR* expression is detected in the ovaries, being mainly increased in the secondary sex organs. Thus, AKH/AKHR may indeed play a central role in oogenesis by regulating mechanisms involved in the metabolic control of energy stores in the form of lipids, carbohydrates or proteins (Oryan et al. 2018). Further, in the oriental fruit fly *Bactrocera dorsalis*, the lipids stored in the fat body seem to be used to promote the maturation of the oocyte via AKH signaling (Hou et al. 2017). However, long-term exposure of the fruit fly *D. melanogaster* to high-fat diet results in elevated *AKH* transcript levels that coincide with an increase in TAG levels in mated flies and impaired fecundity (Liao et al. 2021). Therefore, in this case, an increase in AKH signaling would not have positive effects on reproduction (Lee and Park 2004). Hence, the role of AKH in reproductive events cannot be generalized and definitely needs more attention, which might unveil some interesting new aspects of insect reproduction.

The agonists of insulin signaling, designated insulin-related peptides (IRPs) or most commonly ILPs, have been identified in a wide variety of insect species. They represent the ultimate case of structural and functional conservation of hormones between invertebrates and vertebrates. In the post-genomic era, multiple ILP genes have been reported, ranging from one or two in locusts, eight in *A. aegypti* and *D. melanogaster*, and 38 in the silk moth *Bombyx mori* (Chowański et al. 2021). Insect ILPs are primarily produced by several pairs of medial neurosecretory cells (MNCs) in the brain and, to a lesser extent, by other tissues such as the midgut, ovaries, and fat body, in a tissue- and stage-specific manner (Wu and Brown 2006). Upon secretion, ILPs signaling initiates by binding to insulin receptors (InRs) belonging to the receptor tyrosine kinases (RTK) family. The activation of these receptors regulates two intracellular pathways, the PI3K/Akt/FOXO cascade (phosphatidylinositol-3 kinase/protein kinase B/Fork head box transcription factor) and the Ras/MAPK pathway (mitogen-activated protein kinase). Moreover, some actors of the insulin superfamily can also bind and trigger receptors of the leucine-rich repeat-containing G protein-coupled receptors (LGR) subfamily of GPCRs (Nässel and Vanden 2016; Veenstra 2020). Since the reproductive process is highly energy demanding, it is expected that ILPs regulate the availability, reallocation, and metabolism of lipids and carbohydrates. The ILPs and their cross-talk with other molecular players regulating insect reproduction have been recently reviewed (Chowański et al. 2021). However, the available literature unraveling the direct link between ILPs and lipid metabolism during egg production is still scanty. It is widely known that insulin signaling cascade can activate the mechanistic target of the rapamycin (mTOR) pathway, primarily stimulating protein synthesis, but also lipid synthesis and further uptake of nutrients (Loewith and Hall 2011). In adult females of *R. prolixus*, insulin/TOR signaling is activated in the fat body and the ovaries only after feeding, after blood feeding, being such meal the primary stimulus to initiate vitellogenesis and maximum egg production (Leyria et al. 2020b). Interestingly, InR deficiency reduces TAG storage in the fat body by decreasing *de novo* lipogenesis activity and the capacity to utilize hemolymph-derived fatty acids; as a consequence, InR-deficient females exhibit smaller ovaries and a marked reduction in oviposition (Silva-Oliveira et al. 2021; Leyria et al. 2021).

Sterol regulatory-element binding proteins (SREBPs) are transcription factors from the helix-loop-helix leucine zipper family that regulate the expression of genes in charge of lipid synthesis and uptake (Jeon and Osborne 2012). The PI3K/Akt/mTOR pathway, activated by insulin, regulates lipid biosynthesis via SREBP expression (Porstmann et al. 2008). In mammals, SREBP elicits the expression of numerous genes involved in lipid synthesis and uptake, including the LDLR (Shao and Espenshade 2012). In *D. melanogaster*, SREBP regulates lipid uptake and accumulation in oocytes by regulating the expression of LpR (Sieber and Spradling 2015). In ovaries of *R. prolixus* females, *SREBP* and *LpR* genes are upregulated simultaneously when the insulin/TOR signaling is activated (Leyria et al. 2020b); thus, via insulin signaling, SREBP could also be involved in lipid accumulation by oocytes during vitellogenesis, possibly stimulating *LpR* gene expression. Additionally, in *A. aegypti*, the blood meal promotes egg production by prompting the release of ILPs from MNCs in the brain (Wu and Brown 2006). In this insect model, insulin signaling is triggered after feeding, leading to an increase in lipid storage for egg maturation (Brown et al. 2008). In addition to the blood meal, the regulation of ilp7 and ilp8 by the microRNA miR-277 is necessary for the metabolism of lipids and the development of the ovarian tissue (Ling et al. 2017). In tsetse flies, the growing larva is nurtured by lipid-rich, milk-like secretions. Interestingly, the high lipid content of this milk-like substance is controlled during larvigenesis (i.e., lactating period), oogenesis, and embryogenesis (non-lactating periods) not only by AKH, as discussed above, but also by ILP and JH signaling (Baumann et al. 2013). In *D. melanogaster*, the loss of the insulin signaling antagonist phosphatase and tensin homolog (PTEN) induces the formation of highly enlarged lipid droplets in the ovaries. This phenotype is caused by the accumulation of activated Akt in the cytoplasm and leads to a much higher expression of the lipid storage droplet-2 protein (Lsd-2), the fruit fly homolog to the vertebrate lipid-storage protein perilipin (Vereshchagina and Wilson 2006).

The insulin signaling also interacts with other pathways to regulate lipid metabolism during reproduction. Recently, functional evidence using *A. aegypti* as a model demonstrated that serotonin signaling, via the Aa5HT2B receptor, modulates different ILPs which will adjust the metabolism in the fat body and consequently, the supply of lipid-derived energy for rapid egg development (Ling and Raikhel 2018). In addition, the interplay between insulin pathway and JH and ecdysteroid signaling has also been reported, as discussed later.

5.2 Endocrine Control of Lipid Metabolism by Juvenile Hormones and Ecdysteroids

Insect reproduction is governed mainly by two critical lipid endocrine factors: JHs and ecdysteroids. Juvenile hormones are a group of sesquiterpenoids synthesized primarily by CA in response to different physiological stimuli. Similar to other sesquiterpenoid hormones such as farnesoic acid and methyl farnesoate, JHs are relatively non-polar, lipophilic, and insoluble in aqueous media. Diverse pathways have evolved in the downstream process of sesquiterpenoid production to promote the synthesis of JHs with different structures, depending on the insect species (Tsang et al. 2020). Juvenile hormone homologs are involved in numerous aspects of development and reproduction, such as metamorphosis, caste determination, and ovarian development. These functions are performed mainly via the Methoprene-tolerant (Met) protein, a transcription factor belonging to the basic helix-loop-helix Per/Arnt/Sim (bHLH-PAS) family and reported as the intracellular JH receptor. Binding of JH to Met triggers Met dimerization with Taiman (Tai), a bHLH-PAS protein that acts as co-activator, and the resulting complex induces the transcription of several target genes (Tsang et al. 2020; Riddiford 2020). One of the best-known JH-Met actions on lipid metabolism during reproduction is promoting lipid accumulation in the oocyte by promoting the synthesis of lipid carrier proteins in the fat body, mainly Vg and Lp. In addition, JH induces the formation of large gaps in the follicular epithelium (patency), enabling lipoproteins to reach the oocyte surface (Raikhel and Dhadialla 1992; Wu et al. 2021). Patency is mediated via JH membrane receptors whose identity is still unknown; however, depending on the insect species, either a GPCR or a TKR seems to be promising candidates (Li et al. 2019). Interestingly, the relationship between the nuclear receptor Met and nongenomic receptors of JH has been reported. In the mosquito *A. aegypti*, JH signaling triggered via RTK activation stimulates the phosphorylation of Met to enhance transcriptional regulation activity (Ojani et al. 2016).

On the other hand, Ec is a steroid hormone produced mostly by the prothoracic gland during the larval stages and by the ovaries during the adult stage. This molecule is secreted as a prohormone with low biological activity and is converted into the more biologically active form 20-hydroxyecdysone (20E) in peripheral tissues. In insects, biotransformation of cholesterol into Ec involves hydroxylation and oxidation reactions, led by cytochrome P450 enzymes encoded by the Halloween genes (Igarashi et al. 2018). The 20E interacts mainly with the ecdysone receptor heterodimeric complex composed by the ecdysone receptor (EcR) and ultraspiracle (USP), which in turn activates the transcription of a gene cascade. This functional heterodimeric complex controls the expression of Ec-responsive genes by binding to specific promoter regions, the ecdysone response elements. The signaling triggers a series of processes, such as vitellogenesis, and the coordination of developmental transitions, such as larval molting and metamorphosis (Swevers 2019; Song and Zhou 2020).

The silencing of Met or EcR genes by RNA interference (RNAi) experiments affected several aspects of reproduction, including hatching, fecundity, and oviposition, making these genes an attractive target for insect pest management, not only by RNAi but also other approaches (Qiao et al. 2021; Aathmanathan et al. 2020; Zhou et al. 2020; Thomas et al. 2020; Husain et al. 2020).

Generally, 20E and JHs have contrasting roles in insect development and physiology regulation; while 20E signaling triggers developmental and physiological changes, JH is able to alter the action of 20E in a stage-specific fashion (Liu et al. 2018). In adult insects, 20E and JH regulate female reproduction in different ways, but a proper balance between both is crucial for egg development (Khalid et al. 2021). In evolutionarily primitive hemimetabolous insects such as *Locusta migratoria* and *Blattella germanica*, as well as in *R. prolixus,* one of the oldest models used to set up the bases of insect physiology studies, JH acts independently of 20E to stimulate reproduction. In some holometabolous insects, such as *Tribolium castaneum*, JH controls Vg synthesis in the fat body, and 20E participates in oocyte maturation. In *D. melanogaster*, YPs synthesis in the female fat body is induced mostly by 20E, whereas YPs synthesis by ovarian follicle cells is exclusively under JH control (Bownes et al. 1987; Bownes 2020). In the dipteran *A. aegypti*, 20E along with ILPs stimulate Vg expression and oocyte maturation after a blood meal (Hansen et al. 2014). In lepidopterans, the hormones governing successful reproduction depend on the species (Wu et al. 2021); for instance, in the pantropical insect pest *Maruca vitrata*, two endocrine signals work as gonadotropins, cooperatively promoting egg development. In this sense, while ILPs stimulate previtellogenic development, JH mediates vitellogenesis by activating Vg expression (Al Baki et al. 2019).

In *A. aegypti*, JH and 20E play critical but opposite roles in lipid metabolism regulation during reproduction by binding to the Met and EcR receptors, respectively. Briefly, every gonadotrophic cycle occurs in two phases in female mosquitoes: the pre-feeding phase, which is controlled by JH and prepares the tissues for egg maturation, and the post-blood meal phase, which is mainly regulated by 20E and results in rapid egg development (Hansen et al. 2014). Wang et al. (2017) have reported that the downregulation of the nuclear receptor hepatocyte nuclear factor 4 (HNF4), an ortholog of a well-conserved subfamily of nuclear receptors that contains two mammalian receptors, HNFα and HNFγ, promotes the downregulation of transcripts related to TAG breakdown, β-oxidation enzymes, and TAG accumulation. Interestingly, the expression of the *HNF4* gene is also downregulated by JH/Met at the end of the post-eclosion phase and activated by 20E/EcR and TOR during the post-blood meal phase.

In *D. melanogaster*, ecdysteroids work at two levels to support reproduction. On the one hand, in synergism with the insulin signaling on SREBP regulation, ecdysteroids locally produced by the ovaries promote SREBP-mediated lipid uptake by regulating the expression of LpR, and on the other hand, ecdysteroids act at the CNS level, where stimulate female feeding, promoting high levels of whole-body TAG storage (Sieber and Spradling 2015). Fed females of the malaria mosquito *Anopheles gambiae* allocate nutritional resources to oogenesis only after being inseminated. This process is regulated by the interaction between 20E and the female mating-induced stimulator of oogenesis (MISO) protein, being 20E transferred by males to females during copulation. This interaction translates into an increase of Lp expression, thus enabling oocyte development via lipid accumulation in the ovaries (Baldini et al. 2013).

The transcriptional regulation of *VgR* by JH has been established in several species (Chen et al. 2004; Jing et al. 2021; Miao et al. 2020); however, the evidence on LpR is scarce. In the hematophagous triatomine *Dipetalogaster maxima*, topical administration of JH III induces the expression of *Lp* and *LpR* genes in the fat body and the ovaries and prompt the uptake of Lp and Vg by oocytes (Ramos et al. 2021). Moreover, hormone manipulation experiments consisting of JH analog Methoprene administration showed that the amount of lipids stored in the ovaries of *A. aegypti* during the previllogenic resting stage reflects the nutritional status of the female and that the abundance of *LpR* and lipid accumulation-related mRNA transcripts in the ovaries is regulated by JH (Clifton and Noriega 2012). In *B. germanica*, JH treatment increases LpR protein levels, especially in the fat body, but not those of LpR mRNA, suggesting that LpR expression is post-transcriptionally regulated by JH (Ciudad et al. 2007). However, in *Nilaparvata lugens* and *D. melanogaster*, 20E is the endocrine factor leading to an increase of LpR transcript level (Sieber and Spradling 2015; Lu et al. 2018a).

In *C. bowringi*, photoperiod manipulation determines whether the insect becomes reproductively active or enters diapause. A deficiency in the JH pathway leads to diapause; hence when the JH signaling increases, ovarian development is induced, most likely via an increase in VgR expression. At the same time, the fat body is stimulated to increment Vg production and inhibit lipids synthesis. Therefore, JH acts through Met signaling, facilitating the development of the reproductive system and inhibiting diapause by suppressing lipid synthesis (Tian et al. 2021; Liu et al. 2016, 2019). Remarkably, *Krüppel homolog 1* (*Kr-h1*), one of the best-known JH early-inducible genes, acts as a key mediator by triggering the expression of triacylglycerol lipases to suppress lipid storage and block diapause (Guo et al. 2021b). In *L. migratoria*, Kr-h1 expression via JH-Met signaling also has implications in lipid metabolism during reproduction, since Kr-h1 depletion causes a severe reduction of lipid accumulation in the primary oocytes. In addition, the depletion of Kr-h1 impairs oocyte maturation and ovarian growth, showing the same phenotype of Met-silenced and JH-deprived females (Song et al. 2014).

In *T. castaneum*, nuclear receptors and 20E response genes are required for vitellogenesis and oogenesis. Recently, it has been reported that E93, a member of the helix-turn-helix (HTH) transcription factor family and one of the primary 20E response genes, indirectly regulates reproduction by suppressing genes involved in the metabolism of nutrient supplies, including lipids, as soon as the adult emerges. Therefore, E93 ensures appropriate storage of nutrients to support vitellogenesis, which is initiated 72 h after adult emergence (Eid et al. 2020). In *N. lugens*, JH significantly increases both *VgR* mRNA and protein levels (Lu et al. 2015). Moreover, the most interesting finding in this insect species is that Bmm suppression increases JH esterase enzyme expression, thus regulating the rate of JH degradation. This result was supported by a significant reduction in Vg, Met, Tai, Kr-h1, and VgR transcript levels. Therefore, a new mechanism of lipid homeostasis to achieve successful reproduction was proposed, in which Bmm-mediated lipolysis interacts with the JH signaling pathway to activate vitellogenesis and oocyte maturation (Lu et al. 2018a). On the contrary, Bmm transcription is downregulated in tsetse flies after treatment with the JH analog Methoprene or insulin. Hence, JH signaling induced lipid synthesis in this insect and reduces Bmm expression by suppressing FOXO via insulin signaling activation (Baumann et al. 2013).

Furthermore, in the same fashion YPs act as ecdysteroid carrier protein, Lp not only works as a lipid shuttle among tissues but also as a binding protein. For instance, in the triatomine *D. maxima*, Lp acts as a carrier for a cathepsin D-like peptidase (DmCatD), an enzyme involved in vitellin proteolysis in the ovarian tissue that is synthesized in the fat body as a yolk protein precursor, facilitating its internalization by the oocytes (Leyria et al. 2018). In *B. germanica*, Lp is able to bind JH with high affinity and specificity, regulating JH titers in the hemolymph (Sevala et al. 1997; Schal et al. 1998). Overall, there is a close relationship between lipoproteins and endocrine players necessary to achieve proper reproduction.

Eicosanoid short-acting lipid hormones are oxygenated metabolites of three C20 polyunsaturated fatty acids. Usually, they are biosynthesized from arachidonic acid released from membrane phospholipids by the activation of phospholipase A2. Upon release, arachidonic acid follows two enzymatic oxygenation pathways: one involving cyclooxygenases (COX), which generates prostaglandins and thromboxanes; and other involving lipoxygenases (LOX), producing lipoxins and leukotrienes. Once synthesized, eicosanoids are released into the hemolymph, where they can participate in either autocrine or paracrine mechanisms by the activation of cell surface receptors (Stanley and Kim 2019). Prostaglandins have been involved in promoting egg-laying behavior in some, but not all, insect species (Stanley 2006). Two relevant works on *R. prolixus* support the role of eicosanoids as lipid metabolites regulating oogenesis. In them, the authors showed that products of the COX pathway lead to a reduction in the uptake of yolk proteins by developing oocytes (Medeiros et al. 2002), while products of the LOX pathway promote yolk uptake by modulating the levels of cAMP on the ovarian follicle (Medeiros et al. 2004). In *D. melanogaster*, prostaglandins play a crucial role in eggshell formation since mutations in *pxt*, which encodes a putative *Drosophila* cyclooxygenase, are associated with changes in the timing of chorion gene expression and eggshell defects (Tootle et al. 2011).

6 Female Reproductive System and Its Role in Oogenesis

In a general view, the female reproductive system of insects is composed of a pair of ovaries, oviducts, spermathecas, accessory glands, and the bursa copulatrix or vagina. Each ovary is subdivided into ovarioles, which are the functional units of the reproductive system where new oocytes are produced. Developing oocytes are arranged in an ontogenic series (i.e., from less to more developed) from anterior to posterior location in the ovariole. Mature oocytes are then used to produce the first eggs. A typical insect ovary may contain several ovarioles, generally aligned parallel to one another. The number of ovarioles varies across species, individuals of the same population, and even between the ovaries of a single individual (Hodin 2009; Church et al. 1950). When an insect is actively reproducing, the ovary develops eggs and may nearly fill the abdomen. Success in egg production relies on the uptake of nutrients by immature oocytes. This process is accomplished mainly via receptor-mediated endocytic activity in the oocyte membrane, allowing the accumulation of vast amounts of proteins, mostly Vg, along with lipids and carbohydrates. These resources provide the essential nutrients required for embryonic development (Wu et al. 2021). In oocytes, TAG, the predominantly stored lipid, phospholipids, and cholesterol are the major energetic supply for the developing embryo as well as a source for the synthesis of new membranes. Nevertheless, the capacity of insect oocytes to produce lipids by *de novo* synthesis appears to be poor (Ziegler and Van Antwerpen 2006; Ziegler 1997). Given the relevance of lipids to achieve successful reproduction in insects, the research on their uptake and deposition in insect oocytes is prolific (Fruttero et al. 2017a); below are summarized the main events involved in these processes.

6.1 Lipid Uptake and Accumulation by the Insect Oocyte

Lipid content and composition in the oocyte are central in determining egg quality and offspring survival. Lipids are one of the major constituents of insect oocytes, with TAG as the most abundant lipid class (Ziegler and Van Antwerpen 2006; Fan et al. 2002). For example, in the oocytes of *R. prolixus*, TAG represents near 60% of the neutral lipids (Santos et al. 2011). Further, the oocyte has a very limited capacity for *de novo* lipid synthesis. In the mosquito *A. aegypti* and the tobacco hornworm *Manduca sexta*, about 1% of ovarian lipids come from the synthetic activity of the oocyte (Ziegler 1997; Kawooya et al. 1988), the rest is derived from the circulating lipoproteins Lp and Vg from the fat body (Fruttero et al. 2017a). In mosquitoes, lipid transport from the fat body to growing oocytes takes place within 30–36 h post blood meal. Interestingly, during the first vitellogenic cycle, female mosquitoes use the lipids accumulated throughout larval development to supply the oocytes and then switch to use lipids from the blood meal to restore their fat body lipid reserves (Pinch et al. 2021; Briegel et al. 2003). Additionally, in blood-fed *R. prolixus* females, the incorporation of fatty acids into the ovaries occurs very fast (Santos et al. 2011).

Although Vg has a lower lipid content than Lp, it is taken up by the oocytes in high quantities during vitellogenesis, representing a relevant source for lipid storage. Studies conducted in *M. sexta* indicated that approximately 10% of ovarian lipids are derived from Lp and Vg endocytosis (Canavoso et al. 2001; Ziegler and Van Antwerpen 2006), which is mediated by receptors of the LDLR gene superfamily. The *LDLR* genes are highly conserved, and their modular organization in structural domains has been well documented (Go and Mani 2012). Although LpR and VgR belong to the same superfamily, structural and functional differences exist between them; for instance, LpR has a single lipid-binding domain, and VgR has two. In addition, although VgR is almost exclusively expressed in the ovarian tissue, LpR has been detected in the brain, midgut, and fat body, among other tissues (Tufail and Takeda 2009).

In most species studied to date, including the cockroaches *B. germanica* and *Leucophaea maderae* as well as the mosquito *A. aegypti*, LpR is expressed in ovarian tissue throughout the gonotrophic cycle (Chen et al. 2004; Ciudad et al. 2007; Tufail and Takeda 2009). In *L. maderae*, LpR levels are low during pre-vitellogenesis and high during the vitellogenic period (Tufail and Takeda 2009). In addition, the two isoforms of LpR described in *B. germanica* display distinctive expression patterns: the L isoform levels are high during pre-vitellogenesis and low afterward, while the S isoform expression is low at early vitellogenesis, increases during mid-vitellogenesis, and decreases afterward (Ciudad et al. 2007). In *D. maxima*, the *LpR* gene in the ovarian tissue was significantly upregulated after a blood meal, judging by the high LpR mRNA levels found during early vitellogenesis in comparison with those detected in pre-vitellogenesis (unfed females), late vitellogenesis, and post-vitellogenesis (when the egg-laying has finished) (Ramos et al. 2021). Moreover, in the closely related species *R. prolixus*, the ability of the ovary to take up fatty acids from Lp through a non-endocytic pathway increases soon after the blood meal and reaches its maximum around the second-day post-feeding (Santos et al. 2011). Overall, a relationship between lipid accumulation capacity by endocytic and non-endocytic pathways and the blood meal is strongly suggested. During tsetse fly pregnancy, the knockdown of the *Lp* gene results in a reduction of lipid levels in the hemolymph, a delay in the development of oocytes, and an extension in the larval gestation period. However, the suppression of LpR does not affect lipid levels in the hemolymph nor the duration of oogenesis, although it extends larval development length (Benoit et al. 2011). Thus, Lp/LpR signaling functions as the main shuttle of lipids from the midgut and fat body to the ovaries and milk gland supplying resources for the development of oocytes development and the nourishment of the larvae, respectively. Interestingly, once in the milk gland, lipids seem to be transferred to the developing larvae by a non-Lp-mediated pathway (Benoit et al. 2011).

In the cockroaches *P. americana*, *L. maderae*, and *B. germanica*, as well as in *R. prolixus*, the *VgR* gene is expressed during all the stages of the gonotrophic cycle, but the highest levels are found during pre-vitellogenesis (Ciudad et al. 2006; Tufail and Takeda 2005; Tufail et al. 2007). However, in the triatomine *D. maxima*, the VgR mRNA levels in the ovaries rise from pre-vitellogenesis to vitellogenesis, reaching its maximum at day six after blood feeding (Ramos et al. 2021). At the protein level, VgR in cockroaches reaches maximum expression during vitellogenesis, when the receptor is functionally most needed (Ciudad et al. 2006; Tufail and Takeda 2005; Tufail et al. 2007). In the mosquito *A. aegypti*, the VgR mRNA levels in the ovaries display a different profile, increasing considerably from one-day post-eclosion and during the vitellogenic period, peaking at 24 h post blood meal (Cho and Raikhel 2001). Concerning subcellular distribution, the available data from *D. melanogaster* and *P. americana* indicate that LpR and VgR are distributed in the cytoplasm of the oocyte during pre-vitellogenesis and migrate to the plasma membrane during vitellogenesis, when the endocytosis of yolk protein precursors is highly active (Tufail and Takeda 2005, 2009; Schonbaum et al. 2000).

The lipid transfer particle (LTP) is a very high-density lipoprotein of ~1.23 g/ml first described in the hemolymph of larval *M. sexta* by Ryan et al. (1986). It catalyzes the transfer of lipids between Lp of different densities and from Lp to tissues and vice versa (Liu and Ryan 1991). In the larval *B. mori*, the injection of an anti-LTP antibody decreased lipid accumulation in the oocytes by ~75% (Jouni et al. 2003). The LTP from *B. mori* is encoded by two genes, *apoLTP-II/I* and *apoLTP-III*, being the former a member of the large lipid transfer protein gene family (Yokoyama et al. 2013). Interestingly, LTP is a ligand of LpR that mediates endocytosis of Lp and also stabilizes the lipoprotein in the membrane extracellular domain (Rodríguez-Vázquez et al. 2015; Parra-Peralbo and Culi 2011). Thus, in an intricate scenario of cooperation, LpR recruits LTP to the oocyte plasma membrane, LTP binds Lp and receives their lipid cargo which, in turn, will be transferred to the oocyte cytoplasm (Rodríguez-Vázquez et al. 2015).

There are other membrane proteins that participate in Lp-mediated lipid transfer. In vitellogenic females of the kissing bug *Panstrongylus megistus*, the subunit beta of the ATP synthase (β-ATPase) complex located in the plasma membrane of the oocyte and follicular cells operates as a Lp docking receptor (Fruttero et al. 2017b). In this species, the interaction of Lp and β-ATPase facilitates lipid transfer to the oocytes, and the blocking of this interaction by an anti-β-ATPase antibody significantly impairs that process. It appears that the enzymatic activity of β-ATPase is not necessary for Lp binding (Fruttero et al. 2017b). The role of β-ATPase in lipid transfer was also demonstrated in the midgut and fat body of fifth instar nymphs of the same species (Fruttero et al. 2014, 2019).

Lipoprotein lipases (LPL) are multifunctional enzymes that hydrolyze neutral lipids present in lipoproteins. In vertebrates, the LPL attached to the endothelial heparan sulfated proteoglycans hydrolyses the TAG core of circulating very-low-density lipoproteins (VLDL) and chylomicrons, rendering glycerol and fatty acids available for its use in muscles and adipose tissue (Wang and Eckel 2009). In insects, the research on LPL has been centered in a few species, and major findings have been reviewed elsewhere (Fruttero et al. 2017a; Ziegler and Van Antwerpen 2006). Regarding insect reproduction, soluble LPL from oocytes and eggs of *M. sexta* seems to be involved in lipid depletion of endocytosed Lp and Vg, but to date, no more progress has been made in this area (Kawooya et al. 1988; Haruo et al. 1977; Chinzei et al. 1981; Van Antwerpen and Law 1992).

In *N. lugens*, pancreatic lipase-related protein 2 (PLRP2), which in mammals is critical for the digestion and absorption of dietary fats, is highly expressed in developing oocytes and poorly expressed in laid eggs. PLRP2 has been involved in energy metabolism during reproduction. In fact, its downregulation causes inhibition of lipid uptake by the ovaries, promoting a nutritional imbalance in developing oocytes, and leading to failure of egg maturation and development (Xu et al. 2017).

Although *de novo* fatty acid synthesis by the ovaries is very limited, transcriptomic analyses revealed that lipogenic enzymes such as fatty acid synthase (FAS) and acetyl CoA carboxylase (ACC) are upregulated in *R. prolixus* ovaries before feeding, coincidently with an upregulation of the fatty acid biosynthesis pathway during the previtellogenic stage (Leyria et al. 2020a, b). The massive endocytosis of lipoproteins during vitellogenesis is an energy-dependent process; thus, lipid reserves locally synthetized from non-vitellogenic ovaries could play a critical role in supporting the energetic demands of the growing oocyte at the beginning of vitellogenesis, as suggested in mosquitoes (Clifton and Noriega 2012). Further, ACC has also been detected in purified lipid droplets of *M. sexta* ovaries but in trace amounts (Soulages et al. 2012). In *A. aegypti*, FAS and ACC transcript levels increase in ovaries after a blood meal, and a deficiency of both enzymes leads to decreased lipid biosynthesis and fecundity. Interestingly, only ACC seems to be required to produce specialized eggshell lipids from *de novo* fatty acid biosynthesis (Alabaster et al. 2011). More recently, through ultra-high-resolution mass spectrometry combined with specific isotopes, *de novo* synthesis of TAG in the ovaries of mosquitoes was confirmed (Tose et al. 2021).

In addition to the fat body, the dynamics of lipids in the insect midgut influence ovarian development and reproduction. For instance, stearoyl-CoA desaturase (SCD) plays a key role in converting saturated to unsaturated fatty acids, which are necessary to maintain cell membrane fluidity. Loss of function of SCD1 in *A. coluzzii* adult female mosquitoes severely compromises lipid synthesis in midgut cells, which become depleted of lipid droplets, thus reducing lipid shuttling into the hemolymph leading to underdeveloped ovaries (Ferdous et al. 2021).

In insects, long-chain cuticular hydrocarbons act as barriers to prevent water loss and serve as pheromones or pheromone precursors (Schal et al. 1998, 2003). Insect hydrocarbons are produced by oenocytes from the abdominal integument and are accumulated in the hemolymph and the cuticle (Blomquist and Ginzel 2021). In *B. germanica*, the hydrocarbons comprise about 4.85% of egg lipids; they are taken up by growing oocytes via Lp and stored in lipid droplets until use (Fan et al. 2002). Maternal hydrocarbons are not metabolized during embryogenesis and aid to waterproof the cuticle of the juveniles (Fan et al. 2008). In the cockroach *Diploptera punctata*, the hydrocarbons synthesized by the embryo are just 1%; therefore, the female must have contributed with about 99% of the hydrocarbons found in embryos (Youngsteadt et al. 2005). However, in the triatomine *Triatoma infestans*, there is also an internal contribution of hydrocarbons, arising from the egg itself (Juárez 1994). The transport of hydrocarbons through the hemolymph is accomplished by Lp, LTP, or both; indeed, in several insects, this lipid class may constitute up to 40–50% of the lipoprotein lipid cargo (Sevala et al. 1999; Blacklock and Ryan 1994; Takeuchi and Chino 1993). Recently, it has been reported that LpR regulates adult fecundity and the accumulation of cuticular hydrocarbon in the pea aphid *Acyrthosiphon pisum* (Qiao et al. 2021).

Once in the oocytes, lipids are stored in intracellular organelles called lipid droplets. They are at the center of lipid and energy homeostasis and are composed of a core of neutral lipids, mainly TAG, surrounded by a monolayer of phospholipids and proteins that segregates the hydrophobic core from the aqueous cytosol (Olzmann and Carvalho 2019). During the last years, the view on lipid droplets shifted from passive reservoirs of lipids to dynamic organelles with complex roles on lipid metabolism, participating in intracellular signaling, controlling intracellular lipid trafficking, and capturing proteins involved in transcriptional regulation from other cellular compartments (Ibayashi et al. 2021; Dejgaard and Presley 2021). In insects and other metazoans, the most important physiological function of adipocyte lipid droplets is to store lipids and supply them to circulation or other tissues. On the contrary, the function of non-adipocyte lipid droplets, such as those in insect follicular cells, is to keep lipids to fulfill their own needs rather than to cover the demands of the whole body (Suzuki et al. 2011). The pioneer study of Roth and Porter (1964) showed that, in the ovaries of *A. aegypti*, the nurse cells had occasional lipid droplets, whereas the oocyte cytoplasm displayed an abundance of multilobed lipid droplets (Roth and Porter 1964). Perilipins (PLIN) play a major role in the assembly and transport of lipid droplets and TAG metabolism. In insects, only two members of the PLIN family have been described, PLIN1 (or Lsd-1) and PLIN2 (or Lsd-2) (Miura et al. 2002). PLIN1 is a major regulator of lipolysis, while PLIN2 is mainly needed when the synthesis of diacylglycerol and TAG is active disregard of whether the insect is accumulating or using lipids (Tansey et al. 2004; Chen et al. 2017). In *D. melanogaster*, lipid levels are decreased in early embryos of Lsd-2 mutant females, with a concomitant reduction of their progeny hatching rate. These findings reveal that Lsd-2 has a maternal effect, probably because maternal Lsd-2 is needed to guarantee optimal lipid deposition in the mature egg for embryogenesis. It is also possible that maternal expression of Lsd-2 might be needed in the embryo to ensure the proper use of the lipid pool during its development (Teixeira et al. 2003). In the ovaries of *M. sexta*, Lsd-1 is found in very low levels, consistent with the notion that ovaries and developing oocytes are accumulating rather than mobilizing lipids. The structural apolipoproteins of Lp are also found in ovarian lipid droplets, suggesting that Lp not only mediates TAG transfer to lipid droplets but also serves as an important structural component of these organelles in ovaries (Soulages et al. 2012). In *B. mori*, it is also conceivable that the transport of TAG from the lipid droplets to the embryo involves Lp (Yamahama et al. 2008).

During the early-vitellogenic stage of the dragonfly *Orthetrum chrysis*, the lipid droplets appear in the peripheral region of the oocytes, suggesting the incorporation of yolk lipids from the hemolymph through the follicular epithelium (Tembhare and Thakare 1975). In addition, *Matsucoccus matsumurae* oocytes accumulate lipids at the early-vitellogenic stage rather than proteins; in consequence, by the end of vitellogenesis, the oocyte cytoplasm is packed with abundant lipid droplets and fewer yolk bodies (Liu et al. 2014). However, in the mosquito *Culex quinquefasciatus*, the accumulation of lipids in vitellogenic oocytes was observed alongside the formation of yolk bodies (Cardoso et al. 2010). It has been reported that in the ovaries of the triatomines *R. prolixus, P. megistus*, and *D. maxima*, lipid droplets are numerous, large, and homogeneously distributed throughout the vitellogenic oocytes, filling the spaces among yolk bodies, in a distribution expected for a period of lipid accumulation at a high rate (Huebner and Anderson 1972; Fruttero et al. 2011; Leyria et al. 2014). In contrast, during *D. melanogaster* oogenesis, female flies synthesize lipid droplets in the nurse cells of the egg chambers, which are subsequently transferred to the oocyte to supply energy and membrane building blocks for the developing embryo. Therefore, an impairment of maternal lipid mobilization causes developmental arrest and death of the embryo (Kühnlein 2012).

After complete lipid packaging in oocytes, the lipid droplets will be gradually consumed throughout the embryogenesis process, changing their shape, volume, and location within the ooplasm. The main actors and pathways discussed in this section are integrated in Fig. 1.

6.2 Lipid Utilization During the Embryogenic Process

The TAG accumulated in oocytes can be broken down into fatty acids by lipases when needed, which in turn enter the mitochondria for β-oxidation to provide energy for successful embryogenesis. Lipids also play a crucial role as structural components of cellular membranes and signaling molecules. Moreover, lipid droplets work as protein-sequestration sites; for instance, in *D. melanogaster*, the lipid droplets deposited in oocytes are able to sequestrate large amounts of histones H2A, H2Av, and H2B, allowing embryos to build up extranuclear stores of histones to provide them for chromatin assembly in times of high demand during embryogenesis. Overexpression of histones can cause embryonic lethality; thus, sequestration on lipid droplets maintains maternally provided histones inactive until they are required (Cermelli et al. 2006). Lipid droplets also retain Jabba, a histone-docking protein that protects histones from degradation (Stephenson et al. 2021). Histones H2B and H4 have been found in the ovaries of *M. sexta*; however, this tissue does not function as a storage nor source site of histone H2A (Soulages et al. 2012). In mammals, adipose triglyceride lipases play critical roles in TAG mobilization. The *D. melanogaster* embryos lacking Bmm lipase activity are lethal, showing a phenotype with increased TAG in late embryogenesis (Grönke

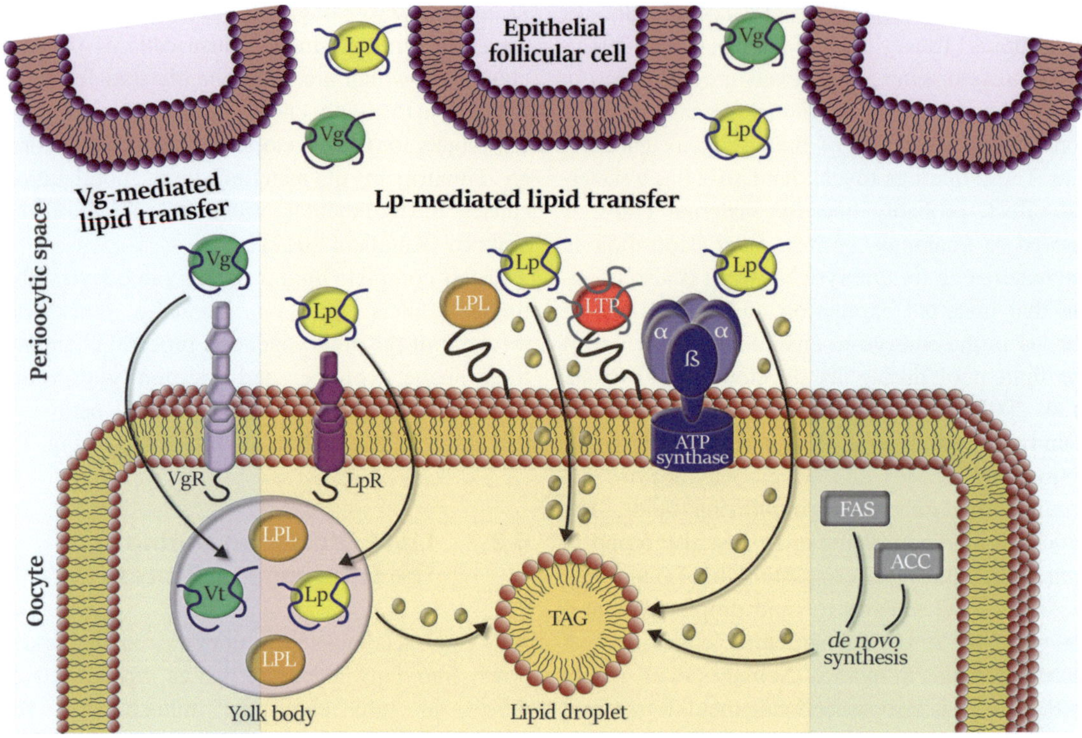

Fig. 1 Schematic processes involved in lipid accumulation by oocytes. The buildup of lipid stores in insect oocytes is mostly accomplished via specialized circulating lipid carriers. Channels between follicular epithelial cells facilitate the passage of the main lipoproteins, lipophorin (Lp) and vitellogenin (Vg), from the hemolymph to the perioocytic space. In the plasma membrane domain of the oocyte, Vg binds its endocytic receptor, VgR, to be taken up and stored in the yolk bodies as vitellin (Vt). The Vg lipid cargo is relocated from the yolk bodies to the lipid droplets by an unknown mechanism. The bulk of neutral lipids in oocytes is supplied mainly by Lp via non-endocytic pathways through its classic docking mechanism. To a lesser extent, lipid storage in oocytes can also be achieved by endocytosis-mediated Lp through the endocytic receptor LpR. The LpR also contributes to the non-endocytic mechanism promoting the recruitment of lipid transfer particles (LTP) to the oocyte plasma membrane, thus stabilizing Lp to facilitate lipid transfer. In addition, the non-endocytic mechanism of Lp involves the β subunit of the ATP synthase complex (β-ATPase) and the membrane-bound lipoprotein lipase (LPL). Although *de novo* fatty acid synthesis by oocytes is minimal, it has been reported to work via lipogenic enzymes such as fatty acid synthase (FAS) and acetyl CoA carboxylase (ACC). Depending on the insect species, the metabolic pathways depicted in the figure may contribute to a different extent to the lipid reserves of the oocyte

et al. 2005). Interestingly, similar phenotypes have been reported for Lsd-2 mutants (Teixeira et al. 2003). Nevertheless, the direct interaction of Lsd-2 with any lipase, which could explain this phenotype, has not been reported (Kühnlein 2012). In *N. lugens* and *G. morsitans* females, *Bmm* gene silencing or inhibition of the lipase activity leads to retarded ovarian development and impaired egg-laying or reproductive capacity (Attardo et al. 2012; Lu et al. 2018b).

Another mechanism regulating lipids availability during embryogenesis is autophagy, a process involved in the degradation of whole organelles, including lipid droplets (Singh et al. 2009). In *D. melanogaster*, autophagosomes might be involved in supplying nutrients to the developing embryo by catabolizing lipid droplets, as these organelles have been found inside autophagosomes (Kuhn et al. 2015). The contribution of autophagy to lipid metabolism during reproduction has been covered elsewhere (Fruttero et al. 2017a).

In the viviparous cockroach *D. punctata* and the mosquito *C. quinquefasciatus*, the fuel used during embryogenesis is almost entirely lipids (Stay and Coop 1973; Van Handel 1993). However, in the tropical butterfly *Bicyclus anynana* glycogen is used largely for embryonic development, while a much more extensive part of lipids is transferred to hatchlings (Geister et al. 2009). In another lepidopteran insect, *Antheraea mylitta,* total lipids are depleted by about 60% during embryogenic development, mostly during the later developmental stages where histogenesis, tissue differentiation, and growth occur (Pant et al. 1979). Further, in *B. mori*, lipid storage is used during late embryogenesis, coinciding with tracheal development in the developing caterpillars (Yamahama et al. 2008). In *R. prolixus*, about 40–50% of TAG reserves are used during embryogenesis; the rest, which is internalized by the embryo probably during dorsal closure, is slowly used by nymphs in an efficient and controlled mechanism (Santos et al. 2011). In agreement, in the milkweed bug *Oncopeltus fasciatus*, 90% of the lipids stored in the egg remain present before hatching (Babcock and Rutschky 1961). In neonate larvae of *A. gambiae* and *A. aegypti*, fatty acids and phospholipids accumulated in the oocyte during vitellogenesis play different functions: the fatty acids distributed along the sides of the body of larval neonates provide energy for their initial movements when seeking food, and the phospholipids emulsify the ingested food facilitating its assimilation (Atella and Shahabuddin 2002).

In *N. lugens*, a new mechanism for lipid droplet content utilization has been reported. RNAi-mediated depletion of 26S proteasome non-ATPase regulatory subunits (NIRpns) decreases the proteolytic activity of the proteasome and hampers the transcription of lipases and Vg genes in the fat body and the ovaries of adult females, disrupting lipid metabolism in both tissues and reducing TAG in the ovaries. Further, impaired lipid utilization in oocytes likely due to lipases downregulation has also been reported (Wang et al. 2021).

Ecdysteroids are supplied maternally into the eggs as physiologically inactive conjugated ecdysteroid phosphates. In *B. mori* oocytes, these inactive lipid-derived hormones synthesized from cholesterol bind to vitellin to form a complex that will be stored in yolk bodies (Yamada and Sonobe 2003). Then, during early embryogenesis, an ecdysteroid-phosphate phosphatase enzyme converts the conjugated ecdysteroids into the active form 20E (Sonobe and Yamada 2004). As shown in other insect species, the resulting free 20E in *B. mori* has a role in morphogenesis at an early embryonic stage in which the prothoracic glands are still not differentiated (Bullière et al. 1979; Lagueux et al. 1984; Warren et al. 1986; Scalia et al. 1987; Yamada et al. 2005).

Thus far, we have reviewed the most important insights of lipid homeostasis regulating insect reproduction. In particular, we have shown how the CNS/CA/CC-fat body-ovary axis is essential for the insect to produce viable eggs. In the next section, we will summarize some findings on lipid metabolism occurring below the CNS/CA/CC-fat body-ovary axis that are equally significant for successful reproduction but that have been less explored.

7 Lipid Signaling in Accessory Glands: Exploring the Role on Spermathecae and Cement Glands

The female insect spermatheca takes, preserves, and releases sperm to fertilize eggs and offers an appropriate environment that guarantees the long-term viability of sperm (Pascini and Martins 2017). Recently, it has been suggested that in *Apis mellifera* the spermatheca of new-laying honeybee queens use lipids and lipid-like molecules, including fatty acyls and glycerophospholipids, phenol lipids, and sterol lipids, to maintain and protect the sperm for long-term storage (Liu et al. 2020). Moreover, prostaglandins appear to be involved in the egg-laying activity of the *Teleogryllus commodus* cricket either by acting directly on muscles or indirectly via the hormonal system.

Prostaglandins are categorized as physiologically active eicosanoids, a group of lipid-derived molecules that are local mediators. They seem to be synthesized in the spermatheca of newly mated females from the prostaglandin synthase complex and a substrate transferred from males. Thus, the route of prostaglandin movement is from the spermathecae to its target sites (Loher 1979). Coincidently, in *B. mori, L. migratoria*, and *T. infestans,* the prostaglandins transferred from male to female during mating are involved in egg-laying behavior (Settya and Rmaiah 1980; Lange 1984; Brenner and Bernasconi 1989). Interestingly, it has been proposed that in *B. mori* prostaglandin-mediated signaling is vital to maintain physiological homeostasis in ovarian follicles (Machado et al. 2007).

In insects, secretions from the female accessory glands play important functions in insemination, supporting the transfer of the sperm into the spermatheca during copulation. They also function as a lubricant, facilitating the transport of the oocytes through an ovipositor, and provide protection from diverse biotic and abiotic stressors (Leopold and Degrugillier 1973; Lococo and Huebner 1980; Degrugillier 1985; Marchini et al. 1991; Lung et al. 2001; Rosetto et al. 2003; Poiani 2006). Therefore, a lipophilic substance is essential for at least the first two roles described above. Accordingly, a morphological and ultrastructural study of the accessory glands in the genital tract of female house crickets revealed that a single gland cell is subdivided in a basal region containing the nucleus and the machinery devoted to protein synthesis and lipid production and an apical region, mostly composed of protein-transferring vesicles, lipid droplets, microfilaments, and a layer of microvilli (Sturm 2012). Indeed, the lipophilic nature of the gland secretion was confirmed since lipids represented about 65% of its content.

8 Pheromones and Their Role in Insect Reproduction

Pheromones are chemical signals used for communication between members of the same species. In insects, lipid pheromones influence survival, reproduction, and social organization. Several lipids present in insect cuticles, produced mainly by the specialized secretory cells oenocytes, function as pheromones and communicate specific information. Fatty acids arising from *de novo* synthesis, the diet, or both, are the most important source for pheromone synthesis (Yew and Chung 2015). From a general point of view, pheromones are essential for successful reproduction because they promote chemical signals used to recognize members of the same species, as aphrodisiacs or attractants of potential mates, and to indicate their reproductive status and fitness (Blomquist et al. 2020). The sensitivity and selectivity of the pheromone-responsive neurons are largely based on specific pheromone receptors residing in their ciliary membrane (Fleischer and Krieger 2018). The first active pheromone identified in *B. mori* was bombykol, which announces the availability and location of insect females (Butenandt 1959). In many dipteran species, pheromones found on the cuticular surface are used as attractants and aphrodisiacs that influence mate choice and induce courtship (Wicker-Thomas 2007). In addition, in eusocial insects such as wasps, bees, ants, and termites, pheromones are essential for the establishment of a social hierarchy as well as for suppression of reproduction in workers (Yew and Chung 2015; Blomquist et al. 2020).

9 Function of Lipids in Spermatogenesis

Compared to its female counterpart, the male reproductive system has received less attention, not only in the volume of publications but also in the number of insect species investigated. The study of the role of lipids in insect spermatogenesis has relied heavily on model organisms such as *D. melanogaster* (Thanh et al. 2020). This trend is due to the fact that *D. melanogaster* and humans not only share genes and gene families related to metabolism but also because their organ systems devoted to lipid storage, uptake, and processing are analogous. Moreover, *D. melanogaster* offers the possibility to work with multiple forward and reverse genetics tools for the discovery of roles in regulation and metabolism in the context of the whole organism, rather than in isolated cells (Wang and Huang 2012; Vrablik and Watts 2013).

The spermatogenesis process can be essentially described as the production of mature spermatozoa from germline stem cells (GSC) which will undergo mitosis to form spermatocytes and, later on, meiosis and cytokinesis to produce spermatids. After several elongation steps, individualization and coiling, mature sperms will emerge (Dallai 2014). In this context, testes are lipid-rich organs where complex membrane reorganizing processes occur, and tight lipid homeostasis is required.

The maintenance of *D. melanogaster* male GSCs relies on mitochondrial fusion. The presence of dysfunctional mitochondria in GSCs leads to a reduction in fatty acid oxidation, causing an accumulation of excess fatty acids in the cytoplasm and resulting in positive-feedback-mediated lipid droplet accumulation. Under these circumstances, GSCs lose their ability to self-renew, undergo differentiation, and thus, spermatogonial differentiation is inhibited (Sênos Demarco et al. 2019). Additionally, there are other types of cells, cyst cells, which encapsulate GSCs to support their development. Cyst cells develop autophagy to regulate lipid levels. Defects in this process promote lipid accumulation and the loss of early cyst cells, thus inhibiting sperm production (Sênos Demarco et al. 2020). Another key gene involved in cyst cell lipid homeostasis is the phosphatidylcholine lysophospholipase gene, *Swiss cheese*, which participates in lipid droplet formation and consequently regulates cyst cell function (Melentev et al. 2022). Moreover, in cyst cells, silencing the *noa* gene which encodes ELOVL6, a member of the fatty acid elongase family that synthesizes very-long-chain fatty acids, led to defects in the spermatogenesis individualization process (Jung et al. 2007). The relevance of elongases and very-long-chain fatty acids in the development of sperms has also been observed in loss-of-function experiments with a mutant *bond* gene encoding a different ELOVL and causing a male sterility phenotype (Szafer-Glusman et al. 2008).

Furthermore, it has been demonstrated that CDP-diacylglycerol synthase (CDS), a crucial enzyme in lipid biosynthesis, plays a significant role in fine-tuning lipid composition during spermatogenesis. It acts by controlling the amounts of signaling lipids, and reduced expression of CDS changes lipid composition, leading to male sterility (Laurinyecz et al. 2016).

The fatty acid metabolism and spermatogenesis were first linked when *scully* (*scu*), an essential gene of *D. melanogaster*, was characterized. *Scu* is a homolog of the mammalian mitochondrial type II L-3-hydroxyacyl-CoA dehydrogenase involved in the beta-oxidation of short-chain fatty acids. In addition, two *scu* mutants have been reported to show abnormal lipid accumulation that led to a significant reduction in the size of testes and spermatocyte degeneration (Torroja et al. 1998).

The requirement of lysophospholipid acyltransferase for normal spermatogenesis was evidenced when three gene homologs of the membrane-bound *O*-acyltransferase domain containing 1 (MBOAT1) enzyme, *Oys*, *Nes*, and *Frj*, presented a combined effect on male fertility, generating a phenotype with defective spermatid individualization (Steinhauer et al. 2009). This phenotype can be caused by abnormally increased levels of saturated fatty acids of several phospholipids.

The regulation of the phosphatidylinositol pathway is carried out by several kinases and phosphatases, and defects in the involved genes including *SigD*, *Sktl*, *fwd*, and *giotto* interfere in proper sperm development (Fabian et al. 2010; Giansanti et al. 2006; Brill et al. 2000; Gatt and Glover 2006; Polevoy et al. 2009). The traffic of cholesterol is also implicated in spermatogenesis as it has been described that Niemann-Pick type C (NPC) and Oxysterol binding protein (OSBP) mutants present phenotypes with dysfunction in spermatid individualization (Wang et al. 2011; Ma et al. 2010).

Overall, several studies discussed in this section strengthen the value of *Drosophila* as a powerful genetic tool and point to the relevance of phosphatidylinositol, fatty acids, and cholesterol for proper spermatogenesis.

10 Conclusions and Future Directions

Insects not only shaped the earliest terrestrial ecosystems about hundreds of million years ago but nowadays, they remain among the most important members of the animal kingdom. In fact, a significant part of these attributes resides in their adaptation skills and the remarkable reproductive capacity. Although at first glance, reproduction seems like a succession of events beginning with mating, followed by oocyte development, and ending with the laying of eggs, females engage in an intricate task, coordinating endocrine and nutritional signals to provisioning eggs with resources to generate hundreds of living and healthy offspring to accomplish successful reproduction (Fig. 2). Lipids are essential energy stores and play a pivotal role by signaling processes accounting for insect reproduction.

Through the brain-CA/CC-fat body-ovary axis, during oogenesis, females manage different ways to build up lipid stockpiles in developing oocytes, which in turn will be broken down to sustain embryogenesis. Lipid-derived pheromones released by females also influence reproduction, affecting successful mating. The integration of genomics and metabolomic advances from a few insect models have allowed the identification and the know-how of key regulatory hormones and neuropeptides driving a well-adjusted lipogenic and lipolytic cascade of events in reproductive females. Although most of these events are part of conserved pathways, due to the extraordinary diversity of insect species and their many reproductive strategies, such notable findings would need to be refined since an endocrine or nutritional signal may elicit dissimilar effects on target tissues, be operative in different

Fig. 2 (continued) supply to improve their energetic status to regulate the onset of reproductive maturity and egg production. *Bombyx mori* and *Aedes aegypti* were used as models to show pheromone and nutrition participation, respectively. The main hormones involved in lipogenesis and lipolysis with a high impact on successful reproduction are insulin-like peptide/insulin-related peptide (ILP/IRP), adipokinetic hormone (AKH), juvenile hormone (JH), and ecdysteroids. ILP/IRP and AKH are peptide hormones synthesized mainly in the brain and *corpora cardiaca* (CC), respectively; JHs and ecdysone (Ec) are produced in the *corpora allata* (CA) and the ovaries, respectively. The integrated model shows the main events involved in lipid homeostasis to accomplish successful reproduction, and it was established on the references described in each corresponding section. Each finding is linked to the insect model used for its discovery (*gray silhouette*). In a general view, the fat body senses hormonal and nutritional stimuli that initiate major biochemical and ultrastructural tissue changes committed to egg production, whose success is based mainly on the uptake of nutrients by the oocytes. In the spermatheca, lipids and lipid-like molecules are used mostly to maintain and protect the sperm for long-term storage. Eicosanoids short-acting lipid hormones constitute an important category of locally active mediators involved in the egg-laying process. Lipid-like secretions from the female accessory glands support the transfer of the sperm into the spermatheca and lubricate the oviducts to facilitate the transport of eggs. After oviposition, the lipid accumulated by the oocytes, largely triacylglycerol, is broken down into fatty acids to provide energy for embryogenesis, serving as structural components of cellular membranes and as signaling molecules. In several insects, part of the maternal lipid accumulated into eggs is used by larva/nymphs in an efficient and controlled mechanism to provide energy for their initial movements when seeking food. The tissues in the figure are schematically drawn and do not match any particular species. *Purple triangular arrows*, tissues from which hormones and peptide hormones or proteins are synthesized and released; *purple circular arrows*, tissues where hormones and peptide hormones or proteins play a role; *green triangular arrows*, stimulation; *red lines*, inhibition. Bmm, Brummer lipase; COX, cyclooxygenases; GP3 pathway, glycerol-3-phosphate pathway; HNF4, hepatocyte nuclear factor 4; LOX, lipoxygenases; Lp, lipophorin; LpR, lipophorin receptor; Lsd2, lipid storage droplet-2 protein; SREBP, sterol regulatory-element binding proteins; TAG, triacylglycerol; Vg, vitellogenin; VgR, vitellogenin receptor; YPs, yolk proteins

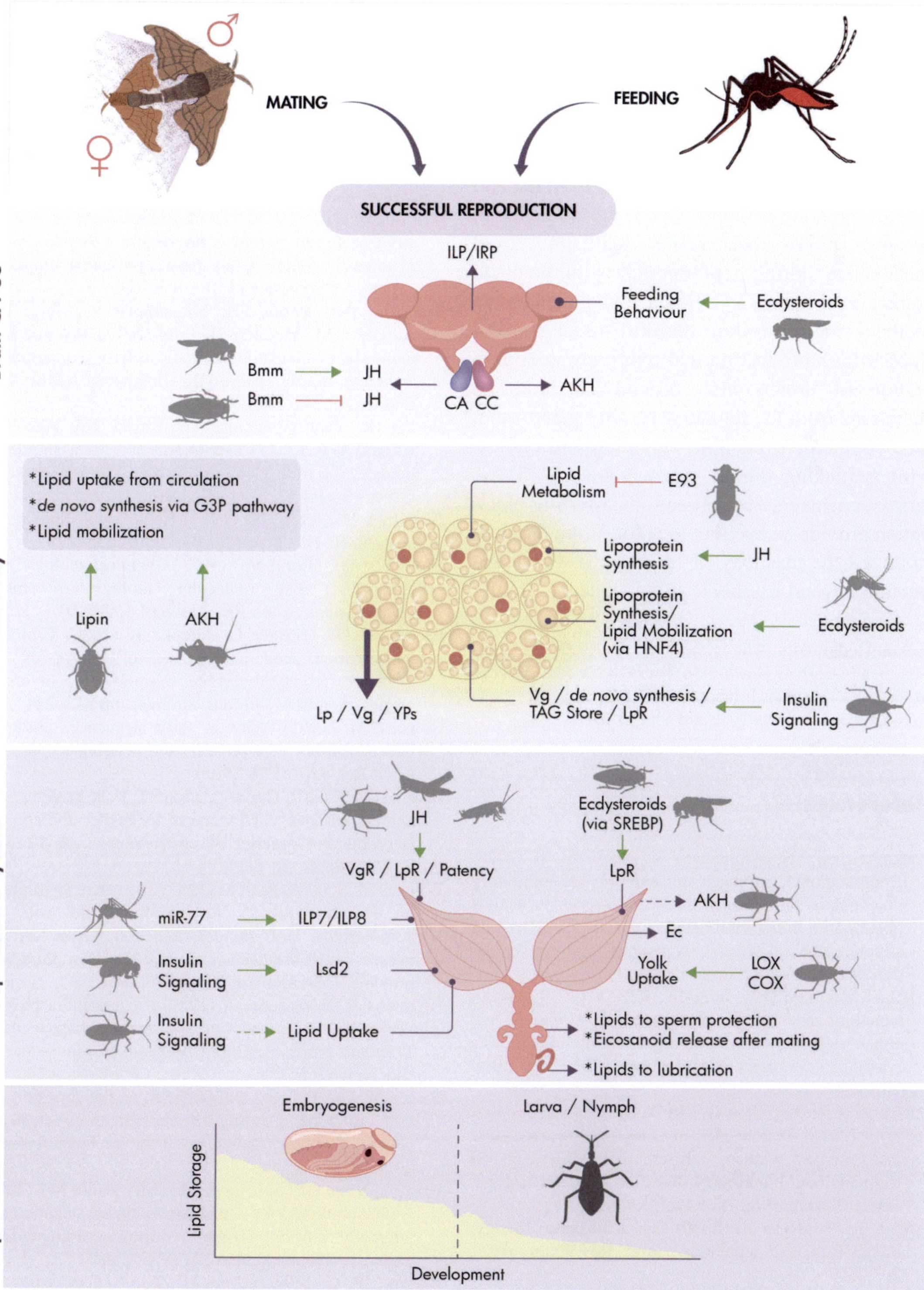

Fig. 2 The central nervous system/*corpora allata*/ corpora cardiaca-fat body-ovary axis and its implication in the lipid dynamics for successful reproduction: main findings in an integrated model. The lipid pheromones released by female insects are used mainly to promote mating. Once mated, adult female insects use dietary

physiological temporal windows or even in a wide range of concentrations to accomplish lipid homeostasis during reproduction.

Therefore, caution should be exerted when drawing general conclusions as more species are studied, and molecular and analytic methods are improved. Emerging evidence linking JH, Ec, and insulin signaling pathways opens the challenge to uncover novel mechanisms regulating lipid metabolism during reproduction with practical applications on more rational and nonchemical methods for controlling harmful insect species. Successful reproduction and a thriving next generation of insects also depend on lipid-like secretions from the female accessory glands and, as a counterpart, on lipid signaling and cholesterol trafficking during spermatogenesis. These largely uncharacterized functions awaiting elucidation provide an exciting area for future research to unveil the plasticity of lipids on insect reproduction beyond its classic role on energy supply.

Acknowledgments The authors thank Dr. David Rojas Marquez (@darwid_illustration) for drawing Fig. 2. Work in the L.E.C and L.L.F laboratory is supported by grants from CONICET, FONCyT, and SECyT-UNC.

References

Aathmanathan VS, Arumugam V, Krishnan M (2020) Computational approach to explore the inhibitory potential of biologically derived compounds against *Spodoptera litura* vitellogenin receptor (VgR) using structure based virtual screening and molecular dynamics. J Biomol Struct Dyn 1–7

Al Baki MA, Lee DW, Jung JK, Kim Y (2019) Insulin signaling mediates previtellogenic development and enhances juvenile hormone-mediated vitellogenesis in a lepidopteran insect, *Maruca vitrata*. BMC Dev Biol 19(1):14

Alabaster A, Isoe J, Zhou G, Lee A, Murphy A, Day WA et al (2011) Deficiencies in acetyl-CoA carboxylase and fatty acid synthase 1 differentially affect eggshell formation and blood meal digestion in *Aedes aegypti*. Insect Biochem Mol Biol 41(12):946–955

Arrese EL, Soulages JL (2010) Insect fat body: Energy, metabolism, and regulation. Annu Rev Entomol 55: 207–225

Atella GC, Shahabuddin M (2002) Differential partitioning of maternal fatty acid and phospholipid in neonate mosquito larvae. J Exp Biol 205(23): 3623–3630

Attardo GM, Benoit JB, Michalkova V, Yang G, Roller L, Bohova J et al (2012) Analysis of lipolysis underlying lactation in the tsetse fly, *Glossina morsitans*. Insect Biochem Mol Biol 42(5):360–370

Babcock KL, Rutschky CW (1961) Lipids in insect eggs: a review with new evidence from the milkweed bug, *Oncopeltus fasciatus* (Hemiptera, Lygaeidae). Ann Entomol Soc Am 54(2):156–164.

Baldini F, Gabrieli P, South A, Valim C, Mancini F, Catteruccia F (2013) The interaction between a sexually transferred steroid hormone and a female protein regulates oogenesis in the malaria mosquito *Anopheles gambiae*. PLoS Biol 11(10):e1001695

Baumann AA, Benoit JB, Michalkova V, Mireji PO, Attardo GM, Moulton JK et al (2013) Juvenile hormone and insulin suppress lipolysis between periods of lactation during tsetse fly pregnancy. Mol Cell Endocrinol 372(1–2):30–41

Benoit JB, Yang G, Krause TB, Patrick KR, Aksoy S, Attardo GM (2011) Lipophorin acts as a shuttle of lipids to the milk gland during tsetse fly pregnancy. J Insect Physiol 57(11):1553–1561

Blacklock BJ, Ryan RO (1994) Hemolymph lipid transport. Insect Biochem Mol Biol 24(9):855–873

Blomquist GJ, Ginzel MD (2021) Chemical ecology, biochemistry, and molecular biology of insect hydrocarbons. Annu Rev Entomol 66:45–60

Blomquist GJ, Tittiger C, Jurenka R (2020) Cuticular hydrocarbons and pheromones of arthropods. In: Wilkes H (ed) Hydrocarbons, oils and lipids: diversity, origin, chemistry and fate. Springer, pp 213–244

Bownes M (1992) Why is there sequence similarity between insect yolk proteins and vertebrate lipases? J Lipid Res 33(6):777–790

Bownes M (2020) The regulation of Yolk protein gene expression and vitellogenesis in higher diptera. In: Reproductive biology of invertebrates, vol 12. Part B. CRC Press, pp 111–144

Bownes M, Scott A, Blair M (1987) The use of an inhibitor of protein synthesis to investigate the roles of ecdysteroids and sex-determination genes on the expression of the genes encoding the drosophila yolk proteins. Development 101(4):931–941

Brenner RR, Bernasconi A (1989) Prostaglandin biosynthesis in the gonads of the hematophagus insect *Triatoma infestans*. Comp Biochem Physiol – Part B Biochem 93(1):1–4

Briegel H, Gut T, Lea AO (2003) Sequential deposition of yolk components during oogenesis in an insect, *Aedes aegypti* (Diptera: Culicidae). J Insect Physiol 49(3): 249–260

Brill JA, Hime GR, Scharer-Schuksz M, Fuller MT (2000) A phospholipid kinase regulates actin organization and intercellular bridge formation during germline cytokinesis. Development 127(17):3855–3864

Brown MR, Clark KD, Gulia M, Zhao Z, Garczynski SF, Crim JW et al (2008) An insulin-like peptide regulates egg maturation and metabolism in the mosquito *Aedes aegypti*. Proc Natl Acad Sci U S A 105(15):5716–5721

Bullière D, Bullière F, de Reggi M (1979) Ecdysteroid titres during ovarian and embryonic development in *Blaberus craniifer*. Wilhelm Roux's Arch Dev Biol 186(2):103–114

Butenandt A (1959) Über den Sexual-Lockstoff des Seidenspinners *Bombyx mori*: Reindarstellung und Konstitution. Zeitschrift Für Naturforschung B, J Chem Sci 14b:283–284

Canavoso LE, Jouni ZE, Karnas KJ, Pennington JE, Wells MA (2001) Fat metabolism in insects. Annu Rev Nutr 21:23–46

Cardoso AF, Cres RL, Moura AS, de Almeida F, Bijovsky AT (2010) *Culex quinquefasciatus* vitellogenesis: morphological and biochemical aspects. Mem Inst Oswaldo Cruz 105(3):254–262

Cazzamali G, Saxild NPE, Grimmelikhuijzen CJP (2002) Molecular cloning and functional expression of a *Drosophila* corazonin receptor. Biochem Biophys Res Commun 298(1):31–36

Cermelli S, Guo Y, Gross SP, Welte MA (2006) The lipid-droplet proteome reveals that droplets are a protein-storage depot. Curr Biol 16(18):1783–1795

Chen ME, Lewis DK, Keeley LL, Pietrantonio PV (2004) cDNA cloning and transcriptional regulation of the vitellogenin receptor from the imported fire ant, *Solenopsis invicta* Buren (Hymenoptera: Formicidae). Insect Mol Biol 13(2):195–204

Chen Y, Rui BB, Tang LY, Hu CM (2015) Lipin family proteins - Key regulators in lipid metabolism. Ann Nutr Metab 66(1):10–18

Chen X, Firdaus SJ, Howard AD, Soulages JL, Arrese EL (2017) Clues on the function of *Manduca sexta* perilipin 2 inferred from developmental and nutrition-dependent changes in its expression. Insect Biochem Mol Biol 81:19–31

Chinzei Y, Chino H, Wyatt GR (1981) Purification and properties of vitellogenin and vitellin from *Locusta migratoria*. Insect Biochem 11(1):1–7

Cho KH, Raikhel AS (2001) Organization and developmental expression of the mosquito vitellogenin receptor gene. Insect Mol Biol 10(5):465–474

Chowański S, Walkowiak-Nowicka K, Winkiel M, Marciniak P, Urbański A, Pacholska-Bogalska J (2021) Insulin-like peptides and cross-talk with other factors in the regulation of insect metabolism. Front Physiol 12:701203

Church SH, De Medeiros BAS, Donoughe S, Márquez Reyes NL, Extavour CG (1950) Repeated loss of variation in insect ovary morphology highlights the role of development in life-history evolution. Proc R Soc B Biol Sci 2021(288):20210150

Ciudad L, Piulachs MD, Bellés X (2006) Systemic RNAi of the cockroach vitellogenin receptor results in a phenotype similar to that of the *Drosophila* yolkless mutant. FEBS J 273(2):325–335

Ciudad L, Bellés X, Piulachs M-D (2007) Structural and RNAi characterization of the German cockroach lipophorin receptor, and the evolutionary relationships of lipoprotein receptors. BMC Mol Biol 8(1):53

Clifton ME, Noriega FG (2012) The fate of follicles after a blood meal is dependent on previtellogenic nutrition and juvenile hormone in *Aedes aegypti*. J Insect Physiol 58(7):1007–1019

Coelho HSL, Atella GC, Moreira MF, Gondim KC, Masuda H (1997) Lipophorin density variation during oogenesis in *Rhodnius prolixus*. Arch Insect Biochem Physiol 35(3):301–313

Dallai R (2014) Overview on spermatogenesis and sperm structure of Hexapoda. Arthropod Struct Dev 43(4): 257–290

Degrugillier ME (1985) In vitro release of house fly, *Musca domestica* l. (Diptera : Muscidae), acrosomal material after treatments with secretion of female accessory gland and micropyle cap substance. Int J Insect Morphol Embryol 14(6):381–391

Dejgaard SY, Presley JF (2021) Interactions of lipid droplets with the intracellular transport machinery. Int J Mol Sci 22(5):2776

Downer RGH, Matthews JR (1976) Patterns of lipid distribution and utilisation in insects. Integr Comp Biol 16(4):733–745

Eid DM, Chereddy SCRR, Palli SR (2020) The effect of E93 knockdown on female reproduction in the red flour beetle, *Tribolium castaneum*. Arch Insect Biochem Physiol 104(4):e21688

Fabian L, Wei HC, Rollins J, Noguchi T, Blankenship JT, Bellamkonda K et al (2010) Phosphatidylinositol 4,5-bisphosphate directs spermatid cell polarity and exocyst localization in *Drosophila*. Mol Biol Cell 21(9):1546–1555

Fan Y, Chase J, Sevala VL, Schal C (2002) Lipophorin-facilitated hydrocarbon uptake by oocytes in the German cockroach *Blattella germanica* (L.). J Exp Biol 205(6):781–790

Fan Y, Eliyahu D, Schal C (2008) Cuticular hydrocarbons as maternal provisions in embryos and nymphs of the cockroach *Blattella germanica*. J Exp Biol 211(4): 548–554

Ferdous Z, Fuchs S, Behrends V, Trasanidis N, Waterhouse RM, Vlachou D et al (2021) *Anopheles coluzzii* stearoyl-CoA desaturase is essential for adult female survival and reproduction upon blood feeding. PLoS Pathog 17(5):e1009486

Fleischer J, Krieger J (2018) Insect pheromone receptors – Key elements in sensing intraspecific chemical signals. Front Cell Neurosci 12:425

Fruttero LL, Frede S, Rubiolo ER, Canavoso LE (2011) The storage of nutritional resources during vitellogenesis of *Panstrongylus megistus* (Hemiptera: Reduviidae): the pathways of lipophorin in lipid delivery to developing oocytes. J Insect Physiol 57(4):475–486

Fruttero LL, Demartini DR, Rubiolo ER, Carlini CR, Canavoso LE (2014) β-chain of ATP synthase as a lipophorin binding protein and its role in lipid transfer in the midgut of *Panstrongylus megistus* (Hemiptera: Reduviidae). Insect Biochem Mol Biol 52(1):1–12

Fruttero LL, Leyria J, Canavoso LE (2017a) Lipids in insect oocytes: From the storage pathways to their multiple functions. In: Kloc M (ed) Oocytes results and problems in cell differentiation. Springer, pp 403–434

Fruttero LL, Leyria J, Ramos FO, Stariolo R, Settembrini BP, Canavoso LE (2017b) The process of lipid storage in insect oocytes: The involvement of β-chain of ATP synthase in lipophorin-mediated lipid transfer in the chagas' disease vector *Panstrongylus megistus* (Hemiptera: Reduviidae). J Insect Physiol 96:82–92

Fruttero LL, Leyria J, Moyetta NR, Ramos FO, Settembrini BP, Canavoso LE (2019) The fat body of the hematophagous insect, *Panstrongylus megistus* (Hemiptera: Reduviidae): Histological features and participation of the β-chain of ATP synthase in the lipophorin-mediated lipid transfer. J Insect Sci 19(4):16

Gäde G (2004) Flight or fight-the need for adipokinetic hormones. Int Congr Ser 1275:134–140

Gäde G, Auerswald L (2003) Mode of action of neuropeptides from the adipokinetic hormone family. Gen Comp Endocrinol 132(1):10–20

Gäde G, Marco HG (2013) AKH/RPCH peptides. In: Kastin A (ed) Handbook of biologically active peptides. Elsevier Inc., pp 185–190

Gäde G, Šimek P, Clark KD, Marco HG (2013) Five functional adipokinetic peptides expressed in the corpus cardiacum of the moth genus *Hippotion* (Lepidoptera, Sphingidae). Regul Pept 184:85–95

Gatt MK, Glover DM (2006) The *Drosophila* phosphatidylinositol transfer protein encoded by vibrator is essential to maintain cleavage-furrow ingression in cytokinesis. J Cell Sci 119(11):2225–2235

Geister TL, Lorenz MW, Hoffmann KH, Fischer K (2009) Energetics of embryonic development: Effects of temperature on egg and hatchling composition in a butterfly. J Comp Physiol B Biochem Syst Environ Physiol 179(1):87–98

Giansanti MG, Bonaccorsi S, Kurek R, Farkas RM, Dimitri P, Fuller MT et al (2006) The class I PITP giotto is required for *Drosophila* cytokinesis. Curr Biol 16(2):195–201

Go GW, Mani A (2012) Low-density lipoprotein receptor (LDLR) family orchestrates cholesterol homeostasis. Yale J Biol Med 85(1):19–28

Gondim KC, Oliveira PL, Masuda H (1989) Lipophorin and oögenesis in *Rhodnius prolixus*: Transfer of phospholipids. J Insect Physiol 35(1):19–27

Gondim KC, Atella GC, Pontes EG, Majerowicz D (2018) Lipid metabolism in insect disease vectors. Insect Biochem Mol Biol 101:108–123

Grönke S, Mildner A, Fellert S, Tennagels N, Petry S, Müller G et al (2005) Brummer lipase is an evolutionary conserved fat storage regulator in *Drosophila*. Cell Metab 1(5):323–330

Guo S, Tian Z, Zhu F, Liu W, Wang XP (2021a) Lipin modulates lipid metabolism during reproduction in the cabbage beetle. Insect Biochem Mol Biol 139:103668

Guo S, Wu QW, Tian Z, Zhu L, King-Jones K, Zhu F et al (2021b) Krüppel homolog 1 regulates photoperiodic reproductive plasticity in the cabbage beetle *Colaphellus bowringi*. Insect Biochem Mol Biol 134: 103582

Hansen KK, Stafflinger E, Schneider M, Hauser F, Cazzamali G, Williamson M et al (2010) Discovery of a novel insect neuropeptide signaling system closely related to the insect adipokinetic hormone and corazonin hormonal systems. J Biol Chem 285(14): 10736–10747

Hansen M, Flatt T, Aguilaniu H (2013) Reproduction, fat metabolism, and life span: What is the connection? Cell Metab 17(1):10–19

Hansen IA, Attardo GM, Rodriguez SD, Drake LL (2014) Four-way regulation of mosquito yolk protein precursor genes by juvenile hormone-, ecdysone-, nutrient-, and insulin-like peptide signaling pathways. Front Physiol 5:103

Haruo C, Downer RGH, Koui T (1977) The role of diacylglycerol-carrying lipoprotein I in lipid transport during insect vitellogenesis. Biochim Biophys Acta (BBA)/Lipids Lipid Metab 487(3):508–516

Hodin J (2009) She shapes events as they come: plasticity in female insect reproduction. In: Whitman D, Ananthakrishnan T (eds) Phenotypic plasticity of insects: mechanisms and consequences. Science Publishers, pp 423–475

Hou QL, Chen EH, Jiang HB, Wei DD, Gui SH, Wang JJ et al (2017) Adipokinetic hormone receptor gene identification and its role in triacylglycerol mobilization and sexual behavior in the oriental fruit fly (*Bactrocera dorsalis*). Insect Biochem Mol Biol 90:1–13

Huebner E, Anderson E (1972) A cytological study of the ovary of *Rhodnius prolixus*. II. Oocyte differentiation. J Morphol 137(4):385–415

Husain M, Rasool KG, Tufail M, Aldawood AS (2020) Molecular characterization, expression pattern and RNAi-mediated silencing of vitellogenin receptor gene in almond moth, *Cadra cautella*. Insect Mol Biol 29(4):417–430

Ibayashi M, Aizawa R, Mitsui J, Tsukamoto S (2021) Homeostatic regulation of lipid droplet content in mammalian oocytes and embryos. Reproduction 162(6):R99–R109

Igarashi F, Ogihara MH, Iga M, Kataoka H (2018) Cholesterol internalization and metabolism in insect prothoracic gland, a steroidogenic organ, via lipoproteins. Steroids 134:110–116

Isoe J, Hagedorn HH (2007) Mosquito vitellogenin genes: Comparative sequence analysis, gene duplication, and the role of rare synonymous codon usage in regulating expression. J Insect Sci 7:1–49

Jeon T, Osborne TF (2012) SREBPs: Metabolic integrators in physiology and metabolism. Trends Endocrinol Metab 23(2):65–72

Jing YP, Wen X, Li L, Zhang S, Zhang C, Zhou S (2021) The vitellogenin receptor functionality of the migratory locust depends on its phosphorylation by juvenile hormone. Proc Natl Acad Sci U S A 118(37): e2106908118

Jouni ZE, Takada N, Gazard J, Maekawa H, Wells MA, Tsuchida K (2003) Transfer of cholesterol and diacylglycerol from lipophorin to *Bombyx mori* ovarioles *in vitro*: Role of the lipid transfer particle. Insect Biochem Mol Biol 33(2):145–153

Juárez P (1994) Hydrocarbon biosynthesis in *Triatoma infestans* eggs. Arch Insect Biochem Physiol 25(3): 193–206

Jung A, Hollmann M, Schäfer MA (2007) The fatty acid elongase NOA is necessary for viability and has a somatic role in *Drosophila* sperm development. J Cell Sci 120(16):2924–2934

Kawooya JK, Osir EO, Law JH (1988) Uptake of the major hemolymph lipoprotein and its transformation in the insect egg. J Biol Chem 263(18):8740–8747

Khalid MZ, Ahmad S, Ngegba PM, Zhong G (2021) Role of endocrine system in the regulation of female insect reproduction. Biology (Basel) 10(7):614

Kuhn H, Sopko R, Coughlin M, Perrimon N, Mitchison T (2015) The Atg1-Tor pathway regulates yolk catabolism in *Drosophila* embryos. Dev 142(22):3869–3678

Kühnlein RP (2012) Thematic review series: Lipid droplet synthesis and metabolism: From Yeast to man. Lipid droplet-based storage fat metabolism in *Drosophila*. J Lipid Res 53(8):1430–1436

Lagueux M, Hoffmann JA, Goltzené F, Kappler C, Tsoupras G, Hetru C et al (1984) Ecdysteroids in ovaries and embryos of *Locusta migratoria*. In: Hoffmann J, Porchet M (eds) Biosynthesis, metabolism and mode of action of invertebrate hormones proceedings in life sciences. Springer, Berlin/ Heidelberg, pp 168–180

Lange AB (1984) The transfer of prostaglandin-synthesizing activity during mating in *Locusta migratoria*. Insect Biochem 14(5):551–556

Laurinyecz B, Péter M, Vedelek V, Kovács AL, Juhász G, Maróy P et al (2016) Reduced expression of CDP-DAG synthase changes lipid composition and leads to male sterility in *Drosophila*. Open Biol 6(1):50169

Lee G, Park JH (2004) Hemolymph sugar homeostasis and starvation-induced hyperactivity affected by genetic manipulations of the adipokinetic hormone-encoding gene in *Drosophila melanogaster*. Genetics 167(1): 311–323

Lee JM, Hatakeyama M, Oishi K (2000) A simple and rapid method for cloning insect vitellogenin cDNAs. Insect Biochem Mol Biol 30(3):189–194

Leopold RA, Degrugillier ME (1973) Sperm penetration of housefly eggs: evidence for involvement of a female accessory secretion. Science 181(4099):555–557

Leyria J, Fruttero LL, Aguirre SA, Canavoso LE (2014) Ovarian nutritional resources during the reproductive cycle of the hematophagous *Dipetalogaster maxima* (Hemiptera: Reduviidae): focus on lipid metabolism. Arch Insect Biochem Physiol 87(3):148–163

Leyria J, Fruttero LL, Ligabue-Braun R, Defferrari MS, Arrese EL, Soulages JL et al (2018) DmCatD, a cathepsin D-like peptidase of the hematophagous insect *Dipetalogaster maxima* (Hemiptera: Reduviidae): Purification, bioinformatic analyses and the significance of its interaction with lipophorin in the internalization by developing oocytes. J Insect Physiol 105:28–39

Leyria J, Orchard I, Lange AB (2020a) Transcriptomic analysis of regulatory pathways involved in female reproductive physiology of *Rhodnius prolixus* under different nutritional states. Sci Rep 10(1):11431

Leyria J, Orchard I, Lange AB (2020b) What happens after a blood meal? A transcriptome analysis of the main tissues involved in egg production in *Rhodnius prolixus*, an insect vector of chagas disease. PLoS Negl Trop Dis 14(10):e0008516

Leyria J, Orchard I, Lange AB (2021) The involvement of insulin/ToR signaling pathway in reproductive performance of *Rhodnius prolixus*. Insect Biochem Mol Biol 130:103526

Leyria J, Orchard I, Lange AB (2022) Impact of JH signaling on reproductive physiology of the classical insect model, *Rhodnius prolixus*. Int J Mol Sci 23(22):13832

Leyria J, Guarneri AA, Lorenzo MG, Nouzova M, Noriega FG, Benrabaa SAM et al (2023) Effects of mating on female reproductive physiology in the insect model, *Rhodnius prolixus*, a vector of the causative parasite of Chagas disease. PLoS Negl Trop Dis 17(9):e0011640

Li K, Jia QQ, Li S (2019) Juvenile hormone signaling – a mini review. Insect Sci 26(4):600–606

Liao S, Amcoff M, Nässel DR (2021) Impact of high-fat diet on lifespan, metabolism, fecundity and behavioral senescence in *Drosophila*. Insect Biochem Mol Biol 133:103495

Ling L, Raikhel AS (2018) Serotonin signaling regulates insulin-like peptides for growth, reproduction, and metabolism in the disease vector *Aedes aegypti*. Proc Natl Acad Sci U S A 115(42):E9822–E9831

Ling L, Kokoza VA, Zhang C, Aksoy E, Raikhel AS (2017) MicroRNA-277 targets insulin-like peptides 7 and 8 to control lipid metabolism and reproduction in *Aedes aegypti* mosquitoes. Proc Natl Acad Sci U S A 114(38):E8017–E8024

Liu H, Ryan RO (1991) Role of lipid transfer particle in transformation of lipoprotein in insect oocytes. Biochim Biophys Acta (BBA)/Lipids Lipid Metab 1085(1):112–118

Liu W, Xie Y, Dong J, Xue J, Tian F, Wu J (2014) Ultra- and microstructure of the female reproductive system of *Matsucoccus matsumurae*. Arthropod Struct Dev 43(3):243–253

Liu W, Li Y, Zhu L, Zhu F, Lei CL, Wang XP (2016) Juvenile hormone facilitates the antagonism between adult reproduction and diapause through the methoprene-tolerant gene in the female *Colaphellus bowringi*. Insect Biochem Mol Biol 74:50–60

Liu S, Li K, Gao Y, Liu X, Chen W, Ge W et al (2018) Antagonistic actions of juvenile hormone and 20-hydroxyecdysone within the ring gland determine developmental transitions in *Drosophila*. Proc Natl Acad Sci U S A 115(1):139–144

Liu W, Guo S, Sun D, Zhu L, Zhu F, Lei CL et al (2019) Molecular characterization and juvenile hormone-regulated transcription of the vitellogenin receptor in the cabbage beetle *Colaphellus bowringi*. Comp Biochem Physiol – Part A Mol Integr Physiol 229: 69–75

Liu Z, Liu F, Li G, Chi X, Wang Y, Wang H et al (2020) Metabolite support of long-term storage of sperm in the spermatheca of honeybee (*Apis mellifera*) queens. Front Physiol 11:574856

Lococo D, Huebner E (1980) The ultrastructure of the female accessory gland, the cement gland, in the insect *Rhodnius prolixus*. Tissue Cell 12(3):557–580

Loewith R, Hall MN (2011) Target of rapamycin (TOR) in nutrient signaling and growth control. Genetics 189(4): 1177–1201

Loher W (1979) The influence of Prostaglandin E2 on oviposition in *Teleogryllus commodus*. Entomol Exp Appl 25(1):107–109

Lorenz MW, Anand AN (2004) Changes in the biochemical composition of fat body stores during adult development of female crickets, *Gryllus bimaculatus*. Arch Insect Biochem Physiol 56(3):110–119

Lorenz MW, Gäde G (2009) Hormonal regulation of energy metabolism in insects as a driving force for performance. Integr Comp Biol 49(4):380–392

Lu K, Shu Y, Zhou J, Zhang X, Zhang X, Chen M et al (2015) Molecular characterization and RNA interference analysis of vitellogenin receptor from *Nilaparvata lugens* (Stål). J Insect Physiol 73:20–29

Lu K, Chen X, Li Y, Li W, Zhou Q (2018a) Lipophorin receptor regulates *Nilaparvata lugens* fecundity by promoting lipid accumulation and vitellogenin biosynthesis. Comp Biochem Physiol – Part A Mol Integr Physiol 219–220:8–37

Lu K, Zhou J, Chen X, Li W, Li Y, Cheng Y et al (2018b) Deficiency of brummer impaires lipid mobilization and JH-mediated vitellogenesis in the brown planthopper, *Nilaparvata lugens*. Front Physiol 9(OCT):1535

Lung O, Kuo L, Wolfner MF (2001) Drosophila males transfer antibacterial proteins from their accessory gland and ejaculatory duct to their mates. J Insect Physiol 47(6):617–622

Ma Z, Liu Z, Huang X (2010) OSBP- and FAN-mediated sterol requirement for spermatogenesis in *Drosophila*. Development 137(22):3775–3784

Machado E, Swevers L, Sdralia N, Medeiros MN, Mello FG, Iatrou K (2007) Prostaglandin signaling and ovarian follicle development in the silkmoth, *Bombyx mori*. Insect Biochem Mol Biol 37(8 SPEC. ISS):876–885

Majerowicz D, Gondim KC (2013) Insect lipid metabolism: insights into gene expression regulation. In: Mandal S (ed) Recent trends in gene expression. Nova Science Publishers, pp 147–189

Marchini D, Bernini LF, Marri L, Giordano PC, Dallai R (1991) The female reproductive accessory glands of the medfly *Ceratitis capitata*: Antibacterial activity of the secretion fluid. Insect Biochem 21(6):597–605

Medeiros MN, Oliveira DMP, Paiva-Silva GO, Silva-Neto MAC, Romeiro A, Bozza M et al (2002) The role of eicosanoids on *Rhodnius* heme-binding protein (RHBP) endocytosis by *Rhodnius prolixus* ovaries. Insect Biochem Mol Biol 32(5):537–545

Medeiros MN, Mendonça LH, Hunter AL, Paiva-Silva GO, Mello FG, Henze IP et al (2004) The role of lipoxygenase products on the endocytosis of yolk proteins in insects: Participation of cAMP. Arch Insect Biochem Physiol 55(4):178–187

Melentev PA, Sharapenkov EG, Surina NV, Ivanova EA, Ryabova EV, Sarantseva SV (2022) *Drosophila* lysophospholipase gene swiss cheese is required for survival and reproduction. Insects 13(1):14

Melo ACA, Valle D, Machado EA, Salerno AP, Paiva-Silva GO, Cunha E Silva NL et al (2000) Synthesis of vitellogenin by the follicle cells of *Rhodnius prolixus*. Insect Biochem Mol Biol 30(7):549–557

Miao L, Zhang N, Jiang H, Dong F, Yang X, Xu X et al (2020) Involvement of Two Paralogous Methoprene-Tolerant Genes in the Regulation of Vitellogenin and Vitellogenin Receptor Expression in the Rice Stem Borer, *Chilo suppressalis*. Front Genet 11:609

Miura S, Gan JW, Brzostowski J, Parisi MJ, Schultz CJ, Londos C et al (2002) Functional conservation for lipid storage droplet association among perilipin, ADRP, and TIP47 (PAT)-related proteins in mammals, *Drosophila*, and *Dictyostelium*. J Biol Chem 277(35): 32253–32257

Nässel DR, Vanden BJ (2016) Insulin/IGF signaling in *Drosophila* and other insects: Factors that regulate production, release and post-release action of the insulin-like peptides. Cell Mol Life Sci 73(2):271–290

Ojani R, Liu P, Fu X, Zhu J (2016) Protein kinase C modulates transcriptional activation by the juvenile hormone receptor methoprene-tolerant. Insect Biochem Mol Biol 70:44–52

Olzmann JA, Carvalho P (2019) Dynamics and functions of lipid droplets. Nat Rev Mol Cell Biol 20(3):137–155

Orchard I, Lange AB (2024) The neuroendocrine and endocrine systems in insect – Historical perspective and overview. Mol Cell Endocrinol 580:112108

Oryan A, Wahedi A, Paluzzi JPV (2018) Functional characterization and quantitative expression analysis of two GnRH-related peptide receptors in the mosquito, *Aedes aegypti*. Biochem Biophys Res Commun 497(2): 550–557

Pant R, Kumar S, Singh SD (1979) Changes in carbohydrates and lipids during embryonic development of *Antheraea mylitta* (Lepidoptera). J Biosci 1(1):27–33

Parra-Peralbo E, Culi J (2011) Drosophila lipophorin receptors mediate the uptake of neutral lipids in oocytes and imaginal disc cells by an endocytosis-independent mechanism. PLoS Genet 7(2):e1001297

Pascini TV, Martins GF (2017) The insect spermatheca: an overview. Zoology 121:56–71

Patel H, Orchard I, Veenstra JA, Lange AB (2014) The distribution and physiological effects of three evolutionarily and sequence-related neuropeptides in *Rhodnius prolixus*: Adipokinetic hormone, corazonin and adipokinetic hormone/corazonin-related peptide. Gen Comp Endocrinol 195:1–8

Pennington JE, Wells MA (2002) Triacylglycerol-rich lipophorins are found in the dipteran infraorder Culicomorpha, not just in mosquitoes. J Insect Sci 2:15

Pinch M, Mitra S, Rodriguez SD, Li Y, Kandel Y, Dungan B et al (2021) Fat and happy: Profiling mosquito fat body lipid storage and composition post-blood meal. Front Insect Sci 1:693168

Poiani A (2006) Complexity of seminal fluid: A review. Behav Ecol Sociobiol 60(3):289–310

Polevoy G, Wei HC, Wong R, Szentpetery Z, Kim YJ, Goldbach P et al (2009) Dual roles for the Drosophila PI 4-kinase Four wheel drive in localizing Rab11 during cytokinesis. J Cell Biol 187(6):847–858

Porstmann T, Santos CR, Griffiths B, Cully M, Wu M, Leevers S et al (2008) SREBP Activity is regulated by mTORC1 and contributes to Akt-dependent cell growth. Cell Metab 8(3):224–236

Qiao JW, Fan YL, Bai TT, Wu BJ, Pei XJ, Wang D et al (2021) Lipophorin receptor regulates the cuticular hydrocarbon accumulation and adult fecundity of the pea aphid *Acyrthosiphon pisum*. Insect Sci 28(4): 1018–1032

Raikhel AS, Dhadialla TS (1992) Accumulation of yolk proteins in insect oocytes. Annu Rev Entomol 37(1): 217–251

Raikhel AS, Brown MR, Belles X (2005) Hormonal control of reproductive processes. In: Gilbert LI (ed) Comprehensive molecular insect science. Elsevier, pp 433–491

Ramos FO, Leyria J, Nouzova M, Fruttero LL, Noriega FG, Canavoso LE (2021) Juvenile hormone mediates lipid storage in the oocytes of *Dipetalogaster maxima*. Insect Biochem Mol Biol 133:103499

Reiff T, Jacobson J, Cognigni P, Antonello Z, Ballesta E, Tan KJ et al (2015) Endocrine remodelling of the adult intestine sustains reproduction in drosophila. Elife 4 (JULY 2015):e06930

Riddiford LM (2020) *Rhodnius*, golden oil, and Met: A history of juvenile hormone research. Front Cell Dev Biol 8:679

Rodríguez-Vázquez M, Vaquero D, Parra-Peralbo E, Mejía-Morales JE, Culi J (2015) *Drosophila* lipophorin receptors recruit the lipoprotein LTP to the plasma membrane to mediate lipid uptake. PLoS Genet 11(6):e1005356

Roma GC, Bueno OC, Camargo-Mathias MI (2010) Morpho-physiological analysis of the insect fat body: a review. Micron 41(5):395–401

Rosetto M, Belardinelli M, Fausto AM, Marchini D, Bongiorno G, Maroli M et al (2003) A mammalian-like lipase gene is expressed in the female reproductive accessory glands of the sand fly *Phlebotomus papatasi* (Diptera, Psychodidae). Insect Mol Biol 12(5):501–508

Roth TF, Porter KR (1964) Yolk protein uptake in the oocyte of the mosquito *Aedes agypti* L. J Cell Biol 20(2):313–332

Roy S, Saha TT, Zou Z, Raikhel AS (2018) Regulatory pathways controlling female insect reproduction. Annu Rev Entomol 63:489–511

Ryan RO, Prasad SV, Henriksen EJ, Wells MA, Law JH (1986) Lipoprotein interconversions in an insect, *Manduca sexta*. Evidence for a lipid transfer factor in the hemolymph. J Biol Chem 261(2):563–568

Santos R, Rosas-Oliveira R, Saraiva FB, Majerowicz D, Gondim KC (2011) Lipid accumulation and utilization by oocytes and eggs of *Rhodnius prolixus*. Arch Insect Biochem Physiol 77(1):1–16

Scalia S, Sbrenna-Micciarelli A, Sbrenna G, Morgan ED (1987) Ecdysteroid titres and location in developing eggs of *Schistocerca gregaria*. Insect Biochem 17(1): 227–236

Schal C, Sevala VL, Young HP, Bachmann JAS (1998) Sites of synthesis and transport pathways of insect hydrocarbons: cuticle and ovary as target tissues. Am Zool 38(2):382–393

Schal C, Fan Y, Blomquist GJ (2003) Regulation of pheromone biosynthesis, transport, and emission in cockroaches. In: Blomquist GJ (ed) Insect pheromone biochemistry and molecular biology. Elsevier Inc., pp 283–322

Schonbaum CP, Perrino JJ, Mahowald AP (2000) Regulation of the vitellogenin receptor during *Drosophila melanogaster* oogenesis. Mol Biol Cell 11(2):511–521

Sênos Demarco R, Uyemura BS, D'Alterio C, Jones DL (2019) Mitochondrial fusion regulates lipid homeostasis and stem cell maintenance in the *Drosophila* testis. Nat Cell Biol 21(6):710–720

Sênos Demarco R, Uyemura BS, Jones DL (2020) EGFR Signaling Stimulates Autophagy to Regulate Stem Cell Maintenance and Lipid Homeostasis in the *Drosophila* Testis. Cell Rep 30(4):1101–1116.e5

Settya BNY, Rmaiah TR (1980) Effect of prostaglandins and inhibitors of prostaglandin biosynthesis on oviposition in the silkmoth *Bombyx mori*. Indian J Exp Biol 18(5):539–541

Sevala VL, Bachmann JAS, Schal C (1997) Lipophorin: A hemolymph juvenile hormone binding protein in the German cockroach, *Blattella germanica*. Insect Biochem Mol Biol 27(7):663–670

Sevala V, Shu S, Ramaswamy SB, Schal C (1999) Lipophorin of female *Blattella germanica* (L.): Characterization and relation to hemolymph titers of juvenile hormone and hydrocarbons. J Insect Physiol 45(5): 431–441

Shao W, Espenshade PJ (2012) Expanding roles for SREBP in metabolism. Cell Metab 16(4):414–419

Sieber MH, Spradling AC (2015) Steroid signaling establishes a female metabolic state and regulates SREBP to control oocyte lipid accumulation. Curr Biol 25(8):993–1004

Silva-Oliveira G, De Paula IF, Medina JM, Alves-Bezerra-M, Gondim KC (2021) Insulin receptor deficiency reduces lipid synthesis and reproductive function in the insect *Rhodnius prolixus*. Biochim Biophys Acta - Mol Cell Biol Lipids 1866(2):158851

Singh R, Kaushik S, Wang Y, Xiang Y, Novak I, Komatsu M et al (2009) Autophagy regulates lipid metabolism. Nature 458(7242):1131–1135

Skowronek P, Wójcik Ł, Strachecka A (2021) Fat body—multifunctional insect tissue. Insects 12(6):547

Song J, Zhou S (2020) Post-transcriptional regulation of insect metamorphosis and oogenesis. Cell Mol Life Sci 77(10):1893–1909

Song J, Wu Z, Wang Z, Deng S, Zhou S (2014) Krüppel-homolog 1 mediates juvenile hormone action to promote vitellogenesis and oocyte maturation in the migratory locust. Insect Biochem Mol Biol 52(1):94–101

Sonobe H, Yamada R (2004) Ecdysteroids during early embryonic development in silkworm *Bombyx mori*: Metabolism and functions. Zoolog Sci 21(5):503–516

Soulages JL, Wells MA (1994) Lipophorin: The structure of an insect lipoprotein and its role in lipid transport in insects. Adv Protein Chem 45:371–415

Soulages JL, Firdaus SJ, Hartson S, Chen X, Howard AD, Arrese EL (2012) Developmental changes in the protein composition of *Manduca sexta* lipid droplets. Insect Biochem Mol Biol 42(5):305–320

Stanley D (2006) Prostaglandins and other eicosanoids in insects: biological significance. Annu Rev Entomol 51: 25–44

Stanley D, Kim Y (2019) Prostaglandins and other eicosanoids in insects: Biosynthesis and biological actions. Front Physiol 9(FEB):1927

Stay B, Coop A (1973) Developmental stages and chemical composition in embryos of the cockroach, *Diploptera punctata*, with observations on the effect of diet. J Insect Physiol 19(1):147–171

Steinhauer J, Gijón MA, Riekhof WR, Voelker DR, Murphy RC, Treisman JE (2009) *Drosophila* lysophospholipid acyltransferases are specifically required for germ cell development. Mol Biol Cell 20(24):5224–5235

Stephenson RA, Thomalla JM, Chen L, Kolkhof P, White RP, Beller M et al (2021) Sequestration to lipid droplets promotes histone availability by preventing turnover of excess histones. Dev 148(15):dev199381

Stryapunina I, Itoe MA, Trinh Q, Vidoudez C, Du E, Mendoza L et al (2024) Precise coordination between nutrient transporters ensures fertility in the malaria mosquito *Anopheles gambiae*. PLoS Genet 20(1): e1011145

Sturm R (2012) Morphology and ultrastructure of the accessory glands in the female genital tract of the house cricket, *Acheta domesticus*. J Insect Sci 12:99

Sun J, Hiraoka T, Dittmer NT, Cho KH, Raikhel AS (2000) Lipophorin as a yolk protein precursor in the mosquito, *Aedes aegypti*. Insect Biochem Mol Biol 30(12):1161–1171

Suzuki M, Shinohara Y, Ohsaki Y, Fujimoto T (2011) Lipid droplets: size matters. J Electron Microsc (Tokyo) 60(SUPPL. 1):S101–S116

Swevers L (2019) An update on ecdysone signaling during insect oogenesis. Curr Opin Insect Sci 31:8–13

Szafer-Glusman E, Giansanti MG, Nishihama R, Bolival B, Pringle J, Gatti M et al (2008) A role for very-long-chain fatty acids in furrow ingression during cytokinesis in *Drosophila* spermatocytes. Curr Biol 18(18):1426–1431

Takeuchi N, Chino H (1993) Lipid transfer particle in the hemolymph of the American cockroach: Evidence for its capacity to transfer hydrocarbons between lipophorin particles. J Lipid Res 34(4):543–551

Tansey JT, Sztalryd C, Hlavin EM, Kimmel AR, Londos C (2004) The central role of perilipin A in lipid metabolism and adipocyte lipolysis. IUBMB Life 56(7): 379–385

Tarone AM, McIntyre LM, Harshman LG, Nuzhdin SV (2012) Genetic variation in the Yolk protein expression network of *Drosophila melanogaster*: Sex-biased negative correlations with longevity. Heredity (Edinb). 109(4):226–234

Teixeira L, Rabouille C, Rørth P, Ephrussi A, Vanzo NF (2003) *Drosophila* Perilipin/ADRP homologue Lsd2 regulates lipid metabolism. Mech Dev 120(9): 1071–1081

Tembhare DB, Thakare VK (1975) The histological and histochemical studies on the ovary in relation to vitellogenesis in the dragonfly, *Orthetrum chrysis* Selys (Libellulidae: Odonata). Zeitschrift fur Mikroskopisch-Anatomische Forsch - Abteilung 2 89 (1):108–127

Terra WR, Barroso IG, Dias RO, Ferreira C (2019) Molecular physiology of insect midgut. Adv In Insect Phys 56:117–163

Thanh MT, Pham TLA, Tran BD, Nguyen YDH, Kaeko K (2020) *Drosophila* model for studying the link between lipid metabolism and development. Front Biosci – Landmark 25(1):147–158

Thomas DS, Varma M, Manoharan C, Rao GP, Vijayan K, Mishra RK et al (2020) Molecular characterization and functional analysis of the vitellogenin receptor from eri silkworm, *Samia ricini*. Comp Biochem Physiol Part – B Biochem Mol Biol 242:110417

Tian Z, Guo S, Li JX, Zhu F, Liu W, Wang XP (2021) Juvenile hormone biosynthetic genes are critical for regulating reproductive diapause in the cabbage beetle. Insect Biochem Mol Biol 139:103654

Tootle TL, Williams D, Hubb A, Frederick R, Spradling A (2011) *Drosophila* eggshell production: Identification of new genes and coordination by Pxt. PLoS One 6(5): e19943

Toprak U (2020) The role of peptide hormones in insect lipid metabolism. Front Physiol 11:434

Torroja L, Ortuño-Sahagún D, Ferrús A, Hämmerle B, Barbas JA (1998) Scully, an essential gene of *Drosophila*, is homologous to mammalian mitochondrial type II L-3-hydroxyacyl-CoA dehydrogenase/amyloid-β peptide- binding protein. J Cell Biol 141(4): 1009–1017

Tose LV, Weisbrod CR, Michalkova V, Nouzova M, Noriega FG, Fernandez-Lima F (2021) Following de novo triglyceride dynamics in ovaries of *Aedes aegypti* during the previtellogenic stage. Sci Rep 11(1):9636

Tsang SSK, Law STS, Li C, Qu Z, Bendena WG, Tobe SS et al (2020) Diversity of Insect Sesquiterpenoid Regulation. Front Genet 11:1027

Tufail M, Takeda M (2005) Molecular cloning, characterization and regulation of the cockroach vitellogenin receptor during oogenesis. Insect Mol Biol 14(4): 389–401

Tufail M, Takeda M (2008) Molecular characteristics of insect vitellogenins. J Insect Physiol 54(12): 1447–1458

Tufail M, Takeda M (2009) Insect vitellogenin/lipophorin receptors: Molecular structures, role in oogenesis, and regulatory mechanisms. J Insect Physiol 55(2):88–104

Tufail M, Takeda M (2012) Hemolymph lipoproteins: role in insect reproduction. In: Hemolymph proteins and functional peptides: recent advances in insects and other arthropods, pp 3–19

Tufail M, Hatakeyama M, Takeda M (2001) Molecular evidence for two vitellogenin genes and processing of vitellogenins in the American cockroach, *Periplaneta americana*. Arch Insect Biochem Physiol 48(2):72–80

Tufail M, Bembenek J, Elgendy AM, Takeda M (2007) Evidence for two vitellogenin-related genes in *Leucophaea maderae*: The protein primary structure and its processing. Arch Insect Biochem Physiol 66(4):190–203

Tufail M, Nagaba Y, Elgendy AM, Takeda M (2014) Regulation of vitellogenin genes in insects. Entomol Sci 17(3):269–282

Ugrankar R, Liu Y, Provaznik J, Schmitt S, Lehmann M (2011) Lipin Is a Central Regulator of Adipose Tissue Development and Function in *Drosophila melanogaster*. Mol Cell Biol 31(8):1646–1656

Van Antwerpen R, Law JH (1992) Lipophorin lipase from the yolk of *Manduca sexta* eggs: Identification and partial characterization. Arch Insect Biochem Physiol 20(1):1–12

Van Der Horst DJ, Rodenburg KW (2010) Lipoprotein assembly and function in an evolutionary perspective. Biomol Concepts 1(2):165–183

Van der Horst DJ, Van Marrewijk WJA, Diederen JHB (2001) Adipokinetic hormones of insect: release, signal transduction, and responses. Int Rev Cytol 211:179–240

Van der Horst DJ, Roosendaal SD, Rodenburg KW (2009) Circulatory lipid transport: Lipoprotein assembly and function from an evolutionary perspective. Mol Cell Biochem 326(1–2):105–119

Van Handel E (1993) Fuel metabolism of the mosquito (*Culex quinquefasciatus*) embryo. J Insect Physiol 39(10):831–833

Van Heusden MC, Erickson BA, Pennington JE (1997) Lipophorin levels in the yellow fever mosquito, *Aedes aegypti*, and the effect of feeding. Arch Insect Biochem Physiol 34(3):301–312

Veenstra JA (2020) Arthropod IGF, relaxin and gonadulin, putative orthologs of Drosophila insulin-like peptides 6, 7 and 8, likely originated from an ancient gene triplication. PeerJ 2020(8):e9534

Vereshchagina N, Wilson C (2006) Cytoplasmic activated protein kinase Akt regulates lipid-droplet accumulation in *Drosophila* nurse cells. Development 133(23): 4731–4735

Vrablik TL, Watts JL (2013) Polyunsaturated fatty acid derived signaling in reproduction and development: Insights from *Caenorhabditis elegans* and *Drosophila melanogaster*. Mol Reprod Dev 80(4):244–259

Wang H, Eckel RH (2009) Lipoprotein lipase: From gene to obesity. Am J Physiol – Endocrinol Metab 297(2): E271–E288

Wang C, Huang X (2012) Lipid metabolism and *Drosophila* sperm development. Sci China Life Sci 55(1):35–40

Wang C, Ma Z, Scott MP, Huang X (2011) The cholesterol trafficking protein NPC1 is required for *Drosophila* spermatogenesis. Dev Biol 351(1):146–155

Wang X, Hou Y, Saha TT, Pei G, Raikhel AS, Zou Z (2017) Hormone and receptor interplay in the regulation of mosquito lipid metabolism. Proc Natl Acad Sci U S A 114(13):E2709–E2718

Wang W, Yang RR, Peng LY, Zhang L, Yao YL, Bao YY (2021) Proteolytic activity of the proteasome is required for female insect reproduction. Open Biol 11(2):200251

Warren JT, Steiner B, Dorn A, Pak M, Gilbert LI (1986) Metabolism of ecdysteroids during the embryogenesis of *Manduca sexta*. J Liq Chromatogr 9(8):1759–1782

White MA, Bonfini A, Wolfner MF, Buchon N (2021) *Drosophila melanogaster* sex peptide regulates mated female midgut morphology and physiology. Proc Natl Acad Sci U S A 118(1):e2018112118

Wicker-Thomas C (2007) Pheromonal communication involved in courtship behavior in Diptera. J Insect Physiol 53(11):1089–1100

Wojda I, Cytryńska M, Zdybicka-Barabas A, Kordaczuk J (2020) Insect defense proteins and peptides. In: Hoeger U, Harris J (eds) Vertebrate and invertebrate respiratory proteins, lipoproteins and other body fluid proteins subcellular biochemistry. Springer, pp 81–121

Wu Q, Brown MR (2006) Signaling and function of insulin-like peptides in insects. Annu Rev Entomol 51:1–24

Wu LT, Hui JHL, Chu KH (2013) Origin and evolution of yolk proteins: expansion and functional diversification of large lipid transfer protein superfamily. Biol Reprod 88(4):102

Wu Z, Yang L, He Q, Zhou S (2021) Regulatory mechanisms of vitellogenesis in insects. Front Cell Dev Biol 8:593613

Xu L, Huang HJ, Zhou X, Liu CW, Bao YY (2017) Pancreatic lipase-related protein 2 is essential for egg hatching in the brown planthopper, *Nilaparvata lugens*. Insect Mol Biol 26(3):277–285

Yamada R, Sonobe H (2003) Purification, kinetic characterization, and molecular cloning of a novel enzyme ecdysteroid-phosphate phosphatase. J Biol Chem 278(29):26365–26373

Yamada R, Yamahama Y, Sonobe H (2005) Release of ecdysteroid-phosphates from egg yolk granules and their dephosphorylation during early embryonic development in silkworm, *Bombyx mori*. Zoolog Sci 22(2): 187–198

Yamahama Y, Seno K, Hariyama T (2008) Changes in lipid droplet localization during embryogenesis of the silkworm, *Bombyx mori*. Zoolog Sci 25(6):580–586

Yew JY, Chung H (2015) Insect pheromones: An overview of function, form, and discovery. Prog Lipid Res 59:88–105

Yokoyama H, Yokoyama T, Yuasa M, Fujimoto H, Sakudoh T, Honda N et al (2013) Lipid transfer particle from the silkworm, *Bombyx mori*, is a novel member of the apoB/large lipid transfer protein family. J Lipid Res 54(9):2379–2390

Youngsteadt E, Fan Y, Stay B, Schal C (2005) Cuticular hydrocarbon synthesis and its maternal provisioning to embryos in the viviparous cockroach *Diploptera punctata*. J Insect Physiol 51(7):803–809

Zhao X, Karpac J (2020) The *Drosophila* midgut and the systemic coordination of lipid-dependent energy homeostasis. Curr Opin Insect Sci 41:100–105

Zhou C, Bin YX, Yang H, Long GY, Jin DC (2020) Effects of sublethal concentrations of insecticides on the fecundity of *Sogatella furcifera* (Hemiptera: Delphacidae) via the regulation of vitellogenin and its receptor. J Insect Sci 20(5):1–9

Ziegler R (1997) Lipid synthesis by ovaries and fat body of *Aedes aegypti* (Diptera: Culicidae). Eur J Entomol 94(3):385–391

Ziegler R, Ibrahim MM (2001) Formation of lipid reserves in fat body and eggs of the yellow fever mosquito, *Aedes aegypti*. J Insect Physiol 47(6):623–627

Ziegler R, Van Antwerpen R (2006) Lipid uptake by insect oocytes. Insect Biochem Mol Biol 36(4 SPEC. ISS):264–272

Jimena Leyria is Assistant Professor at the Universidad Nacional de Córdoba (Argentina) and member of the National Scientific Research Council (CONICET). She received her bachelor's degree in Biochemistry in 2008, and her Ph.D. in Chemical Sciences in 2015, at the Universidad Nacional de Córdoba. During 2018–2013 Dr. Leyria performed her post-doctoral training as fellow in the Department of Biology, at the University of Toronto Mississauga (Canada). Dr. Leyria was awarded two trainee research fellowships at the Universidade Federal do Rio Grande do Sul (Brazil) and another at the Institut de Biologia Evolutiva-UPF-CSIC (Spain). Her field of expertise is centered on insect reproductive physiology, mainly in how tissues communicate with one another to promote successful reproduction.

Leonardo Luis Fruttero is an Assistant Professor in the Department of Clinical Biochemistry and the Center for Research in Clinical Biochemistry and Immunology (CIBICI), Faculty of Chemical Sciences, Universidad Nacional de Córdoba (Argentina). He graduated as Biologist in 2005 and obtained the Ph.D. degree in Chemical Sciences (discipline: Biochemistry) at the Universidad Nacional de Córdoba. Dr. Fruttero performed his postdoctoral training at the Brain Institute, Pontifícia Universidade Católica do Rio Grande do Sul, Brazil, from 2014 to 2017. Since 2018, he has been a member of the National Scientific and Technical Research Council staff in the Assistant category. Current and previous fields of interest of Dr. Fruttero involve insect biochemistry and physiology with focus on lipid and protein metabolism, reproduction, and the effects of entomotoxic proteins on the immune system.

Lilian Etelvina Canavoso is an Associate Professor at the Department of Clinical Biochemistry and the Center for Research in Clinical Biochemistry and Immunology (CIBICI), Faculty of Chemical Sciences, Universidad Nacional de Córdoba (Argentina). She graduated as a Biochemist and obtained the Ph.D. degree in Chemical Sciences (discipline: Biochemistry) in 1995 at the Universidad Nacional de Córdoba. Dr. Canavoso performed her postdoctoral training at the Department of Biochemistry and Center for Insect Science, University of Arizona, USA, from 1998 to 2000. Since 2004, she has been a member of the National Scientific and Technical Research Council staff in the Independent category. Current and previous fields of interest of Dr. Canavoso involve insect biochemistry and physiology with focus on lipid and protein metabolism, reproductive biology, and immunity.

Adv Exp Med Biol (2026) 1494: 345–365
https://doi.org/10.1007/5584_2024_849

Published online: 19 March 2025

Insect Flight and Lipid Metabolism: Beyond the Classic Knowledge

Leonardo L. Fruttero, Jimena Leyria, and Lilián E. Canavoso

Abstract

Insects are the most successful animal group by various ecological and evolutionary metrics, including species count, adaptation diversity, biomass, and environmental influence. This book delves into the underlying reasons behind insects' dominance on Earth. Lipids play pivotal roles in insect biology, serving as fuel for physiological processes, signaling molecules, and structural components of biomembranes and providing waterproofing against dehydration, among other functions. The study of insect flight has been instrumental in advancing our understanding of insect metabolism, with the migratory locust (*Locusta migratoria*) and the tobacco hornworm (*Manduca sexta*) serving as prominent models. Throughout the 1980s and 1990s, numerous studies shed light on the role of adipokinetic hormone (AKH), a crucial neuropeptide in lipid mobilization, to support the extraordinary energy demands of insect flight. Remarkably, AKH was the first identified peptide hormone in insects. These pioneering works linking lipids and flight laid the groundwork for subsequent research characterizing the physiological roles of other neuroendocrine factors in energy substrate mobilization across diverse insect species. However, in the omics era, one may be surprised by the limited understanding of the complex cascade of events governing lipid supply to insect flight muscles. Thus, this chapter aims to provide a concise overview of the evolutionary significance of insect flight, emphasizing key advancements that expand our classical knowledge in this field. Ultimately, we hope this chapter serves as a modest tribute to the pioneering researchers of one of the most captivating areas in insect biology, inspiring further exploration into the myriad roles of lipids in insect biology.

Keywords

Adipokinetic hormone · AKH/corazonin-related peptide · Energetic homeostasis · Fat body · Flight muscle · Insect flight · Juvenile hormone · Lipid metabolism · Lipophorin · Octopamine

L. L. Fruttero, J. Leyria, and L. E. Canavoso (✉)
Departamento de Bioquímica Clínica, Facultad de Ciencias Químicas, Universidad Nacional de Córdoba, Córdoba, CP, Argentina

Centro de Investigaciones en Bioquímica Clínica e Inmunología (CIBICI), Consejo Nacional de Investigaciones Científicas y Técnicas (CONICET), Córdoba, Argentina
e-mail: lfruttero@unc.edu.ar; jimena.leyria@unc.edu.ar; lilian.canavoso@unc.edu.ar

1 Introduction

In insects, flight activity demands a tremendous amount of energy, primarily due to aerodynamical constraints arising from their small size, as well as the relatively low efficiency in converting chemical energy to mechanical energy, which typically ranges from 10% to 20% (Josephson 2007; Klowden 2013). In vertebrates, the increase in muscle respiration rates reach up to fivefold from rest to maximum exertion. However, this pales in comparison to insects, which can exhibit a staggering increase from 50 to 100-fold. Despite this apparent disparity, the muscular structure of vertebrates and insects shares notable similarities. In both groups, skeletal muscles consist of elongated, multinucleated muscle fibers. Remarkably, most of the genes expressed in the muscles of examined insect species have homologs in vertebrate organisms (Cripps 2006; Klowden 2013; Song et al. 2024).

The operation of flight in insects is predominantly aerobic, facilitated by the high mitochondrial density in the muscle cells and the abundant oxygen supply delivered through the extensive network of tracheae (Josephson 2007; Klowden 2013). As discussed in detail in this chapter, the ATP reserves within the muscle are only sufficient to initiate and sustain flight for a few seconds; the remainder must be supplied via the hemolymph in the form of the disaccharide trehalose, diacylglycerol (DAG), and proline, with variations depending on the species and the type of flight (short versus long distance) (Canavoso et al. 2001). This intricate process is finely coordinated by hormones and requires integration among the central nervous system (CNS) and associated endocrine glands, the fat body, and the flight muscle itself. In this chapter, we attempt to provide a comprehensive review of the classical knowledge regarding lipid metabolism in the context of insect flight, complemented with the latest updates emerged in this post-genomic era.

2 Evolutive Significance of Insect Flight

Insects are the first animal group to develop flight and this ability, along with their small size, have played fundamental roles in their evolutive radiation of species. Flight confers significant advantages to the insects in order to find food, localize mates, colonize new habitats, and escape predation. Their small size allows them to occupy niches not available to larger animals. Nevertheless, being small makes flying more difficult by aerodynamic reasons, mainly because there is an inverse relationship between the size of the animal and the wing beat frequency required to flight (Iwamoto 2011; Cao and Jin 2020).

In vertebrates, wings are homologous to forelimbs; however, in insects, wings are novel structures. Despite continuing controversy, two main hypotheses arise to explain the origin of insect wings. The first proposes that wings originated from thoracic winglets or paranotal lobes, initially serving for thermoregulation, then as airfoils for gliding, and finally as flapping wings (Tomoyasu et al. 2017). The second hypothesis, known as the articulated gill theory, posits that wings originated from winglets that derived from gills (Tomoyasu et al. 2017). According to this theory, when the ancestor left the water, these gills transitioned into locomotor appendices. While this hypothesis suggests an aquatic origin for winged insects, it lacks substantial support. However, molecular data provide evidence for a closer ancestral relationship between insects and marine crustaceans, lending support to the articulated gill theory (Alexander 2018).

The origins of flight muscles trace back to thoracic muscles that were later adapted for this new function. For example, the thoracic muscles of orthopterans and blattodeans exhibit greater similarity to those of the "primitive" flightless silverfishes from the order Thysanura than to the muscles of the flying hymenopterans (such as bees) or dipterans (such as flies). This suggests that the muscles from cockroaches and locusts have undergone less divergence from the hypothetical primitive flyer (Josephson 2007; Klowden 2013).

3 Anatomic and Physiological Considerations

The muscles involved in flight, responsible for the oscillatory movement of the wings, can be classified into direct and indirect categories. Direct muscles are inserted at the base of the wing or in the wing articulation, while indirect muscles induce wing movements by changing the position and shape of the thorax (Fig. 1). One evolutionary trend observed in insects is the increasing importance of indirect muscles in powering flight (Josephson 2007). In the Odonata, considered primitive flyers, essentially all flight muscles (wing depressors) are direct. In locusts, flight movement is generated by both direct and indirect muscles, while in flies and bees, all power muscles present are indirect. Power muscles alter flight stability and steering, while control muscles influence wing stroke amplitude and angle of attack (Josephson 2007). Another evolutionary pattern in insects is the reduction in the number of power muscles and the increase in the number of control muscles (Iwamoto 2011; Bullard and Pastore 2019).

Based on the neural control of contraction, muscles can be classified as synchronous and asynchronous. Synchronous muscles are found in what are considered to be the relatively primitive flight muscles of orthopterans, lepidopterans, and odonates. The synchronous muscles found in insects closely resemble any striated muscle (Iwamoto 2011): they receive a neural input, and each contraction is accompanied by evoked muscle action potentials. Thus, the depolarization impulse from the motoneuron triggers action potentials in the muscle fibers. This action potential initiates the release of calcium from the sarcoplasmic reticulum, which in turn diffuses into the cells and binds to the control sites on the contractile machinery, turning on the contractile activity (Gordon and Dickinson 2006). The contraction ends when the calcium is sequestered back to the reticulum in an ATP-driven pumping process. In this way, synchronous muscles maintain a 1:1 ratio between electrical activity and contraction (Fig. 1) (Josephson 2007; Cao and Jin 2020).

Asynchronous muscles are also triggered by action potentials that cause calcium release from the sarcoplasmic reticulum, but once activated, they can contract in an oscillatory manner when connected to a resonant load such as the wings and thorax of the insect. Thus, the alternating contraction and relaxation of the dorsal vertical and dorsal longitudinal muscles induce deformations in the thorax roof that lead to the up- and downstroke movements of the wings (Fig. 1). The oscillatory frequency that asynchronous muscles can achieve is vastly superior to that of synchronous muscles; therefore, flight is more powerful. The asynchronous muscle system operates in such a way that the stretch of the muscle induces its contraction, thus being independent of individual action potential. Asynchronous muscles are most efficient due to lower activation costs and metabolic expenditures of calcium cycling (Josephson 2007; Cao and Jin 2020). Insects belonging to Coleoptera, Diptera, Hymenoptera, and Hemiptera orders possess this type of muscle. These orders have developed smaller wings, allowing them to occupy niches unavailable to larger insects.

4 Energetic Homeostasis of Insect Flight

As previously mentioned, high levels of oxygen are transported from the external environment to the mitochondria of the flight muscles via spiracles and trachea. Muscle cells, with 30–40% of their mass constituted by mitochondria (Iwamoto 2011), exhibit high rates of ATP hydrolysis and glycolytic flux, as indicated by oxygen consumption measurements (Wigglesworth and Lee 1982). During flight, oxygen consumption can increase up to 100-fold compared to resting states (Fig. 2) (Sacktor 1976; Song et al. 2024). For instance, in *Periplaneta americana*, the rate increases from 6 to 600 μl of O_2/g body weight/min. In *Schistocerca gregaria*, it rises from 10 to 500 μl of O_2/g body weight/min; in *Metopsilus procellus*, from 12 to 1700 μl of O_2/g body weight/min; and in *Apis mellifera*, from 53 to 1000 μl of O_2/g body weight/min (Beenakkers et al. 1984). However, unlike vertebrates, the reserves of respiratory fuels in insect flight muscles are sufficient only to initiate flight, lacking the capacity to sustain prolonged flight activity (Goldsworthy and Wheeler 1986). Consequently, flight muscles must obtain additional fuels from alternative sources. To accomplish that, flight muscles of most insect species rely on the uptake of fuels from the hemolymph following

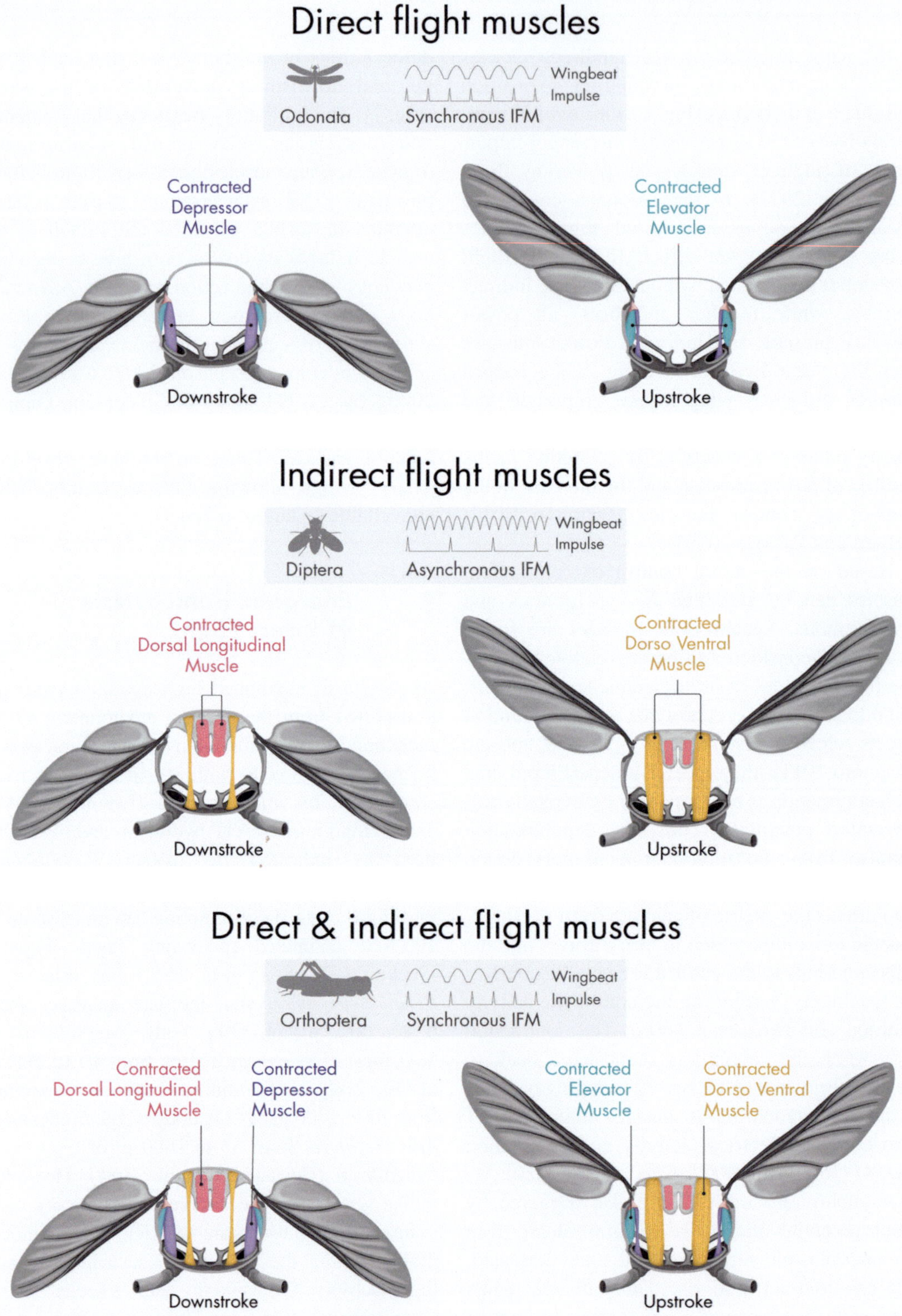

Fig. 1 Schematic diagram illustrating the mechanism of wing movement. Each panel demonstrates a combination of various types of flight muscles (direct vs. indirect) and modes of contraction (synchronous vs. asynchronous), accompanied by corresponding action diagrams and representative insect groups for each type. (Original figure is created by the authors based on the references cited in the corresponding sections)

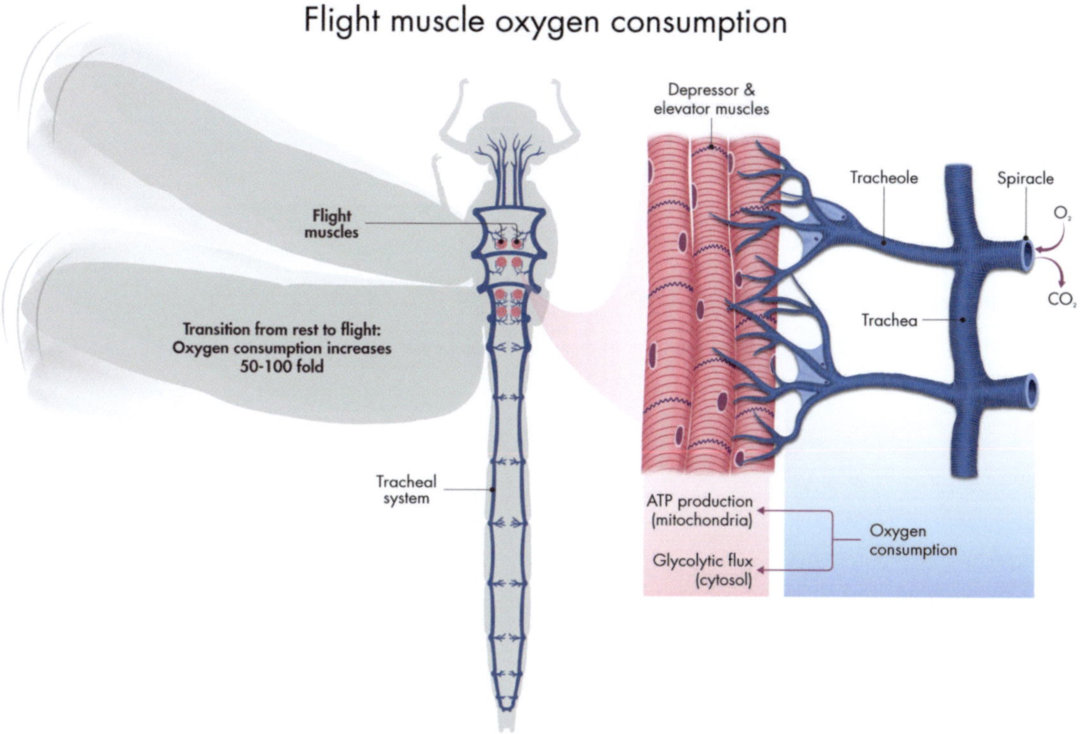

Fig. 2 Schematic diagram illustrating flight muscle oxygen consumption. High levels of oxygen are transported from the external environment to the mitochondria of the flight muscles via the tracheal system. Flight muscle cells exhibit high rates of ATP hydrolysis and glycolytic flux during flight, resulting in an increase in oxygen consumption of up to 50–100-fold compared to resting states

nutrient mobilization, primarily facilitated by the fat body. Depending on the species, insects utilize carbohydrates, lipids, proline, or combinations of them as flight fuels (Fig. 3). The oxidation of glucose yields approximately 30–32 ATP molecules per molecule. Similarly, the oxidation of proline generates around 25–30 ATP molecules per molecule. In contrast, lipids serve as a more energy-dense fuel, with the complete oxidation of fatty acids producing about 100–120 ATP molecules per fatty acid molecule, depending on the type of fatty acid. The total ATP production during flight must align with the energy demand determined by the metabolic rate (Bursell 1981; Mookerjee et al. 2017). For example, due to dietary specialization, honeybees predominantly utilize carbohydrates, while in the blood-feeding tsetse fly and the mosquitoes *Aedes aegypti*, flight muscles use proline as a carbon source to enhance the concentration of Krebs cycle intermediates (Olembo and Pearson 1982; Scaraffia and Wells 2003; Suarez et al. 2005). In addition, in some lepidopterans, such as the tobacco hornworm, lipids represent the main fuel for flight (Kammer and Heinrich 1978; Ziegler and Schulz 1986). Moreover, a combination of carbohydrates and lipids is predominantly utilized by migratory insects and the species with high-lipid diets. For example, muscle contraction during the initial phase of flight in the desert locust *S. gregaria* and the kissing bug *Panstrongylus megistus* is sustained by carbohydrates. However, as flight progresses, the flight fuel gradually shifts from carbohydrates to lipids as the predominant source of energy (Mayer and Candy 1969; Canavoso et al. 2003).

Endocrine signals play a pivotal role in coordinating the supply and mobilization of fuels to sustain the energy requirements for flight

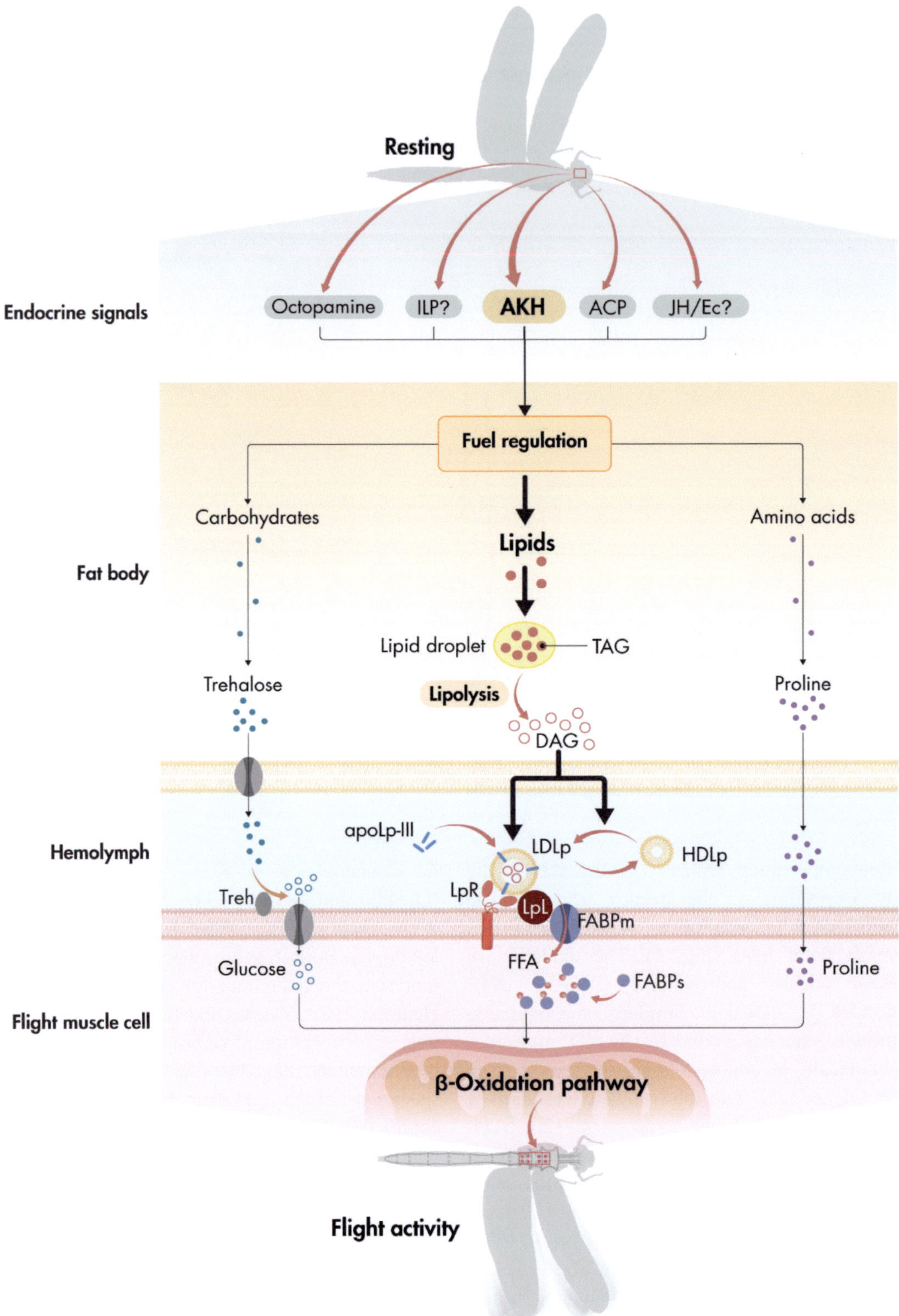

Fig. 3 Simplified scheme illustrating fuel regulation during flight. Endocrine signals play a pivotal role in coordinating the supply and mobilization of fuels to sustain the energy requirements for flight activity. The main hormone involved in nutrient mobilization in the fat body is AKH, with ACP and octopamine also involved

activity (Goldsworthy and Wheeler 1986; Lorenz and Gäde 2009). In this regard, the main hormones involved in nutrient mobilization in the fat body are adipokinetic hormone (AKH) (Van der Horst 2003), AKH/corazonin-related peptide (ACP) (Hou et al. 2021), octopamine (Mentel et al. 2003; Pflüger and Duch 2011), and juvenile hormones (JH) (Whitfield et al. 2006; Shegelski et al. 2021). However, octopamine and AKH signaling are capable of managing energy storage in the flight muscles themselves (Orchard et al. 1993; Van der Horst 2003; Alves-Bezerra et al. 2016). Therefore, the success of short- and long-distance flights is a consequence of a coordinated system of multiple endocrine signals that integrate energy metabolism and utilization for flight (Fig. 3). In the following sections, we focus on endocrine signals modulating lipid availability as the primary fuel for active insect flight, highlighting the role of the fat body as the supplier of nutrients, the hemolymph as the vehicle for lipid transport, and the subsequent lipid uptake and energy generation by flight muscles.

4.1 Endocrine Signals in Flight: The Roles of AKH, AKH/ACP, Octopamine, JH, and Others

Insect AKH peptides are considered the equivalent of the mammalian hormone glucagon, as they induce rapid mobilization of carbohydrates and lipids in the fat body, in the form of trehalose and DAG as intermediates, respectively (Van der Horst 2003). In other words, AKH stimulates catabolic reactions while inhibiting biosynthetic ones. The initiation of flight activity induces the release of AKH, which is synthesized in the *corpus cardiacum* (CC), a neuroendocrine gland physically associated with the insect brain and physiologically analogous to the pituitary gland of mammals (Fig. 3) (Van der Horst et al. 2001). In insects, AKH comprises a family of peptides with around 60 members, all of which are small peptides with 8–11 amino acids. However, so far, only one AKH receptor has been identified. This receptor belongs to the G protein–coupled receptor (GPCR) family and is structurally and evolutionarily related to the gonadotropin-releasing hormone receptors found in vertebrates (Gäde and Marco 2013). One of the primary functions of AKH peptides is to control lipid mobilization in the fat body, an organ serving as an interchange center of materials and energy, remotely integrated with the CNS and associated endocrine glands. Upon binding to its specific GPCR on target cells, AKH triggers a coordinated signal transduction pathway that ultimately results in nutrient mobilization. In vertebrates, mobilization of triacylglycerol (TAG) stored in adipose tissue is enabled by the phosphorylation of several key proteins, including hormone-sensitive lipase (HSL) and lipid droplet proteins belonging to PAT (perilipin/ADRP/TIP47) family (Zechner et al. 2009; Van der Horst and Ryan 2017). In the specific context of lipid metabolism and flight activity, AKH peptides increase intracellular

Fig. 3 (continued) in this process. JH, Ec, and ILPs have been proposed to contribute to nutrient mobilization to sustain flight. Depending on the species, insects utilize different nutrients as flight fuels (e.g., threalose and/or proline), with lipids being the most relevant. The fat body supplies DAG to the circulating HDLp, which incorporates several molecules of apoLp-III, decreasing its density to shift into LDLp. Once at the surface of a muscle cell, DAG-rich LDLp is hydrolyzed to generate free fatty acids, glycerol, and free apoLp-III, leading to DAG-depleted HDLp, which can be reloaded with lipids, acting as the classic reusable lipid shuttle. In flight muscle, the presence of a LpR on the flight muscle membrane may act as a docking site to bind lipophorin to the membrane in the vicinity of the LpL. After the generation of fatty acids, they traffic across the membrane via a putative FABPm into the aqueous cytosol, where they bind FABPs and are subsequently directed to the mitochondria, initiating the β-oxidation pathway cascade, generating the energy required for flight activity. *ACP* AKH/corazonin-related peptide, *apoLp-III* apolipophorin III, *AKH* adipokinetic hormone, *DAG* diacylglycerol, *Ec* ecdysteroids, *FABPm* plasma membrane fatty acid–binding protein, *FABPs* cytoplasmic fatty acid–binding proteins, *ILP* insulin-like peptides, *JH* juvenile hormone, *LpL* lipoprotein lipase, *LpR* lipophorin receptor, *TAG* triacylglycerol, *Treh* trehalase, *FFA* free fatty acids

cAMP levels in target cells, thereby activating a cAMP-dependent protein kinase A which phosphorylates downstream proteins involved in the activation of lipolysis such as triglyceride lipase (TGL) and the perilipin 1/lipid storage droplet 1 (Lsd-1), as discussed below (Arrese and Soulages 2010; Van der Horst and Ryan 2017; Toprak 2020). Interestingly, the TGL identified in *M. sexta* is conserved among insects, displays significant phospholipase A1 activity but shows no homology to the main vertebrate adipose TAG lipase (ATGL) (Arrese and Soulages 2010; Van der Horst and Ryan 2017). In *D. melanogaster*, activation of lipolysis also relies on lipid droplet-associated TAG lipase Brummer (bmm), a homolog of human ATGL (Grönke et al. 2005). The double mutant flies for bmm and AKHR showed extreme obesity and impaired lipid mobilization (Grönke et al. 2005). As a result of the lipolytic events orchestrated by AKH, in locusts and other insect species that rest on fat body TAG stores to power flight, the breakdown of TAG releases *sn*-1,2-DAG into the hemolymph, which will be subsequently transported by LDLp to the flight muscles (Fig. 3) (Van der Horst 1990; Van der Horst and Ryan 2017). The pathways underlying the stereospecific synthesis of *sn*-1,2-DAG remain poorly understood, partly due to technical challenges in isolating or activating TGL. Readers interested in exploring the potential mechanisms involved in *sn*-1,2-DAG generation are encouraged to refer to excellent reviews on the topic (Ryan and Van Der Horst 2000; Van der Horst and Ryan 2017).

The AKH peptides are subjected to circadian rhythm regulation. In crickets, major physical activity, including flight, typically occurs during the scotophase. Coincidentally, the concentration of lipid in the hemolymph is significantly higher during early scotophase than early photophase. This increase appears to depend on circulating AKH levels in the hemolymph, as peptide titers positively correlate with the rise in lipid concentration (Lorenz and Gäde 2009; Fassold et al. 2010). Hence, high rate of lipid mobilization prior to increased physical activity prepares the insects for flight initiation in nocturnal environments. Notably, the timing of lipid appearance in the hemolymph has been linked to the rhythm of fat body sensitivity to the hormone, which, in turn, is influenced by the number of available AKH receptors on target cell membranes (Das et al. 1993). Therefore, the release of AKH peptides and its downstream signaling cannot be simplified to a simple model; rather, it relies on a complex interplay of signals where multiple factors must coordinate to promote lipid mobilization.

It is intriguing that flight muscles of some hemipterans possess abundant lipid droplets, suggesting that such lipid stores could potentially serves as a significant fuel source for flight, especially during the initial stages of flight activity (Cockbain 1961; Ashhurst and Luke 1968; Ward et al. 1982; Cavagnari et al. 2000; Oliveira et al. 2006). In the triatomines *Rhodnius prolixus* and *Triatoma infestans,* a significant decrease in the muscle lipid content occurs during the first hour of flight (Ward et al. 1982). Similar results were reported in *P. megistus* (Canavoso et al. 2003). This lipid mobilization appears to depend on the expression of AKH receptor in flight muscles, which enhances responsiveness of this tissue to AKH. Nevertheless, the extent to which the lipid content within the muscle contributes to flight activity remains unknown (Alves-Bezerra et al. 2016). Although the primary focus of AKH signaling studies has been its role in promoting flight metabolism, as discussed earlier in this chapter, those peptides have also been associated with a neuromodulatory function during flight. Stimulatory effects of AKH on insect muscle activity have been demonstrated in *Pyrrhocoris apterus*, *D. melanogaster*, *Gryllus bimaculatus,* and *P. americana,* where hormone injections or gene manipulation techniques significantly increase locomotory activity (Kodrík 2008).

In locusts, a novel flight modulator, ACP, has been identified, playing a crucial role in sustaining prolonged flight by enhancing lipid utilization in flight muscles. The ACP peptide falls structurally intermediate between AKH and corazonin. Its receptor is widespread among insects, suggesting its relevance as a pivotal insect signaling system. However, the importance of this signaling pathway remained unclear for a significant period

(Hansen et al. 2010). Originally isolated from the storage lobes of the CC in locust, the ACP peptide was initially named locust hypertrehalosemic hormone due to its ability to increase hemolymph trehalose levels in cockroaches, although not in locust (Siegert 1999). A longstanding, unanswered question is how the flight muscles of locust shift from carbohydrate to lipid oxidation even though they maintain significant carbohydrate stores even after 30 min of flight. Recently, it has been reported that ACP signaling facilitates long-term flight by promoting intracellular fatty acid transport to the mitochondria, thereby regulating the expression of fatty acid–binding proteins (FABP) in locust flight muscles, marking the transition to lipid utilization as fuel for flight (Hou et al. 2021).

Juvenile hormone is a unique sesquiterpenoid hormone that primarily regulates insect metamorphosis and reproduction (Truman 2019; Leyria 2024); however, in some species, it also mediates polyphenisms and other environmentally regulated life history events (Riddiford 2020). Because of its lipidic nature, JH easily enters the cell and binds to a nuclear receptor known as Methoprene tolerant (Met), which belongs to an ancient family of basic helix-loop-helix Per/Arnt/Sim (bHLH-PAS) transcription factors. This molecular event triggers the formation of a heterodimer between Met and Tai (Taiman), another bHLH-PAS protein. Upon ligand-induced dimerization, the Met-Tai complex initiates its downstream signaling pathway (Leyria 2024). Additionally, JH has the capability to interact with a proposed membrane receptor, whose identity has not yet been characterized but appears to vary depending on the insect species (Riddiford 2020). Both in vivo and in vitro experiments suggest that JH membrane signaling can activate the GPCR-RTK pathway independently of JH intracellular signaling. This rapid and reversible hormonal response initiates the phospholipase C (PLC)-protein kinase C (PKC) pathway, regulating various cellular processes, including "patency" and protein synthesis in insect reproductive organs (Davey and Huebner 1974; Yamamoto et al. 1988; Liu et al. 2015; Bai and Palli 2016; Gao et al. 2022). Juvenile hormone has been reported as a metabolic rate accelerator in bumble bees, promoting a reduction in lipid levels and flight muscle mass in queens (Shpigler et al. 2021). In the honey bee *A. mellifera*, treatment with the JH-analog methoprene reduces flight span, as well as the total number and duration of successful flights performed throughout the lifetime of the bees (Whitfield et al. 2006). However, in the grasshopper *Melanoplus sanguinipes*, prolonged flight induces an increase in circulating JH titers (Min et al. 2004). The involvement of JH signaling in lipid metabolism has been extensively discussed in the chapter titled "Lipids in insect reproduction: where, how and why" in this book. Nevertheless, the link between lipid metabolism and JH signaling in the context of insect flight remains elusive. Such JH signaling could occur through direct action of the hormone on the CNS or CC, serving as a neuromodulator, or by influencing other signaling pathways, or by promoting a general metabolic effect. Interestingly, a transcriptomic analysis performed on the strong-flying female beetle, *Dendroctonus ponderosae*, reveals an upregulation in the expression of the JH epoxide hydrolase gene, responsible for deactivating JH, which in turn reduces JH signaling (Shegelski et al. 2021).

Ecdysteroids (Ec), derived from dietary cholesterol or plant sterols, induce molting across arthropods. In insects, they primarily originate from the prothoracic glands but can also be synthesized by other tissues like adult gonads. Ecdysone, secreted by the prothoracic glands, is converted to 20-hydroxyecdysone (20E) in peripheral tissues; 20E is typically the active form (Truman 2019). In the same transcriptomic analysis performed on *D. ponderosae*, the expression of the ecdysone 20-monooxygenase gene, which catalyzes the synthesis of 20E, is upregulated in strong flying females, suggesting an enhancement of ecdysone signaling (Shegelski et al. 2021). 20-hydroxyecdysone exerts its effects through a specific nuclear receptor heterodimer complex consisting of the ecdysone receptor (EcR) and ultraspiracle (USP). Interestingly, their respective orthologs in mammals, like the retinoid X receptor (RXR), play critical roles in the regulation of lipid metabolism (Chawta et al. 2001). Hence, it has been proposed that ecdysone signaling may also be involved in lipid mobilization to sustain flight in female beetles (Shegelski et al. 2021).

Octopamine, the insect counterpart of mammal noradrenaline, is a biogenic amine widely distributed within the nervous system of invertebrates. It functions as a neurotransmitter, neurohormone, or neuromodulator (Orchard et al. 1993). Octopamine signaling pathway is mediated through GPCRs located on postsynaptic or target tissue membranes, where activation of the pathway induces unique changes in cAMP or Ca^{2+} levels (Wu et al. 2014). Research has shown that during prolonged flight, carbohydrates initially serve as the primary fuel for flight muscles before transitioning to lipid oxidation. It has been suggested that the octopaminergic system is deactivated in flight power muscles to facilitate the shift from glycolysis to lipid oxidation (Mentel et al. 2003; Pflüger and Duch 2011). However, octopamine signaling, primarily studied in locusts, serves multiple functions (Orchard et al. 1993). For example, as a neurotransmitter, octopamine regulates the release of AKH peptide hormones by modulating intracellular cAMP levels through synaptic connections. As a neurohormone, octopamine may mobilize lipids and carbohydrates through its adipokinetic activity. Furthermore, octopamine might stimulate oxidative metabolism within the muscle itself, likely leading to the modulation of twitch tension amplitude resulting in increased power output during flight activity (Orchard et al. 1993).

While insulin signaling has been widely implicated in wing polymorphism (Zhang et al. 2019), the connection between insulin and lipid metabolism during insect flight has not be fully established. In the fruit fly *D. melanogaster*, the transcription factor forkhead box class O (foxO), a downstream target of insulin signaling, becomes activated in the indirect flight muscle to regulate both autonomous and systemic lipid homeostasis (Zhao and Karpac 2017). When insulin signaling decreases or is absent, foxO undergoes dephosphorylation, becomes activated, and translocates to the nucleus to induce the expression of target genes. In this scenario, foxO can promote lipolysis by upregulating lipases, as reported by Wang et al. (2011), which in turn may supply lipids to induce or sustain the insect's migration to a more nutritionally favorable environment.

4.2 Fat Body as a Supplier of Lipid for Flight

The fat body is a multifunctional tissue distributed throughout the body of the insect, filling the space left by the remaining organs. It is composed of loosely arranged cells called adipocytes that are bathed by the hemolymph (Roma et al. 2010). This particular arrangement allows the fat body to quickly sense the changes in the physiological state of the insect and respond accordingly to meet its metabolic needs. The fat body serves as the main site of intermediate metabolism and the synthesis of hemolymph proteins. It is also the major tissue of lipid and glycogen accumulation and metabolism (Arrese et al. 2001; Arrese and Soulages 2010). This tissue, analogous to the liver and adipose tissue of vertebrates, undergoes significant morphological and biochemical transformations to support various physiological processes, including oogenesis, metamorphosis, fasting, and flight (Arrese and Soulages 2010; Van Der Horst and Rodenburg 2010; Gondim et al. 2018). The fat body specializes in storing large amounts of lipids, primarily as TAG, which serve as fuel for flight (Canavoso et al. 2001).

The energy-demanding activity of flight stimulates the release of hormones, primarily AKH and octopamine (Orchard et al. 1993; Van der Horst 2003). These processes have been extensively studied in moths and locusts. In *M. sexta*, both flight induction and AKH injection result in increased lipase activity (Arrese and Wells 1997; Arrese and Soulages 2010). Adipokinetic hormone induces a fourfold increase in cAMP-dependent protein kinase A (PKA), making PKA-mediated phosphorylation a crucial event for lipolysis in the fat body (Arrese and Wells 1994). Additionally, AKH induces changes in the phosphorylation of the lipid droplet protein Lsd-1, stimulating lipolysis (Arrese and Wells 1994; Patel et al. 2004,

2005). This process leads to the production of DAG, which is then exported from the fat body to the hemolymph for circulation via lipophorin (Fig. 3) (Arrese and Soulages 2010). In the locust fat body, AKH stimulates the production of cAMP and the generation of inositol phosphates, which ultimately mobilize intracellular Ca^{2+} stores (Asher et al. 1984), triggering extracellular Ca^{2+} influx via a store-operated calcium entry mechanism. These signaling pathways mobilize TAG through hormonal activation of lipase, thereby stimulating the release of lipids into the hemolymph (Vroemen et al. 1998; Van der Horst and Ryan 2017). On the other hand, the role of octopamine in the fat body during flight has been extensively studied in locusts (Orchard et al. 1982). It has been proposed that fluctuations in octopamine levels trigger corresponding changes in fat body cAMP, driving the initial release of lipids within the first 10 min of flight. Around 15 min, octopaminergic synapses in the glandular lobe activate, prompting the release of AKHs. These hormones further increase fat body cAMP, initiating a second phase of lipid mobilization. This coordinated process ensures the availability of energy-rich molecules, which are transported to the flight muscles to sustain the high-energy demands of flight (for details, see Orchard et al. 1993).

4.3 Hemolymph as a Vehicle for Transporting Lipids

Lipids are transported through the hemolymph of insects and other arthropods via lipophorin, which serves as the principal lipoprotein in circulation. This versatile particle possesses a lipid content ranging between 30% and 50% of the total protein-lipid complex. Structurally, lipophorin exhibits an outer amphiphilic region comprising phospholipids, free cholesterol, and apolipoproteins, and an internal lipophilic core housing predominantly nonpolar lipids such as DAG, cholesterol esters, and hydrocarbons (Van der Horst et al. 2009). Notably, although DAG serves as the primary transported lipid, the lipid composition of lipophorin varies across species and in response to the insect's physiological state. Lipophorin is categorized based on its buoyant density into high-density lipophorin (HDLp, ~1.15 g/ml) and low-density lipophorin (LDLp, ~1.03 g/ml). Apolipophorin II and I, with molecular weights ranging from ~230–250 kDa to ~70–85 kDa, respectively, serve as structural, non-exchangeable apolipoproteins. The apolipoproteins are encoded by the gene *apolipophorin II/I*. Once synthesized as an apolipophorin II/I precursor in the fat body, it undergoes cleavage by furin, giving mature apolipophorin I and II (Canavoso et al. 2001). Apolipophorin III (apoLp-III), a small exchangeable apolipoprotein (~18 kDa), plays a crucial role during energy-demanding long-distance flight in insects. In long-flying insects such as *L. migratoria* and *M. sexta*, lipid mobilization from the fat body occurs during flight and in response to AKH release, leading to the incorporation of additional DAG into HDLp. Several molecules of apoLp-III associate with the particle, stabilizing the newly formed LDLp. Consequently, during long-distance flight, lipophorin cycles between the fat body, loading itself with DAG, and the flight muscles, unloading its lipid cargo with the assistance of flight-muscle lipoprotein lipases that catalyze the delipidation process. Once delipidated, LDLp is reformed back into HDLp (Fig. 3) (Van der Horst et al. 2002).

4.4 Lipid Uptake by Flight Muscle

In general, with the exception of some hemipteran species (Cavagnari et al. 2000), insect muscles do not store lipids. Consequently, external sources of energy must be transported to and taken up by the flight muscles so that the flight activity can be sustained (Ashhurst and Luke 1968; Ward et al. 1982; Cavagnari et al. 2000; Canavoso et al. 2003). The mechanisms involved in lipid uptake by insect flight muscles were primarily elucidated in the 1980s and 1990s, utilizing models such as *L. migratoria* or *M. sexta* (Van der Horst and Ryan 2017). Using locust muscle cells, Van Antwerpen et al. (1998) demonstrated that lipophorin was only located in the wider spaces of the extracellular matrix, suggesting that lipophorin is not internalized by these cells. In *M. sexta* and *L. migratoria*, upon the onset of flight, the fat body supplies DAG to the circulating HDLp, decreasing its density to become LDLp. Once at the surface of a muscle cell, DAG-rich LDLp is hydrolyzed to generate free fatty acids, glycerol, and free apoLp-III, generating DAG-depleted HDLp, which can be reloaded with lipids, acting as a classic reusable lipid shuttle (Fig. 3) (Downer and Chino 1985; Van Heusden et al. 1991). Wheeler et al. (1984) reported that muscle preparations could hydrolyze lipids associated with lipoproteins at higher rates, suggesting that components of those cells may aid lipolysis. This led to the postulation, for the first time, of the presence of a lipoprotein lipase working in the unloading of lipids into the flight muscles, which was later confirmed to be directly anchored to the cell membrane of the flight muscle cells (Goldsworthy and Wheeler 1986; Wheeler et al. 1986). On the other hand, the fact that LDLp was a much better substrate than HDLp suggested that apoLp-III might have an activating effect on the lipoprotein lipase (Van Heusden et al. 1986). However, efficient enzyme activity also requires close proximity of the substrate to the enzyme. Yoichi (1987) postulated the presence of a lipophorin receptor (LpR) on the flight muscle membrane, which may act as docking site for the binding of lipophorin to the membrane in the vicinity of the lipoprotein lipase. Nonetheless, it was not ruled out that the lipoprotein lipase itself might also bind its own lipoprotein substrate (Van Antwerpen et al. 1990).

After the generation of fatty acids, they undergo trafficking across the membrane of flight muscle cells into the aqueous cytosol and subsequently to the mitochondria, where they are converted into their CoA-esters to initiate the β-oxidation pathway cascade, thereby generating the energy required for flight activity. It is widely acknowledged that integral membrane proteins facilitate the transport of fatty acids from the plasma membrane of flight muscle cells into the cytosol; however, the identity of these proteins remains elusive. In several insect species, three main types of fatty acid–binding proteins have been reported in various tissues, which could also facilitate the entry of lipids into flight muscle cells: plasma membrane fatty acid–binding protein (FABPm), which act as an integral membrane protein; fatty acid translocase (FAT/CD36); and fatty acid transport proteins (FATPs), representing a large family of multifunctional proteins (Kaczmarek and Boguś 2021). Once the fatty acid passes through the plasma membrane into the cytoplasm, cytoplasmatic FABPs play a central role in intracellular lipid trafficking to different compartments, including the mitochondria, to generate the energy necessary for optimal flight activity (Fig. 3). Fatty acid–binding proteins are found ubiquitously in numerous vertebrate and invertebrate tissues. Specifically in insects, several FABPs have been identified and characterized in flight muscle cells (Haunerland and Chisholm 1990; Smith et al. 1992; Van der Horst et al. 1993; Cavagnari et al. 2000; Sim and Denlinger 2009; Caccia et al. 2012; Martínez-Barnetche et al. 2018). Conclusive evidence regarding the requirement of FABPs for sustained muscle activity has recently emerged. In the desert locust *S. gregaria*, RNAi assays targeting FABPs have demonstrated that their absence prevents sustained flight activity (Rajapakse et al. 2019). The authors suggested that FABPs may serve as a sink for fatty acids in the cytosol, maintaining a concentration gradient across the plasma membrane and acting as a vehicle for the rapid transport of fatty acid through the cytosol.

4.5 Energy Generation for Flight

After hydrolysis of LDLp-transported DAG by lipoprotein lipase in flight muscle cells, the resulting fatty acids are ultimately oxidized in the mitochondria, generating ATP to meet the extraordinary energy demand required for flight (Fig. 3). Mitochondrial ATP supply involves specific signaling pathways regulated by various factors, which have recently been thoroughly documented in insects (Mesquita et al. 2021). To enter the mitochondria, fatty acids must first be activated by their conversion into CoA-ester derivatives through the action of a fatty acyl-CoA ligase enzyme. Activated fatty acids are then transported to the mitochondria in a process facilitated by transport proteins belonging to the family of acyl-CoA-binding proteins (ACBPs). Acyl-CoA-binding proteins non-covalently and reversibly bind acyl-CoA with high affinity and specificity, thereby protecting it from degradation (Neess et al. 2015). While identification of ACBPs in several insect species suggests their occurrence in insect muscles (Burton et al. 2005; Ohnishi et al. 2006; Ryuda et al. 2011; Leyria et al. 2020), their specific role in relation to flight activity is still unknown. Since the mitochondrial membrane is impermeable to fatty acyl-CoA esters, shuttle mechanisms are required for their import into the mitochondrial lumen. Carnitine acyltransferases facilitate this process by catalyzing the reversible transfer of an acyl chain from acyl-CoA esters to carnitine, forming acyl-carnitine esters. These esters are then transported into the mitochondrial matrix in exchange for free carnitine. Another carnitine acyltransferase, located on the inner mitochondrial membrane, subsequently transfers the fatty acyl chain back onto CoA (Di Lisa et al. 1995). The activity of carnitine acyltransferase has been documented in the flight muscles of some insect species (Beenakkers et al. 1967; Crabtree and Newsholme 1972; Hansford and Johnson 1976). Additionally, the gene encoding carnitine acyltransferase has been reported in the genome of *D. melanogaster* and *R. prolixus* (Jackson et al. 1999; Majerowicz et al. 2017; Leyria et al. 2020). Interestingly, a carnitine-independent pathway of fatty acid oxidation has also been demonstrated in the flight muscles of the southern armyworm moth, *Prodenia eridania* (Stevenson 1968), suggesting the involvement of alternative pathways in regulating the entry of fatty acyl-CoA esters into the mitochondria. More recently, similar results were obtained for *R. prolixus*, where CPT1 knockdown did not affect lipid oxidation (De Paula et al. 2023). The β-oxidation within the organelle involves sequential reactions like those described in mammals. While several genes associated with fatty acid β-oxidation have been identified in insects (Lee et al. 2012; Jones et al. 2015; Majerowicz et al. 2017; Bretscher and O'Connor 2020; Leyria et al. 2020), to our knowledge, few functional assays using these genes have been reported to confirm their specific roles in lipid metabolism for flight. For instance, it was recently demonstrated that the knockdown of mitochondrial trifunctional protein A subunit (HADHA), a β-oxidation enzyme, decreased flight capacity in *R. prolixus* (Aredes et al. 2022). Migratory insects such as *L. migratoria* and the cotton bollworm *Helicoverpa armigera* require substantial energy to sustain flight muscle activity for extended periods. Interestingly, genes encoding β-oxidation enzymes exhibited differential expression, providing evidence of adaptations in this pathway that enhance long-distance flight capacity (Wang et al. 2014; Jones et al. 2015). In this context, maintaining a balance between energy stored in the fat body, energy provided by the hemolymph, and β-oxidation within the flight muscle cells is crucial for the long-distance flight of certain insects. For instance, gregarious locusts, which undertake long-distance migratory flights, accumulate higher amounts of TAG stores in their fat bodies compared to solitary locusts, thereby enhancing the sustainability of long-distance flight (Ayali et al. 1996; Du et al. 2022).

The process of autophagic degradation of lipid droplets, referred to as lipophagy, is primarily regulated by lipases and classic autophagy (ATG)-related proteins. In recent years, this autophagic process has been demonstrated as an important player in cellular lipid metabolism, providing an alternative pathway for the delivery of fatty acids to the mitochondria (Schulze et al. 2017; Khawar et al. 2019). Lipophagy has been extensively investigated using *D. melanogaster* as a model organism (Maruzs et al. 2019). However, its role in lipid metabolism within flight muscles has recently been explored in the triatomine *R. prolixus*. Silencing of two essential ATGs, Atg6/Beclin1 and Atg8/LC3, led to the accumulation of TAG in larger lipid droplets in the fat body, resulting in decreased forced flight rates and reduced modulation of proteins involved in lipid metabolism (Santos-Araujo et al. 2020). This finding also highlights the connection between the availability of nutritional resources stored in the fat body and the flight activity of nonmigratory insects.

5 Other Processes Involved in Insect Flight

Flight activity confers an advantage to insects in dispersing and foraging for food and mates, among other functions. However, since flight is one of the most energy-consuming processes, a significant portion of their energy budget is invested in building, maintaining, and operating the flight system. Consequently, any reallocation of energy toward other physiological processes such as immune system activation, reproduction, and adapting to environmental changes may impact flight activity. Lymphatic filariasis, commonly known as elephantiasis, is a parasitic disease transmitted by mosquitoes infected with worm larvae species, as the filarial nematode *Brugia malayi* (Lourens and Ferrell 2019). Recently, it has been reported that *B. malayi* infection has a detrimental impact on *A. aegypti*'s flight distance, average speed, and maximum speed. Since immune responses are energetically costly, activation of immune system usually causes the decline in flight due to nutrient reallocation to cover the energy expense needed for immune system. Indeed, there is a significant decline in lipid content in *B. malayi*–infected mosquitoes (Somerville et al. 2019).

Diapause is a programmed developmental arrest accompanied by physiological adaptations aimed at ensuring survival during adverse environmental conditions. During overwintering, adult inseminated females of the mosquito *Culex pipiens* undergo reproductive diapause, utilizing about 60–85% of their lipid reserves accumulated during autumn. Remarkably, such reserves are sufficient to sustain survival for over 500 days under typical cellar conditions (Rozsypal et al. 2021). However, during this period, mosquitoes may transition from a standard metabolic rate to a much higher rate in response to disturbances such as air currents or predator attacks, thereby activating their flight capability to seek out a new optimal microhabitat. Given that flight activity is a significantly energy-consuming process, the autumnal energy reserves may be depleted, potentially limiting the winter survival of these insects (Rozsypal et al. 2021).

In insects, successful reproduction is also associated with a significant energetic cost and often constrained by resource availability. There is a well-established trade-off between flight capability and reproduction, where one process may have priority over the other. Essentially, the energy expended during flight reduces the available resources for oogenesis (Kennedy 1961; Sappington and Showers 1992; Zera and Harshman 2001). In *R. prolixus*, the transfer of lipids from lipophorin to the ovaries is partially halted during flight (Oliveira et al. 2006). Conversely, in certain insect species, flight muscles undergo histolysis to facilitate ovarian growth. For example, female crickets like *G. bimaculatus* and adults of the rice leaf roller *Cnaphalocrocis medinalis* utilize energy-rich substrates derived from degraded flight muscles, including lipids, to fuel oogenesis. However, it is worth noting that flight muscle histolysis can only provide a limited fraction of the substrates required for egg production (Lorenz 2007; Sun et al. 2013).

Genetic alterations have been observed to drive the specialization of phenotypes for alternate life histories in certain insects. For instance, in the cricket species *Gryllus firmus*, distinct genotypes have been described: one capable of flight and another flightless. Interestingly, the flight-capable genotype exhibits a heightened biosynthesis of total lipids, including TAG, which are primarily directed toward somatic tissue during early adulthood. Conversely, the flightless cricket genotype produces higher amounts of phospholipids, crucial for egg development, which are preferentially allocated to the ovaries (Zhao and Zera 2002).

6 Conclusion and Future Directions

Flight holds profound significance in the evolutionary narrative of insects, serving as a cornerstone of their success and ecological dominance.

This remarkable ability has empowered insects to colonize diverse habitats across the globe, establishing them as the most prolific and diverse animal group on Earth. The extraordinary long-distance flights undertaken by small insects such as moths and locusts defy traditional aerodynamic principles, captivating the interest of the scientific community and driving rapid advancements in our understanding of insect flight dynamics. Moreover, the biochemical and physiological insights gained from studying insect flight have sparked considerable interest in their application to pest control strategies. Within this framework, lipids emerge as pivotal players in sustaining prolonged flight.

During flight, the endocrine system orchestrates a finely tuned mechanism involving various components such as AKH and octopamine to efficiently mobilize TAG reserves stored in the fat body through the action of lipases. The resulting DAG is then loaded into circulating HDLp, where several molecules of apoLp-III associate to form LDLp. This DAG-rich lipophorin is transported to the flight muscle, where lipoprotein lipase, along with other necessary, uncharacterized components of the cell membrane, facilitates the unloading of LDLp lipid cargo. Consequently, regenerated HDLp resumes its cycle to meet the energetic demands of flight. Additionally, emerging evidence suggests that ACP, octopamine, JH, 20E, and even AKH may play pivotal roles in modulating energy production and consumption through lipid or carbohydrate metabolism within the flight muscle itself. Nevertheless, further research is fundamental to fully elucidate the comprehensive role of the endocrine system during flight. The main factors involved in lipid metabolism during flight are summarized in Table 1.

As emphasized throughout this book, it is important to approach generalizations about insects with caution, given the vast diversity

Table 1 Summary of the most relevant factors involved in lipid metabolism during flight

Factor(s)	Function in the context of flight metabolism	Main site(s) of action	References
Octopamine	Regulation of AKH release, mobilization of lipids and carbohydrates, and stimulation of the oxidative metabolism	*Corpus cardiacum*/fat body/flight muscles	Orchard et al. (1993), Pflüger and Duch (2011) and Wu et al. (2014)
Insulin-like peptides (ILP)	Putative functions in lipid mobilization	Not characterized	Toprak (2020)
Adipokinetic hormone (AKH)	Mobilization of carbohydrates and lipids. Upon binding to its receptor, AKH triggers a signal pathway that leads to the activation of lipolysis	Fat body	Toprak (2020)
Forkhead box class O (FoxO)	In flight muscle is required to limit systemic AKH signaling, thus promoting lipid synthesis and accumulation in fat storage tissues, as well as maintaining circulating lipid levels	Flight muscles	Zhao and Karpac (2017)
AKH/corazonin-related peptide (ACP)	Enhances lipid utilization during prolonged flight by regulating the expression of fatty acid–binding proteins (FABP)	Flight muscles	Zhou and Nagata (2021)
Juvenile hormone (JH)	Putative functions as neuromodulator or by promoting general metabolism	Central nervous system/*corpus cardiacum*/other organs	Min et al. (2004), Whitfield et al. (2006), Shegelski et al. (2021) and Shpigler et al. (2021)
Ecdysteroids (Ec)	Putative functions in lipid mobilization	Not characterized	Shegelski et al. (2021)

(continued)

Table 1 (continued)

Factor(s)	Function in the context of flight metabolism	Main site(s) of action	References
Lipophorin (Lp)	Lipid transport and transfer	Fat body/flight muscle/other organs	Van der Horst et al. (2002) and Van der Horst and Ryan (2017)
Lipoprotein lipase (LPL)	Delipidation of lipophorin	Flight muscle	Van Der Horst and Rodenburg (2010)
Plasma membrane fatty acid–binding protein (FABPm, putative)	Entry of lipids into flight muscle cells	Flight muscle	Kaczmarek and Boguś (2021)
Fatty acid–binding protein (FABP)	Intracellular fatty acid transport	Flight muscle	Kaczmarek and Boguś (2021)

among species. Much of our current understanding is derived from studies on a limited number of model insects and from classic research published decades ago. Therefore, there is significant potential for further exploration by employing modern experimental designs and techniques, as well as by studying a broader range of species. In today's rapidly changing world, characterized by climate change and the emergence of resistance, new challenges in insect control are constantly arising. It is evident that this information will be essential for developing innovative strategies to address the consequences of these evolving scenarios.

Acknowledgments The authors thank Dr. David Rojas Marquez (@darwid_illustration) for drawing the figures. Work in the laboratory is supported by grants from CONICET, FONCyT, and SECyT-UNC. L.L.F, J.L., and L.E.C are members of the National Scientific and Technical Research Council (CONICET-Argentina).

References

Alexander DE (2018) A century and a half of research on the evolution of insect flight. Arthropod Struct Dev 47: 1–6. https://doi.org/10.1016/j.asd.2017.11.007

Alves-Bezerra M, De Paula IF, Medina JM et al (2016) Adipokinetic hormone receptor gene identification and its role in triacylglycerol metabolism in the blood-sucking insect *Rhodnius prolixus*. Insect Biochem Mol Biol 69:51–60. https://doi.org/10.1016/j.ibmb.2015.06.013

Aredes DS, De Paula IF, Santos-Araujo S, Gondim KC (2022) Silencing of mitochondrial trifunctional protein A subunit (HADHA) increases lipid stores, and reduces oviposition and flight capacity in the vector insect *Rhodnius prolixus*. Front Insect Sci 2:885172. https://doi.org/10.3389/finsc.2022.885172

Arrese EL, Soulages JL (2010) Insect fat body: energy, metabolism, and regulation. Annu Rev Entomol 55: 207–225. https://doi.org/10.1146/annurev-ento-112408-085356

Arrese EL, Wells MA (1994) Purification and properties of a phosphorylatable triacylglycerol lipase from the fat body of an insect, *Manduca sexta*. J Lipid Res 35: 1652–1660. https://doi.org/10.1016/s0022-2275(20)41163-0

Arrese EL, Wells MA (1997) Adipokinetic hormone-induced lipolysis in the fat body of an insect, *Manduca sexta*: synthesis of *sn*-1,2-diacylglycerols. J Lipid Res 38:68–76. https://doi.org/10.1016/s0022-2275(20)37276-x

Arrese EL, Canavoso LE, Jouni ZE et al (2001) Lipid storage and mobilization in insects: current status and future directions. Insect Biochem Mol Biol 31:7–17. https://doi.org/10.1016/S0965-1748(00)00102-8

Asher C, Moshitzky P, Ramachandran J, Applebaum SW (1984) The effects of synthetic locust adipokinetic hormone on dispersed locust fat body cell preparations: cAMP induction, lipid mobilization, and inhibition of protein synthesis. Gen Comp Endocrinol 55:167–173. https://doi.org/10.1016/0016-6480(84)90098-4

Ashhurst DE, Luke BM (1968) Lipid inclusions in the flight muscles of belostomatid water-bugs. Z Zellforsch Mikrosk Anat 92:210–214. https://doi.org/10.1007/BF00335652

Ayali A, Pener MP, Sowa SM, Keeley LL (1996) Adipokinetic hormone content of the *corpora cardiaca* in gregarious and solitary migratory locusts. Physiol Entomol 21:167–172. https://doi.org/10.1111/j.1365-3032.1996.tb00851.x

Bai H, Palli SR (2016) Identification of G protein-coupled receptors required for vitellogenin uptake into the oocytes of the red flour beetle, *Tribolium castaneum*. Sci Rep 6:27648. https://doi.org/10.1038/srep27648

Beenakkers AMT, Dewaide JH, Henderson PT, Lutgerhorst A (1967) Fatty acid oxidation and some participating enzymes in animal organs. Comp

Biochem Physiol 22:675–682. https://doi.org/10.1016/0010-406X(67)90761-X

Beenakkers AMT, Van der Horst DJ, Van Marrewijk WJA (1984) Insect flight muscle metabolism. Insect Biochem 14:243–260. https://doi.org/10.1016/0020-1790(84)90057-X

Bretscher H, O'Connor MB (2020) The role of muscle in insect energy homeostasis. Front Physiol 11:580687. https://doi.org/10.3389/fphys.2020.580687

Bullard B, Pastore A (2019) Through thick and thin: dual regulation of insect flight muscle and cardiac muscle compared. J Muscle Res Cell Motil 40:99–110. https://doi.org/10.1007/s10974-019-09536-8

Bursell E (1981) The role of proline in energy metabolism. In: Downer RGH (ed) Energy metabolism in insects. Springer, Boston, pp 135–154

Burton M, Rose TM, Færgeman NJ, Knudsen J (2005) Evolution of the acyl-CoA binding protein (ACBP). Biochem J 392:299–307. https://doi.org/10.1042/BJ20050664

Caccia S, Grimaldi A, Casartelli M et al (2012) Functional analysis of a fatty acid binding protein produced by *Aphidius ervi* teratocytes. J Insect Physiol 58:621–627. https://doi.org/10.1016/j.jinsphys.2011.12.019

Canavoso LE, Jouni ZE, Karnas KJ et al (2001) Fat metabolism in insects. Annu Rev Nutr 21:23–46. https://doi.org/10.1146/annurev.nutr.21.1.23

Canavoso LE, Stariolo R, Rubiolo ER (2003) Flight metabolism in *Panstrongylus megistus* (Hemiptera: Reduviidae): the role of carbohydrates and lipids. Mem Inst Oswaldo Cruz 98:909–914. https://doi.org/10.1590/S0074-02762003000700009

Cao T, Jin JP (2020) Evolution of flight muscle contractility and energetic efficiency. Front Physiol 11:1038. https://doi.org/10.3389/fphys.2020.01038

Cavagnari BM, Scaraffia PY, Haller JF et al (2000) Presence of a fatty acid-binding protein and lipid stores in flight muscles of *Dipetalogaster maximus* (Hemiptera: Reduviidae). J Med Entomol 37:938–944. https://doi.org/10.1603/0022-2585-37.6.938

Chawta A, Repa JJ, Evans RM, Mangelsdorf DJ (2001) Nuclear receptors and lipid physiology: opening the x-files. Science 294:1866–1870. https://doi.org/10.1126/science.294.5548.1866

Cockbain AJ (1961) Fuel utilization and duration of tethered flight in *Aphis fabae* scop. J Exp Biol 38:163–174. https://doi.org/10.1242/jeb.38.1.163

Crabtree B, Newsholme EA (1972) The activities of lipases and carnitine palmitoyltransferase in muscles from vertebrates and invertebrates. Biochem J 130: 697–705. https://doi.org/10.1042/bj1300697

Cripps RM (2006) The contributions of genetics to the study of insect flight muscle function. In: Nature's versatile engine: insect flight muscle inside and out. Springer, New York, pp 2–15

Das S, Meier OW, Woodring J (1993) Diel rhythms of adipokinetic hormone, fat body response, and haemolymph lipid and sugar levels in the house cricket. Physiol Entomol 18:233–239. https://doi.org/10.1111/j.1365-3032.1993.tb00593.x

Davey KG, Huebner E (1974) The response of the follicle cells of *Rhodnius prolixus* to juvenile hormone and antigonadotropin *in vitro*. Can J Zool 52:1407–1412. https://doi.org/10.1139/z74-178

De Paula IF, Santos-Araujo S, Majerowicz D et al (2023) Knockdown of carnitine palmitoyltransferase I (CPT1) reduces fat body lipid mobilization and resistance to starvation in the insect vector *Rhodnius prolixus*. Front Physiol 14:1201670. https://doi.org/10.3389/fphys.2023.1201670

Di Lisa F, Barbato R, Menabò R, Siliprandi N (1995) Carnitine and carnitine esters in mitochondrial metabolism and function. In: De Jong JW, Ferrari R (eds) The carnitine system: a new therapeutical approach to cardiovascular diseases. Springer, Dordrecht, pp 21–38

Downer RGH, Chino H (1985) Turnover of protein and diacylglycerol components of lipophorin in insect haemolymph. Insect Biochem 15:627–630. https://doi.org/10.1016/0020-1790(85)90124-6

Du B, Ding D, Ma C et al (2022) Locust density shapes energy metabolism and oxidative stress resulting in divergence of flight traits. Proc Natl Acad Sci USA 119:e2115753118. https://doi.org/10.1073/pnas.2115753118

Fassold K, El-Damanhouri HIH, Lorenz MW (2010) Age-dependent cyclic locomotor activity in the cricket, *Gryllus bimaculatus*, and the effect of adipokinetic hormone on locomotion and excitability. J Comp Physiol A Neuroethol Sens Neural Behav Physiol 196:271–283. https://doi.org/10.1007/s00359-010-0513-9

Gäde G, Marco HG (2013) The invertebrate AKH/RPCH family: updated chapter. In: Kastin A (ed) Handbook of biologically active peptides, 2nd edn. Elsevier Inc., Amsterdam, pp 185–190

Gao Y, Liu S, Jia Q et al (2022) Juvenile hormone membrane signaling phosphorylates USP and thus potentiates 20-hydroxyecdysone action in *Drosophila*. Sci Bull 67:186–197. https://doi.org/10.1016/j.scib.2021.06.019

Goldsworthy GJ, Wheeler CH (1986) The endocrine control of flight metabolism in locusts. In: Danthanarayana W (ed) Insect flight. Springer, Berlin/Heidelberg, pp 49–59

Gondim KC, Atella GC, Pontes EG, Majerowicz D (2018) Lipid metabolism in insect disease vectors. Insect Biochem Mol Biol 101:108–123. https://doi.org/10.1016/j.ibmb.2018.08.005

Gordon S, Dickinson MH (2006) Role of calcium in the regulation of mechanical power in insect flight. Proc Natl Acad Sci USA 103:4311–4315. https://doi.org/10.1073/pnas.0510109103

Grönke S, Mildner A, Fellert S et al (2005) Brummer lipase is an evolutionary conserved fat storage regulator in *Drosophila*. Cell Metab 1:323–330. https://doi.org/10.1016/j.cmet.2005.04.003

Hansen KK, Stafflinger E, Schneider M et al (2010) Discovery of a novel insect neuropeptide signaling system closely related to the insect adipokinetic hormone and corazonin hormonal systems. J Biol Chem 285:10736–10747. https://doi.org/10.1074/jbc.M109.045369

Hansford RG, Johnson RN (1976) Some aspects of the oxidation of pyruvate and palmitoylcarnitine by moth (*Manduca sexta*) flight muscle mitochondria. Comp Biochem Physiol Part B Biochem 55:543–551. https://doi.org/10.1016/0305-0491(76)90015-8

Haunerland NH, Chisholm JM (1990) Fatty acid binding protein in flight muscle of the locust, *Schistocerca gregaria*. Biochim Biophys Acta (BBA)/Lipids Lipid Metab 1047:233–238. https://doi.org/10.1016/0005-2760(90)90521-X

Hou L, Guo S, Wang Y et al (2021) Neuropeptide ACP facilitates lipid oxidation and utilization during long-term flight in locusts. elife 10:e65279. https://doi.org/10.7554/eLife.65279

Iwamoto H (2011) Structure, function and evolution of insect flight muscle. Biophysics (Oxf) 7:21–28. https://doi.org/10.2142/biophysics.7.21

Jackson VN, Cameron JM, Zammit VA, Price NT (1999) Sequencing and functional expression of the malonyl-CoA-sensitive carnitine palmitoyltransferase from *Drosophila melanogaster*. Biochem J 341:483–489. https://doi.org/10.1042/0264-6021:3410483

Jones CM, Papanicolaou A, Mironidis GK et al (2015) Genomewide transcriptional signatures of migratory flight activity in a globally invasive insect pest. Mol Ecol 24:4901–4911. https://doi.org/10.1111/mec.13362

Josephson RK (2007) Comparative physiology of insect flight muscle. In: Vigoreaux JO (ed) Nature's versatile engine: insect flight muscle inside and out, 1st edn. Springer, New York, pp 35–43

Kaczmarek A, Boguś M (2021) The metabolism and role of free fatty acids in key physiological processes in insects of medical, veterinary and forensic importance. PeerJ 9:e12563. https://doi.org/10.7717/peerj.12563

Kammer AE, Heinrich B (1978) Insect flight metabolism. Adv In Insect Phys 13:133–228. https://doi.org/10.1016/S0065-2806(08)60266-0

Kennedy JS (1961) A turning point in the study of insect migration. Nature 189:785791. https://doi.org/10.1038/189785a0

Khawar MB, Gao H, Li W (2019) Autophagy and lipid metabolism. In: Qin Z (ed) Autophagy: biology and diseases, Advances in experimental medicine and biology. Springer Singapore, pp 359–374

Klowden MJ (2013) Physiological systems in insects: third edition, 3rd edn. Academic

Kodrík D (2008) Adipokinetic hormone functions that are not associated with insect flight. Physiol Entomol 33:171–180. https://doi.org/10.1111/j.1365-3032.2008.00625.x

Lee SH, Lee SK, Paik D, Min KJ (2012) Overexpression of fatty-acid-β-oxidation-related genes extends the lifespan of *Drosophila melanogaster*. Oxidative Med Cell Longev:854502. https://doi.org/10.1155/2012/854502

Leyria J (2024) Endocrine factors modulating vitellogenesis and oogenesis in insects: an update. Mol Cell Endocrinol 587:112211. https://doi.org/10.1016/J.MCE.2024.112211

Leyria J, Orchard I, Lange AB (2020) What happens after a blood meal? A transcriptome analysis of the main tissues involved in egg production in *Rhodnius prolixus*, an insect vector of chagas disease. PLoS Negl Trop Dis 14:e0008516. https://doi.org/10.1371/journal.pntd.0008516

Liu P, Peng HJ, Zhu J (2015) Juvenile hormone-activated phospholipase C pathway enhances transcriptional activation by the methoprene-tolerant protein. Proc Natl Acad Sci USA 112:E1871–E1879. https://doi.org/10.1073/pnas.1423204112

Lorenz MW (2007) Oogenesis-flight syndrome in crickets: age-dependent egg production, flight performance, and biochemical composition of the flight muscles in adult female *Gryllus bimaculatus*. J Insect Physiol 53:819–832. https://doi.org/10.1016/j.jinsphys.2007.03.011

Lorenz MW, Gäde G (2009) Hormonal regulation of energy metabolism in insects as a driving force for performance. Integr Comp Biol 49:380–392. https://doi.org/10.1093/icb/icp019

Lourens GB, Ferrell DK (2019) Lymphatic filariasis. Nurs Clin North Am 54:181–192. https://doi.org/10.1016/j.cnur.2019.02.007

Majerowicz D, Calderón-Fernández GM, Alves-Bezerra M et al (2017) Lipid metabolism in *Rhodnius prolixus*: lessons from the genome. Gene 596:27–44. https://doi.org/10.1016/j.gene.2016.09.045

Martínez-Barnetche J, Lavore A, Beliera M et al (2018) Adaptations in energy metabolism and gene family expansions revealed by comparative transcriptomics of three Chagas disease triatomine vectors. BMC Genomics 19:296. https://doi.org/10.1186/s12864-018-4696-8

Maruzs T, Simon-Vecsei Z, Kiss V et al (2019) On the fly: recent progress on autophagy and aging in *Drosophila*. Front Cell Dev Biol 7:140. https://doi.org/10.3389/fcell.2019.00140

Mayer RJ, Candy DJ (1969) Changes in energy reserves during flight of the desert locust, *Schistocerca gregaria*. Comp Biochem Physiol 31:409–418. https://doi.org/10.1016/0010-406X(69)90022-X

Mentel T, Duch C, Stypa H et al (2003) Central modulatory neurons control fuel selection in flight muscle of migratory locust. J Neurosci 23:1109–1113. https://doi.org/10.1523/jneurosci.23-04-01109.2003

Mesquita RD, Gaviraghi A, Gonçalves RLS et al (2021) Cytochrome c oxidase at full thrust: regulation and biological consequences to flying insects. Cells 10:470. https://doi.org/10.3390/cells10020470

Min KJ, Jones N, Borst DW, Rankin MA (2004) Increased juvenile hormone levels after long-duration flight in the grasshopper, *Melanoplus sanguinipes*. J Insect Physiol 50:531–537. https://doi.org/10.1016/j.jinsphys.2004.03.009

Mookerjee SA, Gerencser AA, Nicholls DG, Brand MD (2017) Quantifying intracellular rates of glycolytic and oxidative ATP production and consumption using extracellular flux measurements. J Biol Chem 292: 7189–7207. https://doi.org/10.1074/jbc.M116.774471

Neess D, Bek S, Engelsby H et al (2015) Long-chain acyl-CoA esters in metabolism and signaling: role of acyl-CoA binding proteins. Prog Lipid Res 59:1–25. https://doi.org/10.1016/j.plipres.2015.04.001

Ohnishi A, Hull JJ, Matsumoto S (2006) Targeted disruption of genes in the *Bombyx mori* sex pheromone biosynthetic pathway. Proc Natl Acad Sci USA 103: 4398–4403. https://doi.org/10.1073/pnas.0511270103

Olembo NK, Pearson DJ (1982) Changes in the contents of intermediates of proline and carbohydrate metabolism in flight muscle of the tsetse fly *Glossina morsitans* and the fleshfly *Sarcophaga tibialis*. Insect Biochem 12:657–662. https://doi.org/10.1016/0020-1790(82)90053-1

Oliveira GA, Baptista DL, Guimarães-Motta H et al (2006) Flight-oogenesis syndrome in a blood-sucking bug: biochemical aspects of lipid metabolism. Arch Insect Biochem Physiol 62:164–175. https://doi.org/10.1002/arch.20132

Orchard I, Carlisle JA, Loughton BG et al (1982) *In vitro* studies on the effects of octopamine on locust fat body. Gen Comp Endocrinol 48:7–13. https://doi.org/10.1016/0016-6480(82)90031-4

Orchard I, Ramirez JM, Lange AB (1993) A multifunctional role for octopamine in locust flight. Annu Rev Entomol 38:227–249. https://doi.org/10.1146/annurev.en.38.010193.001303

Patel R, Soulages JL, Wells MA, Arrese EL (2004) cAMP-dependent protein kinase of *Manduca sexta* phosphorylates but does not activate the fat body triglyceride lipase. Insect Biochem Mol Biol 34:1269–1279. https://doi.org/10.1016/j.ibmb.2004.08.008

Patel RT, Soulages JL, Hariharasundaram B, Arrese EL (2005) Activation of the lipid droplet controls the rate of lipolysis of triglycerides in the insect fat body. J Biol Chem 280:22624–22631. https://doi.org/10.1074/jbc.M413128200

Pflüger HJ, Duch C (2011) Dynamic neural control of insect muscle metabolism related to motor behavior. Physiology 26:293–303. https://doi.org/10.1152/physiol.00002.2011

Rajapakse S, Qu D, Ahmed AS et al (2019) Effects of FABP knockdown on flight performance of the desert locust, *Schistocerca gregaria*. J Exp Biol 222: jeb203455. https://doi.org/10.1242/jeb.203455

Riddiford LM (2020) *Rhodnius*, golden oil, and Met: a history of juvenile hormone research. Front Cell Dev Biol 8:679. https://doi.org/10.3389/fcell.2020.00679

Roma GC, Bueno OC, Camargo-Mathias MI (2010) Morpho-physiological analysis of the insect fat body: a review. Micron 41:395–401. https://doi.org/10.1016/j.micron.2009.12.007

Rozsypal J, Moos M, Rudolf I, Košťál V (2021) Do energy reserves and cold hardiness limit winter survival of *Culex pipiens*? Comp Biochem Physiol Part A Mol Integr Physiol 255:110912. https://doi.org/10.1016/j.cbpa.2021.110912

Ryan RO, Van Der Horst DJ (2000) Lipid transport biochemistry and its role in energy production. Annu Rev Entomol 45:233–260. https://doi.org/10.1146/annurev.ento.45.1.233

Ryuda M, Tsuzuki S, Matsumoto H et al (2011) Identification of a novel gene, *Anorexia*, regulating feeding activity via insulin signaling in *Drosophila melanogaster*. J Biol Chem 286:38417–38426. https://doi.org/10.1074/jbc.M111.267344

Sacktor B (1976) Biochemical adaptations for flight in the insect. Biochem Soc Symp 41:111–131

Santos-Araujo S, Bomfim L, Araripe LO et al (2020) Silencing of ATG6 and ATG8 promotes increased levels of triacylglycerol (TAG) in the fat body during prolonged starvation periods in the Chagas disease vector *Rhodnius prolixus*. Insect Biochem Mol Biol 127: 103484. https://doi.org/10.1016/j.ibmb.2020.103484

Sappington TW, Showers WB (1992) Reproductive maturity, mating status, and long-duration flight behavior of *Agrotis ipsilon* (Lepidoptera: Noctuidae) and the conceptual misuse of the oogenesis-flight syndrome by entomologists. Environ Entomol 21:677–688. https://doi.org/10.1093/ee/21.4.677

Scaraffia PY, Wells MA (2003) Proline can be utilized as an energy substrate during flight of *Aedes aegypti* females. J Insect Physiol 49:591–601. https://doi.org/10.1016/S0022-1910(03)00031-3

Schulze RJ, Sathyanarayan A, Mashek DG (2017) Breaking fat: the regulation and mechanisms of lipophagy. Biochim Biophys Acta Mol Cell Biol Lipids 1862:1178–1187. https://doi.org/10.1016/j.bbalip.2017.06.008

Shegelski VA, Evenden ML, Huber DPW, Sperling FAH (2021) Identification of genes and gene expression associated with dispersal capacity in the mountain pine beetle, *Dendroctonus ponderosae* Hopkins (Coleoptera: Curculionidae). PeerJ 9:e12382. https://doi.org/10.7717/peerj.12382

Shpigler HY, Magory Cohen T, Ben-Shimol E et al (2021) Juvenile hormone functions as a metabolic rate accelerator in bumble bees (*Bombus terrestris*). Horm Behav 136:105073. https://doi.org/10.1016/j.yhbeh.2021.105073

Siegert KJ (1999) Locust *corpora cardiaca* contain an inactive adipokinetic hormone. FEBS Lett 447:237–240. https://doi.org/10.1016/S0014-5793(99)00299-9

Sim C, Denlinger DL (2009) Transcription profiling and regulation of fat metabolism genes in diapausing adults of the mosquito *Culex pipiens*. Physiol Genomics 39: 202–209. https://doi.org/10.1152/physiolgenomics.00095.2009

Smith AF, Tsuchida K, Hanneman E et al (1992) Isolation, characterization, and cDNA sequence of two fatty acid-binding proteins from the midgut of *Manduca sexta* larvae. J Biol Chem 267:380–384. https://doi.org/10.1016/s0021-9258(18)48505-5

Somerville AGT, Gleave K, Jones CM, Reimer LJ (2019) The consequences of *Brugia malayi* infection on the flight and energy resources of *Aedes aegypti* mosquitoes. Sci Rep 9:18449. https://doi.org/10.1038/s41598-019-54819-2

Song F, Yan Y, Sun J (2024) Energy consumption during insect flight and bioinspiration for MAV design: a review. Comput Biol Med 170:108092. https://doi.org/10.1016/j.compbiomed.2024.108092

Stevenson E (1968) The carnitine-independent oxidation of palmitate plus malate by moth flight-muscle mitochondria. Biochem J 110:105–110. https://doi.org/10.1042/bj1100105

Suarez RK, Darveau CA, Welch KC et al (2005) Energy metabolism in orchid bee flight muscles: carbohydrate fuels all. J Exp Biol 208:3573–3579. https://doi.org/10.1242/jeb.01775

Sun BB, Jiang XF, Zhang L et al (2013) Methoprene influences reproduction and flight capacity in adults of the rice leaf roller, *Cnaphalocrocis medinalis* (GUENÊE) (Lepidoptera: Pyralidae). Arch Insect Biochem Physiol 82:1–13. https://doi.org/10.1002/arch.21067

Tomoyasu Y, Ohde T, Clark-Hachtel C (2017) What serial homologs can tell us about the origin of insect wings. F1000Research 6:268. https://doi.org/10.12688/f1000research.10285.1

Toprak U (2020) The role of peptide hormones in insect lipid metabolism. Front Physiol 11:434. https://doi.org/10.3389/fphys.2020.00434

Truman JW (2019) The evolution of insect metamorphosis. Curr Biol 29:R1252–R1268. https://doi.org/10.1016/j.cub.2019.10.009

Van Antwerpen R, Beekwilder J, Van Heusden MC et al (1990) Interaction of lipophorin with the plasma membrane of locust flight muscles. Biol Chem Hoppe Seyler 371:159–165. https://doi.org/10.1515/bchm3.1990.371.1.159

Van Antwerpen R, Salvador K, Tolman K, Gentry C (1998) Uptake of lipids by developing oocytes of the hawkmoth *Manduca sexta*. The possible role of lipoprotein lipase. Insect Biochem Mol Biol 28:399–408. https://doi.org/10.1016/S0965-1748(98)00012-5

Van der Horst DJ (1990) Lipid transport function of lipoproteins in flying insects. Biochim Biophys Acta Lipids Lipid Metab 1047:195–211. https://doi.org/10.1016/0005-2760(90)90518-3

Van Der Horst DJ (2003) Insect adipokinetic hormones: release and integration of flight energy metabolism. Comp Biochem Physiol B Biochem Mol Biol 136:217–226. https://doi.org/10.1016/S1096-4959(03)00151-9

Van Der Horst DJ, Rodenburg KW (2010) Lipoprotein assembly and function in an evolutionary perspective. Biomol Concepts 1:165–183. https://doi.org/10.1515/bmc.2010.012

Van der Horst DJ, Ryan RO (2017) Lipid transport. In: Reference module in life sciences, 1st edn. Elsevier, pp 1–24

Van der Horst DJ, Van Doorn JM, Passier PCCM et al (1993) Role of fatty acid-binding protein in lipid metabolism of insect flight muscle. Mol Cell Biochem 123:145–152. https://doi.org/10.1007/BF01076486

Van der Horst DJ, Van Marrewijk WJA, Diederen JHB (2001) Adipokinetic hormones of insect: release, signal transduction, and responses. Int Rev Cytol 211:179–240. https://doi.org/10.1016/S0074-7696(01)11019-3

Van der Horst DJ, Van Hoof D, Van Marrewijk WJA, Rodenburg KW (2002) Alternative lipid mobilization: the insect shuttle system. Mol Cell Biochem 239:113–119. https://doi.org/10.1023/A:1020541010547

Van der Horst DJ, Roosendaal SD, Rodenburg KW (2009) Circulatory lipid transport: lipoprotein assembly and function from an evolutionary perspective. Mol Cell Biochem 326:105–119. https://doi.org/10.1007/s11010-008-0011-3

Van Heusden MC, Van der Horst DJ, Van Doorn JM et al (1986) Lipoprotein lipase activity in the flight muscle of *Locusta migratoria* and its specificity for haemolymph lipoproteins. Insect Biochem 16:517–523. https://doi.org/10.1016/0020-1790(86)90029-6

Van Heusden MC, Van der Horst DJ, Kawooya JK, Law JH (1991) *In vivo* and *in vitro* loading of lipid by artificially lipid-depleted lipophorins: evidence for the role of lipophorin as a reusable lipid shuttle. J Lipid Res 32:1789–1794. https://doi.org/10.1016/s0022-2275(20)41633-5

Vroemen SF, Van Der Horst DJ, Van Marrewijk WJA (1998) New insights into adipokinetic hormone signaling. Mol Cell Endocrinol 141:7–12. https://doi.org/10.1016/S0303-7207(98)00079-3

Wang B, Moya N, Niessen S et al (2011) A hormone-dependent module regulating energy balance. Cell 145:596–606. https://doi.org/10.1016/j.cell.2011.04.013

Wang X, Fang X, Yang P et al (2014) The locust genome provides insight into swarm formation and long-distance flight. Nat Commun 5:2957. https://doi.org/10.1038/ncomms3957

Ward JP, Candy DJ, Smith SN (1982) Lipid storage and changes during flight by triatomine bugs (*Rhodnius prolixus* and *Triatoma infestans*). J Insect Physiol 28:527–534. https://doi.org/10.1016/0022-1910(82)90033-6

Wheeler CH, Van der Horst DJ, Beenakkers AMT (1984) Lipolytic activity in the flight muscles of *Locusta migratoria* measured with haemolymph lipoproteins as substrates. Insect Biochem 14:261–266. https://doi.org/10.1016/0020-1790(84)90058-1

Wheeler CH, Boothby KM, Goldsworthy GJ (1986) CL-proteins and the regulation of lipoprotein lipase activity in locust flight muscle. Biol Chem Hoppe Seyler 367:1127–1133. https://doi.org/10.1515/bchm3.1986.367.2.1127

Whitfield CW, Ben-Shahar Y, Brillet C et al (2006) Genomic dissection of behavioral maturation in the honey bee. Proc Natl Acad Sci USA 103:16068–16075. https://doi.org/10.1073/pnas.0606909103

Wigglesworth VB, Lee WM (1982) The supply of oxygen to the flight muscles of insects: a theory of tracheole physiology. Tissue Cell 14:501–518. https://doi.org/10.1016/0040-8166(82)90043-X

Wu SF, Xu G, Qi YX et al (2014) Two splicing variants of a novel family of octopamine receptors with different signaling properties. J Neurochem 129:37–47. https://doi.org/10.1111/jnc.12526

Yamamoto K, Chadarevian A, Pellegrini M (1988) Juvenile hormone action mediated in male accessory glands of drosophila by calcium and kinase C. Science 239: 916–919. https://doi.org/10.1126/science.3124270

Yoichi H (1987) Characterization of lipophorin receptor in locust flight muscles. Biochim Biophys Acta (BBA)/ Lipids Lipid Metab 919:58–63. https://doi.org/10.1016/0005-2760(87)90217-7

Zechner R, Kienesberger PC, Haemmerle G et al (2009) Adipose triglyceride lipase and the lipolytic catabolism of cellular fat stores. J Lipid Res 50:3–21. https://doi.org/10.1194/JLR.R800031-JLR200

Zera AJ, Harshman LG (2001) The physiology of life history trade-offs in animals. Annu Rev Ecol Syst 32: 95–126. https://doi.org/10.1146/annurev.ecolsys.32.081501.114006

Zhang CX, Brisson JA, Xu HJ (2019) Molecular mechanisms of wing polymorphism in insects. Annu Rev Entomol 64:297–314. https://doi.org/10.1146/annurev-ento-011118-112448

Zhao X, Karpac J (2017) Muscle directs diurnal energy homeostasis through a myokine-dependent hormone module in *Drosophila*. Curr Biol 27:1941–1955. https://doi.org/10.1016/j.cub.2017.06.004

Zhao Z, Zera AJ (2002) Differential lipid biosynthesis underlies a tradeoff between reproduction and flight capability in a wing-polymorphic cricket. Proc Natl Acad Sci USA 99:16829–16834. https://doi.org/10.1073/pnas.262533999

Zhou Y, Nagata S (2021) Chapter 83—adipokinetic hormone/corazonin-related peptide. In: Ando H, Ukena K, Nagata S (eds) Handbook of hormones, 2nd edn. Academic, pp 759–760

Ziegler R, Schulz M (1986) Regulation of lipid metabolism during flight in *Manduca sexta*. J Insect Physiol 32:903–908. https://doi.org/10.1016/0022-1910(86)90106-X

Leonardo L. Fruttero is an Assistant Professor in the Department of Clinical Biochemistry and the Center for Research in Clinical Biochemistry and Immunology (CIBICI), Faculty of Chemical Sciences, Universidad Nacional de Córdoba (Argentina). He graduated as Biologist in 2005 and obtained his Ph.D. degree in Chemical Sciences (discipline: Biochemistry) in 2011 at the Universidad Nacional de Córdoba. Dr. Fruttero performed his postdoctoral training at the Brain Institute, Pontifícia Universidade Católica do Rio Grande do Sul, Brazil, from 2014 to 2017. Since 2018, he has been a member of the National Scientific and Technical Research Council staff, currently in the Adjunct category. Current and previous field of interest of Dr. Fruttero involve insect biochemistry and physiology with focus on lipid and protein metabolism, reproduction and the effects of entomotoxic proteins on the immune system.

Jimena Leyria is an Assistant Professor at the Universidad Nacional de Córdoba (Argentina) and member of the National Scientific Research Council (CONICET). She received her bachelor's degree in Biochemistry in 2008 and her Ph.D. in Chemical Sciences in 2015 at the Universidad Nacional de Córdoba. During 2018–2023, Dr. Leyria performed her postdoctoral training as a fellow in the Department of Biology at the University of Toronto Mississauga (Canada). Dr. Leyria was awarded two trainee research fellowships at the Universidade Federal do Rio Grande do Sul (Brazil) and another at the Institut de Biologia Evolutiva-UPF-CSIC (Spain). Her field of expertise is centered on insect reproductive physiology, mainly in how tissues communicate with one another to promote successful reproduction.

Lilián E. Canavoso is an Associate Professor the Department of Clinical Biochemistry and the Center for Research in Clinical Biochemistry and Immunology (CIBICI), Faculty of Chemical Sciences, Universidad Nacional de Córdoba (Argentina). She graduated as a Biochemist and obtained her Ph.D. degree in Chemical Sciences (discipline: Biochemistry) in 1995 at the Universidad Nacional de Córdoba. Dr. Canavoso performed her postdoctoral training at the Department of Biochemistry and Center for Insect Science, University of Arizona, USA, from 1998 to 2000. Since 2004, she has been a member of the National Scientific and Technical Research Council staff in the Independent category. Current and previous fields of interest of Dr. Canavoso involve insect biochemistry and physiology with focus on lipid and protein metabolism, reproductive biology, and immunity.

Adv Exp Med Biol (2026) 1494: 367–388
https://doi.org/10.1007/5584_2024_810

Published online: 15 June 2024

Circadian Control of Lipid Metabolism

Christian Wegener, Kelechi M. Amatobi,
Ayten Gizem Ozbek-Unal, and Agnes Fekete

Abstract

To ensure optimum health and performance, lipid metabolism needs to be temporally aligned to other body processes and to daily changes in the environment. Central and peripheral circadian clocks and environmental signals such as light provide internal and external time cues to the body. Importantly, each of the key organs involved in insect lipid metabolism contains a molecular clockwork which ticks with a varying degree of autonomy from the central clock in the brain. In this chapter, we review our current knowledge about peripheral clocks in the insect fat body, gut and oenocytes, and light- and circadian-driven diel patterns in lipid metabolites and lipid-related transcripts. In addition, we highlight selected neuroendocrine signaling pathways that are or may be involved in the temporal coordination and control of lipid metabolism.

C. Wegener (✉) and K. M. Amatobi
Neurobiology and Genetics, Theodor-Boveri-Institute, Biocenter, University of Würzburg, Würzburg, Germany
e-mail: christian.wegener@biozentrum.uni-wuerzburg.de

A. G. Ozbek-Unal and A. Fekete
Pharmaceutical Biology, Julius-von-Sachs-Institute, Biocenter, University of Würzburg, Würzburg, Germany
e-mail: ayten_gizem.oezbek@stud-mail.uni-wuerzburg.de; agnes.fekete@uni-wuerzburg.de

Keywords

Circadian clock · Fat body · Feeding · Gut · Metabolite cycling · Neuropeptide hormones · Oenocytes · Period · Peripheral clock

Circadian clocks are endogenous timers that allow animals to anticipate daily changes in the environment and to adapt their physiology and behavior accordingly. Circadian clocks temporally organize a multitude of body processes across different scales, from the molecular to the cellular and systemic physiological level all the way up to behavior. In particular, circadian clocks help metabolic organs to adjust in time for periods of high energy need (e.g., periodic phases of increased physical activity) or food intake. The clocks also help to set optimum time windows for anabolic and catabolic processes. For instance, mice with defective clock genes in the adipose tissue or the whole body show various metabolic phenotypes including increased body fat known as hyperlipidemia (see Gooley 2016; Li et al. 2020).

Circadian clocks are based on molecular expression loops that oscillate endogenously with a period close to 24 h (circadian is derived from the latin circa ("around") and *(cotti)dianus* ("daily")). External Zeitgeber (a German term, literally translates to "time givers") such as light or temperature changes are able to entrain circadian clocks in order to align their phase to daily changes in the environment.

Insects, especially the fruit fly *Drosophila melanogaster*, are key models in circadian research. In fact, the first clock genes were identified in *Drosophila*, with homologs later on found in mammals including humans (see Patke et al. 2020). The fruit fly with its sophisticated genetic toolbox is also increasingly used as an experimental model to understand the basics of how body clocks control the metabolism of lipids and other metabolites. Arguably, much research in *Drosophila* in this area is biomedically motivated and aims to better understand the metabolic syndrome associated with circadian disruption in our modern societies (Abbott et al. 2020; Lee et al. 2021). Yet, there is also a clear direct relevance to understand the circadian control of lipid metabolism in insects. For instance, mechanisms that cause higher body mass or optimized lipid composition may actually be beneficial for the development of insects as an environment-friendly food source. Understanding how the clock affects lipid metabolism will also help to understand and predict changes in reproduction, feeding, and metabolism during man-made global change with its artificial light-at-night or geographic shifts to higher latitudes with altered photoperiods (Bozinovic et al. 2011; Sunday et al. 2012; McLay et al. 2018; Fleury et al. 2020).

In this chapter, we review our current knowledge on the diel rhythmicity of lipid metabolism in insects, especially the fruit fly *Drosophila*. We first review the circadian system composed of central clocks in the brain and peripheral clocks in relevant tissues, then we describe the observed clock- and light-driven diel changes in the lipidome and related transcriptome. This is followed by a closer look into the connection between feeding and temporal patterns in lipid metabolism. Finally, we address the still underexplored role of the neuroendocrine system in timing diel changes in lipid metabolism.

1 The Circadian System, and Peripheral Clocks in Organs Involved in Lipid Metabolism

1.1 The Molecular Clock

The molecular clockwork is based on interlinked transcription-translation feedback loops (TTFLs) that lead to a circadian (daily) cycling of clock gene mRNA and protein. In *Drosophila*, the primary TTFL (*per/tim* feedback loop) consists of the clock genes *period* (*per*), *timeless* (*tim-d*), *Clock* (*Clk*), and *cycle* (*cyc*) and their protein products PER, TIM, CLK, and CYC (Fig. 1). The environmental photoperiod can entrain this core TTFL via light signals from the eyes and via the blue light-sensitive protein *Drosophila* Cryptochrome (dCRY) which in many cells is co-expressed with the core clock genes (Fig. 1). The *per/tim* loop is linked via *Clk* with a second TTFL that drives cyclic *Clk* expression (the *Clk* feedback loop) and comprises the genes *Clk*, *cyc*, *vrille* (*vri*), and *Pdp1* (Fig. 1, for recent reviews see e.g. Jarabo and Martin 2017; Patke et al. 2020). The two feedback loops are under further regulation by miRNA and other modulators such as CLOCKWORK-ORANGE (CWO) which modulate *Clk/cyc* transcription (Tataroglu and Emery 2015).

In other insects, the composition and the exact function of the involved clock proteins may vary. For example, the ancestral clock as found in many species, including butterflies, uses dCRY plus a mammalian type Cryptochrome (mCRY), while other species like bees and the linden bug have lost dCRY-d and only use mCRY (Yuan et al. 2007; Tomioka and Matsumoto 2015; Beer and Helfrich-Förster 2020; Kotwica-Rolinska et al. 2022; Deppisch et al. 2023). Further, Hymenopterans appear to have lost the *tim-d* gene (Tomioka and Matsumoto 2015; Beer and Helfrich-Förster 2020; Kotwica-Rolinska et al. 2022). In these cases, entrainment of the central clock to external light-dark cycles occurs exclusively via the eyes as is true for mammals.

1.2 The Central Clock in the Brain

On the neuronal level, the central clock in insects consists of a bilateral network of clock gene-expressing neurons in the central brain, with connection to the accessory medulla, a small neuropile of the optic lobes which is regarded as the circadian pacemaker center that also receives light input from the eyes (Helfrich-Förster et al. 1998; Stengl and Arendt 2016; Li et al. 2018; Beer and Helfrich-Förster 2020). In the brain of

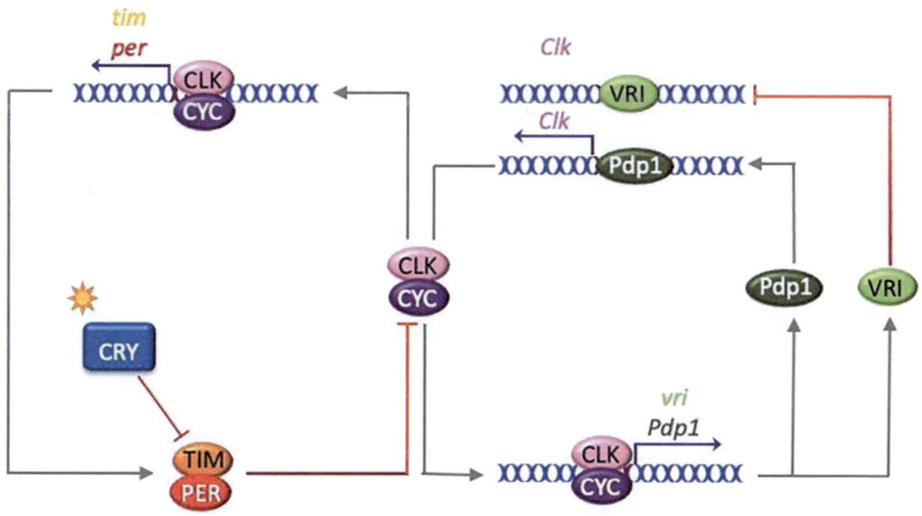

Fig. 1 Simplified sketch of the core transcription-translation feedback loops (TTFL) in the fruit fly *Drosophila*. The expression of *per* and *tim* is induced by binding of a CLK/CYC heterodimer on regulatory elements (E-box). The protein products PER and TIM heterodimerize and inhibit the activity of the CLK/CYC dimer which leads to declining levels of PER and TIM which in turn starts the cycle again. Light changes the phase of the oscillation by activating dCRY which leads to degradation of TIM and delays the inhibition of CLK/CYC. This feedback mechanism is reinforced by a second TTFL that drives cyclic expression of *Clk* via *Pdp1* which is delayed by VRI. *tim* timeless, *per* period, *Clk* Clock, *Cyc* Cycle, *CRY* cryptochrome, *vri* vrille, *Pdp1* Par-domain protein 1

Drosophila, the central clock network consists of around 150 lateral (LN) and dorsal (DN) neurons that are highly interconnected. Many of these clock cells receive light input via the accessory medulla (see Helfrich-Förster 2017; Top and Young 2018; Li et al. 2018; Reinhard et al. 2023). Among the *Drosophila* clock neurons, the lateral neurons (LNs) are best characterized and are regarded as pacemaker neurons. The LNs can be divided into a small and large subset (s- and lLNvs) and express the neuropeptide pigment-dispersing factor (PDF). PDF signaling plays an important role in maintaining temporal alignment within the central clock, and additionally also represents a major clock output factor (Helfrich-Förster and Homberg 1993; Renn et al. 1999; Park et al. 2000; Shafer et al. 2008; Shafer and Yao 2014; Liang et al. 2017). PDF, though also expressed outside the clock, is strongly associated with the central clock throughout the insects (Stengl 1994; Helfrich-Förster et al. 1998; Sehadová et al. 2003; Shafer and Yao 2014; Beer et al. 2018).

1.3 Peripheral Clocks

An oscillatory expression of clock genes is also found in peripheral tissues. In insects, these peripheral clocks comprise, among others, pivotal organs for lipid metabolism: fat body, gut, and oenocytes (Ito and Tomioka 2016). In insects with photosensitive dCRY, the peripheral clocks can (potentially) be directly synchronized by light. This is different in insects without dCRY, and the situation in these species is more like that in mammals which peripheral clocks cannot be directly entrained by light (Ito and Tomioka 2016). Moreover, peripheral clocks show a variable degree of autonomy in insects. While some clocks (e.g., in the Malpighian tubules that have a renal function) are autonomous, others are phase-setted by the central clock (e.g., the oenocyte clock). In line, while the molecular oscillations in some peripheral insect clocks dampen under constant conditions, they remain robust in others (Plautz et al. 1997; Versteven et al. 2020).

1.3.1 The Peripheral Clock in the Fat Body

The insect fat body serves similar functions as the liver and adipose tissue, vertebrate organs known to contain peripheral clocks (Zwighaft et al. 2016; Friedrichs et al. 2018). Pioneering work showed early on that the insect fat body expresses clock proteins. Immunostaining demonstrated PER and TIM expression in the fat body of *Drosophila* and the moths *Antheraea pernyi* and *Spodoptera littoralis* (Sauman et al. 1996; Giebultowicz et al. 2001; Karpowicz et al. 2013; Suszczynska et al. 2017). Labeling of the *Drosophila* fat body was also observed by *per*- or *tim*-promotor-driven expression of GFP (Kaneko and Hall 2000). PER immunostaining did not change in intensity and cytoplasmic location during 24 h in the silkmoth fat body (Sauman et al. 1996). In contrast, *Drosophila* PER and TIM were found to translocate to the nucleus and to cycle with a peak in the second half of the night (Giebultowicz et al. 2001; Xu et al. 2008). This is in line with cycling *tim* transcripts in the fat body which peaked around Zeitgeber time (ZT) 16 (Xu et al. 2008). A similar cycling was also reported for VRI (Xu et al. 2008) (see Fig. 1).

Diel oscillations of clock gene expression in the *Drosophila* fat body were recently confirmed by a heroic circadian transcriptomic analysis (Litovchenko et al. 2021). This analysis supports a robust cycling of *per*, *tim*, *Clk*, *cry*, *vri*, *clockwork orange* (*cwo*), and *Pdp1* in the fat body under light-dark (LD) and constant (DD) conditions. In addition, five uncharacterized genes were found to cycle in a circadian manner and hence may represent further components of the molecular clock in the fat body (Litovchenko et al. 2021). In LD 12: 12 and DD, *tim*, *per*, *vri*, and *Pdp1* transcript levels increased during the second half of the light phase and peaked until the middle of the night with a trough during midday. As expected (Fig. 1), *Clk*, *cry*, and *cwo* cycled in anti-phase (Litovchenko et al. 2021). The temporal pattern of clock gene transcript oscillations was roughly in phase with that observed in the brain, gut, and Malpighian tubules, except for *Pdp1* levels in the brain which lacked a clear rhythmicity (Litovchenko et al. 2021).

The grade of autonomy of the fat body clock has so far not been tested directly, but existing data suggests that the fat body may represent an autonomous clock under LD conditions that dampens in amplitude in DD. In *Drosophila*, cycling of *tim* and *per* was lost after 6 days in DD under *ad libitum* feeding, but was restored and entrained by time-restricted feeding (TRF) (Xu et al. 2011). Depending on the time of TRF, the phase of *tim* and *per* cycling in the fat body was shifted and hence misaligned to the phase of the TRF-insensitive molecular clock in the brain (Xu et al. 2011), suggesting clock autonomy, as well as food entrainability similar to the situation in the liver and other mammalian peripheral clocks (Damiola et al. 2000). In LD, the cycling of *per* transcripts in the fat body is unaltered in flies with an impaired central clock (Erion et al. 2016), suggesting that the light cycle is able to drive *per* oscillations, most likely via dCRY. In fact, dCRY levels in the fat body are strongly reduced by light (Agrawal et al. 2017). This suggests at least for flies with a transparent cuticle that light may act as a dominant Zeitgeber for the fat body clock, similar to other peripheral clocks in the flies that are entrainable by light in a dCRY-dependent manner (Plautz et al. 1997; Giebultowicz et al. 2001; Harper et al. 2017).

In the fat body of the cotton leafworm *S. littoralis*, *per*, *tim*, *vri*, and *m-cry* transcripts were cycling in LD, with peaks during the night (Suszczynska et al. 2017). In line, PER protein showed a rhythmic nuclear translocation. The rhythmic cycling of *per*, *tim*, *cry1*, and *cry2* persisted in DD and hence appears to be driven by the circadian clock. Interestingly, *Clk* transcript levels did not vary in a diel or circadian fashion, and *vri* stopped cycling in DD. Clock gene expression (*per*, *Pdp1*) has also been confirmed by RT-PCR in the linden bug *Pyrrhocoris apterus*, yet no diel cycling of transcript levels was found under short or long days (Dolezel et al. 2008). These findings suggest that *per/tim* expression in the fat body may be driven by species-specific mechanisms with varying robustness across insects.

1.3.2 The Peripheral Clock in the Gut

The insect gut contains a further peripheral clock involved in lipid metabolism. In the midgut epithelium of larvae of the silkmoth *Antheraea pernyi*, PER-immunoreactivity showed cyclic expression and cytoplasm-to-nucleus translocation, with a peak at the second half of the dark phase (Sauman et al. 1996; Sauman and Reppert 1998). *per* and *tim* are also expressed in the anterior stomach and midgut of crickets with a robust cycling in DD (Uryu and Tomioka 2010). In the cotton leafworm *S. littoralis*, clock genes are expressed in the larval midgut: *tim, vri, cry1,* and *cry2* were found to cycle in a diel fashion in LD, while *per* and *clk* transcripts were arrhythmic (Suszczynska et al. 2017). Similarly, PER did not undergo nuclear transition but remained in the cytoplasm (Suszczynska et al. 2017). In DD, only *tim* and *dcry* continued to show significant cycling. None of the investigated clock gene transcripts cycled in a diel fashion in the gut of the linden bug *Pyrrhocoris apterus* (Bajgar et al. 2013b). These findings suggest that clock genes are generally expressed in the digestive system across the insects, yet the degree to which they are periodically expressed may vary.

In *Drosophila*, the gut was first suggested to contain a peripheral clock by immunostaining against the PER and TIM proteins (Siwicki 1988; Giebultowicz et al. 2001) and expression of *per*- or *tim*-promotor driven markers (Liu et al. 1988; Kaneko and Hall 2000) in all major segments, including esophagus, midgut, and hindgut. PER and TIM were found to cycle and translocate from the cytoplasm to the nucleus, as higher nuclear staining intensities occurred late at night (ZT22) compared to mid-day (ZT8) (Giebultowicz et al. 2001). The gut consists of four major cell types: intestinal stem cells, enteroblasts, enterocytes, and enteroendocrine cells. Except for the enteroendocrine cells, PER and TIM cycle in all of them with the same phase in anterior and posterior compartments in the gut of *Drosophila* (Parasram et al. 2018).

The ticking of the molecular clock in the fruit fly gut was comprehensively characterized and confirmed by a recent circadian transcriptomic analysis (Litovchenko et al. 2021). As in the fat body, the main clock genes *per, tim, Clk, cry, vri, cwo*, and *Pdp1* showed robust and strong cycling in the midgut under LD and DD conditions. Oscillations of the individual clock genes were in phase with the oscillations in the fat body, brain, and Malpighian tubules (Litovchenko et al. 2021). Cycling of a *per::GFP* reporter and clock genes in the gut (except for *Pdp1*) was lost in *d-cry*^{01} mutants with ad libitum access to food in LD and DD (Parasram et al. 2018; Litovchenko et al. 2021). Restricted feeding at CT 0-3 and CT 9-12 affected PER cycling in *d-cry*^{01} mutants but not wild-type flies (Parasram et al. 2018). This argues for a dominant role of dCRY-dependent light entrainment, and an effect of feeding on clock gene expression in absence of light entrainment as described for the fat body. This assumption is supported by the strong light-induced degradation of a CRY reporter protein in the midgut (Agrawal et al. 2017).

Ligation and transplantation experiments in the silkmoth larva suggest that nuclear PER translocation in the gut is dependent on regular food intake or neuronal connection to the brain (Sauman and Reppert 1998). In crickets, cycling of *per* in the midgut persisted in LD upon removal of the optic lobes, the location of the central clock in the brain, while the *tim* level was strongly decreased and did not cycle (Uryu and Tomioka 2010). In insects, the gut clock may thus be partially dependent on a functional brain clock, but further research is required to define the actual degree of autonomy.

1.3.3 The Peripheral Clock in the Oenocytes

The oenocytes are of ectodermal origin and are specialized for lipid synthesis and metabolism. They synthesize very long-chain fatty acids and hydrocarbons that are required for cuticular waterproofing and cuticular contact pheromones (Makki et al. 2014). First evidence for the presence of a molecular clock in oenocytes was found in the blood-feeding bug *Rhodnius prolixus,* based on cycling expression of ecdysteroid receptors (Vafopoulou and Steel 2006). Yet, the so far only clear demonstration of a ticking clock in oenocytes comes from *Drosophila* (Krupp et al. 2008). In the oenocytes of wild-type flies in LD and DD, transcripts of *tim*, *per*, and *Clk* cycle in a diel fashion in phase with the other peripheral clocks and the central clock in the brain. In per^{01} mutants in DD, the cycling is abolished for *per* and *tim,* but persists for *Clk* with a shifted acrophase (Krupp et al. 2008). The free-running period of the oenocyte clock appears to be long (25.5–26.5 h, Krupp et al. 2013). The neuropeptide pigment dispersing factor (PDF) is a key signal within the central clock, and also a major clock output factor (Shafer and Yao 2014). Signaling from the central nervous system is not required for rhythmic clock gene expression in the oenocytes, yet PDF seems to provide phase information to the oenocytes (Krupp et al. 2013). Collectively, these results suggest that the oenocyte clock is a slave to the central clock (Ito and Tomioka 2016).

2 Diel and Circadian Changes in Lipid Metabolites

2.1 Circadian Lipidomics in Insects

The specifying feature of rhythmic lipids used here is that their mono- or biphasic temporal profile shows a period length close to 24 h under a light:dark or temperature cycle. In principle, this temporal profile with daily re-occurrence (rhythmicity) can be caused by an endogenous oscillator (circadian oscillation) or by another exogenous factor (e.g., daily changes in light and temperature, food availability) that does not involve an endogenous pacemaker. In chronobiology, the first case of an endogenous oscillator would be called clock-driven (or entrained), and the second case of an exogenous rhythmic factor would be called masking (Mrosovsky 1999). If the observed diel changes ware entirely light-driven, we would observe a loss of rhythmicity in DD (Fig. 2). Diel rhythms are driven by circadian clocks when their oscillation and periodicity i) continues under constant conditions, and ii) is impaired upon disruption of the molecular clock (Fig. 2). Circadian clock-driven metabolite levels can be masked by different environmental factors that may or may not change in time (Fig. 2). For instance, diacylglycerol (DAG) levels in the hemolymph show clear rhythmicity in flies kept continuously on sugar-only medium, but are strongly weakened or disappear when flies were kept on full medium (Amatobi et al. 2023).

To investigate whether temporal oscillations in lipid levels are driven by light or the circadian clock, experimental designs including high-frequency sampling are required (Fig. 2). Due to the small body size of insects, metabolite levels can hardly be determined over time in the same individual. Therefore, values obtained for each time point typically originate from measurements of different individuals, which can lead to weak rhythmicity scores if variation in individual metabolite levels are higher than or similar to daily changes in metabolite levels. To minimize this difficulty, several individuals can be pooled in one sample and the number of replicates can be increased. To identify clock-driven components, metabolite profiles of wild-type animals can be compared to profiles from animals with an impaired molecular clock (Fig. 2). Samples are taken at defined time points, and metabolites are identified that differ in their oscillatory pattern between the two genotypes. Impairment of the endogenous clock may result in either a loss or dampening of metabolic rhythmicity, altered periodicity or a phase shift of the peaks, especially under constant conditions (Fig. 2). General differences in metabolite levels between clock-impaired animals and wild-type controls may hint to pleiotropic effects of the clock on the core metabolism, resulting in metabotypes with constitutively up- or downregulated metabolic pathways.

The current state-of-the-art approach to identify oscillating lipids is to sample total lipid extracts at defined time points. Then, the samples are analyzed by either high-resolution mass spectrometry using untargeted lipidomics, or by tandem mass spectrometry using shotgun lipidomics

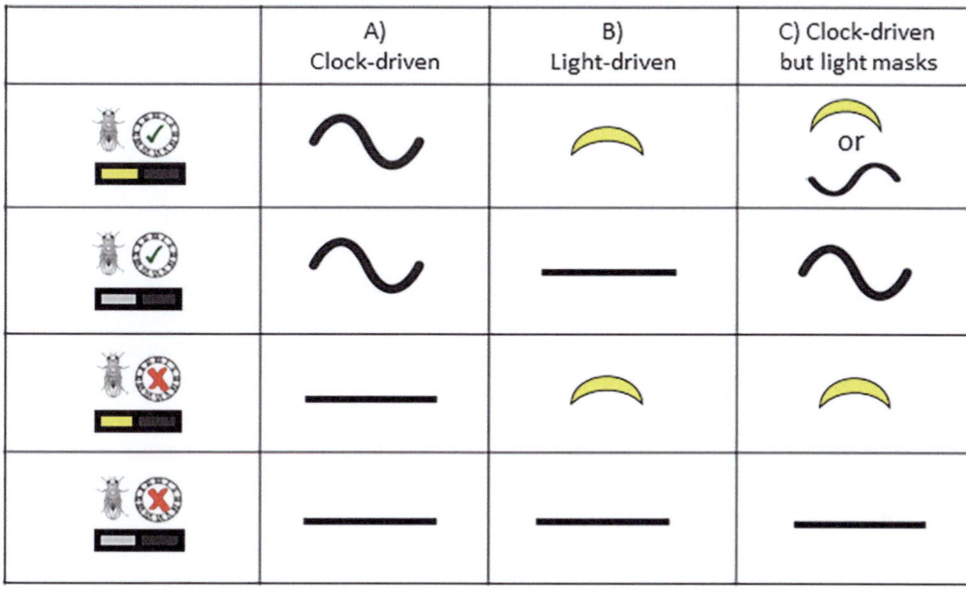

Fig. 2 Schematic overview of the rhythmicity of metabolic processes driven by either circadian clocks or light under different conditions. The first column indicates the condition (LD or DD, flies with a functional or impaired clock). (**a**) depicts the typical rhythmicity of clock-driven processes: daily rhythms are driven by circadian clocks (indicated by sinusoid curve) when rhythmicity observed in LD persists under constant conditions (DD), but is lost in clock mutants. (**b**) Light-driven diel oscillations (indicated by yellow curve) disappear in DD independent of the presence of a functional clock. (**c**) For many lipids, both the clock and light seem to drive diel oscillations. This can lead to complete masking, diminished amplitude, or shifted phase of the clock-driven rhythm in LD

(see chapter "Insect Lipidomics" by Musselman). For data analysis, different multivariate and univariate statistical approaches can be combined to filter out those metabolites which show time-dependent changes in their levels. After statistical analysis, however, manual inspection is typically required to ensure that metabolite levels show clear oscillation patterns. A more straightforward approach uses rhythmicity-detecting algorithms such as JTK_CYCLE (Hughes et al. 2010) or RAIN (Thaben and Westermark 2014) to filter out metabolites with diel rhythmicity. Sampling frequency and duration strongly determines the quality of the automatic detection of cycling metabolites; sampling every 2 h for at least 2 days was shown to decrease the number of false positive results (Wu et al. 2014).

So far, only few comprehensive studies that analyze rhythmic lipid metabolites in insects have been published, all of them in *Drosophila* (Gogna et al. 2015; Rhoades et al. 2018; Schäbler et al. 2020). Gogna et al. (2015) showed that levels of fatty acids and sterols peak at the end of the dark phase when wild-type flies are sampled every 2 h across a 24 h period and acetonitrile extracts are analyzed by nuclear magnetic resonance spectroscopy. Since the flies were kept under light:dark and temperature cycles that strongly promoted synchronization, it is difficult to assess whether the unimodal oscillations in fatty acid and sterol were driven by light:dark and/or temperature cycle or by the circadian clock.

Rhoades et al. analyzed heads and bodies (thorax plus abdomen) of white-eyed w^{1118} control and per^{01} clock mutant flies under LD and DD conditions (Rhoades et al. 2018). Blight-Dyer extraction was performed on flies collected every 2 h over a 48 h period and the aqueous phases were analyzed by hydrophilic interaction liquid chromatography coupled to a triple quadrupole mass spectrometer. Since this extraction and separation protocol is optimized to separate polar metabolites, only a limited number of lipids were detectable. Among the detectable polar lipids, five acylcarnitines (stearoyl-, lauroyl-, hydroxytetradecanoyl-, hydroxyhexadecanoyl-, hydroxyhexadecenoyl-carnitine) showed oscillation in control flies uniquely in DD. In contrast, Schäbler et al. detected weak daily oscillations in levels of eight acylcarnitines (acetyl-, propyonyl-, butyryl-, palmitoyl-, palmytoeoyl-, stearoyl-, oleoyl- and octadecadienyl-carnitine) in bodies of CantonS wild-type flies in LD (Schäbler et al. 2020). In DD, short-chain acylcarnitinines kept their rhythmicity, while long-chain acylcarnitines remained low on a constant level (Schäbler et al. 2020). In bodies of *per*^{01} mutants, the daily oscillation of short- and long-chain acylcarnitines in LD persisted with a weaker rhythmicity (Schäbler et al. 2020).

Similar to what was found in bodies, all detectable acylcarnitines strongly oscillated also in fly heads; short-chain acylcarnitines (from acetyl- to butyryl-carnitine) peaked at the end of the night whereas middle- and long-chain acylcarnitines (from lauroyl till octadecanoyl-cartinine) peaked at ZT4 in CantonS wild-type flies in LD (Schäbler et al. 2020). As in bodies, the cycling of short-chain acylcarnitines in the wild type persisted in DD, whereas long-chain acylcarnitines stopped cycling. In heads of *per*^{01} clock mutant flies, the daily rhythmicity persisted but was weakened in LD. These results suggest that light is a main contributor to daily oscillations of the acylcarnitine pool in head and bodies of *Drosophila*. However, absolute levels of acylcarnitines were significantly lower in isogenic *per*^{01} compared to Canton-S wild-type flies (Schäbler et al. 2020) suggesting that *per* has an impact on acylcarnitine metabolism in a non-circadian manner. Difference between the studies could be explained by differences in the extraction protocol and genotype (Schäbler et al. analyzed acylcarnitines in methanolic extracts of wild-type *Drosophila*; Rhoades et al. analyzed the aqueous phases of Bligh-Dyer extracts in *white* flies) or the number of replicates ($n = 2$ in (Rhoades et al. 2018) and $n = 7$ in (Schäbler et al. 2020)). Collectively, both studies identify acylcarnitines as robust and prominent rhythmic lipid metabolites in *Drosophila* with oscillations driven by both photophase and the clock. This is in line with an analysis of mammalian chronometabolomic studies that identified acylcarnitines as one of the few significantly rhythmic metabolites that were replicated across studies (Rhoades et al. 2017), suggesting that rhythmic acylcarnitine cycling is a shared metabolic feature between flies and mammals.

Acylcarnitines are essential for the transport of fatty acids across the inner mitochondrial membrane, a prerequisite for mitochondrial β-oxidation of activated fatty acid. Further, acylcarnitines are involved in removal of excess acyl groups and in the modulation of intracellular coenzyme A homeostasis. Acylcarnitine levels thus may be used as a proxy for altered mitochondrial metabolism (reviewed in (Reuter and Evans 2012). The different ratio of acylcarnitine and carnitine between CantonS and *per*^{01} flies may thus indicate that mitochondrial lipid metabolism is dependent on the *per* gene (Schäbler et al. 2020), similar to the situation in mice (Neufeld-Cohen et al. 2016).

2.2 Impact of Circadian Clocks on Di- and Triacylglycerol Metabolism

Triacylglycerols (TAG) are key storage molecules in all insect species. In *Drosophila*, there is conflicting information whether central clock neurons and the peripheral clock in the fat body affect storage lipid levels in the fat body.

On one hand, TAG levels in the fat body are elevated in flies after downregulation of *Clk* or expression of a dominant negative allele of *Clk* in PDF-expressing neurons including the abdominal PDF neurons that do not express a functional molecular clock (DiAngelo et al. 2011). On the other hand, overexpression of *tim* in PDF neurons did not affect fat body TAG levels (DiAngelo et al. 2011). *Clk*^{Jrk}, *cyc*^{01}, *tim*^{01}, and *per*^{0} mutants with body-wide clock impairment showed wild-type TAG levels in the fat body (DiAngelo et al. 2011). Significant differences in the pool of TAGs in the body were also not detected between CantonS wild-type and *per*^{01} mutants, although

TAG levels were slightly lower and lipids were remodeled in heads of the clock mutant (Schäbler et al. 2020). Interestingly, the composition of TAG species was different in flies with a non-functional clock: medium-chain saturated TAGs were lowered, whereas long-chain polyunsaturated TAGs were elevated in *per^{01}* flies compared to CantonS controls at ZT6 and ZT18 (Schäbler et al. 2020). In contrast, lower levels of TAGs were found in whole *tim^{01}* clock mutant flies compared to *Berlin-K* wild-type controls (Seay and Thummel 2011). An impact of the circadian clock on lipid storage under dietary restriction was demonstrated by measuring TAG turnover after feeding flies with radioactively labeled glucose (Katewa et al. 2016). Both the synthesis and break-down of TAGs were lower in *per^{01}* and *tim^{01}* mutants, and associated with reduced starvation resistance on a nutrition-depleted diet. A comparison between the different studies (DiAngelo et al. 2011; Seay and Thummel 2011; Katewa et al. 2016; Schäbler et al. 2020) is difficult, as different body parts were sampled in different sexes of flies that were reared on different diets.

While the question whether the clock affects TAG levels remains open, there is good evidence indicating that TAG levels show daily oscillations. Two groups of rhythmic TAGs were identified using LC-TOF MS in whole bodies of wild-type female flies fed with a yeast-reduced diet (Katewa et al. 2016). TAGs such as TAG 36: 0, 38:1, and 39:1 were found to be rhythmic reaching maximum levels at night, while some like TAG 42:0, 44:0, 42:1, and 46:2 peaked during the day. A third group of TAG species was arrhythmic (TAG 44:1, 46:1). In line with this study, we identified daily oscillations in different TAG-species in CantonS wild-type flies in LD with *ad libitum* access to a rich standard medium (Schäbler et al. 2020). Out of the 59 profiled TAGs in body samples, almost half were rhythmic. Of these, two-thirds had a relative amplitude of >5%. In head samples, only 10 out of 53 of the quantified TAG-species were rhythmic with a relative amplitude of >5%. Interestingly, all TAGs identified as rhythmic in the head were also rhythmic in the body samples. Based on their acrophase, the rhythmic TAGs in bodies of the wild type can be divided into two groups (Fig. 3). One group had a maximum around the transition from day to night (ZT10-ZT14), while the other peaked during the night (around ZT16). The first group with a maximum around ZT14 consisted primarily of TAG species with odd-numbered fatty acids and a total chain length <48 carbon atoms with up to two double-bonds (Schäbler et al. 2020). The second group with a maximum around ZT16 contained TAGs with a total chain length of >48 carbon atoms or poly-unsaturated TAGs (>2 double bounds). Total levels of TAGs did not oscillate in head or body samples, possibly caused by a phase extinction of the two profiles due to opposite phases (DiAngelo et al. 2011; Seay and Thummel 2011; Schäbler et al. 2020). In conclusion, the daily rhythmicity of TAG species in *Drosophila* seems to be lipid-species specific and independent of the protein content of the food (Katewa et al. 2016; Schäbler et al. 2020).

Levels of diacylglycerols (DAGs), precursors as well as break-down products of TAGs, were lower in *per^{01}* mutant flies compared to CantonS wild-type flies (Schäbler et al. 2020). This finding supports a role of the circadian clock in TAG metabolism. In addition, DAGs showed weak daily oscillations in heads and bodies of wild-type flies in LD kept on rich standard medium (Schäbler et al. 2020). Out of the ten detected DAG-species, seven oscillated with a maximum in the early morning at ZT2. DAGs are the main transport form of lipids in the hemolymph. Surprisingly, significant daily oscillations of the DAGs in the hemolymph were not found in CantonS flies in LD kept on yeast-containing standard medium. Oscillations became, however, readily detectable on a 5% sugar medium (Amatobi et al. 2023). Under this dietary-restricted condition, hemolymph DAGs in wild-type flies showed maximum levels in the morning at ZT1 and in the evening at ZT10 which were diminished in *per^{01}* mutant flies in LD. The bimodal profile changed to unimodal with a maximum at ZT7 in the middle of the subjective light phase when wild-type flies were kept in DD or food was restricted for 12 h during the light or dark phase in LD12:12 (Amatobi et al. 2023).

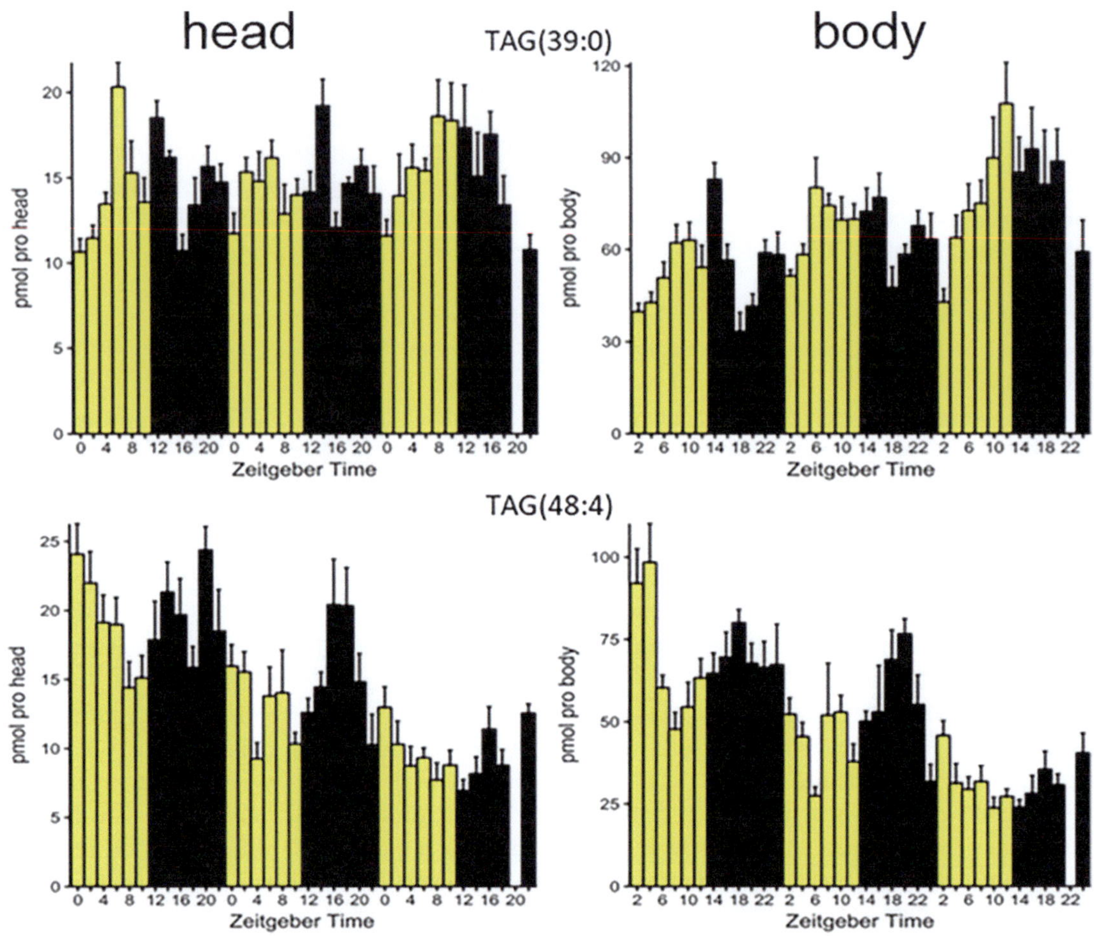

Fig. 3 Time-dependent levels of two TAG-species in wild-type *Drosophila* under LD conditions. CantonS wild-type flies were sampled every 2 h over 3 days and TAGs were analyzed in total lipid extracts of fly heads and bodies using liquid chromatography coupled to mass spectrometry. TAGs with fatty acyls containing odd number of carbon, like TAG(39:0), peaked in the evening (top panels), whereas polyunsaturated TAGs containing long-chain fatty acyls, like TAG(48:4), peaked in the middle of the dark phase. The two TAG-species were defined as number of carbon (first number) and number of double bond (second number) of the three fatty acyls

Diel oscillations of lipids were also found in the hemolymph of other insects and crustaceans. For example, daily rhythmicity of total hemolymph lipids was observed in the cricket *Acheta domestica* (Das et al. 1993), the cockroach *Periplaneta americana* (Turner and Acree 1967), and the prawn *Penaeus japonicas* (Vazquez-Boucard et al. 1989). The juvenile hormone (a lipid) titer is suggested to be regulated by the circadian clock in the cricket *Gryllus firmus* since its level in the hemolymph oscillated in a daily manner, persisted in DD, was temperature-compensated and was phase-shifted by changes in the time of photophase (Zera and Zhao 2009). Daily rhythmicity in hemolymph juvenile hormone titers over the day is also reported for honey bee foragers (Elekonich et al. 2001).

3 Circadian Transcriptomics of Lipid Metabolism

The circadian clock influences lipid metabolism by regulating the transcription of genes related to lipid synthesis, transport, and breakdown (Krishnaiah et al. 2017). For example, in caterpillars of the moth *S. frugiperda*, transcripts for the lipid transporter protein Apolipophorin-III are oscillating with a peak around ZT 4 (Suszczynska et al. 2017). Of note, the finding that fat body, midgut, and oenocytes contain a functional molecular clock does not necessarily imply that the same lipid metabolism-related genes are clock-controlled in all these tissues. In a combinatorial manner, tissue-specific regulatory factors act in conjunction with clock gene-encoded transcription factors to ensure the transcription of tissue-specific sets of genes (Meireles-Filho et al. 2014; Litovchenko et al. 2021) (Fig. 4). Even if the transcription of a given gene is clock-controlled across different tissues, cell/tissue-specific mRNA transcripts may exist due to alternative splicing (Brown et al. 2014). Furthermore, cyclic mRNA expression does not necessarily translate into rhythmic changes in the level or activity of proteins, and cell/tissue-specific post-transcriptional processes may exist.

Only few studies in *Drosophila* have measured circadian transcriptome-wide changes in gene expression in metabolic tissues, including the fat body (Xu et al. 2011; Litovchenko et al. 2021) and midgut (Litovchenko et al. 2021).

We mined the data of cycling transcripts from the most comprehensive study (Litovchenko et al. 2021) for genes associated with lipid metabolism. The study identified 1757 tissue-specific cycling transcripts by genome-wide expression analysis, out of which 67 transcripts were linked to lipid metabolism processes. Based on the adjusted p-values of a JTK_CYCLE (Hughes et al. 2010) analysis carried out by Litovchenko and colleagues, we compiled a list of the strongest lipid-associated cycling transcripts (adj.p $<$ 1.0E-03) (Table 1).

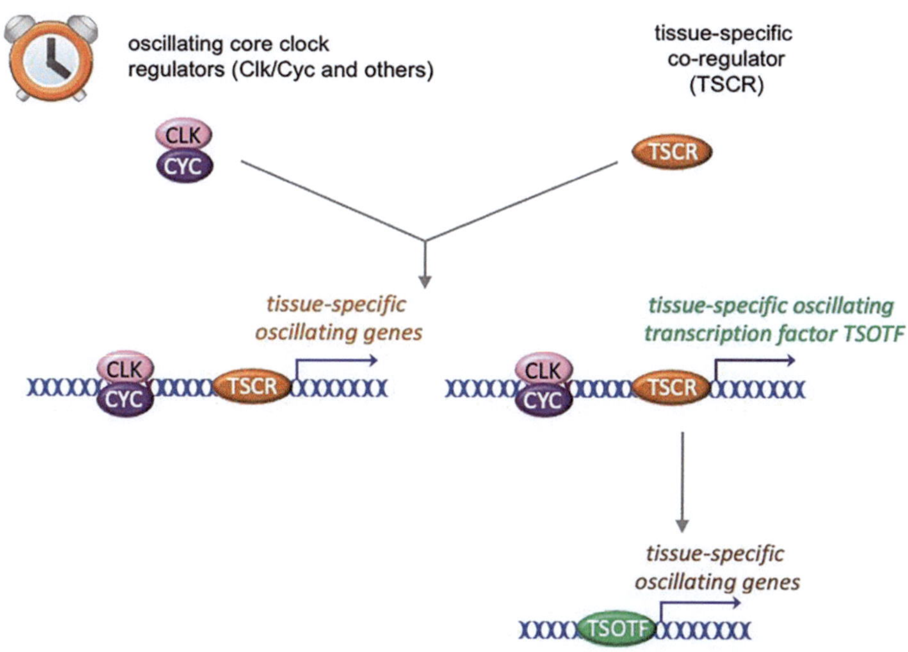

Fig. 4 A general mechanism of how peripheral clocks can induce tissue-specific oscillation of genes (based on (Meireles-Filho et al. 2014; Litovchenko et al. 2021). Tissue-specific oscillations of gene expression are achieved by combined action of core clock regulators (CLK/CYC is given as example) and tissue-specific co-regulators (TSCR). In principle, the co-activation can directly lead to oscillating gene expression (left) or indirectly via cyclic expression of non-clock tissue-specific transcription factors (TSOTF, right) which in turn drive rhythmic expression of downstream tissue-specific genes

Table 1 List of the most strongly cycling transcripts after 1 day in DD (adj.p < 1.0E-03 in JTK_CYCLE analysis) associated with lipid metabolism in w^{1118} flies, sorted after adj.p as reported by Litovchenko and colleagues (Litovchenko et al. 2021)

Gene	Tissue	BH.Q	ADJ.P	PER	LAG	AMP	Class	(predicted) function
CG17562	Fat body	1.38E-14	3.82E-18	26	3	0.34	FA	Fatty-acyl-CoA reductase
Indy	Fat body	6.94E-13	4.82E-16	26	21	0.29	NL	Plasma membrane transporter for Krebs cycle intermediates
CG11407	Brain	8.35E-12	2.22E-15	24	21	0.23	FA	Fatty-acyl-CoA reductase
CG18609	Fat body	5.32E-12	5.17E-15	26	23	0.30	FA	Fatty acid elongase
CG9459	Fat body	6.40E-12	7.10E-15	26	20	0.25	FA/SPL	Fatty acid elongase and sphingolipid biosynthesis
CG17560	Fat body	2.13E-11	2.66E-14	26	4	0.38	FA	Long-chain fatty-acyl-CoA reductase
CG9458	Fat body	1.10E-10	1.84E-13	26	19	0.26	FA	Fatty acid elongase
Spin	Fat body	2.45E-08	4.76E-11	28	15	0.30	SPL	Sphingolipid transporter
CG6432	Fat body	2.92E-08	6.50E-11	26	14	0.22	FA	Short-chain fatty acid-CoA ligase
CG6660	Fat body	3.98E-08	9.94E-11	26	21	0.25	FA/SPL	Fatty acid elongase, sphingolipid biosynthesis
Indy	Gut	4.45E-07	4.19E-10	26	19	0.20	NL	Plasma membrane transporter for Krebs cycle intermediates
CG31523	Fat body	8.19E-07	3.00E-09	26	19	0.24	FA/SPL	Fatty acid elongase activity, sphingolipid biosynthesis
Inos	Fat body	9.80E-07	3.67E-09	24	10	0.20	PL	Inositol-3-phosphate synthase
CG17896	Fat body	1.05E-06	4.09E-09	28	0	0.26	FA	Malonate-semialdehyde dehydrogenase, fatty-acyl-CoA binding
Pdgy	Fat body	1.07E-06	4.31E-09	28	22	0.28	FA	Acyl-CoA synthetase
NLaz	Gut	5.15E-06	7.27E-09	26	22	0.26	Others	Extracellular lipid binding protein of the Lipocalin family
CG14688 /Phyhd1	Brain	3.31E-05	3.08E-08	24	21	0.24	FA	Phytanoyl-CoA dioxygenase
CG14893	Fat body	6.40E-06	3.29E-08	28	15	0.20	FA	Fatty-acyl-CoA reductase
Oys	Fat body	1.63E-05	1.02E-07	24	6	0.21	PL	Lysophospholipid acyltransferase
CG13091	Fat body	2.56E-05	1.81E-07	28	26	0.18	FA	Fatty-acyl-CoA reductase
CG9394	Gut	6.55E-05	1.87E-07	24	9	0.23	PL	Glycerophosphocholine phosphodiesterase
CG12268	Gut	6.55E-05	1.96E-07	24	2	0.22	FA	Fatty-acyl-CoA reductase
CG4020	Fat body	2.89E-05	2.09E-07	24	14	0.20	FA	Fatty-acyl-CoA reductase
Fabp	Fat body	4.08E-05	3.17E-07	28	5	0.12	FA	Fatty acid binding protein
Npc2f	Gut	9.04E-05	3.48E-07	24	23	0.18	Others	Sterol binding and transport
Lace	Gut	9.04E-05	3.83E-07	24	22	0.14	SPL	Sphingolipid biosynthesis
Sws	Fat body	1.00E-04	9.03E-07	28	17	0.20	PL	Lysophospholipase
CG10237	Gut	1.97E-04	1.11E-06	24	15	0.10	PL	Phosphatidylinositol bisphosphate binding
Kst	Gut	3.62E-04	2.84E-06	24	6	0.22	PL	Phospholipid binding
Hydr2	Brain	1.94E-03	3.87E-06	24	14	0.19	Others	Acylglycerol lipase
CG11151	Gut	4.74E-04	4.02E-06	22	13	0.18	Others	Sterol binding
CG14688 /Phyhd1	Gut	7.36E-04	6.70E-06	24	19	0.15	FA	Phytanoyl-CoA dioxygenase
CG13833	Fat body	6.47E-04	8.26E-06	26	21	0.10	Others	Oxidoreductase
CG3376	Fat body	6.66E-04	8.60E-06	26	19	0.17	SPL	Acid sphingomyelinase
Mdy	Fat body	9.93E-04	1.39E-05	24	20	0.14	NL	Diacylglycerol O-acyltransferase
inaE	Fat body	1.11E-03	1.63E-05	28	26	0.10	NL	Diacylglycerol lipase

(continued)

Table 1 (continued)

Gene	Tissue	BH.Q	ADJ.P	PER	LAG	AMP	Class	(predicted) function
CG3603	Gut	1.58E-03	1.88E-05	22	11	0.13	FA	Steroid dehydrogenase
Nes	Fat body	1.48E-03	2.32E-05	22	2	0.10	PL	Lysophospholipid acyltransferase
Lace	Fat body	1.81E-03	3.04E-05	28	19	0.12	SPL	Serine C-palmitoyltransferase
CG8834	Gut	2.37E-03	3.16E-05	26	1	0.14	FA	Long-chain-fatty-acid—CoA ligase
CG2781/ Elovl7	Gut	2.64E-03	3.69E-05	24	19	0.11	FA	Long-chain fatty acid elongase
CG9743	Gut	5.51E-03	9.26E-05	26	4	0.13	FA	Stearoyl-CoA-14-desaturase
Vib	Fat body	4.62E-03	9.69E-05	28	1	0.17	PL	Phosphatidylinositol transfer protein
CG6295	Gut	5.95E-03	1.07E-04	28	13	0.20	Others	TAG lipase, phospholipase
CG3699	Gut	6.70E-03	1.29E-04	24	4	0.14	FA	Steroid dehydrogenase
Taz	Fat body	6.51E-03	1.55E-04	24	20	0.15	SPL	Phosphatidylcholine:Cardiolipin O-linoleoyltransferase
CG5009	Fat body	7.69E-03	1.91E-04	28	5	0.17	FA	Acyl-CoA oxidase
CG8093	Gut	1.38E-02	3.66E-04	24	10	0.08	Others	Triacylglycerol lipase
CG8306	Gut	1.46E-02	4.07E-04	28	16	0.09	FA	Fatty-acyl-CoA reductase
Fabp	Gut	1.56E-02	4.51E-04	24	9	0.12	FA	Fatty acid binding protein
Spin	Brain	1.38E-01	7.87E-04	26	12	0.16	SPL	Sphingolipid transporter
Gpdh	Gut	2.25E-02	8.04E-04	24	22	0.13	NL/PL	Glycerol-3-phosphate dehydrogenase
sxe2	Fat body	1.41E-01	6.75E-03	22	4	0.17	FA	Triacylglycerol lipase
Mdy	Gut	1.40E-01	9.37E-03	26	23	0.17	NL	Diacylglycerol O-acyltransferase
hnf4	Fat body	2.33E-01	1.29E-02	28	17	0.09	Others	Nuclear receptor
hnf4	Brain	1.00E+00	4.63E-02	28	2	0.12	Others	Nuclear receptor

Further cycling genes with an adj.p value >1.0E-03 were selected based on their reported functions (see text). Functional annotation/prediction is based on the DAVID bioinformatics database (Huang et al. 2009) and manual curation in FLYBASE

BH.Q Benjamini-Hochberg q-value, *PER* estimated period of the gene, *LAG* acrophase of the gene (time of max peak), *AMP* amplitude of the gene. CLASS: *FA* fatty acids, *NL* neutral lipids, *PL* phospholipids, *SPL* sphingolipids

The data in Table 1 and Fig. 5 show that the expression of several genes associated with long-chain fatty acid metabolism, fatty-acyl-CoA reductase activity, and fatty acid β-oxidation is highly oscillating in DD. The next large groups of cycling genes were involved in phospholipid biosynthesis, sphingolipid biosynthesis, neutral lipid (TAG and DAG) regulation, and steroid hormone metabolism. These findings clearly show that there are many oscillating genes/transcripts in fat body, gut, or brain that may affect lipid metabolism but are yet to be properly explored. The following part of this section provides some ideas toward cycling genes that have possible roles in lipid metabolism.

The transcripts related to lipid biosynthesis tend to peak at the end of the night (>ZT20), while there is no clear pattern in peak time for other transcripts (e.g., for fatty-acyl-CoA reductases). A microarray study confirms the prominent cycling of long-chain fatty acid elongase transcripts and its late acrophase (Xu et al. 2011). Further evidence for a role of the circadian clock in the regulation of lipid-related genes comes from a study that used chronic circadian misalignment in *Drosophila* (Boomgarden et al. 2019). After two weeks of daily 4 h phase delays in photophase, out of the six genes differentially expressed compared to age-matched controls, three are involved in lipid metabolism: CG4830 (fatty acid ligase),

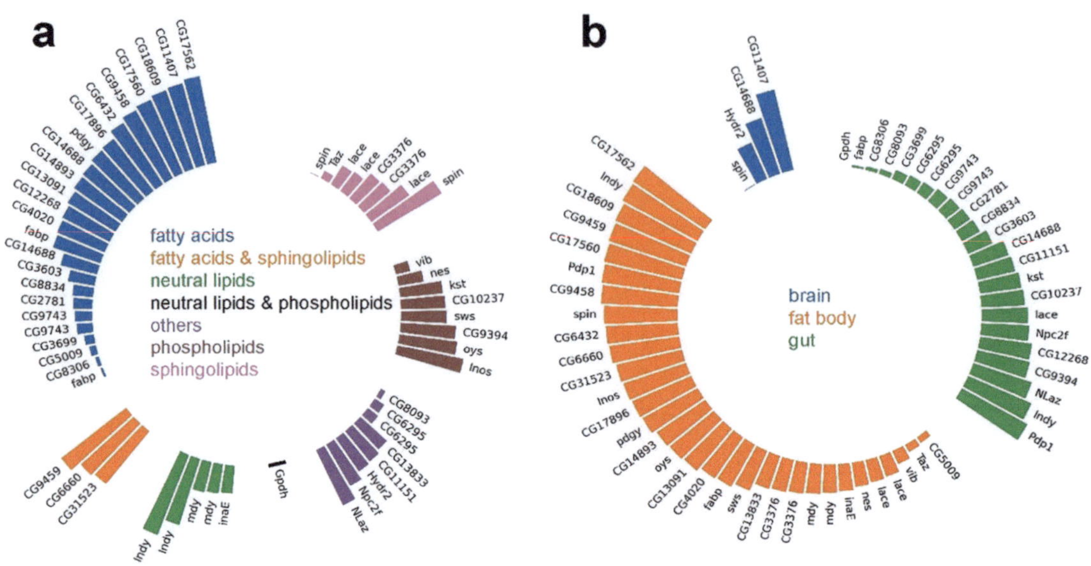

Fig. 5 Classification of cycling genes based on lipid classes and organs. Length of the bar is relative to oscillation strength (adj.p). (**a**) Classification based on lipid classes: FA-related genes (blue), FA-SPL-related genes (orange), NL-related genes (green), NL-PL—related genes (black), other related genes (purple), PL-related genes (brown), SPL-related genes (pink). (**b**) Classification based on organs: fat body-related genes (orange), gut-related genes (green), brain-related genes (blue)

Acox57D-d (an acyl-CoA oxidase), and CG18609, a fatty acid elongase gene that showed very strong rhythmic expression in the fat body and is also expressed in the oenocytes (Gutierrez et al. 2007, Table 1). Fatty acid elongases and fatty acyl-CoA reductases play an important role in the formation of cuticular hydrocarbons by the oenocytes (Blomquist and Ginzel 2021). Their roles in the fat body and midgut may include a support of cuticular hydrocarbon production by providing very long-chain fatty acids or fatty alcohols as precursors to the oenocytes. In line with a circadian regulation of fatty acid elongases and fatty acyl-CoA reductases, the cuticular hydrocarbon profile of *Drosophila* undergoes diel changes (Kent et al. 2007).

The diel and circadian changes in TAG and DAG metabolites (see Section 2) correlate well with the finding of cycling transcripts involved in TAG/DAG synthesis and breakdown (including TAG lipases, glycerol-3-phosphate dehydrogenase, and diacylclycerol acyltransferases). Yet, given that acetylcarnitines are among the most prominent and stable cycling whole-body metabolites (see Section 2) there is a remarkable sparsity of cycling genes related to acylcarnitine synthesis or transport in the fat body or midgut. This extends to genes encoding enzymes involved in β-oxidation (including dehydrogenases and thiolases). Importantly, however, the cycling transcripts identified by Litovchenko et al. (2021) include the nuclear receptor gene *hnf4*, which showed weak rhythmicity. HNF4 is a strong positive master regulator for the various steps in the β-oxidation pathway and is activated by starvation or availability of exogenous long-chain fatty acids (Palanker et al. 2009). It seems thus possible that HNF4 integrates internal metabolic cues and clock input to optimize β-oxidation not in a strict time-dependent fashion, but in an integrated manner that takes account to internal needs. Alternatively, genes related to carnitine/β-oxidation may be rhythmically expressed in lipid-catabolizing organs such as muscles.

4 Feeding and Circadian Lipid Metabolism

The availability of food sources depends on the time of the day and the temporal pattern of foraging activity. It is therefore to be expected that feeding occurs rhythmically at specific times of the day. Nevertheless, feeding rhythmicity has been thoroughly demonstrated and analyzed in only a few insect species, including moth caterpillars (*S. frugiperda* (Suszczynska et al. 2017)), honey and bumble bees (Moore et al. 1989; Stelzer et al. 2010), and mosquitoes (see Meireles-Filho and Kyriacou 2013; Rund et al. 2016). Arguably, the daily pattern and circadian control of feeding is best characterized in *Drosophila*. When given constant access to food, adult flies preferably feed in the morning and/or evening around lights on and off—at the times of their morning or evening peak of locomotor activity (Xu et al. 2008; Seay and Thummel 2011; Ro et al. 2014; Schäbler et al. 2020; Niu et al. 2021; Fulgham et al. 2021). Under constant conditions, this daily rhythmicity in food consumption is lost in various clock mutants including Clk^{jrk} (Xu et al. 2008), per^{01} (Ro et al. 2014), and tim^{01} (Seay and Thummel 2011). These results show that feeding rhythmicity in *Drosophila* is under control of the circadian clock. Yet, also light can drive feeding rhythmicity, as feeding behavior is still rhythmic in these clock mutant flies under LD (Xu et al. 2008; Seay and Thummel 2011; Schäbler et al. 2020).

As feeding is intimately linked to metabolism, the interesting question arises to which extent the time of feeding contributes to daily oscillations in lipid metabolite levels. First studies suggested that feeding does not influence overall TAG levels as these do not show daily rhythmicity (DiAngelo et al. 2011; Seay and Thummel 2011). Lipidomics, however, revealed that some TAG species in fact show daily fluctuations (Schäbler et al. 2020), yet it is unclear whether their temporal pattern is related to feeding activity. In contrast, a clear relation between feeding and lipid levels exists for DAGs, phosphoethanolamines, and phosphocholines in the hemolymph (Amatobi et al. 2023). Their levels peaked at the beginning and end of the light phase in LD, or in the middle of the subjective photophase in DD if flies were kept on lipid-free medium. On lipid-containing food, however, the bimodal oscillation patterns became much weaker, and were no longer classified as rhythmic according to JTK_CYCLE (Amatobi et al. 2023). These results indicate that the endogenous rhythm of DAGs, phosphoethanolamines, and phosphocholines in the hemolymph can be hidden by lipid uptake during feeding.

Feeding also interacts directly with peripheral clocks, at least in the *Drosophila* fat body. In DD, when food was restricted to times at which flies normally eat very little, time-restricted feeding altered the phase of clock gene expression and downstream metabolic cycling (Xu et al. 2011). Though not directly demonstrated yet for genes involved in lipid metabolism, it is conceivable that time-restricted feeding can also phase-shift expression of these genes and subsequently alter lipid metabolism and rhythmicity. The rhythmic transcription of the TAG lipase-encoding gene *sxe2* is regulated by insulin-signaling from the brain and lost in clock mutants (Barber et al. 2016). In per^{01} flies, rhythmicity can however be restored by time-restricted feeding presumably due to restored rhythmic insulin signaling (Barber et al. 2016).

5 Central Control of Daily Changes in Lipid Metabolism

Despite the solid knowledge on the general (neuro)endocrine regulation of lipid metabolism ((Lehmann 2018), chapter "Endocrine Control of Lipid Metabolism" by Krishnan and Bednárová), we still know little about the signaling pathways that modulate diel changes in lipid metabolism and underlie the communication between the relevant body clocks.

5.1 Central to Peripheral Clock Communication

As outlined in 1.1, the fat body of *Drosophila* contains an autonomous clock under LD conditions. Yet, *per* cycling in the fat body is lost in DD when the central clock is impaired specifically in the ventral LN clock neurons (Erion et al. 2016). Moreover, expression of a dominant negative *Clk* allele or *Clk*-RNAi in PDF-expressing neurons including the small ventral LN clock neurons induced an increase in stored TAG levels in the fat body. Loss of PDF however did not affect overall TAG levels (DiAngelo et al. 2011). Moreover, disruption of the molecular clock in neuropeptide F (NPF)-expressing clock neurons (subset of dorsal LNs) disrupted the expression of cytochrome P450 genes in the fat body under LD (Erion et al. 2016). This suggests as yet unidentified and PDF-independent communication pathways between the central clock in the brain and the fat body which partially are masked by photic entrainment of the fat body under LD conditions (Sehgal 2016). As the fat body is not innervated, the neuroendocrine system is likely to play a role. For example, the transcription of *sxe2*, a TAG lipase that cycles in the fat body independently of the fat body clock, may be under control of central clock neurons that use insulin-like peptides as their downstream neuroendocrine outputs (Erion et al. 2016).

A similar situation applies for the digestive tract, for which pioneering transplantation experiments by Sauman described in 1.2 (Sauman and Reppert 1998) hint to a possible involvement of the neuroendocrine system. The expression of clock genes in the gut of the linden bug is not cycling in a diel fashion, yet is regulated in a day length-dependent manner by juvenile hormone (JH) from the endocrine corpora allata (Bajgar et al. 2013a, b). Under long days, JH signaling is active and promotes expression of *Pdp1* in a *Clk* and *cyc*-dependent manner, which in turn suppresses expression of *m-cry2*. JH also increases the expression of several other gut genes, including a lipase and an esterase. RNAi-mediated downregulation of *Pdp1* showed that the JH-induced activation of these metabolic enzymes depends on the presence of *Pdp1* which is downstream of JH and its receptor METHOPRENE TOLERANT (Bajgar et al. 2013b). Under short day conditions, linden bugs go into an energy-saving diapause state, *m-cry* expression is upregulated and inhibits *Pdp1* in the gut. This leads to upregulation of diapause genes and downregulation of the lipase, esterase and other *Pdp1*-regulated genes. Though not directly informative in terms of circadian changes in lipid metabolism, the linden bug example shows that endocrine factors are able to affect clock gene expression which in turn can exert effects on lipid (and other) metabolism.

The most detailed study on central clock-to-peripheral clock communication in metabolic regulation comes from the oenocytes. As mentioned in 1.3, PDF signaling defines the phase of clock gene cycling and a downstream metabolic gene *desaturase 1* in the oenocytes which are not innervated (Krupp et al. 2013)). This PDF signaling requires both the PDF-expressing ventral lateral clock neurons (LN_vs) in the brain, and the neuroendocrine abdominal neurons (PDF_{ab}) in the ventral ganglion which lack a functional molecular clock (Krupp et al. 2013).

5.2 Neuroendocrine Control of Diel Changes in Lipid Metabolism

At least two of the cyclic transcripts in Table 1 are targets of hormonal regulation (see Heier et al. 2021). *pudgy* (*pdgy*) encodes an acyl-CoA synthetase that is positively regulated by the forkhead box transcription factor FOXO which is involved in the regulation of the insulin signaling pathway. FOXO in turn is inhibited by insulin signaling. The lipogenic diacylglycerol O-acyltransferase encoded by *moody* (*mdy*) is inhibited by adipokinetic hormone (AKH).

Drosophila insulin-like peptides (DILPs) produced by neuroendocrine insulin-producing cells in the pars intercerebralis of the brain (see chapter "Endocrine Control of Lipid Metabolism" by Krishnan and Bednárová) provide important lipogenic signals to the fat body. The insulin-producing cells receive inputs from DN1 dorsal clock neurons and show circadian electrical activity (Barber et al. 2016). This suggests that DILPs may be hormonally released in a diel fashion to modulate the circadian activity of the fat body and other organs involved in lipid metabolism.

AKH from AKH-producing endocrine cells (APC) in the glandular part of the corpora cardiaca is a key signal for the mobilization of stored carbohydrates and lipids from the fat body (see chapter "Endocrine Control of Lipid Metabolism" by Krishnan and Bednárová). AKH signaling is largely triggered on demand, for example during extended periods of starvation or physical activity. Some evidence, however, speaks for a circadian modulation of AKH-to-fat body signaling. Crickets and the linden bug *Pyrrhocoris apterus* show a diel variation in the hemolymph level of AKH, total lipids, and carbohydrates. Moreover, the adipokinetic effect of AKH on the fat body is time-dependent, suggesting a circadian modulation of AKH receptor signaling (Das et al. 1993; Maxová et al. 2001; Kodrík et al. 2003). At least in *P. apterus*, these variations appear to be clock-dependent as they persist in DD for AKH hemolymph titer and fat body sensitivity (Kodrik et al. 2005). Curiously, the AKH hemolymph titer phase-shifts in DD compared to LD in parallel with locomotor activity, while the lipid titer oscillations stay in the same phase (Kodrik et al. 2005). These findings seem to suggest that, if at all, circadian modulation of AKH signaling takes place on the level of AKH receptor expression in the fat body rather than AKH release. At least, the tight correlation between locomotor activity and AKH hemolymph levels shows that it is not trivial to discern to which extend the observed diel changes in AKH signaling are due to a direct effect of the clock onto APCs, or are indirectly caused (e.g., clock-driven locomotor activity induces a higher requirement of mobilized energy to supply muscles which in turn triggers increased AKH signaling).

In *Drosophila*, the APCs express a variety of G-protein-coupled receptors for biogenic amines and neuropeptides, including the PDF receptor. Targeted genetic manipulations of this receptor in the APCs affected lifespan under starvation and locomotor activity level (Braco et al. 2022). Effects on lipid metabolism have not been tested yet. The APCs are not innervated by PDF-expressing clock neurons. This suggests that the PDF-expressing abdominal neurons rather than PDF-expressing clock neurons are the source of PDF input to the APCs. This situation is similar to that of the oenocytes, and it would be very interesting to test whether this input is controlled by the central clock or plays a role in temporal alignment of lipid metabolism in the fat body and oenocytes. The situation is complicated by an AKH-dependent FOXO-mediated remodeling of the dorsal projections of PDF-expressing s-LNv clock neurons (He et al. 2020). As for the crickets and *Pyrrhocoris*, it is difficult to discern direct from indirect circadian effects in the fruit fly. For example, daily activity -which is under control of the central clock in the brain (Shafer and Yao 2014) – regulates APC activity via the muscle-borne cytokine Upd2 in a FOXO-dependent fashion. Furthermore, this signalng pathway is important to maintain the diel lipid oscillations in the hemolymph (Zhao and Karpac 2017). In turn, AKH fine-tunes daily activity patterns not only under starvation (Lee and Park 2004; Isabel et al. 2005), but also under *ad libitum* feeding conditions (Pauls et al. 2021).

6 Conclusion and Future Perspectives

During the last decade, the research community has started to provide deeper insights into the daily organization of lipid metabolism and its control by the circadian clock and light in insects. Many shared features such as the presence of peripheral clocks in metabolic organs, or the robust diel cycling of acylcarnitines between insects and mammals, show that insects such as the genetically amenable fruit fly can be excellent models to decipher principles and benefits of temporally controlled lipid metabolism and its importance for human metabolic disorders.

What is obvious from this chapter: we are only at the beginning in this exciting field, and many more metabolomic, transcriptomic, and functional studies are required. For example, we know little about the significance of individual peripheral clocks. For instance, what are the metabolic and systemic consequences of a disrupted fat body clock, if any? How is rhythmicity in lipid metabolism altered by age, and what are the consequences of impaired lipid rhythmicity on aging? There is also still a lot of work ahead until we understand the molecular mechanisms of the clock/light-lipid metabolism interaction. For instance, it will be important to find out which relevant regulatory factors, enzymes and pathways are under clock and light control.

We also need more research on the role and significance of (neuro)endocrine signaling in the temporal coordination of peripheral clocks and lipid metabolism. New insights into the temporal structure and control of insect lipid metabolism will have some translational impact in the field of biomedicine or animal sciences, for example with regard to metabolic diseases related to circadian disruption or improved rearing of insects as a healthy and protein-rich food source.

Nonetheless, and not at least it would be important and exciting to extend research in the field to a broader range of insect species, as insects represent a huge taxon with stunning adaptations in life style, biochemistry, and physiology. For instance, how is the temporal structure of lipid metabolism organized in dietary specialists like aphids that permanently feed on strongly sugar-enriched plant saps, or in species that live under (near) constant conditions like cave beetles or ground-dwelling cicada larvae? From a conservation angle, it seems relevant to ask to which extent artificial light-at-night is disrupting lipid metabolism in insects as this may contribute to the currently ongoing decline in insect abundance due to man-made environmental changes.

Acknowledgments The authors would like to thank the Deutsche Forschungsgemeinschaft (DFG) for generous support of the original work on the topic (lipid measurements were performed with a UPLC-Synapt HDMS funded by the DFG project 179877739 (INST 93/712-1 FUGG) at the Metabolomics Core Unit, University of Würzburg; SFB1047 "Insect timing" project A4 (to AF and Martin J Mueller) and B2 (to CW)), and the Graduate School of Life Sciences (GSLS) of the University of Würzburg and the Turkish Ministry of Education for personal stipends (to KMA and AGOU). We are much indebted to the two unknown reviewers for their valuable comments.

References

Abbott SM, Malkani RG, Zee PC (2020) Circadian disruption and human health: a bidirectional relationship. Eur J Neurosci 51:567–583. https://doi.org/10.1111/ejn.14298

Agrawal P, Houl JH, Gunawardhana KL, Liu T, Zhou J, Zoran MJ, Hardin PE (2017) Drosophila CRY entrains clocks in body tissues to light and maintains passive membrane properties in a non-clock body tissue independent of light. Curr Biol 27:2431–2441.e3. https://doi.org/10.1016/j.cub.2017.06.064

Amatobi KM, Ozbek-Unal AG, Schäbler S, Deppisch P, Helfrich-Förster C, Mueller MJ, Wegener C, Fekete A (2023) The circadian clock is required for rhythmic lipid transport in *Drosophila* in interaction with diet and photic condition. J Lipid Res 100417. https://doi.org/10.1016/j.jlr.2023.100417

Bajgar A, Dolezel D, Hodkova M (2013a) Endocrine regulation of non-circadian behavior of circadian genes in insect gut. J Insect Physiol 59:881–886. https://doi.org/10.1016/j.jinsphys.2013.06.004

Bajgar A, Jindra M, Dolezel D (2013b) Autonomous regulation of the insect gut by circadian genes acting downstream of juvenile hormone signaling. Proc Natl Acad Sci 110:4416–4421. https://doi.org/10.1073/pnas.1217060110

Barber AF, Erion R, Holmes TC, Sehgal A (2016) Circadian and feeding cues integrate to drive rhythms of physiology in drosophila insulin-producing cells. Genes Dev 30:2596–2606. https://doi.org/10.1101/gad.288258.116

Beer K, Helfrich-Förster C (2020) Model and non-model insects in chronobiology. Front Behav Neurosci 14. https://doi.org/10.3389/fnbeh.2020.601676

Beer K, Kolbe E, Kahana NB, Yayon N, Weiss R, Menegazzi P, Bloch G, Helfrich-Förster C (2018) Pigment-dispersing factor-expressing neurons convey circadian information in the honey bee brain. Open Biol 8:170224. https://doi.org/10.1098/rsob.170224

Blomquist GJ, Ginzel MD (2021) Chemical ecology, biochemistry, and molecular biology of insect hydrocarbons. Annu Rev Entomol 66:45–60. https://doi.org/10.1146/annurev-ento-031620-071754

Boomgarden AC, Sagewalker GD, Shah AC, Haider SD, Patel P, Wheeler HE, Dubowy CM, Cavanaugh DJ (2019) Chronic circadian misalignment results in reduced longevity and large-scale changes in gene expression in drosophila. BMC Genomics 20:14. https://doi.org/10.1186/s12864-018-5401-7

Bozinovic F, Calosi P, Spicer JI (2011) Physiological correlates of geographic range in animals. Annu Rev Ecol Evol Syst 42:155–179. https://doi.org/10.1146/annurev-ecolsys-102710-145055

Braco JT, Nelson JM, Saunders CJ, Johnson EC (2022) Modulation of metabolic hormone Signaling via a circadian hormone and biogenic amine in Drosophila melanogaster. Int J Mol Sci 23:4266. https://doi.org/10.3390/ijms23084266

Brown JB, Boley N, Eisman R, May GE, Stoiber MH, Duff MO, Booth BW, Wen J, Park S, Suzuki AM, Wan KH, Yu C, Zhang D, Carlson JW, Cherbas L, Eads BD, Miller D, Mockaitis K, Roberts J, Davis CA, Frise E, Hammonds AS, Olson S, Shenker S, Sturgill D, Samsonova AA, Weiszmann R, Robinson G, Hernandez J, Andrews J, Bickel PJ, Carninci P, Cherbas P, Gingeras TR, Hoskins RA, Kaufman TC,

Lai EC, Oliver B, Perrimon N, Graveley BR, Celniker SE (2014) Diversity and dynamics of the drosophila transcriptome. Nature 512:393–399. https://doi.org/10.1038/nature12962

Damiola F, Minh NL, Preitner N, Kornmann B, Fleury-Olela F, Schibler U (2000) Restricted feeding uncouples circadian oscillators in peripheral tissues from the central pacemaker in the suprachiasmatic nucleus. Genes Dev 14:2950–2961. https://doi.org/10.1101/gad.183500

Das S, Meier OW, Woodring J (1993) Diel rhythms of adipokinetic hormone, fat body response, and haemolymph lipid and sugar levels in the house cricket. Physiol Entomol 18:233–239. https://doi.org/10.1111/j.1365-3032.1993.tb00593.x

Deppisch P, Kirsch V, Helfrich-Förster C, Senthilan PR (2023) Contribution of cryptochromes and photolyases for insect life under sunlight. J Comp Physiol A Neuroethol Sens Neural Behav Physiol. https://doi.org/10.1007/s00359-022-01607-5

DiAngelo JR, Erion R, Crocker A, Sehgal A (2011) The central clock neurons regulate lipid storage in drosophila. PLoS One 6:e19921. https://doi.org/10.1371/journal.pone.0019921

Dolezel D, Zdechovanova L, Sauman I, Hodkova M (2008) Endocrine-dependent expression of circadian clock genes in insects. Cell Mol Life Sci 65:964–969. https://doi.org/10.1007/s00018-008-7506-7

Elekonich MM, Schulz DJ, Bloch G, Robinson GE (2001) Juvenile hormone levels in honey bee (Apis mellifera L.) foragers: foraging experience and diurnal variation. J Insect Physiol 47:1119–1125. https://doi.org/10.1016/S0022-1910(01)00090-7

Erion R, King AN, Wu G, Hogenesch JB, Sehgal A (2016) Neural clocks and neuropeptide F/Y regulate circadian gene expression in a peripheral metabolic tissue. elife 5. https://doi.org/10.7554/eLife.13552

Fleury G, Masís-Vargas A, Kalsbeek A (2020) Metabolic implications of exposure to light at night: lessons from animal and human studies. Obesity (Silver Spring) 28 (Suppl 1):S18–S28. https://doi.org/10.1002/oby.22807

Friedrichs M, Kolbe I, Seemann J, Tsang AH, Cherradi L, Klein J, Oster H (2018) Circadian clock rhythms in different adipose tissue model systems. Chronobiol Int 35:1543–1552. https://doi.org/10.1080/07420528.2018.1494603

Fulgham CV, Dreyer AP, Nasseri A, Miller AN, Love J, Martin MM, Jabr DA, Saurabh S, Cavanaugh DJ (2021) Central and peripheral clock control of circadian feeding rhythms. J Biol Rhythm 07487304211045835. https://doi.org/10.1177/07487304211045835

Giebultowicz JM, Ivanchenko M, Vollintine T (2001) Organization of the insect circadian system: spatial and developmental expression of clock genes in peripheral tissues of Drosophila melanogaster. In: Denlinger DL, Giebultowicz JM, Saunders DS (eds) Insect timing: circadian rhythmicity to seasonality. Elsevier Science B.V, Amsterdam, pp 31–42

Gogna N, Jagdish Singh V, Sheeba V, Dorai K (2015) NMR-based investigation of the Drosophila melanogaster metabolome under the influence of daily cycles of light and temperature. Mol BioSyst 11:3305–3315. https://doi.org/10.1039/C5MB00386E

Gooley JJ (2016) Circadian regulation of lipid metabolism. Proc Nutr Soc 75:440–450. https://doi.org/10.1017/S0029665116000288

Gutierrez E, Wiggins D, Fielding B, Gould AP (2007) Specialized hepatocyte-like cells regulate drosophila lipid metabolism. Nature 445:275–280

Harper REF, Ogueta M, Dayan P, Stanewsky R, Albert JT (2017) Light dominates peripheral circadian oscillations in Drosophila melanogaster during sensory conflict. J Biol Rhythm 32:423–432. https://doi.org/10.1177/0748730417724250

He Q, Du J, Wei L, Zhao Z (2020) AKH-FOXO pathway regulates starvation-induced sleep loss through remodeling of the small ventral lateral neuron dorsal projections. PLoS Genet 16:e1009181. https://doi.org/10.1371/journal.pgen.1009181

Heier C, Klishch S, Stilbytska O, Semaniuk U, Lushchak O (2021) The drosophila model to interrogate triacylglycerol biology. Biochimica et Biophysica Acta (BBA) – Molecular and Cell Biology of Lipids 1866:158924. https://doi.org/10.1016/j.bbalip.2021.158924

Helfrich-Förster C (2017) The drosophila clock system. In: Kumar V (ed) Biological timekeeping: clocks, rhythms and behaviour. Springer India, New Delhi, pp 133–176

Helfrich-Förster C, Homberg U (1993) Pigment-dispersing hormone-immunoreactive neurons in the nervous system of wild-type Drosophila melanogaster and of several mutants with altered circadian rhythmicity. J Comp Neurol 337:177–190

Helfrich-Förster C, Stengl M, Homberg U (1998) Organization of the circadian system in insects. Chronobiol Int 15:567–594

Huang DW, Sherman BT, Lempicki RA (2009) Bioinformatics enrichment tools: paths toward the comprehensive functional analysis of large gene lists. Nucleic Acids Res 37:1–13. https://doi.org/10.1093/nar/gkn923

Hughes ME, Hogenesch JB, Kornacker K (2010) JTK_CYCLE: an efficient non-parametric algorithm for detecting rhythmic components in genome-scale datasets. J Biol Rhythm 25:372–380. https://doi.org/10.1177/0748730410379711

Isabel G, Martin JR, Chidami S, Veenstra JA, Rosay P (2005) AKH-producing neuroendocrine cell ablation decreases trehalose and induces behavioral changes in drosophila. American Journal of Physiology - Regulatory Integrative Comparative Physiology 288:R531–R538

Ito C, Tomioka K (2016) Heterogeneity of the peripheral circadian systems in Drosophila melanogaster: a review. Front Physiol 7:8. https://doi.org/10.3389/fphys.2016.00008

Jarabo P, Martin FA (2017) Neurogenetics of drosophila circadian clock: expect the unexpected. J Neurogenet 31:250–265. https://doi.org/10.1080/01677063.2017.1370466

Kaneko M, Hall JC (2000) Neuroanatomy of cells expressing clock genes in drosophila: transgenic manipulation of the period and timeless genes to mark the perikarya of circadian pacemaker neurons and their projections. J Comp Neurol 422:66–94

Karpowicz P, Zhang Y, Hogenesch JB, Emery P, Perrimon N (2013) The circadian clock gates the intestinal stem cell regenerative state. Cell Rep 3:996–1004. https://doi.org/10.1016/j.celrep.2013.03.016

Katewa SD, Akagi K, Bose N, Rakshit K, Camarella T, Zheng X, Hall D, Davis S, Nelson CS, Brem RB, Ramanathan A, Sehgal A, Giebultowicz JM, Kapahi P (2016) Peripheral circadian clocks mediate dietary restriction-dependent changes in lifespan and fat metabolism in drosophila. Cell Metab 23:143–154. https://doi.org/10.1016/j.cmet.2015.10.014

Kent C, Azanchi R, Smith B, Chu A, Levine J (2007) A model-based analysis of chemical and temporal patterns of cuticular hydrocarbons in male Drosophila melanogaster. PLoS One 2:e962. https://doi.org/10.1371/journal.pone.0000962

Kodrík D, Socha R, Syrova Z (2003) Developmental and diel changes of adipokinetic hormone in CNS and haemolymph of the flightless wing-polymorphic bug, Pyrrhocoris apterus (L.). J Insect Physiol 49:53–61

Kodrik D, Socha R, Zdenka S, Zemek R (2005) The effect of constant darkness on the content of adipokinetic hormone, adipokinetic response and walking activity in macropterous females of Phyrrhocoris apterus (L.). Physiol Entomol 30:248–255

Kotwica-Rolinska J, Chodáková L, Smýkal V, Damulewicz M, Provazník J, Wu BC-H, Hejníková M, Chvalová D, Doležel D (2022) Loss of timeless underlies an evolutionary transition within the circadian clock. Mol Biol Evol 39:msab346. https://doi.org/10.1093/molbev/msab346

Krishnaiah SY, Wu G, Altman BJ, Growe J, Rhoades SD, Coldren F, Venkataraman A, Olarerin-George AO, Francey LJ, Mukherjee S, Girish S, Selby CP, Cal S, Er U, Sianati B, Sengupta A, Anafi RC, Kavakli IH, Sancar A, Baur JA, Dang CV, Hogenesch JB, Weljie AM (2017) Clock regulation of metabolites reveals coupling between transcription and metabolism. Cell Metab 25:1206. https://doi.org/10.1016/j.cmet.2017.04.023

Krupp JJ, Billeter J-C, Wong A, Choi C, Nitabach MN, Levine JD (2013) Pigment-dispersing factor modulates pheromone production in clock cells that influence mating in drosophila. Neuron 79:54–68. https://doi.org/10.1016/j.neuron.2013.05.019

Krupp JJ, Kent C, Billeter J-C, Azanchi R, So AK-C, Schonfeld JA, Smith BP, Lucas C, Levine JD (2008) Social experience modifies pheromone expression and mating behavior in male Drosophila melanogaster. Curr Biol 18:1373–1383. https://doi.org/10.1016/j.cub.2008.07.089

Lee G, Park JH (2004) Hemolymph sugar homeostasis and starvation-induced hyperactivity affected by genetic manipulations of the adipokinetic-hormone-encoding gene in Drosophila melanogaster. Genetics 167:311–323

Lee Y, Field JM, Sehgal A (2021) Circadian rhythms, disease and chronotherapy. J Biol Rhythm 07487304211044301. https://doi.org/10.1177/07487304211044301

Lehmann M (2018) Endocrine and physiological regulation of neutral fat storage in drosophila. Mol Cell Endocrinol 461:165–177. https://doi.org/10.1016/j.mce.2017.09.008

Li M-T, Cao L-H, Xiao N, Tang M, Deng B, Yang T, Yoshii T, Luo D-G (2018) Hub-organized parallel circuits of central circadian pacemaker neurons for visual photoentrainment in drosophila. Nat Commun 9:4247. https://doi.org/10.1038/s41467-018-06506-5

Li Y, Ma J, Yao K, Su W, Tan B, Wu X, Huang X, Li T, Yin Y, Tosini G, Yin J (2020) Circadian rhythms and obesity: timekeeping governs lipid metabolism. J Pineal Res 69:e12682. https://doi.org/10.1111/jpi.12682

Liang X, Holy TE, Taghert PH (2017) A series of suppressive signals within the drosophila circadian neural circuit generates sequential daily outputs. Neuron 94:1173–1189. https://doi.org/10.1016/j.neuron.2017.05.007

Litovchenko M, Meireles-Filho ACA, Frochaux MV, Bevers RPJ, Prunotto A, Anduaga AM, Hollis B, Gardeux V, Braman VS, Russeil JMC, Kadener S, Dal Peraro M, Deplancke B (2021) Extensive tissue-specific expression variation and novel regulators underlying circadian behavior. Sci Adv 7:eabc3781. https://doi.org/10.1126/sciadv.abc3781

Liu X, Lorenz L, Yu QN, Hall JC, Rosbash M (1988) Spatial and temporal expression of the period gene in Drosophila melanogaster. Genes Dev 2:228–238

Makki R, Cinnamon E, Gould AP (2014) The development and functions of Oenocytes. Annu Rev Entomol 59:405–425. https://doi.org/10.1146/annurev-ento-011613-162056

Maxová A, Kodrík D, Zemek R, Socha R (2001) Diel changes in adipokinetic response and walking activity of Pyrrhocoris apterus (Heteroptera) in relation to physiological status and wing dimorphism. Eur J Entomol 98:433–438. https://doi.org/10.14411/eje.2001.063

McLay LK, Nagarajan-Radha V, Green MP, Jones TM (2018) Dim artificial light at night affects mating, reproductive output, and reactive oxygen species in Drosophila melanogaster. J Exp Zool A Ecol Integr Physiol 329:419–428. https://doi.org/10.1002/jez.2164

Meireles-Filho ACA, Bardet AF, Yáñez-Cuna JO, Stampfel G, Stark A (2014) Cis-regulatory requirements for tissue-specific programs of the circadian clock. Curr Biol 24:1–10. https://doi.org/10.1016/j.cub.2013.11.017

Meireles-Filho ACA, Kyriacou CP (2013) Circadian rhythms in insect disease vectors. Mem Inst Oswaldo Cruz 108:48–58. https://doi.org/10.1590/0074-0276130438

Moore D, Siegfried D, Wilson R, Rankin MA (1989) The influence of time of day on the foraging behavior of the honeybee, Apis mellifera. J Biol Rhythm 4:305–325. https://doi.org/10.1177/074873048900400301

Mrosovsky N (1999) Masking: history, definitions, and measurement. Chronobiol Int 16:415–429

Neufeld-Cohen A, Robles MS, Aviram R, Manella G, Adamovich Y, Ladeuix B, Nir D, Rousso-Noori L, Kuperman Y, Golik M, Mann M, Asher G (2016) Circadian control of oscillations in mitochondrial rate-limiting enzymes and nutrient utilization by PERIOD proteins. PNAS 113:E1673–E1682. https://doi.org/10.1073/pnas.1519650113

Niu M, Zhang X, Li W, Wang J, Li Y (2021) dFRAME: a video recording-based analytical method for studying feeding rhythm in drosophila. Front Genet 12:1863. https://doi.org/10.3389/fgene.2021.763200

Palanker L, Tennessen JM, Lam G, Thummel CS (2009) Drosophila HNF4 regulates lipid mobilization and beta-oxidation. Cell Metab 9:228–239. https://doi.org/10.1016/j.cmet.2009.01.009

Parasram K, Bernardon N, Hammoud M, Chang H, He L, Perrimon N, Karpowicz P (2018) Intestinal stem cells exhibit conditional circadian clock function. Stem Cell Rep 11:1287–1301. https://doi.org/10.1016/j.stemcr.2018.10.010

Park JH, Helfrich-Förster C, Lee G, Liu L, Rosbash M, Hall JC (2000) Differential regulation of circadian pacemaker output by separate clock genes in drosophila. Proc Natl Acad Sci U S A 97:3608–3613

Patke A, Young MW, Axelrod S (2020) Molecular mechanisms and physiological importance of circadian rhythms. Nat Rev Mol Cell Biol 21:67–84. https://doi.org/10.1038/s41580-019-0179-2

Pauls D, Selcho M, Räderscheidt J, Amatobi KM, Fekete A, Krischke M, Hermann-Luibl C, Ozbek-Unal AG, Ehmann N, Itskov PM, Kittel RJ, Helfrich-Förster C, Kühnlein RP, Mueller MJ, Wegener C (2021) Endocrine signals fine-tune daily activity patterns in drosophila. Curr Biol 31:4076–4087.e5. https://doi.org/10.1016/j.cub.2021.07.002

Plautz J, Kaneko M, Hall JC, Kay SA (1997) Independent photoreceptive circadian clocks throughout drosophila. Science 278:1632–1635

Reinhard N, Fukuda A, Manoli G, Derksen E, Saito A, Möller G, Sekiguchi M, Rieger D, Helfrich-Förster C, Yoshii T, Zandawala M (2023) Synaptic and peptidergic connectomes of the drosophila circadian clock. bioRxiv. https://doi.org/10.1101/2023.09.11.557222

Renn SC, Park JH, Rosbash M, Hall JC, Taghert PH (1999) A PDF neuropeptide gene mutation and ablation of PDF neurons each cause severe abnormalities of behavioral circadian rhythms in drosophila. Cell 99:791–802

Reuter SE, Evans AM (2012) Carnitine and Acylcarnitines. Clin Pharmacokinet 20

Rhoades SD, Nayak K, Zhang SL, Sehgal A, Weljie AM (2018) Circadian- and light-driven metabolic rhythms in Drosophila melanogaster. J Biol Rhythm 33:126–136. https://doi.org/10.1177/0748730417753003

Rhoades SD, Sengupta A, Weljie AM (2017) Time is ripe: maturation of metabolomics in chronobiology. Curr Opin Biotechnol 43:70–76. https://doi.org/10.1016/j.copbio.2016.09.007

Ro J, Harvanek ZM, Pletcher SD (2014) FLIC: high-throughput, continuous analysis of feeding Behaviors in drosophila. PLoS One 9:e101107. https://doi.org/10.1371/journal.pone.0101107

Rund SSC, O'Donnell AJ, Gentile JE, Reece SE (2016) Daily rhythms in mosquitoes and their consequences for malaria transmission. Insects 7:14. https://doi.org/10.3390/insects7020014

Sauman I, Reppert SM (1998) Brain control of embryonic circadian rhythms in the silkmoth Antheraea pernyi. Neuron 20:741–748. https://doi.org/10.1016/s0896-6273(00)81012-0

Sauman I, Tsai T, Roca AL, Reppert SM (1996) Period protein is necessary for circadian control of egg hatching behavior in the silkmoth Antheraea pernyi. Neuron 17:901–909. https://doi.org/10.1016/s0896-6273(00)80221-4

Schäbler S, Amatobi KM, Horn M, Rieger D, Helfrich-Förster C, Mueller MJ, Wegener C, Fekete A (2020) Loss of function in the drosophila clock gene period results in altered intermediary lipid metabolism and increased susceptibility to starvation. Cell Mol Life Sci 4939–4956. https://doi.org/10.1007/s00018-019-03441-6

Seay DJ, Thummel CS (2011) The circadian clock, light, and Cryptochrome regulate feeding and metabolism in drosophila. J Biol Rhythm 26:497–506. https://doi.org/10.1177/0748730411420080

Sehadová H, Sauman I, Sehnal F (2003) Immunocytochemical distribution of pigment-dispersing hormone in the cephalic ganglia of polyneopteran insects. Cell Tissue Res 312:113–125. https://doi.org/10.1007/s00441-003-0705-5

Sehgal A (2016) Control of metabolism by central and peripheral clocks in drosophila. In: Christen Y (ed) Sassone-Corsi P. A Time for Metabolism and Hormones. Springer, Cham (CH)

Shafer OT, Kim DJ, Dunbar-Yaffe R, Nikolaev VO, Lohse MJ, Taghert PH (2008) Widespread receptivity to neuropeptide PDF throughout the neuronal circadian clock network of drosophila revealed by real-time cyclic AMP imaging. Neuron 58:223–237. https://doi.org/10.1016/j.neuron.2008.02.018

Shafer OT, Yao Z (2014) Pigment-dispersing factor signaling and circadian rhythms in insect locomotor activity. Curr Opin Insect Sci 1:73–80. https://doi.org/10.1016/j.cois.2014.05.002

Siwicki K (1988) Antibodies to the period gene product of drosophila reveal diverse tissue distribution and rhythmic changes in the visual system. Neuron 1:141–150. https://doi.org/10.1016/0896-6273(88)90198-5

Stelzer J, Stanewsky R, Chittka L (2010) Circadian foraging rhythms of bumblebees monitored by radio-frequency identification. J Biol Rhythm 25:257–267. https://doi.org/10.1177/0748730410371750

Stengl M (1994) Pigment-dispersing hormone-immunoreactive neurons in the cockroach Leucophaea maderae share properties with circadian pacemaker neurons. J Comp Physiol A 175:203–213

Stengl M, Arendt A (2016) Peptidergic circadian clock circuits in the Madeira cockroach. Curr Opin Neurobiol 41:44–52. https://doi.org/10.1016/j.conb.2016.07.010

Sunday JM, Bates AE, Dulvy NK (2012) Thermal tolerance and the global redistribution of animals. Nature Clim Change 2:686–690. https://doi.org/10.1038/nclimate1539

Suszczynska A, Kaniewska MM, Bebas P, Giebultowicz JM, Kotwica-Rolinska J (2017) Circadian regulation of caterpillar feeding and growth. J Insect Physiol 101:113–122. https://doi.org/10.1016/j.jinsphys.2017.07.009

Tataroglu O, Emery P (2015) The molecular ticks of the drosophila circadian clock. Curr Opin Insect Sci 7:51–57. https://doi.org/10.1016/j.cois.2015.01.002

Thaben PF, Westermark PO (2014) Detecting rhythms in time series with RAIN. J Biol Rhythm 29:391–400. https://doi.org/10.1177/0748730414553029

Tomioka K, Matsumoto A (2015) Circadian molecular clockworks in non-model insects. Curr Opin Insect Sci 7:58–64. https://doi.org/10.1016/j.cois.2014.12.006

Top D, Young MW (2018) Coordination between differentially regulated circadian clocks generates rhythmic behavior. Cold Spring Harb Perspect Biol 10:a033589. https://doi.org/10.1101/cshperspect.a033589

Turner RB, Acree F (1967) The effect of photoperiod on the daily fluctuation of haemolymph hydrocarbons in the American cockroach. J Insect Physiol 13:519–522. https://doi.org/10.1016/0022-1910(67)90063-7

Uryu O, Tomioka K (2010) Circadian oscillations outside the optic lobe in the cricket Gryllus bimaculatus. J Insect Physiol 56:1284–1290. https://doi.org/10.1016/j.jinsphys.2010.04.010

Vafopoulou X, Steel CGH (2006) Ecdysteroid hormone nuclear receptor (EcR) exhibits circadian cycling in certain tissues, but not others, during development in Rhodnius prolixus (Hemiptera). Cell Tissue Res 323:443–455

Vazquez-Boucard C, Galois R, Ceccaldi HJ (1989) Circadian variations in hemolymph lipids and lipoproteins and in hepatopancreatic lipids of the shrimp Penaeus japonicus. Arch Int Physiol Biochim 97:87–93. https://doi.org/10.3109/13813458909075052

Versteven M, Ernst K-M, Stanewsky R (2020) A robust and self-sustained peripheral circadian oscillator reveals differences in temperature compensation properties with central brain clocks. iScience 23:101388. https://doi.org/10.1016/j.isci.2020.101388

Wu G, Zhu J, Yu J, Zhou L, Huang JZ, Zhang Z (2014) Evaluation of five methods for genome-wide circadian gene identification. J Biol Rhythm 29:231–242. https://doi.org/10.1177/0748730414537788

Xu K, DiAngelo JR, Hughes ME, Hogenesch JB, Sehgal A (2011) The circadian clock interacts with metabolic physiology to influence reproductive fitness. Cell Metab 13:639–654. https://doi.org/10.1016/j.cmet.2011.05.001

Xu K, Zheng X, Sehgal A (2008) Regulation of feeding and metabolism by neuronal and peripheral clocks in drosophila. Cell Metab 8:289–300. https://doi.org/10.1016/j.cmet.2008.09.006

Yuan Q, Metterville D, Briscoe AD, Reppert SM (2007) Insect cryptochromes: gene duplication and loss define diverse ways to construct insect circadian clocks. Mol Biol Evol 24:948–955. https://doi.org/10.1093/molbev/msm011

Zera AJ, Zhao Z (2009) Morph-associated JH titer diel rhythm in Gryllus firmus: experimental verification of its circadian basis and cycle characterization in artificially selected lines raised in the field. J Insect Physiol 55:450–458. https://doi.org/10.1016/j.jinsphys.2008.11.012

Zhao X, Karpac J (2017) Muscle directs diurnal energy homeostasis through a Myokine-dependent hormone module in drosophila. Curr Biol 27:1941–1955.e6. https://doi.org/10.1016/j.cub.2017.06.004

Zwighaft Z, Reinke H, Asher G (2016) The liver in the eyes of a chronobiologist. J Biol Rhythm 31:115–124. https://doi.org/10.1177/0748730416633552

Adv Exp Med Biol (2026) 1494: 389–418
https://doi.org/10.1007/5584_2024_811

Published online: 29 June 2024

Lipid Metabolism in Insect Vectors of Diseases

Juan R. Girotti and Gustavo M. Calderón-Fernández

Abstract

According to the World Health Organization vector-borne diseases account for more than 17% of all infectious diseases, causing more than 700,000 deaths annually. Vectors are organisms that are able to transmit infectious pathogens between humans, or from animals to humans. Many of these vectors are hematophagous insects, which ingest the pathogen from an infected host during a blood meal, and later transmit it into a new host. Malaria, dengue, African trypanosomiasis, yellow fever, leishmaniasis, Chagas disease, and many others are examples of diseases transmitted by insects.

Both the diet and the infection with pathogens trigger changes in many metabolic pathways, including lipid metabolism, compared to other insects. Blood contains mostly proteins and is very poor in lipids and carbohydrates. Thus, hematophagous insects attempt to efficiently digest and absorb diet lipids and also rely on a large de novo lipid biosynthesis based on utilization of proteins and carbohydrates as carbon source. Blood meal triggers essential physiological processes as molting, excretion, and oogenesis; therefore, lipid metabolism and utilization of lipid storage should be finely synchronized and regulated regarding that, in order to provide the necessary energy source for these events. Also, pathogens have evolved mechanisms to hijack essential lipids from the insect host by interfering in the biosynthesis, catabolism, and transport of lipids, which pose challenges to reproduction, survival, fitness, and other insect traits.

In this chapter, we have tried to collect and highlight the current knowledge and recent discoveries on the metabolism of lipids in insect vectors of diseases related to the hematophagous diet and pathogen infection.

Keywords

Blood digestion · Blood-feeding insects · Insect lipids uptake · Insecticide resistance · Lipid synthesis and pathogen infection · Lipophorin · Parasitic diseases · Pathogen attachment and development · Pheromone production

1 Introduction

Insect-borne diseases are illnesses caused by viruses, bacteria, uni- and multicellular parasites that are transmitted by insects. Some of the most

Permissions We confirm that no permissions are needed related to Figs. 1 and 2, and Table 1 of the manuscript.

J. R. Girotti (✉) and G. M. Calderón-Fernández (✉)
Instituto de Investigaciones Bioquímicas de La Plata (CONICET-UNLP), Facultad de Ciencias Médicas, Universidad Nacional de La Plata, La Plata, Argentina
e-mail: jgirotti@med.unlp.edu.ar; gcalderonf@med.unlp.edu.ar

Table 1 Some insect-borne diseases according to WHO (2020)

Vector	Genus	Disease caused	Pathogen type	Pathogen
Mosquitos	*Aedes*	Chikungunya	Virus	Chikungunya virus
		Dengue	Virus	Dengue virus
		Lymphatic filariasis	Multicellular parasite	*Wuchereria bancrofti* and *Brugia* sp.
		Yellow fever	Virus	Yellow fever virus
		Zika	Virus	Zika virus
	Anopheles	Lymphatic filariasis	Multicellular parasite	*Wuchereria bancrofti* and *Brugia* sp.
		Malaria	Unicellular parasite	*Plasmodium* sp.
	Culex	Japanese encephalitis	Virus	Japanese encephalitis virus
		Lymphatic filariasis	Multicellular parasite	*Wuchereria bancrofti* and *Brugia* sp.
		West Nile fever	Virus	West Nile virus
Black flies	*Simulium* and others	Onchocerciasis	Multicellular parasite	*Onchocerca volvulus*
Fleas	*Xenopsylla*	Black death	Bacterium	*Yersinia pestis*
	Tunga	Tungiasis	Flea	*Tunga penetrans*
Lice		Louse-borne relapsing fever	Bacterium	*Borrelia recurrentis*
		Typhus fever	Bacterium	*Rickettsia* sp.
Sand flies	*Lutzomyia* and others	Leishmaniasis	Unicellular parasite	*Leishmania* sp.
		Sandfly fever	Virus	Sandfly fever virus
Stable flies	*Stomoxys*	Meningitis	Bacterium	*Enterobacter sakazakii*
		Sleeping sickness	Unicellular parasite	*Trypanosoma brucei*
Triatomine bugs	*Triatoma* and others	Chagas disease	Unicellular parasite	*Trypanosoma cruzi*
Tsetse flies	*Glossina*	Sleeping sickness	Unicellular parasite	*Trypanosoma brucei*

important diseases transmitted by insects are detailed in Table 1.

They cause about 700,000 deaths every year, showing higher prevalence in tropical and subtropical areas, and affecting mostly the poorest people. They not only affect public health but also strongly impede economic development through medical costs and loss of productivity of sick people. Disease's distribution is determined by complex demographic and environmental factors. Changing land use patterns and increased international travel and trade have brought more people into contact with insect vectors, while climate change and other environmental factors have spread their distribution to new areas (WHO 2020).

Vectors of these diseases are bloodsucking insects, which firstly ingest the disease-producing pathogens from an infected human or animal during blood meal, and then they feed on an uninfected host, transmitting the pathogen. During its transit through the insect digestive system and other body parts, the pathogen usually undergoes cycles of differentiation and replication.

It is considered that blood feeding has evolved independently many times among hematophagous arthropods (Ribeiro 1995). This feeding behavior allowed the coevolution of many pathogens, which changed their ways of transmission and started to use arthropods for this purpose. Both blood feeding and the presence of a pathogen inside different body parts change several metabolic pathways, including lipid metabolism. Viruses in general, and many other pathogens as well, have a reduced genome,

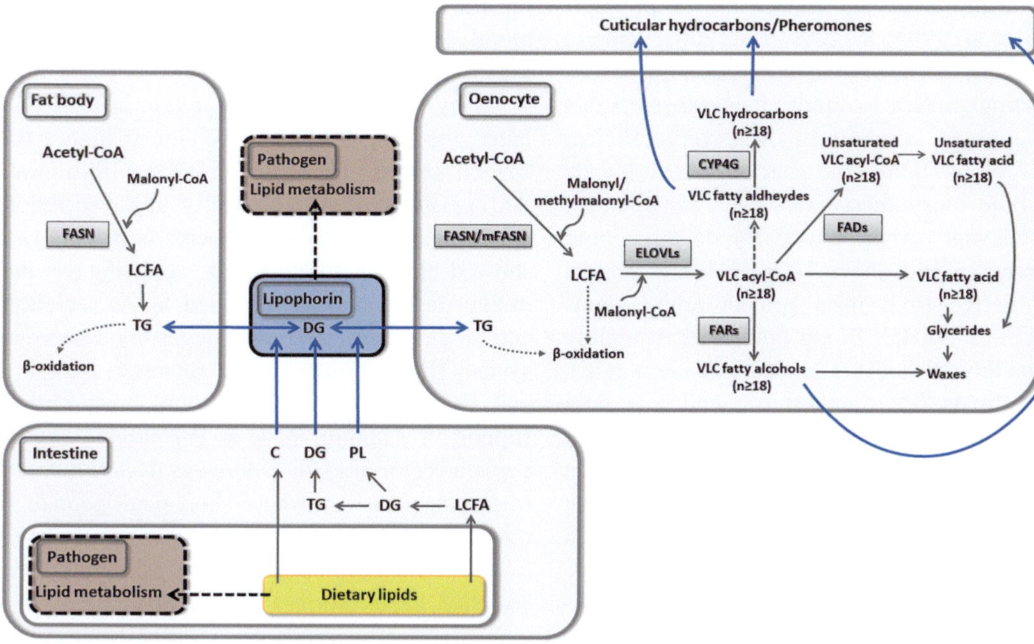

Fig. 1 Relevant pathways of lipid synthesis (gray arrows) and transport (blue arrows) in the fat body, the integument oenocytes and the midgut in insect vectors of diseases. Lipids are synthesized mostly in the fat body, integument and midgut and transported through the hemocoel by lipophorin and other lipoproteins. Pathogens may develop completely within the digestive tract or may traverse the gut lining and undergo some developmental stages within the vector hemocoel, depending on the species. Pathogens hijack lipids from their host (dashed arrows) to fulfill their own lipid metabolism and also trigger alterations in their lipid metabolic pathways, which normally resulted in a decreased reproductive and survival fitness of the insect. Abbreviations: *FASN* fatty acid synthase, *mFASN* microsomal fatty acid synthase, *ELOVL* acyl-CoA elongase, *FAR* acyl-CoA reductase, *FAD* acyl-CoA desaturase, *CYP4G* cytochrome P 450 4G, *LCFA* long chain fatty acid, *VLC*, very long chain, *C* cholesterol, *DG* diglycerides, *PL* phospholipids, *TG* triacylglycerols

which makes them unable to synthesize the complete suite of lipids they require for replication. Thus, intervention in host lipid composition and metabolism is expected in order to compensate for their lipid requirements (Fig. 1).

In this chapter, we will address the current knowledge on lipid metabolism in insects that are vectors of diseases.

2 Blood Digestion, Lipid Intake, and Pathogen Dynamics Within the Digestive Tract

Hematophagous insects do not feed regularly, instead they ingest sporadically large amounts of blood (3–15 times their own weight, depending on the species). Thus, their lipid metabolism is adapted to this feeding behavior and the nutrients the blood contains.

2.1 Blood Digestion

Vertebrate blood is composed of 85% protein, and it is very poor in lipids (less than 1%) (Geering and Freyvogel 1975; Langley et al. 1987a). Therefore, these insects produce and secrete a large amount of peptidases together with minor amounts of lipases and glucosidases in order to digest blood. Mosquitoes and hematophagous flies have an alkaline protein digestion with secretion of trypsin-like enzymes while triatomines have developed an acidic digestion secreting cathepsin-like proteinases. Finally, aminopeptidases and carboxypeptidases produce amino acids from the endopeptidase products (Jahan et al. 1999; Sanders et al. 2003; Ribeiro et al. 2014). In view of the high metabolic cost of converting protein to lipids, it is expected that blood-sucking insects rely on a very efficient use of blood lipids and carbohydrates for de novo lipid biosynthesis. However, protein may be a relevant carbon source for de novo lipid biosynthesis, and about 14% of *Aedes aegypti* lipids derive from meal protein amino acids (Zhou et al. 2004a). Blood lipids consist mainly in triacylglycerols (TG), cholesterol esters (CE), phospholipids (PL), cholesterol, and free fatty acids (FFA) (Wood et al. 1964; Langley et al. 1987a). Experiments using blood meal that contains radiolabeled TG and FFA showed that FFA are rapidly absorbed and transported through midgut epithelium while TG and PL are hydrolyzed. Triacylglycerol hydrolysis may follow two major pathways: complete hydrolysis releasing 3 FFA and glycerol, and partial hydrolysis releasing FFA and monoglycerides (MG) or diglycerides (DG) (Canavoso et al. 2001). As a result of blood lipids digestion, mostly labeled FFA are observed in the midgut of the triatomines *Rhodnius prolixus* and *Panstrongylus megistus* (Canavoso et al. 2004; Grillo et al. 2007) and the tsetse fly *Glossina morsitans* (Langley et al. 1987a) with nearly zero or very low amount of labeled MG and DG, suggesting that the former mechanism seems to clearly predominate. Fifteen transcripts coding for different lipases, mostly TG lipases, were detected in a *R. prolixus* midgut transcriptome and proteome (Ribeiro et al. 2014). In addition, transcripts coding for genes related to phospholipid digestion (inositol polyphosphate phosphatase and secretory phospholipase A2) were detected in salivary glands of *Anopheles gambiae* larvae (Neira Oviedo et al. 2009), and the activity of sphingomyelinase and phospholipase A2 and phospholipase C have been characterized in the mosquitoes *Aedes albopictus* and *Culex quinquefasciatus* and in the stable fly *Stomoxys calcitrans* (Spates et al. 1990; Abdul Rahim et al. 2018). Lipase activity seems to be regulated by the blood meal in hematophagous insects. A significant increase in TG and PL lipases activity was observed about 15 h after blood ingestion in *Ae. aegypti* (Geering and Freyvogel 1975), and about 24 h in tsetse flies (Langley et al. 1987a) and triatomines; in the latter, the enzyme activity continued higher for several days (Canavoso et al. 2004; Grillo et al. 2007). The precise regulation of lipase secretion is not clear but early experiments in *Ae. aegypti* showed that air-filling-based stretching of the midgut resulted in an increased lipase secretion and cholinesterase activity, suggesting a nervous system stimulation of lipase secretion (Geering and Freyvogel 1975). This could have further support in a finding made in *Periplaneta americana*, a vector of many pathogens from viruses to metazoan parasites, where short neuropeptide F released by midgut paraneurons has been implicated in lipase regulation by inhibiting their secretion in starved insects (Mikani et al. 2012).

2.2 Lipid Intake

Free fatty acids released from TG, PL, and CE are transported into the enterocytes. How this transport occurs is yet unknown but may involve membrane fatty acid transporter proteins (FATPs) similar to those identified in mammals. Transcript expression of a FATP showed a significant increase 12–24 h after blood meal in *Ae. aegypti* as well as many genes from enterocytes lipid metabolism (Sanders et al. 2003). Five putative FATP genes have been identified in the *R. prolixus* genome and two of them showed a significantly higher expression in the midgut 2–10 days after a blood meal (Alves-Bezerra et al. 2016). However, despite their identification in mosquitos and triatomines, the transport capacity of fatty acids through the enterocyte membrane is still pending to be characterized.

As blood lipid digestion in the midgut lumen takes place, a rapid rise in the amount of DG is seen within the enterocytes. Subsequently, the amount of DG falls whereas there is an increase in the amount of PL and TG (Langley et al. 1987a) (Fig. 1). Feeding *S. calcitrans* flies with blood labeled with C18 fatty acid evidenced the formation of lipid droplets within the enterocytes about 3–4 h after blood meal; radioactivity was incorporated mostly in TG and PL followed by FFA, DG and minor amounts of MG (Lehane 1977). Lipid droplets and accumulation of TG were also observed within the enterocytes of *G. morsitans*, *Anopheles coluzzii*, and the triatomines *R. prolixus* and *P. megistus* (Langley et al. 1987a; Canavoso et al. 2004; Grillo et al. 2007; Ferdous et al. 2021) suggesting that lipids are stored for a short time previous to be released onto lipophorin. The glycerol-3-phosphate pathway and/or the monoacylglycerol pathway are involved in the synthesis of TG and PL (see Sect. 5). As other insect species, hematophagous insects are unable to synthesize sterols de novo. Thus, they must incorporate them from blood, either from free cholesterol or from the hydrolysis of CE (Canavoso et al. 2001). About 20% of the cholesterol present in the gut lumen (derived either from free cholesterol or CE) is incorporated in *G. morsitans* (Langley et al. 1987a). Cholesterol diffuses from the gut lumen to the enterocyte membrane and then is transported by sterol carrier proteins (SCP) or Niemann–Pick type C1 proteins through the enterocyte cytoplasm until it is loaded into lipophorin for subsequent hemolymphatic transport (Entringer et al. 2021). Inhibition of SCP has been shown to reduce survival in *Ae. aegypti*, *An. gambiae*, and other mosquito species, and has been proposed as an alternative mosquito control strategy (Perera and Wijerathna 2019).

DG and minor amounts of PL and cholesterol are mostly loaded onto lipophorin for subsequent lipid transport to fat body and other tissues through hemolymph (Atella et al. 1995) (Fig. 1). These DG come from the minor amount absorbed from the gut lumen, those biosynthesized by the enterocytes, and the resulting from the hydrolysis of TG accumulated within the enterocytes, which is mediated by a TG lipase (Grillo et al. 2007) (Fig. 1). In triatomines, not all the lipids absorbed by the gut epithelium are loaded onto lipophorin for subsequent delivery to other body organs. Midgut labeling with radioactive C16 fatty acid at 10 days after blood meal has shown that, in contrast to previous days, most of the PL synthesized by the enterocytes were used to form the perimicrovillar membrane while minor amounts were loaded onto lipophorin (Bittencourt-Cunha et al. 2013). The perimicrovillar membrane is a non-chitinous matrix that is positioned between blood and midgut epithelium and acts as a physical and physiological barrier for the blood components and the enzymes involved in the digestive process, in a similar fashion to the peritrophic matrix of most insect species. It consists of a PL bilayer, together with glycoproteins and lipoproteins, and it is very noticeable 10 days after blood meal in *R. prolixus* and *Triatoma infestans* (Alves et al. 2007).

2.3 Pathogen Dynamics within the Insect Vector Digestive Tract

The release of the digestive enzymes together with hemolysins, agglutinins, and many antimicrobial compounds from the salivary glands has profound effects on parasite survival within the digestive tract. The antimicrobial peptide trialysin from *T. infestans* has been shown to lyse *T. cruzi* trypomastigotes (Schaub 2021), and in *R. prolixus*, an 80% reduction in the number of trypomastigotes is detected in the midgut 24 h after blood ingestion, and a few trypomastigotes after 96 h suggesting that the digestive process kills a large number of parasites (Ferreira et al. 2016). Despite this, 1 week after blood ingestion parasites are able to multiply by three their number (Schaub 2021). Many glycoproteins of the perimicrovillar membrane were shown to be essential for the former attachment of *Trypanosoma cruzi* epimastigotes during the first period of parasite infection (Alves et al. 2007). Development of pathogens within the digestive tract is not innocuous to their insect vectors. Several studies have shown that *T. cruzi* affects adversely many life traits of triatomines. Body weight, adult fertility, survival, and development time were affected in *R. prolixus* (Fellet et al. 2014; Peterson et al. 2015) and *Mepraia spinolai* (Botto-Mahan 2009). There are no clear explanations for these effects however, *T. cruzi* is able to alter the lipid metabolism in its insect vectors, diminishing their lipid stores (see Sect. 5). Thus, it is very probable that trypanosomatids hijack lipids and other nutrients for their own development, decreasing their availability to triatomines (Fig. 1). And this may be extensive to virus-infecting mosquitoes. It has been recently shown that infection with dengue virus can alter the dynamics of lipids in the midgut of *Ae. aegypti* (Chotiwan et al. 2018). And similar results have been obtained using whole mosquitoes as well (Koh et al. 2020). In particular, a significant increase in many PL and acyl glycerides was observed in both studies. Although lipid biochemistry has been studied in mosquitoes for several years, very little is known about the relationship between virus and mosquito host and the alteration of intracellular lipids that supports infection capacity. The midgut is the primary site of viral infection and replication, preceding systemic dissemination, and transmission. At the peak of viral replication (7–11 days after blood meal) there was a marked and significant increase in several PL such as phosphatidylethanolamine (PE), phosphatidylserine (PS), sphingomyelin (SM), and cardiolipin (CL). This increase is consistent with the requirements for dengue virus replication within mosquito cells, as the virus induces a massive expansion of intracellular membranes required for the assembly and function of viral replication complexes and after that the assembly and release of enveloped virions. At about day 7 there was an increase in TG, DG, and MG, which might be a result of transportation of these lipids from storage tissues to support the demand for lipids during infection. Many FFA and fatty acyl derivatives, which can be incorporated into PL, acyl glycerides, SM, and other complex lipids that participate in energy storage, immune response, and have structural roles in membranes, showed significant increased or decreased levels indicating alterations in immunomodulation or malfunction of fatty acid oxidation (Chotiwan et al. 2018). Thus, the alteration caused by the pathogen in the lipid repertoire of the mosquito midgut, and other organs as well, may finally lead to compromise the insect reproductive fitness and other physiological parameters.

3 Insect Lipids and Pathogen Attachment and Development

The life cycle of most pathogens includes several phases of differentiation and multiplication that take place within different body parts of their insect vectors. In general, some pathogens restrict their transit to the insect digestive tract, whereas some others start their development within the digestive tract, then colonize the insect hemolymph and finally migrate to the salivary glands where they are ready to infect another vertebrate host. For the pathogen of animal trypanosomiasis, *Trypanosoma vivax*, the proboscis of *Glossina* sp. is the only site of development. *T. vivax* ingested by the tsetse fly gets attached to the fly's labrum and labium and develops directly into epimastigotes. In contrast, *Trypanosoma brucei*, the causative agent of African trypanosomiasis, passes from the mouthparts to the midgut where they develop into short trypomastigotes. Finally, they transform into metacyclic trypomastigotes, the infective form to humans (Vickerman and Tetley 1990). Trypomastigotes of *T. cruzi*, the causative agent of Chagas disease, are ingested by triatomines from infectious vertebrate blood and develop to epimastigotes within the midgut and rectum. Epimastigotes and spheromastigotes multiply outstandingly mostly in the rectum where they go through metacyclogenesis developing to infective metacyclic trypomastigotes. Finally, infective trypomastigotes are released with insect feces (Schaub 2021). The trypanosomatid *Leishmania*, causative agent of Leishmaniasis, develops entirely within the digestive tract of sand flies of the genera *Lutzomyia* and *Phlebotomus*. After being ingested, promastigotes present in the vertebrate blood travel from the mouthparts to the midgut where they develop to amastigotes and promastigotes. Then, they come back to the esophagus and mouthparts where they develop to metacyclic infective promastigotes (Killick-Kendrick 1990). Differently from the previously mentioned trypanosomatids, the apparently non-pathogenic *Trypanosoma rangeli* moves from the digestive tract of the insect to the hemolymph, to finally complete its development within the salivary glands (Garcia et al. 2012). Similar to *T. rangeli*, the many *Plasmodium* species that produce malaria start their development in the *Anopheles* female midgut after ingestion of gametocyte-infected blood from a human. Once in the mosquito midgut lumen, female and male *Plasmodium* gametocytes transform into gametes, which subsequently fertilize to form a zygote. Within 12 to 20 h after blood ingestion, zygotes differentiate into ookinetes, which invade and traverse the midgut epithelium. After reaching the subepithelial basal space, zygotes transform into oocysts, which over a period of 10 to 18 days generates thousands of sporozoites. Sporozoites finally are released into the hemocoel and migrate to mosquito salivary glands from where they are released when the mosquito bites a new host (Yang et al. 2021).

Within the insect vector midgut, the pathogens adhere to either the perimicrovillar membrane, the epithelium surface, or the peritrophic matrix, depending on the insect species. The biochemistry of pathogen adhesion within the midgut has been studied in detail. It involves a variety of sulfated glycoproteins (e.g., cruzipain), glycoinositol phospholipids and lipophosphoglycan in trypanosomatids (Hall et al. 2020; Schaub 2021), and a family of thrombospondin-related adhesive proteins in *Plasmodium* (Zieler et al. 1999; Dinglasan et al. 2007). *Plasmodium* also seems to interact with PL of the mosquito midgut epithelium. Yang et al. (2021) showed that a phospholipid flippase expressed in the ookinete surface is involved in the uptake of phosphatidylcholine from the midgut epithelial cells and is essential for midgut invasion. Depletion of this flippase impairs the ability of the pathogens to traverse the epithelium finally leading to the destruction of the ookinetes inside the midgut cells. The way in which the pathogens are attached to their insect host is a topic of interest, since adhesion is thought to be a prerequisite for differentiation into the infective stage (Vickerman and Tetley 1990; Dinglasan et al. 2007; Schaub 2021). The mouthparts and rectum of the insects have a cuticle lining. The bulk of the insect cuticle consists of the polysaccharide chitin embedded in a complex matrix of many cuticle proteins, and is covered by a lipid layer, mostly a blend of hydrocarbons (HCs), FFA, fatty alcohols, TG, and waxes; non-polar long and very long carbon chains clearly predominate (Lockey 1988; Juárez and Calderón Fernández 2007; Balabanidou et al. 2019). Chitin has been proposed to be the natural substrate for binding of many trypanosomatids; however, this molecule is not exposed over the cuticle surface; gold-labeling and specific staining methods found this polysaccharide not in the cuticle surface but in the deeper region (procuticle) (Schmidt et al. 1998). Therefore, the parasites must attach to the lipid layer that covers the cuticle while they differentiate and multiply in these areas. As previously outlined, many studies have deeply examined the biochemistry of the attachment to parts of the digestive tract lined by an epithelium, but very few analyzed the adhesion to those parts lined by lipids. *T. vivax* undergoes many cycles of multiplication while the parasites are attached in clusters to the cuticle lining of the food canal in the proboscis of the tsetse fly. The adhesion should be strong enough to avoid being removed during subsequent feedings of the fly. Observations based mostly on electron microscopic images showed that the parasite uses its flagellum to adhere to the cuticle surface and form a hemidesmosome-like structure along the attachment region of the flagellar membrane (Vickerman and Tetley 1990). Hemidesmosomes in the flagellum membrane were also reported for *Leishmania* in the sites where the parasite gets attached to areas of the foregut and hindgut lined by cuticle, such as the esophageal valve or pylorus. A distance of about 20–25 nm between the flagellum membrane and the cuticle seems to be necessary for the hemidesmosome to develop (Killick-Kendrick 1990). The deepest studies on this topic have been carried out by Schaub and co-workers relative to the biochemistry of the attachment of *T. cruzi* to the digestive tract of triatomines (Kleffmann et al. 1998; Schmidt et al. 1998; Schaub 2021). Examination of the cuticle in the rectum of *T. infestans* showed that it follows the general pattern found in other insect body parts. The outermost layer that lines the lumen of the rectum was sensitive to extraction by organic solvents, being completely removed after immersion of rectal samples in n-hexane. Also, it was largely lost upon standard fixation for electron microscopy, but it was retained with fixation methods appropriate for lipids. *T. cruzi* epimastigotes were attached by the flagellum to the lipid lining of the rectum and formed a hemidesmosome-like structure in the zone where the membrane adheres to the rectum (Schmidt et al. 1998). Interestingly, they were able to attach to plastic and other hydrophobic materials, to HC droplets, and even to the surface of pieces taken from the wings (Kleffmann et al. 1998; Schmidt et al. 1998). Attachment to the lipid layer of the rectum has been reported to be relevant for development of attached epimastigotes into infective trypomastigotes (Schaub 2021).

Thus, lipids from the cuticle play an important role in the life cycle and differentiation of the pathogens within their insect vector. However, more studies, including molecular techniques, are needed to identify the receptors in the pathogen surface involved in the attachment to cuticle lipids, and correspondingly, how lipids trigger the differentiation of the pathogens once they are attached.

4 Modifications in Composition and Function of Lipid Transport Proteins Related to Hematophagy, Insect Vector Development and Pathogen Infection

Lipid transport through the insect body is carried out by lipophorin, a lipid-protein complex that makes lipids soluble in the hemolymph. As in other insect species, lipophorin is synthesized in the fat body, where the genes coding for lipophorin apoproteins (apoLp-I, apoLp-II, and apoLp-III) are specifically expressed. ApoLp-I and apoLp-II arise through post-translational cleavage of a common precursor protein coded by a single gene while apoLp-III is coded by a distinct gene. Lipophorin biosynthesis and structure is thoroughly addressed in Chapter "Lipophorin the Lipid Shuttle".

4.1 Lipophorin Composition and Its Changes Related to Blood Ingestion, Reproduction, and Other Physiological Aspects

Although DG is in general the main lipid transported, lipid composition of lipophorin greatly varies among hematophagous insects, even between closely related species. In *Ae. aegypti* lipophorin contains unusually higher amounts of PL (~32%) and TG (~41%), together with lower amounts of DG (~7%) compared to most insects (Van Heusden et al. 1998), whereas in *An. gambiae*, HCs reach about 50% of the lipid content, followed by an unidentified lipid (~ 30%), cholesterol and PL (~ 10% each), CE (~ 2%), and low amounts of TG, FFA, and DG (~ 1% each) (Atella et al. 2006).

Lipophorin composition has also been extensively studied in two of the major vectors of Chagas disease, *T. infestans* and *R. prolixus*. In the former lipophorin is about 53% protein and 47% lipid, and its lipid composition consists of about 38.7% DG, 22.7% HCs, 13.0% PL, 12.2% an unidentified lipid, 6.8% TG, 4.8% cholesterol, and minor amounts of FFA and CE (Fichera and Brenner 1982) whereas in the latter consists of about 40% HCs, 35% DG, 10% cholesterol, 9% PL, and minor amounts of TG, FFA, CE, and MG (Majerowicz et al. 2013). Hydrocarbon (HC) content mostly depends on the insect species; high amounts of HCs were reported in the cuticle and eggshell of triatomines, therefore, a large HC transport through the insect body is expected.

Physiological parameters such as the amount and density of lipophorin may vary in hematophagous insects related to the feeding source, developmental cycle, vitellogenesis and egg formation, and feeding status (Gondim et al. 1989; Coelho et al. 1997; Van Heusden et al. 1997; Marinotti et al. 2006). In female samples of the mosquitoes *Ae. aegypti* and *An. gambiae* the blood meal triggers an increase in the amount of lipophorin 24–36 h after feeding probably because egg development induced by blood feeding may need an increased rate of lipid transport to the ovaries. The three- to four-time increase in the lipophorin level after the blood meal indicates that there is a need for an increase in lipid transport and that the mosquitoes respond to this by synthesizing more lipophorin and secreting it into the hemolymph (Van Heusden et al. 1997). In contrast, lipophorin amount remains quite constant in the triatomine *R. prolixus*, showing a slight decrease 5 days after blood meal (Gondim et al. 1989). Regarding the density of lipophorin, in mosquitoes, it reaches a peak 24–36 h after blood meal and then tends to decrease (Van Heusden et al. 1997) whereas in the triatomine *R. prolixus* it shows a marked decrease at days 2–3 after blood meal, then increases abruptly to reach a peak by day 10 after blood meal and by day 18 it has decreased reaching similar values to those previous to the blood meal. This dynamics in the density of lipophorin suggests its loading with lipids from the blood meal shortly after feeding followed by the unloading at the ovaries, fat body, integument, and midgut around day 4 to day 14, and then a new loading with lipids probably from the fat body to supply insect necessities in subsequent days (Gondim et al. 1989; Coelho et al. 1997). Two very high-density lipoproteins (VHDLs), which additionally contribute to lipid transport, have been biochemically characterized in *T. infestans* (Fichera and Brenner 1982; Rimoldi et al. 1989). In contrast to lipophorin, these lipid transporters carry higher amounts of FFA and fatty alcohols, which comprises about 74% of the lipid content, followed by minor amounts of DG, cholesterol, TG, and HCs. Besides their role in lipid transport, lipophorin and VHDLs are of interest in insect vectors of diseases because they transport pyrethroids and other contact insecticides to the target sites after the insecticides have penetrated the integument tissue (Dulbecco et al. 2015). In addition to lipophorin and VHDLs, there are hemolymphatic proteins called lipid transfer particles that participate in the transfer of lipids between lipophorin and other lipoproteins. In *R. prolixus*, they were related to transfer of PL and DG between lipophorin units (Golodne et al. 2001).

4.2 Lipophorin Receptors

In order for lipids to be loaded from and unloaded to different tissues, lipophorin interacts with cells through lipophorin receptors located in the plasma membrane. Lipophorin receptors were identified and/or characterized in many insect vectors of diseases such as the mosquitoes *Ae. aegypti*, *C. quinquefasciatus*, and *An. gambiae* (Seo et al. 2003; Kumar and Paily 2011), the tsetse fly *G. morsitans* (Benoit et al. 2011), and the Chagas disease vectors *T. infestans* and *P. megistus* (Calderón-Fernández et al. 2017; Fruttero et al. 2017). Two isoforms were identified in *Ae. aegypti* (Seo et al. 2003) and *T. infestans* (Calderón-Fernández et al. 2017); in the former, one isoform is specifically expressed in the fat body while the other in the ovary, also showing specific expression times (Seo et al. 2003). Unexpectedly, a β subunit of the ATP synthase with an ectopic localization in the plasma membrane of cells has been reported to be involved in the binding of lipophorin in the midgut, ovary, and fat body of *P. megistus* (Fruttero et al. 2014, 2017, 2019). Very recently, the localization of β-ATP synthase in the plasma membrane of fat body cells has been corroborated in *R. prolixus*, another vector of Chagas disease and an insect model for physiological studies, as well as its role in lipid metabolism through taking part in the interaction between lipophorin and fat body cells (Almeida-Oliveira et al. 2023). Furthermore, the knockdown of both β and α ATP synthase subunit genes affected lipid metabolism and reduced drastically the fecundity of females.

4.3 Lipid Transport Related to Pathogen Infection

An intriguing issue in vector-borne diseases is how the pathogen affects the insect lipid transport and why. The uptake of insect lipids by the pathogen while they are transported was seen many years ago in mosquitos and triatomines (Fig. 1). *T. rangeli* has been shown to uptake the whole lipophorin of *R. prolixus*, transferring a wide range of PL and neutral lipids for its own use, mostly PE, PC, DG, and FFA, and minor amounts of cholesterol and TG (Folly et al. 2003). Similarly, oocysts of *Plasmodium gallinaceum* located in the basal part of the midgut epithelium are able to uptake and internalize the lipophorin particles of *Ae. aegypti* (Atella et al. 2009). In contrast, sporozoites derived from these oocysts, which can be found in the mosquito hemolymph, uptake and transfer PL, and probably other lipids as well, from the circulating lipophorin (Atella et al. 2009). Lipid uptake by malaria pathogens is not restricted to its insect vector. It has been seen during human infection that *Plasmodium falciparum* actively uptakes PL precursors from human blood (FFA, choline, serine) in order to synthesize its own PL, and also can change the PL composition of the erythrocytes they infect (Vial et al. 1999). In *Ae. aegypti* and *An. gambiae*, lipophorin transcript levels are strongly induced by the blood meal (Van Heusden et al. 1998; Marinotti et al. 2005), which is consistent with the necessity of an increased lipid transport, and knockdown of lipophorin gene led to abolition of egg development in the females' ovaries of *Ae. aegypti* (Cheon et al. 2006). Cheon et al. (2006) showed in *Ae. aegypti* that ingestion of blood infected with *P. gallinaceum* caused a strong increase in lipophorin gene expression about 24 after the blood meal, the time when ookinetes invade the midgut wall. They also observed that silencing of the lipophorin gene led to a 90% decrease in oocyst formation, suggesting that pathogens need lipophorin to fulfill the high demand of lipids for membranes during oocyst development. Later on, Rono et al. (2010), using a model of *An. gambiae* infected with *Plasmodium berghei*, showed that the reduction in pathogen survival after lipophorin gene knockdown is indirect and is mediated by reduced levels of vitellogenin, which in turn increased the expression of the antiparasitic thioester-containing protein TEP1. They concluded that the reduction in the number of oocysts after lipophorin silencing is not due to a nutritional dependence of the pathogen on lipids provided by lipophorin but is a consequence of TEP1 activity. However, as it has been demonstrated that *P. gallinaceum* oocysts incorporate lipophorin particles of *Ae. aegypti* (Atella et al. 2009), nutritional dependence might be also possible. Competition for cholesterol and other lipids between *R. prolixus* and the parasites *T. cruzi* and *T. rangeli* has been considered as a factor that impairs the insect reproductive parameters (Fellet et al. 2014). Competition for lipid and other nutrients is not restricted to pathogens. A very recent work in *Glossina fuscipes* showed that the endosymbiotic bacteria *Spiroplasma* makes the fly more resistant to *Trypanosoma* infection but also impairs fly reproduction. *Spiroplasma* competes for many nutrients with the fly, and a higher expression of a lipophorin-related apolipoprotein has been observed in *Spiroplasma*-infected tsetse flies, suggesting a higher lipid mobilization (Son et al. 2021). Major qualitative and quantitative changes in the cuticle hydrocarbons (CHC) of *Anopheles albimanus* infected with *Plasmodium* have been reported (Claudio-Piedras et al. 2021), in particular, a decrease in the number of CHC in infected insects. As lipophorin transports a variety of lipids, the authors suggested that the pathogen effects on lipophorin may have affected the lipid composition of the cuticle.

On the other hand, lipophorin plays a remarkable role in the special way of larval development in the tsetse fly *G. morsitans* and related species. Reproduction in tsetse flies has shifted from oviparity, as happens in most insects, to viviparity, which includes intrauterine development and nourishment of the progeny over the duration of larval development (Tobe and Langley 1978). This mode of reproduction is derived in large alterations in the birth canal, mainly by developing a uterus to accommodate the developing larvae and a milk gland to generate nourishment for the intrauterine progeny. The primary nutrients released by the milk gland are lipids, proteins, and amino acids (Cmelik et al. 1969). Lipids released by the milk gland are not synthesized in the gland itself, instead they derive from those produced and stored in the fat body or are acquired directly from blood feeding. Tsetse flies are obligate hematophagous insects that acquire lipids from the blood meal. Dietary lipids are absorbed from the midgut lumen and loaded onto lipophorin at the midgut/hemolymph interface. Loaded lipophorin travels through the hemolymph and unloads lipids at the fat body for storage, ovaries for oogenesis, and the milk gland for milk production (Tobe and Langley 1978; Langley et al. 1981; Benoit et al. 2011). The gene coding for lipophorin apoproteins I and II are expressed in the fat body of both adult flies and developing larvae and showed higher expression levels during larval development (Benoit et al. 2011). Partial deletion of lipophorin expression resulted in a delayed oocyte and embryonic formation, and a 20% increase in the abortion rate (Benoit et al. 2011), reflecting the relevance of lipophorin transport for this special way of larval development.

5 Lipid Metabolism in Insect Organs Related to Blood Intake, Development, and Pathogen Infection

Although the midgut is very active in lipid metabolism especially during blood digestion, the fat body and the integument of hematophagous insects are the major organs involved in the biosynthesis and catabolism of fatty acids, acyl glycerides, and other lipids in order to support many physiological processes such as molting, starvation, flight, metamorphosis, and reproduction. In fact, a bidirectional lipid metabolic coupling between the fat body and the integument oenocytes has been evidenced in *Drosophila melanogaster* indicating that these organs work in similar fashion to the adipose tissue and the liver in mammals (Gutierrez et al. 2007).

5.1 Fatty Acid Biosynthesis

The biosynthesis, utilization, and catabolism of fatty acids is the central core of lipid metabolism in both the fat body cells and the integument oenocytes (Juárez and Calderón Fernández 2007; Saraiva et al. 2021). As it has been previously mentioned, the blood meal provides little amounts of lipids, therefore fatty acids should be synthesized from other sources such as protein and carbohydrates. For example, *Ae. aegypti* can use 50% of the glucose incorporated with the diet for lipid synthesis (Zhou et al. 2004b). During long starvation periods the de novo synthesis of fatty acids in the fat body and the integument, and their subsequent utilization as an energy source in the same and other organs can be very relevant especially in order to maintain the homeostasis. For instance, some triatomines have been shown to survive about six months or longer without feeding by relying on their lipid reserves (Lima-Neiva et al. 2017).

Long-chain fatty acids (usually C16) are synthesized by the consecutive activity of the acetyl-CoA carboxylase (ACC), which catalyzes the synthesis of malonyl-CoA from acetyl-CoA, followed by the fatty acid synthase (FASN), a multidomain enzyme that extends an acetyl-CoA primer through an iterative series of reactions using malonyl-CoA or methylmalonyl-CoA to add two carbon units each cycle; methyl-branched long-chain fatty acids will be obtained by adding methylmalonyl-CoA instead of malonyl-CoA in some specific cycles of chain elongation. In contrast to mammals, which have two ACC genes with distinct localization and function, a single ACC gene has been identified in *G. morsitans* (Goldring and Read 1994), *Ae. aegypti* (Alabaster et al. 2011), *T. infestans* (Calderón-Fernández et al. 2017), *R. prolixus* (Majerowicz et al. 2017; Saraiva et al. 2021) as well as other non-disease transmitting insects (Parvy et al. 2012). Measures of enzyme activity in *G. morsitans* showed that it reaches the highest values about 6 days after blood meal. The highest level was detected in the fat body although the integument also showed a noticeable value reflecting the relevant role of this organ in lipid metabolism (Goldring and Read 1994). The highest ACC transcript and/or protein expression levels were observed in the fat body in all the species examined. They usually rise about 20 h after blood meal and reach different peak levels depending on the species, suggesting that carbohydrates and proteins from the blood are being used for de novo fatty acid biosynthesis in order to increase lipid reserves. However, *R. prolixus* fat body showed the highest ACC gene and protein expressions before feeding, decreasing in the next days after feeding and then increasing again. Despite this, de novo lipid synthesis was low in unfed insects and shortly after feeding and increased about 5 days after blood meal (Saraiva et al. 2021).

Surprisingly, a high ACC protein expression was detected in the ovary of *R. prolixus* females at day 4 after blood meal (Saraiva et al. 2021) in contrast to *Ae. aegypti* where no protein expression was found by Western blotting (Alabaster et al. 2011). This might indicate some endogenous lipogenic activity in the ovary of *R. prolixus*, instead of largely relying on lipids from the fat body and the integument. Depletion of ACC gene expression in females of both insects resulted in a decreased lipid synthesis, reduced oviposition, defects in eggshell formation, and a strongly reduced egg hatching; a delayed blood digestion and reduced survival were also observed in *R. prolixus* (Alabaster et al. 2011; Moraes et al. 2022).

In contrast to ACC, three or more FASNs were found in most insects. In *Ae. aegypti*, *An. gambiae*, *R. prolixus*, and *T. infestans* three FASN genes were identified (Sim and Denlinger 2009; Calderón-Fernández et al. 2017; Majerowicz et al. 2017; Grigoraki et al. 2020). FASN1 is expressed in the fat body and most somatic cells while FASN2 and FASN3 are expressed in the integument oenocytes. The biochemical characterization of FASN2 and FASN3, located in the microsomal and cytosolic fractions (respectively) of the oenocytes, was reported in *T. infestans* (Juarez et al. 1996). The analysis of insect genomes in *R. prolixus*, *An. gambiae*, and other insect species shows a close localization of FASN1 and FASN2 genes and the presence of FASN gene fragments in the same contig, suggesting that the proliferation of these genes has been the result of gene duplication and gene expansion, followed by acquisition of specific tissue expression and function (Majerowicz et al. 2017; Moriconi et al. 2019). The role of the integument FASNs, but not the fat body FASN, in the formation of cuticle lipids was first demonstrated in the Chagas disease vector *T. infestans* (Juárez and Brenner 1989). FASN activity in the integument oenocytes is coupled to the action of acyl-CoA elongases (ELOVL), acyl-CoA reductases (FAR), and many other enzymes involved in the biosynthesis of HCs and other very long chain lipids that will form part of the cuticle and eggs (Juárez and Brenner 1989; Juárez 1994a). Additionally, FASN2 and FASN3 are able to accept either malonyl-CoA or methylmalonyl-CoA as elongating unit and therefore they can synthesize both straight and methyl-branched fatty acids, in contrast to FASN1 that only accept malonyl-CoA. Recently, the genomic structure of the three FASN genes of *R. prolixus* has been characterized, as well as the relevance of FASN3 in synthesizing the bulk of the methyl-branched lipids that have the major role in the barrier properties of the insect cuticle (Moriconi et al. 2019). Noticeably high transcript levels of ACC, FASNs, and cytochrome P450 4G were detected in starved *R. prolixus* and *T. infestans* (Calderón-Fernández, unpublished; Dulbecco et al. 2020; Saraiva et al. 2021), indicating that during starvation the insects continue a very active de novo biosynthesis of fatty acids from proteins and carbohydrates derived from residual amounts of blood still being digested.

FASN activity of insect hosts is used by the dengue virus for its replication during mosquito cell infection. By using C75, an inhibitor of FASN, the virus replication was significantly reduced in *Ae. aegypti* cells indicating that fatty acid biosynthesis is important for virus viability (Perera et al. 2012).

Desaturation of fatty acids is also very important in fatty acid metabolism as it provides unsaturated fatty acids essential for maintaining cell membrane fluidity and other metabolic functions. Stearoyl-CoA desaturase is the enzyme involved in the formation of $\Delta 9$ unsaturated palmitoleic and oleic fatty acids. Studies in *T. infestans* have shown that the activity of $\Delta 9$ desaturase is higher in the fat body than the integument, as the latter tissue synthesized mostly saturated fatty acids from C14 up to C34 by the successive action of FASN and acyl-CoA elongases (Juárez and Brenner 1989; González and Brenner 1999). The $\Delta 9$ desaturase activity in the fat body increased significantly as a result of successive blood meals or administration of the carbohydrate trehalose giving further support to the use of amino acids and carbohydrates from the blood meal for de novo fatty acid biosynthesis (González and Brenner 1999). The essential role of the stearoyl-CoA desaturase in the general physiology and metabolism of hematophagous insects has been recently reported in the malaria vector *An. coluzzii*. The knockdown of this gene caused drastic effects in the midgut epithelial integrity, egg development, and metabolic homeostasis by disturbing the regulation of many genes involved in lipid, protein and carbohydrate metabolism as well as other metabolic pathways (Ferdous et al. 2021).

5.2 Fatty Acid Catabolism

Fatty acid oxidation is essential for the energy metabolism of vectors of diseases, especially in the species that undergo long periods of starvation or need a sustainable source of energy for flight, although this process has been scarcely studied in these insects. For instance, lipid catabolism has been shown to contribute to the production of proline, the major substrate for tsetse fly *G. morsitans* flight metabolism (Bursell 1977). In contrast, the triatomine *P. megistus* uses carbohydrates during the first 45 min of the flight after which it starts using lipids as energy fuel (Canavoso et al. 2003). Another relevant aspect in hematophagous insects is the proper elimination of surplus nitrogen and other toxic products of blood digestion where lipid oxidation provides the necessary energy for this essential process (Langley 1977). The expression of genes involved in β-oxidation has been analyzed in *Ae. aegypti* females after a blood meal (Wang et al. 2017), in *Culex pipiens* adults undergoing diapause (Sim and Denlinger 2009), and *R. prolixus* females after a blood meal as well as during starvation (Arêdes et al. 2022). In *Ae. aegypti*, transcripts of genes coding for the enzymes that transport the fatty acids into the mitochondria (carnitine O-palmitoyl and -octanoyltransferases) increased in abundance at 18–36 h after blood meal, whereas those of the key enzymes of the β-oxidation pathway (acyl-CoA dehydrogenase, enoyl-CoA hydratase, 3-hydroxyacyl-CoA dehydrogenase, and β-ketoacyl-CoA thiolase) showed a significant increase at 36 h after blood meal (Wang et al. 2017). These genes also showed an increase in *C. pipiens* adults leaving the diapause state, where the insects use energy from lipid catabolism (Sim and Denlinger 2009). In *R. prolixus*, the hydroxyacyl-CoA dehydrogenase trifunctional multienzyme complex subunit alpha gene (containing the long-chain enoyl-CoA hydratase and the long-chain 3-hydroxyacyl-CoA dehydrogenase domains) had higher transcript levels in the flight muscle and ovary during starvation conditions (21 days after the last blood meal) compared to the fat body and midgut, whereas it showed an increase in its expression at about 48 h after blood meal in the fat body and the flight muscle, and then decreased again (Arêdes et al. 2022). Majerowicz et al. (2017) reported the absence of a long-chain acyl-CoA dehydrogenase in the genome of the triatomine *R. prolixus*, although genes coding for short-medium chain acyl-CoA dehydrogenase were detected in a *T. infestans* integument transcriptome (Calderón-Fernández et al. 2017) and in the *R. prolixus* genome as well. Thus, the genomic evidence suggests that triatomines have acyl-CoA dehydrogenases with a broad range of acyl chain length substrates, although the biochemical corroboration is still pending. The relevance of the β-oxidation pathway in essential aspects of insect vector biology has been evidenced after silencing the expression of the hydroxyacyl-CoA dehydrogenase trifunctional multienzyme complex subunit alpha gene in *R. prolixus* females. Silenced insects showed accumulation of TG in their fat body during starvation, and diminished oviposition and flight capacities (Arêdes et al. 2022).

5.3 Glycerolipid Biosynthesis

Metabolic pathways of fatty acids need these components to be activated to acyl-CoAs, which is mediated by long-chain acyl-CoA synthetases. Genes coding for these enzymes have been characterized in detail only in *R. prolixus* (Alves-Bezerra et al. 2016). From up to 20 genes described, two (ACS1 and ACS2) were found to activate long-chain fatty acids from 12 to 20 carbons. ACS2 gene was more abundantly expressed in all organs analyzed than ACS1, and almost exclusively in the fat body. Both genes showed different expression patterns related to the blood meal; the expression increased shortly after blood meal in the midgut and several days later in the fat body, indicating how fatty acid metabolism progresses along different organs. ACSL2 knockdown in this insect vector resulted in 2.5-fold increase in TG content in the fat body, and 90% decrease in fatty acid β-oxidation, together with a longer insect lifespan, delayed digestion, and a noticeably reduced oviposition and egg hatching. Thus, indicating that ACS2 is mainly involved in channeling fatty acids for β-oxidation.

As it has been previously mentioned in this chapter and other chapters as well, TG are the main lipids accumulated by tissues as energy source, thus the biosynthesis and catabolism of TG are essential in lipid metabolism. There are two routes for DG and TG biosynthesis, the monoacylglycerol and the glycerol-3-phosphate pathways. In the monoacylglycerol pathway, DG synthesis from MG and acyl-CoA is catalyzed by the monoacylglycerol acyltransferase (MGAT or MOGT), followed by a subsequent acylation of DG catalyzed by the diacylglycerol transferase (DGAT) to produce TG. In the glycerol-3-phosphate pathway, a glycerol-3-phosphate is acylated to lysophosphatidic acid by the glycerol-3-phosphate acyltransferase (GPAT), then lysophosphatidic acid is acylated to phosphatidic acid catalyzed by the 1-acyl-sn-glycerol-3-phosphate O-acyltransferase (AGPAT), phosphatidic acid is next converted to DG by the phosphatidate phosphohydrolase (PAP), and finally, DG is acylated to TG by DGAT. The glycerol-3-phosphate pathway is widely detected in most tissues and animal species while the monoacylglycerol pathway has been demonstrated to take place mostly in mammals' midgut, and it is restricted to specific tissues of some species in insects (Kühnlein 2012; Soulages et al. 2015). The precise route for DG and TG biosynthesis in insect vectors of diseases has been studied in detail mostly in triatomines, and very recently has been depicted in the mosquito *Ae. aegypti* by Vial et al. (2021) related to PL biosynthesis. It has been early reported in *T. infestans* that the synthesis of TG in the fat body (as well as DG) involved exclusively the glycerol-3-phosphate pathway (Rimoldi et al. 1988), and this finding was later confirmed and extended to other tissues (midgut, ovaries) in *P. megistus* and *R. prolixus* (Canavoso et al. 2004; Alves-Bezerra and Gondim 2012). In the latter species the formation of TG by the glycerol-3-phosphate pathway and the genes being involved were characterized in detail (Alves-Bezerra and Gondim 2012). A putative MGAT gene has been identified in the *R. prolixus* genome (RPRC000305) (Majerowicz et al. 2017); however, the lack of biochemical evidence in this insect for the monoacylglycerol pathway suggests that this gene may have lost its function or may code for an uncharacterized acyltransferase (Alves-Bezerra and Gondim 2012; Majerowicz et al. 2017).

5.4 Mosquitoes

After hatching and before pupation there are four aquatic larval stages in mosquitoes, which feed mostly in bacteria and microorganisms from submerged surfaces. In *Ae. aegypti*, *An. gambiae*, and *C. pipiens* larvae an increasing lipogenesis has been reported while development progresses, with a peak of lipid synthesis during the fourth larval stage. Lipid content, mostly TG, was linearly correlated with body size. Protein and to a lesser extent carbohydrate synthesis followed a similar pattern. A six-fold increase in the lipid content is observed between second- and fourth-stage larvae (Timmermann and Briegel 1999). Mosquito survival and vectorial capacity are related to their size and the amount of energy reserves. The nutritional status of larvae will be manifested as differences in body size and reserves of lipid, protein, and glycogen in teneral adults. Thus, lipid storage accumulated during larval stages significantly contributes to different aspects of adult mosquito life cycle and fitness, such as reproductive success, size, and formation of reserves. Adult females that developed from larvae with higher TG content survived longer and laid a larger number of eggs. Conversely, nutritionally stressed mosquitoes emerge as small individuals that die shortly upon emergence if deprived of a carbohydrate source (Takken et al. 2013).

Feeding *Ae. aegypti* larvae with labeled C18 fatty acid allowed to follow the metabolic fate of the lipid reserves accumulated during the larval stages. This procedure showed that at the beginning of the female first gonotrophic cycle about 76% of the label was retained in lipids (mostly TG), while about 24% were catabolized for subsequent biosynthesis of proteins and amino acids (17%), and carbohydrates (~ 7%), indicating that larval dietary fatty acids can be efficiently carried over into adult females as TG stores. At the end of the first gonotrophic cycle, there were no significant changes in the percentage of proteins and carbohydrates formed from larval lipids, but about 27% of the stored lipids were derived to make egg lipids (~ 13%) and were catabolized for insect requirements such flying, digestion, excretion, and other physiological processes (~ 14%) (Zhou et al. 2004b). These results reflect that during the first gonotrophic cycle, *Ae. aegypti* females use a substantial amount of lipid reserves derived from the larval diet to provide the eggs with lipids. The authors also showed that the residual amount of reserves in 4–6 day-starved female *Ae. aegypti* was 69–86% of the teneral proteins and 16–46% of the teneral lipids, suggesting that lipids derived from larval stages were the main component of energy reserves used by adult mosquitoes before the first blood meal. Consequently, transcripts of genes coding for FASNs were at higher levels during the first 24 h after pupation, as well as many genes from TG catabolism and the β-oxidation pathway (Wang et al. 2017). After pupation, adult mosquitoes have been seen to feed on different carbohydrate sources (Barredo and DeGennaro 2020). Plant nectar and other non-flower sugars are common nutrition sources for adult mosquitoes of both sexes, especially for males, which feed exclusively from plant nectar and require frequent sugar intake for survival. The energy obtained for carbohydrates is stored as glycogen and TG. While glucose, fructose, and sucrose are their preferred sugars, others than arabinose, lactose, and cellulose have proven to be toxic to *Ae. aegypti*. *An. gambiae* females without sugar availability take larger blood meals to achieve their fitness potential and complete egg maturation. *Ae. aegypti* and *An. gambiae* are anautogenous species therefore they must have blood meals in order to develop eggs. Thus, the expression of egg yolk precursor proteins required for egg maturation is initiated when mosquitoes take a blood meal. Despite this, sugar feeding before a blood meal is directly invested in egg production through the synthesis of egg lipids and proteins (Briegel et al. 2002; Zhou et al. 2004b). Feeding *Ae. aegypti* females at the beginning of the gonotrophic cycle with labeled glucose (D-[U-^{14}C]-glucose) showed that radioactivity was incorporated mostly in lipids (~ 50%), followed by glycogen (~ 35%), proteins (~ 8%), sugars (~ 5%), and amino acids (~ 2%). So, it is interesting to note that lipids (mostly TG) constituted the major energy reserves in adult mosquitoes, derived from glycolysis and subsequent conversion of pyruvate to acetyl-CoA (Zhou et al. 2004b). Feeding the insects with glucose differently labeled (D-[3,4-^{14}C]-glucose) resulted in about 24% of the radioactivity incorporated in lipids, which suggested to the authors that the pentose phosphate pathway is also important in mosquitoes for fatty acid synthesis and acyl glycerides production. From this 24% of labeled lipids, about 17% was reserved as maternal lipid stores, 4% transferred to the eggs, and 3% catabolized at the end of the first gonotrophic cycle (Zhou et al. 2004b). Briegel et al. (2002) studied the lipid metabolism during six gonotrophic cycles in female *Ae. aegypti* of two different sizes. Female mosquitoes transfer most of the lipid reserves obtained previously to blood ingestion (larval reserves and those formed by sugar feeding) to the ovaries (~ 80%). After a blood meal, about one third of the blood proteins are used to synthesize lipids. And after oviposition, there is a net synthesis of lipids, which come from sugar meals the insect takes before the next blood meal. Therefore, measuring the lipid content of females previous to blood meal, after blood meal, and after oviposition during several gonotrophic cycles shows that the first cycle depends almost exclusively on teneral lipids, and only after the second cycle sugar-derived and blood-derived lipids contributed to yolk formation. And although those results vary according to female size and age (for a more detailed explanation see Briegel et al. 2002), the evidence suggests that blood meals contribute a little to lipid eggs. Furthermore, females deprived of sugar meal but not blood meal were able to obtain only 21% of egg lipids from blood. Thus, blood meals seem to mostly contribute with proteins for the female and eggs, and trigger the physiological processes related to lipid metabolism and mobilization, and to the beginning of egg production. After each blood meal there is a stepwise sequential increase in transcript levels of lipid metabolism genes in the fat body. FASN transcripts increase 3–6 h after blood meal reaching the highest levels at 18 h, suggesting that blood meal induces an initial activation of lipid biosynthesis. After that, transcripts of genes coding for several lipases (including TG lipases) increase, suggesting the onset of TG degradation and the beginning of the moment where TG catabolism predominates over biosynthesis (Wang et al. 2017). Juvenile hormone promotes lipid biosynthesis while ecdysone TG catabolism and β-oxidation through regulation of the hepatocyte nuclear factor 4 (HNF4). Thus, HNF4 mediates lipid metabolism regulation in reproducing female *Ae. aegypti* mosquitoes (Wang et al. 2017). The ability to synthesize lipids predisposes an organism to function as a host to pathogens that have lost or lack this trait, and viruses rely exclusively on the mosquito and mammal host to fulfill their membrane and lipid requirements. In this sense, dengue virus induces significant rearrangements in the membrane architecture of infected *Ae. aegypti* cells. Virus infection in midgut and other cells has been observed to induce a reconfiguration of the mosquito phospholipidome; in particular, increased amounts of several PL are detected (Perera et al. 2012). An upregulation of phospholipase A2 accounted for the increased lysophospholipids amounts in dengue virus-infected mosquitoes (Perera et al. 2012). Additionally, a downregulation of AGPAT1, which is involved in PL biogenesis has been recently reported (Vial et al. 2021). The evidence suggests that, together with hijacking of lipids for its benefit, the virus triggers de PL remodeling in mosquito cells while actively inhibiting the de novo PL biosynthesis (Vial et al. 2021). Unexpectedly, lipid metabolism and body size derived from nutritional status have shown to affect the number of *Plasmodium* pathogens that develop into oocysts in *Anopheles* species. Small mosquitoes developed fewer oocysts than larger ones, but this is not merely related to the amount of blood ingested. Instead, competition between the insect and pathogens for lipid nutrients seems to be involved (Takken et al. 2013). An interesting aspect regarding the lipid metabolism in mosquitoes is how population density might affect lipid reserves. A recent study measured the TG amount in *Ae. aegypti* fourth-stage larvae exposed to three crowding conditions (Silva et al. 2021). TG content was about 10–20μg/larvae. Surprisingly, higher larvae population densities triggered a significant increase in TG content and this effect persisted in adult females from the more crowded condition. The authors also analyzed the TG mobilization from the fat body of adult females as a result of starvation and found that fat bodies of females from the more crowded condition released TG reserves more slowly. These findings led the authors to suggest that *Ae. aegypti* is able to react rapidly to a potential lack of nutrient availability by accumulating more lipid reserves, and also may tolerate better a longer starvation period by releasing slowly their lipid reserves. Despite these studies, the whole lipid profile of mosquitoes' larvae and its dynamic through development is currently unknown.

5.5 Tsetse Flies

The vectors of sleeping sickness or African trypanosomiasis are obligate hematophagous insects with a diet very rich in proteins and poor in lipids or carbohydrates. Thus, their metabolism is largely based on amino acid utilization. Together with the energy requirements, additional metabolic demands are related in both sexes to flight, post-emergence somatic growth, and in the female for the uterine nourishment of the larva, while in the male accessory gland fluid is accumulated only after blood meals. As it has been previously commented, the amino acid proline has a central role in provision of energy for flight. A high synthesis of TG from amino acids during the 24 h following ingestion of a blood meal is reported in the fat body of male *G. morsitans* (Langley 1977). After lipid reserves have been replenished there is a first lipid oxidation phase, in which proline reserves are built up. Then, a second lipid oxidation phase uses the lipid reserves as the sole input to produce energy, as the residual blood meal has been exhausted. Thus, tsetse flies are unusual in that direct oxidation of fatty acids to provide energy is not a prominent feature of their metabolism, as it is used for the incorporation of lipid carbon into the proline molecule (Bursell 1977; Langley 1977). In tsetse flies, the larvae undergo a viviparous way of development within the female reproductive tract. Langley and Bursell (1980) studied the contribution of the fat body to the product released by the milk gland. Administration of labeled C16 fatty acid showed that the fat body uses fatty acids to synthesize mostly TG and minor amounts of PL as lipid stores that will be later used to provide lipids for the milk gland. Interestingly, some radioactivity was also found in proline reflecting the involvement of lipid carbon in the resynthesis of proline from alanine to provide energy for flight. During early pregnancy (from day 2 to 5 of larval development), the proportion of radioactivity incorporated into TG and PL was greater than in proteins whereas in late pregnancy (from day 6 to 10) the pattern was reversed. Conversely, incorporation rates into the lipid fraction of the milk gland were low during early pregnancy but increased during late pregnancy. There was a massive transfer of lipid from the fat body to the milk gland during late pregnancy indicating that although both the fat body and milk gland are able to synthesize lipids and/or amino acids, the bulk of milk gland lipid is derived from the fat body. Thus, a major proportion of the lipid component of the milk is synthesized during early pregnancy by the fat body and subsequently transferred to the gland, whereas the bulk of the protein component is supplied more directly from the digestion of blood meals. The lipid content of the milk gland secretion comprises about 48% of the total content (52% was protein and amino acids) and includes mostly TG, PL, and cholesterol with minor amounts of HCs and CE (Cmelik et al. 1969). About 15 mg of lipids are transferred from the female to the larva each gonotrophic cycle (Cmelik et al. 1969; Denlinger and Ma 1974). Radioactive label of female lipid was later detected in larval fat body and integument (Langley and Bursell 1980) indicating the major organs involved in lipid metabolism in larvae. Larval lipid content reached about 3.5 mg comprising about 12% of the total larval composition; from this lipid amounts, 80% were lipids found in the gut (Langley 1977). During the first 5 days of pupariation (hardening of the larval cuticle) there was an abrupt fall of DG, some of which are converted to TG, while PL also declines in amount, which has been associated with a higher rate of histolysis. Between days 5 and 15 no lipid is consumed but thereafter, there is an increase in fatty acid oxidation related to the developing pharate adult. PL increases slightly from day 5 onwards, associated with the development of the fat body, flight muscle, and other adult organs. The noticeable fall of glycogen stores indicates that carbohydrates were the main fuel used to fulfill the energy requirements of pupariation and metamorphosis (Langley 1977).

5.6 Triatomines

Triatomines are mostly obligate hematophagous insects. Adults usually have large blood meals followed by extended periods of starvation, whereas for nymphs a single full engorgement is sufficient to trigger the molting to the next nymph stage or the metamorphosis to the adult stage. Following the de novo lipid biosynthesis in fifth-stage nymphs by injection of radiolabeled acetate showed that after a blood meal the highest percent of recovery of the injected radioactivity was ~35% in fat body, ~ 6% in hemolymph, and ~ 3.5% in the integument. Both the fat body and the integument accumulated the radioactivity mostly as TG (~ 54% and 66%, respectively); in the fat body TG were followed by DG (25%), PL (18.5%), and FFA (2.5%) whereas in the integument they were followed by FFA (17%), DG (10%), HCs (3.5%), and minor amounts of waxes and PL (Juárez and Brenner 1989). The different pattern of lipid label between the fat body and the integument seems to indicate their specific roles, the former involved in building energy reserves to fulfill the physiology requirements of the insect organs while the latter mostly in the formation of very long chain lipids or specific compounds (i.e., fatty alcohols) for the cuticle and other body organs. Radioactivity of hemolymph was distributed mostly in DG (~ 67%), PL (~ 21%), and HCs (11%) indicating its role in lipid transport and distribution (Juárez and Brenner 1989). Similar experiments in females showed a noticeable lipid labeling in the ovaries (79% TG, 8% PL, 7% FFA, 3% DG, and 3% HCs) indicating the contribution of both the fat body and the integument to the lipid composition of female reproductive organs (Juárez and Brenner 1989). Although most HCs of insect eggs are synthetized by the integument and subsequently transported to the ovaries by lipophorin, Juárez (1994a) showed for the first time in *T. infestans* that the eggs have active the complete set of enzymes necessary to synthetize some amount of their own straight and methyl-branched HCs. Radiolabeled FFA used to follow the lipid accumulation by fat body in *T. infestans* and *R. prolixus* showed that during the first 45 min there was a period of rapid and linear uptake of the label (about 30% of the total label), followed by a slower uptake that after more than 200 min did still not reach the maximum. In the first 30 min, the fat body incorporated the label in the FFA fraction but later the label was found in TG (~ 60%), PL (~ 20%), and DG (~ 8%) (Rimoldi et al. 1988; Pontes et al. 2008). Free fatty acid incorporation by the fat body rises abruptly from day 2 to day 4 after blood meal when it reaches the maximum and then starts to decrease to reach basal values by day 10 (Pontes et al. 2008). Triacylglycerol content of the fat body increased from about 57μg at day 2 after blood meal to 250–270μg from day 4 to day 13 (that is a five-fold increase), then decreased to about 30μg at day 20, which is the same amount that have unfed insects (Pontes et al. 2008). The very significant increases in FFA incorporation and TG content observed between day 2 and 4 after blood meal are coincident with the moment of higher digestion rates seen in the midgut and subsequent lipid transfer to the fat body mediated by lipophorin (see Sect. 4). Correspondingly, the constantly higher amounts of TG observed between day 4 and day 13 after blood meal reflect similar rates of lipid biosynthesis and their secretion to other organs such as those of the reproductive, excretory, and integumentary systems to provide them with the necessary energy fuel to carry out their physiological functions. Measures of lipid consumption in adults *T. infestans*, *Dipetalogaster maximus*, and *P. megistus* showed that after 30 days of starvation lipid storage of the fat body had fallen about 50% in the first two species, and about 70% in the latter (Canavoso et al. 1998). Regarding the interaction with *T. cruzi*, it has been reported that the parasite depends on active lipid endocytosis from the triatomine host to fulfill the growth requirements. For instance, *T. cruzi* is able to internalize cholesterol and stores it together with TG, PL, DG, and FFA in lipid droplets (Pereira et al. 2018). Although the parasite develops entirely within the digestive tract and never leaves this compartment, the infection alters several lipid metabolic features of the fat body. A downregulation of FASN, DGAT, and lipid storage droplet-associated TG lipase genes was observed in *T. cruzi*-infected *R. prolixus* males, which led to a significant decrease in the amount of acyl glycerides (MG, DG, and TG), CE, and PL (Sousa et al. 2021). It is not clear whether this may be a result of an insect health compromise resulting from parasite infection or may reflect an evolutionary strategy to diminish the insect lipid storage and therefore make lipids more available for the parasite to uptake them.

6 Lipid Metabolism and Pheromone Production

Chemical communication is probably the oldest and widespread form of communication in living organisms. Semiochemicals are the small organic compounds used to transmit intra and interspecific chemical messages (Francke and Schulz 1999; Wyatt 2013). In insects, these chemical cues allow them to evaluate their environment, to detect food sources, potential habitats, toxins, prey, predators, and pathogens, to kin recognition, and to attract the opposite sex for mating (Regnier 1971; Blomquist et al. 2018). Pheromones are a sub-group among semiochemicals, a chemical or blend of chemicals released by an organism that causes a specific behavioral or physiological reaction in one or more individuals of the same species. Therefore, pheromones play critical roles in intraspecific communication (Karlson and Luscher 1959; Nordlund and Lewis 1976). A wide range of compounds have been identified as insect pheromones, including HCs, epoxy HCs, alcohols, acetates, aldehydes, and ketones (Teal and Tumlinson 1987; Francke and Schulz 1999); its chemical structure and functions are very diverse. Pheromones can be classified on the basis of the interaction they mediate, such as alarm, aggregation, sex recognition, and many others that affect insect behavior or physiology (Regnier and Law 1968; Blomquist et al. 2018). Also, they can be short or long-range, or act on contact (Wyatt 2013) (Fig. 1).

6.1 Pheromones of Diptera

In insect vectors, several types of pheromones have been described. One of the first studies on this issue was performed in tsetse flies. In *G. morsitans* it was observed that a single HC component of the female cuticle can elicit copulatory responses from males. The compound was isolated and identified as a methyl-substituted alkane, the 15,19,23-trimethylheptatriacontane, which on the basis of observed fly behavior and its very low volatility was considered as a contact sex recognition pheromone (Langley et al. 1975; Carlson et al. 1978; Langley et al. 1987b). Subsequent studies in other members of the genus have shown the existence of different HCs with sexual pheromone function, in *Glossina austeni* the identified compound was the 15,19-dimethyltritriacontane (Huyton et al. 1980) and in *Glossina pallidipes* the 13,23-dimethylpentatriacontane (Carlson et al. 1984). The first evidence for an oviposition aggregation pheromone in an insect vector was in the mosquito genus *Culex* (Osgood 1971), and later the compound was identified by mass spectrometry and microderivatizations as erythro-6-acetoxy-5-hexadecanolide (Laurence and Pickett 1982). Behavioral observations showed that egg rafts attracted gravid females of *Culex tarsalis*, *C. pipiens molestus*, and *C. quinquefasciatus* to lay eggs at the same oviposition sites (McCall and Cameron 1995). In larval stages of *Ae. aegypti* the n-heneicosane, a C21 straight-chain HC, has been reported to act as an oviposition aggregation pheromone (Mendki et al. 2000). In a recent study it has been shown that n-heneicosane could also be involved in interspecific competition for oviposition sites, acting as a repellent to *Ae. albopictus*. Both species have similar behavior and ecological niches (González et al. 2014). In the genus *Anopheles*, it has been found that males release significantly higher amounts of 3-hydroxy-2-butanone, 6-methyl-5-hepten-2-one, octanal, nonanal, and decanal during swarming. These swarming odors are aggregation pheromones that attract individuals to the swarm enhancing mating success (Mozūraitis et al. 2020). More recently, it has been shown that *Anopheles* males release large amounts of the HC heptacosane during swarm formation, which enhances the interaction between males and courting females; the production of this compound seems to be guided by clock genes, light, and temperature (Wang et al. 2021). The sandfly *Lutzomyia longipalpis* is apparently a complex of sibling species (Lanzaro et al. 1993;

Uribe 1999) where females are attracted to a specific chemical signal released by males. Studies carried out in Brazil with vector populations from different localities suggest that the complex consists of four different species, and also that the species can be distinguished based on the terpenes released by males. The identified terpenes were an homoterpene (1S,3S,7R)-3-methyl-α-himachalene (Jacobina, state of Bahia), another homoterpene (S)-9-methylgermacrene-B (Lapinha Cave, state of Minas Gerais), and two diterpenes, cembrene 1 and cembrene 2 (Sobral, state of Ceará and Jaíbas, state of Minas Gerais) (Hamilton et al. 2004).

6.2 Pheromones of Hemiptera

Regarding triatomines, several compounds that mediate different behaviors have been discovered. For example, it has been reported in *T. infestans* that two cuticle fatty acids (octadecanoic and hexacosanoic acid) trigger aggregation and arrestment behavior (Lorenzo Figueiras et al. 2009), whereas female-specific fatty alcohols (eicosanol and docosanol) are involved in sexual recognition by males and in eliciting the copulatory behavior (Cocchiararo-Bastias et al. 2011). Some volatile compounds of feces (acetic acid, 3-methylbutanoic acid, hexanoic acid, acetamide, and 2,3-butanediol) together with a contact factor deposited on the substrate have shown to induce aggregation of conspecifics in *T. infestans*, *P. megistus,* and *Triatoma brasiliensis* (Mota et al. 2014). Also, volatile compounds released from the metasternal glands of *T. infestans*, *T. brasiliensis*, *Triatoma dimidiata,* and *R. prolixus* seem to promote male orientation to females, mating and male aggregation around mating pairs (Lorenzo et al. 2014).

6.3 Pheromone Metabolism

Pheromone biosynthetic pathways from a number of species have been determined during the last four decades, and this work is making great progress with the use of the powerful tools of molecular biology (Stanley-Samuelson et al. 1988; Blomquist et al. 2018). Most of these investigations were performed in non-vector insects, with an emphasis on sex and aggregation pheromones in insect models of the orders Lepidoptera, Coleoptera, Blattodea, and Diptera. The studies were mainly done on these species due to their economic significance, because they were easy to rear and produced relatively large amount of pheromones, and mainly because studies on pheromone biosynthesis evolved from research on biochemical systems related to the metabolism of fatty acids and HCs (Tillman et al. 1999; Wicker-Thomas et al. 2015). These studies have shown that most pheromones result from modifying metabolic products of the fatty acid and isoprenoid synthesis pathways. Insects require only a few specific enzymes to synthesize these compounds and with a few exceptions, they are synthesized de novo in pheromone gland tissue (Stanley-Samuelson et al. 1988; Tillman et al. 1999; Jurenka et al. 2017). For example, the sex pheromones of many Diptera are HCs derived from fatty acids. Its biosynthesis begins with the synthesis of long-chain fatty acid precursors (saturated, unsaturated, or methyl-branched), these fatty acids are elongated to very long-chain fatty acids, then they are converted to aldehyde and at the end of the metabolic pathway they are transformed into HCs by loss of the carbonyl group in the form of carbon dioxide (Wicker-Thomas et al. 2015). Other mechanisms to produce further modifications include the use of desaturases to introduce double bonds (Bjostad et al. 1987; Knipple and Roelofs 2003; Lassance et al. 2010), highly specific shortening of the fatty acid chain (Plettner et al. 1996; Plettner et al. 1998), or the elongation of polyunsaturated dietary fatty acids followed by loss of the carboxyl carbon atom produces HCs and HC-derived pheromones (Tillman et al. 1999; Millar 2010; Ginzel and Blomquist 2016). Once a specific chain-length pheromone intermediate is produced and has the proper double bonds and methyl branches, the carbonyl carbon atom is modified to form a functional group. Most components of oxygenated pheromones are acetate esters, alcohols, and aldehydes. The production of these components requires in the case of alcohols the reduction of a fatty acyl precursor that is catalyzed by a fatty acyl reductase (Moto et al. 2003; Lassance et al. 2010), for the formation of aldehydes it requires the oxidation of primary alcohols by action of an oxidase (Teal and Tumlinson 1987; Fang et al. 1995) and in acetate esters it is used acetyl-CoA:fatty alcohol acetyltransferases (Morse and Meighen 1987; Jurenka and Roelofs 1989). A detailed overview of the most recent investigations can be found in Blomquist et al. 2018.

In the last decade, based on the aforementioned studies, progress has been made in the knowledge of the routes of pheromone synthesis in insect vectors. For instance, a transcriptomic analysis carried out in one member of the *L. longipalpis* complex allowed to determine the genes expressed in the pheromone gland and flanking tissues and consequently, generate the first sequence catalog for a specialized tissue where the pheromone (S)-9-methylgermacrene-B is produced (González-Caballero et al. 2013).

7 Lipids and Insecticide Resistance

Insecticides are the most important components to control insect vectors of diseases worldwide. Among the various insecticides used for this purpose, pyrethroids have been widely used for many years because they are low toxic, cheap, and long-lasting compared to others such as organophosphate and organochlorine compounds. However, the worldwide development of insecticide resistance, especially to pyrethroids, has become a great concern for the vector-borne disease control programs. There are three major mechanisms involved in insecticide resistance: i) decreased sensitivity of the target site of the insecticide, ii) increased insecticide detoxification, and iii) decreased cuticle penetration (Liu 2012). Contact insecticides are lipophilic compounds that must penetrate the cuticle to finally reach the target tissues. The biosynthesis and metabolism of cuticle lipids have been demonstrated to be related to the latter mechanism. Two studies in the 1990s were the first to show the relationship between cuticle lipids and a diminished insecticide penetration. A multiresistant strain of the diamondback moth (*Plutella xylostella*) showed a significantly slower penetration of the pyrethroid fenvalerate, compared to a susceptible strain. Conversely, the insecticide penetration was enhanced after solvent-washing the cuticle lipids in both the resistant and susceptible strains (Motoyama et al. 1990). In samples of the Chagas disease vector *T. infestans* where the formation of very long chain lipids (HCs, TG, fatty alcohols, very-long-chain fatty acids) had been inhibited, topical application of the organophosphate fenitrothion increased insect mortality by 40%. A raise of 30% in mortality was also observed by applying the insecticide after a solvent-based removal of the cuticle lipids (Juárez 1994b). After these early studies, physiological adaptive changes in the amount of CHC related to insecticide resistance were detected for the first time in *T. infestans*, with an increase of 56% in the amount of CHC in pyrethroid-resistant samples compared to susceptible ones (Pedrini et al. 2009) (Fig. 2). This was correlated to a significant decrease (~50%) in the penetration of deltamethrin through the cuticle of resistant specimens (Juárez et al. 2010). A 29% increase in the CHC content and diminished pyrethroid penetration were later described in resistant samples of the malaria vector *An. gambiae* (Balabanidou et al. 2016). Quantitative real-time PCR analyses showed that a cytochrome P450 4G gene, which codes for the enzyme involved in the last step of HC formation, was overexpressed in both *An. gambiae* (Balabanidou et al. 2016) and *T. infestans* (Calderón-Fernández et al. 2017). Furthermore, microarray and transcriptomic analyses have evidenced that many genes from the biosynthesis pathway of CHC and other cuticle lipids such as acyl-CoA elongases and acyl-CoA reductases were differentially expressed in pyrethroid-resistant specimens of *T. infestans* and *An. coluzzii* (Toé et al. 2015; Calderón-Fernández et al. 2017), although functional analyses to elucidate their contribution to the cuticle factor of insecticide resistance are still pending. Similarly, a transcriptomic analysis evidenced that about 200 genes from lipid metabolism (lipid transport, acyl glycerides and phospholipid metabolism,

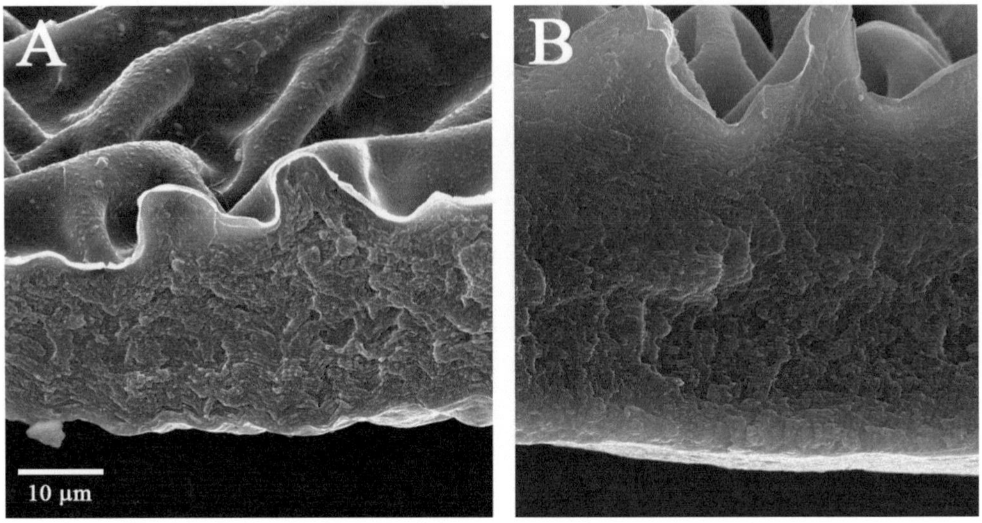

Fig. 2 Scanning electron microscopy images of the cuticle of pyrethroid-susceptible (**a**) and pyrethroid-resistant (**b**) samples of *T. infestans*. Images show transversal cuts from the second abdominal tergite. Insecticide-resistant samples showed a 56% increase in the amount of cuticle hydrocarbons and a thicker cuticle compared to susceptible insects together with a diminished deltamethrin penetration (Pedrini et al. 2009; Juárez et al. 2010). The CYP4G106 gene, involved in the last step of hydrocarbon biosynthesis, was shown to be significantly overexpressed in resistant insects (Calderón-Fernández et al. 2017)

etc.) were differentially expressed in mosquitoes of the *C. pipiens* complex related to resistance to pyrethroid, propoxur, and dichlorvos insecticides (Li et al. 2016) although there were no further studies exploring the significance of this finding. Another unanswered question is how fast insect vectors can increase the production of protective cuticle lipids after being exposed to insecticides and the extent of this induction. Studies in the Asian tiger mosquito *Ae. albopictus* have shown that this dengue vector can enter diapause in response to unfavorable climatic conditions, and that this dormancy state implies an increase in the total lipid content. Lipid accumulation is carried out through downregulation of genes involved in lipid catabolism (lipases, acetyl-CoA synthetase, acyl-CoA dehydrogenases, etc.) simultaneous to upregulation of genes involved in lipid synthesis and storage (lipid storage droplet protein 2, Δ (9)-desaturase, fatty acyl-CoA elongase); the latter involved in the accumulation of TG, unsaturated FFA and CHC needed to resist desiccation (Reynolds et al. 2012). Thus, much work is still needed to fully understand the contribution of lipids to the insecticide resistance of insect vectors of diseases, which is an emerging worldwide concern.

8 Conclusions and Future Directions

Many features of lipid metabolism and other metabolic pathways in insect vectors of diseases are currently unknown. And these insects have certain common features (such as hematophagy), most of their physiological traits regarding development, reproduction, etc. are very different between them making it difficult to generalize. Many aspects of the lipid metabolism of these insects related to blood digestion, specific use of the lipid reserves, role of lipids in insecticide resistance and as chemical signals, and especially how the pathogens affect the lipid metabolism and to what extent, are still mostly misunderstood. We think we are not wrong if we say that reading about the topics of this chapter leaves us with more questions than answers. And that reflects that there is so much research to do on this topic. And much of this research should focus on understanding the function and regulation of genes, pathways, and components of lipid metabolism not only as a contribution to basic science but also related to using them as a way of vector control and elimination.

References

Abdul Rahim NA, Othman M, Sabri M, Stanley DW (2018) A midgut digestive phospholipase A2 in larval mosquitoes, *Aedes albopictus* and *Culex quinquefasciatus*. Enzyme Res 2018:9703413. https://doi.org/10.1155/2018/9703413

Alabaster A, Isoe J, Zhou G, Lee A, Murphy A, Day WA, Miesfeld RL (2011) Deficiencies in acetyl-CoA carboxylase and fatty acid synthase 1 differentially affect eggshell formation and blood meal digestion in *Aedes aegypti*. Insect Biochem Mol Biol 41:946–955. https://doi.org/10.1016/j.ibmb.2011.09.004

Almeida-Oliveira F, Santos-Araujo S, Carvalho-Kelly LF, Macedo-Silva A, Meyer-Fernandes JR, Gondim KC, Majerowicz D (2023) ATP synthase affects lipid metabolism in the kissing bug *Rhodnius prolixus* beyond its role in energy metabolism. Insect Biochem Mol Biol 158:103956. https://doi.org/10.1016/j.ibmb.2023.103956

Alves CR, Albuquerque-Cunha JM, Mello CB, Garcia ES, Nogueira NF, Bourguingnon SC, de Souza W, Azambuja P, Gonzalez MS (2007) *Trypanosoma cruzi*: attachment to perimicrovillar membrane glycoproteins of *Rhodnius prolixus*. Exp Parasitol 116:44–52. https://doi.org/10.1016/j.exppara.2006.11.012

Alves-Bezerra M, Gondim KC (2012) Triacylglycerol biosynthesis occurs via the glycerol-3-phosphate pathway in the insect *Rhodnius prolixus*. Biochim Biophys Acta 1821:1462–1471. https://doi.org/10.1016/j.bbalip.2012.08.002

Alves-Bezerra M, Klett EL, De Paula IF, Ramos IB, Coleman RA, Gondim KC (2016) Long-chain acyl-CoA synthetase 2 knockdown leads to decreased fatty acid oxidation in fat body and reduced reproductive capacity in the insect *Rhodnius prolixus*. Biochim Biophys Acta 1861:650–662. https://doi.org/10.1016/j.bbalip.2016.04.007

Arêdes DS, De Paula IF, Santos-Araujo S, Gondim KC (2022) Silencing of mitochondrial trifunctional protein a subunit (HADHA) increases lipid stores, and reduces oviposition and flight capacity in the vector insect *Rhodnius prolixus*. Front Insect Sci 2. https://doi.org/10.3389/finsc.2022.885172

Atella GC, Gondim C, Masuda H (1995) Loading of lipophorin particles with phospholipids at the midgut of *Rhodnius prolixus*. Arch Insect Biochem Physiol 30: 337–350. https://doi.org/10.1002/arch.940300404

Atella GC, Silva-Neto MAC, Golodne DM, Arefin S, Shahabuddin M (2006) *Anopheles gambiae* lipophorin: characterization and role in lipid transport to developing oocyte. Insect Biochem Mol Biol 36: 375–386. https://doi.org/10.1016/j.ibmb.2006.01.019

Atella GC, Bittencourt-Cunha PR, Nunes RD, Shahabuddin M, Silva-Neto MAC (2009) The major insect lipoprotein is a lipid source to mosquito stages of malaria parasite. Acta Trop 109:159–162. https://doi.org/10.1016/j.actatropica.2008.10.004

Balabanidou V, Kampouraki A, MacLean M, Blomquist GJ, Tittiger C, Juárez MP, Mijailovsky SJ, Chalepakis G, Anthousi A, Lynd A, Antoine S, Hemingway J, Ranson H, Lycett GJ, Vontas J (2016) Cytochrome P450 associated with insecticide resistance catalyzes cuticular hydrocarbon production in *Anopheles gambiae*. Proc Natl Acad Sci USA 113: 9268–9273. https://doi.org/10.1073/pnas.1608295113

Balabanidou V, Kefi M, Aivaliotis M, Koidou V, Girotti JR, Mijailovsky SJ, Juárez MP, Papadogiorgaki E, Chalepakis G, Kampouraki A, Nikolaou C, Ranson H, Vontas J (2019) Mosquitoes cloak their legs to resist insecticides. Proc Biol Sci 286: 20191091. https://doi.org/10.1098/rspb.2019.1091

Barredo E, DeGennaro M (2020) Not just from blood: mosquito nutrient acquisition from nectar sources. Trends Parasitol 36:473–484. https://doi.org/10.1016/j.pt.2020.02.003

Benoit JB, Yang G, Krause TB, Patrick KR, Aksoy S, Attardo GM (2011) Lipophorin acts as a shuttle of lipids to the milk gland during tsetse fly pregnancy. J Insect Physiol 57:1553–1561. https://doi.org/10.1016/j.jinsphys.2011.08.009

Bittencourt-Cunha PR, Silva-Cardoso L, de Oliveira GA, da Silva JR, da Silveira AB, Kluck GEG, Souza-Lima M, Gondim KC, Dansa-Petretsky M, Silva CP, Masuda H, da Silva Neto MAC, Atella GC (2013) Perimicrovillar membrane assembly: the fate of phospholipids synthesized by the midgut of *Rhodnius prolixus*. Mem Inst Oswaldo Cruz 108:494–500. https://doi.org/10.1590/S0074-0276108042013016

Bjostad LB, Wolf WA, Roelofs WL (1987) Pheromone biosynthesis in lepidopterans: desaturation and chain shortening. In: Blomquist GJ (ed) Prestwich, G.D. Elsevier, Pheromone Biochemistry, pp 77–120. https://doi.org/10.1016/B978-0-12-564485-3.50008-7

Blomquist GJ, Tittiger C, Jurenka R (2018) Cuticular hydrocarbons and pheromones of arthropods. In: Wilkes H (ed) Hydrocarbons, oils and lipids: diversity, origin, chemistry and fate. Springer International Publishing, Cham, pp 1–32. https://doi.org/10.1007/978-3-319-54529-5_11-1

Botto-Mahan C (2009) *Trypanosoma cruzi* induces life-history trait changes in the wild kissing bug *Mepraia spinolai*: implications for parasite transmission. Vector Borne Zoonotic Dis 9:505–510. https://doi.org/10.1089/vbz.2008.0003

Briegel H, Hefti M, DiMarco E (2002) Lipid metabolism during sequential gonotrophic cycles in large and small female *Aedes aegypti*. J Insect Physiol 48:547–554. https://doi.org/10.1016/s0022-1910(02)00072-0

Bursell E (1977) Synthesis of proline by fat body of the tsetse fly (*Glossina morsitans*): metabolic pathways. Insect Biochem 7:427–434. https://doi.org/10.1016/S0020-1790(77)90068-3

Calderón-Fernández GM, Moriconi DE, Dulbecco AB, Juárez MP (2017) Transcriptome analysis of the *Triatoma infestans* (Hemiptera: Reduviidae) integument. J Med Entomol 54:1531–1542. https://doi.org/10.1093/jme/tjx151

Canavoso LE, Bertello LE, de Lederkremer RM, Rubiolo ER (1998) Effect of fasting on the composition of the fat body lipid of *Dipetalogaster maximus*, *Triatoma infestans* and *Panstrongylus megistus* (Hemiptera: Reduviidae). J Comp Physiol B 168:549–554. https://doi.org/10.1007/s003600050176

Canavoso LE, Jouni ZE, Karnas KJ, Pennington JE, Wells MA (2001) Fat metabolism in insects. Annu Rev Nutr 21:23–46. https://doi.org/10.1146/annurev.nutr.21.1.23

Canavoso LE, Stariolo R, Rubiolo ER (2003) Flight metabolism in *Panstrongylus megistus* (Hemiptera: Reduviidae): the role of carbohydrates and lipids. Mem Inst Oswaldo Cruz 98:909–914. https://doi.org/10.1590/s0074-02762003000700009

Canavoso LE, Frede S, Rubiolo ER (2004) Metabolic pathways for dietary lipids in the midgut of hematophagous *Panstrongylus megistus* (Hemiptera: Reduviidae). Insect Biochem Mol Biol 34:845–854. https://doi.org/10.1016/j.ibmb.2004.05.008

Carlson DA, Langley PA, Huyton P (1978) Sex pheromone of the tsetse fly: isolation, identification, and synthesis of contact aphrodisiacs. Science 201:750–753. https://doi.org/10.1126/science.675256

Carlson DA, Nelson DR, Langley PA, Coates TW, Davis TL, Leegwater-Van Der Linden ME (1984) Contact sex pheromone in the tsetse fly *Glossina pallidipes* (Austen) identification and synthesis. J Chem Ecol 10:429–450. https://doi.org/10.1007/BF00988090

Cheon H-M, Shin SW, Bian G, Park J-H, Raikhel AS (2006) Regulation of lipid metabolism genes, lipid carrier protein lipophorin, and its receptor during immune challenge in the mosquito *Aedes aegypti*. J Biol Chem 281:8426–8435. https://doi.org/10.1074/jbc.M510957200

Chotiwan N, Andre BG, Sanchez-Vargas I, Islam MN, Grabowski JM, Hopf-Jannasch A, Gough E, Nakayasu E, Blair CD, Belisle JT, Hill CA, Kuhn RJ, Perera R (2018) Dynamic remodeling of lipids coincides with dengue virus replication in the midgut of *Aedes aegypti* mosquitoes. PLoS Pathog 14:e1006853. https://doi.org/10.1371/journal.ppat.1006853

Claudio-Piedras F, Recio-Tótoro B, Cime-Castillo J, Condé R, Maffei M, Lanz-Mendoza H (2021) Dietary and *plasmodium* challenge effects on the cuticular hydrocarbon profile of *Anopheles albimanus*. Sci Rep 11:11258. https://doi.org/10.1038/s41598-021-90673-x

Cmelik SHW, Bursell E, Slack E (1969) Composition of the gut contents of third-instar tsetse larvae (*Glossina morsitans* Westwood). Comp Biochem Physiol 29:447–453. https://doi.org/10.1016/0010-406X(69)91764-2

Cocchiararo-Bastias LM, Mijailovsky SJ, Calderon-Fernández GM, Figueiras ANL, Juárez MP (2011) Epicuticle lipids mediate mate recognition in *Triatoma infestans*. J Chem Ecol 37:246–252. https://doi.org/10.1007/s10886-011-9927-2

Coelho HS, Atella GC, Moreira MF, Gondim KC, Masuda H (1997) Lipophorin density variation during oogenesis on *Rhodnius prolixus*. Arch Insect Biochem Physiol 35:301–313. https://doi.org/10.1002/(SICI)1520-6327(199705)35:3<301::AID-ARCH4>3.0.CO;2-W

Denlinger DL, Ma WC (1974) Dynamics of the pregnancy cycle in the tsetse *Glossina morsitans*. J Insect Physiol 20:1015–1026. https://doi.org/10.1016/0022-1910(74)90143-7

Dinglasan RR, Alaganan A, Ghosh AK, Saito A, van Kuppevelt TH, Jacobs-Lorena M (2007) *Plasmodium falciparum* ookinetes require mosquito midgut chondroitin sulfate proteoglycans for cell invasion. Proc Natl Acad Sci USA 104:15882–15887. https://doi.org/10.1073/pnas.0706340104

Dulbecco AB, Mijailovsky SJ, Girotti JR, Juárez MP (2015) Deltamethrin binding to *Triatoma infestans* (Hemiptera: Reduviidae) lipoproteins. Analysis by solvent Bar microextraction coupled to gas chromatography. J Med Entomol 52:1254–1259. https://doi.org/10.1093/jme/tjv141

Dulbecco AB, Moriconi DE, Lynn S, McCarthy A, Juárez MP, Girotti JR, Calderón-Fernández GM (2020) Deciphering the role of *Rhodnius prolixus* CYP4G genes in straight and methyl-branched hydrocarbon formation and in desiccation tolerance. Insect Mol Biol 29:431–443. https://doi.org/10.1111/imb.12653

Entringer PF, Majerowicz D, Gondim KC (2021) The fate of dietary cholesterol in the kissing bug *Rhodnius prolixus*. Front Physiol 12. https://doi.org/10.3389/fphys.2021.654565

Fang N, Teal PEA, Tumlinson JH (1995) Characterization of oxidase(s) associated with the sex pheromone gland in *Manduca sexta* (L.) females. Arch Insect Biochem Physiol 29:243–257. https://doi.org/10.1002/arch.940290303

Fellet MR, Lorenzo MG, Elliot SL, Carrasco D, Guarneri AA (2014) Effects of infection by *Trypanosoma cruzi* and *Trypanosoma rangeli* on the reproductive performance of the vector *Rhodnius prolixus*. PLoS One 9:e105255. https://doi.org/10.1371/journal.pone.0105255

Ferdous Z, Fuchs S, Behrends V, Trasanidis N, Waterhouse RM, Vlachou D, Christophides GK (2021) *Anopheles coluzzii* stearoyl-CoA desaturase is essential for adult female survival and reproduction upon blood feeding. PLoS Pathog 17:e1009486. https://doi.org/10.1371/journal.ppat.1009486

Ferreira RC, Kessler RL, Lorenzo MG, Paim RMM, Ferreira LDL, Probst CM, Alves-Silva J, Guarneri AA (2016) Colonization of *Rhodnius prolixus* gut by *Trypanosoma cruzi* involves an extensive parasite killing. Parasitology 143:434–443. https://doi.org/10.1017/S0031182015001857

Fichera LE, Brenner RR (1982) Biochemistry of the developmental cycle of *Triatoma infestans* (Vinchuca). VI. Identification and lipid composition of hemolymph lipoproteins of adult males. Acta Physiol Lat Am 32: 21–29

Folly E, Cunha e Silva NL, Lopes AHCS, Silva-Neto MAC, Atella GC (2003) *Trypanosoma rangeli* uptakes the main lipoprotein from the hemolymph of its invertebrate host. Biochem Biophys Res Commun 310:555–561. https://doi.org/10.1016/j.bbrc.2003.09.038

Francke W, Schulz S (1999) Pheromones. In: Barton D, Nakanishi K, Meth-Cohn O (eds) Comprehensive natural products chemistry. Elsevier, Oxford, pp 197–261. https://doi.org/10.1016/B978-0-08-091283-7.00052-7

Fruttero LL, Demartini DR, Rubiolo ER, Carlini CR, Canavoso LE (2014) β-Chain of ATP synthase as a lipophorin binding protein and its role in lipid transfer in the midgut of *Panstrongylus megistus* (Hemiptera: Reduviidae). Insect Biochem Mol Biol 52:1–12. https://doi.org/10.1016/j.ibmb.2014.06.002

Fruttero LL, Leyria J, Ramos FO, Stariolo R, Settembrini BP, Canavoso LE (2017) The process of lipid storage in insect oocytes: the involvement of β-chain of ATP synthase in lipophorin-mediated lipid transfer in the chagas' disease vector *Panstrongylus megistus* (Hemiptera: Reduviidae). J Insect Physiol 96:82–92. https://doi.org/10.1016/j.jinsphys.2016.10.014

Fruttero LL, Leyria J, Moyetta NR, Ramos FO, Settembrini BP, Canavoso LE (2019) The fat body of the hematophagous insect, *Panstrongylus megistus* (Hemiptera: Reduviidae): histological features and participation of the β-chain of ATP synthase in the Lipophorin-mediated lipid transfer. J Insect Sci 19(4). https://doi.org/10.1093/jisesa/iez078

Garcia ES, Castro DP, Figueiredo MB, Azambuja P (2012) Parasite-mediated interactions within the insect vector: *Trypanosoma rangeli* strategies. Parasit Vectors 5:105. https://doi.org/10.1186/1756-3305-5-105

Geering K, Freyvogel TA (1975) Lipase activity and stimulation mechanism of esterases in the midgut of female *Aedes aegypti*. J Insect Physiol 21:1251–1256. https://doi.org/10.1016/0022-1910(75)90095-5

Ginzel MD, Blomquist GJ (2016) Insect hydrocarbons: biochemistry and chemical ecology. In: Cohen E, Moussian B (eds) Extracellular composite matrices in arthropods. Springer International Publishing, Cham, pp 221–252. https://doi.org/10.1007/978-3-319-40740-1_7

Goldring JP, Read JS (1994) Insect acetyl-CoA carboxylase: enzyme activity during adult development and after feeding in the tsetse fly, *Glossina morsitans*. Comp Biochem Physiol Biochem Mol Biol 108:27–33. https://doi.org/10.1016/0305-0491(94)90160-0

Golodne DM, Van Heusden MC, Gondim KC, Masuda H, Atella GC (2001) Purification and characterization of a lipid transfer particle in *Rhodnius prolixus*: phospholipid transfer. Insect Biochem Mol Biol 31:563–571. https://doi.org/10.1016/s0965-1748(00)00161-2

Gondim KC, Oliveira PL, Masuda H (1989) Lipophorin and oögenesis in *Rhodnius prolixus*: transfer of phospholipids. J Insect Physiol 35:19–27. https://doi.org/10.1016/0022-1910(89)90032-2

González MS, Brenner RR (1999) Fatty acid delta9-desaturation in the *Triatoma infestans* fat body: response to food and trehalose administrations. Lipids 34:1199–1205. https://doi.org/10.1007/s11745-999-0472-7

González PV, González Audino PA, Masuh HM (2014) Electrophysiological and behavioural response of *Aedes albopictus* to n -heinecosane, an ovipositional pheromone of *Aedes aegypti*. Entomol Exp Appl 151: 191–197. https://doi.org/10.1111/eea.12184

González-Caballero N, Valenzuela JG, Ribeiro JMC, Cuervo P, Brazil RP (2013) Transcriptome exploration of the sex pheromone gland of *Lutzomyia longipalpis* (Diptera: Psychodidae: Phlebotominae). Parasit Vectors 6:56. https://doi.org/10.1186/1756-3305-6-56

Grigoraki L, Grau-Bové X, Carrington Yates H, Lycett GJ, Ranson H (2020) Isolation and transcriptomic analysis of *Anopheles gambiae* oenocytes enables the delineation of hydrocarbon biosynthesis. elife 9. https://doi.org/10.7554/eLife.58019

Grillo LAM, Majerowicz D, Gondim KC (2007) Lipid metabolism in *Rhodnius prolixus* (Hemiptera: Reduviidae): role of a midgut triacylglycerol-lipase. Insect Biochem Mol Biol 37:579–588. https://doi.org/10.1016/j.ibmb.2007.03.002

Gutierrez E, Wiggins D, Fielding B, Gould AP (2007) Specialized hepatocyte-like cells regulate *drosophila* lipid metabolism. Nature 445:275–280. https://doi.org/10.1038/nature05382

Hall AR, Blakeman JT, Eissa AM, Chapman P, Morales-García AL, Stennett L, Martin O, Giraud E, Dockrell DH, Cameron NR, Wiese M, Yakob L, Rogers ME, Geoghegan M (2020) Glycan–glycan interactions determine *Leishmania* attachment to the midgut of permissive sand fly vectors. Chem Sci 11:10973–10983. https://doi.org/10.1039/D0SC03298K

Hamilton JGC, Brazil RP, Maingon R (2004) A fourth chemotype of *Lutzomyia longipalpis* (Diptera: Psychodidae) from Jaibas, Minas Gerais State, Brazil. J Med Entomol 41:1021–1026. https://doi.org/10.1603/0022-2585-41.6.1021

Huyton PM, Langley PA, Carlson DA, Coates TW (1980) The role of sex pheromones in initiation of copulatory behaviour by male tsetse flies, *Glossina morsitans morsitans*. Physiol Entomol 5:243–252. https://doi.org/10.1111/j.1365-3032.1980.tb00232.x

Jahan N, Docherty PT, Billingsley PF, Hurd H (1999) Blood digestion in the mosquito, *Anopheles stephensi*: the effects of *plasmodium yoelii nigeriensis* on midgut enzyme activities. Parasitology 119:535–541. https://doi.org/10.1017/s0031182099005090

Juárez P (1994a) Hydrocarbon biosynthesis in *Triatoma infestans* eggs. Arch Insect Biochem Physiol 25:193–206. https://doi.org/10.1002/arch.940250303

Juárez P (1994b) Inhibition of cuticular lipid synthesis and its effect on insect survival. Arch Insect Biochem Physiol 25:177–191. https://doi.org/10.1002/arch.940250302

Juárez P, Brenner RR (1989) Fatty acid biosynthesis in the integument tissue of *Triatoma infestans*. Comp Biochem Physiol B 93:763–772. https://doi.org/10.1016/0305-0491(89)90043-6

Juárez MP, Calderón Fernández G (2007) Cuticular hydrocarbons of triatomines. Comp Biochem Physiol A Mol Integr Physiol 147:711–730. https://doi.org/10.1016/j.cbpa.2006.08.031

Juarez MP, Ayala S, Brenner RR (1996) Methyl-branched fatty acid biosynthesis in *Triatoma infestans*. Insect Biochem Mol Biol 26:599–605. https://doi.org/10.1016/S0965-1748(96)00021-5

Juárez MP, Pedrini N, Girotti JR, Mijailovsky SJ (2010) Pyrethroid resistance in Chagas disease vectors: the case of *Triatoma infestans* cuticle. Resist Pest Manag Newsl 19:59–61

Jurenka RA, Roelofs WL (1989) Characterization of the acetyltransferase used in pheromone biosynthesis in moths: specificity for the Z isomer in tortricidae. Insect Biochem 19:639–644. https://doi.org/10.1016/0020-1790(89)90098-X

Jurenka RA, Blomquist GJ, Schal C, Tittiger C (2017) Biochemistry and molecular biology of pheromone production. In: Reference module in life sciences. Elsevier. https://doi.org/10.1016/B978-0-12-809633-8.04037-1

Karlson P, Luscher M (1959) "Pheromones": a new term for a class of biologically active substances. Nature 183:55–56. https://doi.org/10.1038/183055a0

Killick-Kendrick R (1990) The life-cycle of *Leishmania* in the sandfly with special reference to the form infective to the vertebrate host. Ann Parasitol Hum Comp 65(Suppl 1):37–42. https://doi.org/10.1051/parasite/1990651037

Kleffmann T, Schmidt J, Schaub GA (1998) Attachment of *Trypanosoma cruzi* epimastigotes to hydrophobic substrates and use of this property to separate stages and promote metacyclogenesis. J Eukaryot Microbiol 45:548–555. https://doi.org/10.1111/j.1550-7408.1998.tb05115.x

Knipple DC, Roelofs WL (2003) Molecular biological investigations of pheromone desaturases. In: Vogt R (ed) Blomquist, G.J. Elsevier, Insect Pheromone Biochemistry and Molecular Biology, pp 81–106. https://doi.org/10.1016/B978-012107151-6/50006-2

Koh C, Islam MN, Ye YH, Chotiwan N, Graham B, Belisle JT, Kouremenos KA, Dayalan S, Tull DL, Klatt S, Perera R, McGraw EA (2020) Dengue virus dominates lipid metabolism modulations in *Wolbachia*-coinfected *Aedes aegypti*. Commun Biol 3:518. https://doi.org/10.1038/s42003-020-01254-z

Kühnlein RP (2012) Thematic review series: lipid droplet synthesis and metabolism: from yeast to man. Lipid droplet-based storage fat metabolism in *Drosophila*. J Lipid Res 53:1430–1436. https://doi.org/10.1194/jlr.R024299

Kumar BA, Paily KP (2011) Up-regulation of lipophorin (Lp) and lipophorin receptor (LpR) gene in the mosquito, *Culex quinquefasciatus* (Diptera: Culicidae), infected with the filarial parasite, *Wuchereria bancrofti* (Spirurida: Onchocercidae). Parasitol Res 108:377–381. https://doi.org/10.1007/s00436-010-2075-8

Langley PA (1977) Physiology of tsetse flies (*Glossina* spp.) (Diptera: Glossinidae): a review. Bull Entomol Res 67:523–574. https://doi.org/10.1017/S0007485300006933

Langley PA, Bursell E (1980) Role of fat body and uterine gland in milk synthesis by adult female *Glossina morsitans*. Insect Biochem 10:11–17. https://doi.org/10.1016/0020-1790(80)90033-5

Langley PA, Pimley RW, Carlson DA (1975) Sex recognition pheromone in tsetse fly *Glossina morsitans*. Nature 254:51–53. https://doi.org/10.1038/254051a0

Langley PA, Bursell E, Kabayo J, Pimley RW, Trewern MA, Marshall J (1981) Haemolymph lipid transport from fat body to uterine gland in pregnant females of *Glossina morsitans*. Insect Biochem 11:225–231. https://doi.org/10.1016/0020-1790(81)90100-1

Langley PA, Ogwal LM, Felton T, Stafford K (1987a) Lipid digestion in the tsetse fly, *Glossina morsitans*. J Insect Physiol 33:981–986. https://doi.org/10.1016/0022-1910(87)90011-4

Langley PA, Huyton PM, Carlson DA (1987b) Sex pheromone perception by males of the tsetse fly, *Glossina morsitans morsitans*. Physiol Entomol 12:425–433. https://doi.org/10.1111/j.1365-3032.1987.tb00769.x

Lanzaro GC, Ostrovska K, Herrero MV, Lawyer PG, Warburg A (1993) *Lutzomyia longipalpis* is a species complex: genetic divergence and interspecific hybrid sterility among three populations. Am J Trop Med Hyg 48:839–847. https://doi.org/10.4269/ajtmh.1993.48.839

Lassance J-M, Groot AT, Liénard MA, Antony B, Borgwardt C, Andersson F, Hedenström E, Heckel DG, Löfstedt C (2010) Allelic variation in a fatty-acyl reductase gene causes divergence in moth sex pheromones. Nature 466:486–489. https://doi.org/10.1038/nature09058

Laurence BR, Pickett JA (1982) Erythro-6-Acetoxy-5-hexadecanolide, the major component of a mosquito oviposition attractant pheromone. J Chem Soc Chem Commun 59. https://doi.org/10.1039/c39820000059

Lehane MJ (1977) Transcellular absorption of lipids in the midgut of the stablefly, *Stomoxys calcitrans*. J Insect Physiol 23:945–954. https://doi.org/10.1016/0022-1910(77)90121-4

Li C, Guo X, Zhang Y, Dong Y, Xing D, Yan T, Wang G, Zhang H, Zhao T (2016) Identification of genes involved in pyrethroid-, propoxur-, and dichlorvos-insecticides resistance in the mosquitoes, *Culex pipiens* complex (Diptera: Culicidae). Acta Trop 157:84–95. https://doi.org/10.1016/j.actatropica.2016.01.019

Lima-Neiva V, Gonçalves TCM, Bastos LS, Gumiel M, Correia NC, Silva CC, Almeida CE, Costa J (2017) Biology of *Triatoma sherlocki* (Hemiptera: Reduviidae) under laboratory conditions: biological cycle and resistance to starvation. J Med Entomol 54:831–836. https://doi.org/10.1093/jme/tjw249

Liu N (2012) Pyrethroid resistance in insects: genes, mechanisms, and regulation. In Perveen F (ed) Insecticides–advances in integrated pest management. InTech, pp 457–468. https://doi.org/10.5772/28737

Lockey KH (1988) Lipids of the insect cuticle: origin, composition and function. Comp Biochem Physiol Part B Comp Biochem 89:595–645. https://doi.org/10.1016/0305-0491(88)90305-7

Lorenzo Figueiras AN, Girotti JR, Mijailovsky SJ, Juárez MP (2009) Epicuticular lipids induce aggregation in Chagas disease vectors. Parasit Vectors 2:8. https://doi.org/10.1186/1756-3305-2-8

Lorenzo MG, Vidal DM, Zarbin PHG (2014) Control of neglected disease insect vectors: future prospects for the use of tools based on behavior manipulation-interference. J Braz Chem Soc 25:1799–1809. https://doi.org/10.5935/0103-5053.20140202

Majerowicz D, Cezimbra MP, Alves-Bezerra M, Entringer PF, Atella GC, Sola-Penna M, Meyer-Fernandes JR, Gondim KC (2013) *Rhodnius prolixus* lipophorin: lipid composition and effect of high temperature on physiological role. Arch Insect Biochem Physiol 82: 129–140. https://doi.org/10.1002/arch.21080

Majerowicz D, Calderón-Fernández GM, Alves-Bezerra-M, De Paula IF, Cardoso LS, Juárez MP, Atella GC, Gondim KC (2017) Lipid metabolism in *Rhodnius prolixus*: lessons from the genome. Gene 596:27–44. https://doi.org/10.1016/j.gene.2016.09.045

Marinotti O, Nguyen QK, Calvo E, James AA, Ribeiro JMC (2005) Microarray analysis of genes showing variable expression following a blood meal in *Anopheles gambiae*. Insect Mol Biol 14:365–373. https://doi.org/10.1111/j.1365-2583.2005.00567.x

Marinotti O, Capurro M d L, Nirmala X, Calvo E, James AA (2006) Structure and expression of the lipophorin-encoding gene of the malaria vector, *Anopheles gambiae*. Comp Biochem Physiol B Biochem Mol Biol 144:101–109. https://doi.org/10.1016/j.cbpb.2006.01.012

McCall PJ, Cameron MM (1995) Oviposition pheromones in insect vectors. Parasitol Today 11:352–355. https://doi.org/10.1016/0169-4758(95)80192-8

Mendki M, Ganesan K, Prakash S, Suryanarayana M, Malhotra R, Rao K, Vaidyanathaswamy R (2000) Heneicosane: an oviposition-attractant pheromone of larval origin in *Aedes aegypti* mosquito. Curr Sci 78: 1295–1296

Mikani A, Wang Q-S, Takeda M (2012) Brain-midgut short neuropeptide F mechanism that inhibits digestive activity of the American cockroach, *Periplaneta americana* upon starvation. Peptides 34:135–144. https://doi.org/10.1016/j.peptides.2011.10.028

Millar J (2010) Polyene hydrocarbons, epoxides, and related compounds as components of lepidopteran pheromone blends. In: Blomquist GJ, Bagneres A-G (eds) Insect hydrocarbons: biology, biochemistry, and chemical ecology. Cambridge University Press, Cambridge, pp 390–447. https://doi.org/10.1017/CBO9780511711909.019

Moraes B, Braz V, Santos-Araujo S, Oliveira IA, Bomfim L, Ramos I, Gondim KC (2022) Deficiency of acetyl-CoA carboxylase impairs digestion, lipid synthesis, and reproduction in the kissing bug *Rhodnius prolixus*. Front Phys 13. https://doi.org/10.3389/fphys.2022.934667

Moriconi DE, Dulbecco AB, Juárez MP, Calderón-Fernández GM (2019) A fatty acid synthase gene (FASN3) from the integument tissue of *Rhodnius prolixus* contributes to cuticle water loss regulation. Insect Mol Biol 28:850–861. https://doi.org/10.1111/imb.12600

Morse D, Meighen E (1987) Biosynthesis of the acetate ester precursor of the spruce budworm sex pheromone by an acetyl CoA: fatty alcohol acetyltransferase. Insect Biochem 17:53–59. https://doi.org/10.1016/0020-1790(87)90143-0

Mota T, Vitta ACR, Lorenzo-Figueiras AN, Barezani CP, Zani CL, Lazzari CR, Diotaiuti L, Jeffares L, Bohman B, Lorenzo MG (2014) A multi-species bait for Chagas disease vectors. PLoS Negl Trop Dis 8: e2677. https://doi.org/10.1371/journal.pntd.0002677

Moto K, Yoshiga T, Yamamoto M, Takahashi S, Okano K, Ando T, Nakata T, Matsumoto S (2003) Pheromone gland-specific fatty-acyl reductase of the silkmoth, *Bombyx mori*. Proc Natl Acad Sci USA 100: 9156–9161. https://doi.org/10.1073/pnas.1531993100

Motoyama N, Suganuma T, Maekoshi Y (1990) Biochemical and physiological characteristics of insecticide resistance in diamondback moth. In: Talekar NS (ed) Diamondback moth and other crucifer pests: proceedings of the second international workshop, Tainan, Asian Vegetable Research and Development Center, pp 411–418

Mozūraitis R, Hajkazemian M, Zawada JW, Szymczak J, Pålsson K, Sekar V, Biryukova I, Friedländer MR, Koekemoer LL, Baird JK, Borg-Karlson A-K, Emami SN (2020) Male swarming aggregation pheromones increase female attraction and mating success among multiple African malaria vector mosquito species. Nat Ecol Evol 4:1395–1401. https://doi.org/10.1038/s41559-020-1264-9

Neira Oviedo M, Ribeiro JMC, Heyland A, VanEkeris L, Moroz T, Linser PJ (2009) The salivary transcriptome of *Anopheles gambiae* (Diptera: Culicidae) larvae: a microarray-based analysis. Insect Biochem Mol Biol 39: 382–394. https://doi.org/10.1016/j.ibmb.2009.03.001

Nordlund DA, Lewis WJ (1976) Terminology of chemical releasing stimuli in intraspecific and interspecific interactions. J Chem Ecol 2:211–220. https://doi.org/10.1007/BF00987744

Osgood CE (1971) An oviposition pheromone associated with the egg rafts of *Culex tarsalis*. J Econ Entomol 64: 1038–1041. https://doi.org/10.1093/jee/64.5.1038

Parvy J-P, Napal L, Rubin T, Poidevin M, Perrin L, Wicker-Thomas C, Montagne J (2012) *Drosophila melanogaster* acetyl-CoA-carboxylase sustains a fatty acid-dependent remote signal to waterproof the respiratory system. PLoS Genet 8:e1002925. https://doi.org/10.1371/journal.pgen.1002925

Pedrini N, Mijailovsky SJ, Girotti JR, Stariolo R, Cardozo RM, Gentile A, Juárez MP (2009) Control of pyrethroid-resistant Chagas disease vectors with entomopathogenic fungi. PLoS Negl Trop Dis 3: e434. https://doi.org/10.1371/journal.pntd.0000434

Pereira MG, Visbal G, Costa TFR, Frases S, de Souza W, Atella G, Cunha-E-Silva N (2018) *Trypanosoma cruzi* epimastigotes store cholesteryl esters in lipid droplets after cholesterol endocytosis. Mol Biochem Parasitol 224:6–16. https://doi.org/10.1016/j.molbiopara.2018.07.004

Perera H, Wijerathna T (2019) Sterol carrier protein inhibition-based control of mosquito vectors: current knowledge and future perspectives. Can J Infect Dis Med Microbiol 2019:1–6. https://doi.org/10.1155/2019/7240356

Perera R, Riley C, Isaac G, Hopf-Jannasch AS, Moore RJ, Weitz KW, Pasa-Tolic L, Metz TO, Adamec J, Kuhn RJ (2012) Dengue virus infection perturbs lipid homeostasis in infected mosquito cells. PLoS Pathog 8:e1002584. https://doi.org/10.1371/journal.ppat.1002584

Peterson JK, Graham AL, Dobson AP, Chávez OT (2015) *Rhodnius prolixus* life history outcomes differ when infected with different *Trypanosoma cruzi* I strains. Am J Trop Med Hyg 93:564–572. https://doi.org/10.4269/ajtmh.15-0218

Plettner E, Slessor KN, Winston ML, Oliver JE (1996) Caste-selective pheromone biosynthesis in honeybees. Science 80-.). 271:1851–1853. https://doi.org/10.1126/science.271.5257.1851

Plettner E, Slessor KN, Winston ML (1998) Biosynthesis of mandibular acids in honey bees (*Apis mellifera*): *De novo* synthesis, route of fatty acid hydroxylation and caste selective β-oxidation. Insect Biochem Mol Biol 28:31–42. https://doi.org/10.1016/S0965-1748(97)00079-9

Pontes EG, Leite P, Majerowicz D, Atella GC, Gondim KC (2008) Dynamics of lipid accumulation by the fat body of *Rhodnius prolixus*: the involvement of lipophorin binding sites. J Insect Physiol 54:790–797. https://doi.org/10.1016/j.jinsphys.2008.02.003

Regnier FE (1971) Semiochemical – structure and function. Biol Reprod 4:309–326. https://doi.org/10.1093/biolreprod/4.3.309

Regnier FE, Law JH (1968) Insect pheromones. J Lipid Res 9:541–551. https://doi.org/10.1016/S0022-2275(20)42699-9

Reynolds JA, Poelchau MF, Rahman Z, Armbruster PA, Denlinger DL (2012) Transcript profiling reveals mechanisms for lipid conservation during diapause in the mosquito, *Aedes albopictus*. J Insect Physiol 58: 966–973. https://doi.org/10.1016/j.jinsphys.2012.04.013

Ribeiro JM (1995) Blood-feeding arthropods: live syringes or invertebrate pharmacologists? Infect Agents Dis 4:143–152

Ribeiro JMC, Genta FA, Sorgine MHF, Logullo R, Mesquita RD, Paiva-Silva GO, Majerowicz D, Medeiros M, Koerich L, Terra WR, Ferreira C, Pimentel AC, Bisch PM, Leite DC, Diniz MMP, da S G V Junior JL, Da Silva ML, Araujo RN, Gandara ACP, Brosson S, Salmon D, Bousbata S, González-Caballero N, Silber AM, Alves-Bezerra M, Gondim KC, Silva-Neto MAC, Atella GC, Araujo H, Dias FA, Polycarpo C, Vionette-Amaral RJ, Fampa P, Melo ACA, Tanaka AS, Balczun C, Oliveira JHM, Gonçalves RLS, Lazoski C, Rivera-Pomar R, Diambra L, Schaub GA, Garcia ES, Azambuja P, Braz GRC, Oliveira PL (2014) An insight into the transcriptome of the digestive tract of the bloodsucking bug, *Rhodnius prolixus*. PLoS Negl Trop Dis 8:e2594. https://doi.org/10.1371/journal.pntd.0002594

Rimoldi OJ, González MS, Peluffo RO, Brenner RR (1988) Lipid metabolism in fat body of the hematophagous insect: *Triatoma Infestans*. Comp Biochem Physiol Part B Comp Biochem 90:235–241. https://doi.org/10.1016/0305-0491(88)90067-3

Rimoldi OJ, Soulages JL, González SM, Peluffo RO, Brenner RR (1989) Purification and properties of the very high-density lipoprotein from the hemolymph of adult *Triatoma infestans*. J Lipid Res 30:857–864. https://doi.org/10.1016/S0022-2275(20)38311-5

Rono MK, Whitten MMA, Oulad-Abdelghani M, Levashina EA, Marois E (2010) The major yolk protein vitellogenin interferes with the anti-plasmodium response in the malaria mosquito *Anopheles gambiae*. PLoS Biol 8(7):e1000434. https://doi.org/10.1371/journal.pbio.1000434

Sanders HR, Evans AM, Ross LS, Gill SS (2003) Blood meal induces global changes in midgut gene expression in the disease vector, *Aedes aegypti*. Insect Biochem Mol Biol 33:1105–1122. https://doi.org/10.1016/s0965-1748(03)00124-3

Saraiva FB, Alves-Bezerra M, Majerowicz D, Paes-Vieira-L, Braz V, Almeida MGMD, Meyer-Fernandes JR, Gondim KC (2021) Blood meal drives *de novo* lipogenesis in the fat body of *Rhodnius prolixus*. Insect Biochem Mol Biol 133:103511. https://doi.org/10.1016/j.ibmb.2020.103511

Schaub GA (2021) An update on the knowledge of parasite-vector interactions of chagas disease. Res Rep Trop Med 12:63–76. https://doi.org/10.2147/RRTM.S274681

Schmidt J, Kleffmann T, Schaub GA (1998) Hydrophobic attachment of *Trypanosoma cruzi* to a superficial layer of the rectal cuticle in the bug *Triatoma infestans*. Parasitol Res 84:527–536. https://doi.org/10.1007/s004360050443

Seo S-J, Cheon H-M, Sun J, Sappington TW, Raikhel AS (2003) Tissue- and stage-specific expression of two lipophorin receptor variants with seven and eight ligand-binding repeats in the adult mosquito. J Biol Chem 278:41954–41962. https://doi.org/10.1074/jbc.M308200200

Silva ERMN, Santos LV, Caiado MS, Hastenreiter LSN, Fonseca SRR, Carbajal-de-la-Fuente AL, Carvalho MG, Pontes EG (2021) The influence of larval density on triacylglycerol content in *Aedes aegypti* (Linnaeus)

(Diptera: Culicidae). Arch Insect Biochem Physiol 106:e21757. https://doi.org/10.1002/arch.21757

Sim C, Denlinger DL (2009) Transcription profiling and regulation of fat metabolism genes in diapausing adults of the mosquito *Culex pipiens*. Physiol Genomics 39: 202–209. https://doi.org/10.1152/physiolgenomics.00095.2009

Son JH, Weiss BL, Schneider DI, Dera KM, Gstöttenmayer F, Opiro R, Echodu R, Saarman NP, Attardo GM, Onyango M, Abd-Alla AMM, Aksoy S (2021) Infection with endosymbiotic *Spiroplasma* disrupts tsetse (*Glossina fuscipes fuscipes*) metabolic and reproductive homeostasis. PLoS Pathog 17: e1009539. https://doi.org/10.1371/journal.ppat.1009539

Soulages JL, Wu Z, Firdaus SJ, Mahalingam R, Arrese EL (2015) Monoacylglycerol and diacylglycerol acyltransferases and the synthesis of neutral glycerides in *Manduca sexta*. Insect Biochem Mol Biol 62:194–210. https://doi.org/10.1016/j.ibmb.2014.09.007

Sousa G, de Carvalho SS, Atella GC (2021) *Trypanosoma cruzi* affects Rhodnius prolixus lipid metabolism during acute infection. Front Trop Dis 2. https://doi.org/10.3389/fitd.2021.737909

Spates GE, Bull DL, Chen AC (1990) Hydrolysis of sphingomyelin and phosphatidylcholine by midgut homogenates of the stable fly. Arch Insect Biochem Physiol 14:1–12. https://doi.org/10.1002/arch.940140102

Stanley-Samuelson DW, Jurenka RA, Cripps C, Blomquist GJ, de Renobales M (1988) Fatty acids in insects: composition, metabolism, and biological significance. Arch Insect Biochem Physiol 9:1–33. https://doi.org/10.1002/arch.940090102

Takken W, Smallegange RC, Vigneau AJ, Johnston V, Brown M, Mordue-Luntz AJ, Billingsley PF (2013) Larval nutrition differentially affects adult fitness and *plasmodium* development in the malaria vectors *Anopheles gambiae* and *Anopheles stephensi*. Parasit Vectors 6:345. https://doi.org/10.1186/1756-3305-6-345

Teal PEA, Tumlinson JH (1987) The role of alcohols in pheromone biosynthesis by two noctuid moths that use acetate pheromone components. Arch Insect Biochem Physiol 4:261–269. https://doi.org/10.1002/arch.940040404

Tillman JA, Seybold SJ, Jurenka RA, Blomquist GJ (1999) Insect pheromones – an overview of biosynthesis and endocrine regulation. Insect Biochem Mol Biol 29:481–514. https://doi.org/10.1016/s0965-1748(99)00016-8

Timmermann SE, Briegel H (1999) Larval growth and biosynthesis of reserves in mosquitoes. J Insect Physiol 145(5):461–470. https://doi.org/10.1016/s0022-1910(98)00147-4

Tobe SS, Langley PA (1978) Reproductive physiology of *Glossina*. Annu Rev Entomol 23:283–307. https://doi.org/10.1146/annurev.en.23.010178.001435

Toé KH, N'Falé S, Dabiré RK, Ranson H, Jones CM (2015) The recent escalation in strength of pyrethroid resistance in *anopheles coluzzii* in West Africa is linked to increased expression of multiple gene families. BMC Genomics 16:146. https://doi.org/10.1186/s12864-015-1342-6

Uribe S (1999) The status of the *Lutzomyia longipalpis* species complex and possible implications for *Leishmania* transmission. Mem Inst Oswaldo Cruz 94:729–734. https://doi.org/10.1590/s0074-02761999000600005

Van Heusden MC, Erickson BA, Pennington JE (1997) Lipophorin levels in the yellow fever mosquito, *Aedes aegypti*, and the effect of feeding. Arch Insect Biochem Physiol 34:301–312. https://doi.org/10.1002/(SICI)1520-6327(1997)34:3<301::AID-ARCH5>3.0.CO;2-X

Van Heusden MC, Thompson F, Dennis J (1998) Biosynthesis of *Aedes aegypti* lipophorin and gene expression of its apolipoproteins. Insect Biochem Mol Biol 28: 733–738. https://doi.org/10.1016/s0965-1748(98)00068-x

Vial HJ, Eldin P, Martin D, Gannoun L, Calas M, Ancelin ML (1999) Transport of phospholipid synthesis precursors and lipid trafficking into malaria-infected erythrocytes. Novartis Found Symp 226:74–83.; discussion 82-8. https://doi.org/10.1002/9780470515730.ch6

Vial T, Marti G, Missé D, Pompon J (2021) Lipid interactions between Flaviviruses and mosquito vectors. Front Physiol 12. https://doi.org/10.3389/fphys.2021.763195

Vickerman K, Tetley L (1990) Flagellar surfaces of parasitic protozoa and their role in attachment. In: Bloodgood RA (ed) Ciliary and flagellar membranes. Springer, Boston, MA, pp 267–304. https://doi.org/10.1007/978-1-4613-0515-6_11

Wang X, Hou Y, Saha TT, Pei G, Raikhel AS, Zou Z (2017) Hormone and receptor interplay in the regulation of mosquito lipid metabolism. Proc Natl Acad Sci USA 114:E2709–E2718. https://doi.org/10.1073/pnas.1619326114

Wang G, Vega-Rodríguez J, Diabate A, Liu J, Cui C, Nignan C, Dong L, Li F, Ouedrago CO, Bandaogo AM, Sawadogo PS, Maiga H, Alves E Silva TL, Pascini TV, Wang S, Jacobs-Lorena M (2021) Clock genes and environmental cues coordinate *anopheles* pheromone synthesis, swarming, and mating. Science 371:411–415. https://doi.org/10.1126/science.abd4359

Wicker-Thomas C, Garrido D, Bontonou G, Napal L, Mazuras N, Denis B, Rubin T, Parvy J-P, Montagne J (2015) Flexible origin of hydrocarbon/pheromone precursors in *Drosophila melanogaster*. J Lipid Res 56:2094–2101. https://doi.org/10.1194/jlr.M060368

Wood P, Imaichi K, Knowles J, Michaels G, Kinsell L (1964) The lipid composition of human plasma chylomicrons. J Lipid Res 5:225–231. https://doi.org/10.1016/S0022-2275(20)40242-1

World Health Organization (2020) Factsheet: vector-borne diseases [WWW Document]. URL https://www.who.int/news-room/fact-sheets/detail/vector-borne-diseases (accessed 31 January, 2022)

Wyatt TD (2013) Pheromones and animal behavior, 2nd edn. Cambridge University Press, Cambridge. https://doi.org/10.1017/CBO9781139030748

Yang Z, Shi Y, Cui H, Yang S, Gao H, Yuan J (2021) A malaria parasite phospholipid flippase safeguards midgut traversal of ookinetes for mosquito transmission. Sci Adv 7. https://doi.org/10.1126/sciadv.abf6015

Zhou G, Flowers M, Friedrich K, Horton J, Pennington J, Wells MA (2004a) Metabolic fate of [14C]-labeled meal protein amino acids in *Aedes aegypti* mosquitoes. J Insect Physiol 50:337–349. https://doi.org/10.1016/j.jinsphys.2004.02.003

Zhou G, Pennington JE, Wells MA (2004b) Utilization of pre-existing energy stores of female *Aedes aegypti* mosquitoes during the first gonotrophic cycle. Insect Biochem Mol Biol 34. https://doi.org/10.1016/j.ibmb.2004.05.009

Zieler H, Nawrocki JP, Shahabuddin M (1999) *Plasmodium gallinaceum* ookinetes adhere specifically to the midgut epithelium of *Aedes aegypti* by interaction with a carbohydrate ligand. J Exp Biol 202:485–495

Dr. Juan R. Girotti graduated in Biochemistry from the Universidad Nacional de La Plata (UNLP) in 2005. He obtained a doctorate degree at the Facultad de Ciencias Exactas (UNLP) in 2010. He is associate professor of Organic Chemistry at the Facultad de Ciencias Naturales (UNLP) since 2012 and researcher at the Instituto de Investigaciones Bioquímicas de La Plata (INIBIOLP) as a member of the Consejo Nacional de Investigaciones Bioquímicas y Técnicas since 2013. He has expertise in lipid analysis with research interests in biochemistry of insects and pathogenic fungi.

Dr. Gustavo M. Calderón-Fernández graduated in Biological Sciences from the Universidad Nacional de La Plata (UNLP) in 2003. He obtained a doctorate degree in Natural Sciences from the UNLP in 2009. He is associate professor of Animal Histology and Embryology at the Facultad de Ciencias Naturales, UNLP since 2011 and member of the Consejo Nacional de Investigaciones Científicas y Técnicas (CONICET) research career since 2009 at the Instituto de Investigaciones Bioquímicas de La Plata (INIBIOLP). He has expertise in lipid analysis with research interests in biochemistry and molecular biology of cuticle lipids in insect vectors of Chagas disease.

Adv Exp Med Biol (2026) 1494: 419–443
https://doi.org/10.1007/5584_2024_833

Published online: 16 November 2024

Insect Lipid Metabolism in the Presence of Symbiotic and Pathogenic Viruses and Bacteria

Bertanne Visser and Mathilde Scheifler

Abstract

Insects, like most animals, have intimate interactions with microorganisms that can influence the insect host's lipid metabolism. In this chapter, we describe what is known so far about the role prokaryotic microorganisms play in insect lipid metabolism. We start exploring microbe-insect lipid interactions focusing on endosymbionts, and more specifically the gut microbiota that has been predominantly studied in *Drosophila melanogaster*. We then move on to an overview of the work done on the common and well-studied endosymbiont *Wolbachia pipientis*, also in interaction with other microbes. Taking a slightly different angle, we then look at the effect of human pathogens, including dengue and other viruses, on the lipids of mosquito vectors. We extend the work on human pathogens and include interactions with the endosymbiont *Wolbachia* that was identified as a natural tool to reduce the spread of mosquito-borne diseases. Research on lipid metabolism of plant disease vectors is up and coming and we end this chapter by highlighting current knowledge in that field.

Keywords

Aedes aegypti · Cholesterol · *Culex pipiens* · Diapause · Fat content · Fatty acids · Lipogenesis · Phospholipids · Phytophagous insects · *Serratia* · *Spiroplasma* · Triacylglycerols

1 Introduction

All insects harbor a diverse and extensive microbial community, referred to as the microbiota (i.e., the assemblage of microorganisms—bacteria, fungi, viruses, archaea, and protists—associated with a defined host or environment; Berg et al. 2020). The diversification and evolution of insects are closely tied to their symbiotic interactions with microorganisms that may be mutualistic, commensal, or parasitic (Cornwallis et al. 2023; Janson et al. 2008). Bacterial symbionts represent the largest part of the microbiota that can be located either on the surface of the host's body, i.e., ectosymbionts, or reside inside the host's body, i.e., endosymbionts. In insects, endosymbionts are primarily present in

B. Visser
Evolution and Ecophysiology Group, Department of Functional and Evolutionary Entomology, University of Liège – Gembloux Agro-Bio Tech, Gembloux, Belgium
e-mail: bertanne.visser@uliege.be

M. Scheifler (✉)
Institut de Biologie de l'École Normale Supérieure (IBENS), École Normale Supérieure, CNRS, INSERM, Université PSL, Paris, France

Evolution and Ecophysiology Group, Department of Functional and Evolutionary Entomology, University of Liège – Gembloux Agro-Bio Tech, Gembloux, Belgium
e-mail: mathilde.scheifler@gmail.com

the gut or in specialized cells called bacteriocytes (Baumann et al. 2006). Insects can also function as vectors for disease-causing microbes, such as dengue virus (DENV) transmitted by mosquitoes causing dengue fever in humans or plant viruses transmitted by phloem-sucking insects that can have a large effect on crops (e.g., beet, turnip, etc.). Both symbiotic and pathogenic microorganisms can have substantial effects on many different aspects of the insect host's biology.

Symbiotic bacteria are known for a plethora of effects on insect hosts. The insect microbiota, for example, can affect *i)* the host's immune system and protection against various predators, parasites, disease vectors, or pathogens; *ii)* communication and behavior among individuals of the same or of different species; *iii)* host mating preferences and reproductive systems; *iv)* host life histories and fitness-related traits (e.g., development, lifespan, fecundity); and *v)* host resilience to environmental disturbances (e.g., pesticides) (Douglas 2015; Engel and Moran 2013; Engl and Kaltenpoth 2018; Zhang et al. 2022a, b). Notwithstanding these important functions, the provisioning of essential nutrients for the insect host seems to be a primary task of gut microorganisms. Many microorganisms provide nutrients that the insect cannot synthesize, such as amino acids, B vitamins, or sterols (Douglas 2015). The bacterial endosymbiont *Buchnera aphidicola*, for example, is of primary importance for aphid development and adult life by providing essential amino acids, and in return aphids provide a stable and nutrient-rich environment (Douglas et al. 2001). Another well-known insect endosymbiont, *Wolbachia pipientis*, has been shown to supply B vitamins to its host, the bedbug *Cimex lectularius* (Hosokawa et al. 2010; Newton and Rice 2020). Nutrient provisioning by bacteria can compensate for nutrient-poor diets, aids the digestion of recalcitrant food components (e.g., degradation of cellulose in plant cell walls), and supply essential amino acids, metabolic compounds, or nutrients (Engel et al. 2012; Hu et al. 2018; Jing et al. 2020; Russell et al. 2014; Sannino et al. 2018; Tokuda et al. 2018).

Regarding nutritional interactions, symbiotic bacteria were already found to have a major impact on lipid metabolism in humans (Xu et al. 2022). For example, changes in gut bacterial communities are related to metabolic diseases, such as obesity, cardiovascular disease, and type 2 diabetes (Depommier et al. 2019; Liu et al. 2021a, b; Wang et al. 2022). Relatively little is known, however, about the role played by symbiotic microorganisms in insect lipid metabolism. Considering how microorganisms affect key metabolic interactions is important, because more than 10% of insect species rely on obligate bacterial symbionts for survival or reproduction, and many more microorganisms are facultatively associated with insects (Hilgenboecker et al. 2008; Sazama et al. 2017; Weinert et al. 2015; Wernegreen 2002). Recent work on human pathogens, mainly DENV, has, however, revealed major lipid metabolic adjustments in the insect vector incited by the virus that are of importance for viral propagation (Chotiwan et al. 2018; Perera et al. 2012; Tongluan et al. 2017).

Lipids have also been implicated in immune responses of insects, which has already been reviewed extensively (Wrońska et al. 2023; Barletta et al. 2016), and falls beyond the scope of this chapter. We set out to unite research aimed at understanding the role of prokaryotic symbiotic or pathogenic microorganisms on insect lipid metabolism. We focus on prokaryotes, i.e., bacteria and viruses, to be able to set forth and identify commonalities and differences in the ways insect host/vector lipid metabolism is affected.

2 The Impact of Symbiotic Microorganisms on Host Insect Lipid Metabolism

2.1 Influence of the Gut Microbiota

There is growing evidence that the gut microbiota plays a key role in the regulation of insect fat storage. Most studies to date have focused on the impact of gut microbiota and microbe interactions on fat metabolism of the vinegar fly *Drosophila melanogaster* (Fig. 1), which is an

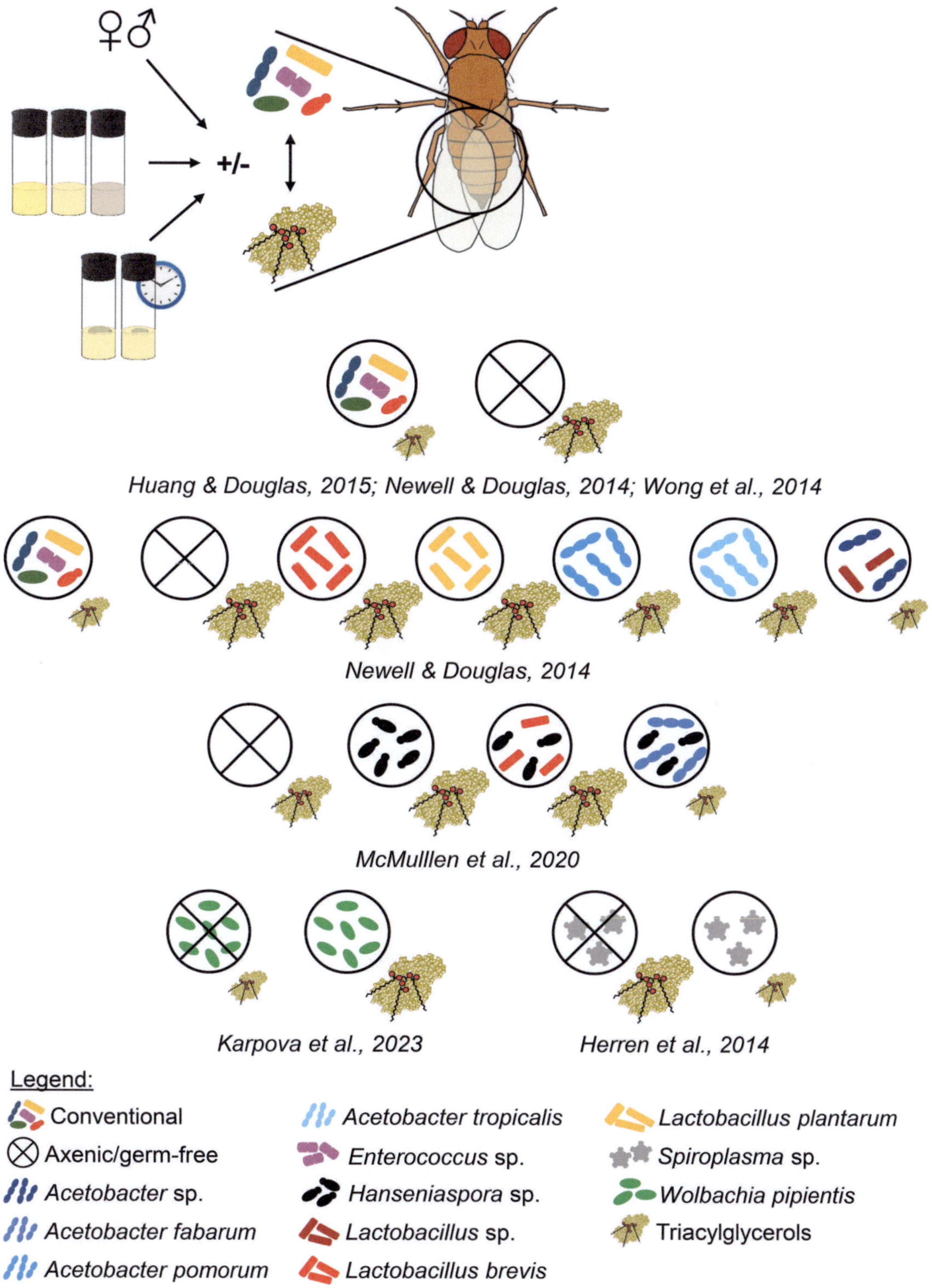

Fig. 1 Summary of microbe effects on *Drosophila melanogaster* triacylglycerol levels (i.e., storage lipids) with quantities represented by the size of the triacylglycerol graphic. A legend of microbial species is provided at the bottom of the graph. Triacylglycerol levels were compared between axenic (germ-free), mono-infected (bacteria-only or fungi-only), dual-infected (bacteria-bacteria and fungi-bacteria interactions), and conventional flies. All relevant references are provided in the figure

emerging model system in the field (Douglas 2019; Erkosar et al. 2013). Generally, *D. melanogaster* deprived of the entire microbiota (i.e., axenic/germ-free individuals) had a higher triacylglycerol content than individuals with microbiota (Huang and Douglas 2015; Newell and Douglas 2014; Wong et al. 2014; but see Ridley et al. 2012 and Henry et al. 2020 who found no difference in fat content between axenic and control *D. melanogaster flies*). The lack of bacteria that usually utilize host gut nutrients could explain the higher triacylglycerol content in axenic *D. melanogaster* flies. Overall, the multitude of studies comparing axenic and microbiota-containing *D. melanogaster* show a range of different results on triacylglycerol content, which can be explained by variation in host-related factors, such as host sex and feeding rate, as well as composition of the diet (e.g., sugar: yeast ratio, nutrient-poor or rich diet) and how each factor interacts with the microbiota and other factors (Huang and Douglas 2015; McMullen et al. 2020; Wong et al. 2014).

Interspecific bacterial interactions can lead to substantial differences in triacylglycerol content of *D. melanogaster,* where both laboratory and wild populations have relatively low-diversity gut microbiota. The *D. melanogaster* gut microbiota is commonly dominated by bacteria in the family *Acetobacteraceae* (mainly represented by the genus *Acetobacter*) and the order *Lactobacillales* (mainly represented by the genus *Lactobacillus*) (Adair et al. 2018; Chandler et al. 2011, 2012; Wong et al. 2011). Both mutualistic and antagonistic associations between *Acetobacter* and *Lactobacillus* have been found in *D. melanogaster*, depending on the bacterial species involved (Consuegra et al. 2020; McMullen et al. 2020; Sommer and Newell 2019). The impact of bacterial taxa (i.e., a total of five *Acetobacter* and *Lactobacillus* species) on *D. melanogaster* triacylglycerol content was assessed by comparing single-, dual-, or multi-species infections (compared to both axenic and conventional flies) (Newell and Douglas 2014). Combinations of bacterial taxa and effects on triacylglycerol content showed that *i)* dual-microbe infected *D. melanogaster* individuals generally had a lower triacylglycerol content than axenic and single-microbe infected individuals; *ii)* bacterial effects on *D. melanogaster* triacylglycerol levels are microbe-specific and dependent on interactions, e.g., mono-infection by *Lactobacillus brevis* and *Lactobacillus plantarum* did not lead to different triacylglycerol levels, only in interaction with *Acetobacter* were levels significantly lower; and *iii)* bacterial interactions are essential to restore the natural insect phenotype (i.e., similar to untreated flies). Newell and Douglas (2014) also highlighted that *Acetobacter tropicalis* abundance is promoted by the colonization of *L. brevis* in *D. melanogaster*. High *A. tropicalis* cell density, in turn, decreased fly triacylglycerol content in a dose-dependent manner (Newell and Douglas 2014). *Drosophila melanogaster* triacylglycerol content is thus mediated by the composition of the gut microbiota, bacterial abundance, and bacterial interactions.

The capacity of some bacteria, such as *Acetobacter* or *Lactobacillus*, to reduce *D. melanogaster* fat content (confirmed by Bozkurt et al. 2023) has been attributed to several, not mutually exclusive, processes. First, the bacteria can reduce host triacylglycerol levels via the consumption of dietary glucose, e.g., *Lactobacillus* produces lactate via the consumption of glucose, the latter being a substrate for acetyl-CoA synthesis that, in turn, is the precursor for fatty acid and subsequent fat synthesis in insects (Huang and Douglas 2015; Sommer and Newell 2019). Second, microorganisms can modulate host nutritional signaling pathways. For example, the increased production of acetic acid by *Acetobacter pomorum,* in response to the production of lactate by *Lactobacillus*, was shown to increase *D. melanogaster*'s insulin levels resulting in reduced adult fat content (Shin et al. 2011). Third, *Lactobacillus* can modulate the TOR (Target of Rapamycin) signaling pathway that also affects insulin signaling (Storelli et al. 2011). Fourth, metabolic models predicted a high release rate of succinate from *Drosophila* gut bacteria (Ankrah et al. 2021), impacting the citric acid cycle by reducing citrate levels available for fatty acid synthesis (Zhang et al. 2022a, b).

Bacteria can also interact with other microorganisms, such as fungi, affecting triacylglycerol levels differently depending on the interactions considered (Bozkurt et al. 2023; McMullen et al. 2020). For example, Bozkurt et al. 2023 showed a positive correlation between the abundances of *Acetobacter persici*, *A. pomorum* and Basidiomycota in *D. melanogaster*, as well as a negative correlation between these microbial taxa and triacylglycerol levels. In contrast, the fungus *Hanseniaspora uvarum* (order Saccharomycetales), also part of the *D. melanogaster* gut microbiota (Chandler et al. 2012), showed antagonistic interactions with *L. brevis* and *Acetobacter fabarum*. When *H. uvarum* is present, there is a negative effect on the abundance of *L. brevis*. The abundance of both *H. uvarum* and *A. fabarum* decreases when present together (McMullen et al. 2020). For the *H. uvarum*-*A. fabarum* interaction, a negative correlation was also observed between *D. melanogaster* triacylglycerol content and acetic acid that varied significantly with the presence of both *A. fabarum* and *H. uvarum*, consistent with previous studies (Newell and Douglas 2014; Sommer and Newell 2019). *Drosophila melanogaster* associated with both *A. fabarum* and *H. uvarum* had high acetic acid levels, but interestingly, triacylglycerol levels were also significantly elevated in flies only infected by the yeast *H. uvarum* (compared to axenic flies). *Hanseniaspora uvarum* was hypothesized to be a producer of acetic acid, like some other fungi (Bueno et al. 2020; Jolly et al. 2014). Interactions between *A. fabarum* and *H. uvarum* could modulate the concentration of acetic acid, reducing triacylglycerol synthesis (McMullen et al. 2020). Taken together, these results demonstrate the key role of the gut microbiota and microbial fermentation products, such as acetic acid, on the nutritional status of *D. melanogaster*, particularly with respect to fat accumulation.

In species other than *Drosophila,* only little progress has been made so far, and contrasting results have been reported regarding insect fat metabolism and fat content. In the aphid *Acyrthosiphon pisum*, axenic individuals showed increased triacylglycerol levels, in line with findings in *D. melanogaster* (Rahbé et al. 1994). In contrast, lower fat content was reported for adults of three fruit fly species, *Ceratitis capitata*, *Bactrocera tryoni*, and *Anastrepha fraterculus,* following antibiotic treatment (Ben-Yosef et al. 2008; Goane et al. 2022; Nguyen et al. 2021). Similar to findings in *D. melanogaster*, the fat content of the other fruit flies was affected by interactions between microbiota, diet, and sex (Ben-Yosef et al. 2008; Nguyen et al. 2021).

In the fruit fly *Bactrocera dorsalis*, a genomic study comparing gene expression of antibiotic-treated and control individuals revealed upregulation of *i)* fatty acid synthesis genes (e.g., fatty acid synthase (*fas*), acetyl-CoA carboxylase), *ii)* genes encoding triacylglycerol catabolism (e.g., lipases, fatty acid hydroxylase), and *iii)* downregulation of genes involved in fatty acid beta-oxidation (e.g., enoyl-CoA hydratase), suggesting a general increase of free fatty acids in the axenic insect (Xie et al. 2023). Downregulation of genes involved in lipid storage (i.e., vitellogenin) and transport (i.e., lipophorins), as well as a decrease in lipid content of the host's fat body have also been reported in *Aedes aegypti* axenic mosquitoes (Romoli et al. 2021). It has remained unclear how and why the expression of fatty acid and triacylglycerol metabolic genes changes depending on gut microbiota presence. One proposed hypothesis is that lipolysis facilitated by endosymbiotic bacteria increases the availability of different lipid types for the insect host. When no bacteria are present, the host insect is forced to start synthesizing different lipid types, while reducing fat storage (due to lower quantities of available precursors; Goane et al. 2022).

Gnotobiotic insects (i.e., insects associated with specific bacterial strains) have also been used in systems other than *Drosophila* to decipher the role of bacterial strains on host fat metabolism and fat content. In the red palm weevil *Rhynchophorus ferrugineus*, for example, a significant reduction in triacylglycerol content was reported in germ-free larvae compared to untreated larvae (Habineza et al. 2019). Introduction of the bacterium *Enterobacter cloacae* into germ-free *R. ferrugineus* larvae partially restored triacylglycerol levels, but no effect was found for

Lactococcus lactis (Habineza et al. 2019). Another study reported that gnotobiotic *Ae. aegypti* mosquitoes associated with *Flavobacterium* or *Paenibacillus* showed higher triacylglycerol levels compared to control mosquitoes, while Enterobacteriaceae and *Lysobacter* had no impact (Giraud et al. 2022). *Enterobacter cloacae* is known to synthetize various carbohydrate-modifying and glycolytic enzymes (e.g., cellulases, trehalases, glucosidases; Habineza et al. 2019), while Flavobacteria are chitinase producers (McBride et al. 2009), suggesting that bacteria other than *Acetobacter* and *Lactobacillus* can play a role in nutrient acquisition of other insect host species.

2.2 Box 1: The Influence of Microbes on Insect Fat Metabolism Associated with Diapause

Many insects have adapted to seasonal changes and low food availability by entering diapause. Diapause is a genetically and hormonally determined program that depends on various environmental signals (e.g., photoperiod, temperature) allowing insects to anticipate pending unfavorable conditions (Denlinger 2002; Denlinger et al. 2012). Diapause may occur during any stage of the insect's life cycle (e.g., embryonic, larval, pupal, or adult), depending on the insect species, and is characterized by reduced metabolic and behavioral activity (Hahn and Denlinger 2011). By delaying development from several weeks up to years, insects can synchronize their life cycle to match with periods suitable for growth, development, and reproduction. In addition to metabolic depression during diapause, insects typically increase energy stores by accumulating fat prior to diapause (Enriquez and Visser 2023). Fat reserves constitute an important source of energy to meet metabolic demands during and post-diapause (Hahn and Denlinger 2007).

Only few studies have explored the role of bacterial symbionts on fat accumulation during diapause, despite the importance of facultative and obligatory diapause for many insects (Hahn and Denlinger 2011). An exception is the work by Didion et al. (2021) that confirmed the critical importance of microbiota for diapause preparation in the mosquito *Culex pipiens*. Diapausing mosquitoes with a low bacterial load had 50% fewer fat reserves, associated with a lower dry mass and lower survival rate. In the parasitic wasp *Nasonia vitripennis*, triacylglycerol levels of diapausing larvae (ranging from 1 to 6 months of diapause) were significantly correlated with changes in microbiota composition (Dittmer and Brucker 2021). This correlation was rather weak, however, when compared to the effects of temperature and quantities of other nutrient types, such as glycerol or glucose (Dittmer and Brucker 2021).

Under laboratory conditions, Liu et al. (2016) investigated the link between gut bacterial symbionts and the metabolic shift from protein synthesis to triacylglycerol accumulation in a vegetable pest beetle, *Colaphellus bowringi,* that occurs when females enter diapause. Gut microbiota composition was slightly different between diapausing and non-diapausing individuals: positive correlations were found between diapause preparation and abundances of *Proteobacteria* (e.g., *Serratia* sp., *Sphingomonas* sp.) and *Firmicutes* (e.g., *Lactococcus* sp.), while a negative correlation was found with the abundance of *Bacteroidetes* (e.g., *Flavobacterium* sp.; but see Didion et al. (2021) that found no difference between the microbiota of diapausing and non-diapausing *C. pipiens* mosquitoes). Based on similar findings on the regulation of obesity by microbiota in mammals (Ley et al. 2005), higher abundances of *Proteobacteria* and *Firmicutes* may affect insect fat accumulation. In a more recent study, the endosymbiont *Wolbachia* appeared to reduce the lipid content (estimated by cholesterol amounts) of the parasitoid wasp *Trichogramma brassicae*, leading to a lower percentage of diapausing individuals (Rahimi-Kaldeh et al. 2019).

The above studies highlight an important functional role of the microbiota in insect diapause, although evidence of host-microbiota interactions during insect diapause remains scarce. As diapause is controlled by the insect's endocrine system, the microbiota is expected to interact closely with the host's hormonal signaling pathways. More work is now needed to increase our understanding of how microbe-insect interactions affect diapause, and more generally how microbes affect host fat metabolism under low-temperature stress (Lv et al. 2023; Raza et al. 2020).

2.3 The Particular Case of the Endosymbiont *Wolbachia*

Wolbachia pipientis is one of the most widespread heritable bacterial endosymbionts harbored by insects, filarial nematodes, crustaceans, and mites (Serbus et al. 2008), infecting at least 65% of all known insect species (Hilgenboecker et al. 2008; Zug and Hammerstein 2012). *Wolbachia* is present in host germ line and somatic tissues, such as the fat body, salivary glands, or hemolymph (Dobson et al. 1999; Pietri et al. 2016), and can affect a wide variety of the insect host's biological functions. In mutualistic interactions, *Wolbachia* can confer fitness advantages, such as protection against viruses, resistance to heat stress or increasing learning ability, immunity, and life history traits (Arai et al. 2019; Cao et al. 2019; Farahani et al. 2017; Faria et al. 2018; Gruntenko et al. 2017; Maistrenko et al. 2016; Mazzucco et al. 2020). *Wolbachia* can also be parasitic and is perhaps best known for manipulating host reproduction in favor of its vertical transmission and spread within insect populations. *Wolbachia* can reduce population sizes, distort population sex ratios through male-killing or feminization of genetic males, induce parthenogenesis or cause cytoplasmic incompatibility (i.e., mating between individuals differing in *Wolbachia* infection status result in embryonic mortality; Charlat et al. 2001) (Dittmer and Bouchon 2018; Hurst et al. 1999; Poinsot et al. 2003; Stouthamer et al. 1999).

Host insect nutrient metabolism appears to be strongly influenced by *Wolbachia*. In *D. melanogaster*, *Wolbachia* affects fatty acid profiles, particularly the odd-chain fatty acid fraction (Molloy et al. 2016; Scheitz et al. 2013). Insects cannot synthesize odd-chain fatty acids. In *D. melanogaster* females, odd-chain fatty acids are likely supplied by *Wolbachia*, where odd-chain fatty acid levels are positively correlated to *Wolbachia* abundance ((Molloy et al. 2016; Scheitz et al. 2013). Odd-chain fatty acids have been found in the insect cuticle and body extracts in a wide variety of insect orders: Diptera (Kaczmarek et al. 2020; Sato et al. 2020), Hymenoptera (Pickett et al. 2000; Stanley-Samuelson et al. 1990), Hemiptera (Bashan et al. 2002; Cakmak et al. 2007a), Coleoptera (Howard and Stanley-Samuelson 1990; Nikolova et al. 2000), Neuroptera (Cakmak et al. 2007b), and Lepidoptera (Akinnawo and Ketiku 2000; Gołębiowski et al. 2010). The widespread occurrence among insects demonstrates that odd-chain fatty acids are fairly common components of insect lipids, although the proportion of odd-chain fatty acids is low compared to even-chain fatty acids. Odd-chain fatty acids can have a role in membrane stability and structure, as they have been found in the phospholipid fraction (Howard and Stanley-Samuelson 1990; Sato et al. 2020). Odd-chain fatty acids have also been found in the triacylglycerol fraction, and used for fat storage (Cakmak et al. 2007a, b). Odd-chain fatty acids can be acquired either by ingestion of symbiotic microorganisms that synthesize them (e.g., bacteria, yeast; Park et al. 2020; Řezanka and Sigler 2009) or synthesized de novo. In *D. melanogaster*, Sato et al. (2020) observed no significant difference in odd-chain fatty acid content between conventional and germ-free flies, suggesting that the microbiota was not involved. Instead, the incorporation of isotopic labels into the odd-chain fatty acids of *D. melanogaster* suggested de novo synthesis (Sato et al. 2020).

An increase in triacylglycerols was observed in *D. melanogaster* flies infected either with the *w*MelPlus, *w*Mel, or *w*MelCS45 *Wolbachia* strain compared to uninfected flies (Karpova et al. 2023). Contrasting results on the effect of *Wolbachia* on host fat metabolism have, however, been reported within and between mosquito species. *Wolbachia* infection led to a decrease in triacylglycerol levels in *Ae. aegypti* (*w*Mel strain) and *Aedes fluviatilis* (*w*Aflu) (Conceição et al. 2021; Koh et al. 2020). Infection of *Ae. aegypti* with *w*Aflu further led to decreased lipid droplet size in the cytoplasm of mosquito cells (Conceição et al. 2021). In *Aedes albopictus*, *w*Mel *Wolbachia* infection decreased diacylglycerol levels by 32% compared to uninfected mosquitoes, while a 17% increase in triacylglycerols was observed in *w*MelPop-infected mosquitoes (Molloy et al. 2016). Overall, *Wolbachia* effects on various lipid types depend on host and *Wolbachia*-related factors (e.g., host species or genotype, *Wolbachia* strain; Molloy et al. 2016), as was already shown for other metabolic pathways (e.g., dopamine metabolism; Gruntenko et al. 2017).

In *Drosophila* and several mosquito species, changes in lipid types other than fatty acids and triacylglycerols were observed in the presence of *Wolbachia* (Conceição et al. 2021; Koh et al. 2020; Molloy et al. 2016). In *Ae. albopictus*, *w*Mel and *w*MelPop *Wolbachia* infection resulted in *i)* a decrease in various sphingolipids (mostly ceramides), as well as phosphatidylcholines, phosphatidylethanolamines, and diacylglycerols, and *ii)* an increase in phosphatidylglycerols and phosphatidylinositols in the host (Molloy et al. 2016). *Wolbachia* infection was also shown to differently affect *Ae. albopictus* lipids depending on the *Wolbachia* strain (i.e., either *w*Mel or *w*MelPop). Ceramide levels, for example, decreased 62% in *Ae. albopictus* infected with the *w*Mel *Wolbachia* strain compared to uninfected mosquitoes, while a decrease of only 20% was observed in mosquitoes infected with the *w*MelPop strain (Molloy et al. 2016). A mean decrease in sphingomyelins of 35% was reported in *w*Mel-infected *Ae. albopictus*, while sphingomyelins increased by 28% in *w*MelPop-infected *Ae. albopictus*. *Aedes aegypti* infected with the same *w*Mel *Wolbachia* strain also revealed a reduction of phosphatidylethanolamines and more complex forms of ceramides (e.g., glucosylceramides) (Koh et al. 2020). As sphingolipids and phospholipids play a major structural role in cell membranes (e.g., complex assembly in lipid rafts), depletion of these lipids was hypothesized to affect host membrane fluidity, curvature, and structure. Changes in the host membrane can facilitate *Wolbachia* colonization within the host (Molloy et al. 2016).

Variation in lipid levels may be related to the dependency of *Wolbachia* on the host insect for lipids. A genome sequencing study indeed revealed that a *Wolbachia* strain (*w*Mel) associated with *D. melanogaster* lost many key metabolic pathways, including pathways for fatty acid and cholesterol metabolism (Wu et al. 2004). Cholesterol is the dominant sterol in most insects, and a vital component for cell membrane stability, hormone regulation, and insect development (Behmer and Nes 2003; Jing and Behmer 2020). The *w*Mel strain thus depends completely on the host to supply fatty acids and cholesterol for its survival and proliferation (Caragata et al. 2017; Zhang et al. 2021). Like some other intracellular bacteria, *Wolbachia* resides in a host-derived vacuole (Cho et al. 2011) within tissues of insects (Dobson et al. 1999; Hughes et al. 2011; Pietri et al. 2016). *Wolbachia* is restricted to the host's Golgi-related vesicles near the endoplasmic reticulum, a site of active nutrient synthesis (Cho et al. 2011). Close positioning next to a lipid-enriched organelle allows *Wolbachia* to acquire nutrients, such as amino acids or lipids, by subverting, modifying (e.g., lipid composition), and redistributing the endoplasmic reticulum of the host to colonize the host cell at a high density (Fattouh et al. 2019).

Only few studies have so far examined the hypothesis that an essential requirement for lipids leads *Wolbachia* to manipulate host lipid metabolism. In adult *Ae. aegypti* infected by the *w*Mel and *w*MelPop *Wolbachia* strain, a decrease of 25.6% and 27.7% in total cholesterol levels was observed, respectively. A reduction in total cholesterol level suggests that *Wolbachia* may use host cellular lipids (Caragata et al. 2014). *Wolbachia* seems to compete for host cholesterol, a pattern already reported for other intracellular bacteria (e.g., *Ehrlichia chaffeensis*, *Anaplasma phagocytophilum*, *Brucella abortus*; Lin and Rikihisa 2003; Watarai et al. 2002). With *Wolbachia* being located in Golgi-related vesicles, where high membrane biogenesis and cholesterol sequestration typically occur, the bacterium has direct access to nutrients metabolized by the insect host (Cho et al. 2011; Howe and Heinzen 2006).

Recent studies highlighted that *Wolbachia* can affect gene expression of host metabolic pathways, including fat metabolism. *Wolbachia* first seems to act on the host's insulin/insulin-like-growth factor pathway (Currin-Ross et al. 2021; Ikeya et al. 2009). Whether *Wolbachia* actively regulates the insulin signaling pathway, however, remains a matter of debate, as both positive and negative regulation have been reported (Currin-Ross et al. 2021; Ikeya et al. 2009). Moreover, genes underlying host fatty acid synthesis (e.g., *fas*) were further found to be upregulated in *Wolbachia*-infected *D. melanogaster* larval stages (*w*Mel *Wolbachia* strain; Zheng et al. 2011), as well as in adult *D. melanogaster* (*w*Mel; Dou et al. 2021) and mosquitoes (*w*Mel and *w*MelPop; Rancès et al. 2012; Wimalasiri-Yapa et al. 2023), suggesting a role for *Wolbachia* in modulating the expression of host genes involved in fat metabolism.

2.4 Endosymbionts Other Than *Wolbachia* Can Also Alter Host Fat Metabolism

Some endosymbionts appear to compete with the host insect for lipids. In *Spiroplasma poulsonii*-infected *D. melanogaster* flies, for example, a significant decrease in circulating lipids, specifically diacylglycerols and sterols, was reported in host hemolymph (compared to *S. poulsonii*-free flies) (Herren et al. 2014). The bacterium *S. poulsonii* subverts and utilizes diglycerides contained in host hemolymph lipoprotein particles (i.e., an important hemolymph lipid carrier; Sieber and Thummel 2012) prior to the arrival of diacylglycerols at the fat body, resulting in lower triacylglycerol levels (as triacylglycerol synthesis and storage in the fat body largely depends on host hemolymph diacylglycerols) (Herren et al. 2014). Proliferation of *S. poulsonii* was also found to be limited by the availability of host hemolymph lipids (Herren et al. 2014). The use of host lipids by *S. poulsonii* was confirmed in a parasitic wasp, *Leptopilina boulardi,* parasitizing *D. melanogaster*. Parasitic wasps depend on a single host insect to complete development and obtain sufficient nutritional resources to fuel life (Scheifler et al. 2024). The presence of *S. poulsonii* led to direct competition with *L. boulardi* for *D. melanogaster* hemolymph lipids (Paredes et al. 2016). In the *D. melanogaster-S. poulsonii-L. boulardi* interaction, competition for lipids underlies the protective role of *S. poulsonii* for *D. melanogaster* larvae by reducing developmental success of the parasitic wasp (Paredes et al. 2016).

Other endosymbionts, such as *Serratia*, are beneficial to the insect by enhancing host fatty acid metabolism. *Serratia symbiotica*-infected aphids (*A. pisum*), for example, up-regulated the expression of genes involved in fatty acid and fat synthesis, such as *fas* and diacylglycerol-o-acyltransferase, resulting in higher triacylglycerol levels in the aphid fat body (Zhou et al. 2021). In the silkworm *Bombyx mori* fed with the symbiont *Bacillus subtilis*, changes in insect gut microbiota composition were correlated with shifts in glycerophospholipid and sphingolipid composition in the host's hemolymph (Li et al. 2022). The abundance of *Enterococcus* was, for example, negatively correlated with some lysophosphatidylcholines and lysophosphatidylethanolamines and positively correlated with some phosphocholines, suggesting a role of *Enterococcus* in the glycerophospholipid metabolism of the host *B. mori*.

3 Interference of Disease-Vector Lipid Metabolism by Human Pathogenic Microbes

3.1 Mosquito-Vector Lipid Metabolism Upon Infection with Human Pathogenic Viruses

Arthropod-borne viruses (arboviruses), such as DENV, West Nile virus, Chikungunya virus, and Zika virus (ZIKV), can cause major health problems for humans with hundreds of millions of infections leading to serious diseases and deaths (Bhatt et al. 2013; Fauci and Morens 2016; Guzman et al. 2010). Like many other viruses, the DENV cycle is initiated with the attachment of the virus to a targeted host cell through the interaction between viral surface proteins and receptor molecules on the host cell surface (Cruz-Oliveira et al. 2015). The internalization of the virus within the infected cell involves receptor-mediated endocytosis (Mosso et al. 2008). Viral genomic RNA is then released into the cytoplasm of the host cell and translated into proteins required for RNA replication and viral particle assembly (Vial et al. 2021). Virus replication is dependent on three cellular pathways: autophagy (e.g., degradation of substrates, such as proteins or lipid droplets), actin polymerization and remodeling (e.g., vesicular trafficking), and fatty acid biosynthesis (Tongluan et al. 2017). Arboviruses are enveloped by lipids derived from the insect vector, leading the scientific community to hypothesize that viral entry, replication, assembly, and release occur in the host's cellular membranes. This led to a surge of studies on the ways in which viruses can manipulate insect vector lipid metabolism.

The fundamental role insect vector lipids play in the virus life cycle has mainly been investigated using flaviviruses, including DENV (see Ratnayake et al. 2023 and Vial et al. 2021 for recent reviews; but see Liu et al. 2021a, b for an investigation of the mechanisms by which FAS is affected following Classical Swine Fever virus infection). Analysis of the vector's fatty acid biosynthesis pathway revealed that the FAS enzyme is essential for DENV replication (Perera et al. 2012; Tongluan et al. 2017). DENV infection induces upregulation of the *fas* gene leading to de novo fatty acid synthesis, and relocalization of the enzyme FAS to sites of DENV replication (Tongluan et al. 2017). Alterations in de novo fatty acid synthesis and the role played by various *Ae. aegypti* fatty acid synthases (aaFAS) were recently investigated by Chotiwan et al. (2022). Seven distinct orthologues of human *fas* were identified, five of which produced transcripts. In females, only aaFAS1 showed high expression in both sugar-fed and blood-fed females, where diet does not seem to play a substantial role. To better understand the role played by the other *fas* orthologues, Chotiwan et al. (2022) knocked down aaFAS1 to determine if aaFAS2, aaFAS3, and aaFAS5 transcription could compensate for significantly reduced aaFAS1 transcription. While the other aaFASs showed a two-fold increase in transcription, aaFAS1 transcription remained higher following knockdown, suggesting that the other aaFASs may not be able to compensate for aaFAS1 function. Knockdown of aaFAS1 further led to a reduction of DENV replication in both mosquito Aag2 cell line and midguts, suggesting that aaFAS1 is required for DENV replication.

Interestingly, in mosquitoes, a significant increase in fat content was observed during the early stages of DENV infection, especially with higher abundances of glycerolipids, including mono-, di-, and triacylglycerols as well as other lipid types, such as glycerophospholipids, sphingolipids, or sterols (Chotiwan et al. 2018; Perera et al. 2012). Mosquito (i.e., *Aedes* sp.) fat content subsequently decreased after a few days. Increased de novo fatty acid synthesis, as well as increased transport of stored fat, suggests that these processes may be required for virus replication, dissemination, and survival during the initial stages of infection (Chotiwan et al. 2018; Perera et al. 2012).

Newly synthesized lipids are redistributed to sites of viral replication, mainly near the insect vector's endoplasmic reticulum membrane. Incorporation of different lipid types can then modify vector membrane structure, i.e., fluidity, permeability, and curvature, altering the functionality of the endoplasmic reticulum to the benefit of virus replication (Vial et al. 2021). DENV translation, replication, and assembly indeed require vector cell endoplasmic reticulum membranes that could affect the synthesis of phospholipids, critical cell membrane components. Vial et al. (2019) used high-resolution mass spectrometry to understand how phospholipid metabolism is affected in *Ae. aegypti* cells, midguts, and whole mosquitoes at various times post-infection. Phospholipidomics first revealed that aminophospholipids, including phosphatidylethanolamine (PE), phosphatidylcholine (PC), and phosphatidylserine (PS), increased at the beginning of the DENV viral cycle, but decreased as time passed. Acylglycerol phosphate acyltransferase (AGPAT) is the rate-limiting enzyme involved in the synthesis of phospholipids (generating phosphatidic acid, a precursor for more complex phospholipids). In *Ae. aegypti,* five AGPAT isoforms were identified, with AGPAT1 being downregulated upon DENV infection (at different times depending on the level of organization, either cell, tissue, or whole organism). Vial et al. (2019) then set out to test whether AGPAT1 regulation is involved in the reconfiguration of the phospholipidome. RNA interference on mosquito cells, used to temporarily knock down *agpat1* and thus mimicking DENV infection, revealed an increase in aminophospholipids. Knockdown of *agpat1* indeed also increased DENV production. The instrumental role of *agpat1* for phospholipid remodeling was confirmed by supplementation of ethanolamine in cells with knocked down *agpat1* expression. Ethanolamine is used in the synthesis of PEs and the presence of ethanolamine in the mosquito cell medium partially restored the observed increase in aminophospholipids in DENV-infected cells. In mosquitoes, knockdown of *agpat1* led to an increase in DENV infection through the consumption or redirection of aminophospholipids.

In a follow-up study, Vial et al. (2020) set out to determine how DENV reconfigures aminophospholipids in mosquitoes, but also how aminophospholipid reconfiguration affects virus proliferation. In the first set of experiments, Vial et al. (2020) knocked down several genes involved in de novo phospholipid synthesis and monitored changes in the phospholipidome. In addition, DENV-infected mosquito Aag2 cells were supplemented with phospholipid precursors to partly restore de novo synthesis. Newly synthesized phospholipids were indeed found to be antiviral, but DENV can inhibit de novo synthesis and initiate phospholipid remodeling to modulate and create a more proviral environment. In a stable isotope tracing experiment using different labeled precursors, Vial et al. (2020) then showed that DENV induces remodeling early on during infection (0–24 h), after which de novo phospholipid synthesis takes place. To test the negative effect of de novo phospholipid synthesis in vivo, mosquitoes were fed an infected blood meal with increased levels of phospholipid precursors. When fed lower precursor levels, DENV was able to increase phospholipid reconfiguration for its own benefit, but reconfiguration was not sufficient at higher precursor concentrations. When DENV-induced remodeling is inhibited by de novo phospholipid synthesis, viral replication (rather than attachment, internalization, or translation) is reduced. Phospholipids were also found to be the main lipid type affected when *Ae. albopictus* cells were infected with ZIKV (Melo et al. 2016).

Cholesterol appears to be essential for the fusion of the lipid envelope of the viral particle with the vector membranes, allowing DENV release and replication (Blanc et al. 2011; Caragata et al. 2013, 2014; Carro and Damonte 2013). In *Ae. aegypti,* sterol carrier protein 2 (SCP-2), involved in cholesterol binding and transport, is essential for cellular cholesterol homeostasis and of importance for DENV production (Fu et al. 2015). Knockdown of SCP-2 indeed reduced DENV production in mosquito Aag2 cells. Further studies with mosquito Aag2 cells revealed that DENV reduced protein expression of low-density lipoprotein receptor-related protein 1 (LRP-1), increasing cholesterol levels and stimulating viral replication (Tree et al. 2019). In mosquitoes, however, low-density-lipoproteins contained in human blood inhibited DENV replication during an early stage of viral infection following a blood meal (also for ZIKV; Wagar et al. 2017). Vertebrate lipids thus seem to have contradictory effects on DENV. To test how DENV responds to low levels of vertebrate lipids, Marten et al. (2022) created cell lines mimicking mosquitoes "feeding" on blood (i.e., provided a normal, control, cell culture medium) or not (i.e., a lipid-depleted medium only). Lipid-depleted cells contained less cholesterol, but similar intracellular lipid levels compared to control cells, despite being smaller and showing reduced proliferation. Mosquito cells thus appear to overcome chronic lipid depletion by reducing lipolysis and increasing de novo lipid synthesis, including fatty acids synthesis. Similar amounts of DENV were found in both cell lines, meaning that mosquito cellular lipid metabolism compensates for a lipid-depleted environment without affecting DENV infection. Cholesterol was also found to play a critical role in alphavirus (e.g., Semliki Forest virus and Sindbis virus) entry and exit in vector cells (Lu et al. 1999).

3.2 Interactions Between *Wolbachia*, Arboviruses, and Lipids

During the past decade, considerable progress has been made in developing novel methods to combat the spread of insect disease vectors, including mosquitoes, and consequently virus transmission. A promising strategy is the use of *Wolbachia* to control and limit arboviral transmission in animals, because *Wolbachia* infection can protect against viral infections (Pimentel et al. 2021). For example, *Wolbachia* can significantly reduce viral load, replication, and transmission of several natural pathogenic RNA viruses associated with the *Drosophila* genus (e.g., Nora virus or Drosophila C virus; Teixeira et al. 2008). A similar effect has also been observed for arthropod-borne viruses, such as West Nile virus or Chikungunya virus, with *Wolbachia* presence generally lowering host insect mortality rate (Glaser and Meola 2010; Hedges et al. 2008; Teixeira et al. 2008). A growing number of studies have, however, suggested that *Wolbachia* can differentially affect viral replication and transmission depending on the insect host species, host strain, and *Wolbachia* strain (Caragata et al. 2013; Hussain et al. 2013; Reyes et al. 2021). For example, replication of West Nile virus in *Ae. aegypti* mosquitoes is significantly reduced by infection with the *w*MelPop *Wolbachia* strain, but no effect was reported for the *w*Mel strain (Hussain et al. 2013).

Wolbachia blocks viral replication and transmission by priming the host's immune system (Angleró-Rodríguez et al. 2017; Bian et al. 2010; Pan et al. 2012, 2018) and/or competing with the virus for host cellular resources, such as amino acids or lipids (Caragata et al. 2013; Moreira et al. 2009). The hypothesis that modification of host metabolic pathways rather than host immune pathways forms the basis for *Wolbachia* pathogen-blocking abilities finds more empirical support. As both the virus and the bacterium are dependent on host lipids for survival and propagation, there can be extreme competition for host lipids, particularly cholesterol. Caragata et al. (2013) tested the influence of a standard, intermediate, or high-cholesterol diet on the ability of *Wolbachia*-infected *D. melanogaster* to resist Drosophila C virus. An increase in cholesterol availability via the enriched diet increased virus replication and reduced the protective effect of *Wolbachia* in a dose-dependent manner. The virus titer was indeed higher in cholesterol-enriched media, leading to earlier death of the flies. An increase in viral replication following cholesterol supplementation was also reported for *Ae. albopictus* and *Ae. aegypti*, suggesting that competition for cholesterol can also play a role in these model systems (Geoghegan et al. 2017; Schultz et al. 2017).

For *Ae. aegypti*, an increase in stored cholesterol (i.e., esterified cholesterol levels) with localized accumulation of lipid droplets in the fat body and a decrease of free cholesterol levels (i.e., potential regulators of lipid transport) were found in *Wolbachia*-infected mosquitoes, suggesting that intracellular cholesterol trafficking may be perturbed (Geoghegan et al. 2017). In *Ae. albopictus*, the abundance of other lipid types, such as sphingolipids, phosphatidylcholines, and diacylglycerols (used by bacteria to enter the cell and activate mechanisms required for bacterial dissemination; Lafont and van der Goot 2005), also decreased following *Wolbachia* infection in DENV-infected *Ae. albopictus* mosquitoes (Molloy et al. 2016). *Wolbachia* and arboviruses may thus compete for multiple lipid types, not only cholesterol. Gene expression studies support these findings, because several genes involved in fatty acid and lipid metabolism, including *fas,* acetyl-coA carboxylase, or sterol-coA desaturase, were downregulated in the presence of *Wolbachia* (Geoghegan et al. 2017; Teramoto et al. 2019). *Wolbachia*-induced metabolic changes, including increased cholesterol storage near viral replication sites, as well as disruption of vesicular trafficking, may thus reduce energy availability needed for viral replication, thereby blocking viral proliferation and transmission (Geoghegan et al. 2017; Schultz et al. 2017, 2018).

The manner by which *Wolbachia* regulates lipid metabolism in the presence of viruses has remained largely unclear. Haqshenas et al. (2019) revealed, however, downregulation of insulin receptor abundance and phosphorylation levels in *Wolbachia*-infected lines, associated with a reduction of DENV and ZIKV proliferation. Inhibition of the insulin receptor revealed that ZIKV and DENV replication is reduced in a dose-dependent manner, suggesting a key role of insulin receptor kinase activity in virus replication. *Wolbachia* may thus reduce insulin receptor phosphorylation and kinase activity, decreasing virus replication (Haqshenas et al. 2019). Insulin was already linked to the activation of the insect host's immune system (Reyes et al. 2021), but further investigation into the underlying mechanisms is needed. Interestingly, here cholesterol could also play a role, as cholesterol is known to affect regulation of the insulin-receptor signaling pathway (Sánchez-Wandelmer et al. 2009).

Wolbachia could become a promising tool for regulating arthropod-borne virus transmission (Ant et al. 2023; Ogunlade et al. 2021). Two recent studies have, however, reported that DENV infection in mosquitoes led to a distinct lipid profile when compared to mosquitoes carrying *Wolbachia* (Koh et al. 2020; Manokaran et al. 2020). This could suggest that DENV and *Wolbachia* may use different lipid types and may not be in competition for lipids. Edenborough et al. (2021) suggested that the intra-thoracic DENV infections used in Koh et al. (2020) could inhibit the effects of *Wolbachia* and may not represent the virus-*Wolbachia* relationship in a natural infection (Fraser et al. 2017). A comprehensive view on the impact of *Wolbachia* and interactions with other microorganisms at the cellular and molecular level is now necessary to fully understand the mechanistic basis of *Wolbachia*-arbovirus interference of lipid metabolism.

4 Plant Pathogen Effects on Insect Vector Fat Metabolism

Plant pathogens represent a major threat to plant populations. In agricultural systems, plant pathogens can reduce yield and affect the quality of agricultural production. Plant pathogens indeed induce significant losses in crops worldwide, representing a major issue for global food security (Fones et al. 2020; Ristaino et al. 2021). Plant viruses can manifest in a variety of symptoms, such as yellowing, spots, necrosis, and distortions of plant structures (Jiang and Zhou 2023). Most plant viruses depend on insect vectors for their survival and transmission, typically phytophagous hemipterans (e.g., aphids, whiteflies, psyllids, leafhoppers, grasshoppers) that use their piercing, sucking mouthparts to feed on plant sap from which the virus is taken up (Hogenhout et al. 2008; Nault 1997). The insect vector then transmits the virus by subsequently feeding on sap from healthy plants.

Plant viruses are generally transmitted by insects via three modes: non-persistent, semi-persistent, and persistent (Nault 1997; Wu et al. 2022). Transmission modes differ in the time during which the insect vector can harbor the virus, ranging from minutes to hours (i.e., non-persistent), days (i.e., semi-persistent), or longer (i.e., persistent; some insects are infected during their entire life and the virus can even be transmitted to insect offspring) (Ng and Falk 2006). Non-persistent and semi-persistent viruses are mainly retained by the insect vector's stylet and foregut, respectively, while persistent viruses infect insect gut cells and are then released in the hemocoel to invade insect tissues and organs (e.g., salivary glands, reproductive system) (Hogenhout et al. 2008; Ng and Falk 2006). The persistent mode of transmission is further categorized as propagative or circulative, depending on whether the location of viral replication is in the insect body or not, respectively (Hogenhout et al. 2008).

Plant viruses have a range of effects on insect vectors by modifying, for example, insect-plant preference/choice, population growth, feeding behavior, or fitness-related traits that may in turn affect survival and transmission of the virus (Blanc and Michalakis 2016; Bosque-Pérez and Eigenbrode 2011; Colvin et al. 2006; Ingwell et al. 2012; Mauck et al. 2012; Stafford et al. 2011). Only little information is available so far on the effects of plant viruses on fat metabolism of insect vectors. Ghodoum Parizipour et al. (2021) investigated the effect of three luteoviruses (i.e., persistent circulative viruses), pea enation mosaic virus (PEMV), bean leafroll virus (BLRV), and barley yellow dwarf virus-PAV (BYDV-PAV) that cause considerable economic losses to cereal and legume fields, on the fatty acid profiles and fat content of the aphid vectors, *A. pisum*, *Aphis fabae*, and *Rhopalosiphum padi*, respectively. Fatty acid profiles differed between infected and uninfected insects in all virus-aphid interactions. In both *A. pisum*-PEMV and *A. fabae*-BRLV interactions, myristic acid (C14:0) quantities increased while an increase in palmitic acid (C16: 0) was reported in *A. fabae*-BRLV and *R. padi*-BYDV-PAV associations. An increase of linoleic acid (C18:2), as well as a decrease of capric (C10:0) and oleic acid (C18:1), were also observed in the *A. pisum*-PEMV, *A. fabae*-BRLV, and *R. padi*-BYDV-PAV interactions, respectively, highlighting specific fatty acid changes depending on the virus-aphid interaction. Infection of *A. fabae* individuals by BRLV further led to a reduction of aphid fat content, while no changes in fat content were reported for the other two virus-aphid interactions (Ghodoum Parizipour et al. 2021). In another virus-aphid vector interaction involving the turnip yellows virus (TuYV) (i.e., a persistent circulative virus, one of the most important viruses infecting cultivated Brassicaceae, e.g., lettuce, broccoli, etc...) and *Myzus persicae*, virus infection also led to a reduction in fat content (Joffrey et al. 2018).

Direct and/or indirect effects of plant viruses have been proposed to explain changes in fatty acid profiles and fat content in *A. fabae* and *M. persicae* infected with BLRV and TuYV respectively. For example, direct immune responses involving lipids, including fatty acids, can protect the insect vector against virus infection (Wrońska et al. 2023). Viruses can also negatively affect plant physiology and quality, decreasing plant biomass and photosynthetic activity, in turn affecting the insect vector (Joffrey et al. 2018). Fat metabolism of the insect vector feeding from the plant sap could be negatively affected due to the lower quantity of nutrients synthetized by the plant (e.g., amino acids). Positive effects of plant virus infection on insect vector fat metabolism have also been reported. The white-backed planthopper *Sogatella furcifera*, vector of the southern rice black-streaked dwarf virus (i.e., a persistent, propagative virus) showed a significant increase in myristic (C14:0), oleic (C18:1), and palmitoleic acid (C16:1) levels in infected individuals (Zhang et al. 2018). Moreover, the small brown planthopper *Laodelphax striatellus,* infected by the maize Iranian mosaic virus (i.e., a persistent propagative virus), harbored more fat that uninfected individuals (Moeini and Tahmasebi 2019). Effects of the maize Iranian mosaic virus on *L. striatellus* fat content was further found to be stage- (i.e., nymph or adult) and sex-specific, where adults and females accumulated more fat. Lipids, including fatty acids, play a key role during viral replication (Konan and Sanchez-Felipe 2014; Lorizate and Krausslich 2011). For persistent propagative viruses, viral replication occurs in the insect tissues/organs; hence increasing and/or modifying insect fat content and fatty acid levels during infection would allow the virus to use insect lipids for replication and dissemination. Finally, an increase of fat storage generally improves insect fitness (Arrese and Soulages 2010; Scheifler et al. 2024, Box 1), allowing the insect to colonize new host plants and, thereby, improve virus transmission.

Plant pathogens other than viruses were also found to affect fat metabolism of insect vectors, including the bacterial pathogen associated with citrus greening disease, *Candidatus Liberibacter asiaticus* (CLAS), for which the Asian citrus phyllid *Diaphorina citri* is the main vector. A proteomic study on *D. citri* adults, infected by CLAS, reported an upregulation of proteins involved in fatty acid beta-oxidation (e.g., enoyl-coA hydratase, acyl-CoA dehydrogenase; Ramsey et al. 2015), while another study found upregulation of *fas* and vitellogenin (i.e., proteins involved in lipid transport) upon infection (Kruse et al. 2018). No change in fatty acid composition was observed between uninfected and infected *D. citri* adults, yet more palmitoleic (C16:1), palmitic (C16:0), linoleic (C18:2), and stearic acid (C18:0) were found in infected nymphs compared to infected adults, suggesting that variation in fatty acid composition is stage-specific (Killiny and Jones 2018). There are thus contrasting results for fat metabolic responses of *D. citri*. Taken together, insect vector fat metabolic responses to plant pathogens are highly dependent on the insect vector-pathogen-host plant interaction considered.

Another topic that has received some attention is the impact of plant viruses and plant physiology and quality on higher trophic levels. Many parasitoids infect vectors of plant pathogens, and virus infection is expected to affect parasitoid performance. Joffrey et al. (2018) studied the effects of TuYV on a plant-aphid-parasitoid interaction, involving the aphid *M. persicae* and the parasitoid *Aphidius colemani*. Reduced photosynthetic activity and lower biomass in TuYV-infected plants led to a decrease in both body size and fat content of *M. persicae* adults. Smaller and leaner aphid adults used as hosts for the parasitoid *A. colemani* led to concomitant decreases in adult parasitoid body size, fat content, and fitness (i.e., lower egg numbers) (Joffrey et al. 2018). No differences were found in host and parasitoid body size and fat content in the aphid *A. fabae,* the parasitoid *Lysiphlebus fabarum* on beets infected with Beet yellows virus (Albittar et al. 2019). Fat storage is particularly important for parasitoids, because most species do not accumulate fat as adults (Visser et al. 2010; Visser et al. 2023; Scheifler et al. 2024). When the amount or quality of fat that can be carried over from the host is reduced due to plant pathogens, there might be negative consequences for parasitoids, a level higher up the trophic food chain. The complexity of these interactions should be studied more carefully to anticipate potential issues in agricultural systems both due to plant disease and complications in biocontrol.

5 Conclusion and Perspectives

Considering the gut microbiota, research on *D. melanogaster* has revealed that individual microbe effects on lipid metabolism appear to be strongly influenced by the metabolic activities of other co-occurring microbes. The complexity of these interactions and their impact on lipid metabolism in general must, therefore, be studied using community-based approaches (rather than mono or dual-infections; Gurung et al. 2019). Furthermore, microbiota composition differs between the sexes in several insect species, suggesting different types of interactions between male and female insect hosts and their respective microbiota (Chen et al. 2016; Fransen et al. 2017; Tang et al. 2012). Metabolic and physiological differences or requirements between the sexes could also explain why interactions between gut microbiota and host fat metabolism are sex-specific, e.g., females require more resources for egg production, mainly lipids. Future work should consider how diet composition and host-related traits, such as genotype and sex, can affect the resident microbiota (Newell and Douglas 2014; Ridley et al. 2012). Such analyses could then be extended to other insect species.

Microbes also seem to play a role in insect recognition and communication. Hertaeg et al. (2021) recently showed that endosymbiotic bacteria can alter the cuticular hydrocarbon (CHC, derived from long-chain fatty acids) composition in the aphid *A. fabae*. CHC profiles depend on the host insect's genetic background, as well as the endosymbiont strain present, which in turn impacts aphid interactions with other insects, such as ants (Hertaeg et al. 2021). We are only beginning to understand the role microbes play in lipid metabolism of insect hosts, but lipid-mediated traits, such as chemical communication, can further affect interspecific insect-insect interactions, also in species other than *A. fabae*.

Wolbachia can have widely different effects on the insect host, including lipid metabolism;

hence *Wolbachia*-insect interactions remain complex to interpret. Modifications of insect fat metabolism and other lipid types appear to depend on insect species, insect host-related traits, and *Wolbachia* strain (Koh et al. 2020; Molloy et al. 2016). Factors other than *Wolbachia* presence should be considered when studying the impact of *Wolbachia* on host fat metabolism, for example, *Wolbachia* density that can vary in response to biotic (Padde et al. 2023; Pascar et al. 2023; Serbus et al. 2008) and abiotic factors (e.g., temperature; Padde et al. 2023; Mouton 2004). We know very little about the mechanism by which *Wolbachia* can modulate insect host fat metabolism. If we want to uncover more about the intricate interplay between *Wolbachia* and insect metabolism, one could investigate lipid-related gene transcription in both the insect host and the *Wolbachia* strain under study. Such a gene-based approach allows for finding correlative data on regulatory and target genes used or exploited by both interacting partners. Once candidate gene regulators and targets have been identified, gene knockdown approaches, such as RNA interference or CRISPR-Cas9, can be used to find a functional link leading to lipid-related phenotypic effects.

Studying the nutritional interplay between symbionts and insect hosts, particularly lipids, is also highly relevant for preventing and managing major public health threats, including vector-borne viruses such as DENV and Chikungunya virus. *Wolbachia* is a promising tool for regulating insect disease vector transmission (Ant et al. 2023; Ogunlade et al. 2021) as *Wolbachia* competes with viruses for multiple host lipid types (Geoghegan et al. 2017; Molloy et al. 2016). A comprehensive overview of the role of fat in *Wolbachia* virus-blocking mechanisms is needed to promote efficient and sustainable virus control in mosquitoes.

Intricate biochemical work on the way in which pathogenic arboviruses manifest within insect mosquito vectors has led to major advancements in our understanding of lipid-virus-mosquito interactions (Vial et al. 2019, 2020, 2021). Viruses critically rely on an array of different lipid types, including fatty acids, phospholipids, and cholesterol, each fulfilling a discrete function for different viral stages. Research on plant pathogen effects on vector lipid metabolism has so far led to varying results, and if lipids are affected, only relatively simple estimates of bulk fat content have been estimated. Lipid effects on vectors could thus be due to indirect effects of infected plants or be a consequence of the viral infection itself. We propose that the research field concerned with plant pathogen-vector interactions draws parallels with the work on pathogenic arboviruses, as the mechanisms by which viruses manipulate and utilize host insect vector lipids may be similar. The use of isotope tracing, precursor supplementation, and genomic interference mechanisms may increase the resolution with which plant pathogen effects can be studied in insect vectors.

Research on the effects of microbes on insect lipid metabolism is up and coming, and we can expect microbes to play unexpected roles in host insect metabolism. The nutritional role lipids play for host insects, microbes or both often remains to be fully elucidated. The repeated evolution of endosymbioses has led to recurrent environmental compensation, where resource provisioning by the insect host has led to genome reduction and trait loss in microbes (Ellers et al. 2012). The loss of fatty acid synthesis pathways in some *Wolbachia* strains is an excellent example of an evolved evolutionary dependence on an insect host (Wu et al. 2004). We can hypothesize that intricate mechanisms to optimize the host environment have evolved in lipid-dependent endosymbionts, for example by stimulating the synthesis of fatty acids or other lipid types by the host. When considering interactions between coexisting microbes, dependence can also evolve when a microbial species provides a common resource, or public good, that is exploited by the community of microbes, also referred to as the Black Queen Hypothesis (Morris 2015). No examples have yet come to light regarding lipids as a public good of microbial origin, but nutrient metabolic interactions could be investigated using recently developed tools, such as NetMet, to predict the metabolic capacities of interacting microbes (Tal et al. 2020). Alternatively, microbes can provide certain nutrients or precursors that are required by the host insect. A well-known example is vitamin B, where different variants are produced by a range of microbes associated with distinct insect species (Serrato-Salas and Gendrin 2023). Regarding various lipids, some microbes, including *Wolbachia*, can synthesize biotin, a co-factor required for acetyl coenzyme A, which is a central intermediary precursor for fatty acid synthesis. We have yet to explore how the synthesis of lipid precursors contributes to lipid dynamics between insect host and symbiont(s).

Acknowledgments BV and MS were supported by the Fonds National de Recherche Scientifique. We would like to thank Thomas Enriquez for making the *D. melanogaster* cartoon used in Fig. 1.

References

Adair KL, Wilson M, Bost A, Douglas AE (2018) Microbial community assembly in wild populations of the fruit fly *Drosophila melanogaster*. ISME J 12:959–972. https://doi.org/10.1038/s41396-017-0020-x

Akinnawo O, Ketiku AO (2000) Chemical composition and fatty acid profile of edible larva of *Cirina forda* (Westwood). Afr J Biomed Res 3:93–96

Albittar L, Ismail M, Lohaus G, Ameline A, Visser B, Bragard C et al (2019) Bottom-up regulation of a tritrophic system by beet yellows virus infection: consequences for aphid-parasitoid foraging behaviour and development. Oecologia 191:113–125. https://doi.org/10.1007/s00442-019-04467-0

Angleró-Rodríguez YI, MacLeod HJ, Kang S, Carlson JS, Jupatanakul N, Dimopoulos G (2017) *Aedes aegypti* molecular responses to zika virus: modulation of infection by the Toll and Jak/Stat immune pathways and virus host factors. Front Microbiol 8:2050. https://doi.org/10.3389/fmicb.2017.02050

Ankrah NYD, Barker BE, Song J, Wu C, McMullen JG, Douglas AE (2021) Predicted metabolic function of the gut microbiota of *Drosophila melanogaster*. mSystems 6:e01369–e01320. https://doi.org/10.1128/mSystems.01369-20

Ant TH, Mancini MV, McNamara CJ, Rainey SM, Sinkins SP (2023) *Wolbachia*-virus interactions and arbovirus control through population replacement in mosquitoes. Pathog Glob Health 117:245–258. https://doi.org/10.1080/20477724.2022.2117939

Arai H, Hirano T, Akizuki N, Abe A, Nakai M, Kunimi Y et al (2019) Multiple infection and reproductive manipulations of *Wolbachia* in *Homona magnanima* (Lepidoptera: Tortricidae). Microb Ecol 77:257–266. https://doi.org/10.1007/s00248-018-1210-4

Arrese EL, Soulages JL (2010) Insect fat body: energy, metabolism, and regulation. Annu Rev Entomol 55:207–225. https://doi.org/10.1146/annurev-ento-112408-085356

Barletta ABF, Alves LR, Nascimento Silva MCL, Sim S, Dimopoulos G, Liechocki S et al (2016) Emerging role of lipid droplets in *Aedes aegypti* immune response against bacteria and dengue virus. Sci Rep 6:19928. https://doi.org/10.1038/srep19928

Bashan M, Akbas H, Yurdakoc K (2002) Phospholipid and triacylglycerol fatty acid composition of major life stages of sunn pest, *Eurygaster integriceps* (Heteroptera: Scutelleridae). Comp Biochem Physiol B Biochem Mol Biol 132:375–380. https://doi.org/10.1016/S1096-4959(02)00045-3

Baumann P, Moran NA, Baumann L (2006) Bacteriocyte-associated endosymbionts of insects. In: Dworkin M, Falkow S, Rosenberg E, Schleifer KH, Stackebrandt E (eds) The prokaryotes. Springer, New York, pp 403–438. https://doi.org/10.1007/0-387-30741-9_16

Behmer ST, Nes WD (2003) Insect sterol nutrition and physiology: a global overview. Adv In Insect Phys 31:1–72. https://doi.org/10.1016/S0065-2806(03)31001-X

Ben-Yosef M, Jurkevitch E, Yuval B (2008) Effect of bacteria on nutritional status and reproductive success of the Mediterranean fruit fly *Ceratitis capitata*. Physiol Entomol 33:145–154. https://doi.org/10.1111/j.1365-3032.2008.00617.x

Berg G, Rybakova D, Fischer D, Cernava T, Vergès M-CC, Charles T et al (2020) Microbiome definition re-visited: old concepts and new challenges. Microbiome 8:103. https://doi.org/10.1186/s40168-020-00875-0

Bhatt S, Gething PW, Brady OJ, Messina JP, Farlow AW, Moyes CL et al (2013) The global distribution and burden of dengue. Nature 496:504–507. https://doi.org/10.1038/nature12060

Bian G, Xu Y, Lu P, Xie Y, Xi Z (2010) The endosymbiotic bacterium *Wolbachia* induces resistance to dengue virus in *Aedes aegypti*. PLoS Pathog 6:e1000833. https://doi.org/10.1371/journal.ppat.1000833

Blanc S, Michalakis Y (2016) Manipulation of hosts and vectors by plant viruses and impact of the environment. Curr Opin Insect Sci 16:36–43. https://doi.org/10.1016/j.cois.2016.05.007

Blanc M, Hsieh WY, Robertson KA, Watterson S, Shui G, Lacaze P et al (2011) Host defense against viral infection involves interferon mediated down-regulation of sterol biosynthesis. PLoS Biol 9:e1000598. https://doi.org/10.1371/journal.pbio.1000598

Bosque-Pérez NA, Eigenbrode SD (2011) The influence of virus-induced changes in plants on aphid vectors: insights from luteovirus pathosystems. Virus Res 159:201–205. https://doi.org/10.1016/j.virusres.2011.04.020

Bozkurt B, Terlemez G, Sezgin E (2023) Basidiomycota species in *Drosophila* gut are associated with host fat metabolism. Sci Rep:13. https://doi.org/10.1038/s41598-023-41027-2

Bueno E, Martin KR, Raguso RA, Mcmullen JG, Hesler SP, Loeb GM et al (2020) Response of wild spotted wing *Drosophila* (*Drosophila suzukii*) to microbial volatiles. J Chem Ecol 46:688–698. https://doi.org/10.1007/s10886-019-01139-4

Cakmak O, Bashan M, Bolu H (2007a) The fatty acid compositions of predator *Piocoris luridus* (Heteroptera: Lygaeidae) and its host *Monosteria unicostata* (Heteroptera: Tingidae) reared on almond. Insect Sci 14:461–466. https://doi.org/10.1111/j.1744-7917.2007.00174.x

Cakmak O, Bashan M, Satar A (2007b) Total lipid and fatty acid compositions of *Lertha sheppardi* (Neuroptera: Nemopteridae) during its main life stages. Biologia 62:774–780. https://doi.org/10.2478/s11756-007-0147-8

Cao LJ, Jiang W, Hoffmann AA (2019) Life history effects linked to an advantage for *w*Au *Wolbachia* in *Drosophila*. Insects 10:126. https://doi.org/10.3390/insects10050126

Caragata EP, Rancès E, Hedges LM, Gofton AW, Johnson KN, O'Neill SL et al (2013) Dietary cholesterol modulates pathogen blocking by *Wolbachia*. PLoS Pathog 9:e1003459. https://doi.org/10.1371/journal.ppat.1003459

Caragata EP, Rancès E, O'Neill SL, McGraw EA (2014) Competition for amino acids between *Wolbachia* and the mosquito host, *Aedes aegypti*. Microb Ecol 67: 205–218. https://doi.org/10.1007/s00248-013-0339-4

Caragata EP, Pais FS, Baton LA, Silva JBL, Sorgine MHF, Moreira LA (2017) The transcriptome of the mosquito *Aedes fluviatilis* (Diptera: Culicidae), and transcriptional changes associated with its native *Wolbachia* infection. BMC Genomics 18:6. https://doi.org/10.1186/s12864-016-3441-4

Carro AC, Damonte EB (2013) Requirement of cholesterol in the viral envelope for dengue virus infection. Virus Res 174:78–87. https://doi.org/10.1016/j.virusres.2013.03.005

Chandler JA, Morgan Lang J, Bhatnagar S, Eisen JA, Kopp A (2011) Bacterial communities of diverse *Drosophila* species: ecological context of a host–microbe model system. PLoS Genet 7:e1002272. https://doi.org/10.1371/journal.pgen.1002272

Chandler JA, Eisen JA, Kopp A (2012) Yeast communities of diverse *Drosophila* species: comparison of two symbiont groups in the same hosts. Appl Environ Microbiol 78:7327–7336. https://doi.org/10.1128/AEM.01741-12

Charlat S, Calmet C, Merçot H (2001) On the *mod resc* model and the evolution of *Wolbachia* compatibility types. Genetics 159:1415–1422. https://doi.org/10.1093/genetics/159.4.1415

Chen B, Teh B-S, Sun C, Hu S, Lu X, Boland W et al (2016) Biodiversity and activity of the gut microbiota across the life history of the insect herbivore *Spodoptera littoralis*. Sci Rep 6:29505. https://doi.org/10.1038/srep29505

Cho K-O, Kim G-W, Lee O-K (2011) *Wolbachia* bacteria reside in host Golgi-related vesicles whose position is regulated by polarity proteins. PLoS One 6:e22703. https://doi.org/10.1371/journal.pone.0022703

Chotiwan N, Andre BG, Sanchez-Vargas I, Islam MN, Grabowski JM, Hopf-Jannasch A et al (2018) Dynamic remodeling of lipids coincides with dengue virus replication in the midgut of *Aedes aegypti* mosquitoes. PLoS Pathog 14:e1006853. https://doi.org/10.1371/journal.ppat.1006853

Chotiwan N, Brito-Sierra CA, Ramirez G, Lian E, Grabowski JM, Graham B et al (2022) Expression of fatty acid synthase genes and their role in development and arboviral infection of *Aedes aegypti*. Parasit Vectors 15:233. https://doi.org/10.1186/s13071-022-05336-1

Colvin J, Omongo CA, Govindappa MR, Stevenson PC, Maruthi MN, Gibson G et al (2006) Host-plant viral infection effects on arthropod-vector population growth, development and behaviour: management and epidemiological implications. Adv Virus Res 67:419–452. https://doi.org/10.1016/S0065-3527(06)67011-5

Conceição CC, da Silva JN, Arcanjo A, Nogueira CL, de Abreu LA, de Oliveira PL et al (2021) *Aedes fluviatilis* cell lines as new tools to study metabolic and immune interactions in mosquito-*Wolbachia* symbiosis. Sci Rep 11:19202. https://doi.org/10.1038/s41598-021-98738-7

Consuegra J, Grenier T, Akherraz H, Rahioui I, Gervais H, da Silva P et al (2020) Metabolic cooperation among commensal bacteria supports *Drosophila* juvenile growth under nutritional stress. iScience 23:101232. https://doi.org/10.1016/j.isci.2020.101232

Cornwallis CK, van't Padje A, Ellers J, Klein M, Jackson R, Kiers ET et al (2023) Symbioses shape feeding niches and diversification across insects. Nat Ecol Evol 7:1022–1044. https://doi.org/10.1038/s41559-023-02058-0

Cruz-Oliveira C, Freire JM, Conceição TM, Higa LM, Castanho MARB, Da Poian AT (2015) Receptors and routes of dengue virus entry into the host cells. FEMS Microbiol Rev 39:155–170. https://doi.org/10.1093/femsre/fuu004

Currin-Ross D, Husdell L, Pierens GK, Mok NE, O'Neill SL, Schirra HJ et al (2021) The metabolic response to infection with *Wolbachia* implicates the insulin/insulin-like-growth factor and hypoxia signaling pathways in *Drosophila melanogaster*. Front Ecol Evol 9: 623561. https://doi.org/10.3389/fevo.2021.623561

Denlinger DL (2002) Regulation of diapause. Annu Rev Entomol 47:93–122. https://doi.org/10.1146/annurev.ento.47.091201.145137

Denlinger DL, Yocum GD, Rinehart JP (2012) Hormonal control of diapause. Insect endocrinology. Elsevier, pp 430–463. https://doi.org/10.1016/B978-0-12-384749-2.10010-X

Depommier C, Everard A, Druart C, Plovier H, van Hul M, Vieira-Silva S et al (2019) Supplementation with *Akkermansia muciniphila* in overweight and obese human volunteers: a proof-of-concept exploratory study. Nat Med 25:1096–1103. https://doi.org/10.1038/s41591-019-0495-2

Didion EM, Sabree ZL, Kenyon L, Nine G, Hagan RW, Osman S et al (2021) Microbiome reduction prevents lipid accumulation during early diapause in the northern house mosquito, *Culex pipiens pipiens*. J Insect Physiol 134:104295. https://doi.org/10.1016/j.jinsphys.2021.104295

Dittmer J, Bouchon D (2018) Feminizing *Wolbachia* influence microbiota composition in the terrestrial isopod *Armadillidium vulgare*. Sci Rep 8:6998. https://doi.org/10.1038/s41598-018-25450-4

Dittmer J, Brucker RM (2021) When your host shuts down: larval diapause impacts host-microbiome interactions in *Nasonia vitripennis*. Microbiome 9:85. https://doi.org/10.1186/s40168-021-01037-6

Dobson SL, Bourtzis K, Braig HR, Jones BF, Zhou W, Rousset F et al (1999) *Wolbachia* infections are distributed throughout insect somatic and germ line tissues. Insect Biochem Mol Biol 29:153–160. https://doi.org/10.1016/S0965-1748(98)00119-2

Dou W, Miao Y, Xiao J, Huang D (2021) Association of *Wolbachia* with gene expression in *Drosophila* testes. Microb Ecol 82:805–817. https://doi.org/10.1007/s00248-021-01703-0

Douglas AE (2015) Multiorganismal insects: diversity and function of resident microorganisms. Annu Rev Entomol 60:17–34. https://doi.org/10.1146/annurev-ento-010814-020822

Douglas AE (2019) Simple animal models for microbiome research. Nat Rev Microbiol 17:764–775. https://doi.org/10.1038/s41579-019-0242-1

Douglas AE, Minto LB, Wilkinson TL (2001) Quantifying nutrient production by the microbial symbionts in an aphid. J Exp Biol 204:349–358. https://doi.org/10.1242/jeb.204.2.349

Edenborough KM, Flores HA, Simmons CP, Fraser JE (2021) Using *Wolbachia* to eliminate dengue: will the virus fight back? J Virol 95:0220320. https://doi.org/10.1128/JVI.02203-20

Ellers J, Toby Kiers E, Currie CR, McDonald BR, Visser B (2012) Ecological interactions drive evolutionary loss of traits. Ecol Lett 15:1071–1082. https://doi.org/10.1111/j.1461-0248.2012.01830.x

Engel P, Moran NA (2013) The gut microbiota of insects – diversity in structure and function. FEMS Microbiol Rev 37:699–735. https://doi.org/10.1111/1574-6976.12025

Engel P, Martinson VG, Moran NA (2012) Functional diversity within the simple gut microbiota of the honey bee. Proc Natl Acad Sci USA 109:11002–11007. https://doi.org/10.1073/pnas.1202970109

Engl T, Kaltenpoth M (2018) Influence of microbial symbionts on insect pheromones. Nat Prod Rep 35:386–397. https://doi.org/10.1039/C7NP00068E

Enriquez T, Visser B (2023) The importance of fat accumulation and reserves for insect overwintering. Curr Opin Insect Sci 60:101118. https://doi.org/10.1016/j.cois.2023.101118

Erkosar B, Storelli G, Defaye A, Leulier F (2013) Host-intestinal microbiota mutualism: "learning on the fly". Cell Host Microbe 13:8–14. https://doi.org/10.1016/j.chom.2012.12.004

Farahani HK, Ashouri A, Zibaee A, Abroon P, Alford L, Pierre J-S et al (2017) Early life nutritional quality effects on adult memory retention in a parasitic wasp. Behav Ecol 28:818–826. https://doi.org/10.1093/beheco/arx042

Faria VG, Martins NE, Schlötterer C, Sucena É (2018) Readapting to DCV infection without *Wolbachia*: frequency changes of *Drosophila* antiviral alleles can replace endosymbiont protection. Genome Biol Evol 10:1783–1791. https://doi.org/10.1093/gbe/evy137

Fattouh N, Cazevieille C, Landmann F (2019) *Wolbachia* endosymbionts subvert the endoplasmic reticulum to acquire host membranes without triggering ER stress. PLoS Negl Trop Dis 13:e0007218. https://doi.org/10.1371/journal.pntd.0007218

Fauci AS, Morens DM (2016) Zika virus in the Americas – yet another arbovirus threat. N Engl J Med 374:601–604. https://doi.org/10.1056/NEJMp1600297

Fones HN, Bebber DP, Chaloner TM, Kay WT, Steinberg G, Gurr SJ (2020) Threats to global food security from emerging fungal and oomycete crop pathogens. Nat Food 1:332–342. https://doi.org/10.1038/s43016-020-0075-0

Fransen F, van Beek AA, Borghuis T, Meijer B, Hugenholtz F, van der Gaast-de JC et al (2017) The impact of gut microbiota on gender-specific differences in immunity. Front Immunol 8:754. https://doi.org/10.3389/fimmu.2017.00754

Fraser JE, De Bruyne JT, Iturbe-Ormaetxe I, Stepnell J, Burns RL, Flores HA et al (2017) Novel *Wolbachia*-transinfected *Aedes aegypti* mosquitoes possess diverse fitness and vector competence phenotypes. PLoS Pathog 13:e1006751. https://doi.org/10.1371/journal.ppat.1006751

Fu Q, Inankur B, Yin J, Striker R, Lan Q (2015) Sterol carrier protein 2, a critical host factor for dengue virus infection, alters the cholesterol distribution in mosquito Aag2 cells. J Med Entomol 52:1124–1134. https://doi.org/10.1093/jme/tjv101

Geoghegan V, Stainton K, Rainey SM, Ant TH, Dowle AA, Larson T et al (2017) Perturbed cholesterol and vesicular trafficking associated with dengue blocking in *Wolbachia*-infected *Aedes aegypti* cells. Nat Commun 8:526. https://doi.org/10.1038/s41467-017-00610-8

Ghodoum Parizipour MH, Tahmasebi A, Shahriari AG, Khashman M, Hemmati F (2021) Luteoviruses affected energy reserves and fatty acid composition of their aphid vectors. J Phytopathol 169:376–386. https://doi.org/10.1111/jph.12993

Giraud É, Varet H, Legendre R, Sismeiro O, Aubry F, Dabo S et al (2022) Mosquito-bacteria interactions during larval development trigger metabolic changes with carry-over effects on adult fitness. Mol Ecol 31: 1444–1460. https://doi.org/10.1111/mec.16327

Glaser RL, Meola MA (2010) The native *Wolbachia* endosymbionts of *Drosophila melanogaster* and *Culex quinquefasciatus* increase host resistance to west nile virus infection. PLoS One 5:e11977. https://doi.org/10.1371/journal.pone.0011977

Goane L, Salgueiro J, Medina Pereyra P, Arce OEA, Ruiz MJ, Nussenbaum AL et al (2022) Antibiotic treatment reduces fecundity and nutrient content in females of *Anastrepha fraterculus* (Diptera: Tephritidae) in a diet dependent way. J Insect Physiol 139:104396. https://doi.org/10.1016/j.jinsphys.2022.104396

Gołębiowski M, Boguś MI, Paszkiewicz M, Stepnowski P (2010) The composition of the free fatty acids from *Dendrolimus pini* exuviae. J Insect Physiol 56:391–397. https://doi.org/10.1016/j.jinsphys.2009.11.009

Gruntenko NE, Ilinsky YY, Adonyeva NV, Burdina EV, Bykov RA, Menshanov PN et al (2017) Various *Wolbachia* genotypes differently influence host *Drosophila* dopamine metabolism and survival under heat stress conditions. BMC Evol Biol 17:252. https://doi.org/10.1186/s12862-017-1104-y

Gurung K, Wertheim B, Falcao SJ (2019) The microbiome of pest insects: it is not just bacteria. Entomol Exp Appl 167:156–170. https://doi.org/10.1111/eea.12768

Guzman MG, Halstead SB, Artsob H, Buchy P, Farrar J, Gubler DJ et al (2010) Dengue: a continuing global threat. Nat Rev Microbiol 8:S7–S16. https://doi.org/10.1038/nrmicro2460

Habineza P, Muhammad A, Ji T, Xiao R, Yin X, Hou Y et al (2019) The promoting effect of gut microbiota on growth and development of red palm weevil, *Rhynchophorus ferrugineus* (Olivier) (Coleoptera: Dryophthoridae) by modulating its nutritional metabolism. Front Microbiol 10:1212. https://doi.org/10.3389/fmicb.2019.01212

Hahn DA, Denlinger DL (2007) Meeting the energetic demands of insect diapause: nutrient storage and utilization. J Insect Physiol 53:760–773. https://doi.org/10.1016/j.jinsphys.2007.03.018

Hahn DA, Denlinger DL (2011) Energetics of insect diapause. Annu Rev Entomol 56:103–121. https://doi.org/10.1146/annurev-ento-112408-085436

Haqshenas G, Terradas G, Paradkar PN, Duchemin J-B, McGraw EA, Doerig C (2019) A role for the insulin receptor in the suppression of dengue virus and zika virus in *Wolbachia*-infected mosquito cells. Cell Rep 26:529–535.e3. https://doi.org/10.1016/j.celrep.2018.12.068

Hedges LM, Brownlie JC, O'Neill SL, Johnson KN (2008) *Wolbachia* and virus protection in insects. Science 322: 702. https://doi.org/10.1126/science.1162418

Henry Y, Overgaard J, Colinet H (2020) Dietary nutrient balance shapes phenotypic traits of *Drosophila melanogaster* in interaction with gut microbiota. Comp Biochem Physiol A Mol Integr Physiol 241: 110626. https://doi.org/10.1016/j.cbpa.2019.110626

Herren JK, Paredes JC, Schüpfer F, Arafah K, Bulet P, Lemaitre B (2014) Insect endosymbiont proliferation is limited by lipid availability. eLife 3:e02964. https://doi.org/10.7554/eLife.02964

Hertaeg C, Risse M, Vorburger C, De Moraes CM, Mescher MC (2021) Aphids harbouring different endosymbionts exhibit differences in cuticular hydrocarbon profiles that can be recognized by ant mutualists. Sci Rep 11:19559. https://doi.org/10.1038/s41598-021-98098-2

Hilgenboecker K, Hammerstein P, Schlattmann P, Telschow A, Werren JH (2008) How many species are infected with *Wolbachia*? A statistical analysis of current data. FEMS Microbiol Lett 281:215–220. https://doi.org/10.1111/j.1574-6968.2008.01110.x

Hogenhout SA, Ammar E-D, Whitfield AE, Redinbaugh MG (2008) Insect vector interactions with persistently transmitted viruses. Annu Rev Phytopathol 46:327–359. https://doi.org/10.1146/annurev.phyto.022508.092135

Hosokawa T, Koga R, Kikuchi Y, Meng X-Y, Fukatsu T (2010) *Wolbachia* as a bacteriocyte-associated nutritional mutualist. Proc Natl Acad Sci USA 107: 769–774. https://doi.org/10.1073/pnas.0911476107

Howard RW, Stanley-Samuelson DW (1990) Phospholipid fatty acid composition and arachidonic acid metabolism in selected tissues of adult *Tenebrio molitor* (Coleoptera: Tenebrionidae). Ann Entomol Soc Am 83:975–981. https://doi.org/10.1093/aesa/83.5.975

Howe D, Heinzen RA (2006) *Coxiella burnetii* inhabits a cholesterol-rich vacuole and influences cellular cholesterol metabolism. Cell Microbiol 8:496–507. https://doi.org/10.1111/j.1462-5822.2005.00641.x

Hu Y, Sanders JG, Łukasik P, D'Amelio CL, Millar JS, Vann DR et al (2018) Herbivorous turtle ants obtain essential nutrients from a conserved nitrogen-recycling gut microbiome. Nat Commun 9:964. https://doi.org/10.1038/s41467-018-03357-y

Huang J-H, Douglas AE (2015) Consumption of dietary sugar by gut bacteria determines *Drosophila* lipid content. Biol Lett 11:20150469. https://doi.org/10.1098/rsbl.2015.0469

Hughes GL, Koga R, Xue P, Fukatsu T, Rasgon JL (2011) *Wolbachia* infections are virulent and inhibit the human malaria parasite *Plasmodium falciparum* in *Anopheles aambiae*. PLoS Pathog 7:e1002043. https://doi.org/10.1371/journal.ppat.1002043

Hurst GDD, Jiggins FM, von der Schulenburg JHG, Bertrand D, West SA, Goriacheva II et al (1999) Male–killing *Wolbachia* in two species of insect. Proc R Soc Lond B Biol Sci 266:735–740. https://doi.org/10.1098/rspb.1999.0698

Hussain M, Lu G, Torres S, Edmonds JH, Kay BH, Khromykh AA et al (2013) Effect of *Wolbachia* on replication of West Nile virus in a mosquito cell line and adult mosquitoes. J Virol 87:851–858. https://doi.org/10.1128/JVI.01837-12

Ikeya T, Broughton S, Alic N, Grandison R, Partridge L (2009) The endosymbiont *Wolbachia* increases insulin/IGF-like signalling in *Drosophila*. Proc R Soc Lond B Biol Sci 276:3799–3807. https://doi.org/10.1098/rspb.2009.0778

Ingwell LL, Eigenbrode SD, Bosque-Pérez NA (2012) Plant viruses alter insect behavior to enhance their spread. Sci Rep 2:578. https://doi.org/10.1038/srep00578

Janson EM, Stireman JO, Singer MS, Abbot P (2008) Phytophagous insect-microbe mutualisms and adaptive evolutionary diversification. Evolution 62:997–1012. https://doi.org/10.1111/j.1558-5646.2008.00348.x

Jing X, Behmer ST (2020) Insect sterol nutrition: physiological mechanisms, ecology, and applications. Annu Rev Entomol 65:251–271. https://doi.org/10.1146/annurev-ento-011019-025017

Jiang T, Zhou T (2023) Unraveling the mechanisms of virus-induced symptom development in plants. Plan Theory 12:2830. https://doi.org/10.3390/plants12152830

Jing T-Z, Qi F-H, Wang Z-Y (2020) Most dominant roles of insect gut bacteria: digestion, detoxification, or essential nutrient provision? Microbiome 8:38. https://doi.org/10.1186/s40168-020-00823-y

Joffrey M, Chesnais Q, Spicher F, Verrier E, Ameline A, Couty A (2018) Plant virus infection influences bottom-up regulation of a plant-aphid-parasitoid system. J Pest Sci 91:361–372. https://doi.org/10.1007/s10340-017-0911-7

Jolly NP, Varela C, Pretorius IS (2014) Not your ordinary yeast: non-*Saccharomyces* yeasts in wine production uncovered. FEMS Yeast Res 14:215–237. https://doi.org/10.1111/1567-1364.12111

Kaczmarek A, Wrońska AK, Kazek M, Boguś MI (2020) Metamorphosis-related changes in the free fatty acid profiles of *Sarcophaga* (Liopygia) *argyrostoma* (Robineau-Desvoidy, 1830). Sci Rep 10:17337. https://doi.org/10.1038/s41598-020-74475-1

Karpova EK, Bobrovskikh MA, Deryuzhenko MA, Shishkina OD, Gruntenko NE (2023) *Wolbachia* effect on *Drosophila melanogaster* lipid and carbohydrate metabolism. Insects 14:357. https://doi.org/10.3390/insects14040357

Killiny N, Jones SE (2018) Metabolic alterations in the nymphal instars of *Diaphorina citri* induced by *Candidatus* Liberibacter asiaticus, the putative pathogen of huanglongbing. PLoS One 13:e0191871. https://doi.org/10.1371/journal.pone.0191871

Koh C, Islam MN, Ye YH, Chotiwan N, Graham B, Belisle JT et al (2020) Dengue virus dominates lipid metabolism modulations in *Wolbachia*-coinfected *Aedes aegypti*. Commun Biol 3:518. https://doi.org/10.1038/s42003-020-01254-z

Konan KV, Sanchez-Felipe L (2014) Lipids and RNA virus replication. Curr Opin Virol 9:45–52. https://doi.org/10.1016/j.coviro.2014.09.005

Kruse A, Ramsey JS, Johnson R, Hall DG, MacCoss MJ, Heck M (2018) *Candidatus* Liberibacter asiaticus minimally alters expression of immunity and metabolism proteins in hemolymph of *Diaphorina citri*, the insect vector of huanglongbing. J Proteome Res 17:2995–3011. https://doi.org/10.1021/acs.jproteome.8b00183

Lafont F, van der Goot FG (2005) Bacterial invasion via lipid rafts. Cell Microbiol 7:613–620. https://doi.org/10.1111/j.1462-5822.2005.00515.x

Ley RE, Bäckhed F, Turnbaugh P, Lozupone CA, Knight RD, Gordon JI (2005) Obesity alters gut microbial ecology. Proc Natl Acad Sci USA 102:11070–11075. https://doi.org/10.1073/pnas.0504978102

Li G, Zheng X, Zhu Y, Long Y, Xia X (2022) *Bacillus* symbiont drives alterations in intestinal microbiota and circulating metabolites of lepidopteran host. Environ Microbiol 24:4049–4064. https://doi.org/10.1111/1462-2920.15934

Lin M, Rikihisa Y (2003) *Ehrlichia chaffeensis* and *Anaplasma phagocytophilum* lack genes for lipid A biosynthesis and incorporate cholesterol for their survival. Infect Immun 71:5324–5331. https://doi.org/10.1128/IAI.71.9.5324-5331.2003

Liu W, Li Y, Guo S, Yin H, Lei C-L, Wang X-P (2016) Association between gut microbiota and diapause preparation in the cabbage beetle: a new perspective for studying insect diapause. Sci Rep 6:38900. https://doi.org/10.1038/srep38900

Liu B-N, Liu X-T, Liang Z-H, Wang J-H (2021a) Gut microbiota in obesity. World J Gastroenterol 27:3837–3850. https://doi.org/10.3748/wjg.v27.i25.3837

Liu Y-Y, Liang X-D, Liu C-C, Cheng Y, Chen H, Baloch AS et al (2021b) Fatty acid synthase is involved in classical swine fever virus replication by interaction with NS4B. J Virol:95. https://doi.org/10.1128/JVI.00781-21

Lorizate M, Krausslich H-G (2011) Role of lipids in virus replication. Cold Spring Harb Perspect Biol 3:a004820–a004820. https://doi.org/10.1101/cshperspect.a004820

Lu YE, Cassese T, Kielian M (1999) The cholesterol requirement for Sindbis virus entry and exit and characterization of a spike protein region involved in cholesterol dependence. J Virol 73:4272–4278. https://doi.org/10.1128/JVI.73.5.4272-4278.1999

Lv W-X, Cheng P, Lei J-J, Peng H, Zang C-H, Lou Z-W et al (2023) Interactions between the gut microcommunity and transcriptome of *Culex pipiens* pallens under low-temperature stress. Parasit Vectors 16:12. https://doi.org/10.1186/s13071-022-05643-7

Maistrenko OM, Serga SV, Vaiserman AM, Kozeretska IA (2016) Longevity-modulating effects of symbiosis: insights from *Drosophila*–*Wolbachia* interaction. Biogerontology 17:785–803. https://doi.org/10.1007/s10522-016-9653-9

Manokaran G, Flores HA, Dickson CT, Narayana VK, Kanojia K, Dayalan S et al (2020) Modulation of acyl-carnitines, the broad mechanism behind *Wolbachia*-mediated inhibition of medically important flaviviruses in *Aedes aegypti*. Proc Natl Acad Sci USA

117:24475–24483. https://doi.org/10.1073/pnas.1914814117

Marten AD, Tift CT, Tree MO, Bakke J, Conway MJ (2022) Chronic depletion of vertebrate lipids in *Aedes aegypti* cells dysregulates lipid metabolism and inhibits innate immunity without altering dengue infectivity. PLoS Negl Trop Dis 16:e0010890. https://doi.org/10.1371/journal.pntd.0010890

Mauck K, Bosque-Pérez NA, Eigenbrode SD, De Moraes CM, Mescher MC (2012) Transmission mechanisms shape pathogen effects on host–vector interactions: evidence from plant viruses. Funct Ecol 26:1162–1175. https://doi.org/10.1111/j.1365-2435.2012.02026.x

Mazzucco R, Nolte V, Vijayan T, Schlötterer C (2020) Long-term dynamics among *Wolbachia* strains during thermal adaptation of their *Drosophila melanogaster* hosts. Front Genet 11:482. https://doi.org/10.3389/fgene.2020.00482

McBride MJ, Xie G, Martens EC, Lapidus A, Henrissat B, Rhodes RG et al (2009) Novel features of the polysaccharide-digesting gliding bacterium *Flavobacterium johnsoniae* as revealed by genome sequence analysis. Appl Environ Microbiol 75:6864–6875. https://doi.org/10.1128/AEM.01495-09

McMullen JG, Peters-Schulze G, Cai J, Patterson AD, Douglas AE (2020) How gut microbiome interactions affect nutritional traits of *Drosophila melanogaster*. J Exp Biol 223:jeb227843. https://doi.org/10.1242/jeb.227843

Melo CFOR, de Oliveira DN, de Olivea Lima E, Guerreiro TM, Esteves CZ, Beck RM et al (2016) A lipidomics approach in the characterization of zika-infected mosquito cells: potential targets for breaking the transmission cycle. PLoS One 11:e0164377. https://doi.org/10.1371/journal.pone.0164377

Moeini P, Tahmasebi A (2019) Maize Iranian mosaic virus infection promotes the energy sources of its insect vector, *Laodelphax striatellus*. J Appl Entomol 143:271–276. https://doi.org/10.1111/jen.12585

Molloy JC, Sommer U, Viant MR, Sinkins SP (2016) *Wolbachia* modulates lipid metabolism in *Aedes albopictus* mosquito cells. Appl Environ Microbiol 82:3109–3120. https://doi.org/10.1128/AEM.00275-16

Moreira LA, Iturbe-Ormaetxe I, Jeffery JA, Lu G, Pyke AT, Hedges LM et al (2009) A *Wolbachia* symbiont in *Aedes aegypti* limits infection with dengue, chikungunya, and *Plasmodium*. Cell 139:1268–1278. https://doi.org/10.1016/j.cell.2009.11.042

Morris JJ (2015) Black Queen evolution: the role of leakiness in structuring microbial communities. Trends Genet 31:475–482. https://doi.org/10.1016/j.tig.2015.05.004

Mosso C, Galván-Mendoza IJ, Ludert JE, del Angel RM (2008) Endocytic pathway followed by dengue virus to infect the mosquito cell line C6/36 HT. Virology 378:193–199. https://doi.org/10.1016/j.virol.2008.05.012

Mouton (2004) PhD thesis. Diversité et densité bactériennes dans les symbioses à infections multiples: régulation et effets sur l'hôte. Cas des associations *Wolbachia*-insectes

Nault LR (1997) Arthropod transmission of plant viruses: a new synthesis. Ann Entomol Soc Am 90:521–541. https://doi.org/10.1093/aesa/90.5.521

Newell PD, Douglas AE (2014) Interspecies interactions determine the impact of the gut microbiota on nutrient allocation in *Drosophila melanogaster*. Appl Environ Microbiol 80:788–796. https://doi.org/10.1128/AEM.02742-13

Newton ILG, Rice DW (2020) The Jekyll and Hyde symbiont: could *Wolbachia* be a nutritional mutualist? J Bacteriol 202:e00589–e00519. https://doi.org/10.1128/JB.00589-19

Ng JCK, Falk BW (2006) Virus-vector interactions mediating nonpersistent and semipersistent transmission of plant viruses. Annu Rev Phytopathol 44:183–212. https://doi.org/10.1146/annurev.phyto.44.070505.143325

Nguyen B, Dinh H, Morimoto J, Ponton F (2021) Sex-specific effects of the microbiota on adult carbohydrate intake and body composition in a polyphagous fly. J Insect Physiol 134:104308. https://doi.org/10.1016/j.jinsphys.2021.104308

Nikolova N, Rezanka T, Nikolova-Damyanova B (2000) Fatty acid profiles of main lipid classes in adult *Chrysomela vigintipunctata* (Scopoli) (Coleopterai: Chrysomelidae). Z Naturforsch C 55:661–666. https://doi.org/10.1515/znc-2000-7-828

Ogunlade ST, Meehan MT, Adekunle AI, Rojas DP, Adegboye OA, McBryde ES (2021) A review: *Aedes*-borne arboviral infections, controls and *Wolbachia*-based strategies. Vaccine 9:32. https://doi.org/10.3390/vaccines9010032

Padde JR, Lu Q, Long Y, Zhang D, Hou M, Chen L et al (2023) The impact of environmental and host factors on *Wolbachia* density and efficacy as a biological tool. Decod Infect Trans 1:100006. https://doi.org/10.1016/j.dcit.2023.100006

Pan X, Zhou G, Wu J, Bian G, Lu P, Raikhel AS et al (2012) *Wolbachia* induces reactive oxygen species (ROS)-dependent activation of the toll pathway to control dengue virus in the mosquito *Aedes aegypti*. Proc Natl Acad Sci USA 109:E23–E31. https://doi.org/10.1073/pnas.1116932108

Pan X, Pike A, Joshi D, Bian G, McFadden MJ, Lu P et al (2018) The bacterium *Wolbachia* exploits host innate immunity to establish a symbiotic relationship with the dengue vector mosquito *Aedes aegypti*. ISME J 12:277–288. https://doi.org/10.1038/ismej.2017.174

Paredes JC, Herren JK, Schüpfer F, Lemaitre B (2016) The role of lipid competition for endosymbiont-mediated protection against parasitoid wasps in *Drosophila*. MBio 7:e01006–e01016. https://doi.org/10.1128/mBio.01006-16

Park Y, Ledesma-Amaro R, Nicaud J-M (2020) De novo biosynthesis of odd-chain fatty acids in *Yarrowia lipolytica* enabled by modular pathway engineering.

Front Bioeng Biotechnol 7:484. https://doi.org/10.3389/fbioe.2019.00484

Pascar J, Middleton H, Dorus S (2023) *Aedes aegypti* microbiome composition covaries with the density of *Wolbachia* infection. Microbiome 11:255. https://doi.org/10.1186/s40168-023-01678-9

Perera R, Riley C, Isaac G, Hopf-Jannasch AS, Moore RJ, Weitz KW et al (2012) Dengue virus infection perturbs lipid homeostasis in infected mosquito cells. PLoS Pathog 8:e1002584. https://doi.org/10.1371/journal.ppat.1002584

Pickett KM, McHenry A, Wenzel JW (2000) Nestmate recognition in the absence of a pheromone. Insect Soc 47:212–219. https://doi.org/10.1007/PL00001705

Pietri JE, DeBruhl H, Sullivan W (2016) The rich somatic life of *Wolbachia*. Microbiol Open 5:923–936. https://doi.org/10.1002/mbo3.390

Pimentel AC, Cesar CS, Martins M, Cogni R (2021) The antiviral effects of the symbiont bacteria *Wolbachia* in insects. Front Immunol 11:626329. https://doi.org/10.3389/fimmu.2020.626329

Poinsot D, Charlat S, Merçot H (2003) On the mechanism of *Wolbachia*-induced cytoplasmic incompatibility: confronting the models with the facts. BioEssays 25: 259–265. https://doi.org/10.1002/bies.10234

Rahbé Y, Delobel B, Febvay G, Chantegrel B (1994) Aphid-specific triglycerides in symbiotic and aposymbiotic *Acyrthosiphon pisum*. Insect Biochem Mol Biol 24:95–101. https://doi.org/10.1016/0965-1748(94)90127-9

Rahimi-Kaldeh S, Bandani A, Ashouri A (2019) Does *Wolbachia* change diapause and energy reserves of *Trichogramma brassicae* in response to light wavelengths? J Agr Sci Tech 21:1173–1182

Ramsey JS, Johnson RS, Hoki JS, Kruse A, Mahoney J, Hilf ME et al (2015) Metabolic interplay between the Asian citrus psyllid and its *Profftella* symbiont: an Achilles' heel of the citrus greening insect vector. PLoS One 10:e0140826. https://doi.org/10.1371/journal.pone.0140826

Rancès E, Ye YH, Woolfit M, McGraw EA, O'Neill SL (2012) The relative importance of innate immune priming in *Wolbachia*-mediated dengue interference. PLoS Pathog 8:e1002548. https://doi.org/10.1371/journal.ppat.1002548

Ratnayake OC, Chotiwan N, Saavedra-Rodriguez K, Perera R (2023) The buzz in the field: the interaction between viruses, mosquitoes, and metabolism. Front Cell Infect Microbiol 13:1128577. https://doi.org/10.3389/fcimb.2023.1128577

Raza MF, Wang Y, Cai Z, Bai S, Yao Z, Awan UA et al (2020) Gut microbiota promotes host resistance to low-temperature stress by stimulating its arginine and proline metabolism pathway in adult *Bactrocera dorsalis*. PLoS Pathog 16:e1008441. https://doi.org/10.1371/journal.ppat.1008441

Reyes JIL, Suzuki Y, Carvajal T, Muñoz MNM, Watanabe K (2021) Intracellular interactions between arboviruses and *Wolbachia* in *Aedes aegypti*. Front Cell Infect Microbiol 11:690087. https://doi.org/10.3389/fcimb.2021.690087

Řezanka T, Sigler K (2009) Odd-numbered very-long-chain fatty acids from the microbial, animal and plant kingdoms. Prog Lipid Res 48:206–238. https://doi.org/10.1016/j.plipres.2009.03.003

Ridley EV, Wong AC-N, Westmiller S, Douglas AE (2012) Impact of the resident microbiota on the nutritional phenotype of *Drosophila melanogaster*. PLoS One 7:e36765. https://doi.org/10.1371/journal.pone.0036765

Ristaino JB, Anderson PK, Bebber DP, Brauman KA, Cunniffe NJ, Fedoroff NV et al (2021) The persistent threat of emerging plant disease pandemics to global food security. Proc Natl Acad Sci USA:118. https://doi.org/10.1073/pnas.2022239118

Romoli O, Schönbeck JC, Hapfelmeier S, Gendrin M (2021) Production of germ-free mosquitoes via transient colonisation allows stage-specific investigation of host–microbiota interactions. Nat Commun 12:942. https://doi.org/10.1038/s41467-021-21195-3

Russell CW, Poliakov A, Haribal M, Jander G, van Wijk KJ, Douglas AE (2014) Matching the supply of bacterial nutrients to the nutritional demand of the animal host. Proc R Soc Lond B Biol Sci 281:20141163. https://doi.org/10.1098/rspb.2014.1163

Sánchez-Wandelmer J, Dávalos A, Herrera E, Giera M, Cano S, de la Peña G et al (2009) Inhibition of cholesterol biosynthesis disrupts lipid raft/caveolae and affects insulin receptor activation in 3T3-L1 preadipocytes. Biochim Biophys Acta Biomembr 1788:1731–1739. https://doi.org/10.1016/j.bbamem.2009.05.002

Sannino DR, Dobson AJ, Edwards K, Angert ER, Buchon N (2018) The *Drosophila melanogaster* gut microbiota provisions thiamine to its host. MBio 9:e00155–e00118. https://doi.org/10.1128/mBio.00155-18

Sato A, Ohhara Y, Miura S, Yamakawa-Kobayashi K (2020) The presence of odd-chain fatty acids in *Drosophila* phospholipids. Biosci Biotechnol Biochem 84: 2139–2148. https://doi.org/10.1080/09168451.2020.1790337

Sazama EJ, Bosch MJ, Shouldis CS, Ouellette SP, Wesner JS (2017) Incidence of *Wolbachia* in aquatic insects. Ecol Evol 7:1165–1169. https://doi.org/10.1002/ece3.2742

Scheifler M, Wilhelm L, Visser B (2024) Lipid metabolism in parasitoids and parasitized hosts. In: Musselman L, Toprak U (eds) Insect lipid metabolism. Advances in experimental medicine and biology. Springer, New York. https://doi.org/10.1007/5584_2024_812

Scheitz CJF, Guo Y, Early AM, Harshman LG, Clark AG (2013) Heritability and inter-population differences in lipid profiles of *Drosophila melanogaster*. PLoS One 8:e72726. https://doi.org/10.1371/journal.pone.0072726

Schultz MJ, Isern S, Michael SF, Corley RB, Connor JH, Frydman HM (2017) Variable inhibition of zika virus replication by different *Wolbachia* strains in mosquito cell cultures. J Virol 91:339–356. https://doi.org/10.1128/JVI.00339-17

Schultz MJ, Tan AL, Gray CN, Isern S, Michael SF, Frydman HM et al (2018) *Wolbachia* wStri blocks zika virus growth at two independent stages of viral replication. MBio 9:e00738–e00718. https://doi.org/10.1128/mBio.00738-18

Serbus LR, Casper-Lindley C, Landmann F, Sullivan W (2008) The genetics and cell biology of *Wolbachia* – host interactions. Annu Rev Genet 42:683–707. https://doi.org/10.1146/annurev.genet.41.110306.130354

Serrato-Salas J, Gendrin M (2023) Involvement of microbiota in insect physiology: focus on B vitamins. MBio 14:e02225–e02222. https://doi.org/10.1128/mbio.02225-22

Shin SC, Kim S-H, You H, Kim B, Kim AC, Lee K-A et al (2011) *Drosophila* microbiome modulates host developmental and metabolic homeostasis via insulin signaling. Science 334:670–674. https://doi.org/10.1126/science.1212782

Sieber MH, Thummel CS (2012) Coordination of triacylglycerol and cholesterol homeostasis by *DHR96* and the *Drosophila* LipA homolog *magro*. Cell Metab 15:122–127. https://doi.org/10.1016/j.cmet.2011.11.011

Sommer AJ, Newell PD (2019) Metabolic basis for mutualism between gut bacteria and its impact on the *Drosophila melanogaster* host. Appl Environ Microbiol 85:e01882–e01818. https://doi.org/10.1128/AEM.01882-18

Stafford CA, Walker GP, Ullman DE (2011) Infection with a plant virus modifies vector feeding behavior. Proc Natl Acad Sci USA 108:9350–9355. https://doi.org/10.1073/pnas.1100773108

Stanley-Samuelson DW, Howard RW, Akre RD (1990) Nutritional interactions revealed by tissue fatty acid profiles of an obligate *Myrmecophilous* predator, *Microdon albicomatus*, and its prey, *Myrmica incompleta*, (Diptera: Syrphidae) (hymenoptera: Formicidae). Ann Entomol Soc Am 83:1108–1115. https://doi.org/10.1093/aesa/83.6.1108

Storelli G, Defaye A, Erkosar B, Hols P, Royet J, Leulier F (2011) *Lactobacillus plantarum* promotes *Drosophila* systemic growth by modulating hormonal signals through TOR-dependent nutrient sensing. Cell Metab 14:403–414. https://doi.org/10.1016/j.cmet.2011.07.012

Stouthamer R, Breeuwer JAJ, Hurst GDD (1999) *Wolbachia pipientis*: microbial manipulator of arthropod reproduction. Ann Rev Microbiol 53:71–102. https://doi.org/10.1146/annurev.micro.53.1.71

Tal O, Selvaraj G, Medina S, Ofaim S, Freilich S (2020) NetMet: a network-based tool for predicting metabolic capacities of microbial species and their interactions. Microorganisms 8:840. https://doi.org/10.3390/microorganisms8060840

Tang X, Adler PH, Vogel H, Ping L (2012) Gender-specific bacterial composition of black flies (Diptera: Simuliidae). FEMS Microbiol Ecol 80:659–670. https://doi.org/10.1111/j.1574-6941.2012.01335.x

Teixeira L, Ferreira Á, Ashburner M (2008) The bacterial symbiont *Wolbachia* induces resistance to RNA viral infections in *Drosophila melanogaster*. PLoS Biol 6: e1000002. https://doi.org/10.1371/journal.pbio.1000002

Teramoto T, Huang X, Armbruster PA, Padmanabhan R (2019) Infection of *Aedes albopictus* mosquito C6/36 cells with the *w*Melpop strain of *Wolbachia* modulates dengue virus-induced host cellular transcripts and induces critical sequence alterations in the dengue viral genome. J Virol 93:e00581–e00519. https://doi.org/10.1128/JVI.00581-19

Tokuda G, Mikaelyan A, Fukui C, Matsuura Y, Watanabe H, Fujishima M et al (2018) Fiber-associated spirochetes are major agents of hemicellulose degradation in the hindgut of wood-feeding higher termites. Proc Natl Acad Sci USA 115:E11996–E12004. https://doi.org/10.1073/pnas.1810550115

Tongluan N, Ramphan S, Wintachai P, Jaresitthikunchai J, Khongwichit S, Wikan N et al (2017) Involvement of fatty acid synthase in dengue virus infection. Virol J 14:28. https://doi.org/10.1186/s12985-017-0685-9

Tree MO, Londono-Renteria B, Troupin A, Clark KM, Colpitts TM, Conway MJ (2019) Dengue virus reduces expression of low-density lipoprotein receptor-related protein 1 to facilitate replication in *Aedes aegypti*. Sci Rep 9:6352. https://doi.org/10.1038/s41598-019-42803-9

Vial T, Tan W-L, Wong Wei Xiang B, Missé D, Deharo E, Marti G et al (2019) Dengue virus reduces AGPAT1 expression to alter phospholipids and enhance infection in *Aedes aegypti*. PLoS Pathog 15:e1008199. https://doi.org/10.1371/journal.ppat.1008199

Vial T, Tan W-L, Deharo E, Missé D, Marti G, Pompon J (2020) Mosquito metabolomics reveal that dengue virus replication requires phospholipid reconfiguration via the remodeling cycle. Proc Natl Acad Sci USA 117:27627–27636. https://doi.org/10.1073/pnas.2015095117

Vial T, Marti G, Missé D, Pompon J (2021) Lipid interactions between flaviviruses and mosquito vectors. Front Physiol 12:763195. https://doi.org/10.3389/fphys.2021.763195

Visser B, Le Lann C, Den Blanken FJ, Harvey JA, van Alphen JJM, Ellers J (2010) Loss of lipid synthesis as an evolutionary consequence of a parasitic lifestyle. Proc Natl Acad Sci USA 107:8677–8682. https://doi.org/10.1073/pnas.1001744107

Visser B, Le Lann C, Hahn DA, Lammers M, Nieberding CM, Alborn HT et al (2023) Many parasitoids lack adult fat accumulation, despite fatty acid synthesis: a discussion of concepts and considerations for future research. Curr Res Insect Sci 3:100055. https://doi.org/10.1016/j.cris.2023.100055

Wagar ZL, Tree MO, Mpoy MC, Conway MJ (2017) Low density lipopolyprotein inhibits flavivirus acquisition in *Aedes aegypti*. Insect Mol Biol 26:734–742. https://doi.org/10.1111/imb.12334

Wang L, Wang S, Zhang Q, He C, Fu C, Wei Q (2022) The role of the gut microbiota in health and cardiovascular diseases. Mol Biomed 3:30. https://doi.org/10.1186/s43556-022-00091-2

Watarai M, Makino S, Michikawa M, Yanagisawa K, Murakami S, Shirahata T (2002) Macrophage plasma membrane cholesterol contributes to *Brucella abortus* infection of mice. Infect Immun 70:4818–4825. https://doi.org/10.1128/IAI.70.9.4818-4825.2002

Weinert LA, Araujo-Jnr EV, Ahmed MZ, Welch JJ (2015) The incidence of bacterial endosymbionts in terrestrial arthropods. Proc R Soc Lond B Biol Sci 282:20150249. https://doi.org/10.1098/rspb.2015.0249

Wernegreen JJ (2002) Genome evolution in bacterial endosymbionts of insects. Nat Rev Genet 3:850–861. https://doi.org/10.1038/nrg931

Wimalasiri-Yapa BMCR, Huang B, Ross PA, Hoffmann AA, Ritchie SA, Frentiu FD et al (2023) Differences in gene expression in field populations of *Wolbachia*-infected *Aedes aegypti* mosquitoes with varying release histories in northern Australia. PLoS Negl Trop Dis 17:e0011222. https://doi.org/10.1371/journal.pntd.0011222

Wong CNA, Ng P, Douglas AE (2011) Low-diversity bacterial community in the gut of the fruitfly *Drosophila melanogaster*. Environ Microbiol 13:1889–1900. https://doi.org/10.1111/j.1462-2920.2011.02511.x

Wong CAN, Dobson AJ, Douglas AE (2014) Gut microbiota dictates the metabolic response of *Drosophila* to diet. J Exp Biol 217:1894–1901. https://doi.org/10.1242/jeb.101725

Wrońska AK, Kaczmarek A, Boguś MI, Kuna A (2023) Lipids as a key element of insect defense systems. Front Genet 14:1183659. https://doi.org/10.3389/fgene.2023.1183659

Wu M, Sun LV, Vamathevan J, Riegler M, Deboy R, Brownlie JC et al (2004) Phylogenomics of the reproductive parasite *Wolbachia pipientis w*Mel: a streamlined genome overrun by mobile genetic elements. PLoS Biol 2:E69. https://doi.org/10.1371/journal.pbio.0020069

Wu W, Shan H-W, Li J-M, Zhang C-X, Chen J-P, Mao Q (2022) Roles of bacterial symbionts in transmission of plant virus by hemipteran vectors. Front Microbiol 13:805352. https://doi.org/10.3389/fmicb.2022.805352

Xie J, Cai Z, Zheng W, Zhang H (2023) Integrated analysis of miRNA and mRNA expression profiles in response to gut microbiota depletion in the abdomens of female *Bactrocera dorsalis*. Insect Sci 30:443–458. https://doi.org/10.1111/1744-7917.13091

Xu Z, Jiang W, Huang W, Lin Y, Chan FKL, Ng SC (2022) Gut microbiota in patients with obesity and metabolic disorders – a systematic review. Genes Nutr 17:2. https://doi.org/10.1186/s12263-021-00703-6

Zhang T, Feng W, Ye J, Li Z, Zhou G (2018) Metabolomic changes in *Sogatella furcifera* under southern rice black-streaked dwarf virus infection and temperature stress. Viruses 10:344. https://doi.org/10.3390/v10070344

Zhang H-B, Cao Z, Qiao J-X, Zhong Z-Q, Pan C-C, Liu C et al (2021) Metabolomics provide new insights into mechanisms of *Wolbachia*-induced paternal defects in *Drosophila melanogaster*. PLoS Pathog 17:e1009859. https://doi.org/10.1371/journal.ppat.1009859

Zhang FQ, McMullen JG, Douglas AE, Ankrah NYD (2022a) Succinate: a microbial product that modulates *Drosophila* nutritional physiology. Insect Sci 29:315–318. https://doi.org/10.1111/1744-7917.12905

Zhang X, Zhang F, Lu X (2022b) Diversity and functional roles of the gut microbiota in lepidopteran insects. Microorganisms 10:1234. https://doi.org/10.3390/microorganisms10061234

Zheng Y, Wang J-L, Liu C, Wang C-P, Walker T, Wang Y-F (2011) Differentially expressed profiles in the larval testes of *Wolbachia* infected and uninfected *Drosophila*. BMC Genomics 12:595. https://doi.org/10.1186/1471-2164-12-595

Zhou X, Ling X, Guo H, Zhu-Salzman K, Ge F, Sun Y (2021) *Serratia symbiotica* enhances fatty acid metabolism of pea aphid to promote host development. Int J Mol Sci 22:5951. https://doi.org/10.3390/ijms22115951

Zug R, Hammerstein P (2012) Still a host of hosts for *Wolbachia*: analysis of recent data suggests that 40% of terrestrial arthropod species are infected. PLoS One 7:e38544. https://doi.org/10.1371/journal.pone.0038544

Adv Exp Med Biol (2026) 1494: 445–477
https://doi.org/10.1007/5584_2024_812

Published online: 09 July 2024

Lipid Metabolism in Parasitoids and Parasitized Hosts

Mathilde Scheifler, Léonore Wilhelm, and Bertanne Visser

Abstract

Parasitoids have an exceptional lifestyle where juvenile development is spent on or in a single host insect, but the adults are free-living. Unlike parasites, parasitoids kill the host. How parasitoids use such a limiting resource, particularly lipids, can affect chances to survive and reproduce. In part 1, we describe the parasitoid lifestyle, including typical developmental strategies. Lipid metabolism in parasitoids has been of interest to researchers since the 1960s and continues to fascinate ecologists, evolutionists, physiologists, and entomologists alike. One reason of this interest is that the majority of parasitoids do not accumulate triacylglycerols as adults. Early research revealed that some parasitoid larvae mimic the fatty acid composition of the host, which may result from a lack of de novo triacylglycerol synthesis. More recent work has focused on the evolution of lack of adult triacylglycerol accumulation and consequences for life history traits. In part 2 of this chapter, we discuss research efforts on lipid metabolism in parasitoids from the 1960s onwards. Parasitoids are also master manipulators of host physiology, including lipid metabolism, having evolved a range of mechanisms to affect the release, synthesis, transport, and take-up of lipids from the host. We lay out the effects of parasitism on host physiology in part 3 of this chapter.

Keywords

Bracovirus · Fat · Fitness · Host-parasitoid interaction · Parasitic wasp · Polydnavirus · Symbiosis · Teratocyte · Venom

1 Introduction

There are many intricacies when it comes to the fat metabolism of parasitoids. Parasitoids have a unique lifestyle, where development takes place inside or on a single host (usually another insect or an arthropod), but the adults are free-living (Godfray 1994) (Fig. 1). During development, the parasitoid consumes the host and its nutritional resources, sometimes only partly or in its entirety in successive steps (Cuny and Poelman 2022), ultimately leading to death of the host. Parasitoidism thus constitutes a parasitic relationship between a parasitoid and its host that differs from parasites in the extent to which the host is harmed, because parasitoids kill their host (Lafferty and Kuris 2002). The parasitoid lifestyle evolved repeatedly in insects, including independent occurrences in beetles, flies,

M. Scheifler, L. Wilhelm, and B. Visser (✉)
Evolution and Ecophysiology Group, Department of Functional and Evolutionary Entomology, Gembloux Agro-Bio Tech, University of Liège, Gembloux, Belgium
e-mail: mathilde.scheifler@uliege.be; bertanne.visser@uliege.be

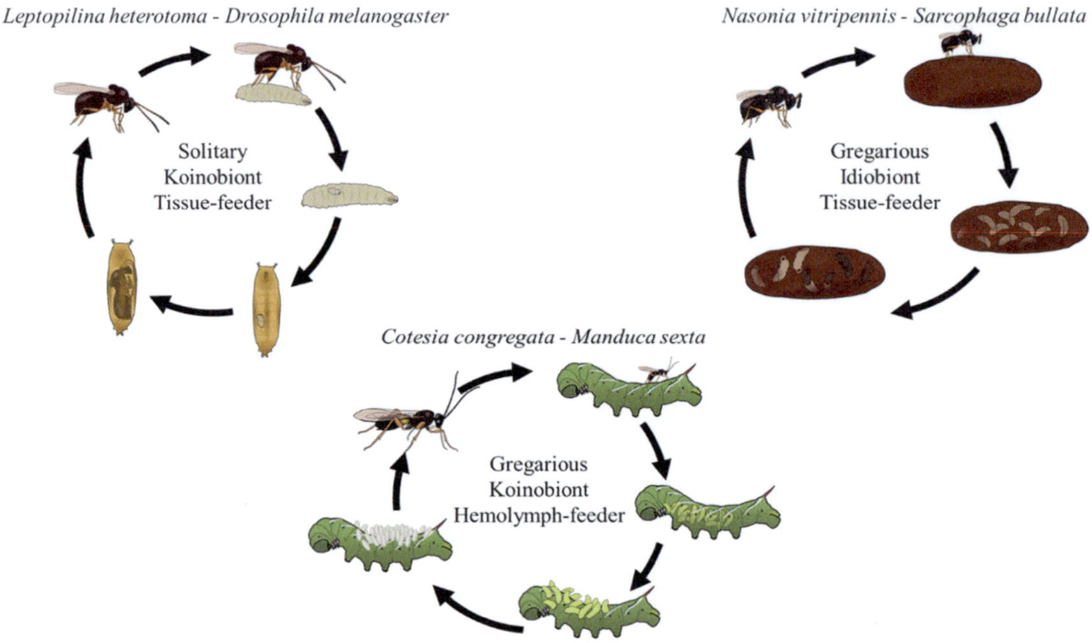

Fig. 1 The parasitoid life cycle showing the different effects on host development (idiobiont versus koinobiont), as well as parasitoid oviposition (solitary versus gregarious), and feeding strategies (hemolymph versus tissue-feeder). Represented species include the amber wasp *Leptopilina heterotoma* on the host *Drosophila melanogaster*, the parasitoid *Cotesia congregata* on the host *Manduca sexta,* and the parasitoid *Nasonia vitripennis* on the host *Sarcophaga bullata*

butterflies, and lacewings (Eggleton and Belshaw 1992, 1993), but Hymenoptera (bees, ants, wasps, and sawflies) is the most speciose order when it comes to parasitoids (and potentially species in general, see Forbes et al. 2018). The large breadth of parasitoid species worldwide, the unique lifestyle, and the plethora of strategies used by parasitoids to parasitize their hosts make them valuable and interesting biological model systems (Hoddle et al. 1998; Liu et al. 2015; Matthews et al. 2009; Quicray et al. 2023; Werren and Loehlin 2009; Whitfield et al. 2017). This is true not only from a basic, fundamental scientific perspective, but also for the applied sciences, because parasitoids play a key role in regulating both natural and agricultural insect communities (i.e., as biocontrol agents against insect pests; Jervis 2005).

This chapter starts with an overview of research done on lipid metabolism of insect parasitoids, from earlier works in the 1960s to the most recent developments in part 2. Host fatty acid composition and fat content, as well as the ability of parasitoids to manipulate host lipids and availability, is of critical importance for both immature and adult parasitoids. Part 3 focuses on the different mechanisms by which parasitoids can manipulate the fat metabolism of their hosts.

2 Fatty Acid Synthesis and Fat Accumulation in Parasitoids

Parasitoids have been of particular interest to biologists regarding lipid metabolism. There has, however, been some recent debate between researchers studying parasitoid lipid metabolism, mainly in terms of semantics (Visser et al. 2023). To avoid confusion about definitions and terminology related to lipid metabolism, the use of stricter definitions that emphasize the difference between the processes of fatty acid synthesis and triacylglycerol/fat accumulation has recently been proposed (Visser et al. 2023). This distinction is important, because these two processes are not synonymous with one another: fatty acids can be synthesized even if triacylglycerols are not accumulated. The main interest of evolutionary ecologists studying lipid metabolism in parasitoids has been focused on the accumulation of triacylglycerols in adults, because energy stored in the form of fat reserves can have a major impact on life histories and fitness (see Box 1 for a brief overview of the link between fat content and life histories in parasitoids). Parasitoids represent a curious case where triacylglycerols are generally not accumulated in response to superfluous feeding, unlike other animals that will readily accumulate triacylglycerols under the same nutritional conditions (Visser et al. 2010; Visser and Ellers 2008). While previously referred to as the "lack of lipogenesis" or "lack of lipid synthesis", this phenomenon is now referred to as the "lack of adult triacylglycerol/fat accumulation" in parasitoids (Visser et al. 2023), with fat being used synonymously with triacylglycerol. We continue part 2 of this chapter with a chronological account of the work done on lipid metabolism, with the main focus on fatty acid synthesis and fat accumulation in insect parasitoids.

Box 1: Survival of the Fattest: Stored Triacylglycerol Levels Impact Longevity, Reproduction, and Other Fitness-Related Traits in Parasitoids

Triacylglycerols are the most comprehensive form in which energy can be stored, both in terms of the storage space needed within a cell and the higher caloric content per unit of weight (Arrese and Soulages 2010). Oxidation of triacylglycerols further releases twice as much water compared to glycogen (i.e., another major metabolite for energy storage). Taking all this into consideration, it is not surprising that fat plays a critical role for insect life histories, including parasitoids (Jervis et al. 2008). We use earlier work on the *Drosophila*-parasitizing braconid wasp *Asobara tabida* as a case study to reveal the close link between fat reserves and life history traits (see Colinet et al. 2006; Giron and Casas 2003a; Le Lann et al. 2014; Luo et al. 2010; Muller et al. 2017; Sheng et al. 2019 for similar findings in other parasitoid species).

Under laboratory conditions, *A. tabida* females can emerge with 0.04 mg or ~ 20% total body fat (Visser et al. 2010), but triacylglycerol levels never exceed those at emergence for the remainder of adulthood (Ellers 1996; Le Lann et al. 2014; Visser et al. 2010), as depicted in Box Fig. 1a. Unlike most other insects (and animals) that rapidly build up triacylglycerol stores when fed a surplus of sugars (see Box Fig. 1b for *Drosophila melanogaster* as an example; Visser et al. 2010), triacylglycerol content in *A. tabida* (without access to hosts) decreases quickly during the first week of life, and then more steadily at a rate of ~0.004 mg per week thereafter (for comparison: when starved, *A. tabida* triacylglycerol-use is ~0.004 mg per day). Moreover, *A. tabida* strains with higher fat content also live longer (Ellers 1996). Fat reserves are thus correlated with adult survival and used to fuel adult life.

Among other metabolic roles, a critical job of the fat body is to store lipids. Fat bodies can become so hypertrophied with lipids that they may fill much of an insect's abdomen. Triacylglycerol levels are correlated with body size in arthropods in general (Lease and Wolf 2011), as well as in *A. tabida* (Ellers 1996; Ellers et al. 1998). Larger, fatter females also have more eggs in their ovarioles (Ellers et al. 1998), and *A. tabida* females can emerge with ~160 yolk-poor (i.e., hydropic) eggs (Carton et al. 1986; Le Lann et al. 2014). Many more eggs can be produced during life (i.e., synovigeny; Box Fig. 2; Jervis et al. 2001), with realized fecundity ranging between 580 and 630 eggs when hosts are available

(continued)

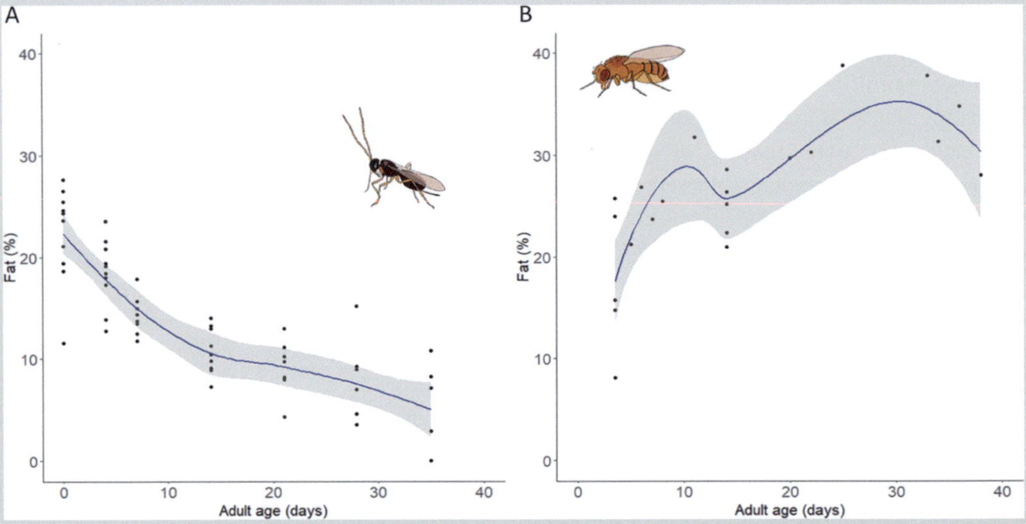

Box Fig. 1 The proportion of fat (in %) in adult *Asobara tabida* (**a**), and *Drosophila melanogaster* (**b**) throughout life. Based on data from Ellers (1996) and redrawn from Service (1987)

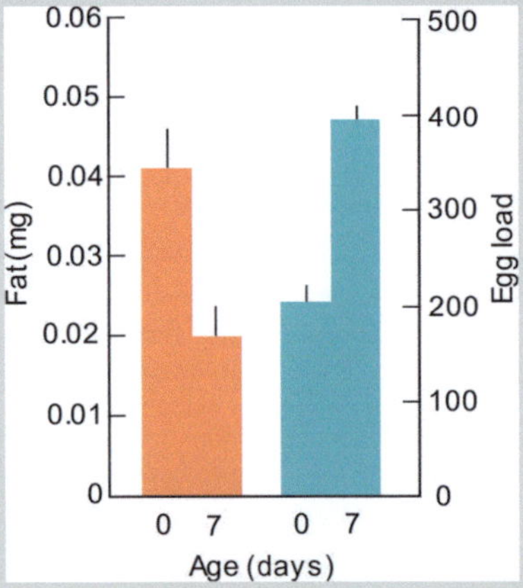

Box Fig. 2 Amount of fat (mean + 1 s.e.) and egg load (mean + 1 s.e.) of *A. tabida* females originating from a population in Kos, Greece, at emergence and 7 days after emergence (with access to food). Redrawn from Ellers and van Alphen (1997)

in excess (Ellers and van Alphen 1997). Similar to lipid-use in non-ovipositing females (Box Fig. 1a), allocation of fat reserves toward reproduction is highest during the first week of life. If the energetic cost of maintenance is similar between ovipositing and non-ovipositing females, then ~25 ng fat is allocated into each egg during the first week of life (based on data of Ellers 1996, and Ellers and van Alphen 1997 for the same population). In *A. tabida,* once fat has been used for reproduction, these reserves cannot be used anymore for other functions (in contrast to some other parasitoid species that can resorb eggs; Jervis et al. 2001). Limiting fat reserves can, therefore, lead to so-called trade-offs in life history traits, because energy can be invested either into reproduction or maintenance/survival (Ellers 1996) or early versus late reproduction (Seyahooei et al. 2020).

Field experiments using a release-recapture approach revealed that dispersal of laboratory-reared *A. tabida* females is size-dependent (Ellers et al. 1998). Larger, fatter females can disperse over larger distances (>15 m) compared to smaller females. Wild-caught *A. tabida* were generally smaller than laboratory-reared females, and size decreased as the season progressed (from July to September). Larger wild-caught females burned more fat than smaller females and carried more eggs at the time of capture. Another study

(continued)

examined the size of field-caught *A. tabida* females over several months (June to October) (Ellers et al. 2001). Female size varied, but larger, fatter females were generally captured at the start and the end of the field season. This pattern may be explained by the differences in temperature throughout the field season, with higher temperatures being reached during summer leading to an increase in metabolic activity and lipid oxidation. Alternatively, fat females have a selective advantage when entering and emerging from diapause early and late in the season, as large fat stores are required to survive months at low environmental temperatures (Ellers and van Alphen 2002).

2.1 Early Studies on Parasitoid Larval Fatty Acid and Triacylglycerol Synthesis

Parasitoid fatty acid metabolism started to receive some attention in the late 1960s. Several researchers took an interest in hymenopteran parasitoids, because unlike other insects (e.g., dipterans and lepidopterans; Barlow 1964, 1965, 1966), some wasps did not seem to have their own characteristic fatty acid composition. Rather than a species-specific qualitative and quantitative fatty acid composition, several wasp species seemed to duplicate the fatty acid composition characteristic of their host. Bracken and Barlow (1967) were the first to investigate this intriguing phenomenon in the ichneumonid parasitoid *Exeristes comstockii*. Using unparasitized larvae of various hosts (the dipteran host *Lucilia sericata,* the lepidopteran host *Galleria mellonella*, and the sawfly host *Neodiprion sertifer* (basal Hymenoptera)) that show substantial interspecific quantitative differences in fatty acids, *E. comstockii* larvae readily duplicated the distinctive fatty acid composition of each host. The host-specific fatty acid composition of *E. comstockii* remained unchanged throughout pupation and into adulthood, meaning that the parasitoid duplication phenomenon is not stage-specific. Similar findings were obtained for another ichneumonid parasitoid, *Itoplectis conquisitor*, when reared on the lepidopteran hosts *G. mellonella* and *Ostrinia nubilalis* (Thompson and Barlow 1970), and in two parasitoid tachinid flies (Delobel and Pageaux 1981).

To determine the generality of duplicated fatty acid compositions in parasitoids, Thompson and Barlow (1972a) tested several other parasitoid species at the larval stage, including aphidids and braconids, as well as species from the superfamily Chalcidoidea, including pteromalids and eulophids. All icheumonids tested (n = 7) had similar fatty acid compositions as their hosts, and the same was true for the pteromalid *Spalangia cameroni* and the eulophid *Dahlbominus fuscipennis.* A subsequent study with 30 species from 5 families of parasitic Hymenoptera revealed that while most ichneumonids had duplicated fatty acid compositions, this was not a general pattern for this group and duplication occurred also in species from other families (Thompson and Barlow 1974). Duplication of the fatty acid composition was suggested to be related to host range of the parasitoid, because most (although not all) species with duplicated compositions were generalists able to develop on a wide range of hosts. It should be noted, however, that in most experiments by Thompson and collaborators unparasitized hosts were compared to parasitoids. When fatty acid compositions differ between hosts and parasitoids, these changes could also result from manipulation of the host's fat metabolism by the parasitoid (see part 3 of this chapter). Furthermore, although in line with sampling expectations at the time, replication is rather low in the early works of Thompson and colleagues. Hence some caution is needed when interpreting their results.

The question arises how and why are some parasitoids duplicating the fatty acid composition of their hosts, but others are not? One explanation is that the parasitoids readily feed on the fat body of the host and incorporate those fatty acids directly into their own fat stores without contributing de novo synthesized fatty acids themselves. Thompson and Barlow (1972b) tested this hypothesis using *E. comtockii* reared on the lepidopteran host *G. mellonella* and the dipteran host *L. sericata*. By injecting radio-labeled acetate (^{14}C-1-acetate) into the hosts, and rearing the parasitoid on both host species, it became apparent that *E. comstockii* larvae synthesized (as well as desaturated and elongated) fatty acids, with palmitate (C16:0) being the main synthetic product. While de novo fatty acid synthesis is clearly taking place in *E. comstockii,* fatty acids also originate from direct incorporation of host fat. This was demonstrated by the presence of eicosenoic acid (C20:1) that was synthesized de novo by *G. mellonella* only and was also present in the fatty acid fraction of *E. comstockii* (but without radioactivity and thus without fatty acid synthesis by the parasitoid).

If host fatty acid composition largely determines that of the parasitoid, what happens if the host is taken out of the equation altogether? Thompson and Barlow (1976) did the test: they reared larvae of another ichneumonid that also duplicates the fatty acid composition of its host, *Exeristes roborator*, on a fatty acid-free artificial medium. Without any host, the parasitoid larvae readily synthesized fatty acids de novo with a composition that did not mimic that of any of its hosts. The developmental environment is strikingly different for parasitoids reared on an artificially defined fatty acid-free medium compared to a natural, developing host insect that is rich in lipids. Triacylglycerols indeed appeared to be less toxic for *E. roborator* to consume than fatty acids (Thompson 1977), which makes sense considering that developing hosts typically contain substantial triacylglycerol quantities compared to free fatty acids. If host acylglycerols can be used by the parasitoid with relative ease, perhaps there is no need to invest in energetically costly triacylglycerol synthesis by the parasitoid itself. The study of Jones et al. (1982) compared triacylglycerol synthesis between ichneumonid species that duplicated the host's fatty acid composition, *E. roborator* and *I. conquisitor*, with species that have their own characteristic fatty acid composition irrespective of that of the host, i.e., the ichneumonids *Aphaereta pallipes, Hyposoter exigua,* and the chalcid *Brachymeria lasus*. What they found is that *E. roborator* and *I. conquisitor* did not incorporate glycerophosphate into acylglycerols, meaning that the de novo triacylglycerol pathway (also known as the Kennedy pathway) was not active. *Aphaereta pallipes* and *H. exigua*, that did not duplicate the host's fatty acid composition, readily incorporated glycerophosphate. Interestingly, all parasitoids were able to use the monoacylglycerol pathway, where monoacylglycerols are sequentially acylated by acyl-CoA to form di- and subsequently triacylglycerols (i.e., by monoacylglycerol transferase, and diacylglycerol acyltransferase, respectively; Fig. 1 in Visser et al. 2023). For *E. roborator*, 75% of triacylglycerols were formed from diacylglycerols, while this was 97% for *I. conquisitor*. The enzymes of the monoacylglycerol pathway further appeared to be substrate-specific in *E. roborator*, meaning that some fatty acid thioesters are more readily used to form triacylglycerols.

Overall, the work of Thompson and colleagues has shed an exceptional light on the mechanistic basis of the duplication of host fatty acid compositions in some parasitoids. When high levels of triacylglycerols are available in a fat-rich host, partial catabolism of triacylglycerols into diacylglycerols (the main form in which lipids are transported through the hemolymph; Soulages and Wells 1994; Turunen 1979) can then facilitate a fast and relatively inexpensive biochemical means to synthesize new triacylglycerols for storage in the parasitoid larvae. The parasitoid larvae thus do not use the de novo triacylglycerol pathway, but rather the monoacylglycerol pathway to accomplish this. The similarity in composition between host and parasitoid is thought to result from acyltransferase specificity favoring fatty acids that are similar to that of the host.

2.2 The Lack of Adult Triacylglycerol Accumulation in Adult Parasitoids

The storage of fat reserves in periods of food abundance is one of the most conserved metabolic responses across all domains of life (Birsoy et al. 2013). Fat is a key energy substrate fueling insect life, including behavior and other components of fitness (i.e., survival, reproduction; Box 1) (Arrese and Soulages 2010). Although adult parasitoids use dietary carbohydrates to meet short-term energetic demands, adult parasitoids show an extraordinary physiological response to sugar feeding, unlike other insects. During the 1990s and 2000s, adults of several parasitoid species were found to not appreciably convert excess carbohydrates into long-term storage in the form of fat. For example, the adult fat content of *Asobara tabida* (Ichneumonoidea) was highest at emergence and declined rapidly with age, despite continuous access to sugar (see Box 1; Ellers 1996). Similar findings were obtained also for other species of various superfamilies: Ichneumonoidea (*Ventura canescens* and *Diadegma insulare;* Casas et al. 2003; Lee et al. 2004), Cynipoidea (*Leptopilina heterotoma*; Eijs et al. 1998), and Chalcidoidea (*Nasonia vitripennis*; *Eupelmus vuilletti*; Giron and Casas 2003b; Rivero and West 2002), demonstrating that this extraordinary physiological phenotype was more common in parasitic Hymenoptera.

Lack of adult fat accumulation was proposed to be an evolutionary consequence of the parasitoid lifestyle (Visser and Ellers 2008). Efficiently utilizing a single host insect and manipulating host nutrient content (see part 3 on host manipulation) generally leads to a lipid-rich environment for developing parasitoid larvae. Under such conditions, where larval host fat content is high, adult fat accumulation is no longer necessary (leading to relaxed selection on adult fat accumulation; Lahti et al., 2009) or too costly to maintain (i.e., leading to selection against adult fat accumulation). Visser and Ellers (2008), therefore, hypothesized that parasitoids lost the ability for adult fat accumulation. A study using a comparative approach with more than 90 insect species indeed revealed that (1) loss of fat accumulation is ancestral in parasitic Hymenoptera; (2) the loss of adult fat accumulation coincided with or followed the evolution of the parasitoid lifestyle; and (3) there is parallel evolution, as the loss of fat accumulation evolved repeatedly and independently in parasitoid flies (Diptera), beetles (Coleoptera), and wasps (Hymenoptera; Visser et al. 2010). There were some exceptions, however, because several generalist parasitoid species did accumulate fat as adults, including *L. heterotoma*, *Pteromalus puparum,* and *Gelis agilis* (Visser et al. 2010). A reason why adult generalist synthesize fat is that manipulation of host fat content is difficult when many species can serve as potential hosts. When a generalist then develops on a fat-poor host, fat accumulation in adults is critical for survival and reproduction (Visser et al. 2010).

An important question is which mechanism (s) underlies the loss of fat accumulation in adult parasitoids? There can be several ways in which the fat accumulation phenotype was lost: (1) the gene (s) underlying fatty acid or triacylglycerol synthesis and accumulation were lost (i.e., not present in the genome anymore; as in the yeast *Malassezia globose*; Xu et al. 2007); (2) the gene(s) have accumulated mutations in the coding regions, leading to non-functionality; (3) the gene(s) remain present, but are silenced through regulatory processes or mutations in regulatory regions. Consequently, either insufficient quantities of fatty acids and triacylglycerols are produced by adult parasitoids or accumulation itself is hampered. The loss or non-functionality of key genes in the fatty acid and triacylglycerol metabolism pathways is unlikely, however, because many genes (e.g., fatty acid synthase *fas*, acetyl-CoA carboxylase *acc*, glycerol-3-phosphate-acyltransferase *gpat*) involved in the conversion of carbohydrates into triacylglycerols, are also essential for the synthesis of other lipid classes and are part of other key metabolic pathways (e.g., pyruvate metabolism, citrate cycle, phospholipids).

The first study on the molecular mechanisms and transcriptional profiles underlying the lack of adult fat accumulation in parasitoids was performed with the chalcid *N. vitripennis*.

Genome analysis indeed revealed that coding sequences of *fas* and *acc* did not contain any stop codons, mutations, or signs of genetic damage (Visser et al. 2012). But contrary to findings in other animals, no effect of continuous access to sugar was found on the transcription levels of *fas* or *acc* (Visser et al. 2012), suggesting that fatty acid synthesis is not taking place. The same was found for *gpat*, involved in the early steps of acylglycerol synthesis (and part of the de novo triacylglycerol synthesis pathway). Genes involved in the monoacylglycerol pathway, e.g., *dgat,* did not respond to continuous access to sugar either. Functionality of *fas* and *acc*, as well as other genes and their enzyme products, were also confirmed in several other parasitoid wasp species (Kraaijeveld et al. 2019; Lammers et al. 2019; Prager et al. 2019; Visser et al. 2021). Presence and functionality of fat-related genes suggest that changes in gene expression, rather than structural genetic changes are involved in the lack of adult fat accumulation (Visser et al. 2012).

For transcriptomic studies on fat metabolism, it is essential to know how fat metabolic phenotypes are affected. In the case of Visser et al. (2012), absence of fatty acid synthesis and fat accumulation in adult *N. vitripennis* were determined using stable isotope tracking methods (of deuterium into fatty acids of the neutral lipid fraction) and bulk fat extractions (comparing fat quantities between emerged and fed wasps), respectively. No incorporation of stable isotopes was found in fatty acids of the neutral lipid fraction, indicating that fatty acid synthesis did not take place in *N. vitripennis*. This was confirmed by quantitative PCR measurements of gene transcripts of key fatty acid synthesis genes, e.g., *fas*, *acc*, and *dgat*. In contrast, the honeybee *Apis mellifera* that readily synthesizes and stores fat as adult, did incorporate stable isotopes into the neutral fat fraction, illustrating that the method could indeed measure fatty acids that were synthesized and incorporated into stored triacylglycerols. In *N. vitripennis,* no adult fat accumulation was detected because fat quantities decreased significantly during life. Even though adult fat did not accumulate, intermediary metabolites involved in fat metabolism could still be synthesized. For example, Ruther et al. (2021) found that several parasitoid wasp species could synthesize fatty acids, and in the case of *N. vitripennis*, utilize these fatty acids in triacylglycerols and eggs (Multerer et al. 2022). However, no increase in bulk triacylglycerol stores was observed (Ruther et al. 2021). This means that even though fatty acids are synthesized and used to form some amount of triacylglycerols, *N. vitripennis* still lacks adult triacylglycerol accumulation.

To further understand the (lack of) fat accumulation phenotype observed in parasitoids, Visser et al. (2017) compared larval and adult fatty acid synthesis between *D. melanogaster*, showing typical and significant fat accumulation after feeding (Box Fig. 1b), a parasitoid that lacked fat accumulation, *E. vuilletti*, and two parasitoids that readily accumulate fat as adults, *Gelis areator* and *G. agilis.* In adults, fatty acid synthesis (of C16:0) was indeed very high for species that accumulate fat, while for *E. vuilletti* that does not accumulate fat, no fatty acid synthesis was detected. The same patterns were found when fatty acid synthesis was analyzed in the larvae of *D. melanogaster, E. vuilletti, G. agilis*, and *G. areator*. There thus seems to be concurrence in fatty acid synthesis phenotypes between larvae and adults.

Work on fat metabolism in parasitoids has had two distinct spurts in terms of studies, one between the 1960s and 1980s focused on the similarity of fatty acid compositions between hosts and parasitoid larvae; the other starting in the 2000s and still ongoing to understand why adult parasitoids do not accumulate fat. It may well be that despite the slightly different interests and focus, both phenomena result from the same underlying mechanism(s) and evolved in a similar way. Only one more recent study has so far compared fatty acid compositions and fat accumulation strategies, using adults of a rose gall wasp community, including the parasitoids *Orthopelma mediator* and *Pteromalus bedeguaris* (Visser et al. 2013). The gall wasp *Diplolepis rosae* is attacked by *O. mediator,* while *P. bedeguaris* can act as a primary parasitoid on *D. rosae* or as a secondary hyperparasitoid on other primary parasitoids of

D. rosae, including *O. mediator*. Both *O. mediator* and *P. bedeguaris* did not accumulate fat as adults, and only the fatty acid composition of *O. mediator* was considerably different from its main host *D. rosae*. *Orthopelma mediator* is an ichneumonid and *P. bedeguaris* a chalcid, both with a very limited host range. Fat accumulation strategy does thus not seem to be related to mimicking of the host's fatty acid composition, as *O. mediator* has a different fatty acid composition than its host *D. rosae*. The similar fatty acid composition of the more specialized *P. bedeguaris* suggests that copying of the host's fatty acid composition is not inherently linked to host breadth (as suggested in Barlow 1964; Bracken and Barlow 1967). The rose gall system may, however, not be ideal for evaluating the link between fat accumulation, host breadth, and fatty acid compositions due to this system's particular ecological niche. More work is thus needed to determine whether the lack of fat accumulation coincides with mimicking of the fatty acid composition of the host.

2.3 More Complex Adult Parasitoid Fatty Acid Synthesis and Fat Accumulation Phenotypes

While the majority of parasitoids do not accumulate fat as adults, for some parasitoid species repeated experiments hinted at more complicated patterns. For example, Moiroux et al. (2010) proposed that the ability of adult parasitoid wasps to accumulate fat was closely tied to geographic location and local environmental conditions. To test this, four geographically distinct *Leptopilina boulardi* populations were collected. Different adult fat accumulation phenotypes were found: two populations accumulated fat, while the two other populations did not (Moiroux et al. 2010). These observations could be related to genetic divergence between populations, as the two populations that accumulated fat were genetically closer to each other than to populations that did not (Seyahooei et al. 2011; Visser et al. 2017).

Like Moiroux et al. (2010), a large-scale study on the ability for fat accumulation of field-caught *L. heterotoma* populations and other *Leptopilina* species also revealed contrasting adult fat accumulation phenotypes (Visser et al. 2018). These differences were found to be related to the fat content of the *D. melanogaster* host strain used. Indeed, parasitoids emerging from a lean host contained a lower quantity of fat, leading to fat accumulation in adults, while no fat accumulation was observed for parasitoids emerging from a fatty host with a high amount of fat (Visser et al. 2018). A more recent study with several *L. heterotoma* populations confirmed that this species can switch fatty acid synthesis and fat accumulation on or off depending on the host's fat content: these wasps generally start synthesizing and accumulating fat on lean *Drosophila* larvae (Visser et al. 2021). Variation in fat accumulation strategies in adult *L. heterotoma* is plastic, meaning that a single genotype can generate different fatty acid synthesis and fat accumulation phenotypes depending on environmental conditions (Fig. 2). What is now needed is to test also other parasitoids for plasticity of fatty acid synthesis and fat accumulation, particularly considering consequences for adult life histories.

3 Lipid Metabolism in Parasitized Hosts

The parasitoid's unique lifestyle has led to the evolution of intricate mechanisms to manipulate host metabolism. A parasitoid relies on only a single host to complete development and to obtain sufficient nutrients to fuel adult life. There is thus an incredible advantage for the parasitoid to "hijack" the host's metabolism for its own benefit. Considering that many parasitoids do not accumulate substantial fat quantities as adults (see part 2) and fat is of key importance for parasitoid life histories and fitness (see Box 1), manipulating host lipid metabolism so that host lipids become more accessible or available for the parasitoid has a clear adaptive value.

Before exploring how host lipid metabolism can be manipulated by the parasitoid, there are some parasitoid-specific traits that have a large impact on host manipulation. Parasitoids show tremendous diversity related to their mode of

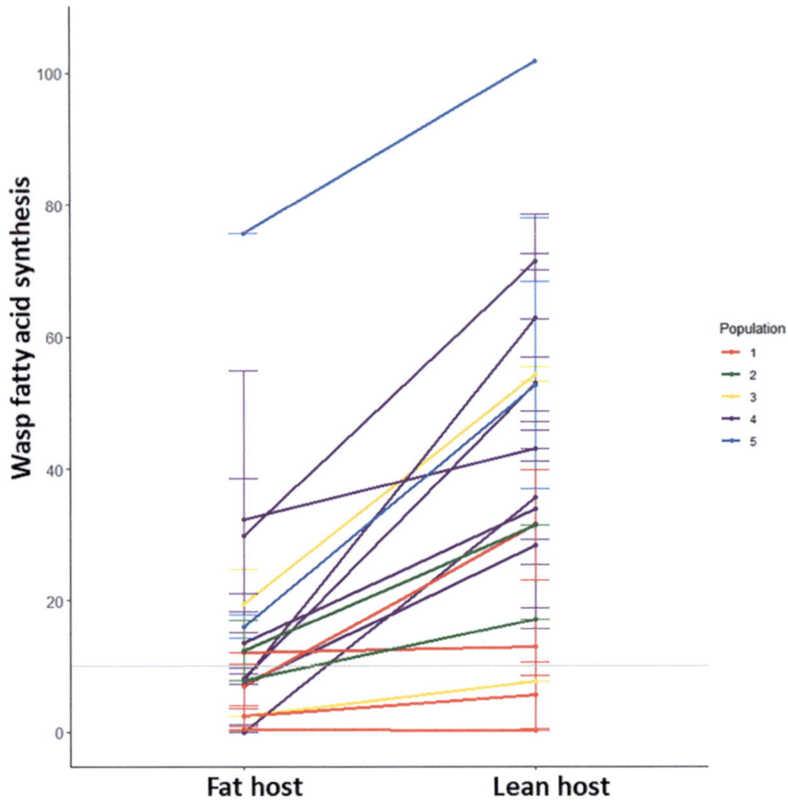

Fig. 2 Fatty acid synthesis (as % deuterium incorporation) of *Leptopilina heterotoma* families (sharing 75% of their genome) originating from different populations (1–5). Reaction norms reveal that some families show plasticity in fatty acid synthesis (synthesis on lean hosts, no synthesis on fat hosts), while other families constitutively synthesize fatty acids or lack fatty acid synthesis (under the gray line). Variation in the slope of the reaction norms suggests that there is genetic variation for plasticity of fatty acid synthesis, meaning that plasticity itself can evolve in response to selection. Redrawn from Visser et al. (2021)

life and general biology (Fig. 1; Godfray 1994; Quicke 1997). An important distinction can be made, for example, between parasitoid species that arrest host development, idiobionts, and species that allow the host to continue feeding and growing, koinobionts. For idiobionts, resource availability for the parasitoid is largely fixed close to the time of oviposition, while for koinobionts host resources generally keep on accumulating, at least for some time while the parasitoid is developing. Among koinobionts, a further distinction can be made between parasitoid species that stop host development prematurely, reducing final host body size, and parasitoid species that prolong host development, increasing final host body size (Cuny and Poelman 2022). Several studies have indeed reported that host food consumption is reduced following parasitism by koinobiont parasitoids (Kaeslin et al. 2005; Morales et al. 2007; Pruijssers et al. 2009; Shi et al. 2015; Thompson 1982a; Thompson 1983), while some parasitized hosts feed longer but remain smaller compared to unparasitized hosts (Thompson and Redak 2008). The extent to which a parasitoid can affect host development can also be dependent on the environment (i.e., phenotypic plasticity). The braconid *Meteorus pulchricornis*, for example, increases final size of the small lepidopteran host *Plutella xylostella* by 30%, while final size of the larger lepidopteran host *Mythimna separata* is increased by 95% (Harvey et al. 2010). The host species can thus have a major impact on resource levels and resource availability for koinobiont parasitoids, which in turn can have major consequences for fatty acid synthesis and fat accumulation of the parasitoid (see part 2.2; Visser et al. 2021).

Parasitoids are highly efficient in carrying over resources from the host, which for some species can mount to >90% of the host's body mass (Harvey et al. 2009). An increase in host fat availability and content can have a major impact on both parasitoid larval development and survival, as well as adult fitness (Rivers et al. 1998). When more fat can be carried over from the host, the parasitoid has more energetic reserves available for allocation into fitness-related traits (see Box 1). For parasitoids in which complete consumption of host tissues is required, developing on larger hosts can be detrimental if it leads to overfeeding (Harvey 1996; Harvey and Strand 2002). Indeed, many parasitoids are so-called "tissue-feeders", where most or all host tissues are consumed during the parasitoid's development (Fig. 1). Within the superfamily Ichneumonoidea, all gregarious (i.e., with multiple offspring emerging from a single host) koinobiont endoparasitoids (e.g., *Microplitis* sp. and *Cotesia* sp., as well as the family Chelonidae) have, however, evolved the ability to feed mostly on host hemolymph (Harvey and Malcicka 2016). These "hemolymph-feeders" initially only feed on hemolymph and part of the fat body of the host but exit the host during the last larval stage to pupate externally (Fig. 1). The adaptive significance of hemolymph feeding is that a wider range of host developmental stages and sizes can be parasitized, including hosts that are much larger than the parasitoid itself.

An important question is whether hemolymph or tissue-feeding koinobionts have evolved different strategies to manipulate host metabolism. We could expect that for tissue-feeders, increasing host fat body lipid content is more important, while for hemolymph-feeders an increase or steady flow of hemolymph lipids can increase the efficiency of scavenging from the host. This could be tested using host-parasitoid systems where the same host species is attacked by multiple parasitoids that differ in feeding strategy. Koinobiont microgastrine braconids and campopleginе ichneumonids may be ideal systems for testing the evolution of host manipulation strategies in parasitoids, e.g., *P. xylostella* parasitized by *Diadegma semiclausum* (tissue-feeder) and *Cotesia vestalis* (hemolymph-feeder), *Pieris brassicae* parasitized by *Hyposoter ebeninus* (tissue-feeder) and *Cotesia glomerata* (hemolymph-feeder), *Spodoptera littoralis* parasitized by *Hyposoter didymator* (tissue-feeder) and *Cotesia marginiventris* (hemolymph-feeder). Comparing host manipulation strategies of hemolymph versus tissue-feeders developing on the same host offers a unique opportunity to increase our understanding of the mechanisms underlying host manipulation.

3.1 The Effects of Parasitism and the Developing Parasitoid(s) on Host Fat Metabolism

The easiest, and sometimes only, way to determine how host metabolism is affected by parasitism is to compare unparasitized with parasitized hosts. Overall, when parasitized, host lipid levels can increase, remain stable, or decrease for a variety of reasons (see Table 1). For example, parasitism of the locust *Chortoicetes terminifera* by the parasitoid fly *Trichopsidea oestracea* led to a steep increase in overall lipid content, although the mechanism has remained unclear (Horwood and Hales 1991). In another parasitoid fly, *Blepharipa sericariae*, the developing larvae were found to secrete a peptide that inhibits lipid transport in the host silkworm, *Philosamia cynthia prieri* (Hayakawa 1987). *Blepharipa sericariae* eggs are consumed by the host during larval feeding, and after parasitoid hatching the parasitoid larvae remain in the second instar until the following spring when the larvae molt and start feeding on the host's pupal tissues. Lipid release from the host fat body into the hemolymph was reduced by 50–70%, and lipid uptake by lipophorin (a blood protein used for lipid transport) was inhibited by ~60% through the action of a parasitoid-secreted peptide (Hayakawa 1986). Similar results were obtained when the locust *Locusta migratoria* was parasitized, with a 50% inhibition of diacylglycerol release (Hayakawa 1987). This finding supports the idea that lipid uptake and transport in the hemolymph, which typically entails the transport of diacylglycerols in insects, is inhibited (Turunen 1979; but see also Ford and van Heusden (1994) who identified a lipophorin transporting triacylglycerols in *Aedes aegypti*). Considering it takes about a year for *B. sericariae* to complete its development, inhibition of lipid transport by lipophorin conserves the triacylglycerol stores of

Table 1 Overview of studies looking at the effect of parasitism on host lipid levels (mainly triacylglycerols) in the whole body, fat body, and/or hemolymph. A distinction is made between the effects of parasitism in general, the venom (with or without the egg, i.e., envenomation), polydnaviruses (with or without the egg, i.e., pseudoparasitization or injection), and teratocytes (with or without the egg, i.e., injection)

Parasitoid species	Parasitoid family	G/S	I/K	Gen/Sp	Ecto/ Endo	Host stage attacked	H/T	Host species	Host order	Host treatment	Host whole body	Host fat body	Host hemolymph	References
Parasitism in general														
Apanteles galleriae	Braconidae	S	K	Gen	Endo	L	T	*Achoria grisella*	Lepidoptera	Par	Sim	–	–	Nurullahoğlu et al. (2004)
Cotesia congregata	Braconidae	G	K	Sp	Endo	L	H	*Manduca sexta*	Lepidoptera	Par	Low	–	–	Thompson and Redak (2008)
Cotesia flavipes	Braconidae	G	K	Sp	Endo	L	H	*Diatraea saccharalis*	Lepidoptera	Par	–	Sim	Low	Salvador and Cônsoli (2008)
								Diatraea flavipennella	Lepidoptera	Par	–	Sim	Low	dos Passos et al. (2019)
Glyptapanteles liparidis	Braconidae	G	K	Sp	Endo	L	H	*Lymantria dispar*	Lepidoptera	Par	Low	–	Sim	Bischof and Ortel (1996)
Cardiochiles nigriceps	Braconidae	S	K	Sp	Endo	L	H	*Heliothis virescens*	Lepidoptera	Par	Sim	–	Sim	Barras et al. (1970)
Hyposoter exigua	Ichneumonidae	G	K	Gen	Endo	L	H	*Trichoplusia ni*	Lepidoptera	Par	Low	–	–	Thompson (1982b)
Trichomalopsis apanteloctena	Pteromalidae	S	I	Gen	Ecto	P	T	*Cotesia kariyai*	Hymenoptera	Par	Low	–	–	Nakamatsu and Tanaka (2004a)
Trichopsidea oestracea	Nemestridae	S	K	Sp	Endo	N	T	*Chortoicetes terminifera*	Orthoptera	Par	Hig	–	–	Horwood and Hales (1991)
Venom														
Bracon nigricans	Braconidae	G	I	Gen	Ecto	L	T	*Spodoptera littoralis*	Lepidoptera	Env	–	Low	Low	Becchimanzi et al. (2017, 2020)
Habrobracon brevicornis	Braconidae	G	I	Gen	Ecto	L	T	*Galleria mellonella*	Lepidoptera	Env	–	Low	Hig	Kryukova et al. (2021)
Lysiphlebia japonica	Braconidae	S	K	Gen	Endo	A	T	*Aphis gossypii*	Hemiptera	Par	Hig	–	–	Xueke et al. (2017)
Euplectrus separatae	Eulophidae	G	K	Sp	Ecto	L	T	*Mythimna separata*	Lepidoptera	Par	–	Low	Hig	Nakamatsu and Tanaka (2003)

								Mythimna separata	Lepidoptera	Env	–	Low	Hig	Nakamatsu and Tanaka (2003), Nakamatsu and Tanaka (2004a, b)
Nasonia vitripennis	Pteromalidae	G	I	Gen	Ecto	P	T	*Sarcophaga bullata*	Diptera	Env	Hig	Hig	Hig	Rivers and Denlinger (1994, 1995)
								Sarcophaga bullata	Diptera	Env	–	–	Hig	Rivers et al. (1998)
								Sarcophaga bullata	Diptera	Par	–	Hig	–	Rivers and Denlinger (1995)
								Sarcophaga bullata	Diptera	Env	–	Hig	–	Rivers and Denlinger (1995)
								Sarcodexia sternodontus	Diptera	Env	–	Low	Low	Rivers and Denlinger (1995)
								Phormia regina	Diptera	Env	–	Low	Hig	Rivers and Denlinger (1995)
								Musca domestica	Diptera	Env	–	Sim	Hig	Rivers and Denlinger (1995)
								Sarcophaga bullata	Diptera	Par	–	Hig	Hig	Rivers and Yoder (1996)
Trichomalopsis apanteloctena	Pteromalidae	S	I	Gen	Ecto	P	T	*Cotesia kariyai*	Hymenoptera	Env	Sim	–	–	Nakamatsu and Tanaka (2004a)
Muscidifurax zaraptor	Pteromalidae	S	I	Gen	Ecto	P	T	*Sarcophaga bullata*	Diptera	Env	–	–	Hig	Rivers et al. (1998)
Trichomalopsis near Americana	Pteromalidae	G	I		Ecto	P	T	*Sarcophaga bullata*	Diptera	Env	–	Low	Hig	Rivers et al. (1998)
Pteromalus puparum	Pteromalidae	G	I	Gen	Endo	P	T	*Pieris rapae*	Lepidoptera	Par	–	Hig	Hig	Wang et al. (2020a)

(continued)

Table 1 (continued)

Parasitoid species	Parasitoid family	G/S	I/K	Gen/Sp	Ecto/ Endo	Host stage attacked	H/T	Host species	Host order	Host treatment	Host whole body	Host fat body	Host hemolymph	References
Scleroderma sichuanensis	Bethylidae	G	I	Gen	Ecto	P	T	*Tenebrio molitor*	Coleoptera	Par	–	Low	Low	Zhuo et al. (2016)
										Env	–	Sim	Low	Zhuo et al. (2016)
Teratocytes														
Meteorus pulchricornis	Braconidae	S	K	Gen	Endo	L	T	*Mythimna separata*	Lepidoptera	Par	–	Low	–	Suzuki and Tanaka (2007)
Cotesia kariyai	Braconidae	G	K	Sp	Endo	L	H	*Mythimna separata*	Lepidoptera	Par	–	Low	–	Nakamatsu et al. (2002)
Microplitis croceipes	Braconidae	S	K	Sp	Endo	L	H	*Heliothis virescens*	Lepidoptera	Par	–	Low	–	Zhang et al. (1997)
										Inj	–	Low	–	Zhang et al. (1997)
Dinocampus coccinellae	Braconidae	S	K	Gen	Endo	L/A	T	*Hippodamia convergens*	Coleoptera	Par	–	Low	–	Sluss (1968)
Dinocampus coccinellae	Braconidae	S	K	Gen	Endo	L/A	T	*Coccinella septempunctata*	Coleoptera	Par	–	Low	–	Gopalapillai et al. (2005)
Polydnavirus														
Microplitis demolitor	Braconidae	S	K	Gen	Endo	L	H	*Chrysodeixis includens*	Lepidoptera	Inj	Low	–	–	Pruijssers et al. (2009)
Cotesia vestalis	Braconidae	S	K	Sp	Endo	L	H	*Plutella xylostella*	Lepidoptera	Par	Low	Low	–	Wang et al. (2021)
		S	K	Sp	Endo	L	H			Ps	Low	Low	–	Wang et al. (2021)
Chelonus inanitus	Braconidae	S	K	Gen	Endo	E/L	H	*Spodoptera littoralis*	Lepidoptera	Par	Hig	–	–	Kaeslin et al. (2005)
										Ps	Hig	–	–	Kaeslin et al. (2005)

G Gregarious, *S* solitary, *I* Idiobiont, *K* Koinobiont, *Gen* Generalist, *Sp* Specialist, *Ecto* Ectoparasitoid, *Endo* Endoparasitoid, *A* Adult, *E* Egg, *L* Larva, *N* Nymph, *P* Pupa, *H* Hemolymph-feeder, *T* Tissue-feeder, *Env* Envenomation, *Inj* Injection, *Par* Parasitization, *Ps* Pseudoparasitization, *Hig* Higher, *Low* Lower, *Sim* Similar, NB: *Trichomalopsis apanteloctena* is a hyperparasitoid the host's fat body. *Blepharipa sericariae* thus prevents the locust host from mobilizing and using lipids. Preventing lipid mobilization by the host is needed for the developing parasitoid to be able to complete its development in spring. The parasitoid fly *T. oestracea* takes a similar time to develop as *B. sericariae*; hence both parasitoid flies have optimized host use, either by increasing or conserving the lipid stores of their respective hosts.

So far, most work on host manipulation has been done on laboratory-reared hymenopteran parasitoids. Unlike the dipteran parasitoids described above, many hymenopteran parasitoids complete their development within several weeks. Major host physiological changes can already be brought about within a short timespan, including a decrease in lipid levels. There are two, not mutually exclusive, reasons why lipid levels are lower in parasitized compared to unparasitized hosts: the host is not able to develop its own fat body (Dahlman 1970) or host and parasitoid compete for lipid resources, with both species consuming and utilizing lipids.

The koinobiont parasitoid *H. exiguae* feeds mainly on lepidopteran host hemolymph (*Trichoplusia ni*) during the first 8 days of development (when the host molts into the third and fourth instar), after which the larvae exit the host to pupate externally (Thompson 1982b). Parasitized larvae had a lower total triacylglycerol content compared to unparasitized larvae near the end of parasitoid development. The reason that parasitized hosts do not get as fat as unparasitized hosts is that the parasitoid is consuming the host's fat or affecting the host's ability to feed and accumulate fat. When comparing unparasitized starved *T. ni* hosts with parasitized *T. ni* hosts, the physiological state in terms of fat content is very similar. In contrast to a starved host, however, a parasitized host still has access to food (at least in this system, where host development continues), which means that host and parasitoid are in direct competition for lipids (Dahlman and Greene 1981). Lipids of parasitized *T. ni* were, however, not depleted completely, suggesting that the parasitoid utilizes resources in such a way that the host does not die prematurely (which would also lead to death of the parasitoid). The above studies contribute to our general understanding of how lipid metabolism of the host is affected following parasitism, including the investigation of rare field-collected hosts that are typically more difficult to study (see Table 1). Experiments focusing solely on the effect of parasitism can, however, be confounded by other factors that can affect host metabolism, such as venom, teratocytes, and mutualistic viruses, which will be discussed in more detail in the following sections.

3.2 Venom-Induced Changes in Host Lipid Metabolism

All female Hymenoptera produce venom in a specialized venom-gland that is a part of the reproductive system (Pennacchio and Strand 2006; Poirié et al. 2014). The venom of parasitoids is injected into the host and consists of both proteinaceous and non-proteinacious compounds (Moreau and Asgari 2015). The venom of ecto and endoparasitoids seems to serve different functions, for the former mainly inducing host paralysis and for the latter mainly interfering with the host's immune system. For all parasitoids, nutrient acquisition during development is critical for survival, investment in costly metamorphosis, and to fuel (at least part of) adult life. In this subsection, we will focus solely on the effects of parasitoid venom on host lipid metabolism.

3.2.1 Venom-Induced Alterations in Host Lipid Metabolism

Venom generally leads to an increase in host lipid levels either in the whole body, the fat body, or the hemolymph (see Table 1). There are some exceptions, however, where host lipid levels were lower, or no changes were observed. For example, in parasitized *S. littoralis,* transmission electron microscopy revealed that the host fat body rapidly released its content (glycogen and lipids) through cell vacuolization and reabsorption (Becchimanzi et al. 2017). Lipid mobilization was aided by hemocytes surrounding the fat body and increased cathepsin L activity. Hemolymph titers of glycerolipids decreased during 48 h after parasitization, probably because the host's tissues require fat for ongoing metabolic activities, albeit reduced. For the coleopteran *Tenebrio molitor* parasitized by the bethylid *Scleroderma sichuanensis,* fat body and hemolymph lipid content also decreased following envenomation and parasitism (Zhuo et al. 2016). This decrease could be due both to consumption of host resources by the parasitoid and the host's own requirement for lipids to stay alive. The host fat body was degraded following parasitism, but envenomation alone did not alter the appearance of the fat body. This suggests that rupture of the fat body cannot be brought about by venom alone.

Host manipulation requires fine-tuned physiological interactions between parasitoid and host that can be highly species-specific. For example, the parasitoid *N. vitripennis* is highly polyphagous, being able to parasitize more than 60 different host species (Desjardins et al. 2010). Yet, despite its wide host range, *N. vitripennis* prefers to oviposit on the fly *Sarcophaga bullata* (Desjardins et al. 2010). Rivers and Denlinger (1995) looked at the effects of parasitism by *N. vitripennis* on four distinct fly species, including *S. bullata, Phormia regina, Musca domestica,* and *Sarcodexia sternodontus.* Only in the host *S. bullata* were marked increases in both fat body and hemolymph lipids observed (Rivers and Denlinger 1995). For *P. regina* and *M. domestica* hemolymph lipids also increased following parasitism but lipid levels in the host fat body did not increase. For the host fly *S. sternodontus* both fat body and hemolymph lipids decreased. For the host *S. bullata* both envenomation and parasitism led to increased fat content, which could result from active fatty acid synthesis and fat accumulation by the host. Parasitism by the wasp *Lysiphlebus japonica* of the aphid *Aphis gossypii* led to upregulation of almost all genes in the glycerolipid pathway, including diacylglycerol acyltransferase that produces triacylglycerols from diacylglycerols, revealing that venom likely induces lipogenesis in hosts (Zhang et al. 2015).

Zooming in on the interaction between *N. vitripennis* and *S. bullata,* the elevation in host hemolymph lipids depended on the location where oviposition occurred on the host pupa. A posterior sting, where parasitism occurs under natural conditions, led to a steeper increase in lipids compared to an anterior sting (Rivers and Yoder 1996). *Nasonia vitripennis* larvae developing on posterior-oviposited hosts also contained more fat than anterior-oviposited hosts, suggesting indeed that lipid availability depends on the location of the parasitoid's sting. Elevation of host hemolymph lipids was also associated with the number of developing parasitoid larvae (Rivers and Yoder 1996). A higher number of eggs laid by the parasitoid led to a greater increase in host hemolymph lipid content. A similar finding was obtained for another gregarious parasitoid, *Trichomalopsis* near *americana* (Rivers et al. 1998). An increase in host hemolymph lipids when more eggs are laid suggests that parasitoid venom increases nutrient content of the host in such a manner that competition between multiple offspring and the host can be avoided. Resource availability is thought to be more restricted for idiobionts that arrest the host's development, where nutrients contained in the host are not altered. Overall, the body of work on *N. vitripennis* suggests that idiobiont parasitoids may not be limited in host lipid resources as substantial increases in host hemolymph and fat body lipid content is brought about by the parasitoid's venom (Rivers and Yoder 1996).

Only few researchers investigated both the composition of the venom and the effects of venom on host lipid metabolism. Wang et al. (2020b) characterized the lipases contained in *P. puparum* venom. Overall, parasitism led to a decrease of triacylglycerols and several phospholipids (e.g., sphingomyelin, phosphatidylcholine etc.) in the host fat body, whereas triacylglycerols and phospholipids increased in the hemolymph (see Table 1). The increase of triacylglycerols in the host hemolymph was concurrent with a decrease in diacylglycerols. In *P. pupurum* venom, diacylglycerol acyltransferase (DGAT2), catalyzing the last step of triacylglycerol synthesis from diacylglycerols, is not present. The venom does, however, contain multiple lipases (some with missing catalytic triads, potentially involved in lipid binding and transport), which suggests that the host's enzymatic machinery facilitates the conversion of di- to triacylglycerols. In the fat body, increasing triacylglycerol levels were mainly observed for highly unsaturated triacylglycerols, while triacylglycerols with fewer double bonds decreased. An increase in unsaturation generally increases triacylglycerol solubility. There was, however, no difference in unsaturation levels of triacylglycerols in the hemolymph. Hence it is unclear what role the unsaturation plays in the fat body (i.e., higher solubility does not lead to increased transport and presence of unsaturated triacylglycerols in the hemolymph for use by the parasitoid larva). Desaturases were not found in the venom of *P. puparum,* but a desaturase was found to be upregulated in the venom glands (Wang et al. 2020a). It thus remains unclear whether the wasp's venom or the host is responsible for the observed changes in triacylglycerol saturation levels.

The decrease of some phospholipids in the fat body and increase in the hemolymph of the host *Pieris rapae* suggest that destruction of the fat body and fat body cell membranes ensues quickly after parasitism by *P. puparum* (Wang et al. 2020b). Parasitized hosts also had an increased cholesteryl ester content in the hemolymph, suggesting that cholesterol transport increased. Increased cholesteryl ester content was also observed in Dufour's gland (i.e., part of the anatomy of the ovipositor) suggesting that cholesteryl esters may be derived from the venom. Lipases with potential cholesteryl esterase function have been identified from the salivary glands of developing *P. puparum* larvae (Wang et al. 2020b). Cholesteryl esterase hydrolyzes cholesteryl esters to form cholesterol, which may allow the developing parasitoid to acquire essential sterols (that insects cannot synthesize). Sterols can subsequently serve important functions as hormone-precursors, signaling molecules, and components of cell membranes, and were found to increase egg viability at least in the wasp *E. vuilletti* (Mondy et al. 2006).

3.2.2 Lipid-Related Parasitoid Venom Components

Venom components related to lipid metabolism have been identified in 23 different parasitoid species (see Table 2). The function of venom enzymes regarding lipid metabolism can be divided into four different categories: lipid catabolism, transport, synthesis, and storage (see Table 2). When venom is injected, even the host's enzymes may participate in freeing lipids for the developing parasitoid(s). Cathepsin of the host *S. littoralis*, for example, contributes to degradation of the host's fat body following parasitization by *Bracon nigricans* (Becchimanzi et al. 2017). On a cellular level, phospholipases play a key role for increasing nutrient transfer from the cytosol to the hemolymph by disintegrating cells to release their content. Various phospholipases have been identified in parasitoid venom that differ in their specific site of action. Phospholipase A1, for example, disrupts phospholipid packaging of cell membranes, leading to cell lysis, while phospholipase A2 hydrolyzes glycerophospholipids in membranes to release free fatty acids and lysophospholipids (Perez-Riverol et al. 2019). Phospholipases can indeed be part of a complex pathway affecting the host's lipid metabolism. The venom of *N. vitripennis*, for example, modifies cell membrane permeability leading to an influx of Na^+ in the cell (Danneels et al. 2010; Rivers et al. 2002). An increase in Na^+ can subsequently activate phospholipase C, leading to an increase in inositol-3-phosphate (a signaling molecule) and the release of Ca^{2+} from the mitochondrion. Calcium in turn activates phospholipase A2, stimulating fatty acid synthesis (Rivers et al. 2002). Within parasitoid venom, phospholipases thus play an important role in making lipids available for parasitoid offspring (see Table 2).

Once lipids are released from the fat body, lipids need to be transported to the developing parasitoid through the hemolymph. Several hemolymph transport enzymes are contained within the venom, including apolipophorin (e.g., Liu et al. 2018; Scieuzo et al. 2021; see Table 2). In addition to this more typical enzyme involved

Table 2 Overview of genes and/or enzymes involved in lipid metabolism found in the venom of different parasitoid species. We did not distinguish between different homologs or orthologs. For each enzyme, a potential function is indicated based on functional studies in animals, including humans

Enzyme	Function	Species		
Lipid catabolism				
Carboxylesterase	Degradation of triacylglycerols, mainly long-chain triacylglycerols (Deng et al. 2021)	*Anisopteromalus calandrae*[1]	*Microplitis mediator*[13]	*Torymus sinensis*[24]
		Bracon nigricans[4]	*Ooencyrtus telenomicida*[16]	
		Hyposoter didymator[9]	*Psyttalia lounsburyi*[19]	
Cathepsin (D, L, J)	Digestive enzymes (e.g. degradation of the fat body) (Becchimanzi et al. 2017; Cristofoletti et al. 2003; Yang et al. 2020)	*B. nigricans*[3]	*Microctonus hyperodae*[12]	*Toxoneuron nigriceps*[25]
		Leptopilina heterotoma[11]	*M. mediator*[13]	*T. sinensis*[24]
		Microctonus aethiopoides[12]	*O. telenomicida*[16]	
Enolase	Mediates host tisue degradation (Falabella et al. 2009; Grossi et al. 2016)	*M. mediator*[13]	*Psyttalia concolor*[19]	*Tetrastichus brontispae*[23]
		O. telenomicida[16]	*P. lounsburyi*[19]	*T. nigriceps*[25]
Enoyl-CoA hydratase	Metabolizing fatty acids in beta-oxidation to produce both acetyl-CoA and ATP	*B. nigricans*[4]	*O. telenomicida*[16]	
Fatty acid binding protein	Fatty acid import, storage, and export (also cholesterol and phospholipids) (Furuhashi and Hotamisligil 2008)	*Diversinervus elegans*[7]	*M. mediator*[13]	
Lipase (3, A, H)	Digestion, transport, processing of dietary lipids (Wang et al. 2020b)	*B. nigricans*[4]	*O. telenomicida*[16]	*Pteromalus puparum*[22]
		Chelonus inanitus[5]	*M. aethiopoides*[12]	*P. lounsburyi*[19]
		Leptopilina boulardi[10]	*M. hyperodae*[12]	*T. sinensis*[24]
		L. heterotoma[11]	*Nasonia vitripennis*[14-15]	
		M. mediator[13]	*Pimpla hypochondriaca*[17]	
Low-density lipoprotein receptor	Low-density lipoprotein, mediating endocytosis of vitellogenin and lipophorin	*Aphidius ervi*[2]	*N. vitripennis*[14-15]	*P. puparum*[22]
		M. mediator[13]	*O. telenomicida*[16]	*T. sinensis*[24]
Low-density lipoprotein receptor-like venom protein	Central role in cholesterol and other lipoprotein metabolism (Scieuzo et al. 2021)	*A. calandrae*[1]	*N. vitripennis*[14-15]	*O. telenomicida*[16]
Phospholipase (A1, A2, B, C)	Hydrolyses phospholipid substrates at specific ester bonds (Richmond and Smith 2011)	*B. nigricans*[4]	*L. heterotoma*[11]	*P. concolor*[19]
		Cotesia chilonis[6]	*M. mediator*[13]	*P. lounsburyi*[19]
		Eupelmus orientalis[8]	*O. telenomicida*[16]	*T. nigriceps*[25]
		D. elegans[7]	*Pimpla turionellae*[17]	

(continued)

Table 2 (continued)

Enzyme	Function	Species		
Vitellogenin receptor	Low-density lipoprotein receptor that transports lipids into a recipient cell	*M. aethiopoides*[12]	*O. telenomicida*[16]	
Lipid transport (in the hemolymph)				
Annexin	Ca^{2+}-dependent lipid binding protein that could be involved in membrane transport processes	*L. heterotoma*[11]	*O. telenomicida*[16]	
		M. mediator[13]	*P. concolor*[19]	
Apolipophorin	Hemolymph lipid transport (Weers and Ryan 2006)	*B. nigricans*[4]	*O. telenomicida*[16]	*T. sinensis*[24]
		D. elegans[7]	*M. mediator*[13]	
		L. heterotoma[10]	*T. brontispae*[23]	
Apolipoprotein D-like	Lipid transport processes in the insect hemolymph (Scieuzo et al. 2021)	*M. mediator*[13]	*O. telenomicida*[16]	*T. sinensis*[24]
Calreticulin	Chaperoning and regulation of Ca^{2+} homeostasis in the endoplasmic reticulum lumen	*A. calandrae*[1]	*M. mediator*[13]	
		C. chilonis[6]	*N. vitripennis*[14-15]	*P. lounsburyi*[19]
		H. didymator[9]	*O. telenomicida*[16]	*T. brontispae*[23]
		M. hyperodae[12]	*P. puparum*[20,22]	*T. sinensis*[24]
		M. aethiopoides[12]	*P. concolor*[19]	*T. nigriceps*[25]
Odorant binding protein	Solubilizing and carrying free fatty acids released by lipases (Ishida et al. 2013; Pelosi et al. 2018)	*A. calandrae*[1]	*M. mediator*[13]	*T. brontispae*[23]
		B. nigricans[4]	*N. vitripennis*[14-15]	*T. sinensis*[24]
		C. inanitus[5]	*O. telenomicida*[16]	
		L. heterotoma[11]	*P. puparum*[21-22]	
Lipid synthesis				
3-oxoacyl-ACP reductase	Fatty acid synthesis and polyunsaturated fatty acid synthesis	*B. nigricans*[4]		
Fatty acid synthase	Catalyzing the de novo synthesis of fatty acids	*A. ervi*[2]	*M. mediator*[13]	*T. brontispae*[23]
		D. elegans[7]	*O. telenomicida*[16]	
n-acetyllactosaminide beta-n-acetylglucosaminyltransferase	Glycosphingolipid synthesis	*L. heterotoma*[11]	*O. telenomicida*[16]	*T. nigriceps*[25]
Phosphatidate phosphatase	Conversion of phosphatidate to diacylglycerols	*D. elegans*[7]	*O. telenomicida*[16]	*T. brontispae*[23]
Lipid storage				
Adipocyte plasma membrane-associated protein-like	Maturation of adipocytes and their capacity to store lipids (Sarjeant and Stephens 2012)	*O. telenomicida*[15]	*T. sinensis*[24]	
Insulin-like growth factor-binding protein	Regulation of lipid metabolism, lipid accumulation, adipocyte differentiation (Pan et al. 2021; Baxter 2023; Kim 2013)	*L. heterotoma*[10]	*O. telenomicida*[16]	
		M. mediator[12]	*T. sinensis*[24]	

(continued)

Table 2 (continued)

Enzyme	Function	Species		
Regucalcin	Ca^{2+} signaling, lipid accumulation in adipocytes (Doğan et al. 2021)	*B. nigricans*[3]	*M. mediator*[13]	*O. telenomicida*[16]

[1]Perkin et al. (2015), [2]Colinet et al. (2014), [3]Becchimanzi et al. (2017), [4]Becchimanzi et al. (2020), [5]Vincent et al. (2010), [6]Teng et al. (2017), [7]Liu et al. (2017), [8]Doury et al. (1997), [9]Dorémus et al. (2013), [10]Colinet et al. (2013), [11]Heavner et al. (2013), [12]Crawford et al. (2008), [13]Lin et al. (2019), [14]de Graaf et al. (2010), [15]Sim and Wheeler (2016), [16]Cusumano et al. (2018), [17]Dani et al. (2005), [18]Uçkan et al. (2006), [19]Mathé-Hubert et al. (2016), [20]Zhu et al. (2010), [21]Wang et al. (2015), [22]Yan et al. (2016), [23]Liu et al. (2018), [24]Scieuzo et al. (2021), [25]Laurino et al. (2016)

in lipid transport, there have been several reports of odorant binding proteins being part of the parasitoid venom (e.g., in *N. vitripennis, P. puparum*, and 8 other species; see Table 2). Most volatiles are lipophilic, and odorant-binding proteins typically serve for the transport of odorant molecules (e.g., pheromones) to olfactory receptors. In the parasitoid venom, odorant binding proteins are hypothesized to play a role as fatty acid and fatty acid ester carriers, as was found in several other insects (e.g., the ant *Camponotus japonicus*; Ishida et al. 2013; the blowfly *P. regina*; González et al. 2009). Existing odorant binding proteins thus seem to have acquired new functions.

The venom of some parasitoid species also contains enzymes that are involved in lipid synthesis, including fatty acids, glycosphingolipids, and diacylglycerols (e.g., Colinet et al. 2014; Heavner et al. 2013; see Table 2). So far, no clear explanation has been proposed as to why the venom would contain enzymes involved in lipogenesis. When hosts are lipid-poor, there could be an advantage for the parasitoid to increase the host's lipogenesis by injecting enzymes involved in lipid synthesis contained in the venom. The aphid *Acyrtosiphon pisum* and the scale insect *Parasaissetia nigra,* for example, are plant sap-sucking insects, a nutritional resource that is expected to contain substantial carbohydrate resources, but not many lipids. Lipid synthesis enzymes present in the venom can then utilize precursors, such as carbohydrates, from the host to increase lipid content and availability. The presence of fatty acid and diacylglycerol synthesizing enzymes in the parasitoid venom may aid the developing parasitoid in obtaining sufficient lipids to complete development and fuel adult life.

Three enzymes implicated in host adipocyte maturation and/or lipid storage were found in the venom of several parasitoid species (e.g., Scieuzo et al. 2021; Sim and Wheeler 2016; see Table 2). At the time of oviposition, the parasitoid is still in the egg or early larval stage, a time at which absorption of nutrients may be relatively little (compared to later developmental stages). For example, during the early stages of parasitism, the braconids *Aphidius ervi* and *Toxoneuron nigriceps* absorb nutrients through the epidermis (Caccia et al. 2005; Grimaldi et al. 2006). In parasitoid offspring in general, some time may be needed to develop a fully functioning gut and absorption of nutrients through the epidermis or the anal vesicle in early larval stages may be more common (Edson and Vinson 1977). Storage of large fat reserves by the parasitoids is also expected to take some time, with fat droplets becoming clearly visible only during later larval instars (e.g., in *E. vuilletti* and *Gelis* sp., Pers. Obs.). When fat precursors accumulate in the hemolymph of the host, initial fat storage in host adipocytes can provide a reserve to be consumed by the developing parasitoid at a later time. Increased fat storage in the preferred host *S. bullata* was indeed found following parasitism by *N. vitripennis* (Rivers and Denlinger 1994, 1995). Parasitoid venom thus contains many enzymes involved in releasing, transporting, synthesizing, and storing fat by the host. What is still mostly lacking are studies identifying the specific functional role of identified venom components on host lipid metabolism.

3.3 Polydnaviruses Increase Host Lipid Availability for the Developing Parasitoid

Polydnaviruses are viruses associated with endoparasitoid wasps (i.e., the primary hosts), persisting as proviruses (i.e., DNA sequences integrated in the host genome) in every germ and somatic cell of the wasp. The proviral genome is composed of both core genes necessary for viral replication and virulence genes that, after amplification and excision from the wasp's genome, form viral particules (i.e., virions). Viral replication occurs only in calyx cells (that are part of the wasp's reproductive tract) during the parasitoid pupal and adult stage. The polydnavirus life cycle starts with virions that are first released from calyx cells to accumulate in the lumen of the parasitoid's reproductive tract where eggs are stored. Eggs, containing both the proviral genome and virions, are then laid by the parasitoid in the host during oviposition along with parasitoid venom (Strand and Burke 2013). After virion injection, the virus integrates into the host genome via a second domain present on the viral DNA termed the host integration motif. Virulence genes are then transcribed in host cells until wasp development is completed and the adult parasitoid has emerged from the host (see Fig. 1 from Strand and Burke 2012).

Polydnaviruses can be grouped into two distinct genera: Bracoviruses, associated with the braconid family, and Ichnoviruses, associated with the ichneumonid family (Strand and Burke 2013). Braco- and Ichnoviruses each have a distinct morphology of the virion (that enters the secondary host, which is the host of the parasitoid) and an independent evolutionary origin (Strand and Burke 2012). The virus participates in the parasitization process, affecting the host's immune system (i.e., to prevent the host from killing the wasp's offspring), host growth, and metabolism (Strand and Burke 2013). During parasitoid oviposition, the bracovirus of the wasp *T. nigriceps*, for example, releases several viral protein tyrosine phosphatases in the host *Heliothis virescens's* body that disrupt the prothoracic gland function of the host and inhibits host metamorphosis (Falabella et al. 2006; but see also Strand and Burke 2015 for other examples). The following subsection focuses on the effects polydnaviruses have on (secondary) host lipid metabolism.

The braconid *Chelonus inanitus* is an endoparasitoid that injects both venom and a bracovirus along with the egg. Kaeslin et al. (2005) disentangled the role of the *C. inanitus* venom, bracovirus, and developing parasitoid on the fat body of the host *S. littoralis*. Separating the effects of venom and bracovirus from developing parasitoids is possible when comparing parasitized hosts with unparasitized hosts, but also using pseudoparasitized hosts, where the eggs within the mother are killed using x-rays prior to oviposition. Pseudoparasitized hosts thus receive the venom and the bracovirus, but the parasitoid larva does not hatch. Venom proteins disappear within 1-2 days after parasitization, while the polydnavirus remains throughout parasitoid development. The parasitoid larva, along with polydnavirus, and potential early effects of venom cause an accumulation of whole-body lipids during development (see Table 1). During the last host larval instar, lipid content was significantly higher in parasitized hosts than in unparasitized larvae, meaning that the parasitoid larva itself also plays a major role in increasing host fat accumulation.

In a recent study, Wang et al. (2021) determined which *C. vestalis* parasitoid-associated factor led to the decrease of lipid levels in the host moth *P. xylostella. Cotesia vestalis* injects venom and bracoviruses and forms teratocytes derived from the embryonic membrane. Wang et al. (2021) used both parasitized and pseudoparasitized *P. xylostella* hosts, thereby removing the effect of teratocytes (as teratocytes are derived from developing parasitoid offspring) and the developing parasitoid. Following parasitization and pseudoparasitization, host whole-body triacylglycerol levels decreased, as did hemolymph fat levels.

Injection of venom alone did not result in any changes; yet a similar reduction in lipids was observed when only the bracovirus was injected. The reduction in host lipids can be due to alterations in lipid absorption and synthesis. Parasitized *P. xylostella* indeed showed reduced formation of neutral lipid droplets in the gut, suggesting that changes in host lipid absorption and synthesis underlie the decrease in whole-body lipids. Transcriptomics further led to the identification of several bracovirus genes that could be involved in manipulating host lipid metabolism (Wang et al. 2021). Expression of one of these genes, *CvBV 9-2*, was indeed found to be responsible for reducing triacylglycerol levels in parasitized larvae by increasing the expression of a tachykinin gene (*PxTk*) in the host gut, suppressing lipogenesis.

3.4 Parasitoid-Derived Teratocytes Increase Fat Availability for the Parasitoid

Teratocytes are specialized cells derived from the dissociation of the cellular membrane surrounding the parasitoid embryo during its development that are released in the host's hemolymph during parasitoid hatching (Strand 2014). Teratocytes are produced by some subfamilies within Braconidae (i.e., Microgastrinae, Meteorinae, Euphorinae, Aphidiinae) and Platygastridae (i.e., Scelioninae, Telenominae, Teleasinae), all of which are endoparasitoids (Dahlman 1990; Strand 2014). Teratocyte-like cells have also been reported in the Ichneumonidae (Rouleux-Bonnin et al. 1999) and Chalcidoidea (Pedata et al. 2003; Strand 1986). The number of teratocytes in parasitoids is species-specific, and can range from 10 (e.g., *Telenomus heliothidis*, Platygastridae; Strand et al. 1988) to more than 1,000 (e.g., *M. pulchricornis*, Braconidae; Suzuki and Tanaka 2007) (Strand 2014). Teratocytes help to disrupt host growth, inhibit host metamorphosis, and also seem to play a role in evading the host's immune system (Ali et al. 2013; Dahlman et al. 2003; Strand 2014). Teratocytes further aid in nutrient acquisition for the developing parasitoid(s), particularly lipids (Falabella et al. 2000, 2005; Nakamatsu et al. 2002; Qin et al. 2000; Suzuki and Tanaka 2007).

Ultrastructure studies revealed that once released in the host's hemolymph, teratocytes show both morphological and metabolic changes (Pennacchio et al. 1994; Strand et al. 1986; Volkoff and Colazza 1992; Zhang et al. 1994), e.g., teratocyte size greatly increases (de Buron and Beckage 1997; Strand and Wong 1991; Volkoff and Colazza 1992). To promote nutrient exchange between the teratocyte's intracellular and extracellular space, teratocytes exhibit long microvilli on their surface (to increase the surface for absorption/secretion), as well as large exosome-like spherical vesicles (containing lipids and other nutrients; Hotta et al. 2001; Salvia et al. 2019; Shelby et al. 2014; Sluss 1968). An abundant rough endoplasmic reticulum, numerous mitochondria, and an extensive vacuolization are observed in the teratocyte cytoplasm (de Buron and Beckage 1997; Gerling and Orion 1973; Sluss 1968; Volkoff and Colazza 1992). Teratocytes further do not divide after being released, but often become highly polyploid associated with an increase of the nuclear area. This polyploidization seems to stimulate the synthesis of proteins of the teratocytes (Gerling and Orion 1973; Hotta et al. 2001; Strand and Wong 1991). In the insect fat body, DNA polyploidy caused by juvenile hormone stimulation was indeed found to increase the transcription of vitellogenin (Dittmann et al. 1989; Nair et al. 1981; Hotta et al. 2001). These characteristics show that teratocytes are specialized cells, able to metabolically interact with other close cells (Dahlman and Bradleigh Vinson 1993; Salvia et al. 2019; Sluss 1968).

Teratocytes supply nutrients to the developing parasitoid by digesting the host's fat body during early parasitoid larval stages when mouth parts are not yet formed. In the host-parasitoid system *Pseudaletia separata-Cotesia kariyai*, triacylglycerol levels of the host fat body decreased 6 days after parasitism but increased in the parasitoid's second instar larva from the 7th day (Nakamatsu et al. 2002). The increased lipase activity in the gut parasitoid larva, as well as the presence of lipid granules in the parasitoid midgut, confirmed the ingestion of host lipids by the parasitoid (Nakamatsu et al. 2002). Interestingly, teratocytes were attached to the host fat body and locally released collagenases (i.e., enzymes that break down the collagen sheath surrounding the host's fat body) to disrupt the host fat body matrix and release fat body cells (Nakamatsu et al. 2002). Teratocytes of other parasitoid species, such as the braconids *Microplitis mediator* or *Microplitis pulchricornis*, seem to play a similar role in disrupting and digesting the host fat body to secure parasitoid survival and development (Qin et al. 2000; Suzuki and Tanaka 2007).

Teratocytes release several other enzymes that can enhance host fat body digestion until complete consumption by the parasitoid larva: a teratocyte-specific carboxylesterase, assumed to be involved in the hydrolysis of host lipids (*Dinocampus coccinellae*; Gopalapillai et al. 2005), enolases and lipases (*A. ervi*, *Microplitis demolitor, D. coccinellae*; Burke and Strand 2014; Falabella et al. 2009; Kadono-Okuda et al. 1998), as well as cathepsin (Burke and Strand 2014). These lipid-catabolic enzymes have also been found in the venom of some parasitoid species (e.g., Dorémus et al. 2013; Perkin et al. 2015; see Table 2). In the parasitoid *T. nigriceps*, teratocytes produced a chitinase during the last larval stage of the parasitoid. This chitinase was hypothesized to be part of the enzymes that help the parasitoid larva's egression by breaking down the host cuticle (Cônsoli et al. 2005).

In addition to fat body disruption, teratocytes also facilitate the transport of lipids from the host hemolymph to the developing larvae. For example, teratocytes of *A. ervi* produce an extracellular fatty acid binding protein that transports fatty acids in the host's hemolymph (Falabella et al. 2000, 2005; Pennacchio et al. 1999). This protein showed a high affinity for C14-C18 saturated fatty acids, oleic acid (C18:1), and a longer chain polyunsaturated fatty acid (arachidonic acid; C20:4) (Falabella et al. 2005). Immunolocalization revealed that the fatty acid binding protein was distributed around lipid particles abundantly present in the hemolymph of the parasitized host, but also in the external epidermal layer and the midgut lumen of parasitoid larvae (Caccia et al. 2012; Falabella et al. 2005). Altogether these findings suggest that (1) fatty acids can be absorbed by the epidermal epithelium of the developing parasitoid, as had previously been found for amino acids and sugars (Caccia et al. 2012) and (2) fatty acid binding protein transports key fatty acids in the host hemolymph to the growing parasitoid larva, which can subsequently be absorbed by the parasitoid and stored as triacylglycerols (Caccia et al. 2012). Similar lipid transport proteins were found in parasitoid venom, such as annexin, apolipophorins, and calreticulin (Crawford et al. 2008; Lin et al. 2019; see Table 2) (Burke and Strand 2014).

A decrease in teratocyte number during later stages of parasitoid development has been observed in several parasitoid species (de Buron and Beckage 1997; Gopalapillai et al. 2005; Kadono-Okuda et al. 1995; Suzuki and Tanaka 2007; Volkoff and Colazza 1992). The number of teratocytes decreases due to the teratocyte undergoing programmed cell death, as evidenced by the appearance of multiple bleb structures (i.e., teratocyte anatomical deformations resulting from the enlargement or coalescence of microvilli; de Buron and Beckage 1997; Zhang et al. 1994) on the teratocycte membrane (de Buron and Beckage 1997; Hotta et al. 2001). Another factor contributing to the declining teratocyte numbers late in parasitoid development is that teratocytes are progressively consumed by the parasitoid larva (e) (Kadono-Okuda et al. 1995; Strand and Wong 1991). Teratocytes produce proteins that can be released in the host's hemolymph for disrupting the host fat body but can also store a high abundance of proteins (e.g., glycoproteins, vitellogenin, amino-acids) as well as lipids (i.e., lipid droplet) that can constitute an additional source of nutrients for successful parasitoid development (de Buron and Beckage 1997; Gopalapillai et al. 2005; Kadono-Okuda et al. 1998; Okuda and Kadono-Okuda 1995). On the contrary, no decrease in teratocyte number was observed during later stages of parasitism of other parasitoids, such as *C. kariyai*, suggesting that the teratocytes are not consumed immediately by the parasitoid and may have another potential role in host regulation or parasitoid development at a later stage (Hotta et al. 2001; Suzuki and Tanaka 2007). Teratocytes produced by some parasitoid wasps are important specialized cells that use a variety of enzymes to disrupt the host's fat body. The release of host fat cells transported from the host to the parasitoid aids parasitoid development and survival.

4 Conclusions and Future Perspectives

Parasitoids are fascinating creatures, particularly regarding lipid metabolism. Parasitoid larvae can mimic the fatty acid composition of the host, because there is little to no de novo triacylglycerol synthesis. The adults of many parasitoid species do not accumulate fat at all, except for some polyphagous species that typically develop on fat-poor hosts. More studies are now needed to determine how host fatty acid composition, host breadth, and the ability to synthesize triacylglycerols are related in parasitoids. Such an endeavor should start with a replication of the work of Barlow and Jones (1981), and Jones et al. (1982), in larvae and adult parasitoids, using tracers to identify if and when the Kennedy pathway for de novo synthesis of triacylglycerols is activated or not. The number of host species a parasitoid can parasitize was found to play a role, where typically specialists mimic the host fatty acid composition, while generalists do not. Generalists were also found to accumulate fat in more recent studies (Visser et al. 2010). To test how host breadth and host fatty acid composition affect parasitoid fatty acid synthesis and fat accumulation, a comparative approach using specialists and generalists developing on the same hosts could be used. For example, the parasitoid guild associated with *Drosophila* contains both specialists and generalists developing on distinct hosts, including *D. melanogaster* and *D. simulans.* Another interesting system to use is the *Nasonia* species complex, with *N. vitripennis* being an extreme generalist (but preferring and manipulating lipid synthesis only of *S. bullata*), and *Nasonia giraulti* and *Nasonia longicornis* that are restricted to hosts in the genera *Protocalliphora* and *Sarcophaga.*

More recently, fatty acid synthesis and fat accumulation were found to vary in response to the fat content of the host and is thus plastic, in the wasp *L. heterotoma*. Plasticity of fatty acid synthesis and fat accumulation may be more common, also in other parasitoid species, but this remains to be explicitly tested on a large scale. More information about genotype-level responses to host fat content in diverse parasitoid species allows to make inferences about the evolution of plasticity and potential consequences for life histories. The latter is particularly relevant for species that are used as important natural enemies in agro-ecosystems. The finding that fat synthesis is plastic can lead to many other interesting avenues for future research. For example, gaining a deeper understanding of the ecological conditions favoring or selecting against more or less plastic phenotypes in natural populations in insects in general. We can further continue to dig into the mechanisms underlying fatty acid synthesis and fat accumulation (and the lack thereof) by experimentally manipulating parasitoid phenotypes (e.g., by changing host fat content; Enriquez et al. 2022). Several other research directions focus on parasitoid lipids, including symbiotic interactions with bacteria, such as *Wolbachia*, as well as the use of parasitoids as a model resisting obesity (being able to switch fat synthesis off when being fat and continuing to feed, as highlighted by Visser et al. 2023). The plethora of future research lines shows that, despite the considerable research effort into parasitoid lipid metabolism since the 1960s, there is still a great diversity of research opportunities that can and hopefully will be pursued.

Parasitoids are masters in host manipulation with the sheer number of mechanisms by which host lipid metabolism can be affected as proof. The diversity of parasitoids and thus host manipulation strategies may seem daunting to try and elucidate because most responses are host and parasitoid-species specific. Using hosts and parasitoids that share an evolutionary history is, therefore, essential to further our understanding of host manipulation in a biologically meaningful way. *Pteromalus puparum* is one of the few species with which complementary studies have been performed to understand the entire process of host manipulation, from physiology to genes and gene diversification (Wang et al. 2020b, 2021). Extending such thorough investigation to other systems (i.e., hemolymph-feeders, koinobionts), also in a comparative context, will certainly enrich our understanding of host manipulation (see Sect. 3). There is also much to learn from parasitoid host manipulation strategies, even for our own benefit. For example, some venom components can be used in biological control of insect pests (Danneels et al. 2010; Moreau and Asgari 2015). Virulence factors associated with teratocytes and polydnaviruses have also been proposed for use in transgenic plants, where virulence genes involved in manipulation of the host are integrated in the plant genome to increase plant resistance to pest attack (Merlin et al. 2021; Kim et al. 2016). Parasitoid venom components were suggested as potential pharmaceuticals against allergies, blood clotting, and as an antibiotic against microbial infections (Moreau and Asgari 2015). Parasitoids can thus inspire the development of new technologies, perhaps even beyond insect pest control.

Acknowledgments We are grateful to Jeff Harvey for providing insights into the biology of several parasitoid species and Jacintha Ellers for providing the original data used to make Box Fig. 1a. We would further like to thank Thomas Enriquez for making the cartoons used in Fig. 1. MS and BV were supported by the Fonds National de Recherche Scientifique.

References

Ali MR, Seo J, Lee D, Kim Y (2013) Teratocyte-secreting proteins of an endoparasitoid wasp, *Cotesia plutellae*, prevent host metamorphosis by altering endocrine signals. Comp Biochem Physiol A Mol Integr Physiol 166:251–262. https://doi.org/10.1016/j.cbpa.2013.06.028

Arrese EL, Soulages JL (2010) Insect fat body: energy, metabolism, and regulation. Annu Rev Entomol 55: 207–225. https://doi.org/10.1146/annurev-ento-112408-085356

Barlow JS (1964) Fatty acids in some insect and spider fats. Can J Biochem 42:1365–1374. https://doi.org/10.1139/o64-148

Barlow JS (1965) Effects of diet on the composition of body fat in *Agria affinis* (Fallén). Can J Zool 43:337–340. https://doi.org/10.1139/z65-032

Barlow JS (1966) Effects of the diet on the composition of body fat in *Musca domestica* L. Can J Zool 44:775–779. https://doi.org/10.1139/z66-077

Barlow JS, Jones D (1981) A comparative study of transacylation in three insect species. Can J Zool 59: 1141–1147. https://doi.org/10.1139/z81-159

Barras DJ, Joiner RL, Vinson SB (1970) Neutral lipid composition of the tobacco budworm, *Heliothis virescens* (Fab.), as affected by its habitual parasite, *Cardiochiles nigriceps* viereck. Comp Biochem Physiol 36:775–783. https://doi.org/10.1016/0010-406X(70)90532-3

Baxter RC (2023) Signaling pathways of the insulin-like growth factor binding proteins. Endocr Rev 44:753–778. https://doi.org/10.1210/endrev/bnad008

Becchimanzi A, Avolio M, Di Lelio I, Marinelli A, Varricchio P, Grimaldi A et al (2017) Host regulation by the ectophagous parasitoid wasp *Bracon nigricans*. J Insect Physiol 101:73–81. https://doi.org/10.1016/j.jinsphys.2017.07.002

Becchimanzi A, Avolio M, Bostan H, Colantuono C, Cozzolino F, Mancini D et al (2020) Venomics of the ectoparasitoid wasp *Bracon nigricans*. BMC Genomics 21:34. https://doi.org/10.1186/s12864-019-6396-4

Birsoy K, Festuccia WT, Laplante M (2013) A comparative perspective on lipid storage in animals. J Cell Sci 126:1541–1552. https://doi.org/10.1242/jcs.104992

Bischof C, Ortel J (1996) The effects of parasitism by *Glyptapanteles liparidis* (Braconidae: Hymenoptera) on the hemolymph and total body composition of gypsy moth larvae (*Lymantria dispar*, Lymantriidae: Lepidoptera). Parasitol Res 82:687–692. https://doi.org/10.1007/s004360050186

Bracken GK, Barlow JS (1967) Fatty acid composition of *Exeristes comstockii* (Cress) reared on different hosts. Can J Zool 45:57–61. https://doi.org/10.1139/z67-007

Burke GR, Strand MR (2014) Systematic analysis of a wasp parasitism arsenal. Mol Ecol 23:890–901. https://doi.org/10.1111/mec.12648

Caccia S, Leonardi MG, Casartelli M, Grimaldi A, de Eguileor M, Pennacchio F et al (2005) Nutrient absorption by *Aphidius ervi* larvae. J Insect Physiol 51:1183–1192. https://doi.org/10.1016/j.jinsphys.2005.06.010

Caccia S, Grimaldi A, Casartelli M, Falabella P, de Eguileor M, Pennacchio F et al (2012) Functional analysis of a fatty acid binding protein produced by *Aphidius ervi* teratocytes. J Insect Physiol 58:621–627. https://doi.org/10.1016/j.jinsphys.2011.12.019

Carton Y, Bouletreau M, van Alphen JJM, van Lenteren JC (1986) The Drosophila parasitic wasps. In: Ashburner M, Carson HL, Thompson JN (eds) The genetics and biology of Drosophila (volume 3). Academic, London, pp 348–394

Casas J, Driessen G, Mandon N, Wielaard S, Desouhant E, van Alphen J et al (2003) Energy dynamics in a parasitoid foraging in the wild. J Anim Ecol 72:691–697. https://doi.org/10.1046/j.1365-2656.2003.00740.x

Colinet H, Hance T, Vernon P (2006) Water relations, fat reserves, survival, and longevity of a cold-exposed parasitic wasp *Aphidius colemani* (Hymenoptera: Aphidiinae). Environ Entomol 35:228–236. https://doi.org/10.1603/0046-225X-35.2.228

Colinet D, Deleury E, Anselme C, Cazes D, Poulain J, Azema-Dossat C et al (2013) Extensive inter- and intraspecific venom variation in closely related parasites targeting the same host: the case of *Leptopilina* parasitoids of *Drosophila*. Insect Biochem Mol Biol 43:601–611. https://doi.org/10.1016/j.ibmb.2013.03.010

Colinet D, Anselme C, Deleury E, Mancini D, Poulain J, Azéma-Dossat C et al (2014) Identification of the main venom protein components of *Aphidius ervi*, a parasitoid wasp of the aphid model *Acyrthosiphon pisum*. BMC Genomics 15:342. https://doi.org/10.1186/1471-2164-15-342

Cônsoli FL, Brandt SL, Coudron TA, Vinson SB (2005) Host regulation and release of parasitism-specific proteins in the system *Toxoneuron nigriceps–Heliothis virescens*. Comp Biochem Physiol B Biochem Mol Biol 142:181–191. https://doi.org/10.1016/j.cbpc.2005.07.002

Crawford AM, Brauning R, Smolenski G, Ferguson C, Barton D, Wheeler TT et al (2008) The constituents of *Microctonus sp.* parasitoid venoms. Insect Mol Biol 17:313–324. https://doi.org/10.1111/j.1365-2583.2008.00802.x

Cristofoletti PT, Ribeiro AF, Deraison C, Rahbé Y, Terra WR (2003) Midgut adaptation and digestive enzyme distribution in a phloem feeding insect, the pea aphid *Acyrthosiphon pisum*. J Insect Physiol 49:11–24. https://doi.org/10.1016/S0022-1910(02)00222-6

Cuny MAC, Poelman EH (2022) Evolution of koinobiont parasitoid host regulation and consequences for indirect plant defence. Evol Ecol 36:299–319. https://doi.org/10.1007/s10682-022-10180-x

Cusumano A, Duvic B, Jouan V, Ravallec M, Legeai F, Peri E et al (2018) First extensive characterization of the venom gland from an egg parasitoid: structure, transcriptome and functional role. J Insect Physiol 107:68–80. https://doi.org/10.1016/j.jinsphys.2018.02.009

Dahlman DL (1970) Trehalose levels in parasitized and nonparasitized tobacco hornworm, *Manduca sexta* larvae. Ann Entomol Soc Am 63:615–617

Dahlman DL (1990) Evaluation of teratocyte functions: an overview. Arch Insect Biochem Physiol 13:159–166. https://doi.org/10.1002/arch.940130303

Dahlman DL, Bradleigh Vinson S (1993) Teratocytes: developmental and biochemical characteristics. In: Beckage NE, Thompson SN, Federici BA (eds) Parasites and pathogens of insects, 1st edn. Elsevier, London, pp 145–165. https://doi.org/10.1016/B978-0-08-091649-1.50012-8

Dahlman DL, Greene JR (1981) Larval hemolymph protein patterns in tobacco hornworms parasitized by *Apanteles congregatus*. Ann Entomol Soc Am 74:130–133. https://doi.org/10.1093/aesa/74.1.130

Dahlman DL, Rana RL, Schepers EJ, Schepers T, DiLuna FA, Webb BA (2003) A teratocyte gene from a parasitic wasp that is associated with inhibition of insect growth and development inhibits host protein synthesis. Insect Mol Biol 12:527–534. https://doi.org/10.1046/j.1365-2583.2003.00439.x

Dani MP, Edwards JP, Richards EH (2005) Hydrolase activity in the venom of the pupal endoparasitic wasp, *Pimpla hypochondriaca*. Comp Biochem Physiol B Biochem Mol Biol 141:373–381. https://doi.org/10.1016/j.cbpc.2005.04.010

Danneels EL, Rivers DB, de Graaf DC (2010) Venom proteins of the parasitoid wasp *Nasonia vitripennis*: recent discovery of an untapped pharmacopee. Toxins (Basel) 2:494–516. https://doi.org/10.3390/toxins2040494

de Buron I, Beckage NE (1997) Developmental changes in teratocytes of the braconid wasp *Cotesia congregata* in larvae of the tobacco hornworm, *Manduca sexta*. J Insect Physiol 43:915–930. https://doi.org/10.1016/S0022-1910(97)00056-5

de Graaf DC, Aerts M, Brunain M, Desjardins CA, Jacobs FJ, Werren JH et al (2010) Insights into the venom composition of the ectoparasitoid wasp *Nasonia vitripennis* from bioinformatic and proteomic studies. Insect Mol Biol 19:11–26. https://doi.org/10.1111/j.1365-2583.2009.00914.x

Delobel B, Pageaux JF (1981) Influence de l'alimentation sur la composition en acides gras totaux de diptères tachinaires. Entomol Exp Appl 29:281–288. https://doi.org/10.1111/j.1570-7458.1981.tb03070.x

Deng Y, Kim BY, Lee KY, Yoon HJ, Wan H, Li J et al (2021) Lipolytic activity of a carboxylesterase from bumblebee (*Bombus ignitus*) venom. Toxins (Basel) 13:239. https://doi.org/10.3390/toxins13040239

Desjardins CA, Perfectti F, Bartos JD, Enders LS, Werren JH (2010) The genetic basis of interspecies host preference differences in the model parasitoid *Nasonia*. Heredity (Edinb) 104:270–277. https://doi.org/10.1038/hdy.2009.145

Dittmann F, Kogan PH, Hagedorn HH (1989) Ploidy levels and DNA synthesis in fat body cells of the adult mosquito, *Aedes aegypti*: the role of juvenile hormone. Arch Insect Biochem Physiol 12:133–143. https://doi.org/10.1002/arch.940120302

Doğan C, Hänniger S, Heckel DG, Coutu C, Hegedus DD, Crubaugh L et al (2021) Characterization of calcium signaling proteins from the fat body of the Colorado potato beetle, *Leptinotarsa decemlineata* (Coleoptera: Chrysomelidae): implications for diapause and lipid metabolism. Insect Biochem Mol Biol 133:103549. https://doi.org/10.1016/j.ibmb.2021.103549

Dorémus T, Urbach S, Jouan V, Cousserans F, Ravallec M, Demettre E et al (2013) Venom gland extract is not required for successful parasitism in the polydnavirus-associated endoparasitoid *Hyposoter didymator* (Hym. Ichneumonidae) despite the presence of numerous novel and conserved venom proteins.

Insect Biochem Mol Biol 43:292–307. https://doi.org/10.1016/j.ibmb.2012.12.010

dos Passos EM, Wanderley-Teixeira V, Porto ALF, Marques EJ, Teixeira ÁAC, Silva FSO (2019) *Cotesia flavipes* Cameron (Hymenoptera: Braconidae) alters the nutrients in the hemolymph, fat body, and cytochemistry of *Diatraea flavipennella* box (Lepidoptera: Crambidae) hemocytes. Semin Cienc Agrar 40: 539–554. https://doi.org/10.5433/1679-0359.2019v40n2p539

Doury G, Bigot Y, Periquet G (1997) Physiological and biochemical analysis of factors in the female venom gland and larval salivary secretions of the ectoparasitoid wasp *Eupelmus orientalis*. J Insect Physiol 43:69–81. https://doi.org/10.1016/S0022-1910(96)00053-4

Edson KM, Vinson SB (1977) Nutrient absorption by the anal vesicle of the braconid wasp, *Microplitis croceipes*. J Insect Physiol 23:5–8. https://doi.org/10.1016/0022-1910(77)90101-9

Eggleton P, Belshaw R (1992) Insect parasitoids: an evolutionary overview. Philos Trans R Soc B 337:1–20. https://doi.org/10.1098/rstb.1992.0079

Eggleton P, Belshaw R (1993) Comparisons of dipteran, hymenopteran and coleopteran parasitoids: provisional phylogenetic explanations. Biol J Linn Soc 48:213–226. https://doi.org/10.1006/bijl.1993.1015

Eijs IEM, Ellers J, van Duinen G-J (1998) Feeding strategies in drosophilid parasitoids: the impact of natural food resources on energy reserves in females. Ecol Entomol 23:133–138. https://doi.org/10.1046/j.1365-2311.1998.00117.x

Ellers J (1996) Fat and eggs: an alternative method to measure the trade-off between survival and reproduction in insect parasitoids. Neth J Zool 46:227–235. https://doi.org/10.1163/156854295X00186

Ellers J, van Alphen JJM (1997) Life history evolution in *Asobara tabida*: plasticity in allocation of fat reserves to survival and reproduction. J Evol Biol 10:771–785. https://doi.org/10.1046/j.1420-9101.1997.10050771.x

Ellers J, van Alphen JJM (2002) A trade-off between diapause duration and fitness in female parasitoids. Ecol Entomol 27:279–284. https://doi.org/10.1046/j.1365-2311.2002.00421.x

Ellers J, van Alphen JJM, Sevenster JG (1998) A field study of size-fitness relationships in the parasitoid *Asobara tabida*. J Anim Ecol 67:318–324. https://doi.org/10.1046/j.1365-2656.1998.00195.x

Ellers J, Bax M, van Alphen JJM (2001) Seasonal changes in female size and its relation to reproduction in the parasitoid *Asobara tabida*. Oikos 92:209–314. https://doi.org/10.1034/j.1600-0706.2001.920213.x

Enriquez T, Lievens V, Nieberding CM, Visser B (2022) Pupal size as a proxy for fat content in laboratory-reared and field-collected *Drosophila* species. Sci Rep 12:12855. https://doi.org/10.1038/s41598-022-15325-0

Falabella P, Tremblay E, Pennacchio F (2000) Host regulation by the aphid parasitoid *Aphidius ervi*: the role of teratocytes. Entomol Exp Appl 97:1–9. https://doi.org/10.1046/j.1570-7458.2000.00710.x

Falabella P, Perugino G, Caccialupi P, Riviello L, Varricchio P, Tranfaglia A et al (2005) A novel fatty acid binding protein produced by teratocytes of the aphid parasitoid *Aphidius ervi*. Insect Mol Biol 14: 195–205. https://doi.org/10.1111/j.1365-2583.2004.00548.x

Falabella P, Caccialupi P, Varricchio P, Malva C, Pennacchio F (2006) Protein tyrosine phosphatases of *Toxoneuron nigriceps* bracovirus as potential disrupters of host prothoracic gland function. Arch Insect Biochem Physiol 61:157–169. https://doi.org/10.1002/arch.20120

Falabella P, Riviello L, De Stradis ML, Stigliano C, Varricchio P, Grimaldi A et al (2009) *Aphidius ervi* teratocytes release an extracellular enolase. Insect Biochem Mol Biol 39:801–813. https://doi.org/10.1016/j.ibmb.2009.09.005

Forbes AA, Bagley RK, Beer MA, Hippee AC, Widmayer HA (2018) Quantifying the unquantifiable: why Hymenoptera, not Coleoptera, is the most speciose animal order. BMC Ecol 18:21. https://doi.org/10.1186/s12898-018-0176-x

Ford PS, van Heusden MC (1994) Triglyceride-rich lipophorin in *Aedes aegypti* (Diptera: Culicidae). J Med Entomol 31:435–441. https://doi.org/10.1093/jmedent/31.3.435

Furuhashi M, Hotamisligil GS (2008) Fatty acid-binding proteins: role in metabolic diseases and potential as drug targets. Nat Rev Drug Discov 7:489–503. https://doi.org/10.1038/nrd2589

Gerling D, Orion T (1973) The giant cells produced by *Telenomus remus* (Hymenoptera: Scelionidae). J Invertebr Pathol 21:164–171. https://doi.org/10.1016/0022-2011(73)90197-3

Giron D, Casas J (2003a) Mothers reduce egg provisioning with age. Ecol Lett 6:273–277. https://doi.org/10.1046/j.1461-0248.2003.00429.x

Giron D, Casas J (2003b) Lipogenesis in an adult parasitic wasp. J Insect Physiol 49:141–147. https://doi.org/10.1016/S0022-1910(02)00258-5

Godfray HCJ (1994) Parasitoids: behavioral and evolutionary ecology, vol 67. Princeton University Press. https://doi.org/10.2307/j.ctvs32rmp

González D, Zhao Q, McMahan C, Velasquez D, Haskins WE, Sponsel V et al (2009) The major antennal chemosensory protein of red imported fire ant workers. Insect Mol Biol 18:395–404. https://doi.org/10.1111/j.1365-2583.2009.00883.x

Gopalapillai R, Kadono-Okuda K, Okuda T (2005) Molecular cloning and analysis of a novel teratocyte-specific carboxylesterase from the parasitic wasp, *Dinocampus coccinellae*. Insect Biochem Mol Biol 35:1171–1180. https://doi.org/10.1016/j.ibmb.2005.05.010

Grimaldi A, Caccia S, Congiu T, Ferrarese R, Tettamanti G, Rivas-Pena M et al (2006) Structure and function of the extraembryonic membrane persisting around the larvae of the parasitoid *Toxoneuron nigriceps*. J Insect Physiol 52:870–880. https://doi.org/10.1016/j.jinsphys.2006.05.011

Grossi G, Grimaldi A, Cardone RA, Monné M, Reshkin SJ, Girardello R et al (2016) Extracellular matrix degradation via enolase/plasminogen interaction: evidence for a mechanism conserved in Metazoa. Biol Cell 108:161–178. https://doi.org/10.1111/boc.201500095

Harvey JA (1996) *Venturia canescens* parasitizing *Galleria mellonella* and *Anagasta kuehniella*: is the parasitoid a conformer or regulator? J Insect Physiol 42: 1017–1025. https://doi.org/10.1016/S0022-1910(96)00069-8

Harvey JA, Malcicka M (2016) Nutritional integration between insect hosts and koinobiont parasitoids in an evolutionary framework. Entomol Exp Appl 159:181–188. https://doi.org/10.1111/eea.12426

Harvey JA, Strand MR (2002) The developmental strategies of endoparasitoid wasps vary with host feeding ecology. Ecology 83:2439–2451. https://doi.org/10.2307/3071805

Harvey JA, Wagenaar R, Bezemer TM (2009) Interactions to the fifth trophic level: secondary and tertiary parasitoid wasps show extraordinary efficiency in utilizing host resources. J Anim Ecol 78:686–692. https://doi.org/10.1111/j.1365-2656.2008.01516.x

Harvey JA, Sano T, Tanaka T (2010) Differential host growth regulation by the solitary endoparasitoid, *Meteorus pulchricornis* in two hosts of greatly differing mass. J Insect Physiol 56:1178–1183. https://doi.org/10.1016/j.jinsphys.2010.03.018

Hayakawa Y (1986) Inhibition of lipid transport in insects by a factor secreted by the parasite, *Blepharipa sericariae*. FEBS Lett 195:122–124. https://doi.org/10.1016/0014-5793(86)80144-2

Hayakawa Y (1987) Inhibition of lipid transport in insects by a parasitic factor. Comparat Biochem Physiol Part B Comparat Biochem 87:279–283. https://doi.org/10.1016/0305-0491(87)90140-4

Heavner ME, Gueguen G, Rajwani R, Pagan PE, Small C, Govind S (2013) Partial venom gland transcriptome of a *Drosophila* parasitoid wasp, *Leptopilina heterotoma*, reveals novel and shared bioactive profiles with stinging Hymenoptera. Gene 526:195–204. https://doi.org/10.1016/j.gene.2013.04.080

Hoddle MS, van Driesche RG, Sanderson JP (1998) Biology and use of the whitefly parasitoid *Encarsia formosa*. Annu Rev Entomol 43:645–669. https://doi.org/10.1146/annurev.ento.43.1.645

Horwood MA, Hales DF (1991) Fat body changes in a locust, *Chortoicetes terminifera* (Walker) (Orthoptera: Acrididae), parasitized by a nemestrinid fly. Arch Insect Biochem Physiol 17:53–63. https://doi.org/10.1002/arch.940170107

Hotta M, Okuda T, Tanaka T (2001) *Cotesia kariyai* teratocytes: growth and development. J Insect Physiol 47:31–41. https://doi.org/10.1016/S0022-1910(00)00089-5

Ishida Y, Ishibashi J, Leal WS (2013) Fatty acid solubilizer from the oral disk of the blowfly. PLoS One 8:e51779. https://doi.org/10.1371/journal.pone.0051779

Jervis MA (2005) Insects as natural enemies. Springer, Dordrecht. https://doi.org/10.1007/978-1-4020-2625-6

Jervis MA, Heimpel GE, Ferns PN, Harvey JA, Kidd NAC (2001) Life-history strategies in parasitoid wasps: a comparative analysis of 'ovigeny'. J Anim Ecol 70: 442–458. https://doi.org/10.1046/j.1365-2656.2001.00507.x

Jervis MA, Ellers J, Harvey JA (2008) Resource acquisition, allocation, and utilization in parasitoid reproductive strategies. Annu Rev Entomol 53:361–385. https://doi.org/10.1146/annurev.ento.53.103106.093433

Jones D, Barlow JS, Thompson SN (1982) *Exeristes*, *Itoplectis*, *Aphaereta*, *Brachymeria*, and *Hyposoter* species: *In vitro* glyceride synthesis and regulation of fatty acid composition. Exp Parasitol 54:340–351. https://doi.org/10.1016/0014-4894(82)90043-1

Kadono-Okuda K, Sakurai H, Takeda S, Okuda T (1995) Synchronous growth of a parasitoid, *Perilitus coccinellae*, and teratocytes with the development of the host, *Coccinella septempunctata*. Entomol Exp Appl 75:145–149. https://doi.org/10.1111/j.1570-7458.1995.tb01920.x

Kadono-Okuda K, Weyda F, Okuda T (1998) *Dinocampus (=Perilitus) coccinellae* teratocyte-specific polypeptide: its accumulative property, localization and characterization. J Insect Physiol 44:1073–1080. https://doi.org/10.1016/S0022-1910(98)00063-8

Kaeslin M, Pfister-Wilhelm R, Lanzrein B (2005) Influence of the parasitoid *Chelonus inanitus* and its polydnavirus on host nutritional physiology and implications for parasitoid development. J Insect Physiol 51:1330–1339. https://doi.org/10.1016/j.jinsphys.2005.08.003

Kim H-S (2013) Role of insulin-like growth factor binding protein-3 in glucose and lipid metabolism. APEM 18: 9–12. https://doi.org/10.6065/apem.2013.18.1.9

Kim E, Kim Y, Yeam I, Kim Y (2016) Transgenic expression of a viral cystatin gene CpBV-CST1 in tobacco confers insect resistance. Environ Entomol 45:1322–1331. https://doi.org/10.1093/ee/nvw105

Kraaijeveld K, Neleman P, Mariën J, de Meijer E, Ellers J (2019) Genomic resources for *Goniozus legneri*, *Aleochara bilineata* and *Paykullia maculata*, representing three independent origins of the parasitoid lifestyle in insects. G3 Genes Genomes Genet 9:987–991. https://doi.org/10.1534/g3.119.300584

Kryukova NA, Mozhaytseva KA, Rotskaya UN, Glupov VV (2021) *Galleria mellonella* larvae fat body disruption (Lepidoptera: Pyralidae) caused by the venom of *Habrobracon brevicornis* (Hymenoptera: Braconidae). Arch Insect Biochem Physiol 106:e21746. https://doi.org/10.1002/arch.21746

Lafferty KD, Kuris AM (2002) Trophic strategies, animal diversity and body size. Trends Ecol Evol 17:507–513. https://doi.org/10.1016/S0169-5347(02)02615-0

Lahti DC, Johnson NA, Ajie BC, Otto SP, Hendry AP, Blumstein DT et al (2009) Relaxed selection in the wild. Trends Ecol Evol 24:487–496. https://doi.org/10.1016/j.tree.2009.03.010

Lammers M, Kraaijeveld K, Mariën J, Ellers J (2019) Gene expression changes associated with the evolutionary loss of a metabolic trait: lack of lipogenesis in parasitoids. BMC Genomics 20:309. https://doi.org/10.1186/s12864-019-5673-6

Laurino S, Grossi G, Pucci P, Flagiello A, Bufo SA, Bianco G et al (2016) Identification of major *Toxoneuron nigriceps* venom proteins using an integrated transcriptomic/proteomic approach. Insect Biochem Mol Biol 76:49–61. https://doi.org/10.1016/j.ibmb.2016.07.001

Le Lann C, Visser B, Mériaux M, Moiroux J, van Baaren J, van Alphen JJM et al (2014) Rising temperature reduces divergence in resource use strategies in coexisting parasitoid species. Oecologia 174:967–977. https://doi.org/10.1007/s00442-013-2810-9

Lease HM, Wolf BO (2011) Lipid content of terrestrial arthropods in relation to body size, phylogeny, ontogeny and sex. Physiol Entomol 36:29–38. https://doi.org/10.1111/j.1365-3032.2010.00767.x

Lee JC, Heimpel GE, Leibee GL (2004) Comparing floral nectar and aphid honeydew diets on the longevity and nutrient levels of a parasitoid wasp. Entomol Exp Appl 111:189–199. https://doi.org/10.1111/j.0013-8703.2004.00165.x

Lin Z, Wang R-J, Cheng Y, Du J, Volovych O, Han L-B et al (2019) Insights into the venom protein components of *Microplitis mediator*, an endoparasitoid wasp. Insect Biochem Mol Biol 105:33–42. https://doi.org/10.1016/j.ibmb.2018.12.013

Liu T-X, Stansly PA, Gerling D (2015) Whitefly parasitoids: distribution, life history, bionomics, and utilization. Annu Rev Entomol 60:273–292. https://doi.org/10.1146/annurev-ento-010814-021101

Liu N-Y, Wang J-Q, Zhang Z-B, Huang J-M, Zhu J-Y (2017) Unraveling the venom components of an encyrtid endoparasitoid wasp *Diversinervus elegans*. Toxicon 136:15–26. https://doi.org/10.1016/j.toxicon.2017.06.011

Liu N-Y, Xu Z-W, Yan W, Ren X-M, Zhang Z-Q, Zhu J-Y (2018) Venomics reveals novel ion transport peptide-likes (ITPLs) from the parasitoid wasp *Tetrastichus brontispae*. Toxicon 141:88–93. https://doi.org/10.1016/j.toxicon.2017.11.008

Luo S, Li J, Liu X, Lu Z, Pan W, Zhang Q et al (2010) Effects of six sugars on the longevity, fecundity and nutrient reserves of *Microplitis mediator*. Biol Control 52:51–57. https://doi.org/10.1016/j.biocontrol.2009.09.002

Mathé-Hubert H, Colinet D, Deleury E, Belghazi M, Ravallec M, Poulain J et al (2016) Comparative venomics of *Psyttalia lounsburyi* and *P. concolor*, two olive fruit fly parasitoids: a hypothetical role for a GH1 β-glucosidase. Sci Rep 6:35873. https://doi.org/10.1038/srep35873

Matthews RW, González JM, Matthews JR, Deyrup LD (2009) Biology of the parasitoid melittobia (Hymenoptera: Eulophidae). Annu Rev Entomol 54:251–266. https://doi.org/10.1146/annurev.ento.54.110807.090440

Merlin BL, Pino LE, Peres LEP, Prataviera F, Ortega EMM, Cônsoli FL (2021) Beyond host specificity: the biotechnological exploitation of chitolectin from teratocytes of *Toxoneuron nigriceps* to control non-permissive hosts. J Pest Sci 94:713–727. https://doi.org/10.1007/s10340-020-01290-y

Moiroux J, le Lann C, Seyahooei MA, Vernon P, Pierre J-S, van Baaren J et al (2010) Local adaptations of life-history traits of a *Drosophila* parasitoid, *Leptopilina boulardi*: does climate drive evolution? Ecol Entomol 35:727–736. https://doi.org/10.1111/j.1365-2311.2010.01233.x

Mondy N, Corio-Costet M-F, Bodin A, Mandon N, Vannier F, Monge J-P (2006) Importance of sterols acquired through host feeding in synovigenic parasitoid oogenesis. J Insect Physiol 52:897–904. https://doi.org/10.1016/j.jinsphys.2006.03.007

Morales J, Medina P, Viñuela E (2007) The influence of two endoparasitic wasps, *Hyposoter didymator* and *Chelonus inanitus*, on the growth and food consumption of their host larva *Spodoptera littoralis*. BioControl 52:145–160. https://doi.org/10.1007/s10526-006-9026-4

Moreau SJM, Asgari S (2015) Venom proteins from parasitoid wasps and their biological functions. Toxins (Basel) 7:2385–2412. https://doi.org/10.3390/toxins7072385

Muller D, Giron D, Desouhant E, Rey B, Casas J, Lefrique N et al (2017) Maternal age affects offspring nutrient dynamics. J Insect Physiol 101:123–131. https://doi.org/10.1016/j.jinsphys.2017.07.011

Multerer M-T, Wendler M, Ruther J (2022) The biological significance of lipogenesis in *Nasonia vitripennis*. Proc Biol Sci 289:20220208. https://doi.org/10.1098/rspb.2022.0208

Nair KK, Chen TT, Wyatt GR (1981) Juvenile hormone-stimulated polyploidy in adult locust fat body. Dev Biol 81:356–360. https://doi.org/10.1016/0012-1606(81)90300-6

Nakamatsu Y, Tanaka T (2003) Venom of ectoparasitoid, *Euplectrus sp.* near *plathypenae* (Hymenoptera: Eulophidae) regulates the physiological state of *Pseudaletia separata* (Lepidoptera: Noctuidae) host as a food resource. J Insect Physiol 49:149–159. https://doi.org/10.1016/S0022-1910(02)00261-5

Nakamatsu Y, Tanaka T (2004a) Food resource use of hyperparasitoid *Trichomalopsis apanteloctena* (Hymenoptera: Pteromalidae), an idiobiotic ectoparasitoid. Ann Entomol Soc Am 97:994–999. https://doi.org/10.1603/0013-8746(2004)097[0994:FRUOHT]2.0.CO;2

Nakamatsu Y, Tanaka T (2004b) Venom of *Euplectrus separatae* causes hyperlipidemia by lysis of host fat body cells. J Insect Physiol 50:267–275. https://doi.org/10.1016/j.jinsphys.2003.12.005

Nakamatsu Y, Fujii S, Tanaka T (2002) Larvae of an endoparasitoid, *Cotesia kariyai* (Hymenoptera:

Braconidae), feed on the host fat body directly in the second stadium with the help of teratocytes. J Insect Physiol 48:1041–1052. https://doi.org/10.1016/S0022-1910(02)00192-0

Nurullahoğlu Z, Kan U, Sak O, Ergi E (2004) Total lipid and fatty acid composition of *Apanteles galleriae* and its parasitized host. Ann Entomol Soc Am 97:1000–1006. https://doi.org/10.1603/0013-8746(2004)097[1000:TLAFAC]2.0.CO;2

Okuda T, Kadono-Okuda K (1995) *Perilitus coccinellae* teratocyte polypeptide: evidence for production of a teratocyte-specific 540 kDa protein. J Insect Physiol 41:819–825. https://doi.org/10.1016/0022-1910(95)00009-J

Pan J, Cen L, Zhou T, Yu M, Chen X, Jiang W et al (2021) Insulin-like growth factor binding protein 1 ameliorates lipid accumulation and inflammation in nonalcoholic fatty liver disease. J Gastroenterol Hepatol 36:3438–3447. https://doi.org/10.1111/jgh.15627

Pedata PA, Garonna AP, Zabatta A, Zeppa P, Romani R, Isidoro N (2003) Development and morphology of teratocytes in *Encarsia berlesei* and *Encarsia citrina*: first record for Chalcidoidea. J Insect Physiol 49:1063–1071. https://doi.org/10.1016/j.jinsphys.2003.08.003

Pelosi P, Zhu J, Knoll W (2018) Odorant-binding proteins as sensing elements for odour monitoring. Sensors 18:3248. https://doi.org/10.3390/s18103248

Pennacchio F, Strand MR (2006) Evolution of developmental strategies in parasitic Hymenoptera. Annu Rev Entomol 51:233–258. https://doi.org/10.1146/annurev.ento.51.110104.151029

Pennacchio F, Vinson SB, Tremblay E (1994) Morphology and ultrastructure of the serosal cells (teratocytes) in *Cardiochiles nigriceps* Viereck (Hymenoptera: Braconidae) embryos. Int J Insect Morphol Embryol 23:93–104. https://doi.org/10.1016/0020-7322(94)90003-5

Pennacchio F, Fanti P, Falabella P, Digilio MC, Bisaccia F, Tremblay E (1999) Development and nutrition of the braconid wasp, *Aphidius ervi* in aposymbiotic host aphids. Arch Insect Biochem Physiol 40:53–63. https://doi.org/10.1002/(SICI)1520-6327(1999)40:1<53:AID-ARCH6>3.0.CO;2-J

Perez-Riverol A, Lasa AM, dos Santos-Pinto JRA, Palma MS (2019) Insect venom phospholipases A1 and A2: roles in the envenoming process and allergy. Insect Biochem Mol Biol 105:10–24. https://doi.org/10.1016/j.ibmb.2018.12.011

Perkin LC, Friesen KS, Flinn PW, Oppert B (2015) Venom gland components of the ectoparasitoid wasp, *Anisopteromalus calandrae*. J Venom Res 6:19–37

Poirié M, Colinet D, Gatti JL (2014) Insights into function and evolution of parasitoid wasp venoms. Curr Opin Insect Sci 6:52–60. https://doi.org/10.1016/j.cois.2014.10.004

Prager L, Bruckmann A, Ruther J (2019) *De novo* biosynthesis of fatty acids from α-D-glucose in parasitoid wasps of the *Nasonia* group. Insect Biochem Mol Biol 115:103256. https://doi.org/10.1016/j.ibmb.2019.103256

Pruijssers AJ, Falabella P, Eum JH, Pennacchio F, Brown MR, Strand MR (2009) Infection by a symbiotic polydnavirus induces wasting and inhibits metamorphosis of the moth *Pseudoplusia includens*. J Exp Biol 212:2998–3006. https://doi.org/10.1242/jeb.030635

Qin Q, Gong H, Ding T (2000) Two collagenases are secreted by Teratocytes from *Microplitis mediator* (Hymenoptera: Braconidae) cultured in vitro. J Invertebr Pathol 76:79–80. https://doi.org/10.1006/jipa.2000.4950

Quicke DL (1997) Parasitic wasps. Springer, London

Quicray MA, Wilhelm L, Enriquez T, He S, Scheifler M, Visser B (2023) The *Drosophila*-parasitizing wasp *Leptopilina heterotoma*: a comprehensive model system in ecology and evolution. Ecol Evol 13:e9625. https://doi.org/10.1002/ece3.9625

Richmond GS, Smith TK (2011) Phospholipases A1. Int J Mol Sci 12:588–612. https://doi.org/10.3390/ijms12010588

Rivero A, West S (2002) The physiological costs of being small in a parasitic wasp. Evol Ecol Res 4:407–420

Rivers DB, Denlinger DL (1994) Redirection of metabolism in the flesh fly, *Sarcophaga bullata*, following envenomation by the ectoparasitoid *Nasonia vitripennis* and correlation of metabolic effects with the diapause status of the host. J Insect Physiol 40:207–215. https://doi.org/10.1016/0022-1910(94)90044-2

Rivers DB, Denlinger DL (1995) Venom-induced alterations in fly lipid metabolism and its impact on larval development of the ectoparasitoid *Nasonia vitripennis* (Walker) (Hymenoptera: Pteromalidae). J Invertebr Pathol 66:104–110. https://doi.org/10.1006/jipa.1995.1071

Rivers DB, Yoder JA (1996) Site-specific effects of parasitism on water balance and lipid content of the parasitic wasp *Nasonia vitripennis* (Hymenoptera: Pteromalidae). Eur J Entomol 93:75–82

Rivers DB, Pagnotta MA, Huntington ER (1998) Reproductive strategies of 3 species of ectoparasitic wasps are modulated by the response of the fly host *Sarcophaga bullata* (Diptera: Sarcophagidae) to parasitism. Ann Entomol Soc Am 91:458–465. https://doi.org/10.1093/aesa/91.4.458

Rivers DB, Rocco MM, Frayha AR (2002) Venom from the ectoparasitic wasp *Nasonia vitripennis* increases Na+ influx and activates phospholipase C and phospholipase A2 dependent signal transduction pathways in cultured insect cells. Toxicon 40:9–21. https://doi.org/10.1016/S0041-0101(01)00132-5

Rouleux-Bonnin F, Renault S, Rabouille A, Periquet G, Bigot Y (1999) Free serosal cells originating from the embryo of the wasp *Diadromus pulchellus* in the pupal body of parasitized leek-moth, *Acrolepiosis assectella*. Are these cells teratocyte-like? J Insect Physiol 45:479–484. https://doi.org/10.1016/S0022-1910(98)00149-8

Ruther J, Prager L, Pokorny T (2021) Parasitic wasps do not lack lipogenesis. Proc R Soc B Biol Sci 288:20210548. https://doi.org/10.1098/rspb.2021.0548

Salvador G, Cônsoli FL (2008) Changes in the hemolymph and fat body metabolites of *Diatraea saccharalis* (Fabricius) (Lepidoptera: Crambidae) parasitized by *Cotesia flavipes* (Cameron) (Hymenoptera: Braconidae). Biol Control 45:103–110. https://doi.org/10.1016/j.biocontrol.2007.12.007

Salvia R, Grimaldi A, Girardello R, Scieuzo C, Scala A, Bufo SA et al (2019) *Aphidius ervi* teratocytes release enolase and fatty acid binding protein through Exosomal vesicles. Front Physiol 10:715. https://doi.org/10.3389/fphys.2019.00715

Sarjeant K, Stephens JM (2012) Adipogenesis. Cold Spring Harb Perspect Biol 4:a008417–a008417. https://doi.org/10.1101/cshperspect.a008417

Scieuzo C, Salvia R, Franco A, Pezzi M, Cozzolino F, Chicca M et al (2021) An integrated transcriptomic and proteomic approach to identify the main *Torymus sinensis* venom components. Sci Rep 11:5032. https://doi.org/10.1038/s41598-021-84385-5

Service PM (1987) Physiological mechanisms of increased stress resistance in *Drosophila melanogaster* selected for postponed senescence. Physiol Zool 60:321–326

Seyahooei M, van Alphen JJM, Kraaijeveld K (2011) Genetic structure of *Leptopilina boulardi* populations from different climatic zones of Iran. BMC Ecol 11:4. https://doi.org/10.1186/1472-6785-11-4

Seyahooei MA, Kraaijeveld K, Bagheri A, van Alphen JJM (2020) Adult size and timing of reproduction in five species of *Asobara* parasitoid wasps. Insect Sci 27: 1334–1345. https://doi.org/10.1111/1744-7917.12728

Shelby KS, Habibi J, Puttler B (2014) An ultrastructural and fluorescent study of the teratocytes of *Microctonus aethiopoides* loan (Hymenoptera: Braconidae) from the hemocoel of host Alfalfa Weevil, *Hypera postica* (Gyllenhal) (Coleoptera: Curculionidae). Psyche (Camb MA) 2014:1–9. https://doi.org/10.1155/2014/652518

Sheng S, Zhang X, Zheng Y, Jiao W, Zhou Y, Liao C et al (2019) Effect of six sugars on the longevity, oviposition performance and nutrition accumulation in an endoparasitoid, *Meteorus pulchricornis* (Hymenoptera: Braconidae). J Asia Pac Entomol 22:263–268. https://doi.org/10.1016/j.aspen.2019.01.010

Shi M, Dong S, Li M, Yang Y, Stanley D, Chen X (2015) The endoparasitoid, *Cotesia vestalis*, regulates host physiology by reprogramming the neuropeptide transcriptional network. Sci Rep 5:8173. https://doi.org/10.1038/srep08173

Sim AD, Wheeler D (2016) The venom gland transcriptome of the parasitoid wasp *Nasonia vitripennis* highlights the importance of novel genes in venom function. BMC Genomics 17:571. https://doi.org/10.1186/s12864-016-2924-7

Sluss R (1968) Behavioral and anatomical responses of the convergent lady beetle to parasitism by *Perilitus coccinellae* (Schrank) (Hymenoptera: Braconidae). J Invertebr Pathol 10:9–27. https://doi.org/10.1016/0022-2011(68)90259-0

Soulages JL, Wells MA (1994) Lipophorin: the structure of an insect lipoprotein and its role in lipid transport in insects. Adv Protein Chem 45:371–415. https://doi.org/10.1016/S0065-3233(08)60644-0

Strand MR (1986) The physiological interactions of parasitoids with their hosts and their influence on reproductive strategies. In: Waage J, Greahead D (eds) Insect Parasitoids. Academic Press, London, pp 97–136

Strand MR (2014) Teratocytes and their functions in parasitoids. Curr Opin Insect Sci 6:68–73. https://doi.org/10.1016/j.cois.2014.09.005

Strand MR, Burke GR (2012) Polydnaviruses as symbionts and gene delivery systems. PLoS Pathog 8:e1002757. https://doi.org/10.1371/journal.ppat.1002757

Strand MR, Burke GR (2013) Polydnavirus-wasp associations: evolution, genome organization, and function. Curr Opin Virol 3:587–594. https://doi.org/10.1016/j.coviro.2013.06.004

Strand MR, Burke GR (2015) Polydnaviruses: from discovery to current insights. Virol 479-480:393–402. https://doi.org/10.1016/j.virol.2015.01.018

Strand MR, Wong EA (1991) The growth and role of *Microplitis demolitor* teratocytes in parasitism of *Pseudoplusia includens*. J Insect Physiol 37:503–515. https://doi.org/10.1016/0022-1910(91)90027-W

Strand MR, Meola SM, Vinson SB (1986) Correlating pathological symptoms in *Heliothis virescens* eggs with development of the parasitoid *Telenomus heliothidis*. J Insect Physiol 32:389–402. https://doi.org/10.1016/0022-1910(86)90052-1

Strand MR, Vinson SB, Nettles WC, Xie ZN (1988) *In vitro* culture of the egg parasitoid *Telenomus heliothidis*: the role of teratocytes and medium consumption in development. Entomol Exp Appl 46:71–78. https://doi.org/10.1111/j.1570-7458.1988.tb02269.x

Suzuki M, Tanaka T (2007) Development of *Meteorus pulchricornis* and regulation of its noctuid host, *Pseudaletia separata*. J Insect Physiol 53:1072–1078. https://doi.org/10.1016/j.jinsphys.2007.06.006

Teng Z-W, Xiong S-J, Xu G, Gan S-Y, Chen X, Stanley D et al (2017) Protein discovery: combined transcriptomic and proteomic analyses of venom from the endoparasitoid *Cotesia chilonis* (Hymenoptera: Braconidae). Toxins (Basel) 9:135. https://doi.org/10.3390/toxins9040135

Thompson SN (1977) Lipid nutrition during larval development of the parasitic wasp, *Exeristes*. J Insect Physiol 23:579–583. https://doi.org/10.1016/0022-1910(77)90051-8

Thompson SN (1982a) Immediate effects of parasitization by the insect parasite, *Hyposoter exiguae* on the nutritional physiology of its host, *Trichoplusia ni*. J Parasitol 68:936–941. https://doi.org/10.2307/3281009

Thompson SN (1982b) Effects of the insect parasite, *Hyposoter exiguae* on the total body glycogen and lipid levels of its host, *Trichoplusia ni*. Comp Biochem Physiol Part B: Comp Biochem 72:233–237. https://doi.org/10.1016/0305-0491(82)90040-2

Thompson SN (1983) The nutritional physiology of *Trichoplusia ni* parasitized by the insect parasite, *Hyposoter exiguae*, and the effects of parallel-feeding. Parasitology 87:15–28. https://doi.org/10.1017/S0031182000052380

Thompson SN, Barlow JS (1970) The change in fatty acid composition pattern of *Itoplectis conquisitor* (Say) reared on different hosts. J Parasitol 56:845–846

Thompson SN, Barlow JS (1972a) Influence of host fatty acid composition on that of Ichneumonoid and Chalcidoid wasps. J Parasitol 58:836–839. https://doi.org/10.2307/3278331

Thompson SN, Barlow JS (1972b) Synthesis of fatty acids by the parasite *Exeristes comstockii* (Hymenop.) and two hosts, *Galleria mellonella* (Lep.) and *Lucilia sericata* (dip.). Can J Zool 50:1105–1110. https://doi.org/10.1139/z72-147

Thompson SN, Barlow JS (1974) The fatty acid composition of parasitic Hymenoptera and its possible biological significance. Ann Entomol Soc Am 67: 627–632. https://doi.org/10.1093/aesa/67.4.627

Thompson SN, Barlow JS (1976) Regulation of lipid metabolism in the insect parasite, *Exeristes roborator* (Fabricius). J Parasitol 62:303–306. https://doi.org/10.2307/3279292

Thompson SN, Redak RA (2008) Parasitism of an insect *Manduca sexta* L. alters feeding behaviour and nutrient utilization to influence developmental success of a parasitoid. J Comp Physiol B 178:515–527. https://doi.org/10.1007/s00360-007-0244-6

Turunen S (1979) Digestion and absorption of lipids in insects. Comp Biochem Physiol A Physiol 63:455–460. https://doi.org/10.1016/0300-9629(79)90171-3

Uçkan F, Ergin E, Rivers DB, Gençer N (2006) Age and diet influence the composition of venom from the endoparasitic wasp *Pimpla turionellae* L. (Hymenoptera: Ichneumonidae). Arch Insect Biochem Physiol:177–187. https://doi.org/10.1002/arch.20154

Vincent B, Kaeslin M, Roth T, Heller M, Poulain J, Cousserans F et al (2010) The venom composition of the parasitic wasp *Chelonus inanitus* resolved by combined expressed sequence tags analysis and proteomic approach. BMC Genomics 11:693. https://doi.org/10.1186/1471-2164-11-693

Visser B, Ellers J (2008) Lack of lipogenesis in parasitoids: a review of physiological mechanisms and evolutionary implications. J Insect Physiol 54:1315–1322. https://doi.org/10.1016/j.jinsphys.2008.07.014

Visser B, le Lann C, den Blanken FJ, Harvey JA, van Alphen JJM, Ellers J (2010) Loss of lipid synthesis as an evolutionary consequence of a parasitic lifestyle. Proc Natl Acad Sci USA 107:8677–8682. https://doi.org/10.1073/pnas.1001744107

Visser B, Roelofs D, Hahn DA, Teal PEA, Mariën J, Ellers J (2012) Transcriptional changes associated with lack of lipid synthesis in parasitoids. Genome Biol Evol 4: 752–762. https://doi.org/10.1093/gbe/evs065

Visser B, van Dooremalen C, Vázquez Ruiz A, Ellers J (2013) Fatty acid composition remains stable across trophic levels in a gall wasp community. Physiol Entomol 38:306–312. https://doi.org/10.1111/phen.12035

Visser B, Willett DS, Harvey JA, Alborn HT (2017) Concurrence in the ability for lipid synthesis between life stages in insects. R Soc Open Sci 4:160815. https://doi.org/10.1098/rsos.160815

Visser B, Hance T, Noël C, Pels C, Kimura MT, Stökl J et al (2018) Variation in lipid synthesis, but genetic homogeneity, among *Leptopilina* parasitic wasp populations. Ecol Evol 8:7355–7364. https://doi.org/10.1002/ece3.4265

Visser B, Alborn HT, Rondeaux S, Haillot M, Hance T, Rebar D et al (2021) Phenotypic plasticity explains apparent reverse evolution of fat synthesis in parasitic wasps. Sci Rep 11:7751. https://doi.org/10.1038/s41598-021-86736-8

Visser B, Le Lann C, Hahn DA, Lammers M, Nieberding CM, Alborn HT et al (2023) Many parasitoids lack adult fat accumulation, despite fatty acid synthesis: A discussion of concepts and considerations for future research. Curr Res Insect Sci 3:100055. https://doi.org/10.1016/j.cris.2023.100055

Volkoff N, Colazza S (1992) Growth patterns of teratocytes in the immature stages of *Trissolcus basalis* (Woll.) (Hymenoptera: Scelionidae), an egg parasitoid of *Nezara viridula* (L.) (Heteroptera: Pentatomidae). Int J Insect Morphol Embryol 21:323–336. https://doi.org/10.1016/0020-7322(92)90027-K

Wang L, Zhu JY, Qian C, Fang Q, Ye GY (2015) Venom of the parasitoid wasp *Pteromalus puparum* contains an odorant binding protein. Arch Insect Biochem Physiol 88:101–110. https://doi.org/10.1002/arch.21206

Wang J, Jin H, Schlenke T, Yang Y, Wang F, Yao H et al (2020a) Lipidomics reveals how the endoparasitoid wasp *Pteromalus puparum* manipulates host energy stores for its young. Biochim Biophys Acta Mol Cell Biol Lipids 1865:158736. https://doi.org/10.1016/j.bbalip.2020.158736

Wang J, Song J, Fang Q, Yao H, Wang F, Song Q et al (2020b) Insight into the functional diversification of lipases in the Endoparasitoid *Pteromalus puparum* (Hymenoptera: Pteromalidae) by genome-scale annotation and expression analysis. Insects 11:227. https://doi.org/10.3390/insects11040227

Wang Y, Wu X, Wang Z, Chen T, Zhou S, Chen J et al (2021) Symbiotic bracovirus of a parasite manipulates host lipid metabolism via tachykinin signaling. PLoS Pathog 17:e1009365. https://doi.org/10.1371/journal.ppat.1009365

Weers PMM, Ryan RO (2006) Apolipophorin III: Role model apolipoprotein. Insect Biochem Mol Biol 36: 231–240. https://doi.org/10.1016/j.ibmb.2006.01.001

Werren JH, Loehlin DW (2009) The parasitoid wasp *Nasonia*: an emerging model system with haploid male genetics, vol 4. Cold Spring Harb Protoc, p pdb. emo134. https://doi.org/10.1101/pdb.emo134

Whitfield JB, Austin AD, Fernandez-Triana JL (2017) Systematics, biology, and evolution of microgastrine parasitoid wasps. Annu Rev Entomol 63: 389–406. https://doi.org/10.1146/annurev-ento-020117-043405

Xu J, Saunders CW, Hu P, Grant RA, Boekhout T, Kuramae EE et al (2007) Dandruff-associated *Malassezia* genomes reveal convergent and divergent virulence traits shared with plant and human fungal pathogens. Proc Natl Acad Sci 104:18730–18735. https://doi.org/10.1073/pnas.0706756104

Xueke G, Shuai Z, Junyu L, Limin L, Lijuan Z, Jinjie C (2017) Lipidomics and RNA-Seq study of lipid regulation in *Aphis gossypii* parasitized by *Lysiphlebia japonica*. Sci Rep 7:1364. https://doi.org/10.1038/s41598-017-01546-1

Yan Z, Fang Q, Wang L, Liu J, Zhu Y, Wang F et al (2016) Insights into the venom composition and evolution of an endoparasitoid wasp by combining proteomic and transcriptomic analyses. Sci Rep 6:19604. https://doi.org/10.1038/srep19604

Yang H, Zhang R, Zhang Y, Liu Q, Li Y, Gong J et al (2020) Cathepsin-L is involved in degradation of fat body and programmed cell death in *Bombyx mori*. Gene 760: 144998. https://doi.org/10.1016/j.gene.2020.144998

Zhang D, Dahlman DL, Järlfors UE, Southgate HH, Wiley SP (1994) Ultrastructure of *Microplitis croceipes* (Cresson) (Braconidae: Hymenoptera) teratocytes. Int J Insect Morphol Embryol 23:173–187. https://doi.org/10.1016/0020-7322(94)90016-7

Zhang D, Dahlman DL, Järlfors UE (1997) Effects of *Microplitis croceipes* Teratocytes on host haemolymph protein content and fat body proliferation. J Insect Physiol 43:577–585. https://doi.org/10.1016/S0022-1910(96)00118-7

Zhang S, Luo J-Y, Lv L-M, Wang C-Y, Li C-H, Zhu X-Z et al (2015) Effects of *Lysiphlebia japonica* (Ashmead) on cotton-melon aphid *Aphis gossypii* Glover lipid synthesis. Insect Mol Biol 24:348–357. https://doi.org/10.1111/imb.12162

Zhu J-Y, Yin Ye G, Fang Q, Hu C (2010) Alkaline phosphatase from venom of the endoparasitoid wasp, *Pteromalus puparum*. J Insect Sci 10:14. https://doi.org/10.1673/031.010.1401

Zhuo Z-H, Yang W, Xu D-P, Yang C-P, Yang H (2016) Effects of *Scleroderma sichuanensis* Xiao (Hymenoptera: Bethylidae) venom and parasitism on nutritional content regulation in host *Tenebrio molitor* L. (Coleoptera: Tenebrionidae). Springerplus 5:1017. https://doi.org/10.1186/s40064-016-2732-1

Adv Exp Med Biol (2026) 1494: 479–499
https://doi.org/10.1007/5584_2024_848

Published online: 08 April 2025

Lipid Properties and Metabolism in Response to Cold

Thomas Enriquez and Nicholas M. Teets

Abstract

Temperature directly shapes insect physiology and has a preponderant effect on life history traits. Winter conditions in temperate and polar regions are especially challenging for insects. Extremely low temperatures can indeed compromise insect survival by promoting freezing of body fluids, but mild cold temperatures above 0 °C (i.e., chilling) can also lead to complex and severe physiological dysregulations. Among physiological damages due to freezing and chilling, insect lipids are one of the primary targets. As low temperatures tend to rigidify phospholipid bilayers, membrane functions are compromised in the cold. Lipid rigidification due to cold also decreases the accessibility of fat stores for metabolic enzymes, and therefore their availability for basal metabolism. These deleterious effects, combined with low food availability in winter, result in substantial nutritional challenges for overwintering insects. Consequently, lipid modifications such as homeoviscous adaptation of cell membranes, fluidity maintenance of fat reserves, cuticular lipid accumulation, and production of antifreeze glycolipids are essential components of the physiological response to cold stress. The aim of the present chapter is to present the physiological challenges caused by low temperatures, the lipid modifications linked with cold tolerance in insects, and the molecular regulation of lipid metabolism during cold exposure.

Keywords

Chilling · Cold tolerance · Energy reserves · Freezing · Homeoviscous adaptation · Overwintering

1 Basic Principles of Cold Tolerance and Chapter Overview

Temperature has a preponderant effect on life by acting from the molecular scale (shaping protein confomration, directing chemical reaction rates, etc. Hochachka and Somero 1984) to the highest levels of biological organizations (shaping fundamental niches of species and population dynamics; Clark et al. 2003; Magnuson et al. 2015). For small-bodied ectotherms like insects, low temperature is a significant challenge (Overgaard and MacMillan 2017; Teets et al. 2023; Sinclair et al. 2003). Most insects are poikilotherms, meaning that their body temperature is directly

T. Enriquez
Evolution and Ecophysiology Group, Department of Functional and Evolutionary Entomology, Gembloux Agro-Bio Tech, University of Liège, Liège, Belgium

N. M. Teets (✉)
Department of Entomology, University of Kentucky, Lexington, KY, USA
e-mail: n.teets@uky.edu

influenced by their surrounding environmental temperature (Potter et al. 2013). Their physiology and life history traits are therefore intimately linked to temperature. Consequently, insect species distributions are highly related to thermal tolerance, and in many cases, distribution is most closely linked to the lowest temperatures they face in their habitat (Bale 2002; Addo-Bediako et al. 2000; Kellermann et al. 2012; Kimura 2004). Historically, insects are divided between "freeze intolerant" species that succumb to ice formation in their body and "freeze tolerant" insects that survive extracellular ice formation (Lee 1989, 2010; Salt 1961). Most insects are freeze intolerant, but even in the absence of ice formation, low temperature creates distinct physiological challenges. Several extensive reviews outline the mechanisms of cold injury (Overgaard and MacMillan 2017; Teets et al. 2023; Rozsypal 2022; Hayward et al. 2014; MacMillan and Sinclair 2011); in brief, low temperature slows down biochemical reactions, directly changes the properties of macromolecules, and leads to a general loss of homeostasis, among other injuries.

In this chapter, we focus on the central role of lipid metabolism in cold tolerance. The fluidity of phospholipid bilayers, which are the major component of cell and organelle membranes, is dictated by temperature, with membranes being fluid at permissive temperatures and rigidifying with decreasing temperatures. We cover the effects of cold on biophysical properties of membranes, and the consequences of these changes, in Sect. 2. Given the central role of membranes in low temperature injury, insects have a suite of adaptations to maintain membrane function in the cold (termed "homeoviscous adaptation"), and we outline these adaptations in Sect. 3. Beyond direct effects on biological membranes, cold exposure is energetically challenging, and lipids play a key role in sustaining insects during the winter and fueling development and reproduction when favorable conditions return. Thus, in Sect. 4, we highlight the energetic challenges of cold stress and the importance of lipid metabolism in fueling insect responses to cold. Further, beyond these well-established roles in membrane properties and energetics, in some species, lipids have unique functions in the cold, and these are detailed in Sect. 5. Finally, advances in molecular biology and high throughput sequencing have advanced knowledge of the molecular regulation of lipid metabolism at low temperature, and these are summarized in Sect. 6.

2 Effects of Cold Temperatures on Lipids and Physiological Consequences in Insects

2.1 Cell Membranes and Their Biophysical Properties at Low Temperature

Cell membranes are essential to organismal organization and physiology by providing cells and organelles with a physical barrier that helps regulate water, solute, and ion diffusion. Biological membranes consist of a glycerophospholipid (GPL) bilayer in which many molecules (mainly proteins) are embedded (Gennis 1989). GPLs are composed of a glycerol residue covalently linked to two nonpolar (hydrophobic) fatty acids and one esterified phosphate group, and this phosphate group is linked to a hydrophilic, polar headgroup that can vary in composition (Fig. 1a). In insects, the main polar groups are choline (forming phosphatidylcholines—PC) and ethanolamine (forming phosphoethanolamines—PE), with additional moieties present in lower abundance, such as phosphatidylglycerol, phosphatidylinositol, or phosphatidylserine (Hammad et al. 2011; Jones et al. 1992). Fatty acid chains are also diverse, as they can change in length and unsaturation ratio. Some GPLs also have one of the fatty acid chains cleaved, forming lysophospholipids (LPLs; e.g., lysophosphatidylcholine and lysophosphatidylethanolamine).

The formation of a GPL bilayer is mainly due to a hydrophobic effect: due to their hydrophobic nature, fatty acid chains are excluded from water and aggregate together within the bilayer structure, while the hydrophilic polar headgroups interact with water molecules and form the membrane surfaces (Gennis 1989; Singer and Nicolson 1972). Other intermolecular interactions

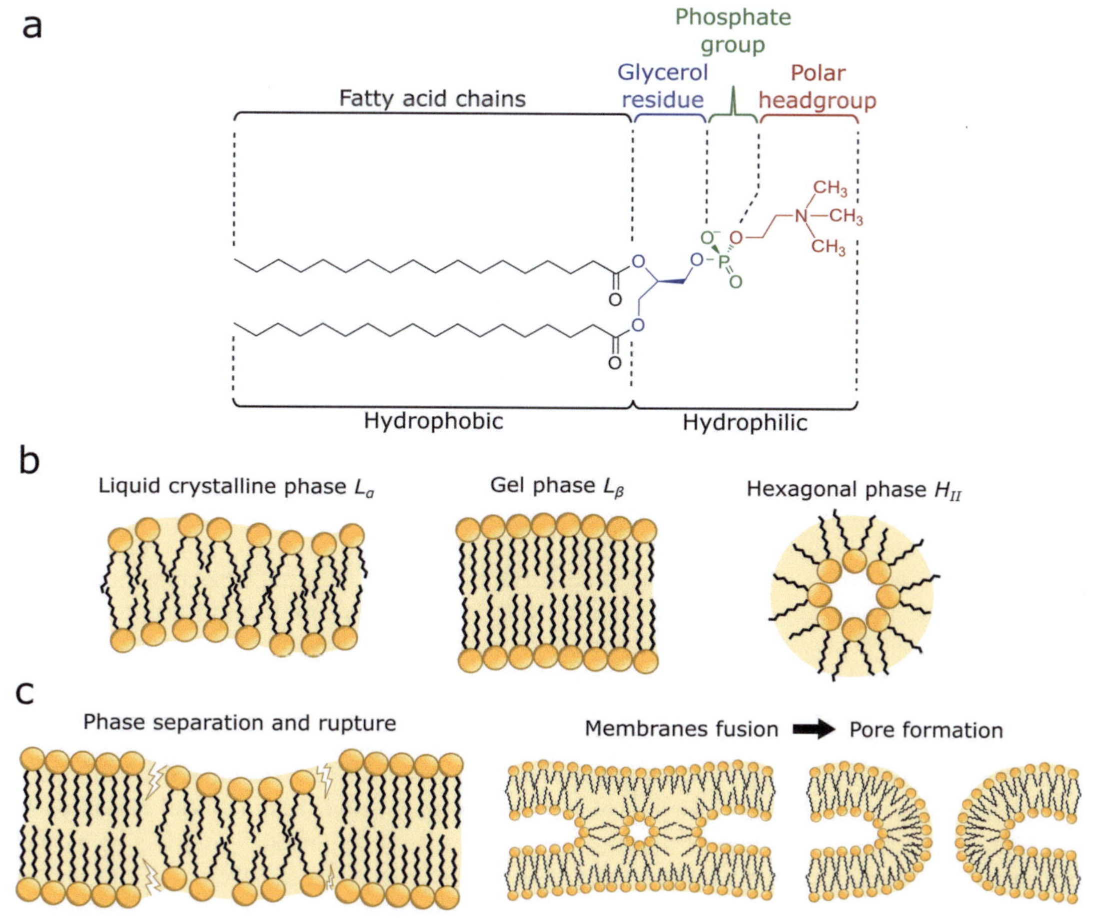

Fig. 1 Glycerophospholipid (GPL) organization and thermal properties. (a) GPLs are composed of two nonpolar (hydrophobic) fatty acids covalently linked to a glycerol residue that contains the esterified phosphate group. This phosphate group is linked to a hydrophilic, polar headgroup which can vary in composition (here a choline is represented). (**b**) Alteration of biophysical properties of biological membranes by temperature. At permissive temperature, the GPL bilayer shows a disordered packing giving the membrane a fluid state, known as the liquid crystalline phase (L_α). At low temperatures, GPLs are more tightly organized, resulting in a rigid gel phase (L_β). Conversely, high temperatures allow lipids to show a loose organization which can lead to a transition into a hexagonal phase (H_{II}). The H_{II} phase can also be provoked by conditions of low humidity. (**c**) Alteration of membrane integrity due to low temperatures. Changes of phases within the bilayer can induce zones of phase transition that can provoke a lateral separation of the phospholipid bilayer at the interface between gel and fluid domains. The H_{II} phase can provoke membrane fusion, which initiate pore formation. These alterations can compromise the permeability of membranes and cause fluid and solutes to leak from the cells

act between GPLs, such as van der Waals forces or hydrogen bonding, but these are relatively minor factors in comparison to hydrophobic forces (Gennis 1989). Membrane bilayers are fluid structures, and the lipids rotate around their own axis, diffuse laterally within their layer, and also move from one layer to another, a movement known as transbilayer flip-flop (Gennis 1989). The fluidity of the membrane is essential for the functioning of enzymes and other proteins embedded in the bilayer, but these molecular movements are highly temperature dependent. Thus, the biophysical properties of cell membranes, defined by their viscosity (fluidity) and also their phase state, can be significantly impacted by low temperatures (Hazel 1989).

At permissive temperatures, fatty acid carbon chains can show isomeric conformations such as gauche rotamers (i.e., a 60° rotation around a carbon-carbon single bond; Kučerka et al. 2011) that result in a disordered packing of GPLs, giving the bilayer a fluid state referred to as the liquid crystalline phase (L_{α}; Fig. 1b). When temperature decreases, fatty acid chains tend to assume a transconformation, which requires less energy, at the expense of rotamers forms. This conformational change leads to a more compact, tight organization of GPLs and therefore promotes a rigidification of the bilayer (Fernandez-Puente et al. 1994), which can eventually lead to the transition to a rigid, highly ordered gel phase ($L\beta$; Fig. 1b; Hazel 1995). Conversely, at high temperatures, the molecular geometry of lipids can change to "conical shape," which in extreme cases can promote a phase transition to a highly disorganized inverted hexagonal phase (H_{II}; Fig. 1b), which is incompatible with the integrity of the GPL bilayer (Hazel 1995). This H_{II} phase can in turn promote membranes fusion, which could initiate pore formation (Jahn et al. 2003), compromising the permeability of membranes and causing fluid and solutes to leak from the cell (Fig. 1c; Rozsypal 2022). In addition to high temperature, transition to H_{II} phase can also be induced by low hydration of molecules, a condition which can be caused by ice formation in freeze tolerant insects, for instance (Rozsypal 2022).

2.2 Physiological Consequences of Lipid Structure Rigidification

Cold-induced losses of membrane integrity can lead to cascading effects on homeostasis. Membrane rigidification and GPL phase perturbation can greatly compromise the activity of transmembrane proteins such as ion channels or transporters (MacMillan and Sinclair 2011; Hazel 1989; Cossins 1994), which alters membrane ionic permeability. Furthermore, the modification of membrane fluidity can cause nonuniform phase changes within the bilayer, resulting in phase transition zones that can lead to a lateral separation of the GPL bilayer at the interface between gel and fluid domains, which can compromise membrane permeability to fluids and solutes (Clerc and Thompson 1995; Koštál 2010). Further, cold exposure may disrupt the association between lipids and proteins in lipid rafts, although this possible consequence has not been explored. In addition to its effects on membrane lipids, cold directly impacts the activity of embedded proteins such as Na+/K+-ATPase pump, leading to cell depolarization (Andersen et al. 2022).

The effects of membrane rigidification on ion homeostasis are particularly problematic in excitable tissues such as nerves and muscles. Ion dysregulation exacerbates temperature-dependent cell depolarization and perturbs action potentials in muscular and nervous cells. In addition, membrane rigidification and depolarization result in the dysfunction of synaptic Ca^{2+} channels, which control neurotransmitter releases (Fig. 2; Findsen et al. 2016). The combined effects of these dysregulations lead to a loss of neuromuscular system coordination, contributing to chill coma induction (Fig. 2; MacMillan and Sinclair 2011). Furthermore, membrane depolarization is a step of apoptosis initiation (Bortner et al. 2001), so loss of water and ionic homeostasis can promote death by apoptosis in insect cells subjected to cold stress (Yi et al. 2007).

Mitochondrial activity is also compromised by low temperatures (Colinet et al. 2017; Kukal et al. 1989). While few studies have directly assessed the effects of low temperature on mitochondria, prolonged cold stress causes declines in the rate of mitochondrial respiration and ATP synthesis, although these declines can be prevented with cold acclimation (Colinet et al. 2017). Also, for cold-susceptible species, low temperatures primarily affect complex I respiration and increase the utilization of alternative energy substrates within the mitochondria (Jørgensen et al. 2023). Together, these changes in mitochondrial function can lead to a deficit in ATP production at the organismal level (Fig. 2; Colinet 2011; Coulson et al. 1992; Dollo et al. 2010) and also lead to the

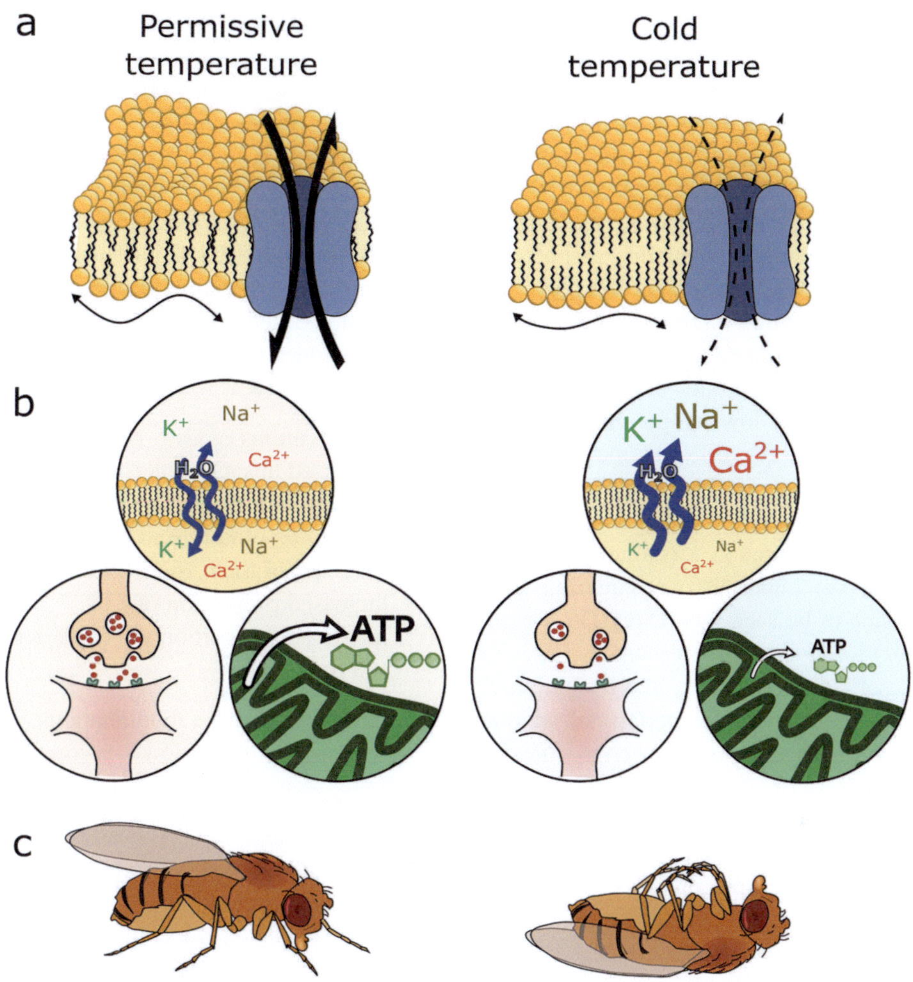

Fig. 2 Summary of low temperature effects on lipids and their physiological consequences. (**a**) Cold temperatures promote the rigidification of membranes, impeding their biological functions, such as the activity of transmembrane proteins. Stored lipids in lipid droplet (triacylglycerides, TAG) fluidity also decreases at cold. (**b**) These alterations lead to cascading effects such as dysregulation of the balance between active and passive ionic and water transfer across the membrane, which can use a loss of hydric and ionic homeostasis. These physiological modifications, combined with membrane rigidification, can lead to membrane depolarization, perturbating action potential and release of neurotransmitters. Rigidification of mitochondrial double membranes compromises their activity. Alteration of mitochondrial function, together with the low accessibility of TAG by hydrolases due to lipid droplets rigidification, can lead to a deficit in ATP production. (**c**) All these dysregulations lead to a global loss of homeostasis, which can result in neuromuscular dysfunction leading to chill coma or death

accumulation of reactive oxygen species, leading to oxidative stress (Lopez-Martinez et al. 2008). Reactive oxygen species can cause oxidative damage to proteins, nucleic acids, and lipids, leading to mutations and damage in tissues (Imlay 2003). In lipids, reactive oxygen species can cause lipid peroxidation, and the end products of lipid peroxidation, including malondialdehyde and 4-hydroxynonenal, can cause membrane damage and further oxidative damage by reacting with and damaging other macromolecules (Ayala et al. 2014).

While empirical evidence is lacking, the effects of cold on mitochondrial function are likely related to changes in mitochondrial membrane properties. Mitochondria contain an inner and outer membrane embedded with proteins, and low temperatures can cause rigidification of these membranes. As is the case with cell membrane transporters discussed above, these changes in mitochondrial membrane properties influence the permeability and activity of the associated enzymes. Furthermore, the activity of the enzymes themselves is directly reduced at low temperatures, which further contributes to the reduced rates of turnover. However, the influence of low temperature on mitochondrial function is still an open area of investigation. In particular, how low temperatures influence mitochondrial lipids, and whether homeoviscous adaptation (see below) in mitochondria is an important overwintering adaptation remains to be seen (see Davy De Virville et al. (2002) for an example of plants).

Outside of cell membranes, low temperature–induced changes in the properties of storage lipids can have important consequences. Triacylgycerols (TAGs) are typically the most abundant lipids in animals, including insects (Williams et al. 2011). Much of these TAGs are stored in cytoplasmic lipid droplets (Arrese and Soulages 2010), and cold-induced rigidification of these droplets may reduce their access to metabolic enzymes and ability to be mobilized as an energy source. For example, triolein, an abundant TAG in many insects, becomes solid at 5 °C (Hagemann et al. 1972), meaning this energy source is likely unusable at low temperatures. Thus, while much of the focus on homeovesicous adaptation (see below) has been on membrane lipids, the ability to maintain storage lipids in a liquid state at low temperature may be just as important.

3 Lipid Modifications Linked with Insect Cold Tolerance

One of the most conserved acclimatory responses of ectotherms to low temperatures is homeoviscous adaptation of cell membranes (Hazel 1995). The term "homeoviscous adaptation" incorporates all modifications of the GPL bilayer composition that adjust membrane fluidity and improve its function at different temperatures (Hazel 1995; Sinensky 1974). Such adjustments not only maintain membrane fluidity but also its phase, a phenomenon known as homeophasic adaptation (Hazel and Eugene Williams 1990). Several changes in GPLs have been identified that impact membrane fluidity, including changes at the headgroup level, in the composition of fatty acid chains or in the nature of lipids embedded in the bilayer. These adjustments can be categorized as follows:

(i) Modifications of GPL headgroups
(ii) Increased unsaturation (or polyunsaturation) of fatty acid carbon chains
(iii) Shortening of fatty acid carbon chains length (increase of the C16/C18 ratio)
(iv) Increased proportion of LPL
(v) Increased proportion of cholesterol in the cell membrane

In addition to these modifications, in Sect. 3 (vi), we discuss the role of cryoprotectants in stabilizing membranes at low temperature.

(i) Modifications of GPL headgroups
GPL polar headgroup modification, typically an increase of PE proportion at the expense of PC, is a common adjustment for homeoviscous adaptation. PC has a cylindrical shape, which gives bilayers a tight and compact organization (therefore, poor fluidity), while PEs have a conical shape that disorganizes GPLs and increases membrane fluidity (Fig. 3). Downer and Kallapur (1981) demonstrated with wide angle X-ray diffraction that mitochondria acclimated to 31 °C (which have a high PE/PC ratio) keep membranes in the fluid L_α state when exposed to 0 °C, while mitochondria acclimated to 45 °C (which have a low PE/PC ratio) have membranes that transition to the rigid L_β phase at 0 °C. The ratio between PC and PE therefore greatly influences membrane fluidity, and for each range of temperature, there is an optimal ratio for proper fluidity and functioning (Hazel 1995). Accordingly, increased

PE/PC ratio in cell membranes is a typical modification linked with adaptation to cold, acclimation, diapause, or winter season in insects (Colinet et al. 2016; Cooper et al. 2012, 2014; Goto and Katagiri 2011; Hodková et al. 1999; Koštál et al. 2011; Michaud and Denlinger 2006; Overgaard et al. 2008; Tomčala et al. 2006; Trenti et al. 2022). However, in some cases, changes in the PE/PC ratio are sex specific (e.g., in *Drosophila suzukii*; Enriquez and Colinet 2019a), or no modification occurs even when cold tolerance increases, such as following rapid cold hardening (i.e., a rapid acclimation process known to increase cold tolerance in insects; MacMillan et al. 2009). In the freeze tolerant larvae of the gall fly *Eurosta solidaginis*, the proportion of PC increases at the expense of PE during winter, opposite to what typically occurs (Pruitt and Lu 2008). In that case, the ordering effect of PC in membrane bilayer could prevent the transition to the H_{II} phase that can occur due to dehydration of the cell during extracellular ice formation. In addition to PE, other GPL headgroups can replace PC in the bilayer, including phosphatidylglycerol or phosphatidylserine, which increase during cold acclimation in *D. suzukii* (Enriquez and Colinet 2019a).

(ii) Increased unsaturation (or polyunsaturation) of fatty acid carbon chains

A major mechanism of homeoviscous adaptation in response to low temperatures is increased unsaturation (or polyunsaturation) of fatty acid carbon chains in GPL and TAG (Fig. 3). Unsaturation of fatty acids is due to desaturase enzymes which insert a *cis* double bond, causing a 30° curvature in the carbon chain. Unsaturated GPLs have a largely conical shape, which increases the fluidity of membrane and TAG stored in lipid droplets. Brankatschk et al. (2018) indeed showed that *Drosophila melanogaster* larvae with high level of unsaturated fatty acids were characterized by increased membrane fluidity at cold (determined from C-Laurdan emission measurements of liposomes prepared from lipid extracts of larvae). In other drosophilids, unsaturation of GPL fatty acids decreases the melting point temperature of membranes in cold acclimated flies, which could participate to decreasing their coma onset temperature (Slotsbo et al. 2016).

Similarly, the transition temperature of TAG decreases in diapausing or cold-adapted *Drosophila* flies (Ohtsu et al. 1993) and in *Cydia pomonella* moths collected in winter (Rozsypal et al. 2014). Increases in unsaturation or polyunsaturation of GPL and TAG are commonly linked with diapause, acclimation, acclimatization, or adaptation to cold in insects and hexapods (Michaud and Denlinger 2006; Trenti et al. 2022; Enriquez and Colinet 2019a; Ohtsu et al. 1993; Bashan and Cakmak 2005; Bennett et al. 1997; Joanisse and Storey 1996a; Koštál and Šimek 1998; Thiry and Hoffmann 1986; van Dooremalen and Ellers 2010) and are likely required to sustain energy metabolism at low temperatures by maintaining lipid storage droplets in a liquid state. While lipid unsaturation has clear benefits at low temperature, double bond fatty acids are highly susceptible to oxidative stress (Storey 1996). Thus, given the increased risk of oxidative stress during cold exposure that is discussed above (Lopez-Martinez et al. 2008; Joanisse and Storey 1996b), there may be a cost to producing unsaturated fatty acids in the winter. Increased antioxidant activities in some insects during winter (Joanisse and Storey 1996b) may therefore protect these sensitive structures during cold exposures.

While many studies have reported increased levels of lipid saturation in the winter, these changes are not universal. In Drosophilidae, several studies reported no major changes in unsaturation of fatty acid (Colinet et al. 2016; Overgaard et al. 2008; MacMillan et al. 2009; Ohtsu et al. 1993), while others reported increased unsaturation and

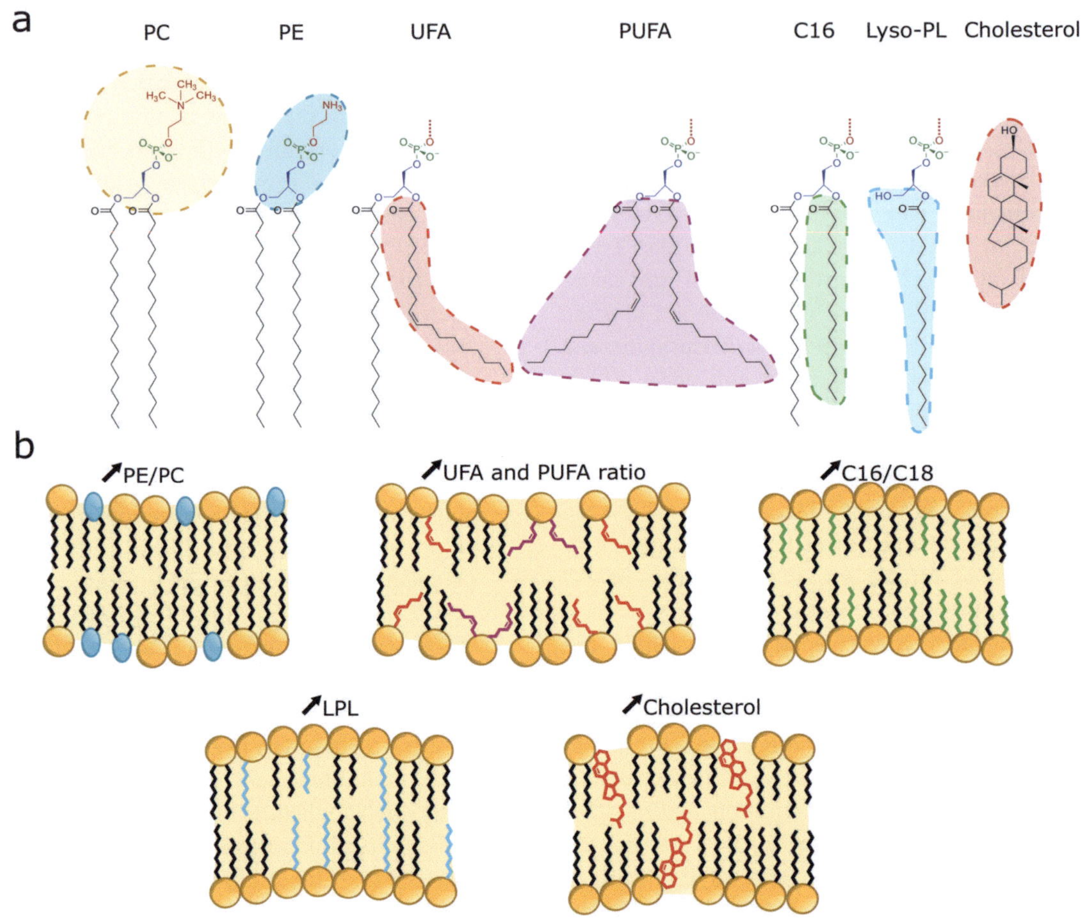

Fig. 3 Modifications of lipid structures linked with homeoviscous adaptation of cell membranes. (**a**) Overview of lipid structure modifications. (**b**) Effects on glycerophospholipid (GPL) organization. PE has a conical shape, contrary to PC that shows a cylindrical shape. Increased proportion of PE at the expense of PC increases space between GPLs, disordering the organization of GPL bilayer, increasing membrane fluidity. Similarly, insertion of *cis* double bonds in fatty acid (FA) carbon chains disorders the organization of the bilayer. An increased proportion of C16 FA in GPLs modifies non-covalent interactions between GPLs, which tends to increase membrane fluidity. LPL has one of the FAs cleaved. This cleavage gives GPLs an "inverted cone" shape which participates in the disorganization of GPL bilayer. Insertion of cholesterol molecules into the bilayer tends to decrease membrane fluidity when PLs are organized in the fluid L_α phase, but conversely increase fluidity when PLs are organized in a rigid L_β phase. Furthermore, the organizing capacities of cholesterol molecules could also prevent membrane phase separation. *PC* phosphocholine, *PE* phosphoethanolamine, *UFA* unsaturated fatty acid, *PUFA* polyunsaturated fatty acid, *LPL* lyso-phospholipid

polyunsaturation in response to cold (Cooper et al. 2012, 2014; Enriquez and Colinet 2019a; Goto et al. 2010; Overgaard et al. 2005, 2006). In other insect taxa, examples of a lack of fatty acid unsaturation due to low temperatures can also be found. For instance, in diapausing eggs of *Aedes albopictus*, the unsaturation of GPLs and TAGs does not change in comparison with non-diapausing eggs (Batz and Armbruster 2018), and similarly no changes are observed in cold-acclimated *Sarcophaga similis* (Goto and Katagiri 2011). No major changes in unsaturation are found in *Locusta migratoria* following cold acclimation (Bayley et al. 2020; Gerber et al. 2021), nor in cold acclimated prepuae of the moth *Cymbalophora pudica* (Koštál and Šimek 1998). Finally, during winter, unsaturation of fatty acid can even decrease, for instance, in thoracic muscles from *Pyrrhocoris apterus* (Hodková et al. 1999).

(iii) Shortening of fatty acid carbon chains length (increase of the C16/C18 ratio)
A widespread response to low temperatures is the shortening of fatty acid carbon chain length: the position and length of carbon chains influence hydrophobic effects between the two fatty acids of GPL. Consequently, shortening of carbon chains reduces the hydrophobic effect and van der Waals forces between fatty acid chains (Colinet et al. 2017) and therefore increases membrane fluidity (Fig. 3). Increased cold tolerance due to fluctuating thermal regimes, adaptation to cold, diapause, or acclimation is often correlated with a shortening of fatty acid carbon chains length in insects, characterized in an increase of the C16/C18 ratio (Colinet et al. 2016; Hodková et al. 1999; Trenti et al. 2022; Thiry and Hoffmann 1986; Overgaard et al. 2006; Ohtsu et al. 1998, 1999; Bahrndorff et al. 2007; Bashan et al. 2002). However, fatty acid chain length does not change in cold acclimated *S. similis* (Goto and Katagiri 2011) nor differ between tropical and subpolar drosophilid species (Slotsbo et al. 2016), and the C16/C18 ratio even decreases in response to acclimation in *D. suzukii* (Enriquez and Colinet 2019a).

(iv) Increased proportion of LPL
In response to cold, the number of LPLs tends to increase in cell membranes (Fig. 3). LPLs are GPLs in which one of the fatty acids is cleaved by the action of phospholipase A. This cleavage gives LPL an "inverted cone" shape, which disrupts the tight organization of GPL, increasing membrane fluidity. An increased proportion of LPLs within the GPL bilayer is known to play a role in the responses to low temperatures during thermal fluctuations in the model species *Drosophila melanogaster* (Colinet et al. 2016), during cold acclimation in *D. suzukii* (Enriquez and Colinet 2019a) and during seasonal acclimatization in the bug *Pyrrhocoris apterus* (Koštál et al. 2013). Also, in *Ceratitis capitata* flies, individuals showing the capacity to quickly recover after a chill coma are characterized by an increased proportion of LPLs (Pujol-Lereis et al. 2016).

(v) Increased proportion of cholesterol in the cell membrane
After GPLs, cholesterol is the most abundant lipid within animal cell membranes. Generally, cholesterol decreases the fluidity of membranes in the L_α liquid crystalline phase while increasing fluidity in the L_β gel phase (Miao et al. 2002). At higher temperatures, cholesterol molecules reduce the rotamerization capabilities of their surrounding GPLs, which tends to reduce bilayer fluidity (Crockett 1998; Yang et al. 2016). This rigidification capacity could give cholesterol a role in maintaining membrane fluidity during heat stress. However, by interacting with fatty acyl chains of GPLs, cholesterol could also prevent membrane phase separation and leakage caused by low temperatures (Crockett 1998; Mocé et al. 2010). Insects cannot synthetize cholesterol de novo, but Shreve et al.2007) showed that *D. melanogaster* fed medium supplemented with cholesterol resulted in an increased proportion of cholesterol in cell membranes, which in turn correlated with increased cold tolerance and cold acclimation capacity. However, follow-up work by Allen et al. (2024) determined that the effect of cholesterol on cold tolerance is trait- and sex dependent, calling into question whether cholesterol plays a universal role in cold tolerance.
In other species, evidence for a role of cholesterol in cold tolerance is inconclusive. In the gall fly *E. solidaginis,* hemolymph cholesterol levels increase during winter

acclimatization, but the proportion of cholesterol embedded in membranes from Malpighian tubules does not change (Yi and Lee 2005) Similarly, in the woolly bear caterpillar *Pyrrhactia Isabella,* cholesterol levels remain stable during cold acclimation in cells from Malpighian tubules and midgut but decrease in the hemolymph and fat body cell membranes (Yi and Lee Jr 2016). As insect sterols originate from their diet, other sterols such as phytosterols can replace the role of cholesterol in insect membranes, as in *D. melanogaster* where ergosterol is an abundant membrane sterol (Rietveld et al. 1999). In the firebug *P. apterus*, winter is characterized by an increased concentration of phytosterols, but it is not clear which proportion of these sterols is incorporated in cell membranes, and what role they play in insect cold tolerance (Koštál et al. 2013). Thus, there is still a lack of knowledge on the function of cholesterol or other sterols in the responses to temperature changes in insects, which should be tackled in future studies.

(vi) Role of cryoprotectants in protecting membrane lipids

While direct modification of lipid composition is one way to preserve membrane function at low temperatures, low molecular weight cryoprotectants can also protect membranes from cold-induced damage. Cryoprotectants are one of the widest studied adaptations to cold, going back to early studies of RW Salt in the mid-1900s (Salt 1957, 1961). Originally, it was thought that cryoprotectants function colligatively and prevent or limit ice formation, but it is now appreciated that cryoprotectants have specific non-colligative functions independent of their concentration (Toxopeus et al. 2019). One of these non-colligative roles appears to be protecting cell membranes from cold injury. For example, trehalose and proline stabilize lipid membranes at low temperatures by binding to head groups of membrane phospholipids, which helps stabilize membranes during periods of cell shrinkage that can occur when insects are chilled or frozen (Yancey 2005). In larvae of *Chymomyza costata,* cold acclimation that enhances freeze tolerance prevents dissociation of fat body membranes, and this effect can be replicated ex vivo by adding specific combinations of croprotectants to fat bodies from non-acclimated larvae (Kučera et al. 2022). Thus, in addition to the extensive membrane modifications described above, low molecular weight cryoprotectants can also contribute to maintaining membrane structure and function at low temperatures.

4 Lipids as Energetic Resources During Cold Stress and Acclimation

Overwintering is a challenging time for insects, as they are faced with the dual threats of abiotic environmental stress (e.g., low temperature, lack of water, limited oxygen, etc.) and lack of food availability (Overgaard and MacMillan 2017; Hahn and Denlinger 2011). The inability of most insects to eat in the winter puts significant energy strain on them, and the role of lipid metabolism in coping with the energetic challenges of overwintering is covered in Chapter "Lipid Metabolism in Diapause". Here, we will address the role of lipid metabolism in supporting the energetic costs of cold acclimation and recovery from cold stress. This section is not intended to be an exhaustive review but will instead focus on general principles and key examples.

The energetic costs of cold stress can be attributed to several factors. First, the bulk movements of water and ions required to restore homeostasis after cold stress are energetically costly, and indeed, recovery from cold stress is often accompanied by elevated metabolic rate relative to conditions before the cold stress (Lalouette et al. 2011; MacMillan et al. 2012). Second, many of the molecular processes involved in resisting or repairing cold-induced damage to macromolecules are energetically costly. For example, heat shock proteins (despite their name) are ubiquitously upregulated during recovery from cold stress (Joplin et al. 1990; Sinclair et al. 2007; Rinehart et al. 2007), and the energetic costs of activating the heat shock response reduce energy availability for reproduction (Krebs and Loeschcke 1994). Beyond heat shock proteins, cold acclimation often involves wholesale changes in transcript abundance; for example, cold acclimation in *D. melanogaster* results in nearly one third of the transcriptome to be differentially regulated (MacMillan et al. 2016). These large-scale changes in gene expression likely carry energetic costs, although the extent to which lipid metabolism provides energy inputs for cold acclimation has not been quantified. Cold acclimation can also cause compensatory changes in body composition. For example, *D. melanogaster* reared at 18 °C have significantly higher lipid reserves than those reared at 30 °C (Klepsatel et al. 2016); so they may be able to accommodate the energetic costs of cold acclimation.

The energetic costs of cold injury are often reflected by a depletion in lipid reserves during or after cold stress. In *D. melanogaster,* exposure to 0 °C beyond 24 h rapidly depletes TAGs, such that less than half the initial reserves are present after 48 h of cold exposure (Chen and Walker 1994). In the same species, exposure to 0 °C for 4 h, which is nonlethal, leads to ~25% reduction in TAG content, and these deficits are permanent and still apparent 10 days after cold exposure (Klepsatel et al. 2016). However, another study in *D. melanogaster* indicated that TAG depletion after cold exposure is only transient (Marshall and Sinclair 2010), and it is unclear why these discrepancies exist. Outside of *D. melanogaster,* similar depletions in lipid reserves following cold stress have been observed in the parasitic wasps *Aphidius colemani* and *A. ervi* (Colinet et al. 2006; Ismail et al. 2010), as well as the freeze-tolerant midge *Belgica antarctica* (Teets et al. 2011, 2019). Similarly, in the freeze-tolerant fly *Chymomyza costata,* recovery from both chilling and freezing causes large increases in metabolic rate, and while lipid content was not directly measured, calculations indicate that as much as half of total lipid reserves could be depleted during recovery (Štětina et al. 2018). In freeze-tolerant larvae of *E. solidaginis,* repeated freeze-thaw cycles promote the conversion of long-chain TAGs to acetylated TAGs (see details below), which causes a net reduction in energy density and corresponds to a concomitant decrease in spring reproductive output (Marshall and Sinclair 2018). This depletion of long-chain TAGs results in decreased egg production in the spring (Marshall and Sinclair 2018), indicating that lipid depletion in response to cold stress can have lasting fitness consequences. However, direct measures of fitness-related traits (e.g., behavior, reproductive output, etc.) are seldom incorporated into energetic studies of cold tolerance. Thus, additional work is needed to quantify the fitness effects of energy depletion caused by cold stress.

It is worth noting that lipid depletion following cold stress is not universal; see references (Ciancio et al. 2021; Chowanski et al. 2015) for examples. Some species primarily use glycogen or other metabolites as an energy source (e.g., Han and Bauce 1993) and thus may not experience any lipid depletion in the cold. Further, as outlined by Sinclair and Marshall (2018), it is likely that life history may influence lipid metabolism after cold stress, as species less able to replenish reserves and unable to feed and replenish lipid reserves after winter may preferentially use other energy sources so that they can reserve lipids for reproduction.

For insects that do experience lipid depletion in response to cold stress, several factors can influence the magnitude of depletion. First, and perhaps most intuitively, the severity and duration of cold stress influence rates of lipid usage. For example, in *B. antarctica,* lipid reserves decrease steadily as the total duration frozen increases, such that the amount of depletion after five cycles of 12 h freezing and 12 h thawing (i.e., 60 h total frozen) is the same as the amount of depletion observed after a continuous 60 h of freezing (Teets et al. 2011). In *E. solidaginis,* lipid metabolism is influenced by both the freezing temperature and the number of freezing events experienced, indicating that the severity of cold exposure and the number of threshold-crossing events can both influence lipid utilization (Marshall and Sinclair 2018). Further, for freeze-tolerant insects, whether they are frozen or supercooled during a cold event can influence lipid metabolism. In *E. solidaginis,* metabolic rate is lower in frozen larvae than supercooled larvae at the same temperature (Irwin and Lee 2002), presumably leading to less energy drain when larvae are frozen. Similarly, in the freeze-tolerant moth *Pyrrharctica isabella,* larvae that overwinter above the snow are more likely to freeze, and frozen larvae consume less lipid, even when accounting for the temperature difference (Marshall and Sinclair 2012). However, in *B. antarctica,* the opposite was observed, in which larvae that inoculatively frozen have higher rates of lipid depletion than those that are supercooled at the same temperature (Teets et al. 2011). Finally, cold acclimation can influence the energetic costs of cold exposure. In *B. antarctica,* 60 h of freezing at −5 °C leads to ~20% depletion of lipid stores in summer-collected larvae (Teets et al. 2011), but winter-acclimatized larvae experience no detectable lipid depletion after 2 weeks of freezing at −5 °C (Teets et al. 2020). Thus, the precise consequences of freezing vs. supercooling on lipid metabolism can vary both across and within species, depending on the exact cold conditions and acclimation state.

While energy drain throughout the winter is a significant challenge for many insects (see Chapter "Lipid Metabolism in Diapause"), conditions that reduce lipid storage could leave less energy available to cope with cold stress (Sinclair 2015). In the case of *D. suzukii,* starvation prior to cold stress that reduces lipid reserves concurrently raises CT_{min} and lengthens chill coma recovery time but improves survival of acute cold shock (De Ro et al. 2021), indicating that distinct cold tolerance traits are differentially affected by energy depletion. These results are consistent with other work showing that dietary composition influences distinct cold tolerance traits in unpredictable ways, and that the precise effect of diet varies across genotypes (Littler et al. 2021). In antlions (Order: Neuroptera) and wormlions (Order: Diptera), starvation has no effect on cold tolerance, although individuals with higher lipid content have faster chill coma recovery time (Scharf et al. 2016). Thus, the relationship between lipid content and cold tolerance appears to be complex and requires further investigation across a wider range of species.

5 Other Functions of Lipids in Relation to Cold Stress

In addition to their roles in homeoviscous adaptation and as energy sources for fueling cold stress responses, lipids play some other roles in cold stress. Most of these minor roles have not been studied extensively, so it is unclear how commonplace they are across species. One potential role for lipids is to serve as a precursor for cryoprotectant production. The glycerol motif of TAG is a potent cryoprotectant, and indeed, glycerol is perhaps the most common low molecular weight cryoprotectant among insects (Storey 1997). Thus, liberation of glycerol by lipases could increase cryoprotectant titers in the bloodstream. However, insects predominantly use glycogen as a precursor for glycerol synthesis (Storey and Storey 2012), as a single glycogen molecule can yield thousands of glycerol molecules, while TAG can only produce a single glycerol molecule. While lipids do not appear to be a major source of glycerol, there are other cases where they can be converted to cryoprotective molecules. For example, trehalose and proline, two potent cryoprotectants, can be derived from fatty acids (Arrese and Soulages 2010; McDougall and Steele 1988), although the extent to which these processes operate in the context of cold tolerance is uncertain.

Another class of lipids that may be important for overwintering insects is cuticular lipids. The insect cuticle contains both structural lipids and a wax layer of free lipids that serves an important waterproofing function (Lockey 1988) and may also serve as a barrier to prevent inoculative freezing. Cold stress is often accompanied by limited water availability, and when insects are in chill coma, they are at increased risk of desiccation due to an inability to drink water (which may be frozen anyway) and an inability to select humid microhabitats (Sinclair et al. 2013). Increased cuticular lipids are observed as part of the diapause program in several species (Benoit and Denlinger 2007; Urbanski et al. 2010; Ala-Honkola et al. 2020), although it appears that the primary benefit is to increase resistance to desiccation during dormancy. Similarly, in the high-elevation species *Drosophila nepalensis,* cold acclimation increases cuticular hydrocarbon deposition (Parkash and Lambhod 2021), although it is once again presumed that this response is to promote cross-tolerance to desiccation. In the larvae of the beetle *Dendroides canadensis*, seasonal changes in the epicuticular composition are linked with increased resistance to ice nucleation (Olsen et al. 1998). Thus, in several species, cuticular lipids appear to be involved in the overwintering program, although it is unclear whether they always play a role in cold tolerance.

Finally, some insects have evolved novel classes of lipids that play an important role in cold hardiness. In *E. solidaginis,* acetylated TAGs make up a staggering 36% of the total lipid pool in overwintering larvae (Marshall et al. 2014). Most animals rely exclusively on long-chain TAGs, and acetylated TAGs have only been observed in extremely trace amounts in animals. Larvae of *E. solidaginis* often overwinter above the snowpack in goldenrod stems, which provide little insulation, and are thus among the most freeze-tolerant insects that have been described (routinely surviving below −55 °C; Lee et al. 1995). Thus, it is proposed that acetylated TAGs, which have lower viscosity than long-chain TAGs, may be more accessible for energy metabolism at extremely low temperature (Marshall et al. 2014). Furthermore, these lipids accumulate in droplets in the fat body, and the fat body of *E. solidaginis* is one of a few known insect tissues that can survive internal ice formation (Bennett and Lee 1997), suggesting that these unique lipids may play a role in intracellular freezing tolerance. The freeze-tolerant beetle, *Upis ceramboides,* also synthesizes a novel xylomannan glycolipid that consists of fatty acid and saccharide components (although it is not clear if the lipid and saccharide are covalently linked), and this compound produces significant thermal hysteresis activity (Walters et al. 2009). This compound was the first nonprotein thermal hysteresis agent to be isolated, and it has since been found in several other insect species, indicating that glycolipids may play a previously unappreciated role in regulating ice formation in insects (Walters et al. 2011).

6 Molecular Regulation of Lipid Metabolism During Cold Exposure

While the molecular regulation of metabolism in relation to diapause is well described in some species (e.g., Hahn and Denlinger 2011; Sinclair and Marshall 2018; Reynolds et al. 2012), the regulation of lipid metabolism during cold acclimation and in response to stress has received less attention. Nonetheless, select studies have identified some of the key lipid metabolism enzymes and genes that directly respond to low temperature. Some of the changes in lipid metabolism discussed above are supported by changes at the molecular level, and we will detail those changes in this section. Our goal is not to provide a comprehensive overview of the biochemistry of lipid metabolism, as such information can be found elsewhere in this book. Rather, we provide select examples of genes and proteins that have been directly linked to cold stress responses.

As discussed above, homeoviscous adaptation is a prevalent response to low temperatures in insects. Desaturases catalyze the formation of double bonds in polyunsaturated fatty acids and thus play an important role in homeoviscous adaptation. Indeed, several studies have shown that desaturases likely play an important role in mediating responses to cold stress (reviewed by Cossins et al. 2002). In the cricket *Acheta domesticus,* progressively decreasing temperatures lead to a 66% increase in Δ^{12} desaturase activity (Batcabe et al. 2000), the specific enzyme responsible for linoelic acid biosynthesis, and accordingly, increases in linoleic acid are sometimes observed in overwintering insects (Rozsypal et al. 2014). At the transcript level, desaturase genes are upregulated during cold acclimation in both the onion maggot *Delia antiqua* (Kayukawa et al. 2007) and *D. suzukii* (Enriquez and Colinet 2019b). Also, in *D. melanogaster,* a desaturase gene is upregulated during recovery from cold stress (Zhang et al. 2011), suggesting desaturases play roles in both preparatory processes and direct response to cold stress. While functional studies on desaturase enzymes in the cold are lacking in insects, in the worm *Caenorhabditis elegans* combined suppression of two desaturase genes reduces the degree of cold acclimation (Murray et al. 2007). However, the effect size of this manipulation is small, suggesting that other non-desaturase processes are responsible for the majority of protection at low temperature.

Further, while changes in gene expression likely contribute to desaturation over longer time periods, in some cases, lipid modifications occur quite quickly, suggesting posttranslational modification. For example, in both *Sarcophaga crassipalpis* and *Drosophila melanogaster,* a few hours of rapid cold hardening can result in membrane desaturation (Michaud and Denlinger 2006; Overgaard et al. 2005). Further, in *Sarcophaga bullata,* direct measurements of membrane fluidity using 31P solid-state NMR indicate that membrane fluidity increases within 2 h at 0 °C (Lee et al. 2006), which is likely too fast (and at too low of a temperature) to support major reorganization of biochemical machinery. Indeed, in *S. bullata*, conditions that elicit rapid cold hardening do not cause any detectable changes in transcript abundance (Teets et al. 2012), suggesting that the existing biochemical machinery is responsible for rapid membrane reorganization observed at low temperature. However, further work is needed to distinguish mechanisms involved in longer-term cold hardening (i.e., acclimation and diapause) from those that occur on shorter time scales.

At the biochemical level, changes in the activities of select lipid metabolism enzymes accompany seasonal acquisition of cold tolerance. Two gall-forming insects, the moth *Epiblema scudderiana* and the fly *E. solidaginis,* have opposite changes in lipid metabolism, with the freeze avoidant *E. scudderiana* depleting lipid reserves over the winter while the freeze-tolerant *E. solidaginis* preserves lipids (Joanisse and Storey 1996a). These changes are consistent with changes in enzyme activity in *E. solidaginis*, as enzymes involved in fat oxidation (hydroxyl-CoA dehydrogenase, carnitine-palmitoyl transferase, and acetoacetyl-CoA thiolase) all decrease in *E. solidaginis* over the winter, likely to preserve lipid reserves. However, in *E. scudderiana,* enzyme activities are more variable and not as well correlated with lipid reserves. Further, in *E. solidaginis,* cold hardy larvae have higher activity of AMP-activated protein kinase (AMPK) (Rider et al. 2011), a signaling molecule that shifts metabolism from ATP consumption to ATP conservation during hypometabolic states. Activation of AMPK also suppresses lipid biosynthesis, and its activation in the winter is consistent with reduced activity of enzymes associated with lipid biosynthesis (e.g., ATP -citrate lyase and malic enzyme) (Joanisse and Storey 1996a). AMPK is also phosphorylated in direct response to cold stress in the beetle *Tribolium castaneum* (Jiang et al. 2019), suggesting this enzyme may play a general role in regulating lipid metabolism at low temperatures.

Select studies have also identified other molecular players involved in lipid metabolism at low temperature. Adipokinetic hormone (AKH) is one of the primary neuroendocrine regulators of lipid metabolism in insects, and while a direct role for this hormone in cold stress has not been identified, genes encoding its receptor (*AkhR*) are activated by low temperatures. *AkhR* transcripts are upregulated in both *D. suzukii* after 9 days of cold acclimation at 10 °C (Enriquez and Colinet 2019b) and after three days of decreased temperatures (from 34 to 25 ° C) in brood of *Apis melifera* (Ramirez et al. 2021), suggesting increased sensitivity to AKH may be required to maintain lipid metabolism at low temperatures. AKH is also involved in regulating supercooling capacity in some insects. In the beetle *Ceruchus piceus,* AKH signaling is shut down in the winter, and exogenous application reduces supercooling capacity by causing the mobilization of ice nucleating lipoproteins

(Xu et al. 1990). In contrast, juvenile hormone (JH) causes a reduction in ice nucleating activity, indicating that the opposing actions of JH and AKH regulate circulation of ice nucleating lipoproteins. In the rice stem borer *Chilo suppressalis,* cold acclimation leads to the upregulation of two transcripts encoding lipase genes (Ma et al. 2020), potentially to liberate glycerol for cryoprotection. In addition to these genes involved in preparatory cold acclimation processes, genes involved in glycerolipid metabolism are rapidly upregulated during recovery from cold shock in the flesh fly *Sarcophaga bullata* (Teets et al. 2012). Finally, during rapid cold hardening, a short-term cold acclimation response that does not require changes in gene expression, lipid storage droplet proteins are rapidly phosphorylated after 2 h of 0 °C (Teets and Denlinger 2016), which would promote lipolysis. However, the functional significance of this phosphorylation event is unclear. Together, the above studies suggest that lipid metabolism is regulated at several levels during cold stress, but work is needed in additional taxa to establish the generalizability of these results.

7 Conclusion and Perspectives

Lipid metabolism plays a central role for overwintering insects, and the various functions are summarized in Fig. 4. In addition to its well-established role for long-term energy storage during diapause (See Chapters 12 and Hahn and Denlinger 2011), lipid metabolism contributes in many other ways to survival at low temperature. Changes in cell membrane lipid composition facilitate homeoviscous adaptation, which permits membrane function and fluidity at low temperatures (Sinensky 1974). Homeoviscous adaptation is commonly observed among overwintering insects, although the manner in which it is achieved can vary from species to species (i.e., changes in headgroups, fatty acid composition, and/or cholesterol content), suggesting multiple evolutionary origins for

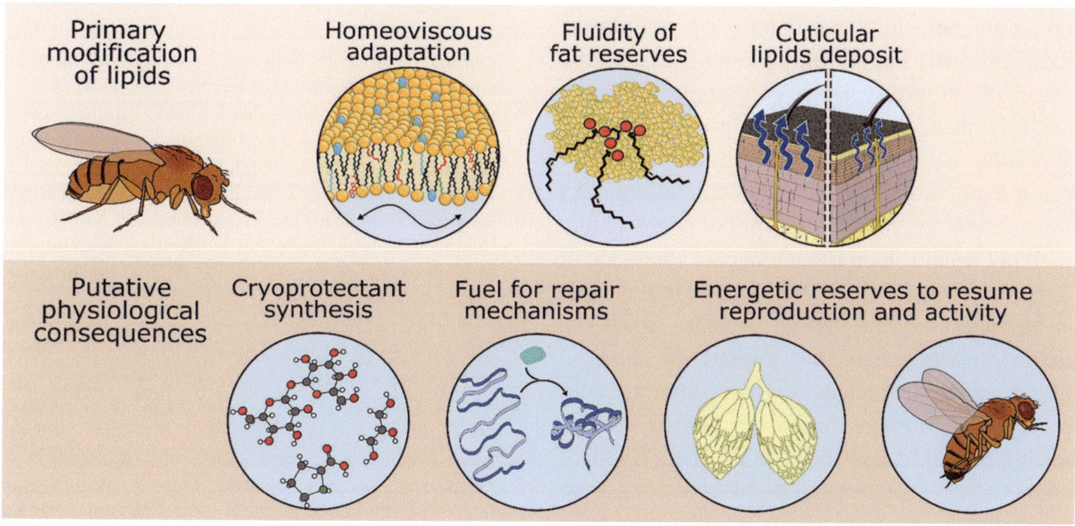

Fig. 4 Modification of lipids during winter and physiological repercussions. Maintenance of lipid structures fluidity allows membrane functioning and guarantees the availability of lipid reserves at cold temperatures. Increased cuticular lipid deposits could protect from dehydration during long-term cold exposures. Lipid reserves can serve as precursors for the synthesis of cryoprotectant molecules, such as saccharose, proline, or glycol, and as fuel for the activation of the repairing machinery, such as production of heat shock proteins, for instance. All these adjustments help in the maintenance of cellular homeostasis. Lipid reserves are also of major importance at the end of winter, as they will allow insects to resume their activity and reproduce

homeoviscous adaptation. Lipids also play an important role in fueling energetically costly stress defense mechanisms in the cold, and indeed, the frequency and severity of cold stress during winter can have a significant impact on energy reserves available when the growing season resumes. Some especially cold-adapted species have also developed novel classes of lipids that function as antifreezes or that can be mobilized as energy sources during periods of freezing. Finally, the expansion of molecular research in recent years has supported an essential role for lipid metabolism in cold stress, as gene expression related to lipid metabolism is involved in both cold acclimation and recovery from cold stress. However, as with much of insect biology, diversity in study systems and methodology makes it challenging to draw general conclusions, so carefully designed, phylogenetically informed studies of lipid metabolism during cold stress are needed in future research.

While lipid metabolism plays a central role in cold tolerance, many research areas require further investigation. First, despite the nearly ubiquitous observation of homeoviscous adaptation, the molecular regulation of membrane modifications is largely unexplored. The expression of desaturase genes is commonly linked to cold tolerance, but direct manipulation of the expression or function of desaturates has not been attempted in insects, to our knowledge. As discussed above, membrane lipid changes play a relatively small role in inducible cold tolerance in *C. elegans* (Murray et al. 2007), and similar studies are necessary in insects to tease apart the relative importance of the myriad biochemical and physiology changes that accompany cold acclimation. Further, while correlative studies indicate that lipid can serve as an energy source to fuel the energetic demands of cold stress, the potential role of lipid transport has not been assessed. In the flesh fly, *Sarcophaga bullata,* brief cold exposure causes phosphorylation of lipid storage droplet proteins (Teets and Denlinger 2016), which in vitro leads to the activation of fat body lipases (Arrese et al. 2008), which suggests that rapid and reversible signaling events could be dynamically regulating lipid metabolism in the cold. Finally, lipid metabolic responses to cold are highly variable across different insect species; so phylogenetically informed studies that address the evolutionary genetics of lipid responses to cold can help identify processes that are essential for low temperature survival.

References

Addo-Bediako A, Chown SL, Gaston KJ (2000) Thermal tolerance, climatic variability and latitude. Proc R Soc Lond Ser B Biol Sci 267(1445):739–745

Ala-Honkola O et al (2020) Diapause affects cuticular hydrocarbon composition and mating behavior of both sexes in *Drosophila montana*. Insect Sci 27(2): 304–316

Allen MC et al (2024) Effects of a high cholesterol diet on chill tolerance are highly context-dependent in *Drosophila*. J Therm Biol 119:103789

Andersen MK, Robertson RM, MacMillan HA (2022) Plasticity in Na+/K+-ATPase thermal kinetics drives variation in the temperature of cold-induced neural shutdown of adult *Drosophila melanogaster*. J Exp Biol 225(24):jeb244923

Arrese EL, Soulages JL (2010) Insect fat body: energy, metabolism, and regulation. Annu Rev Entomol 55: 207–225

Arrese EL et al (2008) Function and structure of lipid storage droplet protein 1 studied in lipoprotein complexes. Arch Biochem Biophys 473(1):42–47

Ayala A, Muñoz MF, Argüelles S (2014) Lipid peroxidation: production, metabolism, and signaling mechanisms of malondialdehyde and 4-hydroxy-2-nonenal. Oxidative Med Cell Longev 2014(1):360438

Bahrndorff S et al (2007) Differences in cold and drought tolerance of high arctic and sub-arctic populations of *Megaphorura arctica* Tullberg 1876 (Onychiuridae: Collembola). Cryobiology 55(3):315–323

Bale JS (2002) Insects and low temperatures: from molecular biology to distributions and abundance. Proc R Soc Lond Ser B Biol Sci 357(1423):849–861

Bashan M, Cakmak O (2005) Changes in composition of phospholipid and triacylglycerol fatty acids prepared from prediapausing and diapausing individuals of *Dolycoris baccarum* and *Piezodorus lituratus* (Heteroptera: Pentatomidae). Ann Entomol Soc Am 98(4):575–579

Bashan M, Akbas H, Yurdakoc K (2002) Phospholipid and triacylglycerol fatty acid composition of major life stages of sunn pest, *Eurygaster integriceps* (Heteroptera: Scutelleridae). Comp Biochem Physiol B Biochem Mol Biol 132(2):375–380

Batcabe JP et al (2000) Effects of developmental age, ambient temperature, and dietary alterations on delta (12) desaturase activity in the house cricket, *Acheta domesticus*. Arch Insect Biochem Physiol 44(3): 112–119

Batz ZA, Armbruster PA (2018) Diapause-associated changes in the lipid and metabolite profiles of the Asian tiger mosquito, *Aedes albopictus*. J Exp Biol 221(Pt 24):jeb189480

Bayley JS et al (2020) Cold acclimation increases depolarization resistance and tolerance in muscle fibers from a chill-susceptible insect, *Locusta migratoria*. Am J Phys Regul Integr Comp Phys 319(4):R439–R447

Bennett VA, Lee RE (1997) Modeling seasonal changes in intracellular freeze-tolerance of fat body cells of the gall fly *Eurosta solidaginis* (Diptera, Tephritidae). J Exp Biol 200(1):185–192

Bennett VA, Pruitt NL, Lee RE (1997) Seasonal changes in fatty acid composition associated with cold-hardening in third instar larvae of *Eurosta solidaginis*. J Comp Physiol B 167(4):249–255

Benoit JB, Denlinger DL (2007) Suppression of water loss during adult diapause in the northern house mosquito, *Culex pipiens*. J Exp Biol 210(2):217–226

Bortner CD, Gomez-Angelats M, Cidlowski JA (2001) Plasma membrane depolarization without repolarization is an early molecular event in anti-Fas-induced apoptosis. J Biol Chem 276(6):4304–4314

Brankatschk M et al (2018) A temperature-dependent switch in feeding preference improves *Drosophila* development and survival in the cold. Dev Cell 46(6): 781–793.e4

Chen C-P, Walker VK (1994) Cold-shock and chilling tolerance in *Drosophila*. J Insect Physiol 40(8): 661–669

Chowanski S et al (2015) Cold induced changes in lipid, protein and carbohydrate levels in the tropical insect *Gromphadorhina coquereliana*. Comp Biochem Physiol A Mol Integr Physiol 183:57–63

Ciancio JJ et al (2021) Cold tolerance, water balance, energetics, gas exchange, and diapause in overwintering brown marmorated stink bugs. J Insect Physiol 128:104171

Clark RA et al (2003) North Sea cod and climate change—modelling the effects of temperature on population dynamics. Glob Chang Biol 9(11):1669–1680

Clerc SG, Thompson TE (1995) Permeability of dimyristoyl phosphatidylcholine/dipalmitoyl phosphatidylcholine bilayer membranes with coexisting gel and liquid-crystalline phases. Biophys J 68(6):2333–2341

Colinet H (2011) Disruption of ATP homeostasis during chronic cold stress and recovery in the chill susceptible beetle (*Alphitobius diaperinus*). Comp Biochem Physiol A Mol Integr Physiol 160(1):63–67

Colinet H, Hance T, Vernon P (2006) Water relations, fat reserves, survival, and longevity of a cold-exposed parasitic wasp *Aphidius colemani* (Hymenoptera: Aphidiinae). Environ Entomol 35(2):228–236

Colinet H et al (2016) Uncovering the benefits of fluctuating thermal regimes on cold tolerance of *Drosophila* flies by combined metabolomic and lipidomic approach. Biochim Biophys Acta 1861(11): 1736–1745

Colinet H, Renault D, Roussel D (2017) Cold acclimation allows *Drosophila* flies to maintain mitochondrial functioning under cold stress. Insect Biochem Mol Biol 80:52–60

Cooper BS et al (2012) In a variable thermal environment selection favors greater plasticity of cell membranes in *Drosophila melanogaster*. Evolution 66(6):1976–1984

Cooper BS, Hammad LA, Montooth KL (2014) Thermal adaptation of cellular membranes in natural populations of *Drosophila melanogaster*. Funct Ecol 28(4):886–894

Cossins AR (1994) Temperature adaptation of biological membranes. Portland Press, London

Cossins AR et al (2002) The role of desaturases in cold-induced lipid restructuring. Biochem Soc Trans 30: 1082–1086

Coulson S, Fisher J, Bale J (1992) 31P NMR investigation of the energy charge of the housefly Musca domestica (Diptera: Muscidae) during rapid cold hardening and cold shock. CryoLetters 13:183–192

Crockett EL (1998) Cholesterol function in plasma membranes from ectotherms: membrane-specific roles in adaptation to temperature. Am Zool 38(2):291–304

Davy De Virville J et al (2002) Homeoviscous and functional adaptations of mitochondrial membranes to growth temperature in soybean seedlings. Plant Cell Environ 25(10):1289–1297

De Ro M et al (2021) Effect of starvation on the cold tolerance of adult *Drosophila suzukii* (Diptera: Drosophilidae). Bull Entomol Res 111(6):694–704

Dollo VH, Yi S-X, Lee RE (2010) High temperature pulses decrease indirect chilling injury and elevate ATP levels in the flesh fly, *Sarcophaga crassipalpis*. Cryobiology 60:351–353

Downer RGH, Kallapur VL (1981) Temperature-induced changes in lipid composition and transition temperature of flight muscle mitochondria of *Schistocerca gregaria*. J Therm Biol 6(4):189–194

Enriquez T, Colinet H (2019a) Cold acclimation triggers lipidomic and metabolic adjustments in the spotted wing drosophila *Drosophila suzukii* (Matsumara). Am J Phys Regul Integr Comp Phys 316(6):R751–R763

Enriquez T, Colinet H (2019b) Cold acclimation triggers major transcriptional changes in *Drosophila suzukii*. BMC Genomics 20(1):413

Fernandez-Puente L et al (1994) Temperature and chain length effects on bending elasticity of phosphatidylcholine bilayers. Europhys Lett (EPL) 28(3):181–186

Findsen A, Overgaard J, Pedersen TH (2016) Reduced L-type Ca2+ current and compromised excitability induce loss of skeletal muscle function during acute cooling in locust. J Exp Biol 219(15):2340–2348

Gennis RB (1989) Biomembranes: molecular structure and function, Springer advanced texts in chemistry. Springer, New York

Gerber L et al (2021) Cold acclimation preserves hindgut reabsorption capacity at low temperature in a chill-susceptible insect, *Locusta migratoria*. Comp Biochem Physiol A Mol Integr Physiol 252:110850

Goto SG, Katagiri C (2011) Effects of acclimation temperature on membrane phospholipids in the flesh fly *Sarcophaga similis*. Entomol Sci 14(2):224–229

Goto SG et al (2010) Fatty acids of membrane phospholipids in *Drosophila melanogaster* lines showing rapid and slow recovery from chill coma. Biochem Biophys Res Commun 391(2):1251–1254

Hagemann JW, Tallent WH, Kolb KE (1972) Differential scanning calorimetry of single acid triglycerides: effect of chain length and unsaturation. J Am Oil Chem Soc 49(2):118–123

Hahn DA, Denlinger DL (2011) Energetics of insect diapause. Annu Rev Entomol 56:103–121

Hammad LA et al (2011) Profiling and quantification of *Drosophila melanogaster* lipids using liquid chromatography/mass spectrometry. Rapid Commun Mass Spectrom 25(19):2959–2968

Han E-N, Bauce E (1993) Physiological changes and cold hardiness of spruce budworm larvae, *Choristoneura fumiferana* (clem.), during pre-diapause and diapause development under laboratory conditions. Can Entomol 125(6):1043–1053

Hayward SAL, Manso B, Cossins AR (2014) Molecular basis of chill resistance adaptations in poikilothermic animals. J Exp Biol 217(1):6–15

Hazel JR (1989) Cold adaptation in ectotherms: regulation of membrane function and cellular metabolism. In: Wang LCH (ed) Animal adaptation to cold. Springer Berlin Heidelberg, Berlin/Heidelberg, pp 1–50

Hazel JR (1995) Thermal adaptation in biological membranes: is homeoviscous adaptation the explanation? Annu Rev Physiol 57(1):19–42

Hazel JR, Eugene Williams E (1990) The role of alterations in membrane lipid composition in enabling physiological adaptation of organisms to their physical environment. Prog Lipid Res 29(3):167–227

Hochachka PW, Somero GN (1984) Biochemical adaptation. Princeton University Press

Hodková M et al (1999) Seasonal changes in the phospholipid composition in thoracic muscles of a heteropteran, *Pyrrhocoris apterus*. Insect Biochem Mol Biol 29(4):367–376

Imlay JA (2003) Pathways of oxidative damage. Ann Rev Microbiol 57:395–418

Irwin JT, Lee RE (2002) Energy and water conservation in frozen vs. supercooled larvae of the goldenrod gall fly, *Eurosta solidaginis* (Fitch) (Diptera: Tephritidae). J Exp Biol 292(4):345–350

Ismail M et al (2010) Physiological costs of cold exposure on the parasitoid *Aphidius ervi*, without selection pressure and under constant or fluctuating temperatures. BioControl 55(6):729–740

Jahn R, Lang T, Südhof TC (2003) Membrane fusion. Cell 112(4):519–533

Jiang H et al (2019) Transcriptional and post-translational activation of AMPKα by oxidative, heat, and cold stresses in the red flour beetle, *Tribolium castaneum*. Cell Stress Chaperones 24(6):1079–1089

Joanisse DR, Storey KB (1996a) Fatty acid content and enzymes of fatty acid metabolism in overwintering cold-hardy gall insects. Physiol Zool 69(5):1079–1095

Joanisse DR, Storey KB (1996b) Oxidative stress and antioxidants in overwintering larvae of cold-hardy goldenrod gall insects. J Exp Biol 199(7):1483–1491

Jones HE et al (1992) Lipid composition of subcellular membranes from larvae and prepupae of *Drosophila melanogaster*. Lipids 27(12):984–987

Joplin KH, Yocum GD, Denlinger DL (1990) Cold shock elicits expression of heat-shock proteins in the flesh fly, *Sarcophaga crassipalpis*. J Insect Physiol 36(11): 825–834

Jørgensen LB et al (2023) Balanced mitochondrial function at low temperature is linked to cold adaptation in *Drosophila* species. J Exp Biol 226(8):jeb245439

Kayukawa T et al (2007) Upregulation of a desaturase is associated with the enhancement of cold hardiness in the onion maggot, *Delia antiqua*. Insect Biochem Mol Biol 37(11):1160–1167

Kellermann V et al (2012) Phylogenetic contstraints in key functional traits behind species' climate niches: patterns of desiccation and cold resistance across 95 *Drosophila* species. Evolution 66(11):3377–3389

Kimura MT (2004) Cold and heat tolerance of drosophilid flies with reference to their latitudinal distributions. Oecologia 140(3):442–449

Klepsatel P et al (2016) Thermal stress depletes energy reserves in *Drosophila*. Sci Rep 6(1):33667

Košťál V (2010) Cell structural modifications in insects at low temperatures. In: Denlinger DL, Lee RE (eds) Low temperature biology of insects. Cambridge University Press, Cambridge, pp 116–140

Košťál V, Šimek P (1998) Changes in fatty acid composition of phospholipids and triacylglycerols after cold-acclimation of an aestivating insect prepupa. J Comp Physiol B Biochem Syst Environ Physiol 168(6):453–460

Košťál V et al (2011) Long-term cold acclimation extends survival time at 0°C and modifies the metabolomic profiles of the larvae of the fruit fly *Drosophila melanogaster*. PLoS One 6(9):10

Košťál V et al (2013) Seasonal changes in minor membrane phospholipid classes, sterols and tocopherols in overwintering insect, *Pyrrhocoris apterus*. J Insect Physiol 59(9):934–941

Krebs RA, Loeschcke V (1994) Costs and benefits of activation of the heat-shock response in *Drosophila melanogaster*. Funct Ecol 8(6):730–737

Kučera L et al (2022) A mixture of innate cryoprotectants is key for freeze tolerance and cryopreservation of a drosophilid fly larva. J Exp Biol 225(8):jeb243934

Kučerka N, Nieh M-P, Katsaras J (2011) Fluid phase lipid areas and bilayer thicknesses of commonly used phosphatidylcholines as a function of temperature.

Biochim Biophys Acta—Biomembr 1808(11): 2761–2771

Kukal O, Duman JG, Serianni AS (1989) Cold-induced mitochondrial degradation and cryoprotectant synthesis in freeze-tolerant arctic caterpillars. J Comp Physiol B 158(6):661–671

Lalouette L et al (2011) Metabolic rate and oxidative stress in insects exposed to low temperature thermal fluctuations. Comp Biochem Physiol A Mol Integr Physiol 158(2):229–234

Lee RE (1989) Insect cold-hardiness—to freeze or not to freeze—how insects survive low temperatures. Bioscience 39(5):308–313

Lee RE (2010) A primer on insect cold tolerance. In: Denlinger DL, Lee RE Jr (eds) Low temperature biology of insects. Cambridge University Press, Cambridge, pp 3–34

Lee RE et al (1995) Cryobiology of the freeze-tolerant gall fly *Eurosta solidaginis*: overwintering energetics and heat shock proteins. Clim Res 5(1):61–67

Lee RE et al (2006) Rapid cold-hardening increases membrane fluidity and cold tolerance of insect cells. Cryobiology 52(3):459–463

Littler AS, Garcia MJ, Teets NM (2021) Laboratory diet influences cold tolerance in a genotype-dependent manner in *Drosophila melanogaster*. Comp Biochem Physiol A Mol Integr Physiol 257:110948

Lockey KH (1988) Lipids of the insect cuticle: origin, composition and function. Comp Biochem Physiol B Comp Biochem 89(4):595–645

Lopez-Martinez G et al (2008) High resistance to oxidative damage in the Antarctic midge *Belgica antarctica*, and developmentally linked expression of genes encoding superoxide dismutase, catalase and heat shock proteins. Insect Biochem Mol Biol 38(8):796–804

Ma W et al (2020) A chromosome-level genome assembly reveals the genetic basis of cold tolerance in a notorious rice insect pest, *Chilo suppressalis*. Mol Ecol Resour 20(1):268–282

MacMillan HA, Sinclair BJ (2011) Mechanisms underlying insect chill-coma. J Insect Physiol 57(1):12–20

MacMillan HA, Guglielmo CG, Sinclair BJ (2009) Membrane remodeling and glucose in *Drosophila melanogaster*: a test of rapid cold-hardening and chilling tolerance hypotheses. J Insect Physiol 55(3): 243–249

MacMillan HA et al (2012) Reestablishment of ion homeostasis during chill-coma recovery in the cricket *Gryllus pennsylvanicus*. Proc Natl Acad Sci USA 109(50):20750–20755

MacMillan HA et al (2016) Cold acclimation wholly reorganizes the *Drosophila melanogaster* transcriptome and metabolome. Sci Rep 6(1):28999

Magnuson JJ, Crowder LB, Medvick PA (2015) Temperature as an ecological resource. Am Zool 19(1): 331–343

Marshall KE, Sinclair BJ (2010) Repeated stress exposure results in a survival-reproduction trade-off in *Drosophila melanogaster*. Proc R Soc Lond Ser B Biol Sci 277(1683):963–969

Marshall KE, Sinclair BJ (2012) Threshold temperatures mediate the impact of reduced snow cover on overwintering freeze-tolerant caterpillars. Naturwissenschaften 99(1):33–41

Marshall KE, Sinclair BJ (2018) Repeated freezing induces a trade-off between cryoprotection and egg production in the goldenrod gall fly, *Eurosta solidaginis*. J Exp Biol 221(15):jeb177956

Marshall KE et al (2014) Seasonal accumulation of acetylated triacylglycerols by a freeze-tolerant insect. J Exp Biol 217(9):1580–1587

McDougall GE, Steele JE (1988) Free fatty acids as a source of energy for trehalose synthesis in the fat body of the American cockroach (*Periplaneta americana*). Insect Biochem 18(6):591–597

Miao L et al (2002) From lanosterol to cholesterol: structural evolution and differential effects on lipid bilayers. Biophys J 82(3):1429–1444

Michaud MR, Denlinger DL (2006) Oleic acid is elevated in cell membranes during rapid cold-hardening and pupal diapause in the flesh fly, *Sarcophaga crassipalpis*. J Insect Physiol 52(10):1073–1082

Mocé E et al (2010) Use of cholesterol in sperm cryopreservation: present moment and perspectives to future. Reprod Domest Anim 45(s2):57–66

Murray P et al (2007) An explicit test of the phospholipid saturation hypothesis of acquired cold tolerance in *Caenorhabditis elegans*. Proc Natl Acad Sci USA 104(13):5489

Ohtsu T et al (1993) Cold adaptations in *Drosophila*. Qualitative changes of triacylglycerols with relation to overwintering. J Biol Chem 268(3):1830–1834

Ohtsu T, Kimura MT, Katagiri C (1998) How *Drosophila* species acquire cold tolerance. Eur J Biochem 252(3): 608–611

Ohtsu T, Katagiri C, Kimura MT (1999) Biochemical aspects of climatic adaptations in *Drosophila curviceps, D. immigrans, and D. albomicans* (Diptera: Drosophilidae). Environ Entomol 28(6):968–972

Olsen TM et al (1998) Factors contributing to seasonal increases in inoculative freezing resistance in overwintering fire-colored beetle larvae *Dendroides canadensis* (Pyrochroidae). J Exp Biol 201(10): 1585–1594

Overgaard J, MacMillan HA (2017) The integrative physiology of insect chill tolerance. Annu Rev Physiol 79: 187–208

Overgaard J et al (2005) Changes in membrane lipid composition following rapid cold hardening in *Drosophila melanogaster*. J Insect Physiol 51(11): 1173–1182

Overgaard J et al (2006) Reorganization of membrane lipids during fast and slow cold hardening in *Drosophila melanogaster*. Physiol Entomol 31(4):328–335

Overgaard J et al (2008) Effects of acclimation temperature on thermal tolerance and membrane phospholipid composition in the fruit fly *Drosophila melanogaster*. J Insect Physiol 54(3):619–629

Parkash R, Lambhod C (2021) Plastic changes in cold and drought tolerance of *Drosophila nepalensis* correlate with sex-specific differences in body melanization, cuticular lipid mass, proline accumulation, and seasonal abundance. Comp Biochem Physiol A Mol Integr Physiol 258:110985

Potter KA, Arthur Woods H, Pincebourde S (2013) Microclimatic challenges in global change biology. Glob Chang Biol 19(10):2932–2939

Pruitt NL, Lu C (2008) Seasonal changes in phospholipid class and class-specific fatty acid composition associated with the onset of freeze tolerance in third-instar larvae of *Eurosta solidaginis*. Physiol Biochem Zool 81(2):226–234

Pujol-Lereis LM et al (2016) Chill-coma recovery time, age and sex determine lipid profiles in *Ceratitis capitata* tissues. J Insect Physiol 87:53–62

Ramirez L et al (2021) Fast weight recovery, metabolic rate adjustment and gene-expression regulation define responses of cold-stressed honey bee brood. J Insect Physiol 128:104178

Reynolds JA et al (2012) Transcript profiling reveals mechanisms for lipid conservation during diapause in the mosquito, *Aedes albopictus*. J Insect Physiol 58(7): 966–973

Rider MH et al (2011) AMP-activated protein kinase and metabolic regulation in cold-hardy insects. J Insect Physiol 57(11):1453–1462

Rietveld A et al (1999) Association of sterol- and glycosylphosphatidylinositol-linked proteins with *Drosophila* raft lipid microdomains. J Biol Chem 274(17):12049–12054

Rinehart JP et al (2007) Up-regulation of heat shock proteins is essentail for cold survival during insect diapause. Proc Natl Acad Sci USA 104(27): 11130–11137

Rozsypal J (2022) Cold and freezing injury in insects: an overview of molecular mechanisms. Eur J Entomol 119(1):43–57

Rozsypal J et al (2014) Seasonal changes in the composition of storage and membrane lipids in overwintering larvae of the codling moth, *Cydia pomonella*. J Therm Biol 45:124–133

Salt R (1957) Natural occurrence of glycerol in insects and its relation to their ability to survive freezing. Can Entomol 89(11):491–494

Salt RW (1961) Principles of insect cold-hardiness. Annu Rev Entomol 6:55

Scharf I et al (2016) The effect of fasting and body reserves on cold tolerance in 2 pit-building insect predators. Curr Zool 63(3):287–294

Shreve SM, Yi S-X, Lee RE (2007) Increased dietary cholesterol enhances cold tolerance in *Drosophila melanogaster*. CryoLetters 28(1):33–37

Sinclair BJ (2015) Linking energetics and overwintering in temperate insects. J Therm Biol 54:5–11

Sinclair BJ, Marshall KE (2018) The many roles of fats in overwintering insects. J Exp Biol 221(Suppl 1): jeb161836

Sinclair BJ et al (2003) Insects at low temperatures: an ecological perspective. Trends Ecol Evol 18(5): 257–262

Sinclair BJ, Gibbs AG, Roberts SP (2007) Gene transcription during exposure to, and recovery from, cold and desiccation stress in *Drosophila melanogaster*. Insect Mol Biol 16(4):435–443

Sinclair BJ et al (2013) Cross-tolerance and cross-talk in the cold: relating low temperatures to desiccation and immune stress in insects. Integr Comp Biol 53(4): 545–556

Sinensky M (1974) Homeoviscous adaptation—homeostatic process that regulates viscosity of membrane lipids in *Escherichia coli*. Proc Natl Acad Sci USA 71(2):522–525

Singer SJ, Nicolson GL (1972) The fluid mosaic model of the structure of cell membranes: cell membranes are viewed as two-dimensional solutions of oriented globular proteins and lipids. Science 175(4023):720–731

Slotsbo S et al (2016) Tropical to subpolar gradient in phospholipid composition suggests adaptive tuning of biological membrane function in drosophilids. Funct Ecol 30(5):759–768

Štětina T et al (2018) Recovery from supercooling, freezing, and cryopreservation stress in larvae of the drosophilid fly, *Chymomyza costata*. Sci Rep 8(1):4414

Storey KB (1996) Oxidative stress: animal adaptations in nature. Braz J Med Biol Res 29(12):1715–1733

Storey KB (1997) Organic solutes in freezing tolerance. Comp Biochem Physiol A Physiol 117(3):319–326

Storey KB, Storey JM (2012) Insect cold hardiness: metabolic, gene, and protein adaptation. Can J Zool 90(4): 456–475

Teets NM, Denlinger DL (2016) Quantitative phosphoproteomics reveals signaling mechanisms associated with rapid cold hardening in a chill-tolerant fly. J Proteome Res 15(8):2855–2862

Teets NM et al (2011) Survival and energetic costs of repeated cold exposure in the Antarctic midge, *Belgica antarctica*: a comparison between frozen and supercooled larvae. J Exp Biol 214(5):806–814

Teets NM et al (2012) Combined transcriptomic and metabolomic approach uncovers molecular mechanisms of cold tolerance in a temperate flesh fly. Physiol Genomics 44(15):764–777

Teets NM et al (2019) Rapid cold hardening protects against sublethal freezing injury in an Antarctic insect. J Exp Biol 222(15):jeb206011

Teets NM et al (2020) Changes in energy reserves and gene expression elicited by freezing and supercooling in the Antarctic midge, *Belgica antarctica*. Insects 11(1):18

Teets NM, Marshall KE, Reynolds JA (2023) Molecular mechanisms of winter survival. Annu Rev Entomol 68(1):319–339

Thiry E, Hoffmann KH (1986) Chemical composition and fatty acids of lipids in an Antarctic beetle (*Hydromedion sparsutum*, Perimylopidae) and an African beetle (*Pachnoda marginata*, Scarabaeidae). Comp Biochem Physiol B 84(3):387–392

Tomčala A et al (2006) Seasonal acquisition of chill tolerance and restructuring of membrane glycerophospholipids in an overwintering insect: triggering by low temperature, desiccation and diapause progression. J Exp Biol 209(20): 4102–4114

Toxopeus J, Kostal V, Sinclair BJ (2019) Evidence for non-colligative function of small cryoprotectants in a freeze-tolerant insect. Proc R Soc Lond Ser B Biol Sci 286(1899):20190050

Trenti F et al (2022) Insect cold-tolerance and lipidome: membrane lipid composition of two chironomid species differently adapted to cold. Cryobiology 106: 84–90

Urbanski JM et al (2010) The molecular physiology of increased egg desiccation resistance during diapause in the invasive mosquito, *Aedes albopictus*. Proc R Soc Lond Ser B Biol Sci 277:2683–2692

van Dooremalen C, Ellers J (2010) A moderate change in temperature induces changes in fatty acid composition of storage and membrane lipids in a soil arthropod. J Insect Physiol 56(2):178–184

Walters KR et al (2009) A nonprotein thermal hysteresis-producing xylomannan antifreeze in the freeze-tolerant Alaskan beetle *Upis ceramboides*. Proc Natl Acad Sci USA 106(48):20210–20215

Walters KR et al (2011) A thermal hysteresis-producing xylomannan glycolipid antifreeze associated with cold tolerance is found in diverse taxa. J Comp Physiol B 181(5):631–640

Williams CM et al (2011) Triacylglyceride measurement in small quantities of homogenised insect tissue: comparisons and caveats. J Insect Physiol 57(12): 1602–1613

Xu L, Neven LG, Duman JG (1990) Hormonal control of hemolymph lipoprotein ice nucleators in overwintering freeze-susceptible larvae of the stag beetle *Ceruchus piceus*: adipokinetic hormone and juvenile hormone. J Comp Physiol B 160(1):51–59

Yancey PH (2005) Organic osmolytes as compatible, metabolic and counteracting cytoprotectants in high osmolarity and other stresses. J Exp Biol 208(15): 2819–2830

Yang S-T et al (2016) The role of cholesterol in membrane fusion. Chem Phys Lipids 199:136–143

Yi S-X, Lee RE (2005) Changes in gut and Malpighian tubule transport during seasonal acclimatization and freezing in the gall fly *Eurosta solidaginis*. J Exp Biol 208(10):1895–1904

Yi S-X, Lee RE Jr (2016) Cold-hardening during long-term acclimation in a freeze-tolerant woolly bear caterpillar, *Pyrrharctia isabella*. J Exp Biol 219(1):17–25

Yi S-X, Moore CW, Lee RE (2007) Rapid cold-hardening protects *Drosophila melanogaster* from cold-induced apoptosis. Apoptosis 12(7):1183–1193

Zhang J et al (2011) Divergent transcriptomic responses to repeated and single cold exposures in *Drosophila melanogaster*. J Exp Biol 214(23):4021–4029

Thomas Enriquez is an ecologist, specialized in entomology, ecophysiology, and behavioral ecology. His main topic of interest is the physiological mechanisms linked with cold tolerance plasticity in insects. He has most recently worked in the Bertanne Visser laboratory at the University of Gembloux in Belgium as a postdoctoral fellow investigating the effects of maternal behavior and environment on the plasticity of fat synthesis in parasitoids.

Nicholas M. Teets is an Associate Professor in Entomology at the University of Kentucky, where he and his colleagues investigate adaptations to environmental stress in insects. Work in the Teets Lab includes fundamental mechanistic studies as well as practical applications of insect genetics and environmental physiology, spanning model systems like fruit flies to insects with unique adaptations like Antarctic insects.

Adv Exp Med Biol (2026) 1494: 501–518
https://doi.org/10.1007/5584_2024_813

Published online: 15 June 2024

Fatty Acid Origin of Insect Pheromones

Russell Jurenka

Abstract

Pheromones are utilized to a great extent in insects. Many of these pheromones are biosynthesized through a pathway involving fatty acids. This chapter will provide examples where the biosynthetic pathways of fatty acid-derived pheromones have been studied in detail. These include pheromones from Lepidoptera, Coleoptera, and Hymenoptera. Many species of Lepidoptera utilize fatty acids as precursors to pheromones with a functional group that include aldehydes, alcohols, and acetate esters. In addition, the biosynthesis of hydrocarbons will be briefly examined because many insects utilize hydrocarbons or modified hydrocarbons as pheromones.

Keywords

Pheromones · Fatty alcohols · Fatty aldehydes · Fatty acetate esters · Desaturases · Fatty acyl reductase · Hydrocarbons

1 Introduction

Pheromones are used extensively in chemical communication in the Insecta. The term pheromone indicates that the chemical signals are secreted by one individual of a species to affect the behavior of another individual of the same species. Although it was known for a long time that insects are using pheromones as chemical communication signals (Fabre 1966; Lintner 1882) it was much later when the first pheromone was identified. The pheromone where the chemical structure was first identified was from the female silkworm moth, *Bombyx mori*, and consisted of (*E,Z*)-10,12-hexadecadien-1-ol (bombykal) (Butenandt et al. 1959). According to Pherobase.com, pheromones and other semiochemicals have currently been identified in over 3,000 insect species. This chapter will concentrate only on those pheromones that are derived through fatty acid biosynthetic pathways and where the biosynthesis of the pheromone has been studied in detail. For information on the biosynthesis of other types of pheromones, such as the monoterpenoids that are produced through the isoprenoid pathway, see Keeling et al. (2021) and Tholl (2021). In addition, some pheromones are derived through modification of host-derived compounds. The major groups of pheromones derived from fatty acids occurs in the Lepidoptera, Coleoptera, and Hymenoptera. In addition, many insects utilize hydrocarbons and their derivatives as pheromone components, therefore the biosynthesis of hydrocarbons is briefly presented in this chapter.

R. Jurenka (✉)
Iowa State University, Iowa, USA
e-mail: rjurenka@iastate.edu

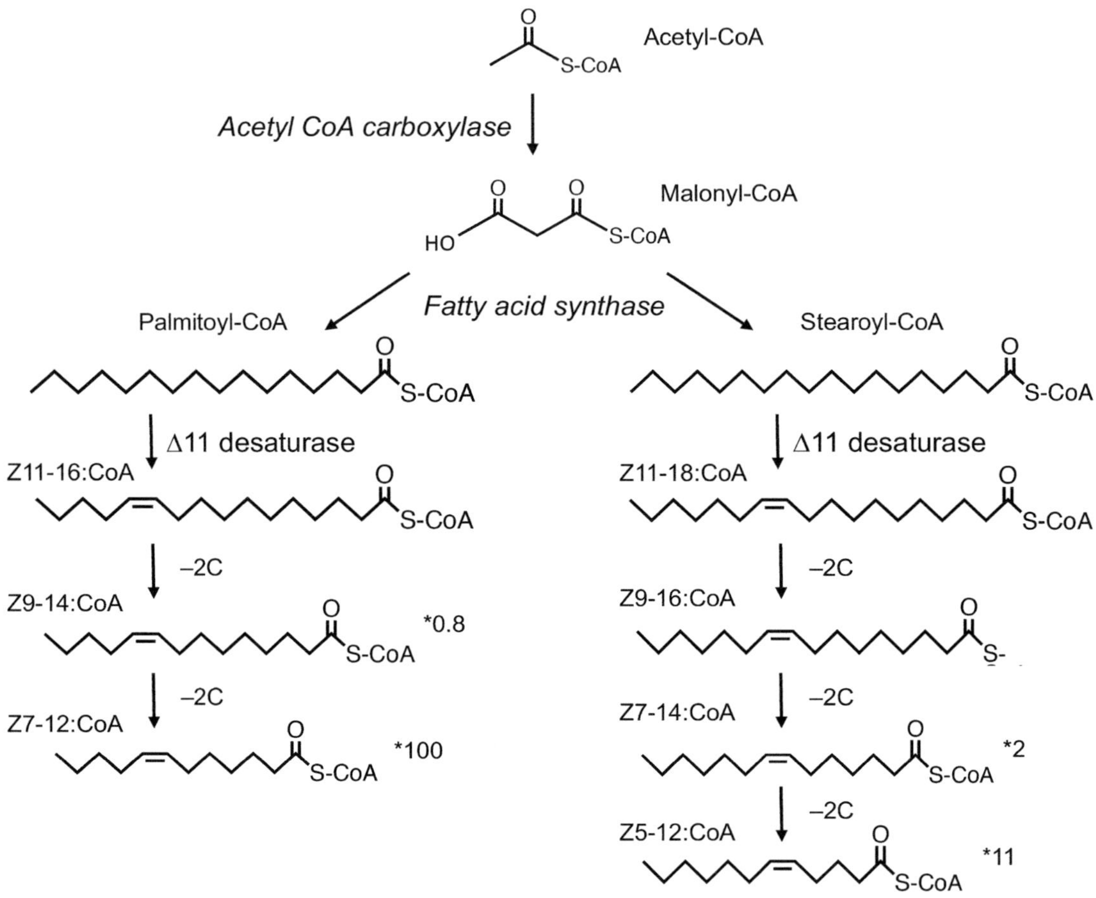

Fig. 1 Biosynthetic pathway of the pheromone blend of the cabbage looper, *Trichoplusia ni*. The asterisk indicates the CoAs that are reduced to an alcohol and acetylated to form acetate esters. The number next to the asterisk indicates the approximate ratio of acetate esters found in the pheromone gland (Bjostad et al. 1984). In addition, 12:CoA and Δ11–12:CoA are precursors to the minor components 12:OAc and 11–12:OAc, respectively

2 Lepidoptera

Many moth sex pheromones contain 10–18 carbon atoms with one or more degrees of unsaturation and a functional group consisting of an acetate ester, alcohol, or aldehyde (Löfstedt et al. 2016). These are the Type I pheromones and are all derived through fatty acid biosynthesis (Jurenka 2021). The first study demonstrating the de novo biosynthesis of a pheromone was in 1978 where injection of radiolabeled acetate was incorporated into the pheromone of female cabbage looper moths (Jones and Berger 1978). The radiolabeled carbons were incorporated throughout the pheromone molecule, (Z)-7-dodecen-1-yl acetate (Z7–12:OAc),[1] indicating a fatty acid biosynthetic pathway. The position of the double bond and the carbon chain length indicated a desaturase acting on a fatty acid precursor. Based on radiolabel experiments it was demonstrated that the major pheromone component Z7–12:OAc was biosynthesized involving a desaturase and chain-shortening enzymes (Bjostad and Roelofs 1983). Figure 1 illustrates the deduced biosynthetic pathway for the major

[1] Shorthand notation of pheromone and precursor fatty acids is as follows. The Z or E denotes double bond configuration, the next number the double bond position, the number after the – indicates number of carbons in the chain, letters after the: denotes a functional group. OAc = acetate ester, Ald = aldehyde, CoA = coenzyme A, OH = alcohol, me = methyl ester.

pheromone components and the minor components in the cabbage looper moth. Since then a number of biosynthetic pathways have been determined in a variety of moths. For a detailed analysis of how female moth pheromones are produced see Jurenka (2021) and Allison and Cardé (2016).

In moths, it is primarily the female that produces the sex pheromone for attraction of the male. These are long-range pheromone molecules that are released in a typical "calling" behavior for attracting males down-wind (Allison and Cardé 2016). The females produce the pheromone in an exocrine gland that is usually associated with the cuticle. In most females this gland is formed from modified epidermal cells located at the posterior tip of the abdomen although in a few moths it can be found in other parts of the body. For the Type I pheromones, the female pheromone gland usually produces the pheromone de novo starting from acetate through fatty acid biosynthesis followed by modifications involving various enzymes. These include desaturases, limited chain-shortening enzymes, fatty acid reductases, oxidases, and acetyltransferases.

2.1 Fatty Acid Synthesis

The pheromone gland contains the ubiquitous acetyl-CoA carboxylase and fatty acid synthetase that are required for the biosynthesis of saturated fatty acids. Currently, no direct enzymatic studies have been carried out on the fatty acid synthesis enzymes from moth pheromone glands, they are probably similar to the enzymes found in other cell types of the insect, like the fat body. Transcriptome studies indicate that the pheromone gland has both acetyl-CoA carboxylase and fatty acid synthetase (Dou et al. 2019; Grapputo et al. 2018; Jurenka 2021). The sequences of both enzymes are the same as that found in other tissues such as the fat body.

The enzymatic reaction of acetyl-CoA carboxylase uses acetyl-CoA, bicarbonate, and ATP to produce malonyl-CoA. The malonyl-CoA is a substrate for fatty acid synthetase in the elongation cycle to produce palmitoyl-CoA and stearoyl-CoA. Labeling studies using labeled acetate or glucose indicate that pheromone glands produce primarily palmitic and stearic acids (Tang et al. 1989; Foster and Anderson 2012). A study by Foster and Anderson (2012) demonstrated that uniformly labeled C13-glucose fed to female *Chloridia* (*Heliothis*) *virescens* moths was broken down to acetate and then incorporated into fatty acids in the pheromone gland. Foster and Anderson (2012) found enrichment in both palmitic and stearic acids. Higher incorporation was found in stearic acid indicating that fatty acid synthetase produces both fatty acids but higher amounts of the 18-carbon fatty acids. Some moths can utilize palmitic and stearic acids directly to produce 16- and 18-carbon pheromone molecules, but in many moths the carbon chain is shortened to 14, 12, and 10 carbons. These shorter chain fatty acid precursors are produced through limited chain-shortening enzymes.

2.2 Limited Chain-Shortening Enzymes

The chain shortening of long-chain fatty acids to specific shorter chain lengths has been described in insects (Stanley-Samuelson et al. 1988). The first studies demonstrating a pheromone biosynthetic pathway in moths determined that limited chain shortening was occurring in the pheromone gland (Bjostad and Roelofs 1983; Bjostad et al. 1984). A combination of desaturation and limited chain shortening can produce the major and minor pheromone components found in the pheromone gland of female *T. ni* (Fig. 1). A combination of desaturation and limited chain shortening can produce a variety of monounsaturated fatty acid precursors. These fatty acid precursors can be utilized to form specific molecules and used in a pheromone blend of components (Fig. 2).

Limited chain shortening of longer chain length fatty acids is thought to be similar to a ß-oxidation pathway but is located in peroxisomes. The enzymes acyl-CoA oxidase and 3-oxoacyl-CoA thiolase are involved in the chain-shortening process. The acyl-CoA oxidase is a multifunctional enzyme containing enoyl-CoA hydratase and

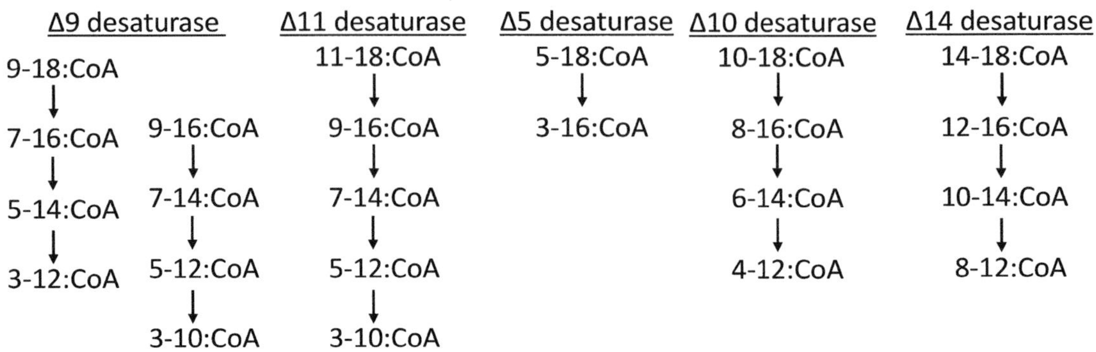

Fig. 2 Desaturation and limited chain shortening can produce a variety of monounsaturated acyl-CoA precursors to moth pheromone components. The down arrow indicates chain shortening by 2 carbons. Further modification of the acyl-CoAs by a reductase to a fatty alcohol followed by an acetyltransferase or oxidase can produce a large variety of moth pheromones in species-specific ratios

3-hydroxyacyl-CoA dehydrogenase activities (van den Bosch et al. 1992). The longer chain fatty acids are shortened by 2 carbons for every round via the enzyme complex. Limited chain shortening was first demonstrated by Bjostad and Roelofs (1983; Bjostad et al. 1984) using fatty acid precursors where the terminal methyl carbon was either carbon 13 or 14. The labeled fatty acid precursors were applied to pheromone glands and incorporation into pheromone molecules was determined. The terminal position of the labeled carbon indicated that the longer chain precursor had to be shortened in order to end up in the pheromone molecule.

A mutant line of cabbage loopers, *T. ni*, was found to have an increased amount of a minor component, Z9–14:OAc over the major component Z7–12:OAc (Haynes and Hunt 1990). The increased amount of Z9–14:OAc indicates that the precursor fatty acid Z11–16:CoA was chain shortened by only one round (Fig. 1). This was confirmed in a cell-free enzyme assay using pheromone glands from the mutant cabbage loopers (Jurenka et al. 1994). The in vitro enzyme assay included microsomes from pheromone glands, acyl-CoA derivatives with the terminal carbon deuterium labeled, and NAD as a cofactor. Results demonstrated that only one round of chain shortening was occurring in the mutant line. In wild-type moths Z11–16:CoA was chain shortened by two rounds to Z7–12:CoA and Z11–18:CoA was chain shortened by three rounds to Z5–12:OAc (Jurenka et al. 1994).

A recent study has demonstrated limited chain shortening using two genes encoding acyl-CoA oxidases identified from *Lobesia botrana* pheromone glands and heterologously expressed in Sf9 cells (Ding et al. 2021). When Z11–14:Me was used as a substrate, a significant amount of Z9–12:acid was extracted from the Sf9 cells. This study indicates that these two acyl-CoA oxidases are involved in the limited chain-shortening reactions found in moth pheromone glands. With the identification of these acyl-CoA oxidases, the molecular mechanism behind the limited chain-shortening reactions occurring in pheromone glands still needs further investigation. An important question is why, in some cases, only one round of chain shortening occurs, while in others, two or three rounds take place.

2.3 Desaturases

Double bonds are inserted into fatty acid chain by desaturases and several have been characterized that are specifically expressed in moth pheromone glands. The desaturases can introduce double bonds into saturated or monounsaturated fatty acids. The desaturases are named based on where the double bond is introduced. So far, Δ5 (Foster and Roelofs 1996), Δ6 (Wang et al.

2010), Δ9 (Löfstedt and Bengtsson 1988; Martinez et al. 1990), Δ10 (Foster and Roelofs 1988), Δ11 (Bjostad and Roelofs 1981, 1983), Δ13 (Villorbina et al. 2003), and Δ14 (Zhao et al. 1990) desaturases have been identified. A variety of pheromone fatty acid precursors can be produced through the action of desaturases and limited chain shortening (Fig. 2). Some of the fatty acid precursors could be produced in two different ways. For example, Δ9 fatty acid precursors could be produced with a Δ9 desaturase or by chain shortening of a Δ11 precursor made with a Δ11 desaturase. The Δ5 desaturase could produce Z5 precursors or they could be produced by two rounds of chain-shortening Δ9 precursors.

Identification of the genes encoding desaturases led to their functional heterologous expression which conclusively identifies the type of desaturase. The first desaturase identified was a Δ11 desaturase from the cabbage looper, *T. ni* (Knipple et al. 1998). Genes encoding desaturases were identified based on sequence similarity to known mammalian desaturases (Knipple and Roelofs 2003). The highly conserved core region contains three histidine-rich sequence motifs that coordinate the two iron ions in the active center of the desaturase protein. Functional expression utilized a yeast strain that lacks an endogenous desaturase and the yeast could grow without added monounsaturated fatty acids by adding the gene from insect pheromone glands. If the desaturase was functional the yeast would multiply and start producing monounsaturated fatty acids. Extraction of the fatty acids and analysis of the methyl esters by GC/MS would help determine which chain-length fatty acid was used as a substrate and where the double bond was introduced. Using this mutant yeast line a number of desaturases have been functionally expressed and identified from moth pheromone glands.

Based on the functional expression and identification of resulting active desaturation, several phylogenetic relationships are apparent (Fig. 3). The largest group are the Δ11 desaturases that include the Δ10, Δ6, and terminal desaturases. The Δ9 desaturases can be divided into those that prefer longer chain length fatty acid precursors and those that prefer palmitate over stearate and vice versa. The few Δ14 desaturases that have been identified are more basal. Multifunctional desaturases that produce Δ11 double bonds can act sequentially to form conjugated dienes or other dienes. A desaturase with Δ11 activity was found in *Spodoptera exigua* that had Δ12 activity (Xia et al. 2019). It was known previously that Z9, E12–14:OAc is made through Δ12 desaturation (Jurenka 1997) but functional expression of the Δ11 desaturase from *S. exigua* demonstrated that the same enzyme introduces the Δ12 double bond. In this case, the multifunctional Δ11 desaturase produces Z11–16:CoA which is chain shortened to Z9–14:CoA and then the same Δ11 desaturase introduces the E12 double bond to make Z9, E12–14:CoA. The latter is then reduced and acetylated to make Z9, E12–14:OAc (Xia et al. 2019).

Primitive moths also utilize pheromones that are produced using a Δ11 desaturase (Liénard et al. 2008). In the moth, *Lampronia capitella* (Prodoxidae), the Δ11 desaturase produces Z11–16:CoA which is chain shortened to Z9–14:CoA and the same Δ11 desaturase introduces another double bond to make Z9, Z11–14:CoA. The presence of a Δ11 desaturase in this primitive moth lineage indicates that Δ11 desaturases evolved early in the evolution of moths (Liénard et al. 2008). A Δ11 desaturase was also found in the butterfly, *Bicyclus anyanna* (Liénard et al. 2014). The latter two studies indicate that the Δ11 desaturase gene family is conserved throughout Lepidoptera.

2.4 Fatty Acyl Reductase

Once a specific chain length and unsaturated acyl:CoA intermediate precursor is produced in the pheromone biosynthetic pathway, a functional group replaces CoA on the carbonyl carbon. The functional group in the majority of Type I pheromone molecules are acetate esters, aldehydes, or alcohols. The next step in the biosynthetic pathway to form all of these functional groups is the formation of a primary alcohol. The primary alcohol is

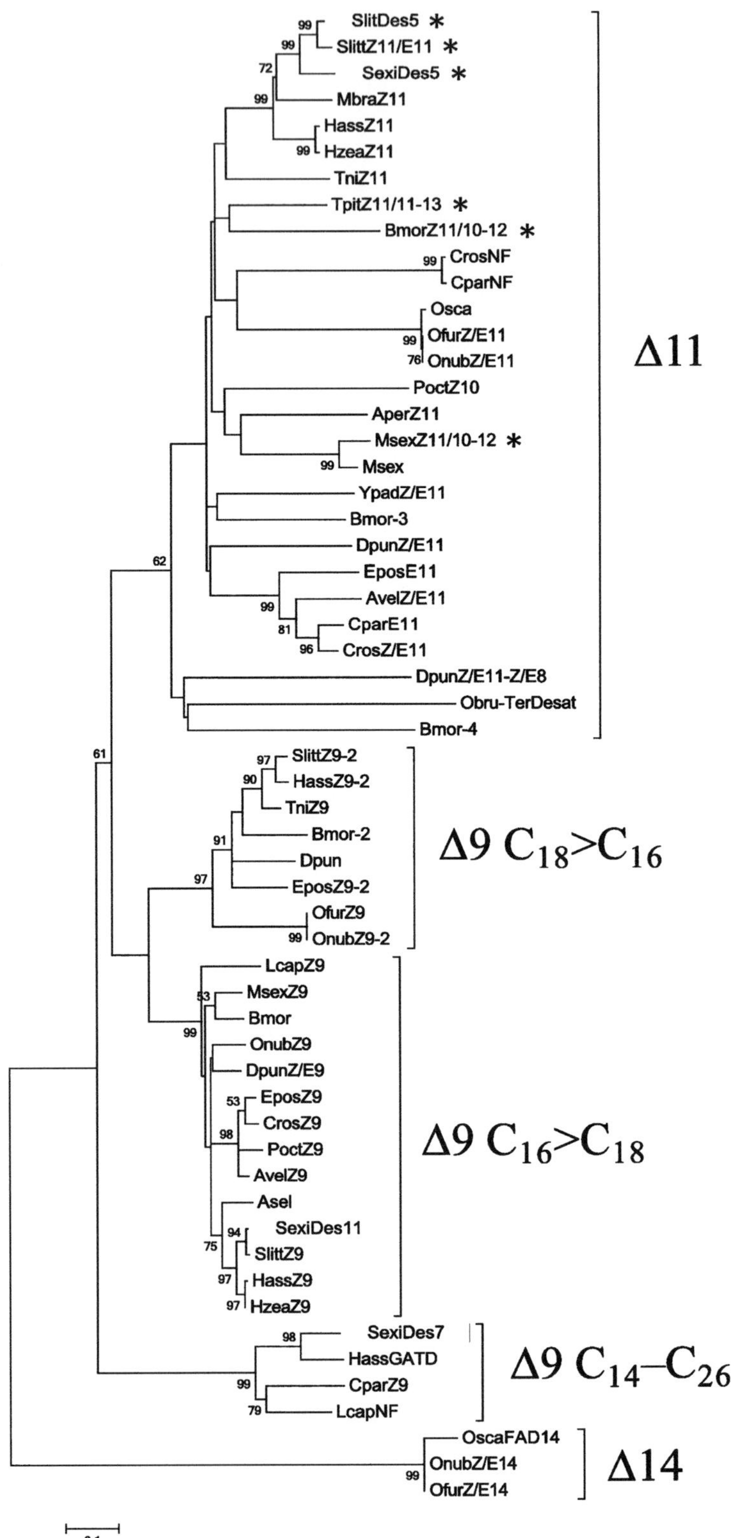

Fig. 3 Phylogenetic relationship of fatty-acyl-CoA desaturases identified from moth pheromone glands. The specificity of the desaturase families is indicated. Bifunctional desaturases are indicated with an asterisk. Modified from Xia et al. (2019)

catalyzed by a pheromone gland-specific fatty acyl CoA reductase (pgFAR). The first pgFAR was identified from the silkworm moth, *B. mori*, that produced bombykol, E10, Z12–16:OH (Moto et al. 2003). The gene was expressed in yeast cells and supplemented with E10, Z12–16: acid. After incubation, the bombykol was extracted. Since then, a number of pgFAR genes have been cloned and functionally expressed in yeast cells (Hagström et al. 2013; Holkenbrink et al. 2020) and plants (Löfstedt and Xia 2021; Xia et al. 2022). Pheromone gland-specific FARs are phylogenetically aligned in one group (Fig. 4).

In some cases the pgFAR was shown to have broad substrate specificity as was shown with three species of *Yponomeuta* (Liénard et al. 2010). Broad substrate specificity was also found in the pgFARs from five species of heliothines (Dou et al. 2020; Hagström et al. 2012), *Spodoptera littoralis* (Carot-Sans et al. 2015), and *Agrotis segetum* (Ding and Löfstedt 2015). In *Spodoptera exigua*, one pgFAR was found to

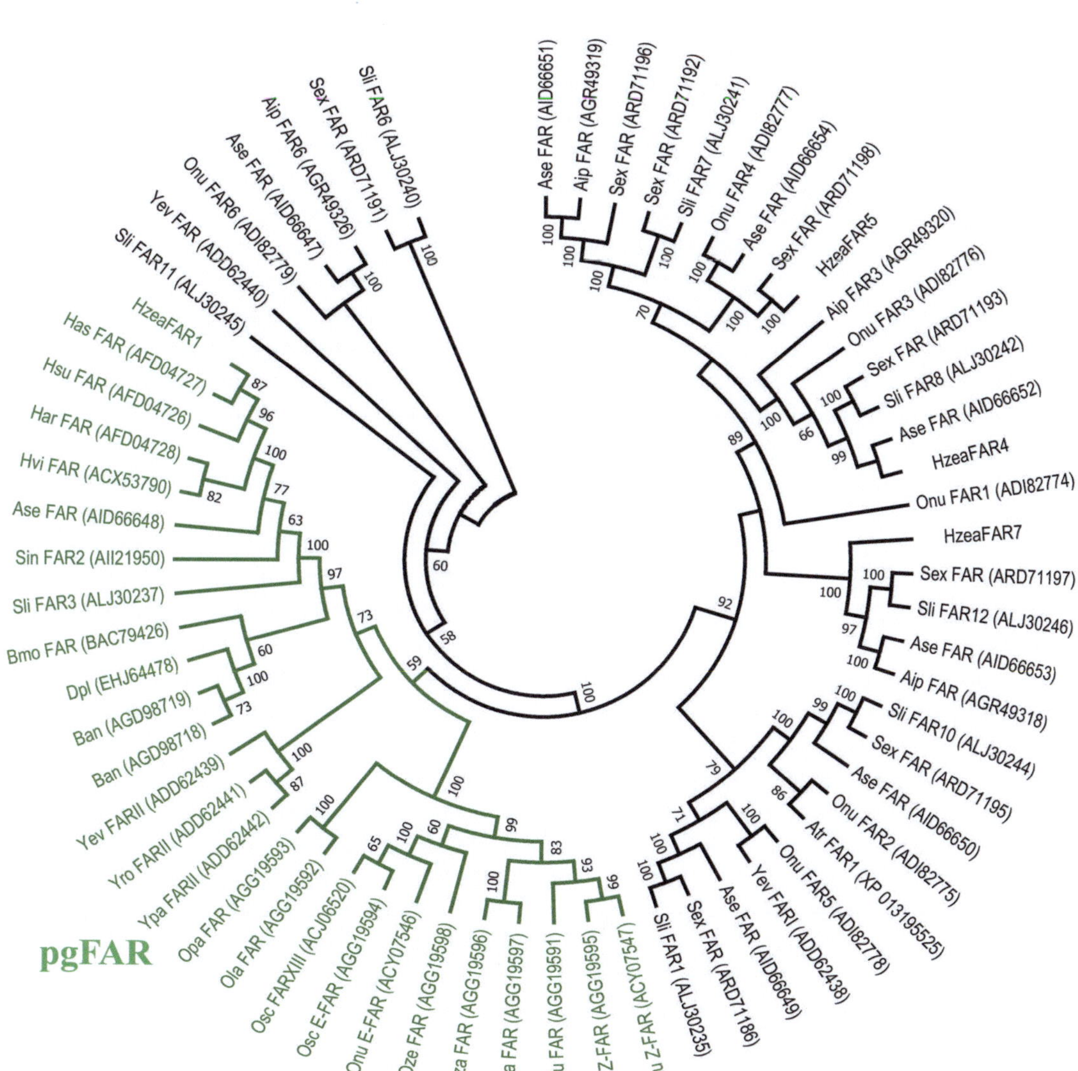

Fig. 4 Phylogenetic relationship of fatty acyl reductases from Lepidoptera. Pheromone gland-specific FARs are highlighted in green. Modified from Dou et al. (2020)

have broad activity for 14 carbon acyl CoAs, while another was specific for 16 carbon acyl CoAs (Antony et al. 2016). In the case of pgFARs from *Ostrinia* moths, they were found to highly specific for certain substrates (Lassance et al. 2013).

In *Ostrinia nubilalis*, there are two populations or strains, one using Z11–14:OAc and the other using E11–14:OAc as the main pheromone components. In both strains, the biosynthetic pathway involves a common Δ11 desaturase that produces Z and E11–14:CoA precursors in a ratio of about 70/30 Z/E (Foster 2004; Roelofs et al. 1987). In order to produce strain-specific acetate ester isomers it was thought that the pgFARs were specific for the Z or E isomer. The pgFARs were cloned from each population and heterologous expression determined the specificity of the reductase (Lassance et al. 2010). Each population had a strain-specific pgFAR that preferentially reduces the Z isomer or E isomer to the corresponding alcohol. Thus, the two strains became isolated due to the production of the Z or E pheromone isomers. Another study compared the functional expression of pgFARs from 8 species of *Ostrinia* and found that most pgFARs were specific for certain fatty acyl precursors to produce the alcohol in the proper ratio for a particular species (Lassance et al. 2013). This indicates that the pgFAR of *Ostrinia* moths is a driving factor in the speciation of the genera.

In some moths, the main pheromone component is a primary alcohol and thus further modification is not required. In most moths, the alcohol is utilized by another enzyme to form a different functional group which includes acetyltransferases and oxidases.

2.5 Acetyltransferase

After the pgFAR produces a primary alcohol another enzyme, an acetyl-CoA:fatty alcohol acetyltransferase will produce acetate esters that are used as the primary pheromone component. The enzyme has been characterized in vitro in several tortricid moths (Morse and Meighen 1987a, b; Jurenka and Roelofs 1989). It uses the primary alcohol and acetyl-CoA as substrates to produce fatty acetate ester. The enzyme found in tortricid moths is specific for Z isomers of monounsaturated fatty alcohols. In contrast, no apparent specificity was found in several noctuid moths (Bestmann et al. 1987; Teal and Tumlinson 1987). Although this enzyme has been characterized in vitro as a membrane-bound enzyme found in pheromone glands, the gene encoding this enzyme has not been identified (Jurenka and Roelofs 1989; Jurenka 2021).

2.6 Fatty Alcohol Oxidase

Aldehydes are produced by the oxidation or dehydrogenation of fatty alcohols (Rizzo 2014). The oxidase has not been identified at the protein or genetic level despite numerous transcriptome studies conducted on moth pheromone glands (Jurenka 2021). It has been characterized by applying alcohols directly to intact or excised pheromone glands of species that utilize only aldehydes as pheromone components (Fang et al. 1995; Hoskovec et al. 2002; Luxova and Svatos 2006; Teal and Tumlinson 1988). This is a unique enzyme in that it is connected to the cuticle and is active in hexane and buffer but not other organic solvents (Teal and Tumlinson 1988; Luxova and Svatos 2006). The substrate specificity has been investigated and shown to convert a variety of primary alcohols to aldehydes (Luxova and Svatos 2006; Kiyota et al. 2011). Location of aldehydes and alcohols in the pheromone gland was investigated in *Chloridea* (*Heliothis*) *virescens* (Foster and Anderson 2019). Fatty alcohols were found inside the gland and the fatty aldehyde pheromones were found on the cuticle overlaying the gland. These results taken together indicate that the oxidase is found in the cuticle and fatty alcohols are produced to generate fatty aldehydes that will be released once the pheromone gland is exposed.

3 Coleoptera

Currently, the Coleoptera consists of the largest order of described insect species. This huge diversity is also reflected in the variations in the types of pheromones that have been identified from beetles. Identification of the biosynthetic pathways for producing beetle pheromones has concentrated on the mevalonate and fatty acid pathways. A considerable amount of information about bark beetle pheromone biosynthesis and the mevalonate pathway is available (Keeling et al. 2021; Tittiger and Blomquist 2016, 2017). Pheromones were demonstrated to be produced through fatty acid biosynthetic pathways, including cucujolide II through X, buibuilactone, japonilure, and *exo*-brevicomin (Seybold and Vanderwel 2003).

The biosynthesis of the sex pheromones buibuilactone and japonilure was investigated in the scarab beetles, *Anomala cuprea* and *Anomala osakama* (Leal et al. 1999). The biosynthesis of buibuilactone in female *A. cuprea* starts with the Δ9 desaturation of palmitate and then a stereospecific hydroxylation on C8 (Fig. 5). Two rounds of chain shortening produce a Z5-4OH-12:acid that undergoes cyclization to the lactone (*R*)-buibuilactone. The biosynthesis of japonilure occurs in the same manner except that the starting fatty acid is stearate (Leal et al. 1999). The *R* and *S* isomers are produced through the stereoselective introduction of the hydroxy group on C8. Different scarab beetles produce either the *R* or *S* isomers. *A. osakana* produces (*S*)-japonilure and several scarabs including *Popillia japonica* produce (*R*)-japonilure (Leal 2001). (*R*)-Japonilure has been identified in several species of *Popillia* and *Anomala* in the Scarabaeidae and two species in the Cetoniidae (pherobase.com). Buibuilactone is primarily found in several species of *Anomala*.

Several species of flat bark beetles in the families Silvanidae and Laemophloeidae produce macrocyclic lactones called cucujolides (Oehlschlager et al. 1988). These are aggregation pheromones attracting both sexes. The cucujolides II through X (Oehlschlager et al. 1988; Hötling et al. 2014) are produced through a fatty acid biosynthetic pathway (Fig. 6a). Cucujolide II, produced by the rusty grain beetle, *Cryptolestes ferrugineus*, and the merchant grain beetle, *Oryzaephilus mercator*, starts with oleate followed by three rounds of chain shortening to Z3–12:CoA, followed by hydroxylation on C11 and cyclization to form the lactone. The other cucujolides could be produced from palmitic or linoleic acids as starting fatty acids (Vanderwel et al. 1990). Macrolactones found in other insects are probably produced in a similar biosynthetic pathway starting with fatty acids (Schulz and Hötling 2015).

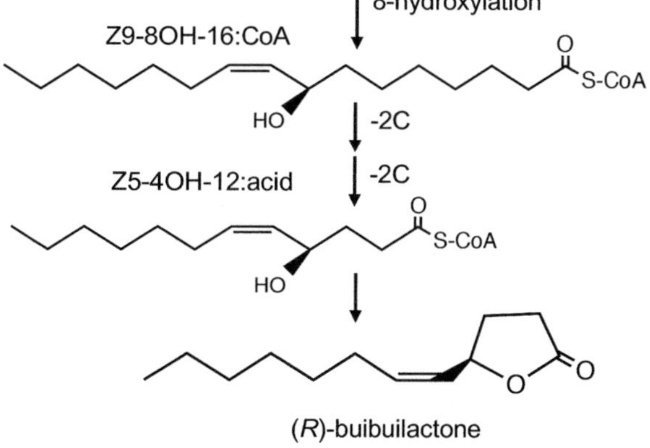

Fig. 5 Biosynthesis of (*R*)-buibuilactone in scarab beetles

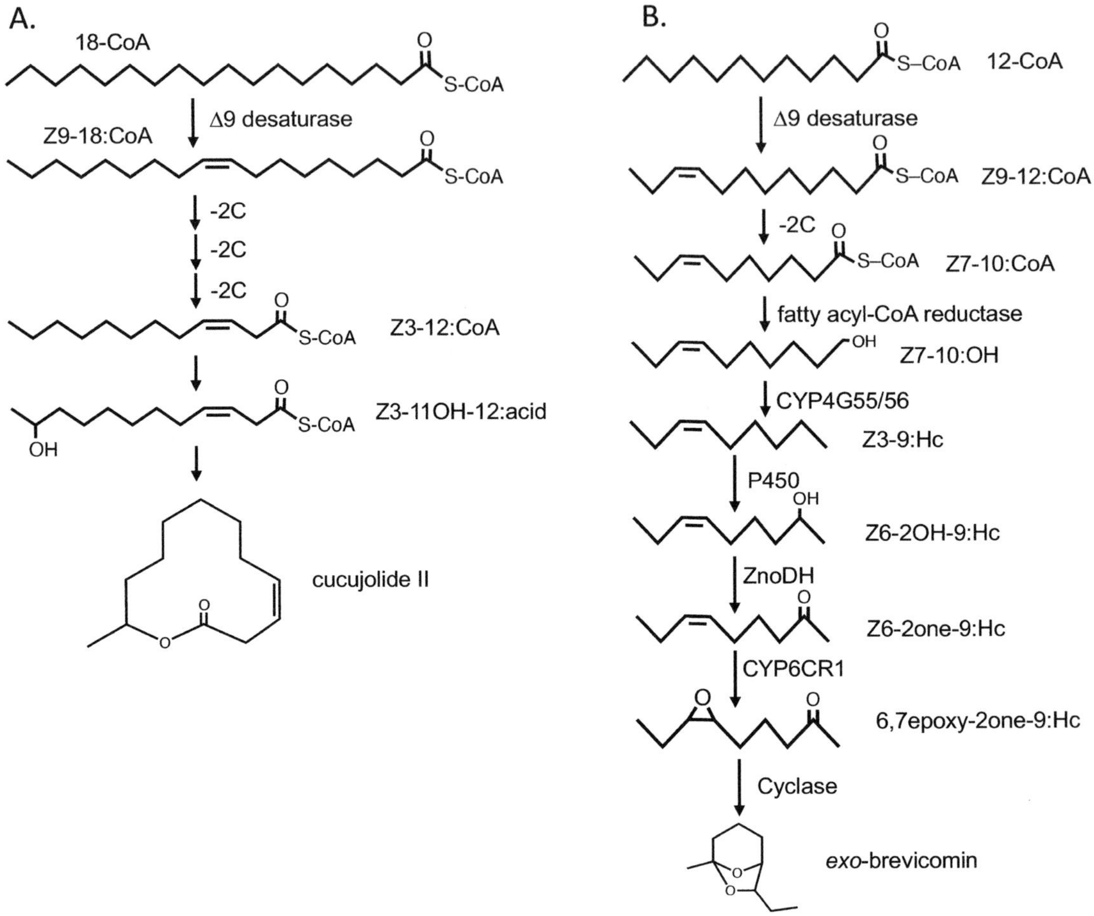

Fig. 6 Biosynthesis of cucujolide II (**a**) and *exo*-brevicomin (**b**) in grain beetles and *Dendroctonus* bark beetles, respectively. −2C: chain shortening by 2 carbons; ZnoDH: (*Z*)-non-6-en-2-ol dehydrogenase. The cucujolide II pathway is based on Vanderwel et al. (1990) and the *exo*-brevicomin pathway is based on Song et al. (2014a, b)

Many bark beetles produce an aggregation pheromone through a mevalonate pathway (Keeling et al. 2021). In *Dendroctonus* bark beetles, males produce a pheromone that synergizes with the female aggregation pheromone. The biosynthesis has been demonstrated in *Dendroctonus ponderosae* (Song et al. 2014a, b) (Fig. 6b). The pathway most likely starts with Δ9 desaturation of dodecanoyl-CoA to produce a ω-3 12C precursor followed by chain shortening to the ω-3 10C CoA. Although this was not indicated in Song et al. (2014b) it is the most likely way to make a ω-3 10C fatty acid. Then a fatty acyl-CoA reductase produces the alcohol which is used by CYP4G55 or 56 to make the 9C hydrocarbon (MacLean et al. 2018). Hydroxylation by an undescribed P450 is followed by (Z)-non-6-en-2-ol dehydrogenase (ZnoDH) to make the ketone. Another P450, CYP6CR1, produces the epoxide (Song et al. 2014a) and the epoxy-ketone is cyclized to *exo*-brevocomin.

4 *Drosophila melanogaster* Acetate Ester

Several types of pheromones are involved in chemical communication in *Drosophila melanogaster* (Montagne and Wicker-Thomas

2021). Hydrocarbons are the main sex pheromones in *Drosophila* and are produced through fatty acid precursors as discussed in Sect. 6. Other pheromones are involved in mating and are produced through fatty acid biosynthesis. Three fatty acids, methyl laurate, methyl myristate, and methyl palmitate were found to be involved in mate attraction in both sexes (Dweck et al. 2015). Another compound involved in aggregation and courtship is Z11–18:OAc (vaccenyl acetate). It is produced in the male ejaculatory bulb and transferred to females during mating (Cortot et al. 2022; Guiraudie-Capraz et al. 2007). This pheromone is produced by chain elongation of Z9–16:CoA to Z11–18:CoA by a specific elongase (Chertemps et al. 2006) and then followed by a reductase and an acetyltransferase.

5 Hymenoptera – Honey Bees and *Nasonia*

5.1 *Apis mellifera* – Honey Bees

The honey bee, *Apis mellifera*, lives in a highly social colony that, in part, is maintained by the use of pheromones. The queen and workers produce a variety of pheromones from various glands, and most studies focused on the biosynthesis of mandibular gland pheromones. Other pheromone components from other glands have been identified and are of obvious fatty acid origin (see Mumoki and Crewe 2021 for a recent review of honey bee pheromones). The tergal gland contains oleic and stearic acids (Wossler and Crewe 1999) and the Dufour's gland contains hydrocarbons and wax esters (Katzav-Gozansky et al. 1997). The developing larvae also produce pheromones that are found on their cuticles and consist of a blend of fatty acid esters (Le Conte et al. 1989).

The biosynthesis of the queen and worker mandibular gland compounds that originate from fatty acids has been studied (Fig. 7). Both queens and workers can produce the compounds shown in Fig. 7 in addition to other compounds not derived from stearic acid (Keeling et al. 2001). Queens primarily produce compounds originating from ω-1 hydroxylation of stearic acid and workers produce compounds originating from ω hydroxylation of stearic acid (Plettner et al. 1996, 1998). Workers primarily produce 10-hydroxydec-2-enoic acid (10-HDA) and 10-hydroxydecanoic acid (10-HDAA). These are the major acids found in Royal Jelly and can be produced at about 5–10 μg per worker (Keeling et al. 2001; Yusuf et al. 2015). The queen produces larger amounts (100–200 μg) of 9-oxo-2-decanoic acid (9-ODA) and lower amounts (50–100 μg) of 9-hydroxydec-2-enoic acid (9-HDA) (Keeling et al. 2001; Yusuf et al. 2015).

The biosynthetic pathway starts with hydroxylation of stearic acid on the terminal, ω carbon, or the penultimate, ω-1 carbon. This hydroxylation reaction is presumably catalyzed by cytochrome P450s (Plettner et al. 1998). Recently, a specific P450, CYP6AS8, was identified in worker mandibular glands and was shown to be involved in the production of 10-HDA after RNAi knockdown of *CYP6AS8* (Wu et al. 2020). Although other P450s have been implicated in the hydroxylation reactions based on differential expression in queens and workers (Malka et al. 2009; Wu et al. 2017; Fujita et al. 2018). After hydroxylation, four rounds of chain shortening occur to produce 10C fatty acids. The 10C acids are then dehydrogenated at the 2 position followed by oxidation to give diacids and 9-ODA. Based on the amounts of components found in queen or worker mandibular glands, caste-specific hydroxylation and chain shortening occur. This has been confirmed with labeling studies (Plettner et al. 1998) and transcriptomic studies (Malka et al. 2009; Wu et al. 2017; Fujita et al. 2018).

5.2 Nasonia

The wasp genus *Nasonia* (Pteromalidae) is a parasitoid of pupae of higher flies. The males of all four species in the genera utilize sex pheromones to attract virgin females. The sex pheromone is deposited on the surfaces of host pupae to arrest

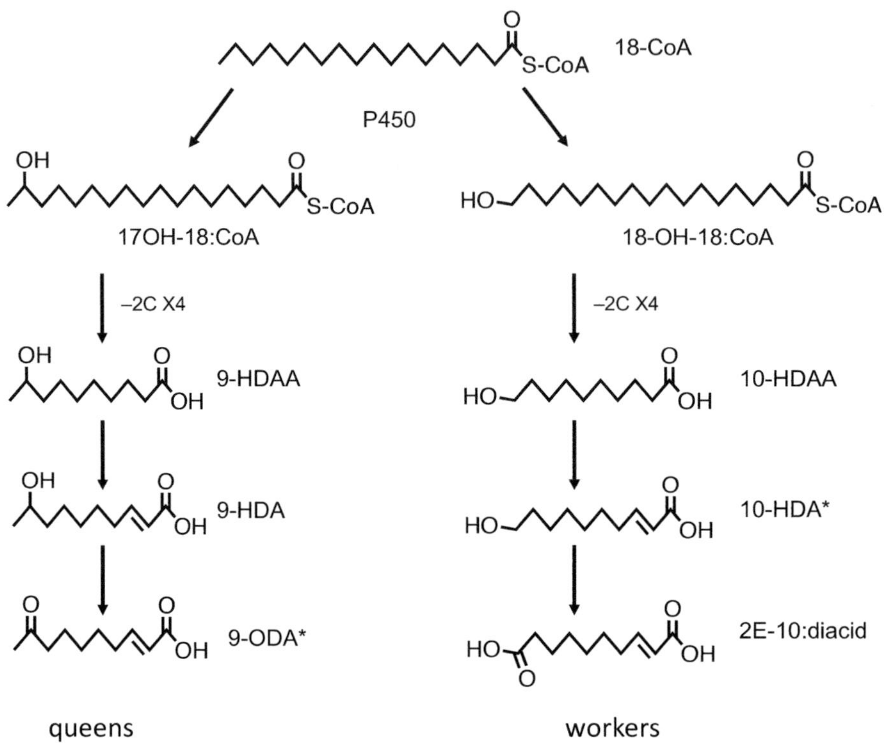

Fig. 7 Biosynthetic pathways for the production of the main fatty acid-derived pheromone components in the honey bee mandibular glands. The primary queen pheromones are on the left and the worker pheromones are on the right. The most abundant acids are marked with an asterisk. HDA: hydroxydec-2-enoic acid; HDAA: hydroxydecanoic acid; ODA: oxo-2-decanoic acid. Based on Plettner et al. (1996, 1998)

emerging females which are then courted by males. The sex pheromone arrestant consists of (4*R*,5*S*)-5-hydroxy-4-decanolide (5-OH-10:lactone) and 4-methylquinazoline. In *Nasonia vitripennis* the (4*R*,5*R*) isomer is also found. The biosynthetic pathway for production of the lactone has been extensively studied. A detailed account of *Nasonia* pheromone biosynthesis can be seen in Ruther et al. (2021).

The biosynthesis of the main pheromone component 5-OH-10:lactone starts with fatty acid biosynthesis. The interesting finding is that linoleic acid is a precursor and males can make linoleic acid de novo (Blaul et al. 2014). This finding indicates a Δ12 desaturase. Based on gene identification of potential desaturases (Wang et al. 2015), two potential Δ12 desaturases were functionally expressed in yeast cells (Semmelmann et al. 2019). Both gene products were characterized as Δ12 desaturases with one having a higher expression in males. Recent research, however, demonstrated that *Nasonia* females can also make linoleic acid de novo as well as other parasitic Hymenoptera (Broschwitz et al. 2021). It seems that the Δ12 desaturase is widely distributed in at least 5 families of parasitic Hymenoptera. Also of interest is the phylogenetic relationship of the Δ12 desaturases with other insect Δ12 desaturases that have been functionally identified from the red flour beetle and house cricket. It seems that the insect Δ12 desaturases do not share a common evolutionary origin, but they have evolved separately in different taxa (Semmelmann et al. 2019). The Δ12 desaturases are not closely related to the multifunctional Δ11 desaturases from Lepidoptera that can also introduce a second double bond on the methyl end of fatty acids.

The next step in the biosynthesis of the *Nasonia* pheromone is to make the epoxy on the

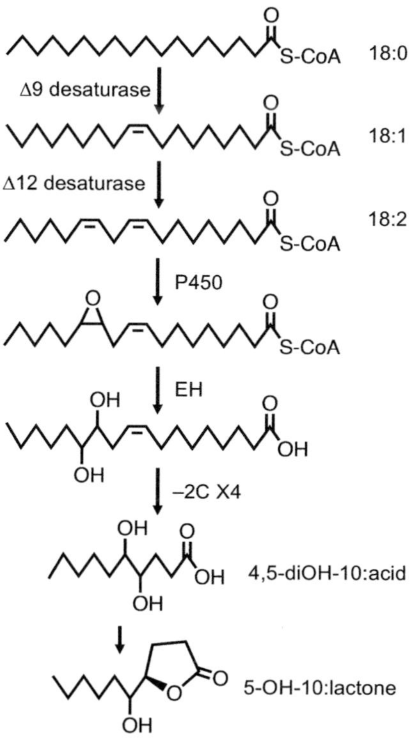

Fig. 8 Proposed biosynthetic pathway of the major sex pheromone components in *Nasonia vitripennis* males. After the formation of linoleic acid (18:2) by the sequential action of a Δ9 and Δ12 desaturase, an unidentified P450 enzyme makes the 12,13 epoxide. The epoxide hydrolase (EH) opens the epoxide to the dihydroxy which is followed by 4 rounds of chain shortening (–2C X4). The resulting dihydroxydecanoic acid undergoes cyclization to form the lactone

Δ12 double bond presumably using a P450 enzyme (Fig. 8). The configuration can be either *RR* or *RS*. Next, an identified epoxide hydrolase opens up the epoxide to two hydroxy groups (Abdel-Latief et al. 2008). Then, four rounds of limited chain shortening produce a 10-carbon dihydroxy acid. The final step is the cyclization to form the lactone. This could be an enzymatic step or possibly it may occur spontaneously (Ruther et al. 2021).

6 Hydrocarbons

The cuticular surface of most insects contains a complex mixture of hydrophobic lipids (Blomquist 2010; Blomquist and Ginzel 2021; Holze et al. 2021). Hydrocarbons are found on the cuticular surface of most insects where their primary function is to prevent desiccation. An additional function is in chemical communication where cuticular hydrocarbons can be used as volatile or nonvolatile semiochemicals. For reviews on the role of hydrocarbons in insect communication see Blomquist and Bagnères (Blomquist and Bagneres 2010), Ginzel et al. (2021), and Blomquist and Ginzel (2021). Hydrocarbons are biosynthesized through a pathway involving fatty acid biosynthesis, desaturases, chain elongation, reductase, and a unique cytochrome P450 reaction to remove the carbonyl carbon (Ginzel et al. 2021).

Hydrocarbons are thought to be exclusively biosynthesized in oenocytes, which are specialized cells associated with epidermal or fat body cells in the abdomen. They are then transported to the cuticle or pheromone gland by lipophorin (Schal et al. 1998). The hydrocarbon can be released without modification or it can be modified with P450s to form epoxy pheromones (Jurenka et al. 2003).

In the oenocytes, fatty acid synthase can make saturated fatty acids or preformed saturated or unsaturated fatty acids can be taken up by oenocytes and elongated. If methyl-branched hydrocarbons are produced, a specific fatty acid synthase is involved (Pei et al. 2019; Yang et al. 2020). The 16- or 18-carbon fatty acids are chain-elongated 2 carbons at a time using elongases (Ng et al. 2015; Dembeck et al. 2015; Li et al. 2019; Pei et al. 2021). The elongase lengthens the fatty acid to a specific chain length that is required before producing the hydrocarbon.

The next step is the reduction of the long-chain fatty acyl-CoA to a primary alcohol. This is catalyzed by a fatty acyl reductase as described in Sect. 2.4 in the biosynthesis of moth pheromones. Oenocytes contain a large number of reductases, some of which are involved in producing primary alcohols (Finet et al. 2019). The primary alcohol is then utilized by a specific cytochrome P450 in the CYP4G family to decarbonylate the alcohol to the hydrocarbon that is one carbon shorter (Ginzel et al. 2021). Based on genome sequences, members of the CYP4G family are found in all insects (Feyereisen 2020). CYP4G1 and CYP4G2 from

D. melanogaster were the first hydrocarbon-producing P450s functionally expressed (Qiu et al. 2012). The decarbonylation mechanism first involves the oxidation of the fatty alcohol to an aldehyde and then the aldehyde is decarbonylated to the hydrocarbon (MacLean et al. 2018). Hydrocarbons usually have an odd number of carbons due to the majority of fatty acids having an even number of carbons. A few hydrocarbon pheromones could have an even number of carbons and these are produced by an α-oxidation reaction to produce an odd-numbered fatty acid (Goller et al. 2007). The odd-numbered fatty acid is then decarbonylated to the even chain length hydrocarbon by the just described CYP4G P450.

Many insects utilize hydrocarbons as pheromones without modifications (Löfstedt et al. 2016; Blomquist and Bagneres 2010). Some insects will produce hydrocarbons with epoxy groups with the epoxidation step occurring in the pheromone gland. The epoxidation is catalyzed by a specific cytochrome P450 that is usually found in the pheromone gland (Rong et al. 2014, 2019a, b).

7 Conclusion

This chapter gave a review of the biosynthesis of insect pheromones that originate with a fatty acid biosynthetic pathway. The majority of research has occurred in the orders Lepidoptera, Coleoptera, and Hymenoptera. Pheromones from other orders of insects are most likely biosynthesized through a fatty acid pathway. The pathways outlined in this chapter provide a foundation for an understanding of how other pheromones could be biosynthesized using fatty acids. The modifications that occur through "normal" fatty acid metabolism include limited chain shortening and modification of the functional group to form specific types of pheromones. As more pheromones are identified, it should become apparent if they are biosynthesized through a fatty acid pathway.

Further research is required to identify some key biosynthetic enzymes involved in producing insect pheromones. The enzymes that are involved in making acetate esters and aldehydes from primary fatty alcohols have not been identified at the genetic level. Identification of the genes encoding these enzymes has proved to be elusive despite attempts to functionally express likely candidates (Xia et al. 2022). Further research on the chain-shortening enzymes will yield insights into the mechanisms behind the occurrence of single versus multiple rounds of chain shortening in the production of pheromone precursors. Additionally, more P450 enzymes involved in hydrocarbon biosynthesis and oxygen modifications will be identified. The discovery of pheromones in orders where few have been identified will offer new avenues for exploring biosynthetic pathways in future research.

References

Abdel-Latief M, Garbe LA, Koch M, Ruther J (2008) An epoxide hydrolase involved in the biosynthesis of an insect sex attractant and its use to localize the production site. Proc Natl Acad Sci USA 105:8914–8919

Allison JD, Cardé RT (2016) Pheromone communication in moths: evolution, behavior, and application. University of California Press, Oakland, California, p 401

Antony B, Ding B-J, Moto K, Aldosari SA, Aldawood AS (2016) Two fatty acyl reductases involved in moth pheromone biosynthesis. Sci Rep 6:29927

Bestmann HJ, Herrig M, Attygalle AB (1987) Terminal acetylation in pheromone biosynthesis by *Mamestra brassicae* L. (Lepidoptera: Noctuidae). Experientia 43:1033–1034

Bjostad LB, Roelofs WL (1981) Sex pheromone biosynthesis from radiolabeled fatty acids in the redbanded leafroller moth. J Biol Chem 256:7936–7940

Bjostad LB, Roelofs WL (1983) Sex pheromone biosynthesis in *Trichoplusia ni*: key steps involve delta-11 desaturation and chain-shortening. Science 220:1387–1389

Bjostad LB, Linn CE, Du J-W, Roelofs WL (1984) Identification of new sex pheromone components in *Trichoplusia ni*, predicted from biosynthetic precursors. J Chem Ecol 10:1309–1323

Blaul B, Steinbauer R, Merkl P, Merkl R, Tschochner H, Ruther J (2014) Oleic acid is a precursor of linoleic acid and the male sex pheromone in *Nasonia vitripennis*. Insect Biochem Mol Biol 51:33–40

Blomquist GJ (2010) Biosynthesis of cuticular hydrocarbons. In: Blomquist GJ, Bagneres A-G (eds)

Insect hydrocarbons: biology. Biochemistry and Chemical Ecology. Cambridge University Press, Cambridge, pp 35–52

Blomquist GJ, Bagneres A-G (2010) Insect hydrocarbons: biology, biochemistry, and chemical ecology. Cambridge University Press

Blomquist GJ, Ginzel MD (2021) Chemical ecology, biochemistry, and molecular biology of insect hydrocarbons. Annu Rev Entomol 66:45–60

Broschwitz B, Prager L, Pokorny T, Ruther J (2021) De novo biosynthesis of linoleic acid is widespread in parasitic wasps. Arch Insect Biochem Physiol 107: e21788

Butenandt A, Beckman R, Stamm D, Hecker E (1959) Uber den Sexuallockstoff des Seidenspinner Bombyx mori, Reidarstellung und Konstitution. Z Naturforsch 14B:283–284

Carot-Sans G, Muñoz L, Piulachs MD, Guerrero A, Rosell G (2015) Identification and characterization of a fatty acyl reductase from a *Spodoptera littoralis* female gland involved in pheromone biosynthesis. Insect Mol Biol 24:82–92

Chertemps T, Duportets L, Labeur C, Ueyama M, Wicker-Thomas C (2006) A female-specific desaturase gene responsible for diene hydrocarbon biosynthesis and courtship behaviour in *Drosophila melanogaster*. Insect Mol Biol 15:465–473

Cortot J, Farine J-P, Cobb M, Everaerts C, Ferveur J-F (2022) Factors affecting the biosynthesis and emission of a *drosophila* pheromone. J Exp Biol 225:jeb244422

Dembeck LM, Böröczky K, Huang W, Schal C, Anholt RRH, Mackay TFC (2015) Genetic architecture of natural variation in cuticular hydrocarbon composition in *Drosophila melanogaster*. elife 4:e09861

Ding B-J, Löfstedt C (2015) Analysis of the *Agrotis segetum* pheromone gland transcriptome in the light of sex pheromone biosynthesis. BMC Genomics 16: 711

Ding B-J, Xia Y-H, Wang H-L, Andersson F, Hedenström E, Gross J et al (2021) Biosynthesis of the sex pheromone component (E,Z)-7,9-dodecadienyl acetate in the European grapevine moth, *Lobesia botrana*, involving Δ11 desaturation and an elusive Δ7 desaturase. J Chem Ecol 47:248–264

Dou X, Liu S, Ahn S-J, Choi M-Y, Jurenka R (2019) Transcriptional comparison between pheromone gland-ovipositor and tarsi in the corn earworm moth *Helicoverpa zea*. Comp Biochem Physiol Part D Genomics Proteom 31:100604

Dou X, Zhang A, Jurenka R (2020) Functional identification of fatty acyl reductases in female pheromone gland and tarsi of the corn earworm, *Helicoverpa zea*. Insect Biochem Mol Biol. 116:103260

Dweck HK, Ebrahim SA, Thoma M, Mohamed AA, Keesey IW, Trona F et al (2015) Pheromones mediating copulation and attraction in *drosophila*. Proc Natl Acad Sci USA 112:E2829–E2835

Fabre J-H (1966) The insect world of J. Henri Fabre. In: Teale EW (ed) Translation of Alexander Teixeira de Mattos. Dodd, Mead, New York, 333 p

Fang N, Teal PEA, Tumlinson JH (1995) Characterization of oxidase(s) associated with the sex pheromone gland in *Manduca sexta* (L.) females. Arch Insect Biochem Physiol 29:243–257

Feyereisen R (2020) Origin and evolution of the CYP4G subfamily in insects, cytochrome P450 enzymes involved in cuticular hydrocarbon synthesis. Mol Phylogenet Evol 143:106695

Finet C, Slavik K, Pu J, Carroll SB, Chung H (2019) Birth-and-death evolution of the fatty acyl-CoA reductase (FAR) gene family and diversification of cuticular hydrocarbon synthesis in *drosophila*. Genome Biol Evol 11:1541–1551

Foster SP (2004) Fatty acid and sex pheromone changes and the role of glandular lipids in the Z-strain of the European corn borer, *Ostrinia nubilalis* (Hübner). Arch Insect Biochem Physiol 56:73–83

Foster SP, Anderson KG (2012) Synthetic rates of key stored fatty acids in the biosynthesis of sex pheromone in the moth *Heliothis virescens*. Insect Biochem Mol Biol 42:865–872

Foster SP, Anderson KG (2019) Production and distribution of aldehyde and alcohol sex pheromone components in the pheromone gland of females of the moth *Chloridea virescens*. J Chem Ecol 45:9–17

Foster SP, Roelofs WL (1988) Sex pheromone biosynthesis in the leafroller moth *Planotortrix excessana* by Δ10 desaturation. Arch Insect Biochem Physiol 8:1–9

Foster SP, Roelofs WL (1996) Sex pheromone biosynthesis in the tortricid moth, *Ctenopseustis herana* (Felder & Rogenhofer). Arch Insect Biochem Physiol 33:135–147

Fujita T, Kozuka-Hata H, Hori Y, Takeuchi J, Kubo T, Oyama M (2018) Shotgun proteomics deciphered age/division of labor-related functional specification of three honeybee (*Apis mellifera* L.) exocrine glands. PLoS One 13:e0191344

Ginzel MD, Tittiger C, MacLean M, Blomquist GJ (2021) Hydrocarbon pheromone production in insects. In: Blomquist GJ, Vogt RG (eds) Insect pheromone biochemistry and molecular biology, 2nd edn. Elsevier, London, pp 205–235

Goller S, Szöcs G, Francke W, Schulz S (2007) Biosynthesis of (3Z,6Z,9Z)- 3,6,9-octadecatriene: the main component of the pheromone blend of *Erannis bajaria*. J Chem Ecol 33:1505–1509

Grapputo A, Thrimawithana AH, Steinwender B, Newcomb RD (2018) Differential gene expression in the evolution of sex pheromone communication in New Zealand's endemic leafroller moths of the genera *Ctenopseustis* and *Planotortrix*. BMC Genomics 19:94

Guiraudie-Capraz G, Pho DB, Jallon J-M (2007) Role of the ejaculatory bulb in biosynthesis of the male pheromone cis-vaccenyl acetate in *Drosophila melanogaster*. Integr Zool 2:89–99

Hagström ÅK, Liénard MA, Groot AT, Hedenström E, Löfstedt C (2012) Semi–selective fatty acyl reductases from four heliothine moths influence the specific pheromone composition. PLoS One 7:e37230

Hagström ÅK, Wang H-L, Liénard MA, Lassance J-M, Johansson T, Löfstedt C (2013) A moth pheromone brewery: production of (Z)-11-hexadecenol by heterologous co-expression of two biosynthetic genes from a noctuid moth in a yeast cell factory. Microb Cell Factories 12:125

Haynes KF, Hunt RE (1990) A mutation in pheromonal communication system of cabbage looper moth, *Trichoplusia ni*. J Chem Ecol 16:1249–1257

Holkenbrink C, Ding B-J, Wang H-L, Dam MI, Petkevicius K, Kildegaard KR et al (2020) Production of moth sex pheromones for pest control by yeast fermentation. Metab Eng 62:312–321

Holze H, Schrader L, Buellesbach J (2021) Advances in deciphering the genetic basis of insect cuticular hydrocarbon biosynthesis and variation. Heredity 126:219–234

Hoskovec M, Luxová A, Svatoš A, Boland W (2002) Biosynthesis of sex pheromones in moths: stereochemistry of fatty alcohol oxidation in *Manduca sexta*. Tetrahedron 58:9193–9201

Hötling S, Haberlag B, Tamm M, Collatz J, Mack P, Steidle JLM et al (2014) Identification and synthesis of macrolide pheromones of the grain beetle *Oryzaephilus surinamensis* and the frog *Spinomantis aglavei*. Chem – A Eur J 20:3183–3191

Jones IF, Berger RS (1978) Incorporation of [1-14C] acetate into cis-7-dodecenyl acetate, a sex pheromone in the cabbage looper *Trichoplusia ni*. Environ Entomol 7:666–669

Jurenka RA (1997) Biosynthetic pathway for producing the sex pheromone component (Z,E)-9,12-tetradecadienyl acetate in moths involves a delta-12 desaturase. Cell Mol Life Sci 53:501–505

Jurenka RA (2021) Lepidoptera: female sex pheromone biosynthesis and its hormonal regulation. In: Blomquist GJ, Vogt RG (eds) Insect pheromone biochemistry and molecular biology, 2nd edn. Elsevier, London, pp 13–88

Jurenka RA, Roelofs WL (1989) Characterization of the acetyltransferase used in pheromone biosynthesis in moths: specificity for the Z isomer in Tortricidae. Insect Biochem 19:639–644

Jurenka RA, Haynes KF, Adlof RO, Bengtsson M, Roelofs WL (1994) Sex pheromone component ratio in the cabbage looper moth altered by a mutation affecting the fatty acid chain-shortening reactions in the pheromone biosynthetic pathway. Insect Biochem Mol Biol 24:373–381

Jurenka RA, Subchev M, Abad J-L, Choi M-Y, Fabriàs G (2003) Sex pheromone biosynthetic pathway for disparlure in the gypsy moth, *Lymantria dispar*. Proc Natl Acad Sci USA 100:809–814

Katzav-Gozansky T, Soroker V, Hefetz A, Cojocaru M, Erdmann DH, Francke W (1997) Plasticity of caste-specific Dufour's gland secretion in the honey bee (*Apis mellifera* L.). Naturwissenschaften 84:238–241

Keeling CI, Otis GW, Hadisoesilo S, Slessor KN (2001) Mandibular gland component analysis in the head extracts of *Apis cerana* and *Apis nigrocincta*. Apidologie 2001(32):243–252

Keeling CI, Tittiger C, MacLean M, Blomquist GJ (2021) Pheromone production in bark beetles. In: Blomquist GJ, Vogt RG (eds) Insect pheromone biochemistry and molecular biology, 2nd edn. Elsevier, London, pp 123–162

Kiyota R, Arakawa M, Yamakawa R, Yasmin A, Ando T (2011) Biosynthetic pathways of the sex pheromone components and substrate selectivity of the oxidation enzymes working in pheromone glands of the fall webworm, *Hyphantria cunea*. Insect Biochem Mol Biol 41:362–369

Knipple DC, Roelofs WL (2003) Molecular biological investigations of pheromone desaturases. In: Blomquist GJ, Vogt RG (eds) Insect pheromone biochemistry and molecular biology. Academic Press, New York, pp 81–106

Knipple DC, Rosenfield C, Miller SJ, Liu W, Tang J, Ma PWK et al (1998) Cloning and functional expression of a cDNA encoding a pheromone gland-specific acyl-CoA Δ11-desaturase of the cabbage looper moth, *Trichoplusia ni*. Proc Natl Acad Sci USA 95:15287–15292

Lassance J-M, Groot AT, Lienard MA, Antony B, Borgwardt C, Andersson F et al (2010) Allelic variation in a fatty-acyl reductase gene causes divergence in moth sex pheromones. Nature 466:486–489

Lassance J-M, Liénard MA, Antonya B, Qian S, Fujii T, Tabata J et al (2013) Functional consequences of sequence variation in the pheromone biosynthetic gene pgFAR for *Ostrinia* moths. Proc Natl Acad Sci USA 110:3967–3972

Le Conte Y, Arnold G, Trouiller J, Masson C, Chappe B, Ourisson G (1989) Attraction of the parasitic mite *Varroa* to the drone larvae of honey bees by simple aliphatic esters. Science 245:638–639

Leal WS (2001) Molecules and macromolecules involved in chemical communication of scarab beetles. Pure Appl Chem 73:613–616

Leal WS, Zarbin PHG, Wojtasek H, Ferreira JT (1999) Biosynthesis of scarab beetle pheromones. Eur J Biochem 259:175–180

Li D-T, Chen X, Wang X-Q, Moussian B, Zhang C-X (2019) The fatty acid elongase gene family in the brown planthopper, *Nilaparvata lugens*. Insect Biochem Mol Biol 108:32–43

Liénard MA, Strandh M, Hedenström E, Johansson T, Löfstedt C (2008) Key biosynthetic gene subfamily recruited for pheromone production prior to the extensive radiation of Lepidoptera. BMC Evol Biol 8:270

Liénard MA, Hagström ÅK, Lassance J-M, Löfstedt C (2010) Evolution of multicomponent pheromone signals in small ermine moths involves a single fatty-acyl reductase gene. Proc Natl Acad Sci USA 107:10955–10960

Liénard MA, Wang H-L, Lassance J-M, Löfstedt C (2014) Sex pheromone biosynthetic pathways are conserved between moths and the butterfly *Bicyclus anynana*. Nat Commun 5:3957

Lintner JA (1882) A new principle in protection from insect attack. West NY Hortic Soc Proc 27:52–66

Löfstedt C, Bengtsson M (1988) Sex pheromone biosynthesis of (E,E)-8,10-dodecadienol in codling moth *Cydia pomonella* involves E9 desaturation. J Chem Ecol 14:903–915

Löfstedt C, Xia Y-H (2021) Biological production of insect pheromones in cell and plant factories. In: Blomquist GJ, Vogt RG (eds) Insect pheromone biochemistry and molecular biology, 2nd edn. Elsevier, London, pp 89–122

Löfstedt C, Wahlberg N, Millar JG (2016) Evolutionary patterns of pheromone diversity in Lepidoptera. In: Allison JD, Cardé RT (eds) Pheromone communication in moths: evolution, behavior, and application. University of California Press, Oakland, California, pp 43–78

Luxova A, Svatos A (2006) Substrate specificity of membrane-bound alcohol oxidase from the tobacco hornworm moth (*Manduca sexta*) female pheromone glands. J Mol Catal B Enzym 38:37–42

MacLean M, Nadeau J, Gumea T, Tittiger C, Blomquist GJ (2018) Mountain pine beetle (*Dendroctonus ponderosae*) CYP4Gs convert long and short chain alcohols and aldehydes to hydrocarbons. Insect Biochem Mol Biol 102:11–20

Malka O, Karunker I, Yeheskel A, Morin S, Hefetz A (2009) The gene road to royalty differential expression of hydroxylating genes in the mandibular glands of the honeybee. FEBS J 2009(276):5481–5490

Martinez T, Fabriàs G, Camps F (1990) Sex pheromone biosynthetic pathway in *Spodoptera littoralis* and its activation by a neurohormone. J Biol Chem 265:1381–1387

Montagne J, Wicker-Thomas C (2021) *Drosophila* pheromone production. In: Blomquist GJ, Vogt RG (eds) Insect pheromone biochemistry and molecular biology, 2nd edn. Elsevier, London, pp 163–181

Morse D, Meighen EA (1987a) Pheromone biosynthesis: enzymatic studies in Lepidoptera. In: Prestwich GD, Blomquist GJ (eds) Pheromone biochemistry. Academic Press, Orlando, FL, pp 121–158

Morse D, Meighen EA (1987b) Biosynthesis of the acetate ester precursors of the spruce budworm sex pheromone by an acetyl CoA: fatty alcohol acetyltransferase. Insect Biochem 17:53–59

Moto K, Yoshiga T, Yamamoto M, Takahashi S, Okano K, Ando T et al (2003) Pheromone gland-specific fatty-acyl reductase of the silkmoth, *Bombyx mori*. Proc Natl Acad Sci USA 100:9156–9161

Mumoki FN, Crewe RM (2021) Pheromone communication in honey bees (*Apis mellifera*). In: Blomquist GJ, Vogt RG (eds) Insect pheromone biochemistry and molecular biology, 2nd edn. Elsevier, London, pp 183–204

Ng WC, Chin JSR, Tan KJ, Yew JY (2015) The fatty acid elongase *bond* is essential for *Drosophila* sex pheromone synthesis and fertility. Nat Commun 6:9263

Oehlschlager AC, Pierce AM, Pierce HD, Borden JH (1988) Chemical communication in cucujid grain beetles. J Chem Ecol 14:2071–2098

Pei X-J, Chen N, Bai Y, Qiao J-W, Li S, Fan Y-L et al (2019) BgFAS1: a fatty acid synthase gene required for both hydrocarbon and cuticular fatty acid biosynthesis in the German cockroach, *Blattella germanica* (L.). Insect Biochem Mol Biol 112:103203

Pei XJ, Fan YL, Bai Y, Bai TT, Schal C, Zhang ZF, Chen N, Li S, Liu TX (2021) Modulation of fatty acid elongation in cockroaches sustains sexually dimorphic hydrocarbons and female attractiveness. PLoS Biol 19:e3001330

Plettner E, Slessor KN, Winston ML, Oliver JE (1996) Caste-selective pheromone biosynthesis in honeybees. Science 1996(271):1851–1853

Plettner E, Slessor KN, Winston ML (1998) Biosynthesis of mandibular acids in honey bees (*Apis mellifera*): de novo synthesis, route of fatty acid hydroxylation and caste selective β-oxidation. Insect Biochem Mol Biol 28:31–42

Qiu Y, Tittiger C, Wicker-Thomas C, Le Goff G, Young S, Wajnberg E, Fricaux T, Taquet N, Blomquist GJ, Feyereisen R (2012) An insect-specific P450 oxidative decarbonylase for cuticular hydrocarbon biosynthesis. Proc Natl Acad Sci USA 109:14858–14863

Rizzo WB (2014) Fatty aldehyde and fatty alcohol metabolism: review and importance for epidermal structure and function. Biochim Biophys Acta Mol Cell Biol Lipids 1841:377–389

Roelofs W, Glover T, Tang X-H, Sreng I, Robbins P, Eckenrode C et al (1987) Sex pheromone production and perception in European corn borer moths is determined by both autosomal and sex-linked genes. Proc Natl Acad Sci USA 84:7585–7589

Rong Y, Fujii T, Katsuma S, Yamamoto M, Ando T, Ishikawa Y (2014) CYP341B14: a cytochrome P450 involved in the specific epoxidation of pheromone precursors in the fall webworm *Hyphantria cunea*. Insect Biochem Mol Biol 54:122–128

Rong Y, Fujii T, Ishikawa Y (2019a) CYPs in different families are involved in the divergent regio-specific epoxidation of alkenyl sex pheromone precursors in moths. Insect Biochem Mol Biol 108:9–15

Rong Y, Fujii T, Naka H, Yamamoto M, Ishikawa Y (2019b) Functional characterization of the epoxidase gene, Li_epo1 (CYP341B14), involved in generation of epoxyalkene pheromones in the mulberry tiger moth *Lemyra imparilis*. Insect Biochem Mol Biol 107:46–52

Ruther J, Bruckmann A, Hofferberth J (2021) Pheromone biosynthesis in *Nasonia*. In: Blomquist GJ, Vogt RG (eds) Insect pheromone biochemistry and molecular biology, 2nd edn. Elsevier, London, pp 237–267

Schal C, Sevala V, Cardé RT (1998) Novel and highly specific transport of a volatile sex pheromone by hemolymph lipophorin in moths. Naturwissenschaften 85: 339–342

Schulz S, Hötling S (2015) The use of the lactone motif in chemical communication. Nat Prod Rep 32:1042–1066
Semmelmann F, Kabeya N, Malcicka M, Bruckmann A, Broschwitz B, Straub K et al (2019) Functional characterisation of two Δ12-desaturases demonstrates targeted production of linoleic acid as pheromone precursor in *Nasonia*. J Exp Biol 222:jeb.201038
Seybold SJ, Vanderwel D (2003) Biosynthesis and endocrine regulation of pheromone production in the Coleoptera. In: Blomquist GJ, Vogt RG (eds) Insect pheromone biochemistry and molecular biology. Elsevier, London, pp 135–200
Song M, Delaplain P, Nguyen TT, Liu X, Wickenberg L, Jeffrey C, Blomquist GJ, Tittiger C (2014a) Exo-Brevicomin biosynthetic pathway enzymes from the mountain pine beetle, *Dendroctonus ponderosae*. Insect Biochem Mol Biol 53:73–80
Song M, Gorzalski A, Nguyen TT, Liu X, Jeffrey C, Blomquist GJ, Tittiger C (2014b) Exo-Brevicomin biosynthesis in the fat body of the mountain pine beetle, *Dendroctonus ponderosae*. J Chem Ecol 40: 181–189
Stanley-Samuelson DW, Jurenka RA, Cripps C, Blomquist GJ, de Renobales M (1988) Fatty acids in insects: composition, metabolism and biological significance. Arch Insect Biochem Physiol 9:1–33
Tang JD, Charlton RE, Jurenka RA, Wolf WA, Phelan PL, Sreng L et al (1989) Regulation of pheromone biosynthesis by a brain hormone in two moth species. Proc Natl Acad Sci USA 86:1806–1810
Teal PEA, Tumlinson JH (1987) The role of alcohols in pheromone biosynthesis by two noctuid moths that use acetate pheromone components. Arch Insect Biochem Physiol 4:261–269
Teal PEA, Tumlinson JH (1988) Properties of cuticular oxidases used for sex pheromone biosynthesis by *Heliothis zea*. J Chem Ecol 14:2131–2145
Tholl D (2021) Biosynthesis of terpene pheromones in hemiptera/stink bugs. In: Blomquist GJ, Vogt RG (eds) Insect pheromone biochemistry and molecular biology, 2nd edn. Elsevier, London, pp 269–284
Tittiger C, Blomquist GJ (2016) Pheromone production in pine bark beetles. Adv Insect Physiol 60:235–264
Tittiger C, Blomquist GJ (2017) Pheromone biosynthesis in bark beetles. Curr Opin Insect Sci 24:68–74
van den Bosch H, Schutgens RBH, Wanders RJA, Tager JM (1992) Biochemistry of peroxisomes. Annu Rev Biochem 61:157–197
Vanderwel D, Pierce HD, Oehlschlager AC, Borden JH, Pierce AM (1990) Macrolide (Cucujolide) biosynthesis in the rusty grain beetle, *Cryptolestes ferrugineus*. Insect Biochem 20:567–572
Villorbina G, Rodriguez S, Camps F, Fabrias G (2003) Comparative sex pheromone biosynthesis in *Thaumetopoea pityocampa* and *T. Processionea*: a rationale for the phenotypic variation in the sex pheromone within the genus *Thaumetopoea*. Insect Biochem Mol Biol 33:155–161
Wang H-L, Liénard MA, Zhao C-H, Wang C-Z, Löfstedt C (2010) Neofunctionalization in an ancestral insect desaturase lineage led to rare Δ6 pheromone signals in the Chinese tussah silkworm. Insect Biochem Mol Biol 40:742–751
Wang X, Werren JH, Clark AG (2015) Genetic and epigenetic architecture of sex-biased expression in the jewel wasps *Nasonia vitripennis* and *giraulti*. Proc Natl Acad Sci USA 112:E3545–E3554
Wossler TC, Crewe RM (1999) The releaser effects of the tergal gland secretion of queen honey bees (*Apis mellifera*). J Insect Behav 12:343–351
Wu Y, Zheng H, Corona M, Pirk C, Meng F, Zheng Y et al (2017) Comparative transcriptome analysis on the synthesis pathway of honey bee (*Apis mellifera*) mandibular gland secretions. Sci Rep 7:4530
Wu Y, Zheng Y, Li-Byarlay H, Shi Y, Wang S, Zheng H et al (2020) CYP6AS8, a cytochrome P450, is associated with the 10-HDA biosynthesis in honey bee (*Apis mellifera*) workers. Apidologie 51:1202–1212
Xia Y-H, Zhang Y-N, Ding B-J, Wang H-L, Löfstedt C (2019) Multi-functional desaturases in two *Spodoptera* moths with Δ11 and Δ12 desaturation activities. J Chem Ecol 45:378–387
Xia Y-H, Ding B-J, Dong S-L, Wang H-L, Hofvander P, Löfstedt C (2022) Release of moth pheromone compounds from *Nicotiana benthamiana* upon transient expression of heterologous biosynthetic genes. BMC Biol 20:80
Yang Y, Zhao X, Niu N, Zhao Y, Liu W, Moussian B, Zhang J (2020) Two fatty acid synthase genes from the integument contribute to cuticular hydrocarbon biosynthesis and cuticle permeability in *Locusta migratoria*. Insect Mol Biol 29:555–568
Yusuf A, Pirk CW, Crewe R (2015) Mandibular gland pheromone contents in workers and queens of *Apis mellifera adansonii*. Apidologie 46:559–572
Zhao C, Löfstedt C, Wang X (1990) Sex pheromone biosynthesis in the Asian corn borer *Ostrinia furnicalis* (II): biosynthesis of (E)- and (Z)-12-tetradecenyl acetate involves Δ14 desaturation. Arch Insect Biochem Physiol 15:57–65

Adv Exp Med Biol (2026) 1494: 519–547
https://doi.org/10.1007/5584_2024_822

Published online: 29 October 2024

Lipid Metabolism as a Target Site in Pest Control

Umut Toprak, Emre İnak, and Ralf Nauen

Abstract

Lipid metabolism is essential to insect life as insects use lipids for their development, reproduction, flight, diapause, and a wide range of other functions. The central organ for insect lipid metabolism is the fat body, which is analogous to mammalian adipose tissue and liver, albeit less structured. Various other systems including the midgut, brain, and neural organs also contribute functionally to insect lipid metabolism. Lipid metabolism is under the control of core lipogenic [e.g. acetyl-CoA-carboxylase (ACC), fatty acid synthase (FAS), perilipin 2 (LSD2)], and lipolytic (lipases, perilipin 1) enzymes that are primarily expressed in the fat body, as well as hormones [insulin-like peptides (ILP), adipokinetic hormone (AKH)], transcription factors (SREBPs, foxO, and CREB), secondary messengers (calcium) and post-translational modifications (phosphorylation). Essential roles of the fat body, together with the fact that proper coordination of lipid metabolism is critical for insects, render lipid metabolism an attractive target site in pest control. In the current chapter, we focus on pest control tactics that target insect lipid metabolism. Various classes of traditional chemical insecticides [e.g. organophosphates, pyrethroids, neonicotinoids, and chitin synthesis inhibitors (Sects. 2.1 and 2.2)] have been shown to interfere with lipid metabolism, albeit it is not their primary site of action. However, the discovery of "lipid biosynthesis inhibitors", tetronic and tetramic acid derivatives commonly known as ketoenols (Sect. 2.3), was a milestone in applied entomology as they directly target lipid biosynthesis, particularly in sucking pests. Spirodiclofen, spiromesifen, and spirotetramat targeting ACC act against various insect and mite pests, while spiropidion and spidoxamat have been introduced to the market only recently. Efforts have concentrated on the development of chemical alternatives, such as hormone agonists and antagonists (Sect. 2.4), dsRNA-based pesticides that depend on RNA interference, which have great potential in pest control (Sect. 2.5) and other eco-friendly alternatives (Sect. 2.6).

Keywords

Fat body · Lipid metabolism · Pest management

U. Toprak (✉) and E. İnak
Faculty of Agriculture, Department of Plant Protection Ankara, Molecular Entomology Lab, Ankara University, Ankara, Turkey
e-mail: utoprak@agri.ankara.edu.tr

R. Nauen (✉)
Bayer AG, Crop Science Division, Monheim, Germany
e-mail: ralf.nauen@bayer.com

1 Introduction

Lipids are one of the primary energy sources and have essential roles in the life of many organisms. They also serve as structural components of cell membranes and signalling elements in the cell. Insects use lipids to supply energy during many physiological life events, such as growth, reproduction, flight, embryo development, diapause, and starvation, as well as to synthesize epicuticular wax, sex pheromones, and defence molecules (Toprak et al. 2020). Although lipids are present in many insect tissues, they are primarily present in the fat body, which serves as an analogous system to the mammalian adipose tissue and liver. The fat body has different cell types; however, adipocytes and hepatocyte-like oenocytes are the primary cells related to lipid metabolism. In both, lipids are stored as triglycerides in special cytoplasmic compartments known as "lipid droplets" (LDs).

The primary role of the fat body is to store and release energy (Arrese and Soulages 2010; Toprak et al. 2020). It also has other essential roles; for example, it synthesizes hormones related to physiology while also acting as a target site for those synthesized by the central nervous system. As the fat body is directly in contact with the organs and haemolymph, it also circulates the metabolites, while secreting many proteins, such as antimicrobial peptides, pattern recognition molecules, lipophorins, vitellogenins, anti-freeze proteins, and detoxification enzymes, into the haemolymph (Arrese and Soulages 2010). Overall, such essential tasks make fat body lipid metabolism an attractive target for pest control.

Lipid metabolism is under the control of various lipogenic [e.g. acetyl-CoA-carboxylase (ACC), fatty acid synthase (FAS), and perilipin 2 (LSD2)] and lipolytic (lipases and perilipin 1) genes, as well as transcription factors (SREBPs, foxO, and CREB), secondary messengers (calcium), hormones [insulin-like peptides (ILP) and adipokinetic hormone (AKH)], and post-translational modifications (phosphorylation) (Toprak 2020; Toprak et al. 2014b, 2020). Various insecticides have been shown to interfere with the metabolism of lipids, albeit not primarily. However, insecticides specifically targeting insect lipid metabolism (lipid biosynthesis inhibitors) have also been developed. As chemicals often have negative outcomes on the environment and non-target organisms if not applied according to label recommendations, efforts have focused on the development of environmentally friendly strategies. One promising strategy is dsRNA-based biologicals based on RNA interference (RNAi) targeting specific mRNAs in the host organism. Below, we first present traditional insecticides and chitin synthesis inhibitors that also impact lipid metabolism and then focus on specific lipid biosynthesis inhibitors. In the final section, we introduce eco-friendly alternatives that target lipid metabolism, such as hormone agonists and antagonists, dsRNA-based insecticides, and finally other eco-friendly alternatives including lipid derivatives, inert dust, botanicals, and microbials.

2 Pest Control Strategies Targeting Lipid Metabolism

An RNAi screen covering 78% of the *Drosophila melanogaster* (Diptera: Drosophilidae) genome revealed that 30% of the genes involved in lipid metabolism are essential, as their silencing leads to lethality (Pospisilik et al. 2010). This suggests that pest control strategies targeting lipid metabolism and its genetic factors might have great potential.

2.1 Traditional Insecticides

Various studies have shown an interaction between lipids and traditional insecticides, which might be considered as an indirect effect. For example, *D. melanogaster* larvae exposed to the pyrethroid permethrin exhibited disruption in lipid metabolism and an increase in free fatty acids, lysolipids, and glycerophosphoalcohols as the degradation products of lipid peroxides by phospholipases (Brinzer et al. 2015). In another study, cuticle analysis by electron microscopy and characterization of lipid extracts in the African malaria mosquito, *Anopheles gambiae* (Diptera: Culicidae), that is resistant to another pyrethroid, deltamethrin, showed that these insects have a thicker epicuticle, the lipid layer of the cuticle, with higher cuticular hydrocarbon content compared to a susceptible strain (Balabanidou et al. 2016). This represents solid evidence indicating a critical role for cuticular hydrocarbons, and specifically for lipids, in resistance against insecticides (Balabanidou et al. 2016). Interestingly, in the deltamethrin-resistant strain of *A. gambiae*, two P450 enzymes were also found to be highly abundant in oenocytes, which is known to secrete cuticular hydrocarbons; however, a functional link between P450 and synthesis of epicuticular lipids was not detected (Balabanidou et al. 2016). In addition, transcriptomic analyses in resistant strains of *Anopheles* and *Aedes* mosquito populations revealed differential expression of genes involved in the biosynthesis of lipid components and fatty acid metabolism (Jones et al. 2013; Toé et al. 2015). On the other hand, studies on the resistance mechanisms against the fumigant phosphine revealed a link between lipid metabolism and resistance. Two enzymes related to lipid metabolism and energy homeostasis, dihydrolipoamide dehydrogenase and fatty acid desaturase, have been involved in the evolution of phosphine resistance (Schlipalius et al. 2018). In another study, the application of the organophosphate insecticide triazophos together with dsRNA specific to the gene encoding acyl-coenzyme A oxidase, the enzyme associated with intracellular oxidation of fatty acids in fat cells and the initial enzyme of the β-oxidation system, led to a reduction in the levels of total lipid and various fatty acids, and fecundity in females of brown planthopper *Nilaparvata lugens* (Hemiptera: Delphacidae) in comparison to only triazophos application (Liu et al. 2016). Another organophosphate insecticide, phoxim, disrupts the activity and expression of various genes involved in insulin signalling in *Bombyx mori* (Lepidoptera: Bombycidae) midgut (Li et al. 2016). In particular, phoxim exposure inhibits the phosphorylation of serine-threonine protein kinase (AKT), which may be the reason for decreased larval body weight (Li et al. 2016). A recent study also revealed major differences between fatty acid profiles of *Leptinotarsa decemlineata* (Coleoptera: Chrysomelidae) populations collected from an organic production field and from a commercially managed field that had a history of systemic neonicotinoid use (Clements et al. 2020). *L. decemlineata* populations collected from a commercially managed field displayed elevated levels of Δ^6-desaturase (Clements et al. 2020). On the other hand, the genes encoding traditional detoxification enzymes, such as cytochrome P450s (Pang et al. 2016), glutathione-S-transferases (Enayati et al. 2005) or carboxylesterases (Jackson et al. 2013; Feng et al. 2018), are also expressed in the fat body, further supporting the notion that the fat body as an organ and lipid metabolism as a central mechanism of biological events could contribute to the evolution of insecticide resistance. Notably, various other genes involved in lipid metabolism, such as those encoding fatty acid synthase (FAS), acyl-CoA synthetase (ACS), and fatty acid binding proteins (FABP), or transcription factors, such as HR96, respond to insecticide applications (Majerowicz and Gondim 2013). In the mosquito *A. gambiae*, *FAS*, *ACS*, and *FABP* are all up-regulated in response to permethrin. On the other hand, loss of *HR96* in *D. melanogaster* leads to an increase in susceptibility against insecticides (King-Jones et al. 2006). In summary, it has been shown that several neurotoxic insecticides from different chemical and mode of action (MoA) classes can impair lipid metabolism, and elevated lipid levels might be one of the factors contributing to the evolution of resistance. However, more detailed studies are needed to uncover the causal link between lipid metabolism and insecticides primarily targeting other sites.

Another major aspect of insecticide resistance is related to the fitness cost that causes high energetic loss in resistant populations when compared to the susceptible counterparts (Kliot and Ghanim 2012). Increased energy accumulation and mobilization might reduce the fitness cost of resistance

(Lopes et al. 2010), suggesting an indirect role for lipids in such events. Indeed, enhanced activity of lipases has been reported in a permethrin-resistant population of *Sitophilus zeamais* (Coleoptera: Curculionidae) (Araújo et al. 2008). However, the knowledge on the correlation between energy accumulation/mobilization and reduced fitness cost is still limited.

2.2 Chitin Synthesis Inhibitors

Several studies have shown that chitin biosynthesis inhibitors interfering with chitin synthase might affect lipid metabolism, for example, the chitin biosynthesis inhibitor, hexaflumuron, reduces lipid levels in the Asiatic rice stem borer, *Chilo suppressalis* (Lepidoptera: Crambidae) (Mirhaghparast et al. 2015). However, hexaflumuron did not significantly affect the level of glucose and lipid content in the predator beetle, *Hippodamia variegata* (Coleoptera: Coccinellidae) (Alimohamadi et al. 2014). Such selectivity might be an advantage for the application of these insecticides together with biological control agents; however, further studies are required.

2.3 Lipid Biosynthesis Inhibitors

2.3.1 Development and Market Value of Lipid Biosynthesis Inhibitor Insecticides/Acaricides

Lipid biosynthesis inhibitor insecticides/acaricides are a good example of how studies on different pesticide classes can inspire each other. Initially, lipid biosynthesis inhibitors were developed as herbicides (aryloxyphenoxypropionates and cyclohexandiones) to control grassy weeds in agricultural areas (Devine 1997). An indication shift led to the discovery and development of a new insecticide/acaricide chemical class with the same MoA, i.e. "inhibiting lipid biosynthesis" (IRAC Group 23), also known as tetronic and tetramic acid derivatives or ketoenols (Nauen et al. 2003; Nauen 2005; Jeschke et al. 2019) (Fig. 1). This group of pesticides has been shown to inhibit acetyl-CoA carboxylase (ACC) (Nauen 2005; Lümmen et al. 2014). ACC has also been investigated as an attractive target for drug discovery against human diseases, such as obesity and cancer (Tong and Harwood Jr 2006).

Spirodiclofen, spiromesifen, and spirotetramat are globally registered compounds inhibiting the biosynthesis of lipids in insects/mites. The first-

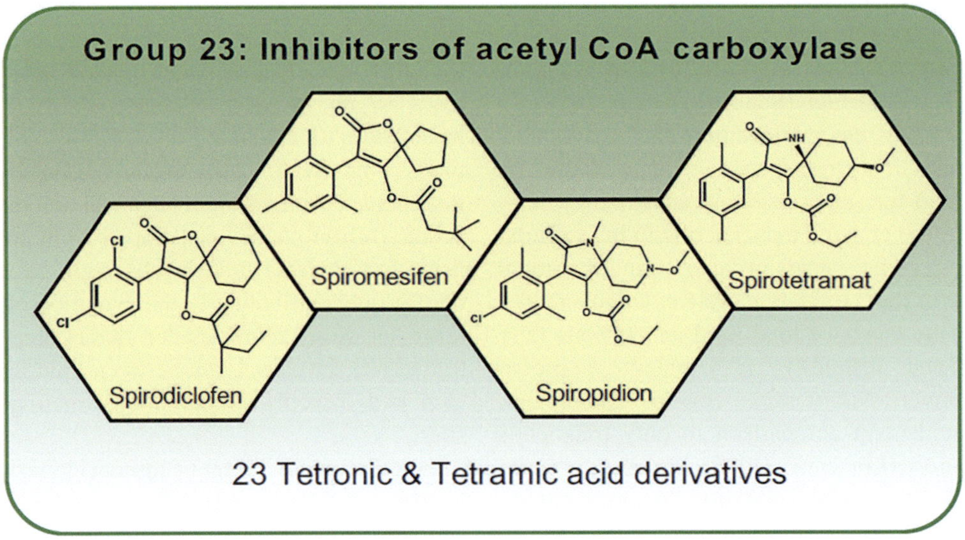

Fig. 1. Tetronic- and tetramic acid derivatives (ketoenols) are acetyl-CoA carboxylase inhibitors classified in mode of action (MoA) group 23 by the Insecticide Resistance Action Committee (IRAC). (Sparks et al. 2020)

generation ketoenol compound, spirodiclofen (launched in 2002), almost exclusively targets mites by contact, whereas the second-generation ketoenol, spiromesifen (launched in 2004), also targets whiteflies and shows some translaminar activity when applied to plant leaves. The third and latest generation represented by spirotetramat (launched in 2008) targets a broad spectrum of sucking pests and was shown to be ambimobile in plants once taken up by the leaf lamina (Nauen et al. 2008) (Fig. 2). In addition, the discovery of a new member of tetramic acids, spiropidion, has been recently reported and has similar properties to spirotetramat (Muehlebach et al. 2020). A fifth ketoenol insecticide, spidoxamat, is close to market introduction and has received its first registration in Cambodia in 2023 (Nauen, unpublished). The commercial success of these compounds in global pesticide markets has been shown by their market value, which reached 652 million USD in 2018 (Sparks et al. 2020). Spirodiclofen has initially led the market and was the most sold acaricide in 2010 (Van Leeuwen et al. 2015); however, spirotetramat has been the most successful ketoenol from a commercial point of view to date (Sparks et al. 2020).

2.3.2 Potential Targets and Mode of Action of Lipid Biosynthesis Inhibitors

Lipid biosynthesis inhibitors target immature stages of sucking insects and mites, such as aphids, psyllids, whiteflies, and spider mites (Bretschneider et al. 2003; Jeschke et al. 2019). In addition, spirodiclofen-treated female adult mites failed to deposit eggs before death (Marčić et al. 2012). Spirotetramat and spiromesifen also lead to decreased fecundity, e.g. in *Aphis gossypii* (Hemiptera: Aphididae) (Gong et al. 2016) and *D. melanogaster*, respectively (Kissoum et al. 2020). Ketoenol treated immature stages of insects and mites usually die when they molt or reach the next quiescent stage, respectively (Nauen 2005; Marčić et al. 2011). The reason underlying this phenomenon may be the requirement for lipids during non-feeding periods, such as molting and hibernation (Toprak et al. 2020).

Ketoenols target ACC (EC 6.4.1.2), an enzyme discovered in 1958 (Wakil et al. 1958); it is a key enzyme in fatty acid biosynthesis (Wakil et al. 1983). ACC is a biotin-dependent enzyme that catalyses the ATP-dependent carboxylation of acetyl-CoA to malonyl-CoA through two distinct

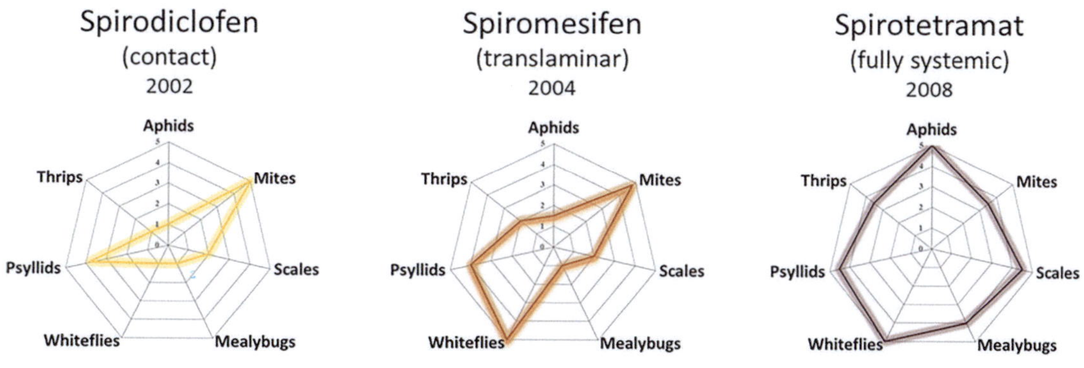

Fig. 2 Spectrum of sucking pest groups targeted by first, second and third generation ketonol insecticides represented by spirodiclofen (launched in 2002), spiromesifen (launched in 2004) and spirotetramat (launched in 2008), respectively. The other two registered ketoenol insecticides, spiropidion and spidoxamat share a pest spectrum similar to spirotetramat

half-reactions: biotin carboxylase and carboxyltransferase (Wakil et al. 1983). In the first half-reaction, biotin carboxylase catalyses the Mg-ATP-dependent carboxylation of the N1 atom of the biotin cofactor using bicarbonate as the CO_2 donor, forming carboxy-biotinyl-ACC. In the second step, the carboxy group is transferred from biotin to acetyl-CoA as acceptor substrate, generating malonyl-CoA (Tong 2013). ACC-inhibitor herbicides bind to the carboxyltransferase domain of the multi-domain plastid ACC in grasses (Linda et al. 2010). Similarly, Lümmen et al. (2014) showed that spirotetramat-enol, the activated metabolite of spirotetramat, binds to the carboxyltransferase domain of *Tetranychus urticae* (Acari: Tetranychidae) ACC. However, mutations that confer herbicide resistance at carboxyltransferase domain did not affect the inhibition of *T. urticae* ACC by spirotetramat-enol, suggesting that herbicides and insecticide/acaricides have different binding modes/sites despite the same MoA (Lümmen et al. 2014).

Inhibition of ACC leads to decreased total lipids in spirodiclofen-treated *T. urticae* female adults (Nauen et al. 2003; Nauen 2005). Spiromesifen reduces both total carbohydrates and glycogen levels in *D. melanogaster* pupae while increasing lactate dehydrogenase, an important glycolytic enzyme involved in carbohydrate metabolism (Kissoum and Soltani 2016). Other studies reported reduced total lipid and protein levels in different dipterans, such as *Culex* and *Drosophila* species (Bensafi-Gheraibia et al. 2013; Bouabida et al. 2017a, b). Specifically, decreased carbohydrate and glycogen levels have been reported from the ovaries of spiromesifen-exposed *D. melanogaster* (Kissoum et al. 2020). Spiromesifen and spirotetramat reduce the total amount of lipids and change the fatty acid profile and levels of specific fatty acids, like palmitic acid, in cotton leaf worm, *Spodoptera littoralis* larvae (Abdel-Fatah et al. 2019). On the other hand, such effects have also been reported in beneficial insects; for example, spirodiclofen leads to lipid depletion in the predatory beetle, *Hippodamia variegata* (Coleoptera: Coccinellidae) (Alimohamadi et al. 2014). Ketoenols might also lead to other alterations in the lipid pathways, as Marnett (1999) reported increased malondialdehyde, a naturally occurring product of lipid peroxidation, which reflects the process of detoxification.

Suppression of ACC activity might have other outcomes on the physiology of target organisms; for example, ACC inhibition in spirotetramat-treated *A. gossypi* led to incomplete ecdysis resulting in death in line with down-regulation of cuticle protein genes (Pan et al. 2015). In another study, proteomic analysis of spirotetramat-resistant *A. gossypi* revealed up-regulation of various genes encoding ACC, heat shock protein 70, ubiquitin-conjugating enzyme, FAS, and UDP-glucose 6-dehydrogenase (Xi et al. 2015). Altered levels of target site proteins are indeed a commonly known mechanism for insecticide resistance (Bass et al. 2014; Lee et al. 2015). Thus, increased amounts of ACC in resistant strains may allow insects to survive, as this would limit the physiological defects due to low ACC activity (Xi et al. 2015). In addition, the reduction in *ACC* mRNA by dsRNA significantly increased the spirotetramat toxicity and impaired cuticle formation in *A. gossypii* (Wei et al. 2016). However, spirodiclofen resistance in *T. urticae* and *Panonychus ulmi (Acari: Tetranychidae)* was not found to be related to the altered expression of *ACC* (Demaeght et al. 2013; Bajda et al. 2015). *ACC* was found to be up-regulated in spirodiclofen-exposed *Panonychus citri* (Yu et al. 2015), which is perhaps linked to a transcriptional mechanism compensating for low ACC activity. The authors speculated that the competition between spirodiclofen and acetyl-CoA may cause an accumulation of acetyl-CoA, resulting in the up-regulation of *ACC* (Yu et al. 2015). Indeed, regulation of *ACC* expression has been associated with two microRNAs, *miR-276* and *miR-3016*, at a post-transcriptional level, resulting in low levels of spirotetramat resistance in *A. gossypi* (Wei et al. 2016). Recently, it was shown that spirotetramat impairs the lipid-based eggshell and cuticle barriers, resulting in increased xenobiotic sensitivity of adults and eggs of *D. melanogaster* (Yang et al. 2023a). Similarly, reduced fitness parameters, such as lower egg number, increased egg desiccation, together with a weak cuticle barrier, after spirotetramat exposure has been reported in *D. suzukii* (Yang et al. 2023b). In brief, the regulation of lipid metabolism in ketoenol-resistant pests, as well as the effects of sublethal concentrations of lipid biosynthesis inhibitors, needs further investigation.

2.3.3 Resistance and Resistance Mechanisms Against Lipid Biosynthesis Inhibitors

Failures in the success of lipid biosynthesis inhibitor insecticides/acaricides have been reported from different pests, including spider mites and whiteflies, due to the evolution of resistance (Khajehali et al. 2011; Karatolos et al. 2012; Bielza et al. 2019; İnak et al. 2019; Lueke et al. 2020; De Rouck et al. 2023; Müller et al. 2024). The mechanisms underlying the resistance have vital importance to the design of resistance management strategies (Mnzava et al. 2015). Arthropods can develop resistance through two main mechanisms; decreasing the amount of pesticide that can reach the target site by detoxification enzymes (pharmacokinetic mechanisms) or altering the target site of the pesticide by acquiring mutations (pharmacodynamic mechanisms) (Li et al. 2007; Van Leeuwen and Dermauw 2016). The latter is quite critical in terms of pesticide resistance management as target site mutations might be considered as "dead-ends" when fixed in a population and lacking fitness costs, resulting in various levels of inactivity of the pesticides depending on the mutation. The first ACC target site mutation, E645K, linked to resistance against lipid biosynthesis inhibitors was reported between biotin carboxylase (BC) and biotin-carboxyl carrier (BCC) domains in *Trialeurodes vaporariorum* (Hemiptera: Aleyrodidae) ACC (Karatolos et al. 2012). That mutation, however, was located outside the carboxyltransferase domain, the binding site of ketoenols, and the resistance levels in this study were relatively low (Karatolos et al. 2012). In a later study with Greek field samples of *T. vaporariorum*, its association with ketoenol resistance was not confirmed (Kapantaidaki et al. 2018) and attempts to introduce the mutation in *Drosophila* using CRISPR/Cas9 genome editing failed as it was homozygous lethal (Douris et al. 2020; Lueke et al. 2020). Several other amino acid substitutions have been reported in ACC of spirotetramat- and spirodiclofen-resistant *A. gossypi* and *T. urticae* populations, respectively (Pan et al. 2017; Wybouw et al. 2019); however, those described in *A. gossypii* require functional validation, whereas transgenic *Drosophila* bearing the *T. urticae* ACC mutation A1079T did not exhibit resistance to spirodiclofen (Lueke et al. 2020). Recently, an A2083V (*Bemisia tabaci* numbering) mutation in a highly conserved region of the carboxyltransferase domain of *B. tabaci* ACC was reported to be associated with high levels of ketoenol resistance and has been functionally validated by CRISPR/Cas9 genome editing in *Drosophila* lines (Lueke et al. 2020). It is interesting to note that ketoenol resistance conferred by the ACC A2083V target site mutation is dominant, i.e. heterozygotes express a fully resistant phenotype (Lueke et al. 2020). The same amino acid substitution at the equivalent position (A2666V) has also been reported from a spirotetramat-resistant *Myzus persicae* (Hemiptera: Aphididae) population collected in Australia (Singh et al. 2021; Umina et al. 2022; for review, see Bass and Nauen 2023). More recently, a F1656L mutation in carboxyltransferase domain of ACC was uncovered in spiromesifen-resistant, field-collected *T. urticae* populations (İnak et al. 2022); however, the role of this mutation in resistance needs to be further elucidated. Müller et al. (2024) recently detected the same dominant A2083V mutation described in B. tabaci in another whitefly species, Aleyrodes proletella, collected in several places in Europe.

A number of studies showed the involvement of detoxification enzymes, such as cytochrome P450 monooxygenases (P450s) and carboxyl/cholinesterases (CCEs), in ketoenol resistance (Rauch and Nauen 2002; Demaeght et al. 2013; Bajda et al. 2015; Peng et al. 2016; Pan et al. 2018; Wybouw et al. 2019; Wei et al. 2020; Pan et al. 2020; İnak et al. 2022, 2023). More specifically, the metabolization of spirodiclofen by CYP392E10, and overexpression of an alternative allele of *CCE04*$^{SR\text{-}VP}$ leading to spirodiclofen sequestration were documented in *T. urticae*

(Demaeght et al. 2013; Wei et al. 2020). A recent study demonstrated the shared overexpression of detoxification genes in separate populations of *T. urticae* selected by spiromesifen and spirodiclofen. This revealed candidate genes that play a role in cross-resistance (İnak et al. 2023). A transcriptomics study with spirodiclofen-susceptible and -resistant European red mites (*P. ulmi*) revealed differential expression of several detoxification genes, including P450s, and suggested their involvement in ketoenol resistance (Bajda et al. 2015). In ketoenol resistant strains of B. tabaci, synergism assays did not lead to significant susceptibility, suggesting that metabolic resistance does not have a critical role in resistance to ketoenols in *B. tabaci* (Bielza et al. 2019).

Considering resistance ratios based on toxicity tests, the resistance against lipid biosynthesis inhibitors seems to be less widespread throughout the world when compared to other pesticide groups (Table 1). Recent studies, however, indicate increased resistance ratios in various target pests. Therefore, rotation of pesticides with different MoA should be favoured, and it is strongly recommended to follow the IRAC "MoA treatment windows" approach when implementing resistance management strategies (Bass and Nauen 2023). More monitoring and genotyping studies, especially in target pests with a history of rapid resistance evolution, are needed to obtain a deeper understanding about the spread of resistance and status of lipid biosynthesis inhibitors to design area-wide control strategies.

2.4 Hormone Agonists and Antagonists

The insect endocrine system orchestrating growth and development is an important target for pest control. Juvenile hormone (JH) agonists, such as pyriproxyfen, methoprene, and fenoxycarb, or 20-hydroxyecdysone (ecdysterone or 20E) agonists, such as bisacylhydrazines, have been used against pests for a long time (Dhadialla et al. 1998). Several studies also indicated the effect of such insecticides on lipids. For example, pyriproxyfen, a JH analogue, reduces lipid levels in sunn pest, *Eurygaster integriceps* (Hemiptera: Scutelleridae) (Zibaee et al. 2011) and Colorado potato beetle, *L. decemlineata* (Fotouhi et al. 2015). Similarly, Ghasemi et al. (2010) reported a significant reduction in the total protein and lipid content of *Plodia interpunctella* (Lepidoptera: Pyralidae), further indicating the negative effect of juvenile hormone analogues on lipid metabolism.

Another potentially interesting approach to interfere with insect growth and development is the use of peptide hormone agonists/antagonists targeting lipid metabolism (Fónagy 2006). A limited number of studies have examined this potential, and insect glucagon, AKH, and diapause hormone (DH) appear to be promising candidates in this regard. AKH has been shown to increase the efficacy of synthetic insecticides (Kodrík et al. 2015; Plavšin et al. 2015), as well as of entomopathogenic bacteria, fungi, and nematodes (Goldsworthy et al. 2005; Mullen and Goldsworthy 2006; Ibrahim et al. 2017). Several studies have examined the MoA of DH and reported a promising potential also for DH agonists and antagonists (Zhang et al. 2011, 2015a). DH is involved in the termination of pupal diapause in *Helicoverpa zea* (Lepidoptera: Noctuidae), and synthetic DH agonists are more potent than the native hormone, leading to premature termination of diapause, as well as an inhibition in the entrance into pupal diapause (Zhang et al. 2011). Notably, the incorporation of a dihydroimidazole trans-proline mimetic motif into a DH agonist allowed the conversion of the agonist into an antagonist that blocks the termination of diapause (Zhang et al. 2011). In summary, DH analogues and antagonists possess a promising potential to disrupt insect diapause as a tool for pest management (Zhang et al. 2015a), albeit the overall value of such strategies to control highly destructive crop pests remains to be shown in applied studies.

2.5 RNA Interference

RNA interference (RNAi) has been a powerful tool for functional genomic studies. The important difference that distinguishes RNAi studies in

Table 1 Resistance developed by field-populations against lipid biosynthesis inhibitors

	Origin of strains	Resistance ratio	Mechanism of resistance	References
Spiromesifen				
Trialeurodes vaporariorum	UK, Germany, Türkiye	Up to 26-fold	E645K between BC and BCC domain	Karatolos et al. (2012)
	Finland	Up to 11.5-fold		Ovčarenko et al. (2014)
	Europe (mainly Greece)	Up to 59-fold	Other than E645K	Kapantaidaki et al. (2018)
Tetranychus urticae	Brazil	Up to 81.5% survival rate at 6.96 mg/L	–	Sato et al. (2016)
	Türkiye	> 54.8% mortality at FD[a]	–	İnak et al. (2019)
		Up to 142-fold	Metabolic resistance, F1656L mutation in CT domain	İnak et al. (2022)
		>30.2-fold (selection)	Metabolic resistance, target-site overexpression	İnak et al. (2023)
	Holland	> 31% mortality at FD[a]		Khajehali et al. (2011)
		Up to 19-fold	P450 (CYP392E10)	Demaeght et al. (2013)
Bemisia tabaci	Israel	Up to 2-fold		Kontsedalov et al. (2009)
	California, Arizona	Up to 29-fold		Prabhaker et al. (2008)
	Spain	Up to 7.1-fold		Fernández et al. (2009)
	Florida	Up to 3.5 fold		Mann et al. (2012)
	India	Up to 32-fold		Roy et al. (2019)
	Brazil	Up to 6096		Dângelo et al. (2018)
	Spain	>38,000-fold		Bielza et al. (2019)
	Greece	>1500-fold		Stavrakaki et al. (2023)
	Brazil, Spain, Australia, selected strain	>640-fold	A2083V in CT domain	Lueke et al. (2020)
Panonychus ulmi	Iran	Up to 22-fold	P450, CCE	Badieinia et al. (2020)
Aleyrodes proletella	Europe (various countries)	>72 fold	A2084V in CT domain	Müller et al., (2024)
Spirodiclofen				
Panonychus ulmi	Europe	>90% mortality 55/63 populations at 25 mg/L		Kramer and Nauen (2011)
	Germany	>7000-fold (selection)	P450 (CYP385C1)	Bajda et al. (2015)
			CCE (TuCCE66-TuCCE13)	
	Iran	Up to 217-fold	P450, CCE	Badieinia et al. (2020)

(continued)

Table 1 (continued)

	Origin of strains	Resistance ratio	Mechanism of resistance	References
Tetranychus urticae	Belgium			
	Greece	Up to 12-fold		Ilias et al. (2012)
	Belgium	274-fold	A1079T outside CT domain	Wybouw et al. (2019)
			D384Y in CPR	
		Up to 680-fold	P450 (CYP392E10)	Demaeght et al. (2013)
		274-fold	CCE (CCE04)	Wei et al. (2020)
	Italy	Up to 1.9-fold	Metabolic resistance	Tirello et al. (2012)
	Greece	>20-fold	T1158N, T2315L outsite CT domain	Papapostolou et al. (2021)
			Metabolic resistance	
	Türkiye	>11.5-fold (selection)	Metabolic resistance	İnak et al. (2023)
Panonychus citri	California	2.5-fold		Ouyang et al. (2012)
	China	Up to 127-fold		Hu et al. (2010)
		712-fold	P450 (CYP385C10)	Cheng et al. (2022)
Bemisia tabaci	Greece	Up to threefold		Ilias et al. (2012)
Spirotetramat				
Bemisia tabaci	California, Arizona	Up to 14-fold		Prabhaker et al. (2014)
	China	Up to 79-fold for eggs		Peng et al. (2017)
		Up to 224.7-fold for eggs		Ji et al. (2023)
		Up to 543-fold		
	Spain	Up to 130-fold		Bielza et al. (2019)
	Brazil, Spain, Australia and selected strain	Up to 93-fold	A2083V in CT domain	Lueke et al. (2020)
	Australia	Up to 2.1-fold		Hopkinson and Pumpa (2019)
	Türkiye	Up to 309-fold		Yükselbaba and Ali (2022)
Aphis gossypii	China	579-fold	P450 (CYP6A2)	Peng et al. (2016)
		Selected 58 generations		
		Selected strain	P450 (CYP6A2)	Pan et al. (2015)
		441-fold for adults	Heat shock protein 70	
		11.9-fold for nymphs	UDP-glucuronosyltransferase (ugt 58, ugt-50)	

(continued)

Table 1 (continued)

	Origin of strains	Resistance ratio	Mechanism of resistance	References
		–	CCE	Gong et al. (2016)
		Selected strain	P450 (CYP380C6)	Pan et al. (2018)
		Selected strain	Acetyl-coA carboxylase (ACC)	Xi et al. (2015)
		578-fold for adults	Heat shock protein 7	
		14.9-fold for nymphs	Ubiquitin-conjugating enzyme	
			Fatty acid synthase	
			UDP-glucose 6-dehydrogenas	
		Selected strain	miR-276, miR-3016-modulated	Wei et al. (2016)
			Acetyl-CoA carboxylase overexpression	
		12.4-fold	UGT344M2	Pan et al. (2020)
Myzus persicae	Global	>90% mortality at 0.5 ppm		
		0% mortality at 0.5 ppm for an Australian strain	A2226V in CT domain	Singh et al. (2021)
	Saudi Arabia	Up to 6.68-fold	Metabolic resistance	Sabra et al. (2023)
	Australia	Up to 170-fold	A2226V in CT domain	Umina et al. (2022)
Aleyrodes proletella	Europe (various countries)	>56 fold	A2084V in CT domain	Müller et al. (2024)
Spiropidion				
Bemisia tabaci	Brazil, Spain, Australia, and selected strain	>750 fold	A2083V in CT domain	Lueke et al. (2020)

[a]FD means registered field dose

entomology from those in medicine or other fields is that the outcome of silencing not only provides functional data but also has potential as a pest control strategy (Toprak et al. 2013, 2014a; Zhu and Palli 2020). Thus, the consequences of silencing genes with important roles in insect physiology could be negative or even fatal for the target pest. In addition, RNAi, as a pest management strategy, has been considered safe in terms of environment and non-target organisms (Martinez et al. 2002; Mehlhorn et al. 2021). Indeed, the first commercial sprayable RNAi application to control Colorado potato beetle has been launched very recently (Yan et al. 2024). Due to the many essential roles of the fat body and the notion that lipid metabolism is vital, genes involved in fat body integrity and lipid metabolism are potentially attractive targets for RNAi-mediated pest control.

RNAi targeting lipid-related genes has been investigated in several agricultural and public health pests. Specifically, *FAS, ACC*, and insulin receptors (*InR*) are the most widely silenced genes related to lipid metabolism. In addition to the fact that down-regulation of these genes can result in significant mortality (Broehan et al. 2013; Pei et al. 2019, 2022), many other effects on developmental biology (i.e. reduced fecundity and lipid levels, developmental arrest, and lower egg hatching rate) have been documented upon silencing lipid-related genes (Table 2). Some of the targets that lead to lethality upon silencing are *ACC*, *FAS*, *InR*,

Table 2 RNAi of the genes involved in lipid metabolism

Species	Target gene	Mortality	Stage	Other Effects	Reference
Blattella germanica	Fatty acid synthase (FAS)	Up to 100% after 96 h	Adult	Accelerated water loss	Pei et al. (2019)
Colaphellus bowringi	Fatty acid synthase 2 (FAS2)	<4%	Adult	Suppressed lipid accumulation	Tan et al. (2017)
Culex pipiens	Fatty acid synthase (FAS)		Adult females	Reduced lipid levels Impaired diapause	Sim and Denlinger (2009)
Blattella germanica	Fatty acid synthase (FAS) Fatty acid elongase	Up to 100% at 120 h	Nymphs (N5)	Body weight loss Lack of waterproofing	Pei et al. (2021)
Spodoptera litura	Fatty Acid Synthase (FAS)	45% reduced survival rate	Larvae (L3)	Reduced density of lipid droplets Reduced triglyceride content in the fat body	Song et al. (2022)
Schistocerca gregaria	Fatty acid binding protein (FABP)	<10%	Adult male	Reduced flight period	Rajapakse et al. (2019)
Bombyx mori	Fatty acid transport protein (FATP)		Pupae	Reduced lipid droplet accumulation and triglyceride Reduced bombykol (pheromone) production	Ohnishi et al. (2009)
Blattella germanica	Fatty acid elongase		Nymphs (N5/N6) Adult	Reduced sexual attractiveness	Pei et al. (2021)
Nilaparvata lugens	Fatty acid desaturase		Nymphs (N3/N5)	Reduced fatty acid and triglyceride content Reduced food intake Reduced body mass Reduced fecundity	Ye et al. (2020)
Aedes aegypti	Acetyl-CoA carboxylase (ACC) Fatty acid synthase 1 (FAS1)		Adult females	Reduced fecundity Defects in eggshell formation Delayed blood meal digestion	Alabaster et al. (2011)
Drosophila melanogaster	Acetyl-CoA-Carboxylase (ACC)	~90% (upon silencing in oenocytes)	Larvae (L2/L3)	Reduced triglyceride (upon silencing in the fat body) Defective spiracles (upon silencing in oenocytes)	Parvy et al. (2012)
Rhodnius prolixus	Acyl-CoA binding protein (ACBP)	Up to 100% after >30 days	Adult females		Almeida et al. (2020)
Nilaparvata lugens	Acetyl Co-A carboxylase (ACC)		Nymphs (N5)	Reduced fecundity	Zhang et al. (2015b)
Coccinella septempunctata	Acetyl-CoA carboxylase (ACC) Long-chain fatty acid-CoA ligase also known as Acyl-CoA synthetase long-chain (ACSL) Elongase of very long-chain fatty acids (ELOVL) Very long-chain 3-oxoacyl-CoA reductase		Adult females	Reduced lipid content	Xiang et al. (2021)

(continued)

Table 2 (continued)

Species	Target gene	Mortality	Stage	Other Effects	Reference
Drosophila melanogaster	Acyl coenzyme A: Diacylglycerol acyltransferase (DGAT) = Midway (Mdy)		Adult flies	Reduced triglyceride	Buszczak et al. (2002)
Leptinotarsa decemlineata	Lipid storage droplet protein 2 (LSD2)		Larvae (L4)	Reduced lipid content	Güney et al. (2021)
Drosophila melanogaster	Lipid storage droplet protein 2 (LSD2)		Larvae (L3)	Apoptosis and developmental defects in the wings	Binh et al. (2019)
Acyrthosiphon pisum	Lipophorin receptor (Lr)		Nymphs (N3)	Reduced hydrocarbon content Reduced fecundity Increased susceptibility of aphids to desiccation	Qiao et al. (2021)
Nilaparvata lugens	Lipophorin receptor (Lr)		Virgin females	Reduced triglyceride Retarded development of ovaries Reduced fecundity	Lu et al. (2018a)
Tenebrio molitor	Apolipophorin-III	Up to 80% for adults	Larvae Pupae	Shedding of pupal skin Shorter extension of the elytra Improper folding of hind wings Adult deformation	Patnaik et al. (2015)
Drosophila melanogaster	Lipoprotein (Lpp)			Reduced circulating lipids Inhibition of *Spiroplasma* (endosymbiont) proliferation	Herren et al. (2014)
Acyrthosiphon pisum	Apolipoprotein D		Nymph (N3)	Increased susceptibility to desiccation Reduced life span	Qiao et al. (2020)
Bombyx mori	Lipase		Adult females	Reduced bombykol (pheromone) production	Du et al. (2012)
Nilaparvata lugens	Brummer lipase		Adult females	Obesity Blocked lipid mobilization Retarded ovarian development Reduced fecundity	Lu et al. (2018b) and Zhou et al. (2018)
Sitotroga cerealella	Triacylglycerol lipase (TGL)		Male adults	Reduced apyrene sperm in males	Yan et al. (2021)
Spodoptera exigua	Calcium-independent phospholipase A		Larvae	Developmental retardation and immunosuppression	Sadekuzzaman et al. (2017)
Spodoptera litura	Sterol carrier protein 2/3-oxoacyl-CoA thiolase (SCP-x)		Molting larvae (L5 → L6)	Reduced cholesterol levels in the haemolymph Retarded larval development	Guo et al. (2009)
Rhodnius prolixus	Mitochondrial-like isoform of glycerol-3-phosphate O-acyltransferase (GPAT)			Reduced triglyceride in the posterior midgut and fat body	Alves-Bezerra et al. (2017)

(continued)

Table 2 (continued)

Species	Target gene	Mortality	Stage	Other Effects	Reference
Drosophila	Adenosine diphosphate ribosylation factor/ coat protein I (ARF/COPI)		Drosophila cell line	Large and disperse lipid droplets (fat body specific knock down)	Guo et al. (2008)
Nilaparvata lugens	Myeloid differentiation factor 2 (MD-2)-related lipid-recognition protein	> 50%	Nymphs	Moulting failure Impaired development Reduced fecundity Reduced 20-hydroxyecdysone production in the nymphs Reduced lipid droplet size	Wang et al. (2021)
Nilaparvata lugens	P-loop NTPase		Adult females	Reduced fecundity Impaired ovarian development Reduced lipid and palmitic acid in females	Pang et al. (2017)
Spodoptera litura	Adipokinetic hormone receptor (AKHR)		Larvae (L6)	Increased triglyceride and lipid droplets in the fat body Reduced triglyceride release into haemolymph Increased free fatty acid	Dong et al. (2022)
Bactrocera dorsalis	Adipokinetic hormone receptor (AKHR)		Adults	Reduced fecundity and sexual courtship activity	Hou et al. (2017)
Gryllus bimaculatus	Adipokinetic hormone receptor (AKHR)		Adults	Reduced diglyceride and trehalose in the haemolymph Increased triglyceride in the fat body Reduced locomotor activity	Konuma et al. (2012)
Drosophila melanogaster	Short neuropeptide F (sNPF) Short neuropeptide F receptor (sNPFR)		Adult flies	Reduced triglyceride and starvation resistance, and increased starvation-mediated hyperactivity response after 24 h of starvation (knockdown of sNPF in circadian clock neurons) Increased starvation resistance and reduced starvation-induced hyperactivity response after 24 h of starvation (knockdown of sNPFR in insulin-producing cells-IPCs)	Geo et al. (2019)
Drosophila melanogaster	Short neuropeptide F (sNPF)		Adult flies	Increased carbohydrates and lipid (knock down of sNPF in dorsal lateral peptidergic neurons)	Lee et al. (2008)
Drosophila melanogaster	Corazonin receptor (CrzR)		Adult flies	Reduced triglyceride	Kubrak et al. (2016)

(continued)

Table 2 (continued)

Species	Target gene	Mortality	Stage	Other Effects	Reference
Locusta migratoria	Insulin-like peptide (ILP, gonadulin)		Adult females	Oocyte growth inhibition	Veenstra et al. (2021)
Leptinotarsa decemlineata	Insulin-like peptide (ILP)	50% larval death +50% pupal death of the survived	Larvae	Reduced weight	Fu et al. (2016)
Rhodnius prolixus	Insulin receptor (InR)	Up to 100% after >80 days	Adult females	Increased lifespan Smaller ovaries Reduced fecundity	Silva-Oliveira et al. (2021)
Rhodnius prolixus	Insulin receptor (InR)	Up to 50% after 15 days	Adult females	Reduced survival and fecundity	Leyria et al. (2021)
Blattella germanica	Insulin receptor (InR)		Nymphs (N5/N6) Adult females	Developmental arrest Wing and tegmina deformation Smaller pronotum	Abrisqueta et al. (2014)
Maruca vitrata	Insulin receptor (InR) Serine-threonine protein kinase (AKT) Forkhead box O (foxO) Target of rapamycin (TOR)		Adult females	Reduced ovarian development Hypotrophied ovaries	Al Baki et al. (2019)
Leptinotarsa decemlineata	Insulin receptor substrate gene Chico Phosphoinositide-3-kinase (PI3K)	~80%	Larvae	Reduced food consumption Affected absorption and metabolism of amino acids and sugars Reduced expression of several 20E signalling genes	Deng et al. (2018)
Nilaparvata lugens	Trehalose phosphate synthase (TPS)	60% at 10d	Larvae		Chen et al. (2010)
Leptinotarsa decemlineata	Trehalose phosphate synthase (TPS) Trehalases (TRE)	20–40%	Larvae	Hyperphagia Reduced chitin content (TRE)	Shi et al. (2016)
Heortia vitessoides	Trehalose phosphate synthase (TPS)	40%	Pupae	Deformed larva and pupae	Chen et al. (2020)
Harmonia axyridis	Glucose transporter 4		Adult females	Impaired ovarian development Reduced fecundity	Li et al. (2021)
Culex pipiens	Glycogen synthase ATP-binding cassette transporter Low-density lipoprotein receptor chaperone		Adult females	Reduced glycogen and lipid	Olademehin et al. (2020)
Tribolium castaneum	AMP-activated protein kinase (AMPK)		20-day-old larvae	Increased whole-body triglyceride and glucose Reduced trehalose	Jiang et al. (2020)

(continued)

Table 2 (continued)

Species	Target gene	Mortality	Stage	Other Effects	Reference
Nilaparvata lugens	Calmodulin (CaM)	40–65% at 6d 66–84% at 10d	Nymph	Feduced egg hatching Sterile female formation	Wang et al. (2018)
Leptinotarsa decemlineata	Calmodulin (CaM) Calcineurin B (CaNB) Regucalcin (RgN)	%80 (CaM)	Larvae (L4)	Reduced fat content (CaNB, RgN) Increased weight (RgN) Reduced weight (CaNB)	Doğan et al. (2021)
Drosophila melanogaster	Sarco/endoplasmic reticulum Ca^{2+}-ATPase (SERCA)		Adult flies	Reduced triglyceride and lipid droplet size (adipocytes)	Baumbach et al. (2014a) and Bi et al. (2014)
Drosophila melanogaster	Inositol trisphosphate receptor (IP_3R) Ryanodine receptor (RyR)		Adult flies	Increased triglyceride and lipid droplet size (adipocytes)	Bi et al. (2014)
Drosophila melanogaster	Stromal interaction molecule (STIM)		Adult flies	Obesity Increased triglyceride and lipid droplet size (adipocytes) Impaired lipid mobilization	Xu et al. (2019b) Baumbach et al. (2014b)
Nilaparvata lugens	Hepatocyte nuclear factor 4		Adult females	Reduced egg hatching Developmental arrest in embryo	Cheng et al. (2020)
Leptinotarsa decemlineata	Taiman	~10%	Larvae	Reduced foliage consumption and nutrient absorption Depleted lipid stores	Xu et al. (2019a)
Drosophila melanogaster	Sterol-regulatory element–binding protein (SREBP)		Adult flies	Reduced total body triglyceride, and cell and organ size	Porstmann et al. (2008)
Drosophila melanogaster	Seven-up (Svp)		Larvae	Increased triglyceride in the larval fat body	Musselman et al. (2018)
Rhodnius prolixus	Mitochondrial trifunctional protein A subunit (HADHA)		Adults	Increased triglyceride and larger lipid droplets in the fat body during starvation Reduced fecundity after feeding Lowered forced flight capacity	Arêdes et al. (2022)
Rhodnius prolixus	Autophagy-related genes (ATG6 and ATG8)	Up to %100 after 40 days	Adult females	Reduced survival Lower locomotor and flight activities throughout starvation Increased triglyceride during starvation	Santos-Araujo et al. (2020)

ILP, acyl-CoA binding protein, trehalose phosphate synthase, apolipophorin-III (Table 2) (Parvy et al. 2012; Patnaik et al. 2015; Fu et al. 2016; Shi et al. 2016; Pei et al. 2019; Almeida et al. 2020; Silva-Oliveira et al. 2021; Leyria et al. 2021). As mentioned above, there have been reports on physiological, morphological, and developmental defects, other than lethality, upon silencing. For example, silencing *FAS* in *Culex pipiens* adults leads to a significant reduction in lipid levels, which impairs diapause (Sim and Denlinger 2009). Similarly, lipid accumulation is suppressed by silencing *FAS-2* in female *Colaphellus bowringi* (Coleoptera: Chrysomelidae) adults; however, interestingly, this did not seriously affect the diapause status of adults (Tan et al. 2017). RNAi-mediated knockdown of the gene encoding serine-threonine protein kinase (*AKT*) in the fat body of *Aedes aegypti* (Diptera: Culicidae) inhibits *vitellogenin* expression, an important egg yolk precursor protein in vitellogenesis (Roy et al. 2007). Silencing lipophorin (*Lp*) in *A. aegypti* impairs the development of oocysts of the parasite *Plasmodium* and reduces the number of oocysts by 90% (Cheon et al. 2006). *ACC* silencing leads to a reduction in the fecundity of the brown planthopper, *N. lugens* (Zhang et al. 2015b), and even death when silenced in *Drosophila* oenocytes, but not fatbody tissue (Parvy et al. 2012). More RNAi screens across a range of insects are warranted to further investigate the potential of ACC silencing for pest control purposes. Silencing perilipin 2 (LSD2) leads to significant reduction in lipid content in *L. decemlineata* larvae (Güney et al. 2021), but not death. Genes involved in calcium homeostasis in relation to lipid metabolism are other promising targets for RNAi. For example, silencing calmodulin (*CaM*) impairs molting and ovary development, which leads to female infertility and death, indicating that CaM could be a potential target for pest control (Wang et al. 2018). Similarly, silencing of *CaM* leads to lethality, while silencing phosphatase calcineurin B (*CaNB*) or the senescence marker protein, regucalcin (*RgN*), leads to decreased fat content in *L. decemlineata* larvae (Doğan et al. 2021). Targeting genes involved in cuticular lipid synthesis might be other potential targets for pest control. ATP-binding cassette (ABC) transporters are involved in the transport of cuticular lipids (Balabanidou et al. 2018). Indeed, silencing *TcABCG-4C* gene in the red flour beetle, *Tribolium castaneum* (Coleoptera: Tenebrionidae) resulted in lipid depletion in epicuticle layers and caused 100% mortality when injected into larvae and pre-pupae (Broehan et al. 2013). In summary, targeting lipid metabolism by RNAi definitely has potential for pest control in RNAi-amenable insects, therefore, future studies to explore candidate genes involved in lipid metabolism for sprayable dsRNA-based bioproducts might be quite promising.

2.6 Other Eco-friendly Alternatives

The development of environmentally friendly and effective insecticidal compounds is a major goal in pest management; lipid derivatives offer promising alternatives in this regard. For example, conjugated linoleic acid has insecticidal activity towards the European corn borer, *Ostrinia nubilalis* (Lepidoptera: Pyralidae) (Gereszek et al. 2008). Clements et al. (2019) demonstrated that conjugated linoleic acid increases *L. decemlineata* larval mortality, retards larval development due to anti-feedant effects, and decreases egg viability, which might be related to reduced Δ^9-desaturase activity. In a more recent study on a series of matrine–cholesterol derivatives, two N-phenylsulfonylmatrinic esters had significant insecticidal activity against *Mythimna separate* (Lepidoptera: Noctuidae), while two N-benzylmatrinic esters exhibited aphicidal activity against *Aphis citricola* (Hemiptera: Aphididae) (Xu et al. 2020). In a subsequent study, oximinoether derivatives of cholesterol-containing isoxazoline/isoxazole fragments showed growth inhibitory activity on *M. separate*, oral toxicity on *Plutella xylostella* (Lepidoptera: Plutellidae), and aphicidal activity on *A. citricola*, further indicating the potential of cholesterol derivatives in pest control (Xu et al. 2021).

Another eco-friendly alternative within the frame of this chapter are the "diatomaceous earth (DE)" products that are mainly composed of amorphous hydrated silicates (more than 80% silicon dioxide) (Golob 1997; Korunić 1998). These products are also known as inert dusts and are considered within the scope of physical control (Subramanyam and Roesli 2000). The DE products possess a high absorption capacity with a primary contact effect on insects. Although the MoA of DE products is not fully understood, they likely absorb lipid molecules in the wax layer of the epicuticle, leading to water loss, desiccation, and eventually death (Ebeling 1971; Korunić

1998; Prasantha et al. 2015). Most of the studies on the insecticidal potential of DE products have been conducted against stored product pests (Athanassiou et al. 2004; Ferizli and Beris 2005; Ertürk et al. 2020; Mortazavi et al. 2020; Atay et al. 2023). Further studies are required to better understand the effect of DE products on cuticular lipids and their specific MoA.

Several boron compounds that are relatively less toxic on non-targets have been shown to be effective against various stored product beetle pests with a significant reduction in triglyceride levels in parallel to the down-regulation of *ACC* and *FAS* (Atay et al. 2024; Ertürk et al. 2024). On the other hand, botanical insecticides (plant extracts or derivatives), such as azadirachtin and pyrethrins, are also promising synthetic chemical alternatives due to their relatively lower risk on mammals, other non-targets, and the environment. Botanical insecticides generally target the insect nervous, endocrine, and respiratory systems (Souto et al. 2021); however, various studies also indicated their interference with insect lipid metabolism. For example, grayanotoxin I, a diterpenoid commonly found in plants of the Ericaceae family, was found to significantly suppress survival rate, growth and development, and pupation of *Spodoptera litura* (Lepidoptera: Noctuidae), while targeting genes involved in lipid metabolism, in particular, lipid synthesis, lipolysis, lipid trafficking, and fat body development (Zhou et al. 2023). Additionally, the haemolymph-free fatty acid content and fat body lipids were notably reduced in response to grayanotoxin application (Zhou et al. 2023). In another study, the essential oil of *Piper aduncum* (Piperaceae) was found to deplete fat body lipid content, leading to the inhibition of development and reproduction in the brown stink bug, *Euschistus heros* (Hemiptera: Pentatomidae) (Cossolin et al. 2019). The total midgut lipid content was found to be reduced in *S. littoralis* larvae treated with garlic and lemon oils, while the lipase levels decreased (Ali et al. 2017). Similarly, corn leaf pieces immersed with citronella oil from *Cymbopogon winterianus* led to reduced lipid content, leading to diminished reproduction (Silva et al. 2016). It is noteworthy that the interaction between insect lipid metabolism and botanical insecticides may not be specific; therefore, further studies are necessary to address this interaction.

Another safe pest management strategy depends on the use of entomopathogens, which might also affect lipid metabolism. For example, the budded form of baculoviruses amplifies primarily in the fat body suggesting the virus is likely to interfere with lipid metabolism (Toprak et al. 2005, 2006). Various studies also indicated the accumulation of lipid droplets with increased whole-body total lipid levels in parallel to the up-regulation of de novo lipogenesis genes in response to viral and bacterial infections (Wrońska et al. 2023). Another example is the entomopathogenic fungi; the spores of such fungi attach to the wax layer of the epicuticle in the host and germinate, therefore targeting insect cuticular lipids for successful penetration (Keyhani 2018). Additionally, internal lipids also act as an antifungal barrier and are targeted by fungal enzymes (e.g. lipases), proteins, and secondary metabolites (Keyhani 2018; Wrońska et al. 2023). Notably, prostaglandins and other eicosanoids, the oxygenated metabolites of certain polyunsaturated fatty acids, mediate cellular immunity against entomopathogenic fungi and bacteria, thereby indicating lipid metabolism as a target site for such microorganisms (Stanley 2006; Kim and Stanley 2021). In brief, a combination of environmentally friendly pest control tactics targeting lipid metabolism might be a promising approach.

3 Conclusion

This chapter provides a concise overview of the potential of pathways and genes orchestrating lipid metabolism for pest management by the application of pesticides and other potential control methods interfering with growth and development. Various classes of chemicals (e.g. organophosphates, pyrethroids, neonicotinoids, and chitin synthesis inhibitors) have been shown to affect lipid metabolism by secondary effects, whereas the primary MoA is linked to molecular target sites not directly involved in lipid metabolism. The discovery and development of the insecticide/acaricide class of "lipid biosynthesis inhibitors" has been a milestone in pest control due to their specific effect on ACC, directly involved in fatty acid biosynthesis. The market introduction of the ketoenol insecticides spirodiclofen, spiromesifen, and spirotetramat between 2002 and 2008, and more recently, spiropidion and spidoxamat, provided chemical options to control and manage some of the world's most destructive sucking pests. As the search for chemical alternatives is likely to continue, other approaches such as RNAi provide a new opportunity to control pests by interfering with growth and development. Sprayable dsRNA-based bioinsecticides/bioacaricides principally have the potential to complement chemical pest control in the future, at least in those pests amenable to RNAi. Efficacy of such biologicals depends on various factors; however, the selection of an appropriate target gene is the first and most essential step for development, followed by formulation development facilitating uptake and distribution. Genes encoding proteins involved in insect lipid metabolism, such as ACC, might have great potential in this regard. Although studies in the last two decades have provided significant progress in our understanding of the genetic and biochemical background of insect lipid metabolism, there are still many questions or variables that have yet to be fully understood or resolved. RNAi and genome editing technology will help to improve our understanding of the functions of a broad range of genes involved in lipid metabolism, which might eventually lead to the development of new bioinsecticides targeting insect lipid metabolism or the discovery of novel target sites selectively druggable by synthetic insecticides.

References

Abdel-Fatah RM, Mohamed SM, Aly AA, Sabry AKH (2019) Biochemical characterization of spiromesifen and spirotetramat as lipid synthesis inhibitors on cotton leaf worm, *Spodoptera littoralis*. Bull Natl Res Cent 43(1):65. https://doi.org/10.1186/s42269-019-0107-9

Abrisqueta M, Süren-Castillo S, Maestro JL (2014) Insulin receptor-mediated nutritional signalling regulates juvenile hormone biosynthesis and vitellogenin production in the German cockroach. Insect Biochem Mol Biol 49:14–23. https://doi.org/10.1016/j.ibmb.2014.03.005

Al Baki MA, Lee DW, Jung JK, Kim Y (2019) Insulin signaling mediates previtellogenic development and enhances juvenile hormone-mediated vitellogenesis in a lepidopteran insect. Maruca Vitrata BMC Dev Biol 19(1):1–14. https://doi.org/10.1186/s12861-019-0194-8

Alabaster A, Isoe J, Zhou G, Lee A, Murphy A, Day WA, Miesfeld RL (2011) Deficiencies in acetyl-CoA carboxylase and fatty acid synthase 1 differentially affect eggshell formation and blood meal digestion in *Aedes aegypti*. Insect Biochem Mol Biol 41(12):946–955. https://doi.org/10.1016/j.ibmb.2011.09.004

Ali AM, Mohamed DS, Shaurub ESH, Elsayed AM (2017) Antifeedant activity and some biochemical effects of garlic and lemon essential oils on *Spodoptera littoralis* (Boisduval) (Lepidoptera: Noctuidae). J Entomol Zool Stud 5(3):1476–1482

Alimohamadi N, Samih M, Izadi H, Shahidi Noghabi S (2014) Developmental and biochemical effects of hexaflumuron and spirodiclofen on the ladybird beetle, *Hippodamia variegata* (Goeze) (Coleoptera: Coccinellidae). J Crop Prod 3(3):335–344

Almeida MG, Arêdes DS, Majerowicz D, Færgeman NJ, Knudsen J, Gondim KC (2020) Expression of acyl-CoA-binding protein 5 from *Rhodnius prolixus* and its inhibition by RNA interference. PLoS One 15(1): e0227685. https://doi.org/10.1371/journal.pone.0227685

Alves-Bezerra M, Ramos I, Paula IF, Maya-Monteiro CM, Klett EL, Coleman RA, Gondim KC (2017) Deficiency of glycerol-3-phosphate acyltransferase 1 decreases triacylglycerol storage and induces fatty acid oxidation in insect fat body. Biochim Biophys Acta (BBA), Mol Cell Biol Lipids 1862(3):324–336. https://doi.org/10.1016/j.bbalip.2016.12.004

Araújo RA, Guedes RNC, Oliveira MGA, Ferreira GH (2008) Enhanced activity of carbohydrate-and lipid-metabolizing enzymes in insecticide-resistant populations of the maize weevil. Sitophilus Zeamais Bull Entomol Res 98(4):417–424. https://doi.org/10.1017/S0007485308005737

Arêdes DS, De Paula IF, Santos-Araujo S, Gondim KC (2022) Silencing of mitochondrial trifunctional protein a subunit (HADHA) increases lipid stores, and reduces oviposition and flight capacity in the vector insect *Rhodnius prolixus*. Front Insect Sci 2:885172. https://doi.org/10.3389/finsc.2022.885172

Arrese EL, Soulages JL (2010) Insect fat body: energy, metabolism, and regulation. Annu Rev Entomol 55: 207–225. https://doi.org/10.1146/annurev-ento-112408-085356

Atay T, Alkan M, Ertürk S, Toprak U (2023) Individual and combined effects of α-pinene and a native diatomaceous earth product on control of stored product beetle pests. J Asia Pac Entomol 26(4):102149. https://doi.org/10.1016/j.aspen.2023.102149

Atay T, Ertürk S, Alkan M, Kordali Ş, Yılmaz F, Barış A, Ghanbari S, Doğan C, Toprak U (2024) Boron compounds are effective on *Tribolium castaneum* (Coleoptera: Tenebrionidae): reduced lipogenesis and induced weight loss. Arch Insect Biochem Physiol 115:e22098. https://doi.org/10.1002/arch.22098

Athanassiou CG, Kavallieratos NG, Andris NS (2004) Insecticidal effect of three diatomaceous earth formulations against adults of *Sitophilus oryzae* (Coleoptera:Curculionidae) and *Tribolium confusum* (Coleoptera: Tenebrionidae) on oat, rye and triticale. J Econ Entomol 97:2160e2167. https://doi.org/10.1093/jee/97.6.2160

Badieinia F, Khajehali J, Nauen R, Dermauw W, Van Leeuwen T (2020) Metabolic mechanisms of resistance to spirodiclofen and spiromesifen in Iranian populations of *Panonychus ulmi*. Crop Prot 105166. https://doi.org/10.1016/j.cropro.2020.105166

Bajda S, Dermauw W, Greenhalgh R, Nauen R, Tirry L, Clark RM, Van Leeuwen T (2015) Transcriptome profiling of a spirodiclofen susceptible and resistant strain of the European red mite *Panonychus ulmi* using strand-specific RNA-seq. BMC Genomics 16(1):974. https://doi.org/10.1186/s12864-015-2157-1

Balabanidou V, Kampouraki A, MacLean M, Blomquist GJ, Tittiger C, Juárez MP, Mijailovsky SJ, Chalepakis G, Anthousi A, Lynd A, Antoine S, Hemingway J, Ranson H, Lycett GJ, Vontas J (2016) Cytochrome P450 associated with insecticide resistance catalyzes cuticular hydrocarbon production in *Anopheles gambiae*. Proc Natl Acad Sci USA 113(33): 9268–9273. https://doi.org/10.1073/pnas.1608295113

Balabanidou V, Grigoraki L, Vontas J (2018) Insect cuticle: a critical determinant of insecticide resistance. Curr Opin Insect Sci 27:68–74. https://doi.org/10.1016/j.cois.2018.03.001

Bass C, Nauen R (2023) The molecular mechanisms of insecticide resistance in aphid crop pests. Insect Biochem Mol Biol 156:103937. https://doi.org/10.1016/j.ibmb.2023.103937

Bass C, Puinean AM, Zimmer CT, Denholm I, Field LM, Foster SP, Gutbrod O, Nauen R, Slater R, Williamson MS (2014) The evolution of insecticide resistance in the peach potato aphid, *Myzus persicae*. Insect Biochem Mol Biol 51:41–51. https://doi.org/10.1016/j.ibmb.2014.05.003

Baumbach J, Hummel P, Bickmeyer I, Kowalczyk KM, Frank M, Knorr K, Kühnlein RP (2014a) A drosophila in vivo screen identifies store-operated calcium entry as a key regulator of adiposity. Cell Metab 19(2): 331–343. https://doi.org/10.1016/j.cmet.2013.12.004

Baumbach J, Xu Y, Hehlert P, Kühnlein RP (2014b) Gαq, Gγ1 and Plc21C control drosophila body fat storage. J Genet Genomics 41(5):283–292. https://doi.org/10.1016/j.jgg.2014.03.005

Bensafi-Gheraibia H, Menail AH, Soltani N (2013) Activity of a lipid synthesis inhibitor (spiromesifen) in *Drosophila melanogaster*: lipid levels and peroxydation, and effects on offspring. Bull Soc Zool Fr 138(1/4):189–199

Bi J, Wang W, Liu Z, Huang X, Jiang Q, Liu GY, Huang X (2014) Seipin promotes adipose tissue fat storage through the ER Ca^{2+}-ATPase SERCA. Cell Metab 19(5):861–871. https://doi.org/10.1016/j.cmet.2014

Bielza P, Moreno I, Belando A, Grávalos C, Izquierdo J, Nauen R (2019) Spiromesifen and spirotetramat resistance in field populations of *Bemisia tabaci* Gennadius in Spain. Pest Manag Sci 75(1):45–52. https://doi.org/10.1002/ps.5144

Binh TD, Pham TL, Men TT, Dang TT, Kamei K (2019) LSD-2 dysfunction induces dFoxO-dependent cell death in the wing of *Drosophila melanogaster*. Biochem Biophys Res Commun 509(2):491–497. https://doi.org/10.1016/j.bbrc.2018.12.132

Bouabida H, Tine-djebbar F, Tine S, Soltani N (2017a) Activity of a lipid synthesis inhibitor (spiromesifen) in *Culiseta longiareolata* (Diptera: Culicidae). Asian Pac J Trop Biomed 7(12):1120–1124. https://doi.org/10.1016/j.apjtb.2017.10.015

Bouabida H, Tine-Djebbar F, Tine S, Soltani N (2017b) Activity of spiromesifen on growth and development of *Culex pipiens* (Diptera: Culicidae): toxicological, biometrical and biochemical aspects. J Entomol Zool Stud 5:572–577

Bretschneider T, Benet-Buchholz J, Fischer R, Nauen R (2003) Spirodiclofen and spiromesifen–novel acaricidal and insecticidal tetronic acid derivatives with a new mode of action. CHIMIA Int J Chem 57(11):697–701. https://doi.org/10.2533/000942903777678588

Brinzer RA, Henderson L, Marchiondo AA, Woods DJ, Davies SA, Dow JA (2015) Metabolomic profiling of permethrin-treated *Drosophila melanogaster* identifies a role for tryptophan catabolism in insecticide survival. Insect Biochem Mol Biol 67:74–86. https://doi.org/10.1016/j.ibmb.2015.09.009

Broehan G, Kroeger T, Lorenzen M, Merzendorfer H (2013) Functional analysis of the ATP-binding cassette (ABC) transporter gene family of *Tribolium castaneum*. BMC Genomics 14(1):6. https://doi.org/10.1186/1471-2164-14-6

Buszczak M, Lu X, Segraves WA, Chang TY, Cooley L (2002) Mutations in the midway gene disrupt a drosophila acyl coenzyme a: diacylglycerol acyltransferase. Genetics 160(4):1511–1518. https://doi.org/10.1093/genetics/160.4.1511

Chen J, Zhang D, Yao Q, Zhang J, Dong X, Tian H, Chen J, Zhang W (2010) Feeding-based RNA interference of a trehalose phosphate synthase gene in the brown planthopper, *Nilaparvata lugens*. Insect Mol Biol 19(6):777–786. https://doi.org/10.1111/j.1365-2583.2010.01038.x

Chen JX, Lyu ZH, Wang CY, Cheng J, Lin T (2020) RNA interference of a trehalose-6-phosphate synthase gene reveals its roles in the biosynthesis of chitin and lipids in *Heortia vitessoides* (Lepidoptera: Crambidae).

Insect Sci 27(2):212–223. https://doi.org/10.1111/1744-7917.12650

Cheng Y, Li Y, Li W, Song Y, Zeng R, Lu K (2020) Effect of hepatocyte nuclear factor 4 on the fecundity of *Nilaparvata lugens*: insights from RNA interference combined with transcriptomic analysis. Genomics 112(6):4585–4594. https://doi.org/10.1016/j.ygeno.2020.08.002

Cheng LY, Hou DY, Sun QZ, Yu SJ, Li SC, Liu HQ, Cong L, Ran C (2022) Biochemical and molecular analysis of field resistance to spirodiclofen in *Panonychus citri* (McGregor). Insect 13(11):1011. https://doi.org/10.3390/insects13111011

Cheon H-M, Woon S, Bian G, Park J-H, Raikhel AS (2006) Regulation of lipid metabolism genes, lipid carrier protein lipophorin, and its receptor during immune challenge in the mosquito *Aedes aegypti*. J Biol Chem 281: 8426–8435. https://doi.org/10.1074/jbc.M510957200

Clements J, Groves RL, Cava J, Barry CC, Chapman S, Olson JM (2019) Conjugated linoleic acid as a novel insecticide targeting the agricultural pest *Leptinotarsa decemlineata*. PLoS One 14(11):e0220830. https://doi.org/10.1371/journal.pone.0220830

Clements J, Olson JM, Sanchez-Sedillo B, Bradford B, Groves RL (2020) Changes in emergence phenology, fatty acid composition, and xenobiotic-metabolizing enzyme expression is associated with increased insecticide resistance in the Colorado potato beetle. Arch Insect Biochem Physiol 103(3):e21630. https://doi.org/10.1002/arch.21630

Cossolin J, Pereira M, Martínez LC, Turchen LM, Fiaz M, Bozdoğan H, Serrão JE (2019) Cytotoxicity of Piper aduncum (Piperaceae) essential oil in brown stink bug *Euschistus heros* (Heteroptera: Pentatomidae). Ecotoxicology 28(7):763–770. https://doi.org/10.1007/s10646-019-02072-8

Dângelo RAC, Michereff-Filho M, Campos MR, Da Silva PS, Guedes RNC (2018) Insecticide resistance and control failure likelihood of the whitefly *Bemisia tabaci* (MEAM1; B biotype): a Neotropical scenario. Ann Appl Biol 172(1):88–99. https://doi.org/10.1111/aab.12404

Demaeght P, Dermauw W, Tsakireli D, Khajehali J, Nauen R, Tirry L, Vontas J, Lümmen P, Van Leeuwen T (2013) Molecular analysis of resistance to acaricidal spirocyclic tetronic acids in *Tetranychus urticae*: CYP392E10 metabolizes spirodiclofen, but not its corresponding enol. Insect Biochem Mol Biol 43(6): 544–554. https://doi.org/10.1016/j.ibmb.2013.03.007

Deng P, Xu QY, Fu KY, Guo WC, Li GQ (2018) RNA interference against the putative insulin receptor substrate gene Chico affects metamorphosis in *Leptinotarsa decemlineata*. Insect Biochem Mol Biol 103:1–11. https://doi.org/10.1016/j.ibmb.2018.10.001

De Rouck S, İnak E, Dermauw W, Van Leeuwen T (2023). A review of the molecular mechanisms of acaricide resistance in mites and ticks. Insect Biochem Mol Biol 159:103981. https://doi.org/10.1016/j.ibmb.2023.103981

Devine MD (1997) Mechanisms of resistance to acetyl-coenzyme a carboxylase inhibitors: a review. Pestic Sci 51(3):259–264. https://doi.org/10.1002/(SICI)1096-9063(199711)51:3<259::AID-PS644>3.0.CO;2-S

Dhadialla TS, Carlson GR, Le DP (1998) New insecticides with ecdysteroidal and juvenile hormone activity. Annu Rev Entomol 43:545–569. https://doi.org/10.1146/annurev.ento.43.1.545

Doğan C, Hänniger S, Heckel DG, Coutu C, Hegedus DD, Crubaugh L, Groves RL, Amutkan Mutlu D, Suludere Z, Bayram Ş, Toprak U (2021) Characterization of calcium signaling proteins from the fat body of the Colorado potato beetle, *Leptinotarsa decemlineata* (Coleoptera: Chrysomelidae): implications for diapause and lipid metabolism. Insect Biochem Mol Biol 133:103549. https://doi.org/10.1016/j.ibmb.2021.103549

Dong X, Chen J, Xu R, Li X, Wang Y, Pan X, Zhang C, Li Y, Wang F, Li C (2022) Molecular identification and lipid mobilization role of adipokinetic hormone receptor in *Spodoptera litura* (F.). Bull Entomol Res 112(6): 758–765. https://doi.org/10.1017/S0007485322000141

Douris V, Denecke S, Van Leeuwen T, Bass C, Nauen R, Vontas J (2020) Using CRISPR/Cas9 genome modification to understand the genetic basis of insecticide resistance: drosophila and beyond. Insect Biochem Mol Biol 167:104595. https://doi.org/10.1016/j.pestbp.2020.104595

Du M, Yin X, Zhang S, Zhu B, Song Q, An S (2012) Identification of lipases involved in PBAN stimulated pheromone production in *Bombyx mori* using the DGE and RNAi approaches. PLoS One 7(2):e31045. https://doi.org/10.1371/journal.pone.0031045

Ebeling W (1971) Sorptive dusts for pest control. Annu Rev Entomol 16:123e158. https://doi.org/10.1146/annurev.en.16.010171.001011

Enayati AA, Ranson H, Hemingway J (2005) Insect glutathione transferases and insecticide resistance. Insect Mol Biol 14:3–8. https://doi.org/10.1111/j.1365-2583.2004.00529.x

Ertürk S, Atay T, Toprak U, Alkan M (2020) The efficacy of different surface applications of wettable powder formulation of Detech® diatomaceous earth against rice weevil, *Sitophilus oryzae* L. (Coleoptera: Curculionidae). J Stored Prod Res 89:101725. https://doi.org/10.1016/j.jspr.2020.101725

Ertürk S, Atay T, Alkan M, Kordali Ş, Yılmaz F, Ghanbari S, Doğan C, Toprak U (2024) Boron compounds are effective on *Sitophilus granarius* (Coleoptera: Curculionidae) and *Rhyzopertha dominica* (Coleoptera: Bostrichidae). J Stored Prod Res 107: 102337. https://doi.org/10.1016/j.jspr.2024.102337

Feng X, Li M, Liu N (2018) Carboxylesterase genes in pyrethroid resistant house flies, *Musca domestica*. Insect Biochem Mol Biol 92:30–39. https://doi.org/10.1016/j.ibmb.2017.11.007

Ferizli AG, Beris G (2005) Mortality and F1 progeny of the lesser grain borer, *Rhyzopertha Dominica* (F), on wheat treated with diatomaceous earth: effects of rate, exposure period and relative humidity. Pest Manag Sci 61:1103–1109. https://doi.org/10.1002/ps.1094

Fernández E, Grávalos C, Haro PJ, Cifuentes D, Bielza P (2009) Insecticide resistance status of *Bemisia tabaci* Q-biotype in South-Eastern Spain. Pest Manag Sci 65(8):885–891. https://doi.org/10.1002/ps.1769

Fónagy A (2006) Insect neuropeptides and their potential application for pest control. Acta Phytopathol Entomol Hung 41:137–152. https://doi.org/10.1556/APhyt.41.2006.1-2.13

Fotouhi K, Fazel MM, Kavousi A (2015) Effects of pyriproxyfen on bioenergetic resources of *Leptinotarsa decemlineata* (say) (Coleoptera: Chrysomelidae). Turkish J Entomol 39(1):11–22. https://doi.org/10.16970/ted.14717

Fu KY, Zhu TT, Guo WC, Ahmat T, Li GQ (2016) Knockdown of a putative insulin-like peptide gene LdILP2 in *Leptinotarsa decemlineata* by RNA interference impairs pupation and adult emergence. Gene 581(2):170–177. https://doi.org/10.1016/j.gene.2016.01.037

Geo A, Pathak H, Kujur AE, Sudhakar SR, Kannan NN (2019) Short neuropeptide F regulates the starvation 1 mediated enhanced locomotor activity in drosophila. BioRxiv 764688. https://doi.org/10.1101/764688

Gereszek LJ, Coats JR, Beitz DC (2008) Effects of dietary conjugated linoleic acid on European corn borer (Lepidoptera: Crambidae) survival, fatty acid profile, and fecundity. Ann Entomol Soc Am 101(2):430–438. https://doi.org/10.1603/0013-8746(2008)101[430:EODCLA]2.0.CO;2

Ghasemi A, Sendi J, Ghadamyari M (2010) Physiological and biochemical effect of pyriproxyfen on Indian meal moth *Plodia interpunctella* (Hübner) (Lepidoptera: Pyralidae). Journal of Plant Protection Research 50(4):416–422. https://doi.org/10.2478/v10045-010-0070-9

Goldsworthy GJ, Opoku-Ware K, Mullen LM (2005) Adipokinetic hormone and the immune responses of locusts to infection. Ann N Y Acad Sci 1040:106–113. https://doi.org/10.1196/annals.1327.013

Golob P (1997) Current status and future perspectives for inert dusts for control of stored product insects. J Stored Prod Res 33:69–79. https://doi.org/10.1016/S0022-474X(96)00031-8

Gong Y, Shi X, Desneux N, Gao X (2016) Effects of spirotetramat treatments on fecundity and carboxylesterase expression of *Aphis gossypii* glover. Ecotoxicology 25(4):655–663. https://doi.org/10.1007/s10646-016-1624-z

Güney G, Toprak U, Hegedus DD, Bayram Ş, Coutu C, Bekkaoui D, Baldwin D, Heckel DG, Hänniger S, Cedden D, Amutkan Mutlu D, Suludere Z (2021) A look into Colorado potato beetle lipid metabolism through the lens of lipid storage droplet proteins. Insect Biochem Mol Biol 133:103473. https://doi.org/10.1016/j.ibmb.2020.103473

Guo Y, Walther TC, Rao M, Stuurman N, Goshima G, Terayama K, Farese RV (2008) Functional genomic screen reveals genes involved in lipid-droplet formation and utilization. Nature 453:657–661. https://doi.org/10.1038/nature06928

Guo XR, Zheng SC, Liu L, Feng QL (2009) The sterol carrier protein 2/3-oxoacyl-CoA thiolase (SCPx) is involved in cholesterol uptake in the midgut of *Spodoptera litura*: gene cloning, expression, localization and functional analyses. BMC Mol Biol 10:102–120. https://doi.org/10.1186/1471-2199-10-102

Herren JK, Paredes JC, Schüpfer F, Arafah K, Bulet P, Lemaitre B (2014) Insect endosymbiont proliferation is limited by lipid availability. eLife 3:e02964. https://doi.org/10.7554/eLife.02964

Hopkinson JE, Pumpa SM (2019) Baseline susceptibility of *Bemisia tabaci* MEAM 1 (Hemiptera: aleyrodidae) in Australia to spirotetramat, cyantraniliprole and dinotefuran, with reference to pyriproxyfen cross-resistance. Austral Entomol 58(4):762–771. https://doi.org/10.1111/aen.12390

Hou QL, Chen EH, Jiang HB, Wei DD, Gui SH, Wang JJ, Smagghe G (2017) Adipokinetic hormone receptor gene identification and its role in triacylglycerol mobilization and sexual behavior in the oriental fruit fly (*Bactrocera dorsalis*). Insect Biochem Mol Biol 90:1–13. https://doi.org/10.1016/j.ibmb.2017.09.006

Hu J, Wang C, Wang J, You Y, Chen F (2010) Monitoring of resistance to spirodiclofen and five other acaricides in *Panonychus citri* collected from Chinese citrus orchards. Pest Manag Sci 66(9):1025–1030. https://doi.org/10.1002/ps.1978

Ibrahim E, Hejníková M, Shaik H, Doležel D, Kodrík D (2017) Adipokinetic hormone activities in insect body infected by entomopathogenic nematode. J Insect Physiol 98:347–355. https://doi.org/10.1016/j.jinsphys.2017.02.009

Ilias A, Roditakis E, Grispou M, Nauen R, Vontas J, Tsagkarakou A (2012) Efficacy of ketoenols on insecticide resistant field populations of two-spotted spider mite *Tetranychus urticae* and sweet potato whitefly *Bemisia tabaci* from Greece. Crop Prot 42:305–311. https://doi.org/10.1002/ps.1978

İnak E, Alpkent YN, Çobanoğlu S, Dermauw W, Van Leeuwen T (2019) Resistance incidence and presence of resistance mutations in populations of *Tetranychus urticae* from vegetable crops in Turkey. Exp Appl Acarol 78(3):343–360. https://doi.org/10.1007/s10493-019-00398-w

İnak E, Alpkent YN, Çobanoğlu S, Toprak U, Van Leeuwen T (2022) Incidence of spiromesifen resistance and resistance mechanisms in *Tetranychus urticae* populations collected from strawberry production areas in Turkey. Crop Prot 160:106049. https://doi.org/10.1016/j.cropro.2022.106049

İnak E, Demirci B, Vandenhole M, Söylemezoğlu G, Van Leeuwen T, Toprak U (2023) Molecular mechanisms of resistance to spirodiclofen and spiromesifen in *Tetranychus urticae*. Crop Prot 172:106343. https://doi.org/10.1016/j.cropro.2023.106343

Jackson C, Liu J-W, Carr P, Younus F, Coppin C, Meirelles T, Lethier M, Pandey G, Ollis DL, Russell RJ, Weik M, Oakeshott J (2013) Structure and function of an insect α-carboxylesterase (αEsterase7) associated with insecticide resistance. Proc Natl Acad Sci USA 110(25):10177–10182. https://doi.org/10.1073/pnas.1304097110

Jeschke P, Fischer R, Nauen R (2019) Inhibitors of lipid synthesis: acetyl-CoA carboxylase inhibitors. In: Jeschke P, Witschel M, Krämer W, Schirmer U (eds) Modern crop protection compounds. Wiley, Ed, pp 1202–1222

Ji Y, Zheng H, Zhang C, Tan X, He C, Fu B, Du T, Liang J, Wei X, Gong P, Liu S, Yang J, Huang M, Yin C, Xue H, Hu J, Du H, Xie W, Yang X, Zhang Y (2023) Dynamic monitoring of the insecticide resistance status of *Bemisia tabaci* across China from 2019–2021. Pest Manag Sci 80:341–354. https://doi.org/10.1002/ps.7763

Jiang H, Zhang N, Ji C, Meng X, Qian K, Zheng Y, Wang J (2020) Metabolic and transcriptome responses of RNAi-mediated AMPKα knockdown in *Tribolium castaneum*. BMC Genomics 21(1):1–13. https://doi.org/10.1186/s12864-020-07070-3

Jones C, Haji K, Khatib B, Bagi J, Mcha J, Devine GJ, Daley M, Kabula B, Ali AS, Majambere S, Ranson H (2013) The dynamics of pyrethroid resistance in *anopheles arabiensis* from Zanzibar and an assessment of the underlying genetic basis. Parasites Vectors 6: 343. https://doi.org/10.1186/1756-3305-6-343

Kapantaidaki DE, Sadikoglou E, Tsakireli D, Kampanis V, Stavrakaki M, Schorn C, Ilias A, Riga M, Tsiamis G, Nauen R, Skavdis G, Vontas J, Tsagkarakou A (2018) Insecticide resistance in *Trialeurodes vaporariorum* populations and novel diagnostics for kdr mutations. Pest Manag Sci 74(1): 59–69. https://doi.org/10.1002/ps.4674

Karatolos N, Williamson MS, Denholm I, Gorman K, Ffrench-Constant R, Nauen R (2012) Resistance to spiromesifen in *Trialeurodes vaporariorum* is associated with a single amino acid replacement in its target enzyme acetyl-coenzyme a carboxylase. Insect Mol Biol 21(3):327–334. https://doi.org/10.1111/j.1365-2583.2012.01136.x

Keyhani NO (2018) Lipid biology in fungal stress and virulence: Entomopathogenic fungi. Fungal Biol 122(6):420–429. https://doi.org/10.1016/j.funbio.2017.07.003

Khajehali J, Van Nieuwenhuyse P, Demaeght P, Tirry L, Van Leeuwen T (2011) Acaricide resistance and resistance mechanisms in *Tetranychus urticae* populations from rose greenhouses in The Netherlands. Pest Manag Sci 67(11):1424–1433. https://doi.org/10.1002/ps.2191

Kim Y, Stanley D (2021) Eicosanoid signaling in insect immunology: new genes and unresolved issues. Genes (Basel) 12(2):211. https://doi.org/10.3390/genes12020211

King-Jones K, Horner M, Lam G, Thummel C (2006) The DHR96 nuclear receptor regulates xenobiotic responses in *Drosophila*. Cell Metab 4(1):37–48. https://doi.org/10.1016/j.cmet.2006.06.006

Kissoum N, Soltani N (2016) Spiromesifen, an insecticide inhibitor of lipid synthesis, affects the amounts of carbohydrates, glycogen and the activity of lactate dehydrogenase in *Drosophila melanogaster*. J Entomol Zool Stud 4(1):452–456

Kissoum N, Bensafi-Gheraibia H, Hamida ZC, Soltani N (2020) Evaluation of the pesticide Oberon on a model organism *Drosophila melanogaster* via topical toxicity test on biochemical and reproductive parameters. Comp Biochem Physiol, Part C: Toxicol Pharmacol 228:108666. https://doi.org/10.1016/j.cbpc.2019.108666

Kliot A, Ghanim M (2012) Fitness costs associated with insecticide resistance. Pest Manag Sci 68(11): 1431–1437. https://doi.org/10.1002/ps.3395

Kodrík D, Plavšin I, Velki M, Stašková T (2015) Enhancement of insecticide efficacy by adipokinetic hormones. In: Montgomery J (ed) Insecticides: occurrence, global threats and ecological impact. Nova Science Publishers, Inc., pp 77–91

Kontsedalov S, Gottlieb Y, Ishaaya I, Nauen R, Horowitz R, Ghanim M (2009) Toxicity of spiromesifen to the developmental stages of *Bemisia tabaci* biotype B. Pest Manag Sci 65(1):5–13. https://doi.org/10.1002/ps.1636

Konuma T, Morooka N, Nagasawa H, Nagata S (2012) Knockdown of the adipokinetic hormone receptor increases feeding frequency in the two-spotted cricket *Gryllus bimaculatus*. Endocrinology 153(7): 3111–3122. https://doi.org/10.1210/en.2011-1533

Korunić Z (1998) Diatomaceous earths, a group of natural insecticides. J Stored Prod Res 34:87e97. https://doi.org/10.1016/S0022-474X(97)00039-8

Kramer T, Nauen R (2011) Monitoring of spirodiclofen susceptibility in field populations of European red mites, *Panonychus ulmi* (Koch)(Acari: Tetranychidae), and the cross-resistance pattern of a laboratory-selected strain. Pest Manag Sci 67(10):1285–1293. https://doi.org/10.1002/ps.2184

Kubrak O, Lushchak O, Zandawala M, Nässel DR (2016) Systemic corazonin signalling modulates stress responses and metabolism in drosophila. Open Biol 6(11):160152. https://doi.org/10.1098/rsob.160152

Lee K, Kwon O, Lee JH, Kwon K, Min K, Jung S, Yu K (2008) Drosophila short neuropeptide F signalling regulates growth by ERK-mediated insulin signalling. Nat Cell Biol 10:468–475. https://doi.org/10.1038/ncb1710

Lee SH, Kim YH, Kwon DH, Cha DJ, Kim JH (2015) Mutation and duplication of arthropod acetylcholinesterase: implications for pesticide resistance and tolerance. Pestic Biochem Physiol 120:118–124. https://doi.org/10.1016/j.pestbp.2014.11.004

Leyria J, Orchard I, Lange AB (2021) The involvement of insulin/ToR signaling pathway in reproductive performance of *Rhodnius prolixus*. Insect Biochem Mol Biol 130:103526. https://doi.org/10.1016/j.ibmb.2021.103526

Li X, Schuler MA, Berenbaum MR (2007) Molecular mechanisms of metabolic resistance to synthetic and natural xenobiotics. Annu Rev Entomol 52:231–253. https://doi.org/10.1146/annurev.ento.51.110104.151104

Li F, Hu J, Tian J, Xu K, Ni M, Wang B, Shen W, Li B (2016) Effects of phoxim on nutrient metabolism and insulin signaling pathway in silkworm midgut. Chemosphere 146:478–485. https://doi.org/10.1016/j.chemosphere.2015.12.032

Li YN, Ren XB, Liu ZC, Ye B, Zhao ZJ, Fan Q, Liu YB, Zhang JN, Li WL (2021) Insulin-like peptide and foxO mediate the trehalose catabolism enhancement during the diapause termination period in the Chinese oak silkworm (*Antheraea pernyi*). Insect 12(9):784. https://doi.org/10.3390/insects12090784

Linda PC, Kim YS, Tong L (2010) Mechanism for the inhibition of the carboxyltransferase domain of acetyl-coenzyme a carboxylase by pinoxaden. Proc Natl Acad Sci USA 107(51):22072–22077. https://doi.org/10.1073/pnas.1012039107

Liu L, Dai R, Yang H, Jin D (2016) Sublethal effects of triazophos on the life table parameters of *Sogatella furcifera* (Hemiptera: Delphacidae). Fla Entomol 99: 292–296. https://doi.org/10.1653/024.099.0221

Lopes KVG, Silva LB, Reis AP, Oliveira MGA, Guedes RNC (2010) Modified α-amylase activity among insecticide-resistant and-susceptible strains of the maize weevil, *Sitophilus zeamais*. J Insect Physiol 56(9):1050–1057. https://doi.org/10.1016/j.jinsphys.2010.02.020

Lu K, Chen X, Li Y, Li W, Zhou Q (2018a) Lipophorin receptor regulates *Nilaparvata lugens* fecundity by promoting lipid accumulation and vitellogenin biosynthesis. Comp Biochem Physiol A Mol Integr Physiol 219: 28–37. https://doi.org/10.1016/j.cbpa.2018.02.008

Lu K, Zhou J, Chen X, Li W, Li Y, Cheng Y, Yan J, You K, Yuan Z, Zhou Q (2018b) Deficiency of Brummer impaires lipid mobilization and JH-mediated vitellogenesis in the brown planthopper, *Nilaparvata lugens*. Front Phys 9:1535. https://doi.org/10.3389/fphys.2018.01535

Lueke B, Douris V, Hopkinson JE, Maiwald F, Hertlein G, Papapostolou KM, Bielza P, Tsagkarakou A, Van Leeuwen T, Bass C, Vontas J, Nauen R (2020) Identification and functional characterization of a novel acetyl-CoA carboxylase mutation associated with ketoenol resistance in *Bemisia tabaci*. Pestic Biochem Physiol 104583. https://doi.org/10.1016/j.pestbp.2020.104583

Lümmen P, Khajehali J, Luther K, Van Leeuwen T (2014) The cyclic keto-enol insecticide spirotetramat inhibits insect and spider mite acetyl-CoA carboxylases by interfering with the carboxyltransferase partial reaction. Insect Biochem Mol Biol 55:1–8. https://doi.org/10.1016/j.ibmb.2014.09.010

Majerowicz D, Gondim KC (2013) Insect lipid metabolism: insights into gene expression regulation. In: Mandal SS (ed) Recent trends in gene expression. Nova Science Publishers, Arlington, pp 147–190

Mann RS, Schuster DJ, Cordero R, Toapanta M (2012) Baseline toxicity of spiromesifen to biotype B of *Bemisia tabaci* in Florida. Fla Entomol 95(1):95–98. https://doi.org/10.1653/024.095.0115

Marčić D, Perić P, Petronijević S, Prijović M, Drobnjaković T (2011) Cyclic ketoenols-acaricides and insecticides with a novel mode of action. Pestic Phytomed 26:185–195. https://doi.org/10.2298/PIF1103185M

Marčić D, Petronijevic S, Drobnjakovic T, Prijovic M, Peric P, Milenkovic S (2012) The effects of spirotetramat on life history traits and population growth of *Tetranychus urticae* (Acari: Tetranychidae). Exp Appl Acarol 56(2):113–122. https://doi.org/10.1007/s10493-011-9500-2

Marnett LJ (1999) Lipid peroxidation-DNA damage by malondialdehyde. Mutat Res-Fund Mol Mech Mutagen 424(1–2):83–95. https://doi.org/10.1016/S0027-5107(99)00010-X

Martinez J, Patkaniowska A, Urlaub H, Lührmann R, Tuschl T (2002) Single-stranded antisense siRNAs guide target RNA cleavage in RNAi. Cell 110(5): 563–574. https://doi.org/10.1016/S0092-8674(02)00908-X

Mehlhorn S, Hunnekuhl VS, Geibel S, Nauen R, Bucher G (2021) Establishing RNAi for basic research and pest control and identification of the most efficient target genes for pest control: a brief guide. Front Zool 18:1–16. https://doi.org/10.1186/s12983-021-00444-7

Mirhaghparast S, Zibaee A, Sendi J, Hoda H, Fazeli-Dinan M (2015) Immune and metabolic responses of *Chilo suppressalis* Walker (Lepidoptera: Crambidae) larvae to an insect growth regulator, hexaflumuron. Pestic Biochem Physiol 125:69–77. https://doi.org/10.1016/j.pestbp.2015.05.007

Mnzava AP, Knox TB, Temu EA, Trett A, Fornadel C, Hemingway J, Renshaw M (2015) Implementation of the global plan for insecticide resistance management in malaria vectors: progress, challenges and the way forward. Malar J 14(1):173. https://doi.org/10.1186/s12936-015-0693-4

Mortazavi H, Toprak U, Emekci M, Bağcı F, Ferizli AG (2020) Persistence of diatomaceous earth, SilicoSec® against three stored grain beetles. J Stored Prod Res 89: 101724. https://doi.org/10.1016/j.jspr.2020.101724

Muehlebach M, Buchholz A, Zambach W, Schaetzer J, Daniels M, Hueter O, Kloer DP, Lind R, Maienfisch P, Pierce A, Pitterna T, Smejkal T, Stafford D, Wildsmith L (2020) Spiro N-methoxy piperidine ring containing aryldiones for thecontrol of sucking insects and mites: discovery of spiropidion. Pest Manag Sci 76(10): 3440–3450. https://doi.org/10.1002/ps.5743

Mullen L, Goldsworthy G (2006) Immune responses of locusts to challenge with the pathogenic fungus *Metarhizium* or high doses of laminarin. J Insect Physiol 52(4):389–398. https://doi.org/10.1016/j.jinsphys.2005.10.008

Müller V, Maiwald F, Haack LC, Köhler H, Nauen R (2024) Mapping and characterization of target-site resistance to cyclic ketoenol insecticides in cabbage whiteflies, Aleyrodes proletella (Hemiptera: Aleyrodidae). Insects 15:178

Musselman LP, Fink J, Maier E, Gatto J, Brent M, Baranski T (2018) Seven-up is a novel regulator of insülin signaling. Genetics 208(4):1643–1656. https://doi.org/10.1534/genetics.118.300770

Nauen R (2005) Spirodiclofen: mode of action and resistance risk assessment in Tetranychid Pest mites. J Pestic Sci 30:272–274. https://doi.org/10.1584/jpestics.30.272

Nauen R, Bretschneider T, Elbert A, Fischer R, Tieman R (2003) Spirodiclofen and spiromesifen. Pestic Outlook 14(6):243–246

Nauen R, Reckmann U, Thomzik J, Thielert W (2008) Biological profile of spirotetramat (Movento®)–a new two-way systemic (ambimobile) insecticide against sucking pest species. Bayer CropSci J 61(2): 245–278

Ohnishi A, Hashimoto K, Imai K, Matsumoto S (2009) Functional characterization of the *Bombyx mori* fatty acid transport protein (BmFATP) within the silkmoth pheromone gland. J Biol Chem 284(8):5128–5136. https://doi.org/10.1074/jbc.M806072200

Olademehin OP, Liu C, Rimal B, Adegboyega NF, Chen F, Sim C, Kim SJ (2020) Dsi-RNA knockdown of genes regulated by foxo reduces glycogen and lipid accumulations in diapausing *Culex pipiens*. Sci Rep 10(1):1–13. https://doi.org/10.1038/s41598-020-74292-6

Ouyang Y, Montez GH, Liu L, Grafton-Cardwell EE (2012) Spirodiclofen and spirotetramat bioassays for monitoring resistance in citrus red mite, *Panonychus citri* (Acari: Tetranychidae). Pest Manag Sci 68(5): 781–787. https://doi.org/10.1002/ps.2326

Ovčarenko I, Lindström L, Saikkonen K, Vänninen I (2014) Variation in mortality among populations is higher for pymetrozine than for imidacloprid and spiromesifen in *Trialeurodes vaporariorum* in greenhouses in Finland. Pest Manag Sci 70(10): 1524–1530. https://doi.org/10.1002/ps.3766

Pan Y, Yang C, Gao X, Peng T, Bi R, Xi J, Xin X, Zhu E, Wu Y, Shang Q (2015) Spirotetramat resistance adaption analysis of *Aphis gossypii* glover by transcriptomic survey. Pestic Biochem Physiol 124: 73–80. https://doi.org/10.1016/j.pestbp.2015.04.007

Pan Y, Zhu E, Gao X, Nauen R, Xi J, Peng T, Wei X, Zheng C, Shang Q (2017) Novel mutations and expression changes of acetyl-coenzyme a carboxylase are associated with spirotetramat resistance in *Aphis gossypii* glover. Insect Mol Biol 26(4):383–391. https://doi.org/10.1111/imb.12300

Pan Y, Chai P, Zheng C, Xu H, Wu Y, Gao X, Xi J, Shang Q (2018) Contribution of cytochrome P450 monooxygenase CYP380C6 to spirotetramat resistance in *Aphis gossypii* glover. Pestic Biochem Physiol 148:182–189. https://doi.org/10.1016/j.pestbp.2018.04.015

Pan Y, Wen S, Chen X, Gao X, Zeng X, Liu X, Tian F, Shang Q (2020) UDP-glycosyltransferases contribute to spirotetramat resistance in *Aphis gossypii* glover. Pestic Biochem Physiol 104565. https://doi.org/10.1016/j.pestbp.2020.104565

Pang R, Chen M, Liang Z, Yue X, Ge H, Zhang W (2016) Functional analysis of CYP6ER1, a P450 gene associated with imidacloprid resistance in *Nilaparvata lugens*. Sci Rep 6:34992. https://doi.org/10.1038/srep34992

Pang R, Qiu J, Li T, Yang P, Yue L, Pan Y, Zhang W (2017) The regulation of lipid metabolism by a hypothetical P-loop NTPase and its impact on fecundity of the brown planthopper. Biochim Biophys Acta Gen Subj 1861(7):1750–1758. https://doi.org/10.1016/j.bbagen.2017.03.011

Papapostolou KM, Riga M, Charamis J, Skoufa E, Souchlas V, Ilias A, Dermauw W, Ioannidis P, Van Leeuwen T, Vontas J (2021) Identification and characterization of striking multiple-insecticide resistance in a *Tetranychus urticae* field population from Greece. Pest Manag Sci 77(2):666–676. https://doi.org/10.1002/ps.6136

Parvy JP, Napal L, Rubin T, Poidevin M, Perrin L, Wicker-Thomas C, Montagne J (2012) *Drosophila melanogaster* acetyl-CoA-carboxylase sustains a fatty acid–dependent remote signal to waterproof the respiratory system. PLoS Genet 8(8):e1002925. https://doi.org/10.1371/journal.pgen.1002925

Patnaik BB, Patnaik HH, Park KB, Jo YH, Lee YS, Han YS (2015) Silencing of apolipophorin-III causes abnormal adult morphological phenotype and susceptibility to L isteria monocytogenes infection in *Tenebrio molitor*. Entomol Res 45(2):116–121. https://doi.org/10.1111/1748-5967.12099

Pei XJ, Chen N, Bai Y, Qiao JW, Li S, Fan YL, Liu TX (2019) BgFas1: a fatty acid synthase gene required for both hydrocarbon and cuticular fatty acid biosynthesis in the German cockroach, *Blattella germanica* (L.). Insect Biochem Mol Biol 112:103203. https://doi.org/10.1016/j.ibmb.2019.103203

Pei XJ, Fan YL, Bai Y, Bai TT, Schal C, Zhang ZF, Chen N, Li S, Liu TX (2021) Modulation of fatty acid elongation in cockroaches sustains sexually dimorphic hydrocarbons and female attractiveness. PLoS Biol 19(7):e3001330. https://doi.org/10.1371/journal.pbio.3001330

Pei XJ, Bai TT, Zhang ZF, Chen N, Li S, Fan YL, Liu TX (2022) Two putative fatty acid synthetic genes of BgFas3 and BgElo1 are responsible for respiratory waterproofing in *Blattella germanica*. Insect Sci 29(1):33–50. https://doi.org/10.1111/1744-7917.12900

Peng T, Pan Y, Yang C, Gao X, Xi J, Wu Y, Huang X, Zhu E, Xin X, Zhan C, Shang Q (2016) Overexpression of CYP6A2 is associated with spirotetramat resistance and cross-resistance in the resistant strain of *Aphis gossypii* glover. Pestic Biochem Physiol 126:64–69. https://doi.org/10.1016/j.pestbp.2015.07.008

Peng Z, Zheng H, Xie W, Wang S, Wu Q, Zhang Y (2017) Field resistance monitoring of the immature stages of the whitefly *Bemisia tabaci* to spirotetramat in China. Crop Prot 98:243–247. https://doi.org/10.1016/j.cropro.2017.04.009

Plavšin I, Stašková T, Šerý M, Smýkal V, Hackenberger B, Kodrík D (2015) Hormonal enhancement of insecticide efficacy in *Tribolium castaneum*: oxidative stress and metabolic aspects. Comp Biochem Physiol, Part C: Toxicol Pharmacol 170:19–27. https://doi.org/10.1016/j.cbpc.2015.01.005

Porstmann T, Santos CR, Griffiths B, Cully M, Wu M, Leevers S, Schulze A (2008) SREBP activity is regulated by mTORC1 and contributes to AKT-dependent cell growth. Cell Metab 8(3):224–236. https://doi.org/10.1016/j.cmet.2008.07.007

Pospisilik J, Schramek D, Schnidar H, Cronin S, Nehme N, Zhang X, Knauf C, Cani PD, Aumayr K, Todoric J, Bayer M, Haschemi A, Puviindran V, Tar K, Orthofer M, Neely GG, Dietzl G, Manoukian A, Funovics M, Prager G, Wagner O, Ferrandon D, Aberger F, Hui CC, Esterbauer H, Penninger J (2010) *Drosophila* genome-wide obesity screen reveals hedgehog as a determinant of brown versus white adipose cell fate. Cell 140(1):148–160. https://doi.org/10.1016/j.cell.2009.12.027

Prabhaker N, Castle SJ, Buckelew L, Toscano NC (2008) Baseline susceptibility of *Bemisia tabaci* B biotype (Hemiptera: Aleyrodidae) populations from California and Arizona to spiromesifen. J Econ Entomol 101(1):174–181. https://doi.org/10.1603/EC13429

Prabhaker N, Castle S, Perring TM (2014) Baseline susceptibility of *Bemisia tabaci* B biotype (Hemiptera: Aleyrodidae) populations from California and Arizona to spirotetramat. J Econ Entomol 107(2):773–780. https://doi.org/10.1603/EC13429

Prasantha BDR, Reichmuth C, Adler C, Felgentreu D (2015) Lipid adsorption of diatomaceous earths and increased water permeability in the epicuticle layer of the cowpea weevil *Callosobruchus maculatus* (F.) and the bean weevil *Acanthoscelides obtectus* (say) (Chrysomelidae). J Stored Prod Res 64:36–41. https://doi.org/10.1016/j.jspr.2015.08.003

Qiao JW, Fan YL, Wu BJ, Wang D, Liu TX (2020) Involvement of apolipoprotein D in desiccation tolerance and adult fecundity of *Acyrthosiphon pisum*. J Insect Physiol 127:104160. https://doi.org/10.1016/j.jinsphys.2020.104160

Qiao JW, Fan YL, Bai TT, Wu BJ, Pei XJ, Wang D, Liu TX (2021) Lipophorin receptor regulates the cuticular hydrocarbon accumulation and adult fecundity of the pea aphid *Acyrthosiphon pisum*. Insect Sci 28(4):1018–1032. https://doi.org/10.1111/1744-7917.12828

Rajapakse S, Qu D, Ahmed AS, Rickers-Haunerland J, Haunerland NH (2019) Effects of FABP knockdown on flight performance of the desert locust, *Schistocerca gregaria*. J Exp Biol 222(21):jeb203455. https://doi.org/10.1242/jeb.203455

Rauch N, Nauen R (2002) Spirodiclofen resistance risk assessment in *Tetranychus urticae* (Acari: Tetranychidae): a biochemical approach. Pestic Biochem Physiol 74(2):91–101. https://doi.org/10.1016/S0048-3575(02)00150-5

Roy S, Hansen I, Raikhel A (2007) Effect of insulin and 20-hydroxyecdysone in the fat body of the yellow fever mosquito, *Aedes aegypti*. Insect Biochem Mol Biol 37(12):1317–1326. https://doi.org/10.1016/j.ibmb.2007.08.004

Roy D, Bhattacharjee T, Biswas A, Ghosh A, Sarkar S, Mondal D, Sarkar PK (2019) Resistance monitoring for conventional and new chemistry insecticides on *Bemisia tabaci* genetic group Asia-I in major vegetable crops from India. Phytoparasitica 47(1):55–66. https://doi.org/10.1007/s12600-018-00707-w

Sabra SG, Abbas N, Hafez AM (2023) First monitoring of resistance and corresponding mechanisms in the green peach aphid, *Myzus persicae* (Sulzer), to registered and unregistered insecticides in Saudi Arabia. Pestic Biochem Physiol 105504. https://doi.org/10.1016/j.pestbp.2023.105504

Sadekuzzaman M, Gautam N, Kim Y (2017) A novel calcium-independent phospholipase A2 and its physiological roles in development and immunity of a lepidopteran insect, *Spodoptera exigua*. Dev Comp Immunol 77:210–220. https://doi.org/10.1016/j.dci.2017.08.014

Santos-Araujo S, Bomfim L, Araripe LO, Bruno R, Ramos I, Gondim KC (2020) Silencing of ATG6 and ATG8 promotes increased levels of triacylglycerol (TAG) in the fat body during prolonged starvation periods in the Chagas disease vector *Rhodnius prolixus*. Insect Biochem Mol Biol 127:103484. https://doi.org/10.1016/j.ibmb.2020.103484

Sato ME, Veronez B, Stocco RS, Queiroz MCV, Gallego R (2016) Spiromesifen resistance in *Tetranychus urticae* (Acari: Tetranychidae): selection, stability, and monitoring. Crop Prot 89:278–283. https://doi.org/10.1016/j.cropro.2016.08.003

Schlipalius D, Tuck A, Jagadeesan R, Nguyen T, Kaur R, Subramanian S, Barrero R, Nayak M, Ebert P (2018) Variant linkage analysis using *de novo* transcriptome sequencing identifies a conserved phosphine resistance gene in insects. Genetics 209(1):281–290. https://doi.org/10.1534/genetics.118.300688

Shi JF, Xu QY, Sun QK, Meng QW, Mu LL, Guo WC, Li GQ (2016) Physiological roles of trehalose in *Leptinotarsa* larvae revealed by RNA interference of trehalose-6-phosphate synthase and trehalase genes. Insect Biochem Mol Biol 77:52–68. https://doi.org/10.1016/j.ibmb.2016.07.012

Silva CT, Wanderley-Teixeira V, Cunha FM, Oliveira JV, Dutra Kde A, Navarro DM, Teixeira ÁA (2016)

Biochemical parameters of *Spodoptera frugiperda* (JE smith) treated with citronella oil (Cymbopogon winterianus Jowitt ex Bor) and its influence on reproduction. Acta Histochem 118(4):347–352. https://doi.org/10.1016/j.acthis.2016.03.004

Silva-Oliveira G, De Paula IF, Medina JM, Alves-Bezerra-M, Gondim KC (2021) Insulin receptor deficiency reduces lipid synthesis and reproductive function in the insect *Rhodnius prolixus*. Biochim Biophy Acta (BBA) Mol Cell Biol Lipids 1866(2):158851. https://doi.org/10.1016/j.bbalip.2020.158851

Sim C, Denlinger D (2009) Transcription profiling and regulation of fat metabolism genes in diapausing adults of the mosquito *Culex pipiens*. Physiol Genomics 39(3): 202–209. https://doi.org/10.1152/physiolgenomics.00095.2009

Singh KS, Cordeiro EM, Troczka BJ, Pym A, Mackisack J, Mathers TC, Duarte A, Legeai F, Robin S, Bielza P, Burrack HJ, Charaabi K, Denholm I, Figueroa CC, Ffrench-Constant RH, Jander G, Margaritopoulos JT, Mazzoni E, Nauen R, Ramírez CC, Ren G, Stepanyan I, Umina PA, Voronova NV, Vontas J, Williamson MS, Wilson ACC, Xi-Wu G, Youn YN, Zimmer CT, Simon JC, Hayward A, Bass C (2021) Global patterns in genomic diversity underpinning the evolution of insecticide resistance in the aphid crop pest *Myzus persicae*. Commun Biol 4(1):1–16. https://doi.org/10.1038/s42003-021-02373-x

Song Y, Gu F, Liu Z, Li Z, Wu F, Sheng S (2022) The key role of fatty acid synthase in lipid metabolism and metamorphic development in a destructive insect pest, *Spodoptera litura* (Lepidoptera: Noctuidae). Int J Mol Sci 23(16):9064. https://doi.org/10.3390/ijms23169064

Souto AL, Sylvestre M, Tölke ED, Tavares JF, Barbosa-Filho JM, Cebrián-Torrejón G (2021) Plant-derived pesticides as an alternative to pest management and sustainable agricultural production: prospects, applications and challenges. Molecules 26(16):4835. https://doi.org/10.3390/molecules26164835

Sparks TC, Crossthwaite AJ, Nauen R, Banba S, Cordova D, Earley F, Ebbinghaus-Kintscher U, Fujioka S, Hirao A, Karmon D, Kennedy R, Nakao T, Popham HJR, Salgado V, Watson GB, Wedel BJ, Wessels FJ (2020) Insecticides, biologics and nematicides: updates to IRAC's mode of action classification-a tool for resistance management. Pestic Biochem Physiol 104587. https://doi.org/10.1016/j.pestbp.2020.104587

Stanley D (2006) Prostaglandins and other eicosanoids in insects: biological significance. Annu Rev Entomol 51: 25–44. https://doi.org/10.1146/annurev.ento.51.110104.151021

Stavrakaki M, Tsagkarakou A, Vontas J, Roditakis E (2023) A multi-year monitoring survey on insecticide resistance for cotton whitefly *Bemisia tabaci*. Entomol Gen 43(3):567–574. https://doi.org/10.1127/entomologia/2023/2134

Subramanyam B, Roesli R (2000) Inert dusts. In: Subramanyam B, Hagstrum DW (eds) Alternatives to pesticides in stored-product IPM. Kluwer Academic Publishers, Dordrecht, p 321e380

Tan QQ, Liu W, Zhu F, Lei CL, Wang XP (2017) Fatty acid synthase 2 contributes to diapause preparation in a beetle by regulating lipid accumulation and stress tolerance genes expression. Sci Rep 7(1):1–11, 40509. https://doi.org/10.1038/srep40509

Tirello P, Pozzebon A, Cassanelli S, Van Leeuwen T, Duso C (2012) Resistance to acaricides in Italian strains of *Tetranychus urticae*: toxicological and enzymatic assays. Exp Appl Acarol 57(1):53–64. https://doi.org/10.1007/s10493-012-9536-y

Toé K, N'Falé S, Dabiré R, Ranson H, Jones C (2015) The recent escalation in strength of pyrethroid resistance in *Anopheles coluzzi* in West Africa is linked to increased expression of multiple gene families. BMC Genomics 16(1):146. https://doi.org/10.1186/s12864-015-1342-6

Tong L (2013) Structure and function of biotin-dependent carboxylases. Cell Mol Life Sci 70(5):863–891. https://doi.org/10.1007/s00018-012-1096-0

Tong L, Harwood HJ Jr (2006) Acetyl-coenzyme a carboxylases: versatile targets for drug discovery. J Cell Biochem 99(6):1476–1488. https://doi.org/10.1002/jcb.21077

Toprak U (2020) The role of peptide hormones in insect lipid metabolism. Front Physiol 11:434. https://doi.org/10.3389/fphys.2020.00434

Toprak U, Bayram Ş, Gürkan MO (2005) Gross pathology of SpliNPVs and alterations in *Spodoptera littoralis* Boisd. (Lepidoptera: Noctuidae) morphology due to baculoviral infection. J Agr Sci 11:65–71. https://doi.org/10.1501/Tarimbil_0000000495

Toprak U, Bayram Ş, Gürkan MO (2006) Comparative biological activities of a plaque-purified variant and a Turkish native isolate of SpliNPV-B against *Spodoptera littoralis* (Lepidoptera: Noctuidae). Pest Manag Sci 62:57–63. https://doi.org/10.1002/ps.1128

Toprak U, Baldwin D, Erlandson M, Gillott C, Harris S, Hegedus D (2013) In vitro and in vivo application of RNA interference for targeting genes involved in peritrophic matrix synthesis in a lepidopteran system. Insect Sci 20(1):92–100. https://doi.org/10.1111/j.1744-7917.2012.01562.x

Toprak U, Coutu C, Baldwin D, Erlandson M, Hegedus D (2014a) Development of an improved RNA interference vector system for *agrobacterium*-mediated plant transformation. Turk J Biol 38:40–47. https://doi.org/10.3906/biy-1304-4

Toprak U, Guz N, Gurkan MO, Hegedus DD (2014b) Identification and coordinated expression of perilipin genes in the biological cycle of sunn pest, *Eurygaster maura* (Hemiptera: Scutelleridae): implications for lipolysis and lipogenesis. Comp Biochem Physiol B Mol Biol 171:1–11. https://doi.org/10.1016/j.cbpb.2014.02.001

Toprak U, Hegedus D, Doğan C, Güney G (2020) A journey into the world of insect lipid metabolism.

Arch Insect Biochem Physiol e21682. https://doi.org/10.1002/arch.21682

Umina PA, Bass C, van Rooyen A, Chirgwin E, Arthur AL, Pym A, Mackisack J, Mathews A, Kirkland L, Kirkland L (2022) Spirotetramat resistance in *Myzus persicae* (Sulzer)(Hemiptera: Aphididae) and its association with the presence of the A2666V mutation. Pest Manag Sci 78(11):4822–4831. https://doi.org/10.1002/ps.7103

Van Leeuwen T, Dermauw W (2016) The molecular evolution of xenobiotic metabolism and resistance in chelicerate mites. Annu Rev Entomol 61:475–498. https://doi.org/10.1146/annurev-ento-010715-023907

Van Leeuwen T, Tirry L, Yamamoto A, Nauen R, Dermauw W (2015) The economic importance of acaricides in the control of phytophagous mites and an update on recent acaricide mode of action research. Pestic Biochem Physiol 121:12–21. https://doi.org/10.1016/j.pestbp.2014.12.009

Veenstra JA, Leyria J, Orchard I, Lange AB (2021) Identification of gonadulin and insulin-like growth factor from migratory locusts and their importance in reproduction in *Locusta migratoria*. Front Endocrinol 12:675. https://doi.org/10.3389/fendo.2021.693068

Wakil SJ, Titchener EB, Gibson DM (1958) Evidence for the participation of biotin in the enzymic synthesis of fatty acids. Biochim Biophys Acta 29(1):225–226. https://doi.org/10.1016/0006-3002(58)90177-x

Wakil SJ, Stoops JK, Joshi VC (1983) Fatty acid synthesis and its regulation. Annu Rev Biochem 52(1):537–579

Wang W, Wan P, Lai F, Zhu T, Fu Q (2018) Double-stranded RNA targeting calmodulin reveals a potential target for pest management of *Nilaparvata lugens*. Pest Manag Sci 74(7):1711–1719. https://doi.org/10.1002/ps.4865

Wang W, Ma Y, Yang RR, Cheng X, Huang HJ, Zhang CX, Bao YY (2021) An MD-2-related lipid-recognition protein is required for insect reproduction and integument development. Open Biol 11(12):210170. https://doi.org/10.1098/rsob.210170

Wei X, Zheng C, Peng T, Pan Y, Xi J, Chen X, Zhang J, Yang S, Gao X, Shang Q (2016) miR-276 and miR-3016-modulated expression of acetyl-CoA carboxylase accounts for spirotetramat resistance in *Aphis gossypii* glover. Insect Biochem Mol Biol 79:57–65. https://doi.org/10.1016/j.ibmb.2016.10.011

Wei P, Demaeght P, De Schutter K, Grigoraki L, Labropoulou V, Riga M, Vontas J, Nauen R, Dermauw W, Van Leeuwen T (2020) Overexpression of an alternative allele of carboxyl/choline esterase 4 (CCE04) of *Tetranychus urticae* is associated with high levels of resistance to the keto-enol acaricide spirodiclofen. Pest Manag Sci 76(3):1142–1153. https://doi.org/10.1002/ps.5627

Wrońska AK, Kaczmarek A, Boguś MI, Kuna A (2023) Lipids as a key element of insect defense systems. Front Genet 14:1183659. https://doi.org/10.3389/fgene.2023.1183659

Wybouw N, Kosterlitz O, Kurlovs AH, Bajda S, Greenhalgh R, Snoeck S, Bui H, Bryon A, Dermauw W, Van Leeuwen T, Clark RM (2019) Long-term population studies uncover the genome structure and genetic basis of xenobiotic and host plant adaptation in the herbivore *Tetranychus urticae*. Genetics 211(4):1409–1427. https://doi.org/10.1534/genetics.118.301803

Xi J, Pan Y, Wei Z, Yang C, Gao X, Peng T, Bi R, Liu Y, Xin X, Shang Q (2015) Proteomics-based identification and analysis proteins associated with spirotetramat tolerance in *Aphis gossypii* glover. Pestic Biochem Physiol 119:74–80. https://doi.org/10.1016/j.pestbp.2015.02.002

Xiang M, Zhang HZ, Jing XY, Wang MQ, Mao JJ, Li YY, Zang LS, Zhang LS (2021) Sequencing, expression, and functional analyses of four genes related to fatty acid biosynthesis during the diapause process in the female ladybird, *Coccinella septempunctata* L. Front Phys 12:706032. https://doi.org/10.3389/fphys.2021.706032

Xu QY, Deng P, Mu LL, Fu KY, Guo WC, Li GQ (2019a) Silencing Taiman impairs larval development in *Leptinotarsa decemlineata*. Pestic Biochem Physiol 160:30–39. https://doi.org/10.1016/j.pestbp.2019.06.013

Xu Y, Borcherding AF, Heier C, Tian G, Roeder T, Kühnlein RP (2019b) Chronic dysfunction of stromal interaction molecule by pulsed RNAi induction in fat tissue impairs organismal energy homeostasis in drosophila. Sci Rep 9(1):1–15. https://doi.org/10.1038/s41598-019-43327-y

Xu J, Sun Z, Hao M, Lv M, Xu H (2020) Evaluation of biological activities, and exploration on mechanism of action of matrine-cholesterol derivatives. Bioorg Chem 94:103439. https://doi.org/10.1016/j.bioorg.2019.103439

Xu H, Zhang K, Lv M, Hao M (2021) Construction of cholesterol oxime ether derivatives containing isoxazoline/isoxazole fragments and their agricultural bioactive properties/control efficiency. J Agric Food Chem 69(29):8098–8109. https://doi.org/10.1021/acs.jafc.1c01884

Yan WH, Wu MY, Shah S, Yao YC, Elgizawy KK, Tang N, Wu G, Yang FL (2021) Silencing the triacylglycerol lipase (TGL) gene decreases the number of apyrene sperm and inhibits oviposition in *Sitotroga cerealella*. Cell Mol Life Sci 79(1):44. https://doi.org/10.1007/s00018-021-04048-6

Yan J, Nauen R, Reitz S, Alyokhin A, Zhang J, Mota-Sanchez D, Kim Y, Palli SR, Rondon SI, Nault BA, Jurat-Fuentes JL, Crossley MS, Snyder WE, Gatehouse AMR, Zalucki MP, Tabashnik BE, Gao Y (2024) The new kid on the block in insect pest management: sprayable RNAi goes commercial. Sci China Life Sci 67(8):1766–1768. https://doi.org/10.1007/s11427-024-2612-1

Yang J, Flaven-Pouchon J, Yang Y, Gehring N, Moussian B (2023a) The greenhouse and field insecticide Spirotetramat differentially affects the surface barrier efficiency in non-target *Drosophila melanogaster*. Entomol Gen 43(2):433–440. https://doi.org/10.1127/entomologia/2022/1732

Yang J, Flaven-Pouchon J, Wang Y, Moussian B (2023b) Spirotetramat reduces fitness of the spotted-wing drosophila. Insect Sci, Drosophila suzukii. https://doi.org/10.1111/1744-7917.13283

Ye W, Zeng J, Hu W, Bustos-Segura C, Noman A, Lou Y (2020) The desaturase gene Nlug-desata2 regulates the performance of the brown planthopper *Nilaparvata lugens* and its relationship with rice. Int J Mol Sci 21(11):4143. https://doi.org/10.3390/ijms21114143

Yu SJ, Tian H, Yang JS, Ding L, Chen F, Li X, Yue J, Liu H, Ran C (2015) Cloning of acetyl CoA carboxylase cDNA and the effects of spirodiclofen on the expression of acetyl CoA carboxylase mRNA in *Panonychus citri*. Entomol Exp Appl 156(1):52–58. https://doi.org/10.1111/eea.12314

Yükselbaba U, Ali IH (2022) Insecticide resistance status of *Bemisia tabaci* (Genn., 1889) (Hemiptera: Aleyrodidae) populations to cyantraniliprole, pyriproxyfen and spirotetramat in Antalya (Türkiye). *Turkish*. J Entomol 46(3):263–274. https://doi.org/10.16970/entoted.1112553

Zhang Q, Nachman RJ, Kaczmarek K, Zabrocki J, Denlinger DL (2011) Disruption of insect diapause using agonists and an antagonist of diapause hormone. Proc Natl Acad Sci USA 108(41):16922–16926. https://doi.org/10.1073/pnas.1113863108

Zhang Q, Nachman R, Denlinger D (2015a) Diapause hormone in the *Helicoverpa/Heliothis* complex: a review of gene expression, peptide structure and activity, analog and antagonist development, and the receptor. Peptides 72:196–201. https://doi.org/10.1016/j.peptides.2015.05.005

Zhang YX, Ge LQ, Jiang YP, Lu XL, Li X, Stanley D, Song QS, Wu JC (2015b) RNAi knockdown of acetyl-CoA carboxylase gene eliminates jinggangmycin-enhanced reproduction and population growth in the brown planthopper, *Nilaparvata lugens*. Sci Rep 5:15360. https://doi.org/10.1038/srep15360

Zhou J, Chen X, Yan J, You K, Yuan Z, Zhou Q, Lu K (2018) Brummer-dependent lipid mobilization regulates starvation resistance in *Nilaparvata lugens*. Arch Insect Biochem Physiol 99(2):e21481. https://doi.org/10.1002/arch.21481

Zhou Y, Wu Y-m, Fan R, Ouyang J, Zhou X-l, Li Z-b, Janjua MU, Li H-g, Bao M-h, He B-s (2023) Transcriptome analysis unveils the mechanisms of lipid metabolism response to grayanotoxin I stress in *Spodoptera litura*. Peer Journal 11:e16238. https://doi.org/10.7717/peerj.16238

Zhu KY, Palli SR (2020) Mechanisms, applications, and challenges of insect RNA interference. Annu Rev Entomol 65(1):293–311. https://doi.org/10.1146/annurev-ento-011019-025224

Zibaee A, Zibaee I, Jalali Sendi J (2011) A juvenile hormone analog, pyriproxifen, affects some biochemical components in the haemolymph and fat bodies of *Eurygaster integriceps* Puton (Hemiptera: Scutelleridae). Pestic Biochem Physiol 100(3):289–298. https://doi.org/10.1016/j.pestbp.2011.05.002

Adv Exp Med Biol (2026) 1494: 549–577
https://doi.org/10.1007/5584_2024_827

Published online: 16 October 2024

Insect Models to Study Human Lipid Metabolism Disorders

Thomas B. Rundell and Thomas J. Baranski

Abstract

Disorders of lipid metabolism such as obesity have become some of the most significant diseases of the twenty-first century. Despite these metabolic diseases affecting more than a third of the population in highly industrialized nations, the mechanisms underlying disease development remain poorly understood. Insect models, such as *Drosophila melanogaster,* offer a means of systematically examining conserved lipid metabolism and its pathology. Over the past several decades, *Drosophila melanogaster* has been used to greatly expand on our knowledge of metabolic disease, often taking advantage of the extensive genetic tools available to researchers. Additionally, *Drosophila melanogaster* has served and will continue to serve as a powerful tool for validating the results of genome-wide approaches to the study of diseases. This chapter explores the advancements of insect models in the study of lipid metabolism disorders as well as highlight opportunities for future areas of research.

Keywords

Drosophila · Fatty acid beta oxidation · Insect models · Lipid disorders · Lipid metabolism · Lipid transport · Obesity

1 Introduction

Lipids are ubiquitous within all metazoan organisms. Lipids serve as the primary component of cell membranes, allowing cellular compartmentalization of function. They serve as the backbone for many hormone-signaling molecules, facilitating endocrine control of an organism. They serve as one of the primary energy storage molecules, providing energy for costly processes like development and reproduction. As such, proper lipid homeostasis is pivotal to an animal's healthy metabolism. Often, severe metabolic disease is incurred when an animal fails to, either through genetic or environmental means, regulate its lipid metabolism. In humans, these diseases can manifest as rare, inborn genetic diseases where the loss of function of a single enzyme can result in severe phenotypes, such as Niemann-Pick (Naureckiene et al. 2000). Lipid metabolism disorders can also result from environmental factors such as overnutrition in the case of type 2 diabetes or dysfunctional regulation as is

T. B. Rundell
Department of Biological Sciences, Binghamton University, Binghamton, NY, USA
e-mail: trundel1@binghamton.edu

T. J. Baranski (✉)
Division of Endocrinology, Metabolism, and Lipid Research, Department of Medicine, Washington University School of Medicine, St Louis, MO, USA
e-mail: baranski@wustl.edu

seen in type 1 diabetes (Abdullah et al. 2010; Ng et al. 2014; Herder and Roden 2011).

Since the mid twentieth century, insect models have increasingly become pivotal in our understanding of metabolism (Arias 2008; Musselman and Kühnlein 2018). In recent decades, the proliferation of new technologies has enabled researchers to model many human diseases with *Drosophila melanogaster*. Here, we will review the contributions that insect models have made to lipid metabolism disorder research, with further notes on opportunities for more insect models.

1.1 The Fly as a Model

It would be difficult to overstate the contributions that *Drosophila melanogaster*, henceforth referred to as "drosophila," has made to the scientific inquiry over the last century. While it would be impossible to give a full history of drosophila research here, it is important to note that many of the advances in genetics during the twentieth century begin with the work of Thomas H Morgan and his students of drosophila. Discovering the process of mutagenesis, chromosomes, and basic gene structure, Morgan's work would serve as a foundation for much of modern genetics (Arias 2008; Musselman and Kühnlein 2018). Headway continued to be made until the turn of the twenty-first century when major breakthroughs such as the full sequencing of the drosophila genome and the development of genetic tools such as RNA interference catapulted drosophila into the realm of functional genetics (Dietzl et al. 2007; Brand and Perrimon 1993).

1.1.1 Advantages of Using an Insect Model

Drosophila has become a particularly powerful tool for understanding human disease thanks to several core attributes. First, of approximately 14,000 genes spread over the 4 chromosomes of the fly genome, it has been estimated that nearly 75% have homologs in the human genome (Reiter et al. 2021). Indeed, an estimated 77% of human genes involved in disease have homologs in the fly (Reiter et al. 2021). The pathways that regulate carbohydrate and lipid metabolism are especially conserved, often displaying 80–90% nucleotide identity in conserved functional domains (Pandey and Nichols 2011). Conservation is often so strong that mammalian homologs can be inserted into the fly genome (or vice versa) and recuse the deleterious effects of a knockout (Manzo et al. 2018; Zagoraiou et al. 2001; Weinmaster et al. 1991).

In addition to gene homology, the fly also possesses significant homology with mammalian tissues, organs, and organ systems (Fig. 1, Table 1). Like the human brain, the drosophila central nervous system (CNS) mediates a wide range of behaviors including feeding behavior. Food is taken in via a proboscis to be absorbed in the digestive system (gut) in a similar fashion to the human gastrointestinal tract. As metabolites are absorbed by a gut like our own if more simplistic, the fly dorsal vessel functions as a rudimentary tube-like heart that pumps nutrient-rich hemolymph around the open circulatory system. In contrast to our mammalian close circulatory system, fly hemolymph makes direct contact with the organs and is a composite of signaling peptides, nutrients, macrophages, and other elements of interstitial fluid. Three major sites of nutrient exchange, metabolic regulation, and detoxification are the fly fat body, oenocytes, and Malpighian tubules, respectively. The fat body and oenocytes together fulfill many of the roles of mammalian adipose tissue (lipid molecule storage) and liver (detoxification and metabolic control). Finally, hemolymph is filtered for purposes of osmoregulation and excretion by the Malpighian tubules, analogous to human kidneys. Adipokinetic hormone (Akh)–producing cells (APCs) release Akh, a hormone similar in function to mammalian glucagon that stimulates metabolite catabolism. Thus, APCs are analogous in function to human pancreatic α-cells. Conversely, insulin-like peptide (Ilp)-producing cells (IPCs) release Ilps into the fly hemolymph to stimulate anabolism much like mammalian insulin.

One of the greatest advantages to insect models, and drosophila especially, is the vast genetic toolkit at researchers' disposal. The great

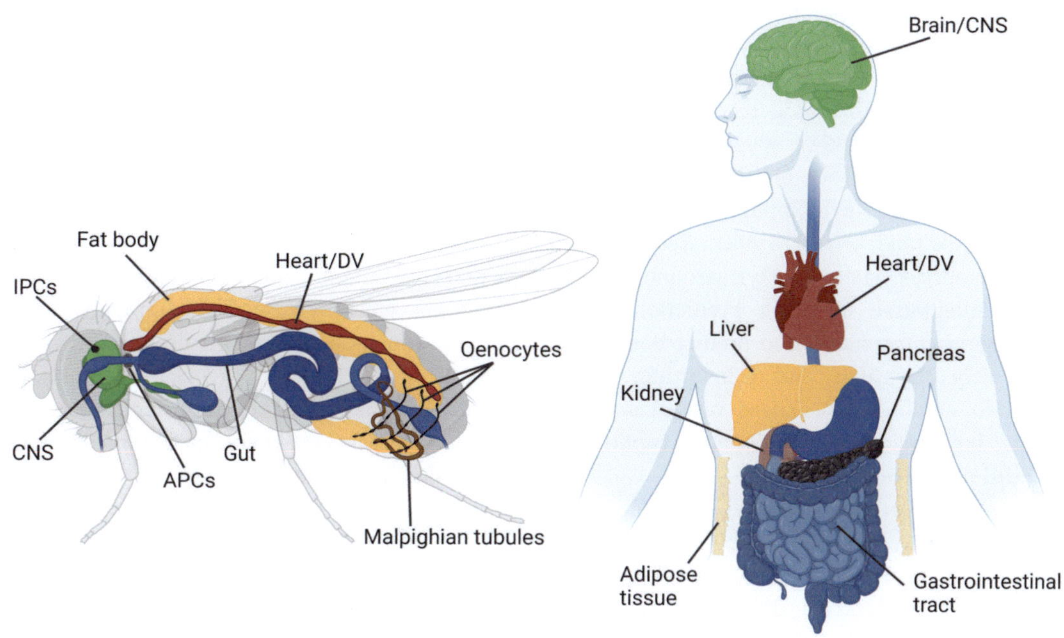

Fig. 1 Homologous organs between the insect *Drosophila melanogaster* and humans. See Table 1. Made with Biorender

Table 1 Tissues and organs homologous between the fly and human

Drosophila melanogaster	Human
Central nervous system	Brain/central nervous system
Adipokinetic hormone–producing cells (APCs)	Pancreatic α-cells
Insulin-like peptide–producing cells (IPCs)	Pancreatic β-cells
Gut	Gastrointestinal tract
Dorsal vessel (heart)	Heart/circulatory system
Fat body	Adipose tissue, liver, immune system
Oenocytes	Liver
Malpighian tubes	Kidneys

workhorse of drosophila genetics has been the bipartite GAL4-UAS system, originally developed by A H Brand and N Perrimon in 1993 (Brand and Perrimon 1993). This system makes use of two manipulated fly genotypes. One line expresses the yeast transcriptional factor GAL4, while the other possesses GAL4's guiding response elements, UAS (Upstream Activation Sequence), upstream of the desired target gene. Mating the two lines together results in progeny that will express the target gene (Brand and Perrimon 1993). This system has proven remarkably mutable for discerning gene function. Researchers have devised means of overexpressing genes of interest by driving a transgene with the GAL4-UAS system. Researchers also utilize RNA interference, whereby GAL4-UAS is used to drive the expression of an RNA molecule complementary to the mRNA of the gene of interest (Dietzl et al. 2007; Brand and Perrimon 1993; Clemens et al. 2000; Kaya-Çopur and Schnorrer 2016). Endogenous mechanisms of double-stranded RNA removal then effectively "knock down" the gene. These techniques, when combined with other strategies of tissue-specific and temporal expression, have made the GAL4-UAS indispensable to modeling research. These tools further benefit from extensive publicly available resources. First, various stock centers, such as the Vienna Drosophila Research Center collection, provide publicly available lines with UAS response elements targeting ~95% of known genes in *Drosophila melanogaster* (Dietzl et al. 2007; Brand and Perrimon 1993; Kaya-Çopur and Schnorrer 2016; Phelps and Brand 1998). Second, free online resources, such as FlyBase and DIOPT (DRSC Integrative Ortholog Prediction Tool), catalog gene information and predict homology, respectively (Drysdale and Consortium the F 2008; Hu et al. 2011). These resources taken together have made *Drosophila* an exceptionally powerful model organism.

In conjunction with RNA interference, the CRISPR/Cas9 system has in recent years been adapted for use in flies (Housden et al. 2015; Gratz et al. 2014, 2015; Bassett et al. 2013; Shah et al. 2013). The use of the CRISPR/Cas9 DNA modification system allows researchers to alter the specific locus of an endogenous gene in the *Drosophila* germ line. The system makes use of a streamlined two-part system derived from *Streptococcus pyogenes* comprising the endonuclease Cas9 and a guide RNA (Housden et al. 2015; Gratz et al. 2014, 2015; Bassett et al. 2013; Shah et al. 2013). The guide RNA is an easily constructed 20 nucleotide long sequence that guides the Cas9 endonuclease to the target site via homologous base pairing. The Cas9 protein then uses its two nuclease domains to cut the DNA at the desired locus in a double-stranded break. This triggers endogenous DNA repair mechanisms that researchers exploit. A supplied donor template consisting of the desired sequence is provided to the cell's own repair mechanism. Homology-directed repair then uses the template to repair the break, including the desired single-nucleotide polymorphism (SNP) or sequence into the organism's germ line. The only requirement for this model is the existence of protospacer adjacent motifs (PAMs) near the target of interest (Housden et al. 2015; Gratz et al. 2014, 2015; Bassett et al. 2013; Shah et al. 2013). Services for protospacer insert are commercially available.

Another advantage an insect model affords researchers, though no less important, is their rapid life cycle and reproduction. Over the course of several days, a single drosophila female may lay hundreds of eggs that will hatch in ~12 days when reared at 25 °C (Barnes et al. 2008). This not only allows for rapid characterization of phenotypes but also allows researchers to maintain extremely large populations of flies. In recent decades, it has also allowed for experimental evolution and evolve-and-resequence paradigms whereby researchers impose selective pressure on populations of drosophila (Burke et al. 2010; Hardy et al. 2015; Turner et al. 2011; Kofler et al. 2012; Kofler et al. 2011; Tobler et al. 2014). As such, researchers can observe evolution in real-time in a multi-organ species.

1.1.2 Model Considerations

Insect models are powerful tools for studying human biology, but these models have several considerations that limit their ability to model aspects of human biology and pathophysiology. While many of the fundamental biochemical processes are conserved between humans and insects, their implementation in the whole organism can often be radically different. For example, both drosophila and humans can suffer from increased "blood" sugar in response to a diet high in sucrose. In the fly, however, the primary sugar of circulation is not glucose but the disaccharide trehalose, and it circulates via an open circulatory system (Sang et al. 2021). Further, some organ systems operate in very different ways in insects. For example, it is difficult to model the complex human disease atherosclerosis, the buildup of fats and cholesterol in arteries, as flies lack blood vessels.

While single genes and their functions may be accurately modeled in the fly with high degrees of conservation to humans, the size constraints of drosophila present logistic problems for system biology research. For example, the small size of drosophila impacts the ability to do elegant physiologic studies like insulin clamps to determine glucose utilization rates and tissue-specific insulin sensitivity. As a result, many complex human diseases must be modeled in piecemeal, addressing one element at a time to elucidate the larger picture (Pandey and Nichols 2011).

1.2 Categorization

Diseases of lipid metabolism are nearly as varied as the diverse functions of lipids themselves. To aid the reader, we have broken these many diseases into two broad categories: polygenic/complex diseases and monogenic diseases. Monogenic diseases are rare conditions that result from mutations in single genes. These variations frequently produce wholly or partially defective proteins, and thus directly affect a biochemical pathway. Their prevalence varies widely; monogenic variants may be as rare as 1 in 10,000 or as common as 1 in 250 live births. Further, phenotype severity may range from a mild increase in risk of cardiovascular disease to lethality within months of birth. Here we have divided monogenic lipid metabolism disorders into three categories based on the metabolic pathway affected: 1. lipid storage/turnover, 2. fatty acid oxidation, and 3. lipid transport. While we have attempted to be thorough, this certainly does not constitute an exhaustive list of monogenic disorders of lipid metabolism. Polygenic disorders, by contrast, are those that result from the complex interaction of genes and environment. While debate continues as to the relative effect size of genetic vs environmental/lifestyle factors, there is an emerging body of literature showing strong associations between disease states and genetic influences. Here we will be discussing major polygenic diseases that include defective lipid metabolism: diabetes mellitus, obesity, and nonalcoholic fatty liver disease.

2 Monogenic Diseases and Their Models

Monogenic diseases of lipid metabolism, dyslipidemias, are a diverse group of rare inherited human disorders marked by disruptions in lipid storage, transport, or oxidation. These inborn diseases are often inherited in an autosomal recessive or X-linked fashion (Rigon et al. 2021): however, some are autosomal dominant. Most dyslipidemias present with a variety of phenotypes depending on the pathway and metabolite affected, ranging from hypercholesterolemia to severe neurological impairment. As such, we will examine each of the disorder categories in turn.

2.1 Lipid Lysosomal Storage Disorders

Lysosomal storage disorders (LSDs) are a group of diseases that affect the function of specific enzymes within lysosomes. These inborn diseases are often very rare, with incidences ranging from 7.5 to 20.5 per 100,000 births (Rigon et al. 2021). They are most frequently inherited in an autosomal recessive fashion, though there are several X-linked and even autosomal dominant diseases within this category (Rigon et al. 2021). LSDs are categorized primarily by the metabolites and genes affected by the inborn error; mucolipidoses, neuronal ceroid lipofuscinoses, sphingolipidoses, and acid lipase disorders are the categories addressed here.

Lysosomes serve a key role in lipid and general metabolic homeostasis. They serve as the site of breakdown for a variety of metabolites, including proteins, nucleic acids, carbohydrates, and lipids (Settembre et al. 2013; Xu and Ren 2015). They are one of the primary destinations for lipoproteins following cellular endocytosis, and where such lipoproteins are processed. Lysosomes also play a role in the autophagy of lipid droplets, thus functioning as a lipid metabolic hub (Rigon et al. 2021; Settembre et al. 2013; Xu and Ren 2015). As such, diseases of lysosomal enzymes are often far-reaching and severe.

Given the diversity of metabolites affected, these diseases present with a wide spectrum of symptoms. The most common phenotypes, however, are defects in the musculature and skeleton, locomotion, and neurological impairment (Settembre et al. 2013; Xu and Ren 2015).

While LSDs have been well characterized as disease states, much work remains to be done. There are currently no cures to LSDs and what gene therapies exist are often limited in their ability to address neurological symptoms (Rigon et al. 2021). Thus, further elucidating these disease states remains important. LSDs have been characterized in a drosophila model, but much potential remains to be explored. Listed in Table 2 (and all subsequent tables) are human LSD and corresponding fly genes. The DIOPT score refers to the number of homology-predicting software tools that endorse a human-fly gene pairing as homologous (Hu et al. 2011). The maximum DIOPT score number for a given human-fly gene pair is 19 (Hu et al. 2011).

2.1.1 Mucolipidoses

Mucolipidoses (MLs) are a group of LSDs that result in the abnormal trafficking of lysosomal enzymes and substrates to the lysosome. While there has been some reclassification in recent years as to what diseases precisely belong in this category, they were named for the observed improper transport and accumulation of mucolipids. Here we will describe types II, III, and IV.

Mucolipidosis types II and III result from a mutation in the gene *GNPTAB* that encodes the alpha and beta subunits of the enzyme N-acetylglucosamine-1-phosphotransferase. This enzyme is responsible for tagging proteins destined for the lysosome with mannose-6-phosphate (Bargal et al. 2000). Without the tag, lysosomal enzymes are poorly trafficked in mammals; however, flies do not appear to use mannose-6-P to target its hydrolases to the lysosome because they lack other required components, such as an uncovering enzyme and a mannose-6-phosphate receptor (Hasanagic et al. 2015). Mucolipidosis type IV results from a mutation in the gene *MCOLN1* that encodes the lysosomal protein mucolipin TRP cation channel 1 (Bargal et al. 2000). This protein is associated with endo- and exocytosis in and out of the lysosome. Thus, in all types of mucolipidosis, fatty substrates that would normally be broken down by the transported lysosomal enzymes accumulate in the lysosomes, causing the disease's pathophysiology (Bargal et al. 2000). While *GNPTAB* does have a predicted fly homolog, *CG8027*, it is yet to be modeled in drosophila. *CG8027* is predicted to have N-acetylglucosamine-1-phosphotransferase activity, but it remains uncharacterized (Drysdale and Consortium the F 2008). *MCOLN1*, by contrast, does have a model in the drosophila gene *Trpml*. By p-element insertion, a mutant *Trpml* fly was generated that captured the most prominent MLIV phenotypes, such as accumulation of fatty molecules, decreased motor control, and neurodegeneration (Venkatachalam et al. 2008). Venkatachalam et al. further show that these phenotypes largely stem from impaired autophagy in a pathway conserved between humans and flies (Venkatachalam et al. 2008).

2.1.2 Neuronal Ceroid Lipofuscinoses

Neuronal ceroid lipofuscinoses (NCLs), also known as Batten disease, are LSDs that result from genes involved in the lysosomal metabolism of lipofuscins. Lipofuscins are protein-lipid molecules frequently found in the lysosomes and are by-products of fatty acid oxidation (Haltia and Goebel 2013). A disease state occurs when an abnormal accumulation of these lipofuscins arises typically in the retina, nerve cells, liver, kidney, and heart muscle (Haltia and Goebel 2013). There are ten NCL types generally recognized, categorized by age presentation and the specific gene mutated. Types 2, 4, 5, and 6 do not have fly homologs, while type 8 does have a homolog, *CG17841*, which is yet to be characterized or modeled (Drysdale and Consortium the F 2008). NCL type 1 has been modeled with the fly homolog *PPT1*; the mutant fly possessed an abnormal accumulation of storage lipofuscins but did not recapitulate other disease phenotypes (Hickey et al. 2006). The NCL type 3 gene, *CLN3*, has a fly homolog by the same name; mutant flies interestingly show increased stress under conditions of oxidative stress, potentially accounting for neurodegeneration (Tuxworth et al. 2011). NCL type 10 has perhaps been most successfully modeled as an NCL disease in flies. Type 10 results from a mutation in the gene *CTSD* which encodes cathepsin D, a protein involved in lysosomal protein turnover by degradation (Myllykangas et al. 2005). The disease has been modeled via a mutant of the homologous drosophila gene *cathD*. Mutant flies captured many

Table 2 Human lipid lysosomal storage disorders and their insect models

Disease	Human gene	Protein encoded	Pathway/function	Fly ortholog	DIOPT score	Reference for model
Mucolipidosis II and III	GNPTAB	N-Acetylglucosamine-1-phosphate transferase subunits alpha and Beta	Vesicular transport of lysosomal enzymes	CG8027	13	
Mucolipidosis IV	MCOLN1	Mucolipin TRP cation channel 1	Lysosomal endo- and exocytosis	Trpml	12	Venkatachalam et al. (2008)
NCL type 1	PPT1	Palmitoyl-protein Thioesterase 1	Fatty Acyl-CoA Biosynthesis	PPT1	14	Hickey et al. (2006)
NCL type 2	TPP1	Tripeptidyl peptidase 1	Lysosomal peptidase	None		
NCL type 3	CLN3	CLN3 Lysosomal/Endosomal transmembrane protein	Lysosomal transport	CLN3	13	Tuxworth et al. (2011)
NCL type 5	CLN5	CLN5 transmembrane ER and Lysosomal protein	Lysosomal transport	None		
NCL type 4 NCL type 6	CLN6	CLN6 transmembrane ER protein	Lysosomal transport	None		
NCL type 7	MFSD8	Major facilitator superfamily domain containing 8	Lysosomal transport	Cln7	14	Mohammed et al. (2017)
NCL type 8	CLN8		Lysosomal transport	CG17841	2	
NCL type 10	*CTSD*	Cathepsin D	Lysosomal protein turnover	cathD	12	Myllykangas et al. (2005)
Niemann-Pick disease type 1C	NPC1	NPC intracellular cholesterol transporter 1	Lysosomal cholesterol and sphingolipids transport	Npc1a-b	14	Huang et al. (2005)
Niemann-Pick disease type 2C	NPC2	NPC intracellular cholesterol transporter 2	Lysosomal cholesterol and sphingolipids transport	Npc2a-h	13	Huang et al. (2007)
Niemann-Pick disease type A/B	SMPD1	Sphingomyelin phosphodiesterase 1	Lysosomal sphingolipid metabolism	CG3376	15	
Gaucher	GBA	Glucosylceramidase beta	Lysosomal sphingolipid metabolism	Gba1a, Gba1b	14	Kawasaki et al. (2017, Davis et al. (2016), Maor et al. (2016), Maor et al. (2013)
Fabry disease	GLA	Galactosidase alpha	Ceramide and sphingolipid metabolism	CG7997	9	Braunstein et al. (2020)
GM1 gangliosidosis	GLB1	Galactosidase beta 1	Lysosomal glycosaminoglycan metabolism	Ect3, Gal	12, 11	Pearson and Crews (2014)
Sandhoff disease	HEXB	Hexosaminidase subunit beta	Lysosomal sphingolipid metabolism	Hexo1, Hexo2, fdl	8, 7, 7	Léonard et al. (2006), Boquet et al. (2000), Cattaneo et al. (2006)
Wolman disease	LIPA (LAL)	Lysosomal acid lipase	Lysosome catalysis of cholesteryl esters and triglycerides	Lip3, Lip4, CG8093, CG31871, mag	13, 12, 12, 12, 9	Sieber and Thummel (2012), Vihervaara and Puig (2008), Sellin et al. (2020)

of the disease phenotypes, showing accumulation of fatty storage materials in the brain and fat body, followed by progressive neurodegeneration (Myllykangas et al. 2005).

2.1.3 Sphingolipidoses

The sphingolipidoses are a large group of LSDs characterized by the accumulation of sphingolipids or their derivatives in the lysosome. These diseases frequently lead to severe neurodegeneration and death, depending on severity (Rigon et al. 2021).

The first subset of sphingolipidoses we will cover is the Niemann-Pick diseases (NPCs). There are three main types: type 1C, type 2C, and type A/B. The genes *NPC1* and *NPC2* encode the proteins NPC intracellular cholesterol transporter 1 and 2, respectively, and control cholesterol transport from the lysosome. Both have been well modeled in drosophila, with multiple homologs for each gene (two homologs for NPC1 and eight homologs for NPC2). Mutants for drosophila *Npc1a* showed larval lethality until dietary supplementation of 7-dehydrocholesterol allowed them to reach adulthood. Such adults showed the accumulation of sterols in the brain, similar to the human phenotype (Huang et al. 2005). Mutants of *Npc2a* exhibited reduced lifespan accompanied by sterol accumulation throughout all body tissues (Huang et al. 2007). Further work remains to be done, however, as numerous other homologs of human *NPC2* have not been well studied in drosophila (Rigon et al. 2021).

Gaucher disease arises from a mutation in the gene *GBA* which encodes the enzyme glucosylceramidase beta. This enzyme is responsible for the hydrolysis of glucosylceramides into free glucose and ceramides in the lysosome. Thus, it plays a key role in cellular membrane turnover (Dandana et al. 2016). Defects in this gene result in the accumulation of glucosylceramides in the lysosome. The disease presents with skeletal, liver, and spleen abnormalities, along with neurological impairment in severe forms. Drosophila possesses two homologs, *Gba1a* and *Gba1b*. Both have been modeled along with a double mutant deficient in both homologs. Both the *Gba1b* and *Gba1a*/*Gba1b* mutants exhibited an increase in glucosylceramide accumulation and decreased locomotion with age (Kawasaki et al. 2017; Davis et al. 2016). Indeed, as Maor et al. have shown, the Gaucher model has been expanded upon to model Parkinson's disease (Maor et al. 2013, 2016). Heterozygous mutants exhibited upregulation of the unfolded protein response (UPR) pathway and other markers of the Parkinson's (Maor et al. 2013, 2016).

Mutations in the gene *GLA* result in Fabry disease, a rare X-linked LSD. *GLA* encodes the enzyme alpha-galactosidase A, which normally breaks down a glycolipid. Mutations that disrupt this function result in an accumulation of ceramides in the heart's endothelium (Aerts et al. 2008). This disease presents with renal and heart impairment, including the thickening of the heart's smooth muscle. The fly homolog *CG7997* of *GLA* has been disrupted to model Fabry disease; mutants show a decrease in locomotion and lifespan. Interestingly, like the Gaucher fly model, this model shows an upregulation in the unfolded protein response pathway (Braunstein et al. 2020).

Next are the gangliosidoses, a group of disorders that result from the accumulation of ganglioside lipids. There are two groups, GM1 and GM2 gangliosides. We will address GM1 and Sandhoff, a GM2 here. GM1 gangliosidosis is typified by symptoms of neurodegeneration, seizures, and muscle atrophy (Okada and O'Brien 1968). This results from defects in *GLB1*, a lysosomal protein that encodes galactosidase beta 1. With reduced or absent galactosidase beta 1 activity, gangliosides are not degraded and accumulate in nerve cells (Okada and O'Brien 1968). The fly has two predicted homologs, *Ect3* and *Gal*; however, neither have been modeled in relation to gangliosidosis (Pearson and Crews 2014).

The GM2 gangliosidosis, Sandhoff disease, is clinically indistinguishable from the more famous GM2 Tay-Sachs. The gene *HEXB* codes for the beta subunit of hexosaminidases (Léonard et al. 2006; Cattaneo et al. 2006). Drosophila possesses three homologs *hexo1, hexo2*, and *fdl*. Of the three, *fdl* may serve as the most promising model for GM2 diseases. Mutations in *fdl* have been shown to dramatically affect brain development, causing the fusion of mushroom body lobes in adults (Boquet et al. 2000).

2.1.4 Lysosomal Acid Lipases

The final lipid LSD we will address here is Wolman disease or lysosomal acid lipase deficiency. The classic LSD, Wolman, results from a defect in the gene *LIPA* which encodes the protein lysosomal acid lipase (LAL). Lacking the enzyme, which breaks down triglycerides and cholesterols, causes the accumulation of fats in the liver, spleen, and blood vessels (Reiner et al. 2014). Human *LIPA* has many homologs in drosophila, multiple of which have been characterized extensively in the fly under a variety of contexts, while others remain unexplored (Sieber and Thummel 2012; Vihervaara and Puig 2008; Sellin et al. 2020). Indeed, it has been shown using a drosophila model of *LIPA* that the acid lipases are regulated by the key insulin-sensitive transcription factor dFOXO (Vihervaara and Puig 2008). Much work remains to be done, however, elucidating acid lipase regulation and its links to other complex diseases.

2.2 Fatty Acid β-Oxidation Disorders

Fatty acid β-oxidation disorders (FAODs) are a group of approximately 20 rare diseases relating to the transport to and oxidation of lipids within the mitochondria (Fig. 2) (Vishwanath 2016; Merritt II et al. 2018). Mutations in key enzymes prevent the body from properly processing either individual species or bulk fatty acids. These diseases are all autosomal recessive, with a collective incidence of 1 in 5000–10,000 live births (Vishwanath 2016; Merritt II et al. 2018). Milder forms present with neonatal cardiomyopathy, liver dysfunction, hypoglycemia, rhabdomyolysis, and skeletal myopathy. Severe varieties frequently lead to hyperammonemia and subsequent death due to the necessary catabolism of proteins for energy (Vishwanath 2016; Merritt II et al. 2018). Here we divide the FAODs into two categories: diseases that result from the dysfunctional carnitine shuttle responsible for importing fatty acids into the mitochondrial matrix, and diseases that result from the dysfunction of β-oxidation proper.

2.2.1 Dysfunctional Carnitine Shuttle

Carnitine is an essential compound required for the transport of long-chain fatty acids into the mitochondria for oxidation via the carnitine shuttle (Fig. 2). Carnitine uptake into the cell is controlled by carnitine transporter 2, encoded by *SLC22A5* (Li et al. 2010). Defects in *SLC22A5* result in systemic primary carnitine deficiency (CDSP). As carnitine cannot enter cells without the transporter, β-oxidation is impaired. Symptoms are often milder, with chronic muscle weakness and cardiomyopathy. The disease is often managed with high doses of carnitine (Li et al. 2010). *SLC22A5* has multiple fly homologs, notably *BalaT, Orct2,* and *Orct* (Engelhart et al. 2020; Han et al. 2017). While some functional phenotyping for these genes has been conducted, they have not been modeled for the purposes of studying CDSP. *BalaT* has been shown to function as a β-alanine transporter and necessary for neurotransmission (Han et al. 2017) (Table 3).

The carnitine shuttle consists of three major players: CPT1, CACT, and CPT2 (Fig. 2). Carnitine palmitoyl transferase 1, encoded by *CPT1A*, generates and transports acyl-bound carnitine across the outer mitochondrial membrane, thereby allowing long-chain fatty acids to pass the first mitochondrial barrier (Strub et al. 2008). Defects in *CPT1A* result in carnitine palmitoyl transferase 1 deficiency. *CPT1A* possesses the fly homolog *whd*, notable for its extremely withered wings (Strub et al. 2008). Researchers have shown that *whd* mutants are hypersensitive to oxidative stress and starvation, capturing a key phenotype of CPT1 deficiency (Strub et al. 2008). The next step of the carnitine shuttle involves the cotransporter carnitine-acylcarnitine translocase (CACT), which transports an acyl-carnitine across the inner mitochondrial membrane into the matrix as it transports a carnitine molecule in the opposite direction (Vishwanath 2016; Merritt II et al. 2018). Defects in the gene *SLC25A20* cause CACT deficiency, a disease

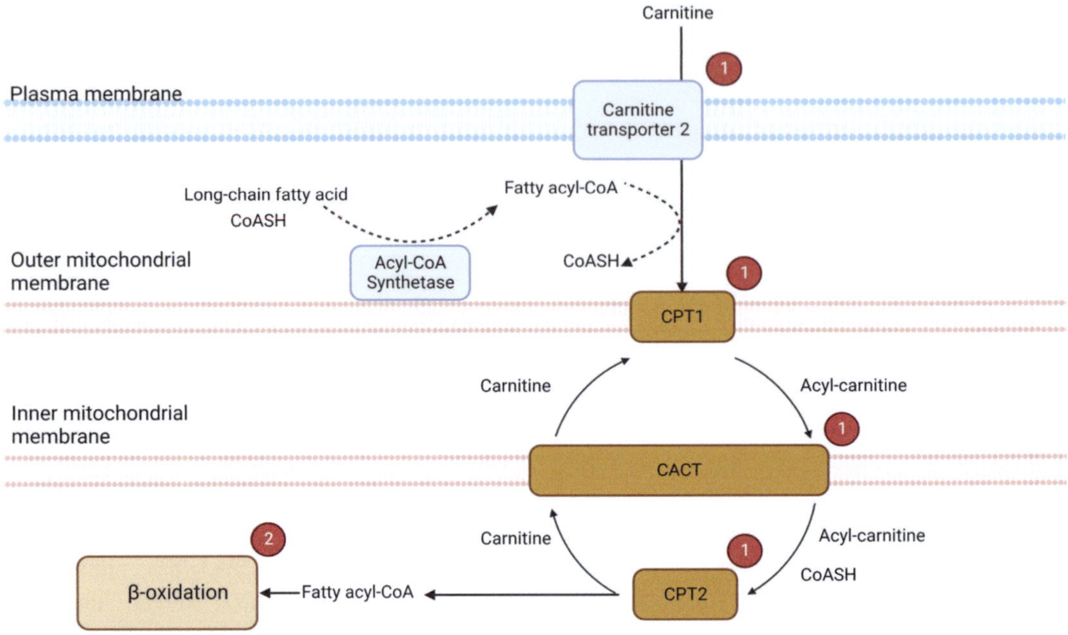

Fig. 2 Sites of mitochondrial β-oxidation disorders. Fatty acid β-oxidation disorders (FAODs) can be roughly divided into two categories: (1) diseases that result from a dysfunctional carnitine shuttle and (2) diseases that result from dysfunctional oxidation enzymes. Made with Biorender

with similar phenotypes to the previous two carnitine-related diseases (Vishwanath 2016; Merritt II et al. 2018). *SLC25A20* has a fly homolog, *colt*, but this gene has not been modeled for the purposes of studying carnitine deficiency diseases. The final member of the carnitine shuttle is carnitine palmitoyl transferase 2 (CPT2) and is encoded by *CPT2*. Defects result in CPT2 deficiency, and this disease has been robustly modeled in drosophila (Manzo et al. 2018; Schulz et al. 2015). Using two p-element insertions into *dCPT2*, Schulz et al. have shown high mutant mortality within 3 days of eclosion, not dissimilar to the high human mortality shortly after birth (Schulz et al. 2015). This phenotype was fully rescued by the insertion of rat CPT2. Interestingly, it has also been shown that alterations in dCPT2 were protective against fly mortality under conditions of amyotrophic lateral sclerosis (Manzo et al. 2018). In this model, researchers previously overexpressed human TAR DNA binding protein 43 (TDP-43) in drosophila, successfully capturing the phenotypes of the neurodegenerative disease in addition to a deficit in the carnitine shuttle and reduced β-oxidation (Manzo et al. 2018; Estes et al. 2011). Through genetic (knocking *dCPT2* down) and dietary (addition of medium-chain fatty acids into the diet) means, researchers were able to rescue reduced fly mortality by bypassing the carnitine shuttle.

2.2.2 Dysfunctional β-Oxidation

A variety of inborn β-oxidation errors exist due to mutation in one of a number of enzymes involved in the mitochondrial oxidation of fatty acids (Vishwanath 2016; Merritt II et al. 2018). Depending on the enzyme affected, the diseases can range from treatment of dietary restrictions to rapid death upon birth (Vishwanath 2016; Merritt II et al. 2018). However, widespread genetic screening has increased the likelihood of a proper diagnosis and thus the proper treatment regime, drastically decreasing mortality (Merritt II et al. 2018). Listed in Table 3 are many of the inborn β-oxidation diseases, several of which do not

Table 3 Human fatty acid β-oxidation disorders and their insect models

Disease	Human gene	Protein encoded	Pathway/function	Fly ortholog	DIOPT score	Reference for model
Systemic primary carnitine deficiency (CDSP)	*SLC22A5*	Solute carrier family 22 member 5	Cellular uptake of carnitine	*BalaT, Orct2, Orct*	7, 7, 7	Engelhart et al. (2020), Han et al. (2017)
Carnitine palmitoyl transferase I deficiency	*CPT1A*	Carnitine palmitoyltransferase 1A	Carnitine-dependent transport of fatty acids	*whd*	13	Strub et al. (2008)
Carnitine-acylcarnitine translocase (CACT) deficiency	*SLC25A20*	Solute carrier family 25 member 20	Transport of acylcarnitines to mitochondrial matrix	*colt*	14	Carillo et al. (2020)
Carnitine palmitoyltransferase II deficiency	*CPT2*	Carnitine palmitoyltransferase 2	Carnitine-dependent transport of fatty acids	*CPT2*	14	Manzo et al. (2018), Schulz et al. (2015)
Very long-chain acyl-coenzyme A dehydrogenase deficiency (VLCADD)	*ACADVL*	Acyl-CoA dehydrogenase very long chain	Lipid β-oxidation	*CG7461*	15	
Long-chain 3-hydroxyacyl-coenzyme A dehydrogenase deficiency (LCHAD)	*HADHA*	Hydroxyacyl-CoA dehydrogenase trifunctional multienzyme complex subunit alpha	Lipid β-oxidation	*Mtpalpha*	15	Kishita et al. (2012)
Medium-chain acyl-CoA dehydrogenase deficiency (MCADD)	*ACADM*	Acyl-CoA dehydrogenase medium chain	Lipid β-oxidation	*Mcad*	14	Course et al. (2018)
Short-chain acyl-coenzyme A dehydrogenase deficiency (SCADD)	*ACADS*	Acyl-CoA dehydrogenase short chain	Lipid β-oxidation	*Arc42*	15	
3-Hydroxyacyl-coenzyme A dehydrogenase deficiency (HADD)	*HADH*	Hydroxyacyl-CoA dehydrogenase	Lipid β-oxidation	*Had2*	2	
2,4 Dienoyl-CoA reductase deficiency	*NADK2*	NAD kinase 2, mitochondrial	Lipid β-oxidation	*CG8080*	12	

possess an insect model despite strong candidate homologs existing.

The diseases long-chain 3-hydroxyacyl-coenzyme A dehydrogenase deficiency (LCHAD) and medium-chain acyl-coenzyme A dehydrogenase deficiency (MCADD) have been modeled in the fly. The gene *HADHA*, responsible for LCHAD, has the fly homolog *Mtpalpha*, which researchers knocked out to study the disease (Kishita et al. 2012). The knockout flies exhibited reduced lifespan and fecundity, defective locomotion, and hypersensitivity to starvation. Flies were unable to mobilize fats for energy production under conditions of starvation (Kishita et al.

2012). *ACADM*, the gene responsible for MCADD, has been implicated in Parkinson's disease, as a target of a crucial kinase *PINK1* (Course et al. 2018).

2.3 Lipid Transport Disorders

Diseases of fatty acid transport are by far the most common inborn errors of lipid metabolism, with some having incidences of 1 in 250 people (Palm et al. 2012; Rodríguez-Vázquez et al. 2015). Symptoms range from relatively mild to severe. These diseases are classified as transport diseases as they affect genes controlling proteins involved in the transport of lipids through the blood. The most common vector of this transport is the lipoprotein, whose metabolism or regulation may be exaggerated (Palm et al. 2012; Rodríguez-Vázquez et al. 2015). Due to the increased complexity of the mammalian cardiovascular system, many of the causal genes for these diseases lack homologs in an insect model. That said, some of the key genes involved in lipid transport are conserved (Palm et al. 2012; Rodríguez-Vázquez et al. 2015).

2.3.1 Hyperlipidemias

The first group of dyslipidemias we will cover here are the hyperlipidemias. These include familial hypercholesterolemia, hyperalphalipoproteinemia, familial hypertriglyceridemia, and sitosterolemia. This family of dyslipidemias is characterized by extremely high levels of cholesterol/low-density lipoprotein (LDL), high-density lipoprotein (HDL), and very low-density lipoproteins/triglycerides, respectively (Brown and Goldstein 1974; Giammanco et al. 2021; Schonfeld et al. 2005; Bruckert and Gallo 2014). Sufferers of these hyperlipidemias are at much higher risk of atherosclerosis; thus cardiovascular disease is the main complication of these diseases (Brown and Goldstein 1974; Schonfeld et al. 2005; Ripatti et al. 2020).

Of the groups of diseases that we have so far covered, this group perhaps offers the greatest room for future exploration. While key genes are conserved between humans and insects, few have been used to model their diseases (see Table 4). These diseases range widely in prevalence, with the symptomatic heterozygote form of familial hypercholesterolemia occurring within 1 in 250 individuals who possess a mutation in *LDLR*, the LDL receptor (Brown and Goldstein 1974). By contrast, mutations for *APOB*, another causal mutation for familial hypercholesterolemia, occur in 1 in 1000 births (Ripatti et al. 2020). Human *APOB* has two homologs in drosophila, both having been characterized for their role in insect lipoprotein metabolism. Researchers have shown that a knockdown of *apolpp*, both whole-animal and cardiomyocyte-specific, resulted in significantly lowered systemic triglyceride levels. A knockdown in cardiomyocytes alone showed an accumulation of dietary lipids in the gut. Further, flies fed a high-fat diet had significantly reduced expression (Lee et al. 2017; Ding et al. 2021). Taken together, these results suggest that *apolpp* expressed in the heart plays a key role in moderating lipid transport (Lee et al. 2017; Ding et al. 2021). Likewise, an *apoltp* knockdown has been shown to cause a significant accumulation of neutral lipids in the midgut (Rodríguez-Vázquez et al. 2015). This further suggests these homologs are involved in the uptake and transport of lipids from the gut.

2.3.2 Familial Hypobetalipoproteinemia

Familial hypobetalipoproteinemia (FHBL) is a rare autosomal inborn error leading to very low levels (below the 5th percentile) of cholesterol-bound LDL in the blood (Schonfeld et al. 2005). FHBL has been associated with several genes, including *APOB*, *SAR1B*, and *MTTP*, all of which have drosophila homologs. These genes are associated with LDL itself (the product of *APOB* being a major component of LDL), or LDL-cholesterol metabolism (Schonfeld et al. 2005). These mutations typically result in reduced LDL or very LDL while remaining clinically asymptomatic, or in extreme disruptions to betalipoprotein metabolism, fatty liver, and fat malabsorption (Schonfeld et al. 2005). Of the three genes, only the *MTTP* has been modeled for a disease deficiency. Kishita et al. have shown drosophila with knockouts in *Mtp* exhibited

Table 4 Human lipid transport disorders and their insect models

Disease	Human gene	Protein encoded	Pathway/function	Fly ortholog	DIOPT score	Reference for model
Familial hypercholesterolemia	*APOB*	Apolipoprotein B	Activated TLR4 signaling and lipoprotein metabolism	*Apoltp, apolpp*	4, 4	Rodríguez-Vázquez et al. (2015), Lee et al. (2017), Ding et al. (2021)
Familial hypercholesterolemia	*LDLR*	Low-density lipoprotein receptor	Cholesterol synthesis/regulation	*LpR2*	11	Rodríguez-Vázquez et al. (2015)
Familial hypercholesterolemia	*LDLRAP1*	Low-density lipoprotein receptor adaptor protein 1	Lipoprotein metabolism and Clathrin-mediated endocytosis	*CG14968*	2	
Familial hypercholesterolemia	*USF1*	Upstream transcription factor 1	Signaling mediated by p38-alpha and p38-beta	*USF*	9	Rodríguez-Vázquez et al. (2015)
Hyperalphalipoproteinemia	*LIPC*	Lipase C, hepatic type	Lipoprotein metabolism and triacylglycerol biosynthesis	*CG6472*	5	
Hyperalphalipoproteinemia	*SCARB1*	Scavenger receptor class B member 1	Mediates cholesterol transfer to and from HDL	*crq*	8	Guillou et al. (2016)
Familial hypertriglyceridemia	*LPL*	Lipoprotein lipase	Lipoprotein metabolism	*CG6271, CG6283*	3, 3	
Familial hypertriglyceridemia	*GPD1*	Glycerol-3-phosphate dehydrogenase 1	Glycerophospholipid biosynthesis and acyl chain remodeling of PE	*Gpdh1*	14	O'brien and Shimada (1974)
Sitosterolemia	*ABCG5*	TP-binding cassette (ABC) transporter	Sterol transport across the cell membrane	*CG11069*	13	
Sitosterolemia	*ABCG8*	ATP-binding cassette subfamily G member 8	Sterol transport	*CG31121*	4	
Hypobetalipoproteinemia	*SAR1B*	Secretion-associated Ras related GTPase 1B	Lipoprotein metabolism	*Sar1*	12	
Hypobetalipoproteinemia	*MTTP*	Microsomal triglyceride transfer protein	Lipoprotein metabolism	*Mtp*	14	Kishita et al. (2012, Palm et al. (2012)

reduced lifespan and locomotion, as well as abnormal fat accumulation and a failure to mobilize fats under starvation conditions (Kishita et al. 2012; Lee et al. 2017). Thus, drosophila seems to be a strong candidate for modeling diseases of lipoprotein deficiency.

3 Polygenic Diseases and Their Models

Polygenic diseases associated with lipid metabolism are a broad category of complex diseases with a variety of causes, including genetic and environmental components. Often, genes thought to be associated with the disease are not causative on their own. Genome-wide association studies (GWAS) have located regions of the genome marked by single-nucleotide polymorphisms (SNPs) that confer increased or decreased risk of metabolic and lipid disorders. The effect size of any SNP is small, though, and the majority of SNPs do not reside within genes and may serve only as markers of regions either containing regulatory elements of neighboring genes or regions regulated in an epigenetic fashion to contribute to disease phenotypes (Vujkovic et al. 2020). Given the complex nature of genetic and environmental influences, these diseases have multifactorial etiologies but typically present with altered lipid homeostasis and an increased risk of cardiovascular disease (Ng et al. 2014; Bullard et al. 2018; Himanshu et al. 2020; Mooradian 2009). We will first discuss candidate genes for the major polygenic disorders that include lipid diseases: type 2 diabetes mellitus, obesity, and nonalcoholic fatty liver (NAFL) disease.

3.1 Diabetes Mellitus

Diabetes mellitus is a group of metabolic diseases primarily resulting in hyperglycemia and dysregulated lipid metabolism. The group consists of type 1 diabetes, an autoimmune disorder; type 2 diabetes, which makes up roughly 90% of cases; mature onset diabetes in youth (MODY); and gestational diabetes, a form of the disease found in at-risk pregnant women (Bullard et al. 2018; Himanshu et al. 2020; Mooradian 2009). In recent decades, diabetes has come to represent a significant health burden in the United States, with 20 million Americans being diagnosed with the disease in 2016 (Bullard et al. 2018). Indeed, diabetes has become a growing problem for the world. The prevalence of diabetes in 1985 was just over 30 million people; in 2015, over 390 million people were diagnosed with type 2 diabetes (Ng et al. 2014). As such, many efforts have been made to model aspects of this disease in drosophila (Alfa and Kim 2016; Álvarez-Rendón et al. 2018).

Diabetes is characterized by high blood sugar, low insulin production, insulin resistance, frequent urination, increased thirst, and metabolic flux often leading to weight loss (Bullard et al. 2018; Himanshu et al. 2020; Mooradian 2009). The disease leads to complications and comorbidities such as increased risk of cardiovascular disease and kidney failure among others. Diabetes also frequently presents with diabetic dyslipidemia, characterized by high concentrations of chylomicrons, VLDL and LDL, and decreased HDL in the blood. This is thought to result from the high free fatty acid flux that accompanies insulin resistance (Mooradian 2009).

In both major types of diabetes mellitus, type 1 diabetes (T1D) and type 2 diabetes (T2D), the ultimate cause is a deficiency in insulin or insulin action (Bullard et al. 2018; Himanshu et al. 2020; Graham and Pick 2017). T1D is an autoimmune disease whereby insulin-producing β-cells in the pancreas are targeted by the immune system. This causes insulin production to be severely limited from a young age. While the cause of this autoimmune disorder is not well understood, it is thought to be due to a combination of environmental and genetic susceptibility (Atkinson et al. 2014). T1D typically first presents in children. T2D, by contrast, results from the complex interaction of diet, lifestyle, and genetics most often in adults. The presence of obesity, especially chronic obesity, lack of physical exercise, and a diet high in fats and sugars are the primary contributing factors in the development of T2D (Abdullah et al. 2010; Herder and Roden 2011; Himanshu et al. 2020). Genetically, T2D is highly polygenic with each associated gene having very low effect sizes (Herder and Roden 2011). Potential genetic causes are often identified as single-nucleotide polymorphisms (SNPs) near or rarely within genes by genome-wide association studies (GWAS) (Vujkovic et al. 2020; Pendse et al. 2013). Often, these SNPs do not mutate the gene in such a way as to severely inhibit function, but are thought to influence function or regulation, potentially in epistasis with other genes (Vujkovic et al. 2020; Pendse et al. 2013). GWAS thus reveal candidate genes that further studies attempt to validate, often in model organisms, by more drastically altering their function and observing the phenotype (Pendse et al. 2013; Alfa and Kim 2016; Park et al. 2014).

3.1.1 Insulin Pathway Homology

The insulin signaling pathway is broadly conserved between humans and insects, thus allowing researchers to model both T1D and T2D (Alfa and Kim 2016; Álvarez-Rendón et al. 2018). In humans, the hormone insulin is produced by Langerhans islet β-cells in the pancreas that is released upon stimulation by spikes of glucose in the blood. In drosophila, this hormonal function is performed by a small family of signaling peptides called *Drosophila* insulin-like peptides (ILPs) (Alfa and Kim 2016; Semaniuk et al. 2018). ILPs are produced primarily by specialized neurons, insulin-producing cells (IPCs) (Fig. 1). There are seven ILPs in drosophila that function in a like manner to insulin; they are thought to have nonredundant function and are sometimes produced by different tissues under unique physiological conditions (Alfa and Kim 2016; Álvarez-Rendón et al. 2018). Interestingly, insect insulin-like peptides are also homologous to human insulin-like growth factors (IGFs) and thus serve the dual function of promoting growth (Alfa and Kim 2016; Álvarez-Rendón et al. 2018).

The insulin receptor (IR) is also evolutionarily conserved between flies and mammals, as well as much of the downstream machinery such as human *IRS1* and drosophila *chico* (Murillo-Maldonado et al. 2011; Alfa and Kim 2016). The critical signaling hub, and downstream target of insulin, in both humans and flies is Akt/PKB. This protein acts via the phosphorylation of target proteins, affecting many anabolic pathways. Akt/PKB stimulates glucose uptake from the blood, stimulates protein synthesis through TOR signaling (a process that serves to couple animal growth with the amount of available nutrients), and inhibits the catabolic controller FOXO. These processes are well conserved in insect models (Álvarez-Rendón et al. 2018). Remarkably, human insulin can bind to and activate the fly insulin receptor, resulting in the phosphorylation of dAkt (Musselman et al. 2011).

3.1.2 Diet-Induced Models of Diabetes

Among the first to use a diet-induced model of T2D in an insect model were Musselman et al. in 2011 (Musselman et al. 2011). Here they used a high-sugar (1 M sucrose) food media to model insulin resistance in a fly model. To this effect, larvae reared on high-sugar food possessed hyperglycemia akin to both diabetic humans and flies with loss of function in the insulin signaling pathway. Indeed, this increased hemolymph glucose concentration persisted despite increased expression of *ilp*s, in particular *ilp2*, suggesting that the flies had become insulin resistant (Musselman et al. 2011). Further supporting this conclusion, insulin-stimulated phospho-Akt was measured as a proxy for insulin resistance (see above). Musselman et al. found that larvae reared on high-sugar food indeed had blunted insulin stimulation of phospho-Akt, again suggesting insulin signaling had been systemically impaired (Musselman et al. 2011). High-sugar-fed larvae also exhibited increased fat storage akin to diabetes and reduced size (as one might expect given ILPs' dual roles of glucose homeostasis and growth signaling) (Musselman et al. 2011). Since then, many other researchers have used a high-sugar paradigm to produce a diabetes-like phenotype in *Drosophila melanogaster*, using not only sucrose but glucose and fructose as well (Graham and Pick 2017; Na et al. 2013; Morris et al. 2012; Musselman et al. 2019; Ecker et al. 2017; Musselman et al. 2018; Lüersen et al. 2019). In addition to diabetes-like phenotypes, the fly lipidome was significantly affected by a high-sugar diet. Flies fed this diet exhibited increased levels of free fatty acids and had significantly larger lipid droplet size, a phenotype observed in obese humans (Musselman et al. 2011). Lipid species were also affected by a high-sugar diet. There was a relative increase in the abundance of shorter triglyceride species in the fat body, as well as an increase in the number of double bonds present in the lipids (Musselman et al. 2013). Limitations persist, however. The diets used to produce the diabetes-like phenotypes have not been fully defined in a chemical fashion that considers the yet poorly understood micronutrient requirements of drosophila (Graham and Pick 2017; Lüersen et al. 2019). Further, factors such as genotype/genetic background, unique environmental conditions, and larval density have been shown to influence levels of circulating glucose, potentially influencing results (Ugrankar et al. 2015). Regardless, as calls for more standardized "diabetes-inducing diets" continue, diet will remain a powerful tool for modeling T2D in insects (Lüersen et al. 2019).

3.1.3 Genetic Models of Diabetes

With regard to genetic models of diabetes, we might first start with forward genetic approaches whereby researchers hunt for genes responsible for the diabetic disease state. As discussed above, complex diseases such as T2D are thought to be largely polygenic with small effect size SNPs contributing to the disease (Vujkovic et al. 2020; Pendse et al. 2013). To date, GWAS are the favored approach for identifying regions of the genome harboring such causal genes. As GWAS have continued to be carried out with larger and larger sample sizes in novel populations, the list of potential causal genes has grown from roughly 30 to some 586 associations (Vujkovic et al. 2020). Recently, Vujkovic et al. conducted GWAS using 1.4 million US military veteran participants (more than 200,000 cases and 1.1 million controls) (Vujkovic et al. 2020). With the benefit of ancestral data, the researchers identified 318 new regions marked by SNPs associated with risk for developing T2D (Vujkovic et al. 2020).

One of the greatest advantages of the fly model, as we have discussed previously, is the plasticity with which to manipulate the fly genome and study its effects. Following GWAS, researchers then can "validate" an association in a drosophila model (Pendse et al. 2013; Alfa and Kim 2016; Park et al. 2014). Many phenotypes such as hyperglycemia and reduced size might serve as confirmation that a gene is indeed related to the pathways affected by T2D. Indeed, researchers have also developed a direct means to measure the insulin-analogous Ilp2 levels to quantify insulin resistance (Park et al. 2014). One such validation study showed that under conditions of high-sugar feeding, disruptions of GWAS-identified genes led to sucrose-dependent toxicity (Pendse et al. 2013). One such gene was *dHHEX,* homologous to human *HHEX*, a transcription factor associated with organ development (Topisirovic et al. 2003). A tissue-specific knockdown of *dHHEX* in the fat body (analogous to the human liver and adipose tissue; see Fig. 1) showed significantly decreased insulin signaling on a high-sugar diet compared to controls, suggesting a heightened diabetes-like phenotype upon loss of function of *dHHEX* (Pendse et al. 2013).

In addition to functional genomic studies that attempt to uncover the genetic basis of diabetes, T1D and T2D diabetes can be modeled by direct manipulation of insulin signaling genes (Table 5). In both cases, the goal is to genetically reduce or ablate insulin signaling. T1D has been modeled successfully using flies with *ilp1–5* simultaneously deleted, thus removing insulin-like peptide function almost entirely (Graham and Pick 2017; Broughton et al. 2005). These flies exhibited significantly increased triglyceride levels, reduced size, and increased circulating sugar, i.e., diabetic-like phenotypes (Graham and Pick 2017; Broughton et al. 2005). Further, by varying the combination of *ilp* deletions, they have shown *ilp2, 3,* and *5* to be the primary *ilps* responsible for sugar uptake in drosophila (Graham and Pick 2017). Likewise, T2D can be modeled in the fly by genetic manipulation of other elements of the insulin signaling pathway (Pendse et al. 2013; Murillo-Maldonado et al.

Table 5 Human polygenic lipid disorders and their drosophila models

	Human gene	Protein encoded	Pathway/function	Fly ortholog	DIOPT score	Reference for model
Type 2 diabetes	HHEX	Hematopoietically expressed Homeobox	Metabolism-related HOX-class transcription factor	HHEX	8	Pendse et al. (2013)
Type 2 diabetes	INS	Insulin	Carbohydrate and lipid metabolism	ilp1–5	2	Graham and Pick (2017), Broughton et al. (2005)
Type 2 diabetes	INSR	Insulin receptor	Carbohydrate and lipid metabolism	InR	10	Murillo-Maldonado et al. (2011)
Type 2 diabetes	IRS1	Insulin receptor substrate 1	Carbohydrate and lipid metabolism	Chico	10	Murillo-Maldonado et al. (2011)
Type 2 diabetes	RHEB	Ras homolog, MTORC1 binding	mTOR signaling	Rheb	12	Murillo-Maldonado et al. (2011)
Obesity	FTO	FTO alpha-ketoglutarate-dependent dioxygenase	FTO obesity variant mechanism and homology directed repair	None		
Obesity	PNPLA2, PNPLA3	*Patatin-like phospholipase domain–containing protein 2, 3*	Triglyceride lipolysis	Bmm	12, 11	Grönke et al. (2005)
Obesity	STIM1, STIM2	Stromal interaction molecule 1, 2	Mediation of Ca^{2+} flux	Stim	13, 10	Baumbach et al. (2014)
Obesity	TUB	*TUB bipartite transcription factor*	Transcriptional regulation	ktub	11	Musselman et al. (2013)
Nonalcoholic fatty liver disease	PNPLA3	*Patatin-like phospholipase domain–containing protein 3*	Triglyceride lipolysis	Bmm, dob	11, 9	Grönke et al. (2005)
Nonalcoholic fatty liver disease	TM6SF2	*Transmembrane 6 superfamily member 2*	Regulator of liver fat metabolism	None		
Nonalcoholic fatty liver disease	GCKR	*Glucokinase regulator*	Regulator of glucokinase in the liver	None		
Nonalcoholic fatty liver disease	HNF4	*Hepatocyte nuclear factor 4*	Hepatocyte differentiation and lipid metabolism regulation	Hnf4	11	Palanker et al. (2009; Almeida-Oliveira et al. (2021)

2011; Jünger et al. 2003). Null mutants of genes functioning downstream of insulin-like peptide signaling such as *InR, chico, Rheb,* and others have been utilized to induce insulin resistance and other diabetic-like phenotypes (Murillo-Maldonado et al. 2011). In this way, the phenotype can be investigated with respect to or absent from lifestyle/diet considerations.

3.1.4 Cell Ablation Model of T1D

The final diabetic modeling method we will discuss here has been used to model T1D in a fashion functionally like the pathophysiology in humans. In 2002, Rulifson et al. demonstrated that the ablation of IPCs, the neuronal cells responsible for insulin-like peptide production, resulted in T1D-like phenotypes in drosophila larvae (Rulifson et al. 2002). Animals exhibited impaired development and growth, as well as increased sugars in the hemolymph. In addition, these phenotypes were rescued by the ectopic expression of *ilp2* (Graham and Pick 2017; Broughton et al. 2005). This method was repeated in adult flies by Haselton et al., showing comparable phenotypes (Haselton et al. 2010). Indeed, they further demonstrated that like human T1D, IPC-ablated flies experienced a spike in glucose following a meal that was cleared far more slowly than control flies (Haselton et al. 2010).

3.2 Obesity

Obesity is a metabolic disease characterized by the accumulation of excess body fat (Ng et al. 2014; Levitsky et al. 2019). This has been classically measured by a body mass index (BMI) of over 30 kg/m^2 (Seidell 1999). Over the past 50 years, obesity has become one of the largest health burdens in the United States and is predicted to continue to rise in prevalence globally. Indeed, in 1980, there were an estimated 857 million obese adults compared to 2.1 billion adults in 2013 (Ng et al. 2014). In the United States, the prevalence of obesity was roughly 35.8% of adult women and 35.5% of adult men in 2010 (Flegal et al. 2012). This has posed not just a significant health crisis, as we will discuss below, but also a significant strain on the financial health burden, with a staggering $147 billion dollars spent on medical costs for obesity and its various sequelae (Finkelstein et al. 2009).

While the primary phenotype of obesity is increased fat storage, it is also accompanied by an array of co-morbidities. Most prominent among these are T2D, cardiovascular disease and other heart health complications, osteoarthritis, some types of cancer, obstructive sleep apnea, and nonalcoholic fatty liver disease. All these together cause obesity to be significantly associated with reduced life expectancy (Bray 2004).

Much like T2D, obesity is caused by the complex interaction of diet, lifestyle, and genetics. The primary cause of obesity is increased food intake (Levitsky et al. 2019; Swinburn et al. 2004). Prevalence of obesity correlates tightly with the availability of high-energy foods (Swinburn et al. 2004). Indeed, developed nations show a higher prevalence and risk of developing obesity (Swinburn et al. 2004). Obesity is also connected to a sedentary lifestyle. As a complex disease, obesity has also been found to have a significant genetic risk contribution. Like T2D, obesity has been shown to be highly polygenic in nature, especially as obesity is an extremely heterogenous disease (Li and Loos 2008; Carmelli et al. 1994; Ichihara and Yamada 2008). GWAS have been conducted using a variety of obesity phenotypic markers, most successfully identifying a SNP in the *FTO*, or fat mass and obesity-associated protein, gene (Turcot et al. 2017). The gene has been shown to have several risk alleles, but adults homozygous for the SNP rs9939609 on average weighed 3 kg more and had a 1.67-fold increased chance of developing obesity in comparison to adults with no copies (Turcot et al. 2017). Much work remains to be done, however, characterizing the roles of other risk SNPs and genes.

3.2.1 Obesity Phenotype Homology

Much like humans, insects primarily store excess energy as glycogen (short term) and triglycerides (long term). In drosophila, both storage molecules are primarily housed in the fat body, analogous to the human adipose tissue and liver (see Fig. 1). The fly fat body is composed of many polyploid cells which store lipids in organelles called lipid droplets an organelle conserved in human adipose tissues (Musselman and Kühnlein 2018; Grönke et al. 2005; Trinh and Boulianne 2013; Gáliková and Klepsatel 2018; Smith et al. 2014). In an obesogenic state, both the size and number of lipid droplets are increased in both humans and flies (Musselman and Kühnlein 2018; Trinh and Boulianne 2013; Gáliková and Klepsatel 2018). Thus, the fly fat body is the central organ under focus for much of the drosophila obesity research.

Analogous to human glucagon release under conditions of fasting, the fly APCs release adipokinetic hormone (Akh) to mobilize lipids (see Fig. 1). The cAMP/PKA pathway that glucagon targets is likewise conserved in insect models. Indeed, many of the fly genetic models of obesity target the downstream targets of the cAMP/PKA pathway (Musselman and Kühnlein 2018; Trinh and Boulianne 2013; Smith et al. 2014). One such gene is the well-studied *bmm* (*brummer*), a target of the catabolic controller Foxo (Grönke et al. 2005). This gene encodes a lipase homologous to *PNPLA2*, a gene encoding human adipose triglyceride lipase (ATGL). Catalyzing the conversion of triglycerides to diglycerides, ATGL thus plays a central role in the mobilization of stored fats for energy needs. Mutant *bmm* flies accordingly accumulate fats as they are unable to begin the process of mobilizing them from the fat body (Grönke et al. 2005).

3.2.2 Diet-Induced Models of Obesity

The most used diet-induced model for obesity in drosophila is the high-sugar diet, like the T2D diet model (Musselman et al. 2011, 2013; Na et al. 2013; Trinh and Boulianne 2013; Gáliková and Klepsatel 2018; Musselman et al. 2016). This diet can be used to generate obesity in both adults and larvae. While a high-sucrose diet predominates, researchers have also used a high-glucose and high-fructose diet to induce similar phenotypes (Musselman et al. 2011, 2013, 2019). The exact chemical composition of these diets varies, but frequently, high-sugar diets possess roughly 30% sugar w/v (Musselman and Kühnlein 2018; Musselman et al. 2011, 2013, 2016; Na et al. 2013; Trinh and Boulianne 2013; Gáliková and Klepsatel 2018). In addition to the desired model of increased fat storage and hyperglycemia, high-sugar fed flies also exhibit delayed larval development and heart function complications such as arrhythmia and fibrosis, reduced fecundity, and generally increased mortality (Musselman et al. 2011; Na et al. 2013; Trinh and Boulianne 2013; Gáliková and Klepsatel 2018). While fats do not comprise a significant proportion of a fruit fly's diet, they, nevertheless, can be artificially fed high-fat diets as a means of inducing obesity-like phenotypes in flies. This is typically administered via coconut oil in a 20–30% fat content food media (Gáliková and Klepsatel 2018; Hong et al. 2016; Schultzhaus et al. 2018; Birse et al. 2010). Like the high-sugar diet, the high-fat diet is typified by increased fat storage, hyperglycemia, cardiac complications such as fat accumulation in the heart, and increased mortality (Gáliková and Klepsatel 2018; Hong et al. 2016; Schultzhaus et al. 2018; Birse et al. 2010).

3.2.3 Genetic Models of Obesity

Like T2D studies, the GWAS remain one of the primary tools for uncovering the genetic component of obesity susceptibility. However, as the clinical definition of obesity is somewhat allometrically problematic, researchers such as Muller et al. have argued that BMI is a suboptimal GWAS criterion (Müller et al. 2018). This follows as they argue that BMI is an oversimplification of the phenotype and cannot capture the complexity that results from the interplay of energy intake and expenditure. Thus, clear insights are muddied. Instead, these researchers suggest other energy traits, such as resting metabolic rate (Müller et al. 2018). Other researchers, however, have argued that BMI is, while imperfect to be sure, an adequate trait for obesity GWAS. Speakman et al., for example, have argued that while indeed BMI is a flawed metric to categorize obesity when considering individuals with extreme muscularity, it is, nevertheless, accurate for most individuals (Speakman et al. 2018). Perhaps most importantly, Speakman et al. argue that BMI remains the cheapest and easiest measure of obesity, whatever its faults. It thus allows for extremely large sample sizes to increase the power of the GWAS (Speakman et al. 2018). This high power, coupled with the ability to screen loci using drosophila, will likely mean BMI retains its place as the preeminent trait for obesity GWAS. Indeed, drosophila has already been used to great effect in testing such GWAS-identified SNPs (Pendse et al. 2013; Williams et al. 2015; Chalmers et al. 2021; Baranski et al. 2018).

In addition to using drosophila to investigate SNPs identified by GWAS, drosophila has been used to directly model obesity. As mentioned above, one common method is to genetically knockout or knockdown key lipases, such as the lipase encoded by *bmm* (Grönke et al. 2005). As the flies are unable to fully mobilize fats, lipids accumulate in the fat body. Mutant *bmm* flies exhibit 10% decreased survival compared to controls, but interestingly show as much as 50% increase in survival under conditions of starvation (Grönke et al. 2005). This surprising result suggests that fat mobilization is under the control of multiple lipases in the fly fat body. Another genetic model is that of *Akh* where the removal of *Akh* function results in significant increases in fat content as well as hemolymph sugar concentrations (Gáliková et al. 2015). As the hormone Akh serves as the primary catabolic signaling molecule, it is perhaps unsurprising that its removal results in excessive storage. By contrast, knockdowns of the *stim* gene in drosophila elicit adiposity without hyperglycemia (Baumbach et al. 2014). *Stim* functions in the store-operated calcium entry pathway and broadly controls lipogenic and lipolytic gene regulation. Through its manipulation, researchers can study the effects of adiposity without the additional factor of hyperglycemia (Baumbach et al. 2014). Interestingly, the knockdown of the drosophila gene *king-tubby*, a homolog of known obesogenic mammalian *tubby*, produces larvae with increased triglyceride storage but *reduced* hemolymph glucose levels when fed a high-sugar diet (Musselman et al. 2013). This model shows that increased lipogenesis may indeed be protective against the negative effects of a high-sugar diet through the increased clearance of glucose from the blood (Musselman et al. 2013).

3.3 Metabolic Dysfunction–Associated Fatty Liver Disease

Metabolic dysfunction–associated fatty liver disease (MAFLD), previously nonalcoholic fatty liver disease (NAFLD), is a pervasive form of fatty liver disease whereby there is an excessive accumulation of fat in the liver. The disease is related to alcoholic fatty liver disease, but with more complex dietary causes than alcohol consumption (Rinella 2015; Younossi et al. 2016). Though rare, MAFLD may progress into metabolic dysfunction–associated steatohepatitis (MASH), formerly nonalcoholic steatohepatitis (NASH), a more severe form of the disease, if the root causes are not addressed (Rinella 2015; Younossi et al. 2016). MAFLD is the most common liver disease in the United States and was estimated to have affected 75–100 million adult Americans in 2017 (Younossi et al. 2016).

In most cases, MAFLD lacks any noticeable symptoms and is only diagnosable via blood tests, but the disease is occasionally accompanied by fatigue and mild jaundice (Tilg et al. 2016). However, in cases where MAFLD does progress into MASH, liver function may be severely impaired or fail, liver cancer may develop, or the patient may develop cardiovascular disease. MAFLD has been associated with several comorbidities including T2D, obesity, impaired insulin sensitivity, and increased risk for developing cardiovascular disease (Tilg et al. 2016).

Like the above diseases, MAFLD results from the combination of diet, lifestyle, and genetic predisposition (Lim et al. 2010; Arquier and Léopold 2007; Wong et al. 2018). The primary causal effect in the development of MAFLD is thought to be increased food intake (Lim et al. 2010; Arquier and Léopold 2007; Wong et al. 2018). In particular, an increase in the consumption of fructose has been linked to the development and worsening of the disease (Lim et al. 2010). Genetically, characterization of MAFLD is not as fully developed as T2D and obesity. That said, there have been strong findings via GWAS as to high-risk loci. Three variants have consistently been identified to be associated with MAFLD: *PNPLA3*, *TM6SF2*, and *GCKR* (Wong et al. 2018; Kienesberger et al. 2009). The SNP rs738409 in *PNPLA3* is missense mutation that occurs in roughly 38% of the population and is associated with a moderate risk of developing MAFLD (Wong et al. 2018; Kienesberger et al. 2009). *PNPLA3* encodes an adipocyte lipase involved in triglyceride mobilization (Kienesberger et al. 2009). The case for *PNPLA3* is further strengthened by its fly homolog *bmm*. The other variants in *TM6SF2* and *GCKR* are associated with lower risk. These two genes regulate liver fat metabolism and glucokinase activity in the liver, respectively (Wong et al. 2018).

3.3.1 MAFLD: A Model with Room to Grow

As has been discussed previously, the primary mammalian organs for the regulation of fat metabolism are the liver and adipose tissue. During conditions of energy excess, adipose tissue and liver cells activate lipogenesis. Likewise, under conditions of starvation, the same sites promote the action of lipases to mobilize fats for lipolysis. Interestingly, under starvation conditions, the liver begins to accumulate fats, some of which are oxidized down to ketone bodies for energy (Arquier and Léopold 2007).

These functions are largely conserved in the drosophila model by the fat body (adipose tissue) and oenocytes (liver) (see Fig. 1). Under conditions whereby the fly mobilizes fats, such as starvation, the oenocytes accumulate lipids (Arquier and Léopold 2007; Gutierrez et al. 2006). Indeed, the activation of lipases such as the lipase encoded by *bmm* is enough to induce the uptake of lipids into the oenocytes (Grönke et al. 2005). Interestingly, fly larvae genetically induced to lack oenocytes were unable to fully mobilize their fat stores under conditions of starvation. Such larvae accumulated more lipid droplets than their control counterparts (Arquier and Léopold 2007; Gutierrez et al. 2006). Thus, Gutierrez et al. have argued that the regulation of lipid metabolism between the oenocyte and fat body is bidirectional, much like the mammalian system (Gutierrez et al. 2006). It is this regulatory link between the fat body and oenocyte that gives drosophila the most promise for modeling MAFLD. In addition to the above functional homology, Arquier and Leopold have shown that there are more than 20 genes that function in the fly oenocytes homologous to human liver cells (Arquier and Léopold 2007). Included in this list is *HNF4*, a nuclear receptor protein primarily involved in liver development. Drosophila *Hnf4* has been shown to play a critical role in lipid mobilization and the stimulation of β-oxidation in larvae (Palanker et al. 2009; Almeida-Oliveira et al. 2021). While a drosophila model of MAFLD has not been fully developed, the crosstalk governing lipid metabolism between the fat body and oenocytes is fertile ground for a better understanding of the nature of the disease.

3.4 Future Outlook

Drosophila has served as an excellent tool for providing insight into the underlying biochemistry and genetics of human lipid metabolism disease. However, many exact mechanisms remain poorly understood. Future studies will likely focus on regulation and the interaction between pathways using a combination of approaches, such as genetics and lipidomics.

3.4.1 GWAS Validation

As GWAS continue to be refined and increase their sample sizes, drosophila will remain a powerful tool for screening candidate genes near SNPs. Drosophila remains a cheap means of a high throughput screen of potentially causal genes. One of the main challenges of identifying causal variants utilizing GWAS is the small effect size common variants have on the disease phenotype (Müller et al. 2018). Recent studies have shown that GWAS on smaller, especially isolated populations, may represent a path to uncovering rare variants with larger effect sizes (Turcot et al. 2017; Speakman et al. 2018; Minster et al. 2016). Recent BWI GWAS conducted on the Samoan population identified a SNP nearby *CREBRF* that occurs in 45% and accounted for 1.36–1.45 kg per copy of the allele (Minster et al. 2016). While extremely common in the Samoan population, the variant is rare in global populations and thus likely to never be identified by GWAS studying global population data. Another challenge remaining to GWAS is reliably detecting rare intronic variants. Typically requiring sequencing to ~30-fold depth, detecting rare noncoding variants is often prohibitively expensive (Speakman et al. 2018). As sequencing technologies improve, however, deep sequencing of exomes and noncoding regions will become more realistic, allowing researchers to identify rare coding and noncoding variants that associate with disease traits. Drosophila will be well poised to model such rare variants as highly specific gene editing tools such as CRISPR/Cas9 continue to improve.

3.4.2 Mechanism-Based Therapeutic Drug Discovery

Drosophila further represents an opportunity for improving upon therapeutic drug discovery. It has been noted that drug discovery of novel therapeutics has declined over the past several decades (Pandey and Nichols 2011). This is partly due to the traditional method of identifying new drug compounds. The traditional drug discovery method typically begins with the identification of a protein that is involved in the malady of interest. It is then screened with a high volume of chemical compounds to identify those that affect the catalytic, binding, or other biochemical functions of the protein in vitro. The compound is then tested in an animal model. This process has an extremely high failure rate due to unforeseen issues in absorption, metabolism, excretion, or toxicity (Pandey and Nichols 2011). This is largely due to the in vitro origins of the process. Drosophila, by contrast, offers an alternative approach to drug discovery. As already discussed, the fly model has the advantages of low cost, short animal lifespan, ease of husbandry, and highly developed genetic tool (Pandey and Nichols 2011). Thus, drosophila can be utilized to screen compounds, albeit with much reduced throughput, and eliminate most of the downsides of the traditional method (Pandey and Nichols 2011).

In addition to screening, drosophila offers opportunities for a more targeted approach. As novel pathways are associated with disease, the fly model allows researchers the ability to target that pathway with known compounds in a novel fashion. For example, Musselman et al. identified that chronically high-sugar-fed flies underwent reduced triglyceride synthesis due to a deficiency in coenzyme-A (CoA) (Musselman et al. 2016). Thus, the lipogenic pathway was targeted by the dietary supplementation of pantothenic acid (vitamin B5, a precursor to CoA). Dietary supplementation of pantothenic acid ameliorated the negative phenotypes associated with chronic high-sugar feeding in addition to the expected increased lipogenesis. Supplemented larvae exhibited the restoration of whole-animal weights, lowered hemolymph glucose, and increased levels of phospho-Akt (thus improved insulin signaling) (Musselman et al. 2016). Taken together, these data suggest that lipogenesis is protective against the negative effects of high-sugar diets by more effectively clearing sugars from the hemolymph. This may provide key insight into the mystery of the metabolically healthy obese (healthy obese sources). In such cases, otherwise healthy individuals, nevertheless, suffer from extreme adiposity for unknown reasons.

3.4.3 Prospects in Insect Lipid Metabolism

Finally, our understanding of human lipid metabolism disorders will invariably be improved as insect lipid metabolism becomes more well understood. Much remains unknown regarding the lipidome, regulation mechanisms, and the interaction between lipid metabolism and other biochemical pathways. For example, the interaction between insulin resistance and lipoproteins remains poorly understood and is of high relevance to our understanding of diabetes mellitus. It has been previously shown in drosophila that lipoproteins interact with insulin signaling via the accumulation of LDL receptor–related proteins on IPCs. Brankatschk et al. have demonstrated that the proteins LRP1 and Megalin cross the blood-brain barrier to accumulate on the surface of IPCs (Brankatschk et al. 2014). However, this connection has not been investigated in light of insulin resistance and disease.

One approach to better understand drosophila lipid metabolism is to make use of GWAS in flies. These studies can be used to identify novel genes involved in a variety of processes, such as lipid droplet dynamics (Fei and Yang 2012). One great advantage of using the fly in GWAS is the ability to manipulate nutritional conditions and measure desired outcomes. In addition, there are multiple catalogs of fly lines that have been metabolically phenotyped, such as the DGRP lines, which have extensive sequencing data available (Mackay et al. 2012; Mackay and Huang 2018). A further opportunity afforded by drosophila is the ability to perform "evolve-and-resequence" experiments on drosophila (Burke et al. 2010; Hardy et al. 2015, 2018; Kofler et al. 2012; Tobler et al. 2014; Kang et al. 2016; Versace et al. 2014). In these paradigms, populations of drosophila are selected under metabolically relevant pressure for numerous generations. Researchers then phenotypically and genetically "track" their populations. In one such experiment, populations of drosophila were adapted to conditions of starvation for over 80 generations (Hardy et al. 2015, 2018). The population exhibited a significant increase in fat storage in comparison to ancestral flies. Indeed, GWAS was conducted on the evolved populations that identified 382 genes, many of which were associated with metabolism and lipid droplet function (Hardy et al. 2015, 2018). Evolve-and-resquence paradigms may offer a powerful means of identifying novel functional and regulatory genes for lipid homeostasis.

4 Conclusion

The insect models have served an invaluable role in furthering the study of metabolic diseases. Drosophila is well poised for modeling diseases of lipid metabolism, as much of the core mechanistic biology is conserved between fruit flies and mammals. With dietary means, researchers can accurately model widely prevalent diseases such as T2D and obesity. This coupled with the well-established genetic tools and prospects for further development of these tools ensures that drosophila will remain at the forefront of lipid metabolic disease research.

References

Abdullah A, Peeters A, de Courten M, Stoelwinder J (2010) The magnitude of association between overweight and obesity and the risk of diabetes: a meta-analysis of prospective cohort studies. Diabetes Res Clin Pract 89(3):309–319

Aerts JM, Groener JE, Kuiper S, Donker-Koopman WE, Strijland A, Ottenhoff R et al (2008) Elevated globotriaosylsphingosine is a hallmark of Fabry disease. Proc Natl Acad Sci USA [Internet] 105(8): 2812–2817. https://pubmed.ncbi.nlm.nih.gov/18287059/. Cited 12 Dec 2021

Alfa RW, Kim SK (2016) Using Drosophila to discover mechanisms underlying type 2 diabetes, pp 365–376

Almeida-Oliveira F, Tuthill BF, Gondim KC, Majerowicz D, Musselman LP (2021) dHNF4 regulates lipid homeostasis and oogenesis in *Drosophila melanogaster*. Insect Biochem Mol Biol 133:103569

Álvarez-Rendón JP, Salceda R, Riesgo-Escovar JR (2018) *Drosophila melanogaster* as a model for diabetes type 2 progression. Biomed Res Int [Internet] 2018. /pmc/articles/PMC5941822/. Cited 4 Dec 2021

Arias AM (2008) *Drosophila melanogaster* and the development of biology in the 20th century. Methods Mol Biol [Internet]. 420:1–25. https://pubmed.ncbi.nlm.nih.gov/18641938/. Cited 25 Jan 2022

Arquier N, Léopold P (2007) Fly foie Gras: modeling fatty liver in Drosophila. Cell Metab 5(2):83–85

Atkinson MA, Eisenbarth GS, Michels AW (2014) Type 1 diabetes. Lancet 383(9911):69–82

Baranski TJ, Kraja AT, Fink JL, Feitosa M, Lenzini PA, Borecki IB et al (2018) A high throughput, functional screen of human Body Mass Index GWAS loci using tissue-specific RNAi *Drosophila melanogaster* crosses. PLOS Genet [Internet] 14(4):e1007222. https://journals.plos.org/plosgenetics/article?id=10.1371/journal.pgen.1007222. Cited 26 Jan 2022

Bargal R, Avidan N, Ben-Asher E, Olender Z, Zeigler M, Frumkin A et al (2000) Identification of the gene causing mucolipidosis type IV. Nat Genet [Internet] 26(1):118–121.https://pubmed.ncbi.nlm.nih.gov/10973263/. Cited 25 Jan 2022

Barnes AI, Wigby S, Boone JM, Partridge L, Chapman T (2008) Feeding, fecundity and lifespan in female *Drosophila melanogaster*. Proc R Soc B Biol Sci [Internet] 275(1643):1675–1683. https://royalsocietypublishing.org/doi/abs/10.1098/rspb.2008.0139. Cited 25 Jan 2022

Bassett AR, Tibbit C, Ponting CP, Liu JL (2013) Highly efficient targeted mutagenesis of Drosophila with the CRISPR/Cas9 system. Cell Rep 4(1):220–228

Baumbach J, Hummel P, Bickmeyer I, Kowalczyk KM, Frank M, Knorr K et al (2014) A Drosophila in vivo screen identifies store-operated calcium entry as a key regulator of adiposity. Cell Metab [Internet] 19(2): 331–343. https://pubmed.ncbi.nlm.nih.gov/24506874/. Cited 26 Jan 2022

Birse RT, Choi J, Reardon K, Rodriguez J, Graham S, Diop S et al (2010) High-fat-diet-induced obesity and heart dysfunction are regulated by the TOR pathway in Drosophila. Cell Metab [Internet] 12(5):533–544. https://pubmed.ncbi.nlm.nih.gov/21035763/. Cited 9 Jan 2022

Boquet I, Hitier R, Dumas M, Chaminade M, Preat T (2000) Central brain postembryonic development in Drosophila: implication of genes expressed at the interhemispheric junction—PubMed [Internet]. J Neurobiol 33–48. https://pubmed.ncbi.nlm.nih.gov/10623899/

Brand AH, Perrimon N (1993) Targeted gene expression as a means of altering cell fates and generating dominant phenotypes. Development 118(2):401–415

Brankatschk M, Dunst S, Nemetschke L, Eaton S (2014) Delivery of circulating lipoproteins to specific neurons in the Drosophila brain regulates systemic insulin signaling. elife 3(October):1–19

Braunstein H, Papazian M, Maor G, Lukas J, Rolfs A, Horowitz M (2020) Misfolding of lysosomal α-galactosidase a in a fly model and its alleviation by the pharmacological chaperone migalastat. Int J Mol Sci [Internet] 21(19):1–20. https://pubmed.ncbi.nlm.nih.gov/33036426/. Cited 12 Dec 2021

Bray GA (2004) Medical consequences of obesity. J Clin Endocrinol Metab [Internet] 89(6):2583–2589. https://academic.oup.com/jcem/article/89/6/2583/2870290. Cited 26 Jan 2022

Broughton SJ, Piper MDW, Ikeya T, Bass TM, Jacobson J, Driege Y et al (2005) Longer lifespan, altered metabolism, and stress resistance in Drosophila from ablation of cells making insulin-like ligands. Proc Natl Acad Sci [Internet] 102(8):3105–3110. https://www.pnas.org/content/102/8/3105. Cited 1 Jan 2022

Brown MS, Goldstein JL (1974) Familial hypercholesterolemia: defective binding of lipoproteins to cultured fibroblasts associated with impaired regulation of 3-hydroxy-3-methylglutaryl coenzyme A reductase activity. Proc Natl Acad Sci U S A [Internet] 71(3):788. /pmc/articles/PMC388099/?report=abstract. Cited 28 Dec 2021

Bruckert É, Gallo A (2014) Familial Hypercholesterolemia. Ochsner J [Internet]14(4):669. /pmc/articles/PMC4295745/. Cited 28 Dec 2021

Bullard KM, Cowie CC, Lessem SE, Saydah SH, Menke A, Geiss LS et al (2018 Mar 30) Prevalence of diagnosed diabetes in adults by diabetes type – United States, 2016. Morb Mortal Wkly Rep [Internet] 67(12):359. /pmc/articles/PMC5877361/. Cited 30 Dec 2021

Burke MK, Dunham JP, Shahrestani P, Thornton KR, Rose MR, Long AD (2010) Genome-wide analysis of a long-term evolution experiment with Drosophila. Nature [Internet] 467(7315):587–590. http://www.nature.com/articles/nature09352. Cited 8 Oct 2019

Carillo MR, Bertapelle C, Scialò F, Siervo M, Spagnuolo G, Simeone M et al (2020) L-Carnitine in Drosophila: a review. Antioxidants [Internet] 9(12):1310. https://www.mdpi.com/2076-3921/9/12/1310/htm. Cited 25 Jan 2022

Carmelli D, Cardon L, Fabsitz R (1994) Clustering of hypertension, diabetes, and obesity in adult male twins: same genes or same environments? – PubMed. Am J Hum Genet [Internet]. 55(3):566–573. https://pubmed.ncbi.nlm.nih.gov/8079995/. Cited 7 Jan 2022

Cattaneo F, Intra J, Matsumoto M, Briani F, Hoshi M, Perotti ME (2006) Identification and expression analysis of *Drosophila melanogaster* genes encoding β-hexosaminidases of the sperm plasma membrane. Glycobiology [Internet] 16(9):786–800. https://academic.oup.com/glycob/article/16/9/786/736245. Cited 13 Dec 2021

Chalmers J, Tung YCL, Liu CH, O'Kane CJ, O'Rahilly S, Yeo GSH (2021) A multicomponent screen for feeding behaviour and nutritional status in Drosophila to interrogate mammalian appetite-related genes. Mol Metab (43):101127

Clemens JC, Worby CA, Simonson-Leff N, Muda M, Maehama T, Hemmings BA et al (2000) Use of double-stranded RNA interference in Drosophila cell lines to dissect signal transduction pathways. Proc Natl Acad Sci [Internet] 97(12):6499–503. https://www.pnas.org/content/97/12/6499. Cited 4 Dec 2021

Course MM, Scott AI, Schoor C, Hsieh CH, Papakyrikos AM, Winter D et al (2018) Phosphorylation of MCAD selectively rescues PINK1 deficiencies in behavior and metabolism. Mol Biol Cell [Internet] 29(10):1219–1227. https://www.molbiolcell.org/doi/abs/10.1091/mbc.E18-03-0155. Cited 20 Dec 2021

Dandana A, Ben Khelifa S, Chahed H, Miled A, Ferchichi S (2016) Gaucher disease: clinical, biological and therapeutic aspects. Pathobiology [Internet] 83(1):13–20. https://pubmed.ncbi.nlm.nih.gov/26588331/. Cited 12 Dec 2021

Davis MY, Trinh K, Thomas RE, Yu S, Germanos AA, Whitley BN et al (2016) Glucocerebrosidase deficiency in Drosophila results in α-synuclein-independent protein aggregation and neurodegeneration. PLOS Genet [Internet] 12(3):e1005944. https://journals.plos.org/plosgenetics/article?id=10.1371/journal.pgen.1005944. Cited 12 Dec 2021

Dietzl G, Chen D, Schnorrer F, Su KC, Barinova Y, Fellner M et al (2007) A genome-wide transgenic RNAi library for conditional gene inactivation in Drosophila. Nature 448(7150):151–156

Ding M, Zheng L, Li QF, Wang WL, Da PW, Zhou M (2021) Exercise-training regulates apolipoprotein B in Drosophila to improve hfd-mediated cardiac function damage and low exercise capacity. Front Physiol (12):873

Drysdale R, Consortium the F (2008) FlyBase. Methods Mol Biol [Internet] 420:45–59. https://link.springer.com/protocol/10.1007/978-1-59745-583-1_3. Cited 8 Dec 2021

Ecker A, Gonzaga TKS, Do N, Seeger RL, dos Santos MM, Loreto JS, Boligon AA et al (2017) High-sucrose diet induces diabetic-like phenotypes and oxidative stress in *Drosophila melanogaster* : protective role of *Syzygium cumini* and *Bauhinia forficata*. Biomed Pharmacother [Internet] 89:605–616. http://www.ncbi.nlm.nih.gov/pubmed/28267671. Cited 8 Oct 2019

Engelhart DC, Azad P, Ali S, Granados JC, Haddad GG, Nigam SK (2020) Drosophila SLC22 orthologs related to OATs, OCTs, and OCTNs regulate development and responsiveness to oxidative stress. Int J Mol Sci [Internet]. 21(6):2002. https://www.mdpi.com/1422-0067/21/6/2002/htm. Cited 25 Jan 2022

Estes PS, Boehringer A, Zwick R, Tang JE, Grigsby B, Zarnescu DC (2011) Wild-type and A315T mutant TDP-43 exert differential neurotoxicity in a Drosophila model of ALS. Hum Mol Genet [Internet] 20(12): 2308–2321. https://pubmed.ncbi.nlm.nih.gov/21441568/. Cited 7 Feb 2022

Fei W, Yang H (2012) Genome-wide screens for gene products regulating lipid droplet dynamics. Methods Cell Biol [Internet] 108:303–316. https://pubmed.ncbi.nlm.nih.gov/22325608/. Cited 20 Jan 2022

Finkelstein EA, Trogdon JG, Cohen JW, Dietz W (2009) Annual medical spending attributable to obesity: payer-and service-specific estimates. Health Aff [Internet] 28(Supplement 1):w822–w831. http://www.ncbi.nlm.nih.gov/pubmed/19635784. Cited 8 Oct 2019

Flegal KM, Carroll MD, Kit BK, Ogden CL (2012) Prevalence of obesity and trends in the distribution of body mass index among US adults, 1999–2010. JAMA [Internet] 307(5):491. http://www.ncbi.nlm.nih.gov/pubmed/22253363. Cited 8 Oct 2019

Gáliková M, Klepsatel P (2018) Obesity and aging in the Drosophila model. Int J Mol Sci [Internet] 19(7). /pmc/articles/PMC6073435/. Cited 4 Dec 2021

Gáliková M, Diesner M, Klepsatel P, Hehlert P, Xu Y, Bickmeyer I et al (2015) Energy homeostasis control in drosophila adipokinetic hormone mutants. Genetics [Internet] 201(2):665–683. https://pubmed.ncbi.nlm.nih.gov/26275422/. Cited 9 Jan 2022

Giammanco A, Noto D, Barbagallo CM, Nardi E, Caldarella R, Ciaccio M et al (2021) Hyperalphalipoproteinemia and beyond: the role of HDL in cardiovascular diseases. Life [Internet] 11(6). /pmc/articles/PMC8235218/. Cited 28 Dec 2021

Graham P, Pick L (2017) Drosophila as a model for diabetes and diseases of insulin resistance 121. /pmc/articles/PMC5444297/. Cited 2021 Dec 30

Gratz SJ, Ukken FP, Rubinstein CD, Thiede G, Donohue LK, Cummings AM, et al. (2014) Highly specific and efficient CRISPR/Cas9-catalyzed homology-directed repair in Drosophila. Genetics [Internet] 196(4): 961–971. http://www.ncbi.nlm.nih.gov/pubmed/24478335. Cited 1 Nov 2019

Gratz SJ, Rubinstein CD, Harrison MM, Wildonger J, O'Connor-Giles KM (2015) CRISPR-Cas9 genome editing in Drosophila. Curr Protoc Mol Biol 2015: 31.2.1–31.2.20

Grönke S, Mildner A, Fellert S, Tennagels N, Petry S, Müller G et al (2005) Brummer lipase is an evolutionary conserved fat storage regulator in Drosophila. Cell Metab 1(5):323–330

Guillou A, Troha K, Wang H, Franc NC, Buchon N (2016) The Drosophila CD36 homologue croquemort is required to maintain immune and gut homeostasis during development and aging. PLOS Pathog [Internet] 2(10):e1005961. https://journals.plos.org/plospathogens/article?id=10.1371/journal.ppat.1005961. Cited 25 Jan 2022

Gutierrez E, Wiggins D, Fielding B, Gould AP (2006) Specialized hepatocyte-like cells regulate Drosophila lipid metabolism. Nature [Internet] 445(7125): 275–280. https://www.nature.com/articles/nature05382. Cited 16 Jan 2022

Haltia M, Goebel HH (2013) The neuronal ceroid-lipofuscinoses: a historical introduction. Biochim Biophys Acta [Internet] 1832(11):1795–1800. https://pubmed.ncbi.nlm.nih.gov/22959893/. Cited 12 Dec 2021

Han Y, Xiong L, Xu Y, Tian T, Wang T (2017) The β-alanine transporter balaT is required for visual neurotransmission in drosophila. elife 6

Hardy CM, Birse RT, Wolf MJ, Yu L, Bodmer R, Gibbs AG (2015) Obesity-associated cardiac dysfunction in starvation-selected *Drosophila melanogaster*. Am J Physiol Integr Comp Physiol [Internet] 309(6):R658–R667. http://www.ncbi.nlm.nih.gov/pubmed/26136533. Cited 8 Oct 2019

Hardy CM, Burke MK, Everett LJ, Han M V., Lantz KM, Gibbs AG (2018) Genome-wide analysis of starvation-selected *Drosophila melanogaster* – a genetic model of obesity. Mol Biol Evol [Internet] 35(1): 50–65. https://academic.oup.com/mbe/article/35/1/50/4371414. Cited 20 Jan 2022

Hasanagic M, Van Meel E, Luan S, Aurora R, Kornfeld S, Eissenberg JC (2015) The lysosomal enzyme receptor protein (LERP) is not essential, but is implicated in lysosomal function in *Drosophila melanogaster*. Biol Open [Internet] ;4(10):1316–1325. https://journals.biologists.com/bio/article/4/10/1316/1218/The-lysosomal-enzyme-receptor-protein-LERP-is-not. Cited 7 Feb 2022

Haselton A, Sharmin E, Schrader J, Sah M, Poon P, Fridell YWC (2010) Partial ablation of adult Drosophila insulin-producing neurons modulates glucose homeostasis and extends life span without insulin resistance. Cell Cycle [Internet] 9(15):3135–3143. https://pubmed.ncbi.nlm.nih.gov/20699643/. Cited 5 Jan 2022

Herder C, Roden M (2011) Genetics of type 2 diabetes: pathophysiologic and clinical relevance. Eur J Clin Invest [Internet] 41(6):679–692. https://onlinelibrary.wiley.com/doi/full/10.1111/j.1365-2362.2010.02454.x. Cited 30 Dec 2021

Hickey AJ, Chotkowski HL, Singh N, Ault JG, Korey CA, MacDonald ME et al (2006) Palmitoyl-protein thioesterase 1 deficiency in *Drosophila melanogaster* causes accumulation of abnormal storage material and reduced life span. Genetics [Internet] 172(4):2379–2390. https://pubmed.ncbi.nlm.nih.gov/16452138/. Cited 12 Dec 2021

Himanshu D, Ali W, Wamique M (2020) Type 2 diabetes mellitus: pathogenesis and genetic diagnosis. J Diabetes Metab Disord [Internet] 19(2):1959–1966. http://www.ncbi.nlm.nih.gov/pubmed/33520871. Cited 30 Dec 2021

Hong SH, Kang M, Lee KS, Yu K. (2016) High fat diet-induced TGF-β/Gbb signaling provokes insulin resistance through the tribbles expression. Sci Rep [Internet] 6. https://pubmed.ncbi.nlm.nih.gov/27484164/. Cited 9 Jan 2022

Housden BE, Valvezan AJ, Kelley C, Sopko R, Hu Y, Roesel C, et al (2015) Identification of potential drug targets for tuberous sclerosis complex by synthetic screens combining CRISPR-based knockouts with

RNAi. Sci Signal [Internet] 8(393):rs9. http://www.ncbi.nlm.nih.gov/pubmed/26350902. Cited 1 Nov 2019

Hu Y, Flockhart I, Vinayagam A, Bergwitz C, Berger B, Perrimon N et al (2011) An integrative approach to ortholog prediction for disease-focused and other functional studies. BMC Bioinformatics [Internet] 12(1): 1–16. https://bmcbioinformatics.biomedcentral.com/articles/10.1186/1471-2105-12-357. Cited 4 Dec 2021

Huang X, Suyama K, Buchanan JA, Zhu AJ, Scott MP (2005) A Drosophila model of the Niemann-Pick type C lysosome storage disease: dnpc1a is required for molting and sterol homeostasis. Development [Internet] 132(22):5115–5124. https://pubmed.ncbi.nlm.nih.gov/16221727/. Cited 12 Dec 2021

Huang X, Warren JT, Buchanan JA, Gilbert LI, Scott MP (2007) Drosophila Niemann-Pick type C-2 genes control sterol homeostasis and steroid biosynthesis: a model of human neurodegenerative disease. Development [Internet] 134(20):3733–3742. https://pubmed.ncbi.nlm.nih.gov/17804599/. Cited 12 Dec 2021

Ichihara S, Yamada Y (2008) Genetic factors for human obesity. Cell Mol Life Sci [Internet] 65(7–8): 1086–1098. https://pubmed.ncbi.nlm.nih.gov/18097636/. Cited 7 Jan 2022

Jünger MA, Rintelen F, Stocker H, Wasserman JD, Végh M, Radimerski T et al (2003) The Drosophila forkhead transcription factor FOXO mediates the reduction in cell number associated with reduced insulin signaling. J Biol [Internet] 2(3). https://pubmed.ncbi.nlm.nih.gov/12908874/. Cited 1 Jan 2022

Kang L, Aggarwal DD, Rashkovetsky E, Korol AB, Michalak P (2016) Rapid genomic changes in *Drosophila melanogaster* adapting to desiccation stress in an experimental evolution system. BMC Genomics 17(1)

Kawasaki H, Suzuki T, Ito K, Takahara T, Goto-Inoue N, Setou M et al (2017) Minos-insertion mutant of the Drosophila GBA gene homologue showed abnormal phenotypes of climbing ability, sleep and life span with accumulation of hydroxy-glucocerebroside. Gene 30(614):49–55

Kaya-Çopur A, Schnorrer F (2016) A guide to genome-wide in vivo RNAi applications in Drosophila. Methods Mol Biol [Internet] 1478:117–143. http://www.ncbi.nlm.nih.gov/pubmed/27730578. Cited 1 Nov 2019

Kienesberger PC, Oberer M, Lass A, Zechner R (2009) Mammalian patatin domain containing proteins: a family with diverse lipolytic activities involved in multiple biological functions. J Lipid Res [Internet] 50(Suppl): S63. /pmc/articles/PMC2674697/. Cited 16 Jan 2022

Kishita Y, Tsuda M, Aigaki T (2012) Impaired fatty acid oxidation in a Drosophila model of mitochondrial trifunctional protein (MTP) deficiency. Biochem Biophys Res Commun 419(2):344–349

Kofler R, Pandey RV, Schlötterer C (2011) PoPoolation2: identifying differentiation between populations using sequencing of pooled DNA samples (Pool-Seq). Bioinformatics 27(24):3435–3436

Kofler R, Betancourt AJ, Schlötterer C (2012) Sequencing of pooled DNA samples (Pool-Seq) uncovers complex dynamics of transposable element insertions in *Drosophila melanogaster*. PLoS Genet 8(1)

Lee S, Bao H, Ishikawa Z, Wang W, Lim HY (2017) Cardiomyocyte regulation of systemic lipid metabolism by the apolipoprotein B-containing lipoproteins in drosophila. PLOS Genet [Internet] 13(1):e1006555. https://journals.plos.org/plosgenetics/article?id=10.1371/journal.pgen.1006555. Cited 28 Dec 2021

Léonard R, Rendić D, Rabouille C, Wilson IBH, Préat T, Altmann F(2006) The Drosophila fused lobes gene encodes an N-Acetylglucosaminidase Involved in N-Glycan processing *. J Biol Chem [Internet] 281(8):4867–4875. http://www.jbc.org/article/S0021925819767829/fulltext. Cited 13 Dec 2021

Levitsky DA, Sewall A, Zhong Y, Barre L, Shoen S, Agaronnik N et al (2019) Quantifying the imprecision of energy intake of humans to compensate for imposed energetic errors: a challenge to the physiological control of human food intake. Appetite (133):337–343

Li S, Loos RJF (2008). Progress in the genetics of common obesity: size matters. Curr Opin Lipidol [Internet] 19(2):113–121. https://pubmed.ncbi.nlm.nih.gov/18388690/. Cited 7 Jan 2022

Li FY, El-Hattab AW, Bawle E V., Boles RG, Schmitt ES, Scaglia F, et al. Molecular spectrum of SLC22A5 (OCTN2) gene mutations detected in 143 subjects evaluated for systemic carnitine deficiency. Hum Mutat [Internet]. 2010 31(8):E1632–E1651. https://onlinelibrary.wiley.com/doi/full/10.1002/humu.21311. Cited 25 Jan 2022

Lim JS, Mietus-Snyder M, Valente A, Schwarz JM, Lustig RH (2010) The role of fructose in the pathogenesis of NAFLD and the metabolic syndrome. Nat Rev Gastroenterol Hepatol [Internet] 7(5):251–264. https://www.nature.com/articles/nrgastro.2010.41. Cited 16 Jan 2022

Lüersen K, Röder T, Rimbach G (2019) *Drosophila melanogaster* in nutrition research – the importance of standardizing experimental diets. Genes Nutr [Internet] 14(1):1–5. https://genesandnutrition.biomedcentral.com/articles/10.1186/s12263-019-0627-9. Cited 4 Jan 2022

Mackay TFC, Huang W (2018) Charting the genotype–phenotype map: lessons from the *Drosophila melanogaster* genetic reference panel, Wiley interdisciplinary reviews: developmental biology, vol 7. Wiley

Mackay TFC, Richards S, Stone EA, Barbadilla A, Ayroles JF, Zhu D et al (2012) The *Drosophila melanogaster* genetic reference panel. Nature [Internet] 482(7384):173–178. http://www.nature.com/articles/nature10811. Cited 22 Sep 2019

Manzo E, O'Conner AG, Barrows JM, Shreiner DD, Birchak GJ, Zarnescu DC (2018) Medium-chain fatty acids, beta-hydroxybutyric acid and genetic modulation of the carnitine shuttle are protective in a drosophila model of ALS based on TDP-43. Front Mol Neurosci 31(11):182

Maor G, Rencus-Lazar S, Filocamo M, Steller H, Segal D, Horowitz M (2013) Unfolded protein response in Gaucher disease: from human to Drosophila. Orphanet J Rare Dis [Internet] 8(1):140. /pmc/articles/PMC3819655/. Cited 12 Dec 2021

Maor G, Cabasso O, Krivoruk O, Rodriguez J, Steller H, Segal D et al (2016) The contribution of mutant GBA to the development of Parkinson disease in Drosophila. Hum Mol Genet [Internet] 25(13):2712–2727. https://pubmed.ncbi.nlm.nih.gov/27162249/. Cited 12 Dec 2021

Merritt JL, II, Norris M, Kanungo S (2018 Apr) Fatty acid oxidation disorders. Ann Transl Med [Internet] 6(24): 181–3. Available from: /pmc/articles/PMC6331364/. Cited 4 Dec 2021

Minster RL, Hawley NL, Su CT, Sun G, Kershaw EE, Cheng H et al (2016) A thrifty variant in CREBRF strongly influences body mass index in Samoans. Nat Genet [Internet] 48(9):1049–1054. https://pubmed.ncbi.nlm.nih.gov/27455349/. Cited 22 Jan 2022

Mohammed A, O'Hare MB, Warley A, Tear G, Tuxworth RI (2017) In vivo localization of the neuronal ceroid lipofuscinosis proteins, CLN3 and CLN7, at endogenous expression levels. Neurobiol Dis (103):123–132

Mooradian AD (2009) Dyslipidemia in type 2 diabetes mellitus. Nat Clin Pract Endocrinol Metab [Internet] 5(3):150–159. https://pubmed.ncbi.nlm.nih.gov/19229235/. Cited 30 Dec 2021

Morris SNS, Coogan C, Chamseddin K, Fernandez-Kim SO, Kolli S, Keller JN et al (2012) Development of diet-induced insulin resistance in adult *Drosophila melanogaster*. Biochim Biophys Acta 1822(8): 1230–1237

Müller MJ, Geisler C, Blundell J, Dulloo A, Schutz Y, Krawczak M et al (2018) The case of GWAS of obesity: does body weight control play by the rules? Int J Obes (Lond) [Internet] 42(8):1395–1405. https://pubmed.ncbi.nlm.nih.gov/29795468/. Cited 9 Jan 2022

Murillo-Maldonado JM, Sánchez-Chávez G, Salgado LM, Salceda R, Riesgo-Escovar JR (2011) Drosophila insulin pathway mutants affect visual physiology and brain function besides growth, lipid, and carbohydrate metabolism. Diabetes [Internet] 60(5):1632–1636. http://flybase.org. Cited 5 Jan 2022

Musselman LP, Kühnlein RP (2018) Drosophila as a model to study obesity and metabolic disease. J Exp Biol [Internet] 121(Suppl 1). www.who.int/. Cited 4 Dec 2021

Musselman LP, Fink JL, Narzinski K, Ramachandran PV, Hathiramani SS, Cagan RL et al (2011) A high-sugar diet produces obesity and insulin resistance in wild-type Drosophila. Dis Model Mech 4(6):842–849

Musselman LP, Fink JL, Ramachandran PV, Patterson BW, Okunade AL, Maier E et al (2013) Role of fat body lipogenesis in protection against the effects of caloric overload in *Drosophila*. J Biol Chem [Internet] 288(12):8028–8042. http://www.jbc.org/lookup/doi/10.1074/jbc.M112.371047. Cited 22 Sep 2019

Musselman LP, Fink JL, Baranski TJ (2016) CoA protects against the deleterious effects of caloric overload in Drosophila1. J Lipid Res [Internet] 57(3):380–387. http://www.jlr.org/article/S0022227520354377/fulltext. Cited 20 Jan 2022

Musselman LP, Fink JL, Grant AR, Gatto JA, Tuthill BF, Baranski TJ (2018) A complex relationship between immunity and metabolism in Drosophila diet-induced insulin resistance. Mol Cell Biol [Internet] 38(2). https://journals.asm.org/doi/abs/10.1128/MCB.00259-17. Cited 4 Jan 2022

Musselman LP, Fink JL, Baranski TJ (2019) Similar effects of high-fructose and high-glucose feeding in a Drosophila model of obesity and diabetes. PLoS One [Internet] 14(5): e0217096. https://journals.plos.org/plosone/article?id=10.1371/journal.pone.0217096. Cited 4 Jan 2022

Myllykangas L, Tyynelä J, Page-McCaw A, Rubin GM, Haltia MJ, Feany MB (2005) Cathepsin D-deficient Drosophila recapitulate the key features of neuronal ceroid lipofuscinoses. Neurobiol Dis [Internet] 19(1–2):194–199. https://pubmed.ncbi.nlm.nih.gov/15837574/. Cited 25 Jan 2022

Na J, Musselman LP, Pendse J, Baranski TJ, Bodmer R, Ocorr K et al (2013) A Drosophila model of high sugar diet-induced cardiomyopathy. Rulifson E, editor. PLoS Genet [Internet] 9(1):e1003175. https://dx.plos.org/10.1371/journal.pgen.1003175. Cited 8 Oct 2019

Naureckiene S, Sleat DE, Lacklan H, Fensom A, Vanier MT, Wattiaux R et al (2000) Identification of HE1 as the second gene of Niemann-Pick C disease. Science 290(5500):2298–2301. https://pubmed.ncbi.nlm.nih.gov/11125141/. Cited 12 Dec 2021

Ng M, Fleming T, Robinson M, Thomson B, Graetz N, Margono C et al (2014) Global, regional and national prevalence of overweight and obesity in children and adults 1980–2013: a systematic analysis. Lancet [Internet]. 384(9945): 766. /pmc/articles/PMC4624264/. Cited 11 Jan 2022

O'brien SJ, Shimada Y (1974). THE α-glycerophosphate cycle in drosophila melanogaster IV. Metabolic, ultrastructural, and adaptive consequences of αGpdh-1 null mutations. J Cell Biol [Internet]. 63(3):864–882. http://rupress.org/jcb/article-pdf/63/3/864/1266802/864.pdf. Cited 25 Jan 2022

Okada S, O'Brien JS (1968) Generalized gangliosidosis: beta-galactosidase deficiency. Science (80-) [Internet] 160(3831):1002–1004. https://www.science.org/doi/abs/10.1126/science.160.3831.1002. Cited 13 Dec 2021

Palanker L, Tennessen JM, Lam G, Thummel CS (2009) Drosophila HNF4 regulates lipid mobilization and β-oxidation. Cell Metab 9(3):228–239

Palm W, Sampaio JL, Brankatschk M, Carvalho M, Mahmoud A, Shevchenko A et al (2012) Lipoproteins in *Drosophila melanogaster* – assembly, function, and influence on tissue lipid composition. PLOS Genet [Internet] 8(7):e1002828. https://journals.plos.org/

plosgenetics/article?id=10.1371/journal.pgen.1002828. Cited 4 Dec 2021

Pandey UB, Nichols CD (2011) Human disease models in *Drosophila melanogaster* and the role of the fly in therapeutic drug discovery. Pharmacol Rev [Internet] 63(2): 411. /pmc/articles/PMC3082451/. Cited 4 Dec 2021

Park S, Alfa RW, Topper SM, Kim GES, Kockel L, Kim SK (2014) A genetic strategy to measure circulating Drosophila insulin reveals genes regulating insulin production and secretion. PLoS Genet [Internet] 10(8). https://pubmed.ncbi.nlm.nih.gov/25101872/. Cited 1 Jan 2022

Pearson JC, Crews ST (2014) Enhancer diversity and the control of a simple pattern of Drosophila CNS midline cell expression. Dev Biol 392(2):466–482

Pendse J, Ramachandran P V, Na J, Narisu N, Fink JL, Cagan RL et al (2013) A Drosophila functional evaluation of candidates from human genome-wide association studies of type 2 diabetes and related metabolic traits identifies tissue-specific roles for dHHEX. BMC Genomics [Internet] 14(1):1–11. https://bmcgenomics.biomedcentral.com/articles/10.1186/1471-2164-14-136. Cited 1 Jan 2022

Phelps CB, Brand AH (1998) Ectopic gene expression in Drosophila using GAL4 system. Methods A Companion Methods Enzymol 14(4):367–379

Reiner Ž, Guardamagna O, Nair D, Soran H, Hovingh K, Bertolini S et al (2014) Lysosomal acid lipase deficiency – an under-recognized cause of dyslipidaemia and liver dysfunction. Atherosclerosis 235(1):21–30

Reiter LT, Potocki L, Chien S, Gribskov M, Bier E (2021) A systematic analysis of human disease-associated gene sequences in Drosophila melanogaster. www.genome.org. Cited 4 Dec 2021

Rigon L, De Filippis C, Napoli B, Tomanin R, Orso G (2021) Exploiting the potential of Drosophila models in lysosomal storage disorders: pathological mechanisms and drug discovery. Biomed [Internet] 9(3):268. https://www.mdpi.com/2227-9059/9/3/268/htm. Cited 4 Dec 2021

Rinella ME (2015) Nonalcoholic fatty liver disease: a systematic review. JAMA [Internet] 313(22): 2263–2273. https://jamanetwork.com/journals/jama/fullarticle/2319168. Cited 15 Jan 2022

Ripatti P, Rämö JT, Mars NJ, Fu Y, Lin J, Söderlund S et al (2020) Polygenic hyperlipidemias and coronary artery disease risk. Circ Genomic Precis Med [Internet] 3(4): e002725. /pmc/articles/PMC7176338/. Cited 28 Dec 2021

Rodríguez-Vázquez M, Vaquero D, Parra-Peralbo E, Mejía-Morales JE, Culi J (2015) Drosophila lipophorin receptors recruit the lipoprotein LTP to the plasma membrane to mediate lipid uptake. PLOS Genet [Internet] 11(6):e1005356. https://journals.plos.org/plosgenetics/article?id=10.1371/journal.pgen.1005356. Cited 28 Dec 2021

Rulifson EJ, Kim SK, Nusse R (2002) Ablation of insulin-producing neurons in flies: growth and diabetic phenotypes. Science [Internet] 296(5570):1118–1120. https://pubmed.ncbi.nlm.nih.gov/12004130/. Cited 5 Jan 2022

Sang J, Dhakal S, Lee Y (2021) Cucurbitacin B suppresses hyperglycemia associated with a high sugar diet and promotes sleep in *Drosophila melanogaster*. Mol Cells [Internet] 4(2):68 /pmc/articles/PMC7941002/. Cited 11 Dec 2021

Schonfeld G, Lin X, Yue P (2005). Familial hypobetalipoproteinemia: genetics and metabolism. Cell Mol Life Sci [Internet] 62(12):1372–1378. https://link.springer.com/article/10.1007/s00018-005-4473-0. Cited 28 Dec 2021

Schultzhaus JN, Bennett CJ, Iftikhar H, Yew JY, Mallett J, Carney GE. (2018) High fat diet alters Drosophila melanogaster sexual behavior and traits: decreased attractiveness and changes in pheromone profiles. Sci Rep [Internet] 8(1). https://pubmed.ncbi.nlm.nih.gov/29599496/. Cited 9 Jan 2022

Schulz JG, Laranjeira A, Van Huffel L, Gärtner A, Vilain S, Bastianen J et al (2015) Glial β-oxidation regulates Drosophila energy metabolism. Sci Rep [Internet] 5(1):1–9. https://www.nature.com/articles/srep07805. Cited 25 Jan 2022

Seidell JC (1999) Obesity: a growing problem. Acta Pædiatrica [Internet] 88(s428):46–50. https://onlinelibrary.wiley.com/doi/full/10.1111/j.1651-2227.1999.tb14350.x. Cited 26 Jan 2022

Sellin J, Fülle JB, Thiele C, Bauer R, Bülow MH (2020) Free fatty acid determination as a tool for modeling metabolic diseases in Drosophila. J Insect Physiol (126):104090

Semaniuk UV, Gospodaryov DV, Feden'ko KM, Yurkevych IS, Vaiserman AM, Storey KB et al (2018) Insulin-like peptides regulate feeding preference and metabolism in Drosophila. Front Physiol 9 (Aug):1083

Settembre C, Fraldi A, Medina DL, Ballabio A (2013) Signals for the lysosome: a control center for cellular clearance and energy metabolism. Nat Rev Mol Cell Biol [Internet] 14(5):283. /pmc/articles/PMC4387238/. Cited 13 Dec 2021

Shah SA, Erdmann S, Mojica FJM, Garrett RA (2013) Protospacer recognition motifs: mixed identities and functional diversity. RNA Biol [Internet] 10(5): 891–899. http://www.ncbi.nlm.nih.gov/pubmed/23403393. Cited 1 Nov 2019

Sieber MH, Thummel CS (2012) Coordination of triacylglycerol and cholesterol homeostasis by DHR96 and the Drosophila LipA homolog magro. Cell Metab 15(1):122–127

Smith WW, Thomas J, Liu J, Li T, Moran TH (2014) From fat fruitfly to human obesity. Physiol Behav [Internet] 0:15. /pmc/articles/PMC4125553/. Cited 4 Dec 2021

Speakman JR, Loos RJF, O'Rahilly S, Hirschhorn JN, Allison DB (2018) GWAS for BMI: a treasure trove of fundamental insights into the genetic basis of obesity. Int J Obes [Internet] 42(8):1524–1531. https://www.nature.com/articles/s41366-018-0147-5. Cited 9 Jan 2022

Strub BR, Parkes TL, Mukai ST, Bahadorani S, Coulthard AB, Hall N et al (2008) Mutations of the withered (whd) gene in *Drosophila melanogaster* confer hypersensitivity to oxidative stress and are lesions of the carnitine palmitoyltransferase I (CPT I) gene. https://doi.org/101139/G08-023 [Internet] 51(6):409–420. https://cdnsciencepub.com/doi/abs/10.1139/G08-023. Cited Dec 2021

Swinburn BA, Caterson I, Seidell JC, James W (2004) Diet, nutrition and the prevention of excess weight gain and obesity. Public Health Nutr [Internet] 7(1a):123–146. https://www.cambridge.org/core/journals/public-health-nutrition/article/diet-nutrition-and-the-prevention-of-excess-weight-gain-and-obesity/6DB071DD5989F7D0096DD70B0EEDC77B. Cited 26 Jan 2022

Tilg H, Moschen AR, Roden M (2016) NAFLD and diabetes mellitus. Nat Rev Gastroenterol Hepatol [Internet] 14(1):32–42. https://www.nature.com/articles/nrgastro.2016.147. Cited 15 Jan 2022

Tobler R, Franssen SU, Kofler R, Orozco-Terwengel P, Nolte V, Hermisson J et al (2014) Massive habitat-specific genomic response in *D. melanogaster* populations during experimental evolution in hot and cold environments. Mol Biol Evol 31(2):364–375

Topisirovic I, Culjkovic B, Cohen N, Perez JM, Skrabanek L, Borden KLB (2003) The proline-rich homeodomain protein, PRH, is a tissue-specific inhibitor of eIF4E-dependent cyclin D1 mRNA transport and growth. EMBO J [Internet] 22(3):689. /pmc/articles/PMC140753/. Cited 5 Jan 2022

Trinh I, Boulianne GL (2013) Modeling obesity and its associated disorders in Drosophila. Physiology [Internet] 28(2):117. /pmc/articles/PMC3768089/. Cited 30 Dec 2021

Turcot V, Lu Y, Highland HM, Schurmann C, Justice AE, Fine RS et al (2017) Protein-altering variants associated with body mass index implicate pathways that control energy intake and expenditure in obesity. Nat Genet 2017 501 [Internet] 50(1):26–41. https://www.nature.com/articles/s41588-017-0011-x. Cited 22 Jan 2022

Turner TL, Stewart AD, Fields AT, Rice WR, Tarone AM (2011) Population-based resequencing of experimentally evolved populations reveals the genetic basis of body size variation in *Drosophila melanogaster*. Gibson G, editor. PLoS Genet [Internet] 7(3): e1001336. http://dx.plos.org/10.1371/journal.pgen.1001336. Cited 8 Oct 2019

Tuxworth RI, Chen H, Vivancos V, Carvajal N, Huang X, Tear G (2011). The Batten disease gene CLN3 is required for the response to oxidative stress. Hum Mol Genet [Internet] 20(10):2037–2047. https://pubmed.ncbi.nlm.nih.gov/21372148/. Cited 12 Dec 2021

Ugrankar R, Berglund E, Akdemir F, Tran C, Kim MS, Noh J, et al. (2015) Drosophila glucome screening identifies Ck1alpha as a regulator of mammalian glucose metabolism. Nat Commun [Internet] 6(1):1–10. https://www.nature.com/articles/ncomms8102. Cited 4 Jan 2022

Venkatachalam K, Long AA, Elsaesser R, Nikolaeva D, Broadie K, Montell C (2008) Motor deficit in a Drosophila model of mucolipidosis type IV due to defective clearance of apoptotic cells. Cell 135(5):838–851

Versace E, Nolte V, Pandey RV, Tobler R, Schlötterer C (2014) Experimental evolution reveals habitat-specific fitness dynamics among Wolbachia clades in *Drosophila melanogaster*. Mol Ecol 23(4):802–814

Vihervaara T, Puig O (2008) dFOXO Regulates transcription of a Drosophila acid lipase. J Mol Biol 376(5): 1215–1220

Vishwanath VA (2016) Fatty acid beta-oxidation disorders: a brief review. Ann Neurosci [Internet]. 23(1):51. /pmc/articles/PMC4934411/. Cited 4 Dec 2021

Vujkovic M, Keaton JM, Lynch JA, Miller DR, Zhou J, Tcheandjieu C et al (2020) Discovery of 318 new risk loci for type 2 diabetes and related vascular outcomes among 1.4 million participants in a multi-ancestry meta-analysis. Nat Genet 2020 [Internet] 52(7): 680–691. https://www.nature.com/articles/s41588-020-0637-y. Cited 30 Dec 2021

Weinmaster G, Roberts VJ, Lemke G. (1991) A homolog of Drosophila Notch expressed during mammalian development. Development [Internet] 113(1): 199–205. https://journals.biologists.com/dev/article/113/1/199/37154/A-homolog-of-Drosophila-Notch-expressed-during. Cited 7 Feb 2022

Williams MJ, Eriksson A, Shaik M, Voisin S, Yamskova O, Paulsson J et al (2015) The obesity-linked gene Nudt3 Drosophila homolog aps is associated with insulin signaling. Mol Endocrinol [Internet] 29(9):1303–1319. https://academic.oup.com/mend/article/29/9/1303/2556759. Cited 9 Jan 2022

Wong VWS, Adams LA, de Lédinghen V, Wong GLH, Sookoian S (2018) Noninvasive biomarkers in NAFLD and NASH – current progress and future promise. Nat Rev Gastroenterol Hepatol 2018 [Internet] 15(8):461–478. https://www.nature.com/articles/s41575-018-0014-9. Cited 15 Jan 2022

Xu H, Ren D (2015) Lysosomal physiology. Annu Rev Physiol [Internet] 77:57. /pmc/articles/PMC4524569/. Cited 13 Dec 2021

Younossi ZM, Koenig AB, Abdelatif D, Fazel Y, Henry L, Wymer M (2016) Global epidemiology of nonalcoholic fatty liver disease – meta-analytic assessment of prevalence, incidence, and outcomes. Hepatology [Internet] 64(1):73–84. https://onlinelibrary.wiley.com/doi/full/10.1002/hep.28431. Cited 15 Jan 2022

Zagoraiou L, Drabek D, Alexaki S, Guy JA, Klinakis AG, Langeveld A, et al. (2001) In vivo transposition of Minos, a Drosophila mobile element, in mammalian tissues. Proc Natl Acad Sci [Internet] 98(20): 11474–11478. https://www.pnas.org/content/98/20/11474. Cited 7 Feb 2022

MIX
Papier aus verantwortungsvollen Quellen
Paper from responsible sources
FSC® C105338

If you have any concerns about our products,
you can contact us on
ProductSafety@springernature.com

In case Publisher is established outside the EU,
the EU authorized representative is:
Springer Nature Customer Service Center GmbH
Europaplatz 3, 69115 Heidelberg, Germany

Printed by Libri Plureos GmbH
in Hamburg, Germany